HUMAN NUTRITION

HUMAN NUTRITION

FOURTEENTH EDITION

Edited by

Catherine **GEISSLER**

Professor Emerita of Human Nutrition, King's College London, UK;
Secretary General of the International Union of Nutritional Sciences,
2013–2022

Hilary J. **POWERS**

Professor Emerita of Nutritional Biochemistry; University of Sheffield, UK

OXFORD
UNIVERSITY PRESS

Great Clarendon Street, Oxford, OX2 6DP,
United Kingdom

Oxford University Press is a department of the University of Oxford.
It furthers the University's objective of excellence in research, scholarship,
and education by publishing worldwide. Oxford is a registered trade mark of
Oxford University Press in the UK and in certain other countries

Thirteenth Edition 2017
Twelfth Edition 2011
Eleventh Edition 2005

Public sector information reproduced under Open Government Licence v3.0
(http://www.nationalarchives.gov.uk/doc/open-government-licence/open-government-licence.htm)

Published in the United States of America by Oxford University Press
198 Madison Avenue, New York, NY 10016, United States of America

British Library Cataloguing in Publication Data

Data available
Library of Congress Control Number: 2023906872

ISBN 978-0-19-886665-7

Printed in the UK by
Bell & Bain Ltd., Glasgow

Preface

Significant advances have been made in our understanding of the links between what we eat and drink and our health, and the mechanisms that can explain these links. However, malnutrition, be it undernutrition, vitamin or mineral deficiencies, or overweight and obesity, continues to pose a serious threat to human health on a global scale.

Access to healthy food can be a struggle for the majority of people in low-income countries and for the most deprived in all populations. The realization that climate change and global warming pose a potentially catastrophic challenge to tackling food security can no longer be ignored. Equally, the way in which food is produced, processed, transported, and consumed is a key influencer of global warming. A greater understanding of the impact of climate change on food security in all its forms, and how in turn food systems influence carbon emissions and global warming, may prove central to making further progress in reducing malnutrition across the world.

This textbook provides a clear, comprehensive account of our current understanding of human nutrition in its broadest sense. The book is arranged into 36 chapters in seven Parts; foods and nutrients, physiology and metabolism, micronutrient function and metabolism, dietary requirements for specific groups, nutrition and disease, assessment of nutritional status, and public health nutrition. The reader is presented first with a description of the composition of foods and food patterns worldwide, before exploring the manner in which nutrients in foods are absorbed and metabolized. The causal links between diet and health or ill-health are explored with reference to specific diseases or conditions, with supporting examples from the research literature. Later chapters offer readers insight into the interplay of global factors which influence food access and ultimately malnutrition across continents. Specific examples are given of nutrition policies and interventions that have proved effective, or which hold promise for improving the health of the world's population.

We are delighted to welcome new authors to our international authorship, all experts in their field, who share with readers an up-to-date view of their specialist subjects. All chapters have been revised and updated to address the most recent research findings and in response to valuable feedback from readers and reviewers. New material is included, for example in the areas of food sustainability, the gut microbiome, dementias, the social impact of alcohol consumption, and the implications of climate change to food security. Emerging trends are highlighted and discussed. Additionally, where relevant, and evidence permits, SARS-CoV-2 (COVID-19) infection is discussed in the context of diet and nutritional status.

Readers will benefit from a substantial online resource of additional material to complement the chapters, and of test questions and answers for each chapter.

The editors are grateful, as ever, to the many authors who have shared their expertise and experience in contributing to this new edition of Human Nutrition.

Catherine Geissler
Hilary J. Powers

New to this edition

- New authors have been added to our international authorship, all experts in their field, providing an up-to-date view of their specialist subjects.

- All chapters have been revised and updated to address the most recent research findings and in response to valuable feedback from readers and reviewers.

- New material is included, for example in the areas of food sustainability, the gut microbiome, dementias, the social impact of alcohol consumption, and the implications of climate change to food security.

- Emerging trends are highlighted and discussed.

- Where relevant, and evidence permits, SARS-CoV-2 (COVID-19) infection is discussed in the context of diet and nutritional status.

- Readers will benefit from a substantial online resource of additional material to complement the chapters, and of test questions and answers for each chapter.

For extended coverage on the topics included in this textbook, in addition to further reading and weblinks for each chapter, visit the dedicated Online Resources at www.oup.com/he/geissler14e/.

Contents

Abbreviations

AA	amino acid, arachidonic acid
ACAORN	Australasian Child and Adolescent Obesity Research Network
ACE	angiotensin-converting enzyme, adverse childhood experiences
ACH	arm–chest–hip index
ACNFP	Advisory Committee on novel foods and processes
AD	Alzheimer's disease
ADH	alcohol dehydrogenase
ADHD	attention deficit hyperactivity disorder
ADI	acceptable daily intake
ADP	adenosine diphosphate
AGEs	advanced glycation end products
AgRP	agouti-related peptide
AI	adequate intake
AIDS	acquired immune deficiency syndrome
ALA	alpha linoleic acid
ALARP	as low as reasonably practicable
ALD	alcoholic liver disease
ALDH	aldehyde dehydrogenase
ALP	atherogenic lipoprotein phenotype
AMC	arm muscle circumference
AMDR	acceptable macronutrient distribution range
AMP	adenosine monophosphate
AN	anorexia nervosa
ANGPLT	angiopoietin-like proteins
ANLS	astrocyte–neuron lactate shuttle
APCs	antigen-presenting cells
APT	atopy patch test
AR	attributable risk
ARBD	alcohol-related birth defect
ARDS	acute respiratory distress syndrome
ARFID	avoidant/restrictive food intake disorder
ART	anti-retroviral therapy, assisted reproductive technology
ASP	animal source protein
ATP	adenosine triphosphate
ATGL	adipose triglyceride lipase
AUDIT	Alcohol Use Disorder Identification Test
BAPEN	British Association for Parental and Enteral Nutrition
BAT	brown adipose tissue
BCAA	branched-chain amino acids (leucine, isoleucine, valine)
BCAT	branched-chain amino transferase enzyme
BCKD	branched-chain keto-acid dehydrogenase
BDA	British Dietetic Association
BDNF	brain-derived neurotrophic factor
BED	binge eating disorder
BER	base excision repair
BF	body fat
BFH	baby-friendly hospital
BIA	bio-impedance analyser
BIE	bioelectrical impedance
BIVA	bioelectrical impedance vector analysis
BMC	bone mineral content
BMD	benchmark dose
BMD	bone mineral density
BMI	body mass index
BMR	basal metabolic rate
BN	bulimia nervosa
bp	base pair
BPL	below poverty level
BSE	bovine spongiform encephalopathy
BSID	Bayley Scores of Infant Development
BSF	biceps skinfold thickness
BV	biological value
cAMP	cyclic adenosine monophosphate
CAC	Codex Alimentarius Commission
CAGE	Cut, Annoyed, Guilty, Eye-Opener
CAP	Common Agricultural Programme
CART	cocaine- and amphetamine-regulated transcript
CAT	computer-assisted tomography
CBT	cognitive behavioural therapy
CCK	cholecystokinin
CCS	continued cephalic stimulation
CD	Crohn's disease, cluster of differentiation
CDC	Centers for Disease Control and Prevention (USA)
cDNA	complementary DNA
CDRR	chronic disease risk reduction intake
cds	coding sequence
CED	chronic energy deficiency
CEL	carboxyl ester lipase
CETP	cholesterol ester transfer protein

CGI	comparative gene identification	DEQAS	vitamin D external quality assurance scheme
CGIAR	Consultative Group on International Agricultural Research	DES	dietary energy supply
CGM	continuous glucose monitoring	DEXA	dual-energy X-ray absorptiometry
CHD	coronary heart disease	DFE	dietary folate equivalents
CI	confidence interval	DHA	docosahexaenoic acid, dehydroascorbic acid
CLA	conjugated linoleic acids		
CMAM	Community Management of Acute Malnutrition	DIT	diet-induced thermogenesis
		DHM	donor human milk
CMPI	cow's milk protein intolerance	DHSS	Department of Health and Social Security (UK)
CMV	cytomegalovirus		
CNS	central nervous system	DIAAS	digestible indispensable amino acid score
CoA	coenzyme A		
CoFID	Composition of Foods Integrated Dataset	DIT	diet-induced thermogenesis, di-iodotyrosine
COMA	Committee On Medical Aspects of Food Policy	DMFT	index of decayed, missing, and filled teeth (permanent)
COPD	chronic obstructive pulmonary disease	Dmft	index of decayed, missing, and filled teeth (deciduous)
COT	Committee on toxicity of chemicals in food, consumer products, and the environment	DNA	deoxyribonucleic acid
		DNL	de novo lipogenesis
		DNMT1	DNA methyltransferase 1
COUP	chicken ovalbumin upstream promoter	DNSBA	Diet and Nutrition Survey of British Adults
COX	cyclo-oxygenase	DoH	Department of Health (UK)
CPR	cephalic phase responses	DOHaD	developmental origins of health and disease
CR	caloric restriction		
CRD	component-resolved diagnostics	DPA	docosapentaenoic acid
CRP	C-reactive protein	DRA	dietary reducing agents
CSF	cerebrospinal fluid	DRI	dietary reference intake (USA)
CSII	continuous subcutaneous infusion	DRV	dietary reference value (UK)
CT	computer-assisted tomography	dUMP	deoxyuridine monophosphate
CUG	catch-up growth	DXA	dual X-ray absorptiometry
CVA	cerebrovascular accident	EAA index	essential amino acid index
CVD	cardiovascular disease	EAR	estimated average daily requirements (UK)
CYP	cytochrome P450 mono-oxygenase		
DA	dopamine	EC DG SANTE	directorate-general of the European Commission
DALY	disability-adjusted life years		
DADS	diallyl disulfide	ECF	extracellular fluid
DAPA	Diet, Anthropometry, and Physical Activity Measurement Toolkit (UK)	ECHO-MRI	quantitative magnetic resonance
		ED	elemental diet, eating disorder
DAS	diallyl sulfide	EDD	estimated date of delivery
DASH	dietary approaches to stop hypertension	EDI	estimated daily intake
		EDNOS	not otherwise specified (eating disorders)
DBP	vitamin D-binding protein		
DBPCFC	double-blind placebo-controlled food challenge	EDTA	ethylene diamine tetra-acetic acid
		EE	energy expenditure
DCCT	Diabetes Control and Complications Trial	EFA	essential fatty acid
		EFS	Expenditure and Food Survey
DDT	Dichlorodiphenyltrichloroethane	EFSA	European Food Safety Authority
DEB	disordered eating behaviours	EGF	epidermal growth factor
DEFRA	Department of the Environment, Food, and Rural Affairs (UK)	EGRAC	erythrocyte glutathione reductase activation coefficient

eLENA	eLibrary of Evidence for Nutrition Actions—WHO	FXr	farnesyl X receptor
ELISA	enzyme-linked immunosorbent assay	GAD	glutamic acid decarboxylase
EMA	European Medicines Agency	GAIN	Global Alliance for Improved Nutrition
ENS	enteric nervous system	GALT	gut-associated lymphatic tissue
EoE	eosinophilic oesophagitis	GAP	good agricultural practices
EPA	eicosapentaenoic acid, Environmental Protection Agency (USA)	GATT	General Agreement on Tariffs and Trade
ER	endoplasmic reticulum, oestrogen receptor	GAVI	Global Vaccine Alliance
		GDP	guanosine diphosphate
ERP	event-related potential	GI	gastrointestinal, glycaemic index
ESPEN	European Society for Parenteral and Enteral Nutrition	GIFT	Global Individual Food Consumption Data Tool
EU	European Union	GINA	Global Database on the Implementation of Nutrition Action
FA	fatty acid		
FABP	fatty acid binding protein	GIP	gastric inhibitory peptide
FACS	fluorescence activated cell sorter	GLP	Good Laboratory Practice
FAD	flavin adenine dinucleotide, fatty acid desaturase	GLP-1	glucagon-like peptide 1
		GM	genetically modified
FAEE	fatty acid ethyl ester	GMB	gut microbiome
FAO	Food and Agriculture Organization of the United Nations	GMOs	genetically modified organisms
		GNR	global nutrition report
FAS	foetal alcohol syndrome	GPx	glutathione peroxidase
FAT	fatty acid translocase	GRAS	generally regarded as safe
FATP	fatty acid transport protein	GWAS	genome-wide association studies
FBS	food balance sheet	GTF	glucose tolerance factor
FBT	family-based therapy	GTP	guanosine triphosphate
FDA	US Food and Drug Administration	HAART	highly active antiretroviral therapy
FDB	familial defective apolipoprotein	H+	hydrogen ions
FDG-PET	fluorodeoxyglucose positron emission tomography	HACCP	Hazard Analysis Critical Control Point
Fe	iron	HATs	histone acetyl transferases
FES	Family Expenditure Survey	HAZ	height-for-age score
FFA	free fatty acid	Hb	haemoglobin
FFM	fat-free mass	HBV	hepatitis B virus
FFQ	food frequency questionnaire	HCG	human chorionic gonadotrophin
FGR	foetal growth restriction	HDACs	histone deacetylases
FH	familial hypercholesterolaemia	HDL	high density lipoproteins
FIES	food insecurity experience scale	HEP	high energy phosphate
FIGLU	formiminoglutamic acid	HFA	height for age
FM	fat mass	HFCS	high-fructose corn syrup
FMN	flavin mononucleotide	HFE	haemochromatosis
FODMAP	fermentable oligosaccharides, disaccharides, monosaccharides, and polyols	HFSS	high in fat, salt, and sugar
		HIC	high-income countries
		HIV	human immunodeficiency virus
FOSHU	foods for specified health use	HL	hepatic lipase
FPG	fasting plasma glucose	HLA	human leucocyte antigen
FPN	ferroportin	HMBs	human milk banks
FSA	Food Standards Agency	HMOs	human milk oligosaccharides
FSIS	Food Safety and Inspection Service (USA)	HMTs	histone methyltransferases
		HPLC	high-performance or high-pressure liquid chromatography
FTO	fat mass and obesity-associated		
FTT	failure to thrive	HPV	human papillomavirus

HSL	hormone-sensitive lipase	LCF	living costs and food
IAA	indispensable amino acid	LCPUFA	long-chain polyunsaturated fatty acid
IAP	intracisternal A particle	LCTG	long-chain triglyceride
IAPP	islet amyloid polypeptide	LDL	low density lipoproteins
IARC	International Agency for Research on Cancer	LIC	low-income countries
IBD	inflammatory bowel disease	LIP	labile intracellular pool
IBS	irritable bowel syndrome	LLNA	large neutral amino acid
ICC	interstitial cells of Cajal	LMICs	low and middle income countries
ICF	intracellular fluid	LMP	last menstrual period
ID	iron deficiency	LNAAs	large neutral amino acids
IDA	iron-deficiency anaemia	LNS	lipid-based nutritional supplement
IDD	iodine deficiency disorder	LOAEL	lowest observed adverse effect level
IDDM	insulin-dependent diabetes mellitus	LOFFLEX	low-fat fibre-limited exclusion diet
IDL	intermediate density lipoproteins	LOX	lipoxygenase
IF	intestinal failure	LPL	lipoprotein lipase
IFG	impaired fasting glucose	LPS	lipopolysaccharide
IFN-γ	interferon-gamma	LRNI	lower reference nutrient intake
IFPRI	International Food Policy Research Institute	LTI	lower threshold intake
Ig	immunoglobulin	MAC	mid-arm circumference
IGF-1	insulin-like growth factor 1	MAFF	Ministry of Agriculture, Fisheries, and Food (UK)
IGT	impaired glucose tolerance	MAM	moderate acute malnutrition
IHD	ischaemic heart disease	MAST	Michigan Alcohol Screening Tool
IHS	Integrated Household Survey	MANTRA	Maudsley model of treatment for adults with anorexia nervosa
IL	interleukin	MARSIPAN	management of really sick patients with anorexia nervosa
ILSI	International Life Sciences Institute	MCT	medium-chain triglycerides
IMP	inosine monophosphate	MDG	Millennium Development Goal
IMTG	intramyocellular triacylglycerol	ME	metastable epiallele
INDDEX	International Dietary Data Expansion Project	MED	minimal erythemal dose
INFOODS	International Network of Food Data Systems	MEOS	microsomal ethanol oxidizing system
IOC	International Olympic Committee	MET	metabolic equivalent of a task
IOM	Institute of Medicine	Mg	magnesium
IOTF	International Obesity Task Force	MHC	major histocompatibility complex
IP$_3$	inositol-1,4,5-triphosphate	MIND	Mediterranean–DASH intervention for neurodegenerative delay
IPM	integrated pest management	MIT	mono-iodotyrosine
IPT	interpersonal psychotherapy	MJ	megajoule
IU	international unit	MLN	mesenteric lymph node
IUGR	intra-uterine growth retardation	MMA	methylmalonic acid
JECFA	Joint Expert Committees on Food Additives	MMC	migrating motor complex
JEMRA	Joint Expert Meeting on Microbial Risk Assessment	MNA	Mini Nutritional Assessment
JMPR	Joint Meetings on Pesticide Residues	MODY	maturity-onset diabetes of the young
K	potassium	MOE	margin of exposure
KEQAS	vitamin K external quality assurance scheme	MPR	minimum protein requirement
LA	linoleic acid	MRI	magnetic resonance imaging
LBM	lean body mass	MRLs	maximum residue levels
LBW	low birthweight	mRNA	messenger RNA
LCAT	lecithin-cholesterol acyltransferase	miRNA	microRNA
LCD	low calorie diet	MRS	magnetic resonance spectroscopy
		MSI	minimum safe intake
		MTCT	mother-to-child transmission
		MTHFR	methylenetetrahydrofolate reductase

MUAC	mid-upper arm circumference
MUFA	mono-unsaturated fatty acid
MUST	Malnutrition Universal Screening Tool
MZ	monozygotic
N4G	nutrition for growth
Nac	nucleus accumbens
NAD	nicotinamide adenine dinucleotide
NAADP	nicotinic acid adenine dinucleotide phosphate
NADP	nicotinamide adenine dinucleotide phosphate
NAFLD	non-alcoholic fatty liver disease
NaHCO$_3$	sodium bicarbonate
NAMs	new approach methodologies
NAS	National Academy of Sciences
NCD	non-communicable disease
NCEP	National Cholesterol Education Program (USA)
NCGS	non-coeliac gluten sensitivity
NCMP	national child measurement programme (UK)
ncRNA	non-coding RNA
NDNS	National Diet and Nutrition Survey
NEAT	non-exercise activity thermogenesis
NEC	necrotizing enterocolitis
NEFA	non-esterified fatty acids
NFS	National Food Survey
NHANES	National Health and Nutrition Examination Survey
NICE	National Institute for Health and Care Excellence
NIH	National Institute of Health (USA)
NK	natural killer
NMN	N-methyl nicotinamide
NMR	nuclear magnetic resonance
NPY	neurotransmitters neuropeptide Y
NPYR	NPY receptor
NOAEL	no adverse effect level
NR-NCD	nutrition-related non-communicable disease
NSP	non-starch polysaccharide
NSAID	non-steroidal anti-inflammatory drug
NTD	neural tube defect
ODA	official development assistance
ODS	Office of Dietary Supplements
Ofcom	UK Office of Communications
OGTT	oral glucose tolerance test
ONL	obligatory nitrogen loss
ONS	oral nutritional supplements, Office of National Statistics
OR	odds ratio
OSFED	other specified feeding and eating disorders

OT	oral tolerance
PA	phase angle
PABA	p-amino benzoic acid
Pi	inorganic phosphate
PAH	polycyclic aromatic hydrocarbon
PAL	physical activity level
PAT	Paddington Alcohol Test
PBM	peak bone mass
PBB	polybrominated biphenyl
PBPK	physiologically based pharmacokinetic
PCA	principal components analysis
PCB	polychlorinated biphenyl
PCFT	proton-coupled folate transporter
PCI	protein C inhibitor
PCOS	polycystic ovary syndrome
PDCAAS	protein digestibility corrected amino acid score
PEPCK	phosphoenol-pyruvate carboxykinase
PEM	protein–energy malnutrition
PEP	phosphoenolpyruvate
PER	protein efficiency ratio
PEU	protein–energy undernutrition
PFAS	perfluoroalkyl substances
PFK	phosphofructokinase
PGA	phenylacetylglutamine
PGC	primordial germ cell
PHE	Public Health England
PHV	peak height velocity
PIVKA	protein induced by vitamin K absence or antagonism
PLP	pyridoxal 5'-phosphate
PLRP	pancreatic lipase-related protein
PMP	pyridoxamine 5'-phosphate
PODs	points of departure
POMC	pro-opiomelanocortin
POU	prevalence of undernourishment
PPAR	peroxisome proliferation activated receptor
PPIs	proton-pump inhibitors
PPU	postprandial protein utilization
PRI	population reference intake (of nutrients)
PRSL	potential renal solute load
P/S ratio	polyunsaturated/saturated fatty acid ratio
PSD	psychosocial deprivation
PSP	plant source protein
PT	preterm
PTH	parathyroid hormone
PTL	pancreatic triacylglycerol lipase
PUFA	polyunsaturated fatty acids
PVAT	perivascular adipose tissue
PVC	polyvinyl chloride

PWS	Prader–Willi syndrome		SFA	saturated fatty acid
PYY	peptide YY		SGA	small for gestational age, subjective global assessment
QPM	quality protein maize			
RAE	retinoid activity equivalent		SGLT	sodium glucose-linked transporter
RAR	retinoic acid receptor		SIBO	small intestinal bacterial overgrowth
RAST	radio-allergosorbent test (food allergy)		SMR	standardized mortality ratio
RBC	red blood cell		SNP	single nucleotide polymorphism
RBP	retinol binding protein		SO	sarcopenic obesity
RCPCH	Royal College of Paediatrics and Child Health		SOD	superoxide dismutase
			SPA	spontaneous physical activity
RCT	randomized controlled trial		SR	sarcoplasmic reticulum (complex)
RDA	recommended daily allowance		T_3	tri-iodothyronine
RDI	recommended daily intake		T_4	tetra-iodothyronine
rDNA	recombinant DNA		TAG	triacylglycerol
RDR	relative dose response		TBK	total body potassium
RE	retinol equivalent		TBW	total body water
REACH	Resource for Advancing Children's Health		TCA cycle	tricarboxylic acid cycle
			TDI	tolerable daily intake
REE	resting energy expenditure		TDP	thiamin diphosphate
RfD	reference dose		TEE	total energy expenditure
RISC	RNA-induced silencing complex		TET	ten-eleven translocation methylcytosine dioxygenases
RMR	resting metabolic rate			
RNA	ribonucleic acid		Tf	transferrin
RNI	reference nutrient intake (UK)		TFFP	trefoil factor family
ROS	reactive oxygen species		tHcy	total homocysteine
RQ	respiratory quotient		TIBC	total iron-binding capacity
RR	relative risk, risk ratio		TMAO	trimethylamine oxide
RS	resistant starch		TMP	thymidine monophosphate
RTF	ready to feed (infant formula)		TNF	tumour necrosis factor
RUTF	ready to use therapeutic foods		TNF-α	tumour necrosis factor alpha
RXR	retinoic X receptor		TLR	toll-like receptors
SAA	sulfur amino acids		TPN	total parenteral nutrition
SAC	S-allylcysteine		TRL	triacylglycerol-rich lipoprotein
SACN	Scientific Advisory Committee on Nutrition		tRNA	transfer RNA
			TSF	triceps skinfold thickness
SADQ	Severity of Alcohol Dependence Questionnaire		TSH	thyroid-stimulating hormone
			TUL	tolerable upper limit
SAM	S-adenosylmethionine, severe acute malnutrition		TWI	tolerable weekly intake
			UC	ulcerative colitis
SAMC	S-allylmercaptocysteine		UCP-1	uncoupling protein 1
SARs	structure–activity relationships		UDP	uridine diphosphate
SB	small bowel		UF	uncertainty factor
SCF	Scientific Committee on Food (EU)		UFED	unspecified feeding and eating disorders
SCFA	short-chain fatty acid			
SDGs	sustainable development goals		UIC	urinary iodine concentration
SCID	severe combined primary immunodeficiencies		UIE	urinary iodine excretion
			UL	upper tolerable intake level
SD	standard deviation		UNHCR	United Nations High Commission for Refugees
SEDU	specialist eating disorder unit			
SENECA	Survey in Europe on Nutrition and the Elderly		UNICEF	United Nations Children's Fund
			UNSCN	United Nations Standing Committee on Nutrition (also SCN)
SEPE	social-economic-political-emotional			

UNU	United Nations University	VO_2max	maximal oxygen uptake
UPF	ultra-processed food	WBC	white blood cell
USDA	United States Department of Agriculture	WBPT	whole-body protein turnover
		WCRF	World Cancer Research Fund
USRDA	United States Recommended Daily Allowance	WFA	weight for age
		WFH	weight for height
UTP	uridine triphosphate	WFP	World Food Programme
UTR	untranslated region	WFS	World Food Summit
VAD	vitamin A deficiency	WHI	Women's Health Initiative
VAS	visual analogue scale, vitamin A supplementation	WHO	World Health Organization
		WHR	waist-to-hip ratio
vCJD	new variant Creutzfeldt–Jakob disease	WHZ	weight-for-height
		WIC	Women, Infants and Children—Food and Nutrition Service, USA
VDR	vitamin D receptor	WIN	Weight Control Information Network
VIMNS	vitamin and mineral nutrition information system	WKS	Wernicke–Korsakoff syndrome
		WOF	World Obesity Federation (formerly IOTF)
VKDB	vitamin K deficiency bleeding	WTO	World Trade Organization
VLBW	very low birthweight	Zn	zinc
VLCD	very low calorie diet		
VLDL	very low density lipoproteins		
VKDP	vitamin K dependent protein		

Contributors

Professor Arne Astrup University of Copenhagen, Denmark

Professor David A. Bender University College, London, UK

Professor Barry Bogin Loughborough University, UK

Dr Lutgarda Bozzetto University of Naples Federico II, Naples, Italy

Dr Kathryn E. Bradbury University of Auckland, NZ

Professor John M. Brameld University of Nottingham, UK

Professor Eric Brunner University College London, UK

Professor Philip C. Calder University of Southampton, UK

Professor Brunella Capaldo University of Naples Federico II, Naples, Italy

Professor Parul Christian Johns Hopkins University, Baltimore, MD, USA

Dr Marc J. Cohen Oxfam America, Washington, DC, USA

Dr Meera Cush Ramboll UK Limited

Dr Abdul G. Dulloo University of Fribourg, Switzerland

Dr Louise Dunford Warwick Medical School, University of Warwick, UK

Dr Ruan Elliott University of Surrey, UK

Dr Charlotte Evans University of Leeds, UK

Professor Catherine Geissler King's College London, UK

Professor Godfrey S. Getz University of Chicago, USA

Professor George Grimble University College London, UK

Dr Bridget Holmes Food and Nutrition Division, FAO, Rome

Professor J. O. Hunter Addenbrooke's Hospital, Cambridge, UK

Dr Yannan Jin Leicester School of Allied Health Sciences, De Montfort University, UK

Professor Timothy J. Key University of Oxford, UK

Professor Mairead E. Kiely University College Cork, Ireland

Professor Anna Lartey University of Ghana, Legon

Dr Faidon Magkos University of Copenhagen, Denmark

Dr Laura Medialdea Marcos Universidad Carlos III, Madrid, Spain. Action Against Hunger, Madrid, Spain

Professor John C. Mathers Newcastle University, UK

Professor Helene McNulty Nutrition Innovation Centre for Food and Health (NICHE), Ulster University, Coleraine, Northern Ireland

Professor Joe D. Millward University of Surrey, Guildford, UK

Dr Victoria Hall Moran University of Central Lancashire, Preston, UK

Dr Annhild Mosdøl The Gjensidige Foundation, Oslo, Norway

Professor Paula Moynihan University of Adelaide, Australia

Dr Bruno Palazzo Nazar Federal University of Rio de Janeiro, Brazil. Maudsley Hospital NHS. London

Ms Claire V. Oldale Gloucestershire Hospitals NHS Foundation Trust, Cheltenham, UK

Dr Saskia Osendarp Micronutrient Forum, Washington DC

Professor Tim Parr University of Nottingham, UK

Dr Vinood B. Patel School of Life Sciences, University of Westminster, London, UK

Professor Kristina Pentieva Nutrition Innovation Centre for Food and Health (NICHE), Ulster University, UK

Associate Professor Aurora Perez-Cornago University of Oxford, UK

Dr Gerda Pot King's College London UK, Wageningen University The Netherlands

Professor Hilary J. Powers University of Sheffield, UK

Professor Victor R. Preedy King's College London, UK

Dr Catherine A. Reardon University of Chicago, USA

Professor Ian R. Reid University of Auckland, New Zealand

Professor Angela A. Rivellese University of Naples Federico II, Naples, Italy

Dr Joseph V. Rodricks Ramboll Environment and Health, Arlington, Virginia, USA

Professor Andy Salter School of Biosciences, University of Nottingham, UK

Professor Thomas A. B. Sanders King's College London, UK

Professor Yves Schutz University of Lausanne and Fribourg, Switzerland

Dr Domenico Sergi University of Ferrara, Italy

Professor Paul Sharp King's College London, UK

Dr Mario Siervo School of Population Health, Curtin University, Western Australia

Professor Stephan Strobel University College London, UK

Professor Janet Treasure Kings College London, UK

Dr Jorn Trommelen Maastricht University, The Netherlands

Dr Luc J. C. van Loon Maastricht University, The Netherlands

Dr Carina Venter University of Colorado, Colorado, USA

Dr Janette Walton Munster Technological University, Ireland

Dr Elizabeth A. Williams University of Sheffield, UK

Dr Lynda M. Williams Rowett Institute of Nutrition and Health, Aberdeen, UK

Professor Parveen Yaqoob University of Reading, UK

PART
1

Food and nutrients

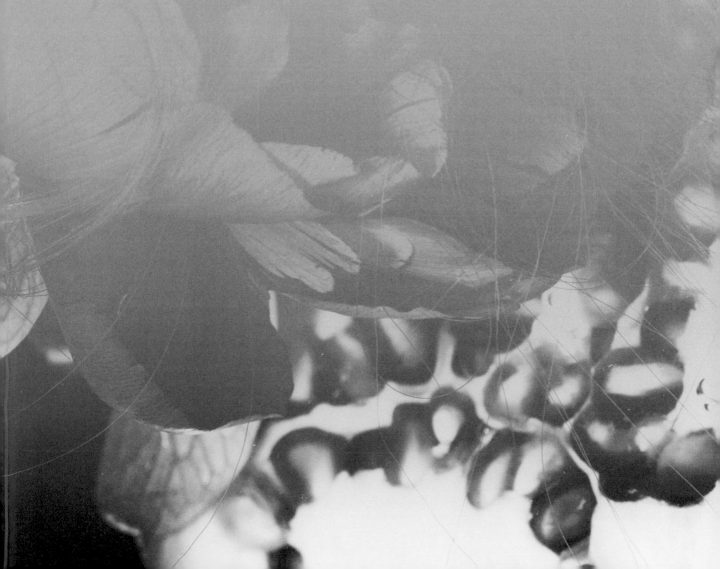

1 Food and nutrient patterns

Janette Walton and Gerda Pot

OBJECTIVES

By the end of this chapter you should be able to:

- identify the main food groups and their contribution to nutrient intakes in Western diets
- understand the social, psychological, geographic, and economic factors determining food choices and dietary patterns
- appreciate the similarities and variability in dietary patterns in different population groups and countries
- be aware of changing trends in dietary patterns over time.

1.1 Introduction

This chapter examines food and nutrient patterns in the context of the major foods and food products in the Western diet and their nutritional importance. Taking the UK as an example of a country with a typical Western diet, it considers the main food groups and identifies the main sources of nutrients. It also outlines the main contributors to the nutrient and non-nutrient content of the UK diet. Another important aspect of this chapter is the exploration of the variations in dietary patterns including the causes of variation such as availability (geographic, trade, demand) as well as economics, food beliefs, and cultural differences. Examples of variations in dietary patterns in population subgroups such as vegetarians, and those defined by religion and region (national and international) are also outlined. This should enable the reader to appreciate the similarities and variability in different population groups and to clarify the social, psychological, and geographical factors influencing food intake patterns. The variability in the consumption of foods, nutrients, and non-nutrients in relation to time (secular trends) and place (geographical differences between developing and developed countries as well as between different developed countries), is described. This highlights the ability of different diets (foods consumed) to provide optimal nutrient intake. The wide diversity in the quantities of foods consumed between different countries and the changes with nutritional transition will serve to illustrate this.

In summary, in this chapter the major food groups in the diet are examined according to their nutrient and non-nutrient contribution, their variability between countries and between subgroups in the population, as well as changing trends in food intake patterns, including novel foods, over time.

1.2 Major food groups in Western diets

Western diets are composed of several food groups collectively providing the nutritional needs of the body. The particular food groups in Western diets that provide all the nutrients and non-nutrients for optimum health include: cereals and cereal products (e.g., bread and breakfast cereals); fruits and vegetables; roots and tubers; milk and other dairy products, meats, fish, eggs and other sources of protein, and fats and oils. The food intake patterns for UK adults are illustrated here as an example of a Western diet (Public Health England 2020) (Table 1.1). How these patterns vary between countries (i.e., geographically) is discussed in Section 1.3, while changes over time are discussed in Section 1.4. The non-nutrients discussed in this chapter are those believed to have a potentially beneficial effect on human health. They include both dietary fibre and phytoprotectants such as flavonoids and phytoestrogens. Other non-nutrients in foods such as contaminants, allergens, and food additives do not have specific nutritional benefits and are not discussed here (see Chapters 2 and 3).

TABLE 1.1 Food groups making important contributions to macronutrient intakes in UK adults (19–64 years)

	%
Carbohydrates	
Cereals & cereal products (pasta, rice, pizza, breads, rolls, and breakfast cereals)	37
Potatoes, potato products, and savoury snacks	9
Fruit (including juices) and nuts	7
Biscuits, cakes, pastries, and puddings	9
Drinks including soft drinks, alcoholic drinks, tea, coffee, and water	6
Sugar, confectionery, and preserves	6
Protein	
Meat and meat products	34
Cereals & cereal products (pasta, rice, pizza, breads, rolls, and breakfast cereals)	24
Milk and milk products	13
Fish and fish dishes	7
Vegetables	6
Potatoes, potato products, and savoury snacks	3
Total fat	
Meat and meat products	22
Cereals & cereal products (pasta, rice, pizza, breads, rolls, and breakfast cereals)	21
Milk and milk products	13
Butter, fat spreads, and oils	9
Potatoes, potato products, and savoury snacks	8
Vegetables (excluding potatoes)	5
Saturated fat	
Meat and meat products	21
Milk and milk products	21
Cereals & cereal products (pasta, rice, pizza, breads, rolls, and breakfast cereals)	21
Butter, fat spreads, and oils	10
Eggs and egg dishes	5
Potatoes, potato products, and savoury snacks	4

Source: Data from NDNS from years 9–11 (combined) of the Rolling Programme (2016/17–2018/19). Public Health England 2020 for carbohydrates, protein, total fat, and saturated fat.

Nutritional importance of food groups in the Western diet

The main food sources of macronutrients, vitamins, and minerals in UK adults are outlined in Tables 1.1–1.3 (see Weblink Section 1.1).

Cereals and cereal products

Cereals are the staple foods for almost all populations in both the developing and developed world. They are all seeds from domesticated members of the grass family and represent the most important plant foods in the human diet for their contribution to energy and carbohydrate intake and many micronutrients. Their contribution to energy intake varies markedly between developing and developed countries. In developing countries such as those in Africa and parts of Asia, cereals may contribute as much as 70 per cent of energy intakes. By contrast, in developed countries such as the UK, cereals and cereal

products provide approximately 30 per cent of the energy intake. In the UK, cereals and cereal products also provide about 20 per cent of fat, 25 per cent of protein and almost 50 per cent of available carbohydrates and also make a significant contribution (40 per cent) to intakes of dietary fibre. The major cereals in the human diet are wheat and rice followed by maize, barley, oats, and rye.

Fruits and Vegetables

These include a wide range of plant families and consist of any edible portion of the plant including roots, leaves, stems, buds, flowers, and fruits. The leaves, stems, buds, flowers, and some fruits (tomatoes, cucumbers, marrows, and pumpkins) are commonly classified as vegetables while the fruits that are sweet are classified as fruits.

Fruits and Vegetables are primarily seen as sources of vitamins and are important contributors to the intake of dietary fibre. Fruits and Vegetables contribute approximately one-third of total fibre intake, being particularly

TABLE 1.2 Food groups making important contributions to vitamin intakes by UK adults (19–64 years)

Vitamins	Main food sources
Thiamin	Cereals & cereal products, meat and meat products, vegetables (especially potatoes)
Riboflavin	Milk and milk products, cereal & cereal products (especially breakfast cereals), meat and meat products
Niacin	Meat and meat products, cereals & cereal products, vegetables (including potatoes)
Vitamin B_6	Meat and meat products, cereals & cereal products, vegetables & potatoes, milk & milk products, beer
Vitamin B_{12}	Milk and milk products, meat & meat products, fish & fish dishes, eggs and egg dishes
Folate	Cereals and cereal products, vegetables & potatoes (mainly vegetables), meat & meat products, milk and milk products, beverages
Vitamin C	Vegetables (including potatoes), fruits (including juices), beverages
Vitamin A (including pre-formed retinol and carotenoids)	Vegetables & vegetable dishes, milk & milk products, meat & meat dishes, fat spreads
Vitamin D	Meat & meat products, eggs, oily fish, cereal products, fat spreads, milk & milk products
Vitamin E	Vegetables and potatoes, cereals & cereal products, meat & meat products, fat spreads

Source: Data from NDNS from years 9–11 (combined) of the Rolling Programme (2016/17–2018/19). Public and from NDNS from years 1, 2, 3, and 4 (combined) of the Rolling Programme (2008/2009–2011/12). Public Health England and Food Standards Agency, 2014 for thiamine, niacin, vitamin B6, vitamin B12, vitamin C, and vitamin E.

rich in soluble fibre components that have been shown to be most effective in lowering serum cholesterol levels and slowing glucose absorption. Fruits and Vegetables tend to be high in potassium and low in sodium. Although quite low in B vitamins (with the exception of folates, for which leafy vegetables are rich sources), they are the most important source of vitamin C. Fruits and vegetables are also important sources of non-nutrients such as the phytoprotectants, carotenoids, and anthocyanins (flavonoids in berries). Fruits and vegetables are versatile in the diet and may be consumed either raw or cooked and can be processed and preserved in many ways such as frozen, dried, bottled, pickled, canned, or conserved with sugar to make jams.

Some of the most commonly consumed Fruits and Vegetables come from the following families: *Brassica* (cabbage, cauliflower, Brussels sprouts, kales); *Allium* (onions, leeks, garlic), *Compositae* (lettuce, endives, chicory), legumes (peas, beans, lentils) and other leafy vegetables including spinach. The legumes consumed most commonly in the UK are peas, green beans, and 'baked beans'.

TABLE 1.3 Food groups making important contributions to mineral intakes by UK adults (19–64 years)

Minerals	Main food sources
Calcium	Milk and milk products, cereals and cereal products, dark green vegetables
Magnesium	Cereals and cereal products, vegetables and potatoes, meat and meat products, beverages, milk and milk products
Potassium	Vegetables and potatoes, meat and meat products, cereals and cereal products, milk and milk products
Sodium	Meat and meat products, cereals and cereal products, milk and milk products
Iron	Cereals and cereal products, meat and meat products, vegetables
Copper	Cereals and cereal products, meat and meat products, vegetables and potatoes
Zinc	Meat and meat products, cereals and cereal products, milk and milk products, vegetables and potatoes
Iodine	Milk and milk products, cereals and cereal products, beverages (including beer), fish, meat and meat products

Source: Data from NDNS from years 9–11 (combined) of the Rolling Programme (2016/17–2018/19) Public Health England 2020 for calcium, magnesium, potassium, sodium, iron, and zinc and from NDNS from years 1, 2, 3, and 4 (combined) of the Rolling Programme (2008/2009–2011/12) Public Health England and Food Standards Agency, 2014 for copper.

Roots and tubers

Roots and tubers are the underground organs of many plants and could be included in the vegetable and fruit group. However, due to differences in their nutrient composition and the important role they play in the Western diet (especially the potato) they are discussed separately.

Root crops: Most roots have a high water content and provide relatively small amounts of energy in the form of carbohydrate, both sugars and starch. They are typically low in protein and fat. Examples of root crops are carrots, beets, and turnips.

Tubers: Tubers are not true roots but rather underground stems that store large quantities of carbohydrate, usually starch. The potato is a stem tuber, native to the Andes and was brought into Europe in the seventeenth century. The nutrient composition of the potato varies according to variety but all contain large amounts of starch. Potatoes also contain significant amounts of vitamin C and their high levels of consumption make them an important source of this vitamin contributing up to 16 per cent of total vitamin C intake in the UK diet.

Meat and meat products

Meat has been a central component of the human diet for a large part of our history and still is the centrepiece of most meals in developed countries. In many developing countries, non-animal-based sources of protein such as legumes are still dominant. In the USA and the UK, the most important meat sources are pigs, sheep, and cattle. In other regions such as India, the Middle East, and Africa, goat and camel are the main meats consumed. Other meat sources include wild animals such as rabbits, deer for venison, and poultry (chicken, ducks, turkey, and geese). In the UK and a number of other countries including China, poultry (chicken) has now become the most popular meat source. Apart from the muscle, other parts of the animal collectively described as offal are also consumed (but to a much lesser extent). The liver, kidneys, brain, and pancreas (sweetbreads) are the most commonly consumed organs. Meat products such as sausages and pork pies count for almost half of total meat consumption in the UK. Conventionally viewed as protein foods, meats as a whole are a major source of protein of high biological value. In developed countries such as the UK, where meat consumption is relatively high, it provides the main source of protein. Meat is also an important source of fat in the diet.

Fish

While fish catches worldwide are on the increase according to the Food and Agriculture Organization (FAO), fish stocks are being depleted due to over-fishing. Fish can be categorized into two main categories: vertebrates (finfish) which include white fish and oily fish, and seafood invertebrates (shellfish) which includes crustaceans and molluscs. Fish are an important source of good quality protein and are low in fat, except for oily fish which provide a very good source of long-chain polyunsaturated fatty acids (PUFA). Fish are also a major source of iodine which has been accumulated from their environment. Also, they may be an important source of calcium (in fish with fine bones). The only vitamin provided in significant quantity is vitamin D where fish provide up to 14 per cent of intake (see Chapter 12). Fish, at such low consumption levels (<30 g/day) is not a major contributor to energy intakes, providing on average just 1 per cent of energy and 4.7 per cent of protein and 1.3 per cent of fat. Compared to many European countries such as Portugal and Spain, consumption of fish in the UK is low at 22g per head per day for adults. Furthermore, recent figures from the NDNS data for 2016 to 2019 found that the average consumption of oily fish is well below the recommended one portion (140g) per week in all age groups.

Eggs

The most widely consumed eggs in the UK diet are hens' eggs. The proteins found in eggs contain the amino acids essential for the chick embryos' complete development. Because of this, eggs are considered the 'reference protein' for biological evaluation of other protein sources in terms of their amino acid patterns. Lipids found in eggs are rich in phospholipids and cholesterol. The fatty acid profile of eggs shows a high proportion of polyunsaturated fatty acids relative to saturated fatty acids (high P:S ratio). A fall in the level of consumption of eggs was observed in Europe and the USA in the late twentieth century primarily due to guidance at that time to limit foods high in cholesterol. With updated dietary guidance now shifted away from restriction of dietary cholesterol, per capita egg consumption is on the rise again in the last 20 years with a 15 per cent increase reported in the USA in the period from 2000 to 2020.

Milk and other dairy products

Cows provide the bulk of all milk consumed in the UK, with goat and sheep making only a minor contribution to overall milk consumption. Milk is an excellent source of many nutrients. The major protein in milk is casein, comprising up to 80 per cent of the protein in cow's milk. Other proteins include lactalbumin and immunoglobulins that are responsible for the transfer of maternal immunity of the young animal for a short period following birth. Milk from ruminant animals, such as the cow, contains a large proportion of short-chain fatty acids produced from the fermentation of carbohydrates in the rumen. Milk and its products are excellent sources of many inorganic nutrients especially calcium and certain vitamins, both fat-soluble and water-soluble. Other dairy products include:

- cheese, which contains many of milk's nutrients in concentrated form
- yoghurt, which is produced from the culturing of a mixture of milk and cream products with the lactic acid-producing bacteria *Lactobacillus bulgaricus* and *Streptococcus thermophilus* and other bacterial cultures (e.g., *Lactobacillus acidophilus, Bifidobacteria*).

The contribution of dairy products to the UK diet, and to that of many other northern European countries, is very important. Dairy products are an important source of calcium and riboflavin especially in children and adolescents where they contribute approximately 35 per cent of total riboflavin intake and 40 per cent of calcium intake in the diets of UK children and adolescents. The recommendation to decrease intakes of full-fat dairy foods should be considered as an attempt to reduce intake or move to reduced-fat varieties and not an attempt to restrict consumption of these foods altogether.

A growing number of consumers are opting for plant-based milk and dairy substitutes, either for medical reasons or as a lifestyle choice, and as a result, availability of these plant-based alternatives is growing. Consumers should be aware that these products are not always fortified with equivalent levels of micronutrients (such as calcium and riboflavin) that would be naturally found in milk and dairy products, and hence may be of lesser nutritional quality.

Fats and oils

These are distinguished by their physical characteristics, with fats being solid at room temperature (due to a high relative concentration of saturated fatty acids) while oils are liquid and usually of plant origin, either from the flesh of the fruit (olive oil) or from the seed (sunflower and linseed). Oils typically have a higher concentration of unsaturated fatty acids. Lipids that are isolated from animal products tend to be solid fats (e.g., butter, lard, and suet). In recent years, there has been an increased consumption of margarine made from highly unsaturated fats (such as sunflower) due to the beneficial effects on serum cholesterol. In many countries, including the UK, margarines are required by legislation to be fortified with vitamins A and D so that they are nutritionally equivalent to butter. Other fat spreads are often fortified also with vitamins A and D.

Food sources of nutrients and health-related non-nutrients

The importance of specific foods as sources of macro- and micronutrients for particular populations depends not only on the level of the nutrient in the food or food product (and its availability to the body) but also on the extent to which the food is consumed (the quantity and the proportion of the population that consumes that food). The way the food is consumed (raw, cooked, or processed) will also influence its nutritional importance. Thus, it is critical to distinguish between the nutritional importance of various foods generally and 'the main contributors' from food sources to nutrient intakes of a particular population. Such information is obtained from individual dietary surveys including diet histories or food records (see Chapter 30). In the UK, a rolling programme of National Diet and Nutrition Surveys (NDNS) is carried out to report dietary intake among the UK population with the most recent survey being carried out in 2018/19 (Public Health England 2020). The main food sources of carbohydrates in the UK are cereal products and potatoes. Biscuits, cakes, and snacks also make a sizeable contribution to carbohydrate intake. Important contributors to protein and fat intakes are meat, fish, and eggs as well as dairy products, while fruits and vegetables make important contributions to a considerable number of micronutrients (vitamins and minerals), dietary fibre, and other non-nutrients (phytoprotectants).

Dietary fibre

Dietary fibre (total fibre) in the diet is now defined to include dietary fibre (indigestible carbohydrate and lignin which are intrinsic and intact in plants) and functional fibre (isolated non-digestible carbohydrates that have beneficial physiological effects) (see Chapter 7). The main contributors to dietary fibre in the diet of adults in the UK are cereals and cereal products and vegetables (including potatoes).

Phytoprotectants

Epidemiological studies indicate that a diet rich in fruits and vegetables has health benefits particularly related to protection against certain chronic diseases such as cancer. Experimental studies are being conducted to help establish causality and to elicit mechanisms for these observations, and this currently focuses on the antioxidant potential of certain nutrients and non-nutrients. The term 'phyto' originates from a Greek word meaning plant. Also referred to as phytochemicals, phytonutrients, and bioactive compounds, they are derived from foods of plant origin and, while not regarded as nutrients, they may be beneficial to health (e.g., reduced risk of certain cancers and cardiovascular disease, and of age-related blindness). Fruits, vegetables, grains, legumes, nuts, and teas are rich sources of phytoprotectants.

Phytoprotectants are extremely varied in their chemical composition, the plants in which they are found, and their putative beneficial effects. While there are tens of thousands of phytoprotectants in plants that have not yet been tested for health benefits, it is likely that the best way to benefit from their potential health benefits is to increase both consumption and variety of plant foods (particularly variety of colour). Some of the common classes of phytoprotectants include: flavonoids (polyphenols) including isoflavones (phytoestrogens), inositol phosphates (phytates), lignans (phytoestrogens), isothiocyanates, indoles, phenols, and sulphides and thiols.

Polyphenols: Polyphenolic compounds, also known as secondary plant metabolites, are natural components of a wide variety of plants and can be classified as flavonoids or non-flavanoids. Much of the total antioxidant activity of fruits and vegetables is related to their phenolic content. Chlorogenic acid and caffeic acid are also important phenolic compounds in our diet. The interest in these phenolic compounds is increasing because they have antioxidant activity *in vitro*. It has recently been shown that chlorogenic acid and caffeic acid are absorbed in humans, which increases the possibility that they might affect health. Food sources rich in polyphenols include onions, apples, tea, red wine, red grapes, grape juice, strawberries, raspberries, blueberries, cranberries, and certain nuts.

The flavonoids quercetin and catechins are the most extensively studied polyphenols in relation to their absorption and metabolism. Experimental studies in animals suggest that dietary quercetin, at relatively high doses, could inhibit the initiation and development of tumours in humans. These results are supported by *in vitro* studies showing that quercetin inhibits the growth of isolated human tumour cell lines. Furthermore, quercetin, the most commonly consumed flavonoid, is reported to exhibit antioxidant activity. A relationship has been reported between quercetin intake and reduced risk of cancer and coronary heart disease in a Finnish population. Further studies on epidemiology, mechanisms, and interventions are needed before firm conclusions on the health protective effects of quercetin can be drawn.

Contribution of macronutrients to energy needs

Energy is required for the body to function, in particular to maintain basal metabolic rate and also for physical activity and for thermogenesis (see Chapter 6). The energy needs of the body are served by the contribution of the three macronutrients: carbohydrate, fat, and protein. Alcohol is the other energy source. Carbohydrate and fat are the primary fuel sources and for this purpose they can be largely used interchangeably. To a large extent the body can synthesize *de novo* the specific carbohydrates and lipids it needs, with the exception of the requirement for small amounts of carbohydrate and n-6 and n-3 fatty acids. Thus a mixture of macronutrients is required as a source of fuel to meet the energy requirements of the body. Defining the optimal mix of energy sources to optimize health and promote longevity is not easy. There are no clinical trials that compare various macronutrient combinations with longevity in humans.

Patterns of consumption of the macronutrients have changed radically from those of our ancient ancestors, for whom the relative contribution to energy has been estimated as 34 per cent protein, 45 per cent carbohydrate, and 21 per cent fat, contrasting to that of a typical current Western diet of 12 per cent protein, 46 per cent carbohydrate and 42 per cent fat. Differences also exist in the energy contribution of these macronutrients between countries; developing countries have a much higher energy contribution from carbohydrates relative to fat than developed countries, where fat constitutes well over a third of energy intake. The acceptable (i.e., healthy) macronutrient distribution ranges (AMDR) are 10–15 per cent of energy for protein, 25–30 per cent of energy for fat and 45–65 per cent of energy for carbohydrate. The recommendations (dietary reference values (DRVs)) for total fat are 33 per cent of total energy or 35 per cent of food energy. Considerable debate exists on the importance of total fat (quantity versus the qualitative composition of the fat) in relation to cardiovascular disease and certain cancers (see Chapter 8).

1.3 Variations in dietary patterns

Developing versus developed countries

Diets in the developing world are very different from those in the developed countries, although there is evidence of convergence between the two. The diets of the former tend to be characterized by lower total calorie intake, more calories from cereals and roots and tubers, less diversity of food groups consumed, fewer animal source foods (and less fat), less processed food, and less food consumed outside the home. However, in more recent decades food consumption patterns in developing countries have started to resemble those of more developed countries, which is described later in the section on changing dietary patterns. Many of these differences are explained by socioeconomic factors such as

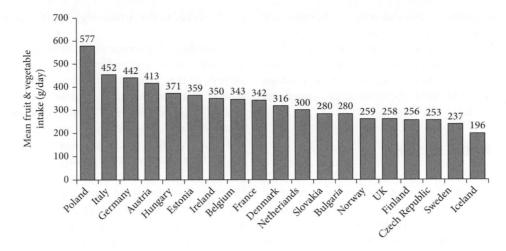

FIGURE 1.1 Fruit and vegetable intake in selected European countries.
Source: Reproduced from European Food Safety Authority (2008). Concise database summary statistics—total population. Available at: https://www.eufic.org/en/healthy-living/article/fruit-and-vegetable-consumption-in-europe-do-europeans-get-enough. The European Food Information Council (EUFIC) www.eufic.org

occupation, education, income, wealth, and residential location, but there is considerable variation in consumption patterns even accounting for such factors (see Weblink Section 1.2).

Europe

Differences in European diets can be examined from two main sources: a questionnaire-based survey carried out by the World Health Organization (WHO) Regional Office for Europe; and food supply data published by the Food and Agriculture Organization of the United Nations (FAO).

Data from the European Food Safety Authority (EFSA) compiling national food consumption data based on dietary surveys showed vegetable consumption tends to be higher in Southern Europe than in Northern Europe (Figure 1.1). Indeed, large variations were found, with up to a twofold variation in mean fruit and vegetable intakes from 196g/d in Iceland to 452 g/day in Italy.

Whereas the intakes of polyunsaturated fatty acids (PUFA) are generally similar for countries in the north and south of Europe, the intakes of monounsaturated (MUFAs) and saturated fatty acids (SFAs) differ markedly. The intakes of SFA as per cent energy are somewhat higher in northern European countries, ranging from 13 per cent to 15 per cent energy compared with 11 per cent to 15 per cent in southern European countries, while intakes of MUFAs as percentage energy are higher in southern European countries (12 to 22 per cent) compared with 12–13 per cent in northern European countries. This difference in MUFA intake may be directly attributed to the higher consumption of olive oil in the southern countries of Europe. For percentage energy

from fats and fatty acid categories in selected European countries see Table 1.4 (Eilander et al. 2015).

One specific diet survey, the SENECA (Survey in Europe on Nutrition and the Elderly, a Concerted Action) project examined cross-cultural variations in intakes of food groups among elderly Europeans in four European countries (Denmark, Netherlands, Switzerland, and Spain). The dietary pattern followed a typical Western diet, with all participants from all four countries consuming milk,

TABLE 1.4 Per cent energy from fatty acid categories in selected EU member countries

Country	SFA (%E)	MUFA (%E)	PUFA (%E)
Northern Europe			
Belgium	15.4	13.7	7.0
Denmark	15.0	12.0	5.2
Finland	13.6	12.6	6.2
Germany	14.4	12.8	6.5
Netherlands	14.4	13.0	6.9
Sweden	13.0	12.8	5.6
Southern Europe			
Greece	13.1	22.3	6.6
Italy	11.2	17.5	4.6
Portugal	14.0	12.4	4.9
Spain	14.8	13	3.9

Source: Eilander et al. (2015) *Eur J Lipid Sci and Technol* **117**, 1370–1377.

grain products, and vegetables and virtually all consuming meat and fruit; however fewer participants were consumers of fish, eggs, and sugar. While variation was found between these food groups, the main variation seen between countries was in the types of foods comprising the food groups and with which meals the foods are consumed (see Weblink Section 1.3).

Variation within the UK

The UK Living Costs and Food Survey (DEFRA 20020) enables comparison of different patterns according to socioeconomic group and region (see Table 1.5a, Table 1.5b, Table 1.6, and Weblink **Tables 1.1 and 1.2.**

There is considerable variation between countries in the UK and more specifically across regions of England with the lowest and highest levels of consumption of particular foods. For example, the highest consumption of fruit is in England and the lowest in Scotland, whereas the highest consumption of meat products and confectionary is in Wales and Scotland and the lowest is in England. The low levels of vegetable and fruit consumption recorded in Scotland is the result of a number of historical and economic factors, and this observation may be linked to the current high levels of cardiovascular disease.

A number of important variations are also evident in relation to income and employment status. Previous surveys of the National Expenditure and Food Survey showed that non-pensioner households with the lowest head of household income (quintile 1 or bottom 20 per cent) per week consumed lower amounts of fresh fruits and fresh vegetables (excluding potatoes), skimmed milk, cheese, fruit juices, breakfast cereals, and alcoholic drinks than those earning the most (quintile 5 or top 20 per cent). However, they consumed more liquid whole milk, eggs, fats and oils, sugar and preserves, fresh potatoes, frozen and canned vegetables, and bread compared to the top income households (Table 1.3) (see Weblink Section 1.4).

Factors determining dietary variations

So far in this chapter we have considered foods and the nutritional contribution of food groups. However, in practice, people do not eat 'food groups' any more than they eat 'nutrients'. They choose foods to eat which usually contain ingredients from a number of the groups, for example pizza, apple pie, beef stew. Food groupings

TABLE 1.5A Food purchase according to income level in the UK, 2015–2017 (grams per person per week, unless otherwise stated)

	Lowest income group (Quintile 1)	Highest income group (Quintile 5)
Milk & cream (ml)	2089	1647
Cheese	110	123
Carcass meat	185	199
Other meat/meat products	826	722
Fish	157	146
Eggs (no.)	2	2
Fats and oils	180	128
Sugar and preserves	144	83
Potatoes	735	581
Total vegetables (excl. potatoes)	1036	1170
Total fruit	1058	1186
Total cereals	1563	1410
Soft drinks (ml)	1546	1529
Alcoholic drinks (ml)	624	758
Confectionery	134	121

Source: Adapted from DEFRA (2019). Gross income quintile group (GIQ) (updated with revised 2016/17 data).

TABLE 1.5B Food purchase according to employment status in the UK, 2016/17–2018/19 (grams per person per week, unless otherwise stated)

	Employed (full time)	Unemployed	Retired*
Milk & cream (ml)	1771	1702	1819
Cheese	124	120	90
Carcass meat	178	126	130
Other meat/meat products	744	706	755
Fish	123	150	127
Eggs (no.)	2	1	2
Fats and oils	125	114	159
Sugar and preserves	91	57	116
Potatoes	599	471	617
Total vegetables (excl. potatoes)	1133	829	988
Total fruit	1102	876	889
Total cereals	1427	1351	1385
Soft drinks (ml)	1455	1347	1502
Alcoholic drinks (ml)	678	752	695
Confectionery	127	94	110

* Min NI pension age.
Source: Adapted from DEFRA (2020) household purchases updated with 2018/19 data.

TABLE 1.6 Changes over time in foods consumed by adults in the UK: A comparison between 1986/7 adult survey (Gregory et al. 1990) and 2000/1 NDNS of adults (Henderson et al. 2002)

More likely to consume in 1986/7	More likely to consume in 2000/1
Wholemeal bread, biscuits, buns, cakes, pastries, & fruit pies	Pasta, rice, & other miscellaneous cereals, white bread
Puddings (including dairy desserts & icecreams)	Breakfast cereals
Milk (in particular whole milk)	Yogurt & fromage frais, semi-skimmed milk
Cheese	Savoury snacks
Eggs & egg dishes	Nuts
Fats & oils (in particular butter)	Salad vegetables
Meat, meat dishes, & meat products	Chicken and turkey
Fish & fish dishes	Oily fish, shellfish
Sugars, preserves & sweet spreads	Ssoft drinks, low-calorie alcoholic drinks (in particular wine)
Tea & water	

Source: J Gregory, K Foster, H Tyler et al. (1990) *The Dietary and Nutritional Survey of British Adults*. London: HMSO; L Henderson, J Gregory, G Swan (2002) *National Diet and Nutrition Survey: Adults Aged 19 to 64 Years. Volume 1: Types and Quantities of Food Consumed*. London: TSO.

are useful tools for nutritionists, especially in relation to public nutrition policy, but they are not entirely satisfactory in picturing the way in which people eat. The way that people think about foods and how they group them together when making choices about what to eat may be very different from the viewpoint of the nutritionist with respect to food groups.

Of all the animal and plant species that are considered safe for human consumption, humans choose from a relatively narrow range of species. Those which may

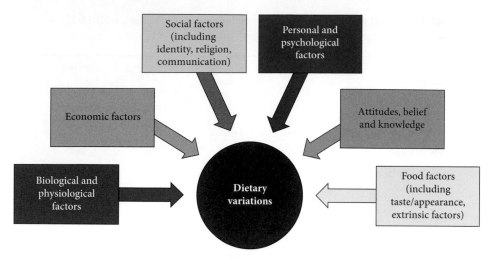

FIGURE 1.2 Factors determining dietary variation.

be considered a delicacy by some groups are rejected as inedible by others. Individuals' food choices at any given time will be influenced not only by what they consider to constitute 'food' and whether it is available (either physically accessible or affordable) but also by what is appropriate according to a variety of sociocultural factors, ideas, beliefs and attitudes, as well as psychological factors and their level of hunger or satiety. The different factors that determine dietary variation are presented in Figure 1.2 and are discussed below.

Biological and physiological factors

Human physiology places few restrictions on food choice. Clearly the quantity of food consumed is limited to some extent by the capacity of the gastrointestinal tract, which results in the necessity to feed regularly, at least once a day and preferably more often. The quality of the diet is also limited, for example, by the fact that humans are monogastric, which renders plants with very high fibre contents impossible to digest.

Although factors such as age, gender, pregnancy, lactation, and activity patterns all affect nutritional requirements, humans are unable to perceive their specific needs and act in response. No convincing evidence has been produced by studies that have attempted to demonstrate 'gustatory sensibility' (the innate ability of an individual to select foods that meet variations in nutritional status with regard to specific nutrients). However, studies with infants fed solely on milk have shown an innate ability to regulate energy intake in early life by adjusting the quantity of milk consumed in response to the level of dilution in order to maintain energy intake. This ability appears to decline when the diet becomes more varied and is easily overridden by other factors influencing the

timing of eating and the amount and types of food consumed in later life. Nonetheless the familiar experience of a decrease in pleasantness in association with increasing consumption of a particular food serves to limit the amount of that food which is eaten. This innate mechanism, called 'sensory specific satiety', serves to ensure that humans eat a varied diet and are therefore more likely to meet their requirements for all nutrients. While this may have served humans well from a survival point of view in earlier times, it is potentially more problematic when surrounded by a huge variety of different, highly palatable, food items.

Humans experience general and non-specific feelings of hunger or satiety as a result of complex physiological processes which are not fully understood but involve an integrated set of feedback mechanisms (see Chapter 4,) Section 4.5). In the past decades there has been growing interest in the possible identification of gene-related determinants of nutritional behaviour. It has been proposed that these might be modulated through sensory sensitivity to specific chemical substances and reflected in taste preferences. However, it appears that while biology determines humans' nutritional needs, it does not substantially affect food choices (see Weblink Section 1.5).

Economic factors

Access to food is a primary determinant of food choice and dietary variation. The availability of food will be influenced by (a) geography, season, and factors such as food preservation and distribution systems which affect physical availability, and (b) the ability of the individual to acquire what is available. These two elements are closely interlinked and the relative importance of either will depend on the situation.

In remote rural areas, with poorly developed food markets, decisions about what to produce have an important effect on what is consumed, both at the village and at household levels. In these cases, decisions about production and consumption cannot be treated as entirely separate. Where markets are stronger, physical access to foods becomes less important and economic constraints assume a greater role. About 20 per cent of the world's 8 billion population participate in the cash economy and about 90 per cent of the world's food consumption occurs where it is produced. In towns and cities people depend almost entirely on purchased foods, while in rural populations people consume around 60 per cent of the food they produce (McMichael 2015).

In situations where food must be purchased, there are two key aspects of the relationship between income and food consumption patterns. The first relates to the overall level of expenditure on food, the second to the types of food consumed. As income levels increase, families spend more on food, although as a proportion of overall expenditures food is likely to decline in importance (Engels' law). This may occur within a fairly narrow range in an industrialized country, such as the UK, where expenditure may range from 15 per cent to 29 per cent of reported expenditure between high income and low-income households. However, in developing countries the differences may be far more dramatic, ranging between 15 and 80 per cent of household expenditure.

One important outcome of this is the increased vulnerability of low-income groups to changes in the price of foodstuffs and other necessities. This may be particularly important to the urban poor in less industrialized countries. If a diet which requires a high level of expenditure is barely adequate nutritionally, then any increase in price which leads to the purchase of reduced quantities of food may have serious nutritional consequences. Income may also be very unreliable. People are likely to be casually employed, or receive small amounts of cash intermittently as a result of different activities, and there is usually only enough money to buy small quantities of food at a time. Low-income families have few opportunities to accumulate savings. As a result it is not possible to take advantage of the economies which are possible through bulk purchase. The urban poor may also be disadvantaged in the range and quality of foodstuffs available to them as well as the price they pay for food.

The cost of utilities—fuel, water, clothing, transport, rent—also affect food choices. The budgets of low-income households leave very little room for manoeuvre and often food is the most flexible element of expenditure. If the price of cooking fuel increases, disposable income goes down and food purchases are affected. This may result in a change in the types of food purchased and a reduction in the amount of food eaten. Some or all of the family members may go without meals. It is commonly a priority to maintain the diet of the 'breadwinner', while women and children are particularly adversely affected by any shortfall.

The second area where income has a marked influence on food intake is in relation to the type of foods eaten. It is an internationally observed phenomenon that as income levels rise, people diversify away from a reliance on cereals and roots/tubers and begin to purchase more animal source foods, fruits, and vegetables. In the UK more affluent households eat more fruits and vegetables, polyunsaturated margarine, low-fat products and carcass meat. In contrast, low-income families eat less fruit and more of the cheaper cuts of meat and meat products such as meat pies and sausages. The use of high-fat foods such as these contributes to the observation that (in a European context), energy-dense diets tend to cost less (Darmon & Drenowski 2008). When for example studying the price per unit energy, it has been shown that grains and sugars are cheaper than fruits and vegetables (Drenowski 2010; Rao et al. 2013).

The prices of one food can affect the consumption of other foods. If two foods are complements (i.e., they are eaten in combination), a price increase in one, will, in the short run, lead to a lower demand for it and a lower demand for its complement. For two foods that are substitutes (e.g., one type of cereal for another), if the price of one increases, people will switch to the substitute, increasing the demand for it. These food prices can be dramatically affected by food policies that subsidize the consumption and production of certain foods, typically not based on nutritional goals (see Chapter 34).

Food prices are not the only prices that affect food consumption and diets. Financial needs and the availability of paid employment may cause men and women to work outside the home. As a result, less food will be prepared at home and more food will be consumed away from home. This has an effect on the types of food that are eaten. Only certain foods are 'portable' and suitable for advance preparation and consumption elsewhere, especially in the absence of refrigeration. Food choices become increasingly influenced by the types of food produced by others and available at times and in places convenient to the work schedule. The tendency to consume food away from home increases in urban areas, associated with a greater concentration of food vendors and restaurants.

The recent COVID-19 pandemic has revealed the particular vulnerabilities of our food system. Due to travel restrictions, foods became less available and many people lost their income leading to even more food insecurity.

Social factors

The human is basically an omnivore who has to learn what to eat, and what is learned is strongly influenced by social factors like culture and religion. This includes learning what is acceptable or not acceptable as food, and appropriate foods for different occasions and different types of people according to age, gender, and social status. Rules about the preparation of food and how it is eaten may also be culturally determined. Food has many important social uses and although the food items associated with these uses may vary from society to society and change over time, these 'non-nutritional' functions persist.

Identity

As part of the socialization process, children grow up learning the conventions of their social, gender, and age group concerning appropriate food choices and the manner in which such foods should be prepared and consumed. In some parts of the world, notably northern India and Pakistan, such conventions may have a marked impact on the diets of women and children, particularly girls, as a result of restrictions on the amount and types of foods which are considered suitable. More generally, the selection of foods that are deemed as inappropriate for a particular situation may have social consequences. The saying *Tell me what you eat, and I will tell you who you are* is attributed to Brillat-Savarin and encapsulates this phenomenon. People's choice of food identifies them in various aspects of social background and may demonstrate whether they do or do not 'fit in' with another social group. The existence of clearly defined 'food rules' may play an important role in reinforcing 'in-group' identity. This may assume considerable importance in the context of religion, where such rules operate to set those with particular beliefs apart from others who do not share their beliefs.

Ethics and religion

Food may play a very special role in the 'living out' of an individual's beliefs. Since food is ultimately incorporated into the body, food choices become part of who we are in a very real sense and as food must be eaten every day it serves as a constant reminder of what we believe. In recent years vegetarianism and veganism have become increasingly popular in the UK (see Chapter 17) and for many people this choice has been based on ethical concerns about the exploitation of farm animals. In contrast to ethical considerations, which operate primarily on an individual level, religious food-rules serve not only to enhance the spiritual life of the individual but also to enhance allegiance to a community of believers.

An individual may use the restriction of food choice or fasting as a means to enhance personal spiritual growth through the rejection of worldliness, or use foods in rituals associated with communication with God/supernatural forces. Following specified food-rules can also be used to express separateness from non-believers and enhance feelings of identity and belongingness with co-religionists. The 2011 UK census showed a culturally diverse population. While the eating habits of the majority, white Christian population are relatively unaffected by their religious affiliation, nearly 5 per cent of the population are Muslim (2.7 million) whose religion requires the following of particular food rules as is the case for the next largest groups, Hindus (816,633), Sikhs (423,158) and Jews (263,346).

Some dietary restrictions are based on direct injunctions from the holy texts of the religions concerned, while others have their origins in the commemoration of particular events in religious history. For example, the prohibition against pork in Judaism is firmly based on Leviticus 11:4, whereas the eating of matzahs by Jewish people at Passover commemorates the deliverance from Egypt, when their ancestors had no time to allow the bread to rise.

Such rules may affect different aspects of food choice. This includes the items which are considered acceptable as food. For example, those that are acceptable to Muslims are described as 'halal' and those that are forbidden, such as pork and alcohol, as 'haram'. Certain foods may be proscribed on particular days—Roman Catholics should not eat meat on Fridays during Lent. The time of day at which food is eaten may also be restricted—Buddhist monks should not eat after midday. Rules may include the preparation of food. Ritual slaughter is important in rendering meat acceptable to both Muslims (halal) and Jews (kosher). In addition, meat and dairy products must be kept separate and no food prepared on the Sabbath in Jewish households. Fasting is important to the followers of many religions and may take different forms, varying in length and the extent of restriction. In the case of the Ramadan fast for Muslims, no food or drink should be taken between sunrise and sunset for one month. In the Greek Orthodox Church it is expected that during the 40 days of Lent, which precede Easter, the faithful will abstain from all animal foods. In contrast, Hindus may 'fast' once or twice a week throughout the year, restricting themselves to 'pure' foods.

Communication

An invitation to share food or drink, in a range of different settings, is widely used to initiate and maintain personal relationships. The actual foods and drinks that are consumed will vary according to the nature of those relationships and are hence instilled with layers of social meaning. What may be considered appropriate when the setting is a casual encounter between old friends will probably be very different from what might be served on a more formal occasion with people who are less well known and whom the host might want to impress. Food can be used as a means of demonstrating status and prestige. Arrangements for feeding guests at a wedding celebration may well be designed to reflect the status of the bride's father rather than being a response to the perceived hunger and nutritional needs of the guests. In addition to cementing relationships through hospitality and demonstrating social status, food is widely used to reward, punish, or influence the behaviour of others. This may range from the use of sweets by parents to reward children for good behaviour to, in the wider world, the giving/withholding of food aid to particular countries, depending on whether the politics of the regime in power are acceptable to the donor government.

Personal and psychological factors

Personal factors, including emotions, personality, self-esteem and self-efficacy, as well as beliefs and attitudes play an important role in shaping eating habits.

Food becomes associated with emotion from the very start of life, when feeding provides a pleasurable experience of comfort, security, and well-being as well as satiety. Throughout childhood, humans learn to associate particular foods with feelings related to the circumstances in which they were eaten. Thus positive feelings of pleasure may be associated with foods given as a reward, or eaten on special occasions with much-loved people whilst negative feelings may be felt with foods used as a punishment or which had to be eaten because of financial hardship. These associations are woven together and colour an individual's response to food and situations in later life (Birch 1980).

As a result, a response to stress, loneliness, and anxiety may be to eat and to choose particular foods that provide comfort through positive associations. Food may be used as a means of feeling secure, and the hoarding of food is often seen as a response to past experiences of food insecurity. All of these elements may play a part in promoting eating habits which lead to overweight and obesity for some individuals. In addition, parents' desire to show love, to spare their children any deprivation which they themselves experienced when growing up, or to deal with guilt that they do not spend enough 'quality time' with their children, may result in overfeeding them.

Food also provides a useful vehicle for demonstrating emotions such as anger and protest and may elicit very powerful feelings in the bystanders. The refusal to eat is a particularly powerful weapon—whether that is in the context of a child refusing to eat because they are angry or attention-seeking or a politically motivated hunger striker.

Even very young children have been shown to associate particular personality characteristics with different body sizes (thin—mean and sneaky, overweight—stupid and lazy) and the media are often criticized for reinforcing these stereotypes. But the evidence linking personality type to food preferences is not at all clear, although some work has suggested that introverts may have more food dislikes than extroverts.

The impact of 'self-esteem' and 'locus of control' on an individual's eating habits is of particular interest in relation to weight control, but is not clearly established. Some studies have pointed to the higher prevalence of overweight and obesity in relatively deprived population groups in North America and Europe, where self-esteem and people's perception of their ability to exert control over their lives may be low. Consequently some health promotion programmes incorporate activities designed to increase self-esteem and 'empower' participants.

Attitude and belief factors

Traditional approaches to understanding health beliefs that underpin eating behaviour have focused on three particular aspects: perception of risks (relevance and seriousness), beliefs that dietary advice is effective, and beliefs that the benefits of adopting a particular eating habit outweigh the costs (both social and financial).

To prevent and treat illness

Later chapters of this book will examine in detail the scientific basis for our understanding of nutritional requirements and how dietary intake can affect health. However, popular beliefs about the links between diet, health, and disease are shaped by their cultural context and consequently vary in different parts of the world. Not only do ideas vary geographically, but they also change over time. The history of medicine (using the term broadly) is not characterized by a linear succession in which old systems of thought are exchanged wholesale for new ones. We can see this clearly in relation to diet. Despite an enormous growth in the understanding of physiology and the origins of disease, popular advice in the early nineteenth century in England was still based on the dietetic works of Hippocrates and ideas about the effects of diet on humoral balance attributed to Galen and Avicenna.

The concept of 'balance' as a key to health is central to these and in relation to diet focuses particularly on the 'heating' and 'cooling' effects attributed to foods. Individuals by nature of their age, gender, and other characteristics are perceived to have a basic tendency towards the hot or cold end of a spectrum and in order to be healthy should select foods which counteract these tendencies and bring them back into a neutral position. Foods are classified as inherently 'heating' or 'cooling' regardless of the actual temperature of the food at the moment when it is eaten, although the same food may be classified differently in different cultures. In addition, some conditions and diseases are associated with an excess of 'cold' or 'heat' and treatment should include foods with the opposite characteristic to restore health. It is usually considered to be particularly important for women to avoid imbalances during menstruation, pregnancy, and lactation. Young children are also considered to be more vulnerable to imbalance and this may affect the foods that parents consider it appropriate to feed them. These ideas about the heating and cooling effects of foods are found in the classical health systems of the Indian subcontinent (Ayurvedic-Hindu and Unani-Islamic, medicine) as well as in traditional Chinese medicine, and are still widely practised in these communities. In addition, this approach was carried by the Spaniards (under the influence of Islamic medicine) to Central and South America. In some instances these beliefs may, in practice, support an improvement in diet for an individual, but in others they may restrict the diet and increase the risk of nutritional inadequacy.

Perceptions of risk

The general public find it very difficult to assess the relative risk of different diet-related health threats, particularly in the face of almost weekly 'food scares'. In addition, food safety concerns have become more pronounced in the past decade, partly as a result of the loss of trust in governments' ability to regulate food supply ('mad cow disease' is the leading example), the new possibilities afforded by new technologies (e.g., biotechnology and genetically modified foods) and partly by the longer food chains, many of which originate in areas that may be out of reach of domestic food safety agencies. These risk perceptions are powerful drivers of food choice and are influenced by trust in government, the food industry, and scientists and also by the accuracy of the information contained in food labels and in media reporting.

Perceptions of effectiveness of dietary advice

Two factors in particular may contribute to public uncertainty. Coverage of nutritional topics in the popular media is frequently adversarial, where a topic is debated by two 'experts' with opposing views. This may serve to confuse people. In addition, the most widespread and serious nutritional problems in Western countries are often themselves complex, multifactorial, and develop over long time periods. Consequently dietary advice may change as scientific understanding unfolds. In addition, dietary change in relation to these health problems does not necessarily result in a clearly visible 'quick-fix' to the problem (as giving vitamin C to someone with scurvy would), producing convincing evidence of the effectiveness of change.

Because eating is a social activity, making changes may involve social costs as well as financial ones. Perceptions of the behaviour that is considered to be the 'norm' in an individual's particular social group and the extent to which they feel bound to comply with that 'norm' and able, if necessary, to go against that norm, may be important determinants of food choice.

Educational factors

Within most societies, people who have more knowledge about nutrition tend to have better diets. However, it is clear that just giving people more information about food and health does not necessarily result in a change to healthier eating habits. For many the practical difficulties of implementing change (access to affordable healthy food) and the social barriers to changing eating patterns (lack of support from family, friends, and neighbours) will result in apparently insurmountable difficulties when the benefits of making changes are weighed up against the costs.

Food factors (taste and appearance)

People are generally cautious when encountering foods for the first time (neophobic) and repeated exposure and consumption of a range of different foods in early childhood plays an important part in establishing a varied diet in later life.

A range of senses (vision, smell, hearing, etc.) contributes to our perception of the appearance, texture, and flavour of foods. Whether or not we like those aspects of the food when it is presented to us will depend in part on our expectations of how that food should be, our memories of past experiences with that food, and the context in which the food is served. Finally our level of hunger will determine whether we accept and consume the food item.

Vision plays a critical role in food acceptance. Although colour is probably the most important visual aspect, other attributes such as gloss (on fruit), translucency (of jelly), size, shape, and appearance of surface texture (of vegetables and baked goods) are all visual attributes which make a major contribution to the consumer's perception of the quality and appeal of a food item.

There are four basic taste qualities, salty, sweet, sour, and bitter, and the experience of taste is also influenced by the smell of food (Cardello 1996). A further taste, umami, was identified in Japan in 1908; it is provided by glutamate, and is abundant in plants and animal proteins, and by 5'-ribonucleotides, including 5'-inosinate (mainly in meat), 5'-guanylate (in plants), and 5'-adenylate (abundant in fish and shellfish) (Yamaguchi & Ninomiya 2000). It appears that infants have an innate preference for sweet tastes. A preference for salty foods does not appear until later in the first year of life and is more susceptible to modification by experience. People learn to like a certain level of saltiness. Taste/food *aversions* are strongly influenced by conditioning—particularly when consumption of a particular food has been followed by illness. Conditioning of taste/food *preferences* seems to be based on a more subtle mixture of positive associations with exposure, other flavours, and satiety. Ageing is often associated with marked loss in sensitivity to taste, which is an important contributory factor to a decline in the enjoyment of food by older people.

The appeal of different textures may also vary throughout life. A crisp and crunchy apple may be much more attractive to a teenager than to an elderly person who has lost all their teeth. Furthermore, the role of texture in food acceptance is highly product-dependent. In the case of meat, celery, and mashed potato, for example, perception of texture makes a major contribution to the overall acceptability of the food. Socially and culturally learned expectations also play an important part in evaluating texture.

Awareness of the sensory attributes of food which appeal to consumers is clearly of major importance to food manufacturers. Insights into how the senses interact in the experience of eating can also have implications for product formulation; for example, consumers of a fruit drink with a deeper colour will report a stronger fruit taste. Foods are also made tastier through the use of fats, sugars, and salts, all items that in modest excess can lead to health problems such as diabetes, hypertension, and some forms of heart disease and cancer.

Extrinsic food factors

Huge amounts of money are spent each year marketing foods in rich and middle-income countries. In the UK, the top 18 spending brands of potato chips, confectionery, and sugary drinks put over £143 million (US$190 million) a year into advertising. Advertising has a very powerful influence on food choice, especially for highly processed and packaged foods. Some groups of the population may be much more susceptible to this influence and there has been particular concern about the impact on children. In experimental situations, young children clearly have high levels of recall of advertisements and are more likely to request advertised products. The extent to which this results in increased consumption will depend on the relationship with the parent or food provider and the extent to which that person is resistant to 'pester power'. Once children are older and have more money to spend they are clearly in a position to make their own purchases. Advertising also contributes to the creation of social norms—projecting images about what 'people do' and offering images of people whom the audience might want to identify with. This has been the focus of extensive debate in relation to the contribution that media images make to the development of eating disorders, particularly anorexia nervosa, in young girls.

The types of foods that are advertised are often high in fat and simple sugars and the emphasis on the promotion of these types of foods in contrast to the low level of marketing of fruits and vegetables is also considered to have a distorting effect on food habits. In less affluent countries it is also the nature of the foods that are promoted (expensive and nutrient-poor) and the distortion that consumption of these may cause to already overstretched food budgets that raises ethical questions when Western food companies are involved. Some countries have introduced restrictions or bans for specific types of foods targeted at children. For example, in 2007 the UK Office of Communications (Ofcom) introduced restriction on TV advertising foods high in fat, salt, and sugar (HFSS) for pre-school or primary school aged children (< 16 years) with further restrictions to be implemented in 2021 and 2022 as part of a new UK-wide policy.

Lastly, seasonality could be of influence. In parts of the world where food markets are not well integrated, local physical availability is a key determinant of food consumption. In rich countries, there are fewer and fewer seasonal food items as foods are sourced from all corners of the globe.

1.4 Changing dietary patterns

Changing dietary patterns: migration

The large-scale rural–urban and international migration seen around the world in the second half of the twentieth century and into the twenty-first century has been accompanied by major changes in eating habits as people have adapted to the demands of their new environments. As people move from the countryside to urban areas they are forced to enter into a cash economy, while at the same time often having few employment-related skills. Lack of cash may severely limit access to food and food choice. Income generation may also often involve several jobs spread over long working hours. This has particular implications for women in relation to food preparation and childcare. Facilities for cooking are often limited and living conditions poor.

Access to familiar foods, economic factors, and food preparation facilities may also contribute to changes in the eating habits of international migrants. But the eating habits of the host community will also have an impact, depending on the extent to which the migrants are exposed to these and the strength of individual factors which may affect their resistance to change. The dietary changes that are made first usually involve the adoption of foods that are convenient, affordable, and do not clash with religious or cultural beliefs. The adoption of ready-to-eat breakfast cereals and the decline in cooking traditional foods at breakfast has been widely observed in migrants to the USA and UK. Breakfast is also a meal with less social significance, whereas traditional dishes and foods are more often retained at evening meals when family members gather to eat. The presence of children in a family can accelerate the incorporation of non-traditional foods and eating patterns and the gradual development of eating patterns that incorporate elements of both the traditional and new food environments (see Weblink Section 1.6).

Changing dietary patterns: supply and demand

In response to changing lifestyle and popular demand (initially among more affluent, more industrialized Western countries, and subsequently the growing middle classes in developing countries) global food supply systems have been transformed, through changes in agriculture, food technology, and transport in the past number of decades.

These trends are linked to two major developments in relation to patterns of diet and health on a global basis. The first of these is the demographic transition—the shift from high fertility and high mortality to low fertility and low mortality which has been associated with increased industrialization. The second is the epidemiological transition, the shift from a pattern of high levels of infectious diseases associated with unreliable food supply, malnutrition, and poor environmental sanitation to a high prevalence of chronic degenerative diseases associated with urban-industrial lifestyles. These two changes underpin the so-called 'nutrition transition' which describes the shift from a high prevalence of undernutrition to a situation where nutrition-related non-communicable diseases (NR-NCDs) predominate. The recent COVID-19 pandemic resulting in movement restrictions of workers, changes in demand of consumers, closure of food production facilities, restricted food trade policies, and financial pressures in the food supply chain has exposed what a large pandemic can do for local access to food supply (Aday & Aday 2020).

At one time, NR-NCDs were referred to as 'diseases of affluence', more commonly found in industrialized Western countries than the developing world. However, it has long been recognized that this name was misleading in higher income countries, where these diseases were in fact more common in lower-income groups, associated with lifestyle and dietary differences. It has also become apparent that NR-NCDs are emerging increasingly among lower- and middle-income groups in less affluent countries.

The dietary changes associated with the nutrition transition are outlined below and commented on in relation to global trends in the following sections of this chapter.

Against a background of urbanization, economic growth, technological changes for work, leisure, food processing, and mass media growth, the changes in dietary and activity patterns can be summarized in three stages (Popkin et al. 2002) including:

- receding famine pattern characterized by: diet (starchy, low variety, low fat, high fibre); labour intensive work/leisure
- degenerative disease pattern characterized by: increased fat, sugar, processed foods; shift in technology of work and leisure
- behavioural change pattern characterized by: reduced fat, increased fruits and vegetables, carbohydrate, fibre; increased activity.

The process by which immigrants adopt the dietary practices of the host country is also called 'dietary acculturation'. Although lifestyle and other factors clearly lie behind these changes in eating habits (Figure 1.3) there are other aspects of the 'globalization' process that need to be recognized as having a powerful part to play in shaping food choice.

Throughout history, foods have 'migrated' and been incorporated into the diets of people thousands of miles away—the tomato and potato, originally from America, transformed the cuisines of Europe. These dietary changes have sometimes had huge social consequences, as in the case of sugar and the slave trade, or the development of the Irish dependence on the potato and subsequent impact of the potato famine. What is different about the changes in the past 50 years is the pace and scale of change. This is in part due to the application of marketing techniques to mould changes in taste. Most cuisines have traditionally included 'fast foods' but the spread of the hamburger has been achieved through systematic and sophisticated marketing strategies (see Weblink Section 1.7).

The current phase of globalization is also marked by concentration at regional, national, and international levels. This applies throughout the food chain including production, processing, and retailing. Decisions in all these areas affect the choices that people can exercise about what they actually eat. A relatively small number of companies dominate the world food markets, affecting every aspect of the route from farm to consumer. Supermarkets place contracts with distant suppliers to enable previously seasonal foods to be available all year round. The UK's food manufacturing sector is also highly concentrated. In 2018, two companies (Associated British Foods and Boparan Holdings) dominated UK food manufacturing, and half the world's top 100 food sector companies are

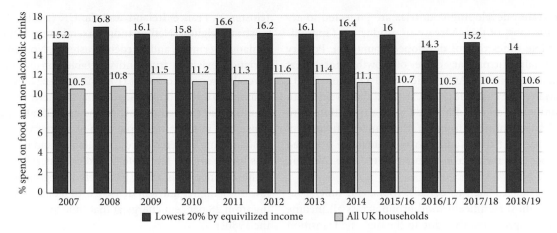

FIGURE 1.3 Information on relative income spend on food and alcoholic drinks and how this has changed over the years in the UK.
Source: Adapted from DEFRA (2020) *UK Household Purchases Updated with 2018/19 Data*. https://www.gov.uk/government/statistical-data-sets/family-food-datasets [accessed April 2021]. Reproduced with permission.

USA-owned. Estimates suggest that the global food industry will come to be dominated by up to 200 groups who account for approximately two-thirds of sales. In the classical market economy, many suppliers compete for the attention of the consumer, responding to what the consumer wants. It has been suggested that in the evolving hypermarket economy, sophisticated systems of contracts, and specifications and tight managerial control enable the retailer rather than the primary food producer or consumer to control the entire supply chain. Selection of foods which are *acceptable* to an individual increasingly takes place in a context where *availability* is substantially influenced by the food industry and food retailers.

Examples of variation in diets over time
Worldwide

On a worldwide basis, the consumption of meat, milk, and eggs varies widely among countries, reflecting differences in food production resources, production systems, income, and cultural factors. Per capita consumption is much higher in developed countries but the current rapid increase in many developing countries is projected to continue (Figure 1.4).

According to the OECD, total meat consumption is projected to increase by 12 per cent between the years 2020 and 2029. Due to high population levels and

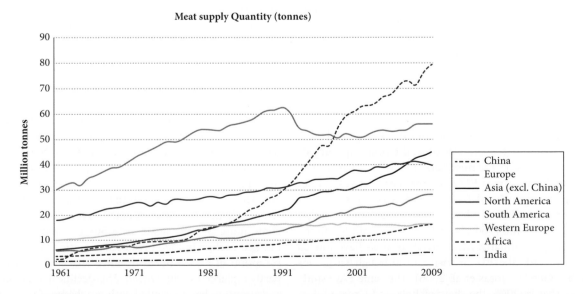

FIGURE 1.4 Trend in meat supply for different countries and continents over the past five decades.
Source: FAO (2012a). Food and Agriculture Organization of the United Nations. Data accessed on 18 October 2015 at http://faostat.fao.org

growth rates, developing countries are projected to account for most of this increase in production (approximately five times that of developed countries). Poultry meat remains the primary driver of growth in total meat production (OECD-FAO Agricultural Outlook 2020). In high income countries it is expected that in line with environmental and health guidelines, a transition from animal-based protein towards alternative sources will be seen together with a more immediate move from red and processed meat to fish and poultry (see Weblink Section 1.8).

Whilst staples such as cereals, roots, and tubers, remain the most significant food group across all income groups, it is projected that due to the ongoing transition in global diets towards a higher share of animal products, staples will decline between the years 2020 and 2029 albeit at different rates for each income group. World trends in the supply of vegetables highlight the regional and temporal variations in the per capita availability of vegetables per year over the past few decades. According to the FAOSTAT food balance data, in 2018, the global annual average per capita vegetable supply was highest in Asia (178 kg), and lowest in South America (54 kg) and Africa (67 kg).

Diets in developing countries are changing rapidly as incomes grow, populations urbanize and become older, and food choice options change. The latter occurs due to transformations in technology (such as global positioning systems and other information and communications technology) and food distribution systems (such as a reduction in the numbers of wholesalers and retailers and longer food chain linkages). The transition in diets is happening most rapidly in China (see Chapter 34) and in middle-income, urbanizing countries such as Brazil, Mexico, Indonesia, and Nigeria. The nutrition transition is causing a relative shift in the causes of disease away from those related to undernutrition only, towards those related to both under- and overnutrition—the double burden of malnutrition.

Europe

As trade and cultural links within the European Union (EU) have strengthened over time, the consumption patterns of the first 15 EU members (EU-15) have been converging. Results from FAO's Global Perspectives Studies Group based on Food Balance Sheets (FBS), show that the diets of the UK and Sweden were much more similar to the USA in the 1960s, but are now more similar to the EU average. Similarly, Mediterranean diets have moved more towards the EU mainstream although variations persist (Garcia-Closas et al. 2006). The same FAO study shows that in 2000, the diets of Italy and Greece (cluster 1) were separate from those of Spain and Portugal

(cluster 2), France, Ireland, and the UK (cluster 3), and Austria, Belgium, Netherlands, Luxembourg, Denmark, Finland, Germany, and Sweden (cluster 4). Notably, wide variation exists in fruit and vegetable consumption in several European countries (EFSA 2008) (see Figure 1.1). Examples of temporal changes in a southern European country may be seen in Italy. In the early 1960s, the diet consumed in the rural population of southern Italy was based mainly on cereals, fresh vegetables, and olive oil. It was low in animal fat, protein, and cholesterol and high in fibre. Since then, there has been a progressive move towards a higher nutrient density with fat intake increasing from 28 per cent energy at that time to 36 per cent in the 1980s.

The UK

Ongoing survey-based estimates of diets are available only for high-income countries. In the UK, the National Food Survey (NFS) reveals 50-year trends in food intakes from 1940 to 1990 (MAFF 1991). This has been followed up with the Expenditure and Food Surveys (EFS), the results of which are being published by the Department for Environment, Food & Rural Affairs (DEFRA) since 2002. From January 2008, the EFS became known as the Living Costs and Food (LCF) module of the Integrated Household Survey (IHS). The National Diet and Nutrition Survey (NDNS) programme began in 1992 as a series of cross-sectional surveys, each covering a different age group: pre-school children (aged 1.5 to 4.5 years); young people (aged 4 to 18 years); adults (aged 19 to 64 years), and older adults (aged 65 years and over). Since 2008 however, the NDNS has been a rolling programme (RP) covering adults and children aged 1.5 years and over. Thus, this makes comparisons between the newer data and that preceding 2008 difficult owing to different methodologies (Table 1.6).

The Second World War and its aftermath had a significant effect on the UK diet. During the war years and in the early post-war years dietary habits were shaped by the prevailing scarcity and restriction of many foods. While many foods such as fruits and meat were restricted, others such as foods rich in starch including potatoes and wholemeal brown bread (known as national bread) were increased. From 1954, consumers were able to return to their pre-war diets, higher in butter, sugar, fresh meat, and white bread. It was not until the 1980s that the brown and wholemeal breads considered as healthier foods become more popular again. The 50-year trend shows a marked decline in total bread consumption. While consumption of brown bread declined sharply, it was only partly replaced by white bread. This decline in total bread consumption has continued with purchases of bread being 14 per cent lower in 2019 than in 2013.

Recent changes in the pattern of consumption of milk products in the UK are reflected in a lower proportion of the population consuming full-fat milk and considerably more now consuming semi-skimmed and skimmed milk as purchases of semi-skimmed milk overtook whole milk in the early 1990s. Whole milk purchases were 19 per cent lower in 2013 than in 2010, equivalent to a reduction of 67 ml per person per week but have started to increase again slightly based on data available to 2019 (DEFRA 2020). Over the same period (2010–2013) purchases of semi-skimmed and skimmed milks were unchanged. However purchases have decreased again by 8 per cent to 2019. Changes in consumption patterns of whole and reduced fat milks may have resulted from dietary guidelines recommending a reduction in total fat and more specifically saturated fat intake. In addition, butter has been partly replaced by margarines and low-fat spreads. These changes are most evident from the early 1980s when lower-fat products became readily available. However, butter purchases had increased steadily over ten years to 2016 to 42 grams per person per week, but are beginning to decrease again with purchases of 33 grams per person per week reported in 2019.

The main changes in meat consumption between 1940 and 1990 were the rise in consumption of beef, lamb, and pork in the early 1950s and the subsequent decline in lamb consumption. This has been accompanied by a significant rise in the consumption of chicken, which was rarely consumed 50 years ago in the UK and has now become the most common form of dietary protein. Overall, fish consumption trends show little change over time from a low intake level and data from 2019 show purchases of 146 grams per person per week similar to that reported since the early 1980s.

Taken together, fruit and vegetable consumption did not change appreciably in the 50 years to the late 1990s in the UK. Vegetable consumption declined slightly but this was offset by an increase in fruit consumption. There was a decline in consumption of brassica vegetables including cabbage, cauliflower, and Brussels sprouts as well as the traditional root vegetables. On the other hand, there has been an increase in salad vegetables and frozen vegetables. The sizeable increase in the consumption of the fruit group since the 1970s may be attributed to fruit juice (now accounting for 75 per cent of fruit products) as well as the increasing year-round availability of fresh fruit. Apples are the most commonly consumed fruit followed by bananas and oranges. Results of the Living Costs and Food Survey (formerly the Expenditure and Food Survey), successor to the Family Expenditure Survey (FES) and the National Food Survey (NFS) and published by DEFRA <https://www.ons.gov.uk/peoplepopulationand community/personalandhouseholdfinances/incomeand wealth/methodologies/livingcostsandfoodsurvey> indicate varying trends in food purchases over the last number of years. People in the UK were buying more fruits and vegetables between 2003/4 and 2006, with quantities of fresh and processed fruit purchased having risen by 10 per cent. Purchases of fruit have now levelled off and are similar in 2019 to purchases in 2013. However, vegetable purchases have increased by 10 per cent in the same period. Food consumption data from the UK NDNS rolling survey has shown a slight increase in fruit and vegetable consumption between 2008–2010 and 2016–2019 for 11–75 year olds with the percentage of the population meeting the 5-a-Day recommendation ranging from 12 per cent in 11–18 year olds to over 40 per cent in adults aged 65–74 years. Household purchases of soft drinks were 3.1 per cent lower in 2013 compared to 2010, a fall of 54 grams per person per week and decreased by a further 9 per cent to 2018 before increasing again in 2019 to levels close to 2013. Within this category, household purchases of 'not low calorie soft drinks' are on a downward trend since 2010, mirrored by an upward trend in 'low calorie soft drinks'. The survey also found a decline in purchases of alcoholic drinks between 2010 and 2019. Purchases of various household foods are on a clear short-term downward trend, including milk and milk products, meat products, potatoes, and bread.

Plant-based alternatives are becoming more popular on the market as consumers are encouraged to move to a more plant-based diet and reduce animal product intake for both health and environmental benefit. With an increasing trend towards flexitarian, vegetarian, and vegan diets, plant protein sources such as soya, mycoprotein, tofu, pulses and legumes are expected to become more popular in certain subgroups of the population.

With the proportion of the population living with overweight and obesity high in Western countries, weight-loss diets have been popular since the early 1960s and many quick fixes and fad diets are marketed widely. Fad diets typically involve a diet plan which is very restrictive either in foods that are allowed (e.g., Atkins, Dukan) or with regard to times that food can be consumed (e.g., 5:2 diet). The intention is for rapid weight loss but the reality is that for most individuals when they go back to eating freely again, they end up putting back on the weight lost and often more. Following the food-based dietary guidelines (Eatwell plate in the UK) together with increasing physical activity should be encouraged for healthy weight balance.

Novel foods

'Novel foods' are foods or food ingredients that do not have a significant history of consumption in the European Union before 1997. A novel food may be defined as: (1) a substance, including a microorganism, that does not have a history of safe use as a food; (2) a food that has been manufactured, prepared, preserved, or packaged by a process that (i) has not been previously applied to that food, and (ii) causes the food to undergo a major change; (3) a food that is derived from a plant, animal, or microorganism that has been genetically modified such that (i) the plant, animal, or microorganism exhibits characteristics that were not previously observed in that plant, animal, or microorganism, (ii) the plant, animal, or microorganism no longer exhibits characteristics that were previously observed in that plant, animal, or microorganism, or (iii) one or more characteristics of the plant, animal, or microorganism no longer fall within the anticipated range for that plant, animal, or microorganism. Foods commercialized in at least one Member State before the entry into force of the Regulation on Novel Foods on 15 May 1997 are on the EU market under the 'principle of mutual recognition', which means that novel foods must undergo a health safety assessment before being placed on the EU market (EU 2015). Only those products considered to be safe for human consumption are authorized for marketing. Companies that want to place a novel food on the EU market need to submit their application in accordance with Commission procedures.

Insects as novel foods

In 2021, dried yellow mealworm (larvae of the beetle *Tenebrio molitor*) was the first insect to be approved as a novel food intended to be used as a whole dried insect in the form of snacks or as a food ingredient, in a number of products (EC2021). It is expected that the insects as novel foods market will continue to grow in the twenty-first century due to the rising cost of animal protein, food insecurity, environmental pressures, population growth, and increasing demand for protein.

Other novel foods

Most of these products are crop plants (e.g., corn, canola (rape), potatoes, and soya bean) that have been genetically modified to improve agronomic characteristics such as crop yield, hardiness and uniformity, insect and virus resistance, and herbicide tolerance. Tomatoes that express delayed ripening characteristics have also been assessed and approved. A few of the products have been modified to intentionally change their composition, for example, canola oil with increased levels of lauric acid. In the UK, the Food Standards Agency has a research programme on the safety of novel foods, with specific emphasis on GM foods. However, other novel foods, including 'functional foods', also fall within the scope of this programme.

Genetically modified foods

Genetically modified foods (GM foods) have ingredients in them that have been modified by a technique called gene technology (see Chapter 34 and Weblink Section 1.9). This technology allows food producers to alter certain characteristics of a food crop by introducing genetic material and proteins from another source. An example of this is a corn plant with a gene that makes it resistant to insect attack. With this technology it is possible to speed up the breeding of new and 'improved' crop varieties and to introduce completely new genetic information, for example, from bacteria or animals into plants. Examples of genetically modified foods that are authorized in the EU market include varieties of genetically modified soya beans, maize, cotton and rapeseed oil (canola). However, several others have been notified to the Commission as being substantially equivalent to traditional varieties according to the 1997 *Novel Food Directive*. In the USA more than 50 new recombinant-DNA (r-DNA) derived foods have been evaluated successfully by the Food and Drug Administration. The USA is the market leader in the total area occupied by genetically modified crops at 68 per cent, while in Europe the figure is close to zero. For almost a decade, genetically modified foods such as corn, soya beans, canola, and tomatoes have been part of the American diet; however this is not the case in Europe. The slower adoption of GM foods in Europe may be due to a lack of trust in government sources with respect to information on food safety, based on previous history in how food safety issues have been handled, for example, the BSE outbreak in Europe in the late 1980s.

Genetic modification and other strategies that target single nutrients promise selective improvement of plant nutrient composition. Although genetically modified (GM) organisms are subject to considerable scrutiny for their potential adverse effects on human health, this technology also has potential ecological and social effects that require careful evaluation. The proponents of GM foods see the benefits of r-DNA technology as a more abundant and economical food supply for the world, continued improvement in nutritional quality (including foods of unique composition for populations whose diets are lacking in essential nutrients), fresh fruits and vegetables with longer shelf-life, foods with reduced allergenicity, and the development of functional foods that may provide certain health benefits. On the other hand, those with particular concerns generally have fears about safety concerns regarding GM food, as well as environmental risks and ethical aspects of using r-DNA technology. These fears should be allayed considerably by a proposed EU directive for the approval, safety evaluation, traceability, and labelling of

GM foods. The general attitude in the UK has traditionally been averse to genetically modified (GM) products; however, a slight shift in attitude towards GM products has recently been reported. Government policy states that, provided it is used safely, GM foods could be a tool with which to address global food security and climate change, and help with sustainable agricultural protection.

Functional foods

Consuming a nutritionally balanced diet was formerly considered as eating an adequate diet to avoid deficiency. But among developed countries, consuming a nutritionally balanced diet has come to mean consuming an optimal diet for promoting health as well as reducing the risk of diet-related chronic diseases. Optimal nutrition focuses on optimizing the quality of the diet in terms of its quantity of nutrients and non-nutrients that favour the maintenance of health. This is where functional foods may have an important role to play since they are considered to have a specific role in relation to disease or the promotion of health. Indeed, a functional food is one claiming to have additional benefits other than nutritional value, for example a margarine that contains a cholesterol-lowering ingredient.

Japan is the birthplace of the term 'functional food'. In the early 1980s three large research programmes were launched and funded by the Japanese government on 'systematic analysis and development of food functions'. A variety of terms related to the Japanese Foods for Specified Health Use (FOSHU) have subsequently appeared. These include such terms as nutraceuticals, designer foods, pharmafoods, medifoods, vitafoods, dietary supplements, and fortified foods. Functional foods must remain foods and they must demonstrate their effects in amounts that are deemed to be normally consumed in the diet. They are not pills or capsules but rather part of a normal food pattern.

A functional food may be a natural food in which one of the components has been naturally enhanced through special growing conditions, a food to which a component has been added to provide benefits, a food from which a component has been removed so that the food has less adverse health effects (e.g., the reduction of saturated fatty acids), a food in which the nature of one or more components has been chemically modified to improve health (e.g., the hydrolysed protein in infant formulas to reduce the likelihood of allergenicity), or a food in which the bioavailability of one or more components has been increased to provide greater absorption of a beneficial component. A functional food may not

necessarily induce a health benefit in all members of the population. The matching of selected food intakes with individual biochemical needs will become an increasingly important factor as our understanding of gene–nutrient interactions develops. Given the potential benefits to health of novel foods (including GM foods and functional foods) and the increasing awareness and interest in healthy eating by the consumer, their consumption levels will inevitably continue to increase in the Western diet.

KEY POINTS

- Typical Western diets are characterized as relatively high in fat and low in carbohydrates when compared to the typical diet of most developing countries.

- While nutrient intakes and even broad food groups may not differ dramatically between developed countries, the foods that contribute to these nutrient intakes do differ markedly and there is no single 'Western diet', or even European diet.

- Factors in dietary variations include those that are: physiological, such as age, gender, pregnancy, lactation, activity; economic; cultural and religious; psychological; beliefs and perceptions, such as risk, benefit; educational; related to food characteristics such as taste, appearance, texture; and extrinsic, such as advertising.

- Diets in developing countries are changing rapidly, driven by urbanization, food distribution and retail technology, increased trade, income growth, and food price policies. Rather than transition from undernutrition through health to overnutrition, many countries are moving to a situation of a double burden of malnutrition.

- While nutrient intakes have remained remarkably consistent in the last 50 years (in the UK), there have been notable changes in food consumption patterns including partial replacement of butter by low-fat spreads and vegetable-based margarines, partial replacement of full-fat milk by low-fat and skimmed milk, and a decrease in vegetables such as swedes, turnips, and brussels sprouts which have been replaced by salad vegetables and mushrooms.

- Novel foods, including genetically modified and functional foods, are increasing in Western diets with a growing emphasis on an optimal diet for promoting health as well as reducing the risk of diet-related chronic diseases.

ACKNOWLEDGEMENTS

The authors would like to acknowledge the contribution of John Kearney to earlier editions of this chapter.

 Be sure to test your understanding of this chapter by attempting multiple choice questions. See the Further Reading and Useful Websites lists for additional material relevant to this chapter.

REFERENCES

Aday S & Aday M S (2020) Impact of COVID-19 on the food supply chain. *Food Qual & Safety* **4**(4), 167–180, https://doi.org/10.1093/fqsafe/fyaa024

Birch L L (1980) Effects of peer models' food choices and eating behaviors on preschoolers' food preferences. *Child Dvelopmnt* **51**(2), 489–496.

Cardello A (1996) The role of human senses in food acceptance. In: *Food Choice, Acceptance and Consumption,* 1–82 (Meiselman H L & MacFie H J H eds). Blackie Academic and Professional, Glasgow.

Darmon N & Drewnowski A (2008). Does social class predict diet quality? *Am J Clin Nutr* **87**(5), 1107–1117.

Department for Environmental Food and Rural Affairs (DEFRA) (2020) *UK Household Purchases Updated with 2018/19 Data.* https://www.gov.uk/government/statistical-data-sets/family-food-datasets (accessed April 2021).

Department for Environmental Food and Rural Affairs (DEFRA) (2019) Gross income quintile group (GIQ) (updated with revised 2016/17 data) https://www.gov.uk/government/statistical-data-sets/family-food-datasets (accessed April 2021).

Drewnowski A (2010) The cost of US foods as related to their nutritive value. *Am J Clin Nutr.* **92**(5), 1181–1188.

Eilander A, Harika R K, Zock P L (2015) Intake and sources of dietary fatty acids in Europe: Are current population intakes of fats aligned with dietary recommendations? *Eur J Lipid Sci Technol.* **117**(9), 1370–1377.

European Commission (2021) *Approval of First Insect as Novel Food.* https://ec.europa.eu/food/safety/novel_food/authorisations/approval-first-insect-novel-food_en (accessed May 2021).

European Food Safety Authority (2008) *Concise Database Summary Statistics: Total Population* (accessed April 2021).

European Parliament and Council of the European Union (2015) Regulation (EU) 2015/2283 of the European Parliament and of the Council of 25 November 2015 on novel foods, amending regulation (EU) No. 1169/2011 of the European Parliament and of the Council and repealing regulation (EC) No. 258/97 of the European Parliament. *J. Eur. Econ. Assoc.* **327**, 1–22.

Garcia-Closas R, Berenguer A, González C A (2006) Changes in food supply in Mediterranean countries from 1961 to 2001. *Publ Hlth Nutr* **9**(1), 53–60.

Gregory J, Foster K, Tyler H, Wiseman M (1990) *The Diet and Nutritional Survey of British Adults.* The Stationery Office, London.

Henderson L, Gregory, J, Swan L (2002) *The National Diet and Nutrition Survey: Adults Aged 19 to 64 Years.* The Stationery Office, London.

McMichael A J, Butler C D, Dixon J (2015). Climate change, food systems and population health risks in their eco-social context. *Publ Hlth.* **129**(10), 1361–1368.

Ministry of Agriculture, Fisheries & Food (1991) *National Food Survey Annual Report On Food Expenditure, Consumption and Nutrient Intakes.* The Stationery Office, London.

OECD/FAO (2020) OECD-FAO *Agricultural Outlook 2020–2029.* OECD Publishing, Paris/FAO, Rome.

Popkin B M, Lu B, Zhai F (2002). Understanding the nutrition transition: Measuring rapid dietary changes in transitional countries. *Publ Hlth Nutr* **5**(6A), 947–953.

Public Health England (2020) *National Diet and Nutrition Survey Rolling Programme Years 9 to 11 (2016/2017 to 2018/2019)* https://www.gov.uk/government/statistics/ndns-results-from-years-9-to-11-2016-to-2017-and-2018-to-2019 (accessed April 2021).

Rao M, Afshin A, Singh G et al. (2013) Do healthier foods and diet patterns cost more than less healthy options? A systematic review and meta-analysis. *BMJ Open* **3**, e004277.

Yamaguchi S & Ninomiya K (2000). Umami and food palatability. *J Nutr* **130**(4S), 921S–6S. doi: 10.1093/jn/130.4.921S. PMID: 10736353.

2

Food and nutrient structure

Yannan Jin and Louise Dunford

OBJECTIVES

By the end of this chapter you should be able to:

- describe the main structures of the macronutrients
- characterize the effects of food processing on nutrient content
- describe the main types of natural toxins, pollutants, and pathogenic agents
- identify the classes and functions of phytoprotectants.

2.1 Introduction

Humans have found that a wide range of plant varieties, animals, and some microbial sources can be consumed as food. Originally, food raw materials were selected on the basis of conveying no harmful effects upon consumption, and it was subsequently found that processing of foods by heat could improve the texture, and in some cases inactivate harmful components within the foods. As understanding of the composition of foods has developed, it has become convenient to classify food components into macronutrients, which are present as bulk components of foods, and micronutrients, which are present at lower levels. Energy is mainly derived from the macronutrients, which comprise carbohydrates, fat, proteins, and alcohol, but both macronutrients and micronutrients provide dietary components that are essential for normal physiological processes. This chapter informs readers about the chemical structures, properties, and functions of a wide range of food components including macronutrients, phytoprotectants, and toxic substances. The effects of processing and storage on these components are discussed.

2.2 Chemical characteristics of macronutrients

For the chemical characteristics of the micronutrients, see Chapters 10–13.

Carbohydrates

Carbohydrates (see Chapter 7) are mainly important as a source of energy (16 kJ/g or 3.8 kcal/g) in the human diet, being converted to glycogen or fat as energy stores. Carbohydrates include simple sugars (i.e., monosaccharides and disaccharides), such as glucose, fructose, and sucrose, plus polymers of sugars, which are termed polysaccharides, of which starch is the most important dietary component. Other polysaccharides include glycogen and cellulose. Many properties of sugars and polysaccharides are very different, but the molecular formulae of all carbohydrates approximate to (CH_2O_n), and the presence of carbon and water in this structure is why the class gained its name. Carbohydrates are synthesized in plants from carbon dioxide and water during photosynthesis.

Sugars

Monosaccharides are the building blocks for oligosaccharides and polysaccharides. Monosaccharides and some oligosaccharides occur as sweet components in foods, with glucose and fructose being the most common monosaccharides, and lactose and sucrose being examples of disaccharides which belong to the class termed oligosaccharides that are sweet and are classed as sugars. Oligosaccharides contain from two to ten monosaccharide units linked together. The sweetness found in the monosaccharides disappears in the higher molecular weight oligosaccharides. Polysaccharides have very different properties from sugars, since they are non-sweet and their solubility in water is normally very low. They are commonly of significance as thickeners or gelling agents in foods.

Glucose occurs in fruits and vegetables, but it is also formed by the hydrolysis of oligosaccharides and polysaccharides during digestion in the small intestine.

Fructose occurs together with glucose in fruits and vegetables and in some sweet foods such as honey. Hydrolysis of sucrose in the small intestine is also a source of fructose.

The chemical structures of monosaccharides comprise chains of between four and six carbon atoms with multiple hydroxyl substituents. Each monosaccharide contains a carbonyl group along the chain, and this may either be at the end of the chain, in which case it corresponds to an aldehyde, for example glucose, or it may be away from the end of the chain, in which case it corresponds to a ketone group as in fructose. The chemical structures can be represented in several ways. The Fischer projection formula (a convention used to depict a stereoformula in two dimensions without destroying the stereochemical information) is often used to show sugar structures. Figure 2.1 shows the orientation of the hydrogen atoms and hydroxyl groups relative to carbon atoms (which are not drawn but which are present at each point where four bonds meet). The ends of the carbon skeleton are behind the plane of the paper with the substituents above the plane of the paper.

In solution, monosaccharides exist as the open-chain form shown in Figure 2.1 in equilibrium with ring forms, which develop when one of the hydroxyl groups forms an intramolecular bond with the carbonyl carbon atom. Since six-membered rings are most stable, glucose exists in solution as an equilibrium mixture of the open-chain form (2 per cent) and the six-membered ring forms termed glucopyranose (98 per cent). When the open-chain form of a sugar closes to form a ring, the hydroxyl group at carbon-1 may be either in an axial position in the chair structure as in α-glucopyranose or in an equatorial position as in β-glucopyranose. These isomers are termed anomers.

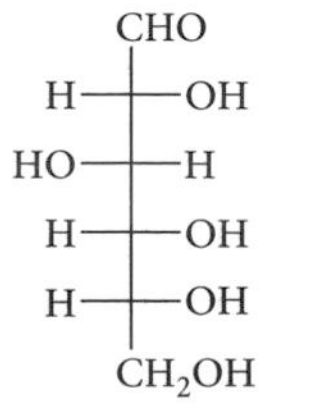

FIGURE 2.1 Fisher projection formula of glucose.

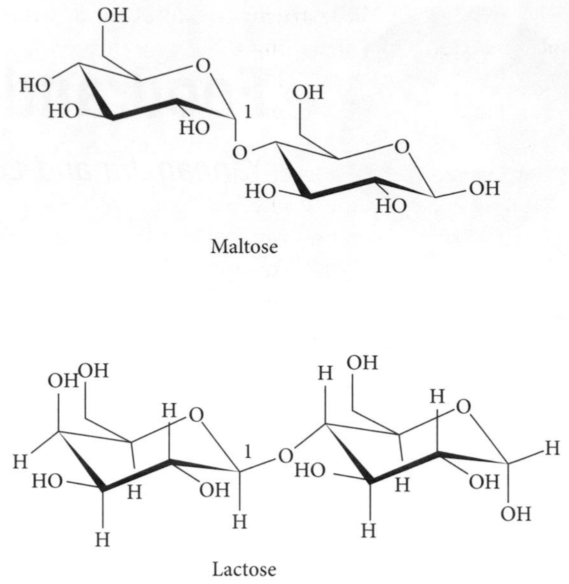

FIGURE 2.2 Structures of maltose with α-1,4 linkage and lactose with β-1,4 linkage.

Monosaccharide units may be linked together by a glycosidic link which is formed when the carbonyl group of one monosaccharide unit links to a hydroxyl group on a second monosaccharide unit forming an acetal in the process. Molecules comprising a small number of monosaccharide units are termed oligosaccharides, with sucrose, maltose, and lactose being common disaccharides, comprising two monosaccharide units. Oligosaccharides vary in whether the monosaccharide units are linked by α-linkages as in maltose or β-linkages as in lactose (Figure 2.2). Raffinose and stachyose are examples of tri- and tetrasaccharides that occur in foods.

Sucrose is well known as the most common sugar used in domestic kitchens and added to processed foods. It is refined on an industrial scale from sugar cane and sugar beet. Hydrolysis of sucrose during digestion provides glucose and fructose.

Lactose is present in cow's milk at about 4.7 per cent of the milk, which corresponds to nearly 40 per cent of the dry matter in milk. Its concentration in human milk is even higher at about 6.8 per cent of the milk or nearly 60 per cent of the dry matter in the milk. Hydrolysis of lactose forms galactose and glucose.

Polysaccharides

Polysaccharides contain large numbers of monosaccharide units linked together. Polysaccharides include starch, glycogen, and cellulose, which are all polymers of glucose. They vary in whether the monosaccharide units are linked by α- or β-linkages, and whether they are linear or branched. These structural features are important since many enzymes including those of significance for the digestion of polysaccharides are selective in terms of which type of linkage is cleaved in the presence of the enzyme.

Starch, which is an important polysaccharide in plant foods including potatoes and bread, is a mixture of two polymers, amylose and amylopectin. Amylose contains glucose units linked by α-linkages between carbon-1 and carbon-4 of successive monosaccharide units, and these linkages cause the molecule to be linear. Amylopectin contains some α-1:6 linkages as well as α-1:4 linkages. The α-1:6 linkages cause the polysaccharide to be branched. Starches from different plants differ in properties due to variations in the ratio of amylose to amylopectin present.

Glycogen is the form in which animals store carbohydrates, and this is similar to amylopectin in having both α-1:4 and α-1:6 linkages between glucose molecules, although there are rather more α-1:6-linkages in glycogen than in amylopectin.

Cellulose is a polymer of glucose with β-1:4 linkages between the glucose units. As a consequence of this structure, cellulose is non-digestible because humans lack enzymes which are capable of hydrolysing polymers of glucose linked by β-glycosidic bonds. In contrast, starch is split by amylases, which are secreted by the salivary glands and by the small intestine, so that starch is initially hydrolysed to disaccharides which can be hydrolysed to monosaccharides for absorption.

Pectins are important polysaccharides for the food industry. They occur in fruits and vegetables including apples and carrots, but they are important for their gelling properties in acid foods with a high sugar content such as jam. Pectins are polymers mainly comprising α-1:4 linked galacturonic acid with some other sugars. The degree of esterification of the polysaccharide has a major effect on the gelling properties.

Fats

Fats (see Chapter 8) are components of foods that are extractable with organic solvents such as hexane or diethyl ether, but are insoluble in water. Dietary fats are converted to energy (37 kJ/g or 8.8 kcal/g), but they are also sources of essential fatty acids. Dietary fatty acids or metabolites are incorporated into phospholipids in cell membranes to provide the membrane structure. Essential fatty acids are converted into prostaglandins and other biologically active compounds described as eicosanoids, which control biochemical reactions within cells.

Fats comprise several classes of chemical compounds, which are termed lipids. The bulk of the fats in food are triacylglycerols (often >95 per cent of the fat), which are often referred to by the more traditional name of triglycerides. These molecules consist of three fatty acid residues esterified to a glycerol backbone (Figure 2.3). The properties of triacylglycerols depend on the structures of the constituent fatty acid residues. The structure of fatty acids consists of a chain of carbon atoms with the required number of hydrogen atoms attached and a carboxylic acid residue at one end of the chain. The most common fatty acids have a chain of 16 or 18 carbon atoms, but fatty acids with between four and 22 carbon atoms occur in food lipids. Fatty acids with even numbers of carbon atoms occur almost exclusively because of the biosynthetic pathway by which they are formed in plants and animals. Fatty acids may be classified according to the number of carbon–carbon double bonds as saturated with 0, monounsaturated with 1, and polyunsaturated with 2–6 double bonds (Figure 2.4).

In polyunsaturated fatty acids, each of the carbon–carbon double bonds is separated from the next double bond by a methylene (CH_2) group. This allows a convenient shorthand nomenclature to be used for fatty acid structures, with the number of carbon atoms in the fatty acid chain followed by a colon, the number of double bonds and then the position of the double bond nearest the methyl end of the fatty acid chain. The position of the carbon of the first double bond (x) counting from the methyl end of the molecule is indicated by n-x. Hence linoleic acid may be represented as 18:2 n-6 (Figure 2.4). An alternative nomenclature is to denote the position of the carbon of the first double bond by ω, so that linoleic acid is denoted as 18:2 ω6.

The most common fatty acids in food lipids are oleic acid (18:1 n-9), palmitic acid (16:0), stearic acid (18:0), linoleic acid (18:2 n-6), and α-linolenic acid (18:3 n-3). The nutritionally important polyunsaturated fatty acids eicosapentaenoic acid (EPA) (20:5 n-3) and docosahexaenoic acid (DHA) (22:6 n-3) are present in fish oil. Linoleic and linolenic acids are important dietary components because humans lack the enzymes required

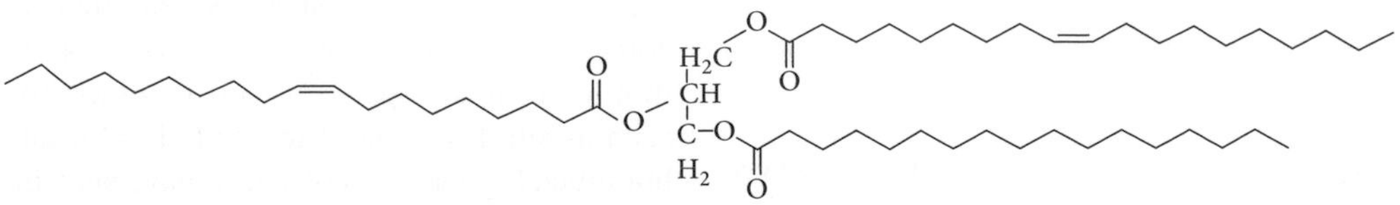

FIGURE 2.3 Typical triacylglycerol structure.

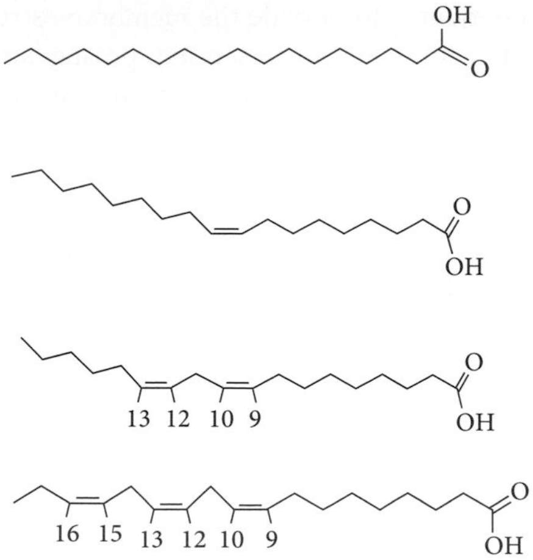

Stearic acid (18:0)

Oleic acid (18:1 n-9)

Linoleic acid (18:2 n-6)

Linolenic acid (18:3 n-3)

FIGURE 2.4 Fatty acid structures.

to introduce double bonds into fatty acids at the n-6 or n-3 positions, but metabolism of linoleic and α-linolenic acids allows the biosynthesis of metabolites including arachidonic acid (20:4 n-6) and eicosapentaenoic acid (20:5 n-3) by desaturation and chain elongation. Although longer-chain metabolites may be formed from 18:3 n-3, the process is not very efficient and the consumption of EPA (20:5 n-3) and DHA (22:6 n-3), which occur in fish oils, has been found to be beneficial in reducing inflammation and risk factors for coronary heart disease including plasma triacylglycerol levels.

The substituents at carbon–carbon double bonds of fatty acids may be on the same side of the bond (*cis*) or on the opposite side of the double bond (*trans*), as shown in Figure 2.5. The configuration of the double bonds in natural plant lipids is exclusively *cis*, although small amounts of *trans*-unsaturated fatty acids are found in animal fats, such as milk fat, due to formation by hydrogenation of polyunsaturated fatty acids in the rumen of the animal. *Trans* fatty acids are also sometimes found in processed foods, because fats containing these fatty acids can be prepared from liquid oils by an industrial process known as hydrogenation and these fats have the correct melting properties for use in foods such as margarine. However, because of concern about the health effects of excessive levels of intake of *trans* fatty acids, manufacturers in Western Europe normally use other methods of preparing fats that do not form *trans* fatty acids. Low levels of *trans* fatty acids are consumed in foods containing animal products such as butter or are present at very low levels in vegetable oils as a consequence of the high temperatures reached during the refining process. Although *cis* and *trans* unsaturated fatty acids are similar in chemical structure (Figure 2.5), the different stereochemistry has important consequences. Polyunsaturated fatty acids containing one or more *trans* double bonds cannot act as essential fatty acids. Excessive dietary intake of *trans* unsaturated fatty acids may lead to increased risk of cardiovascular diseases.

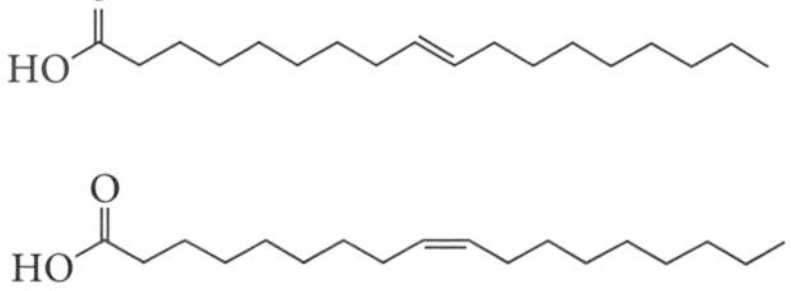

Elaidic acid, 18:1 t-9

Oleic acid, 18:1 c-9

FIGURE 2.5 *Cis* and *trans* fatty acid structures.

Phospholipids, the second major lipid class in foods besides triacylglycerols, are important structural components in biological membranes, and consequently are present in plants and animals consumed as food. The main phospholipids are phosphatidylcholine, phosphatidylethanolamine, phosphatidylserine, phosphatidylinositol, and phosphatidic acid. These are acylglycerol derivatives with fatty acids at positions 1 and 2 of the glycerol molecule, with a phosphoric acid derivative at carbon-3 (Figure 2.6). Phospholipids are added to many foods as they act as emulsifiers, helping to stabilize emulsions such as mayonnaise.

Sterols occur as minor lipid components in biological membranes. Animal tissues contain almost exclusively cholesterol (Figure 2.7), whereas plant tissues contain a mixture of sterols, termed phytosterols, with β-sitosterol being a major component. Other phytosterols include campesterol and stigmasterol. Sterols are extracted with edible oils and fats from plant tissues, and they commonly occur at 1 per cent concentration in edible oils such as sunflower oil. Their solubility in oils is limited but much higher levels of sterols have been incorporated into some functional foods by esterification of the sterols with fatty acids to form steryl esters. A functional food

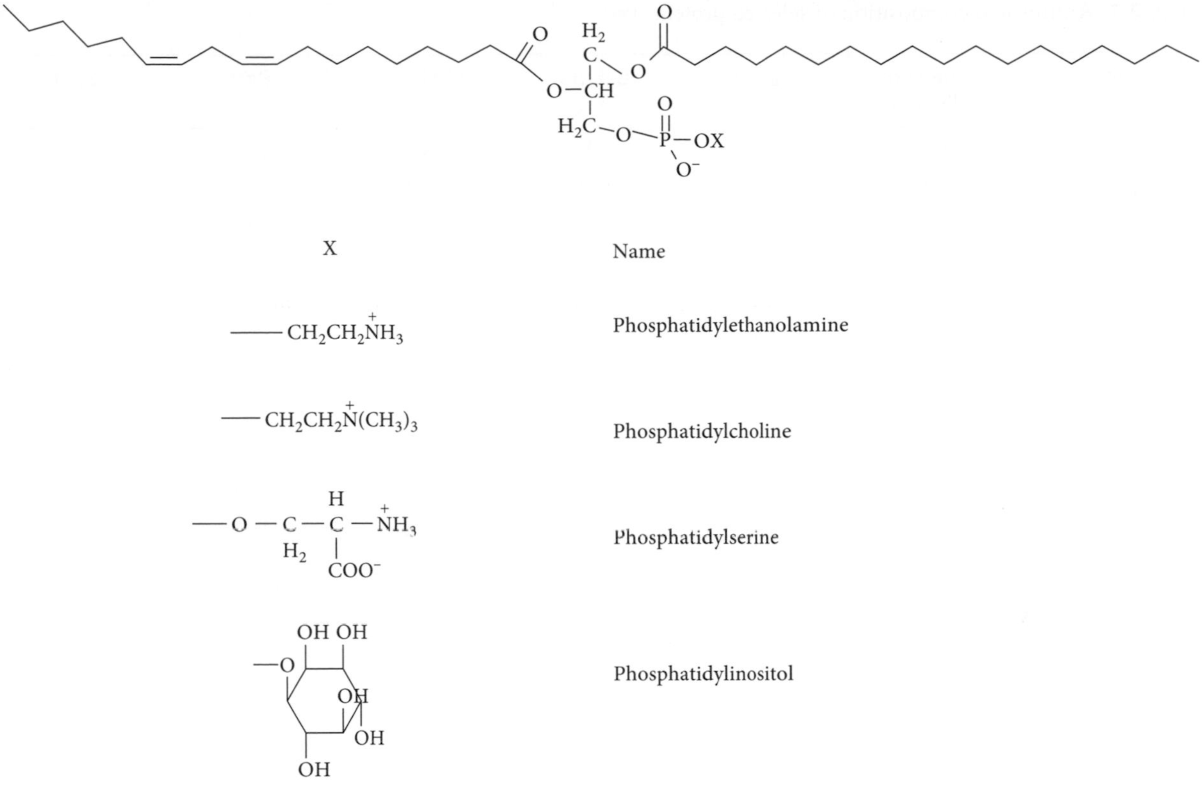

X	Name
—— $CH_2CH_2\overset{+}{N}H_3$	Phosphatidylethanolamine
—— $CH_2CH_2\overset{+}{N}(CH_3)_3$	Phosphatidylcholine
$-O-\underset{H_2}{C}-\overset{H}{C}-\overset{+}{N}H_3$ $\underset{|}{COO^-}$	Phosphatidylserine
	Phosphatidylinositol

FIGURE 2.6 Phospholipid structure.

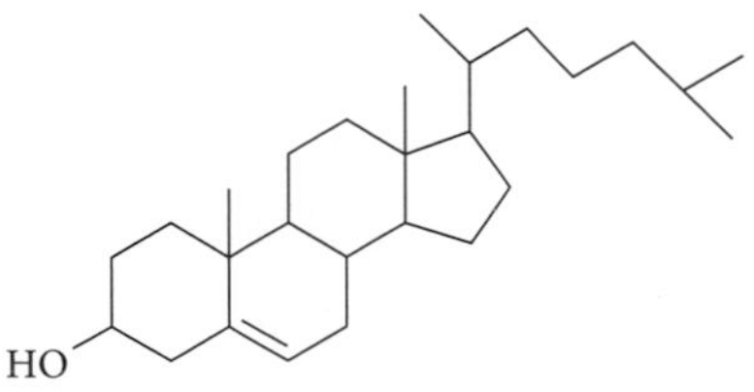

FIGURE 2.7 Cholesterol.

may be defined as a food having health-promoting benefits and/or disease-preventing properties over and above its usual nutritional value. In the case of foods containing steryl esters, the effect is to reduce blood serum cholesterol levels.

Other lipids occur as minor components in foods. These include glycolipids, sphingolipids, mono- and diacylglycerols.

Proteins

Proteins (see Chapter 9) are nitrogen-containing macromolecules that occur in major amounts in foods. Dietary proteins provide energy (17 kJ/g or 4.06 kcal/g) but they are also sources of amino acids that are essential for the synthesis of a wide variety of proteins with important functions including carriers of vitamins, oxygen, and carbon dioxide plus enzymes and structural proteins (Table 2.1). Proteins comprise polymers of amino acids with molecular weights varying from about 5000 up to several million. Twenty amino acids (Figure 2.8) occur in most proteins but additional amino acids, namely hydroxyproline and hydroxylysine, occur in some animal proteins such as gelatin. In proteins, the amino acids are linked together by peptide linkages, which are formed when the carboxylic acid group of one amino acid condenses with the amine group of a second amino acid with the elimination of a water molecule.

Most amino acids have the chemical structure H_2N-C(R)H–COOH, with variation in chemical structure arising from the nature of the substituent R, which may correspond to an aromatic residue as in phenylalanine or an aliphatic residue as in leucine or to a hydrogen atom in the case of glycine (Figure 2.8). In the case of proline, the amino acid contains a secondary amine group in a ring instead of a primary amine group. The common amino acids of nutritional significance in humans comprise alanine, arginine, asparagine, aspartic acid, cysteine, glutamine, glutamic acid, glycine, histidine, isoleucine, leucine, lysine, methionine, phenylalanine, proline, serine,

TABLE 2.1 Amino acid composition of selected proteins (%)

Amino acid	Bovine serum albumin	Casein	Gelatin	Whole egg	Pork	Beef
Alanine	6.3	3.1	11	5.4	15.3	20
Arginine	5.9	3.3	8.8	6.1	–	–
Aspartic acid	10.9	7.6	6.7	10.7	5	0.5
Cystine (0.5)	6.0	0.3	0.0	1.8	–	–
Glutamic acid	16.5	24.5	11.4	12	7.1	8.2
Glycine	1.8	1.9	27.5	3.0	10	4.2
Histidine	4.0	3.8	0.8	2.4	9.3	7.3
Hydroxyproline	0.0	0.0	14.1	0.0	–	–
Isoleucine	2.6	5.6	1.7	5.6	3.8	3.6
Leucine	12.3	9.2	3.3	8.3	9.8	6.8
Lysine	12.8	8.9	4.5	6.3	15.6	11
Methionine	0.8	1.8	0.9	3.2	2.5	3.6
Phenylalanine	6.6	5.3	2.2	5.1	1.9	2.4
Proline	4.8	13.5	16.4	3.8	2.3	–
Serine	4.2	5.3	4.2	7.9	10.8	13.4
Threonine	5.8	4.4	2.2	5.1	1.8	2.0
Tryptophan	–	–	–	1.8	–	–
Tyrosine	5.1	5.7	0.3	4.0	2	3.3
Valine	5.9	6.8	2.6	7.6	1.1	5.3
1-Methylhistidine	–	–	–	–	1.8	8.5

Data from J M Deman (1999) *Principles of Food Chemistry*, 3rd edn; Aspen, Gaithersburg, MD; and F J Francis (2000) *Encyclopedia of Food Science and Technology*, 2nd edn, John Wiley & Sons, New York.

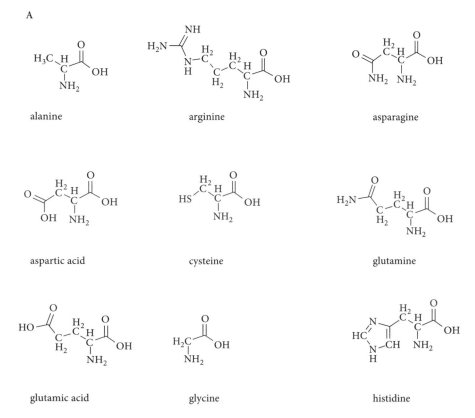

FIGURE 2.8 Amino acid structures.

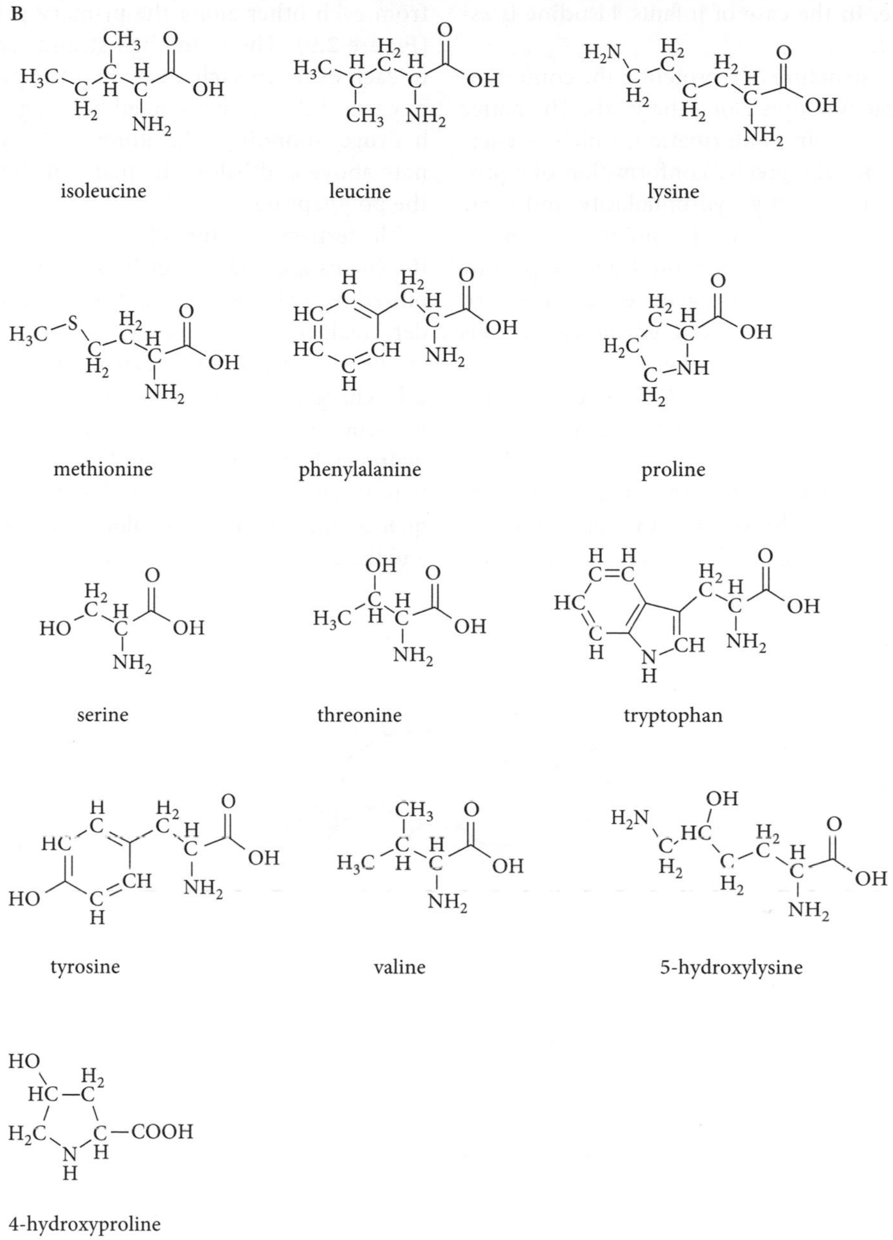

B

isoleucine

leucine

lysine

methionine

phenylalanine

proline

serine

threonine

tryptophan

tyrosine

valine

5-hydroxylysine

4-hydroxyproline

FIGURE 2.8 (*Continued*)

threonine, tryptophan, tyrosine, and valine. These amino acids provide the common polypeptide backbone of proteins, with the identity of the amino acids determining the substituents along the polypeptide chain of the protein. The amino acid sequence of a protein is described as the primary structure of the protein.

Plants can synthesize all the amino acids they need from inorganic nutrients, but animals cannot synthesize the amine group. Animals must consume plants in order to introduce amino acids for their protein synthesis, but they are able to synthesize some amino acids by transamination in which an amine group is shifted from one amino acid to another. Transaminases act with pyridoxal phosphate (the active form of vitamin B6) as a coenzyme to catalyse this reaction. However, isoleucine, leucine, lysine, methionine, phenylalanine, histidine, threonine, tryptophan, and valine cannot be synthesized in this way and these amino acids are described as essential amino acids, because they can only be introduced into the body by eating foods that contain these amino acids. The adult human body can maintain nitrogen equilibrium on a mixture of these amino acids as the sole source of nitrogen. If one or more of the essential amino acids is omitted from the diet, an adult would go into a negative nitrogen balance. In the case of infants, histidine is essential for growth.

The secondary structure of a protein is the conformation that the protein adopts along the y axis. The native protein adopts a certain conformation, which is energetically favourable. The precise conformation of a protein depends on the polarity, hydrophobicity, and steric hindrance of the side-chains of the amino acids in the protein. The α-helix is a conformation that is adopted by many proteins. It has 3.6 amino acid residues per turn with a separation of 0.54 nm between successive coils of the helix (Figure 2.9). The α-helix is stabilized by hydrogen-bonding between the backbone carbonyl group of one residue and the backbone NH group of the fourth residue along the chain.

The β-pleated sheet is another common secondary protein structure. The sheet is formed from several individual β-strands, which are at a distance from each other along the primary protein sequence (Figure 2.9). The individual strands are aligned next to each other in such a way that the peptide carbonyl oxygens interact with neighbouring NH groups by hydrogen-bonding. The atoms of the α-carbons alternate above and below the plane of the main chain of the polypeptide.

The tertiary structure of a protein is established when the chains are folded over into compact structures stabilized by hydrogen bonds, disulphide bridges, and van der Waals forces. Van der Waals forces are weak, non-covalent interactions between atoms that are not electrically charged. The quaternary structure of proteins is the non-covalent association of subunits of proteins due to hydrogen bonds and van der Waals forces. The full protein structure is thus characterized by the amino acid sequence, the conformation along the y axis, the folding, and the association of subunits.

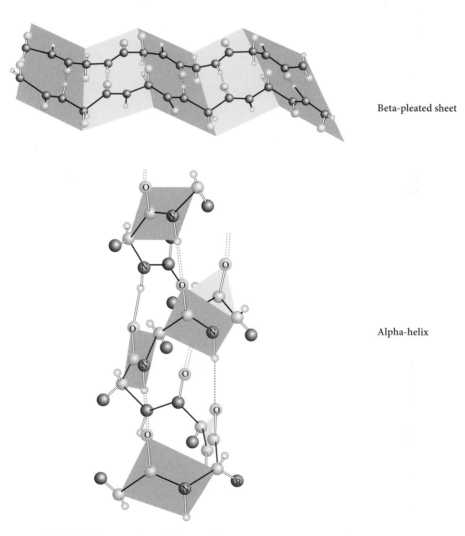

Beta-pleated sheet

Alpha-helix

FIGURE 2.9 Beta pleated sheet and alpha helix protein secondary structures.

Denaturation of proteins is the loss of structure that commonly occurs on heating proteins in solution. It can also be caused by salts, pH, surface effects, or freezing. No covalent bonds are broken, but the change in properties can be very large. Minor degrees of denaturation may be reversible when the denaturing agent is removed, but if denaturation has proceeded further it is commonly irreversible. Enzymes lose activity on denaturation, and the texture of foods may change due to a reduction in protein solubility on denaturation. Thus, the change in eggs on boiling, or the toughening of fish on frozen storage are examples of changes in food texture due to protein denaturation.

Alcohol

Alcohol (see Chapter 10), which is more fully described as ethanol, C_2H_5OH, is formed by the anaerobic fermentation of sugars during manufacture of beer, wines, and spirits. Beer, wines, fortified wines and spirits contain ethanol at concentrations of about 30, 100, 150, and 300 g/l. Ethanol is fully miscible with water and fat. It is absorbed rapidly and metabolized in the liver to yield energy (29.7 kJ/g or 7.1 kcal/g). Many organic molecules are more soluble in aqueous ethanol than in water and the presence of alcohol in a beverage may increase the bioavailability of minor components such as flavonoids from red wine.

Summary of relationship between structure and function

Fats provide a major contribution to the sensory quality of food, and they are also the richest energy source in foods. They are present at high levels in some foods such as butter and margarine, where the fat content is about 80 per cent. Liquid oils provide foods with an oily texture, but semi-solid fats are important in foods such as chocolate, where the melting of the fat as it warms in the mouth is an important contributor to the sensory quality of the product. Fats are important in foods as carriers of the fat-soluble vitamins A, D, E, and K, and they are also important as carriers of oil-soluble flavour compounds. Some liquid oils such as sunflower oil are rich in the essential polyunsaturated fatty acid linoleic acid (18:2 n-6), and some liquid oils also contain α-linolenic acid (18:3 n-3). Fish oils are the main source of the fatty acids eicosapentaenoic acid (20:5 n-3) and docosahexaenoic acid (22:6 n-3), which are considered as beneficial dietary components.

Proteins may be soluble or insoluble in water. If they are soluble they may be important as enzymes and they may contribute to a viscous texture in liquid foods, but when they are insoluble they are normally more important in determining the texture of foods such as in meat. Proteins are sources of the essential amino acids, that is, isoleucine, leucine, lysine, methionine, phenylalanine, histidine, threonine, tryptophan, and valine.

Carbohydrates vary in properties from low molecular weight components that contribute sweetness through to high molecular polysaccharides such as starch that may be important for food texture. If the carbohydrates are not hydrolysed in the small intestine, they are normally important as dietary fibre. Table 2.2 provides the typical values for the protein, fat, and carbohydrate content of selected foods.

Modern trends of macronutrient distribution in diet

The macronutrient profile of diet depends on the quality and quantity of its food composition. Modern diets feature a large diversity in their food compositions, with typical examples of omnivorous, (semi-) vegetarian, and vegan diets, the latter two of which are becoming more and more adopted in younger populations especially in Western countries. As a result, the macronutrient composition varies greatly among groups with different dietary patterns.

Vegetarian and vegan diets (see Chapter 17) have been increasingly drawing interest among nutrition scientists for their evident health benefits and also potential risk of inadequate nutrient intakes. Both of them are plant-based diets and devoid of flesh-based foods, such as red meats, poultry, fish, and sea foods, for example. They are associated with improved blood lipid profile and metabolic function, and reduced risks of certain chronic conditions such as cardiovascular disease. The observed health benefits are attributed to the high diversity of their plant-based diet such as fruits and vegetables, cereals, nuts, beans, pulses and seeds, and reduced intakes of saturated and trans fats. However, those diets are susceptible to inadequate intake of certain types of macronutrients. Studies have shown that vegan diets generally have low total energy intake, which is reflected in their low protein intake. This can be prevented by a regular intake of legumes and soy products to avoid protein deficiencies. In vegetarians' and particularly vegans' diets, intakes of EPA and DHA are commonly lower compared to omnivorous ones. Therefore, both dietary groups are recommended to enhance their intake of alpha-linolenic acid as the endogenous precursor for EPA and DHA from rich sources of walnuts and their oils, hemp, canola, chia, flax, and camelina. More details on the nutritional values of those diets can be found in Chapter 17.

TABLE 2.2 Typical values for protein, fat, and carbohydrate content of selected foods (% of wet weight)

Food	Protein	Fat	Carbohydrate
Apple: eating, raw	0.6	0.5	11.6
Runner beans: raw	1.6	0.4	3.2
Runner beans: boiled	1.2	0.5	2.3
Beef: lean, raw	22.5	4.3	0.0
White bread	8.7	2.1	48.7
Hard cheese	24.9	34.5	0.1
Roast chicken	26.3	12.5	0
Baked cod	21.3	0.4	0
Sweetcorn: kernels, raw	3.4	1.8	8.1
Egg: chicken, white, raw	10.8	Trace	Trace
Haddock, steamed	19.0	0.5	0
Whole milk, pasteurized	3.4	3.6	4.6
Peas, raw	6.9	1.5	11.3
Pork: lean, raw	21.8	4.0	0
Potatoes: new, boiled	1.8	0.1	14.9
Rice: white, basmati, raw	8.1	0.5	83.7
Rice: cooked	2.8	0.4	31.1
Sardines, canned in brine, drained	22.1	9.1	0
White wheat plain flour	9.1	1.4	80.9

Source: Public Health England (2015) *McCance and Widdowson's The Composition of Foods Integrated Dataset* (CoFID). The data is available at: https://www.gov.uk/government/publications/composition-of-foods-integrated-dataset-cofid. Contains public sector information licensed under the Open Government Licence v3.0.

2.3 Effects of food processing and storage

Types of food processing

Foods are processed either to improve their palatability or to extend their lifetime before deterioration reduces their sensory or microbial quality to a level where they can no longer be consumed. The methods vary in the temperatures used and in the contact with water or oil as heat transfer media. Other variables that affect the rate of destruction of nutrients are the presence of oxygen and light, and the pH of the aqueous phase.

Cooking processes such as roasting animals on a fire or boiling in a pot have been used for thousands of years. Preservation processes such as salting or drying in air have also been applied for hundreds of years. Modern industrial processes have been developed to maintain or improve flavour, texture, and nutritional or other quality aspects. Preservation techniques have been improved so as to improve the shelf-life of products, whilst maintaining optimal quality. Common processing methods include baking, frying, boiling, grilling, and microwave heating, each of which results in variable losses of nutrients. For example, when the raw meats of round herring underwent conventional cooking including grilling, boiling, frying, steaming, and microwave heating respectively, about 50 per cent of the vitamin B_{12} content was lost in each cooking process. The loss was found to depend on the temperature and time used in those cooking processes (Nishioka et al. 2011). Losses of vitamin C during the processing of small cylinders of potato fell in the order boiling > frying > baking > microwave cooking (Han et al. 2004). However, the pattern is reversed in the case of –vitamin K. Its level in both potatoes and chards increased by 16–41 per cent after boiling. Similarly, its increase was also observed in broccoli and spinach after microwaving by 1.8–21.2 per cent (Lee et al. 2018).

Baking involves cooking cakes, pastry, potatoes, or fruit in a dry oven at temperatures in the range of 170–230°C. The surface of the food reaches the oven temperature, but in most foods, for example potatoes or fruit, the high water content limits the internal temperature to 100°C or less. Some losses of nutrients such as vitamin C may occur from the surface layer, but the bulk of the nutrients are retained. The loss of thiamin has been widely studied. In the mildly acidic environment of many fermented products including bread, most of the thiamin is retained but if the pH rises above 6, most of the vitamin can be lost. Some of the amino acid lysine can be destroyed by baking due to the non-enzymatic browning reactions between proteins and carbohydrates that contribute to the brown colour and aroma of baked products. Average losses of lysine during breadmaking were reported to be about 15 per cent.

Frying involves cooking in oil at temperatures in the range of 160–200°C. Oil gives better heat transfer than air, so cooking times tend to be shorter than for baking. Losses of nutrients, especially vitamin C may occur, but the food absorbs some oil which in the case of vegetable oils contains vitamin E. Plant foods rich in water such as potatoes tend to absorb more fat than meat does, because the fat is sucked into channels near the surface as the water is lost by evaporation. In potatoes, 80 per cent of the ascorbic acid was converted to dehydroascorbic acid (DHA) during frying for five minutes at 180°C, with no residual ascorbic acid being detectable (Davey et al. 2000). However, DHA is readily reduced by glutathione in cells, and this change should not reduce the bioavailability of ascorbic acid from fried foods. Longer frying periods can cause irreversible losses of ascorbic acid due to oxidation. Also, prolonged frying on potatoes after 12 hours eradicated their antioxidant potential tested *in vitro* (Santos et al. 2018). Frying of meat caused losses of thiamin and riboflavin, with reductions of up to 72 per cent and 55 per cent respectively during the frying of chicken. Frying of round herring caused reduction of the vitamin B_{12} content by up to 62 per cent (Nishioka et al. 2011).

Grilling involves the application of dry heat by radiation from a heating element. Grilling of beef caused losses of 34 per cent vitamin A, 14 per cent vitamin E, 74 per cent thiamin, 49 per cent riboflavin, and 40 per cent nicotinic acid (Gerber et al. 2009). With high temperatures being reached by the surface of the food during grilling, small concentrations of toxic and carcinogenic compounds, such as polycyclic aromatic hydrocarbons (PAHs) are formed. Grilling imposes higher risks of PAHs' generation in animal-based foods compared to its effect on vegetables (Cheng et al. 2019).

Boiling involves cooking in water at about 100°C. It is commonly applied to vegetables and starch-rich foods such as rice or potatoes. When boiled, plant cell walls soften, and starch-rich foods absorb water and swell. Water-soluble vitamins such as vitamin C, thiamin, riboflavin, and minerals leach out of the food and the losses increase with the amount of added water and prolonged cooking time. Mashing of boiled potatoes increases losses of vitamin C by oxidation due to the increased exposure to oxygen. Immersing small amounts of vegetables into rapidly boiling water minimizes losses of vitamin C by rapid denaturation of enzymes that catalyse oxidation of ascorbic acid. Losses of vitamin C during the boiling of spinach were almost 60 per cent, compared to approximately 42 per cent through blanching and 9 per cent through microwaving (Lee et al. 2018). The benefits of steaming broccoli were reported with no significant losses in the vitamin C content whilst boiling resulted in a reduction of 33 per cent of the vitamin (Yuan et al. 2009). Boiling of beef led to losses of 100 per cent thiamin and 83 per cent riboflavin (Gerber et al. 2009). Eggs stored at refrigerated temperature for one week, then boiled, lost approximately 20 per cent of their total free amino acid content, 15 per cent of total carotenoid content and 50 per cent of the antioxidant activity of the egg yolk (Nimalaratne et al. 2016).

Microwave heating involves shorter cooking times than other processing methods, and the fact that the food is heated internally rather than from the surface explains the reduced losses of vitamin C when food is cooked by this method. Other methods, such as baking, have higher temperatures on the surface, where exposure to oxygen is greater and oxidation is more severe, or in the case of boiling allow losses by leaching into the processing water. Retention of thiamin (vitamin B_1), riboflavin (vitamin B_2), and vitamin C in vegetables processed by microwave cooking was also higher than in vegetables that were boiled in water. In a recent study on the effect of cooking methods on free amino acid contents in vegetables, microwave heating was shown to retain 80 per cent of the total free amino acids in cabbage after cooking, comparing to the amount of 70 per cent loss through boiling (Ito et al. 2019).

Canning is a common food preservation technique. The times and temperatures applied in canning must be sufficient to ensure that all pathogens are inactivated or destroyed. The most heat-resistant pathogen normally found in canned food is *Clostridium botulinum*. Heating for 2.5 minutes at 121°C is sufficient to destroy *C. botulinum* spores. For foods with a pH of less than 4.5, a less severe heat treatment may be applied because *C. botulinum* spores are less heat stable and will not grow under these conditions. The amount of oxygen available in the headspace of canned food is small and this limits losses of vitamin C by oxidation. Losses of well over 50 per cent of vitamin C are typical for canned vegetables such as green beans (63 per cent), broccoli (84 per cent), and carrots (88 per cent) (but losses are much less for most fruits due to the stabilizing effects of low pH. For some types of vegetables, a much lesser extent of vitamin C loss has been observed after canning, such as 0.25 per cent in corn and 10 per cent in beets (Rickman et al. 2007). However, losses of vitamin C in canned foods during storage at ambient temperature are small, that is, <15 per cent in one year. Compared to household-cooking using boiling water, canning resulted in significantly decreased contents of protein and total dietary fibre in cooked pulses including kidney beans, white beans, and chickpeas (Margier et al. 2018).

Pasteurization is a mild heat treatment applied to foods to reduce the number of spoilage organisms present and to kill the pathogenic organisms, but it does not render the food sterile. Although commonly associated with milk, it is applied to a wide variety of foods including fruit juices and ice cream mixes. The process involves a temperature–time combination sufficient to kill enough organisms to achieve reduction to an acceptable number, but without unacceptable changes in the flavour or nutrients of the food. For heat treating milk, either a high temperature/short time process, which involves a temperature of 71.7°C for 15 seconds, or a low temperature/longer time process, which involves holding the milk at 62.8°C for 30 minutes, is applied. The milk must then be cooled rapidly to below 10°C. These conditions are sufficient to destroy tuberculosis organisms and other more sensitive pathogenic organisms in the milk. For liquid egg, the pasteurization conditions require heating at 64.4°C for 2.5 minutes prior to cooling to below 3.3°C.

Fermentation is defined as the action of microorganisms or enzymes on food raw materials to cause biochemical changes. Fermentation is widely used as a processing method for improving the nutritional quality, digestibility, safety, or flavour of food. Fermentation is a relatively low-energy process that can improve product life and reduce the need for refrigeration. Foods such as beer, cheese, mushrooms, bread, yoghurt, soy sauce, pepperoni, tempeh, and many others are produced by fermentation. Lactic acid bacteria which convert carbohydrates to lactic acid, thereby lowering the pH, and which also produce flavour compounds, are important in many fermented foods including fermented milk, cheese, meat, and cereal products. Fermentation of cereals and legumes is a common processing method to enhance the protein digestibility and bioavailability of minerals and phytochemicals contained in those foods (Nkhata et al. 2018).

Storage of foods may cause deterioration by aerial oxidation or by enzyme action, which can lead to losses of nutrients and textural changes. Microorganisms may multiply in some foods such as milk and yoghurt unless the storage temperature is kept below 6°C, but dry foods can be kept at higher temperatures because microorganisms will not grow in foods with a water activity below

0.6. The effect of water on microbiological changes is often discussed with reference to the water activity, a_w, where a_w is the ratio of the vapour pressure of water in a food to the saturated vapour pressure of water at the same temperature, the a_w of pure water being 1. In intermediate water foods (jam, dried and salted meats and fish, cakes, dried figs, etc.) where the a_w is 0.70–0.90, mould and yeast spores, or pathogenic bacteria, can survive and then they can grow if the moisture content accidentally rises.

Refrigeration slows down changes in stored foods. A temperature below 5°C is recommended to retard the growth of microorganisms but chemical or enzymatic changes can still occur slowly at these temperatures. Leafy vegetables such as spinach are very vulnerable to vitamin C loss during storage, but reducing the storage temperature from 20°C to 4°C significantly reduces losses (Favell 1998). Storage of fresh eggs at 4°C for six weeks did not significantly change the antioxidant properties of egg yolk with respect to the antioxidant activity and the contents of free amino acids, carotenoids, and malondialdehyde, which was studied as a marker of lipid oxidation in egg yolk (Nimalaratne et al. 2016).

Freezing slows down all chemical and enzymatic changes, but these processes can progress slowly even at −20°C, which is about the temperature in domestic freezers. Blanching, which involves heating briefly in hot water or steam, is commonly applied to vegetables prior to freezing in industrial processes in order to inactivate enzymes such as lipoxygenase that catalyse biochemical deterioration during frozen storage. Vitamin C is mainly lost at the blanching stage both due to thermal degradation and due to leaching into the blanching medium. Losses of 10–40 per cent of the vitamin C throughout the freezing process are typical, but the extent of losses depends on exposure to air during the process and the type of blanching process that is used. Losses are generally lower for steam blanching than for water blanching (Favell 1998).

Drying or dehydration is the removal of water and is traditionally achieved by leaving commodities in the sun. The larger the surface area the faster the drying process but the application of vacuum and heat accelerates the drying process. Drying reduces the activity of enzymes and also inhibits the growth of microorganisms. Industrially, dehydration is applied more widely to potatoes than to fruits and vegetables, and losses of about 75 per cent of the vitamin C can occur. Vitamin C is commonly added to dehydrated potatoes, because of the importance of potatoes as a dietary source of vitamin C.

Freeze-drying is a process where water is removed under vacuum by sublimation from the frozen state of a food. Freeze-dried foods can normally be rehydrated to dissolve rapidly in water or to recover the original texture more closely than foods dried at higher temperatures. Thus, freeze-drying is commonly applied in the manufacture of instant coffee powder because the powder has good solubility in water. Other areas of its application include pharmaceutical products, food supplements, spices, and infant formulae. Freeze-dried food products generally show better organoleptic and nutritional profiles compared to their counterparts processed by conventional drying methods.

Salting is an effective method for preserving foods such as meat because the water activity is reduced. Salting in combination with smoking and drying was the main preservation method used for centuries prior to the twentieth century. Salt-preserved products such as ham and bacon are still important in the Western diet but canning, refrigeration, and freezing have become more common preservation techniques.

Irradiation is a physical method of processing food using ionizing radiation. The radiation generates free radicals, which can react with the DNA of living insects and microorganisms. For food irradiation, irradiation doses of up to 10 kGy are used, although lower doses are often used. Irradiation is permitted in Europe for fruits, vegetables, cereals, fish, shellfish, and poultry, but the main application is herbs and spices, where chemical treatment is often required if irradiation is not used. Low irradiation doses, <1 kGy, inhibit sprouting, sterilize insects and delay ripening. At 1–3 kGy, the numbers of some spoilage microorganisms are reduced, but viruses and spores of sporulating bacteria such as *Clostridium botulinum* are not affected by food irradiation. Thiamin is the most sensitive of the water-soluble vitamins. More than 50 per cent of the thiamin in chicken can be lost by an irradiation dose of 10 kGy at 10°C. Vitamin E is also sensitive to irradiation, with significant losses of vitamin E in irradiated cereals. However, although foods containing vitamins E and A are susceptible to irradiation, the main food sources of these vitamins are not irradiated. Losses of nutrients by irradiation are often no greater than losses by thermal processing methods. In a recent study conducted by Khan et al. (2018), the shelf-life of peaches was extended up to 17 days when treated with gamma irradiation at the dose of 5.0 kGy with no further reduction of moisture contents, crude protein, fat, fibre, and ascorbic acid comparing to untreated controls. Irradiation of selected foods, especially herbs and spices, is a very safe procedure and would be much more widely applied if consumers' concern about the effects of radiation on human tissues was not carried over into prejudice against irradiated foods, where no radiation is retained by the food.

High pressure processing: When high pressures up to 1000 MPa (10^4 bar) are applied to packages of food that are submerged in a liquid, the pressure is distributed instantly and uniformly throughout the food. The high pressure causes destruction of microorganisms and inactivation of some enzymes, although a combination of heating and high pressure is often applied. The process has the potential for providing products with improved flavour, good retention of nutrients, and changed texture compared to products that are processed by heat alone. Fruit juices processed by high pressure processing are now sold in Europe, and other products processed by this technique including jams, fruit jellies, sauces, fruit yoghurts, and salad dressings are on sale in Japan. Losses of ascorbic acid from vegetable juice increase in the order high pressure processing < pasteurization < sterilization. In addition, ascorbic acid in high-pressure-processed foods showed better stability during refrigeration storage as compared to that of thermally processed ones (Tewari et al. 2016).

Additives (including preservatives, flavours, colours, sweeteners, and processing aids)

The use of additives in foods in Europe is strictly controlled by legislation. The legislation aims to protect the health of the consumer and to prevent fraud. Additives are allowed for specific functions (antioxidant, colour, etc.) and are restricted to those listed in the regulations. In most cases additives are only allowed for use in selected foods and at limited levels. Prior to being included in European regulations, the technical need and the safety of an additive must be demonstrated.

Preservatives are substances added to foods to inhibit microbial spoilage and extend storage time. Common foods including meats, cheeses, baked goods, fruit juices, and soft drinks are likely to include preservatives. Even if sterile foods are produced initially by thermal processing, infection with bacteria, fungi, and yeasts can occur in these foods, which are often not consumed at one sitting, and preservatives are required to extend the shelf-life of the products. Sorbic acid, benzoic acid, sulphites, thiabendazole, nitrites, and biphenyl are amongst the substances approved for use as food preservatives. Some food preservatives including benzoates and sulphur dioxide have been identified as causing sensitizing or allergic reactions including chronic urticaria and asthma in susceptible individuals. Nitrites and nitrates, which are used in cured meats and some cheeses, have caused some concern because secondary amines may react with nitrite derivatives to form N-nitroso compounds that are possibly carcinogenic. However, nitrites are formed naturally in saliva within the body and nitrates are ingested in vegetables and water, so it is considered that the possibility of N-nitrosated compounds being formed in treated foods may be of less consequence than other sources of these compounds. Sulphur dioxide is not used in foods containing thiamin because it brings about the destruction of this vitamin in the food. Emerging interests have been drawn by meat and seafood manufacturers in using natural preservatives derived from plants (e.g., grape extract), bacteria (e.g., bacteriocins and organic acids) and animals (e.g., chitosan) to extend product shelf-lives as to their antimicrobial or antioxidant properties.

Flavours added to food may be natural flavour complexes derived from plants and other natural sources by physical processes such as extraction or distillation. A range of essential oils are isolated by these processes and they are widely used for flavouring foods such as peppermint for chewing gums, and orange oil for flavoured drinks. Other flavour compounds are synthesized by controlled chemical synthesis, by transformations using living biological systems, by enzyme-catalysed synthesis, or by the reaction-flavour approach which mimics established food-cooking techniques. Synthetic flavour compounds include vanillin, menthol, methyl salicylate, benzaldehyde, maltol, ethyl maltol, and cinnamaldehyde.

Colours: The classes of natural or nature-identical colourings used for food include carotenoids, chlorophyll, anthocyanins, and betalaines. Besides these, some synthetic compounds are allowed for addition to food. Synthetic food colours can be classified as azo (e.g., sunset yellow FCF), azo-pyrazolone (e.g., tartrazine), triarylmethane (e.g., green S), xanthene (e.g., erythrosine), quinoline (e.g., quinoline yellow), or indigoid compounds (e.g., indigo carmine). Most of the allowed synthetic food colours are water-soluble. Some food colours have been found to cause allergic responses in susceptible individuals. Most concern has been expressed about tartrazine. It has been estimated that about 100,000 people in the USA are sensitive to tartrazine. Symptoms of the allergic response include urticaria, swelling, often of the face and lips, runny nose, and occasionally asthma. The EU requires a mandatory warning to be put on the packaging of any food that contains tartrazine, quinoline yellow, sunset yellow, carmoisine, ponceau 4R, and allura red because of concern that they could make children hyperactive and inattentive.

Sweeteners: Consumers are very interested in low-calorie food and beverage products. Since the sugars present are significant contributors to the calorific content of many foods, the food industry has developed a range of zero or low-calorie sweeteners. Aspartame, saccharin, acesulfame K, cyclamates, neotame, and steviol glycosides (stevia plant extracts) are intense sweeteners, which are approved for use in foods in the EU.

Aspartame is used widely in foods because it is the sweetener that most closely mimics the taste of sucrose. It consists of two amino acids, aspartic acid and phenylalanine, and it is 180 times sweeter than sucrose. Aspartame is quite stable in acid foods but it is less stable in the neutral pH range found in baked foods. Its safety of consumption was reconfirmed by the European Food Safety Authority (EFSA) in 2013. The Acceptable Daily Intake (ADI) is 40 mg/kg body weight (bw) for the general population. People with phenylketonuria, a rare inherited disorder, must avoid foods containing aspartame because they do not metabolize phenylalanine effectively. Neotame is an artificial sweetener that is between 7000–13,000 times sweeter than sucrose. It is chemically related to aspartame but is more stable and is used at lower concentrations.

Saccharin is a very stable, highly water-soluble and cheap food additive, but although it is a high-potency sweetener, that is, 300–500 times sweeter than sucrose, some consumers are sensitive to its bitter and metallic off-tastes. Saccharin cannot be metabolized by the body and therefore it does not give any calories upon consumption. Although concerns about the safety of saccharin were raised following a Canadian study in the 1970s in which bladder tumours were found in the second generation of rats fed with saccharin, there is no evidence that saccharin causes cancer in humans. Some studies have suggested that it was the sodium component of sodium saccharin that caused the bladder tumours in rats at the high level of consumption, and there is no evidence that similar effects occur in humans at realistic consumption levels. Its ADI is 5 mg/kg bw. Acesulfame K is structurally related to saccharin and suffers from similar taste defects. Although it is not as stable as saccharin, it is used together with other sweeteners, especially aspartame, in foods such as diet cola drinks. Its ADI is 9 mg/kg bw. Cyclamates are sodium or calcium salts of cyclamic acid. The taste is considered better than that of saccharin, but the concentration is limited to 400 mg/l in soft drinks. Weak biological effects for the main metabolite of cyclamate, cyclohexylamine, have limited the ADI to 11.0 mg/kg bw.

Steviol glycosides were approved by the EU for use in foods in 2010. The sweetener is extracted from the leaves of the Stevia plant and is 200–300 times sweeter than sucrose. It has a bitter aftertaste and is therefore commonly used in combination with other sweeteners. The ADI is 4 mg/kg bw.

Sucralose was developed as a non-nutritive sweetener in the 1970s. It is a trichlorinated derivative of sucrose with excellent flavour and stability. Its ADI is 15 mg/kg bw.

Sugar alcohols including sorbitol, mannitol, and xylitol have comparable sweetness and about the same calorific content as sucrose, but they are absorbed more slowly from the digestive tract and do not raise postprandial blood sugar and insulin levels; thus they are suitable for sweetening diabetic foods.

Processing aids are substances used to facilitate food processing by acting as chelating agents, enzymes, antifoaming agents, catalysts, solvents, lubricants, or propellants. They are not consumed as food ingredients by themselves, and are used during food processing without the intention that they should be present in the final product. Residues of the processing aids may be present in the finished product and it is a legal requirement that they do not present any health risk to humans. Chymosin, which was developed as a replacement for rennet, the milk-clotting enzyme traditionally used in cheese manufacture, was classified as a processing aid.

Chelating agents such as glycine, ethylenediaminetetraacetic acid (EDTA), and citric acid are widely used in the food industry for stabilizing the chemical, sensory, and nutritional properties of foods. They act by scavenging free metal ions in the food system by forming stable complexes with them, as the latter can deteriorate foods by initiating fat or oil oxidation.

Genetically modified organisms (GMOs)

GMOs are plants, animals, or microorganisms which have had DNA inserted into them by means other than the natural processes of combination of an egg and a sperm or natural bacterial conjugation. The use of GMOs in food has been the subject of intense debate in recent years. GMOs have a number of potential benefits including improved nutritional attributes, for example, reduction of anti-nutritive and allergenic factors, and increase of the vitamin A content in rice to reduce blindness in Southeast Asia. Tomato puree produced from GM tomatoes was widely accepted for several years in the UK but it is no longer available. Advantages claimed for the product included better flavour, consistency, and lower price than the non-GM alternatives. However, concerns about antibiotic resistance, transferring allergenic components between plant species, and the risks of GM crops to the agricultural environment have prevented widespread acceptance of GM foods amongst consumers in the UK. In the EU, it must be clearly labelled if the foods contain GMOs or ingredients produced from GMOs. Examples of widely used GM foods are GM maize and soybeans which are widely used as animal feeds and also in some processed foods such as cooking oils and biscuits in many places worldwide, including America, the UK and other European countries.

2.4 Toxic components formed by processing

The processing of foods is essential to inactivate microorganisms and to allow flavour and texture development. However, processing may lead to losses of nutrients, and it may also lead to the formation of toxic components under certain conditions (see also Chapter 3). A wide range of chemical reactions occur at high temperatures such as those that occur during frying or grilling of food. The following compounds are examples of toxic products that may be formed.

3-Monochloropropane-1, **3-diol (3-MCPD)** is an example of a chemical that may form in foods by the reaction of chloride with lipids. It has been shown to be a carcinogen by laboratory animal studies. 3-MCPD can be formed as a result of industrial processing of foods such as hydrolysed vegetable protein or soya sauce but it may also form during domestic cooking of foods or it may transfer into foods from packaging materials. The European Commission's Scientific Committee on Food (SCF) have proposed a tolerable daily intake of 2 mg/kg body weight.

Acrylamide (2-propenamide) (Figure 2.10) is found in fried and baked goods at levels up to 3 mg/kg. Highest levels are found in crisps, crispbread, chips, and fried potatoes. Acrylamide forms as a by-product of the Maillard reaction, which is a chemical reaction between amino acids (particularly asparagine) and reducing sugars that gives browned food its distinctive flavour. The WHO classifies acrylamide as a probable human carcinogen based on experiments on laboratory animals and effects on humans exposed to high levels through industrial exposure. In 2015 the European Food Standards Agency published a risk assessment which stated that acrylamide in food may increase the risk of cancer, and European food manufacturers are required to have mitigation measures to reduce the presence acrylamide in food.

Polycyclic aromatic hydrocarbons (PAHs) are a group of several hundred related compounds that are present in wood smoke and are detected at low levels in charred meat and in foods exposed to smoke. PAHs are chemically very stable and they are widespread in the environment, since they do not degrade easily. Consequently, a variety of food products that are not smoked including vegetable oils and fish also contain detectable levels of PAHs. Many PAHs are carcinogenic to animals, and carcinogenic effects have been demonstrated in humans following occupational exposure to high levels of PAHs. For the general public, food is the main route of exposure to PAHs. Benzo(a)pyrene (BaP) is the PAH about which there is most concern as a potential carcinogen. European regulatory limits set out the maximum levels of BaP allowed in the foods at most risk of contamination.

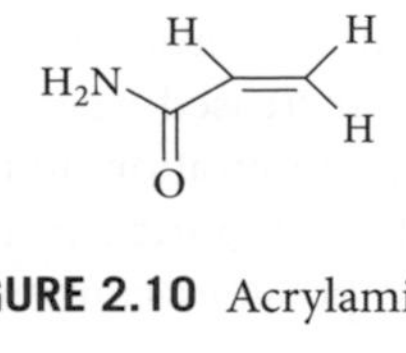

FIGURE 2.10 Acrylamide.

2.5 **Natural food toxins**

Many natural substances are harmful to health when consumed in foods at a sufficient dose (see also Chapter 3). For example, fat-soluble vitamins cause toxic effects when excessive amounts are consumed and these can be fatal at a high level of intake. However, toxins are substances that cause harmful effects when foods are consumed at levels comparable to those which may be eaten by consumers. These may be natural products that accumulate in foods during processing or storage or they may be introduced into plants or animals that are subsequently consumed as foods. Some toxins, for example polychlorinated biphenyls, are cumulative, but other toxins such as glycoalkaloids are completely harmless when consumed repeatedly at sub-toxic doses. Food components may cause toxic effects within hours, days, or longer-term consumption of the food, or they may have mutagenic or carcinogenic effects in which an inheritable change in the genetic information of a cell may lead to cancer or other disease states over a period of years. When there is evidence from animal experiments that food components are toxic, regulatory authorities normally allow the components to be present in foods when the maximum amount consumed is 100 times less than the minimum amount shown to have an adverse effect in animals with due allowance for the body weight of the animal. Close monitoring of foods by government agencies helps to prevent chemical toxins reaching levels at which harmful effects occur, and pathogenic bacteria are much more common causes of human disease than chemical toxins.

Natural plant toxins

Glycoalkaloids: Solanine is the main glycoalkaloid, which commonly occurs with other glycoalkaloids including chaconine at low levels in potatoes. However, high levels of solanine may be found in green potatoes, since conditions which lead to an increase in chlorophyll content also cause increases in solanine levels.

Glycoalkaloid levels of over 200 mg/kg fresh weight may lead to harmful effects. The ingested levels of 2–5 mg/kg body weight can cause toxic symptoms, and it can become fatal when the levels reach 3–6 mg/kg body weight. Solanine acts by inhibiting the enzyme cholinesterase, which catalyses hydrolysis of acetylcholine to acetate and choline. The action of this enzyme is essential for the repolarization of neurons following transmission of a nerve impulse. Increased gastric pain followed by nausea, vomiting, and respiration difficulties, which may cause death, have been reported in individuals following consumption of potatoes containing high levels of glycoalkaloids. Cholinesterase activity recovers within a few hours following ingestion of low doses of glycoalkaloids, so there are no ill-effects of repeated ingestion of small doses.

Marine toxins

The neurotoxin tetrodotoxin occurs in some organs of the puffer fish, which is a culinary delicacy in Japan. Great skill is required by the chef to separate the muscles and testes, which are free of the toxin, from the liver, ovaries, skin, and intestines which are not. Tetrodotoxin is fatal above a dose of 1.5–4.0 mg, whereas the concentration may exceed 30 mg/100 g in the liver and ovaries. The toxin blocks movement of sodium ions across the membranes of nerve fibres, inhibiting transmission of nerve impulses, which causes distressing symptoms that develop into total paralysis and respiratory failure causing death within 6–24 hours. There is no known antidote to the poison, and fatalities continue to occur regularly. Japanese regulatory authorities require chefs to be licensed for cooking the puffer fish, and this has helped to reduce the number of fatalities in recent years.

Paralytic shellfish poisoning is caused by consumption of shellfish such as clams or mussels that have fed on dinoflagellate algae, which reach high concentrations in red tides that develop in seawater. The algal bloom is common in coastal waters close to Europe, North America, Japan, and South Africa. The algae, especially *Gonylaux* spp., produce a toxin, saxitoxin, that accumulates in the flesh of the shellfish. Most shellfish break down or excrete the toxin within three weeks after ingestion ceases, but some species of clams may retain the toxin for several months. Saxitoxin is considered to be fatal at a dose of about 4 mg, with symptoms including numbness of lips, hands, and feet that develop into vomiting, coma, and death. A dose of 1 mg causes mild intoxication.

Mycotoxins

Many species of fungi produce metabolites that are toxic. Mycotoxins are produced by filamentous fungi. Epidemics due to the consumption of rye that had been stored in damp conditions were common in the Middle Ages. The grain was contaminated by the fungus *Claviceps purpurea*, known as ergot, and ergot poisoning was manifested either as gangrenous or convulsive ergotism. Gangrenous ergotism caused severe pain, inflammation and blackening of limbs, and loss of toes and fingers. Convulsive ergotism caused numbness, blindness, paralysis, and convulsions.

The main pharmacologically active compounds in ergot are a series of alkaloid derivatives of lysergic acid. The most important alkaloids are ergotamine, ergonovine, and ergotoxin. Some of these alkaloids have found applications in medicine for treatment of migraine.

Greater care over the harvesting and storage of grain has reduced the occurrence of ergot in grain very considerably, but an outbreak occurred in Ethiopia as recently as 2001. However, concern over the presence of mycotoxins in mouldy food has developed since the 1960s, when it was found that metabolites produced by the fungus *Aspergillus flavus* (aflatoxins) were toxic to animals.

Aflatoxins occur in mouldy grain, soya beans, or nuts and are carcinogenic at very low levels of intake. Toxin-producing fungi usually produce only two or three aflatoxins under a given set of conditions, but fourteen chemically related toxins have been identified. Aflatoxins are a series of bisfuran polycyclic compounds, which vary in the structure of at least one ring or substituent. Aflatoxin B_1 is one of the most potent chemical carcinogens known. A high incidence of liver cancer is induced in rats by feeding diets containing 15 mg/kg aflatoxin B_1. There have been many reports of animals suffering from toxic effects following the consumption of mouldy feeds. In 1960, over 100,000 turkeys died in England with extensive necrosis of the liver after consuming mouldy feed. In 2016 there was an outbreak of acute aflatoxicosis in Tanzania in humans who had consumed contaminated maize.

As well as directly consuming contaminated food, aflatoxins may be ingested via an indirect route. Cattle consuming feed contaminated by aflatoxins excrete milk containing aflatoxin M_1. Aflatoxins may also be transmitted to humans via animals in meat or eggs. When food has been contaminated with mycotoxins, the toxins remain in the food even after the mould has been removed or has died. Aflatoxins and many other mycotoxins are quite stable during normal food processing operations.

Most toxins from other fungi are much less potent than the aflatoxins. Some mushrooms are toxic but sometimes they can be rendered edible by cooking, and only a few species are lethal if consumed. *Amanita muscaria* is a fleshy fungus that grows widely in temperate areas of the world. It causes hallucinogenic effects and has commonly been used as a narcotic or intoxicant. The compounds responsible for the toxic effects are a series of isoxazoles including muscimol.

Pollutants

Pesticides are used to control weeds or insects that would otherwise reduce crop production or cause post-harvest losses. Several hundred pesticides, which correspond to several chemical classes, are used in agriculture. The use of pesticides has made a major contribution to food production and preservation. Pesticides are designed to kill or adversely affect living organisms, and consequently there is in principle a risk to health. Maximum residue levels of many pesticides in foods such as fruit, vegetables, and other plant and animal products are specified in the legal regulations.

Antibiotics used to treat animals may remain as residues in meat. There is concern that this may lead to the development of antibiotic resistance in humans, although it is only one of the mechanisms for the development of antibiotic-resistant organisms in humans.

Hormones are used in the rearing of animals in some countries because they act as active growth promoters. The Scientific Committee on Veterinary Measures relating to Public Health concluded that no ADI could be established for any of these hormones. Although a total ban exists on the use of hormones in raising cattle in the EU, other countries such as the USA and Canada permit their use, and the ban is considered controversial.

Heavy metals have received attention as widespread environmental contaminants and as accidental food contaminants. They enter the environment as a consequence of industrial pollution and they enter the food chain in various ways. The two metals of main concern are mercury and cadmium.

Polychlorinated and polybrominated biphenyls (PCBs and PBBs)

PCBs and PBBs are mixtures of inert molecules, which were previously used as electrical insulators and fire retardants. They are resistant to chemical and biological breakdown and as a consequence of their stability they have become widespread environmental contaminants, which are a potential hazard to human health. PCBs are frequently found at mg/kg levels in fish, poultry, milk, and eggs.

Radioactive fallout

Radioactive isotopes, or radionuclides, are atoms that are unstable and emit energy as radiation when they decay to more stable atoms. Some radionuclides such as carbon-14, uranium-238, and radon-222 occur naturally in the environment due to the effects of cosmic radiation or due to their creation in prehistoric times. However, fallout from nuclear power stations and the use of radioactive materials in industry, medicine, and research have led to increases in the levels of radionuclides in the environment. When living tissue is exposed to ionizing radiation it will absorb some of the radiation's energy and may become damaged, with the development of cancer.

Pathogenic agents

Pathogenic agents are a very common cause of food poisoning. Organisms may be associated with endogenous animal infections transmissible to humans (zoonoses) by consumption of meat or fish. This group includes bacterial, viral, fungal, helminthic, and protozoan species. The second group of organisms are species that are exogenous contaminants of food, which may cause infections in humans. This group includes common food poisoning organisms such as *Salmonella*, *Staphylococcus*, and *Clostridium botulinum*.

Seafood poisoning

Bacterial decomposition of fish that is stored at unacceptably high temperatures or for excessive time is the main cause of seafood poisoning. Seafood poisoning is often called scombroid poisoning because fish of the *Scombroidea* species including mackerel and tuna are widely consumed. However, fish from other species including sardines and herring may also cause outbreaks. Consumption of contaminated fish may cause symptoms to appear within about two hours of the fish being eaten. The main symptoms include pain, vomiting, and diarrhoea. Seafood poisoning has been attributed to the formation of histamine by bacterial decomposition of the amino acid histidine in fish. However, pure histamine has low oral toxicity, and it appears that other components in the fish such as putrescine and cadaverine are important in allowing the toxicity of histamine to be manifested.

Food poisoning organisms

Salmonella is one of the most common intestinal infections in the UK. In recent years, there have been about 8500 reported cases per year. The most common organism involved is *S. enteritidis* followed by *S. typhimurium* but the known number of strains of the bacterium is over 2300. Eggs are the most common vehicle for the transmission of salmonellosis. The common contamination of eggs is due to the fact that hens lay eggs through a passageway called the vent (or cloaca), which is an exit shared by their intestines, where *Salmonella* is commonly present. Other foods of animal origin including unpasteurized milk, poultry, and cheese are also possible sources. The organism is readily inactivated by cooking, but cross-contamination of cooked food by uncooked food is a common pathway for contamination of food in domestic kitchens. Symptoms normally follow ingestion of contaminated food within 6–48 hours. *Salmonella* causes diarrhoea, often with fever and abdominal cramps. The onset may be sudden and there may be nausea and vomiting initially.

Campylobacters are the leading cause of bacterial diarrhoea in humans. Campylobacteriosis commonly affects babies and young children. The illness is a gastrointestinal infection caused by *C. jejuni* or *C. coli*. Symptoms may show themselves as bloody diarrhoea, fever, nausea, and abdominal cramps. The illness normally lasts 2–10 days but some symptoms may persist for several months. The disease is usually self-limiting, so antibiotic treatment is not normally required except in serious cases. Campylobacters occur widely in the intestinal tract of many animals, especially chickens and turkeys. During slaughtering and preparation of raw birds, a large number of birds may become contaminated, and therefore undercooked poultry meat and offal are a major source of infection. Raw milk and poorly treated water supplies are also causes of campylobacter infections.

Verocytotoxin-producing Escherichia coli is an uncommon but important pathogen because serious infections, particularly in children, may result. The illness was first recognized in the early 1980s. Infection may produce mild diarrhoea or a severe or fatal illness. Cattle are the main source of infection, with most cases being associated with the consumption of undercooked beef burgers and similar meat, or raw milk. The infective dose appears to be very low, probably less than ten cells. A temperature of 70°C for two minutes is sufficient to destroy the organism in meat.

Listeria monocytogenes is a potentially dangerous foodborne pathogen. The bacterium occurs in cheese. Hard cheeses do not support the multiplication of the bacterium and numbers less than 20/g do not present a hazard to health in hard cheeses. However, the bacterium can multiply in soft cheeses at refrigeration temperatures, and the occurrence of the bacterium in cheeses should be minimized by the use of Hazard Analysis Critical Control Point (HACCP) systems throughout the whole food chain. *L. monocytogenes* causes very serious illnesses including meningitis and septicaemia. The mortality rate can be as high as 30 per cent, with pregnant women, infants, the elderly, and people who are immune suppressed being most vulnerable.

Clostridium perfringens is a common source of food poisoning, normally from meat that has been cooked and then stored at insufficiently low temperatures. The organism is anaerobic and spores that survive cooking may develop vegetative forms that multiply in the gut and produce an enterotoxin. Symptoms, which include diarrhoea and abdominal pain, normally occur within eight to 24 hours following consumption of the food.

Staphylococcus aureus is a usually harmless part of the microbiota, but can become pathogenic. Foods may be contaminated by carriers, and ingestion of contaminated food may be followed within two to four hours by vomiting and diarrhoea, which may be severe and may lead to collapse due to dehydration.

Clostridium botulinum is an organism found in soils, which may contaminate canned meats and meat pastes. The organism forms heat-resistant spores, which may develop vegetative forms anaerobically if not inactivated by heat. The vegetative forms of the organism produce a toxin that is extremely potent at very low levels. Difficulty in swallowing may develop into paralysis and death. Cases of botulism are rare, since the food industry uses nitrites to prevent anaerobic growth in processed meats.

Viral infections

Viruses have no cellular structure and possess only one type of nucleic acid (either RNA or DNA) wrapped in a protein coat. Consequently, they cannot multiply in foods, but food handlers with dirty hands or dirty utensils may allow foods to become contaminated by viruses, which can subsequently multiply in the intestinal tract by using the host cells' mechanism for replication. Viral infections often have much longer incubation periods of up to several weeks compared to several hours for bacterial infections.

Hepatitis A is a member of the genus *Enterovirus*, which causes symptoms of anorexia, fever, malaise, nausea, and vomiting followed after a few days by liver damage. Exposure to the virus is limited in the developed world due to good public hygiene and sanitation. However, there was a large outbreak of 300,000 cases in Shanghai in 1988 due to consumption of raw clams, and in Delhi in 1955–1956 there were 36,000 cases due to contamination of the mains water supply.

Poliomyelitis is another enterovirus that can be transmitted by contaminated food such as milk. However, there has been a global effort to eradicate the disease through vaccination programmes, and only a few cases occur each year in developing countries.

Norovirus is a highly infectious virus and is the most common cause of gastroenteritis. It can be spread through contaminated food or water, or through the air, and causes about 200,000 deaths per year globally. It is common in both developed and undeveloped countries.

Prions

Transmissible spongiform encephalopathies are a group of progressive neurodegenerative conditions which exist in both animals and humans. They are thought to be caused by prions (proteinaceous infectious particles) which are misfolded proteins which can trigger other proteins to fold abnormally. Prions form amyloids, which are abnormal aggregates of proteins, and these build up in infected tissue and lead to tissue damage and cell death. Prions are very stable and are not destroyed by cooking or disinfectants.

A new form of Creutzveldt–Jakob disease called variant CJD linked to eating meat from cattle with bovine spongiform encephalopathy (BSE) was identified in 1996. It is thought that the cattle were infected by eating meat and bone meal, which included the remains of other animals. Although CJD had been known as a rare disease since the 1920s, vCJD affected younger people more than the earlier form, and there were unusual clinical features. vCJD presents as psychiatric disorders including anxiety, depression, and withdrawal but it develops over a period of months into forgetfulness and memory disturbance. A cerebellar syndrome develops with gait and limb ataxia and eventually death. The disease in cattle was reduced after 1993 by the removal of meat and bone meal from cattle feed concentrates and by the slaughter of large numbers of animals. There have been 173 deaths in the UK from vCJD, with a peak of 28 in the year 2000, but reductions in the annual number of deaths have been recorded since that year, and no cases have been reported in the UK since 2016.

Parasites

Parasites are organisms that live on or inside other organisms, known as hosts. They generally do not kill their host and can live for a long time; however, they can cause considerable harm.

Helminths and nematodes are flatworms and roundworms, which develop as parasites in humans following consumption of contaminated water or food, especially meat or raw salads. Liver flukes, *Fasciola hepatica*, and tapeworms of the genus *Taenia* are the most common helminths. The liver fluke develops as a leaf-like animal in humans, sheep, or cattle. It grows up to 2.5 cm long by 1 cm wide and establishes itself in the bile duct after entering and feeding on the liver. When mature the liver fluke produces eggs, which are excreted in the faeces. Symptoms include fever, tiredness, and loss of appetite with pain and discomfort in the liver region of the abdomen.

Tapeworms include *Taenia solium*, which occurs in pork, and *T. saginata*, which is associated with beef. The mature tapeworm develops in the human intestine and may cause severe symptoms in the young and in individuals weakened by other diseases. Effects may include nausea, abdominal pain, gut irritation, anaemia, and a nervous disorder resembling epilepsy. If gut damage allows eggs to be released in the stomach, the resulting bladder worms may invade the central nervous system with fatal consequences.

Trichinella spiralis is a nematode (roundworm) that causes trichinosis in humans with symptoms of discomfort, fever, and even death. Consumption of raw or poorly cooked infected pork products is the normal source of foodborne illness.

Giardia duodenalis is a microscopic parasite which colonizes and reproduces in the intestines and causes a disease called Giardiasis, which is also known as beaver fever. It is the most common enteric protozoal infection globally. Symptoms include diarrhoea, abdominal pain, and weight loss, although about 10 per cent of people infected are asymptomatic. It is spread through eating or drinking food or water that has been contaminated with faeces containing Giardia duodenalis cysts. Giardiasis can be successfully treated with a nitroimidazole or anti-helminthic medication.

2.6 Phytoprotectants

(For a fuller version of this section with references, see Weblink Section 2.1.)

There is strong evidence that a diet rich in fruits and vegetables can reduce the risk of heart disease and probably some cancers in the human population. However, identification of the individual components of the diet which may confer this protection is still a matter of intense study. Plants and plant extracts have been used for centuries in the treatment of chronic ailments. In a recent systematic review and meta-analysis (Aune 2019), a linear inverse relationship of total fruits and vegetables intake with the risk of coronary artery disease or mortality from stroke was found at the dose range between 0 and 800 g per day. In particular, such associations were established for citrus fruits, apples and pears, green leafy vegetables, salads, and cruciferous vegetables. The protective role of an increased intake of fruits and vegetables up to 600 g per day against total cancer risk was also reported. The benefits were found to be associated with green/yellow vegetables and cruciferous vegetables only. The risk of certain types of cancer appeared to be negatively correlated with the intake of specific groups of fruits and vegetables, although this does not demonstrate a causal relationship. Examples are the relationships between intake of citrus fruits, cruciferous vegetables, and green leafy vegetables and risk of lung cancer; between intake of banana and spinach and risk of colon cancer; as well as between intake of citrus fruits and risk of bladder cancer. Research has identified an array of phytochemicals including organosulphur compounds, carotenoids, terpenes, and polyphenols including flavonoids and phytoestrogens that have been shown to produce beneficial physiological effects such as antimicrobial, induction of carcinogen detoxification, antioxidant, anti-inflammatory, anti-thrombotic, cholesterol-lowering, neuroprotective, and immuno-modulating effects.

Phytoprotectants may inhibit tumour development by scavenging chemical carcinogens to prevent them binding to electron-rich sources in the cell including DNA, RNA, and proteins. They may also act by their effect on phase I enzymes involved in activation of environmental and chemical carcinogens or phase II enzymes involved in detoxification pathways.

Organosulphur compounds found in garlic represent some of the more effective phase I and II enzyme modulators. Diallyl sulphide (DAS), diallyl disulphide (DADS), triallyl sulphide, and diallyl polysulphides are lipid-soluble components in garlic, and S-allylcysteine (SAC) and S-allylmercaptocysteine (SAMC) are among the water-soluble components that are considered important. DAS and DADS from garlic, and isothiocyanates, which occur in cabbage, broccoli, and cauliflower, have been shown to inhibit chemically induced cancers in animals. Sulforaphane, an isothiocyanate found in broccoli, is a potent inducer of phase II enzymes.

Carotenoids are a class of structurally related compounds which are strongly coloured red or orange, and which are found in foods that are coloured red, orange, or green including tomatoes, carrots, and spinach. Carotenoids comprise molecules with hydrocarbon structures including α-, β-, and γ-carotene and lycopene as well as molecules with one or more polar substituents attached to the hydrocarbon backbone, including astaxanthin, zeaxanthin, lutein, and β-cryptoxanthin. Some of the most common carotenoids act as provitamin A, being converted to vitamin A by β-carotene-15, 15'-oxygenase in the intestinal mucosa. Provitamin A activity is restricted to carotenoids such as β-carotene in which at least half of the molecule shares the ring and chain structure of vitamin A. The role of carotenoids in cancer prevention is uncertain. Lycopene does not have provitamin A activity. A number of animal models using tomato products or lycopene have shown strong anticarcinogenic effects against prostate cancer, yet, the latest report of the World Cancer Research Fund International (WCRF 2014) has concluded that the strength of existing evidence on the protective role of lycopene-rich foods against prostate cancer in humans is 'limited-no conclusion'. Processed tomato products that are rich in bioavailable lycopene were shown to be linked with a reduced risk of prostate cancer, the association of which was not found in studies using raw tomatoes. This suggests that the bioavailability of lycopene influences its association with cancer risk.

Polyphenols are defined as classes of plant components, the structure of which includes more than one aromatic ring and several phenolic hydroxyl groups. The term was originally restricted to vegetable tannins, which are water-soluble compounds having molecular weights between 500 and 3000, and which have the ability to precipitate proteins. The theaflavins and thearubigins, which contribute the colour to black tea, are examples of tannins. In recent years, the use of the term polyphenols has been extended to include lower molecular weight

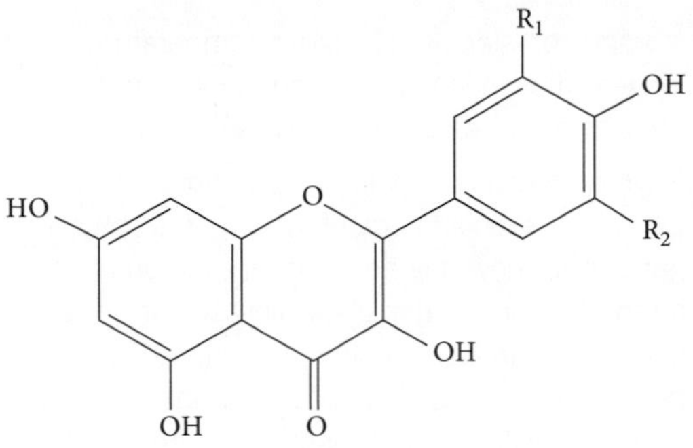

R_1 = H;	R_2 = H:	Kaempferol
R_1 = OH;	R_2 = H:	Quercetin
R_1 = OH;	R_2 = OH:	Myricetin
R_1 = OCH$_3$;	R_2 = H:	Isorhamnetin

FIGURE 2.11 Flavonol structures.

compounds containing at least one aromatic ring and one or more hydroxyl groups. Flavonoids comprise the most important group of low molecular weight polyphenols.

Flavonoids are phenolic compounds, which occur widely in plant tissues, mainly as glycosides. All flavonoids have a C_6-C_3-C_6 structure in which a six-carbon aromatic ring is linked by a three-carbon bridging unit to a second aromatic ring (see Figure 2.11). The nature of the central C_3 unit defines the class of flavonoid. The classes include: the anthocyanins, which contribute the red, purple, or black colours to fruits such as strawberries, plums, and blackcurrants; the flavonols, which occur widely in vegetables including onions, broccoli, and beans; flavanols, which occur in green tea, cocoa, and red wine; flavones, which occur in aubergines and celery; flavanones which occur in citrus fruits, and proanthocyanidins which occur in cocoa, cider, and wine.

The position of the benzenoid substituent on the C-ring divides the flavonoid class into flavonoids (2-position) and isoflavonoids (3-position). Isoflavones in soya products are of current nutritional interest because of their oestrogenic structure. Other flavonoids are of interest as bioactive dietary components because of effects on cell-signalling, cognitive function, or other effects. They also have antioxidant properties. Epidemiological evidence to date suggests a protective role of flavonoids in reducing the risk of different types of cancers including gastric, breast, prostate, and colorectal cancers. Yet, more human studies are needed to confirm this hypothesis.

Phytoestrogens

Phytoestrogens include any plant substance or metabolite that induces biological responses in vertebrates and can mimic or modulate the actions of endogenous oestrogens, usually by binding to oestrogen receptors. Most of the work investigating the properties of phytoestrogens has been on isoflavones, with a little work on prenylated flavonoids, coumestrol, or lignans. Phytoestrogens mimic or block the action of the human hormone oestrogen, although they are much less potent. They are of interest because they may have benefits for prevention of certain cancers such as prostate, breast, and colorectal cancers, cardiovascular and metabolic diseases as well as menopausal symptoms such as osteoporosis and hot flushes in postmenopausal women. However, there are possible links to fertility problems in animals, which raises concerns that similar effects could occur in humans, particularly babies fed soya-based infant formulas.

Isoflavones are present in several legumes mainly including soya beans, green beans, and mung beans with soya beans being the primary human dietary source. The three main isoflavones are genistein, daidzein, and glycitein, which are mainly present as the glycosides genistin, daidzin, and glycitin in raw materials. These glycosides are poorly absorbed on consumption of foods. Isoflavones also occur as malonyl or acetyl esters but these derivatives are commonly hydrolysed during the processing of soya products such as tofu (Song et al. 1998). During the processing which involves the main steps of steaming, cooking, roasting, and microbial fermentation, protease inhibitors are destroyed and the glycoside bond is cleaved to generate absorbable aglycones in processed soya food products. Consequently, the digestibility and bioavailability of isoflavones are increased by processing, which also enhances the shelf-life of products including natto, soya milk, and miso (Zaheer and Akhtar 2015).

Metabolites of isoflavones such as glucuronide and sulphate conjugates are formed by metabolism in the intestine and the liver. Among them, equol, a bacterial metabolite of daidzein, demonstrates the most potent oestrogenic and antioxidant activity *in vitro*.

KEY POINTS

- Carbohydrates include sugars and polysaccharides. They are important as an energy source or as dietary fibre.
- Fats are mainly triacylglycerols. They are a source of the essential fatty acids, linoleic acid and linolenic acid, which are utilized for eicosanoid synthesis.
- Proteins are macromolecules comprising chains of amino acids linked by a peptide bond.
- Macronutrient composition in vegetarian and vegan diets vary largely from the omnivorous diet, accompanied with both health benefits and risk of nutritional inadequacies in certain types of nutrients such as protein, EPA, and DHA.

- Foods are processed either to improve their palatability or to extend their shelf-life, especially by the destruction of microorganisms.

- Food processing operations applied to foods include baking, frying, grilling, boiling, microwave cooking, canning, pasteurization, irradiation, freeze-drying, drying, and high-pressure processing.

- Food additives include preservatives, processing aids, flavours, colours, and sweeteners.

- Growth of pathogenic organisms including bacteria, yeasts, and moulds may cause food poisoning.

- Processing of foods, for example, heat treatment under selected conditions, reduces the levels of many pathogenic organisms, so that the foods may be consumed safely. Cross-contamination of cooked products by unprocessed products during domestic storage, or storage at higher temperatures or for longer times than recommended by manufacturers, are common causes of food poisoning incidents.

- Phytoprotectants include organosulphur compounds, flavonoids, carotenoids, and phytoestrogens. And they may be important in contributing to a reduction in the risk of cardiovascular diseases or cancer by various mechanisms. These include inhibition of cancer by their effect on phase I enzymes that are involved in activation of environmental and chemical carcinogens or phase II enzymes that are involved in detoxification pathways.

Be sure to test your understanding of this chapter by attempting multiple choice questions. See the Further Reading list for additional material relevant to this chapter.

REFERENCES

Aune D (2019) Plant foods, antioxidant biomarkers, and the risk of cardiovascular disease, cancer, and mortality: A review of the evidence. *Adv Nutr* **10**, S404–421.

Cheng J, Zhang X, Ma Y et al. (2019) Concentrations and distributions of polycyclic aromatic hydrocarbon in vegetables and animal-based foods before and after grilling: Implication for human exposure. *Sci of Total Environ* **690**, 965–972.

Davey M W, Van Montague M, Inze D et al. (2000) Plant L-ascorbic acid chemistry, metabolism, bioavailability and effects of processing. *J of Sci of Food and Agric* **80**, 825–860.

Favell D J (1998) A comparison of the vitamin C content of fresh and frozen vegetables. *Food Chem* **62**, 59–64.

Gerber N, Scheeder M R L, Wenk C (2009) The influence of cooking and fat trimming on the actual nutrient intake from meat. *Meat Sci* **81**, 148–154.

Han J S, Kozukue N, Young K-S et al. (2004) Distribution of ascorbic acid in potato tubers and in home processed and commercial potato foods. *J of Agric and Food Chem* **52**, 6516–6521.

Ito H, Kikuzaki H, Ueno H (2019) Effect of cooking methods on free amino acid contents in vegetables. *J Nutr Sci Vitaminol* **65**, 264–271.

Khan Q U, Mohammadzai I, Shah Z et al. (2018) Effect of gamma irradiation on nutrients and shelf life of peach (Prunus persical) stored at ambient temperature. *Open Conf Proc J* **9**, 8–15.

Lee S, Choi Y, Jeong H S et al. (2018) Effect of different cooking methods on the content of vitamins and true retention in selected vegetables. *Food Sci Biotechnol* **27**, 333–342.

Margier M, Georgé S, Hafnaoui N et al. (2018) Nutritional composition and bioactive content of legumes: Characterization of pulses frequently consumed in France and effect of the cooking method. *Nutrients* **10**, 1668.

Nimalaratne C, Schieber A, Wu J P (2016) Effects of storage and cooking on the antioxidant capacity of laying hen eggs. *Food Chem* **194**, 111–116.

Nkhata S G, Ayua E, Kamau E H et al. (2018) Fermentation and germination improve nutritional value of cereals and legumes through activation of endogenous enzymes. *Food Sci Nutr* **6**, 2446–2458.

Nishioka M, Kanosue F, Yabuta Y et al. (2011) Loss of vitamin B(12) in fish (round herring) meats during various cooking treatments. *J Nutr Sci Vitaminol (Tokyo)* **57**, 432–436.

Rickman J C, Barrett D M, Bruh C M (2007) Nutritional comparison of fresh, frozen and canned fruits and vegetables. Part 1. Vitamins C and B and phenolic compounds. *J Sci Food Agric* **87**, 930–944.

Santos C S P, Molina-Garcia L, Cunhaa S C et al. (2018) Fried potatoes: Impact of prolonged frying in monounsaturated oils. *Food Chem* **243**, 192–201.

Song T, Barua K, Buseman G, Murphy P (1998) Soy isoflavone analysis. *AJCN* **68**, 1474S–9S.

Tewari S, Sehrawat R, Nema P K et al. (2016) Preservation effect of high pressure processing on ascorbic acid of fruits and vegetables: A review. *J Food Biochem* **41**, e12319.

World Cancer Research Fund International/American Institute for Cancer Research *Continuous Update Project Report: Diet, Nutrition, Physical Activity, and Prostate Cancer 2014*. Available at: https://www.wcrf.org/wp-content/uploads/2021/02/prostate-cancer-report.pdf

Yuan G F, Sun B, Yuan J et al. (2009) Effects of different cooking methods on health-promoting compounds of broccoli. *J of Zhejiang Univ-Sci* **10**, 580–588.

Zaheer K & Akhtar M H (2015) An updated review of dietary isoflavones: Nutrition, processing, bioavailability and impacts on human health. *Crit Rev Food Sci Nutr*: **13**,0 e-pub ahead of print.

FURTHER READING

Ballantine B, Marrs T, Syversen T (2009) *General and Applied Toxicology*. 3rd edn. Wiley, New York.

Coultate T P (2015) *Food: The Chemistry of its Components*. 6th edn. RSC, London.

Deshpande S S (2002) *Handbook of Food Toxicology*. CRC Press, London.

Fellows P J (2009) *Food Processing Technology: Principles and Practice*. 3rd edn. Woodhead Publishing, Cambridge.

Tiwari B K, Brunton N P, Brennan C (2013) *Handbook of Plant Food Phytochemicals: Sources, Stability and Extraction*. Wiley-Blackwell, New York.

Ryley J, & Kajda P (1994) Vitamins in thermal processing. *Food Chem* **49**, 119–129.

Shahidi F (ed.) (1997) *Antinutrients and Phytochemicals in Food*. American Chemical Society, Washington.

Shibamoto T, & Bjeldanes L F (2009) *Introduction to Food Toxicology*. 2nd edn. Academic Press, San Diego.

USEFUL WEBSITES

British Nutrition Foundation: www.nutrition.org.uk

Institute of Food Science & Technology: www.ifst.org

UK Food Standards Agency Committee on Toxicity 2002: https://cot.food.gov.uk

3 Food safety

Joseph V. Rodricks, Karen Huleback, and Meera Cush

OBJECTIVES

By the end of this chapter you should be able to:

- define food safety and describe the major threats to food safety
- describe the scientific tools and methods used to identify food-related health risks
- explain principles of risk assessment and management
- describe the risk management approaches to ensuring food safety.

3.1 Introduction

The science of food safety is devoted to understanding how substances present in food may have adverse effects on health, and the extent to which human exposures to (or intakes of) those substances must be limited to avoid those effects. Identifying safe exposures is gained by subjecting the evidence on adverse health effects to a process called risk assessment. The programmes used to protect human health are developed and implemented within a process called risk management. The latter is shaped by laws and regulatory and public health policies applicable to food. While there are some national differences in laws and policies, risk assessment and management approaches are generally consistent among nations. Indeed, an international intergovernmental organization called the Codex Alimentarius Commission (CAC) is devoted to achieving a high degree of global harmony in standards for food safety (Codex 1979).

This chapter describes risk assessment and management practices used to ensure the safety of food. It begins with a brief description of the constituents and common contaminants of food that, if not adequately controlled, could cause adverse health effects. It then turns to a discussion of the scientific methods used to identify the types of adverse health effects these substances may

cause. These methods involve the sciences of epidemiology and toxicology. Application of risk assessment to evidence regarding health effects follows and focuses on identifying safe exposure levels. A significant element of risk assessment pertains to uncertainties in scientific understanding and methods for dealing with them. Some typical risk management practices are introduced at this point in the chapter.

The chapter ends with discussion of the international risk management programme of the CAC, its supporting scientific structures, and its relationship to risk management agencies of its member nations. Emerging food safety issues are also described, along with some new methods for studying and assessing food-related risks (Codex 1979).

It should be noted that food safety is concerned with individual constituents and contaminants of food, and not with diets as a whole.

3.2 Constituents and contaminants of food

Chapter 2 of this textbook, 'Food and nutrient structure', provides a description of the natural chemical constituents of food, the types of chemicals intentionally introduced into food, and of the chemical and biological agents that are common contaminants of food. Table 3.1 contains a summary of the main categories of constituents, additives, and contaminants that are referred to in this chapter. Each category presents food safety issues in common with the others and also issues requiring special attention (Rodricks et al. 2020; Kotsonis & Burdock 2013).

It is no exaggeration to say that food is by far the most chemically complex part of the environment to which humans are continuously exposed. The nutrients and immense numbers of natural constituents of food, on their own, provide support for this statement. Substances

TABLE 3.1 Constituents and contaminants of food

Category	Characteristics

Natural constituents
  Nutrients
  Non-nutritive constituents
	Largest and chemically most diverse class

Chemicals produced during food processing and preparation	Reaction products from heating, irradiation, etc.
Substances intentionally introduced	Food additives, pesticides, veterinary drugs, and feed additives
Indirect additives	Substances migrating from food contact surfaces
Contaminants	Substances not expected to be present, or present at greater than expected (natural) levels, either chemical or biological

added to food directly (food and colour additives) and indirectly (pesticides, veterinary drugs, and constituents of food packaging and other food contact materials) add thousands of chemicals to those naturally present. Chemicals produced during food processing, particularly through heating, are coming under increased scrutiny, and the number of substances labelled contaminants, of both natural and industrial origin, is significant. All of these many thousands of food constituents and contaminants can, under conditions of excessive intake, cause adverse health effects, and pose potential food safety concerns. Biological contaminants, especially certain microbial pathogens, cause very large numbers of food poisonings, and so are at the centre of attention of public health and food safety agencies.

Our understanding of this highly complex matter has increased immensely in the past century, and that understanding has been put to good use to achieve a relatively high degree of public health protection. At the same time, it is recognized that the degree of protection achieved varies considerably around the world, and that even the most advanced societies have far from perfect food safety systems. Efforts to improve understanding of food safety risks, and to develop more effective risk management programmes, are thus continuing and pressing societal needs.

3.3 Evolution of efforts to investigate food hazards

Although concerns about food safety have existed throughout human history, there was little systematic, scientific investigation of the issue until the early twentieth century. There was, during very early human history, no doubt much 'trial and error' experimentation with potential sources of food to eliminate those plants and animals that were obviously poisonous. But it was

the revolution in chemical science that began in the mid-nineteenth century that initiated modern concerns about the safety of food. That revolution created the industrial capacity to produce a wide variety of synthetic chemicals, a capacity that continues to expand in the present age.

Efforts to investigate the composition of food, and to find ways to create new varieties of processed foods, also began in the early twentieth century, and food scientists began to rely heavily on products of the ever-expanding chemical industry to aid their efforts. The increasing uses of chemical additives (see Chapter 2 for a review of the various functions of additives in food), together with emerging concerns about lack of sanitation during food processing, led, in the United States, to the enactment of the first federal law directed at creating a regulatory system to ensure the safety of food (IOM 1998). This law, the Pure Food and Drug Act of 1906, set into motion many efforts to investigate health hazards associated with food constituents.

Similar laws emerged in Europe, and eventually in virtually every country. These laws have been updated periodically, generally in response to new scientific understanding of food-related risks, and in many cases the burden of demonstrating the safety of additives has shifted to manufacturers, replacing a system in which government regulators had the burden of demonstrating a substance was harmful before it could be prohibited from addition to food. In the UK, the Food Safety Act of 1990 replaced earlier English and Scottish laws governing food safety. Until recently the UK food industry was governed by a broad range of European legislation (www.food.gov.uk). Following the withdrawal from the European Union (EU), the UK now has its own regulations called the UK Statutory Instruments. For at least the various categories of additives listed in Table 3.1, those intending to introduce a substance into food are required to develop the data necessary to demonstrate safety, and to submit the data to government regulators; additives generally have to be approved for use by regulators, and no substance may be added to food without such an approval. The specific conditions of approved uses are specified in regulations. There are many variations on this basic model for ensuring the safety of substances deliberately introduced into food, and some differences in approach among countries, but to discuss these variations here would be beyond the scope of this chapter.

The model described above for the various categories of intentionally introduced food substances also applies to two other categories of substances: crop protection chemicals, otherwise known as pesticides, and veterinary drugs used in food-producing animals. Residues of chemical pesticides and veterinary drugs may be found in the foods to which they are applied, and in most cases, manufacturers are required to demonstrate the safety of such residues before regulatory approvals of their uses can be gained (Kotsonis & Burdock 2013).

Evidence required

The question of what types of evidence would be necessary to demonstrate the safety of these categories of food substances began to be investigated in the early decades of the twentieth century and continues to be investigated and refined in the present. Those scientists who first took on this work understood that, in sufficiently large intakes (doses), all chemical substances could cause some type of harm to health. They recognized that the type of harm would vary among chemicals, and that its severity would increase with increasing doses, and often with increasing duration of exposure. By the early 1950s, scientists in government agencies responsible for food safety decided that the safety of a food substance could be demonstrated if: (1) its adverse effects on health have been identified; and (2) human intake (dose) of the substance because of its presence in food was only a (specified) small fraction of the dose observed to cause adverse effects. In the section below on risk assessment the evolution of this simple idea about safety into modern concepts of risk assessment will be outlined.

It was evident to those early investigators that it would not be ethical to study the adverse effects of chemicals by deliberately exposing human beings in clinical trials. Thus, toxicity testing of chemicals in experimental animals was born. Animals had been used in various research settings for several centuries, but it was in the 1920s that the systematic uses of experimental animals to identify the hazardous properties of chemicals of all types began to be accepted, and government regulators began in the 1950s to insist on their use to support claims of safety. Animal tests to identify many different manifestations of chemical toxicity have grown in scientific rigour over the past several decades, and results from them

are widely accepted for use in chemical risk assessment. At the same time, it is now widely recognized that there might be better methods for identifying chemical toxicity that do not involve the use of animals; efforts underway to achieve this goal are discussed in the closing section of the chapter.

Epidemiological methods, in particular observational studies, have also come into use to study the adverse effects of chemicals. These methods have been particularly useful for the study of food contaminants. Categories of food substances other than the additives, pesticides, and animal drugs (Table 3.1) present somewhat different challenges for risk assessment and management, to be discussed in later sections (IOM 1998).

3.4 Toxicity and toxicology studies

Toxicity refers to any type of adverse health effect caused by exposure to a chemical. An effect is said to be adverse if it results in damage to the structure or function of the body. Effects may be relatively mild and even reversible, or may be serious and irreversible, but only the serious and irreversible effect is adverse. The effect can occur in any organ of the body (liver, kidney, skin, etc.), can impair whole systems (nervous, immune, endocrine, reproductive, etc.), can lead to cancer in any tissue of the body, or can harm a developing foetus. The particular manifestation of toxicity that is observed is a function of the chemical, the doses used, route (oral, inhalation, dermal) and duration of exposure. Thus, a complete description of a chemical's toxicity is rarely a single effect but includes several types of harm occurring at different doses and exposure durations, and even by different routes. Toxicity is studied in experimental animals in laboratory conditions. The type of toxic effects observed may depend on the test species (e.g., rat, mouse) studied. It can be difficult to interpret which of the many different results from animal tests is most relevant to humans (Kotsonis & Burdock 2013). This is a critical question in the risk assessment process, and approaches to answering it are described in the risk assessment section below.

In risk assessment terminology, manifestations of toxicity are referred to as hazards. The likelihood, or probability that these hazards will occur in humans under specific conditions of exposure to the substance is referred to as risk. Thus, to compensate for the limitations in the test sensitivity associated with the use of relatively small sample sizes, escalating doses are used to examine risk greater than about 10 per cent and up to 100 per cent (i.e., all animals affected). (Rodricks et al. 2020; Brock et al. 2003).

Toxicity protocols

Over the past 75 years, toxicologists have devoted enormous effort to developing protocols to be used to study toxicity, and to defining laboratory practices that should be followed to ensure the reliability of the observations made. Most of these protocols now have worldwide acceptance, and results from their application are used not only to assess the risks of food constituents and contaminants but also to assess risks of chemicals in many product applications, and risks of chemical contaminants of air, water, and soils.

Many protocols are designed to identify the types of toxicity a chemical produces after a single (acute) exposure, and after repeated daily exposures, running from a few weeks to several months (the 90-day subchronic exposure study in rodents is a widely used, highly informative protocol), to a full lifetime (18–24 months in laboratory rodents, chronic). Acute toxicity studies refer to the exposure of animals to single or multiple doses of a test substance within 24 hours by a known route (oral, dermal, or inhalation) and the evaluation of adverse effects that occur. Repeat dose studies are designed to identify the types of toxicity caused by a chemical following repeated administration. All such studies involve a group of control animals, identical in every way to animals receiving the chemical being tested, and several groups of animals receiving doses of increasing size. Causation is a relatively straightforward matter in these controlled experiments; effects observed in treated animals that are of a type not observed in controls are likely to be due to the chemical being tested. The case for causation is strengthened greatly if the severity and frequency of the observed effect is seen to increase as the size of the treatment dose increases. The notion that toxic responses of all kinds increase, in frequency or severity, as doses increase is a firmly established principle, and the 'dose–response relationship' is a key element of the risk assessment process (Rodricks et al. 2020).

The dose–response curve generated from toxicity studies can be used to identify whether there is a threshold dose for adverse effects, that is, a dose that must be exceeded before an adverse effect is produced. For such effects, the highest dose at which no adverse effects are seen is termed the no observed adverse effect level (NOAEL). The lowest dose at which adverse effects are seen, is termed the lowest observed adverse effect level (LOAEL).

For substances that may not have a threshold for effect, such as certain types of carcinogens, the dose that produces a predetermined change in response rate for an adverse effect compared with background is called a benchmark dose (BMD) (Crump 1984). The statistical 95 per cent lower confidence limit of the BMD (BMDL) is used in risk assessment. The NOAEL, LOAEL, and BMDL are often referred to as points of departure (PODs).

In an effort to harmonize the way in which toxicity testing is conducted there exists a set of standard methods for toxicity testing for assessing the potential effects of chemicals on human health and the environment by the OECD Guidelines for the Testing of Chemicals. These toxicity test protocols are in widespread use and can be found at www.oecd.org.

The single and repeat dose protocols are not designed to test any specific hypothesis about a chemical's toxicity. They are simply applied, and observations of toxicity are recorded. Other protocols are designed to test hypotheses—that is, to determine if a chemical causes a specific type of toxicity. Thus, there are several protocols available to test the hypothesis that a chemical causes neurotoxicity, and even to provide information about the specific types of nervous system damage caused. Protocols to study effects on the reproductive process and on the developing foetus and newborn are especially important because the biological systems controlling these processes tend to be very sensitive to external stressors such as chemicals. The study of perturbation to the endocrine system is evaluated from a number of protocols. These studies investigate the effect a chemical has on endocrine organs such as the thyroid and on hormone levels. These effects must be accompanied by histopathological effects on affected tissues and pathology such as reproductive/developmental effects or cancer.

Determining whether a chemical can cause cancer generally requires lifetime exposure studies in animals. Cancer usually begins when a carcinogen (cancer-causing chemical) reaches the cell and causes some kind of genetic damage or alteration in key genes. The path from initial damage to development of an observable malignancy is long and complex, and often involves a continuing series of cellular alterations over the greater part of a lifetime. While genetic changes (genetic instability and mutation) are important in the aetiology of cancer, a number of non-genetic events (tumour-promoting inflammation; sustained proliferative signalling; insensitivity to anti-growth signals; resistance to cell death; replicative immortality; dysregulated metabolism; angiogenesis; tissue invasion and metastasis; and avoiding immune destruction) are also involved (COC Guidance Statement G08—version 2.0 https://www.gov.uk/government/groups/committee-on-carcinogenicity-of-chemicals-in-food-consumer-products-and-the-environment-coc). A variety of *in vitro* test systems are available to identify a chemical's capacity to cause genetic damage and can be useful in supporting the understanding of the biological mechanism of particular cancers of genotoxic origin in whole animals (FDA 1982). Note that the progression from an initial insult to genetic material to a cancerous lesion is not inevitable, because most cells have the capacity to repair damage.

Whether toxicity tests focus on the effects of repeat dosing, or are designed to investigate a chemical's capacity to cause specific types of damage, the procedures used to conduct the test, and the observations to be made, are fully set out in test protocols. The combined skills of toxicologists, pathologists, and statisticians are required to make the observations (clinical findings, tissue pathology, etc.) and to determine which are caused by the chemical being tested and not due to chance alone. Thousands of observations are made during these tests, and detailed pathology work is performed on tissues after the dosing period has concluded and animals are killed. Good Laboratory Practice (GLP) regulations (Directive 2004/9/EC and Directive 2004/10/EC in the EU) define rules and criteria for a quality system designed to ensure harmonization (reliability, reproducibility, consistency, quality, and integrity) of observations made and accuracy of reporting the findings. GLP must be followed by testing laboratories if study results are to be accepted by regulators.

Various types of toxicity investigations usually are undertaken in research settings, and results from them are often published in scientific journals. The studies generally are not focused on the questions that the standardized protocols are designed to address but are directed at understanding the biochemical and physiological processes that underlie the production of toxicity, that is, mechanistic information. Such research studies may also focus on the fate of the chemical in the body, its metabolism, and its excretion (all part of the study of pharmacokinetics/toxicokinetics). Research of these types provide information on a chemical's mechanism of toxic action and can be highly valuable in determining how results from animal studies can be applied to humans (Kotsonis & Burdock 2013; Dourson & Stara 1983).

Finally, as noted earlier, standardized animal tests are generally carried out employing doses greatly in excess of often ≥100 times the doses people would normally be exposed to because of the chemical's presence in food. One important reason for the use of such high doses pertains to the fact that, for purely practical reasons, only relatively small groups of animals, perhaps 50–100 per test group, can be used. Such small sample sizes greatly limit the sensitivity of these tests for toxicity detection. For example, when groups of 50 animals are used in control and test groups, the lowest toxicity response rate that would be statistically significant would be 5 of 50 animals (assuming the observed response in control animals is 0/50, which is not usually the case). Thus, such a test design could not detect risk rates less than 10 per cent (5/50 animals). Toxicity risks of 10 per cent are quite large, and yet these are approximately the smallest risks detectable in animal studies of practical size. Thus, to compensate for the test sensitivity limitations associated with the use of relatively small sample sizes used in toxicity studies, doses are used to examine risks greater than about 10 per cent and up to 100 per cent (all animals affected). These percentages translate to risks (probabilities of 0.1 to 1.0).

3.5 Evidence from epidemiology studies

Clinical trials are important sources of information about the health effects of medicines and certain foods, but cannot, for ethical reasons, be used to study toxicity. Although adverse side effects may sometimes be observed in such trials, they are not major sources of information about chemical toxicity. It is, however, possible to evaluate the effects of exposures that occur in populations that are exposed to chemicals in the ordinary course of their lives, and with studies of appropriate design, determine whether those exposures are associated with adverse health effects. Studies of this kind are referred to as observational epidemiology studies, and most commonly involve studies of occupational and various kinds of environmental exposures. Such studies can be valuable sources of information for certain industrial chemicals that have entered the environment and then come to contaminate food. Information on the adverse health effects of lead, methyl mercury, and arsenic has been developed from both occupational and environmental studies. Note that for metals in particular, we all experience what is referred to as a background exposure, because all occur naturally in soils. In some regions, natural levels of some metals may greatly exceed most natural background levels, and this may create elevated exposures in local populations. Excessive levels may also occur because of certain uses of the metals, for example, historical uses of lead in paint and many other products, and because of localized industrial contamination.

Some important food contaminants that have been the subjects of extensive epidemiological study are listed in Table 3.2, along with some of their important adverse health effects (Rodricks et al. 2020; Kotsonis & Burdock 2013).

Studies in humans can be immensely important for risk assessment. But, as is well known, establishing causal relationships using observational studies is often problematic. Identifying quantitative dose–response relationships is often not possible for lack of reliable and complete information on the exposures experienced by the populations studied. Also, opportunities to conduct such studies are limited.

TABLE 3.2 Some important chemical contaminants of food and health effects found in epidemiology studies

Contaminant*	Example of health effect
Arsenic	Cancer of skin, liver, bladder
Lead	Neurological and cognitive impairments Blood and kidney effects
Methyl mercury	Neurological and cognitive impairments
Chlorinated dioxins	Carcinogenicity
Aflatoxins	Liver cancer

* See Chapter 2 for information on occurrence in food.

3.6 Risk assessment and its relationship to safety

The term 'dose' is an important term in risk assessment and refers to the amount of a chemical taken into the body over a given period of time. In most cases, the period of time of interest is a single day—thus, milligram of chemical/day (mg/day). For purposes of assessing toxic risk, mg/day is divided by body weight of the organism (human, rat, mouse) receiving the chemical. Dose expressed as mg chemical/kg body weight/day is known to be a reliable indicator of a way to compare expected degrees of toxic response across animal species. It is not the only expression of dose in use, but it is the most common.

The dose of a chemical received from the consumption of food is determined by multiplying the concentration of the chemical in food (mg/kg food) by the weight of food consumed per day (kg food/day), and dividing the product by the weight of the exposed organism. The concentration is determined by chemical analysis or may be known because a known amount of the chemical was added to a known amount of food. While dose is the measure typically used in assessing risk of toxicity, most food scientists prefer the term 'daily intake'.

Risk assessment acquired its current meaning and structure in a 1983 report from the United States' National Academy of Sciences (NAS) (*Risk Assessment in the Federal Government: Managing the Process*). The expert NAS committee produced this report in response to the growing need of US government agencies to evaluate the risks of increasing numbers of chemicals present in consumer products, food, air, water, soil, and workplace environments. It became clear to the expert committee asked to examine how this need could best be fulfilled that a highly systematic and consistently applied framework was needed as a guide, and its report contains such a framework. Government agencies almost immediately adopted the framework, and, with some modification in the terms used, EU Member State agencies and the CAC did the same. Risk assessment for chemical substances now proceeds in a similar manner globally (Rodricks et al. 2020; IOM 1998).

Risk is the probability of harm from exposure to a hazard. Risk can be expressed quantitatively, and has values ranging from zero to 1.0 (0 per cent to 100 per cent chance of harm). Determining whether, in any specific circumstance, the probability of harm is absolutely zero (a condition of absolute safety) is scientifically impossible, and thus safety has to be defined as a condition in which there is a small probability of harm. In many cases, probabilities of harm may actually be zero, but no scientific means exists to establish such a condition. In chemical safety assessments, risk is described qualitatively, and safety is typically described only as a condition of 'very small risk' (or in similar terms). Strictly speaking, decisions about how small risks should be to protect human health, and measures to be taken to achieve those risks, fall within the domain of risk management. This risk assessment–risk management distinction was also proposed in the 1983 NAS report and is widely accepted (Rodricks et al. 2020; Brock et al. 2003).

3.7 The conduct of risk assessment

Risk assessment is a four-step process, carried out by experts in toxicology, epidemiology, biostatistics, and scientists who are experts in assessing adverse health effects following human exposures to chemicals. The assessment process relies upon the interpretation of the scientific evidence described earlier. It also involves the issue of certain assumptions (often called 'default assumptions') to fill gaps in scientific knowledge. These various assumptions refer to uncertainties in our knowledge of variability in response to a given dose of a hazardous agent across different species (e.g., rodent–human differences) and across members of the human population. Other sources of variability in response also need to be considered, most especially that relating to sensitive subpopulations such as children (Kotsonis & Burdock 2013; Dourson & Stara 1983).

That these variabilities exist is certain; their magnitudes are not firmly established. Various assumptions are used in risk assessment to account for these variabilities, and they are expressed as Uncertainty Factors (UFs), but also known as safety factors, variability factors, and assessment factors. The typical value ('default values') of 100 (10-fold factor for interspecies extrapolation and a 10-fold factor to cover human variability) has been used for extrapolating experimental data from the NOAEL for the critical effect in animal studies to produce health-based guidance values/reference doses and is widely accepted as precautionary (i.e., they are likely to overestimate true variabilities). Regulatory authorities generally accept this view, but it is not universally accepted in the scientific community (Kotsonis & Burdock 2013; Dourson & Stara 1983).

Four steps in risk assessment

There are essentially four steps that describe risk assessment (Figure 3.1): hazard identification, hazard characterization (dose–response evaluation), exposure assessment, and risk characterization (NRC 1983; IPCS 2009).

Step one is known as Hazard Identification. It involves assembling all of the available toxicology and epidemiology evidence for the substance under review, evaluating the quality of individual studies, and synthesizing from the evidence the key findings. The purpose is to identify all of the toxic effects (hazards) produced by the substance at the tissue and cellular levels, the conditions under which they are produced (dose, duration, species tested, exposure route), and the strength of the scientific evidence supporting each type of hazard. The strength of the evidence is judged by considering factors such as study quality and consistency of the finding across different studies. Data from human experimental or observational and epidemiological studies are preferred, but information obtained from animal studies are typically more readily available. Data from *in vitro* and structure–activity relationships (SARs) may also be used to identify hazards and often used as supporting information. Techniques such as toxicogenomics (i.e., the application of functional genomics technologies such as transcriptomics, proteomics, and metabolomics) are being increasingly used in toxicological risk assessment. While these techniques are becoming more prevalent in the safety assessment of cosmetics in the EU where testing on animals is prohibited, its use in identification of hazards from food is limited. This is primarily due to the fact these techniques have not been incorporated into the guidelines for toxicological studies and are yet to achieve regulatory approval.

Step two is called Dose–Response Evaluation in the USA and Hazard Characterization in the EU and some other countries, and in the CAC. The purpose of this step is to identify the study in which the critical toxic effect has been identified, and the dose–response curve for that effect. In most cases the critical effect is that occurring at the lowest dose. Using dose–response data for the effect observed at lowest dose as the starting point for risk assessment ensures that the risk assessment is protective of other effects occurring at higher doses. For example, liver toxicity caused by chloroform occurs at lower doses than nervous system toxicity caused by chloroform. If the risk assessment reveals the conditions under which liver toxicity is likely avoided, then it can be assumed that under those conditions, nervous system effects are avoided (Rodricks et al. 2020; Brock et al. 2003; Dourson & Stara 1983).

The dose–response curve for the critical effect reveals how its risk increases with increasing dose. The study should also reveal the maximum dose on the curve at which no adverse effect is observed, as measured against effects observed in control animals. The NOAEL (which should properly be called a no-observed-adverse-effect-dose, not a 'level'), is a safe dose for the experimental animal group studied. Toxicity does not begin to be observed until some threshold dose is exceeded; the NOAEL might represent that threshold, although the true threshold could well be at a greater (untested) dose.

Step two of the risk assessment process involves the application of UFs to the point of departure (e.g., NOAEL), to consider variability as described above. The result of this process is the derivation of acceptable daily intake (ADI) or tolerable daily intake (TDI), which are collectively known as health-based guidance values (Figure 3.2).

$$ADI/TDI = POD/UF_1 \times UF_2 \times \ldots UF_n$$

ADIs are derived for chemicals that are intended for use in food while TDIs are derived for chemicals that are unavoidably present in food, such as contaminants. Tolerable usage reflects the view that we are forced to tolerate, and not 'allow', a certain level of contamination. The ADI is a dose (although it is called an intake) likely to pose very little risk and is usually taken to represent a condition of safety. The US Environmental Protection Agency, responsible for the regulation of pesticides in food and for many other chemicals in the environment, refers to the dose derived in this same way as a toxicity reference dose (RfD). Generally, food safety agencies use the ADI nomenclature, or TDI for contaminants.

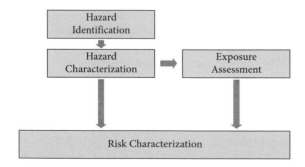

FIGURE 3.1 Stages of toxicological risk assessment.
  * Dose–response assessment in the USA.

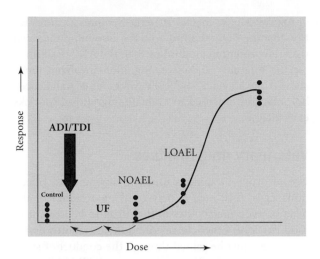

FIGURE 3.2 Derivation of ADI/TDI from animal studies. ADI: acceptable daily intake; NOAEL: no observed adverse effect level; LOAEL: lowest observed adverse effect level; TDI: tolerable daily intake; UF: uncertainty factor.

Step three involves assessment of Human Exposure to the substance of concern. The dose, or intake, of the substance because of its presence in food, is estimated. Human exposure assessment typically focuses on the 'high end' of the range of intakes occurring in the human population or subpopulation (e.g., toddlers, children, adults, elderly, ethnic groups); thus individuals incurring doses at the 95th percentile of the distribution of doses in the population, or even higher percentiles, are the principal focus of risk assessment. The result of Step three is the Estimated Daily Intake (EDI).

Step four is called Risk Characterization. In this step the EDI is compared with the TDI, ADI, or RfD. Safety is said to have been achieved if the EDI is less than these safety measures. If the EDI exceeds the ADI, a risk to human health is suggested or demonstrated, depending on the magnitude of exceedance. The Risk Characterization contains not only a description of the EDI/ADI relationship but also a summary characterization of the chemical hazards, and the associated scientific uncertainties.

3.8 Carcinogen risk assessment

Chemicals that can induce cancer may do so by mechanisms different from those operating in the production of other types of toxic effects. The latter is thought to arise only after a threshold dose has been exceeded (represented conceptually for a population by the ADI). Carcinogens may operate through non-threshold mechanisms, a hypothesis that translates to the proposition that any exposure to a carcinogen increases cancer risk (COC 2018). This proposition does not mean that any exposure

can cause cancer; it means only that the odds of cancer occurring are increased. Those odds may range from the extremely small lifetime probabilities of one-in-a-million or less, for example, to very large (probabilities of one-in-one hundred or more). In the United States, regulatory agencies have adopted methodologies that involve inferences about the dose–response relationship in the range of zero dose up to the dose associated with the minimum cancer risk detectable in experimental animal studies or in epidemiology studies (a lifetime risk of about one-in-ten). In the UK and EU, exposure to genotoxic carcinogens should be as low as reasonably practicable (ALARP). For this reason, substances that do not have a threshold for safety are not permitted to be intentionally added to food or authorized for consumption as food. However, the ALARP principle is employed for genotoxic substances that are unavoidable in the diet (e.g., aflatoxins, polycyclic aromatic hydrocarbons, acrylamide). For such substances a Margin of Exposure (MOE) approach is used to help prioritize risk and support risk management decisions (EFSA 2005). The MOE approach uses a dose that causes a low but measurable response in animals such as the $BMDL_{10}$, often taken from an animal study, and compares this with various dietary intake estimates in humans, considering differences in consumption patterns. $BMDL_{10}$ is the 10 per cent change in the response rate of an adverse effect relative to the response of control group. When using this approach, risk assessors indicate whether the MOE is of high concern, low concern, or unlikely to be of safety concern. An MOE of 10,000 or higher is considered of low concern from a public health point of view with respect to the carcinogenic effect if it is based on a $BMDL_{10}$ as the point of departure (EFSA 2005).

$$MOE = Exposure/BMDL_{10}$$

3.9 Risk management to achieve food safety

Food additives

No substance can be intentionally introduced into food unless it has been shown to have a technical function in the food (see Chapter 2, Section 2.3 on Additives), and unless the EDI resulting from its use in foods does not exceed the ADI established for the additive. In addition, any reliable evidence of carcinogenicity from animal studies would prohibit its addition to food in any amount. Regulations for food additives are generally publicly available and describe the basis for the additive's approval. They also describe specifications for the additive's purity and other relevant physical and chemical characteristics that must be met for legal use of the additive. Regulations typically list specific approved uses, and the amounts of an additive that are not to be exceeded. In cases in which an additive is used in several foods, the EDI from all such uses must not exceed the ADI (Rodricks et al. 2020; FDA 1983; IOM 1997–2002).

Food additives in the EU/UK are heavily regulated with defined sets of data requirements covering toxicological endpoints, absorption and metabolism in the gastrointestinal tract, and chemical/physical parameters. The latter is used for setting specifications for the additive.

GRAS

There is a category of added substances in the United States that has importance for safety. Under federal law, some additives may meet the criteria for what is called GRAS status: Generally Recognized as Safe. Substances having a history of safe use in food prior to the enactment of new food additive laws in the 1950s were defined as GRAS; in effect, lawmakers saw no need to require new animal testing on such substances. In addition, the laws allowed experts in food safety who were outside of the federal regulatory agencies (i.e., in academia or the private sector) to judge whether a food substance could be considered GRAS; this system still exists in the United States, and many food spices and flavouring agents, and substances such as acetic acid, calcium carbonate, lecithin, rennet, and benzoic acid are considered to be GRAS. Although the EU has a system for food additives similar to that of the United States, it has nothing comparable to the GRAS category (https://www.efsa.europa.eu/en).

Regulatory action to prohibit or limit use of an approved additive (or GRAS substance) may be taken whenever new data become available that suggest the approved use is not safe. Regulators have such authority with all categories of substances deliberately introduced into food.

Indirect food additives

Substances present in packaging and other food contact materials are considered to be additives if they are shown to migrate from the material that contacts the food into the food. Many polymers, for example, are used in food packaging materials and thus come into contact with food, and such polymers contain small amounts of monomeric chemicals from which they are synthesized. Polymers are generally quite inert materials, and do not themselves become components of food, but small molecular weight monomers can usually migrate in some amounts. The extent of migration depends upon the solubility characteristics of the monomer or other constituent of the contact material.

Approval of a contact material requires demonstration that the EDI resulting from the amount of migrated material present in food is less than the ADI established for the migrant. Studies of migration involving food-simulating solvents, typically water and water/alcohol mixtures, are used to evaluate migration (Rodricks et al. 2020).

Veterinary drug residues

Use of drugs in animals used for human food (meat, milk, eggs, fish, honey) may result in residues of the drug or even metabolites of that drug to be present in food. As with other classes of additives, approvals for such uses must be gained through the conduct of studies that provide the data necessary to establish ADIs and to estimate EDIs. Carcinogens are generally disallowed, although they may be permitted if EDIs are shown to create lifetime risks of less than one-in-one million (United States), or if research shows that the mechanisms by which they induce cancer require the exceedance of some threshold dose.

Crop protection chemicals (pesticides)

Uses of pesticides in food production are permitted if the residues of pesticides remaining in food lead to EDIs that are less than the ADIs established for these pesticides. Field trials of proposed uses are conducted to study residue levels and their persistence in food, and toxicology studies are conducted to establish Maximum Residue Levels (MRLs). These are the maximum levels of pesticide residue that are protective of health and may remain in food following correct use of the pesticide product in accordance with Good Agricultural Practices (GAP).

Many pesticides, such as Dichlorodiphenyltrichloroethane (DDT) and other chlorinated compounds, were prohibited from use in the 1970s and 1980s, because of safety concerns, in part related to chemical stability and extremely long persistence in the environment. Residues of some of these pesticides can still be found in fatty tissue and blood of humans and other animals, and in certain foods, such substances are considered to be contaminants (Rodricks et al. 2020; Kotsonis & Burdock 2013).

Substances formed by processing

The use of heat in the processing and preparation of foods, as well as smoking and other processes, results in numerous chemical changes (see Chapter 2, Section 2.3). A number of important and potent carcinogens, such as polycyclic aromatic hydrocarbons (PAHs, substances form when any type of organic matter is burned), N-nitrosamines, and the animal carcinogen, acrylamide, are just a few of the food constituents in this category. There is much study of their risks, and effort to find ways to reduce their occurrence. Although these substances are not, strictly speaking, 'contaminants' of food, they present difficult risk management challenges similar to those presented by contaminants (next section). No ideal risk management approach has yet been identified to deal with such substances (www.food.gov.uk).

Contaminants

Substances that are not expected to be present in food are considered contaminants. They are not intrinsic (natural) components of food, and they are not intentionally introduced. They enter food because they are, for one reason or another, present in environments where food is produced. They may be industrial chemicals or produced naturally (fungal and marine toxins). A subcategory of contaminants are substances, particularly metals, that occur naturally in foods because of their natural occurrence in soils and water; if these same materials have been produced industrially, they may contaminate environments at greater than normal 'background' levels, and then become contaminants of foods (see Section 2.5 of Chapter 2).

Unlike intentional additives, contaminants cannot simply be eliminated from foods. Many contaminants have been subjected to extensive epidemiological and toxicological study, and TDIs and cancer-risk estimates have been developed for many of them. But in many cases purely risk-based limits on the level of contaminants allowed in foods would, if 100 per cent enforceable (an impossible goal), result in the destruction of enormous amounts of food because levels of contamination in many cases exceed levels allowable based purely on risk-based goals. Regulators often set limits for such contaminants based on what are thought to be reasonably achievable goals (www.food.gov.uk: see views on arsenic in rice).

Some important food contaminants of industrial origin are listed in Table 3.2. Extensive efforts are underway to improve risk assessments for these substances, to study their effects in human populations, and to identify risk reduction strategies. Similar, though generally less extensive, efforts are underway for important contaminants of natural origin (Kotsonis & Burdock 2013; IARC 2012; NRC 2002).

Chemicals of recent concern are the perfluoroalkyl substances (PFAS) which form a group of approximately 4700 man-made chemicals that remain in use in consumer products. PFAS have also been used in oil-resistant coatings for paper and board products approved for food contact. Due to their extremely persistent nature these substances have been found in food such as fish, fruit, eggs, and egg products. The critical effect for

risk assessment of PFAS is the immune system and substances are known to accumulate in the body. A tolerable weekly intake (TWI) has been established for this group of substances and given that the TWI is set for the critical effect (immune system) it is expected to be protective against the other known adverse effects of PFAS (EFSA 2020). There are work programmes currently underway to replace the use of PFAS in food contact materials (OECD 2020).

Nutrients and other intrinsic components of food

By intrinsic, we refer to substances that are naturally present in food. Among these are substances having nutritional value and huge numbers of natural products. Nutrients obviously play critical roles in maintaining health, and it is also recognized that deficiencies in nutrient intake can have adverse health effects. Thus, adverse effects on health can result from both inadequate and excessive intakes of a food substance. Excessive intakes of nutrients, even essential nutrients, can cause adverse effects, and efforts have been underway for the past 20 years to identify what is called the 'tolerable upper intake level' (UL). In concept, the UL for a nutrient is derived in a manner that is similar to the derivation of an ADI, and is the highest level of daily nutrient intake that is likely to pose no risk of adverse health effects in the general population over a long period of time. ULs are available now for many nutrients (IOM 1997-2002; EFSA 2010). ULs are used, along with measures of nutritional deficiencies, to develop Dietary Reference Intakes (DRIs) in the United States and Canada, and Dietary Reference Values (DRVs) in the EU/UK. DRVs and DRIs are used for planning and assessing diets for healthy people to avoid deficiencies and excess intakes.

Although the many thousands of natural food constituents have not been subjected to extensive study, the toxic properties of some of these substances have been identified. It seems that, under ordinary conditions of food consumption, these substances do not pose significant health risks, but this conclusion must be heavily qualified, considering our lack of data. Moreover, there is now much interest in studying the possible health benefits of many of these natural products (so-called 'bioactives'). Important members of this class are sometimes referred to as phytoprotectants (Chapter 2). There are now widespread research efforts underway to improve understanding of the roles of nutrients and other natural food constituents in chronic disease risks, both increased and decreased.

In 2019, a United States government sponsored review of the health effects of sodium and potassium intakes, both excesses and inadequacies, led to the development of what the expert committee called a Chronic Disease Risk Reduction Intake (CDRR) for sodium (insufficient data were available to develop a CDRR for potassium). The CDRR includes a range of intakes over which reductions of intake will reduce the risk of several interrelated chronic diseases (cardiovascular disease, hypertension, and systolic and diastolic blood pressure); for practical applications it is expressed as the daily intake above which intake reductions will reduce disease risk (for sodium, 2300 mg/day for adults). The UL remains in place, although the committee review of sodium and potassium concluded that there was insufficient evidence to develop ULs for these nutrients. The introduction of the CDRR stemmed from the realization that most chronic disease risks exhibit dose–risk relationships that are different in character from those related to more acute forms of toxicity, where there tend to be relatively sharp divisions between safe and unsafe levels.

Microbial pathogens

Bacteria that, under certain conditions, can produce toxins and thereby cause illness, fall into this category. The problematic species are not infectious agents in the usual sense, that is, they are not transmitted from person-to-person. Rather they can, under some circumstances, contaminate food or water. In some cases, the pathogens grow and produce toxins in foods; in other cases, the pathogen is swallowed, comes to infect the human gastrointestinal tract, and while growing there produces its toxins. In either case, the result can be 'food poisoning' (www.food.gov.uk).

Data regarding disease outbreaks are published by the UK Food Standards Agency and the US Centers for Disease Control and Prevention (CDC). The great majority of these illnesses are sporadic, individual cases, not associated with each other (at least, as far as we know). A small percentage of these illnesses occur in foodborne illness outbreaks, in which two or more people become ill from consuming a common food or foods. Many categories of food are susceptible, and likelihood of contamination depends upon a host of factors. As these numbers show, most foodborne illness cases amount to nothing more than the extreme discomfort associated with vomiting and diarrhoea, but mortality rates can be high especially in certain subpopulations—the very young, the very old, pregnant women, and people with impaired immune functions.

Bacterial (and viral) pathogens can enter the food chain very early; in fact, many foods leaving the farm carry them. But they become dangerous only if foods are mishandled, and conditions for their growth (and for some bacteria, toxin production) are created (Abebe,

Gugsa, & Ahmed 2020). They can be destroyed by heat, by other food processing techniques, and by interventions designed to reduce pathogen contamination at early processing steps, such as in slaughter operations. And, of course, cooking in the home or in restaurants may destroy them. Food preparation in the home and in restaurants also provides opportunities for cross-contamination between contaminated foods (that themselves may later be cooked) to foods such as raw produce that are not cooked and are eaten raw (and contaminated). So, in the entire 'farm-to-table' chain, there are opportunities for both the growth and the destruction of foodborne pathogens (Redmond & Griffith 2003).

The major culprits are certain species and subtypes of *Staphlococcccus aureus*, *Clostridium botulinum*, *Clostridium perfringens*, and *Escherichia coli* O157:H7 and other O157 STECs and non-O157 STECs (Shiga-toxin producing *Escherichia coli*), all of which produce toxins that cause illness. The toxins produced in food are proteins of extreme toxicity; some are vulnerable to destruction by heating (botulinum toxin and the toxin produced by *Clostridium perfringens*) but other toxins (for example, those formed by *Staphlococcus aureus*) are not destroyed by heating. *Campylobacter jejuni* and other species of *Campylobacter*, *Salmonella*, and *Shigella* species, *Listeria monocytogenes*, and *Escherichia coli* non-STEC subtypes cause illness directly and not through toxin production (IOM 1998; Brock et al. 2003; Thorns 2000).

The combined result of research and surveillance innovations has been a much improved understanding of which foods are most likely to be associated with illnesses and outbreaks, the relative impacts of various categories of foods (e.g., raw produce, meat, poultry, shellfish, etc.) on the overall foodborne illness totals, and the effects of various regulatory interventions to reduce incidence and prevalence of certain foodborne pathogens. A leading example of such an intervention is Hazard Analysis and Critical Control Points (HACCP). It is an internationally recognized preventative food safety system in which every step in the manufacture, storage, and distribution of a food product is scientifically analyzed for microbiological, physical, and chemical hazards, and points are identified in the system (often through the use of risk assessment) at which interventions will achieve the greatest reductions in hazard, and therefore the greatest reductions in risk to the ultimate consumer. It forms part of food safety laws internationally and is a way of managing food safety hazards.

According to the European Food Safety Authority (EFSA) and the European CDC, *Campylobacter* is the most common cause of food-borne illness, and *Salmonella* is the second most common cause of illness. *Listeria*, which also has the highest case-fatality rate in

Europe, is the fifth most common cause of illness (www
.food.gov.uk).

Microbial risk assessment models

Successful development of microbial risk assessment
models is a leading priority in public health and regulatory agencies. The risk assessment framework described
earlier for chemical toxicity is applicable to microbial risk
assessment. Once information is available on microbial
hazards, which are for the most part acute (immediately
or relatively quickly observable—occurring within days
or weeks of exposure) conditions resulting from acute
(one-time) exposures, and on their dose (pathogen
count)–response characteristics, it is possible to assess
the risks associated with any dose of interest. Hazard
information for important pathogens is readily available but, as expected, dose–response characteristics are
much harder to obtain. These food-safety hazards have
a characteristic that is not associated with chemicals—
they are living organisms that can reproduce. So, in
their journey from farm to table, organism counts can
increase and, if conditions for growth are unfavourable,
they can decrease.

Microbiologists have developed ways to model microbial growth using assumptions related to the expected
behaviour of organisms under different environmental
conditions. These models are then coupled with dose–
response models with the result that risks (responses) can
be estimated, given a certain degree of knowledge about
initial microbe counts and the environmental conditions
(related to food processing) these microbes encounter as
they move from these initial conditions to the plate.

Most microbial risk assessors hold to the 'no-threshold'
hypothesis for those pathogens that can infect. This is
because a single bacterium, if still viable when it reaches
the gastrointestinal tract, can multiply and cause disease.
Virtually all current guidelines for microbial risk assessment acknowledge this fact and note that, practically
speaking, it is not possible to distinguish between a very
low non-zero risk and a true threshold. Sigmoid (S)-
shaped dose–response models, with no threshold characteristics, are typically proposed for microbial risk assessments (although threshold models are used for those
toxins produced in foods). Microbial risk assessment is
in a state of rapid evolution, and it is exciting to observe
the worldwide scientific dialogue that is underway. At the
same time, public health approaches to detect outbreaks
and to respond to them are improving, and regulatory
actions to put into place practices to minimize food
contamination are becoming ever more sophisticated.
Appropriate controls at many steps in the farm-to-table
pathway are necessary to reduce the heavy burden of
food-pathogen related diseases. It should be emphasized
that advances in this area of food safety occurring in developed countries are not evident in many less developed
countries, where information on foodborne disease burdens is lacking.

Food safety and the microbiome

One of the most dynamic areas of current research
concerns the various roles of the human microbiome
in human disease (see also Chapter 23). Evidence is accumulating that interactions between substances in the
environment and the microbiome can affect risks of a
variety of diseases, in both beneficial and detrimental
ways. There is already a substantial body of literature
regarding the role of diet in microbiome-mediated
health conditions and developing literature on the interactions between pharmaceuticals and other chemicals and the microbiome, with most of the research to
date focusing on the gut microbiome (GMB). Of particular interest to food safety is the question of whether
non-nutritional substances present in foods can affect
the assembly, maturation, and stability of the microbiomes, and whether those effects have consequences for
host disease status. This issue is now the subject of substantial research effort, and may become a significant
element of future food safety risk assessment. A study
of this subject has been published by the US National
Academies of Science, Engineering and Medicine
(https://onlinelibrary.wiley.com/doi/abs/10.1111/
risa.13316).

Genetically modified foods

When reference is made to biotechnology ('biotech') reference is often being made to genetically modified foods
or to food products derived from genetically modified
organisms (GMO). Genetic modification involves the
human manipulation or modification of genetic material
of a plant, microorganism, and animals, by means other
than traditional plant hybridization or animal breeding.
Such changes have been made, among many other changes, to maintain peak flavour of fruits over longer shelf-
lives; to enable certain crop plants to resist the effects of
certain widely used herbicides; and to enable salmon to
grow to market size in a significantly shortened period of
time. Proponents of genetic engineering advocate for the
technology as a valuable tool that will benefit humanity
as we strive to feed more and more people with fewer and
fewer natural resources, thereby increasing food security
worldwide, while at the same time reducing food losses
and food waste and, as well, reducing overall chemical inputs into the environment for crop protection purposes

and reducing the overall environmental impacts of large-scale agriculture.

Critics of genetic modification claim that there are too many unknowns surrounding the man-made manipulation of plant and animal genetic material, and that, furthermore, we can't know all the possible direct or indirect negative consequences of genetic manipulation that could result in unanticipated and unintended health effects or other consequences, such as environmental damage.

Gene editing is a method that allows the DNA of organisms to be changed by removing, adding, or replacing sections of the DNA so that the desired physical traits can be achieved. Currently in the UK GMOs are defined in the Environmental Protection Act 1990 that requires that all gene edit organisms are classified as GMOs irrespective of whether they could be produced by traditional breeding methods (C-528/16 - Confédération paysanne and Others). However, the UK regulator, Department for Environment Food & Rural Affairs (DEFRA), wishes to amend this regulation so that organisms produced by gene editing or by other genetic technologies should not be regulated as GMOs if they could have been produced by traditional breeding methods. This would be an important change in legislation as new approach methodologies means that there is increased likelihood that organisms can be manipulated in this way, but if this potential regulation were to come into force such gene-edited organisms would not have to be subject to the rigorous risk assessment as for GMOs.

3.10 Food allergies and intolerances

Many people are allergic (or hypersensitive) to certain foods and food ingredients (see also Chapter 27). In most cases, the reaction is immediate and reversible, but delayed-onset allergies also occur. These reactions involve the immune system, and a prior exposure to the allergenic agent is required to precipitate an event. Immunoglobins are typically involved, and reactions can be cutaneous (hives, dermatitis, rash), gastrointestinal (nausea, vomiting, diarrhoea), or respiratory (asthma, wheezing, rhinitis). Anaphylactic shock and death can also occur. Many foods have been reported to be allergenic, although the majority of cases involve peanuts and other tree nuts, milk, fish, and shellfish. It appears that most allergenic agents in food are large molecular weight glycoproteins, although closely related foods with similar proteins do not all cause allergies. The reaction is highly individualized, and not similar to toxic responses described earlier. There is no available risk model to establish safe intake levels for individuals who are hypersensitive, and reactions are avoidable only by avoiding foods that are sources of the allergen.

Some types of genetic predisposition may lead to certain forms of food intolerance in individuals; these may resemble allergic reactions, but they are not immune mediated. In many cases, individuals lack certain enzymes necessary for the normal metabolism of food ingredients. Lactose intolerance, for example, results from the lack of the enzyme lactase, and this leads to excessive lactose accumulation in the bowel. Microbial fermentation of lactose in the bowel has an osmotic effect, leading to malabsorption and diarrhoea. Lactose intolerance appears to be less likely in the population in Northern Europe but reaches 90 per cent in Southern Italy and nearly 100 per cent in Southeast Asia. In the UK, lactose intolerance is more common in people of Asian or African-Caribbean descent (https://www.nhs.uk/conditions/lactose-intolerance/). Other intolerances include those to fava bean (favism), chocolate (migraine), and red wine (similar to allergic reactions). 'Asian flush syndrome' is seen in many Asians after alcohol consumption. No risk model is available to develop safe intake levels for people who experience these idiosyncratic reactions.

There are other categories of unusual reactions to certain foods, including anaphylaxis related to histamine content. Food–drug interactions of different kinds are not uncommon, and drug labelling is needed to warn people about them (such as warnings about grapefruit consumption for many drugs). Allergic and idiosyncratic reactions to certain foods and food constituents are unfortunately not manageable except by the individuals who are susceptible to them. Public health and regulatory authorities can only educate and provide warnings.

3.11 Food safety institutions

In the EU, the Directorate-General of the European Commission (EC DG SANTE) is responsible for ensuring that the food that we eat is safe. They do this by having effective control systems and setting standards for food safety. All applications for authorization of a food or food ingredients are presented to the EC in the first instance. The EFSA provides independent scientific advice on all aspects relating to food safety. A broad range of legislation governs the safety of the food industry in EU Member States. Specifically, food additives, novel foods, genetically modified foods, and food contact materials (www.efsa.europa.eu). The EU published the Farm to Fork Strategy (https://ec.europa.eu/food/farm2fork_en) which forms part of the European Green Deal. This strategy aims to achieve a fair, healthy, and environmentally friendly food system in order to make the availability of food sustainable.

In the UK, the Food Standards Agency (FSA) is responsible for food safety and food hygiene in England, Wales, and Northern Ireland. Scotland is represented by Food Standards Scotland. The FSA was established to replace the Food Advisory Committee by an act of parliament on 10 June 1999, which set out its main objective of 'protecting public health in relation to food and the functions that it will assume in pursuit of that aim, and give the FSA the powers necessary to enable it to act in the consumer's interest at any stage in the food production and supply chain'. Following the UK exit from the EU ('Brexit') in 2020, the UK has retained all current EU food regulations and transcribed these into UK law. However, the Northern Ireland Protocol states that any business seeking a new authorization for a regulated food and feed product placed on the Northern Ireland market will have to continue to follow EU rules, although UK legislation will still be applicable. The FSA has in place a number of scientific advisory committees (SACs) which are responsible for providing independent advice based on the best and most recent scientific evidence.

Examples of such committees are Committee on Toxicity of Chemicals in Food, Consumer Products and the Environment (COT) and its sister committees on carcinogenicity (COC) and mutagenicity (COM) and the Advisory Committee on Novel Foods and Processes (ACNFP) (https://sac.food.gov.uk/).

In the United States, the majority of food safety, scientific, and regulatory activities reside in three agencies: the Food and Drug Administration (fda.hhs.gov); the Food Safety and Inspection Service (fsis.usda.gov); and the Environmental Protection Agency (epa.gov). FSIS is responsible for regulatory risk management of food safety with respect to all meat, poultry, and egg products; FDA is responsible for regulatory risk management of all other foods; and EPA is responsible for regulatory risk management of pesticides in foods.

At the international level, the CAC (codexalimentarius.net) develops and publishes food safety and quality standards for all foods circulating in international trade. The CAC is a joint function of two United Nations agencies: the Food and Agriculture Organization (FAO; fao.org) and the World Health Organization (WHO; who.int). Codex standards are voluntary; all of Codex's 187 member countries have the sovereign right to establish their own food safety regulations to protect their own consumers as they see best. Codex standards are the reference standards for the World Trade Organization (WTO), the 'court' that trading countries turn to in cases of trade disputes. Given, however, that adoption of a Codex standard confers the assumption of WTO compliance, and that any country not employing Codex standards must be able to justify its non-Codex standard with rigorous science and risk assessment. How Codex standards are written is obviously of critical importance to countries with significant food trade. The risk management (standard setting) activities of Codex (Codex Alimentarius—the compendium of all Codex standards) are supported by scientific evaluations and risk assessments performed by three panels of independent scientific experts, which are also jointly administered and managed by FAO and WHO (the Joint Expert Committees on Food Additives—JECFA; the Joint Meetings on Pesticide Residues—JMPR; and the Joint Expert Meeting on Microbial Risk Assessment—JEMRA).

Much activity relating to food safety is underway in public health agencies, including the US Centers for Disease Control and Prevention (CDC; cdc.gov) and the International Agency for Research on Cancer (https://www.iarc.who.int/). The International Life Sciences Institute (ilsi.org) brings together scientists from industry, governments, and academia to study important and emerging food safety issues. The Centre for Science in the Public Interest (cspinet.org), and the European Consumer Organization (beuc.org) are among many groups offering manufacturer and consumer advocacy perspectives.

3.12 Monitoring and surveillance

To ensure compliance with food safety standards and to provide data for continuing risk assessments, most country-level agencies monitor the food supply and also conduct surveillance. Much activity of similar natures is conducted at provincial and state levels. Monitoring generally refers to sampling of selected foods to determine whether they follow applicable standards. Regulators can take action to recall from commerce, or to prevent entry into commerce, products found not to comply. Surveillance generally refers to sampling and analyses conducted to identify emerging problems. The websites of food safety institutions, such as those mentioned in the previous section, contain such information.

3.13 Some emerging issues

Efforts to improve toxicity testing of chemicals are underway in the EU, the USA, and elsewhere. Current test methods not only involve the harming of animals but also are exceedingly slow, perhaps requiring four to five years and millions of dollars in cost. Emphasis is placed on increasing the speed with which tests can be conducted, providing data more relevant to human risks, and reducing the use of animals in such testing. High-throughput testing, using a series of *in vitro* assays that involve human cells and cellular constituents (see https://www.epa.gov/chemical-research/exploring-toxcast-data[toxcast]) is now a major research effort. Although substantial toxicity data are available for food constituents and contaminants, there remain substantial gaps in knowledge for many. While our ability to interpret data from high-throughput, *in vitro* testing is still limited, research to allow interpretation is a large enterprise that seems highly likely to continue. In 2016, the EC launched the EU-ToxRisk research project to develop and promote animal-free approaches in toxicology. These approaches include new approach methodologies (NAMs) which would be used for read-across and for screening and prioritization of substances in testing regimes. NAMs include a range of methods that investigate/elucidate mechanistic information (systems biology) *in silico* approaches, *in chemico* and *in vitro* assays, 'high-throughput screening' and 'high-content methods', for example, genomics, proteomics, metabolomics. In addition to these existing conventional methods, newer methods such as physiologically based pharmacokinetic (PBPK) modelling and simulation have been developed to improve understanding of toxic effects from the toxicokinetic and toxicodynamic knowledge. Such methodologies identify key pathways at the molecular level that can be linked with an adverse outcome at the organ/organism level.

Closely linked to this new model for acquiring toxicity information are developments in risk assessment that involve heavy reliance on data relating to the chemical and biological mechanisms underlying the production of toxicity. The *in vitro* assays used in high-throughput testing and designed to provide direct information about the pathways leading from initial exposure to the production of toxicity, and risk assessors are finding ways to use understanding of pathways to develop more accurate measure of human risk. The fact that certain subpopulations, notably children but others as well, may exhibit unusually severe or frequent responses to toxic agents is not new, but identifying risk assessment models that adequately take such subpopulations into account remains a challenge.

New health concerns relating to chemical responses are constantly emerging. Perhaps the capacity of chemicals to interfere with endocrine system function is the most visible of these concerns, but concerns regarding effects on nutrition status, behaviour and learning, and metabolic disorders are emerging. Thus, while there is a strong basis for believing that food systems are considerably safer now than ever before in human history, many significant questions remain. Moreover, while the better-developed countries of the world may have a basis

for believing their systems are relatively secure, it is far from clear that a similar degree of food safety exists in the vast and hugely populous lesser-developed countries (Grace 2015).

3.14 Food and SARS-Cov-2 disease (COVID-19)

The US Centers for Disease Control (CDC) has stated that there is currently no evidence that handling or consuming food is associated with COVID-19. The agency also states that breastfeeding is not a likely source of COVID-19 transmission, and the WHO encourages mothers to continue to breastfeed even if they have been confirmed to have the disease. CDC does have guidance aimed at reducing transmission among workers in the food and agriculture sectors (see CDChttps://www.cdc.gov/ and WHOURL: https://www.who.int websites).

KEY POINTS

- Food safety is focused on the constituents and contaminants of food, and not on the diet as a whole. Methods to identify the hazardous properties (adverse health effects) are well developed and have international acceptance.

- Risk assessments are applied to hazard data and dose–response data to develop measures of safe intakes (doses) of food constituents and contaminants. A food safety risk may exist if safe intake levels are exceeded.

- The safety of substances intentionally introduced into foods must be established. Producers and manufacturers of food products have ultimate responsibility for ensuring their safety. Health-based guidance values such as ADI for intentionally added substances and TDI for unintentional substances (e.g., impurities, contaminants) are set using hazard and dose– response data from epidemiology and/or toxicological studies. MRLs are established for pesticides, generally with limits on their concentrations in foods, so that EDIs are not exceeded for total intake.

- The types of data needed to assess human intakes of different categories of intentionally introduced substances (Table 3.1) varies in relation to their use characteristics, but in all cases safety assurance requires that those EDIs fall below relevant ADIs.

- Although TDIs have been established for some contaminants and process-formed chemicals, these standards may not be strictly risk-based. Factors such as the technical feasibility of achieving various limits may be considered—as low as reasonably practicable (such an approach is generally not taken for intentionally introduced substances).

- Efforts to increase understanding of risks related to natural constituents of food use are increasing, as well as efforts to understand whether such constituents, including nutrients, have effects on the risks of chronic diseases.

- Microbiological risk assessment models microbial growth in foods and, using assumptions on expected behaviour of organisms under different environmental conditions, are then coupled with dose–response models to estimate risks (responses). Most microbial risk assessments use the 'no-threshold' assumption (i.e., that a single bacterium is sufficient to cause disease) and current guidelines note that, practically speaking, it is not possible to distinguish between a very low non-zero risk and a true threshold.

- Food allergies and intolerances are not avoidable by the establishment of health-based guidance values (and tolerances) and individuals having such sensitivities must manage risks by avoiding foods causing those conditions. Food safety institutions provide warnings and education.

- Faster, more risk-relevant, high-throughput toxicity testing techniques are currently the subject of intense research and development interest. These promise much faster results, at greatly reduced cost, and have the additional benefit of avoiding use of test animals. However, they are still under development and have not received international validation or regulatory acceptance, therefore their uptake for assessing food safety is expected to be slow.

- Numerous national and international institutions with public health or regulatory responsibilities are involved in establishing safety standards for food substances and for providing guidance or regulations that establish how such standards are to be met. Many non-governmental organizations focus on emerging safety issues and bring them to public attention. Regulatory authorities monitor foods to determine whether foods in commerce comply with such standards and can remove from commerce foods found to be out of compliance.

 Be sure to test your understanding of this chapter by attempting multiple choice questions.

REFERENCES

Brock W J et al. (2003) Food safety and risk assessment. *Int J Toxicol* **22**, 435–451.

Codex Alimentarius Commission (1979) *Guide to the Safe Use of Food Additives*. Food and Agriculture Organization, Rome.

Committee on Carcinogenicity of Chemicals in Food, Consumer Products and the Environment (COC) (2018) *Hazard Identification and Characterisation: Animal Carcinogenicity Studies*. Public Health England.

Crump K (1984) A new method for determining allowable daily intakes. *Fund App Toxicol* **4**, 854–871.

Dourson M L & Stara J F (1983) Regulatory history and experimental support of uncertainty (safety) factors. *Regul Toxicol Pharmacol* **3**, 224–238.

Abebe E, Gugsa G, Ahmed M (2020) Review on Major Food-Borne Zoonotic Bacterial Pathogens. *Journal of Tropical Medicine*. https://doi.org/10.1155/2020/4674235

European Food Safety Authority (EFSA) (2005) Opinion of the Scientific Committee on a request from EFSA related to A Harmonised Approach for Risk Assessment of Substances Which are both Genotoxic and Carcinogenic. *EFSA J* **282**, 1–31.

European Food Safety Authority (EFSA) (2010) Scientific opinion on principles for deriving and applying dietary reference values. *EFSA J* **8**, 1–30.

European Food Safety Authority (EFSA) (2020) Risk to human health related to the presence of perfluoroalkyl substances in food. *EFSA J* **18**, 1–391.

Food and Drug Administration (FDA) (1982) Policy for regulating carcinogenic chemicals in food and color additives: Advance notice of proposed rulemaking. *Federal Register* **47**, 14464–14470.

Food and Drug Administration (FDA) (1983) *Food Additives Permitted for Direct Addition to Food for Human Consumption. Section 172.804. Aspartame. Title 21*. Code of Federal Regulations, Washington, DC.

Grace D (2015) Food Safety in Developing Countries: An Overview. *International Livestock Research Institute*. DOI: http://dx.doi.org/10.12774/eod_er.oct2015.graced

Institute of Occupational Medicine (IOM) (1998) *Ensuring Safe Food: From Production to Consumption*. National Academy Press, Washington, DC.

Institute of Occupational Medicine (IOM) (1997–2002) Series of 6 volumes: *Dietary Reference Intakes*. National Academy Press, Washington, DC.

International Agency for Research on Cancer (IARC) (2012) *Improving Public Health through Mycotoxin Control*. Lyon, World Health Organization.

International Programme on Chemical Safety (IPCS) (1999) Principles for the Assessment of Risks to Human Health from Exposure to Chemicals. WHO: Environmental Health Criteria 210.

Kotsonis F N & Burdock G A (2013) Food toxicology. In: *Casarett and Doull's Toxicology* 8th edn (C D Klassen ed.). McGraw Hill, New York.

National Research Council (US) Committee on the Institutional Means for Assessment of Risks to Public Health (NRC) (1983) *Risk Assessment in the Federal Government: Managing the Process*. National Academy Press, Washington, DC.

National Research Council (NRC) (2002) *Toxicological Effects of Methylmercury*. National Academy Press, Washington, DC.

OECD (2020) *PFASs and Alternatives in Food Packaging (Paper and Paperboard): Report on the Commercial Availability and Current Uses*. OECD Series on Risk Management No. 58, Environment,Health and Safety,Environment Directorate, OECD.

Redmond E C & Griffith C J (2003) Consumer food handling in the home: a review of food safety studies. *J Food Prot*. Jan;66(1):130-61. doi: 10.4315/0362-028x-66.1.130. PMID: 12540194.

Rodricks J V, Turnbull D, Chowdhury F, Wu F (2020) Food constituents and contaminants. In: *Environmental Toxicants: Human Exposures and their Health Effects*, 4th edn, 47–203 (M Lippmann & G D Leikauf, eds). John Wiley, Hoboken, New Jersey.

Thorns C J (2000) Bacterial food-borne zoonoses. *Rev Sci Tech*. Apr;19(1):226-39. doi: 10.20506/rst.19.1.1219. PMID: 11189717.

FURTHER READING

Hulebak K L, Rodricks J V, Smith DeWaal C (2013) Integration of animal health, food pathogen and foodborne disease surveillance in the Americas. In: *Scientific and Technical Review: Coordinating Surveillance Policies in Animal Health and Food Safety 'from Farm to Fork'*, 529–536 (S Slorach, ed.). World Organization for Animal Health (OIE), Paris.

International Programme on Chemical Safety (IPCS) (2009) *Principles and Methods for the Risk Assessment of Chemicals in Food*. WHO: Environmental Health Criteria 240.

National Research Council (2008) *Science and Decisions: Improving Risk Analysis*. National Academy Press, Washington, DC.

Rodricks J V (2007) *Calculated Risks* 2nd edn. Cambridge University Press, Cambridge.

US Department of Agriculture (2012) *Microbial Risk Assessment Guideline: Pathogenic Microorganisms with Focus on Food and Water*. Prepared by the Interagency Microbiological Risk Assessment Guideline Workgroup. US Department of Agriculture. FSIS Publication No. USDA/FSIS/2012-001; EPA Publication No. EPA/100/J12/001.

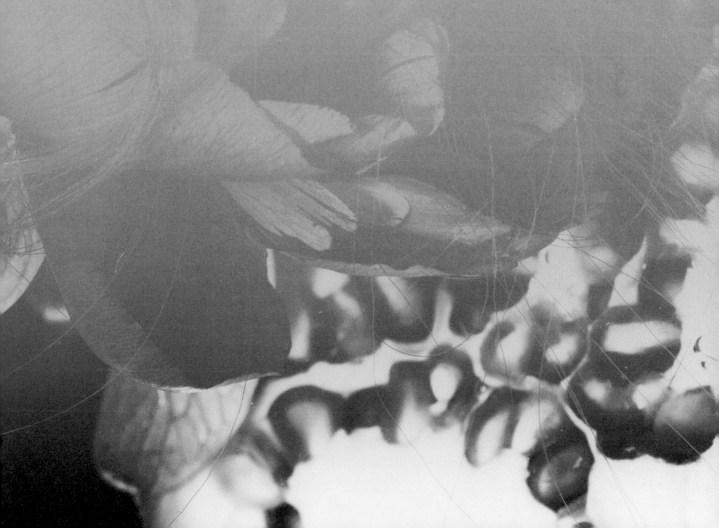

PART 2

Physiology and macronutrient metabolism

4

The physiology of nutrient digestion and absorption

George Grimble

OBJECTIVES

By the end of this chapter, you should:

- be able to understand the most important functions of the various parts of the gastrointestinal tract
- be aware of the main actions of digestive enzymes and the neural and hormonal regulators of digestion
- know the main features of absorption and secretion of specific nutrients, water, and electrolytes
- appreciate that luminal nutrients play a role in the regulation of food intake, through chemosensors which also maintain intestinal integrity and immune defences.

4.1 Introduction

The major components of the diet (starches, sugars, fats, and proteins) must be hydrolysed to their constituent smaller molecules for absorption and metabolism to occur. Starches and sugars are absorbed as monosaccharides; fats are absorbed as free fatty acids and glycerol (plus a small amount of intact triacylglycerol); proteins are absorbed as their constituent amino acids and small peptides. The processes of digestion and absorption occur in the gastrointestinal tract, which is not only a digestive and absorptive tube but also the largest endocrine organ in the body and home to both a large part of the immune system and the colonic microbiome. It is divided into zones anatomically and in terms of substrate digestion and absorption, electrolyte absorption and secretion, metabolism, and neural control. Nutrient absorption can be described by reference to its linear progression along the intestine; specialized compartments with different digestive and absorptive functions follow each other. Digestive and absorptive processes are very efficient because most are duplicated. Thus, impairment of one process by disease does not necessarily lead to complete malabsorption of a particular nutrient. The

digestive and absorptive capacity of the human intestine closely matches the metabolic mass of each individual, just as it matches the metabolic mass of species, small and large, such as the shrew and the whale, respectively (see Weblink Figure 4.1). Clearly, an excessively large intestine would be inefficient because at the extreme, its maintenance cost may exceed the energy value of food ingested. Inefficient digestion also carries penalties. The most rapid period of intestinal growth occurs after birth and if part of the intestine of a neonate is removed by surgery because of gastrointestinal disease, some adaptation of remaining digestive and absorptive capacity can occur but is not great. As a result, intravenous nutrition may become necessary. See Figure 4.1 for a scheme of the intestinal tract.

The gastrointestinal tract is metabolically active, accounting for 10–20 per cent of total energy expenditure even though it is, at most 5 per cent of body weight if the liver is included. It meets its energy requirements from substrates in arterial blood and from the products of digestion in the lumen. Components of this energy expenditure include the effort of chewing (1–2 per cent of the energy content of the food itself), metabolism of one half of dietary protein to meet the amino acid requirements of the gut, and direct oxidation of ~10 per cent of dietary glucose. Ingestion of food leads to increased blood flow around the intestine and to increased transmembrane transport of substrates, water, and ions, all of which are energy-consuming processes.

Intestinal function is modulated by more than 100 gastrointestinal peptide hormones and an enteric nervous system (ENS) that has as many neurons as the spinal cord. Individual neurons are independent of the central nervous system (CNS) but the ENS communicates with the CNS via sympathetic and parasympathetic pathways. This autonomic system controls functions such as contraction, secretion, motility, and mucosal immune defenses and is influenced by central mechanisms, which may set the threshold of autonomic phenomena such as appetite and satiety. In addition, the gut itself signals

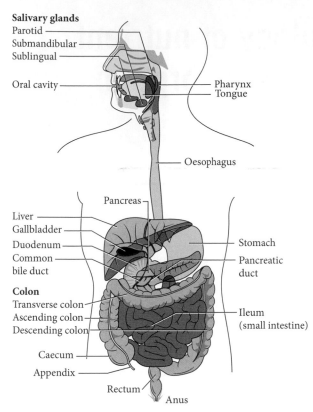

FIGURE 4.1 The gastrointestinal tract.

sensations of hunger and will initiate the chain of events which lead not only to the start of food-seeking behaviours and eating behaviours but also to satiety and cessation of eating. These processes are all modified by disease and by obesity. See Figure 4.2.

4.2 The structure and function of the gastrointestinal tract

The gastrointestinal tract in an adult human is 7–10 metres long, stretching from mouth to anus and, together with the liver comprises about 3 per cent of adult body weight but consumes 10–20 per cent of resting energy expenditure. It is divided into zones that accomplish different digestive and absorptive tasks. The first zone, the mouth and pharynx, leads to the oesophagus, a pipe that enters the stomach via the oesophageal sphincter. Exit from the stomach is governed by the pyloric sphincter that allows food to pass into the small intestine (duodenum, jejunum, and ileum) that is 2.8 to 8.5 metres long and is shorter in women, who have lower average metabolic mass than men. The duodenum (20–30cm long, 5cm diameter) leads to the jejunum and ileum that comprise the upper two-fifths and lower three-fifths of the

small bowel, respectively. There is no strict anatomical differentiation between these zones but the lumen narrows towards the terminal ileum and this reduction in volume per segment length reflects the decreasing luminal fluid loads along the small bowel. Similarly, the jejunal wall is thicker and villi are longer and this correlates with the amount of substrate absorbed in each region. Some of these regional differences are summarized in Table 4.1. The small bowel enters the large bowel via the ileocaecal valve, which, like the oesophageal sphincter, prevents back-flow or reflux of luminal contents. Zones within the colon are defined as the caecum, transverse, and distal colon which terminates in the rectum and anus (see Figures 4.1 and 4.3).

Chewing reduces the particle size of food and increases the surface area available for digestive enzyme action. In addition, it will release the intracellular contents of meat and plants. Some enzymes are secreted with saliva. The stomach is a muscular sac which not only physically reduces the particle size of food but also forms a barrier to ingress of bacteria into the gastrointestinal tract. There is considerable release of enzymes that break down (hydrolyse) lipid and protein.

Stomach emptying occurs when the particle size has been reduced sufficiently by grinding in the antrum. The stomach is therefore a mill, a fermenter, and a pump with built-in particle-size sensing. The thin emulsion of food which enters the small bowel from the stomach is neutralized by duodenal bicarbonate secretions and further digested by enzymes secreted by the pancreas. This process is aided by detergent bile salts released by the gall bladder. During passage along the jejunum, a large amount of water (9 litres) is secreted and then reabsorbed with the products of luminal and brush-border digestion. Efficient reabsorption means that only 60–120 ml of fluid pass the ileocaecal valve each hour, carrying undigested material into the caecum to be metabolized by the numerous bacteria that have established a stable environment there. Colonic fermentation generates short chain fatty acids (SCFA) that stimulate absorption of salts and water to produce a formed stool, which is passed per rectum.

Figure 4.1 summarizes the anatomy of the gastrointestinal tract, and Table 4.1, the main processes that occur in each region (see also Weblink Figure 4.2). The intestinal wall also has zoned anatomy (Figure 4.3) the absorptive surface of which is amplified by three structures.

1. Folds or ridges (rugae) in the intestinal wall increase absorptive area.
2. Villi, finger-like projections 0.5–1.0 mm long covered with mucosal absorptive cells (enterocytes) further increase the absorptive capacity of mammals with higher continuous metabolic rates.

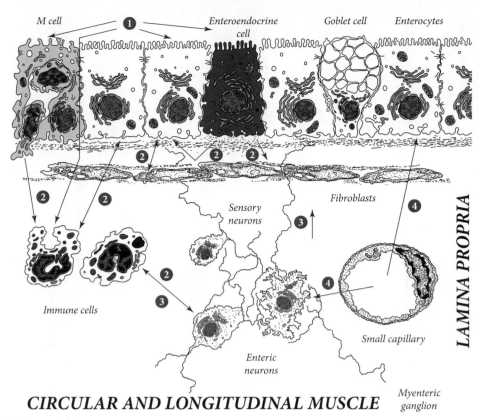

M cell · Enteroendocrine cell · Goblet cell · Enterocytes

Sensory neurons

Fibroblasts

LAMINA PROPRIA

Small capillary

Immune cells

Enteric neurons

Myenteric ganglion

CIRCULAR AND LONGITUDINAL MUSCLE

FIGURE 4.2 The regulation of intestinal absorption and secretion by communication between intestinal cells. (1) Luminal nutrients cause lymphocytes and macrophages to move to the basolateral surface. (2) They secrete cytokine mediators. (3) Signals from sensory neurons and the central nervous system integrated by the enteric nervous system will modulate gut motility and mucosal function. (4) Circulating hormones regulate mucosal function and the cells of lamina propria.
Source: J Pacha (2000) Development of intestinal transport function in mammals. *Physiol Rev* **80**, 1633–1667. Fig. 3, with permission. Copyright © 2000, The American Physiological Society.

3. Enterocytes have further finger-like projections on their luminal surface, known as microvilli and these define the brush-border membrane.

These features increase the absorptive area of the human intestine to 200m², which is about the same area as a singles tennis court. Each villus is supplied by an arteriole and is drained by a venule and a lacteal. The venules carry water-soluble nutrients. The lacteals are part of the lymphatic system and carry the water-insoluble products of fat digestion and absorption to the thoracic duct and then the subclavian vein. This means that most dietary lipids avoid 'first-pass' metabolism by the liver and are metabolized instead by peripheral tissues first.

Blood is supplied abundantly (500 ml/min) to the intestine by arteries that branch from the aorta. The coeliac artery and its branches feed the stomach, pancreas, spleen, and liver whilst the small intestine and large intestine are supplied by the superior mesenteric artery and inferior mesenteric artery, respectively. In the arch

of the superior mesenteric artery, it has been observed that the blood flow to small arteries feeding successive segments of the small bowel increase significantly for 15–30 minutes in step as the main part of the meal moves along. Only one-quarter of the arterial blood supply supplies the submucosa, muscularis, and serosa, the remainder goes to the mucosal layer that has very active metabolism (and needs a good oxygen supply) and where absorbed nutrients are quickly diluted out and removed to the portal vein, thus preventing any high osmotic loads developing. Venules drain via the inferior mesenteric vein (colon) to the splenic vein (spleen) and they join the superior mesenteric vein (small intestine) to form the portal vein, which also receives blood from the gastric vein. The portal vein feeds the liver with nutrients and substances absorbed from the gastrointestinal tract. The liver regulates the supply of nutrients to the periphery through the hepatic vein into the vena cava. In some respects, the intestine behaves like a 'pre-liver' because it:

TABLE 4.1 Regional anatomy of the intestine and sites of nutrient absorption

Region	Functions performed	Mucosal surface	Nutrients digested	Nutrients absorbed	Major site of absorption	Electrolytes absorbed
Mouth	Grinding food to smaller particle size Moistening food (saliva) Initial digestion by lipase and alpha-amylase Initiation of satiety mechanisms	Small folds	Small amount of protein Starch	Small amounts of glucose, peptides, and amino acids	No	No
Stomach	Intestinal defence (e.g., acid secretion) Homogenizing food to smaller particle size Moistening food (gastric secretions) Further enzyme digestion Gastric emptying meters delivery of nutrients to the small intestine Feedback of satiety messages	Rugae and pits	Protein, lipid	Insignificant amounts	No	No
Small intestine	Completion of digestion by pancreatic enzymes Absorption of digestion products of carbohydrate, protein, and fat Absorption of water and electrolytes Absorption of mineral and micronutrients Feedback of satiety messages	Rugae, villi, and microvilli	Protein, lipid, carbohydrate	Amino acids, peptides, fatty acids, glucose, fructose, galactose, glycerol, vitamins	Carbohydrate, fat, protein, water, electrolytes	Sodium, potassium, calcium, magnesium, chloride, phosphate
Colon	Final salvage of water and electrolytes Mucin breakdown Conversion of bilirubin to urobilinogen Cholesterol catabolism Organic acid production ('acetate buffer')	Rugae and pits	Dietary fibre—digested by bacteria and fermented to short-chain fatty acids	Acetate, propionate and butyrate and dicarboxylic acids	SCFA, water, electrolytes	Magnesium and calcium in form of soaps with fatty acids

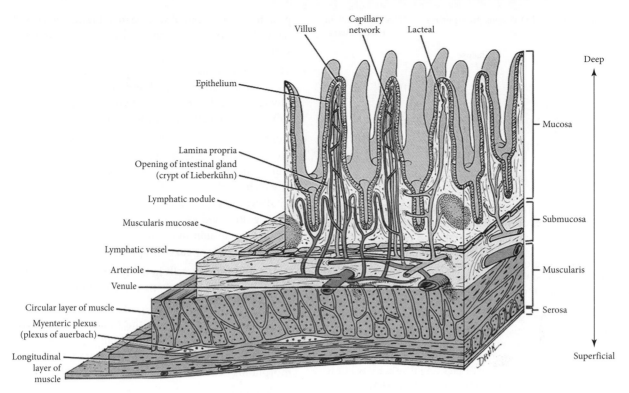

FIGURE 4.3 Intestinal architecture.
The intestine is an organ of digestion and absorption. The villus is covered with absorptive cells called enterocytes. Their outer surface is covered with microvilli, rich in digestive enzymes and nutrient and solute transporters. Folds, villi, and microvilli increase the absorptive area of the intestine approximately 1800-fold to about the size of a doubles tennis court. Water-soluble nutrients are carried to the liver through the venules which drain into the portal vein. Lipids and lipid-soluble vitamins are transported via the lacteals into the lymphatic system.
Reproduced with permission from G J Tortora & N P Anagnostukos (1990) *Principles of Anatomy and Physiology*. Harper & Row, New York.

1. Metabolizes considerable amounts of dietary glucose and amino acids.

2. Transforms and degrades dietary arginine and nucleotides completely.

3. Detoxifies drugs and dietary toxins through the action of mucosal cytochrome P-450 enzymes and the UDP glucosyltransferases and sulphotransferases extrudes toxins back into the gut lumen, via the multidrug resistance transporters (see Figure 4.4).

The gut is therefore a formidable barrier to dietary carcinogens and the continual, rapid shedding of mature enterocytes from the villus tips may explain why tumours of the small intestine are much rarer than those in the large bowel.

Changes in intestinal motility will alter absorptive efficiency. This is because they might increase the contact time between nutrients and the absorptive mucosal surface. In contrast, motility disorders often cause nutrient malabsorption. The importance of gut motility can be judged from the depth of the muscular zone of the

intestine and the size of the enteric nervous system (see Figure 4.2) in which regular nerve plexuses control motility at a local level whilst also responding to signals generated by the presence of luminal nutrients in the lumen and by the vagus nerve.

The intestine defends its integrity by balancing aggressive factors against several protective mechanisms. Aggressive endogenous factors include gastric juice, which is chemically very corrosive (typically pH 1.0), contains aggressive proteolytic enzymes, and is regularly augmented by exogenous factors such as hot foods and, sometimes, alcohol. The stomach will also be exposed to detergent bile acid reflux. Non-steroidal anti-inflammatory drugs are also a potent exogenous aggressive factor. The protective mechanisms are numerous and include the following.

• Gastric mucus contains ~25 per cent phospholipids (by dry weight) which makes it very hydrophobic and damaging agents have limited ability to wet it and thus cause local injury. The gastric glands secrete

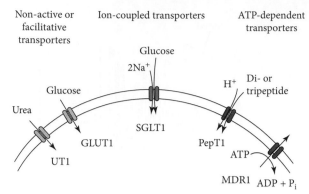

FIGURE 4.4 Transporters and their classification. Three types of transporters exist in the plasma cellular membrane.
Non-active or facilitative transporters (monoporters) move solutes across a membrane 'down' a concentration gradient. Ion-coupled transporters (symporters) move one solute across a membrane 'up' a concentration gradient, using the drive of a coupled solute which is moving 'down' a concentration gradient (a good example is SGLT1). ATP-dependent transporters use the energy derived from hydrolysis of ATP. An example is the multidrug resistance protein (MDR1), responsible for transporting toxic compounds out of cells. PepT1 lacks ATPase activity but is similar to the ATP-dependent transporters.
Adapted from: M A Hediger (1994) Structure, function and evolution of solute transporters in prokaryotes and eukaryotes. *J Exptl Biol* **196**:15–49. Fig. 4. With permission of the Company of Biologists Ltd.

acid through narrow channels in the mucus layer that guide it away from mucosal cells and into the gastric lumen. If the mucus layer remains intact, gastric acid provides an effective bactericidal barrier. The total thickness of mucus varies throughout the length of the gastrointestinal tract (see Weblink Figure 4.3) comprising an inner viscous shear-resistant layer and an outer loosely adherent shear-compliant layer (Allen and Flemstrom 2005). This means that bacteria which penetrate the mucus will be swept off this fragile outer layer by peristaltic waves in the gut. The mucus layer is thickest in the stomach (0.3mm) and colon, regions with the highest luminal concentrations of bacteria. The trefoil factor family (TFFP) peptides secreted with mucin promote restoration of the epithelium and also increase the viscosity of mucus. Bicarbonate is also secreted into this stable mucus gel layer to create a pH gradient from sub-mucus neutrality to luminal pH 1.0. This is the first line of mucosal defense against luminal acid and is also the most important protection against digestion of the gastric (or duodenal) epithelium by pepsin (or pancreatic enzymes).

- The mucosal surface epithelium is protected by continual replacement of cells which are capable of secreting bicarbonate, TFFP, mucins, and defensive proteins. Their luminal surface membrane is enriched with phospholipids which gives a very hydrophobic surface.

- The microvasculature can be repaired rapidly after injury by aggressive factors. Endothelial mediators rapidly cause local arterial dilatation which will provide substrate for bicarbonate production.

- Continuous epithelial cell renewal repairs injury. In the stomach, complete replacement takes 3–7 days and is under the control of several endogenous factors of which chief is the epidermal growth factor receptor (EGF-R). Its expression is triggered by prostaglandin E2 or gastrin and its agonists are transforming growth factor-β or insulin-like growth factor-1. In addition, stress proteins (e.g., survivin) promote mucosal healing.

- About one-third of cells in the intestine comprise the gut-associated lymphatic tissue (GALT) that secretes IgA into the gut lumen.

- Some dietary nutrients in the lumen stimulate G-protein-coupled receptors, normally considered as taste-receptors but which also maintain intestinal morphology and immune defence.

It should therefore not be surprising to find that when patients are fed intravenously and these defence mechanisms are not operating normally, the gut barrier may become compromised.

4.3 Processes of digestion

Digestive processes in the mouth

Chewing grinds the food and mixes it with saliva which contains enzymes that initiate digestion of dietary starch and protein. Saliva not only lubricates food and aids in formation of the bolus to be swallowed but also protects the mucosal surface of the pharynx and oesophagus. Bicarbonate in saliva will neutralize any acid that refluxes back up the oesophagus and thus protects the oesophageal mucosa. Lipid digestion also begins in the mouth, initiated by lipase secreted by the tongue (lingual lipase).

The efficiency of chewing varies through the life-cycle. Infants can efficiently eat only milk and puréed food before they develop teeth, whereas nearly three-quarters of the elderly population in the UK have few teeth because of a lifetime of periodontal disease or dental neglect. The link between tooth loss in the elderly and poor nutrition is complex. Fitting dentures does not always improve nutritional status because dentures require a symmetrical bite if they are not to become displaced and not all people can achieve this because the normal bite is often asymmetrical.

Swallowing transfers a food bolus from the mouth to the oesophagus, and involves contraction and relaxation of at least 14 groups of muscles in about ten seconds in healthy subjects (see Weblink Figure 4.4). It is a complex phenomenon comprising oral, pharyngeal, and oesophageal phases, the first of which is voluntary, the other two being reflex actions which ensure that the airway is protected from the food bolus. Older people may have an abnormal swallowing process, but this may not always predispose to problems if adequate adaptation has occurred. However, a stroke which affects the relevant neurons within the dorsal and ventrolateral medullas, may impair swallowing. Recovery of this (and tube-feeding if swallowing is not re-established) remains a major nutritional challenge in care of elderly people.

Digestive processes in the stomach

Control of gastric secretion

The surface of the mucosa is covered with gastric pits or crypts that are lined with four types of cells.

1. G Cells which secrete the hormone gastrin (which stimulates acid secretion).

2. Parietal cells which secrete hydrochloric acid.

3. Chief cells which secrete pepsinogen, the inactive precursor of pepsin.

4. Mucous cells which secrete the glycoprotein mucin, are found at the neck of the pit.

The bulk of the gastric contents is typically at pH 1.6–2.7 increasing progressively to 5.0 within the strongly adherent mucus layer. At the outlet to the gastric crypt, pH falls to 4.6 and then to pH 3.0 at the base of the crypt. By contrast, the intracellular pH of cells lining the crypt is neutral or slightly alkaline in the deepest parietal cells. This pH gradient is maintained by the directional secretion of H^+ into the gastric lumen and formation of HCO_3^- which neutralizes acid and so protects mucosal cells. The flow of H^+ out of the cells is matched by an influx of K^+, whilst outflow of HCO_3^- is balanced by inflow of Cl^- which will eventually be secreted with H^+ as HCl. This constant pumping of large amounts of ions (typical of all regions of the intestine) means that the parietal cells have very high metabolic rates. Parietal cells carry receptors for three agents that stimulate acid secretion; acetylcholine, gastrin, and histamine.

The cephalic, gastric, and intestinal phases of gastric digestion

The end of the hunger phase of the hunger–satiety cycle is marked by eating, but before this happens, the brain and intestine 'prepare themselves for dinner', that is the

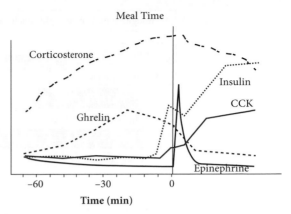

FIGURE 4.5 Time courses of several hormones relative to the start of anticipated meals.
This diagram shows likely changes in hormone profiles in anticipation of eating food during the so-called 'cephalic phase' of the hunger–satiety cycle. Data are approximated from different sources.
Source: M H Tschop, T R Castaneda, S C Woods (2006) The brain is getting ready for dinner. Cell Metab **4**, 257–258. Fig. 1. With permission from Elsevier.

very large amount of nutrients which would otherwise overwhelm fasting metabolism and fasting gastrointestinal function (Smeets et al. 2010) and Figures 4.5 and 4.6). The process follows three well-defined phases.

1. The 'cephalic phase' of eating (sight, aroma, and anticipation of food) stimulates the parasympathetic intestinal nervous system via the vagus nerve, with acetylcholine release near parietal- and G-cells leading to acid and gastrin secretion. Ghrelin and corticosterone concentrations have reached their peak and just before eating starts, there is a small preliminary peak of insulin release (Figure 4.5).

2. The 'gastric phase' is defined by increased acid and pepsinogen secretion in response to stretching of mechanoreceptors by food ingestion. This stimulates gastrin release, which increases secretions.

3. The 'intestinal phase' occurs towards the end of liquidization of food in the stomach.

Receptors in other parts of the intestine inhibit gastric emptying through neural and hormonal pathways. This is known as the 'pyloric brake'. Each of these phases has a counterpart that controls appetite and eating behaviour.

Digestive properties of gastric juice

Dietary lipid (triacylglycerol) and protein are hydrolysed by enzymes secreted in the gastric juice; salivary amylase action continuing until inactivated by low stomach pH. Food lipid is reduced to an emulsion of droplets 10–100 μm in size through the processes of chewing and

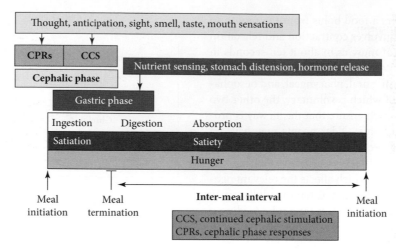

FIGURE 4.6 The hunger–satiety cycle.
The intensity of satiation, satiety, and hunger are represented by the intensity of the grey value. Continued cephalic stimulation (CCS) occurs during ingestion of food. The cephalic phase is the period from anticipation of eating until the last mouthful is swallowed and it comprises cephalic phase responses (CPR). The gastric phase is the time during which food is present in the stomach. Meal initiation (MI) and meal termination (MT) are self-explanatory.
Adapted from: P A Smeets, A Erkner, C de Graaf (2010) Cephalic phase responses and appetite. *Nutr Rev* **68**, 643–655. Fig. 1. With permission.

of mechanical mixing and also the action of gastric lipase. Triacylglycerols form a reverse-emulsion of lipid in water micelles to whose surface gastric lipase binds, releasing palmitic acid and oleic acids from the sn-3 position of triacylglycerols. These fatty acids generate osmotic pressure within the micelle and lead to budding of new, smaller micelles of triacylglycerols. In addition, these fatty acids stimulate the activity of pancreatic lipase in the small intestine. Although gastric lipase hydrolyses only 10–30 per cent of dietary triacylglycerol, it is enough to allow efficient emulsion formation (Figure 4.7). The process is self-limiting as these fatty-acid rich buds inhibit surface-bound lipase.

Proteins are relatively resistant to enzymic hydrolysis until their structure has been denatured by heat (e.g., in cooking) or by low gastric pH. Although this occurs to a considerable extent, some proteins such as ingested lactoferrin, an iron-binding transferrin-like protein in human milk, survive passage through the stomach, intact.

Pepsins comprise a family of seven aspartate proteinases which initiate the process of protein digestion and are stored as inactive zymogens in granules of the chief cells of human gastric mucosa (e.g., pepsinogen A) and secreted as pepsinogens; inactive precursors or zymogens. They contain a pro-segment (i.e., an extra amino acid sequence at the N-terminus of the enzyme) that stabilizes the inactive form and blocks the active site. They are activated at the sub-mucosal layer by gastric acid secretion; low pH unwinds this pro-segment which is cleaved off by the active site of the enzyme itself, thus resulting in active pepsin which may then activate other

pepsinogens. They are denatured at pH >7.0 but operate at higher pH because food rapidly buffers the acid to achieve a pH of ~4.5 (Roberts 2006).

The gastric pepsins are paradoxical because they are endogenous aggressive factors and do not extensively hydrolyse food protein but can reduce its viscosity greatly. However, their action is essential because they hydrolyse protein sequences mainly at the carboxyl side of the aromatic amino acids, releasing relatively large, soluble, oligopeptides and the free aromatic amino acids tyrosine, phenylalanine, and tryptophan. These amino acids stimulate release of gastrin and cholecystokinin (CCK) which cause acid secretion and gastric accommodation (i.e., relaxation) respectively, in addition to release of bile and pancreatic enzymes. Pepsins therefore promote orderly digestion as does the cephalic phase (see 'The cephalic, gastric, and intestinal phases of gastric digestion') and dysregulation of pepsin synthesis may lead to functional dyspepsia (Petersen 2018).

Control of the rate of gastric emptying

Gastric emptying matches food eaten and its composition to the progress made in liquidizing it within the stomach. Simple fluids like water will empty at a rate proportional to gastric volume, whereas solid nutrients will empty at a rate that depends on their energy density and potency in altering the duodenal brake. Furthermore, fat that reaches the ileum exerts a profound inhibitory effect on gastric motility, known as the 'ileal brake'. After a meal, the two parts of the stomach show different motility responses.

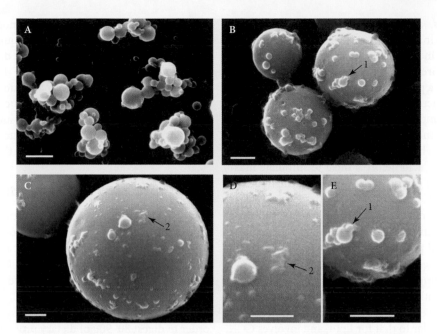

FIGURE 4.7 Hydrolysis of lipid droplets by human gastric lipase
Scanning electron micrographs of lipid droplets during hydrolysis by human gastric lipase (HGL).
The smooth, spherical lipid droplets (A, 0.5–2 μm diameter) become swollen (B, 2.6–6.4 μm) to large droplets
(C, 5.6–10 μm). Note the surface 'buds' (C, arrow 2, enlargement in D and enlargement in E).
Source: Y Pafumi et al. (2002) Mechanisms of inhibition of triacylglycerol hydrolysis by human gastric lipase.
J Biol Chem **277**, 28070–28079, 2002. Fig. 5, with permission. © 1969, Elsevier.

1. Upper stomach (fundus, upper body)—initially, upon feeding this relaxes to accommodate gastric contents. That accommodation avoids an increase in intragastric pressure which, itself, generates the sensation of satiety. Over time the fundus increases its tone progressively, thus forcing its contents to migrate to the antrum. So, the fundus experiences changes in tone rather than phasic contractions. In fact, the ICC-MP or pacemaker cells are absent in the gastric fundus and only present in the mid corpus and antrum.

2. Lower stomach (lower body, antrum)—powerful peristaltic contractions towards the pyloric sphincter.

These combined motility patterns, together with hydrolysis of lipids and proteins, lead to liquidization of food that is released into the duodenum in spurts. The upper stomach acts as a 'pressure-pump'. In the lower stomach, solid food larger than 1–2mm is recycled through the 'antral mill' until it has reached a size small enough to pass the pyloric sphincter.

Gastric emptying is controlled by neural or hormonal signals arising in response to nutrients which activate chemoreceptors in other parts of the gut (see Section 4.5). As described, a feedback loop from lipids in the proximal and distal small bowel inhibits gastric emptying and promotes gastric accommodation by relaxation of the fundus; the effect is strongest with free fatty acids with a chain length >10 carbon atoms. There seem to be four types of mechanisms by which lipids modulate gastric motility.

1. Luminal lipid stimulates release of the regulatory peptide hormones CCK, neurotensin, peptide YY (PYY) and glucagon-like peptide 1 (GLP-1). CCK has local effects on motility.

2. CCK stimulates afferent nerve pathways in the intestine that inhibit gastric activity.

3. The products of lipid absorption, chylomicrons packaged for export to lymph (see Section 4.4) are also sensed by an, as yet, unknown mechanism.

4. Short-chain fatty acids (SCFA) which reflux back into the ileum are sensed, and signal the release of PYY that inhibits gastric motility.

In this way, nutrient overload in the intestinal lumen is avoided and motility is adjusted to permit enough time before the remainder of the meal is released. Even at rest, the stomach is never quiescent. Rhythmic waves of polarization and depolarization of gastric smooth muscle, starting in the corpus and progressing towards the antrum, occur every 20 seconds. These so called slow waves are controlled by pacemaker cells located in the myenteric plexus, also known as interstitial cells of Cajal. These cells not only play a role as gut pacemakers but also facilitate excitatory (cholinergic) and inhibitory (nitric oxide) motor responses, and are distributed throughout the gastric and intestinal smooth muscle layers. Gastric motor dysfunction includes gastroparesis (delayed gastric emptying), impaired gastric relaxation after ingesting a meal (reduced in a proportion of patients with functional dyspepsia and increased gastric compliance in patients with bulimia), and over-rapid gastric emptying (e.g., 'dumping syndrome'). Slow emptying can be treated by prokinetic drugs whilst dumping can be treated by giving 1–2 grams of oleic acid before a meal to stimulate maximum inhibition of gastric motility by both the duodenal and ileal brakes. Currently, drugs for improving gastric accommodation include buspirone (anxiolytic drug, 5-hydroxytryptamine 1A receptor agonist) and clonidine (treats high blood pressure, alpha-2 (α2) adrenergic receptor agonist).

Gastric emptying is also slowed by increased blood glucose concentration and is accelerated by insulin injection that reduces blood glucose concentration. The mechanism is similar to that for lipid, and involves CCK, PYY, GLP-1, and Amylins (co-secreted with insulin). This is an example of a feedback mechanism that matches the amount of nutrient presented for absorption by the small intestine, with the amount already absorbed. Unsurprisingly, delayed gastric emptying is a consequence of poorly controlled diabetes and can be treated with appropriate insulin therapy (e.g., continuous subcutaneous insulin infusion in patients with poor glycaemic control). Gastric emptying is also controlled through inhibition of eating behaviours and this is mediated through a complex ballet of gut hormones (e.g., GLP-1, CCK, PYY, oxyntomodulin, and glucagon), which generally increase satiety.

Finally, gastric distension is a very powerful inhibitory signal that increases feelings of fullness and satisfaction (satiety) and hence counters the stimulatory afferent signals produced by eating tasty food. When the former signal predominates, eating is reduced and the stomach will, on balance, empty.

Digestive processes in the small intestine

Intestinal secretions and their control

Gastrin secreted by the stomach stimulates the secretion of enzymes by acinar cells in the pancreas. As the meal is released by the pylorus, acid-sensing cells in the duodenal mucosa release the hormone secretin, which stimulates water and bicarbonate secretion by pancreatic duct cells. This in turn flushes the pancreatic enzymes into the duodenum via the pancreatic duct. A second hormone, CCK is also released and elicits two responses. 1) The acinar cells of the pancreas release large quantities

of pancreatic enzymes as inactive zymogens. 2) The gallbladder contracts powerfully and squirts bile into the duodenum through the common bile duct.

Although digestion will increase luminal osmolality and secretion of water into the duodenal lumen, the absorption of digestion products reduces osmolality and the water will be reabsorbed. This is an impressive process; it is estimated that 7.5 litres of water are secreted and absorbed in this way, every day, from the small intestine.

Digestive function of intestinal secretions

As for gastric pepsinogens, pancreatic enzymes are secreted as inactive precursors (zymogens), accounting for up to 30 per cent of the protein passing through the gastrointestinal tract with the meal. If these pancreatic enzymes were completely hydrolysed in the lower intestine and the amino acids absorbed, a large amount of pancreatic enzyme would need to be synthesized daily in order to digest 80–90g of dietary protein. Is this the case? On the one hand, patients with an ileostomy (surgical fistula that drains ileal contents) do excrete that amount of partially digested protein and there is a daily rhythm of synthesis of new zymogen granules and their release into the pancreatic ducts. An alternative view is that pancreatic enzymes are absorbed intact and recycled in an enteropancreatic circulation that is analogous to the enterohepatic circulation of bile salts. Compelling arguments for this view are that pancreatic enzymes can be detected in the circulation (usually considered to be of pathological not physiological significance) and that the pancreas does not have the capacity to synthesize such a large amount of secretory enzymes each day. The mechanism of the proposed selective intestinal absorption of pancreatic enzymes is unknown.

The process of export of inactive zymogen precursors is dependent on ATP generation by mitochondria. The process starts in the endoplasmic reticulum before the zymogen is packaged into vesicles budded from the Golgi apparatus. These mature into membrane vesicles or zymogen granules which will eventually fuse with the external cell membrane and release their contents into the pancreatic duct. The process is tightly controlled by accessory proteins and damaged or dysfunctional organelles are removed via autophagy-lysosomal-endosomal pathways. Failure of the process can lead to release of zymogens inside the pancreatic acinar cells and precipitate pancreatitis, with disastrous consequences.

Various pancreatic enzymes hydrolyse proteins (proteases), lipids (lipase, phospholipase), starch (amylase), and nucleic acids (ribonuclease, deoxyribonuclease) together with esterases and two specific proteases, gelatinase and elastase. These enzymes carry out luminal digestion of more highly polymerized substrates of >10 units (e.g., larger maltodextrins) whereas brush-border hydrolases favour shorter oligomers (e.g., maltose—maltopentaose).

Proteases

The pancreatic proteases are either endopeptidases (trypsin, chymotrypsin, and elastase) that cleave internal amino acid bonds or they are carboxypeptidases (A & B) that will sequentially cleave amino acids from the C-terminal of oligopeptides. The endopeptidases have specificities for bonds adjacent to dibasic amino acids (trypsin), hydrophobic amino acids (chymotrypsin), or small neutral amino acids (elastase). The combined actions of these enzymes will reduce dietary proteins to a mixture of free amino acids and peptides with a chain length of 2–8 amino acids.

Like gastric pepsins, pancreatic proteases are secreted as inactive zymogens. Trypsinogen is first converted to trypsin by enterocyte brush-border enteropeptidase which cleaves off a peptide sequence that blocks the active site of trypsin. Whilst trypsin cannot catalyse trypsinogen conversion, it does activate the zymogens of the other major proteases to yield chymotrypsin, elastase, carboxypeptidase A, and carboxypeptidase B. This cascade initiator, enteropeptidase, is most prevalent in the duodenum and then decreases distally; and its level of expression on the membrane depends on the luminal presence of pancreatic enzymes, amino acids, or glucose. The sequence of events which leads enteropeptidase to activate the trypsinogen cascade is still unknown. One hypothesis is that enteropeptidase can be inactivated by protein C inhibitor (PCI) which is a serine protease inhibitor and a member of the serpin superfamily. In this model, serpins contain an exposed reactive centre loop which is recognized by the protease, an enzyme inhibitor complex is formed and the protease then clips off the loop which remains bound to the protease, inhibiting it in the process. Since this inhibitory loop binds reversibly, the enteropeptidase will be activated by dissociation of the serpin fragment. Enteropeptidase is triggered pathologically, in the case of acute necrotizing pancreatitis, but the cause is unclear.

This explanation does not tell us what triggers trypsinogen activation but merely pushes the ultimate cause back up the chain. The alternative scheme for reabsorption of intact pancreatic enzymes invoked a similar group of serpin-like inhibitors, which inactivated these enzymes during passage through the blood stream.

Amylase

Both salivary and pancreatic α-amylases are most active at neutral pH and act as endoglucosidases that have an absolute specificity for α-1,4 glucose linkages with two adjacent α-1,4 linkages. Therefore, amylase will not cleave other glucose polymers such as β-glucans (oats), cellulose (plants), dextran (dental plaque), or lactose or sucrose. The end products of exhaustive starch digestion are therefore maltose, maltotriose, and the α-1,6 branched limit dextrins. No free glucose is released.

Further digestion of the branched limit dextrins can only occur at the brush-border (catalysed by isomaltose or glucoamylase). The chain length of the linear, α-1,4 linked dextrins in the lumen after a starch meal depends on the extent to which α-amylase digestion has gone to completion, but is probably in the range 5–10 glucose units.

Lipases

In addition to salivary and gastric lipases, there are four pancreatic lipases.

- Pancreatic triacylglycerol lipase (PTL)
- Carboxyl ester lipase (CEL)
- Pancreatic lipase-related proteins 1 and 2 (PLRP-1, -2)
- Group 1B phospholipase A2 (sPLA2-1B).

PTL is the most abundant and important in adult life; pancreatic lipase-related proteins 1 and 2 (PLRP-1, PLRP-2) are expressed pre- and perinatally but not in adulthood. PTL is a true lipase which preferentially hydrolyses triacylglycerols which form oil-in-water emulsions. Its binding to the surface of the oil droplet is aided by colipase (another pancreatic protein) and bile salt. The N-terminal of the enzyme has a 'lid' sequence; a highly mobile, hydrophobic structure which, upon lipid binding, will swing aside to reveal the active site and thus allow lipid hydrolysis to occur.

CEL, PLRP, and sPLA2-1B have broad specificity and will hydrolyse phospholipids. Their preference is for emulsions made of micelles formed with bile salts. CEL is activated by bile salts and has a broad specificity towards cholesteryl esters, tri-, di-, and mono-acylglycerols, phospholipids, lysophospholipids, and ceramide. In addition, it will also hydrolyse fat-soluble vitamins and triglycerides. Luminal CEL can be transported into the bloodstream after endocytosis by enterocytes. It associates with low-density lipoprotein and is excreted via the kidney.

Fat digestion comprises the following steps.

1. Partial digestion and emulsification in the stomach.
2. Mixing of triacylglycerols, diacylglycerols, monoacylglycerols, and fatty acids with detergent (bile salts), cholesterol, and phospholipid to form mixed micelles that have a hydrophobic core and hydrophilic outer surface. Their small size and high surface area leads to efficient hydrolysis.

3. Binding of colipase and lipase to the surface of these micelles leads to release of free fatty acids and retinol.

4. Osmotic pressures generated within the micelle by triacylglycerol hydrolysis causes budding of smaller monoacylglycerol- and fatty acid- rich micelles from the surface of these structures. These easily penetrate the unstirred water layer adjacent to the absorptive surface of the enterocyte.

5. This presents lipid substrate for transport across the enterocyte membrane at a much higher concentration than would occur if triacylglycerols arrived there by simple diffusion.

Regulation of small intestinal motility

The intestine mixes and propels gut contents using different combinations of muscle and nerve systems. This system has redundancy because failure of one system (or its deletion in gene knockout animals) does not seem to alter the way the gut moves its contents along.

Intestinal motility can be measured in several ways.

1. Simple transit time of liquid or radio-opaque markers from mouth to anus.

2. High-resolution manometry using a fibre-optic catheter with solid-state pressure sensors and electrodes at <1cm intervals. It can measure propagating pressure waves and electrical patterns.

3. Wireless capsule endoscopy filming gives a complete continuous record of local movement and of intestinal diameter during passage through the intestine. An algorithm picks the best representation of intestine wall/lumen for each frame and the resulting complete 'motility bar' for the intestine identifies specific contractile patterns.

4. MRI of the abdomen allows reconstruction of a 3D bowel image and estimates of volumes at specific locations.

It has been found that networks of interstitial cells of Cajal (ICC) act as pacemakers and propagate rhythmic slow oscillations of depolarization of membrane potential that increase the likelihood of a contraction of smooth muscle in the gut to produce rhythmic peristalsis. The strongest contractions are defined as rhythmic propulsive motor complexes (or migrating motor complexes or giant migrating complexes). In addition, the expression of the pattern of intestinal motility depends critically on whether the subject is fasted, fed, or post-prandial. Several gut pathologies also disorder motility

(e.g., delayed gastric emptying, Chagas disease and megacolon, diabetes, pseudo-obstruction).

The ICC slow waves migrate into adjacent segments of the intestine and force that segment to oscillate at the same frequency. The frequency of gastric oscillations is different to that of the duodenum and the pylorus acts as an electrical 'break' between the two. However, emptying of gastric contents into the duodenum provides a stimulus for duodenal peristalsis, as luminal distension is another mechanism promoting smooth muscle cells contractility.

During the fasting period, a stereotypical pattern of contractility is activated. These patterns, called the migrating motor complex (MMC), start in the antrum and will occur 4–6 hours after a meal. It is a complex entity, comprising four phases that cycle continuously whilst no further food is ingested.

Phase I—inactivity (30–40 minutes)

Phase II—irregular pressure spike activity (30–40 minutes).

Phase III—intense repetitive high amplitude contractions (4–6 minutes)

Phase IV—irregular activity.

This cycle moves down the intestine at 4–6 cm/min before slowing in the terminal ileum. The MMC plays an important role in preventing small bowel bacterial overgrowth, because it moves bacteria, which have refluxed in from the caecum, distally. Feeding initiates irregular activity throughout the small intestine, which resembles phase II of the MMC. It leads to greatly reduced rates of intestinal contraction and rate of movement, and this lengthens transit time in the bowel. The absorption of nutrients from the lumen of the small intestine is increased by the way, in which the mucosa repeatedly dips into the chyme, minimizing the diffusion barrier to absorption. At the same time, villous contractions help lymph and blood flow to carry away absorbed nutrients. These repetitive segmenting contractions are interspersed with erratic motile patterns that move chyme forwards rapidly by 10–30cm before segmenting contractions recommence.

These responses are nutrient dependent. Lipid has particularly potent effects because it generates strong clustered contractions that enhance emulsification. In summary, fasting motility sweeps debris, shed cells, and bacteria down the intestine whereas fed motility enhances digestion and absorption of nutrients. The transition between fasting and fed patterns is modulated by the presence of nutrients in the lumen. After the start of a meal, a small portion of chyme ('the head of the meal') will be rapidly swept down to the distal ileum, where the presence of digested fat evokes the ileal brake leading to a marked reduction in transit rate. In addition, the passage of food into the small bowel stimulates colonic segmental movement, known as the 'gastrocolonic reflex'. This reflex is partially controlled by the cephalic phase of eating and leads to increased churning of colonic contents that increases absorption of nutrients, water, and electrolytes from the colon and will eventually lead to defaecation. The SCFA produced by bacterial fermentation in the colon not only stimulate water absorption, but if they reflux back through the ileo-caecal valve into the ileum will simultaneously inhibit gastric emptying and stimulate peristalsis in the terminal ileum. The net effect is to sweep colonic bacteria from the distal small bowel.

Colonic motility differs and most human studies have used simpler manometric methods, focusing on the so-called high-amplitude (>100 mm Hg) propulsive contraction which is infrequent during each day. Forward-moving contractions are more frequent after a meal and after waking in the morning. Backward-propagating contractions are also frequent and contribute to allow water absorption. One new idea is the 'neuromechanical loop' hypothesis whereby mechanical dilation of the colon activates polarized gut circuits to produce propulsion of the bolus. This activates further distal circuits which produce distension, thus closing the loop (i.e., pressure falls). This system can adapt to a range of consistencies of colonic contents. For example, patients with constipation exhibit increased numbers of backwards contractions whilst both the amplitude and length of propagation of forward propagating contractions is significantly reduced.

Nasogastric tube feeding is associated with diarrhoea. One cause is that the slow rate of nutrient infusion (4.2–6.3 kJ/minute) is insufficient to trigger a normal post-prandial slowing of intestinal transit. It does, however maintain colonic water secretion and hence provokes diarrhoea.

4.4 The absorption and secretion of nutrients

How molecules cross the intestinal mucosal barrier

Absorbed nutrients must cross four barriers to reach the bloodstream:

1. The mucus layer, a thin diffusion barrier in the small intestine;
2. The enterocyte apical membrane—a lipid bilayer, which requires transport proteins for water-soluble molecules;
3. The enterocyte—a metabolic barrier which may metabolize the nutrient;
4. The basolateral membrane—a lipid bilayer which again requires transport proteins.

Most solute absorption occurs across the enterocyte membrane (transcellular) but some occurs via the tight junction between enterocytes (paracellular)

TABLE 4.2 Substrate transporters present in the human intestine and in other tissues

Transport system	Substrates	Transporter protein names	Gene name for solute carrier (SLC) family
Na+-dependent			
A	Small aliphatics	SNAT2	SLC38
N	Gln, His, Asn	SNAT3, SNAT5	SLC38
B⁰,⁺	Ala, Lys, Arg, Orn, Gly	BAT1	SLC7
GLY	Gly	GLYT, GLYT-1a, -1b, GLYT-2, BGT-1, PRO	SLC6
ASC	Small aliphatics and cysteine	ASCT1, ASCT2 SATT	SLC1A4/5
X⁻_AG	Asp, Glu	EAAC1, GLAST, GLT1	SLC1A1/2/3/6/7
y⁺L	Leu, Met	4F2hc	SLC3
y⁺	Gln, homoserine, citrulline	mCAT-1, -2, -2A, CAT-1, -2A, -2B	SLC7
IMINO	Pro	SIT1?	SLC6A20
Glucose transporter	Glucose	SGLT-1, SGLT-2, SAAT1	SLC5

TABLE 4.2 *Continued*

Transport system	Substrates	Transporter protein names	Gene name for solute carrier (SLC) family
Na⁺-independent			
L	Leu, Ileu, Val, Phe		SLC43
b⁰,⁺	Lys, Leu, Trp, Met	rBAT, D2, NBAT	SLC3SLC7
y⁺	Basic	mCAT-1, -2, -2A, CAT-1, -2A, -2B	
y⁺L	Leu, Met	4F2hc	SLC7
x⁻c exchange transporter	Glu exchanges Cys		SLC7A11
	Urea	UT	SLC14
Glucose transporter	Glucose and fructose	GLUT 1-7	SLC2
Proton energized			
Peptide transporter	di- and tripeptides, β-lactams, ACE inhibitors	PEPT-1, PEPT-2	SLC15
Organic anion transporter/ATP-binding cassette transporter			
Multidrug resistance protein	glutathione and glucuronide anionic conjugates	MRP1, MRP2 (cMRP/cMOAT) MRP-3	ABCC1
Multidrug resistance protein	xenobiotics	MDR1/MDR2/p-glycoprotein	ABCB1
Organic anion transporting protein	glutathione, conjugated bile salts	OATP1	SLCO1
Fatty acid transporter			
Fatty acid transport protein	Long-chain fatty acids	FATP1/2/3/4/5/6	SLC27
Monocarboxylic acid transporter	Medium-chain fatty acids	MCT	SLC16
Sodium-coupled monocarboxylate transporter	Short-chain fatty acids	SMCT1/2	SLC5A12

4F2hc (heavy subunit of LAT1), ASCT1, ASCT2, SATT (Alanine/Serine/Cysteine/Threonine Transporter, neutral amino acid transporter), BGT-1 (Betaine/GABA Transporter-1), CAT-1, CAT-2B, CAT-2A, mCAT-1, mCAT-2A, mCAT-2 (Cationic Amino Acid Transporter), cMOAT, MRP1, MRP2, MRP3, cMRP, MDR1, MDR2, p-glycoprotein (Multidrug Resistance Protein), EAAC1 (High-affinity glutamate transporter), FATP1/2/3/4/5/6 (Fatty Acid Transport Protein), GLAST (Glial High Affinity Glutamate Transporter), GLT1 (Glial High Affinity Glutamate Transporter), GLUT 1-7 (Glucose Transporter), GLYT-1a, GLYT-1b, GLYT-2 (Neurotransmitter Transporter, Glycine), LAT1 (large neutral amino acid transporter), MCT (Monocarboxylate Transporter), NBAT, rBAT (Cystine, Dibasic And Neutral Amino Acid Transporters), OATP1(Organic anion transporter/ATP-binding cassette transporter), PEPT-1, PEPT-2 (Peptide Transporter), SAAT1(Is a glucose transporter, not Sodium Couple Amino Acid Transporter), SGLT-1, SGLT-2 (Sodium Glucose-linked Cotransporter), SIT1 (Sodium/Imino-Acid Transporter), SMCT1/2 (Sodium/Monocarboxylate Cotransporter), UT (Urea Transporter).

(Figure 4.8). The classification of transporters is shown in Figure 4.4 and main intestinal transporters are listed in Table 4.2. The method for naming these transporters is complex and outside the scope of this chapter (Broer and Fairweather 2018). Suffice it to say that a transporter may be named after its overall system (e.g., system A, neutral amino acids), after the gene for its transport protein (e.g., SNAT2), or after its SoLute Carrier (e.g., SLC38A2). This example is the transporter for glycine, proline, alanine, serine, glutamine, asparagine, and histidine. Although different transporters carry very different substrates, they share many common structural features. They have regions of hydrophobic amino acids that can fold into helices which, when grouped together like the staves of a barrel, span the membrane and form a 'pore' through which substrates can be transported. Parts of the protein (often bearing a sugar-polymer) are outside the membrane and can act as a signalling receptor to allow other compounds to control the rate of transport of the main substrate.

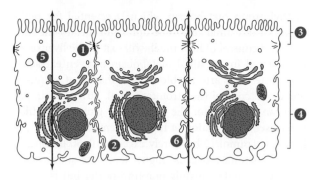

FIGURE 4.8 Intestinal epithelium and routes of solute transport.
Nutrients, salts, and water can cross the mucosal barrier either by going through the cells (5 transcellular pathway) regulated by transporters in the apical (3) and basolateral membranes (4) or by moving between cells (6 paracellular pathway). The tight junctions (1) prevent free movement of water and solutes through the space between cells.
Source: J Pacha (2000) Development of intestinal transport function in mammals. *Physiol Rev* **80**, 1633–1667, 2000. Fig. 2, with permission. Copyright © 2000, The American Physiological Society.

Alternatively, a transport protein may be linked to another regulatory protein that can chaperone the transporter from an intracellular vesicle into the membrane and thus modulate transport capacity. Transport may be either passive, allowing the concentration of the transported nutrient to come to equilibrium across the membrane, or active, permitting a higher concentration to be achieved on one side of the membrane than on the other (Figure 4.4).

These permit the transfer of a solute across the membrane in either direction, down a concentration gradient (so-called 'downhill transport'). Net accumulation of the transported material in the cell can occur as a result of either onward metabolism to a compound that does not cross the membrane (e.g., vitamin B_6 is accumulated intracellularly by phosphorylation to pyridoxal phosphate) or by binding to cytosolic proteins (e.g., iron binds to ferritin).

Active transporters

These transport solutes against a concentration gradient, linked to either direct ATP hydrolysis (P-type transporters) or co-transport of an ion down its concentration gradient (symporters, which transport two solutes in parallel) (Figure 4.4). Direct utilization of ATP involves phosphorylation of the transport protein, which permits it to transport one or more solute molecules in one direction only; solute transport causes dephosphorylation of the protein, so closing the pore.

Symporters commonly utilize a sodium ion gradient across the membrane, although some systems use a hydrogen ion gradient. The ionic gradient is generated by membrane ATPases that pump ions across the membrane. Intestinal absorption of glucose and some amino acids is by sodium-linked symporters.

Digestion and absorption of carbohydrates

The main dietary carbohydrates are starch, lactose, and sucrose, as well as smaller amounts of glucose, sugar alcohols, and fructose. Carbohydrates are only absorbed as monosaccharides, so starch assimilation proceeds in two phases.

As described above, starch digestion by amylase yields a mixture of maltose and isomaltose and glucose oligomers (5–10 glucose units), which require further hydrolysis by brush-border glucosidases to glucose. Disaccharides are hydrolysed to their constituent monosaccharides by disaccharidases on the brush-border of the enterocytes: lactase, trehalase, and a bifunctional enzyme, sucrase/isomaltase.

Glucose and galactose are taken up by the same active (sodium-linked) transporter, sodium glucose-linked transporter 1 (SGLT1) (Figure 4.9), while fructose, some other monosaccharides, and sugar alcohols are carried by passive transporters (GLUT5). This means that only a proportion of fructose and sugar alcohols can be absorbed, and after a large dose much may remain in the lumen, leading to osmotic diarrhoea.

Glucose transport is duplicated. SGLT1 has high affinity for glucose and low transport capacity. Glucose transporter 2 (GLUT2) has high activity but low affinity and is stored intracellularly in the enterocyte and inserted into the membrane in response to glucose absorption via SGLT1. This process is controlled by insulin and dietary amino acids, mediated by protein kinase C activation. This leads to a great increase in transport capacity in response to dietary load. In addition, luminal glucose concentration is sensed by G-protein, gustducin and leads to increased GLUT2 expression. The evidence for a similar effect of artificial sweeteners is conflicting (Ahmad et al. 2020).

Starch and fat are digested and absorbed at different intestinal sites. For starch, this occurs in the duodenum, upper jejunum, and proximal ileum because: 1) they have the highest mucosal expression of sucrase-isomaltase and SGLT1; 2) the rapid appearance of blood glucose after a starch meal fits with this site of absorption; 3) removal of the distal small intestine hardly affects glucose assimilation. By contrast, most fat and fat-soluble vitamin assimilation occurs in the ileum which has the highest transport capacity. Surgical removal of the ileum

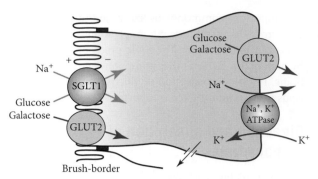

FIGURE 4.9 Glucose transport via both transporter routes. SGLT1 and GLUT2 both transport glucose in the small intestine, but SGLT1 co-transports two Na^+ and ~220 water molecules. The apical transporters are shown in addition to the single basolateral transporter for efflux into the portal circulation. The process is driven by ATP hydrolysis and coupling of glucose, sodium, and water transport is the basis of oral rehydration therapies for secretory diarrhoea (e.g., cholera).

may therefore leave the patient at risk of essential fatty acid and fat-soluble vitamin deficiency.

Digestion and absorption of protein

Gastric and pancreatic proteases hydrolyse dietary proteins to oligopeptides and aminopeptidases in the intestinal mucosa, and pancreatic carboxypeptidases can hydrolyse further to produce free amino acids and di- and tripeptides. The apical membrane of intestinal enterocytes contains a collection of different amino acid transporters, which ensure efficient uptake of all amino acids from the intestinal lumen. These transport systems were originally shown to be stereospecific (L-amino acids >> D-amino acids), specific (only a few chemically related amino acids per carrier system), duplicated (some amino acids transported by more than one carrier system). Some were sodium-linked, others were not. They can be generally categorized as systems for uptake of neutral amino acids, dibasic (cationic) amino acids, acidic (anionic) amino acids, imino acids, and β-amino acids. The same or different transporters are present in the basolateral membrane of the enterocyte and this allows amino acids to be released into the portal circulation for further metabolism by the liver and release into the peripheral circulation.

Returning to the case of System A/SNAT2/SLC38A2, this also transports methionine which also uses another eight different carriers, of which four are on the apical enterocyte surface. This highlights the fact that although System A transports the majority of methionine, it is a precious amino acid also involved in secondary

methyl-group metabolism which is essential for replication of intestinal cells. The uptake of dietary methionine and control of how much will exit to the liver or be sequestered in intestinal metabolism or even extracted from intestinal arterial blood is a matter of exquisite metabolic control (Mastrototaro et al. 2016). A similar case can be made for the amino acid glutamate, which is released by digestion at high concentration in the gut lumen (>1mmol/L) but rarely exceeds 10–50 μmol/L in blood—fortunate, because higher levels are quite neurotoxic. Glutamate has only one transporter but is extensively metabolized in the intestinal to α-ketoglutarate, L-alanine, ornithine, arginine, proline, glutathione, and γ-aminobutyric acid (GABA). It is also a potent signalling molecule involved in the cephalic phase of digestion (cf. monosodium glutamate) and neuroendocrine responses including control of intestinal inflammation. These considerations explain why monosodium glutamate is not toxic despite its potential to be so. Lastly, the aromatic amino acid phenylalanine has two transporters and undergoes little intestinal metabolism so that what is absorbed mostly reaches the liver and peripheral circulation.

Di- and tripeptides are taken up into the enterocyte by peptide transporters (PEPT1 and PEPT2), which are stereospecifically promiscuous (e.g., cyclic peptides) and non-specific. This means that they will transport most, if not all, of the theoretically possible 400 dipeptides and 8000 tripeptides, in addition to β-lactam antibiotics (e.g., penicillin) and valacyclovir, an anti-Herpes drug that has no peptide bond. The absorbed di- and tripeptides are hydrolysed by intracellular peptidases, and released into the portal circulation as free amino acids.

Some relatively large peptides (large enough to elicit antibody formation) enter the bloodstream intact, either by passing between cells or by uptake into mucosal cells. These are normally trapped by the gut-associated lymphoid tissue, but can enter the systemic circulation—this is the basis of food allergy (See Chapter 27).

Digestion and absorption of fat

As described, the mixed micelles transfer their constituent free fatty acids, monoacylglycerols, cholesterol, and bile salts across the enterocyte membrane. Short chain fatty acids enter the villus microcirculation, but most of the fatty acids are re-esterified to triacylglycerol in the mucosal cell, then are packaged into chylomicrons and secreted into the lacteals, and then into the lymphatic system (see Chapter 8).

Cholesterol and fat-soluble vitamins in the hydrophobic core of the micelles are absorbed directly and cholesterol is esterified in the enterocyte before insertion into new chylomicrons. Competition between cholesterol and other sterols/stanols for the acyltransferase probably explains why these compounds taken in dietary spreads will reduce cholesterol absorption, and hence have a hypocholesterolaemic action.

Fatty acids can cross the enterocyte by two routes.

1. Being lipophilic they can dissolve in the membrane and traverse it, by a 'flip-flop' mechanism, as protonated, uncharged fatty acids.
2. Via membrane-bound transport proteins. Candidates include CD36 (Fatty Acid Translocase) and FATP4 (Fatty Acid Transporter Protein 4).

'Flip-flop' permeation of a protonated fatty acid across a simple lipid bilayer membrane is rapid, about 10^3 faster than water and $>10^6$ faster than glucose. The process is rendered unidirectional because inside the cell they are either ionized at higher intracellular pH or incorporated into TAG. However, the internal dissociation step may be rate-limiting. The second proposed route of uptake is via membrane-bound transport proteins. Candidates include CD36 (Fatty Acid Translocase) and FATP4 (Fatty Acid Transporter Protein 4). There has been considerable controversy over the quantitative significance of either route. One intriguing proposal is that most fatty acids cross via 'flip-flop' but the rate-limiting internal desorption step is aided by FATP (Glatz and Luiken 2020) and Figure 4.10.

Within the enterocyte, fatty acids are transferred by intracellular fatty acid binding proteins (FABP) to the nascent lipid droplet where they are re-esterified to triacylglycerol. This lipid droplet enters the endoplasmic reticulum together with cholesterol, phospholipids, fat-soluble vitamins, and apolipoproteins before moving to the Golgi apparatus where the chylomicron matures. Buds from the Golgi apparatus fuse with the enterocyte lateral membrane, leading to exocytosis and the release of chylomicrons into the lymphatic system.

The rate-limiting step in this process is transfer of lipid from the endoplasmic reticulum to the Golgi apparatus and excess dietary lipid may either be oxidized for cellular energy or temporarily stored within the endoplasmic reticulum. The rate of movement of dietary fat into lymph depends on its fatty acid composition. Olive oil appears to be most rapidly absorbed, and cocoa butter and menhaden oil are most slowly absorbed. In malabsorption syndromes such as cystic fibrosis, medium chain triglycerides (MCT, chain length of 8–10 carbon atoms) are used because their shorter chain length and greater water solubility results in uptake into the portal circulation and hence a faster overall rate of macronutrient uptake.

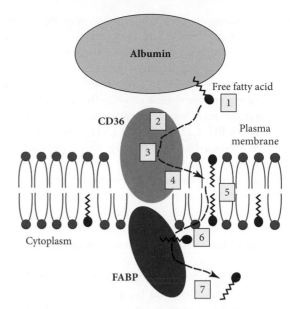

FIGURE 4.10 Sequential steps in fatty acid uptake by cells. Uptake of long-chain fatty acids by cells. 1. Release of fatty acids from (interstitial) albumin. 2. Binding in the hydrophobic cavity of CD36. 3. Guidance of the fatty acid through CD36 and past the unstirred water layer. 4. Exit from CD36 to the phospholipid bilayer. 5. Transmembrane translocation ('flip-flop') of single fatty acids. 6. Desorption of fatty acids from the inner membrane leaf and binding to the interior of FABP anchored to CD36. 7. Diffusion of the fatty acid-FABP complex towards sites of intracellular fatty acid metabolism.
Redrawn from Figure 1 in: J F C Glatz & J J F P Luiken (2020) Time for a detente in the war on the mechanism of cellular fatty acid uptake. *J Lipid Res* **61**, 1300–1303. (With permission.)

Absorption of vitamins

The fat-soluble vitamins A, D, E, and K, and carotenoids (see Chapter 12) are poorly soluble in water and are chaperoned in the circulation by specific transporter proteins or in lipoproteins (e.g., chylomicrons or VLDL/HDL) as they shuttle between tissues which comprise the absorptive, storage, or user organs. They are absorbed in the intestine, from lipid micelles which are formed by the products of fat digestion. During the process of food digestion and emulsification, these vitamins are released and selectively partition into the lipid phase of the luminal contents. It was assumed for a long time that like fatty acids these vitamins were absorbed by passive diffusion across the brush-border membrane of the enterocyte before repackaging into chylomicrons and exported in lymph. Current understanding reveals a more complex and controlled process. In the first place, this class of molecules is also absorbed through the action of a group of scavenger receptors which were previously assumed to have other functions. These molecules recognize and take up macromolecules which have negative charge, such as oxidized low-density lipoprotein. As the name implies, their function is to clean or scavenge undesirable molecules. They are found in the membranes of many tissues (e.g., macrophages in atherosclerotic lesions) and in the enterocyte membrane as well. The most common gut scavenger receptors are Scavenger Receptor Class B type I (SR-BI), CD36 (Cluster Determinant 36) and Niemann-Pick C1-like 1 (NPC1L1). This dual functionality of control molecules is a common theme in metabolic nutrition. Another example would be the activation of vitamin D precursors by enzymes of the cytochrome P450 superfamily which are also associated with detoxification reactions. The current understanding of the assimilation of the fat-soluble vitamins is summarized in Table 4.3.

Vitamin A may be absorbed as β-carotene via SR-B1 and is either exported into lymph in chylomicrons to be stored in the liver or is cleaved to form retinal and then retinol which is chaperoned inside the enterocyte by cellular retinol binding protein type II (CRBPII). It may be exported and bound to retinol binding protein 4 (RBP4) which is stabilized by transthyretin (TTR, formerly known as prealbumin). Retinyl esters require prior hydrolysis before uptake and are then re-esterified inside the enterocyte before packaging in chylomicrons.

Vitamin D is absorbed by passive diffusion and via SRB1, NCP1L1, and CD36 and is excreted at the basolateral membrane by ATP-binding cassette, subfamily A (ABCA1) before packaging into chylomicrons.

Vitamin E is absorbed in the same way by passive diffusion or via SR-B1, or NPC1L1 and is excreted as for vitamin D. Synthetic forms of vitamin E are esterified, natural forms not. The synthetic forms therefore require hydrolysis in the intestinal lumen before absorption.

Vitamin K is assumed to be absorbed in the same way as other fat-soluble vitamins. There is further conversion to other forms of phylloquinones in the gut, whilst the menaquinones are formed by colonic bacteria or are present in some fermented foodstuffs.

The water-soluble vitamins (see Chapter 11) are presented to the gut mucosal surface in several forms, each of which may have its specific transport protein (Table 4.3). Vitamin C is present in the intestinal lumen as both reduced ascorbic acid and oxidized dehydroascorbic acid. Ascorbic acid is absorbed by a sodium-dependent transporter (SVCT1), while dehydroascorbic acid is structurally similar to glucose (from which some mammalian species can synthesize it) and absorbed by the sodium-independent glucose transporters GLUT1, GLUT3, and GLUT4. Some vitamins require dephosphorylation before they can be absorbed (e.g., vitamins B_1, and B_6) and are trapped inside the intestinal cell by rephosphorylation (see Chapter 11). Vitamin B_{12} is absorbed bound to intrinsic factor, a glycoprotein that is secreted by the parietal cells of the gastric mucosa (see Chapter 11). Finally, it should be noted that vitamins are provided by two routes. The small intestine is the site of absorption of dietary vitamins. Patients who have undergone surgical removal of the ileum may often develop deficiency of fat-soluble vitamins. In contrast, the colon is the site of absorption of considerable amounts of vitamin K and most of the water-soluble vitamins (thiamin, riboflavin, niacin, pantothenic acid, pyridoxine, biotin, folate, cyanocobalamin) because they are synthesized there by the colonic microbiota. The species responsible for this are at present unknown but some lactobacilli possess some, but not all parts of the vitamin B_{12} synthetic pathways. This concept will be of great importance in the future for several reasons. First, diet affects the colonic microbiome and thus human health. Second, diseases such as obesity and diabetes alter the microbiome. Third, the increasing adoption of vegetarian or vegan diets in modern times requires scrupulous attention to water-soluble vitamin intake inadequacies which may not be compensated for by colonic biosynthesis. Two recent surveys in Switzerland (Schupbach et al. 2017) and of the worldwide literature (Pawlak et al. 2013) revealed widespread deficiency amongst vegans and vegetarians. The latter concluded that *'with few exceptions, the reviewed studies documented relatively high deficiency prevalence among vegetarians'* (Pawlak et al. 2013). Finally, it is possible that the use of specific prebiotic strains of lactobacilli to produce prefermented foods, which contain naturally synthesized vitamins, may revolutionize dietary patterns in the future.

Digestion and absorption in the colon

Professor Sir Tore Midtvedt was the first person to call the colonic microbiome *'an organ within an organ'* on the basis of its ability to self-regulate.

- It establishes an 'acetate buffer' (about one-tenth the strength of condiment vinegar), which suppresses the growth of some pathogenic bacteria.
- It produces short-chain fatty acids (SCFA) which protect the colon.
- Species such as bacteroides degrade pancreatic enzymes and their low prevalence is associated with inflammatory bowel disease.
- It converts bilirubin (haem breakdown product) to urobilinogen (yellow pigment) which is either reabsorbed and excreted by the kidney or converted to stercobilin (brown pigment).
- It digests mucins shed from the epithelial mucus layer.
- It catabolises and excretes bile salts which are synthesized from cholesterol.

TABLE 4.3 Current understanding of the assimilation of the fat- and water-soluble vitamins in the human intestine

Vitamin name	Dietary form	Absorbed from	Brush-border membrane uptake		Cytoplasmic transport	Basolateral membrane efflux transporter	Transport in circulation
			Passive absorption	Transporter			
Vitamin A	β-carotene	Mixed micelles	Yes	SR-B1	Cleaved to retinal then retinol OR remains intact as β-carotene	?	Lacteals/ chylomicrons
	Retinol	Mixed micelles	Yes	?	CRBPII and some converted to retinyl ester	? but small amount	RBP4
	Retinyl-esters	Mixed micelles	Yes	Hydrolysed before transport	CRBPII and most re-converted to retinyl ester	?	Lacteals/ chylomicrons
Vitamin D	Ergocalciferol (D2) cholecalciferol (D3)	Mixed micelles	Yes	SRBI, NCP1L1, CD36	? None identified yet	ABCA1	Lacteals/ chylomicrons
Vitamin E	Tocopherol	Mixed micelles	Yes	SRBI, NCP1L1	? None identified yet	ABCA1	Lacteals/ chylomicrons
Vitamin K	Phylloquinone, menaquinone	Mixed micelles and from colon	Yes	?	?	Incorporated into chylomicrons	Lacteals/ chylomicrons
Vitamin C	Reduced— Ascorbic acid	Aqueous phase	No	SVCT1	None	SVCT2	Portal vein
	Oxidized— Dehydro-L-ascorbic acid	Aqueous phase	No	GLUT1, GLUT3 and GLUT4	None		Portal vein
Vitamin B1	Thiamine	Aqueous phase of small and large bowel	No	THTR1/ THTR2	none	THTR1	Portal vein
	Thiamine phosphate			Requires dephosphorylation before absorption			
Vitamin B2	Riboflavin	Aqueous phase of small and large bowel	No	RFT1		RFT2	Portal vein
	Flavin mononucleotide			Requires hydrolysis before Flavin can be absorbed			
	Flavin adenine nucleotide						
Vitamin B3	Niacin	Aqueous phase of small and large bowel	No	OAT10?	None		
Vitamin B5	Panthothenate	Aqueous phase	No	SMVT	None	?	Portal vein
Vitamin B6	Pyridoxal, pyridoxine, and pyridoxamine.	Aqueous phase of small and large bowel	No	?	?	?	Portal vein
	Pyridoxal phosphate	Aqueous phase	No	Requires dephosphorylation before absorption			

TABLE 4.3 *Continued*

Vitamin name	Dietary form	Absorbed from	Brush-border membrane uptake		Cytoplasmic transport	Basolateral membrane efflux transporter	Transport in circulation
Vitamin H/ B7	Biotin	Aqueous phase of small and large bowel	No	SMVT	None	?	Portal vein
	Protein-bound biotin	Requires release from dietary proteins by biotinidase before absorption					
Vitamin B9	Folate polyglutamates	Aqueous phase of small bowel	No	Require removal of polyglutamate before absorption as monoglutamate			
	Reduced folate monoglutamate	Aqueous phase of small and large bowel	No	RFC	?	?	Portal vein
	Oxidized folate monoglutamate	Aqueous phase of small and large bowel	No	PCFT	?	?	Portal vein
Vitamin B12	Cobalamin	Aqueous phase	Cobalamin binds to haptocorrin or intrinsic factor released from the gastric mucosa. The complex binds to cubilin on the ileal enterocyte brush-border membrane and is taken up by endocytosis		Digestion of intrinsic factor, release of cobalamin	?	Portal vein

ABCA1 (ATP-binding cassette, sub-family A, member 1), CD36 (Cluster Determinant 36), NCP1L1 (Niemann–Pick C1-like 1), OAT10 (Organic Anion Transporter 10), PCFT (Proton-Coupled Folate transporter), RFC (Reduced Folate Carrier), RFT1 (Riboflavin Transporter 1), SMVT (Sodium-dependent Multivitamin Transporter), SR-BI (Scavenger Receptor Class B, Type I), SVCT1/SVCT2 (Sodium Vitamin C Transporter 1/2), THTR1/THTR2 (Thiamine Transporter 1/2).

It is also organ-like because of the diversity of reactions which are performed by ~10^{14} colonic microbes. Within this microbiome there is considerable cross-talk between species and cross-feeding of substrates (cf. liver periportal and perivenous hepatocytes). In addition, there is considerable two-way cross-talk between host and microbe. Their actions take place in an anaerobic environment, which means that those bacteria which hydrolyse fibre (i.e., soluble oligo-saccharides, di-saccharides, and polyols) or salvage and oxidize proteins and amino acids (which have escaped small bowel assimilation) for energy, must donate protons to an acceptor such as NAD^+ or $NADP^+$. Acetyl CoA is the main 2-carbon product of glycolysis and provides feedstock for direct production of acetate (directly) or of propionate (3-carbons, from succinate) or butyrate (4-carbons, from acetate via

ketogenesis) in the ratio 6:3:1. Butyrate and propionate are both synthesized using energy derived from glycolysis since the complete tricarboxylic acid cycle does not operate in an anaerobic environment. Hydrogen is a prominent end-product of this process and represents a 'sink' of reducing equivalents, which can be expelled as gas.

Although dietary surveys reveal that the UK population consumes ~18g/day of dietary fibre, this is much less than is required to generate the average daily stool weight. What is missing from the calculation is the contribution from other malabsorbed macronutrients (e.g., resistant starch) and from mucins and sloughed intestinal cells (Figure 4.11).

If all of this were oxidized, with H_2 as the end-product, it would produce ~22 litres of gas whereas the average

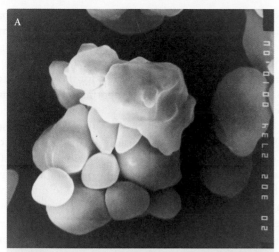

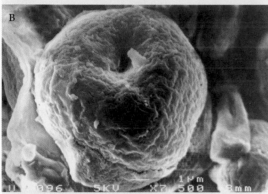

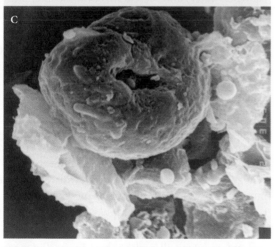

FIGURE 4.11 Resistant starch granules and their digestion in the colon.
Scanning electron micrographs of native high-amylose starch granules sampled in: Top—human ileostomy effluent; Middle—the caecum of a pig; Bottom—the proximal large bowel of a pig. Amylase causes pitting and etching of the starch granules and glucose oligomers that are released will be fermented in the large intestine.
Source: D L Topping & P M Clifton (2001) Short-chain fatty acids and human colonic function: Roles of resistant starch and nonstarch polysaccharides. *Physiol Rev* **81**, 1031–1064.
Fig. 1, with permission. Copyright © 2001,
The American Physiological Society.

daily flatus is ~1 litre. The difference probably reflects further H_2 consumption in 1) secondary hydrogenation of CO_2 to CH_4 and 2) hydrogenation of malabsorbed polyunsaturated fats which enter the colon. Together they will reduce the volume considerably.

SCFA absorption can occur by simple diffusion of protonated species across the apical colonocyte membrane or via a sodium-coupled monocarboxylate transporter (SLC5A8) which significantly stimulates water uptake and electrolyte salvage from the colon, much as sodium-linked glucose transport (for example) does in the small bowel.

Butyrate promotes cell differentiation, induces apoptosis in colon cancer cells and in T cells which are responsible for colonic inflammation. It does this by inhibiting histone deacetylases (HDACi) and activating the Fas receptor-mediated extrinsic death pathway. Both butyrate and propionate inhibit histone deacetylases, a property they share with all of the SCFA up to a chain length of 8-carbons (including the antipsychotic drug, valproate). Acetylation of lysine and arginine residues in histone proteins leads to a looser binding with DNA which allows chromatin expansion, permitting genetic transcription to take place. Because butyrate maintains the transcription of key genes by preventing deacetylation of the histone, it is in effect a transcription factor. These effects are butyrate dose-dependent and the concentration of butyrate is proportional to the amount and type of fermentable fibre in the colon. After a two-week daily enema of butyrate in human subjects, there was increased transcription of genes, which were mainly associated with energy metabolism, increasing fatty acid oxidation, and in protecting against oxidative stress. Butyrate is also rapidly oxidized leading to generation of large amounts of reactive oxygen species by mitochondria which peroxidize long-chain n-3 polyunsaturated fatty acids from fish oil and the products stimulate apoptosis. This occurs at the base of the mucosal crypt and is effective in all stages of tumourigenesis.

The SLC5A8 gene is a tumour suppressor, silenced in colon cancer by DNA methylation. Its product, the SCFA transporter, causes uptake of butyrate and other SCFA into colonic cells where they exert an anti-tumour effect, as described. Two other cell membrane receptors, hydroxycarboxylic acid receptor 2 (HCA2, product of gene GPR109A) and free fatty acid receptor 2 (FFAR2, product of gene GPR43) are agonized by butyrate and modulate fatty acid metabolism and provide protection against inflammation, respectively. This therefore provides a satisfying explanation for the cancer-preventing effects of a diet containing oily fish and high fruit and vegetable intake whose colonic fermentation produces butyrate, propionate, and acetate (Chapkin et al. 2020).

Water and electrolytes

The human small intestine absorbs 6.5 to 7.5 litres of water each day (Figure 4.1). This comes from several sources as indicated in Figure 4.12. How this gets across the membrane is still unclear because the lipid bilayer that surrounds each cell is impermeable to water, and the intestinal mucosal surface is rather hydrophobic. Before describing several hypotheses of water movement across the mucosa, it is worth summarizing the known characteristics of the process. Water absorption is proportional to the amount of substrate and electrolyte that moves across the membrane. For example, SGLT1 moves 2 Na^+ ions and 220 water molecules across the membrane with each cycle of transport of one glucose molecule. Absorption of 180 grams of glucose in the small intestine would result in the uptake of 4.7 litres of water and 116 grams of NaCl. The direction of water movement is governed by solute movement. The mechanisms for secretion and reabsorption of water and electrolytes by the intestine are very efficient. Under normal circumstances, daily secretion of Na^+ and K^+ by the gut is equivalent to a half and two-thirds of these electrolytes present in the human body's extracellular space, respectively. These movements are achieved by coupled transport systems which will excrete HCO_3^- in response to Cl^- uptake, in order to achieve electroneutrality across the mucosal barrier. At more distal sites, Cl^- and HCO_3^- will move in the opposite direction to conserve bicarbonate. Similar mechanisms exist for exchange of Na^+, K^+, and divalent cations. As a result, despite this huge transfer and counter-transfer of water and ions across the small and large intestinal mucosa, faecal outputs are modest and comprise only 200g of water, 5 mmol Na^+, 2 mmol Cl^-, 9 mmol K^+ and approximately 5 mEq HCO_3^-. In cholera, excessive secretion of chloride into the colon is accompanied by water secretion and diarrhoea, leading to dehydration and eventually death, unless treated. Therefore, stimulation of water uptake by glucose (and sodium) is the basis of oral rehydration therapy for the treatment of diarrhoea, which is the most important cause of infant death worldwide (Ofei and Fuchs 2019).

The mechanisms by which the bulk of water crosses epithelial membranes is still a cause of controversy which is explained clearly in this review (Hill 2008). The intestinal mucosa acts as a semipermeable membrane through which water flows either way, in response to differences in osmotic pressure. Luminal nutrient digestion renders the bulk phase hypertonic and water moves from the plasma compartment into the gut lumen. Nutrient absorption, however, renders it less hypertonic because some water is absorbed along with these solutes. In this way, the luminal contents are adjusted to near isotonicity throughout the small bowel. There are several models for this process.

1. Simple osmosis. The osmolality gradient is small (3–30 mOsmol/kg) and this model would mean that enterocytes replaced their entire fluid volume every few seconds.

2. Water moves through 'tight junctions' between enterocytes, the paracellular route. Enterocytes absorb Na^+ (not water) and secrete this into the lateral intercellular space as a hyperosmolar fluid. This is the most popular hypothesis.

3. Specific transporters or aquaporins move water across membranes. These transporters are versatile and can also transport other small molecules such as urea, CO_2, NH_3, and glycerol in addition to water. Their structure is similar to that of other transporters except that the hydrophobic inner surface of the transmembrane-pore is lined with some hydrophilic amino acids, which will selectively allow passage of water. This means that enterocytes will replace their intracellular water volume every few seconds.

4. Water is co-transported with substrates and ions by other substrate transporters (e.g., SGLT1). This means that enterocytes will replace their intracellular water volume every few seconds.

Sugar alcohols, used as sweeteners, such as xylitol, lactitol, and sorbitol, are poorly absorbed and will enter the colon with sufficient water to maintain luminal isotonicity before fermentation and the absorption of SCFA, water, and Na^+. If the colonic fermentation capacity is exceeded, then osmotic diarrhoea ensues because the excess water cannot be absorbed fast enough.

Clinically the synthetic disaccharide lactulose (which is not hydrolysed in the small intestine) and the sugar alcohol lactitol are used as laxatives. Other causes of

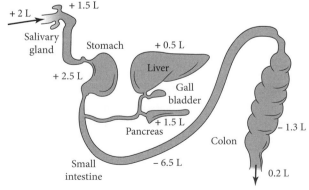

FIGURE 4.12 Water movements in the gut. Dietary intake of water and stool output are modest in comparison to secretion and absorption of water within the human intestine which has large transport capacity. *Source*: T Ma & A S Verkman (1999) Aquaporin water channels in gastrointestinal physiology. *J Physiol (Lond)* **517**, 317–326. Fig. 1. Reproduced with permission. Copyright 2004, John Wiley.

osmotic diarrhoea include dietary fibre such as guar gum, probiotics such as fructose oligosaccharide (i.e., inulin), and beans that contain large quantities of stachyose, all of which are good substrates for bacterial fermentation. The laxative threshold of unacceptable gastrointestinal symptoms for most readily fermented, non-absorbed carbohydrates is about 70g/day, but most people will notice the effects of 40g/day. The FODMAPS (Fermentable Oligo-, Di-, Monosaccharides and Polyols) diet for treatment of irritable bowel syndrome is an attempt to reduce the amount of fermentable substrate in the diet. It is an alternative to the recommendations of NICE (National Institute for Health and Care Excellence), which include smaller and more frequent meals and a reduction of dietary fibre.

Most minerals are absorbed by carrier-mediated diffusion, and are then accumulated by binding to intracellular binding proteins, followed by sodium-dependent transport from the enterocyte into the villous microcirculation. The main system for transport of most divalent metal ions and ferrous iron is the divalent metal-ion transporter 1 (DMT1) whose activity responds to systemic signals of adequate nutritional Zn^{2+}, Mg^{2+} or Fe^{2+} or Fe^{3+} stores. For example, the protein hepcidin inhibits iron transport by binding to the iron export channel (ferroportin) on the enterocyte basolateral membrane. Hepcidin is secreted when iron stores are adequate. Genetic defects of either the intracellular binding proteins or the active transport systems at the basal membrane of the enterocyte can result in mineral deficiency despite an apparently adequate intake. As discussed in Chapter 12 the enterocyte calcium binding protein is induced by vitamin D, and vitamin D deficiency results in much reduced absorption of calcium.

4.5 The role of the gastrointestinal tract in the regulation of feeding

Long-term and short-term mechanisms regulate food intake and energy expenditure, so as to maintain energy balance (Chapter 6). The processes of eating and digesting food trigger a large and orderly array of short-term signals within the time frame of the hunger–satiety cycle (Smeets et al. 2010) Figure 4.6. Several of these have already been described in this chapter and they mostly comprise classic homeostatic control via negative feedback loops.

The mouth

The taste of food, as well as the smell before eating, stimulates secretion of gastric juice and intestinal motility. In addition, the tongue can sense the fatty acids liberated from triacylglycerols by lingual lipase. Combinations of taste and sensation may have additive effects, for example sugar mixed with fat is particularly pleasurable, and salt may be useful in masking bitter flavours that are taste-aversive. The importance of taste in controlling sensations of hunger and satiety is seen in patients receiving long-term tube-feeding who experience constant feelings of hunger although nutritionally replete. The taste of food provides a strong signal to stimulate eating. However, this process operates only when there are sensations of hunger and is suppressed when sensations of satiety become strong. This is known as sensory-specific or conditioned satiety.

The products of digestion and microbial metabolites also act on intestinal chemosensors (aka taste sensors, chemoreceptors) present on the tongue and throughout the intestine. They can detect sweet, sour, bitter, and salty flavours in addition to the more elusive umami (Japanese 旨味 = pleasant savoury taste). Salty and sour tastes are mediated by ion-channels whereas sweet/umami are detected by taste receptor family 1 (TAS1) and bitter taste is recognized by taste receptor family 2 (TAS2). When stimulated, these receptors cause intracellular generation of inositol trisphosphate (IP3) and diacylglycerol (DAG) which trigger Ca^{2+} release from the endoplasmic reticulum into the cytoplasm. This causes a Na^+-mediated cascade within the cell membrane leading to purinergic activation of the associated afferent 'taste' nerve.

These receptors can sense digestion products or bacterial metabolites. For example, the tongue can sense the fatty acids liberated from triacylglycerols by lingual lipase, and other examples include:

- succinate (microbial product)—receptor SUCNR1 triggers defensive type 2 immune inflammatory response
- valine and amino acids (digestion product)—calcium-sensing receptor (CaSR) stimulates GLP-1 secretion
- glutamate (digestion product)—receptor TAS1R1-TAS1R3 (umami) triggers cephalic phase and CCK release
- di- and tripeptides (digestion product)—lysophosphatidic acid receptor 5 (LPA5R) stimulates CCK release, via CaSR receptor stimulates GLP-1 release
- free fatty acids (digestion product)—free fatty acid receptors stimulate anorexigenic hormones, CCK and GLP-1. SCFA inhibit release of orexigenic hormone, ghrelin
- glucose (digestion product)—receptor TAS1R2 stimulates GLP-1 and PYY release.

This is extensively reviewed elsewhere (Steensels & Depoortere 2018).

Before describing the role of intestinal chemo-sensing during digestion and absorption, it should be emphasized that it has great importance in the global business of meal production. For example, non-nutritive sweeteners such as aspartame tell the intestine to mount a hormonal and metabolic response to something which isn't there. This may alter food intake inappropriately and lead to obesity in an extreme case, although the evidence is that it does not (Klaassen et al. 2021). Second, aversive bitter peptides are often released during fermentation or enzyme treatment of foodstuffs and they typically contain hydrophobic amino acids. This is problematic in applications as diverse as production of soy sauce or therapeutic protein hydrolysate-based diets for children with cow's milk intolerance. Not all peptides are bitter, indeed the sweetener aspartame is L-aspartyl-L phenylalanine-Omethyl ester, whereas its L,-D, D,L and D,D isomers are bitter (Temussi 2012). This problem has traditionally been resolved by substituting proteases with different specificity into the process to change the peptide profile. A newer approach is to use small peptides released by hydrolysis of beef or other protein sources to block the TAS2 bitter taste receptor. Third, many drugs stimulate TAS2 and this can lead to inadvertent food aversion (Schiffman 2018). Infusion of a mixture of tastants into the duodenum reduced food consumption and hunger sensations in a following meal (van Avesaat et al. 2015). In this experiment, the tastants were bitter (quinine), sweet (rebaudioside A, a stevioside), and umami (monosodium glutamate). Lastly, all good chefs know that good flavour is the foundation of a delicious meal, which is eagerly anticipated by smell and enjoyed by taste.

The relative speed of feedback loops in relation to eating

The control of energy balance is impressively precise. The adipose tissue stores of an average lean and obese man contain approximately 170,000 kcal and 413,000 kcal, respectively. The process which led to the lean man becoming obese over 30 years involved an imbalance in energy intake over requirements of 20–30 kcal/day, compared to an energy intake of 2800 kcal/day, that is, a ~1 per cent difference. The more important question is not 'Why has this man become obese?' but 'Why is he not five-times the size?' In other words, regulatory mechanisms of eating are exquisitely balanced, and a general view is that obesity results when this regulation is subverted by a plentiful supply of energy-rich, high Glycaemic Index, hyper-palatable, ultra-processed food (UPF), available in the long-term. Epidemiological studies which confirm or challenge this depend critically on the prior definition of UPF. One plausible explanation for an effect has come from a recent trial comparing *ad libitum* ultra-processed

or unprocessed diets consumed for two weeks in a cross-over design (Hall et al. 2019). Each subject acted as their control and UPF consumption led to increased weight because more of it was eaten. This was not because UPF was tastier than the unprocessed diet, nor did it cause hyperinsulinaemia or alter glycaemia. Subjects just ate it faster. This suggests that satiety signals were insufficiently nimble for this rapid rate of ingestion. 'Fast food' is often defined as food that is quick to prepare or that is eaten in large amounts and is high in fat and sugar. The ease of eating it at speed could also be added to that definition and may also help explain why it is obesogenic.

Aversive taste effects

The Cephalic Phase (see 'The cephalic, gastric, and intestinal phases of gastric digestion') is mediated by food odours which lead to liking or aversion. A clinically relevant example relates to eating by older people, because taste sensitivity declines with age (Methven et al. 2012). The sights, sounds, and smells of a ward are also food-aversive for hospital patients (Schiffman 2018). Recent studies have shown that addition of umami tastants to foodstuffs can increase the amount eaten at each meal, and in the long-term this strategy has been shown to increase food intake in cancer patients.

The gastric phase

During eating, food stretches the stomach and induces a complex series of signals that lead to cessation of eating. For example, taking a glass of water with a meal (not afterwards) will induce feelings of satiety whilst eating. Conversely, in rats in which gastric contents are continuously removed during the meal through a fistula ('sham feeding'), the amount of food eaten each meal is greatly increased. The mechanism is due to stretch, not gastric pressure, and works through direct inhibition of the stimulating effect of pleasurable tastes on eating (Figure 4.13A). Signals from taste receptors in the mouth and from gastric stretch receptors are integrated in the parabrachial nucleus of the pons in the brain stem so that one signal will down-regulate the other (Figure 4.13B).

During the gastric phase, several hormones initiate satiety.

1. CCK acts directly on the vagal afferent nerves and on the hypothalamus.

2. Peptide YY (PYY) secretion signals the presence of nutrients in the gut lumen and reduces voluntary food intake as confirmed by PYY infusion studies.

3. Leptin is mainly secreted by adipose tissue and its main function is to regulate long-term food intake

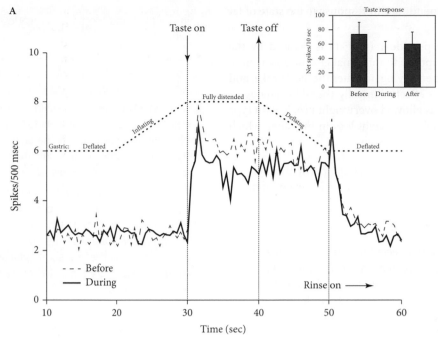

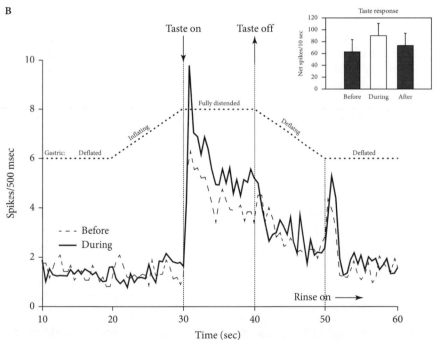

FIGURE 4.13 How the taste of food and gastric distension have opposite effects on appetite.
Taste and gastric distension have opposite effects on appetite measured from electrical activity of the parabrachial nucleus of anaesthetized rats. Taste solutions were applied to the tongue or the stomach and the time course of response before and during gastric distension was measured in A) Inhibitory cells and B) Excitatory cells. Gastric distension enhances satiation and inhibits eating which is activated by taste. These dynamic signals control food intake during feeding. This is like driving a car by pressing the brake and accelerator together!

Source: J P Baird et al. (2001) Integration of gastric distension and gustatory responses in the parabrachial nucleus. *Am J Physiol Regul Integr Comp Physiol* **281**, R1581–R1593. Fig. 4, with permission. Copyright © 2001, The American Physiological Society.

and energy expenditure in response to the state of fat reserves.

4. Glucagon-like peptide-1 (GLP-1) is released by the pancreas in response to intestinal glucose absorption. It slows gastric emptying, initiates the 'ileal brake', and has an anorectic effect. Clinically, the GLP-1 analogue, Semaglutide has allowed overweight people with type 2 diabetes to achieve weight loss in the range which results in diabetes remission.

Gastric distention induces satiety, partly through the release of hormones such as gastrin-releasing peptide.

The small intestine

In addition to stretch receptors, the intestinal mucosa possesses an abundance of receptors for acid and for fatty acids and glucose and amino acids, which will provide information about the contents of the lumen that the brain stem will integrate and use to control eating behaviour (van Avesaat et al. 2015). CCK, released in response to luminal fat leads to powerful inhibition of eating.

Absorbed nutrients are also potent signals that modulate eating behaviour. For example, in adequately nourished subjects, the intravenous infusion of lipid stimulates dopamine activity (which acts as a feeding inhibitor) and increases satiety ratings, feelings of fullness, and reduces the desire to select particular foodstuffs. In contrast, studies in patients who were admitted acutely to hospital (which experience ~40 per cent malnutrition) have shown that fortification of hospital food with fat actually stimulates energy intake. This mechanism thus depends on sensations of hunger that are related to nutritional status. However, these mechanisms can be overridden centrally. An example of this would be the inability to resist the unexpected offer of a plate of strawberries and cream after a particularly heavy meal.

KEY POINTS

- The gastrointestinal tract provides a linear sequence of events resulting in the hydrolysis of dietary carbohydrates, triacylglycerols, and proteins, and the absorption of the products of digestion. This process starts before eating, when salivary and gastric secretions are stimulated and continues when the presence of food in the mouth and stomach stimulates further secretion.
- Gastric emptying is controlled by both the amount of food eaten during a meal, and also nutrients present in the food and the progress made in liquidizing it within the stomach.
- Pancreatic and intestinal secretion is stimulated by hormones secreted in response to the presence of food in the stomach.
- The monosaccharides resulting from carbohydrate digestion, and free amino acids from protein digestion, are absorbed into the hepatic portal vein, and the liver regulates the entry into the peripheral circulation of the products of digestion.
- Amylases in saliva and pancreatic juice catalyse hydrolysis of starch to disaccharides and limit dextrins; disaccharides are hydrolysed by intestinal brush-border enzymes, and monosaccharides are absorbed by active transport (glucose and galactose) or passive transport (other monosaccharides and sugar alcohols).
- Lipases secreted by the tongue, in gastric juice and pancreatic juice catalyse the progressive hydrolysis of triacylglycerol until dietary lipid is emulsified into micelles small enough to be absorbed across the small intestinal lumen. Most absorbed fatty acids are re-esterified in the mucosal cells and absorbed into the lymphatic system in chylomicrons, but medium-chain fatty acids are absorbed in to the hepatic portal vein.
- Proteolytic enzymes are secreted as inactive zymogens. Pepsinogen in the gastric juice is activated by gastric acid and autocatalysis; trypsinogen in the pancreatic juice is activated by intestinal enteropeptidase. Trypsin then activates the other intestinal zymogens.
- Protein digestion begins with the action of endopeptidases, which hydrolyse proteins at specific sites within the molecule, resulting in the formation of a large number of oligopeptides. Exopeptidases then remove amino- and carboxy-terminal amino acids, resulting in free amino acids and di- and tripeptides.
- Free amino acids are absorbed by a variety of group-specific transporters; di- and tripeptides are absorbed by a specific transporter with wide substrate tolerance, and hydrolysed within the intestinal mucosal cells.
- Luminal digestion products bind to specific receptors on intestinal mucosal cells and elicit changes in hormones which alter metabolism (e.g., GLP-1) or which alter eating behaviours, usually by inducing satiety (e.g., CCK, PYY).

ACKNOWLEDGEMENTS

Dr Natalia Zarate-Lopez, Consultant Gastroenterologist, GI Services, University College London Hospitals NHS Foundation Trust, 235 Euston Rd, London NW1 2BU.

 Be sure to test your understanding of this chapter by attempting multiple choice questions.

REFERENCES

Ahmad S Y, Friel J K, Mackay D S (2020) Effect of sucralose and aspartame on glucose metabolism and gut hormones. *Nutr Rev* **78**, 725–746.

Allen A & Flemstrom G (2005) Gastroduodenal mucus bicarbonate barrier: Protection against acid and pepsin. *Am J Physiol Cell Physiol* **288**, C1–19.

Broer S & Fairweather S J (2018) Amino acid transport across the mammalian intestine. *Compr Physiol* **9**, 343–373.

Chapkin R S, Navarro S L, Hullar M A J et al. (2020) Diet and gut microbes act coordinately to enhance programmed cell death and reduce colorectal cancer risk. *Dig Dis Sci* **65**, 840–851.

Glatz J F C & Luiken J J F P (2020) Time for a detente in the war on the mechanism of cellular fatty acid uptake. *J Lipid Res* **61**, 1300–1303.

Hall K D, Ayuketah A, Brychta R et al. (2019) Ultra-processed diets cause excess calorie intake and weight gain: An inpatient randomized controlled trial of ad libitum food intake. *Cell Metab* **30**, 67–77.

Hill A E (2008) Fluid transport: A guide for the perplexed. *J Membr Biol* **223**, 1–11.

Klaassen T, Keszthelyi D, Troost F J et al. (2021) Effects of gastrointestinal delivery of non-caloric tastants on energy intake: a systematic review and meta-analysis. *Eur J Nutr* Published online: 8 Feb. 2021. doi: 10.1007/s00394 021 02485 4

Mastrototaro L, Sponder G, Saremi B et al. (2016) Gastrointestinal methionine shuttle: Priority handling of precious goods. *IUBMB Life* **68**, 924–934.

Methven L, Allen V J, Withers C A et al. (2012) Ageing and taste. *Proc Nutr Soc* **71**, 556–565.

Ofei S Y & Fuchs G J III (2019) Principles and practice of oral rehydration. *Curr Gastroenterol Rep* **21**, 67.

Pawlak R, Parrott S J, Raj S et al. (2013) How prevalent is vitamin B_{12} deficiency among vegetarians? *Nutr Rev* **71**, 110–117.

Petersen K U (2018) Pepsin and its importance for functional dyspepsia: Relic, regulator or remedy? *Dig Dis* **36**, 98–105.

Roberts N B (2006) Review article: human pepsins—their multiplicity, function and role in reflux disease. *Aliment Pharmacol Ther* **24** Suppl 2, 2–9.

Schiffman S S (2018) Influence of medications on taste and smell. *World J Otorhinolaryngol Head Neck Surg* **4**, 84–91.

Schupbach R, Wegmuller R, Berguerand C et al. (2017) Micronutrient status and intake in omnivores, vegetarians and vegans in Switzerland. *Eur J Nutr* **56**, 283–293.

Smeets P A, Erkner A, de Graaf C (2010) Cephalic phase responses and appetite. *Nutr Rev* **68**, 643–655.

Steensels S & Depoortere I (2018) Chemoreceptors in the gut. *Annu Rev Physiol* **80**, 117–141.

Temussi P A (2012) The good taste of peptides. *J Pept Sci* **18**, 73–82.

van Avesaat M, Troost F J, Ripken D et al. (2015) Intraduodenal infusion of a combination of tastants decreases food intake in humans. *Am J Clin Nutr* **102**, 729–735.

FURTHER READING

Benga G (2012) The first discovered water channel protein, later called aquaporin 1: Molecular characteristics, functions and medical implications. *Mol Aspects Med* **33**(5–6), 518–534.

Dermiki M, Prescott J, Sargent L J et al. (2015) Novel flavours paired with glutamate condition increased intake in older adults in the absence of changes in liking. *Appetite* **90**, 108–113.

Duque-Correa M J, Codron D, Meloro C et al. (2021) Mammalian intestinal allometry, phylogeny, trophic level and climate. *Proc Biol Sci* **288**, 20202888.

Hill A E & Shachar-Hill Y (2015) Are aquaporins the missing transmembrane osmosensors? *J Membr Biol* **248**, 753–765.

Lindquist S & Hernell O (2010) Lipid digestion and absorption in early life: An update. *Curr Opin Clin Nutr Metab Care* **13**, 314–320.

Litou C, Psachoulias D, Vertzoni M et al. (2020) Measuring pH and buffer capacity in fluids aspirated from the fasted upper gastrointestinal tract of healthy adults. *Pharm Res* **37**, 42.

Modvig I M, Kuhre R E, Jepsen S L et al. (2021) Amino acids differ in their capacity to stimulate GLP-1 release from the perfused rat small intestine and stimulate secretion by different sensing mechanisms. *Am J Physiol Endocrinol Metab* **320**, E874–E885.

Rothman S, Liebow C, Isenman L (2002) Conservation of digestive enzymes. *Physiol Rev* **82**, 1–18.

Sasaki S (2008) Introduction for special issue for Aquaporin: Expanding the world of aquaporins: New members and new functions. *Pflügers Arch* **456**, 647–664.

Schneider C, O'Leary C E, Locksley R M (2019) Regulation of immune responses by tuft cells. *Nat Rev Immunol* **19**, 584–593.

Spring K R (1999) Epithelial fluid transport: A century of investigation. *News Physiol Sci* **14**, 92–98.

Stevens C E & Hume I D (1998) Contributions of microbes in vertebrate gastrointestinal tract to production and conservation of nutrients. *Physiol Rev* **78**, 393–427.

Tomé D (2018) The roles of dietary glutamate in the intestine. *Ann Nutr Metab* **73** Suppl 5, 15–20.

Wilding J P H, Batterham R L, Calanna S et al. (2021) Once-weekly semaglutide in adults with overweight or obesity. *N Engl J Med* **384**, 989–1002.

Zhang C, Alashi A M, Singh N et al. (2019) Glycated beef protein hydrolysates as sources of bitter taste modifiers. *Nutrients* **11**, 2166.

5 Body size and composition

Mario Siervo

OBJECTIVES

By the end of this chapter you should be able to:

- define terms for components of body composition
- describe their relative size and variation
- describe the main characteristics and functions of body components
- summarize the use of body composition information in nutrition
- describe the main methods for measurement of body fat, fat-free mass, lean body mass, body water, and blood fractions.

5.1 Introduction

Nutrients in the diet are essential to provide the elements from which the human body is built, the energy on which all metabolic activity depends, and cofactors (vitamins and trace elements) which cannot be synthesized within the body. This chapter is concerned with composition of the body, and the information that this gives us about the quality of the diet, lifestyle, and disease status. For example, normal growth in children is impaired without an adequate intake of protein and energy, normal fat stores depend on balancing energy input and output, synthesis of essential components such as haemoglobin or thyroid hormone depends respectively on an adequate intake of iron, or iodine, and important physiological functions such as vision and bone health development depends on adequate intake of vitamin A and vitamin D. The chapter will also discuss the main measurement techniques, their strengths and limitations, and recent advances in assessment of body composition using multi-compartment models and novel diagnostic models.

5.2 Body composition and function

Body composition components: definitions

The science of body composition started more than 2000 years ago when Archimedes gave name to the principle stating that any object, wholly or partially immersed in a fluid, is buoyed up by a force equal to the weight of the fluid displaced by the object and the weight of the displaced fluid is directly proportional to the volume of the displaced fluid. This principle has been applied to calculate body density and derive body composition parameters such as fat mass and fat free mass. However, the earliest measurements of body composition in humans were made to discover if overweight individuals contained an unusually large amount of fat, or of muscle (see section 'Body fat', below). The method used was to measure average body density, since fat has a density of 0.900 g/cm^3 and a typical mixture of non-fat tissues has a density of 1.100 g/cm^3. Therefore density, and other methods that have been calibrated against density, like skinfold thickness yield estimates of *fat* and *fat-free mass*, which together make up total body weight.

Computer-scanning methods recognize tissues, such as adipose tissue, by the amount of energy (e.g., X-radiation) they absorb. Adipose tissue is not pure fat, but approximately 79 per cent fat, 3 per cent protein, and 18 per cent water. Tissue other than adipose tissue is not totally fat-free, since there is lipid in cell membranes to increase membrane fluidity, in nervous tissue to efficiently conduct electric signals, and in muscular tissue to act as energetic reservoir. Therefore, scanning methods measure *adipose tissue* and *lean body mass*, which together make up total body weight. The terms fat-free mass and lean body mass are often used interchangeably, but this is not correct as it may have implications for the interpretation of body composition analysis.

Human cadaver analysis

Our understanding of body composition in human subjects is based on seminal studies, conducted approximately six decades ago, which performed detailed chemical analyses of a small number of human cadavers (~6 bodies). First the fat in these bodies was separated by dissection and extraction with an ether-chloroform mixture, so what remained was fat-free mass. The fat-free body in adults has a fairly constant composition with respect to its water, protein, and potassium content. Table 5.1 summarizes these results. On average, fat-free tissue contains about 72.5 per cent water, 20.5 per cent protein, and 69 mmol of potassium (K) per kg. In Section 5.4 there is further discussion about the methods by which estimates of whole body density, or water, or potassium, combined with the data in Table 5.1, can be used to calculate the fat-free mass of living people. The antique study of anatomy using human cadavers crossed the path of modern imaging techniques in the Visible Man project. This was a scientific initiative coordinated by the National Institute of Health (NIH, USA), which used advanced imaging techniques (magnetic resonance imaging) to obtain

transversal sections of the human body from three human cadavers to provide detailed and high-resolution images on the anatomical and topographical features of tissues and organs.

The concept of a fat-free mass of constant composition is a great help in estimating fat-free mass in living people, but Table 5.1 also shows that this assumption is often an over-simplification. For example, skin and brain have very different chemical compositions. Skin has a very high protein content and low water content (30.0 per cent and 69.4 per cent, respectively) and very little K (23.7 mmol/kg). By contrast, brain has a very low protein and high water content (10.7 per cent and 77.4 per cent, respectively) and a high concentration of K (84.6 mmol/kg). This is because brain has a lot of intracellular water, which contains K, whereas skin has little water, and most of that is extracellular. This detailed information about the chemical composition of individual tissues and organs can only be obtained by chemical analysis of autopsy or biopsy samples. In addition, different physiological (growth, pregnancy, lactation) and pathological conditions (kidney failure, muscular dystrophy, lipodystrophy, cancer, ascites, heart failure) can challenge these assumptions and complicate even more the assessment of the chemical properties of bodily tissues and quantitative assessment of body composition.

TABLE 5.1 The contribution of water and protein to the fat-free weight of adult bodies and in some organs

Age (years)	Water (g/kg)	Protein (g/kg)	Remainder (g/kg)	Potassium	K:N ratio (mmol/kg)
Fat-free whole bodies					
25	728	195	77	71.5	2.29
35	775	165	60	–	–
42	733	192	75	73.0	2.38
46	674	234	92	66.5	1.78
48	730	206	64	–	–
60	704	238	58	66.6	1.75
Mean	724	205	71	69.4	2.05
Selected organs					
Skin	694	300	6	23.7	0.45
Heart	827	143	30	66.5	2.90
Liver	711	176	113	75.0	2.66
Kidneys	810	153	37	57.0	2.33
Brain	774	107	119	84.6	4.96
Muscle	792	192	16	91.2	2.99

Data compiled from various sources.

Water and electrolytes: intracellular and extracellular

Using the data from cadaver analyses we can construct a diagram of the components of the body of a typical normal young adult who weighs 70 kg: this is shown in Figure 5.1. We call it the Reference Man.

For the sake of simplicity we will use rounded numbers, and assume a fat content of 12 kg (17 per cent), so the fat-free mass is 58 kg. By far the largest component of fat-free mass is water (42 kg), which is 72.5 per cent of fat-free weight. Approximately 70 per cent of this water (28 kg) is inside cells, that is, intracellular fluid (ICF), and the remaining 30 per cent (14 kg) is extracellular, or extracellular fluid (ECF).

The total amount of water in the body, and the partition of body water between ICF and ECF, is closely regulated. Body stores of protein, energy, vitamins, and minerals ensure survival for many weeks without dietary intake, but deprivation of water causes death in a few days by multi-organ failure due to electrolyte disorders (see section 'Changes in hydration: diarrhoea, oedema, kwashiorkor, heart or kidney failure', below).

The electrolyte composition of ICF and ECF is shown in Figure 5.2.

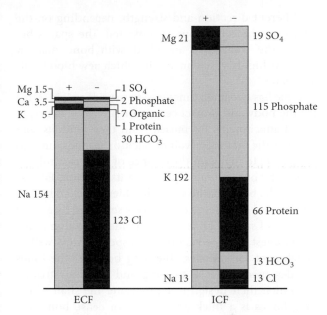

FIGURE 5.2 Electrolyte composition (mEq/kg water) of extracellular fluid (ECF) and intracellular fluid (ICF).

The principal cation in ICF is K, with a small amount of Na and Mg. The principal cation in ECF is Na, and Cl is the principal anion. The total electrical charge of anions and cations is exactly balanced in ICF and in ECF. The total concentration of ions is greater in ICF than ECF, because ICF contains polyvalent cations, particularly proteins, but the osmolality (determined by the concentration of molecules, rather than ions) of the two fluids is identical.

Methods by which we can measure the total amount of water in a living subject, and the proportion of this water that is extracellular, are described in the section 'Total body water (TBW)', below.

Bone: composition, types, density, growth, and turnover

Bone contains a matrix of fibrillar collagen that gives it tensile strength. Packed in an orderly manner around the collagen fibres is the mineral that gives the bone rigidity. This mineral is a complex crystalline calcium phosphate, called hydroxyapatite. Finally, the bone is covered by a vascular fibrous sheath, the periosteum. An adult body contains approximately 1 kg of Ca, of which 99 per cent is in bone. The skeleton also contains approximately 500 g of P, and more than half the collagen in the body, the remaining collagen being in skin, tendons, and fascial sheaths.

Every bone has a dense outer osseous layer which is filled inside by spongy bone arranged with trabeculae

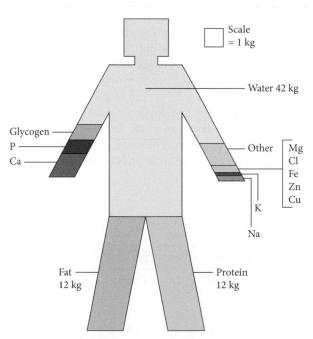

FIGURE 5.1 Diagrammatic representation of the body composition of a normal adult male weighing 70 kg. The contributions of the components to body weight are represented by their area in the diagram: only fat, protein, and glycogen contribute to the energy stores of the body.

of different direction and strength, depending on the stress to which the bone is subjected. The spaces between the trabeculae are filled with bone marrow which is highly vascular, and in which new blood cells are made.

Bone has two main functions: to provide a rigid frame for the body and protect certain organs from injury, to afford attachments for muscles and their tendons, and to act as the main reservoir of calcium, magnesium, and phosphate to maintain the balance of these electrolytes. The shape of the bone depends on its function. For example, the bones of the skull, shoulder blade, and pelvis are flat, and have dense outer layers separated by a small amount of spongy bone. They provide protection to the brain, chest, and pelvic organs, respectively, as well as attachment for muscles. The long bones of the limbs are designed to resist twisting and bending stresses, for which their design is appropriate. The shaft of the long bones is a thick-walled tube of dense bone that has expanded ends to bear the articular surfaces of the shoulder, elbow, wrist, hip, knee, or ankle joints, and to provide anchorage for the powerful muscles that move these joints.

Bone density (g/cm^3) is determined by the quantity of bone per unit volume. This in turn depends on the balance of activity between osteoblasts (cells that stimulate new bone formation) and osteoclasts (cells that cause resorption of bone). Factors that influence this balance are discussed in Chapter 25. It has become possible to measure the bone mineral content of selected bones since the development of quantitative radiological scanning techniques such as Dual X-Ray Absorptiometry (DXA) (see section 'Body fat', below). Normally bone density reaches a maximum around the age of 30 years, and then declines. Peak bone density is greater in men than in women, so women are more liable than men to develop osteoporosis in later life. This is a condition in which the quantity of bone per unit volume is decreased, so the trabeculae become thinner and weaker, and the dense cortical bone becomes porous. (This should not be confused with osteomalacia (or rickets, in children), which is a defect of bone mineralization, and causes deformities because the bones are unduly pliable.) Osteoporosis is a common and important disorder in the elderly, since it predisposes to fractures, especially of the vertebrae, femoral neck and forearm, and it is an important risk factor for physical disability and frailty.

Bone growth does not progress at an equal rate in all parts of the skeleton. The brain (and therefore also the skull that encloses it) is proportionately very large in the foetus, but after birth the limbs grow more rapidly in both length and strength. At birth there are no ossification centres for the wrist bones, or for the ends of the humerus, radius, ulna, or fibula. This is of importance for nutritional assessment, since a child who is chronically undernourished will show delayed development of these ossification centres. If the undernutrition is severe and prolonged the normal growth of long bones will not occur, and adult height will be stunted. The normal changes in body size and composition throughout the lifespan are described further in Section 5.3. The proportion of body weight as adipose tissue, muscle bone and skin is shown in Table 5.2.

The constituents of bone are not static: both the protein and mineral in bone are constantly exchanging with the amino acid and mineral metabolic pools in the body. Whereas most body proteins turn over with a half-life of days or weeks, the turnover of collagen is very slow. Enzymatic hydrolysis of collagen yields a mixture of amino acids that have a low nutritional value, because in the synthesis of collagen proline is hydroxylated, and hydroxyproline cannot be used to synthesize new protein. The turnover rate of collagen can be estimated from the urinary excretion of hydroxyproline, which is typically about 20 mg per day in an adult. This is very little compared with the 10 g of urea nitrogen excreted daily from the catabolism of all the other body proteins in an adult on a normal diet.

TABLE 5.2 Proportion of body weight as adipose tissue, muscle, bone, and skin

	% of body weight		% of adipose tissue free weight	
	Male	Female	Male	Female
Adipose tissue	28.1 ± 6.4	34.6 ± 9.3	–	–
Muscle	37.4 ± 4.9	32.9 ± 6.3	52.0 ± 4.3	50.0 ± 4.4
Bone	14.3 ± 2.0	13.4 ± 2.2	19.9 ± 2.4	20.6 ± 2.6
Skin	5.6 ± 0.6	5.5 ± 0.7	7.8 ± 0.8	8.5 ± 1.2

Source: From data reported by J P Clarys, A D Martin, D T Drinkwater (1984). Gross tissue weights in the human body by cadaver dissection. *Human Biology* **56**, 459–473.

Bone calcium also exchanges with calcium in other organs, especially kidney and gut. These fluxes are controlled by the action of parathyroid hormone, 1,25-dihydroxycholecalciferol (vitamin D, see Chapter 12) and calcitonin. It is essential for normal muscular and neurological function that plasma Ca is maintained in the range of 2.25–2.60 mmol/l. Disorders of bone metabolism are described further in Chapter 25.

Muscle: types, growth, repair, function

In a typical adult, skeletal muscle accounts for approximately 40 per cent of body weight, and another 10 per cent is smooth muscle. The water, protein, and potassium content of muscle is shown in Table 5.1. Since muscle makes up so much of fat-free weight, the composition of muscle is similar to that of the average of all fat-free tissues. However, skeletal muscle differs markedly from other tissues in an important respect: resting muscle has a lower energy consumption per unit weight than tissues such as heart, liver, kidneys, or brain, but during vigorous physical exercise the energy consumption in muscle greatly exceeds the total of all other body tissues. Organs can be ranked according to their energy expenditure per unit of body mass (kg) and in proportion to their relative contribution to resting energy expenditure (REE). Humans can be considered at the top of the phylogenetic evolutionary ladder and the brain is one of the most expensive organs, energetically speaking, as it consumes 1008 kJ/kg/day. The kidneys and the heart have the highest metabolic rate (1848 kJ/kg/day for both) followed by the liver (840kJ/kg/day), skeletal muscle (55 kJ/kg/day), and adipose tissue (19 kJ/kg/day).

There are three types of muscle: skeletal, smooth, and cardiac. Skeletal muscle consists of bundles of individual muscle fibres, each of which ranges from 10 to 80 µm in diameter. In most muscles these fibres extend the whole length of the muscle. Each fibre is activated by a single nerve ending, situated near the middle of the fibre. Each fibre contains several hundred to several thousand myofibrils. Each myofibril contains about 1500 myosin filaments and 3000 actin filaments arranged longitudinally with an overlapping zone between the two types of filament that shows as a darker band on electron micrographs, which causes a striated appearance on the whole muscle. When the muscle contracts the overlap between the actin and myosin fibres increases, and thus the two ends of the muscle, and the bones to which these ends are attached, are pulled towards each other. During muscle relaxation the actin and myosin units slide apart, but even at full relaxation they retain a degree of overlap.

Smooth muscle also consists of actin and myosin units sliding together and apart, but the muscle fibres are far smaller. Smooth muscle fibres are 2 to 5 µm in diameter, and only 50 to 200 µm in length, whereas the diameter of skeletal muscle fibres is up to 20 times greater, and they are thousands of times longer. Smooth muscle does not move joints, but lies in the walls of blood vessels, gut, bile ducts, ureters, and uterus. Contraction and relaxation of the smooth muscle in the walls of these organs alters the diameter of the tubes to control, for example, vascular tone, or may be organized in peristaltic waves so as to move forwards the contents of the tube.

Cardiac muscle, as its name implies, forms the chambers of the heart. Unlike skeletal muscle it is not organized as bundles of individually innervated muscle fibres but is a syncytium of cells fused end-to-end in a latticework. The electrical potential that causes one cell to contract readily passes to adjacent cells, and the whole mass of muscle contracts and relaxes synchronously. Thus, contraction expels the blood within the chamber through an exit valve, and then as it relaxes the chamber refills by permitting blood to flow in through an entry valve.

In skeletal muscle, physiologists distinguish two types of muscle cell: red (slow twitch or Type I) and white (fast-twitch or Type II) fibres. Type I fibres utilize oxygen to generate energy for movement and have a higher density of mitochondria. This makes them dark. Type II fibres use less oxygen, they contain less mitochondria, and they used high-energy stores (i.e., phosphocreatine) to support rapid bursts of movements. In birds, for example, muscles in the breast are designed for the rapid movement required for flight, and breast muscle is white. However, the legs of birds contain mainly red muscle, since legs are required to sustain posture for long periods, but not to move as quickly as wings. In human subjects there is a less marked difference between the fibre types of muscles in different parts of the body, and most muscles in human beings contain a mixture of red and white fibres. The effect of different types of physical training on the size and proportions of red and white fibres is considered in Chapter 18, and fuel selection in muscle in Chapter 6.

Blood: serum, plasma, RBC, WBC, normal values, cell turnover

Blood consists of plasma and cells—red blood cells (RBC), white blood cells (WBC) and platelets. It is primarily a medium for the transport of oxygen, nutrients, and hormones to the tissues, and for the removal of carbon dioxide and other waste products from tissues. In a typical adult the volume of blood is approximately 5 litres, of which about 55 per cent is plasma and 45 per cent the volume of packed cells (the haematocrit). If a sample of blood is withdrawn from a peripheral vein and put in a glass test tube it will solidify as a web of fibrin forms. The fibrin then contracts and squeezes together the trapped cells, so after a few minutes there is a firm dark red mass at the bottom of the tube (a blood clot) and a clear yellowish supernatant fluid (serum). Serum is plasma from which the proteins involved in blood clotting have been removed. If we wish to obtain a sample of plasma it is necessary to put the fresh blood sample into a tube with an anticoagulant, and then to separate the cells by centrifugation.

If blood did not contain red cells (RBC) it would be able to transport only 0.3 ml of oxygen dissolved in 100 ml of plasma. This would be quite inadequate to meet the oxygen requirements of the body tissues, even under conditions of basal metabolism. RBCs are packets of haemoglobin in a rather tough bi-concave cell wall which is freely permeable to oxygen, so the presence of RBCs enables 100 ml of blood to transport up to 20 ml of oxygen.

Red blood cells develop in the bone marrow from stem cells to the reticulocyte stage in about three days. The reticulocyte is so called because, on staining, remnants of nucleic acid appear as a blue network in the cytoplasm. This is the most immature form of red cell normally seen in the peripheral circulation, but normally 99 per cent of red blood cells are fully mature, with no nucleic acid or nucleus. The mature red blood cells survive in the circulation for about 120 days, after which they are destroyed in the reticulo-endothelial system. The haemoglobin is broken down to bilirubin, and the iron is recycled for the synthesis of new haemoglobin for new red blood cells.

White cells are approximately a thousand times less numerous than red cells in the blood. The commonest type of white blood cell is the polymorph, or neutrophil granulocyte, which, like RBCs, is formed in bone marrow. The neutrophil count increases rapidly in response to infection or tissue injury. Neutrophils survive in the peripheral circulation for a very short time: their half-life is estimated at six to eight hours. The next commonest white blood cell is the lymphocyte, which is formed in lymphoid tissue. This cell is involved in immune responses. The remaining white blood cells (monocytes, basophil, and eosinophil granulocytes) occur still more rarely. Platelets are very small cells that are involved in blood clotting and the repair of damaged blood vessels.

Body fat: subcutaneous, intra-abdominal and intra-organ, and brown fat

Body fat has become unfashionable and undesirable by humans: in modern affluent communities it is desirable to be slim, but to our ancestors, and to those now living in subsistence economies, body fat is a valuable store of energy during times of famine, and a thermal insulator when the environment is cold. These valuable characteristics of fat still exist, but in affluent countries (and increasingly in developing countries) the need for

protection from famine and cold is less often required, but excessive fatness is increasingly common.

The typical adult illustrated in Figure 5.1 contains 12 kg of fat, which is 17 per cent of body weight. This degree of fatness is within the healthy range. Usually 90 per cent of this fat is in a layer under the skin (subcutaneous fat), but there is also fat within the abdominal cavity (visceral fat), and a small amount is in the fascial planes between muscles (extra-myocellular fat). Not all of body fat is available as an energy store: even in the bodies of people who have died of starvation there is still about 2 kg of fat remaining which reinforce the concept of essentiality of body fat. A reserve of 10 kg of fat contains 375 MJ (90,000 kcal), equivalent to about four weeks of normal energy requirements assuming a total energy expenditure of ~3000 kcal/day. In fact, people of normal weight on total starvation survive for about ten weeks, because energy expenditure decreases. In severely obese people the fat stores may reach 80 kg: such people can survive a year of starvation, but this is not appropriate treatment for severe obesity. See Chapters 6 and 20 for further discussion of this topic.

The distribution of body fat differs between men and women, and this has metabolic significance. At puberty, women tend to store fat around breast, hip, and thigh regions, whereas men tend to accumulate fat in the abdominal area. If adults become obese they may deposit the excess fat in female (or gynoid) pattern, or the male (android) pattern (see Figure 5.3). This pattern is determined by the action of sex hormones which are probably linked to the genetic background: even when excess fat is lost the characteristic pattern is preserved. *In vitro* measurement of rates of lipolysis in biopsies of adipose tissue have shown similar basal rates of lipolysis in femoral and abdominal adipose tissue in non-pregnant women. However, lipolysis in femoral fat increases during lactation. It seems that femoral fat is conserved in the non-pregnant woman, but made available as an energy source in later stages of pregnancy and lactation.

It has been observed that obesity, and particularly fat deposited in the abdominal cavity (and hence associated with the android body shape), strongly predisposes to the development of metabolic syndrome, type II diabetes, and cardiovascular diseases. This is probably because the intra-abdominal fat is much more sensitive to lipolytic stimuli than subcutaneous fat, and this intra-abdominal lipolysis causes a rapid increase in the free fatty acid flow into the liver via the portal vein. This in turn predisposes to hyperinsulinaemia, hyperlipidaemia, and hypertension. For further discussion of the risk factors associated with abdominal obesity see Chapter 20.

Body fat is synthesized by, and stored in, fat cells or adipocytes. Adipocytes from subcutaneous white fat can be sampled by needle biopsy and studied in the laboratory.

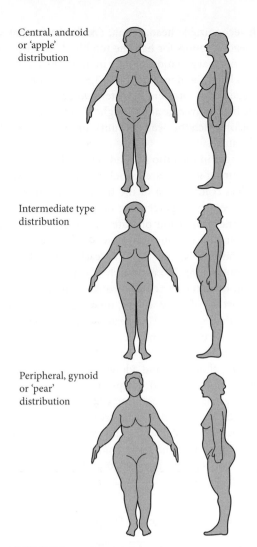

Central, android or 'apple' distribution

Intermediate type distribution

Peripheral, gynoid or 'pear' distribution

FIGURE 5.3 Outline of three obese women with approximately the same Quetelet's index, but differing in pattern of fat distribution.

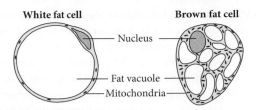

FIGURE 5.4 Diagram to illustrate chief morphological differences between white and brown fat cells. Adapted from J S Garrow (1988). *Obesity and Related Diseases.* Churchill Livingstone, Edinburgh.

They have a nucleus in a thin rind of cytoplasm surrounding a large fat globule (see Figure 5.4). A typical adipocyte in a biopsy sample from a normal-weight person contains about 0.6 μg of fat. If we assume that the 12 kg of fat in the man in Figure 5.1 is stored in adipocytes of this size we can calculate that he has 2.0×10^{10} fat cells in his body. If a biopsy sample is taken from an obese adult with, say, 60 kg of body fat, the average adipocyte is not five times the normal size, but only about 50 per cent larger than normal, so there must be more adipocytes than normal in obese people. This reasoning led to the hypothesis that if a child became obese before a certain age then proliferation of adipocytes was stimulated, and the child would be predisposed to obesity in later life.

We now know the hypothesis is not true, because new adipocytes can develop at any age if the amount of fat to be stored increases. However, the idea stimulated research into the metabolic activity of adipocytes, which has yielded valuable information. We now know that white adipose tissue has two important functions related to energy balance and reproductive function. Aromatase is an enzyme in adipose tissue that catalyses the conversion of androstenedione to oestrogen, and in postmenopausal women it is the only route of oestrogen production. This probably explains the association between obesity and abnormalities of sex steroid hormones, especially in conditions such as polycystic ovarian syndrome.

There has been intense research into the product of the ob gene, which has been knocked out in the obese mutant mouse. This is the hormone leptin. It was discovered in 1994, it is produced by adipocytes, and it is involved in the control of both food intake and energy expenditure. Obese mice are deficient in leptin and infertile, so there was great excitement when it was shown that administration of leptin cured both obesity and infertility in these animals. Unfortunately, studies on obese human subjects have shown that most had abnormally high, rather than low, leptin concentrations (leptin resistance), and therapeutically the administration of leptin has not realized early expectations. However, a very small number of obese subjects may have very low levels of leptin due to a congenital disorder impairing the secretion of leptin. The phenotype of these subjects is characterized by childhood obesity, excessive eating, and other endocrinological abnormalities (hypothyroidism, hypogonadism, puberty failure, hyperinsulinemia). Replacement therapy with recombinant leptin improves most of the metabolic disorders and it reduces dramatically their energy intake.

Since the discovery of leptin, the physiological role of adipose tissue has dramatically changed and it is now considered as an independent metabolically active organ capable of secreting several molecules involved in the control of appetite (leptin), insulin sensitivity (adiponectin, visfatin, omentin), inflammation (tumor necrosis factor alpha, interleukin 6, resistin), angiogenesis (vascular endothelial growth factor), or coagulation (plasminogen activator inhibitor-1, tissue factor). These are just a few examples and the number of proteins and adipokines secreted by adipocytes is continuously expanding. Unlike the white fat cell, the brown fat cell has small fat droplets in a cytoplasm rich in mitochondria in which heat can be generated (see Figure 5.4). It is important to small mammals to be able to generate heat to maintain body temperature in cold environments. Hypothermia is particularly a problem in newborn mammals (including human babies) because their small body mass does not generate enough heat by normal metabolic processes to maintain body temperature, so some extra thermogenic source is required.

A defect in thermogenesis is the cause of obesity in some animal models, so it was thought that defective brown fat might be a cause of human obesity. However, obese human beings have a higher (not lower) total heat production than lean people, and in mammals weighing more than 6 kg the heat generated in normal metabolism is sufficient to maintain body temperature in normal environments, so there is no need for additional thermogenesis.

5.3 Changes in body size and composition

Throughout the lifespan: normal growth and composition

On the first day after fertilization the human embryo is a single cell, approximately 0.15 mm in diameter. After two months of intrauterine life it is about 30 mm long with recognizable head, trunk, and limbs; at that stage the head accounts for half the total body length. By the end of normal gestation, at nine months, the foetus is 500 mm in length and weighs about 3.5 kg: the head is then one-quarter of total length. After two decades of extrauterine life the average adult weighs about 70 kg, is 1.7–1.8 m tall, of which the head accounts for only one-eighth of total stature.

Changes in the ratio of head size to that of the whole body during growth are associated with changes in the chemical composition of the tissues. The embryo contains a very high percentage of water, but with maturation the proportion of water decreases, and there is a shift in the distribution of water from extracellular (with sodium, Na^+, as the chief cation) to intracellular (with potassium, K^+, as the chief cation). As the proportion of water decreases, protein and electrolytes, as a proportion of body weight, increase. Fat is deposited mainly during the last trimester of pregnancy and the first year of extrauterine life. These changes are shown in Table 5.3.

During childhood and adolescence the proportion of water in the body, and the proportion of extracellular to intracellular water, continues to decrease. Increase in total body K, Ca, and P in fat-free tissue reflects the increase in intracellular water and growth in the skeleton. Fat-free mass remains fairly constant in both men and women between the ages of 20 and 65 years, but then

TABLE 5.3 Effect of growth, malnutrition, and obesity on the composition of the body and of fat-free tissue

	Foetus (20–25 weeks)	Premature baby	Full-term baby	Infant (1 year)	Adult man	Malnourished infant	Obese man
Body weight (kg)	0.3	1.5	3.5	20	70	5	100
Water (%)	88	83	69	62	60	74	47
Protein (%)	9.5	11.5	12	14	17	14	13
Fat (%)	0.5	3.5	16	20	17	10	35
Remainder (%)	2	2	3	4	6	2	5
Fat-free weight (kg)	0.30	1.45	2.94	8.0	58	4.5	65
Water (%)	88	85	82	76	72	82	73
Protein (%)	9.4	11.9	14.4	18	21	15	21
Na (mmol/kg)	100	100	82	81	80	88	82
K (mmol/kg)	43	50	53	60	66	48	64
Ca (g/kg)	4.2	7.0	9.6	14.5	22.4	9.0	20
Mg (g/kg)	0.18	0.24	0.26	3.5	0.5	0.25	0.5
P (g/kg)	3.0	3.8	5.6	9.0	12.0	5.0	12.0

Data compiled from various sources.

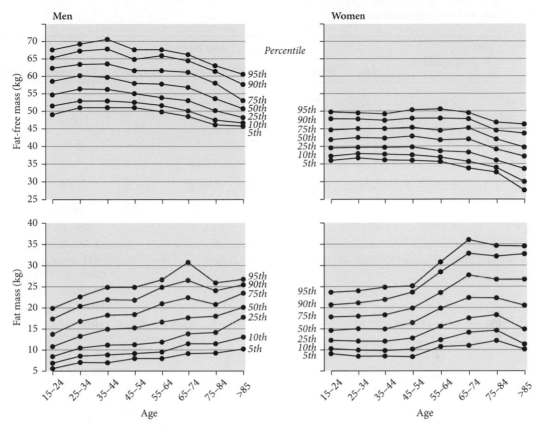

FIGURE 5.5 Percentile changes in fat-free mass and fat mass of Caucasian men and women with advancing age. *Source*: L Genton, D Hans, U G Kyle et al. (2002) Dual X-ray absorptiometry and body composition: Differences between devices and comparison with reference methods. *Nutrition* **18**, 66–70. With permission from Elsevier.

decreases by about 15 per cent in the next two decades (see Figure 5.5).

Throughout adult life there is a trend for fat mass to increase in both men and women, but the increase is more rapid in postmenopausal women with also a change in distribution as more fat is deposited centrally. Changes in body composition with age are shown in Table 5.4 and changes in tissue protein content with age in Table 5.5.

Effects of diet and exercise on body size and composition

The laws of thermodynamics require that if over a given period the energy intake of an individual is 10 MJ (2400 kcal) less than energy output, then the energy stored in the body must be reduced by 10 MJ. That is invariably true. However, a decrease of 10 MJ in energy stores may be achieved by losing 0.27 kg of fat, or 2.4 kg of fat-free tissue, or (more probably) an intermediate amount of weight made up from a mixture of lean and fat tissue. But it does not follow that if the individual increases his energy output by 10 MJ (for example by exercising) and

does not change his energy intake, the above losses will be achieved. As weight is lost, resting energy expenditure also decreases, so for a given energy intake the energy deficit, and hence the rate of weight loss, also decreases, a process called metabolic adaptation. This effect is quite small: about 70 kJ (16 kcal)/day/kg weight lost in men, and about 50 kJ (12 kcal)/day/kg weight lost in women.

A larger rate of weight loss was observed in a study of 108 obese women whose average starting weight was 100.8 kg (SD 23.6) (see Figure 5.6). On a diet supplying 3.4 MJ (800 kcal)/day their average weight loss in three weeks was 5.0 kg, but it was significantly faster (330 g/day) during the first week than during the next two weeks (210 g/day). This is because during the first week on a diet, glycogen and its associated water is lost, with an energy density of 4.2 MJ/kg, but by the second week the tissue lost is a mixture of 75 per cent fat and 25 per cent fat-free mass (FFM), which has an energy density of 30 MJ (7000 kcal)/kg (see Chapter 6). Very low carbohydrate diets cause rapid initial weight loss because they cause early loss of glycogen and the attached water (glycogen:water ratio is ~ 1:3).

During total starvation weight loss is even more rapid, partly because the energy deficit is greater, and partly

TABLE 5.4 Changes in body composition with age

Age	Fat % body weight	g/kg fat-free mass						
		Protein	**Water**	**Sodium**	**Potassium**	**Calcium**	**Magnesium**	**Phosphorus**
Foetus, 1.5 kg	3.5	111	856	105	49.6	354	19.7	126
Foetus, 2.5 kg	7.6	126	837	101	52.1	413	20.4	152
Birth	16.2	137	822	97	50.8	464	20.6	171
1 year	20	170	780	95	58.0	698	24.7	226
4.5 years	22.7	238	697	99.9	65.0	1050	29.6	338
Adult, mean	19.7	205	723	87	69	1064	38.4	375
Adult, range	12.5–27.9	165–238	674–775	78–96	66.5–73.0	912–1240	35.4–40.3	284–452

Source: From data reported by J S Garrow, K Fletcher, D Halliday (1965). Body composition in severe infantile malnutrition. *J Clin Invest* **44**, 417–425.

TABLE 5.5 Changes in tissue protein content with age

	Ratio protein: water		
	Newborn	**2–9 months old**	**Adult**
Skeletal muscle	0.16	0.23	0.27
Adipose tissue	0.11	0.14	0.05
Skin	0.20	0.51	0.48
Bone	0.27	0.30	1.10
Liver	0.18	0.20	0.25
Heart	0.15	0.16	0.17
Mean of tissues reported	0.17	0.26	0.39

Source: From data reported by S J Fomon (1967) Body composition of the reference male infant during the first year of life. *Pediatrics* **40**, 863–870, and S J Fomon, F Hasche, E E Ziegler et al. (1982). Body composition of reference children from birth to age 10 years. *Am J Clin Nutr* **35**, 1169–1175.

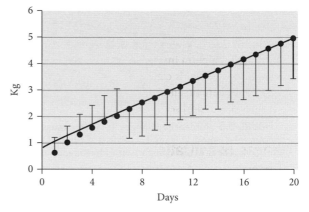

FIGURE 5.6 Average cumulative weight loss in 108 obese women on a reducing diet. Vertical bars indicate 1 SD. Continuous line is the best-line fit for data from days seven to 21.
Source: Adapted from J D Webster, J S Garrow (1989) Weight loss in 108 obese women on a diet supplying 800 kcal/d for 21 d. *Am J Clin Nutr* **50**, 41–45, and reproduced with permission by the *American Journal of Clinical Nutrition* © *Am J Clin Nutr*. American Society for Clinical Nutrition.

because a higher proportion of the weight loss is FFM, to provide amino acids for glucose synthesis. (See Chapter 7, section 'Gluconeogenesis: the synthesis of glucose from non-carbohydrate precursors', and Chapter 9.4 'Amino acid metabolism'. Severely obese patients lose 300 to 500 g/day on prolonged starvation, but about 50 per cent of the weight lost is FFM. Total starvation ceased to be an acceptable treatment for gross obesity because several patients died unexpectedly, due to damage to heart muscle, mostly related to altered electrolytes movements, abnormal acid-base homeostasis, and decreased protein synthesis. In recent years, alternative weight loss approaches (e.g., alternate-day fasting, time-restricted eating, 5:2 diet) have been proposed as a more moderate and safer approach to personalize weight loss and increase adherence to obesity treatments. Results from preliminary studies are encouraging and able to achieve comparable results to hypocaloric diets.

The interaction between diet, exercise, and muscle bulk is controversial. There is no doubt that strenuous isometric exercise causes muscle hypertrophy, and immobilization causes atrophy of muscle (see Chapter 18). It is also true that exercise increases energy expenditure, so it is plausible that in obese people a combination of a low energy intake and exercise should achieve more fat loss, and less loss of FFM, than diet alone. A meta-analysis of 28 publications reporting trials on overweight subjects (BMI 25–29) showed that aerobic exercise without dietary restriction caused a weight loss of 3 kg in 30 weeks in men, and 1.4 kg in 12 weeks in women, but had no effect on FFM. Another systematic review, including 26 cohorts undergoing dietary and behavioural treatment and 29 studies evaluating the effects of bariatric surgery on body composition, has instead observed that the level of negative energy balance is a determinant of FFM loss. In addition, in three randomized controlled trials on obese subjects there was a protective effect of exercise on FFM loss during weight reduction. In summary, a combination of diet and exercise seems to have a protective effect on body composition by preserving fat-free mass. However, more studies are needed to confirm these results, particularly in obese subjects.

The effect of diet and exercise on bone density is even more controversial. Studies have shown that calcium supplementation in the diet increases bone density, but bone density is high in African and other countries where people have a low calcium intake but a high level of physical activity. Bone density is clearly affected by diet, exercise, and no doubt genetic and other factors not yet clarified. For further discussion see Chapter 25.

Changes in hydration: diarrhoea, oedema, kwashiorkor, heart or kidney failure

In the previous section, it has been assumed that there will be no change in the hydration of the body with diet and exercise, but if there is a change in hydration during changes in these factors, there will be weight changes that bear no simple relationship to energy balance.

Total body water is regulated to maintain an osmotic pressure in body fluids of 285 mosmol/kg. If the tonicity of these fluids increases then water intake is stimulated by thirst, and water losses are reduced by secretion of more concentrated urine. If the tonicity decreases, dilute urine is secreted to remove the excess water. Urine osmolality can vary from 50 to 1200 mmol/kg. This regulatory mechanism may be overwhelmed, and dehydration may occur if water losses are very high, as in severe diarrhoea, or with sweat loss in high ambient temperature or during prolonged vigorous exercise. Dehydration may also occur with abuse of diuretic drugs to achieve weight loss, or during the recovery phase of diabetic coma. In severe untreated diabetes mellitus (due to infection, or interruption of insulin administration in a diabetic), the patient becomes dehydrated by two mechanisms.

First, the excretion of large amounts of glucose in urine causes an osmotic diuresis and hence excessive loss of water. Second, ketosis is associated with vomiting and further water loss, and potassium leaches out of cells. Treatment of this condition with insulin only makes the dehydration even worse because correction of the acidosis allows potassium to re-enter the cells and causes a catastrophic reduction in extracellular fluids (including blood volume). In the management of diabetic coma the replacement of fluids (as well as insulin) is essential, and potassium must be replaced judiciously to avoid the dangers of either hyper- or hypokalaemia.

The opposite problem of water excess, indicated by oedema, is much more common than dehydration. The commonest cause in elderly people is congestive heart failure: the heart is unable to pump blood out as fast as it returns from the veins, the veins become engorged and leak fluid into the tissues. A similar situation may arise when the glomeruli in the kidney leak albumin into the urine or the liver is unable to synthesize enough albumin, so the plasma oncotic pressure decreases and fluid passes from the vascular system into the extracellular spaces.

Oedema is also a feature of certain types of malnutrition. In prolonged starvation hepatic albumin synthesis is reduced, on which the oncotic pressure of plasma mainly depends, and oedema is a striking feature of kwashiorkor (see Chapter 31 Weblink 31.6 and Chapter 33). It is particularly important to understand the fluid and electrolyte metabolism of malnourished children who may appear to be dehydrated when in fact they are overhydrated.

A marasmic child does not show obvious oedema, but is terribly wasted with wrinkled skin covering an emaciated body. Such children often have vomiting and diarrhoea, so the temptation is to rehydrate the child with intravenous fluid, but this is much more likely to kill the child than save it because an aggressive rehydration therapy would impose a strong haemodynamic load on a weak cardiac muscle causing heart failure. Measurements of total body water (see section 'Total body water (TBW)', below) in such children show that water accounts for too large a proportion of body weight: the limbs are thin and the skin wrinkled because almost all the muscle and fat has been lost, not because the child is dehydrated.

5.4 Use of size and composition data in nutrition

Assessment of nutritional status: obesity, thinness, muscle wasting, stunting, osteoporosis, anaemia

Clinical nutrition is about understanding the relationship between diet and health and about the correction of nutritional disorders in diseased states to re-establish a normal health. To learn more about these relationships we need reliable measures of health and of diet in either individuals or communities. We may define health in terms of longevity, development limited only by genetic potential, and freedom from disease and disability. All these are measurable with reasonable accuracy, but the weakest link is in reliably measuring habitual diet in free-living people. Chapter 30 discusses methods for assessing diet and body composition, many of which are satisfactory for assessing if an individual has been taking a suitable diet, or if the diet has supplied too little, or too much, of certain nutrients. For example, a simple measurement of weight and height, and perhaps also waist circumference, is adequate to identify those individuals who are obese, too thin, or stunted in growth. But simple anthropometry will not tell us if a malnourished infant is overhydrated or underhydrated (see section 'Changes in hydration: diarrhoea, oedema, kwashiorkor, heart or kidney failure', above). Unless we could measure body composition accurately we would have not learned that total starvation, or treatment with supra-physiological doses of thyroid hormones, is not an appropriate treatment for obesity, although it causes spectacular weight loss: the problem is that the weight lost includes too much lean tissue and too little fat. We would not know the extent and severity of tissue potassium depletion in a comatose diabetic, or a child with kwashiorkor, if we could not measure total body potassium: in such cases serum potassium concentrations are seriously misleading. We could not compare the relative effects of exercise and diet on the progress of osteoporosis if we could not measure bone density.

Many of the methods for measuring body composition discussed in Section 5.5 are research tools that cannot be used in normal clinical practice, but they serve as reference methods against which to calibrate simpler methods. For example, skinfold thickness has been calibrated against body density measured by underwater weighing to give a measure of body fat. Similarly, bioelectrical impedance has been validated against TBW measured by isotope dilution methods (deuterium or oxygen-18) to estimate fat free mass.

For standardization of metabolic rate/ drug dosage/renal clearance

Many physiological measurements give results that are related to body size or composition. For example, glomerular filtration rate is higher in large people than in small people, because they have larger kidneys, although someone who weighs 120 kg does not have kidneys twice as large as someone who weighs 60 kg. Autopsy data showed that kidney weight is well correlated with body surface area, which was typically 1.73 m^2 in normal young men. So now glomerular filtration rates are 'corrected' to a surface area of 1.73 m^2. Energy physiologists

had a similar problem, and unfortunately chose the same (inappropriate) solution. It is said that obese people have a lower resting metabolic rate than lean people, and this is a cause of their obesity. Consider Figure 5.1: this person weighs 70 kg and contains 12 kg of fat and 48 kg of FFM. Suppose he becomes obese and gains 40 kg, of which 30 kg is fat and 10 kg is FFM. His resting metabolic rate (RMR) will increase by a factor of 58/48 (1.21) because RMR is proportional to FFM, and his FFM has increased by 1.21. But his weight has increased by 110/70 (1.57) and his surface area by 2.25/1.80 (1.25). If his obese metabolic rate is compared with baseline values in absolute terms it has *increased* by 21 per cent, if 'corrected' for FFM it is unchanged, if 'corrected' for surface area it has *decreased* by 1.21/1.25 (~3.2 per cent), and if 'corrected' for body weight it has *decreased* by 1.21/1.57 (~23 per cent). The correct calculation depends on the question to be answered. If the question is 'What is the relationship of energy requirements for weight maintenance after weight gain compared with baseline?' the change should be expressed in absolute values (e.g., ~21 per cent). If the question is 'What is the change in metabolism per unit weight of FFM?' the answer is that it has not changed. There is no justification for 'correcting' metabolic rates by either surface area or body weight because they would give an erroneous estimate of energy metabolism due to the confounding effect of non-energy generating compartments (water) included in the correction.

There are situations in which it is necessary to adjust for body size or composition. This particularly applies to calculating doses of drugs given to small children, like in cancer therapy and dialysis,

5.5 Measurement and interpretation of body composition

Anthropometry

Measurements of body composition may be made to assess current nutritional status, or serial measurements may be made to assess change in status. These two purposes require different levels of precision in measurement. For example, if an observer is asked to determine which of two people is more obese a simple measurement of skinfold thickness would serve to rank the subjects correctly. However, if they returned two weeks later having lost some weight, and wanted to know which of them had lost more fat, a repeat of the skinfold measurements would not be adequate, because the error of an estimation of fat from skinfolds is large compared with the amount of fat that people lose in two weeks. Furthermore, many anthropometric measurements are subject to observer bias. If the observer believes subject

A should have lost more fat than B, then a measurement of a skinfold, or circumference, or diameter, may support that prejudice because the observer (consciously or unconsciously) pulls the tape a little tighter on A than on B. (For detailed discussion of the techniques of anthropometry see Chapter 31).

For sensitivity, precision, and objectivity, body weight is the best and most reliable of all anthropometric measurements. Do not trust the slimming club leader who says: 'Your weight has slightly increased since last week, but my measurements show that your fat has decreased, so you must have gained lean tissue.' It is far more likely that the measurement of fat is inaccurate. This is mostly due to the imprecision of the body composition methods commonly used in clinical practice such as skinfold thickness or bioelectrical impedance.

In adults the easiest anthropometric measure of fatness is to measure weight (W, kg) and height (H, m) and calculate W/H^2. The normal range in adults is between 20 and 25 kg/m². This measurement was suggested by the Belgian astronomer Quetelet in 1869, and so is known as Quetelet's index. However, in the USA the same index was proposed in 1972 by Keys, and named body mass index (BMI). The application of this index in the study of obesity and undernutrition is discussed in more detail in Chapters 20 and 31. In growing children the normal range of BMI changes with age, so age-specific standards must be used. See Ellis (2000) and Siervo et al. (2010) for a more detailed overview of this method.

Skinfold thickness and mid-upper arm circumference (MUAC)

In adults the percentage body fat can be estimated by measuring, with special calipers, the sum of thickness of the skinfolds over the biceps, triceps, subscapular, and supra-iliac sites. For a given skinfold thickness the corresponding fat content varies with age and gender: Table 5.6 shows the relationship of skinfold to percentage body fat. Accurate measurement of skinfold thickness requires good technique, which is described in Chapter 31.

Mid-upper arm circumference is a useful tool for assessing the nutritional status of both adults and children

TABLE 5.6 Percentage body fat in men and women related to the sum of four skinfolds (biceps, triceps, subscapular, and suprailiac)

Skinfold (mm)	Men, age (years)				Women, age (years)			
	17–29	30–39	40–49	50+	17–29	30–39	40–49	50+
20	8.1	12.2	12.2	12.6	14.1	17.0	19.8	21.4
30	12.9	16.2	17.7	18.6	19.5	21.8	24.5	26.6
40	16.4	19.2	21.4	22.9	23.4	25.5	28.2	30.3
50	19.0	21.5	24.6	26.5	26.5	28.2	31.0	33.4
60	21.2	23.5	27.1	29.2	29.1	30.6	33.2	35.7
70	23.1	25.1	29.3	31.6	31.2	32.5	35.0	37.7
80	24.8	26.6	31.2	33.8	33.1	34.3	36.7	39.6
90	26.2	27.8	33.0	35.8	34.8	35.8	38.3	41.2
100	27.6	29.0	34.4	37.4	36.4	37.25	39.7	42.6
110	28.8	30.1	35.8	39.0	37.8	38.6	41.0	42.9
120	30.0	31.1	37.0	40.4	39.0	39.6	42.0	45.1
130	31.0	31.9	38.2	41.8	40.2	40.6	43.0	46.2
140	32.0	32.7	39.2	43.0	41.3	41.6	44.0	47.2
150	32.9	33.5	40.2	44.1	42.3	42.6	45.0	48.2
160	33.7	34.3	41.2	45.1	43.3	43.6	45.8	49.2
170	34.5	34.8	42.0	46.1	44.1	44.4	46.6	50.0

Source: From data reported by J V Durnin & J Womersley (1974) Body fat assessed from total body density and its estimation from skinfold thickness: Measurements on 481 men and women aged from 16 to 72 years. *Br J Nutr* **32**(1), 77–97.

in famine conditions. The technique of measurement and calculation is given in Chapter 31.

It has been recognized in the last decade that intra-abdominal fat has a greater influence on the risk of heart disease, metabolic syndrome, and diabetes than an equal weight of fat in subcutaneous sites. Attempts have therefore been made to estimate intra-abdominal fat by measurements of waist circumference, or sagittal diameter. However, these measurements are difficult to make accurately, and computerized scanning techniques, like computer assisted tomography or magnetic resonance imaging, are now generally used to assess intra-abdominal fat. Recent developments of methods for measuring intra-abdominal fat have tried to apply less invasive and less costly methods such as ultrasonography or abdominal bio-electrical impedance. See Ellis (2000) and Siervo et al. (2010) for a more detailed overview of these methods.

Body fat

There is no 'best' method for measuring body composition in living subjects: every method has errors, and some methods require expensive laboratory equipment that would be impossible to use in the field, for example during famine relief or in extreme environments such as high altitude. The first two methods described below estimate FFM and so, by subtraction from total body weight, fat mass (FM). They all require expensive laboratory equipment and cooperative subjects. They are used (preferably in combination) to provide reference values with which simpler methods can be compared and calibrated. They all depend on an assumption, based on the data in Table 5.1, that the fat-free body has a constant density, water content, and potassium content.

Total body water (TBW)

The total amount of water in the body can be measured quite accurately by isotope dilution. The subject is given a known dose of water labelled with deuterium (^{2}H), the stable heavy isotope of hydrogen, and this is allowed to equilibrate with total body water, which takes about three hours. Then the dilution of ^{2}H in a sample representative of body water, such as blood plasma, is measured, and TBW is calculated. Fat (by definition) does not contain any water. If we assume that the fat-free tissues contain 73 per cent of water, then FFM is TBW/0.73. By subtraction of FFM from body weight, the weight of fat in the body is estimated.

The limitations of this approach are that it requires a high-precision isotope-ratio mass spectrometer with a competent operator to measure the concentration of ^{2}H in the equilibrated body water sample, and that in many conditions the assumption that FFM is 73 per cent water is not valid. In patients with oedema caused by heart or kidney disease, in severely obese people, and in severely malnourished people, the assumption that FFM is TBW/0.73 will overestimate FFM and hence underestimate fat. In dehydrated subjects the same assumption will overestimate fat.

It is possible to measure sub-compartments of total body water using a similar dilution principle. Instead of using deuterated water as a tracer, a tracer that distributes only in the vascular space (such as radiolabelled red cells) can be used to measure blood volume. Various compounds (such as thiocyanate, sodium bromide, sodium sulphate, or inulin) that distribute in the extracellular water can be used to measure the volume of extracellular water. These measurements are rarely made in clinical practice. The situation in which it is dangerous to life to have an excessive volume of extracellular water is congestive heart failure, leading to over-filling of the venous circulation, pulmonary oedema, and death. In this case central venous pressure is easier, quicker, and more relevant to measure.

Total body potassium (TBK)

All potassium (K), including that in the human body, contains a natural radioactive isotope (^{40}K), so each gram of K emits 3 gamma rays per second. This radiation is of high energy (1.46 MeV) so most of it emerges from the tissues and can be counted by high sensitivity detectors. However, this level of radiation is low compared with the normal background, which arises mainly from cosmic rays, so the subject being measured, and the detectors, must be screened by a massive shield of lead and steel. With this cumbersome and expensive 'whole-body counter' it is possible to measure total body potassium and, assuming a constant K content in FFM, to calculate FFM, and hence fat mass.

The value of this technique is mainly that the errors in estimates of body composition by TBK and TBW, arising from oedema or dehydration are in opposite directions. With oedema TBK overestimates fat, while TBW underestimates fat; with dehydration the converse applies. However, the practical inconvenience and the limited availability of whole-body counters means that this method is very rarely used today.

Body density

Human fat at body temperature has a density of 0.900 g/cm^3. The remainder of the body (the fat-free mass) is a mixture of water, protein, bone mineral, glycogen, and minor components such as nucleic acids and electrolytes. This mixture has a density of approximately 1.100 g/cm^3 (Keys & Brozek 1953). Therefore, if we know the average density of all the tissues of the body we can cal-culate the ratio of fat to fat-free mass: for example, if the average density was 1.00 g/cm^3 then that person must be 50 per cent fat and 50 per cent FFM. The body shown in Figure 5.1 weighs 70 kg and has 12 kg fat, so his fat content is 17 per cent and his total body density would be 1.06 g/cm^3.

It is easy to measure the weight of a subject accurately, but difficult to measure tissue volume with similar ac-curacy. For example, the 70 kg subject who is 50 per cent fat will have a volume of 70 litres, but if he had only 17 per cent fat his volume would be 66.05 litres. The usual method for measuring body volume is to compare body weight in air and totally immersed in water: the decrease in weight on immersion shows the volume of water dis-placed. However, some of the water is displaced by air trapped in the subject's lungs and gut, and if allowance is not made for this the fat content of the subject will be overestimated. It is difficult to measure this trapped air, so other methods for measuring FFM that do not require total immersion have been developed, such as air dis-placement phlethysmography, commonly known as 'Bod Pod', which is gradually replacing the hydrodensitomet-ric method. A further development of this technique is a smaller device, using the same principle, called Pea-Pod, to measure body density in infants (Garrow et al. 1979; McCrory et al. 1995; Fields et al. 2002; Ma et al. 2004).

Dual-energy X-ray absorptiometry (DEXA)

Unlike the methods discussed so far, this technique measures three compartments of the body: fat, lean soft tissue, and bone mineral. The analysis is based on an X-ray image composed of individual pixels, rather than shades of grey on a conventional X-ray film, just as digital photography differs from conventional film photography. The X-ray source produces rays of two defined energies, ranging from 38 to 140 keV, which depends on the type of instrument being used (Hologic, Lunar, Norland). These pairs of beams are scanned across the area of interest in the subject (which may be the whole body) and the ener-gy spectrum of the emerging beam is analysed. When the beam passes through material of high opacity to X-rays (such as bone mineral) the emerging energy is severely attenuated, but especially the lower energy beam. When the tissue being irradiated is fat there will be very little at-tenuation of either beam, and if the tissue is lean soft tis-sue the attenuation will be intermediate. At each instant the energy emerging is recorded in a pixel relating to that particular beam position, and the information stored in all the pixels is integrated by the computer to provide an estimate of the composition of the tissue scanned. An important check on the validity of DEXA analyses is that the computer makes an independent estimate of total body weight (by summing estimated fat, soft tissue, and bone), and this estimate can be compared with body weight obtained by simple weighing.

The technique was originally devised to measure bone mineral density, which can be done by comparing the en-ergy spectrum in pixels derived from an instant when the beam went through bone with adjacent pixels in which the beam has just missed the bony structure. However, the newer instruments analyse the energy spectrum in pixels where the tissue is a mixture of lean soft tissue and fat and, since the attenuation coefficient of both of these is known, the ratio of lean tissue to fat can be calculated. For a more detailed review of the technique see Kendler et al. (2013).

It is important to note that the 'fat' determined by the methods previously discussed is not the same as the 'adipose tissue' determined by DEXA. With the DEXA technique what is being measured is the ability of a par-ticular core of tissue to absorb X-rays of two energies. If this attenuation matches the values assigned to 'fat' then the computer reports the tissue as fat, but if that tissue was dissected out it would be mainly fat, but also contain some protein and water, because the adipose cells that contain fat also contain some protein and water. By con-trast, methods that depend on measuring body density or water would include the water component of adipose tissue as 'fat-free mass', because it is not fat. It would help if authors would use the terms 'fat' and 'fat-free mass' to describe results that are obtained by methods that measure the chemical composition of the tissue, and 'adipose tissue' and 'lean body mass' to describe results that are obtained by measuring the X-ray absorption of tissues. Unfortunately, many authors use the terms interchangeably.

The DEXA method has been described in some detail because it has become one of the most common methods for measuring body composition in living subjects. The scan can be performed in five to ten minutes, which is much quicker and less demanding on the subject than measurements of density, water, or potassium. It is much more accurate and independent of operator bias than any technique depending on anthropometry. The radia-tion dose to the subject is very low, about equivalent to one day of background radiation. It measures bone min-eral, which can otherwise only be measured by CT scan, which is more expensive and involves a higher radiation dose. Perhaps the most important advantage over all oth-er methods (except CAT and MRI scans) is that it permits analysis of specific regions of the body, in particular the adipose tissue within the abdominal area, and the mus-cle mass in the limbs. This has now become a reference method for the assessment of appendicular skeletal mass and sarcopenia and novel body composition phenotypes such as sarcopenic obesity and cachexia.

Multi-compartment models: a 'gold standard' for measurement of fat and fat-free mass

For research purposes it is useful to have a best, or 'gold standard', estimate of fat and FFM in living subjects against which simpler methods can be compared. The two-compartment models regard body weight as the sum of fat and FFM, and make assumptions about density, water, or potassium content for FFM. If the assumptions are wrong, then the estimate is wrong. To try to avoid this source of error the four-compartment models measure water (see section 'Total body water (TBW)', above) and bone mineral by DEXA (see section 'Dual-energy X-ray absorptiometry (DEXA)', below) and density (see section 'Body density', above). Knowing the water and bone mineral content of the body, the remainder must be either fat or non-mineral, non-water FFM, which means essentially protein and glycogen. Thus, by combining the results of three measurements (water, density, and bone mineral) it is not necessary to assume that FFM has a density of 1.100 or a water content of 73 per cent (Wang et al. 1992). The relation of body weight to body fat, measured by these three methods, is shown in Figure 5.7. Over a wide range of weight (42–132 kg) and fat (9–75 kg) the relationship is linear ($r = 0.960$) with a slope of 1.27. This shows that for every 1 kg extra fat by which an obese woman exceeds a normal-weight control, body weight is (on average) increased by 1.27 kg. The excess weight is not pure fat, but about 75 per cent fat and 25 per cent a mixture of water and protein.

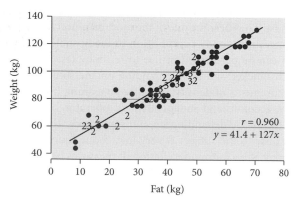

FIGURE 5.7 Relation of body weight to total body fat in a series of 104 women. Body fat was calculated from the mean of estimates by density, water, and potassium in each woman. *Source*: Adapted from J D Webster, R Hesp, J S Garrow (1984) The composition of excess weight in obese women estimated by body density, total body water and total body potassium. *Hum Nutr Clin Pract* 38C, 299–306 and reproduced by permission of Nature Publishing Group (http://www.nature.com).

Body composition assessment by computer imaging techniques

Undoubtedly the development of better techniques for computer analysis of X-ray imaging have led to the most important advances in the measurement of human body composition in the last decade. In conventional radiography a wide parallel beam of X-rays (or gamma rays) is passed through the tissues of a subject, and from the rays that emerge on the other side an image is formed on a sheet of photographic film. Where the tissue is relatively opaque to the X-rays (as in bone) the film is little exposed, and therefore remains clear when the film is developed. Where the tissue is easily penetrated by X-rays (as in lung) the film is dark, and where there is tissue of intermediate density (such as muscle) there are various shades of grey.

In computer-assisted tomography (CT) the X-rays are emitted in a narrow beam from a source that travels in a semi-circle round the subject, and the energy emerging from the body is recorded by a detector which is mounted diametrically opposite the X-ray source. A computer is programmed to analyse the constantly changing output of the detector, and from this information it constructs a picture of the 'slice' of body that has been scanned. Since fat, water, lean tissue, and bone have different absorption characteristics, the computer shows on a video screen a picture of the distribution of these tissues, as if the subject had been cut through at the level of the scan. If serial scans are performed at different levels of the body from head to foot it is possible to build up data on the total volume of different types of tissue, and how these tissues are distributed in different sections of the body. An alternative technique is to use helical computed tomography, in which the radiation source rotates around the supine patient, who lies on a table that travels at a constant speed. Thus the scan, instead of being in a series of parallel slices, is made in a continuous spiral. This technique reduces errors arising from respiratory movements in the subject.

CT scanning has proved very valuable in showing the size and position of abnormal tissue (such as tumours) in the body, but it is expensive and involves a rather high dose of radiation, so it is not suitable for routine estimation of body composition in normal subjects.

Magnetic resonance imaging (MRI) is a relatively new application in body composition research and its application is constantly growing. The MRI utilizes the different electromagnetic properties of tissues to reconstruct high-quality images of the organs inside the body. More specifically this technique exploits the physical properties of the water molecule which carries an asymmetric charge. If a body is placed in a strong magnetic field some hydrogen nuclei will change their orientation in the field but they will flip back when the magnetic field is switched off. The return to their original state will release energy which is then captured, integrated, and converged into images which show the differences in energy state and level of hydration of the tissues. The potential applications of the MRI are large in body composition research due to the high resolution of the images and to the non-radioactive nature of the radiations. The procedure is very expensive but the information on tissue distribution and morphological characteristics of organs can give precious insights into the relationship between body composition and nutritional status in health and disease. A further development of the MRI method is the use of magnetic resonance spectroscopy (MRS), which allows non-invasive and *in vivo* exploration of the molecular composition of tissues by detecting signals from different chemical nuclei within the body. The most common elements to be studied are hydrogen, phosphorus, sodium, and fluorine. More recently, a new technique, called Quantitative Magnetic Resonance (ECHO-MRI), has been developed and applied for the measurement of TBW, lean body mass, and fat mass. This technique relies on the different physical properties of the hydrogen nuclei (relaxation time, spin, and density) in tissues with contrasting chemical composition. The different nuclear magnetic resonance (NMR) signal amplitudes and relaxation times of fat, body-free fluid, and muscle enhance the tissue contrast, which can then be further magnified by application of certain radio frequency sequences. Water can be easily distinguished from fat based on NMR high-resolution spectra. The instrument provides a very quick assessment (less than three minutes) of body composition, and it has a higher upper body weight limit compared with other methods (~250 kg). This is a relatively new technique for the assessment of body composition in human subjects and its use is currently mainly limited to research settings.

The application of medical ultrasonography for the assessment of body composition is primarily organ-specific and the technique is routinely used for example to assess fat deposition or liquid accumulation in organs such as the liver or kidneys. The magnitude of the frequency of the acoustic waves (ultrasound) is associated with resolution and deepness of the measurement. Lower frequencies have lower resolution and travel deeper into the body compared with high frequencies. The sound waves are reflected and produce a reverberating echo signal every time there is a change of density in the tissues. The images formed are on a grey scale, and the intensity of the scale is correlated to the density of the tissues (acoustic impedance). Air has extremely low acoustic impedance, whereas bone tissue is characterized by high values of impedance. Soft issues have intermediate values. Medical ultrasound is a safe, portable, interactive technique used for the diagnostic visualization of subcutaneous tissues and organs. This technique has been used for the assessment of regional distribution of body fat and quantitative determination of the subcutaneous fat layer. Ultrasound has also been applied for the determination of abdominal subcutaneous fat (deeper and superficial) and measurement of intra-abdominal diameter. A moderate agreement with MRI and CT scans (r = 0.62–0.83) was observed in studies assessing the precision and accuracy of this technique for the determination of subcutaneous fat and intra-abdominal diameter. This technique requires highly trained operators to obtain reliable and valid estimates of body composition. The assessment provides more qualitative than quantitative information. The measurements are also affected by standardization of protocols, technical errors, and anatomical irregularities. For a more detailed review of the techniques see Shen et al. (2008) and Siervo et al. (2010).

Neutron activation

If a person is irradiated with a beam of fast neutrons, some elements (notably nitrogen, calcium, chlorine, sodium, carbon, and some of the trace elements) form very short-lived radioactive isotopes. The radiation from these isotopes can be detected, and hence the body content of these elements can be calculated. By this method it is possible to measure, for example, total body calcium. This is an extremely expensive procedure, and involves a significant radiation dose to the subject, so it is only applicable to some very specialized research protocols.

Electrical conductivity

Fat is an electrical (as well as a thermal) insulator, but the FFM is a tissue bathed in an electrolyte solution, and therefore is a good conductor. If a small electric current is passed from the hand to the foot of two subjects of the same weight and height, but with different proportions of fat to FFM, the voltage drop will be greater in the fatter than in the leaner subject, because the higher fat content in the fatter subject offers more resistance to the electric current. Obviously, there are many problems to be overcome before this difference in electrical conductivity between the two hypothetical subjects can reliably be converted into estimates of FFM. Heitmann (1990) validated estimates of FFM obtained with a commercial 'bio-impedance analyser' (BIA) against a four-compartment model using total body water, total body potassium, weight, and height. The standard deviation of the difference between the two methods was 4.36 kg. There are now several commercial BIA instruments; some require the operator to attach electrodes to the hands and feet (tetrapolar method) of the subject while others have electrodes in the base-plates of a stand on machine (leg-to-leg method).

A weakness of whole-body BIA is that it measures the average impedance of an ill-defined electrical path through the arm, trunk, and leg. The cross-sectional area of the limbs is much less than that of the trunk, so the limbs make a disproportionately great contribution to overall impedance and, other things being equal, subjects with long limbs will have a greater impedance, and therefore a smaller estimated FFM, than subjects with shorter limbs. Recent innovations include the development of the method to estimate segmental body composition by assessing electrical conductivity of the limb and truncal sections separately. Also, the frequency of the alternating current affects the ability of the current to penetrate from extracellular to intracellular water. Multi-frequency devices have been developed to measure the specific resistance offered by the extracellular fluid (low frequencies) and intracellular fluid compartments (high frequencies). Other approaches such as the determination of the phase angle (PA) or bioelectrical impedance vector analysis (BIVA) have been developed to provide a semi-quantitative assessment of body cell mass and hydration, based on the advantage of utilizing of raw impedance values. See Kyle et al. (2004) and Norman et al. (2012) for a more detailed description of this method.

Estimating change in body composition

Body composition measurements may be made to find out if a given patient, or population, has an abnormal amount of some component, such as fat, or bone mineral. The techniques described above will usually provide a reliable answer to this question.

The estimate of changes in body compartments is a different matter and therefore the accuracy, the precision, and the technical features of the different body composition methods discussed above have to be taken into account to maximize the likelihood of the detecting significant clinical changes after an intervention. Different techniques differ in their capacity to measure changes in body composition and the integration of this capacity with the precision of the method, the sample size, and the biological variation of the outcome will give an estimate of the likelihood to detect significant differences. For example, a bigger sample size will increase the chance to detect significant changes whereas a decrease in precision or an increase in the biological variation will decrease the sensitivity. The choice of the best method is very much based on the nature of the primary outcome of interest. If, for example, it is important to measure in individuals the composition of changes in body fat, an evaluation of the characteristics of the body composition techniques

available in the study, the sample size, the biological variability of fat mass in that particular population, and the time interval between measurements are objective indicators which have to be considered for a rational selection of the method to be used in the intervention. The four-compartment model of body composition is currently considered the best method to estimate changes in fat mass as it eliminates most of the assumptions made by the individual techniques. The alternative is to perform metabolic balance studies to measure the total intake and output of the element of interest—for example N balance for protein, Ca balance for bone mineral.

Novel diagnostic approaches for the assessment of body composition phenotypes

There is growing recognition that fat mass and lean body mass are closely connected and therefore an integrated, diagnostic approach is needed in order to improve the assessment of nutritional status and increase the sensitivity for disease risk prediction. This concept is based on the established pathogenetic role associated with low FFM (sarcopenia) and excess adiposity (obesity), which has led to the proposal of a new body composition phenotype, namely sarcopenic obesity (SO). Recent results seem to suggest that sarcopenic obesity represents an important risk factor for metabolic and cardiovascular morbidities as it is characterized by the double-burden of a reduced oxidizing capacity (sarcopenia) and increased metabolic load (adiposity). At the moment a universal consensus for its diagnosis does not exist and different procedures, parameters, and cut-off points have been proposed in the literature to define it. Moreover, SO also represents a difficult challenge with regard to treatment. Reducing fat mass with recovery or at least maintaining lean mass, which would be the logical targets of therapeutic intervention in the SO, are difficult to achieve. Both obesity and sarcopenia are associated with the reduction of muscle strength and the impairment of physical performance and therefore with the onset of disability. Furthermore, sarcopenia and obesity are both characterized by metabolic aspects related to insulin resistance and inflammation in particular. SO is more strictly associated to CVD risk factors (e.g., arterial hypertension, carotid arteria intima-media thickness, lipid profile, insulin resistance) than either obesity or sarcopenia. However the studies that investigated the association of SO with known cardiometabolic and/or cardiovascular risk factors, described controversial results depending on the considered population and the criteria used for the definition of SO. For a more detailed review on the topic see Prado et al. (2012) and Batsis et al. (2018).

KEY POINTS

- There are remarkably few data on the chemical composition of the body; these come from analysis of cadavers; a healthy adult male contains about 60 per cent water, 17 per cent protein, and 17 per cent fat.

- The skeleton accounts for 14 per cent of total body weight. Bone contains 99 per cent of total body calcium, which acts as a reservoir to maintain an appropriate plasma concentration of calcium. Bone turnover can be estimated from the urinary excretion of hydroxyproline.

- Skeletal muscle accounts for 40 per cent of body weight, and smooth muscle 10 per cent, but resting muscle accounts for only 22 per cent of basal metabolic rate.

- Fat reserves are in white adipose tissue. White adipocytes contain about 80 per cent triacylglycerol, as a single central droplet surrounded by a thin layer of cytoplasm containing the nucleus.

- Males tend to accumulate fat in the abdomen, while women accumulate subcutaneous fat stores around the breast, hip, and thigh. In obesity, people may accumulate fat in the male or female pattern; this is probably genetically determined. Abdominal adipose tissue is more closely related to diabetes and heart disease than is total body fat.

- Body composition changes throughout life. The water content of the body decreases, and the content of protein and fat increases, through gestation, infancy, and into adolescence.

- Fat-free mass remains relatively constant from age 20 to 65, but then decreases; fat mass increases throughout adult life. Food restriction and starvation result in loss of fat-free mass as well as adipose tissue.

- A variety of conditions can lead to excessive accumulation of body water, resulting in oedema. Even severely wasted undernourished people may be oedematous.

- The two most common methods of assessing body fat are bioelectrical impedance and (becoming increasingly applicable outside research laboratories) dual-energy X-ray absorptiometry (DEXA) and air displacement plethysmography.

- Sarcopenic obesity is a novel body composition phenotype which seems to confer greater risk for cardiometabolic diseases compared to obesity and sarcopenia alone. However, consensus on the diagnostic criteria for sarcopenic obesity has not been reached.

 Be sure to test your understanding of this chapter by attempting multiple choice questions. See the Further Reading list for additional material relevant to this chapter.

REFERENCES

For additional references see Chapter 5 Further Reading.

Batsis J A, Villareal D T (2018) Sarcopenic obesity in older adults: Aetiology, epidemiology and treatment strategies. *Nat Rev Endocrinol* **14**(9), 513–537.

Durnin J V G A & Womersley J (1974) Body fat assessed from total body density and its estimation from skinfold thickness: Measurements on 481 men and women aged from 16 to 72 years. *Br J Nutr* **23**, 77–97.

Ellis K (2000) Human body composition: In vivo methods. *Physiol Rev* **80**, 649–680.

Fields D A, Goran M I, McCrory M A (2002) Body-composition assessment via air-displacement plethysmography in adults and children: A review. *Am J Clin Nutr* **75**, 453–467.

Garrow J S, Stalley S, Diethelm R et al. (1979) A new method for measuring body density of obese adults. *Br J Nutr* **42**, 173–183.

Kendler D L, Borges J L, Fielding R A et al. (2013) The official positions of the International Society for Clinical Densitometry: Indications of use and reporting of DXA for body composition. *J Clin Densitom* **16**(4), 496–507.

Keys A & Brozek J (1953) Body fat in adult man. *Physiol Rev* **33**, 245–325.

Kyle, U G, Bosaeus I, De Lorenzo A et al. (2004) Bioelectrical impedance analysis—part I: review of principles and methods. *Clin Nutr* **23**, 1226–1243.

Kyle, U G, Bosaeus I, De Lorenzo A et al. (2004) Bioelectrical impedance analysis—part II: utilization in clinical practice. *Clin Nutr* **23**, 1430–1453.

Norman K, Stobäus N, Pirlich M et al. (2012) Bioelectrical phase angle and impedance vector analysis: Clinical relevance and applicability of impedance parameters. *Clin Nutr* **31**(6), 854–61.

Prado C M, Wells J C, Smith S R et al. (2012) Sarcopenic obesity: A critical appraisal of the current evidence. *Clin Nutr* **31**(5), 583–601.

Shen W & Chen J (2008) Application of imaging and other noninvasive techniques in determining adipose tissue mass. *Methods Mol Biol* **456**, 39–54.

Siervo M & Jebb S A (2010). Body composition assessment: Theory into practice: Introduction of multicompartment models. *IEEE Eng Med Biol Mag.* **29**(1), 48–59.

FURTHER READING

Chaston T B, Dixon J B, O'Brien P E (2007) Changes in fat free mass during significant weight loss: A systematic review. *Int J Obes* **31**, 743–750.

Donini L M, Busetto L, Bauer J M et al. (2020) Critical appraisal of definitions and diagnostic criteria for sarcopenic obesity based on a systematic review. *Clin Nutr* **39**(8), 2368–2388.

Heitman B L (1990) Evaluation of body fat estimated from body mass index, skinfolds and impedance: a comparative study. *Eur J Clin Nutr* **44**, 831–837.

Keys A & Brozek J (1953) Body fat in adult man. *Physiol Rev* **33**, 245–325.

Ma G, Yao M, Liu Y et al. (2004). Validation of a new paediatric air phlethysmography for assessing body composition in infants. *Am J Clin Nutr* **79**, 653–660.

Marra M, Sammarco R, De Lorenzo A et al. (2019) Assessment of body composition in health and disease using bioelectrical impedance analysis (BIA) and dual energy X-ray absorptiometry (DXA): A critical overview. *Contrast Media Mol Imaging* **2019**, 3548284.

McCrory M A, Gomez T D, Bernauer E M et al. (1995) Evaluation of a new air displacement phlethysmography for measuring human body composition. *Med Sci Sports Exerc* **27**, 1686–1691.

Norgan N G (2005) Laboratory and field measurements of body composition. *Public Health Nutr* **8**, 1108–1122.

Rogalla P, Meii N, Hoksch B et al. (1998) Low-dose spiral computed tomography for measuring abdominal fat volume and distribution in a clinical setting. *Eur J Clin Nutr* **52**, 597–602.

Wang Z, Pierson R Jr, Heymsfield S (1992) The five-level model: A new approach to organizing body-composition research. *Am J Clin Nutr* **56**, 19–28.

6

Energy balance and body weight regulation

Abdul G. Dulloo and Yves Schutz

OBJECTIVES

By the end of this chapter you should be able to:

- understand the concept of metabolizable energy intake
- describe the main factors affecting energy intake
- describe the assessment of energy needs
- explain the components of energy expenditure, their relative size and variability
- describe the classical and modern methods of measurement of energy expenditure
- identify the main types of signals in relation to hunger and satiety
- describe how energy expenditure changes in response to undernutrition and overnutrition
- summarize the components of models of energy intake and energy expenditure.

6.1 Introduction

Understanding how body weight is regulated is still challenging for human research today. It is likely that long-term constancy of body weight is achieved through a highly complex network of regulatory systems through which changes in food intake, body composition, and energy expenditure are interlinked. Failure of this regulation leads either to obesity and its co-morbidities or to protein-energy malnutrition and cachexia in disease states such as anorexia, cancer, and infections. Between these disorders attributed to 'failure of regulation' lies those due to chronic undernutrition because of poverty, war, and famine. Achieving energy balance and weight homeostasis is central to the quality of life. An understanding of how they are achieved requires an appreciation of the following:

i. the basic concepts and principles about the flux of energy transformations through which body weight is regulated,

ii. factors affecting food intake and energy expenditure, which represent the entry and exit in this flux of energy transformations,

iii. the methods for assessing energy expenditure and energy requirements, and

iv. a number of models that have been proposed to explain the regulation of body weight and body composition in humans.

6.2 Basic concepts and principles in human energetics

Energy balance and the laws of thermodynamics

Energy represents the capacity of a system to perform work. It can appear in various forms—light, chemical, mechanical, electrical—all of which can be completely converted to heat. According to the *first law of thermodynamics*, energy cannot be created or destroyed, but can only be transformed from one form into another. Like machines, biological systems depend on the transformation of some form of energy in order to perform work. Whereas plants depend on light energy captured from the sun to synthesize molecules like carbohydrates, proteins, and fats, animals meet their energy needs from chemical energy stored in plants or in other animals. The chemical energy that is derived from foods can be utilized for performing a variety of work; for example, the synthesis of new macromolecules (*chemical work*), in muscular contraction (*mechanical work*), or in the maintenance of ionic gradients across membranes (*bio-electrical work*).

Some of the food energy can also be stored in the body in the form of fat, protein, and glycogen. The term energy balance is described by the following equation:

$$\text{Energy intake} = \text{Energy expenditure} + \Delta\text{Energy stores}$$

Despite the simplicity of this equation, we must recall the temporality and the non-syncronicity of the two factors, that is, energy intake which is discontinuous and energy expenditure which is continuous. Thus, if the total body energy does not change (i.e., Δ Energy stores = 0) over a period of time, this implies that energy expenditure must be equal to energy intake and the individual is hence in a state of *energy balance* over that period of time. If energy intake and energy expenditure are not equal, then a change in body energy content will occur, with *negative energy balance* resulting in the utilization and hence depletion of the body's energy stores (glycogen, fat, and protein) or *positive energy balance* resulting in an increase in body energy stores, primarily as fat. However, changes in energy expenditure can influence energy intake, and vice versa. For example, voluntary energy intake may increase following intense physical activity, while overeating may lead to an increase in energy expenditure. Furthermore, both energy intake and energy expenditure can be influenced by changes in body energy stores; in particular, the loss of body fat stores may trigger compensatory mechanisms to restore it through increased hunger and diminished energy expenditure (see Sections 6.6 and 6.7).

The *second law of thermodynamics*, in biological terms, makes a subtle distinction between the potential energy of food, useful work, and heat. It states that when food is utilized in the body, whether for muscle contraction, synthesis of new tissues, or for maintenance of ionic equilibrium across membranes, these processes must inevitably be accompanied by heat. In thermodynamic terms, some energy is degraded, and such heat energy, which is not available for performing work, is termed 'entropy'. Thus, the conversion of available food energy is not a perfectly efficient process, and about 75 per cent of the chemical energy contained in foods may be ultimately dissipated as heat because of the inefficiency of intermediary metabolism in transforming food energy into a form (e.g., adenosine triphosphate, ATP) which can be used for useful work, whether it be the internal work required to maintain structure and function or external physical work. Thus, in the transformation of food energy to perform work, only 25 per cent of the food energy is actually used to perform work, resulting in the efficiency of work being 25 per cent, as depicted in Figure 6.1. The overall process is referred to as energy expenditure or metabolic rate or heat production. In energy equilibrium, the sum of heat and work is equal to the metabolizable energy intake, that is, the food energy that is available for intermediary metabolism after taking into account the losses in faeces and urine (see Section 6.3 below).

Energy balance and heat balance

Total daily energy intake, total energy expenditure (i.e., total heat production) and total heat loss (i.e., heat dissipation) are depicted schematically in Figure 6.2 within the context of energy balance and heat (thermal) balance, that is, with heat production in equilibrium with heat loss. This equilibrium is determined by the thermostatic 'set-point' of the individual, and is expressed as the core (central) body temperature. When both sides of the heat balance are identical (i.e., total heat production = total heat loss), core body temperature remains unchanged. When there is a gap between the two sides of the heat balance, heat is either gained (+ve gap) or lost (−ve gap) from the body and hence body core temperature increases or decreases respectively.

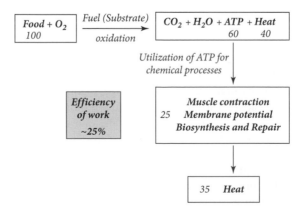

FIGURE 6.1 Energy flux from food fully oxidized (energy metabolizable) to heat.
For every 100 kcal of food energy that is oxidized, 40 kcal is lost as heat during formation of ATP, and another 35 kcal is lost as heat when ATP is used to perform work, resulting in the efficiency of work being 25 per cent.

Heat is lost from the body by four major routes shown here in descending order of importance, namely: (1) radiation (emitted by the body to the environment surrounding it, (2) evaporation (insensible water loss through the skin and respiration), (3) convection (effect of cold or hot winds on the body or use of an external ventilator, for example), and (4) conduction (effect of being in contact with physical objects in the environment with the skin at different temperature from that of the body). When thermal balance is positive, endogenous heat can be stored in the body leading to a rise in core temperature, and vice versa. Exogenous heat can also be gained from the environment when the gradient of the radiative components is negative, for example, the surrounding environment (i.e., walls on which the body radiates) is warmer than the average body surface (skin) temperature.

The mammalian body attempts to maintain core body temperature essentially constant (homeothermy), which requires equilibrium between total heat loss vs. total heat produced. In real life, core body temperature is rarely constant; it has a nycthemeral oscillation, called circadian rhythm, over approximately a 24-hour period. It fluctuates slightly within a small window, since it is one of the best-regulated physiological variables in the body together with pH, arterial pO2, and so on. Since there is no immediate synchronization between total heat production (rapid response) and total heat loss (delayed response) over time, the core body temperature must oscillate slowly over 24 hours, and the thermal balance is not expected to be in equilibrium hour-by-hour, but only when averaged over a 24-hour period. If this is the case, then total 24-hour energy expenditure (i.e., heat production) will be equal to total heat dissipation (i.e., heat loss) over a 24-hour period.

The core temperature fluctuation observed during the day and night (circadian rhythm) is explained by several factors such as (i) the thermogenic effect of acute food ingestion which increases heat production more rapidly than the heat loss response, and (ii) the effect of acute physical activity (i.e., movement- or exercise-related thermogenesis).

New developments in measuring internal temperature inconspicuously and continuously over 24–48 hours (or more) by using small miniature telemetric pill-size sensors are increasingly used for research in the field of thermal control in humans. The capsule is ingested with some water. It progresses, at enteral physiological speeds, through the gut before being expelled in the faeces allowing the calculation of the transit time. The use of these electronic capsules to monitor 24-hour core temperature fluctuations constitutes an interesting new tool for testing the hypothesis that daily and nightly low-core body temperature may be a factor that, through energy conservation, predisposes some people to obesity.

Units of energy

Since all the energy used by the body at rest is ultimately lost as heat, and physical (external) work will also be eventually degraded as heat, the energy that is consumed, stored, and expended is expressed as its heat equivalent. The calorie was originally adopted as the unit of energy in nutrition; it is defined as the amount of heat required to raise the temperature of 1 gram of water by 1°C (from 14.5 to 15.5°C); in nutrition the kilocalorie (= 1000 calories) is used. With the introduction of the SI system, energy is expressed in joules (J). One joule is the energy used when a mass of 1 kilogram (kg) is moved through 1 metre (m) by a force of 1 newton (N). Because one joule is a very small unit of energy, it is more convenient to use kilojoule (kJ), or megajoule (MJ) in nutrition. Rates of energy expenditure (often referred to as metabolic rate) are expressed in kJ or MJ per unit time (kJ/min or MJ/day), which correspond to 10^3 J and 10^6 J, respectively. The conversion of calorie to joules is: 1 calorie = 4.18 J, or 1 kilocalorie (kcal) = 4.18 kilojoule (kJ).

TABLE 6.1 Macronutrient storage in the body, its energy density, and its degree of auto-regulation

Substrate	Form of storage	Associated H2O in tissue pool (%)	Pool size	Tissues	Energy density (kJ/g)	Autoregulation
Carbohydrate	Glycogen	75	Small	Liver Muscle	~4	Accurate
Fat	Triacylglycerols	5	Moderate-large (unlimited)	Adipose tissue	~33	Poor
Protein	Protein	73	Moderate (limited)	Lean tissue	~4	Accurate

Source: Adapted from Y Schutz and J S Garrow (2000) Energy and substrate balance, and weight regulation. In: *Human Nutrition and Dietetics* (10th edn), 137–148 (J S Garrow, W P T James, A Ralph eds), Churchill Livingstone, Edinburgh.

Sources of energy and macronutrient balance

The macronutrients (carbohydrate, fat, protein, and alcohol) are the exogenous sources of energy, so it makes sense to consider energy balance and macronutrient balance together. There is a strong relationship between energy balance and macronutrient balance, and the sum of individual substrate balance (expressed as energy) must be equivalent to the overall energy balance. Thus, it follows that:

exogenous carbohydrate—carbohydrate oxidation = carbohydrate balance

exogenous protein—protein oxidation = protein balance

exogenous lipid—lipid oxidation = lipid balance

exogenous alcohol—alcohol oxidation = alcohol balance (transitorily positive)

total energy intake—total oxidation = energy balance.

Unlike the size of the fat stores, which can increase very considerably, there is a limited capacity for storing protein in fat-free mass and carbohydrate as glycogen in liver and muscles; glycogen, which is stored with water in tissues and has an energy density about 9 times lower (1 kcal/g; i.e., about 4 kJ/g) than fat (triacylglycerols with 9 kcal/g; i.e., about 33 kJ/g) stored in adipose tissue (Table 6.1). It is therefore not surprising that protein and glucose tend to be oxidized more readily than fat. Alcohol, which is not stored in the body as such, is oxidized at an essentially constant rate, during which it spares fat.

Acute substrate imbalance, resulting from acute changes in either substrate intake or substrate oxidation, or both combined, is of paramount importance for our understanding of short term, day-to-day body weight changes.

In positive energy balance, lipids stored in adipose tissue can originate from dietary (exogenous) lipids or from nonlipid precursors, mainly from carbohydrates but also from ethanol, that is, from substrates, which produce acetyl-CoA during their catabolism, and are therefore susceptible to be converted to fatty acids in intermediary metabolism (liver); this process is known as *de novo* lipogenesis (DNL) (Figure 6.3). The net conversion of carbohydrate into fat is a high

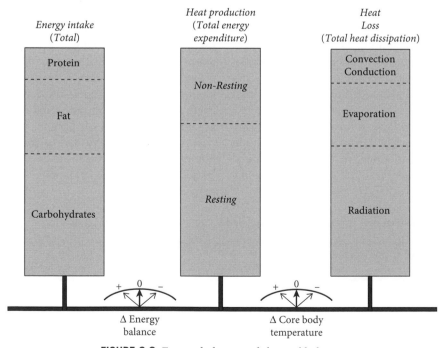

FIGURE 6.2 Energy balance and thermal balance.
Schematic representation of daily energy expenditure from a perspective of thermal balance.
Total heat production, i.e., energy expenditure (EE), and total heat dissipation (heat losses) can be represented
as the two plates of a scale; the difference between the total heat production and total heat losses is
called thermal balance (or heat balance) by analogy to energy balance.
Source: A G Dulloo and Y Schutz (2016) Energy balance and body weight regulation. In: *Human Nutrition* (13th edn), 114–135 (C Geissler & H Powers eds). Churchill Livingstone, Edinburgh.

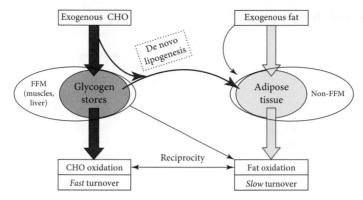

FIGURE 6.3 Dietary fat and carbohydrate (CHO) turnover.
This scheme shows that in certain nutritional circumstances such as after a few days of overfeeding (particularly with excessive carbohydrates), the stimulation of *de novo* lipogenesis represents a sort of 'carbohydrate sink' (i.e., a glycogen overflow pathway) leading to an increase in body fat stores. This process is essential for achieving a new state of carbohydrate balance. In addition, note that a general reciprocity occurs in carbohydrates vs. fat utilization/oxidation, the higher the former, the lower the latter.
Source: Adapted from A G Dulloo & Y Schutz (2011) Energy balance and body weight regulation. In: *Human Nutrition* (12th edn), 87–110 (C Geissler & H Powers eds). Churchill Livingstone, Edinburgh.

energy-requiring process requiring ATP as compared to the direct storage of exogenous fat as body fat. Due to the inefficiency of this process, about 25 per cent of the carbohydrate energy is converted into heat, whereas the storage of dietary triacylglycerols into adipose tissue requires only about 2 per cent of fat energy. Therefore, DNL from carbohydrate would theoretically constitute a protective factor as recruitment of this energetically costly pathway indirectly limits the increase in body fat stores.

In summary, fat-free mass (FFM) is a large compartment (about 80 per cent of body weight in men without obesity) that is capable of dynamically utilizing glucose and fatty acids released from adipose tissue and other tissues. In the size-limited, labile glycogen-water pool (which logically belongs to the FFM compartment) exogenous carbohydrate is stored (mainly in muscles and liver) and is quickly released on metabolic demand thanks to the fast glycogen turnover (Figure 6.3). Carbohydrates may be transformed into fat (by net *de novo* lipogenesis) involving a spilling-over process in the case of continuous metabolic carbohydrate overload. This occurs progressively when the glycogen stores become full.

By contrast, the adipose tissue is a smaller compartment than FFM (about 20 per cent of body weight in men without obesity) but larger than the glycogen–water pool (a few kg). Paradoxically, the adipose tissue oxidizes primarily glucose as fuel substrate rather than fatty acids which are released into the circulation by lipolysis for covering the need of other organs and tissues such as resting muscles and liver. The turnover of fat mass is very slow compared with the glycogen–water pool. It is now recognized that, in addition to the liver, adipose tissue is also a site of *de novo* lipogenesis (see Chapter 8).

6.3 Energy intake

Energy value of foods: the utility of developing the 'Atwater factors'

The traditional way of measuring the energy content of foodstuffs is to use a 'bomb calorimeter' in which the heat produced when a sample of food is combusted (in the presence of oxygen) is measured. When the food is combusted, it is completely oxidized to water, carbon dioxide, and oxides of other elements such as sulphur and nitrogen. The total heat liberated (expressed in kilocalories or kilojoules) represents the *gross energy* value or *heat of combustion* of the food (Figure 6.4). The heat of combustion differs between carbohydrates, proteins, and fats. There are also important differences within each category of macronutrient. The gross energy yield of sucrose, for example, is 16.5 kJ/g, whereas starch yields 17.7 kJ/g. The energy yield of butterfat is 38.5 kJ/g and of lard, 39.6 kJ/g. These values have been rounded off to give 17.3 kJ/g for carbohydrates rich in starch and poor in sugar, 39.3 kJ/g for average fat, and 23.6 kJ/g for mixtures of animal and vegetable proteins. The heat of combustion of pure alcohol is intermediary with 29 kJ/g. Since the density of ethanol is much lower than that of water (0.79 g/100 ml), the energy value per unit volume is 23 kJ/ml.

The gross energy value of foodstuffs, however, does not represent the energy actually available to the body, since no potentially oxidizable substrate can be considered available until it is presented to the cell for oxidation. None of the foodstuffs is completely digested and absorbed; some energy therefore never enters the body and is excreted in faeces. Digestibility of the major

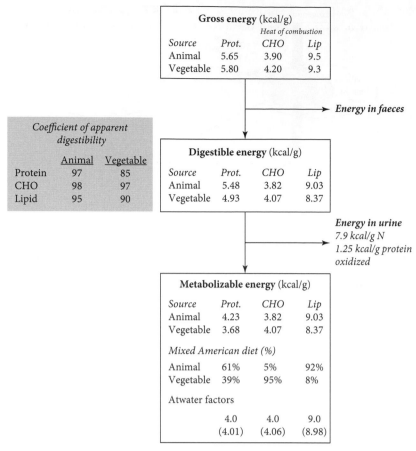

FIGURE 6.4 Physiological energy value of energy yielding nutrients obtained in human. These are shown from the flow of energy originating from food energy content in protein, carbohydrates, and lipids (as gross energy), through that absorbed across the intestines and hence accounting for losses in faeces (digestible energy), and that available for metabolism (as metabolizable energy) after taking into account the losses both in faeces and in urine. *Source*: A G Dulloo & Y Schutz (2011) Energy balance and body weight regulation. In: *Human Nutrition* (12th edn), 87–110 (C Geissler & H Powers eds). Churchill Livingstone, Edinburgh.

foodstuffs, however, is high; on average 97 per cent of ingested carbohydrates, 95 per cent of fats, and 92 per cent of proteins are absorbed from the intestinal lumen. There is a difference between the true and apparent digestibility—the latter includes energy which is excreted in the faeces from sources such as bacteria (microbiota) in the gut and enzyme secretions.

In the body, the tissues are able to oxidize carbohydrate and fat molecules completely to carbon dioxide and water, but the oxidation of protein is not complete, since the end product of protein oxidation results in the formation of residual nitrogenous compounds such as urea and, to a lesser extent, ammonia—both excreted in the urine. Determination of both the heat of combustion and the nitrogenous content in dried (lyophililized) urine indicates that approximately 33.0 kJ/g of urine nitrogen (i.e., 7.9 kcal/g N) which is equivalent to 5.3 kJ/g (i.e., 1.25 kcal/g) of protein, knowing that one g urinary N excreted arises from ~ 6.25 g protein.

This residual energy represents a protein metabolic loss and therefore it must be subtracted from the 'digestible' energy of protein. From these considerations, Atwater has derived (at the beginning of the twentieth century) his factors used for calculating available energy (called metabolizable energy) for the three macronutrients ingested (see Figure 6.4). Note that it is the metabolizable energy value that is quoted in food composition tables, making allowance for the small amount of energy that is lost in faeces and urine after food ingestion. They are physiological *approximations* based on experiments on a limited number of subjects who were fed the typical diet eaten at that time.

Patterns of food intake

Human beings eat food in a discontinuous manner, even under conditions of nibbling, and the amount of food eaten can range from zero to up to 21MJ/day in highly active individuals or during acute episodes of hyperphagia. This contrasts with energy expenditure, which is continuous but variable. This irregularity of food behaviour occurs both within-day and between-days, which explains why there is a 2–3 times greater coefficient of variation for energy intake (15–20 per cent) than for energy expenditure (5–8 per cent). It also explains the difficulty in assessing food intake in order to obtain a representative picture of 'habitual' food (energy) intake. The physiological control of food intake is highly complex (see Section 6.6 below). Eating behaviour varies widely. This makes it extremely difficult to interpret data on food intake, the measurement of which has plagued nutritionists for more than a century (Dhurandhar et al. 2015).

Factors affecting patterns of food intake and energy intake

Since the ultimate function of energy intake is the provision of energy for metabolic processes and performance of work, body size and physical activity are important factors influencing energy intake. Little is known, however, as to how the body's energy needs are sensed, integrated, and translated into eating behaviours. Furthemore, eating is a pleasure which fulfils not only nutritional but also social, cultural, emotional, and psychological needs. The increasing buying power in industrialized society, combined with the intense marketing from food companies, has led to a progressive change in eating behaviour over the past two decades. Food technologists are constantly inventing new foods and flavours, which may not be compatible with sound nutritional guidelines. Many processed foods and snacks are rich in refined sugars and fats, and their high energy density and palatability are conducive to overeating. Apart from snacking and quick eating, a non-exhaustive list of the exogenous factors contributing to a poor control of food intake is given in Table 6.2. Among these factors, it has long been known that the nutrient composition of the diet has marked effects on food intake. Diets which are either very low or very high in protein, as well as those with an unbalanced amino acid mixture, tend to depress food intake. Similarly, low-fat diets cause a reduction in food intake, in part because they are bland and difficult to swallow, and also because of their low energy density, the total bulk may limit energy intake through greater gastric distension and delayed gastric emptying. Furthermore, carbohydrates, and specifically glucose, have been directly implicated in the control of food intake, and it is well established that low blood glucose (hypoglycaemia) stimulates hunger and feeding. By contrast, high-fat diets, in addition to adding palatability to foods, have high energy density and low bulk, which leads to diminished gastric distension and gastric emptying, so retarding the feeling of fullness and the cessation of eating. Furthermore,

TABLE 6.2 Exogenous factors, typically encountered in affluent societies, contributing to a poor control of food intake in humans

1. Large food diversity and high palatability diets
2. Profuse availability of food
3. Television watching (reduced activity, pressure of food advertising)
4. Snacking rather than meal eating
5. Fast rate of eating ('fast foods')
6. High-energy density diets (e.g., high-fat diet)
7. Eating outside home and unsociable eating
8. Technological developments, less activity
9. Reduced physical activity level
10. Urbanization: more access to energy-dense food, less need to walk

Source: Y Schutz and J S Garrow (2000) Energy and substrate balance, and weight regulation. In: *Human Nutrition and Dietetics* (10th edn), 137–148 (J S Garrow, W P T James, A Ralph eds). Churchill Livingstone, Edinburgh.

liquid carbohydrates (e.g., sugar-sweetened beverages) generally produce less satiety than solid forms.

6.4 Energy expenditure

Components of total daily energy expenditure

Daily energy expenditure is often divided into three components: the energy spent for basal metabolism (or basal metabolic rate), the energy spent on physical activity, and the increase in energy expenditure in response to a variety of stimuli (including food, cold, stress, and drugs); the latter thermogenic effects are grouped under the component of energy expenditure referred to as 'thermogenesis'. These three components are depicted in Figure 6.5 (Model A).

Basal metabolic rate (BMR)

For most people, BMR is the largest component (60–75 per cent) of total daily energy expenditure. It corresponds to the energy required for the work of vital functions, namely maintaining ionic equilibrium across cell membranes, cell and protein turnover, respiratory and cardiovascular functions. BMR is measured under standardized conditions, that is, in the morning after an overnight fast (usually 10–12 hours after the last meal), with the person awake and lying comfortably in the supine position at thermoneutrality, and in a state of physical and mental rest.

The most important factor influencing BMR is body size, and in particular the FFM, which is influenced by weight, height, gender, and age. On average, men have

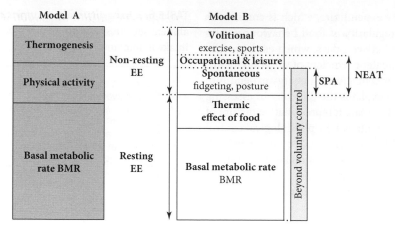

FIGURE 6.5 Components of energy expenditure.
Model A: Energy expenditure (EE) is divided into three components: basal metabolic rate (BMR),
physical activity, and thermogenesis (mostly due to the thermic effects of meals).
Model B: EE is divided into resting EE and non-resting EE. The latter is subdivided into volitional and non-exercise
activity thermogenesis (NEAT), which in turn is subdivided into spontaneous physical activity (SPA) and occupational/
leisure activities. Resting EE comprises all measurements of EE made at rest—basal metabolic rate (BMR), sleeping EE,
and the thermic effect of food—and which are essentially beyond voluntary control (i.e., subconscious).
Non-resting EE is also divided into voluntary and involuntary (subconscious) physical activities.
Source: Adapted from A G Dulloo & Y Schutz (2011) Energy balance and body weight regulation. In: *Human Nutrition* (12th edn),
87–110 (C Geissler & H Powers eds). Churchill Livingstone, Edinburgh.

greater fat-free mass and higher BMR than women of the same age, weight, and height, and older people have lower fat-free mass and BMR than young adults of the same sex, weight, and height. Most, but not all, of these between-group differences in BMR disappear when BMR is expressed as a function of FFM. This is expected as FFM contains tissues and organs which have high relative metabolic activities (per kg organ/tissue) such as liver, kidneys, heart, and to a lesser extent the resting muscles. The lower relative metabolic rate of muscle can be compensated for by a large muscle mass (see section below on 'Fuel metabolism at the level of organs and tissues'). In contrast, the contribution of adipose tissue to BMR is small (< 5 per cent in individuals without obesity). As a result, the most accurate equations for predicting BMR or resting metabolic rate (RMR) have been based on absolute FFM (in kg) or both FFM and fat mass combined. An equation is given below to illustrate the calculation (Nelson et al. 1992)

$$RMR(kJ/d) = 1265 + (93.3 \times FFM(kg)$$
$$(r2 = 0.727, P < 0.001):$$

$$RMR(kcal/d) = 302 + 22.3 \times FFM(kg)$$

Note that RMR is often used interchangeably with BMR. However, RMR may not satisfy all criteria of BMR measurements; for example, when the metabolic rate at rest is measured in the sitting posture rather than supine position and/or without overnight fasting as specified for BMR measurements.

BMR can vary up to ±10 per cent between healthy individuals of the same age, gender, body weight, and FFM, thereby underscoring the importance of genetic factors. The day-to-day intra-individual variability in BMR is low in men (coefficient of variation of 1–3 per cent) but is larger in women (2–5 per cent) because of variations in BMR over the menstrual cycle. In both men and women, BMR is greater than the metabolic rate during sleep by 5–20 per cent, the difference between BMR and sleeping metabolic rate being explained by the effect of arousal. The thyroid hormones play an important role in the maintenance of BMR; the latter is elevated in hyperthyroidism and reduced in hypothyroidism.

BMR is known to be depressed during starvation. Although this is to a large extent explained by the loss in body weight and lean tissues, the fall of BMR is often reported to be lower than predicted from the reduction in body weight or FFM, which could reflect adaptive mechanisms for energy conservation (resulting from starvation-induced reductions in circulating thyroid hormones and in sympathetic nervous system activity). During rapid overfeeding, the evidence that BMR is increased is equivocal, and when it is found to be increased, it is within 5–10 per cent of the excess energy intake.

Energy expenditure due to physical activity

The energy expenditure due to physical activity is determined by the type and intensity of the physical activity and on the time spent in different activities. Physical activity is often considered to be synonymous with 'muscular work' which has a strict definition in physics—force × distance, when external work is performed on the environment. During muscular work (muscle contraction), the muscle produces 3–4 times more heat than mechanical energy (the energy for performing work on the environment). The energy cost of any given physical activity varies both within and between individuals; the latter variation is due to differences in body size and in the speed and dexterity with which an activity is performed. In order to adjust for differences in body size, the energy cost of physical activities is expressed as multiples of BMR or RMR, and referred to as the metabolic equivalent of a task or MET. MET generally ranges from 1–5 for most movement-associated activities of everyday life, but can reach values between 10 and 14 during intense exercise, and 20 or more during maximum work capacity maintained only for a few minutes (test called VO_2max). In heavy manual workers or competition athletes, physical activity can account for up to 70 per cent of daily energy expenditure. For most people in industrialized societies, however, the contribution of physical activity to daily energy expenditure is relatively small (10–15 per cent). In a hospitalized patient in bed, it is even lower. However, in the latter case the RMR itself can increase moderately to substantially due to the disturbed metabolic effect of the disease (for example infection with fever), as well as the level of stress factor incurred.

Energy expenditure in response to various thermogenic stimuli

The component of energy expenditure referred to as 'thermogenesis' is best described by the various forms in which it can exist (Miller 1982), and summarized below:

(i) *Isometric thermogenesis:* This is due to increased muscle tension (without muscle shortening); no physical work is actually done. For example, the differences in energy expenditure in a person who is lying, sitting, or standing are mainly due to changes in muscle tone for posture maintenance. While most physical activities consist of both dynamic and static (isometric) muscle actions, the isometric component is very often essential for the optimal performance of dynamic work given its role in coordinating posture during standing, walking, and most physical activities of everyday life. The multiple expressions of isometric thermogenesis are depicted in Figure 6.6: isometric muscle contraction for posture maintenance can involve skeletal muscle in the upper body, lower body, and in the trunk. Its nature can be classified as voluntary, involuntary, and spontaneous. The magnitude of internal and external load (if any) influences muscle contraction and hence isometric thermogenesis. In a quantitative way, as compared to low-intensity dynamic exercises such as low-power cycling (10 to 50 Watt) generating MET values of

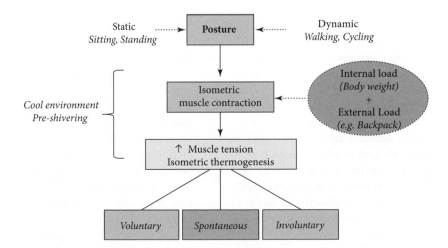

FIGURE 6.6 Multiple expressions of isometric thermogenesis in humans.
Multiple expressions of isometric thermogenesis in humans. Isometric muscle contraction for posture maintenance can involve skeletal muscle in the upper body, lower body, and in the trunk. Its nature can be classified as voluntary, involuntary, and spontaneous. The magnitude of internal and external load (if any) influences muscle contraction and hence isometric thermogenesis. Exposure to cool environment can also trigger an increase in skeletal muscle tension and hence increased (isometric) thermogenesis. *Source*: A G Dulloo, J L Miles-Chan, J P Montani, Y Schutz (2017) Isometric thermogenesis at rest and during movement: A neglected variable in energy expenditure and obesity predisposition. *Obesity Reviews* **18** (Suppl 1), 56–64. With permission from John Wiley.

2 to 4, the intensity of isometric-structured activities (such as leg press, passive lower body vibrations) is closer to 'no-load' cycling (viz. rotation of legs with virtually zero resistance on a stationary bicycle ergometer) which is in the range of 1.2 to 2 METs (Dulloo et al. 2017)

(ii) *Psychological thermogenesis:* The psychological state may influence energy expenditure, as anxiety (or emotional state), anticipation, and acute stress stimulate adrenaline secretion, leading to increased heat production. For example, the energy cost of sitting at ease and sitting playing chess may differ by as much as two-folds—a difference that cannot entirely be attributed to muscular movement. Furthermore, the increase in energy expenditure when pilots are under air traffic control (or take off) is inversely related to their level of experience. Note that heart rate response (versus an appropriate pre-stimuli baseline) constitutes a proxy of this phenomenon, but consider that heart rate is a poor predictor of energy expenditure at rest.

(iii) *Cold-induced thermogenesis:* Human beings rarely need to increase heat production for the purpose of thermal regulation. This is because they are able to seek an equitable environment or wear suitable clothing. At low temperatures, resting metabolic rate, and hence heat production, increases. For example, normal weight women maintained in identical clothing when the temperature in the respiration chamber was lowered from 28 to 22°C increased their 24h heat production by about 7 per cent. It is customary to distinguish between two forms of cold-induced thermogenesis: (i) shivering thermogenesis which involves rhythmic muscle contraction and (ii) non-shivering thermogenesis which is defined as increased heat production not associated with muscle contraction, and is thought to be due to increased sympathetic nervous system activity, particularly in brown adipose tissue (BAT) in small mammals. Non-shivering thermogenesis is inversely correlated with body size, age, and ambient temperature and has been demonstrated in adult human beings chronically exposed to extreme temperatures. Although several lines of evidence are consistent with an important role for the sympathetic nervous system in the regulation of thermogenesis in humans, the quantitative importance of BAT as a site of adaptive thermogenesis in adults has proven to be elusive. However, morphological and scanning studies have raised the possibility that BAT in humans may not be as rare as once believed. Indeed, the use of fluorodeoxyglucose positron emission tomography (FDG-PET) scans have visualized areas of uptake that correspond to BAT, with main depots occurring primarily in the supraclavicular and neck regions, with some additional locations in the axillary and paravertebral regions of healthy adult individuals. These BAT-like depots express uncoupling protein 1 (UCP1)—the unique identifying characteristic of BAT. Furthermore, UCP1-containing adipocytes may also occur within white fat depots and are referred to as beige or brite adipocytes. These findings have regenerated interest into approaches to activate BAT for obesity management.

(iv) *Diet-induced thermogenesis:* Heat production increases following the consumption of a meal, and this thermic effect of food is classically termed 'specific dynamic action'. Heat production also increases on a high plane of nutrition, the so-called 'luxusconsumption'. These two forms of thermogenesis related to food have been regrouped historically under the term 'diet-induced thermogenesis' or DIT (Miller 1982). The magnitude of DIT depends upon a large number of endogenous and exogenous factors: These include gender, body composition, menstrual cycle, aging, obesity, nutritional status, familial and genetic factors.

DIT is often divided into two subcomponents: 1) an *obligatory* and 2) a *facultative* component. The obligatory component is related to the energy costs of absorption and metabolic processing of nutrients or the energy cost of tissue synthesis during overfeeding. The *facultative* component results partly from the sensory aspects of foods (smell and taste) and partly from the stimulation of the sympathetic nervous system; such diet-induced thermogenesis could contribute to adaptive thermogenesis that minimize weight gain during overfeeding (see Section 6.7).

(v) *Drug-induced thermogenesis:* The consumption of caffeine, nicotine, and alcohol may form an integral part of daily life for many people, and all three of these drugs stimulate thermogenesis. A cup of coffee (containing 60–80 mg caffeine) can increase BMR by 5–10 per cent over an hour or two. Oral intake of 100 mg caffeine every two hours during the day or smoking of a packet of 20 cigarettes increase 24-hour energy expenditure by 5 per cent and 15 per cent, respectively. Furthermore, the thermogenic effect of nicotine is potentiated by caffeine. The cessation of elevated thermogenesis induced by nicotine or nicotine and caffeine may be a factor that contributes to the average weight gain of 5 kg within a year after cessation of smoking. Several non-caloric ingredients in food and beverages (often referred to bioactive food ingredients) such as green tea polyphenols and red pepper capsaicins have also been shown to stimulate thermogenesis and fat oxidation in humans (reviewed in Dulloo 2011).

Spontaneous physical activity and non-exercise activity thermogenesis

Since the turn of this century, the term non-exercise activity thermogenesis (NEAT) has been used to define physical activity other than volitional exercise and sports activities. This is depicted in Figure 6.5 (Model B) where non-resting energy expenditure is divided into voluntary and involuntary physical activities. The voluntary physical activity comprises volitional activities such as exercise and sports as well as occupational activities (going to work and performing work duties) and leisure activities (e.g., gardening). The involuntary physical activity comprises spontaneous and subconscious fidgeting and posture maintenance, and is referred to as spontaneous physical activity (SPA). These essentially involuntary movements comprise a larger proportion of thermogenesis induced by isometric work; the actual work done on the environment during SPA is thus very small compared to the total energy spent on such activities. In subjects confined to a respiration chamber, the 24h energy expenditure attributed to SPA (as assessed by radar systems measuring their total durations) was found to vary widely between individuals (range from 400 to 3000 kJ/day), and to be a significant predictor of subsequent weight gain over the following three years.

SPA is thus an important component of NEAT, which is not limited solely to SPA but also includes energy expended for 'voluntary' occupational and leisure-time activities. The potential importance of variations in SPA and NEAT in body weight regulation is discussed below (Section 6.7).

Fuel metabolism at the level of organs and tissues

The energy supplied by the diet is in the form of macronutrients (also called substrates or metabolic fuels): carbohydrates, proteins, fat, and alcohol. The macromolecules (carbohydrates, proteins, and fat) cannot be directly utilized by the tissues as such but must be first broken down into smaller molecules: carbohydrates into monosaccharides, triacylglycerols into fatty acids (+glycerol), and proteins into amino acids. Ethanol (alcohol), which in a strict sense is not considered a macronutrient, but as a toxic substance when drunk in excess, also constitutes a source of energy utilized by the body. The major substrates, which circulate in the blood and are taken up by the tissues to serve as fuels, are shown in Table 6.3. The amount stored in the tissues, the level of exogenous supply, and the metabolic state of the individual determine the relative importance of the utilization of each fuel, with synthesis, breakdown, and utilization of body reserves controlled by the neuro-hormonal system.

TABLE 6.3 Substances which circulate in the blood and are used to supply energy

Fuel	Source
Glucose	Dietary carbohydrate; Glycogen stores; Gluconeogenesis in liver and kidney from lactate amino acids and glycerol
Free fatty acids (FFA)	Dietary fats; Triacylglycerol stores (especially in adipose tissue); Synthesized carbohydrate in liver and adipose tissue, especially after feeding on low-fat diets
Amino acids	Dietary protein; Tissue protein stores; Synthesized from carbohydrates
Ketone bodies (acetoacetate, 3-hydroxybutyrate)	Produced from FFA and some amino acids in liver
Glycerol	Produced from triacylglycerol breakdown
Lactate	Anaerobic glycolysis
Acetate	Gut fermentation of carbohydrates; Produced from FFA in liver and muscle and from ethanol in liver
Ethanol	Dietary intake; Gut fermentation
Fructose	Dietary sucrose
Galactose	Dietary intake, especially as milk lactose

Source: M Elia (2000) Fuel of the tissues. In: *Human Nutrition and Dietetics* (10th edn), 37–59 (J S Garrow, W P T James, A Ralph eds). Churchill Livingstone, Edinburgh.

Metabolic rate at the level of organs and tissues

The heat production of individual tissues and organs can be calculated from the oxygen consumption by measuring blood flow and the arterio-venous difference in oxygen concentration across tissues and organs. Normalized for body mass, adipose tissue has the lowest metabolic rate (approx. 18.8 kJ/kg/day for sub-cutaneous abdominal adipose tissue). In a subject without obesity, adipose tissue contributes 3–5 per cent of the total resting energy expenditure, although it represents 20–30 per cent of body weight. The majority of the heat production (about 60 per cent) comes from active organs such as the liver, kidney, heart, and brain, although they account for only 5 to 6 per cent of total body weight (Table 6.4). The heat production of muscle per unit mass (42 to 63 kJ/kg) is 15–40 times lower that of metabolically active organs, but because of its large size (more than half of the total fat-free mass) it contributes about 20 per cent of total heat production.

The heat production or metabolic rate per kg of organ seems to change little during growth and development. However, the metabolic rate per kg body weight (or per kg fat-free mass) is much greater in young children than in adults. The reduction of metabolic rate with increasing age is mostly due to a change in a proportion of different tissues, and to a lesser extent to a reduction in the metabolic rate per kg body weight (Elia 2000). The larger proportion of metabolically active tissues (brain, liver, heart, kidneys) in infants and children explains their higher metabolic rates compared with adults when expressed in relation to fat-free mass. The contribution of different tissues to body weight and BMR in a 'reference male' is shown in Table 6.4.

Fuels used by different tissues and fuel selection

The main fuels available to tissues are glucose, triacylglycerol, free fatty acids, and ketone bodies; Table 6.5 shows the fuels that can be used by different tissues. Red blood cells are wholly reliant on anaerobic metabolism of glucose, releasing lactate. The brain oxidizes glucose as a source of energy since it cannot utilize fatty acids. Nevertheless, during prolonged fasting and starvation, the production of ketone bodies can progressively take over and can contribute to about two-thirds of the brain's energy needs. The ketone bodies are by-products of the metabolism of endogenous fatty acids resulting from the accelerated lipolysis of adipose tissue during fasting and starvation. Other tissues can utilize a variety of fuels, depending on their availability in the circulation, as well as their hormonal control.

Measurement of energy expenditure

The energy expended by an individual can be assessed by two different techniques: direct and indirect calorimetry. Direct calorimetry is the direct measurement of heat output; it consists of the measurement of heat dissipated by the body through radiation, convection, conduction, and evaporation (Section 6.2). Indirect calorimetry depends on the fact that the heat released by metabolic processes can be calculated from the rate of oxygen consumption

TABLE 6.4 Contribution of different tissues and organs to basal metabolic rate (BMR) of a reference man

	Weight of tissue		Organ/tissue metabolic rate		
	(kg)	(% body weight)	(MJ/kg/day)	(MJ/day)	(% BMR)
Liver	1.8	2.6	0.84	1.51	21
Brain	1.3	2.0	1.00	1.41	20
Heart	0.33	0.5	1.84	0.61	9
Kidney	0.31	0.4	1.84	0.57	8
Muscle	28.0	40	0.054	1.52	22
Adipose tissue	15	21.4	0.019	0.28	4
Miscellaneous by difference e.g., skin, intestine, bone	23.16	33.1	0.049	1.13	16
Whole body	70	100	0.1	7.03	100

Source: M Elia (2000) Fuel of the tissues. In: *Human Nutrition and Dietetics* (10th edn), 37–59 (J S Garrow, W P T James, A Ralph eds). Churchill Livingstone, Edinburgh.

TABLE 6.5 Important fuels utilized by various tissues

Brain	Glucose, ketone bodies
Muscle	Glucose, FFA, ketone bodies (starvation), acetate (after alcohol ingestion), triacylglycerol, branched-chain amino acids
Liver	Amino acids, fattly acids including short chain fatty acids, glucose, alcohol
Kidney	
Cortex	Glucose, FFA, ketone bodies
Medulla	Glucose (glycolysis)
Brown adipose tissue	Mainly FFA
White adipose tissue	Glucose, ? FFA
Gastrointestinal tract	
Small intestine	Glutamine, ketone bodies (starvation), a variety of other fuels in smaller amounts
Large intestine	Short-chain fatty acids, glutamine, glucose, and other fuels in smaller amounts
Red blood cells	Glucose (glycolysis)
Lymphocytes/macrophages	Glutamine, glucose, ? FFA/Ketones

FFA; Free fatty acids

Source: M Elia (2000) Fuel of the tissues. In: *Human Nutrition and Dietetics* (10th edn), 37–59 (J S Garrow, W P T James, A Ralph eds). Churchill Livingstone, Edinburgh.

(VO_2). This is because energy expenditure to maintain electrochemical gradients, support biosynthetic processes, and generate muscular contraction utilizes ATP (adenosine triphosphate), which is formed by oxidative phosphorylation, directly linked to the oxidation of substrates and reduction of oxygen to water (Figure 6.1). It is the rate of ATP utilization that determines the rate of substrate oxidation and therefore oxygen consumption. The energy expenditure per mole of ATP formed can be calculated from the heat of combustion of 1 mole of substrate, divided by the total number of moles of ATP generated in its oxidation. Each mol of ATP formed is accompanied by the release of about the same amount of heat (~75 kJ/mol ATP) during the oxidation of carbohydrates, fats, or proteins, and the consumption of 1 litre of oxygen is equivalent to 20.3 kJ energy expenditure, regardless of the substrate being oxidized (Livesey & Elia 1988). Under conditions of thermal equilibrium in a subject at rest and in postabsorptive conditions, heat production, measured by indirect calorimetry, is identical to heat dissipation, measured by direct calorimetry. This is an obvious confirmation of the first law of thermodynamics, which states that the energy released is ultimately transformed into heat (and external work during exercise) and validates the use of indirect calorimetry. An overview of the classical and modern methods for the assessment of human energy expenditure is provided in Weblink Section 6.1.

Estimations of energy requirements

The energy requirement of an individual is defined by the World Health Organization (WHO) as 'the level of energy intake that will balance energy expenditure when the individual has a body size and composition, and a level of physical activity, consistent with long-term good health'. The energy requirement should also allow the maintenance of economically necessary and socially desirable physical activity. In children and pregnant women, the energy requirement includes the extra energy needs associated with the deposition of tissues during growth in infants and foetus development in pregnancy, consistent with good health. During lactation, the exported energetic nutrients due to the secretion of milk must be added to the energy expenditure measured in the lactating mother.

There are two approaches to assess the energy requirement of people of different age, sex, and physical activity:

(i) indirect: assessment of food intake followed by the calculation of energy intake, which assumes energy equilibrium;

(ii) direct: assessment of total energy expenditure.

The energy needs of a group, in contrast to protein and micronutrient needs, represent the *average* value of the individuals making up that group, since in the latter situation there is a probabilistic cut-off level below which the supply will become insufficient. When possible, energy requirements should be based on estimates of energy expenditure rather than on energy intake for two reasons: (i) the day-to-day intra-individual variability of energy intake is at least 3–4 times greater than total energy expenditure (TEE), the lower level being resting (sleeping) values and 'clamped' by the rather narrow span of possible physical activity levels (see in paragraphs below), as compared to the energy ingestion which can vary from zero to extremely high values, and (ii) the accuracy and precision of energy intake measurement is much lower for energy intake than for energy expenditure. In the former, involuntary 'cheating' and bias exist whereas energy output is based on physiological measurements and does not depend upon the reliability of the subject (for example, misreporting).

The term 'requirements' refers to the 'habitual' or 'usual' requirements over a certain period. From one day to the next, individuals are not expected to maintain energy balance precisely, and hence energy intake and energy expenditure measurements may not give the same values. Because the variability of energy intake is greater than the variability of expenditure, the habitual energy requirements can be best determined from expenditure rather than intake measurements. In addition, it is difficult to measure energy intake accurately without influencing the subject's eating behaviour (Dhurandhar et al. 2015).

In the estimation of daily energy expenditure by the so-called 'factorial' approach, the BMR is first calculated (Ainsworth et al. 2011). The physical activities are broken down into occupational activities (work) and discretionary (i.e., leisure) activities. Occupational activities include salaried and non-salaried chores (such as housework). The energy expenditure for occupational activities will depend on the type of occupation, the time spent in performing the work, and the physical characteristics of the individuals. These activities are classified into *light* (1.7 × BMR), *moderate* (2.2 × BMR for women and 2.7 × BMR for men) and *heavy* (2.8 × BMR for women and 3.8 × BMR for men). Discretionary activities include socially desirable activities (such as the exploratory activities of children and the participation in tasks implying social improvement), exercise for physical and cardiovascular fitness, and optional household tasks. In addition, the BMR is used to estimate the energy cost of sleeping. Finally, the residual time during which there is no clear definition of activity has been taken as BMR × 1.4.

Once the separate components of energy expenditure (sleep, physical activity, and residual time) have been calculated, the total energy requirement can be calculated by summation. It should be noted that when the energy expenditure is calculated *over 24 hours* and categorized into 'light', 'moderate', and 'heavy' work, the value expressed in multiples of BMR is obviously much lower than that calculated during a working task. For example, a group with occupational work classified as 'moderate' activity will have an energy requirement calculated over 24 hours of 1.78 × BMR in men and 1.64 × BMR in women because it includes sleeping hours and residual time, whereas during the actual performance of the given task, the energy expended will be 2.7 × BMR for men and 2.2 × BMR for women.

The rate of total energy expenditure directly assessed in a respiration chamber or by doubly labelled water (see Weblink Section 6.1 on 'Measurements of energy expenditure'), can be expressed as a multiple of some baseline value such as BMR. This approach has been used by an international expert committee (James & Schofield 1990) for calculating the energy requirement by the 'factorial' method. The ratio of TEE and BMR provides a rough index of physical activity (referred to as 'physical activity level' or PAL) but the contribution of the thermic effect of foods represents a small confounding factor. Since the energy cost of a given activity is proportional to body weight, especially for weight-bearing activities, the absolute energy expenditure during weight-bearing activity will be linearly related to body weight. In subjects without obesity, the ratio TEE:BMR ranges from 1.64 to 1.98, based on doubly labelled water measurements of energy expenditure. In the confined condition of a respiratory chamber (without prescribed exercise on a treadmill or bicycle, etc.) the ratio is 1.3 to 1.35, indicating that small discontinuous activities of daily life (washing, moving around, studying, watching TV, etc.) increase energy expenditure above basal energy expenditure by one-third. While the James and Schofield analysis has served a significant role in re-establishing the importance of using BMR to predict human energy requirements, the universal validity and application of their equations have been questioned. Subsequently, a series of more recent equations (Oxford equations) have been developed by Henry (2005) using a dataset of more than 10,000 BMR values, of which >4,000 were obtained from people living in the tropics. A very comprehensive and critical British review of energy requirement (SACN report 2011) has been published, which provides additional useful information and detailed explanations on energy allowances in humans from infants to adulthood, including pregnancy and lactation.

6.5 Timescale of energy balance and body weight variability

It is important to emphasize three cardinal features of energy balance and weight regulation (Figure 6.7).

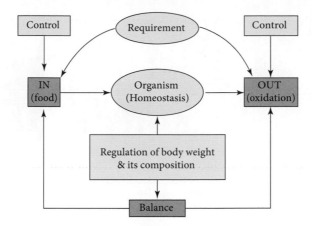

FIGURE 6.7 A simple model of body weight regulation with control systems operating via energy intake and energy expenditure.
In this model, the dynamic aspect of energy balance is regulated by feedback systems loops on both the input side (food intake) and output side (total energy expenditure), the level of the latter determining the energy 'requirement' and hence the energy intake necessary to match energy expenditure. If body weight and body composition of the organism is maintained essentially constant over time (calculated at least over one week, or one month or more), this indicates that input should be equal to output.

(i) In humans, the balance between energy intake and energy expenditure is not achieved on a day-to-day basis. Furthermore, positive energy balance on one day is not spontaneously fully compensated by negative energy balance over the following days. Near equality of intake and expenditure most often occurs over 1–2 weeks. Over months and years, total energy intake and expenditure must be very close in any individual whose body weight has remained relatively constant.

(ii) This matching of long-term energy intake and energy expenditure must be extremely precise since a systematic theoretical error of only 1 per cent between input and output of energy, if persistent, will lead to a theoretical gain or loss of about 10 kg per decade. But this does not occur for most individuals without obesity, whose body weight remains essentially constant with a few kg gain over several decades during aging.

(iii) Even in adults whose body weight appears to be stable over years and decades, there is no 'absolute' constancy of body weight per se. Instead, body weight tends to fluctuate or oscillate around a constant mean value, with small or large deviations from a 'set' or 'preferred' value being triggered by events that are seasonal and/or cultural (weekend parties, holiday seasons), psychological (stress, depression, anxiety, or emotions) and pathophysiological (ranging from minor health perturbations to serious diseases). Indeed, very short-term day-to-day variations in body weight have a standard deviation of about 0.5 per cent, while longitudinal observations over 10–30 years indicate that individuals experience slow trends

and reversal of body weight amounting to between 7 and 20 per cent of mean body weight (Garrow 1974). In order to illustrate this the day-to-day variation in body weight over a four-year period is shown for a young male of normal body weight (Figure 6.8). Despite a large variability in body weight ranging from a few kg upwards and downwards, the final body weight was almost identical to the initial one.

6.6 Control of food intake

Hunger and satiety

Because habitual food intake is difficult to measure and eating behaviour is altered by the experiments themselves, much of the work carried out in human beings has been concerned with short-term studies in relation to hunger and appetite or with short-term satiety and satiation. It is important to differentiate between these terms. Hunger may be defined as a 'demand for energy' (e.g., in response to starvation), while appetite refers to 'a demand for a particular food'. If a person is deprived of food, he becomes hungry, and if he has eaten a lot, he becomes satiated. *Satiation* refers to processes involved in the termination of a meal, and is studied by providing the subjects with test meals and then measuring the amount consumed when the food is freely available. *Satiety* refers to the inhibition of further intake of a food or meal after eating has ended. However, appetite is a powerful but poorly controlled stimulus to eat even when not hungry, and the total energy ingested in a day is determined by the interaction of many endogenous and exogenous factors. Indeed, it is common experience that feeding patterns

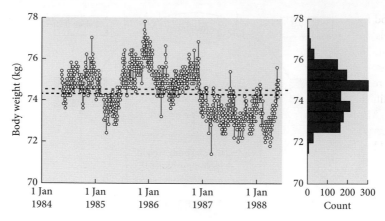

FIGURE 6.8 Day-to-day fluctuations in body weight over a four-year period in a lean young man (BMI = 22.4 kg/m²). The frequency distribution on the right-hand side (with weight intervals of 0.5 kg) indicates the number of times (counts) that subject's body weight was between 71.5 and 77.5 kg. The horizontal broken line represent the mean body of this person over this four-year period.

are influenced strongly by psychological, economic, and social factors (Table 6.2). Even though individuals may feel satiated by one particular food, they can continue to eat when a new food is presented—a phenomenon that is referred to as 'sensory specific satiety'. Conversely, when subjects are presented with a monotonous diet, their intakes are usually low. In certain communities in low-income countries, the major part of energy intake derives from one staple food, which, together with low fat intakes, constitutes a bland and monotonous diet, so that even when supplies are adequate, obesity is rarely seen. These observations suggest that when the psycho-social incentives to eat are removed, human beings can control food intake quite precisely.

Hunger–satiety control centres in the brain

Much of our understanding about centres in the brain that are involved with the control of food intake derives from studies conducted in laboratory animals. Although many areas of the brain play a role in the control of food intake, the identification two subpopulations of hypothalamic neurons, whose activations have opposing effects on food intake, has provided some insights into the functions of the hypothalamic centres. People with damage in the hypothalamus, due to trauma or tumour, often show abnormalities in feeding behaviour and weight regulation.

Hunger–satiety signals from the periphery

The sensations of hunger and satiety result from the central integration of numerous signals originating from a variety of peripheral tissues and organs; these include the gastrointestinal tract, liver, adipose tissue, and skeletal muscle. The hunger–satiety signalling systems that have generated the most interests are outlined below.

Signals from the gastrointestinal tract

The progression of food through the stomach and small intestine initiates a number of sequential peripheral satiety signals—from stretch- and mechano-receptors or from chemoreceptors that respond to the products of digestion (sugars, fatty acids, amino acids, peptides)—which are transmitted via the vagus nerve to the hindbrain for integration. By this pathway, the properties of food influence food intake in the short-term by limiting the size of a single meal and may also affect energy intake in a subsequent meal. Numerous endocrine signals from the gut are believed to exert important influences on food intake, and these include:

(a) cholecystokinin (CCK), released from the small intestine into the circulation in response to food, decreases meal size;

(b) gastric inhibitory peptide (GIP), glucagon-like peptide-1 (GLP-1) and the PYY, the peptide YY(3-36), released after food consumption, reduce appetite;

(c) ghrelin, secreted by the stomach, increases after food deprivation and decreases after food consumption. Ghrelin is the only known circulating factor to increase hunger.

Aminostatic or proteinstatic signals

Dietary protein induces satiety in the short-term, and consumption of low-protein diets leads to increased appetite for protein-containing foods. The aminostatic theory states that food intake is determined by the level of plasma amino acids, possibly related to the regulation of lean body mass, which is rigorously defended against experimental or dietary manipulation. A 'protein-stat' mechanism for the regulation of lean body mass has been proposed by Millward and is supported by human data from the classic Minnesota Starvation Experiment showing the post-starvation hyperphagia is determined by the loss of body fat but also of lean tissue (reviewed in Dulloo 2021).

Glucostatic and glycogenostatic signals

A glucostatic theory for the regulation of feeding behaviour, first proposed by Mayer in the 1950s, postulates that there are chemoreceptors in the hypothalamic satiety centre which would be sensitive to the arteriovenous difference in glucose or to the availability and utilization of glucose. Flatt later (1995) proposed that the control of food intake, via the prevention of hypoglycaemia and maintenance of adequate glycogen levels, primarily serves the maintenance of the carbohydrate balance (see section 'Macronutrient balance model' below).

Lipostatic or adipostat signals

A lipostatic theory of food intake control, first proposed by Kennedy in the 1950s, postulates that substances released from the adipose tissue stores act as satiety signals. Body fat is thus maintained at a set value, and any deviation from this value is detected by the hypothalamus via a circulating metabolite which is related to the size of the fat stores, eliciting compensatory changes in energy intake. The lipostatic hypothesis provides the most plausible explanation of long-term regulation of the fat stores. A major advance occurred in 1995 following the cloning of the ob gene, whose protein product (leptin) is primarily produced by adipocytes. Leptin is released into the circulation and acts on hypothalamic receptors to induce satiety. Rare people with mutations causing complete leptin deficiency show marked hyperphagia and severe obesity—which can be reversed by administration of small doses of leptin. However, circulating leptin is elevated in people with obesity leading to the hypothesis that resistance to the action of leptin is a factor in obesity. As blood leptin concentration varies widely in individuals with the same degree of obesity; subpopulations might have relative leptin deficiency. Insulin also has a

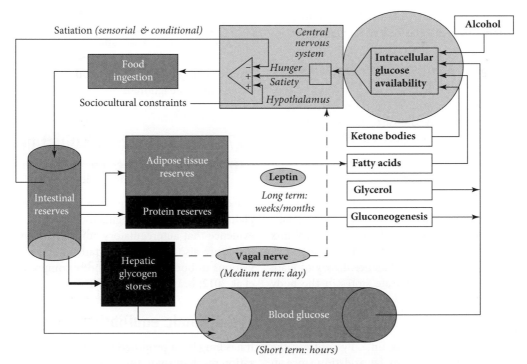

FIGURE 6.9 Sequence of events in food intake control.
The control of food intake, via alterations in the neuro-hormonal system and metabolites exchanging with the various energy reserves—involving the regulation of carbohydrates (as glycogen), fat, and protein—is considered in three phases: short-term (hour-to-hour), medium-term (day-to-day), and long-term (over weeks and months).
Source: Adapted from M Fantino (1994) Contrôle physiologique de la prise alimentaire. In: *Traité de Nutrition Clinique de l'Adulte*, 25–34 (A Basdevant, M Laville, E Lerebours eds) Flammarion, Médecine-Science, Paris.

role as an adiposity signal as it also circulates at levels proportional to body fat content, and acts on the same subpopulations of hypothalamic neurons through which leptin controls satiety.

Impact of peripheral signals on brain higher centres

Although the classic signals from the periphery such as leptin, insulin, gut hormones, and circulating nutrients themselves, act mainly on a few areas of the brain such as specific areas of the hypothalamus and brain stem, there is increasing evidence suggesting that these signals have a much broader influence on brain function. Indeed, leptin and gut hormones do not only act on the 'energy balance' control circuits in the hypothalamus and brain stem but also on cortico-limbic systems involved in cognitive, reward, and executive brain functions important for ingestive and exercise behaviour.

Integrated models of control of food intake and energy expenditure

The various signals from the periphery can be integrated in a model (see Figure 6.9) in which the control of food intake is considered in three phases:

(a) *short-term* (hour-to-hour) blood glucose homeostasis by dampening episodes of hypoglycaemia or hyperglycaemia;

(b) *medium-term* (day-to-day) maintenance of adequate hepatic stores of glycogen;

(c) *long-term* (weeks, months, or years) maintenance of the body's fat reserve (in adipose tissue) and protein reserve compartments, that is, fat mass and fat-free mass.

Macronutrient balance model

The long-term stability of body weight requires not only that energy expenditure is equal to energy intake but also that the composition of the substrate fuel mix which is oxidized corresponds to that which is ingested. According to the macronutrient balance theory of Flatt, since the protein and carbohydrate stores in the body are limited (Table 6.1), they tend to be modulated by an autoregulatory process which allows an increase in their own oxidation in response to an increase in their exogenous supply after a few days lag time. By contrast, the fat stores are not well regulated by fat oxidation since an increase in dietary fat does not promote its own oxidation. In other words, unlike carbohydrate and protein, the fat balance is passively (hence not accurately) regulated since it depends upon the co-ingestion of carbohydrates, which stimulates insulin, an anti-lipolysis hormone. The failure to increase fat oxidation to match an increase in dietary fat intake will lead to a reliance on the carbohydrate reserves, and hence to depletion of the glycogen stores, with postulated negative feedback on energy intake. This energy imbalance would persist until the adipose fat stores build up sufficiently to provide a greater supply of FFA for fat oxidation. When the higher endogenous fat oxidation (resulting from the higher fat mass) matches the higher exogenous intake, the individual would then be both in fat balance and in energy balance equilibrium, but at the cost of a higher percentage of body fat; this is schematized in Figure 6.10. Note that the time required to reach a new equilibrium in fat balance is very long (years) and depends upon the extent to which fat oxidation increases with increased fat storage, the latter being individually determined by endogenous (genetic predisposition) and exogenous factors. Based on the highly significant correlation between body fat mass and resting fat oxidation, a 10 kg of excess body fat would increase fat oxidation by (only) 20g per day (Schutz 2004). Consequently, if an excess fat of about 180 kcal (20g) is ingested at time zero, the gain in body fat required to offset this excess would be approximately 10 kg gain of adipose tissue progressively stored over several years.

This macronutrient-balance concept, which centres upon the need to maintain specific carbohydrate (glycogen) stores as a determinant of appetite, has been challenged (Stubbs et al. 1998). In people fed a very low carbohydrate (high fat) diet to deplete the glycogen stores, appetite did not increase, and fat oxidation increased to meet energy needs. The complex relationships between fat and other constituents of foods in the control of appetite cannot be ignored.

6.7 Autoregulatory adjustments in energy expenditure

The control of food intake is not by itself sufficient to explain long-term weight regulation. There is indeed ample evidence that autoregulatory adjustments in energy expenditure also play an important role in correcting deviations in body weight and body composition (Dulloo et al. 2012; Bray & Bouchard 2020).

The dynamic equilibrium model

It has long been proposed that there is a built-in stabilization mechanism in the overall homeostatic system for body weight regulation. Any imbalance between energy intake and energy requirements (i.e., energy gap) would result in a change in body weight which, in turn, will alter the maintenance energy requirements in a direction that will tend to counter the original imbalance (i.e., diminishing the energy gap) and hence be stabilizing.

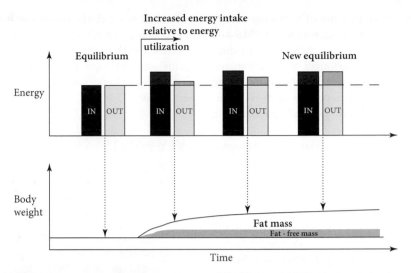

FIGURE 6.10 Effect of a step-increase in energy intake on energy balance and fat balances.
In response to an increase in daily energy and fat intake that is sustained over time, body weight will progessively increase, and this will lead to an increase in energy expenditure (due to the energy cost of extra weight gained and its maintenance energy requirement). The time required to reach a new equilibrium in energy and fat balances is very long (years) and depends upon the extent to which fat oxidation increase with increase fat storage, the latter being individually determined by endogenous (genetic predisposition) and exogenous factors.
Source: Adapted from Y Schutz (2004) Dietary fat, lipogenesis and energy balance. *Physiology & Behaviour*, **83**, 557–564.

There is a built-in negative feedback and the system thus exhibits *dynamic equilibrium*. For example, an increase in body weight will increase energy expenditure, which will then produce a negative energy balance and hence a subsequent decline in body weight. Conversely, a decrease in body weight would result in a lowering of energy expenditure (due to the loss in lean and fat tissues), which will tend to produce a positive balance and hence a subsequent return towards the previous weight. In reality, the homeostatic system is much more complex than this simple effect of 'mass action' since the *efficiency of metabolism* may also alter in response to the alterations in body weight (Stock 1999; Bray & Bouchard 2020).

Variability in adaptive thermogenesis

This notion of altered metabolic efficiency in response to changes in energy intake and/or body weight is strongly supported by the demonstration that in subjects who were deliberately made to maintain body weight at a level within 10 per cent above or 10 per cent below their habitual body weight for several weeks, there was an increase or decrease in 24-hour energy expenditure even after adjusting for changes in body weight and composition (Leibel et al. 1995). These compensatory changes in energy expenditure (referred to as adaptive thermogenesis) reflect alterations in metabolic efficiency that tend to oppose the maintenance of a body weight that is above or below the 'set' or 'preferred' body weight. This above-mentioned study (Leibel et al. 1995) also underscored a large inter-individual variability in the ability to readjust energy expenditure, with some individuals showing little or no evidence for adaptive thermogenesis, while others revealed a marked capacity to decrease or increase energy expenditure through adaptive changes in thermogenesis. The fact that genes play an important role in determining the magnitude of adaptive increase in thermogenesis, and hence contribute to resistance or susceptibility to weight gain and obesity has in fact been established from an overfeeding experiment in identical twins. Conversely, a role for genotype in human variability in the amount of weight loss during therapeutic dieting, in part resulting from variability in adaptive reduction in thermogenesis, has been suggested from studies in which identical twins underwent slimming therapy on a very-low-calorie diet. Thus, in addition to the control of food intake, changes in efficiency of energy utilization, in part through regulated heat production (i.e., adaptive thermogenesis) play an important role in the regulation of body weight, and that the magnitude of these adaptive changes in thermogenesis is strongly influenced by the genetic make-up of the individual, notwithstanding the role of exogenous factors previously discussed.

Adaptive thermogenesis at rest and during movement

Several components of energy expenditure (Figure 6.5) could be contributing to adaptive thermogenesis in the regulation of body weight. Decreases in BMR (after statistical adjustment for reductions in fat-free mass and fat mass) in response to weight loss and increases in mass-adjusted BMR in response to overfeeding have been demonstrated, and hence reflect adaptive changes in thermogenesis in the compartment of resting energy expenditure. Reports of a lower mass-adjusted BMR and a lower thermic effect of food (as a percentage of meal energy) in subjects who were no longer obese, compared to those controls who never developed obesity, may also be suggestive of adaptive thermogenesis in these compartments of energy expenditure (generally measured in resting conditions).

There is also some evidence for adaptive thermogenesis occurring in the compartment of non-resting energy expenditure, that is, in the compartment of physical activity and including NEAT. In particular, the efficiency of skeletal muscle contraction during cycling at low intensity has been shown to be increased after weight loss (i.e., favouring energy conservation) or decreased during overfeeding (i.e., favouring energy dissipation), thereby reflecting adaptive thermogenesis operating during low-intensity physical activities of everyday life. There is also some evidence which suggests that SPA may be increased in response to overfeeding and decreased after weight loss. As SPA is essentially subconscious and hence beyond voluntary control, a change in the *level* or *amount* of SPA in a direction that defends body weight stability also constitutes an auto-regulatory change in energy expenditure (Dulloo et al. 2012).

Overall, any change in metabolic efficiency in the resting or non-resting state that would tend to attenuate the magnitude of energy imbalance or to restore body weight and body composition towards its 'set' or 'preferred' value constitutes an evidence of adaptive changes in thermogenesis. In industrial societies where food is plentiful all year round and physical activity demands are low, subtle variations between individuals in thermogenesis, cumulated over the long-term, can be important in determining constancy of body weight in some individuals and in provoking the drift towards obesity in others. It is relevant as much to the aetiology of obesity as to the inability to sustain weight loss after slimming therapy.

6.8 Integrating energy intake and energy expenditure

To achieve long-term constancy of body weight, compensatory adjustments must occur in both energy intake and energy expenditure, so that unravelling the importance of one or other is difficult in practice, if not impossible. Models of body weight regulation have primarily focused on physiologically induced *autoregulatory* adjustments in energy intake and in energy expenditure, that is, those beyond voluntary control. However, there is certainly some degree of cognitive control, as pointed out

by Garrow, when clothes no longer fit, and changes occur in appearance, exercise tolerance, and general well-being (Garrow 1974). In many individuals, the cognitive (conscious) controls over food intake and energy expenditure can be as important as non-conscious physiological regulation.

KEY POINTS

- In the transformation of the chemical energy in food (i.e., gross energy) to energy available for the body (i.e., metabolizable energy), 5–10 per cent is lost through faeces and urine.
- Energy balance is the difference between metabolizable energy intake and total energy expenditure. It is strongly related to macronutrient balance, and the sum of individual substrate balances, expressed as energy, must mathematically be equivalent to the overall energy balance.
- The matching between energy intake and expenditure is poor over the short-term, but (in most people) it is accurate over the long-term.
- Because the day-to-day variability in energy intake is much greater than that in energy expenditure, the habitual energy requirements is best determined from total energy expenditure.
- The mechanisms underlying long-term energy balance and weight regulation involve both involuntary controls as well as conscious alterations in lifestyle to correct unwanted changes in body weight.
- Modern lifestyles have led to considerable changes in the type and amount of food eaten and in the amount of time spent on physical activity, leading to an environment where matching energy intake and energy expenditure is more difficult to achieve over the long term.
- Adaptive thermogenesis plays a role as regulator of heat production that contributes to homeothermy and to the defence of body weight and body composition.
- Undernutrition leads to a decrease in energy expenditure which results from the loss in body weight (and metabolically active tissues) and from increased efficiency of metabolism (i.e., adaptive reduction in thermogenesis).
- Overnutrition leads to gain in body weight which is often less than predicted because of compensatory increases in energy expenditure through an adaptive increase in thermogenesis; there is a large inter-individual variability in the capacity for adaptive thermogenesis.

 Be sure to test your understanding of this chapter by attempting multiple choice questions. See the Further Reading list for additional material relevant to this chapter.

REFERENCES

Ainsworth B E, Haskell W L, Herrman S D et al. (2011) Compendium of physical activities: A second update of codes and MET values. *Med Sci Sports Exerc* **43**, 1575–1581.

Bray G A & Bouchard C (2020) The biology of human overfeeding: A systematic review. *Obes Rev* **21**, e13040. doi: 10.1111/obr.13040.

Dhurandhar N V, Schoeller D, Brown A W et al. (2015) Energy balance measurement: when something is not better than nothing. *Int J Obes* **39**, 1109–1113.

Dulloo A G (2011) The search for compounds that stimulate thermogenesis in obesity management: From pharmaceuticals to functional food ingredients. *Obes Rev* **12**, 866–883.

Dulloo A G (2021). Physiology of weight regain: Lessons from the classic Minnesota Starvation Experiment on human body composition regulation. *Obes Rev* **22**, Suppl 2, e13189. doi: 10.1111/obr.13189.

Dulloo A G, Jacquet J, Montani J P et al. (2012) Adaptive thermogenesis in human body weight regulation: More of a concept than a measurable entity? *Obes Rev* **13** Suppl 2, 105–121.

Dulloo A G, Miles-Chan J L, Montani J P et al. (2017). Isometric thermogenesis at rest and during movement: A neglected variable in energy expenditure and obesity predisposition. *Obes Rev* **18**, Suppl 1, 56–64.

Elia M (2000) Fuel of the tissues. In: *Human Nutrition and Dietetics*, 10th edn, 37–59 (J S Garrow, W P T James, A Ralph eds). Churchill Livingstone, Edinburgh.

Garrow J S (1974) *Energy balance and Obesity in Man*. North-Holland, Amsterdam.

Henry C J (2005) Basal metabolic rate studies in humans: measurement and development of new equations. *Public Health Nutr* **8**, 1133–1152.

James W P T & Schofield E C (1990) *Human Energy Requirements. A Manual for Planners and Nutritionists*. Oxford University Press, Oxford.

Leibel R L, Rosenbaum M, Hirsch J (1995) Changes in energy expenditure resulting from altered body weight. *N Engl J Med* **332**, 621–628.

Livesey G & Elia M (1988) Estimation of energy expenditure, net carbohydrate utilisation and net fat oxidation and synthesis by indirect calorimetry: Evaluation of errors with special reference to the detailed composition of fuels. *Am J Clin Nutr* **47**, 608–628.

Miller D S (1982) Factors affecting energy expenditure. *Proc Nutr Soc* **41**, 193–202.

Nelson K M, Weinsier R L, Long C L et al. (1992) Prediction of resting energy expenditure from fat-free mass and fat mass. *Am J Clin Nutr* **56**, 848–856.

SACN Report (2011) *Dietary Reference Values for Energy*. The Scientific Advisory Committee on Nutrition report on the DRVs for energy. https://www.gov.uk/government/publications/sacn-dietary-reference-values-for-energy (accessed May 2021).

Schutz Y (2004) Dietary fat, lipogenesis and energy balance. *Physiol Behav* **83**: 557–564.

Stock M J (1999) Gluttony and thermogenesis revisited. *Int J Obes* **23**, 1105–1117.

Stubbs R J (1998) Appetite, feeding behaviour and energy balance in humans. Proceedings of the Nutrition Society **57**, 341–356.

FURTHER READING

Dulloo A G & Schutz Y (2005) Energy balance and body weight regulation. In: *Human Nutrition and Dietetics*, 11th edn, 83–102 (C Geissler & H Powers eds). Churchill Livingstone, Edinburgh.

Dulloo A G and Schutz Y 2011 Energy balance and body weight regulation. In: *Human Nutrition and Dietetics*, 12th edn, 87–110 (C Geissler & H Powers eds). Churchill Livingstone, Edinburgh.

Fantino M (1994) Contrôle physiologique de la prise alimentaire. In: *Traité de Nutrition Clinique de l'adulte*, 25–34 (A Basdevant, M Lavilleand, E Lerebours eds). Flammarion Médecine Sciences, Paris.

Schutz Y & Garrow J S (2000) Energy and substrate balance, and weight regulation. In: *Human Nutrition and Dietetics*, 10th edn. 137–148 (J S Garrow, W P T James, A Ralph eds). Churchill Livingstone, Edinburgh.

7 Carbohydrate metabolism

John M. Brameld, Tim Parr, and David A. Bender

OBJECTIVES

By the end of this chapter you should be able to:

- classify the dietary carbohydrates
- explain the importance of the glycaemic index and glycaemic load
- describe the main functions of glycaemic and non-glycaemic carbohydrates
- describe the main types of dietary fibre (non-starch polysaccharides, resistant starch, resistant oligosaccharides), and their effects on health, including intestinal bacterial fermentation and effects on digestion and absorption of other nutrients
- describe the pathways of carbohydrate metabolism and their regulation
- describe the role of glycogen as a carbohydrate reserve and explain how its synthesis and utilization are regulated
- explain the importance of gluconeogenesis and describe the pathway
- explain the importance of carbohydrate in exercise.

7.1 Introduction

Carbohydrates are not essential nutrients since there is no absolute requirement for a dietary intake, as long as there is an adequate intake of protein to permit *de novo* synthesis of glucose, and of fat as an energy source. Red blood cells are wholly dependent on (anaerobic) glycolysis and so have an absolute requirement for glucose. The central nervous system is largely dependent on glucose but can meet a proportion of its energy needs from ketone bodies. A very low carbohydrate diet, however, results in a chronically increased production and plasma concentrations of the ketone bodies (ketosis) and absence of glycogen stores, with adverse effects on high-intensity work by muscles. Other possible adverse effects of diets very low in carbohydrates are bone mineral loss, hypercholesterolaemia, and increased risk of urolithiasis.

Based on the rate of carbohydrate utilization by the brain, the average requirement for carbohydrate has been estimated to be 100 g/day (20 per cent of energy intake). As little as 10 per cent of energy intake from carbohydrate (50 g/day) will prevent ketosis but there are advantages of higher intakes of carbohydrate, not least in reducing the proportion of energy provided by fat. However, when carbohydrate provides more than about 80 per cent of energy, it is difficult to consume sufficient bulk of food to meet energy requirements.

The main role of dietary carbohydrate is as a metabolic fuel: in average Western diets carbohydrates provide about 40–45 per cent of energy intake; in developing countries, where fat is scarce, 75 per cent or more of energy may come from carbohydrate. It is considered desirable that carbohydrate should provide 50 per cent of energy intake (SACN 2015). Dietary carbohydrate can affect satiety, insulin secretion and glucose homeostasis, lipid metabolism, gut function, and intestinal microflora, so that the amount (and type) of carbohydrate consumed will be important with respect to body weight control, type 2 diabetes mellitus, cardiovascular disease, and cancer.

7.2 The classification of dietary carbohydrates

As shown in Table 7.1, dietary carbohydrates can be divided into three groups: sugars (mono- and disaccharides), oligosaccharides (with 3–9 monomer units), and polysaccharides (with 10 or more monomer units). Polysaccharides can be further divided into starch and non-starch polysaccharides.

There are two different ways of determining the carbohydrate content of foods. The older method is carbohydrate 'by difference', where carbohydrate is calculated as the difference between the total mass of the food and the measured content of water, fat, protein, and minerals.

TABLE 7.1 Classification of food carbohydrates

Class	Subgroup	Principal components
sugars (1–2 monomer units)	monosaccharides	glucose, fructose, galactose
disaccharides	sucrose, lactose, maltose, trehalose	
sugar alcohols (polyols)	sorbitol, mannitol, lactitol, xylitol, erythritol, isomalt, maltitol	
oligosaccharides (3–9 monomer units)	malto-oligosaccharides (α-glucans)	maltodextrins—glycaemic polymers of glucose
fructo-oligosaccharides (including inulin)	non-glycaemic polymers of fructose	
galacto-oligosaccharides	non-glycaemic polymers of galactose	
other oligosaccharides	polydextrose, raffinose, stachyose	
polysaccharides (>10 monomer units)	starch (α-glucan)	amylose, amylopectin, modified starches

Source: Adapted from Asp N-G (1996) Dietary carbohydrates: classification by chemistry and physiology. *Food Chem* **57**, 9–4.

More accurate determination of available carbohydrate (as used in UK tables of food composition since the 1920s) measures sugars and starches specifically. Measurement of carbohydrate by difference not only compounds the errors in determining the other components of the food but includes indigestible carbohydrates (oligosaccharides and non-starch polysaccharides) and non-carbohydrate material, and overestimates both the carbohydrate content and energy yield of the food.

Dietary sugars

Structures of the main dietary sugars are shown in Figure 7.1. Sugars can be considered in two groups: those that are in solution (known as free sugars) and those in plant foods that are contained within cell walls (known as intrinsic sugars). Free sugars will include both sugars naturally present in fruit juice, honey, syrups, and so on, and those added in manufacture and at the table. Lactose naturally present in milk and milk products is excluded from free sugars. Intrinsic sugars will have a lower availability than extrinsic sugars because the cellulose of plant cell walls is not digested. However, what is measured analytically is the total sugar content of the food, after more or less complete disruption of cell walls, and this is what is reported in nutrition labelling and food composition tables.

Free sugars (other than lactose) increase the risk of dental caries, because of bacterial fermentation in the mouth, leading to formation of acids that attack dental enamel. Prospective studies show an increased risk of developing type 2 diabetes mellitus with an increased intake of free sugars. Especially in sweetened drinks, free sugars lead to an increased energy intake, leading to the development of obesity.

SACN (2015) recommended that free sugars should not account for more than 5 per cent of total energy intake (compared with an average in UK of more than 15 per cent). This is because (added) sugars will dilute the nutrient density of foods, and can be readily over-consumed, leading to positive energy balance, overweight, and obesity. Sugar-sweetened drinks are an important source of excess energy intake, contributing to obesity. In populations in which the intake of free sugars is 5 per cent of energy intake or less, the incidence of dental caries is significantly lower than when the intake is above 10 per cent of energy intake.

Fruits also provide fructose, and small amounts of pentose (5-carbon) and other sugars, as well as sugar alcohols (polyols), in which the aldehyde or ketone group has been reduced to a hydroxyl group. Fructose is 40 per cent sweeter-tasting than sucrose, and 75 per cent sweeter than glucose. High-fructose syrups, prepared by hydrolysis of corn starch and isomerization of some of the glucose to fructose, are now widely used in food manufacture, and have been associated with increased synthesis of fatty acids and the development of hyperlipidaemia and obesity.

Glycaemic and non-glycaemic carbohydrates

Nutritionally, carbohydrates can be classified in two broad categories.

- Those that are digested in, and absorbed from, the small intestine, and so provide carbohydrate for metabolism; these are glycaemic carbohydrates—they increase the blood glucose concentration. This group includes sugars, much dietary starch, and dextrins (glucose oligosaccharides resulting from starch digestion).

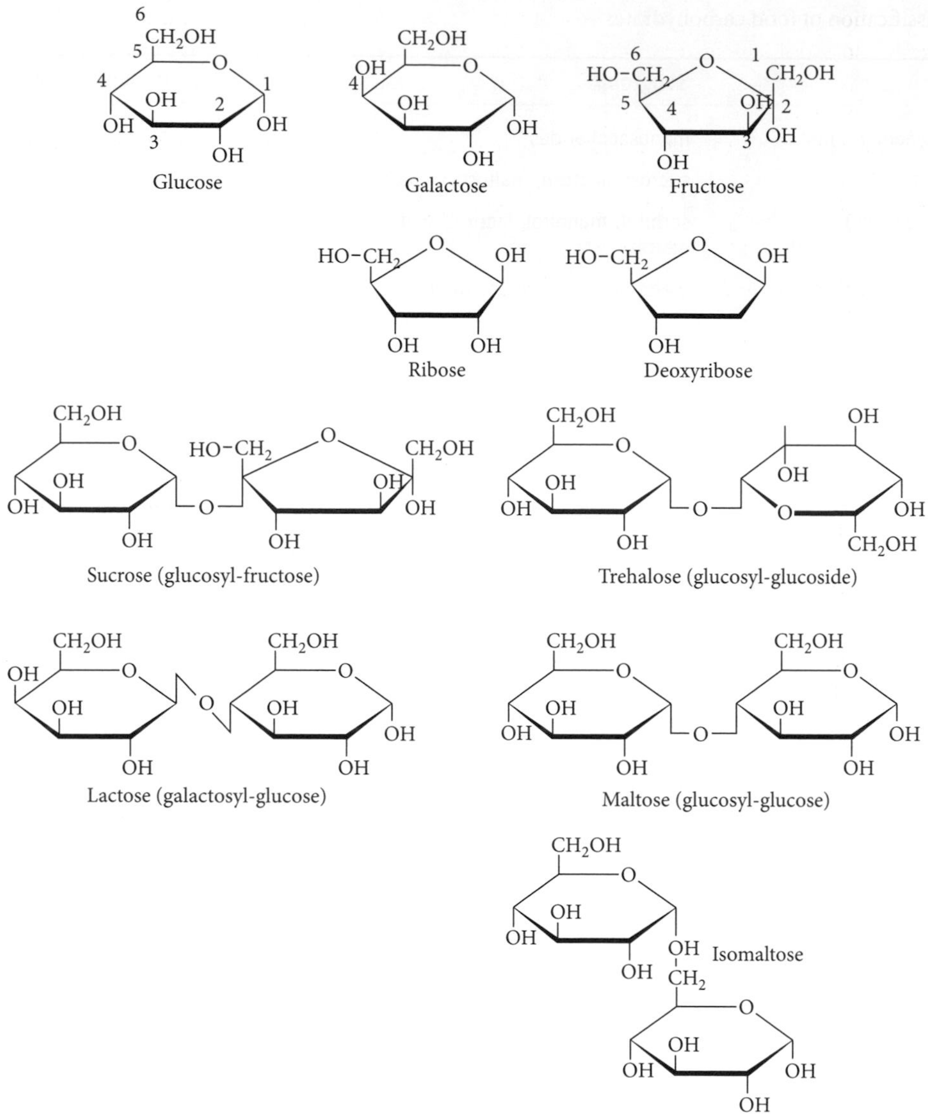

FIGURE 7.1 Structures of the main dietary carbohydrates.

- Those that are not digested in the small intestine, but pass into the large intestine, providing substrates for the colonic microflora; these are non-glycaemic carbohydrates—they do not increase the blood glucose concentration. This group includes non-starch polysaccharides, most oligosaccharides, and a proportion of starch that is resistant to digestion.

Glycaemic carbohydrates, the glycaemic index, and glycaemic load

The main glycaemic carbohydrates are glucose and fructose (monosaccharides), sucrose and lactose (disaccharides), starch and dextrins, and small amounts of other disaccharides (maltose, isomaltose, and trehalose).

The concept of the glycaemic index (GI) of foods was originally introduced to simplify exchanges between carbohydrate foods in the control of diabetes mellitus (McClenaghan 2005). The GI of a carbohydrate-containing food is the extent to which it raises the blood glucose concentration compared with an equivalent amount of a reference carbohydrate. It is calculated from the area under the blood glucose response curve after the consumption of a portion of the food containing 50 g of carbohydrate, divided by the blood glucose response after either 50 g of glucose or a standard food, commonly white bread, containing 50 g of carbohydrate.

Carbohydrate foods with a high GI generally provoke a higher secretion of insulin than those with a low GI, and both the GI of, and insulin response to, dietary carbohydrates are important in the maintenance of glycaemic control in diabetes mellitus. However, amino acids from protein digestion also stimulate insulin secretion, and foods with a low GI may nevertheless provoke a relatively large insulin response. Protein and fat influence the glycaemic response to a meal, as does the amount of liquid taken with the meal.

A food with a low GI but a high carbohydrate content will have a greater effect on blood glucose and insulin secretion than a food with a high GI but a low carbohydrate content. The glycaemic load is the product of the GI of a food multiplied by the amount of carbohydrate in a normal portion. Diets with a high GI or high glycaemic load are associated with increased risk of developing type 2 diabetes mellitus.

The GI of a food is influenced by a number of factors: the nature of the different mono- and disaccharides present; the nature of the starch (the relative amounts of amylose and amylopectin); cooking and processing, which affect both the extent of disruption of plant cell walls and the degree of gelatinization of starch. Intact kernels in cereal products hinder starch digestion, thus lowering the GI. A proportion of the starch and intrinsic sugars in foods may be enclosed by plant cell walls, which are composed of indigestible non-starch polysaccharides, and so is protected against digestion. Other factors lowering the GI are viscous solutions of non-starch polysaccharides (which hinder diffusion), organic acids (which inhibit gastric emptying and interact with starch), and amylase inhibitors. Food factors that affect the glycaemic index are listed in Table 7.2.

Starch is a polymer of glucose, containing a variable, but large, number of glucose units. There are two main forms of starch: amylose, which is a linear polymer with glucose units linked $\alpha1{\rightarrow}4$, and amylopectin, which has a branched structure, where every 30th glucose molecule has links to three others, forming a branch point with an $\alpha1{\rightarrow}6$ glycoside link. On average, 20–25 per cent of dietary starch is amylose, with the remainder as amylopectin. Glycogen, the storage carbohydrate of mammalian tissues, has a similar structure to that of amylopectin, but is more highly branched, with a branch point about every tenth glucose unit.

Uncooked starch is resistant to amylase action, because it is present as small insoluble, partly crystalline, granules. The process of cooking swells the starch granules, resulting in a gel on which amylase can act. The swelling and solubilization of starch by cooking is gelatinization. Thus, uncooked starches have very low GI, which increases with the degree of gelatinization, but falls again as foods stale and the starch undergoes retrogradation (crystallization). Because of the high activity of amylase in the small intestine, fully gelatinized starch may have a GI as high as glucose.

The concentration of blood glucose is determined by three main factors: the rate of intestinal absorption, the net liver uptake or output, and peripheral glucose uptake, which in turn depends upon insulin secretion and the sensitivity of tissues to insulin. With a constant dietary carbohydrate load, there is a range of blood glucose responses between individuals, from low responses through what is defined as impaired glucose tolerance, diabetes mellitus.

The aim of dietary treatment of diabetes mellitus is to avoid both hyperglycaemia and an excessive postprandial rise in blood glucose, which will lead to a large insulin response, with the possibility of reactive hypoglycaemia. A stable blood glucose concentration is also advantageous in relation to satiety and mood, and epidemiological studies suggest that a low glycaemic load may help to reduce the risk of developing type 2 diabetes and cardiovascular disease.

The glycaemic response to one meal may influence the response to the next. Improved glucose tolerance has been reported at lunchtime after a low GI breakfast, and in the morning after a late evening meal with low GI. The colonic fermentation of non-starch polysaccharides may be partly responsible for this, via formation of short-chain fatty acids that stimulate the secretion of anorectic hormones.

Some intervention studies, mainly in people with diabetes, but also in non-diabetics, have demonstrated a reduction in both total and LDL-cholesterol (and an increase in HDL cholesterol) with low GI diets. For further information about diet and diabetes, see Chapter 21.

TABLE 7.2 Food factors influencing the glycaemic index of foods

Structural properties of the food	Properties of the starch	Other factors
Gross structure (e.g., whole cereal grains)	Degree of gelatinization	Viscous solutions of non-starch polysaccharide
Cellular structures (e.g., leguminous seeds)	Amylose: amylopectin ratio	Organic acids (e.g., lactate, propionate)
Granular structure of starch	Crystallization retrogradation (recrystallization of gelatinized starch)	Amylase inhibitors disaccharidase inhibitors

Carbohydrate intake and obesity

Although the average requirement for carbohydrate is only about 10 per cent of energy intake, the consensus is that carbohydrates should provide 50 per cent of energy intake, with free sugars providing no more than 5 per cent. The benefit of increasing starch intake is that it will replace fat, and so help to achieve the goal of 30–35 per cent of energy derived from fat. There is no evidence that total carbohydrate intake (as opposed to free sugar intake) is a factor in development of type 2 diabetes or obesity. Indeed, increased carbohydrate (mainly starch) intake will be beneficial in terms of reducing the prevalence of obesity, since it is easier to overconsume a high fat food than one that is isocaloric but high in starch.

High carbohydrate meals lead to greater short-term satiety than low carbohydrate meals. This is mainly an effect of the bulk of starch and non-starch polysaccharides; sugars in free solution have a low satiety value, and it is easy to overconsume them.

The extent to which there is *de novo* lipogenesis from carbohydrates in human beings consuming a relatively high-fat diet is still unclear. An intake of carbohydrate in excess of energy requirements leads to obesity, but this may be more a matter of positive energy balance, with increased utilization of carbohydrate, and hence decreased utilization of lipids, so sparing adipose tissue fat reserves, rather than lipogenesis from glucose. There is evidence that *de novo* lipogenesis is depressed in response to a high fat diet, and especially by long-chain fatty acids. However, in obesity and insulin resistance, *de novo* lipogenesis accounts for half the palmitate in very low density lipoproteins (VLDL). There is evidence that fructose (including that derived from sucrose) leads to increased fatty acid synthesis and hypertriglyceridaemia.

There seems to be no difference between the efficacy of isocaloric low-fat and low-carbohydrate diets for weight reduction. Very low-carbohydrate diets (providing only 20 g of carbohydrate per day in the early stage) are effective for weight reduction, but much of their effectiveness is the result of nausea and hence loss of appetite caused by ketosis, as well as the energy cost of synthesizing glucose from amino acids, which requires metabolism of fat to provide the ATP required.

Non-glycaemic carbohydrates: non-starch polysaccharides, oligosaccharides, and dietary fibre

For many years, the non-glycaemic carbohydrates were regarded as having no nutritional importance, simply providing bulk in the diet, but now their importance to health is widely recognized. Dietary fibre is defined as all carbohydrates that are neither digested nor absorbed in the small intestine and have a degree of polymerization of three or more monomeric units, plus lignin, as measured by standardized methods.

There are three classes of non-digestible carbohydrate: non-starch polysaccharides (NSP), resistant starch (RS), and resistant oligosaccharides. In the large intestine, all are substrates for anaerobic fermentation by the intestinal microflora, leading to increased bacterial cell mass and metabolites that affect both the colon and peripheral metabolism.

NSP consists of cellulose, a glucose polymer, which is the main component of plant cell walls, and a variety of other polysaccharides of differing monomer composition, molecular size, and structure—hemicelluloses, pectins, gums, and mucilages. NSP can also be classified according to the main monomeric constituents, for example, glucans (glucose oligosaccharides), fructans (fructose oligosaccharides), and galactans (galactose oligosaccharides).

The physico-chemical properties of different components of dietary fibre vary widely, as do their physiological effects. Viscous solutions of NSP affect digestion and reduce the rate of absorption from the small intestine, attenuating postprandial rises in blood glucose and lipids. Insoluble dietary fibre is relatively resistant to fermentation and binds water in the distal large intestine, providing faecal bulk.

RS is starch that escapes digestion in the small intestine and may be a substantial fraction of the total carbohydrate delivered to the colon. Foods with a low GI also often have a high content of RS, for example, beans, but in bread and some breakfast cereals there is a substantial RS content in spite of a high GI for the digestible starch.

Resistant oligosaccharides include polymers of fructose (fructo-oligosaccharides, from onions and artichokes, as well as added inulin and oligofructose), and galactose-containing oligosaccharides from legumes.

7.3 Functions of glycaemic carbohydrates

The most obvious function of glycaemic carbohydrates is as an energy source, yielding 16 kJ (4 kcal)/g. In the fed state, most tissues metabolize glucose as their main metabolic fuel. In addition, liver and muscle synthesize the polysaccharide glycogen in the fed state, as a reserve of carbohydrate for use in the fasting state between meals. Carbohydrate metabolism is discussed in more detail in Section 7.5.

Synthesis of amino acids and other molecules from carbohydrates

The carbon skeletons of the non-essential amino acids are derived from intermediates of carbohydrate metabolism (see Chapter 9). Indeed, as indicated later (Section 7.5, 'The biosynthetic role of glycolysis'), the breakdown of glucose (glycolysis) is used to provide the carbon skeleton for the synthesis of a number of other biomolecules, including amino acids (e.g., serine), glycerol phosphate (for formation of triacylglycerol or TAG), nucleotides and lipids (e.g., phospholipids and sphingolipids). As shown in Figure 7.2, catabolism of the carbon skeletons of several amino acids yield pyruvate, the end-product of glycolysis, and others yield intermediates of the citric acid cycle that can be used for gluconeogenesis. Hence the metabolism of glucose/carbohydrates, amino acids, and lipids are all interconnected via glycolysis and/or the citric acid cycle (see Section 7.6).

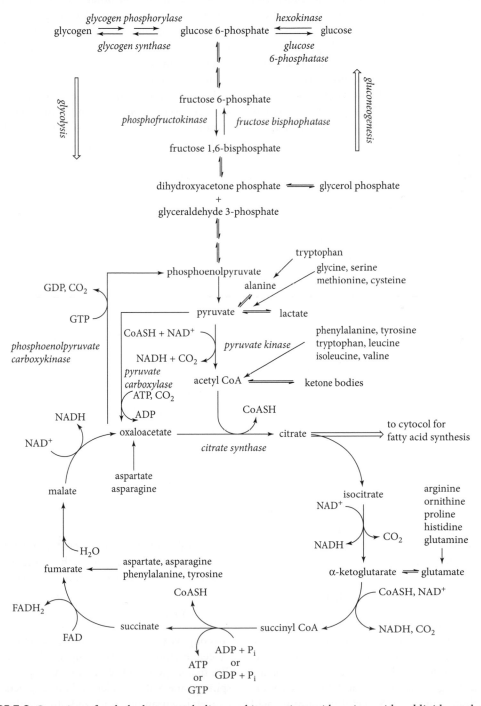

FIGURE 7.2 Overview of carbohydrate metabolism and interactions with amino acid and lipid metabolism.

Synthesis of other sugars with specific functions

The main carbohydrate entering the bloodstream from carbohydrate digestion is glucose, with smaller amounts of fructose and galactose, and traces of other monosaccharides. A number of sugars are required for specific purposes, and these can all be synthesized from glucose. Glycerol phosphate, which is required for esterification of fatty acids to triacylglycerol and phospholipids, arises from dihydroxyacetone phosphate, an intermediate in glycolysis (see Figure 7.2).

There is a requirement for relatively large amounts of two pentoses, ribose for synthesis of RNA and deoxyribose for synthesis of DNA. In addition, ribose is required for the synthesis of ATP and GTP, as well as the coenzymes NAD and NADP. The sugar alcohol, ribitol, is required for the synthesis of the flavin coenzymes derived from vitamin B_2, riboflavin. Ribose is an intermediate in, and can be a product of, the pentose phosphate pathway of glucose metabolism (see section 'The pentose phosphate pathway: an alternative to glycolysis', below). Deoxyribose and ribitol are synthesized by reduction of ribose.

Glucuronic acid is required for the conjugation of bile salts, steroid hormone metabolites, and the metabolites of a variety of xenobiotics (compounds either ingested in the diet or taken as drugs) in order to render them water-soluble for excretion in bile or urine. It is synthesized from glucose, by the oxidation of uridine diphosphate glucose. In animals for which vitamin C is not a dietary essential, glucuronic acid is metabolized onwards to ascorbate; in human beings and other animals for which ascorbate is a dietary essential, glucuronic acid is metabolized to yield the pentose sugar xylulose and its alcohol xylitol, which is further metabolized via the pentose phosphate pathway.

Genetic lack of the enzyme that reduces xylulose to xylitol results in the harmless condition of pentosuria—urinary excretion of relatively large amounts of xylulose. Xylulose is a reducing sugar, and gives a positive result when urine is tested with old-fashioned copper reagents for monitoring glycaemic control in diabetes mellitus, although it does not react with glucose oxidase, which is the basis of dip-stick tests for glucose.

Carbohydrates with special functions

The amino sugar glucosamine is synthesized from fructose 6-phosphate (an intermediate in glycolysis, see section 'Glycolysis: the (anaerobic) metabolism of glucose', below), and is the precursor for synthesis of other amino sugars, including *N*-acetyl-glucosamine, *N*-acetyl-galactosamine, and the 9-carbon sialic acid *N*-acetyl-neuraminic acid.

These amino sugars are important for the synthesis of glycoproteins, glycosaminoglycans, and glycolipids.

Glycoproteins

Glycoproteins are proteins esterified to one or more oligosaccharides containing various sugars and amino sugars; the carbohydrate content of different glycoproteins ranges between 1 and 85 per cent of the mass. They include albumin and other serum proteins, collagen in connective tissue, mucins secreted to protect the intestinal mucosa, immunoglobulins, peptide hormones, and enzymes. In addition, glycoproteins at cell surfaces are important in cell recognition, including the major blood group determinants.

Lectins are proteins that recognize, and bind to, cell surface glycoproteins, and may cause cell agglutination. A number of lectins occur in legumes and other foods, and if not denatured by adequate cooking can cause severe intestinal disorder by binding to intestinal mucosal cell surface glycoproteins.

Glycosaminoglycans and proteoglycans

Glycosaminoglycans are unbranched polysaccharides made up of repeating disaccharide units; one component of the disaccharide is an amino sugar (glucosamine or galactosamine), and the other is normally glucuronic acid. Most glycosaminoglycans are also sulphated, and apart from hyaluronic acid they all occur as proteoglycans, linked to proteins. Major glycosaminoglycans include:

- hyaluronic acid, in synovial fluid, cartilage, and loose connective tissue
- chondroitin sulphate, in cartilage and at the sites of calcification in bone
- keratan and dermatan sulphates, which also occur in cartilage, and have a critical role in maintaining the transparency of the cornea
- heparin, a proteoglycan secreted by the liver that acts as an anticoagulant.

A number of proteoglycans are found at the outer surface of cell membranes, where they function as receptors and mediators of cell–cell communication.

Glycolipids

Glycolipids (a large group of sphingolipids) are found in cell membranes, especially in the nervous system. The major glycolipids consist of a fatty acid esterified to sphingosine with covalently bound glucose, galactose, *N*-acetyl-neuraminic acid, and other amino sugars. Like glycoproteins and proteoglycans, they act as receptors and cell-surface recognition compounds.

7.4 Functions and health benefits of non-glycaemic carbohydrates

Because of the health benefits of dietary fibre discussed below, the recommendation is that total dietary fibre intake should be 30 g per day for people aged over 16; 15 g/day for children aged 2 to 5 years; 20 g/day for children aged 5 to 11 years; and 25 g/day for children aged 11 to 16 years (SACN 2015).

Non-glycaemic carbohydrates (dietary fibre) may influence digestion and absorption in the small intestine and provide faecal bulk and substrates for the intestinal microflora (Anderson et al. 2009). The fermentation products, mainly short-chain fatty acids, are important for the function and health of the colon as well as effects on host metabolism. Insoluble, especially lignified, dietary fibre, as found in cereal bran, is largely resistant to fermentation. It has laxative effects due to binding water in the distal colon. Fermented carbohydrates also contribute to the faecal bulk through an increased bacterial mass. The increase in faecal mass ranges from 1 g per gram of ingested pectin to 6 g per gram of wheat bran fibre.

Fermentation in the large intestine

Carbohydrates that are not digested in the small intestine are substrates for anaerobic fermentation by the colonic microflora. The rate, extent, and site of fermentation in the colon are dependent on both the substrate and host factors—the molecular structure and physical form of the carbohydrates, the mix of bacterial flora, and intestinal transit time. The main products of fermentation are short-chain fatty acids (acetate, propionate, and butyrate), and gases, (carbon dioxide, hydrogen, and methane).

The decrease in pH of the colon contents resulting from the formation of these short-chain fatty acids may reduce the formation of bile salt metabolites that have been implicated in carcinogenesis. A low pH may also enhance colonic absorption of calcium. Butyrate is a metabolic fuel for colonocytes, with effects on cell differentiation and apoptosis that may be protective against cancer. Acetate and propionate are absorbed and metabolized in the liver and other tissues and have effects on carbohydrate and lipid metabolism. Propionate inhibits liver cholesterol synthesis in experimental animals and is a substrate for gluconeogenesis. Resistant starch and oat bran yield more butyrate than other substrates.

Depending on the extent of fermentation and the mixture of short-chain fatty acids produced, fermentation of non-glycaemic carbohydrates results in an energy yield of 4–8 kJ (1–2 kcal)/g.

There are receptors for short-chain fatty acids in intestinal cells, and other tissues. Short-chain fatty acids arising from bacterial fermentation have a satiating effect (Byrne et al. 2015), due to increased secretion of the anorectic gut hormones glucagon-like peptide 1 (GLP-1) and peptide YY (PYY), so reducing food intake. In addition, they lead to increased energy expenditure and decreased lipogenesis, so improving energy homeostasis and reducing obesity. They also increase secretion of leptin, so inhibiting proliferation of adipocytes.

The intestinal microflora contains several hundred different species, some of which are beneficial, while others are potentially pathogenic. The name prebiotics has been coined for non-glycaemic carbohydrates that stimulate the growth or activity of bacteria in the colon that have the potential to improve host health, and out-compete pathogens, and prevent them colonizing the large intestine. Such beneficial micro-organisms are known as probiotics, and there is some evidence that they enhance immune function. Fructo-oligosaccharides (inulin and shorter oligosaccharides) as well as other oligosaccharides, (galacto-oligosaccharides, resistant malto-oligosaccharides, and resistant starch) have prebiotic effects (Bosscher et al. 2006), including increasing the growth of *Bifidobacteria* and *Lactobacillus* spp.

Oligosaccharides and unabsorbed sugar alcohols are fermented rapidly in the proximal colon, and gas formation can lead to abdominal discomfort and flatulence if the intake exceeds some 20 g/day. Resistant starch is fermented more slowly and does not cause flatulence. Lactose may also cause intolerance in some people with low intestinal lactase activity if the intake exceeds 5–10 g/day. Otherwise, unabsorbed lactose can also be regarded as a prebiotic carbohydrate with potentially beneficial effects in stimulating what is perceived as a 'healthy' intestinal microflora (Bode 2009).

Effects of non-starch polysaccharides and oligosaccharides in the upper gastrointestinal tract

Some oligosaccharides inhibit amylase and have been marketed as slimming aids (so-called 'starch blockers'), although there is little evidence of efficacy.

Total and LDL blood cholesterol are reduced by viscous solutions of NSP, such as pectins, gums, and mucilages—especially guar gum, which is a galactomannan. These effects are related to reduced absorption of cholesterol and bile salts, helping to lower the body pool of cholesterol. Viscous solutions of NSP also attenuate the postprandial rise in blood glucose and insulin responses, and thus lower the glycaemic index of foods. This permits improved glycaemic control in diabetes mellitus. They act by forming diffusion barriers due to increased viscosity of the intestinal contents, hindering both amylase action in the lumen and the absorption of monosaccharides. Inhibition of gastric emptying may also contribute.

Foods rich in fibre inhibit the absorption of iron, zinc, and calcium. However, this is mainly due to phytate (inositol hexaphosphate) associated with the fibre in cereal bran and legumes, and can be abolished by the degradation of phytate (e.g., by the action of phytase from yeast in bread making, and phytase secreted in the small intestine). Calcium absorption is enhanced by the acidic products of fibre fermentation.

Non-starch polysaccharides and chronic diseases

There is good evidence that high intakes of fibre (especially cereal fibre and wholegrain cereals, Smith & Tucker 2011) are protective against colorectal cancer, cardiometabolic disease, and type 2 diabetes, as well as increased faecal mass and decreased intestinal transit time. High intakes of β-glucans (from oat bran) lead to decreased circulating total and LDL cholesterol and triacylglycerols, as well as reduced blood pressure. It is difficult to disentangle the effects of fibre from the effects of phytonutrients present in the fruits, vegetables, and whole-grain cereals that are the sources of fibre, or the reduction in fat intake that is implied by a diet rich in carbohydrate and fibre. For this reason it is considered that the recommended intake of dietary fibre be from fibre-rich foods rather than added isolated fibre.

A high intake of fibre increases faecal bulk, both as a result of the water-binding capacity of polysaccharides that are not fermented, and also as a result of the increase in bacterial mass that results from providing additional fermentable substrates. This increased faecal bulk enhances peristalsis and reduces straining on defecation, and so reducing the risk of developing diverticular disease of the colon and haemorrhoids. A high intake of fibre may reduce the risk of colorectal and other cancers by dilution of the intestinal contents (and reduced intestinal transit time), so reducing exposure to, and absorption of, dietary carcinogens. In addition, many potential carcinogens will bind to cellulose and other components of fibre, so preventing their absorption.

When there is plenty of fermentable carbohydrate available, intestinal bacteria will use all the available amino acids for synthesis of bacterial proteins and biomass; when the bacteria have limited carbohydrate available for fermentation they also ferment amino acids, yielding a variety of potentially carcinogenic (and foul smelling) end-products. Perhaps more importantly, the butyrate produced by fermentation of NSP and RS is the preferred fuel for colon enterocytes. It regulates cell growth and induces proliferation and apoptosis, so providing protection against colorectal cancer.

There is evidence from both epidemiological and intervention studies that fibre may be an important factor in combatting obesity. This is mainly the effect on satiety of the greater bulk of food consumed, but the products of bacterial fermentation may also affect satiety. A number of epidemiological studies have shown a lower risk of coronary heart disease with increased intake of whole-grain cereals and fruit and vegetables. This may relate to the lowering of serum cholesterol by viscous solutions of NSP described above, but could also relate to the protective effects of compounds (e.g., phytonutrients) found in whole-grain cereals and fruits. There is epidemiological evidence for a protective effect of fibre against type 2 diabetes. There are two factors involved here. A diet high in fibre is associated with lower incidence of obesity, and obesity (especially abdominal obesity) is a major risk factor in the development of insulin resistance and type 2 diabetes. In addition, viscous solutions of NSP slow absorption and reduce the glycaemic index of foods. Intervention trials show a beneficial effect of viscous NSP, while prospective studies show a beneficial effect of whole-grain cereals.

Carbohydrate malabsorption and disaccharide intolerance

High intakes of sugar alcohols and fructose may cause gastrointestinal upset, since they are only slowly absorbed in the small intestine and after a relatively large intake a significant amount may enter the colon to undergo rapid fermentation, producing relatively large volumes of gases and increasing the osmolality of the intestinal contents, so causing osmotic diarrhoea.

The disaccharide lactulose is an isomer of lactose that is not digested. It is used as a mild osmotic laxative. In the colon it is fermented to 4 mol of lactic acid, so increasing the osmolality of the intestinal contents. It is also used clinically to treat hyperammonaemia, as occurs in liver failure. Here it is the protons formed by fermentation that are important in lowering the pH of the intestinal contents. While most blood ammonia is present as ammonium ions, that cannot cross the intestinal wall, there is equilibrium with a small amount of ammonia (NH_3) that can diffuse into the intestinal lumen. Here it is trapped as ammonium ions because of the lower pH.

Malabsorption of sugars as a result of lack of disaccharidase activity may also cause excessive delivery of readily fermentable substrate to the colon, leading to abdominal bloating, cramps and watery diarrhoea (Swallow 2003). Genetic lack of lactase (alactasia) is rare, but can lead to severe diarrhoea and consequent malnutrition in affected infants (breast milk and most infant milk formula contains about 7 per cent lactose, providing half the energy). Genetic lack of sucrase is also rare, causing diarrhoea and malnutrition when sucrose is introduced into the diet. Lack of sucrase is relatively common among the Inuit (Swallow 2003), but only became a problem as sucrose-containing foods and beverages were introduced—there are no sources of sucrose in the traditional Inuit diet.

In most ethnic groups, apart from people of northern European origin, intestinal lactase is lost gradually through late adolescence and early adult life. Some, but not all, people with low lactase activity experience gastrointestinal symptoms due to fermentation of unabsorbed lactose in the colon. Many adults can drink 1–200 ml of milk (containing 5–10 g of lactose) without symptoms. In yoghurt much of the lactose has been fermented to lactic acid, and cheese is almost lactose free. These products are therefore normally well tolerated.

7.5 The metabolism of glycaemic carbohydrates

Figure 7.2 shows an overview of carbohydrate metabolism. The first stage is glycolysis, which results in the cleavage of glucose to yield 2 mol of the 3-carbon compound pyruvate and a net yield of 7 × ATP under aerobic conditions. Glycolysis is (indirectly) reversible, so that glucose can be synthesized from non-carbohydrate precursors that yield pyruvate (see section 'Gluconeogenesis', below). Pyruvate arising from glycolysis enters the mitochondria and is oxidized and decarboxylated to yield 2.5 × ATP per pyruvate (and hence 5 per glucose) and acetyl CoA, which may undergo one of two fates:

- complete oxidation in the citric acid cycle, yielding 10 × ATP per acetyl CoA (and hence 20 per glucose);
- utilization for synthesis of fatty acids and cholesterol (and hence steroid hormones and the bile acids) by way of formation of citrate that is exported from the mitochondria into the cytosol.

Under anaerobic conditions there is a net yield of only 2 × ATP per glucose from glycolysis, with lactate being synthesized from the pyruvate instead. This anaerobic glycolysis is important in muscle under conditions of maximum exertion, and in red blood cells at all times.

Glycolysis: the (anaerobic) metabolism of glucose

Glycolysis occurs in the cytosol; overall it results in cleavage of the 6-carbon glucose molecule into two 3-carbon units. The key steps in the pathway are:

- two phosphorylation reactions to form fructose-bisphosphate;
- cleavage of fructose-bisphosphate to yield two molecules of triose (3-carbon sugar) phosphate;
- two steps in which phosphate is transferred from a substrate onto ADP, forming ATP (4 × ATP formed per glucose);
- one step in which NAD^+ is reduced to NADH (2 mol of NADH formed per glucose);
- formation of 2 mol of pyruvate per mol of glucose metabolized.

The immediate substrate for glycolysis is glucose 6-phosphate; this may arise from two sources:

- by phosphorylation of glucose, at the expense of ATP, catalysed by hexokinase (and by glucokinase in the liver in the fed state);
- by phosphorolysis of glycogen in liver and muscle to yield glucose 1-phosphate, catalysed by glycogen phosphorylase, using inorganic phosphate as the phosphate donor. Glucose 1-phosphate is readily isomerized to glucose 6-phosphate.

The pathway of glycolysis is shown in Figure 7.3. Although the aim of glucose oxidation is to form ATP by phosphorylation of ADP, there is a modest cost of 2 × ATP to initiate glucose metabolism. There are two steps in which ATP is used, one to form glucose 6-phosphate when glucose is the substrate, and the other to form fructose bisphosphate.

Glucose 6-phosphate is isomerized to fructose 6-phosphate, which is then phosphorylated to fructose bisphosphate. The formation of fructose bisphosphate, catalysed by phosphofructokinase, is the main regulatory step in glucose metabolism. Fructose bisphosphate is then cleaved into two 3-carbon compounds, dihydroxyacetone phosphate and glyceraldehyde 3-phosphate, which are interconvertible. The onward metabolism of these 3-carbon sugars is linked to both the reduction of NAD^+ to NADH, and direct (substrate-level) phosphorylation of ADP to form ATP. The result is the formation of 2 mol of pyruvate from each mol of glucose.

As discussed below in section 'Gluconeogenesis', the reverse of the glycolytic pathway is important as a means of glucose synthesis—the process of gluconeogenesis. Most of the reactions of glycolysis are readily reversible,

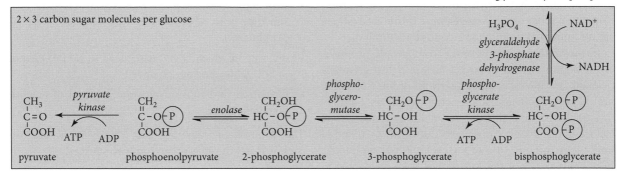

FIGURE 7.3 Glycolysis.

but at three points (the reactions catalysed by hexokinase, phosphofructokinase, and pyruvate kinase) there are separate enzymes involved in glycolysis and gluconeogenesis. The glycolytic pathway also provides a route for the metabolism of fructose, galactose (which is phosphorylated to galactose 1-phosphate and isomerized to glucose 1-phosphate), and glycerol. Some fructose is phosphorylated directly to fructose 6-phosphate by hexokinase, but most is phosphorylated to fructose 1-phosphate by fructokinase. Fructose 1-phosphate is then cleaved to dihydroxyacetone phosphate and glyceraldehyde, which is phosphorylated to glyceraldehyde 3-phosphate by triose kinase. The metabolism of galactose and fructose occurs in the intestinal mucosa and liver, so little fructose or galactose normally reaches the peripheral circulation. Ethanol inhibits the metabolism of galactose, leading to increased plasma levels and galactosuria.

Glycerol, arising from the hydrolysis of triacylglycerols, can be phosphorylated and oxidized to dihydroxyacetone phosphate. Glycerol phosphate, which is important for triacylglycerol synthesis, is formed from dihydroxyacetone phosphate.

Glycolysis under aerobic conditions

Glycolysis occurs in the cytosol, but the oxidation of NADH, linked to phosphorylation of ADP to ATP, occurs inside the mitochondria. The mitochondrial inner membrane is impermeable to NAD, and two substrate shuttles

are used to oxidize cytosolic NADH, transport reduced substrates into the mitochondrion, and reoxidize them at the expense of mitochondrial coenzymes. These are then reoxidized by the electron transport chain, linked to phosphorylation of ADP to ATP. The two substrate shuttles operate as follows:

- The malate–aspartate shuttle involves reduction of oxaloacetate by NADH in the cytosol to form malate, which enters the mitochondria and is reduced back to oxaloacetate (forming NADH). Oxaloacetate is then transaminated to aspartate for export back into the cytosol.

- The glycerol phosphate shuttle involves the reduction of dihydroxyacetone phosphate by NADH in the cytosol to form glycerol 3-phosphate, which enters the mitochondria and is oxidized back to dihydroxyacetone phosphate (reducing FAD). Dihydroxyacetone phosphate is then exported back into the cytosol.

Glycolysis under anaerobic conditions: the Cori cycle

In red blood cells, which lack mitochondria, the NADH formed in glycolysis cannot be reoxidized aerobically. Similarly, under conditions of maximum exertion, for example in sprinting, the rate at which oxygen can be taken up into muscle is not great enough to allow for the reoxidation of all the NADH that is formed. In order to maintain glycolysis and the formation of ATP, NADH is oxidized to NAD$^+$ by the reduction of pyruvate to lactate, catalysed by lactate dehydrogenase.

The resultant lactate is exported from the muscle and red blood cells, and taken up by the liver, where it is used for the resynthesis of glucose. This is the Cori cycle (Figure 7.4), an inter-organ cycle of glycolysis in muscle and gluconeogenesis in liver. The synthesis of glucose from lactate is an ATP (and GTP) requiring process. The oxygen debt after strenuous physical activity is due to an increased rate of energy-yielding metabolism to provide the ATP and GTP that are required for gluconeogenesis from lactate. While most of the lactate will be used for gluconeogenesis, a proportion will have to undergo oxidation to CO_2 in order to provide the ATP and GTP required. Lactate may also be taken up by other tissues, where oxygen availability is not a limiting factor, such as the heart. Here it is oxidized to pyruvate, which is then a substrate for complete oxidation to carbon dioxide and water, via the citric acid cycle (see section 'Oxidation of acetyl CoA', below).

The biosynthetic role of glycolysis

Many tumours have a poor blood supply and hence a low capacity for oxidative metabolism, so that much of their energy-yielding metabolism is anaerobic. Lactate produced by the tumours is exported to the liver for gluconeogenesis. The energy cost of this increased cycling of glucose between anaerobic glycolysis in the tumour and gluconeogenesis in the liver accounts for much of the weight loss (cachexia) that is seen in patients with advanced cancer.

However, in cancer cells the increased rates of glycolysis is used to synthesize other biomolecules needed for

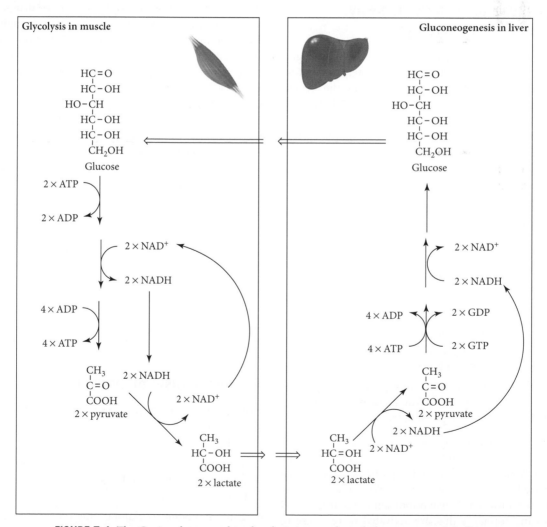

FIGURE 7.4 The Cori cycle: anaerobic glycolysis in muscle and gluconeogenesis in liver.

cell proliferation rather than simply to provide energy. This was originally called the Warburg effect, named after the Nobel Prize winning scientist Otto Warburg who first described the phenomenon whereby cancer cells have high rates of glycolysis even under aerobic conditions, resulting in increased production of lactic acid (Rosenzweig et al. 2018). Subsequent research showed that the increased flux through glycolysis was also associated with an increase in one carbon metabolism, including increased synthesis of serine, nucleotides, and sphingolipids, as well as impacting on methylation reactions and the generation of reducing factors. Whether this reprogramming of metabolism takes place in normal (non-cancerous) cells is currently under investigation, but there are indications that glycolysis might also be used at certain times for the synthesis of biomolecules in other cells and tissues.

The pentose phosphate pathway: an alternative to glycolysis

There is an alternative pathway for the conversion of glucose 6-phosphate to fructose 6-phosphate, the pentose phosphate pathway (sometimes known as the hexose monophosphate shunt), shown in Figure 7.5. Overall, the pathway produces 2 mol of fructose 6-phosphate, 1 mol of glyceraldehyde 3-phosphate and 3 mol of carbon dioxide from 3 mol of glucose 6-phosphate, linked to the reduction of 6 mol of $NADP^+$ to NADPH. The sequence of reactions is as follows:

- 3 mol of glucose are oxidized to yield 3 mol of the 5-carbon sugar ribulose 5-phosphate and 3 mol of carbon dioxide;
- 2 mol of ribulose 5-phosphate are isomerized to yield 2 mol of xylulose 5-phosphate;
- 1 mol of ribulose 5-phosphate is isomerized to ribose 5-phosphate;
- 1 mol of xylulose 5-phosphate reacts with the ribose 5-phosphate, yielding (ultimately) fructose 6-phosphate and erythrose 4-phosphate;
- The other mol of xylulose-5-phosphate reacts with the erythrose 4-phosphate, yielding fructose 6-phosphate and glyceraldehyde 3-phosphate.

This is also the pathway for the synthesis of ribose for nucleotide synthesis, and the source of about half the NADPH required for fatty acid synthesis. Tissues that synthesize fatty acids have a high activity of the pentose phosphate pathway. It is also important in the respiratory burst of macrophages that are activated in response to infection.

The pentose phosphate pathway in red blood cells: favism

NADPH is required in red blood cells to maintain an adequate pool of reduced glutathione, which is used to remove hydrogen peroxide. The tripeptide glutathione (γ-glutamyl-cysteinyl-glycine) is the reducing agent for glutathione peroxidase, which reduces H_2O_2 to H_2O and O_2. Oxidized glutathione (GSSG) is reduced back to GSH by glutathione reductase, which uses NADPH as the reducing agent. Glutathione reductase is a flavin-dependent enzyme, and its activity in red blood cells can be used as an index of vitamin B_2 status (see Chapter 11, Section 11.3).

Partial lack of glucose 6-phosphate dehydrogenase (and hence impaired activity of the pentose phosphate pathway) is the cause of favism; acute haemolytic anaemia with fever and haemoglobinuria. It is precipitated in genetically susceptible people by the consumption of broad beans (fava beans) and a variety of drugs, all of which, like the toxins in fava beans, undergo redox cycling, producing hydrogen peroxide. Infection can also precipitate an attack, because of the increased production of oxygen radicals as part of the macrophage respiratory burst. Favism is one of the commonest genetic defects; an estimated 200 million people worldwide are affected. It is an X-linked condition, and female carriers are resistant to malaria; this advantage presumably explains why defects in the gene are so widespread. Because of the low activity of the pentose phosphate pathway in affected people, there is a lack of NADPH in red blood cells, and hence an impaired ability to remove hydrogen peroxide, which causes oxidative damage to the membrane lipids, leading to a haemolytic crisis. There are two main variants of favism. In the most severe form, there is very low activity of glucose 6-phosphate dehydrogenase in all red blood cells. In the less severe form of the disease, the enzyme is unstable, so it is only older red blood cells that are affected. Other tissues are unaffected because there are mitochondrial enzymes that can provide a supply of NADPH; but red blood cells have no mitochondria.

The metabolism of pyruvate

Pyruvate arising from glycolysis can be metabolized in four different ways, depending on the metabolic state of the body:

- Reduction to lactate under anaerobic conditions (see section 'Glycolysis under anaerobic conditions: the Cori cycle', above).
- Transamination to alanine; this provides a pathway for the indirect utilization of muscle glycogen as a

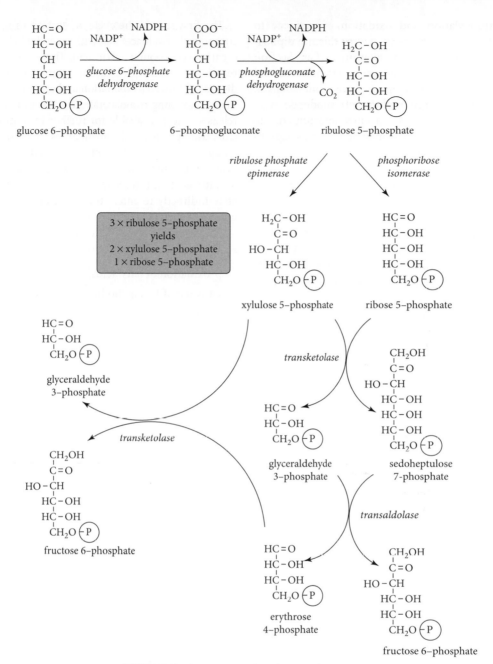

FIGURE 7.5 The pentose phosphate pathway.

source of blood glucose in the fasting state, the glucose-alanine cycle (see section 'Glycogen utilization' below).

- As a substrate for gluconeogenesis (see section 'Gluconeogenesis', below). This also provides a pathway for gluconeogenesis from pyruvate arising from amino acids (see Figure 7.2).

- Oxidation to acetyl CoA, followed by complete oxidation to carbon dioxide and water through the citric acid cycle (see section 'Oxidation of acetyl CoA: the citric acid cycle', below).

The oxidation of pyruvate to acetyl CoA

The first step in the complete oxidation of pyruvate is a reaction in which carbon dioxide is lost, and the resulting 2-carbon intermediate is oxidized to acetate, linked to the reduction of NAD^+ to NADH. Since 2 mol of pyruvate are formed from each mol of glucose, this step represents the formation of 2 mol of NADH, equivalent to 5 × ATP, for each mol of glucose metabolized. The acetate is released from the enzyme esterified to coenzyme A, as acetyl CoA.

The decarboxylation and oxidation of pyruvate to form acetyl CoA requires the coenzyme thiamin diphosphate, which is formed from vitamin B_1 (see Section 11.2). In thiamin deficiency this reaction is impaired, and deficient subjects have impaired glucose metabolism. Especially after a test dose of glucose or moderate exercise they develop high blood concentrations of pyruvate and lactate. In some cases this may be severe enough to cause life-threatening acidosis.

Oxidation of acetyl CoA: the citric acid cycle

The acetate of acetyl CoA undergoes a stepwise oxidation to carbon dioxide and water in a cyclic pathway, called the citric acid cycle or tricarboxylic acid (TCA) cycle, shown in Figure 7.2 (and Figure 7.6). This pathway is sometimes known as the Krebs' cycle, after its discoverer, Sir Hans Krebs. For each mol of acetyl CoA oxidized in this pathway, there is a yield of 10 ATP equivalents, and hence 20 per glucose.

The first step in the cycle is reaction of the 4-carbon compound, oxaloacetate, with acetyl CoA to form a 6-carbon compound, citric acid. There is then a series of reactions in which two carbon atoms are lost as carbon dioxide, followed by oxidation and other reactions, eventually reforming oxaloacetate. The CoA of acetyl CoA is released and is available for further formation of acetyl CoA from pyruvate. Citrate in excess of requirements for energy-yielding metabolism is exported from mitochondria to the cytosol where it is cleaved to yield acetyl CoA for fatty acid and sterol synthesis. The resultant oxaloacetate indirectly re-enters the mitochondrion for further uptake of acetyl CoA.

The citric acid cycle is also involved in the oxidation of acetyl CoA arising from other sources:

- β-oxidation of fatty acids;
- cleavage of ketone bodies in muscle and other extrahepatic tissues;
- oxidation of alcohol;

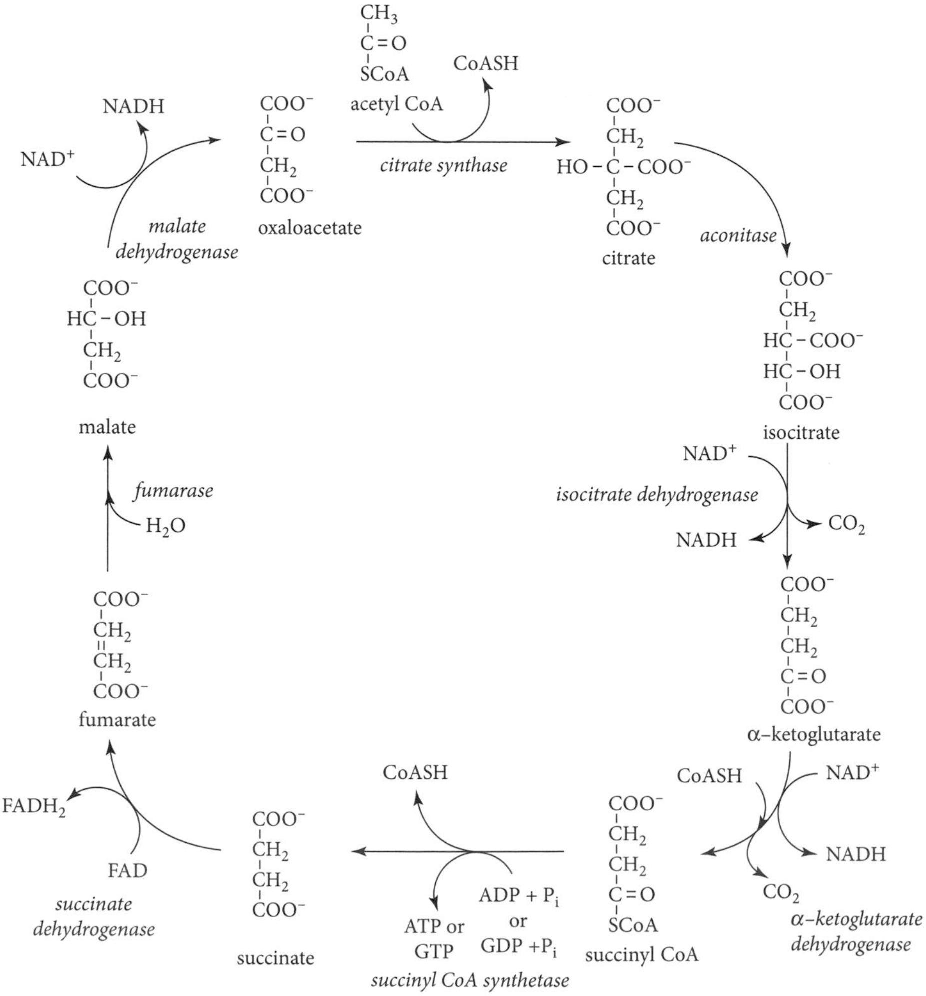

FIGURE 7.6 The citric acid cycle.

- acetate entering from the gut as a result of bacterial fermentation of NSP;
- those amino acids that give rise to acetyl CoA or acetoacetate (the ketogenic amino acids, see Chapter 9).

Citrate is isomerized to isocitrate, followed by two reactions involving both oxidation (linked to reduction of NAD^+ to NADH) and decarboxylation, yielding α-ketoglutarate, then the 4-carbon succinyl moiety of succinyl CoA. Succinyl CoA loses its CoA to yield succinate in a reaction linked to phosphorylation of either GDP or ADP. The GTP-linked reaction occurs only in tissues that are capable of gluconeogenesis, and serves to control the rate of removal of oxaloacetate from the cycle for glucose synthesis.

The sequence of reactions between succinate and oxaloacetate is chemically the same as that involved in the β-oxidation of fatty acids. Oxidation of succinate yields fumarate, linked to the reduction of a flavin. Addition of water across the carbon–carbon double bond of fumarate yields malate, which is oxidized in a reaction linked to reduction of NAD^+, yielding oxaloacetate.

The citric acid cycle as pathway for metabolic interconversion

In addition to its role in oxidation of acetyl CoA, the citric acid cycle is an important central metabolic pathway, providing the link between carbohydrate, fat, and amino acid metabolism. Many of the intermediates can be used for the synthesis of other compounds:

- α-Ketoglutarate and oxaloacetate can give rise to the amino acids glutamate and aspartate respectively (and from these a variety of other non-essential amino acids)
- Oxaloacetate is the precursor for glucose synthesis in the fasting state (see section 'Gluconeogenesis', below)
- Citrate formed in the mitochondria is exported to the cytosol to provide acetyl CoA for fatty acid and sterol synthesis.

If oxaloacetate is removed from the cycle for gluconeogenesis, it must be replaced, since if there is not enough oxaloacetate available to form citrate, the rate of acetyl CoA metabolism, and hence the rate of formation of ATP, will slow down. A variety of amino acids give rise to citric acid cycle intermediates, so replenishing the cycle and permitting the removal of oxaloacetate for gluconeogenesis. In addition, the direct conversion of pyruvate to oxaloacetate by pyruvate carboxylase can be a major source of oxaloacetate to maintain citric acid cycle activity. The removal of oxaloacetate for gluconeogenesis is controlled by phosphoenol-pyruvate carboxykinase (PEPCK) enzymes (see section 'Gluconeogenesis' below) that require GTP to catalyse the decarboxylation and phosphorylation of oxaloacetate to form phosphoenolpyruvate (PEP), the main substrate for gluconeogenesis. In tissues which are active in gluconeogenesis, such as liver and kidney, the major source of GTP in mitochondria is the reaction of succinyl CoA→succinate. If so much oxaloacetate were withdrawn that the rate of cycle activity fell, there would be inadequate GTP to permit further removal of oxaloacetate. In tissues such as brain and heart, which do not carry out gluconeogenesis, this reaction is linked to phosphorylation of ADP rather than GDP.

Glycogen as a carbohydrate reserve

The main carbohydrate reserve in the body is glycogen: some 50–120 g in the liver (depending on whether the person is in the fed or fasting state) and 350–400 g in muscle. Other tissues contain small amounts of glycogen to meet their short-term needs.

Glycogen is a branched polymer of glucose, with essentially the same structure as amylopectin, except that it is more highly branched, with a branch point about every tenth glucose unit. The branched structure of glycogen means that it binds a considerable amount of water within the molecule. In the early stages of food restriction there is depletion of muscle and liver glycogen, with the release and excretion of this trapped water. This leads to an initial rate of weight loss that is very much greater than can be accounted for by catabolism of adipose tissue, and, of course, it cannot be sustained—once glycogen has been depleted the rapid loss of water (and weight) will cease.

Glycogen synthesis

In the fed state, glycogen is synthesized from glucose in both liver and muscle. The reaction is a stepwise addition of glucose units onto the glycogen that is already present. The pathway is shown in Figure 7.7.

Glycogen synthesis involves the intermediate formation of UDP-glucose (uridine diphosphate glucose) by reaction between glucose 1-phosphate and UTP (uridine triphosphate). As each glucose unit is added to the growing glycogen chain, so UDP is released, and must be rephosphorylated to UTP by reaction with ATP. There is thus a significant cost of ATP in the synthesis of glycogen: 2 mol of ATP are converted to ADP + phosphate for each glucose unit added, and overall the energy cost of glycogen synthesis may account for 5 per cent of the energy yield of the carbohydrate stored. Glycogen synthase

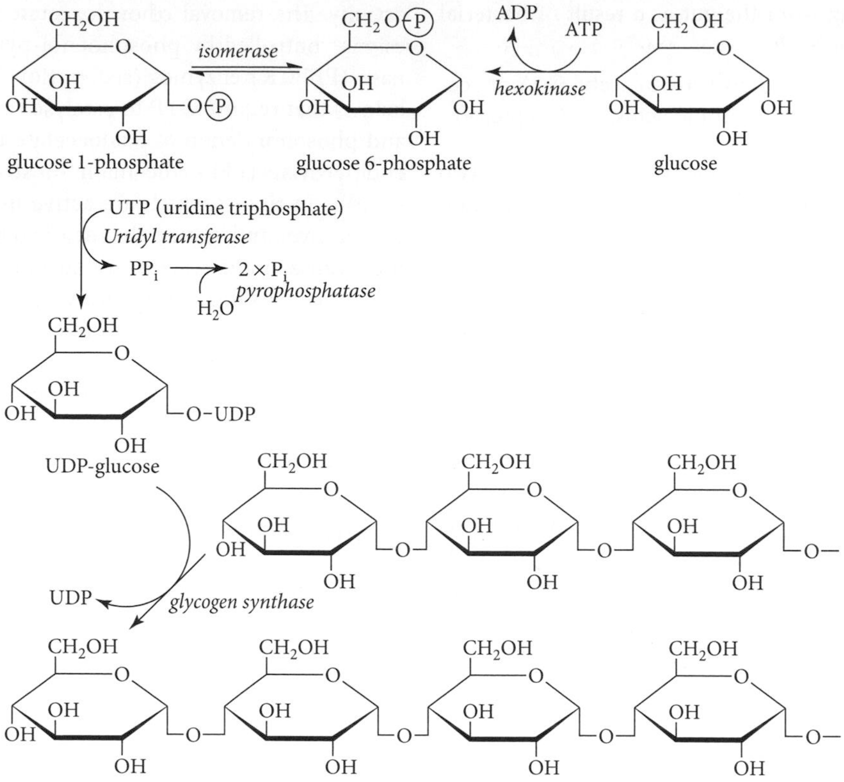

FIGURE 7.7 Glycogen synthesis.

forms only the α1→4 links that form the straight chains of glycogen. The branch points are introduced by the transfer of 6–10 glucose units in a chain from carbon-4 to carbon-6 of the glucose unit at the branch point.

Glycogen utilization

In the fasting state, glycogen is broken down by the removal of glucose units one at a time from the many ends of the molecule. The reaction is a phosphorolysis—cleavage of the glycoside link between two glucose molecules by the introduction of phosphate. The product is glucose 1-phosphate, which is then isomerized to glucose 6-phosphate. In the liver glucose 6-phosphatase catalyses the hydrolysis of glucose 6-phosphate to free glucose, which is exported for use by the brain and red blood cells.

Muscle cannot release free glucose from the breakdown of glycogen since it lacks glucose 6-phosphatase. However, muscle glycogen can be an indirect source of blood glucose in the fasting state. Glucose 6-phosphate from muscle glycogen undergoes glycolysis to pyruvate, which is then transaminated to alanine. Alanine is exported from muscle and taken up by the liver for use as a substrate for gluconeogenesis (see section 'Gluconeogenesis', below). Glycogen phosphorylase stops cleaving α1→4 links four glucose residues from a branch point, and a

debranching enzyme catalyses the transfer of a three glucosyl unit from one chain to the free end of another chain. The α1→6 link is then hydrolysed by a glucosidase, releasing glucose. The branched structure of glycogen means that there are a great many end points at which glycogen phosphorylase can act; in response to stimulation by adrenaline there can be a very rapid release of glucose 1-phosphate from glycogen.

Gluconeogenesis: the synthesis of glucose from non-carbohydrate precursors

The pathway of gluconeogenesis is essentially the reverse of the pathway of glycolysis (see section 'Glycolysis: the (anaerobic) metabolism of glucose', above). However, at three steps there are separate enzymes involved in the breakdown of glucose (glycolysis) and gluconeogenesis. The reactions of pyruvate kinase, phosphofructokinase, and hexokinase cannot readily be reversed (i.e., they have equilibria that are strongly in the direction of the formation of pyruvate, fructose bisphosphate, and glucose 6-phosphate respectively). There are therefore separate enzymes, under distinct metabolic control, for the reverse of each of these reactions in gluconeogenesis.

Pyruvate is converted to phosphoenolpyruvate for glucose synthesis by a two-step reaction, with the intermediate formation of oxaloacetate. Pyruvate is carboxylated to oxaloacetate by pyruvate carboxylase in an ATP-dependent reaction in which the vitamin biotin is the coenzyme. This reaction can also be used to replenish oxaloacetate in the citric acid cycle when intermediates have been withdrawn for use in other pathways. Oxaloacetate then undergoes a phosphorylation reaction, catalysed by either of two PEPCK enzymes, one being the cytosolic isoform, PEPCK-C (encoded by the Pck1 gene) and the other the mitochondrial isoform, PEPCK-M (encoded by the Pck2 gene). They both catalyse the conversion of oxaloacetate to phosphoenolpyruvate (PEP), but in different cell compartments, with PEP then being the main substrate for gluconeogenesis. It is unclear as to the relative importance of the two PEPCK enzymes, although PEPCK-C is the one normally associated with hepatic gluconeogenesis. However, there is increasing evidence that PEPCK-M is upregulated in certain cancers and therefore may play a role in the metabolic reprogramming described earlier (see section 'The biosynthetic role of glycolysis'). It should be noted that PEPCK-M interacts directly with the oxaloacetate within the mitochondria and it is the PEP that then crosses the mitochondrial membrane. By contrast the oxaloacetate in the mitochondria must be converted to malate (by malic enzyme in the mitochondria) to be able to cross the mitochondrial membrane, and then is converted back to oxaloacetate (by malic enzyme in the cytosol) to be a substrate for PEPCK-C.

The only source of GTP in mitochondria is the reaction of succinyl CoA → succinate; this provides a link between citric acid cycle activity and gluconeogenesis, preventing excessive withdrawal of oxaloacetate for gluconeogenesis if citric acid cycle activity would be impaired.

- Fructose bisphosphate is hydrolysed to fructose 6-phosphate by a simple hydrolysis reaction catalysed by the enzyme fructose bisphosphatase.

- Glucose 6-phosphate is hydrolysed to free glucose and phosphate by the action of glucose 6-phosphatase.

The other reactions of glycolysis are readily reversible, and the overall direction of metabolism, either glycolysis or gluconeogenesis, depends mainly on the relative activities of phosphofructokinase and fructose bisphosphatase.

The main substrates for hepatic gluconeogenesis are lactate (from anaerobic glycolysis in muscle or red blood cells), glycerol (from adipose tissue lipolysis), or amino acids (from muscle glycolysis or proteolysis). Lactate, the volatile fatty acid propanoic acid, and many of the products of amino acid metabolism can all be used for gluconeogenesis, since they are sources of pyruvate or intermediates in the citric acid cycle, and hence provide increased oxaloacetate. The requirement for gluconeogenesis from amino acids in order to maintain a supply of glucose for the nervous system and red blood cells explains why there is a considerable loss of muscle in prolonged fasting or starvation, even if there are apparently adequate reserves of adipose tissue to meet energy needs. The liver is the main site of gluconeogenesis, although the kidneys also make a significant contribution to blood glucose in the fasting state. The enzymes for gluconeogenesis have been detected in the small intestinal mucosa, and there is evidence that intestinal gluconeogenesis makes some contribution to blood glucose. Propionate (arising from intestinal bacterial fermentation of carbohydrates) is the main substrate for intestinal gluconeogenesis, which is stimulated by butyrate (also arising from bacterial fermentation).

Substrates that give rise to acetyl CoA directly (alcohol, fatty acids, ketone bodies, and ketogenic amino acids) cannot be substrates for gluconeogenesis since the two carbons added in the reaction to yield citrate are lost in the citric acid cycle, and acetyl CoA cannot provide an increase in the pool of oxaloacetate to act as a precursor for gluconeogenesis. However, the glycerol that arises from the hydrolysis of triacylglycerol in the fasting state can be a substrate for gluconeogenesis, since it is interconvertible with dihydroxyacetone phosphate, an intermediate in glycolysis.

7.6 The control of carbohydrate metabolism and integration with lipid and protein metabolism

Energy expenditure is relatively constant throughout the day, but most food intake normally occurs in two or three meals. There is therefore a need for metabolic regulation to ensure that there is a more or less constant supply of metabolic fuel to tissues, regardless of the variation in intake.

There is a particular need to regulate carbohydrate metabolism since the nervous system is largely reliant on glucose as its metabolic fuel, and red blood cells are entirely so. The plasma concentration of glucose is maintained between 3 and 5.5 mmol/l in the fasting state. If it falls below about 2 mmol/l there is loss of consciousness—hypoglycaemic coma. An excessively high concentration in the fed state can cause hyperglycaemic coma, and prolonged moderate hyperglycaemia leads to the complications of poorly controlled diabetes mellitus.

The plasma glucose concentration is maintained in short-term fasting by the utilization of hepatic glycogen stores, and by releasing free fatty acids from adipose tissues, and later ketone bodies from the liver, which are preferentially used by muscle, so sparing such glucose as is available for tissues that require it. However, the total body content of glycogen would be exhausted within 12–18 hours of fasting if there were no other source of glucose, and in more prolonged fasting gluconeogenesis from amino acids and the glycerol of triacylglycerol are important. The regulation of blood glucose is achieved largely by changes in the rates of glycolysis and gluconeogenesis, as well as changes in the rates of glycogen synthesis and breakdown (Wahren & Ekberg 2007). Both hormonal control (mainly insulin and glucagon) and regulation of carbohydrate metabolism by intermediates of fatty acid metabolism are important.

Hormonal control of carbohydrate metabolism in the fed state

During the 3–4 hours after a meal, there is an ample supply of metabolic fuel entering the circulation from the gut. Glucose from carbohydrate digestion and amino acids from protein digestion are absorbed into the portal circulation, and to a considerable extent the liver controls the amounts that enter the peripheral circulation. Under these conditions, when there is a plentiful supply of glucose, it is the main metabolic fuel for most tissues, and after a meal the RQ (respiratory quotient the ratio of the volume of carbon dioxide formed to oxygen consumed in metabolism) is close to 1.0, indicating that it is mainly glucose that is being oxidized.

The increased concentration of glucose and amino acids in the portal blood stimulates the β-cells of the pancreas to secrete insulin and suppresses the secretion of glucagon by the β-cells of the pancreas. Insulin has five main actions:

- Increased uptake of glucose into muscle and adipose tissue. This is effected by recruitment to the cell surface of glucose transporters (GLUT4) that are in intracellular vesicles in the fasting state

- Stimulation of the synthesis of glycogen from glucose in both liver and muscle, by activation of glycogen synthase, and parallel inhibition of glycogen phosphorylase

- Stimulation of lipogenesis (fatty acid synthesis and TAG synthesis) in adipose tissue by activation of acetyl CoA carboxylase (ACC) and diacylglycerol acyl transferase (DGAT) and parallel inactivation of lipolysis via hormone-sensitive lipase

- Stimulation of amino acid uptake into tissues, leading to increased rates of protein synthesis

- Induction of lipoprotein lipase in muscle and adipose tissue, thereby directing the uptake of fatty acids from triacylglycerol in chylomicrons and very low density lipoprotein.

In the liver, glucose uptake is by carrier-mediated diffusion and metabolic trapping as glucose 6-phosphate, independent of insulin. The uptake of glucose into the liver increases very significantly as the concentration of glucose in the hepatic portal vein increases, and the liver has a major role in controlling the amount of glucose that reaches peripheral tissues after a meal. There are two isoenzymes that catalyse the formation of glucose 6-phosphate in liver.

- Hexokinase has a K_m of approximately 0.15 mmol/l. In the liver, this enzyme is saturated, and therefore acting at its V_{max}, under all conditions. It acts mainly to ensure an adequate uptake of glucose into the liver to meet the demands for liver metabolism. In other tissues the availability of glucose for hexokinase action depends on (insulin-dependent) glucose transport, so that the rate of phosphorylation varies with the intracellular concentration of glucose.

- Glucokinase is an allosteric enzyme, with an apparent K_m of approximately 20 mmol/l. This enzyme will have very low activity in the fasting state, when the concentration of glucose in the portal blood is between 3 and 4 mmol/l. However, after a meal the portal concentration of glucose may well reach 20 mmol/l or higher, and under these conditions glucokinase has significant activity, and there is increased formation of glucose 6-phosphate in the liver. Unlike hexokinase, glucokinase is not inhibited by its product. Most of this additional glucose 6-phosphate will be used for synthesis of glycogen, although some may also be used for synthesis of fatty acids that will be exported in very low density lipoprotein.

In response to increased availability of glucose in the liver, there is increased synthesis of the enzymes for fatty acid synthesis and glycolysis, by binding of the carbohydrate response element binding protein (ChREBP) to the carbohydrate response elements (ChoRE) in the promoter regions of their genes (Abdul-Wahed et al. 2017). In the fasting state, ChREBP is phosphorylated (by protein kinase A (PKA) or AMP Kinase (AMPK)) and located in the cytosol. In response to an increase in the intracellular concentration of glucose (or more likely a metabolite) in the fed state, ChREBP is dephosphorylated by protein phosphatase 2 (also called PP2A), which results in its migration into the nucleus, where it binds to ChoRE sequences on target genes thereby increasing their transcription.

Glucokinase is also expressed in the β-cells of the pancreas, where it acts as a glucose sensor to increase the secretion of insulin. The additional glucose 6-phosphate produced by glucokinase as the blood concentration of glucose increases is metabolized by glycolysis, leading to increased ATP formation. The increase in intracellular ATP closes an ATP/K^+ channel, leading to depolarization and opening of a voltage gated calcium channel. The influx of calcium ions leads to fusion of the insulin secretory granules with the cell membrane and secretion of insulin. Glucokinase expression is decreased in response to a high fat diet, so impairing both insulin secretion and glycaemic control. There is some evidence that glucokinase is also expressed in pancreatic α-cells, where, by a similar mechanism, it leads to reduced secretion of glucagon.

Hormonal control of carbohydrate metabolism in the fasting state

In the fasting or the post-absorptive state (beginning about 4–5 hours after a meal, when the products of digestion have been absorbed), metabolic fuels enter the circulation from the reserves of glycogen, triacylglycerol, and protein laid down in the fed state. Because the brain is largely dependent on glucose, and red blood cells are entirely so, those tissues that can utilize other fuels do so, in order to spare glucose for the brain and red blood cells.

As the concentration of glucose and amino acids in the portal blood falls, so the secretion of insulin by the β-cells of the pancreas decreases and the secretion of glucagon by the α-cells increases. Glucagon has two main actions:

- stimulation of the breakdown of liver glycogen to glucose 1-phosphate, resulting in the release of glucose into the circulation;
- stimulation of the synthesis of glucose from gluconeogenic substrates in liver and kidney (gluconeogenesis, see Section 7.3, under 'Gluconeogenesis').

At the same time, the reduced secretion of insulin results in:

- reduced glucose uptake into muscle;
- a reduced rate of protein synthesis, so that the amino acids arising from protein catabolism are available for gluconeogenesis;
- relief of the inhibition of hormone-sensitive lipase in adipose tissue, leading to release of non-esterified fatty acids and glycerol.

Three further hormones are important for glucose homeostasis. Adrenaline (epinephrine) stimulates glycogenolysis and gluconeogenesis, as well as lipolysis, making fatty acids available as an alternative fuel. Cortisol and growth hormone stimulate gluconeogenesis and lipolysis, and also inhibit glucose uptake, with a resulting blood glucose elevating effect.

Control of glycogen synthesis and utilization

In response to insulin (secreted in the fed state) there is increased synthesis of glycogen (via activation of glycogen synthase) and decreased glycogen breakdown (via inactivation of glycogen phosphorylase). In response to glucagon (secreted in the fasting state) or adrenaline (secreted in response to fear or fright) the opposite occurs via inactivation of glycogen synthase and activation of glycogen phosphorylase, thereby permitting utilization of glycogen reserves. Both effects are mediated by protein phosphorylation and dephosphorylation.

- In response to glucagon or adrenaline, protein kinase is activated (Figure 7.8), which catalyses the phosphorylation of both glycogen synthase (resulting in reduced activity) and glycogen phosphorylase (resulting in increased activity).
- In response to insulin, phosphoprotein phosphatase is activated (Figure 7.8), which catalyses the dephosphorylation of both phosphorylated glycogen synthase (increasing its activity) and phosphorylated glycogen phosphorylase (reducing its activity).

Intracellular metabolites can override this hormonal regulation. Inactive glycogen synthase is allosterically activated by high concentrations of its substrate, glucose 6-phosphate. Active glycogen phosphorylase is allosterically inhibited by ATP, glucose, and glucose 6-phosphate, all of which signal that there is an ample supply of glucose available.

Control of glycolysis: the regulation of phosphofructokinase

The reaction catalysed by phosphofructokinase in glycolysis, the phosphorylation of fructose 6-phosphate to fructose 1,6-bisphosphate, is essentially irreversible. In gluconeogenesis the hydrolysis of fructose 1,6-bisphosphate is catalysed by a separate enzyme, fructose bisphosphatase. Regulation of the activities of these two enzymes therefore determines whether the overall metabolic flux is in the direction of glycolysis or gluconeogenesis.

Inhibition of phosphofructokinase leads to an accumulation of glucose 6-phosphate in the cell; this inhibits

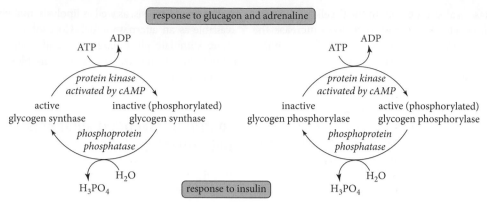

FIGURE 7.8 Hormonal control of glycogen synthesis and utilization.

hexokinase, which has an inhibitory binding site for its product. The result is decreased entry of glucose into the glycolytic pathway. Phosphofructokinase is inhibited by ATP, binding at a regulatory site that is distinct from the substrate-binding site for ATP. This is end-product inhibition, since ATP can be considered to be an end-product of glycolysis. When there is a requirement for increased glycolysis, and hence increased ATP production, this inhibition is relieved, and there may be a 1000-fold or greater increase in glycolytic flux in response to increased demand for ATP. However, there is less than a 10 per cent change in the intracellular concentration of ATP, which would not have a significant effect on the activity of the enzyme. What happens is that as the concentration of ADP begins to increase, so adenylate kinase catalyses the reaction $2 \times ADP \rightarrow ATP + 5'\text{-}AMP$. $5'AMP$ acts as an intracellular signal that energy reserves are low, and ATP formation must be increased. It binds to phosphofructokinase and reverses the inhibition caused by ATP. It also binds to fructose 1,6-bisphosphatase, reducing its activity.

Phosphoenolpyruvate, which is synthesized in increased amounts for gluconeogenesis, and citrate, which accumulates in the cytosol when fatty acids are to be synthesized, both inhibit phosphofructokinase.

Substrate cycling

A priori, it would seem sensible that the activities of opposing enzymes such as phosphofructokinase and fructose 1,6-bisphosphatase should be regulated in such a way that one is active and the other inactive at any time. If both were active at the same time then there would be cycling between fructose 6-phosphate and fructose 1,6-bisphosphate, with hydrolysis of ATP—a so-called futile cycle. What is observed is that both enzymes are indeed active to some extent at the same time, although the activity of one is greater than the other, so there is a net metabolic flux. One function of such substrate cycling is

thermogenesis—hydrolysis of ATP for heat production and maintenance of body temperature.

Substrate cycling also provides a means of increasing the sensitivity and speed of metabolic regulation. The increased rate of glycolysis in response to need for ATP for muscle contraction would imply a more or less instantaneous 1000-fold increase in phosphofructokinase activity if phosphofructokinase were inactive and fructose 1,6-bisphosphatase active. If there is moderate activity of phosphofructokinase, but greater activity of fructose 1,6-bisphosphatase, so that the metabolic flux is in the direction of gluconeogenesis, then a more modest increase in phosphofructokinase activity and decrease in fructose 1,6-bisphosphatase activity will achieve the same reversal of the direction of flux.

Control of the utilization of pyruvate

Pyruvate is at a metabolic crossroads, and in the liver and kidney it can either undergo decarboxylation to acetyl CoA, and hence oxidation in the citric acid cycle, or be carboxylated to provide oxaloacetate for gluconeogenesis. Its metabolic fate is largely determined by the oxidation of fatty acids.

Pyruvate dehydrogenase is inhibited by acetyl CoA, and an increase in the NADH:NAD⁺ ratio in the mitochondrion. The concentration of acetyl CoA will be high when β-oxidation of fatty acids is occurring, and there is no need to utilize pyruvate as a metabolic fuel. Similarly, the NADH:NAD⁺ ratio will be high when there is an adequate amount of metabolic fuel being oxidized in the mitochondrion, so that again pyruvate is not required as a source of acetyl CoA. Under these conditions it will mainly be carboxylated to oxaloacetate for gluconeogenesis.

The regulation of pyruvate dehydrogenase is by phosphorylation. Pyruvate dehydrogenase kinase is activated by acetyl CoA and NADH, and catalyses the phosphorylation of the enzyme to an inactive form. Pyruvate dehydrogenase phosphatase acts continually to dephosphorylate the inactive enzyme, so restoring its activity, and maintaining sensitivity to changes in the concentrations of acetyl CoA and NADH.

Control of glucose utilization in muscle

Muscle can use a variety of fuels: plasma glucose; its own reserves of glycogen; triacylglycerol from plasma lipoproteins; plasma non-esterified fatty acids; plasma ketone bodies; triacylglycerol from adipose tissue reserves within the muscle. The selection of metabolic fuel depends on the intensity of work being performed and whether the individual is in the fed or fasting state (Mittendorfer & Klein 2003).

The effect of work intensity on muscle fuel selection

Skeletal muscle contains two muscle fibre types: Type I, slow-twitch oxidative (SO) fibres, mainly used in prolonged, moderate intensity activity, and Type II, fast-twitch fibres, used in short-term high intensity activity (see Chapter 18, Section 18.2).

Intense physical activity requires rapid generation of ATP, usually for a relatively short time. When substrates and oxygen cannot enter the muscle at an adequate rate to meet the demand for aerobic metabolism, muscle depends on anaerobic glycolysis of its glycogen reserves. However, exercise that utilizes anaerobic glycolysis as its main means of energy production cannot be sustained for more than a few (2–3) minutes. As discussed above in section 'Glycolysis under anaerobic conditions: the Cori cycle', this leads to the release of lactate into the bloodstream, which is used as a substrate for gluconeogenesis in the liver, mainly after the exercise has finished. Less intense physical activity is often referred to as aerobic exercise, because it involves mainly type I (red, SO) and type IIA (white, fast oxidative glycolytic or FOG) muscle fibres and there is less accumulation of lactate.

The increased rate of glycolysis for exercise is achieved in three ways.

- As ADP begins to accumulate in muscle, it undergoes the reaction catalysed by adenylate kinase: $2 \times ADP \rightarrow ATP + 5'AMP$. 5'AMP activates phosphofructokinase, so increasing the rate of glycolysis.

- Nerve stimulation of muscle results in an increased cytosolic concentration of calcium ions, and hence activation of calmodulin. Calcium-calmodulin activates glycogen phosphorylase, increasing the rate of formation of glucose 1-phosphate, so providing increased substrate for glycolysis.

- Adrenaline, released from the adrenal glands in response to fear or fright, acts on cell surface receptors, leading to the formation of cAMP, which leads to increased activity of protein kinase, and increased activity of glycogen phosphorylase.

In prolonged aerobic exercise at a relatively high intensity (e.g., cross-country or marathon running), muscle glycogen and endogenous triacylglycerol are the major fuels with a gradual switch from glucose to fatty acid oxidation and a modest contribution from plasma non-esterified fatty acids and glucose. The higher the intensity of aerobic exercise the greater the dependence on muscle glycogen as the predominant fuel. As the exercise continues, and muscle glycogen and triacylglycerol begin to be depleted, so plasma non-esterified fatty acids (from adipose tissue lipolysis) become more important. However, in high intensity exercise there is still a requirement for glucose as the predominant fuel, so the depletion of muscle glycogen is associated with an increased dependence on blood glucose. When blood glucose begins to decline there is a decrease in the intensity of aerobic exercise as the high intensity aerobic exercise cannot be maintained using non-esterified fatty acids as the predominant fuel.

Muscle fuel utilization in the fed and fasting states

Glucose is the main fuel for muscle in the fed state, but in the fasting state glucose is spared for use by the brain and red blood cells; glycogen, fatty acids, and ketone bodies are now the main fuels for muscle. There are four mechanisms involved in the control of glucose utilization by muscle.

- The uptake of glucose into muscle is dependent on insulin. This means that in the fasting state, when insulin secretion is low, there will be little uptake of glucose.

- The activity of glycogen phosphorylase is increased in response to glucagon in the fasting state and the resultant glucose 6-phosphate inhibits hexokinase and hence the utilization of glucose.

- The activity of pyruvate dehydrogenase is reduced in response to increasing concentrations of NADH and acetyl CoA. This means that the oxidation of fatty acids and ketone bodies will inhibit the decarboxylation of pyruvate. Under these conditions the pyruvate that is formed from muscle glycogen by glycolysis will either form lactate (via lactate dehydrogenase) or be transaminated to alanine, both of which are used for gluconeogenesis in the liver.

- ATP inhibits both pyruvate kinase and phosphofructokinase. This means that under conditions where the supply of ATP is more than adequate to meet requirements, the metabolism of glucose is inhibited.

Manipulation of dietary carbohydrate

The manipulation of carbohydrate in the diet facilitates glycogen stores to either be optimized or depleted. Glycogen depletion will decrease blood glucose levels which will lead to the mobilization of fatty acids and the potential production of ketone bodies via ketogenesis, (See Chapter 8, Section 8.4). By manipulating dietary carbohydrate, metabolic capacity can be influenced which then impacts upon exercise performance or weight loss.

Carbohydrate loading and exercise performance

Carbohydrate plays an important role as the major fuel in both aerobic and anaerobic exercise. During aerobic exercise at a relatively high intensity (e.g., marathon running) or very high intensity (e.g., 10 km running) endurance exercise, depletion of muscle glycogen, and an associated decrease in blood glucose concentrations lead to fatigue and therefore an inability to maintain the exercise intensity. It has been demonstrated that when muscle glycogen stores become depleted following exercise to exhaustion, subsequent dietary carbohydrate intake plays a key role in the post-exercise increase in muscle glycogen (Bergstrom et al. 1967). Hence, individuals who consumed high carbohydrate diets were able to supercompensate their muscle glycogen stores, resulting in an increase in the length of time they could subsequently carry out relatively high endurance exercise until exhaustion (see Chapter 18, Section 18.4).

Very low carbohydrate/ketogenic diets for weight loss and exercise performance

Very low carbohydrate/ketogenic diets consist of low carbohydrate, typically 5 per cent of energy intake, and therefore supply 80 per cent of the energy as fat, along with a nutritionally adequate (15 per cent) protein content. Although the diets have an altered macronutrient composition, they do not impose a reduction in energy intake. These diets result in the stimulation of ketone body production. Ketone bodies become elevated in the circulation either during nutrient deprivation or when there is low carbohydrate availability, as experienced during fasting, exercise, or diabetes mellitus. Ketone bodies consist of acetoacetate and beta-hydroxybutyrate,

with the latter predominating in circulation. They are produced by the liver from acetyl CoA following beta-oxidation of fatty acids and therefore are elevated when circulating free fatty acids are increased due to increased adipose tissue lipolysis.

Very low carbohydrate/ketogenic diets can promote weight loss (Gershuni et al. 2018), as they induce lipolysis in adipose tissue and the release of non-esterified fatty acids, which the liver then uses to produce ketone bodies. When these diets are sustained, they result in weight loss but the exact mechanisms by which they achieve this is not clear. A common observation associated with these diets is a reduction in appetite, which is a reported effect of elevated circulating ketone bodies. High protein intakes are also known to induce satiety, thereby reducing appetite and may involve changes in various appetite-controlling hormones. It also has been reported that the switch to ketones as a principal fuel is energetically more wasteful, due to the activation of processes like gluconeogenesis to maintain blood glucose in the early phase of these diets. However, experimental data to support this hypothesis lacking, but there is consistent evidence that very low carbohydrate/ketogenic diets give greater long-term reductions in body weight compared with low fat diets. In addition, these ketogenic diets having favourable effects on blood lipid profiles, lowering serum triglycerides and total cholesterol.

At low to moderate intensity aerobic exercise, fatty acid oxidation makes a significant contribution as a fuel, while for higher intensity exercise, the contribution of fatty acid as a fuel increases with the duration of the exercise. Since very low carbohydrate/ketogenic diets promote the utilization of non-esterified fatty acids and ketones as fuel source, theoretically, an increased use of these substrates as fuel for aerobic exercise might allow endurance exercise to be sustained for longer. Although use of these diets in endurance-trained individuals does appear to decrease fat mass, the effects on performance are inconsistent (Harvey et al. 2019). This is probably because performance at high intensity aerobic exercise is predominantly influenced by the availability of carbohydrate, in the form of glycogen, as a fuel. Anaerobic exercise at very high intensity can only be sustained for short periods (less than two minutes) and is dependent on muscle glycogen stores. Consequently, the very low carbohydrate/ketogenic diets have been shown to have no positive effects on anaerobic exercise performance.

Specific features of dietary fructose metabolism

Fructose does not stimulate the secretion of insulin or leptin, nor does it suppress the secretion of ghrelin, so it does not suppress appetite in the same way as glucose does. This means that a high intake of fructose per se, or arising from sucrose, may be a contributing factor in positive energy balance and the development of obesity (van Buul et al. 2014). High intakes of fructose cause increased plasma concentrations of triacylglycerol. Hepatic fructose metabolism begins with phosphorylation by fructokinase to fructose 1-phosphate, which then enters the glycolytic pathway at the level of triose phosphate. Fructose thus bypasses phosphofructokinase, which is the major control point for glycolysis, so that more enters the pathway than is required for energy-yielding metabolism. The resultant citrate formed from acetyl CoA is used for fatty acid synthesis and lipogenesis and may potentially lead to obesity. It is unclear to what extent normal intakes of fructose are a factor in the development of obesity, as opposed simply to excess energy intake from glucose.

Fructose also increases the turnover of purines, resulting in an increased rate of production of uric acid, and high intakes may be a factor in the aetiology of gout. Because fructose enters glycolysis without control at the level of phosphofructokinase, its metabolism is not regulated by ATP (which inhibits phosphofructokinase), and high intakes of fructose can lead to accumulation of phosphorylated glycolytic intermediates, leading to depletion of inorganic phosphate and reduced formation of ATP.

KEY POINTS

- Carbohydrates, and especially starch, provide the main energy source in the diet, and after a meal it is mainly carbohydrate that provides metabolic fuel to all tissues.

- Dietary carbohydrates can be classified as glycaemic (sugars, starches, and dextrins) or non-glycaemic (non-starch polysaccharides, oligosaccharides, and resistant starch).

- The glycaemic index of a food is the extent to which it raises blood glucose compared with an equivalent amount of a reference carbohydrate. Low GI foods confer benefits in metabolic control of diabetes and may help to reduce risk factors for chronic diseases.

- Non-glycaemic carbohydrates (including resistant starch) are not digested in the small intestine but provide substrate for fermentation by the colonic microflora. They also have a prebiotic action, promoting the growth of *Lactobacillus* and *Bifidobacteria* spp., which inhibit the growth of pathogenic bacteria.

- Glucose is the precursor for synthesis of ribose, deoxyribose, glucuronic acid, and the carbohydrate moieties of complex carbohydrates, including glycoproteins and glycolipids, which have important structural and cell signalling and recognition functions.

- The main monosaccharides are all metabolized by the same pathway, glycolysis, which can operate either aerobically or anaerobically. Under anaerobic conditions lactate is formed and is used in the liver for resynthesis of glucose (gluconeogenesis).

- Under aerobic conditions, pyruvate arising from glycolysis is oxidized to acetyl CoA, which can either be a precursor for fatty acid synthesis or undergo complete oxidation in the citric acid cycle.

- Pyruvate and intermediates of both glycolysis and the citric acid cycle can be used for synthesis of non-essential amino acids and can also be formed by the metabolism of amino acids.

- The main storage carbohydrate in the body is glycogen in liver and muscle; its synthesis in the fed state and utilization in the fasted state are regulated by insulin and glucagon, with glucose being spared for the brain and red blood cells during fasting via regulation of its uptake into, and utilization by, muscle and other tissues.

- The plasma concentration of glucose is regulated within strict limits; failure of this regulation, and poor glycaemic control, leads to the complications of diabetes mellitus.

 Be sure to test your understanding of this chapter by attempting multiple choice questions.

REFERENCES

Abdul-Wahed A, Guilmeau S, Postic C (2017) Sweet sixteenth for ChREBP: Established roles and future goals. *Cell Metab* **26**, 324–341.

Anderson J W, Baird P, Davis R H et al. (2009) Health benefits of dietary fiber. *Nutr Rev* **67**, 188–205.

Asp N-G (1996) Dietary carbohydrates: Classification by chemistry and physiology. *Food Chem* **57**, 9–14.

Bergstrom J, Hermansen L, Hultman E et al. (1967) Diet, muscle glycogen and physical performance. *Acta Physiol Scandin* **71**, 140–150.

Bode L (2009) Human milk oligosaccharides: Prebiotics and beyond. *Nutr Rev* **67** (suppl 2), s183–191.

Bosscher D, van Loo J, Franck A (2006) Inulin and oligofructose as prebiotics in the prevention of intestinal infections and disease. *Nutr Res Rev* **19**, 216–226.

Byrne C S, Chambers E S, Morrison D J et al. (2015) The role of short chain fatty acids in appetite regulation and energy homeostasis. *Intern J Obes* **39**, 1331–1338.

Gershuni V M, Yan S L, Medici V (2018) Nutritional ketosis for weight management and reversal of metabolic syndrome. *Curr Nutr Rep* **7**, 97–106.

Harvey K L, Holcomb L E, Kolwicz S C (2019) Ketogenic diets and exercise performance. *Nutrients* **11**, 2296–2231.

McClenaghan N H (2005) Determining the relationship between dietary carbohydrate intake and insulin resistance. *Nutr Res Rev* **18**, 222–240.

Mittendorfer B & Klein S (2003) Physiological factors that regulate the use of endogenous fat and carbohydrate fuels during endurance exercise. *Nutr Res Rev* **16**, 97–108.

Rosenzweig A, Blenis J, Gomes A P (2018) Beyond the Warburg effect: How do cancer cells regulate one-carbon metabolism? *Front Cell Dev Biol* **6**, 90.

SACN (Scientific Advisory Committee on Nutrition) (2015) *Carbohydrates and Health*. The Stationery Office, London. Available at: https://www.gov.uk/government/publications/sacn-carbohydrates-and-health-report.

Smith C E & Tucker K L (2011) Health benefits of cereal fibre: A review of clinical trials. *Nutr Res Rev* **24**, 118–131.

Swallow D M (2003) Genetic influences on carbohydrate digestion. *Nutrition Research Reviews* **16**, 37–43.

van Buul V J, Tappy L, Brouns F J (2014) Misconceptions about fructose-containing sugars and their role in the obesity epidemic. *Nutr Res Rev* **27**, 119–130.

Wahren J & Ekberg K (2007) Splanchnic regulation of glucose production. *Ann Rev Nutr* **27**, 329–345.

FURTHER READING

Bender D & Cunningham S M C (2021) *Introduction to Nutrition and Metabollism* 6th edn. CRC Press, Boca Raton.

Foster-Powell K, Holt S H A, Brand-Miller J C (2002) International table of glycemic index and glycemic load values. *Am J Clin Nutr* **76**, 5–56.

Frayn K N & Evans R (2019) *Human Metabolism: a Regulatory Perspective* 4th edn. Wiley-Blackwell, Oxford.

Joint FAO/WHO Scientific Update on Carbohydrates in Human Nutrition (2007) A collection of reviews covering the FAO/

WHO Scientific Update on Carbohydrates in Human Nutrition. Available at: https://www.nature.com/collections/zpmfrlxswp.

Rodwell V, Bender D A, Botham K M et al. (eds) (2015) *Harper's Illustrated Biochemistry* 30th edn. McGraw-Hill Education, New York.

WHO (2015) *Guideline: Sugars Intake for Adults and Children*. World Health Organization, Geneva. Available at: http://www.who.int/nutrition/publications/guidelines/sugars_intake/en/

8 Fat metabolism

Philip C. Calder and Parveen Yaqoob

OBJECTIVES

By the end of this chapter you should be able to:

- relate fat structure with function
- describe the main features of metabolism of fat in the fed and fasted states
- explain the time-frame and location of fat metabolism following a meal containing fat
- understand the main regulatory features of fat metabolism
- explain the relationship between fat, carbohydrate, and protein metabolism.

8.1 Introduction

The major form of dietary fat is triacylglycerol (Figure 8.1), which comprises approximately 90–95 per cent of total dietary fat, with cholesterol and phospholipids being the other main components. Fat is a major contributor to total energy intakes in most Western diets, supplying 35–40 per cent of food energy through consumption of 80–100 g of fat per day. Because a gram of fat yields more than twice as much metabolizable energy (9 kcal/g) as a gram of either carbohydrate or protein (4 kcal/g), altering the fat content of a food can profoundly affect its energy density. All fat sources contain mixtures of saturated, monounsaturated, and polyunsaturated fatty acids; although the proportions vary depending on the source (see Chapter 2). The fatty acids present in dietary fat mostly contain an even numbers of carbon atoms and the most abundant have 16 or 18 carbon atoms. Dietary fat is digested in the lumen of the small intestine and the digestion products are packaged into lipoproteins in intestinal epithelial cells before delivery to the lymph and then the bloodstream, where they become available for uptake by tissues (see Chapter 4).

Fat plays diverse roles in human nutrition. It is important as a source of energy, both for immediate utilization by the body and in laying down a storage depot (adipose tissue) for later utilization when food intake is reduced. Dietary fat acts as a vehicle for the ingestion and absorption of fat-soluble vitamins (see Chapter 12). Two fatty acids, linoleic acid and α-linolenic acid, cannot be synthesized in mammalian tissues and are essential in the diet. Linoleic and α-linolenic acids are both 18 carbons in length. Lack of essential fatty acids in animals leads to typical scaling of the skin, growth retardation, and impaired reproduction. Similar skin symptoms have been described in human infants fed artificial milk formula and in patients on long-term intravenous nutrition, but essential fatty acid deficiency is rarely observed in people consuming mixed diets. Essential and non-essential fatty acids are integral components of cell membrane phospholipids and play important structural and functional roles in the cell (Calder 2015). Following cell stimulation, long-chain (20 and 22 carbon chain length) polyunsaturated fatty acids (PUFAs) in membrane phospholipids are the precursors of bioactive lipid mediators including prostaglandins and other eicosanoids, compounds that have local hormone-like effects. Cholesterol is also an essential component of cell membranes and is the precursor for synthesis of adrenocorticoid and sex hormones, bile acids, and vitamin D.

Because most Western diets contain relatively large amounts of fat, the regulation of fat storage in the fed state is essential to ensure that circulating triacylglycerol does not exceed optimal limits and that dietary fat is stored in adipose tissue rather than in non-adipose tissue organs, such as skeletal muscle and liver. Triacylglycerol accumulated in non-adipose tissue organs is known as ectopic fat. Raised fasting and postprandial concentrations of plasma triacylglycerol are recognized risk factors for coronary heart disease; hence a good understanding of the processes through which triacylglycerol concentrations in the circulation are regulated is important. Conversely, in the fasted state the mobilization of triacylglycerol stored in adipose tissue is central to the maintenance of tissue and cellular energy homeostasis, prevents excessive catabolism of protein stores, and aids in the regulation of blood glucose concentration. In addition, the

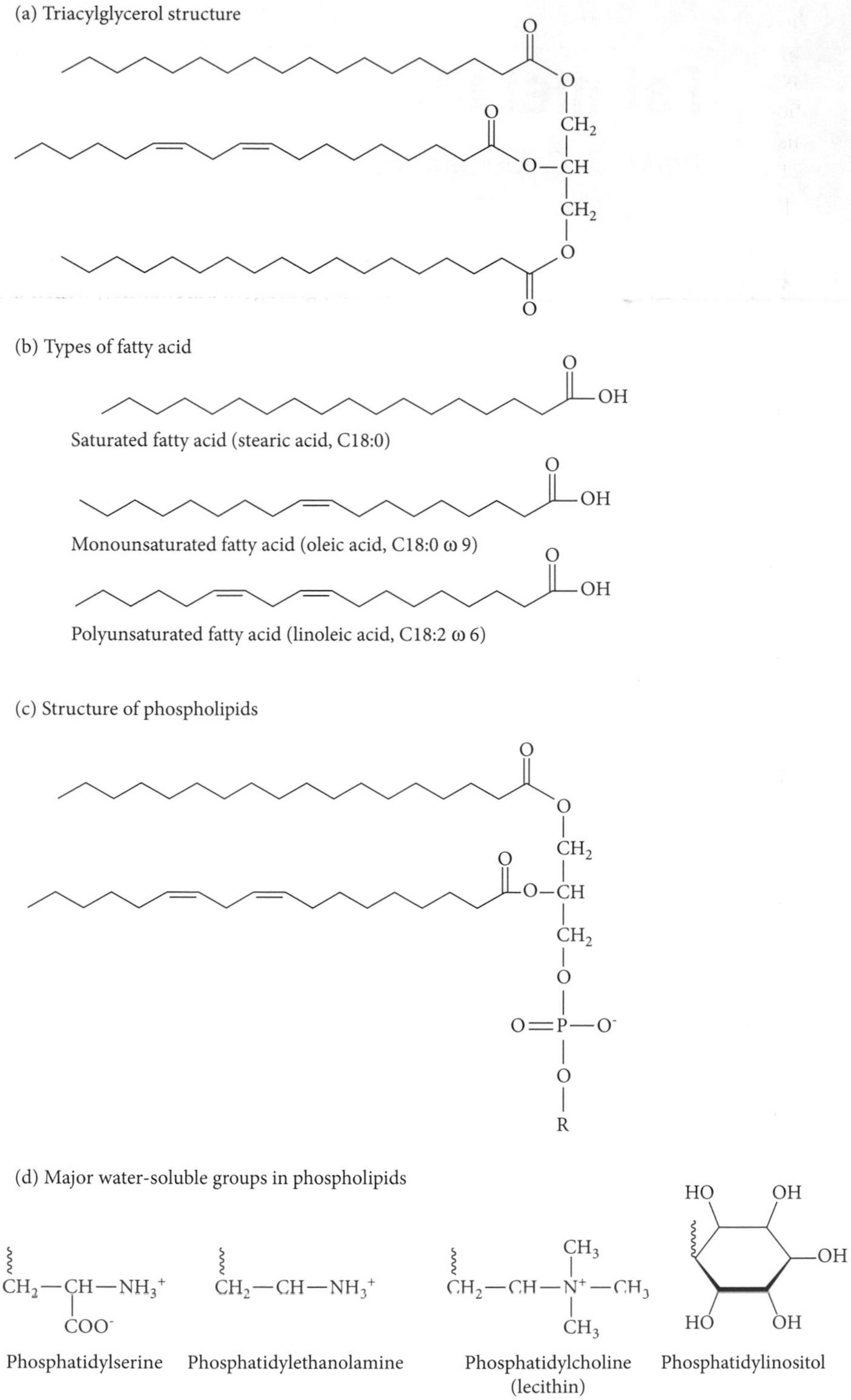

(a) Triacylglycerol structure

(b) Types of fatty acid

Saturated fatty acid (stearic acid, C18:0)

Monounsaturated fatty acid (oleic acid, C18:0 ω 9)

Polyunsaturated fatty acid (linoleic acid, C18:2 ω 6)

(c) Structure of phospholipids

(d) Major water-soluble groups in phospholipids

CH_2—CH—NH_3^+
|
COO^-
Phosphatidylserine

CH_2—CH—NH_3^+
Phosphatidylethanolamine

CH_3
|
CH_2—CH—N^+—CH_3
|
CH_3
Phosphatidylcholine
(lecithin)

Phosphatidylinositol

FIGURE 8.1 Structure of triacylglycerols (TAG) and phospholipids.
Triacylglycerols (a) consist of a glycerol backbone esterified to three fatty acids (b). In phospholipids (c) carbon-3 of glycerol is esterified to phosphate, then to one of a number of water-soluble bases shown in (d).

balance in the type of fat in the diet (whether saturated or unsaturated) can influence circulating cholesterol concentration in terms of the pro-atherogenic low density lipoprotein (LDL)-cholesterol and anti-atherogenic high density lipoprotein (HDL)-cholesterol. Because of their importance in understanding relationships between dietary fat and chronic disease, this chapter pays particular attention to the transport of fat in the body as exogenous (dietary) and endogenous (hepatic) lipoproteins (Section 8.3) and the regulation of fatty acid metabolism at the cellular and whole-body level in the fed and fasted states (Sections 8.4 and 8.5). However, undue emphasis on adverse effects of excess fat intake can obscure the essential structural and functional roles that fat plays in the body (Section 8.2) and this chapter also considers the role of fat as a precursor for the synthesis of specialist molecules (Section 8.2 under 'Specific functions in membranes').

8.2 Functions of dietary fat

Fat performs a range of essential functions within the body, including the provision of energy, structural and specific functional roles in cell membranes, and hormone-like activities.

Energy storage

Due to its energy density, fat, as triacylglycerol (Figure 8.1), is the nutrient of choice to act as a long-term fuel reserve for the organism. The majority of fat is stored as triacylglycerol. The majority of this is stored in specially adapted cells of adipose tissue termed adipocytes, although some fat reserve is also found in other cells in the body, such as liver and muscle cells (where excessive accumulation has pathological consequences). Although there is a relationship between the long-term intake of individual fatty acids and the fatty acid composition of the adipose tissue (e.g., an individual eating a diet rich in PUFAs will have a greater proportion of these in their adipose tissue), fatty acids of triacylglycerol stored in the adipocyte tend to be more saturated than the fatty acids in cell membrane phospholipids. The fatty acid composition of triacylglycerol stored in adipose tissue differs between sites. For example, subcutaneous abdominal adipose tissue contains a higher proportion of saturated fatty acids and lower proportion of monounsaturated fatty acids than subcutaneous adipose tissue on the buttock (gluteal); there are no clear differences in the proportions of PUFAs, suggesting that the differences in fatty acid composition are due to endogenous metabolism rather than selective tissue uptake of specific fatty acids.

Following a meal, ingested fat is transported in the bloodstream to the adipose tissue in lipoproteins called chylomicrons (Section 8.3). The fatty acids are hydrolysed from the triacylglycerol in circulating chylomicrons and other lipoproteins by the enzyme lipoprotein lipase (LPL), taken up by the adipose tissue and re-esterified into triacylglycerol for storage. The uptake of fatty acids from chylomicron-triacylglycerol occurs to a greater extent in subcutaneous adipose tissue on the upper (abdominal) body compared to lower body adipose tissue (gluteal). Not all fatty acids are taken up by adipose tissue: some of the hydrolysed fatty acids escape uptake into adipose tissue and 'spill over' into the systemic circulation. These non-esterified fatty acids may subsequently be taken up by skeletal muscle and the liver. When dietary energy is limited (e.g., after an overnight fast), the triacylglycerol stored in adipose tissue is hydrolysed and non-esterified fatty acids are released from the adipocyte into the circulation, where they are bound to serum albumin. This tightly controlled dynamic process is regulated by the concentration of metabolites (glucose, fatty acids, triacylglycerol) in the blood and by hormones (insulin, glucagon, adrenaline), which are themselves responsive to diet (Section 8.5).

In addition to its role as an energy reserve, subcutaneous adipose tissue is important in the maintenance of body temperature, whereas internal fat (visceral fat) protects the vital organs such as the kidney and spleen. Accumulation of excessive visceral fat (abdominal obesity) is linked with insulin resistance and is a risk factor for heart disease and diabetes (see Chapter 20).

Structural functions: as components of cell membranes

Lipids form an integral part of cell membranes, which create a barrier between a cell and its external environment. Intracellular membranes compartmentalize different areas within the cell, such as the nucleus. The basic structural unit of most biological membranes is phospholipids, which, like triacylglycerol, have fatty acids esterified at carbons-1 and -2 of glycerol. However different from triacylglycerol, in phospholipids carbon-3 of glycerol is esterified to a phosphate group, which in turn is esterified to one of a variety of bases (choline, ethanolamine, serine, inositol); this contributes to the amphiphilic nature of membrane lipids providing both hydrophobic (fat-soluble) and hydrophilic (water-soluble) regions. In mammalian tissues the most common base is choline, and phosphatidylcholine is the main membrane phospholipid (Figure 8.1).

Lipids based on a sphingosine rather than a glycerol backbone (sphingolipids) are also widespread in membranes and are particularly abundant in the brain and nervous system. In sphingomyelin the amino group of the long unsaturated hydrocarbon chain of sphingosine is linked to a fatty acid and the hydroxyl group is esterified to phosphoryl choline, yielding a molecule with a similar conformation to phosphatidylcholine (see Weblink Figure 8.1). Glycolipids contain carbohydrate (see Section 7.3, under 'Glycolipids'). They consist of a sphingomyelin backbone and a fatty acid unit bound to the amino group, with one or more sugars attached to the hydroxyl group. The simplest is cerebroside, which contains a single sugar, either glucose or galactose.

In membranes, phospholipids and sphingolipids arrange themselves in a lipid bilayer with the hydrophobic fatty acid tails facing inwards and the hydrophilic head interacting with the aqueous environment of the cytosol (at the inner face) and the extracellular fluid at the outer face (Figure 8.2). The chain length and degree of unsaturation of the fatty acids within the bilayer have an impact on the physical properties of the membrane, altering membrane fluidity and therefore function. For example, incorporation of long-chain PUFAs into a membrane will increase its fluidity compared to incorporation of shorter chain saturated fatty acids. Dietary fatty acid intake affects membrane composition to some extent. The presence of lipid-soluble antioxidants such as α-tocopherol within the membrane serves to minimize oxidation of the unsaturated fatty acids.

Cholesterol, which is almost entirely absent from plant tissues, is the most common sterol found in animal tissues. It inserts itself into the lipid bilayer where its hydrophobic interactions with fatty acids are essential to maintain membrane structure and fluidity. On a diet rich in PUFAs, an increase in the cholesterol to phospholipid ratio of a membrane serves to maintain its fluidity.

In addition to lipids, membranes contain a variety of proteins: enzymes, receptors, or transporters (Figure 8.2). The protein content is variable and reflects the function of the cell.

Specific functions in membranes

In addition to their relatively non-specific function in membrane structure, membrane lipids have a wide variety of specific roles such as a lung (pulmonary) surfactant, in cell signalling and as precursors of a diverse range of the metabolically active lipid mediators.

Pulmonary surfactant

Each time we exhale, our lungs are prevented from collapsing by a protein–lipid mixture known as pulmonary surfactant. This contains about 85 per cent lipid, dominated by a single compound, dipalmitoylphosphatidylcholine. Surfactant forms a solid film on the alveolar surface as we breathe out, reducing the surface tension and preventing lung collapse. Its importance is evident in newborn infants with acute respiratory distress, an often fatal condition, resulting from an inability to synthesize pulmonary surfactant, leading to respiratory dysfunction.

Cell signalling

Various lipids are involved in cell signalling and the conversion of extracellular signals into intracellular ones. The discovery of the phosphatidylinositol cycle indicated that membrane inositol phospholipids are important mediators of hormone and neurotransmitter action (see Weblink Figure 8.2). The binding of a hormone to membrane receptor proteins activates the enzyme phospholipase C, which hydrolyses the phosphatidylinositol molecule to diacylglycerol and inositol-1,4,5-triphosphate (IP_3). Both these products act as secondary messengers involved in regulation of cellular processes such as smooth muscle contraction, glycogen metabolism (see

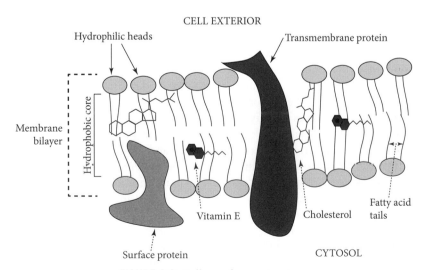

FIGURE 8.2 Cell membrane structure.
In Figure 8.2 the spheres represent the polar heads of phospholipids, which are in contact with the aqueous environment of the extracellular fluid (outside the cell) and the cytosol (inside the cell). The phospholipids arrange themselves in a 'lipid bilayer' with the hydrophobic fatty acid tails facing inwards towards each other. Cholesterol maintains the fluidity of the membrane, whereas vitamin E protects the fatty acids from oxidation. In addition, the membrane contains a range of proteins such as receptors, transporters, and enzymes.

Section 7.5, under 'Glycogen as a carbohydrate reserve'), and cell proliferation and differentiation.

In addition to inositol phospholipids, sphingolipids are important modulators of membrane receptor activity in all stages of the cell cycle including apoptosis (programmed cell death) and in inflammation. Intact membrane sphingolipids act as a ligand for receptors on nearby cells and modulate the activity of receptors and membrane-associated proteins in the same cell. Hydrolysis of sphingolipids can give rise to a variety of second messengers such as ceramide, lactosylceramide, glycosylceramide, and sphingosine, via the sphingomyelin cycle (see Weblink Figure 8.3). Ceramide is the best known of these sphingoid signalling molecules. Activation of membrane-bound sphingomyelinase releases ceramide from membrane sphingolipids. Ceramide activates protein kinases involved in various metabolic processes within the cell, including cell growth and death, and inflammatory responses.

Eicosanoids and other lipid mediators

Membrane unsaturated fatty acids, in particular the 20 and 22 carbon PUFAs, are the precursors of a variety of hormone-like compounds known collectively as eicosanoids (20 carbon) and docosanoids (22 carbon), which mediate a variety of cellular functions including smooth muscle contraction, blood clotting, inflammation, and the immune response (Calder 2020). They act locally to their site of synthesis and are metabolized very rapidly. It is largely this role of fatty acids as precursors of eicosanoids and docosanoids that underlies the essentiality of linoleic and α-linolenic acids, since these two fatty acids, which cannot be synthesized in the body, are the precursors of the 20 and 22 carbon PUFAs. The main precursor for the synthesis of eicosanoids is the omega-6 (n-3) PUFA arachidonic acid (20:4 n-6), which is synthesized from linoleic acid. Arachidonic acid is found in fairly high amounts in membrane phospholipids of many cell types. It can be released from membrane phospholipids by phospholipase A_2 following an appropriate stimulus and is then a substrate for cyclooxygenases (COX), lipoxygenases (LOX), or cytochrome P450 monooxygenases (CYPs) as illustrated in Figure 8.3. These enzymes generate enzymatically oxidized fatty acid products, which are collectively termed eicosanoids. Metabolism by lipoxygenases gives rise to leukotrienes, lipoxins, and hydroxy fatty acids, while metabolism by cyclo-oxygenase gives rise to prostaglandins, thromboxanes, and prostacyclin. More detail on the synthesis of eicosanoids can be found in Weblink Section 8.1(a). The range of biological activities of the eicosanoids is enormous and varies from tissue to tissue. More detail on the actions of eicosanoids can be found in Weblink Section 8.1(b).

Although arachidonic acid is regarded as the main precursor of eicosanoids, a separate family of eicosanoids is derived from the 20-carbon n-3 polyunsaturated fatty acid eicosapentaenoic acid (EPA) (see Figure 8.3). Oily fish and fish oils are rich sources of dietary EPA, which is readily incorporated into biological membranes and can replace arachidonic acid to some degree. This has two consequences. First, the replacement of arachidonic acid in the membranes of eicosanoid-synthesizing cells by EPA results in a decrease in the production of arachidonic acid-derived eicosanoids. Second, there appears to be production of selected EPA-derived eicosanoids. The physiological significance of the n-3 polyunsaturated fatty acid-derived eicosanoids is relatively poorly understood. Some studies have demonstrated that the EPA-derived eicosanoids are less potent than those derived from arachidonic acid. This type of observation has formed the basis of suggestions that the n-3 polyunsaturated fatty acids possess anti-thrombotic, anti-inflammatory, and immunomodulatory properties.

The last decades have seen the discovery of new families of bioactive mediators produced from EPA and another n-3 PUFA, docosahexaenoic acid (DHA) (Figure 8.3). These mediators are termed resolvins, protectins, and maresins. Their synthesis also involves the COX and LOX enzymes in rather complex pathways that are not yet fully understood. Together resolvins, protectins, and maresins are referred to as 'specialized pro-resolving mediators' because their main function appears to be the resolution (switching off) of inflammatory processes (Chiang & Serhan 2020).

8.3 Plasma lipoproteins

As triacylglycerol, cholesterol, and phospholipids are not water-soluble, they cannot be transported free in the blood; transport of these largely hydrophobic compounds occurs using specialized structures known as lipoproteins. Lipoproteins are particles in the circulation whose function is to 'shuttle' lipids to tissues where they are needed. They have a hydrophobic core of triacylglycerol and cholesteryl-esters and a hydrophilic surface consisting of phospholipids and free cholesterol, which interact with the aqueous environment of the blood and lymph (Figure 8.4) Free cholesterol is amphiphilic, as the free hydroxyl group gives the molecule some hydrophilic properties. Lipoproteins also contain specific proteins, called apoproteins, which, in addition to being essential for maintaining the structure and the solubility of the particle, determine how the lipoprotein is metabolized. Apoproteins recognize and interact with specific receptors on the cell surface, and the receptor–lipoprotein complex is internalized into the cell by the

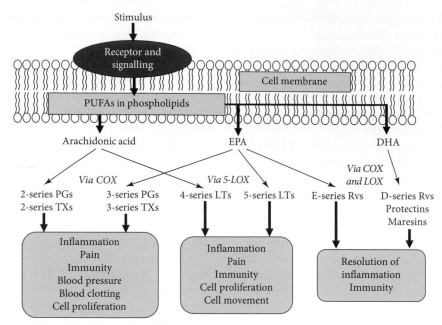

FIGURE 8.3 Overview of the synthesis and functions of eicosanoids and docosanoids. Eicosanoids and docosanoids are oxidized metabolites of 20- and 22-carbon polyunsaturated fatty acids (PUFAs). The PUFA substrate is released from cell membrane phospholipids upon cell stimulation and can then act as a substrate for cyclooxygenase (COX) and lipoxygenase (LOX) enzymes. The COX pathway produces prostaglandins (PGs) and thromboxanes (TXs) from arachidonic acid and eicosapentaenoic acid (EPA). The 5-LOX pathway produces leukotrienes (LTs) from arachidonic acid and EPA. EPA can be metabolized by pathways involving both COX and LOX to produce E-series resolvins (Rvs), while docosahexaenoic acid (DHA) is metabolized to D-series Rvs, protectins, and maresins. Together the resolvins, protectins, and maresins are termed specialized pro-resolving mediators.

process of endocytosis. Apoproteins also determine the activities of a range of proteins, including hydrolysing enzymes (lipases), receptors, and lipid transfer proteins which are involved in all stages of lipoprotein metabolism (Table 8.1). Five main series of apoproteins have been identified: apoA (1, 2), apoB (B48 and B100), apoC (1, 2, 3), apoE and apo(a). However, other classes have been recognized recently (apoD, apoF, apoH, apoL, and apoM), although as yet no functions have been assigned to them.

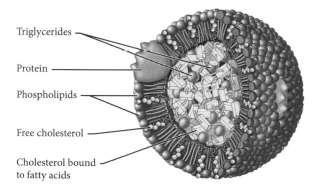

FIGURE 8.4 General structure of a lipoprotein particle.
Source: This figure was published in *Mosby's Medical Dictionary*. Copyright Elsevier (2012). Reproduced with permission.

Lipoprotein classes and their apoproteins

In order to aid description, lipoproteins have traditionally been classified (according to their density) into four main subgroups: chylomicrons, very low-density lipoproteins (VLDLs), LDLs, and HDLs. Recent research has focused on a fifth category of lipoprotein, lipoprotein (a) (Lp(a)), due to its strong independent association with the development of atherosclerosis.

Chylomicrons formed in the gut wall, and VLDLs formed in the liver, are the main transporters of triacylglycerols to the tissues and are referred to as triacylglycerol-rich lipoproteins (TRLs). They are the least dense of the lipoproteins as they have the lowest protein:lipid ratio. ApoB48 and apoB100 represent the main protein component of chylomicrons and VLDLs, respectively, with both proteins encoded for by the same gene; apoB48 represents the N-terminal domain (48 per cent) of apoB100. Only one molecule of apoB protein is present per lipoprotein particle. ApoE and apoC are smaller apoproteins, which are passed from one particle to another, and the transfer of these molecules from HDLs is an important mediator of triacylglycerol metabolism.

The smaller, denser LDLs and HDLs are involved in the transport of cholesterol to and from the cells, with

TABLE 8.1 Major classes of plasma lipoproteins

	Chylomicrons	VLDLs	IDLs	LDLs	HDLs
Density (g/ml)	<0.95	<1.006	1.006–1.019	1.020–1.063	1.064–1.210
Diameter (nm)	75–1200	30–80	25–35	18–25	5–12
M_r (10^3 kDa)	400	10–80	5–10	2.3	0.175–0.36
% protein	1.5–2.5	5–10	15–20	20–25	40–55
% phospholipids	7–9	15–20	22	15–20	20–35
% free cholesterol	1–3	5–10	8	7–10	3–4
% triacylglycerols	84–89	50–65	22	7–10	3–5
% cholesteryl esters	3–5	10–15	30	35–40	12
Electrophoretic mobility	At origin	Pre-beta	Between pre-beta and beta	Beta	Alpha
Major apoproteins	A1, A2, B-48, C1, C2, C3, E	B-100, C1, C2, C3, E	B-100, C3, E	B-100	A1, A2, B-48, C1, C2, C3, D, E
Turnover in plasma	4–5 min	1–3 h	1–3 h	45%/day	4 days

about 70 per cent of total cholesterol present in LDLs. LDLs are derived from the catabolism of VLDLs in the circulation and therefore also contain apoB100 as their main apoprotein. HDLs, which are originally synthesized in the gut and the liver, are responsible for the removal of excess cholesterol from peripheral tissues and its return to the liver, a process called reverse cholesterol transport. HDL particles are relatively small and contain a high protein:lipid ratio (50:50), and are therefore more dense than chylomicrons, VLDLs, or LDLs. The apoA series are the main proteins of HDLs.

Lipoprotein(a) is a small, cholesteryl-ester rich, LDL-like particle containing apo(a) as its characteristic protein. Assembly of Lp(a) appears to occur extracellularly, with apo(a) receiving cholesterol from circulating LDLs. Lp(a) is thought to be highly variable within populations, with levels largely genetically determined by a very polymorphic (variable) gene. Recent evidence suggests that this particle is highly atherogenic.

Although it is useful from a descriptive point of view to divide lipoproteins into five main classes, lipoproteins are a heterogeneous and dynamic group of particles, with apoproteins and lipids continuously moving between them.

The exogenous and endogenous lipoprotein pathway

The exogenous lipoprotein pathway, as the name suggests, distributes fat entering the circulation from outside the body, that is. dietary fat. In contrast, the endogenous pathway distributes fat either synthesized or stored in the liver. These two pathways are shown in Figure 8.5.

The exogenous lipoprotein pathway

As described in Chapter 4, dietary fat is packaged into chylomicrons in the enterocytes of the small intestine. Chylomicrons are secreted into the lymphatic system and then enter the circulation via the thoracic duct. They pass through the lungs and heart relatively unchanged and subsequently rapidly acquire apoC2. This apoprotein is essential for interaction of chylomicrons with lipoprotein lipase (LPL) as they pass through the capillaries of skeletal and heart muscle, mammary gland, and adipose tissue. Unlike LDLs, which can be taken up by cells intact, chylomicrons are too large to move through the capillary wall, so cells cannot take them up directly. Adipocytes, muscle, and mammary cells synthesize and secrete LPL, which is transported and attached to capillary endothelium by the protein GPIHBP1. LPL hydrolyses the triacylglycerol in chylomicrons, and the released fatty acids are subsequently taken up by the tissues. As LPL is a key determinant of plasma triacylglycerol clearance and tissue uptake of fatty acids, its activity needs to be carefully regulated in order to match the rate of uptake of plasma triacylglycerol-derived fatty acids to the needs of tissue and the ability of the tissue to dispose of the fatty acids. LPL activity in adipose tissue is upregulated in response to insulin whilst exercise increases LPL activity in skeletal muscle. The physiological variation in LPL activity in various tissues is primarily achieved via post-translational mechanisms involving two groups of

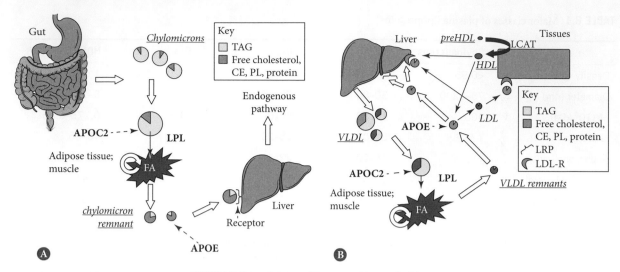

FIGURE 8.5 Pathways of lipoprotein metabolism.

Figure 8.5 (a) shows the exogenous pathway of lipoprotein metabolism. Chylomicrons may pass through the tissue capillary beds several times where they are hydrolysed by lipoprotein lipase (LPL), resulting in a chylomicron remnant which has lost a large portion of its original triacylglycerol (TAG) content. The resultant particle is taken up by the liver by receptor-mediated endocytosis. Figure 8.5 (b) shows the endogenous pathway of lipoprotein metabolism. VLDLs may pass through the tissue capillary beds several times where they are hydrolysed by LPL, resulting in VLDL remnants which have lost a large portion of the original triacylglycerol content. The remnant particle has two possible metabolic fates: (a) it can be further hydrolysed to LDL, which is the main transporter of cholesterol to target tissues; (b) the remnant particle is taken up by the liver by receptor-mediated endocytosis. Excess tissue cholesterol is returned to the liver by reverse cholesterol transport mediated by HDL and lecithin cholesterol acyltransferase (LCAT). *Abbreviations:* LDL-R, LDL receptor; LRP, LDL-R-related protein; LPL, lipoprotein lipase; CE, cholesteryl esters; PL, phospholipids; LCAT, lecithin-cholesterol acyltransferase.

proteins: apolipoproteins (apoC1, apoC2, apoC3, apoA5, and ApoE) and angiopoietin-like proteins (ANGPLT).

The released fatty acids rapidly cross the endothelium into the interstitial space, enter the cell and are immediately re-esterified in order to maintain a concentration gradient. In adipose tissue and mammary gland, the fatty acids are stored as triacylglycerol as an energy reserve and to provide milk fatty acids, whereas in the muscle the fatty acids are used as an immediate source of fuel to provide ATP for muscle contraction. In healthy individuals, the rate at which fatty acids are utilized for fuel in muscle is dependent on the rate of uptake. However, when fatty acid uptake is greater than utilization and in excess of requirements they are stored as triacylglycerol within lipid droplets.

The process of chylomicron hydrolysis is rapid, with 50 per cent of chylomicron triacylglycerol being removed within the first 2–3 minutes of entry into the bloodstream. Following a number of cycles through the tissues and the eventual removal of a large fraction of the triacylglycerol and a portion of cholesterol, and the surface phospholipids and apoproteins (to HDLs), a smaller, more cholesteryl-ester-enriched particle remains. This contains the original apoB48 and the majority of the fat-soluble vitamins. ApoE is acquired from HDLs upon arrival at the liver, and the liver cell takes up the chylomicron remnant by

a receptor-mediated process with the primary receptors. These are thought to be the LDL-receptor (LDL-R, also known as the apoB/apoE receptor) and the LDL-R related protein (LRP).

The endogenous lipoprotein pathway: VLDL metabolism

VLDLs distribute triacylglycerol from the liver to the tissues via the endogenous lipoprotein pathway. The sources of fatty acids for liver triacylglycerol synthesis include fatty acids returned to the liver by chylomicron remnants, LDLs, or HDLs, fatty acids delivered bound to albumin, fatty acids already present within the cytosolic storage pool of hepatocyte, along with fatty acids formed by *de novo* lipogenesis in the liver. This latter source is thought to be small on a typical Western diet, but may become more significant on a high-carbohydrate diet, particularly if the diet is high in sugar-rich foods (e.g., soft drinks and sweets) or when an individual has insulin-resistance. ApoB100 is the main structural and functional protein of VLDLs. There are two distinct sub-classes of VLDLs: $VLDL_1$ is larger and more triacylglycerol-rich than $VLDL_2$ which can either be secreted directly from the liver, or formed by the peripheral hydrolysis of $VLDL_1$. Upon secretion into the bloodstream VLDLs acquire apoC2 from HDLs, then VLDL triacylglycerol is hydrolysed by LPL and the fatty acids are accumulated by tissues in a similar manner to those from chylomicrons. However, chylomicrons are thought to provide a better substrate for LPL and are hydrolysed preferentially when both particles are present in the postprandial (fed) state. In healthy individuals VLDL production is suppressed in response to insulin during the postprandial period and despite suppression there are approximately ten times more VLDL than chylomicron particles.

VLDL remnants, known as intermediate density lipoproteins (IDLs), have two metabolic fates. Approximately 40–50 per cent are taken up by the liver by receptor-mediated endocytosis, the remaining 50–60 per cent lose all surface components except for a layer of phospholipids, free cholesterol and apoB100, and become LDLs, the major carrier of cholesterol in the blood. An increased secretion or delayed clearance of triacylglycerol-rich lipoproteins (VLDLs and chylomicrons) is a significant risk factor for coronary heart disease (see Chapter 19).

The endogenous lipoprotein pathway: LDL metabolism

The role of LDLs is to transport cholesterol to the peripheral tissues and regulate *de novo* synthesis of cholesterol at these sites. On arrival at the cell surface, the apoB100 component of LDLs is recognized by the LDL-receptor. Following internalization of the LDL-receptor complex, the vesicle fuses with lysosomes, which contain a variety of degradative enzymes. The apoB100 protein is hydrolysed to free amino acids and the cholesteryl-esters to free cholesterol. The majority of the LDL-receptor is returned to the cell surface unaltered. The released cholesterol can be used immediately for incorporation into cell membranes or synthesis of steroid hormones. Alternatively, the cholesterol can be re-esterified and stored within the cell. Cellular cholesterol is derived from both extracellular sources (LDLs) and synthesized in the cell. The process of cellular cholesterol metabolism is tightly regulated.

The physiological importance of the LDL-receptor in cholesterol homeostasis is demonstrated in the condition familial hypercholesterolaemia (FH), in which there is an absence or deficiency of functional LDL-receptors. Marked elevations in circulating LDL levels are evident, which leads to deposition of cholesterol in a variety of tissues, including the artery walls, thus contributing to atherogenesis.

A number of additional receptors which recognize LDLs, one class of which is known as the scavenger receptors, have been identified. These receptors, which are present in large numbers on the surface of macrophages, do not bind to native LDLs, but only to LDLs that have been chemically modified, for example, oxidized. Unlike the LDL-receptor, scavenger receptors are not subject to downregulation and therefore macrophages can take up LDL indefinitely until they become lipid laden, when they are known as foam cells. This process forms the basis of the lipid accumulation in the blood vessel wall, which occurs in the development of atherosclerosis and the process is accelerated in people with high circulating LDL levels (see Chapter 19).

Reverse cholesterol transport: HDL metabolism

Excessive accumulation of cholesterol in tissues is toxic as the cell cannot break down cholesterol and in the artery wall it leads to the development of atherosclerosis. This excess cholesterol is transported in HDLs back to the liver, where it can be excreted in the bile, or be transported to other cells via the VLDL–LDL pathway. More than 40 per cent of individuals who have a myocardial infarction (heart attack) have low HDL levels.

Pre-β-HDL is synthesized in the intestine and liver and secreted into the bloodstream as a discoidal shaped pre-HDL particle containing apoA1 and a small amount of phospholipid. The emerging HDL particles gather some surface material (phospholipids and free cholesterol) released following the hydrolysis of chylomicrons and VLDLs by LPL. Nascent HDL particles bind to cell surface receptors and avidly absorb cholesterol from the cell membrane. Lecithin-cholesterol acyltransferase (LCAT) present in HDLs esterifies the cholesterol, allowing it to move to the core of the particle and freeing up space on the surface for more cholesterol. LCAT, which is activated by apoA1, ensures a unidirectional movement of cholesterol from the cell to the HDL particle. Gradually the HDLs accumulate cholesterol and become mature spherical α-HDL particles (HDL_2).

Subsequent movement of this excess cholesterol in HDLs back to the liver is mediated by either a direct or an indirect pathway. In the direct pathway, HDLs take the cholesterol to the liver, although quantitatively this is not the most important route. The majority of cholesterol delivery is achieved via the indirect route, where HDLs transfer their cholesterol to chylomicrons and VLDL remnants, which subsequently transport the cholesterol to the liver. In addition to a role in reverse cholesterol transport, HDLs may also have some additional benefits with respect to the development of atherosclerosis. For example, HDLs inhibit the movement of macrophages (cells which accumulate cholesterol) into the artery wall, are important for maintaining endothelial (cells lining the blood vessels) health and inhibit LDL oxidation.

Genetic and dietary regulation of lipoprotein metabolism

The concentration and composition of circulating lipoproteins are influenced by both genetic and environmental factors, including diet.

Genetic factors influencing circulating lipoprotein concentrations

Rare but major gene defects, such as familial hypercholesterolaemia (FH), cause large increases in lipoprotein concentrations, in particular LDL cholesterol, resulting in high risk of premature myocardial infarction. More common variants (known as single nucleotide polymorphisms (SNPs)) in genes encoding for apolipoproteins, lipases, lipoprotein receptors, and fatty acid transport proteins, have lesser impact on the concentration and composition of circulating lipoproteins, but contribute to a moderately altered risk of cardiovascular disease (CVD). Overall, greater than 50 per cent of the variation in fasting and postprandial lipoprotein concentrations is genetically determined. The adverse impact of these lesser gene defects on lipoprotein concentrations can vary according to background diet. A collection of lipoprotein abnormalities termed the Atherogenic Lipoprotein Phenotype (ALP) (raised plasma triacylglycerol, reduced levels of HDL cholesterol and a high proportion of LDLs in the small, dense atherogenic form), is becoming increasingly common and is linked with the metabolic syndrome, central obesity and ectopic fat accumulation within the liver in individuals who do not consume greater than the recommended intake of alcohol. This is known as non-alcoholic fatty liver disease (NAFLD). NAFLD is associated with an overproduction of VLDL-triacylglycerol, leading to raised plasma triacylglycerol concentrations. The ALP, which increases risk of CVD 3–4 fold, is thought to arise due to complex interactions between diet, adiposity, and a number of SNPs in genes encoding for proteins which regulate lipoprotein metabolism and insulin function. More detail on SNPs and lipoprotein metabolism can be found in Weblink Section 8.2.

Familial hypercholesterolaemia (FH) (type II$_a$ hyperlipidaemia)

Familial hypercholesterolaemia is an autosomal recessive disorder first characterized by the pioneering studies of Goldstein et al. (1983). The condition is characterized by a deficiency or complete absence of functional LDL receptors on the cell surface. As a result, the uptake of cholesterol into the liver and other tissue is impaired,

resulting in markedly elevated levels in the circulation. Heterozygotes typically have circulating cholesterol levels two- to three-fold higher than normal. Homozygotes have almost no receptors for LDL and have up to 10-fold higher cholesterol than the levels in 'normal' individuals. Untreated, more than 85 per cent of males with FH will suffer a myocardial infarction (heart attack) before the age of 60 years and many before the age of 40 years. The condition is usually treated by following a reduced total fat, saturated fat, and cholesterol diet in combination with a combined drug therapy of resins (which sequester bile acid) and statins (which inhibit HMG-CoA reductase, the main enzyme involved in cholesterol synthesis). Such a combined treatment can reduce LDL cholesterol by up to 50 per cent.

Familial defective apolipoproteinaemia

A number of dyslipidaemias are due to an inherited defect in a particular apolipoprotein. Familial defective apolipoprotein B–100 (FDB) results in defective lipoprotein clearance and increased atherogenesis. It is clinically indistinguishable from heterozygous FH. An altered apoB100 structure results in decreased binding of the LDL particle to its receptor and subsequent increases in LDL cholesterol levels. Familial defective apolipoprotein E is characterized by a low affinity of apoE for lipoprotein receptors, resulting in the accumulation of chylomicron and VLDL remnants in the circulation.

Atherogenic lipoprotein phenotype and the metabolic syndrome

The importance of hypertriacylglycerolaemia as a risk factor for coronary heart disease (CHD) is well established (Havel 1994; Castelli 1986). In addition to the ability of triacyglycerol-rich lipoproteins (TRLs) to directly infiltrate and transport cholesterol into the developing atherosclerotic plaque, raised circulating triacylglycerol has been implicated as a major metabolic component of the ALP. As described above, this is a term frequently used to describe a collection of pro-atherogenic lipoprotein abnormalities, which constitute the main dyslipidaemia of the metabolic syndrome (Austin et al. 1990; Ruotolo & Howard 2002). The ALP lipid profile, which occurs in up to 25 per cent of middle-aged males, is associated with a three-fold increased CHD risk. It is characterized by a moderate fasting hypertriacylglycerolaemia (1.5–4.0 mmol/l), exaggerated postprandial triacylglycerol levels, low HDL cholesterol levels (<1.1 mmol/L), and a predominance of the potentially atherogenic small dense LDL-3 particles (>40 per cent of total LDL).When levels of TRLs are high in the circulation there is a net transfer of triacylglycerol from these particles to LDL and HDL in exchange for cholesteryl-ester. This system of 'neutral lipid exchange' is catalysed by cholesteryl-ester transfer protein (CETP) (Weblink Figure 8.4). The cholesterol-enriched TRLs have an increased capacity to transport cholesterol into the artery wall. In addition, the triacylglycerol-enriched LDLs and HDLs make ideal substrates for hepatic lipase (HL), which hydrolyses the relevant particle into small dense LDL-3 and HDL-3. HDL-3 is rapidly removed from the circulation, thereby reducing circulating levels of the protective HDL cholesterol. LDL-3 has greater atherogenic potential than normal LDLs due to the fact that it is less readily removed from the circulation by hepatic LDL receptors, remains in circulation longer than normal LDL particles, and is more prone to oxidation and uptake into macrophages in the arterial wall.

Diet and other environmental factors influencing lipoprotein concentration

Overview of the effect of diet and other environmental factors

Numerous environmental factors (diet, exercise, alcohol consumption, and prescribed drug use) influence both fasting and postprandial lipid levels. Pharmacological therapies such as statins and fibrates lower the blood concentrations of total and LDL cholesterol, and lower triacylglycerol and increase HDL cholesterol, respectively. These changes lower risk of CVD. The evidence suggests a J- or U-shaped relationship between alcohol consumption and CHD, with light to moderate consumption (1–4 units per day, up to 20 g and 30 g of alcohol per day in women and men) being associated with a reduced risk (see also Chapter 10). Alcohol is a lipogenic substrate. Intake of alcohol above recommended limits can result in alcoholic fatty liver disease, which may in part be due to an up-regulation of hepatic *de novo* lipogenesis. Numerous mechanisms have been proposed to explain the protective effect of light alcohol consumption, including a beneficial impact on thrombosis, inflammation, or insulin sensitivity. However, an increase in HDL-cholesterol concentration due to an in increase in apoA1 and A2 is thought to be in large part responsible for the cardioprotective actions of moderate alcohol consumption. A beneficial impact of aerobic exercise on all lipoprotein classes has been reported with the greatest effect evident for triacylglycerol and HDL cholesterol levels. This is likely to be in part attributable to the known effect of exercise on insulin sensitivity as well as increased utilization of circulating triacylglycerol (due to an up-regulation of LPL) as a substrate for oxidation in skeletal muscle.

Effect of the amount and composition of dietary fat

The replacement of saturated fat, in particular lauric acid (c12:0), myristic acid (c14:0), and palmitic acid (c16:0), with n-6 polyunsaturated fat is associated with a reduction in LDL cholesterol concentration whilst the effect of n-3 polyunsaturated fatty acids in reducing LDL cholesterol is less evident. In 2003, Mensink and co-workers (Mensink et al. 2003) published a meta-analysis of 60 randomized controlled trials published between 1970–1998, involving a total of 1672 participants, with intervention periods ranging from 13 to 91 days. The results of the meta-analysis were summarized by a series of predictive equations which indicated that the replacement of 1 per cent of dietary energy from saturated fat, by carbohydrate, monounsaturated fat, or polyunsaturated fat would result in a decrease in LDL cholesterol of 0.036, 0.042, and 0.57 mmol/L respectively. However, it must be noted that replacement of saturated fatty acids also decreases the concentration of beneficial HDL cholesterol. Replacement of a large amount of saturated fatty acids with carbohydrate should be avoided because, although this will lower LDL cholesterol, it will also result in a significant increase in plasma triacylglycerol concentration. The type of carbohydrate consumed may play a role in the effect, for example a diet high in free/added sugars increases plasma concentrations of triacylglycerol (which will in part be mediated by an up-regulation of hepatic *de novo* lipogenesis), total and LDL cholesterol, with less effect on HDL cholesterol concentration. The consumption of soluble dietary fibre and plant sterols/stanols is associated with a significant reduction in LDL-cholesterol which is attributable to the reduced absorption of dietary cholesterol and of bile salts which are derived from cholesterol. The n-3 polyunsaturated fatty acids eicosapentaenoic acid (EPA) and docosahexaenoic acid (DHA) are highly effective hypotriacylglycerolaemic agents, which is more evident in individuals with an elevated (>2 mmol/L) plasma triacylglycerol concentration. Intakes of 3–4 g per day of EPA and DHA decrease fasting and postprandial TAG by 20–35 per cent, through reducing hepatic *de novo* lipogenesis, along with the production and secretion of hepatic triacylglycerol and improved LPL-mediated TRL clearance.

8.4 The routes of intracellular fat metabolism

Fatty acid uptake and activation

Fatty acids are delivered to tissues either in the form of non-esterified fatty acids (NEFAs), bound to serum albumin, or by the hydrolysis of the triacylglycerol component of circulating lipoproteins. The mechanism by which NEFAs are taken up by cells is not clear, but it is likely that both carrier-mediated transport and diffusion are involved. The proteins implicated in fatty acid transport are fatty acid binding proteins (FABPs), fatty acid translocase (FAT), and the fatty acid transport protein (FATP). In addition to their roles in fatty acid uptake by cells, FABPs are also important intracellular carriers of fatty acids, delivering them to subcellular organelles, such as mitochondria, where they can be oxidized to generate ATP. Before fatty acids can take part in metabolic reactions they are esterified to coenzyme A (CoA), forming the thiol ester acyl CoA. The formation of different acyl CoA molecules is catalysed by several acyl CoA synthetases, which differ in their subcellular location and their specificity for fatty acids of different chain length. More detail on the acyl CoA synthetases can be found in Weblink Section 8.3(a).

Fatty acid synthesis

De novo fatty acid synthesis usually signifies an excess of energy-yielding substrates; carbon for fatty acid synthesis is supplied by carbohydrate, or, in some cases, amino acids. As discussed in Section 7.5 under 'The oxidation of pyruvate to acetyl CoA', these precursors are metabolized to acetyl CoA in the mitochondria. There are three major steps in the pathway leading to the synthesis of fatty acids (Figure 8.6):

(i) the transport of acetyl CoA to the cytoplasm;

(ii) the formation of malonyl CoA from acetyl CoA; and

(iii) elongation of the fatty acid chain using malonyl CoA as the donor.

Under conditions which favour fatty acid synthesis, citrate that has been formed in mitochondria by condensation of acetyl CoA and oxaloacetate is transported into the cytosol where it is cleaved by ATP citrate lyase to yield acetyl CoA and oxaloacetate (Figure 8.7). This compartmentalization separates fatty acid synthesis, which occurs in the cytosol, from fatty acid oxidation, which occurs exclusively in mitochondria, and the transport step is therefore crucial in control of fatty acid synthesis. The oxaloacetate re-enters the mitochondria as pyruvate, yielding about half the NADPH required for fatty acid synthesis in the process; the other half comes from the pentose phosphate pathway (see Section 7.5, under 'The pentose phosphate pathway: an alternative to glycolysis').

The malonyl group formed by acetyl CoA carboxylase is transferred onto an acyl carrier protein, and then reacts with the growing fatty acid chain, bound to the central acyl carrier protein of the fatty acid synthase complex. The carbon dioxide that was added to form malonyl CoA

is lost in this reaction. For the first cycle of reactions, the central acyl carrier protein carries an acetyl group, and the product of reaction with malonyl CoA is acetoacetyl ACP; in subsequent reaction cycles, it is the growing fatty acid chain that occupies the central ACP, and the product of reaction with malonyl CoA is a keto-acyl ACP.

This intermediate is reduced then dehydrated to yield a carbon–carbon double bond, which is further reduced to yield a saturated fatty acid chain. Thus, the sequence of chemical reactions is the reverse of that in β-oxidation (see section 'Oxidation of fatty acids', below). For both reduction reactions in fatty acid synthesis, NADPH is the hydrogen donor.

The normal end-product of FAS action is palmitic acid, a saturated, 16-carbon fatty acid, which is cleaved from the complex by an integral thioesterase. However, many membranes contain longer-chain fatty acids, which may be unsaturated. These are formed by elongation and/or desaturation of fatty acids after palmitic acid has been cleaved from the FAS complex. Fatty acid synthesis is increased after consumption of a mixed meal, with peak postprandial fatty acid synthesis occurring 3–4 hours after the peak in plasma insulin concentrations. Individuals who are insulin-resistant, or have NAFLD, have high fasting and postprandial hepatic fatty acid synthesis, as the lipogenic capacity of the liver does not become resistant to insulin and hepatic fatty acid/VLDL-triacylglycerol production may be further enhanced.

Elongation and desaturation of fatty acids

Elongases are enzymes that add carbon atoms to pre-formed fatty acids that either have been synthesized *de novo* or originate from the diet. Two elongation systems exist in many tissues, one in the mitochondria and the other in the endoplasmic reticulum. The mitochondrial system involves the addition of 2-carbon units from acetyl CoA, whereas elongation in the endoplasmic reticulum employs malonyl CoA as the donor. An example of elongation is the formation of the 18-carbon saturated fatty acid stearic acid from palmitic acid.

One of the most important roles of the elongases and desaturases is the conversion of the essential fatty acids, linoleic acid and α-linolenic acid, to their longer-chain derivatives (Figure 8.8). Thus, linoleoyl-CoA undergoes sequential desaturation and elongation to form intermediates of the n-6 family of PUFAs, the key product of which is arachidonic acid. As a result of the sequential nature of these reactions, PUFAs usually contain methylene-interrupted double bonds. A similar series of desaturations and elongations generate the n-3 family of polyunsaturated fatty acids. The derivatives of linoleic and

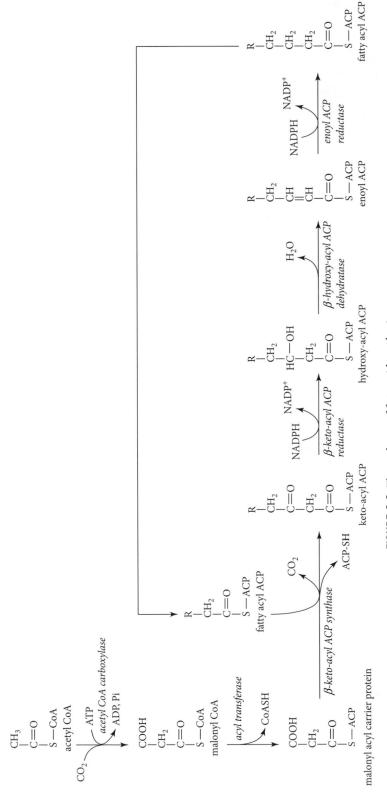

FIGURE 8.6 The pathway of fatty acid synthesis.

Fatty acids are synthesized by the successive addition of two-carbon units from acetyl CoA, followed by reduction. Two key multi-enzyme complexes are responsible for the synthesis of fatty acids from acetyl CoA. The first is acetyl CoA carboxylase, which catalyses the carboxylation of acetyl CoA to malonyl CoA, a three-carbon unit. The second enzyme complex is fatty acid synthase, which catalyses a series of reactions involving the successive addition of two-carbon units to a growing fatty acid chain, using malonyl CoA as the donor of each two-carbon unit. The enzymes are arranged in a series of concentric rings around a central acyl carrier protein (ACP), which carries the growing fatty acid chain from one enzyme to the next. The malonyl group formed by acetyl CoA carboxylase is transferred onto an ACP, and then reacts with the growing fatty acid chain, bound to the central ACP of the fatty acid synthase complex. The carbon dioxide that was added to form malonyl CoA is lost in this reaction. For the first cycle of reactions, the central ACP carries an acetyl group, and the product of reaction with malonyl CoA is acetoacetyl-ACP; in subsequent reaction cycles, it is the growing fatty acid chain that occupies the central ACP, and the product of reaction with malonyl CoA is a keto-acyl-ACP. This intermediate is then reduced to yield a hydroxyl group. In turn, this is dehydrated to yield a carbon–carbon double bond, which is reduced to yield a saturated fatty acid chain. For both reduction reactions in fatty acid synthesis, NADPH is the hydrogen donor.

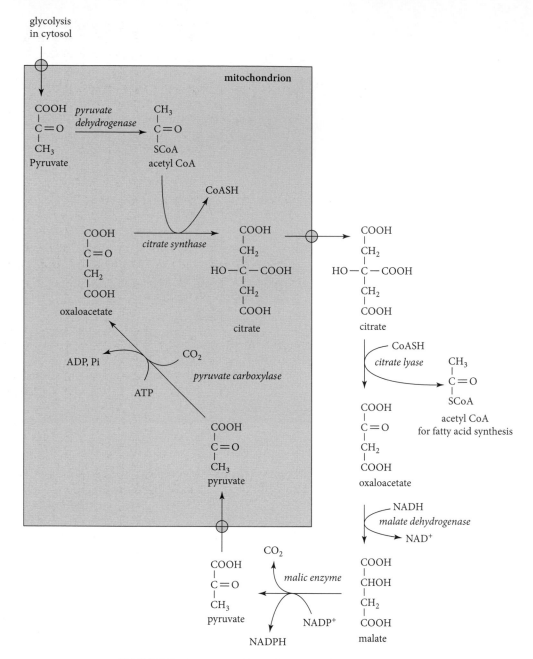

FIGURE 8.7 The source of acetyl CoA for fatty acid synthesis

Fatty acids are synthesized by the successive addition of two-carbon units from acetyl CoA, followed by reduction. Two key multi-enzyme complexes are responsible for the synthesis of fatty acids from acetyl CoA. The first is acetyl CoA carboxylase, which catalyses the carboxylation of acetyl CoA to malonyl CoA, a 3-carbon unit. Its activity is regulated in response to insulin and glucagon. Malonyl CoA is not only the substrate for fatty acid synthesis but also a potent inhibitor of carnitine palmitoyl transferase (see section 'Oxidation of fatty acids' below), so inhibiting the uptake of fatty acids into mitochondria for β-oxidation. The second enzyme complex is fatty acid synthase (FAS), which catalyses a series of reactions involving the successive addition of 2-carbon units to a growing fatty acid chain, using malonyl CoA as the donor of each 2-carbon unit. The enzymes required for fatty acid synthesis form a multi-enzyme complex, arranged in a series of concentric rings around a central acyl carrier protein (ACP), which carries the growing fatty acid chain from one enzyme to the next.

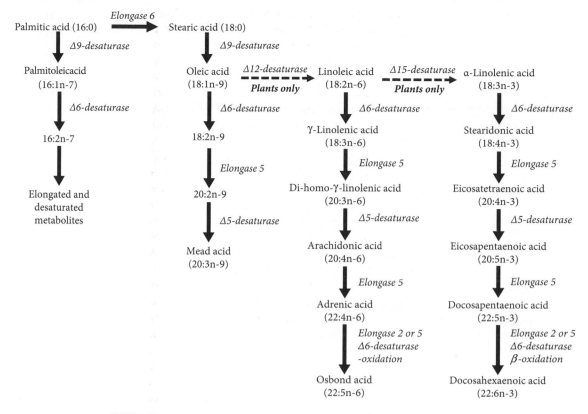

FIGURE 8.8 Overview of the pathways of unsaturated fatty acid synthesis.

α-linolenic acid are often termed 'conditionally essential', since their synthesis is determined by the presence of the essential fatty acid precursors. The extent and regulation of conversion of α-linolenic acid to EPA, DPA, and DHA remains unclear. It appears that in humans, α-linolenic acid can be converted to EPA and DPA, but only very low levels of DHA are synthesized. Women of child-bearing age have a greater capacity to convert α-linolenic acid to EPA, DPA, and DHA than men of a similar age. Importantly, the elongation and desaturation pathways for linoleic and α-linolenic acids share one set of desaturase and elongase enzymes, which means that there is competition between the n-6 and n-3 families of fatty acids.

The elongation and desaturation of oleic acid does not occur to a significant degree in mammalian tissues, However, if essential fatty acid deficiency occurs, oleic acid is desaturated and elongated, usually to mead acid (C20:3 n-9). The presence of mead acid in biological samples is interpreted as a sign of essential fatty acid deficiency.

The factors regulating elongases and some desaturases remain unclear. However, it appears that if a product of elongation or desaturation is present and not essential, then these processes are down-regulated. Insulin and many nutritional factors (such as the composition of dietary fat, the amount of dietary sugars) affect the activity of stearoyl CoA desaturase, the rate-limiting enzyme in the synthesis of monounsaturated fatty acids such as palmitoleic acid (C16:1 n-7) and oleic acid (C18:1 n-9) from palmitic acid (C16:0) and stearic acid (C18:0), respectively.

Oxidation of fatty acids

Fatty acids can undergo oxidation starting at the α-, β- or ϖ-carbon; β-oxidation is the most physiologically important pathway. In β-oxidation, fatty acids are degraded with the sequential liberation of acetyl CoA units. Although mitochondria are the major site for β-oxidation, the peroxisomes also contain the enzymes for this pathway. This additional site is particularly important in the liver, serving to oxidize very long-chain fatty acids to medium-chain products, which are subsequently transported to the mitochondria for complete oxidation. In addition to partial oxidation of long-chain fatty acids, peroxisomes are also the site for the degradation of xenobiotics and eicosanoids.

Once it has entered the mitochondrion, fatty acyl CoA undergoes a repeating series of four reactions, as shown in Figure 8.9, which results in the cleavage of the fatty acid molecule to give acetyl CoA and a new fatty acyl CoA which is two carbons shorter than the initial substrate.

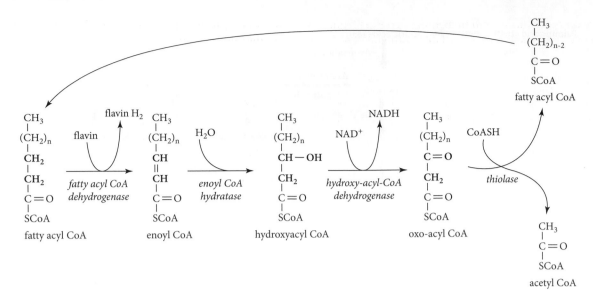

FIGURE 8.9 The pathway of ß oxidation of fatty acids.
The β-oxidation pathway involves four enzymes which act sequentially to cleave two carbons, in the form of acetyl CoA, from the acyl chain.

This new, shorter, fatty acyl CoA is then a substrate for the same sequence of reactions, which is repeated until the final result is cleavage to yield two molecules of acetyl CoA. The reactions of β-oxidation (below), are chemically the same as those in the conversion of succinate to oxaloacetate in the citric acid cycle, and the reverse of those in fatty acid synthesis:

(i) removal of two hydrogens from the fatty acid forms a carbon–carbon double bond—an oxidation reaction which yields a reduced flavin, so for each double bond formed in this way there is a yield of ~2 ATP;

(ii) the newly formed double bond in the fatty acyl CoA then reacts with water, yielding a hydroxyl group—a hydration reaction;

(iii) the hydroxylated fatty acyl CoA undergoes a second oxidation in which the hydroxyl group is oxidized to an oxo group, yielding NADH (equivalent to ~3 ATP);

(iv) the oxo-acyl CoA is then cleaved by reaction with CoA, to form acetyl CoA and the shorter fatty acyl CoA, which undergoes the same sequence of reactions.

More detail on the degradation of fatty acids can be found in Weblink Section 8.3(b).

Regulation of the rate of β-oxidation

The rate of β-oxidation is regulated by two mechanisms, the availability of fatty acids and the rate of utilization of β-oxidation products. The availability of fatty acids is, in turn, dictated by the insulin:glucagon ratio, which, when high, inhibits the breakdown of triacylglycerol in adipose tissue and therefore limits the release of NEFAs from adipose stores. The insulin:glucagon ratio will be high in the fed state, when there is adequate availability of fuel from the ingested food and the release of NEFAs from adipose tissue is therefore not required. In muscle, the rate of β-oxidation is dependent on the plasma NEFA concentration and the energy demand of the tissue. A reduction in energy demand (e.g., when physical activity is low) will lead to accumulation of NADH (which will inhibit the citric acid cycle) and acetyl CoA.

The role of carnitine in fatty acid uptake for oxidation

Fatty acyl CoAs cannot cross the mitochondrial membranes to enter the matrix. On the outer face of the outer mitochondrial membrane, the fatty acid is transferred from CoA onto carnitine, forming acylcarnitine, a process catalysed by the enzyme carnitine palmitoyl transferase 1. The acylcarnitine then enters the inter-membrane space through an acylcarnitine transporter and must then cross the inner mitochondrial membrane (Weblink Figure 8.5). Acylcarnitine can cross only the inner mitochondrial membrane on a counter-transport system which takes in acylcarnitine in exchange for free carnitine being returned to the inter-membrane space. Once inside the mitochondrial inner membrane, acylcarnitine transfers the acyl group onto CoA ready to undergo β-oxidation. This counter-transport system provides regulation of the uptake of fatty acids into mitochondria for oxidation. As long as there is free CoA available in the mitochondrial matrix, fatty acids can be taken up, and the carnitine returned to the outer membrane for uptake of more fatty acids. However, if most of the CoA in the mitochondrion is acylated, then there is no need for further fatty acid uptake immediately, and indeed, it is not possible. Further control is exerted by malonyl CoA (the precursor for fatty acid synthesis) which is a potent inhibitor of carnitine palmitoyl transferase I in the outer mitochondrial membrane. Thus, when malonyl CoA levels are high (e.g., when there is a high insulin:glucagon ratio such as in the fed state) fatty acid β-oxidation is inhibited.

Synthesis of ketone bodies

In the liver, the acetyl CoA formed by β-oxidation is positioned at a crossroad for two important metabolic fates. It can react with either oxaloacetate to form citrate (and hence undergo complete oxidation) or it can form ketone bodies (called ketogenesis). Its fate is determined chiefly by the rate of β-oxidation and the availability of

oxaloacetate. If the rate of β-oxidation is high (as in fasting), then oxaloacetate will be diverted towards gluconeogenesis, so reducing the amount of acetyl CoA entering the citric acid cycle. In this situation, acetyl CoA will be directed towards ketogenesis (Figure 8.10).

Under conditions that favour ketone body synthesis (i.e., during fasting), plasma insulin levels are low and fatty acid β-oxidation to acetyl CoA predominates in the liver and other tissues that are able to oxidize fatty acids. As the liver converts fatty acids to acetyl CoA, the citric acid cycle becomes progressively less able to oxidize the acetyl CoA formed, partly because high amounts of ATP begin to inhibit the activity of the cycle and partly because oxaloacetate is diverted towards gluconeogenesis (see Section 7.5) and so becomes limiting for citric acid cycle activity. This is a situation specific to the liver because of its important role in synthesizing and secreting glucose during fasting and starvation. Acetyl CoA that does not undergo further oxidation is condensed to form the four-carbon compound acetoacetyl CoA, which is further metabolized to form the ketone bodies acetoacetate and β-hydroxybutyrate.

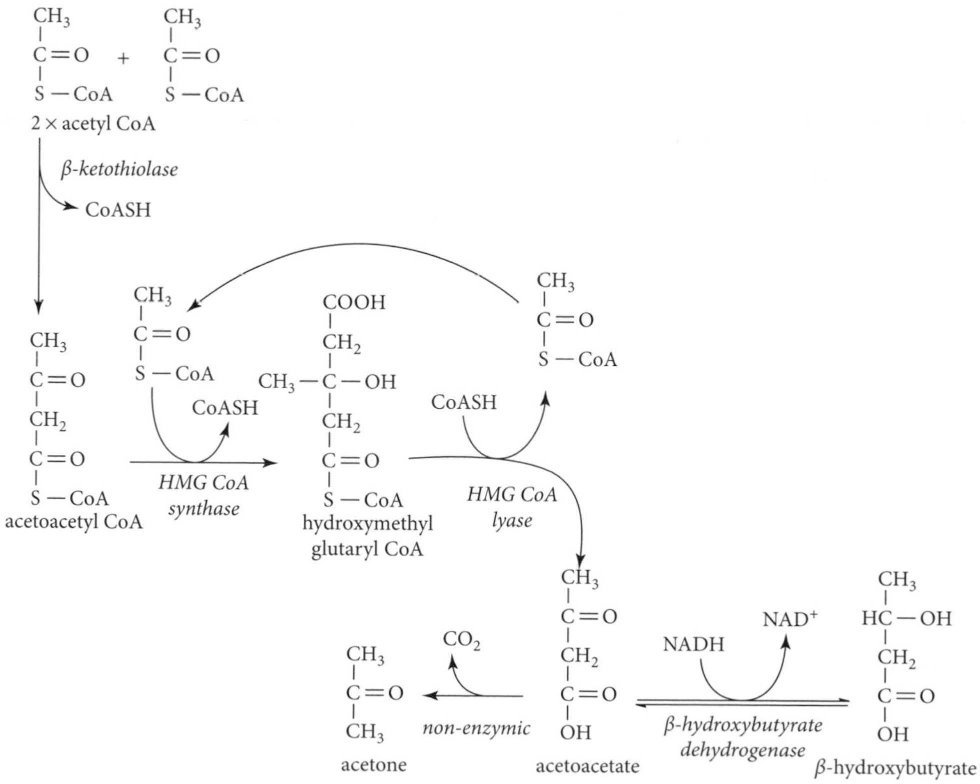

FIGURE 8.10 The pathway of ketone body synthesis.
When the rate of fatty acid oxidation by the liver is high (as in fasting), there is a high likelihood that oxaloacetate will be diverted towards gluconeogenesis and will therefore be present in too low amounts for acetyl CoA to enter the TCA cycle. Acetyl CoA will therefore be directed towards ketogenesis, the first step of which is a condensation reaction joining two molecules of acetyl CoA. There are two ketone bodies, which are the end-products of this pathway, acetoacetate and β-hydroxybutyrate.

Most ketone bodies are converted back into acetyl CoA by muscle and other tissues that are able to use ketone bodies as a fuel (Figure 8.11), and the acetyl CoA is oxidized in the citric acid cycle.

The formation of ketone bodies from fatty acids released by adipose tissue during starvation is extremely important because ketone bodies provide a water-soluble fuel to meet part of the energy requirements of the brain, which cannot oxidize fatty acids, so sparing glucose. Normal levels of circulating ketone bodies in the fed state are approximately 0.01 mmol/L, but they can rise to 0.1 mmol/L after an overnight fast and 6–8 mmol/L following several days of starvation. Excessively high concentrations of ketone bodies (which are acidic) can cause acidosis, inducing coma or death if untreated. This usually only occurs in uncontrolled type 1 diabetes mellitus, when it is termed diabetic ketoacidosis (see Chapter 21).

Synthesis of cholesterol

Cholesterol can be obtained through the diet, but all nucleated cells have the capacity to synthesize cholesterol, with the liver being quantitatively the most important site. About 80 per cent of circulating cholesterol is synthesized in the body. An important function of cholesterol is its structural role in membranes, but it is also important as a precursor for the synthesis of bile acids, steroid hormones, and vitamin D. The precursor for the synthesis of cholesterol is cytosolic acetyl CoA

(Weblink Figure 8.6). The rate limiting step in cholesterol biosynthesis is hydroxymethyl-glutaryl CoA reductase; this is the enzyme inhibited by statins to lower blood cholesterol concentrations. Since high levels of unesterified cholesterol are likely to be undesirable for cells, and cells (other than the liver) are unable to oxidize cholesterol, excess cholesterol is converted into cholesteryl esters by the enzyme acyl CoA cholesterol acyltransferase, which is located on the endoplasmic reticulum. The cholesteryl esters can be stored in lipid droplets within the cytosol; these are commonly observed in steroidogenic tissues such as the adrenal glands and the ovaries. More detail on the synthesis and regulation of intracellular cholesterol can be found in Weblink Section 8.3(c).

Synthesis and utilization of triacylglycerols

Triacylglycerols are both the chief form of dietary fat, and the main form in which fat is stored in the body. Triacylglycerols provide a highly reduced, anhydrous form of metabolic fuel, which can be stored in very large amounts. Whenever energy supply from the diet exceeds the energy expenditure of the body, triacylglycerols are deposited in adipose tissue. In particular cases triacylglycerols may be stored in alternative sites such as muscle, liver, or pancreas, referred to as ectopic fat storage. This generally occurs in obese individuals, with its occurrence

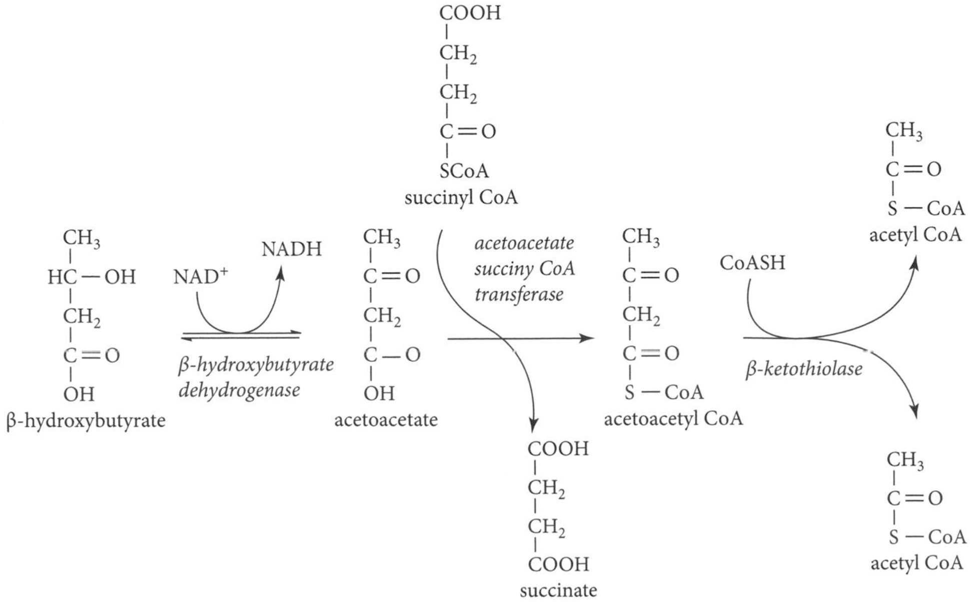

FIGURE 8.11 The pathway of ketone body utilization. The acetyl CoA produced can enter the TCA cycle.

leading to dysregulation of tissue metabolism. Ectopic fat storage is considered to be an underlying cause of insulin resistance and is a risk factor for type 2 diabetes and cardiovascular disease.

White adipose tissue is distributed throughout the body, surrounding many internal organs, and provides a protective subcutaneous layer. The cells within adipose tissue are adipocytes, which are bound together by connective tissue and are supplied by an extensive network of blood vessels. When a fat-containing meal is consumed, adipocytes acquire fat from circulating lipoproteins by hydrolytic breakdown of triacylglycerols by lipoprotein lipase (LPL), releasing fatty acids. Biosynthesis of triacylglycerols involves the esterification of three fatty acids (acyl groups) to a glycerol backbone (Weblink Figure 8.7). It occurs in tissues other than adipose tissue, predominantly liver, enterocytes, and the mammary gland during lactation.

In the reverse situation, when there is a demand for fatty acids for metabolism, triacylglycerols in adipocytes are hydrolysed by adipose triglyceride lipase (ATGL) and also by the enzyme hormone-sensitive lipase (HSL). Removal of one fatty acid by ATGL leaves a diacylglycerol, the next fatty acid is removed by HSL. ATGL and HSL act at the surface of the triacylglycerol droplet. Monoacylglycerol lipase is responsible for the removal of the third fatty acid, so three fatty acids and one glycerol molecule are produced from each molecule of stored triacylglycerol. Along with fatty acids, glycerol also leaves the cell as it cannot be utilized for esterification of fatty acids as adipose tissue does not have the enzyme glycerol kinase which is necessary for this. The regulation of HSL is best understood (Weblink Figure 8.8), whilst less is known about the regulation of ATGL, which was only discovered in 2004. ATGL activity is strongly stimulated by an activator protein—comparative gene identification-58 (CGI-58). The phases of triacylglycerol mobilization are integrated and controlled by the nutritional status of the individual through a number of hormones, the most important of which is insulin (see 'Integration of fat metabolism from the fasted to the fed state at the whole-body level', in Section 8.5 below).

Under conditions where the demand for mobilization of fuel reserves increases, usually signalled by low concentrations of insulin, biosynthetic pathways are inhibited and ATGL and HSL are activated within adipocytes. Once released, the NEFAs are bound to plasma albumin and may be taken up by tissues that are able to utilize fatty acids as a fuel source.

More detail on the synthesis and degradation of triacylglycerols can be found in Weblink Section 8.3(d).

8.5 Integration, control, and dynamics of fat metabolism

Coordinated regulation of fatty acid synthesis and oxidation

In the fed state, when carbohydrate may be used in the synthesis of fatty acids, the concentration of malonyl CoA is raised and this results in inhibition of carnitine palmitoyl transferase 1 (CPT1), which controls the uptake of acyl CoA into mitochondria for oxidation. Hence this situation inhibits β-oxidation of fatty acids. In the fasting state, the reverse situation occurs and CPT1 activity is high, promoting β-oxidation and ketogenesis. Coinciding with this, in the fasting state, a low insulin:glucagon ratio and/or the release of adrenaline inhibits acetyl CoA carboxylase activity (Weblink Figure 8.9), reducing the synthesis of malonyl CoA and relieving the inhibition of CPT. More detail on the regulation of acetyl CoA carboxylase and FAS can be found in Weblink Section 8.4.

Tissues such as muscle, that oxidize fatty acids but do not synthesize them, also have acetyl CoA carboxylase, and produce malonyl CoA in order to control the activity of carnitine palmitoyl transferase I, and thus control the mitochondrial uptake and β-oxidation of fatty acids. Tissues also have malonyl CoA decarboxylase, which acts to remove malonyl CoA and so reduce the inhibition of carnitine palmitoyl transferase I. The two enzymes are regulated in opposite directions in response to insulin, which stimulates fatty acid synthesis and reduces β-oxidation, and glucagon, which reduces fatty acid synthesis and increases β-oxidation.

Fatty acids are the major fuel for red muscle fibres, which are the main type involved in moderate exercise. Children who lack one or other of the enzymes required for carnitine synthesis, and are therefore reliant on a dietary intake, have poor exercise tolerance, because they have an impaired ability to transport fatty acids into the mitochondria for β-oxidation. Provision of supplements of carnitine to the affected children overcomes the problem which is fairly rare.

Integration of fat metabolism with the metabolism of carbohydrate and protein

As described above, fats can circulate in the blood in the form of NEFAs, as triacylglycerol in lipoproteins, and as ketone bodies in the fasted state. In addition to these fat-derived fuels, the circulation provides glucose, lactate, pyruvate, glycerol, and amino acids, which may all be oxidized to generate energy. What determines which of these fuels is oxidized by a tissue at any given time? This question is best answered by considering:

- The ability of tissue to oxidize the fuel; some tissues are anaerobic, or lack mitochondria, and therefore cannot oxidize fatty acids or ketone bodies.

- The availability of fuel; this will be determined by the prevailing conditions. If an individual has been fasting for 18 hours, liver glycogen stores will be largely depleted and the circulating concentration of fatty acids will be high.

In summary, fatty acids are the preferred fuel for oxidation whenever their circulating concentration is high and glucose is spared whenever necessary (see Figure 8.12). When the energy provided by the diet exceeds immediate requirements, excess carbohydrate is preferentially used to replenish liver glycogen stores. Excess amino acids will be oxidized only after satisfying the needs for protein synthesis (see Chapter 9). Any remaining excess of fuel will be used for fatty acid and triacylglycerol synthesis for storage in adipose tissue (see Figure 8.12). All of these processes will be coordinated by changes in the circulating levels of hormones, the most important of which are insulin and glucagon.

Interconversion of fuels and the energy paradox

The following rules regarding interconversion of fuels are absolutely central to understanding the integration of metabolic pathways:

- fatty acids can be made from carbohydrates and amino acids, but cannot be converted to either;

- carbohydrates can be made from amino acids and can be used to make fatty acids and triacylglycerols.

The inability to convert fatty acids to glucose gives rise to what is known as the 'energy paradox'. The basis of this paradox is that the brain requires 500 kcal of water-soluble fuel (usually glucose) per day, yet the chief energy store in the body is fat, not glycogen, and fatty acids cannot be converted to glucose. The energy paradox is dealt with in four ways.

(i) The oxidation of fatty acids (especially by muscle) spares glucose for use by the brain.

(ii) Lipolysis of triacylglycerol during starvation releases glycerol as well as fatty acids and the glycerol can be used as a substrate for gluconeogenesis. Thus, glucose can be synthesized from the glycerol component of triacylglycerol.

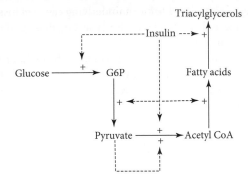

A: DURING MEALS

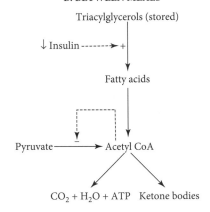

B: BETWEEN MEALS

FIGURE 8.12 Integration of carbohydrate and fat metabolism in the fed and fasted states. When the energy provided by the diet exceeds immediate requirements, excess carbohydrate is preferentially used to replenish liver glycogen stores. Any remaining excess of fuel will most likely be converted into triacylglycerol for storage in adipose tissue (A). All of these processes will be coordinated by changes in the levels of hormones in the blood, the most important of which are insulin and glucagon. Between meals (B), decreasing insulin concentrations allow the hydrolysis of stored triacylglycerol, and fatty acids are released. The oxidation of fatty acids (to generate ATP by many tissues or ketone bodies in the liver) allows the sparing of glucose through the glucose–fatty acid cycle (B).

(iii) Fatty acid oxidation in the liver provides the ATP required for gluconeogenesis.

(iv) Fatty acids can be converted to ketone bodies, a water-soluble fuel which can be used by the brain to meet perhaps one fifth of its energy needs in starvation.

Mechanisms for integration of fat and carbohydrate metabolism

Control of phosphofructokinase (PFK) activity

As discussed in Chapter 7, phosphofructokinase (PFK) catalyses an irreversible and controlling step in glycolysis. Control of this enzyme is key to the integration of the metabolism of fat and carbohydrate. High levels of ATP inhibit PFK and therefore inhibit glycolysis. In tissues that are oxidizing fatty acids (e.g., during fasting or exercise), large amounts of ATP are generated. As a result, the oxidation of fatty acids will prevent oxidation of glucose by inhibiting glycolysis and glucose will be spared for other tissues. This regulatory mechanism is termed the 'glucose–fatty acid cycle'.

Control of pyruvate dehydrogenase (PDH) activity

Pyruvate dehydrogenase catalyses the irreversible oxidation of pyruvate to acetyl CoA. When glucose is freely available, PDH is active and acetyl CoA does not accumulate because it is rapidly used for synthesis of citrate and either complete oxidation in the citric acid cycle or fatty acid synthesis (if carbohydrate is in excess). However, when glucose supplies are diminished and plasma NEFA levels increase as a result of lipolysis in adipose tissue, the oxidation of fatty acids in tissues results in an increase in intracellular acetyl CoA, ATP, and NADH. These inhibit PDH activity, reinforcing the glucose–fatty acid cycle. Thus, oxidizing fatty acids will conserve glucose by inhibiting both PFK and PDH.

Integration of fat metabolism from the fasted to the fed state at the whole-body level

Non-esterified fatty acids

Following an overnight fast, the plasma NEFA concentration is normally about 0.5 mmol/L and the triacylglycerol concentration about 1 mmol/L (largely contributed by VLDLs) in healthy, lean individuals; both these concentrations are higher in individuals with overweight or obesity. NEFAs are released by lipolysis of adipose tissue triacylglycerol by ATGL and HSL and they are taken up by a number of tissues, including skeletal muscle and liver. The regulatory mechanisms which lead to the activation of ATGL and HSL and the reverse process, the esterification of fatty acids in adipose tissue, are key determinants of the plasma concentration of NEFAs. The rate of oxidation of NEFAs by tissues depends mainly on their plasma concentration, so that the higher the concentration, the greater the rate of utilization. The plasma concentration of NEFAs is directly related to the rate of their release from adipose tissue (and therefore activation of the lipases involved versus esterification). Since the key regulatory signal for activation of HSL is a fall in insulin

concentration, the plasma NEFA concentration over the course of a day is normally an inverse reflection of the plasma concentrations of glucose and insulin. In the fasting state the concentrations of glucose and insulin are at their lowest and those of NEFAs highest. The plasma concentration of NEFAs has an upper limit of approximately 2 mmol/L, because above this concentration, the relative proportion of NEFAs which are not bound to albumin increases; NEFAs not bound to albumin will cause significant haemolysis and may also have adverse effects on other tissues, particularly the heart.

Following consumption of a meal (the absorptive phase), the rise in blood glucose concentration stimulates insulin secretion, which suppresses lipase activity within adipose tissue, and the plasma concentration of NEFAs will subsequently fall to <0.1 mmol/L. The concentration at which insulin inhibits the activity of HSL is much lower than the concentration at which it stimulates glucose metabolism. This means that NEFAs fall very rapidly and dramatically very soon after food ingestion. However, the activity of this enzyme is never completely suppressed, even at very high concentrations of insulin. After a meal, the increase in plasma glucose and insulin concentrations will increase the uptake of glucose and of glycolysis within the adipocyte, and, as a result, glycerol-3-phosphate will become available for re-esterification of fatty acids, so any NEFA released by the action of ATGL and hormone sensitive lipase will be re-esterified. After a meal, therefore, release of NEFAs from adipose tissue will be almost completely suppressed in lean, healthy individuals although it remains higher in individuals who are insulin resistant or have NAFLD.

As a result of the fall in NEFA concentration, tissues that were oxidizing fatty acids in the fasted state (and so sparing glucose) will reduce their uptake and oxidation of fatty acids and utilize glucose once more. The increased insulin:glucagon ratio on feeding also leads to a reduction in the synthesis of ketone bodies by the liver, so that plasma levels of ketone bodies fall from overnight fasted values of 0.1–0.2 mmol/L to levels that are almost undetectable. The absorptive phase finally begins to decline after about five hours, the exact time depending on the composition of the meal, allowing insulin concentrations to decline and relaxing the restraint on fat mobilization. In general, a meal containing a significant amount of fat slows absorption.

Triacylglycerol

After an overnight fast, the plasma triacylglycerol concentration is normally about 1 mmol/L, almost all in the endogenous triacylglycerol-rich lipoprotein particles (VLDLs). Consumption of a meal containing fat results in the formation of chylomicrons in the enterocyte and their entry into the bloodstream approximately 3–5 hours after a meal. The postprandial plasma concentration of triacylglycerol can rise to between 1.5 mmol/L and 3.0 mmol/L, depending on the amount of fat in the meal and the metabolic capacity of the individual. The magnitude and duration of the postprandial lipaemic response will depend on the efficiency of the regulatory mechanisms for the disposal and storage of the triacylglycerol. Lipoprotein lipase is upregulated by insulin and will therefore be most active following a meal; at the adipose tissue its activity reaches a peak approximately 3–4 hours after a meal, coinciding with the peak in postprandial plasma triacylglycerol. The hydrolysis of chylomicrons and VLDLs in tissues is shown in Figure 8.13.

Insulin clearly plays a key role in the coordination of all aspects of fat metabolism, since both the hydrolysis of chylomicron-triacylglycerol by LPL and the subsequent uptake of the liberated fatty acids are facilitated by the fact that the activity of ATGL and HSL are suppressed. Furthermore, insulin also promotes the re-esterification of fatty acids to form triacylglycerols for storage. However, adipose tissue is not the only tissue able to utilize the fatty acids released from chylomicron-triacylglycerols by the action of lipoprotein lipase. Skeletal muscle, for

example, uses fatty acids from chylomicrons (or VLDLs) as a source of energy.

During the postprandial period, chylomicron-triacylglycerols represent only a proportion of the total plasma triacylglycerols (perhaps 0.3–0.4 mmol/L after a very fatty meal). This is because the endogenous pathway (VLDL synthesis) is always active and after a meal the hydrolysis of VLDLs is suppressed in favour of hydrolysis of chylomicrons. In addition, not all the NEFAs released from chylomicrons are taken up and re-esterified in adipose tissue; there is some spillover into the systemic circulation and subsequent uptake by the liver can result in them being used as substrates for the synthesis of VLDLs. Thus, a significant proportion of the postprandial lipaemic response is, in fact, contributed by VLDLs. It should also be noted that the duration of elevation of triacylglycerol in blood following a fat-containing meal is quite prolonged. It may be 6–8 hours before concentrations return towards the fasted values and because most people eat fat-containing meals throughout the day, postprandial lipaemia is the normal state. Once chylomicrons have been completely hydrolysed and their remnants removed by the liver, the exogenous pathway ceases and the endogenous pathway once again becomes the dominant route of triacylglycerol metabolism in the body.

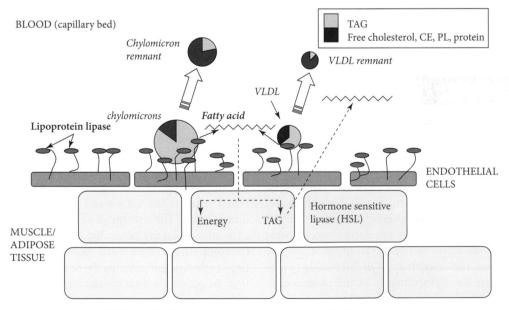

FIGURE 8.13 Hydrolysis of chylomicrons and VLDLs in tissues.
Chylomicrons and VLDLs are hydrolysed in adipose tissue or muscle by lipoprotein lipase, which is attached to the luminal surface of the capillary endothelial cells. The released fatty acids enter the underlying cells and are immediately used for energy or re-esterified into triacylglycerol (TAG), thereby maintaining a concentration gradient. At times of energy deficit, stored TAGs are hydrolysed by hormone-sensitive lipase and the fatty acids released into the circulation.

KEY POINTS

- Fats perform a range of essential functions in the body; they can be stored for later release of energy, they are important structural components of cell membranes, they play roles in cell signalling, are essential for the absorption of fat-soluble vitamins, and are precursors for the synthesis of hormones and other physiological mediators.

- Dietary fats are packaged into chylomicrons in enterocytes within the small intestine and enter the exogenous lipoprotein pathway via the lymph system. The triacylglycerols they carry are hydrolysed by the enzyme lipoprotein lipase, which is found on the surface of endothelial cells lining capillaries. The NEFAs released are taken up for use by the tissues.

- The endogenous pathway of lipoprotein metabolism involves the synthesis of VLDLs by the liver and their subsequent metabolism to LDLs. Chylomicrons and VLDLs are carriers of triacylglycerols, while LDL and HDL transport cholesterol.

- Linoleic and α-linolenic acids are essential because they cannot be synthesized by animal cells. These essential fatty acids give rise to other n-6 and n-3 PUFAs respectively through the action of desaturases and elongases. The metabolism of the essential fatty acids is competitive because the same set of enzymes is shared by the n-6 and n-3 pathways.

- The oxidation of fatty acids is regulated by their availability and their rate of utilization. Fat is stored as triacylglycerol in adipose tissue, which can be mobilized by the action of hormone-sensitive lipase during fasting. Fats can circulate as triacylglycerols in lipoproteins, NEFAs, or ketone bodies (in fasting). Fatty acids are the preferred fuel for oxidation whenever their circulating concentrations are high and glucose is spared whenever necessary.

ACKNOWLEDGEMENTS

The authors would like to acknowledge the contributions of Leanne Hodson, Anne-Marie Minihane, and Christine Williams to earlier editions of this chapter.

 Be sure to test your understanding of this chapter by attempting multiple choice questions. See the Further Reading list for additional material relevant to this chapter.

REFERENCES

Austin M A, King M-C, Vranizan K M et al. (1990) Atherogenic lipoprotein phenotype: A proposed genetic marker for coronary heart disease. *Circulation* **82**, 495–506.

Calder P C (2015) Functional roles of fatty acids and their effects on human health. *J Parent Ent Nutr* **39**(1 Suppl), 18S–32S.

Calder P C (2020) Eicosanoids. *Essays Biochem* **64**, 423–441.

Castelli W P (1986) The triglyceride issue: A view from Framingham. *Am Heart J* **112**, 432–437.

Chiang N & Serhan C N (2020) Specialized pro-resolving mediator network: An update on production and actions. *Essays Biochem* **64**, 443–462.

Goldstein J L, Kita T, Brown M S (1983) Defective lipoprotein receptors and atherosclerosis. *New Engl J Med* **286**, 283–296.

Havel R (1994) McCollum Award Lecture 1993: Triglyceride-rich lipoproteins and atherosclerosis: new perspectives. *Am J Clin Nutr* **59**, 795–799.

Mensink R P, Zock P L, Kester A D et al. (2003) Effects of dietary fatty acids and carbohydrates on the ratio of serum total to HDL cholesterol and on serum lipids and apolipoproteins: A meta-analysis of 60 controlled trials. *Am J Clin Nutr* 77, 1146–1155.

Ruotolo G & Howard B V (2002) Dyslipidaemia of the metabolic syndrome. *Curr Cardiol Rep* **4**, 494–500.

FURTHER READING

Assmann G (1993) *Lipoprotein Metabolism Disorders and Coronary Heart Disease*. MMV Medizin Verlag, Munich.

Betteridge D J, Illingworth D R, Shepherd J (1999) *Lipoproteins in Health and Disease*. Arnold, London.

British Nutrition Foundation (1992) *Unsaturated Fatty Acids: Nutritional and Physiological Significance—Report of the British Nutrition Foundation's Task Force*. Chapman & Hall, London.

Evans R & Frayn K H (2019) *Human Metabolism: A Regulatory Perspective*. Wiley Blackwell Science, Oxford.

Frayn K N (2010) *Metabolic Regulation: A Human Perspective*, 3rd edn. Wiley Blackwell Science, Oxford.

Gunstone F D, Harwood J L, Padley F B (1994) *The Lipid Handbook*, 2nd edn. Chapman & Hall, London.

Gurr M I, Harwood J L, Frayn K N et al. (2016) *Lipids: Biochemistry, Biotechnology and Health*, 6th edn. Wiley-Blackwell, Oxford.

Vance D E & Vance J E (1996) *Biochemistry of Lipids, Lipoproteins and Membranes*. Elsevier, Amsterdam.

Protein metabolism and requirements

Joe D. Millward and Andy Salter

9.1 Introduction

Proteins, polymers of l-α-amino acids linked through peptide bonds, are by far the most diverse of the large macromolecules which provide structure and enable function of the organism. Diversity reflects the chemistry of the side chains of the 21 unique amino acids occurring in proteins (see Figure 9.1). The human proteome comprises ~20,000 primary proteins, each displaying a unique shape determined by the interactions of the amino acid side chains with water and with each other, both within each primary molecule and between different polypeptide chains which form multimeric protein structures. Thus, human proteins comprise a wide range of structures with many different functions (see Table 9.1). This chapter begins with a consideration of proteostasis, the maintenance of an intact proteome, the entire set of expressed proteins within a cell or tissue, through control of protein synthesis and proteolysis, followed by a brief overview of the integration of proteostasis with amino acid metabolism in the whole body and a review of amino acid metabolism as the metabolic setting for a consideration of the nutritional requirement for protein and amino acids. This includes a review of the magnitude of the protein requirements in the context of an adaptive metabolic demands model, focusing on the implications of this model for current values of protein requirements. Protein quality and its assessment is considered in the context of the nutritional classification of amino acids, and the nutritional value of traditional and novel plant proteins is reviewed. Finally, the impact of protein production, and how we are going to sustainably meet the future requirements of the expanding and aging human population is considered.

9.2 Proteostasis

The body protein pool

The body protein pool, about 11kg for a 70kg adult man, ~20 per cent of the fat free mass, is distributed between the cellular mass (75 per cent), extracellular solids (bone, cartilage, tendons, fascia), 23 per cent, and a minor part in the extracellular fluid (2 per cent). Of the cellular mass within the organs, skeletal muscle protein accounts for about 50 per cent in adults.

All intracellular proteins and many extracellular proteins continually turn over, that is, they are degraded by proteolytic enzymes in all cells and replaced by new

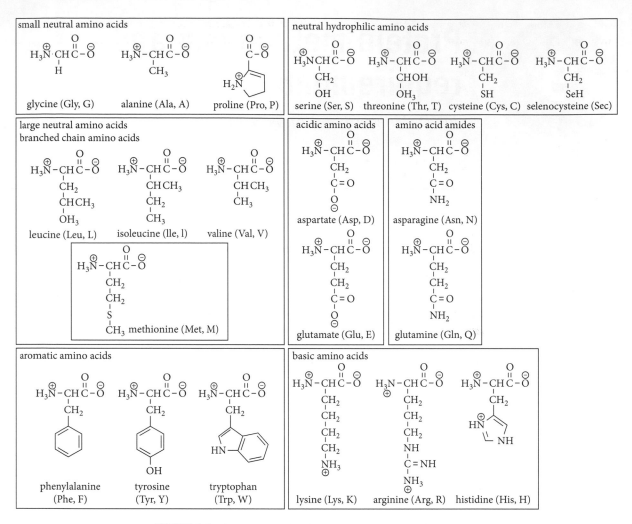

FIGURE 9.1 The amino acids, showing three and one letter codes.
Those on the left are hydrophobic, with those on the right, hydrophilic. Within the protein molecule in general the hydrophobic sequences will tend to avoid water and form the interior of the final structure with the hydrophilic sequences on the outside.

TABLE 9.1 What do proteins do

Function	Examples
Muscular contraction	Myosin, actin, tropomysin, & troponin
Intracellular movement	Kinesin
Structure	Collagens, elastin, actin
Enzymatic catalysis	Hexokinase, citrate synthetase, glutamate dehydrogenase
Transport in blood	B_{12} binding proteins, ceruloplasmin, apolipoproteins
Immunity	Antibodies
Plasma oncotic pressure	Albumin
Storage/sequestration	Ferritin, metallothionein
Peptide hormones	Insulin, glucagon, growth hormone IGF-1
Intercellular signalling molecules	Cytokines
Intracellular signalling molecules	Tyrosine kinase, mTOR
Regulatory proteins	Protein synthesis initiation factors, peptide growth factors

synthesis, a phenomenon first suggested by the French physiologist François Magendie in Paris early in the nineteenth century and subsequently investigated with isotopic tracers in animals and humans in the twentieth century. The term proteostasis is used to describe the maintenance of an intact proteome through strict control of the initial production and folding of a protein, its conformational maintenance, control of abundance and subcellular localization, and disposal by degradation. Proteostasis accounts for a significant part of cellular energy expenditure and is managed by the proteostasis network, which contains in all 1000–2000 protein components. Individual proteins in the nucleus and cytosol, as well as in the endoplasmic reticulum (ER) and mitochondria, turnover at widely differing rates that vary from minutes for some regulatory enzymes, to days or weeks for proteins such as actin and myosin in skeletal muscle, or months for haemoglobin which turns over with the entire red cell during its destruction within the spleen. With the development of proteomic techniques for the separation and analysis of all proteins within cells, and with stable isotope labelling, it is now possible to evaluate the turnover rates of proteins within the entire proteome.

The genetic code and protein synthesis

The primary structure, the polypeptide amino acid sequence, of each of the 20,000 proteins in the human proteome is encoded as the human genome in DNA, mostly in nuclear chromosomes. DNA comprises a double helix of two linear polymers of nucleotides (base + sugar + phosphate), with the backbone comprising the pentose sugar, deoxyribose, linked by phosphate diester bonds. The bases of the four different nucleotides in one strand, the purines, adenine (A), and guanine (G), and the pyrimidines, thymine (T), and cytosine (C), project from this sugar phosphate backbone and bind with specific bases in the second strand of DNA (A with T and G with C), to form the double helix structure as discovered by Watson and Crick. The information for the amino acid sequence of proteins is coded in the form of successive triplet sequences of nucleotides which correspond to each amino acid. Of the two strands one acts as the 'sense' strand, containing the genes, with the other acting as an 'antisense' strand with a sequence of matching anti-codons.

Of the 64 possible mRNA codons, three are used to designate the end of a protein chain, (STOP codons), the other 61 code for the 21 amino acids (Table 9.2). Thus, the genetic code is degenerate, that is, there is more than one codon for each amino acid, three on average, with Met and Trp having only one codon, and Leu, Ser, and Arg having six each. This uneven distribution of codons between amino acids has influenced to some extent the

TABLE 9.2 The genetic code, showing the codons in mRNA

First base	Second base				Third base
	U	C	A	G	
U	Phe	Ser	Tyr	Cys	U
U	Phe	Ser	Tyr	Cys	C
U	Leu	Ser	STOP	STOP*	A
U	Leu	Ser	STOP	Trp	G
C	Leu	Pro	His	Arg	U
C	Leu	Pro	His	Arg	C
C	Leu	Pro	Gln	Arg	A
C	Leu	Pro	Gln	Arg	G
A	Ile	Thr	Asn	Ser	U
A	Ile	Thr	Asn	Ser	C
A	Ile	Thr	Lys	Arg	A
A	Met	Thr	Lys	Arg	G
G	Val	Ala	Asp	Gly	U
G	Val	Ala	Asp	Gly	C
G	Val	Ala	Glu	Gly	A
G	Val	Ala	Glu	Gly	G

* UGA also codes for selenocysteine in a specific context.

amino acid composition of proteins, since in the database of more than 0.5×10^6 known protein sequences, the frequency of occurrence of Leu, Met, and Trp is 9.7 per cent, 2.4 per cent, and 1.1 per cent respectively, with their frequency in some food proteins shown in Table 9.3. However, because of the specific amino acid sequence of the individual proteins most abundant in the various food protein sources, the average amino acid content of plant and animal source proteins in food differs (see Table 9.3). Thus, in muscle proteins (e.g., beef in Table 9.3), Lys is more abundant than would be expected from its two codons, as are Met and Cys in egg albumin, and tryptophan in milk proteins. In cereal grains, the major proteins are the prolamin storage proteins, which contain a high Gln and Pro and low Lys content. These include wheat gluten, barley hordein, rye secalin, maize zein, sorghum kafirin, and as a minor protein, avenin in oats. The unique structure and high content of gluten in wheat enables wheat flour to be made into bread and pasta. Zein in maize is low in Try as well as Lys. The globulin storage proteins in legumes have lower levels of Met and Cys than in cereals or in animal source foods. The implications of these differences in amino acid content for the nutritional quality of food proteins is discussed below.

TABLE 9.3 Amino acid composition of food protein sources (mg/g protein)

	Codons	Egg	Beef	Milk	Soya	Wheat	Potato	Rice	Maize	Mean plant	Mean animal
Leu	6	83	80	104	76	72	61	86	136	86	89
Thr	4	51	46	44	38	29	38	35	36	35	47
Val	4	75	53	51	50	48	51	61	53	53	60
Ile	3	56	51	38	48	36	42	40	37	37	48
Lys	2	62	91	71	65	26	54	39	26	26	75
Phe + Tyr	2.2	46	42	41	43	40	37	46	48	48	43
Cys + Met	2.1	25	20	18	13	22	14	19	15	15	21
Trp	1	18	13	25	13	12	14	13	7	7	18

Transcription involves the copying of the DNA nucleotide sequence of an individual gene on the sense strand to make a matching (antisense), ribonucleotide sequence of a messenger (m)RNA molecule, which, after some processing, moves out of the nucleus into the cytoplasm. The triplet sequences of ribonucleotides in mRNA, U, C, A, and G, are what we identify as the genetic code (see Table 9.2), even though they are in fact the antisense triplets.

Translation involves the assembly of the polypeptide sequence indicated by the codon sequence of the mRNA molecule. As shown in Figure 9.2 it is achieved by

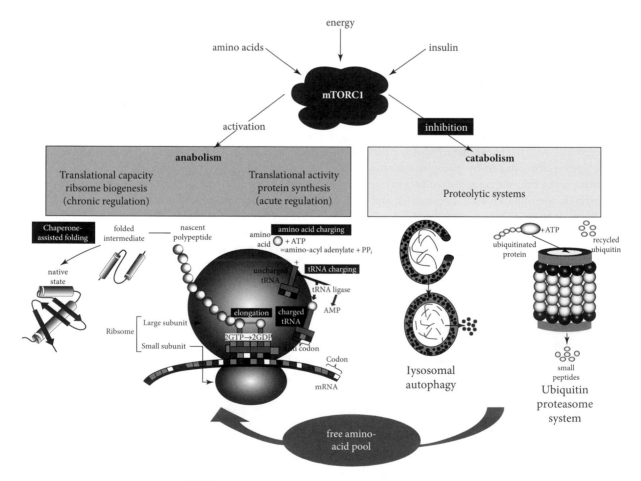

FIGURE 9.2 Proteostasis and its regulation by mTORC1.

ribosomes, large protein-RNA complexes and the smaller amino-acyl transfer RNA (tRNA) molecules. Amino acids are first activated by ATP, then transferred to their specific tRNA, and then aligned by the ribosome with their specific codon on the mRNA, and finally chemically joined to the previous amino acid in the growing polypeptide chain. This is released as the completed protein after the final amino acid is added and a STOP codon is reached. After release the completed polypeptide chain is folded into its active shape by a series of specific chaperone proteins. (See Weblink Section 9.1 for more details of the genetic code and protein synthesis.)

Protein degradation

Multiple proteolytic systems in cells enable turnover, with complex regulatory mechanisms to ensure that turnover is highly selective and to prevent excessive breakdown of cell constituents, but two systems are under nutritional control and are likely to account for a large proportion of whole-body protein turnover (WBPT, see Figure 9.2). The autophagic–lysosomal pathway involves a number of acidic proteases within the membrane-bound lysosome which are capable of degrading any protein or even cellular structures such as mitochondria to amino acids. Proteins can be delivered to a lysosome for degradation by several different mechanisms, including chaperone-mediated autophagy which involves proteins selectively recognized through a specific amino acid sequence motif present on 40–50 per cent of all proteins, and delivered by a cytosolic chaperone to the lysosomal surface, where, upon unfolding, they are internalized through a membrane translocation complex.

Another major system is the ubiquitin–proteasome pathway which degrades only individual proteins. Ubiquitin is a small peptide that is attached to the ε-amino groups of lysine residues in target proteins for proteolysis in a three stage (E1, E2, and E3), ATP-dependent process to mark them for degradation by a very large multicatalytic protease complex, the 26S proteasome. This proteolytic machine first unfolds the ubiquitinated proteins then degrades them to small peptides, which are released and subsequently degraded to amino acids by cytosolic peptidases, with the ubiquitin recycled. The molecular determinants of which proteins are targeted for ubiquitination are defined by the E3 ubiquitin ligase enzymes, of which 500–1000 different species are encoded in the human genome.

The regulation of proteostasis

Proteostasis ensures that the rates of protein synthesis and degradation within cells are balanced precisely to avoid rapid growth or atrophy of cellular mass and is regulated through highly complex multiple signalling systems which exert fine control. Signal transduction involves hormones, acting through receptors on the cell surface and within the nucleus, of which insulin is the most important mediator of nutritional regulation; cytokines, small signalling peptides which do not cross the plasma membrane of cells; and metabolites which provide energy and amino acids, the latter acting both as substrates for and specific activators of protein synthesis. At the centre of the regulatory network in most cells is one complex, mTOR, the mammalian (or mechanistic) Target of Rapamycin. mTOR is a highly conserved serine/threonine kinase which phosphorylates other proteins involved in signal transduction pathways regulating cell growth and metabolism. This occurs in cells in the form of two structurally and functionally distinct multiprotein complexes mTORC1 and mTORC2, the former acting as the control centre for the nutritional regulation of proteostasis. As shown in Figure 9.2, mTORC1 integrates signals from amino acids, the energy status of the cell and growth factor pathways, for example, insulin, all of which change after food intake, promoting protein synthesis and inhibiting proteolysis.

Thus, after a meal providing protein and energy, which induces an amino acidaemia and insulin secretion, protein deposition is mediated by the combination of amino acids, especially Leu, stimulating protein synthesis, and insulin, inhibiting proteolysis via mTORC1. The stimulation of the anabolic responses includes both the chronic induction of ribosomal biogenesis (e.g., by activating ribosomal protein S6 kinase), to increase the translational capacity for protein synthesis, and the acute stimulation of translation of mRNA molecules to synthesize new proteins (e.g., by activating initiation factor 4E-binding protein 1). At the same time the catabolic phase of protein turnover is inhibited with both lysosomal autophagy and the ubiquitin proteasome system inhibited. When the plasma concentrations of amino acids and insulin fall in the postabsorptive state these feeding responses are reversed resulting in a net catabolic state.

The energy cost of proteostasis

Within proteostasis, protein turnover accounts for a major part of the cell's energy expenditure, although the actual cost is uncertain. An approximate estimate is about 5 moles of ATP per mole of peptide bond synthesis (4 moles for translation and 1 additional mole for miscellaneous processes) amounting to about 14 per cent of the BMR. (See Weblink Section 9.2 for more details of the regulation and energy cost of proteostasis.)

The functional importance of protein turnover within proteostasis

Protein turnover is fundamental to proteostasis and some argue that loss of proteostasis is a 'hallmark' of aging, which might involve protein unfolding, oxidative damage, and post-translational modifications and aggregation, resulting in altered rates of protein turnover. In this context age-related diseases and conditions can be associated with the inability of the cell to maintain its healthy proteome and eliminate defective proteins, as observed in neurodegenerative diseases, cardiac dysfunction, cataracts, and sarcopenia. The overall concept is therefore that protein turnover is essential for cellular 'quality control'.

Another perspective is the role of protein turnover within the day-to-day, and hour-to-hour regulation of the overall mass of tissue protein, within the nutritionally sensitive tissues and organs of the splanchnic bed and skeletal muscle. It is the case that in healthy individuals in the postabsorptive state there is a net loss of tissue protein which acts as an energy store providing free amino acids as fuel and to satisfy the obligatory metabolic demand (see Section 9.3) that is, protein degradation > synthesis, resulting in amino acid oxidation and utilization for ATP production either directly or after conversion in the liver to glucose or ketones with the nitrogen excreted as urea. This continues until a meal provides energy and protein to reverse the catabolism and mediate net protein deposition to replace the postabsorptive losses. This diurnal cycling can only occur because of continuing protein degradation and because both protein synthesis and degradation are sensitive to nutrition. In the whole body both proteolysis and protein synthesis are sensitive

to feeding which mediates an inhibition of postabsorptive proteolysis of up to 50 per cent together with an increase in protein synthesis. The continuing intracellular proteolysis even in the postprandial state means that the continual removal and replacement of cellular proteins, as with 'quality control' described above, is absolutely necessary for normal functioning of cellular processes and metabolism.

It is often stated that the reutilization of amino acids during protein turnover is inefficient, with only 70 per cent of amino acids reutilized, and that this accounts for a major part of the nutritional requirement for protein, but there is no evidence for this. In fact, the detailed mechanisms of the way in which mTORC1 and the lysosome interact during mTORC1 activation include quite complex shuttling of amino acids between protein synthesis and proteolysis. The 70 per cent efficiency figure derives from an inadequate understanding of the regulation of oxidative pathways of amino acids which generates a metabolic demand for their dietary replacement. This is discussed further below in relation to the metabolic basis of the protein requirement.

9.3 An overview of whole-body protein and amino acid metabolism

A simplified scheme integrating proteostasis with amino acid metabolism in the whole body in relation to dietary intake and nitrogen excretion is shown in Figure 9.3. The purpose of this is to show the nature of the metabolic demand for dietary protein, which is the basis for the

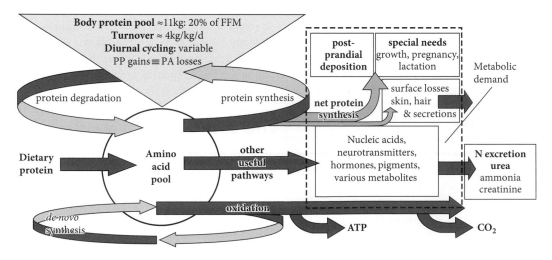

FIGURE 9.3 Simplified scheme for the metabolic demand for amino acids in the context of proteostasis and amino acid metabolism.

dietary requirement for protein, discussed in detail below. The processes shown are turnover, diurnal-cycling in which postabsorptive protein losses are replaced by postprandial deposition, net protein synthesis for special needs, and synthesis of proteins in skin, hair, and various secretions from the body surface. Other useful pathways include amino acids irreversibly transformed to a variety of other compounds. Oxidation is the catabolism of amino acids to urea and ammonia generating ATP and CO_2, while *de novo* formation is the synthesis of amino acids from other amino acids, glucose and other nitrogen sources (e.g., urea and ammonia). Those pathways within the dotted area identified as the metabolic demand are discussed further below.

9.4 **Amino acid metabolism**

Free amino acids occur in both extracellular (ECF) and intracellular (ICF) pools. The ECF pool present in blood plasma and interstitial fluid is of mainly uniform composition throughout the body because of blood flow. The plasma membrane of cells is impermeable to amino acids, but specific transporters mediate amino acid transport which allows rapid exchange between ECF and ICF compartments. The ICF pools within individual tissues and organs differ in their amino acid composition from the ECF pool and vary between organs and tissues but generally all amino acids are maintained at higher concentrations intracellularly. This is apparent in Figure 9.4 which shows amino acid concentrations in the human ECF and skeletal muscle ICF. Many amino acids are highly concentrated within the ICF ($\geq$5–10 fold), especially glutamine. The physiological implications of these differences between amino acids are discussed further below.

Amino acid (AA) transport into cells is achieved by specific amino acid transporters which generally recognize a range of structurally similar AAs as cargo for transport: for example, large neutral AAs, small neutral AAs, acidic or basic AAs (see Figure 9.1). AA transporter expression is tissue specific hence the concentration gradients maintained can vary between tissues. However, in all tissues maintenance of concentration gradients requires energy, and transport systems are therefore linked to the primary active transporter, the Na^+/K^+ pump which maintains the high ICF concentration of K^+ and high ECF concentration of Na^+. There are two main ways the Na^+/K^+ pump can be linked to amino acid transport, each involving the potential energy of the Na^+ concentration gradient energizing secondary and tertiary active transport systems. A secondary H^+/Na^+ exchanger transporter links the inward flow of Na^+ down the concentration gradient to promote the export of protons which will then return to the ICF on a tertiary proton-coupled AA transporter (e.g., Pro, Gly, Ala). Alternatively, a secondary Na^+/AA co-transporter carries both the Na^+ ion and an AA (e.g., Gln) into the cell allowing the high IC concentration of the Gln to develop with the Na^+ pumped out by the Na^+/K^+ pump. This high ICF concentration of Gln can then drive a tertiary active transporter of another AA (e.g., Leu) by exchange.

It is quite clear that in the context of protein turnover in which amino acids are recycled between the protein-bound and free pools, the composition of the protein-bound and free amino acid pools are very different, with leu, for example, the most abundant protein-bound amino acid but one of the least abundant in the free pool. The implication of this is discussed further below.

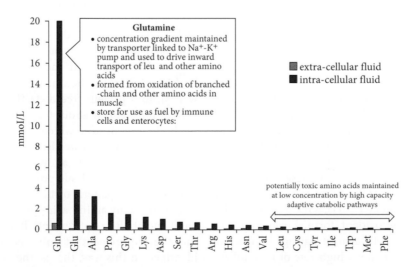

FIGURE 9.4 Free amino acid concentrations in the extra and intracellular pools.

TABLE 9.4 Useful pathways of amino acid consumption

Amino acid	Molecules synthesized
Arginine	Creatine, nitric oxide, proline, polyamines
Aspartate	Purine and pyrimidine nucleotides
Cysteine	Glutathione, taurine
Glutamate	Glutathione, N-acetylglutamate, γ-aminobutyrate
Glutamine	Purine and pyrimidine nucleotides, amino sugars, ammonia transport, cellular fuel
Glycine	Creatine, porphyrins (for haemoglobin and cytochromes), purines, bile acids, glutathione, hippuric acid
Histidine	Histamine, carnosine, anserine
Lysine	Carnitine
Methionine	Creatine, carnitine, choline, acetylcholine, ornithine, putrescine
Serine	Ethanolamine, choline, sphingosine
Tyrosine	Adrenaline, noradrenaline, dopamine, melanin pigments, thyroxine
Tryptophan	Serotonin (5-hydroxytryptamine), melatonin (N-acetyl-5-methoxy tryptamine)

Amino acids as metabolic precursors

Free amino acids are precursors for protein synthesis and for synthesis of most nitrogen-containing compounds in the organism, shown as other useful pathways in Figure 9.3. These range from small molecules like nitric oxide (NO), an important signalling molecule within the vascular system, to the nucleotides which make up the nucleic acids. Some important examples are listed in Table 9.4. These include: *neurotransmitters* serotonin, dopamine, glutamate, and acetyl choline; *hormones* adrenaline, noradrenaline, thyroxine; *inflammatory mediators* histamine; *cofactor precursors* nicotinamide; *antioxidants* (glutathione), and numerous other compounds. These are formed in many cases through unidirectional pathways with the nitrogen either eventually converted to urea after catabolism of the intermediates or excreted directly (as with creatinine). The amounts involved may be quite small as with nitric oxide for example. Arg is the precursor of NO and the total amount of NO synthesized (and degraded) per day may represent less than 1 per cent of the daily Arg intake. In contrast, the synthesis and turnover of glutathione (GSH), a major intracellular thiol and important antioxidant, formed from Glu, Gly, and Cys, that protects cells against damage by reactive oxygen species, accounts for a high rate of Cys utilization. This is especially the case in conditions of oxidative

stress, where the oxidized form of glutathione (GSSG) may leak out of the tissues and some of it may be transformed into mercapturic acid and excreted. When this occurs, there will be an increased demand for Met and/or Cys and some suggest they should be supplemented as part of nutritional therapy of trauma patients. (See Weblink Section 9.3 for more details of the conditionally essential amino acids.)

Urinary creatinine derives from muscle creatine phosphate, the store of high-energy phosphate which can regenerate ATP from ADP. Both creatine and creatine phosphate concentrations are regulated at a specific level but undergo an irreversible spontaneous reaction to form creatinine at a constant rate of about 2 per cent/day. This is no longer useful and leaves muscle to be excreted by the kidneys by very efficient renal clearance. This means that the urinary creatinine excretion rate varies with the skeletal muscle creatine/creatine phosphate pool size, itself a function of skeletal muscle mass. The assumed relationship is 1 g/d of excreted creatinine is equivalent to 18–22 kg of muscle mass. This also means that the amount of creatinine in a single sample of urine can be used to calculate what proportion of the daily urine production the sample represents, enabling the amount of any urinary metabolite in the urine sample (e.g., sodium or iodine) to be calculated as a 24-hour equivalent, in instances where 24-hour urine samples cannot be obtained. In the same way a urinary metabolite concentration can be standardized by expressing it as the metabolite/creatinine concentration. However, variable dietary creatine intakes (from meat), variation in hepatic creatine synthesis, and an influence of physical activity on creatinine formation from creatine, means that the precision of this relationship is low with quite large day-to-day variations in creatinine excretion (≈20 per cent).

The quantitative importance of amino acid utilization as precursors for such compounds as those shown in Table 9.4 is poorly understood. An upper limit can be estimated from the obligatory nitrogen losses (ONL), the total urinary and faecal nitrogen losses from healthy adults after their adaptation (usually over two weeks), to a protein-free diet. This is about 48mg N/kg/d, equivalent to about 0.3g protein/kg/d lost from the body protein store which is assumed to provide for the basal amino acid needs. However, this net loss of tissue protein releases amino acids in proportion to their occurrence in tissue protein and this pattern will not match the pattern of basal amino acid consumption, with many amino acids present in excess of their basal metabolic requirement. It is possible however, that the requirement of only one amino acid can be identified in this way, that is, the amino acid with the highest ratio of metabolic demand for maintenance to tissue protein content. Methionine is believed to be the limiting amino acid, with its metabolic consumption in effect 'driving' the ONL. This means that the over-all metabolic requirements for amino acids in other pathways can be considerably less than their content in 0.3g protein/kg/d.

9.5 Oxidation of amino acids and disposal of nitrogen

The oxidation of amino acids requires the removal of the amino group and its subsequent conversion to urea, and the disposal of the remaining carbon skeletons in energy-generating pathways or as storage as lipids. Although urea production occurs in the liver, there is considerable amino acid metabolism and oxidation in extrahepatic tissues including muscle, brain, kidney, adipose tissues, and the intestine.

Transamination reactions, *de novo* synthesis of amino acids, and removal of the amino nitrogen group

The amino group of most amino acids is very labile and can exchange between amino acids and form new amino acids (*de novo* synthesis) when the keto-acid carbon skeleton can be synthesized from simple metabolic intermediates. The exchange of amino groups with keto-acids involves pyridoxal phosphate-dependent transamination reactions as shown in Figure 9.5. The keto-acids include α-ketoglutarate, pyruvate, 3-phospho hydroxypyruvate, and oxaloacetate, which form Glu, Ala, Ser (via phosphoserine), and Asp respectively.

Through the process of deamination some amino acids can be directly oxidized to their corresponding keto (or α- oxo)-acids thereby releasing ammonia (see Figure 9.6). Four amino acids (Glu, Gly, Ser, and Thr) are deaminated by specific enzymes. Of these, glutamate

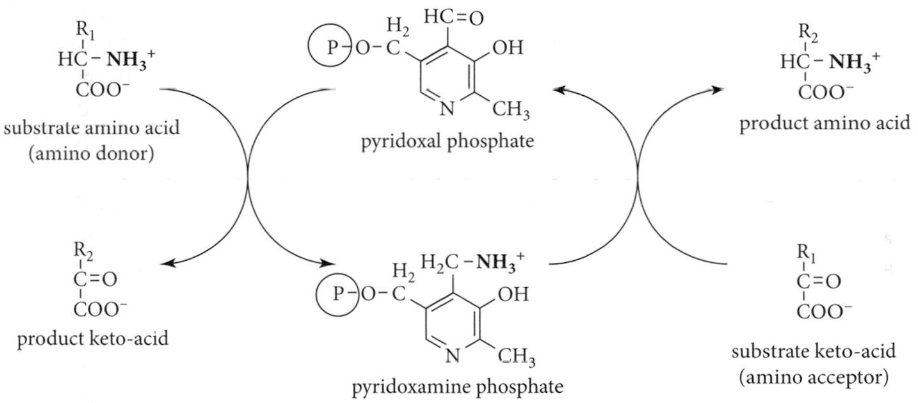

FIGURE 9.5 Transamination of amino acids.

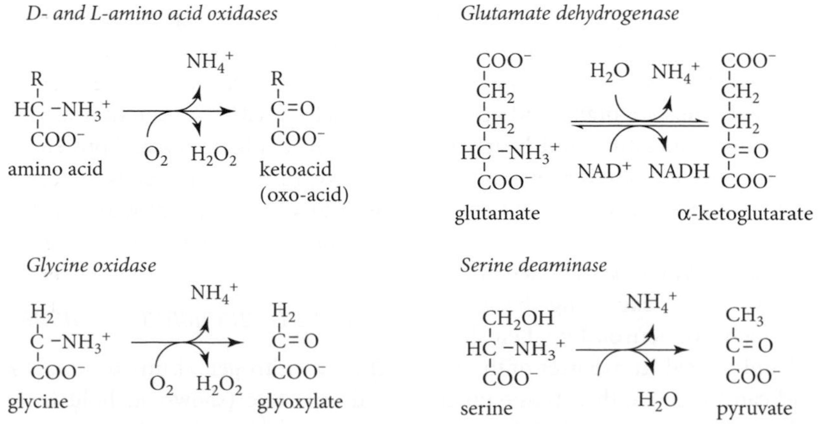

FIGURE 9.6 Deamination of amino acids.

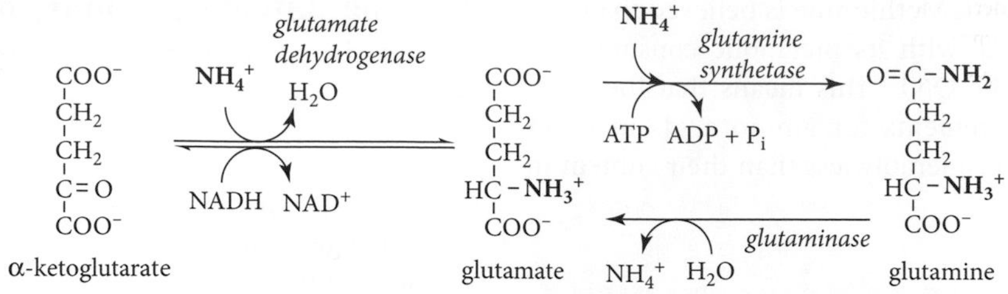

FIGURE 9.7 Synthesis and hydrolysis of glutamine.

dehydrogenase, a mitochondrial enzyme which is present in liver and most tissues, is particularly important, mediating the oxidative deamination of glutamate and releasing ammonia in an NAD/NADH-linked reaction, capturing some energy for subsequent ATP production. Gly is oxidized by glycine oxidase and also by the mitochondrial glycine cleavage system generating tetrahydrofolate, ammonia, and NADH, while Ser, Thr, and His are deaminated by their dehydratase enzymes, liberating ammonia without any NAD/NADH-linked reactions or ATP generation, only heat which is part of the heat increment of feeding.

Gln also plays an important role in nitrogen transport and amino acid oxidation and in the removal of ammonia which is highly toxic. Ammonia is rapidly removed, first by the formation of Glut from α-ketoglutarate by glutamate dehydrogenase, then of Gln by glutamine synthetase, an ATP requiring enzyme acting on Glu (see Figure 9.7). Glutaminase mediates the reverse reaction liberating Glu and ammonia.

Both Glu and Gln play an important role in the oxidation of the branched chain amino acids (BCAAs), which can account for 20 per cent of dietary protein. The BCAAs reversibly transaminate with α-ketoglutarate to produce Glu and their respective α-oxoacids, via the branched-chain amino transferase enzyme (BCAT), which acts on all three BCAAs. Subsequently the irreversible branched-chain keto-acid dehydrogenase, BCKD, acts on the branched-chain α-oxoacids, such as α-oxo isocaproate (OIC from Leu), which is the first step in their oxidation. The enzymes BCAT and BCKD are widespread in peripheral tissues, especially muscle, brain, and adipose tissue, as well as liver, kidney, intestine, and heart. The peripheral location of BCAA metabolism means that they can serve as a source α-amino nitrogen and energy to the periphery.

For example, human brain avidly extracts Leu and releases Gln, with the entry of Leu exceeding that of any other amino acid. Although not neuroactive, Leu shares the same large neutral amino acid transporter of the aromatic amino acids and can influence their tissue uptake and the synthesis of neurotransmitters deriving from them (dopamine from Tyr and serotonin from Try). Leu

is especially important for brain glutamate metabolism, being 'trafficked' between cellular compartments, providing -NH$_2$ groups for Glu synthesis in compartments of high Glu concentration and supplying OIC to act as a sink for -NH$_2$ groups from Glu in compartments of low Glu concentration.

In skeletal muscle, the uptake of BCAA represents ≅ 50 per cent of all amino acids taken up after a meal, with their nitrogen released as Ala and Gln. The Ala released from muscle derives from Glu (by alanine aminotransferase acting on pyruvate) with the Glu deriving from the BCAAs (by BCAT acting on α-KG). Gln can be synthesized as described above (see Figure 9.7), from Glu and ammonia also deriving from Glu by via glutamate dehydrogenase (although the source of the ammonia in muscle is not entirely clear). The released glutamine is mainly taken up by enterocytes in the gastrointestinal tract and in some circumstances by the kidney as discussed below, while the alanine is taken up by the liver so that its ammonia can be made available for urea synthesis.

Urea synthesis

The ornithine cycle of urea synthesis (see Figure 9.8), is the way in which mammals detoxify ammonia liberated from amino acids and excrete it as urea. The urea cycle also serves as a pathway for bicarbonate (HCO_3^-), excretion. The complete oxidation of amino acids yields HCO_3^- and NH_4^+. Air-breathing animals can excrete volatile CO_2 from their lungs, but not HCO_3^-. The kidney, with its usual range of urine production, cannot remove the daily HCO_3^- production associated with oxidation of amino acids from usual protein intakes. Thus, hepatic urea synthesis, which consumes two mol of HCO_3^- and two mol of NH_4^+ per mol urea formed is the major pathway for disposal of HCO_3^-.

Sources of ammonia and aspartate

The two nitrogen atoms of urea derive from ammonia and aspartate (shown in bold in Figure 9.8). Whereas urea production takes place largely within the liver, much of the ammonia used in its synthesis is derived, directly

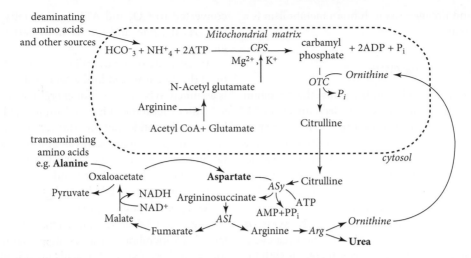

FIGURE 9.8 Urea synthesis by the ornithine cycle enzymes and their distribution between the mitochondrion and cytosol in the liver.
CPS, carbamoyl phosphate synthetase; OTC, ornithine transcarbamylase; Asy, argininosuccinic synthetase; ASl, argininosuccinate lyase; Arg, arginase.

or indirectly, from extrahepatic tissues. Thus, alanine and glutamine reach the liver from the peripheral oxidation of the BCAAs as described above, while ammonia derives from the intestine.

The major immediate source of ammonia for the first step of the urea cycle, the formation of carbamyl phosphate, is glutamate, via the action of mitochondrial glutamate dehydrogenase (see Figure 9.6). The glutamate is derived by transamination from alanine, aspartate, and other transaminating amino acids and by the deamidation of glutamine by glutaminase (see Figure 9.7), another important source of ammonia. Ammonia is also generated by the deamidation of asparagine and the deamination of glycine, serine, threonine, and histidine. Finally, 25 per cent of the ammonia derives via the portal vein from the intestine through amino acid utilization in the intestinal mucosa and from urea hydrolysis by bacterial ureases in the colon. (See Weblink Section 9.4 for more details of intestinal ammonia metabolism.)

The aspartate which provides the second nitrogen molecule in urea derives from the citric acid cycle intermediate oxaloacetate by transamination from alanine or other transaminating amino acids by the abundant aspartate aminotransferase. Aspartate is always present in excess and rapidly replenished as it is used up in the formation of argininosuccinate.

Ammonia detoxication by glutamine synthesis

Apart from urea synthesis the other major pathway for ammonia detoxication in liver is glutamine synthesis by glutamine synthetase, which is anatomically separate from the urea cycle enzymes. Portal blood from the intestine delivered to the liver first encounters the urea cycle system in periportal hepatocytes which will remove intestinal ammonia. However, after this, at the perivenous end of the liver functional unit, a small hepatocyte subpopulation, perivenous scavenger cells, contains no urea cycle enzymes but do contain glutamine synthetase which can eliminate with high affinity any ammonia that was not used by the upstream urea-synthesizing compartment. These cells are of crucial importance for the maintenance of non-toxic ammonia levels in the hepatic vein and throughout the entire body. Up to 25 per cent of the ammonia delivered via the portal vein escapes periportal urea synthesis and is used for glutamine synthesis in these cells which exhibit high-affinity uptake of carbon precursors for glutamine synthesis, mainly α-ketoglutarate (see Figure 9.7). Perivenous glutamine synthesis allows excess NH_4^+ to be transported via the bloodstream to the kidney as glutamine, to be degraded by renal glutaminase with NH_4^+ excreted via the tubules. By intricate intercellular compartmentation in the liver of a more periportal localization of urea synthesis and glutaminase, the regulation of hepatic glutaminase provides a major point of pH control in the organism. In acidosis, both urea synthesis and glutaminase are shut off, decreasing bicarbonate removal, favouring hepatic glutamine formation by the perivenously located glutamine synthetase, and causing increased renal excretion of NH_4^+ instead of urea. The converse holds in alkalosis. Thus, a hepatic-intercellular glutamine cycle between periportal and perivenous cells of the lobule serves a regulatory function in the pH homeostasis of the organism. While such a scheme is clear from animal studies the relationship between pH homeostasis and ureagenesis in human metabolism is a controversial issue.

Regulation of the urea cycle

Experimental *in vitro* studies indicate that the operation of the urea cycle is automatic, controlled entirely by kinetic and thermodynamic factors. The activity of the first enzyme, CPS-1, is irreversible and appears to control the rate of urea production. However, CPS-1 is inactive without a co-factor, N-acetyl glutamate (N-AG), which is an allosteric activator (see Figure 9.8). Because the synthesis of N-AG is stimulated by glutamate, through a substrate effect, and by arginine, many believe N-AG to be a regulator or even overall controller of the urea cycle. Thus, the amount of N-AG in the liver increases with increased amino acid supply and urea production: that is, glutamate stimulates the formation of N-AG which then exerts control over CPS-1. However, others believe that although N-AG activates CPS-1, substrate availability, that is, free NH_3 and ornithine, exerts immediate control.

The energy cost of urea synthesis

It is conventional to describe the urea cycle as consuming 4 moles of ATP, 2 in the synthesis of carbamyl phosphate and 2 associated with the synthesis of argininosuccinate as shown in Figure 9.8. However, this ignores the oxidative steps of the regeneration of aspartate (malate to oxaloacetate which generates NADH), and ammonia generation from glutamate (as shown in Figure 9.6): each would yield 3 moles of ATP. However, the variety of potential sources of ammonia shown in Figure 9.6 and ammonia from intestinal metabolism make it is difficult to describe a precise stoichiometry for the urea cycle. It is clear, however, that from the perspective of the oxidation of amino acids, the final pathways involved in urea production are unlikely to involve an energy cost and therefor ever be limited by energy supply.

The metabolism of amino acid carbon skeletons

The various pathways by which the carbon skeletons of amino acids are converted into useful energy result in one of two end products. Most amino acids are glucogenic (glucose-forming), after conversion to pyruvate, oxaloacetate, α-ketoglutarate, propionyl CoA, succinyl CoA, or fumarate. The ketogenic amino acids leucine and lysine are converted into acetyl CoA which feeds into the tricarboxylic acid cycle and is converted to CO_2 and ATP, or into fatty acid synthesis. Trp, ile, phe, and tyr are both part ketogenic and glucogenic.

An estimate of the overall metabolic fate of a protein meal (beef) consumed in the fasting state (when all of it is assumed to be oxidized for energy), is shown in Figure 9.9. After digestion all luminal amino acids (and probably some dipeptides but these are ignored in the scheme), are taken up into enterocytes and most pass into the portal blood en route to the liver. However, the dicarboxylic amino acids and their amides (Glu, Asp, Gln, and Asn) are a major metabolic fuel for the enterocyte and are completely oxidized, together with Gln (originating from muscle), taken up from arterial blood via the basolateral cell membrane. The resulting alanine and ammonia from these four amino acids are added to the other absorbed amino acids leaving the enterocyte via the portal blood. All this is extracted by the liver, apart from the branched chain amino acids, which will be taken up by muscle. Within the liver the amino acid nitrogen is converted to urea although some glutamate leaves via the hepatic vein ultimately to be taken up by muscle. Because complete oxidation of the amino acid carbon skeletons would produce more ATP than needed by the liver, some is converted to glucose and ketones (acetoacetate), which also leaves via the hepatic vein ultimately taken up by the brain. In muscle the three branched-chain amino acids which escaped uptake by the liver are oxidized together with the glutamate, with the nitrogen leaving the muscle as glutamine to be extracted by enterocytes of the intestinal epithelium, and alanine to be extracted by the liver. A detailed analysis of the stoichiometry of these reactions (not shown here), suggests that consumption of a protein meal containing 110g of protein from beef, of which all of it was oxidized, indicates that of the ATP generated, 70 per cent occurred in brain, 21 per cent in muscle, 8 per cent in enterocytes, and less than 1 per cent in the liver. Thus, the liver was in near balance after using the ATP generated from the amino acid oxidation to drive urea synthesis, gluconeogenesis, and ketogenesis, and would have had to utilize its stored glycogen and/or some fat oxidation for its own metabolic needs. Experiments in which glucose and nitrogen excretion after a protein meal were measured in subjects after taking the drug phlorhizin, which blocks renal glucose reabsorption, indicated a glucose:nitrogen ratio very similar to the stoichiometry calculation. It is interesting that the increase in blood ketones which occurs in subjects on very low-carbohydrate, high-protein diets is often suggested to be an indication that increased fat oxidation is occurring, whereas it is in fact an inevitable reflection of the protein oxidation shown in Figure 9.9.

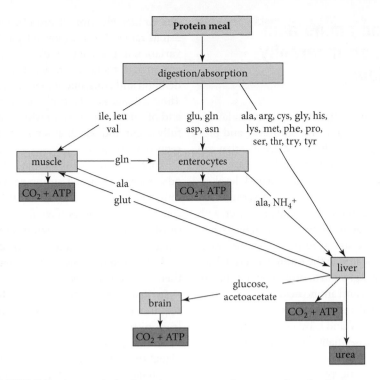

FIGURE 9.9 The metabolic fate of a protein meal consumed in the fasting state.

KEY POINTS: PROTEIN AND AMINO ACID METABOLISM

- Each of the 20,000 proteins within the human proteome is a large polymer of 21 different amino acids with a unique amino acid sequence determined by the gene for that protein and which can be folded to form proteins with many types of functions.

- Proteostasis is the maintenance of the cellular proteome through strict control of the initial production and folding of a protein, its conformational maintenance, control of abundance and subcellular localization, and disposal by degradation, all achieved by a network of 1000–2000 component proteins.

- Within proteostasis the continual catabolism of tissue proteins and their replacement results in a dynamic equilibrium of protein turnover which accounts for a significant part of resting energy expenditure, and there are overall gains and losses of cellular protein according to the acute nutritional state of feeding or fasting, in subjects maintaining overall body weight.

- Proteostasis is controlled by hormones including insulin and nutrients, especially amino acids of which leucine is especially important, acting through the mTORC1 signalling complex.

- The cellular metabolic demand for amino acids includes their role as precursors for tissue protein synthesis and for a wide range of nitrogen (N), containing molecules, and this latter demand drives the obligatory N losses observed in subjects adapted to a protein-free diet. This obligatory metabolic demand, equivalent to about 0.3g/kg/d, is part of the dietary protein requirement.

- Disposal of amino acids involves their oxidation in ATP-forming pathways with N-excretion as urea and/or ammonia after transamination and deamination reactions involving interorgan metabolic pathways between muscle, intestine, liver, and the kidneys, with glutamine and alanine playing an important role. Their carbon skeletons can act as fuels for tissues and when in excess results in formation of glucose, ketones, and in lipid storage.

9.6 Protein and amino acid requirements: an inherently difficult problem

Discussions in the area of nutritional recommendations are seldom without some degree of controversy and protein is an exemplar of the difficulties and debates that can occur. In terms of the dietary protein needs to allow optimal expression of the genome during growth and development and to maintain the adult lean body mass, questions of how much and what sort remain difficult to answer, with a paucity of examples of genuine protein deficiency and with virtually no effective markers of protein status or deficiency in adults. As indicated in the online material dealing with the historical background to our understanding of both protein requirements and amino acid essentiality (see Weblink Section 9.5), the debate about protein needs started in the nineteenth century and is by no means resolved. Also, whilst amino acid essentiality in terms of indispensability and dispensability is resolved (see Table 9.5), debate continues about the concept and clinical importance of conditional dispensability (see Weblink Section 9.3). Finally, the currently accepted values for the requirement values for the indispensable amino acids (IAAs) discussed below are by no means entirely satisfactory.

Part of the difficulties can be related to the marked variation in dietary protein intakes within otherwise healthy populations (see Figure 9.10). As shown this is ≥3-fold but given that food energy intake can also vary markedly due to lifestyle, overall protein intakes can vary over a 5-fold range. Most importantly in response to this wide variation in protein intakes there are adaptive changes in amino acid metabolism which influences the metabolic demand and consequent dietary requirement for protein. The challenge is to identify the protein intake at the lower end of the intake range to which adaptation can successfully occur and which as a result satisfies the criteria of a requirement value.

Another problem is methodological. Balance methods, N or ^{13}C-labelled amino acid oxidation, remain the only option in practice (see FAO/WHO/UNU 2007), and stable-isotope studies have failed to provide completely satisfactory alternative methodologies to nitrogen (N)-balance. Although there is some discussion about *optimal* protein requirements as shown in Figure 9.10, as yet there is no consensus.

In this section the current definition of the protein requirement is defined, followed by estimates obtained from nitrogen balance studies. After this the uncertainties in the values obtained is reviewed, followed by an explanation of these uncertainties in terms of an adaptive metabolic demand model of the protein requirement.

Definition of the protein requirement

In general terms the protein requirement is the amount of dietary protein in an otherwise nutritionally adequate diet, necessary to maintain the desired structure and function of the organism and to provide for the extra needs for growth, reproduction, and lactation. In the

TABLE 9.5 Essentiality/indispensability of the 21 coded amino acids in proteins

Group	Essential/indispensable *no synthesis*	Conditionally essential/indispensable *Limited synthesis at times*	Non-essential/dispensable *Unlimited synthesis*
Branched chain	Leucine Isoleucine Valine		
Aromatic	Phenylalanine Tryptophan	→ Tyrosine	
Sulphur/selenium	Methionine	→ Cysteine	Selenocysteine (from inorganic Se)
Basic	Lysine Histidine	Arginine	
Acidic and amides		Glutamine	Glutamate Aspartate Asparagine
Neutral	Threonine	Glycine Proline	Alanine Serine

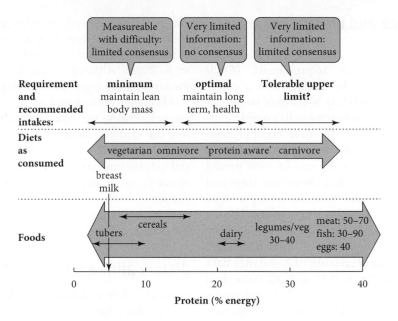

FIGURE 9.10 Protein requirements and intakes.

2007 WHO report (FAO/WHO/UNU 2007), the minimum protein requirement (MPR), was defined as: *the lowest level of dietary protein intake that will balance the losses of nitrogen from the body, and thus maintain the body protein mass, in persons at energy balance with modest levels of physical activity, plus, in children or in pregnant or lactating women, the needs associated with the deposition of tissues or the secretion of milk at rates consistent with good health.* This is conceptually straightforward with N-balance as its focus. Whilst several stable-isotopic approaches based on different paradigms have been described, no consensus has been reached on any specific alternative approach, with some identified as deeply flawed. Nevertheless N-balance studies are very difficult to conduct and can raise difficulties if they are overinterpreted. The measurement of N intakes and losses over a range of intakes, do in theory directly identify the MPR as the intake for nitrogen equilibrium when intake = loss and balance = 0, the value indicated by the ratio of intercept/slope from a linear regression of intakes on balance. However, problems arise when this regression slope is interpreted as an indication of the efficiency of dietary protein utilization and used to predict protein requirements for growth in children and for pregnancy in factorial models. Such values at ≈50 per cent or less, are biologically unreasonable. This is discussed further below.

The nitrogen balance method

The principle of N-balance is that the overall state of whole-body protein metabolism, as judged by a stable lean body mass, reflects the adequacy of dietary intake to meet metabolic demands and is determined as the difference between dietary intake of nitrogenous compounds (mainly protein, but also nucleic acids and small amounts of other compounds) and the output of nitrogenous compounds from the body, with zero N-balance indicating the adequacy of the protein intake. Nitrogen constitutes 16 per cent of most proteins, and the protein content of foods is calculated as N × 6.25, although for some foods with an unusual amino acid composition other factors can be used.

The output of N from the body is largely in the urine and faeces, but significant amounts may also be lost in sweat and shed skin cells. The urinary output is mainly urea, with small amounts of other end products of amino acid, purine, and pyrimidine metabolism and very small amounts of free amino acids with variable amount of ammonium ions, to neutralize excessively acid urine.

Three states of nitrogen balance can be defined.

1. *Nitrogen balance or equilibrium*: meaning an adult in good health consuming a healthy diet with a stable body weight and body protein pool, and with daily N losses = N intake.

2. *Positive nitrogen balance*: such as during child growth or pregnancy when total body protein is increasing and during lactation with milk protein secretion is occurring so that N losses < N intake.

3. *Negative nitrogen balance*: such as in response to trauma, infection, starvation, or an inadequate protein intake when total body protein is decreasing and N losses > N intake.

Nitrogen balance and current values of the MPR

Currently accepted protein requirements (FAO/WHO/UNU 2007), derive for adults from a meta-analysis of all eligible published N-balance studies in which the subjects were fed nutritionally complete diets containing at least three different levels of protein, usually in the sub-maintenance to maintenance range, for preferably two weeks. At the end of each period all N-losses were measured over about four days and compared with the N-intake from the food provided and balance calculated as intake minus losses. The intake for equilibrium, the MPR, is calculated as intercept/slope from linear regression of the balance data for each individual. The regression of N-balance data on intake is shown in Figure 9.11 with all individual balance points from 19 studies with 235 individuals studied at ≥3 intakes assembled for the meta-analysis by Rand, Pellett, and Young (2003). The intercept, N-balance at zero N intake should indicate the obligatory nitrogen loss and the intake for N equilibrium is indicated by the intercept/slope. In practice the median value of the MPR was obtained from the distribution of each individual subject regression intercept/slope values from the 235 subjects and this is shown in Figure 9.12. Sub-group analysis showed no differences according to gender, type of diet (animal or plant protein), or age, although numbers in subgroups were

small. For the elderly, N-balance studies conducted since the meta-analysis have indicated no effect of age confirming the results of the meta-analysis (Campbell et al. 2008). Although there have been several calls for revising the requirement values for the elderly towards higher values based on reports of higher protein intakes slowing the rate of development of sarcopenia, the evidence base for this is minimal (Millward 2012a). For children and other special groups a factorial calculation was used based on needs for maintenance from the adult data, assumed to be the same per kg of body weight for all groups and life stages, and the needs for protein deposition during growth, pregnancy, and lactation.

Dietary allowances

These are reference intakes for individuals or population groups which meet the dietary requirements of most individuals within the population, that is, the intakes are associated with a very low risk of deficiency (with deficiency defined as an intake < requirement) (FAO/WHO/UNU 2007). The average or median protein requirement, 0.66g/kg/d for men and women of all ages, is the mid-point (median) of the range of individual requirements. The safe individual intake (or recommended nutrient intake, RNI, in the UK), lies at the upper end of the range, the 97.5th percentile of the distribution, nominally the average+2SD and is associated with a very low (<2.5 per

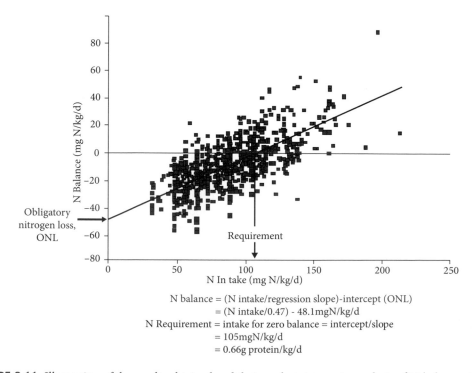

N balance = (N intake/regression slope)-intercept (ONL)
= (N intake/0.47) - 48.1mgN/kg/d
N Requirement = intake for zero balance = intercept/slope
= 105mgN/kg/d
= 0.66g protein/kg/d

FIGURE 9.11 Illustration of the results obtained and their analysis in a meta-analysis of N-balance studies.

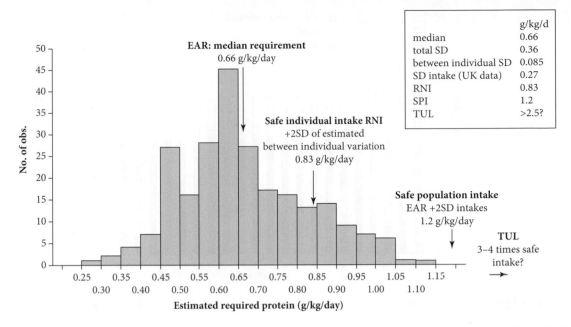

	g/kg/d
median	0.66
total SD	0.36
between individual SD	0.085
SD intake (UK data)	0.27
RNI	0.83
SPI	1.2
TUL	>2.5?

FIGURE 9.12 Distribution of individual protein requirements from nitrogen balance studies and values for EAR, RNI, safe population intakes, and the tolerable upper limit.

cent) risk of deficiency. In fact the requirements values assembled for the meta-analysis shown in Figure 9.12, exhibited a very marked variability, reflecting many factors. However, the true interindividual variability was estimated to be much less than the overall variability, so that the safe individual intake or RNI, 0.83g/kg/d, is lower than would be expected from simple inspection of the results shown in Figure 9.12.

The safe intake for a *population* needs to take into account the distribution of both intakes and requirements and is usually greater than the RNI, approximating to the average requirement +2SD of intakes or + 3–5SD of requirements, see (FAO/WHO/UNU 2007, chapter 3). In the UK the distribution of protein intakes reported in the national diet and nutrition survey, NDNS, indicates an SD of intake of 0.27g/kg/d, that is, 2–3 times that of the between-individual requirement resulting in a safe population intake of 1.2g/kg/d. The UK estimated average daily requirements (EAR) can be used to judge the adequacy of population intakes in that the prevalence of deficiency approximates to the proportion of the population with intakes <EAR (this is called the 'cut-point' method). The current intake within the UK adult population (a mean value of 1.28g/kg/d, range 0.69–2.72g/kg/d after correcting for underreporting by trimming for unrealistically low energy intakes), indicates no protein deficiency.

Figure 9.12 shows that a tolerable upper limit (TUL) has not been identified for protein. However, most accept that intakes of 3–4 times the safe individual intake (e.g., ~35 per cent protein energy), can be consumed without obvious harm.

Protein requirements for growth and special needs: the semi-factorial model for the MPR

During child growth and pregnancy the MPR is calculated as maintenance, the same as the adult MPR per kg body weight, plus special needs. These are daily rates of protein deposition from analyses of body composition during child growth, pregnancy, or as rates of breast milk protein secretion, adjusted by an efficiency factor for dietary protein-utilization. These efficiency factors were derived from the slopes of N-balance studies reported either for adults or for the special groups: that is, 0.58 (normal child growth), 0.70 (child catch-up growth), 0.42 (pregnancy), and 0.47 (lactation, the value from the mean slope of all adults in the meta-analysis as shown in figure 9.11). Figure 9.13 shows the safe allowances for infants, children, adolescents, and adults in terms of maintenance, growth, and 2SD calculated from the weighted mean of the coefficient of variation values for maintenance (12 per cent) and growth (43 per cent).

For pregnancy the total of the requirements for maintenance and for the dietary requirement for protein retention in the products of conception and in the maternal tissues associated with a gestational weight gain of 13.8 kg were calculated as an additional 1, 9, and 31 g/day protein in the first, second, and third trimesters, respectively.

Lactation requirements were derived from rates of milk produced by well-nourished women exclusively breastfeeding their infants during the first six months

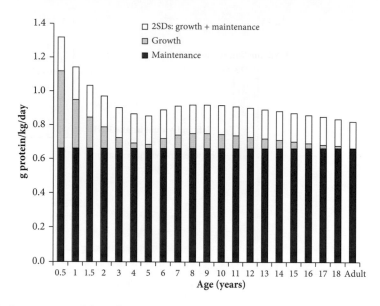

FIGURE 9.13 Components of the safe protein allowances for infants, children, adolescents, and adults.

postpartum and partially breastfeeding in the second six months postpartum. The mean concentrations of protein and non-protein nitrogen in human milk give average milk protein outputs. The additional dietary requirement to meet this were calculated at 19–20g/day during the first six months and 12.5g/d after that, assuming an efficiency of dietary protein utilization of 47 per cent.

Methodological explanations of the apparent variation in the minimum protein requirement

The apparently simple, but laborious, N-balance approach is beset with several serious problems. Balance is the small difference between two large amounts and can seldom be precisely measured. Systematic errors (an overestimation of intake because subjects might eat less than their provided test meals, and an underestimation of the losses), mean that balance may be overestimated, with weight-stable adults appearing to gain N. Often individual balance curves are not linear so that curve fitting rather than linear regression has to be used to predict the intake for zero balance. In extreme cases with a shallow slope at equilibrium the problem of accounting for all losses becomes critical, to the extent that the values for N equilibrium vary markedly according to the predicted (i.e., unmeasured), integumental losses. In many of the early balance studies no corrections for such losses were made and intakes for equilibrium were underestimated, often quite markedly.

Body protein equilibrium can be influenced by intakes of other nutrients, especially energy, and ensuring that energy intakes are sufficient is difficult with very few studies paying attention to this. Excess energy intake leads to weight and lean tissue gain and positive N-balance. Too little energy intake results in weight-loss and negative N-balance because dietary protein is oxidized as an energy source. Thus, the MPR is a function of the state of energy balance. In fact, on the basis of what is known about the influence of energy balance on N-balance, an error of only ±10 per cent in estimating energy needs is equivalent to about 85 per cent of the estimated true interindividual variance in the studies in Figure 9.12.

Adaptation of protein metabolism to a change in intake is recognized so that multilevel balance studies usually involve measurements over at least three to four days at the end of a two-week period at each intake level. However, data suggests that a new equilibrium takes considerably more than two weeks to achieve. Incomplete adaptation from the habitual to the lower test diet intake results in excess N excretion, so that the intake for equilibrium is overestimated. This would suggest that some of the observed MPR variability between studies shown in Figure 9.12 reflects incomplete adaptation to the test diets. How these uncertainties in terms of energy imbalance and inadequate adaptation to the reduced protein intakes during the balance trial could account for some of the apparent variation in reported protein requirement values is suggested in Figure 9.14.

Underestimation of efficiency of utilization and overestimation of the requirements for special needs

Historically, the efficiency factors used within factorial models for the protein requirement were decided somewhat arbitrarily at 0.7. However, the WHO 2007

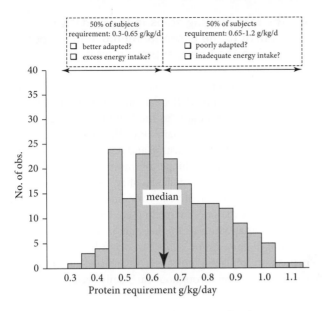

FIGURE 9.14 Factors which could account for the variation in the apparent protein requirement.

report adopted values chosen from the slopes of linear regressions of the N-balance studies reported for the different population groups, which were much lower and varied between the different population groups in which the factorial model was deployed (from 0.42 to 0.7). Thus, the requirements for growth, pregnancy, and lactation are higher (in the case of pregnancy markedly so), than they would be if more physiologically realistic values were used. This was not discussed in the report so that there is uncertainty and controversy about these requirement values (Millward 2012b). Furthermore, the fact that the adult MPR, at 0.66g/kg/d was much greater than the ONL (~0.3g protein/kg/d), indicating an apparent efficiency of utilization of dietary proteins of 47 per cent was also not discussed. However, metabolic studies of amino acid oxidation, postprandial protein utilization, and nitrogen excretion during the diurnal cycle of feeding and fasting have explained why the MPR is substantially greater than the ONL. This work has established a different model of the protein requirement based on the concept of an adaptive metabolic demand (Millward 2003).

Metabolic demand for amino acids: obligatory and adaptive components

The demand for amino acids is to maintain cellular function and tissue protein at appropriate levels (maintenance), together with provision for any special needs: growth, rehabilitation, pregnancy, and lactation, as shown in Figure 9.3.

The metabolic demand is supplied from the free amino acid pool which is regulated within narrow concentration limits and which is supplied from three sources: dietary protein after digestion and absorption, proteolysis during tissue protein turnover, and *de novo* synthesis of dispensable amino acids from their carbon skeletons by either transamination from amino acids present in excess of metabolic needs or from ammonia, from the various sources discussed above. Removal of free amino acids from the amino acid pool occurs during their use as substrates for protein turnover and the other pathways shown of the metabolic demand, and by oxidation of surplus intake during the transient increases in free amino acids after a protein meal. In weight-stable subjects protein synthesis and proteolysis are in balance so turnover does not exert a significant net metabolic demand. This means that removal of amino acids to meet the metabolic demand involves their utilization by several irreversible pathways and these are either obligatory or adaptive.

Obligatory demand

At maintenance, as shown in Figure 9.3, there is some net protein synthesis, a small amount of protein loss as skin, hair, and secretions and replacement of postabsorptive losses, and other irreversible metabolic transformations of individual amino acids to non-protein products, listed in Table 9.4. This demand results in eventually nitrogenous end-products, mainly urea and other compounds in urine, faeces, or sweat.

These diverse obligatory demands for amino acids for maintenance represent an essential but probably quite small intrinsic metabolic demand, assumed empirically to be equal to the protein-equivalent of the obligatory nitrogen loss (ONL), the sum of all N losses from the body observed in adult subjects fed a protein-free diet after 7–14 days of adaptation, ~48 mg N/kg/d or about 0.3 g protein/kg/d tissue protein mobilized. In fact, the obligatory demand may be less than the ONL because it includes a mixture of amino acids with a profile which is unknown but unlikely to match that of the tissue protein mobilized to meet the demand, because not all the amino acids mobilized serve useful functions. Thus, feeding selective amino acids, such as threonine, tryptophan, and methionine lowers the ONL. Any special needs for growth or pregnancy also constitute an obligatory metabolic demand. (See Weblink Section 9.6 for more details of the obligatory demand.)

Adaptive demand

This represents oxidative losses of amino acids not obviously associated with useful utilization, that is, the oxidation pathway shown in Figure 9.3. The rate is set by the habitual protein intake, which in turn sets the activity and capacity levels of pathways of oxidative catabolism. It can be identified as a demand because this oxidation of amino acids occurs continuously, day and night, and changes only slowly, over several weeks, when the habitual protein intake changes (Millward 2003). This adaptive and therefore variable demand is one reason why protein requirements have proved so difficult to determine in practice. It is not entirely understood and may have evolved because of the slow growth-rate and body weight stability of humans consuming diets which provide protein, which is often considerably in excess of minimum needs. Four aspects of human protein and amino acid metabolism and behaviour help to explain the adaptive metabolic demand.

1. Some amino acids are potentially toxic

The branched chain, aromatic, and sulphur amino acids (SAAs) are toxic at high concentrations in the blood and tissues, and inborn errors of metabolism in children associated with their oxidative disposal result in severe pathologies. Figure 9.4 shows their very low concentrations in blood and muscle tissue even though three of them, Leu, Val, and Ileu, are the most abundant amino acids in tissue and food proteins (see Table 9.3). This means that after a meal, the potentially toxic amino acids must be rapidly disposed of by high-capacity oxidative pathways or by deposition as tissue protein.

2. There are no protein stores

Although there is a small capacity for the liver protein mass to increase with feeding, it is not possible to expand the skeletal muscle protein pool by depositing protein from food. Muscle myofibres are encased in an inelastic connective tissue framework which determines myofibre volume and this must be remodelled by mechanotransduction following bone lengthening to allow muscle growth. In the adult, after epiphyseal fusion of the long bones, height growth stops with muscle mass fixed at its phenotypical size which is a function of height (Millward 1995, 2021). No further growth can occur unless muscle is subject to passive stretch through resistance exercise. This means that dietary protein in excess of the minimum demands cannot be simply stored as larger muscles.

3. Humans are usually daytime meal eaters

Human eating behaviour solves the problem of meal protein disposal. The diurnal sleep–wake cycle throughout the lifecycle, with long periods between meals, especially during sleep, results in a feeding–fasting cycle with a substantial postabsorptive state, tissue protein loss, amino acid oxidation, and urea excretion, creating a space into which excess protein can be deposited after a meal. This repletes postabsorptive losses until the regulated maximum size of the muscle mass is reached.

4. The rate limiting steps in the oxidative catabolism of the essential amino acids appear to adapt to the habitual protein intake

This is the least understood feature: that is, how the capacity or Vmax for oxidative catabolism of the potentially toxic amino acids adapts to the habitual protein intake and is, to a large extent, maintained throughout the diurnal cycle of feeding and fasting and does not change until a sustained period of a higher or lower protein intake occurs. Thus, the acute regulation of amino acid oxidation appears to be less sensitive to food intake than indicated by studies of the rate-controlling enzymes. This ensures that after adaptation to the habitual protein intake, disposal of dietary amino acids can occur both by protein deposition, replacing postabsorptive losses, and by amino acid oxidation throughout each daily cycle.

Consequences of the adaptive demand for daily balance

The consequences of these features of protein and amino acid metabolism are shown in Figure 9.15 in terms of body protein balance regulation during the diurnal cycle. Figure 9.15a shows the two components of amino acid oxidative losses which comprise the metabolic demand occurring continuously throughout each 24 hours of the diurnal cycle. This is measurable as the hourly rate of postabsorptive $^{13}C_1$-leucine oxidation as a tracer for total amino acid oxidation or nitrogen excretion, scaled up to 24 hours. In the postabsorptive state the demand is met from tissue protein resulting in a negative balance as tissue protein is lost. After feeding, the metabolic demand is met by the dietary protein which also repletes the tissue protein pool and this allows the essential amino acids to be disposed of without excessive increases in the free amino acid pool. Arbitrary responses to three meals are shown with very brief between-meal periods of negative balance which might occur. Any dietary protein in excess of the demand or amino acids not matching the amino acid pattern of the demand will be oxidized and these are shown as meal-dependent amino acid losses, that is, an increase in amino acid oxidation above the postabsorptive rate. The extent of this increase will determine the true efficiency of protein utilization. Clearly in practice, because protein intake varies from day to day, so will overall 24 hour N-balance, but for any habitual dietary pattern of protein intake above the minimal level, the adaptive metabolic demand will ensure overall balance.

Figure 9.15b shows how the body protein content, mainly a function of muscle mass, is regulated at its phenotypical level throughout the diurnal cycle with two days indicated at each of three levels of habitual protein

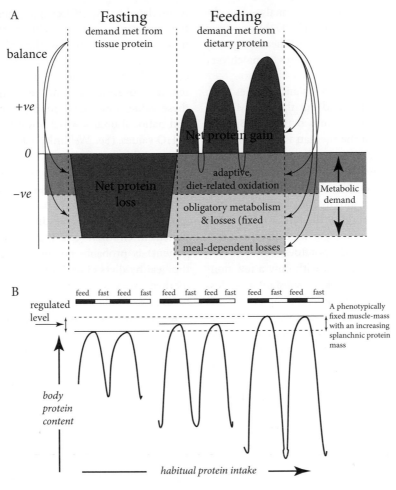

FIGURE 9.15 Body protein balance regulation during the diurnal cycle.
A (Top). Gains and losses in relation to the metabolic demand over 24 hours.
B (Bottom). Influence of adaptation to increasing protein intakes on diurnal gains and losses of protein in relation
to the regulated size of the skeletal muscle mass.

intake (Millward 1995, 2021). With an increased habitual protein intake level after complete adaptation, the amplitude of the postabsorptive losses and repletion increases. Although there is a phenotypically fixed muscle mass there may be a small increasing splanchnic protein mass associated with digestion absorption and metabolism of the increasing protein intakes. The increasing postabsorptive losses resulting from the adaptive increases in the capacity of the amino acid oxidation pathways with increasing habitual dietary protein intakes, means that to maintain muscle at its phenotypic regulated mass, increasing deposition to replete muscle protein will be observed in the postprandial state.

When protein intake changes to a lower habitual intake, mobilization of tissue protein occurs with a negative N-balance for as long as it takes oxidative losses to adapt to the lower intake. This was previously identified as the *labile protein reserves* but no specific protein stores

were ever identified, and it is now believed to originate from both skeletal muscle and the splanchnic organs. Change to a new higher intake poses a problem for the immediate disposal of dietary amino acids in excess of the adapted metabolic demand. This is partially solved by the appetite mechanism in which protein is the most satiating nutrient, especially when consumption exceeds habitual intakes.

Implications of the adaptive metabolic demand model of the protein requirement

By definition, the adaptive metabolic demand model sets the protein requirement at the habitual intake within an adaptive range. The practical problem involves both the lower limit of that range and the timescale of adaptation. While a maximum of one to two weeks was adopted in most of the multilevel balance studies analysed in the WHO 2007 report (FAO/WHO/UNU 2007), long-term balance studies in the 1960s indicated a time scale greatly in excess of this was necessary for N-losses to match very low intakes.

Incomplete adaptation in the N-balance studies from which the MPR is derived would put the true MPR in the lower range of observed values between 0.3g/kg/d (the ONL), and 0.65g/kg/d the median value of all studies. However, in practice protein intakes as low as this are rare and probably not possible from otherwise nutritionally complete diets. For example a 70 kg young adult male consuming either cereal or starchy roots, tubers, or fruit staples as sole sources of energy at the rate of 1·75 × BMR (183 kJ or 43·7 kcal)/kg), would consume digestible protein intakes of 0·82g/kg/d from potatoes, between 0.76 and 1·80 g/kg per d from cereals, with only a few non-cereal staples providing intakes substantially less than 0·50 g/kg per d: for example, Ethiopian banana (Ensete) 0·36, plantain 0·35 or cassava 0·23 g/kg per d. Furthermore, such diets are seldom consumed as sole sources of energy with other constituents providing protein. Even if they were they would almost certainly be nutritionally limiting in several key micronutrients and minerals that would impact on most indicators of nutritional status. This means that the actual value of the true MPR becomes of academic rather than practical interest with no reason not to continue with the currently accepted EAR of 0.65g/kg/d. However, there are other important consequences of this metabolic model that are of practical importance.

The slope of the N-balance regression line markedly underestimates the efficiency of protein utilization.

This is a quite complicated conceptual issue but is of great importance for the factorial model used to derive the protein requirements for children and for pregnancy and lactation. Most N-balance studies have involved proteins of high digestibility and biological value, so that the low efficiency of utilization (≈50 per cent), is not biologically plausible. Furthermore, direct studies of the efficiency of utilization of proteins with ^{13}C-1 leucine oxidation balance studies have shown that the true efficiency of utilization of dietary protein is as would be expected, that is, 95–100 per cent for milk and high quality proteins and lower (60–70 per cent) for wheat (Millward 2003). The shallow slope (low efficiency) of the N-balance regression is explained in terms of the adaptive metabolic demand, which results in increasing oxidative losses and consequent metabolic demand, that is, >ONL or intercept, with increasing intakes, that is, the N-balance slope

is unrelated to the efficiency of utilization. The slope of the regression is only of value in terms of defining the intake for equilibrium (i.e., the requirement = intercept/slope = intake for zero balance). This means that the special needs for children, pregnancy, and lactation are all markedly overestimated in the current WHO report and these values have been adopted by other international and national bodies which invariably tend to adopt the WHO values. (See Weblink Section 9.7 for more details.)

For children (see Figure 9.13), the dietary intake to meet the demands for growth is experimentally determined lean-tissue protein gain adjusted by an efficiency factor 0.58, the slope of N-balance studies in children. In practical terms the overestimation of the growth component in children (and that of the maintenance component) is probably not of great importance because the healthy diets of children usually provide protein in marked excess of their needs. For pregnancy and lactation (see Table 9.6), the extra protein intakes derive from calculations in which the protein deposition during gestational weight gain, for which there is a consensus, is adjusted by an efficiency factor of only 0.42, deriving from N-balance studies in pregnant women. These values in Table 9.6 are almost double previous reports in which much higher efficiency values were used. Given reports of increased neonatal deaths with high protein supplements, the report recommended that the higher intake during pregnancy should consist of normal food, rather than commercially prepared high-protein supplements. In fact, during the third trimester an extra 31 g.d^{-1} of protein would represent an extra 3.6 MJ of a mixed diet assuming it contains 15 per cent protein energy, considerably more additional energy than is recommended at this stage of pregnancy by any agency.

TABLE 9.6 Extra protein requirements for pregnancy and lactation

	Safe intake (g/day)	Additional energy requirement (kJ/day)	Protein:energy ratio
Pregnancy trimester			
1	1	375	0.04
2	10	1200	0.11
3	31	1950	0.23
Lactation			
First 6 months	19	2800	0.11
After 6 months	13	1925	0.11

Given the concern for adverse influences of overweight and obesity on pregnancy outcomes and given that pregnant women often have successful pregnancy outcomes without any increase in food intake, it is likely that this amount of additional food would result in excessive weight gain. On this basis these recommendations are arguably unsafe.

Implications of adaptation for nutrition policy

In general, protein requirement values serve two purposes.

One is as a basis for prescription, that is, advice on safe diets through recommending appropriate dietary intakes. The adaptive model described here implies a much lower, but difficult to define requirement. Formulation of policy will inevitably and correctly be most concerned with satisfying the upper range of demands for protein and, where there is uncertainty, including positive margins of error. In this case it is unwise to adopt an adaptive model and reduce the requirement. Indeed, an adaptive model does not mean that protein is an unimportant nutrient for the maintenance of human health and well-being, but that indicators other than balance or maintenance of body composition need to be identified. Unfortunately, it remains the case that there is mixed evidence about the overall long-term health effects of protein (FAO/WHO/UNU 2007) with no overall consensus. This results in a dilemma for those attempting to frame dietary guidelines. It is probably wise to retain current values for children and adults until it becomes possible to quantify the benefits (and any risks) of protein intakes within the adaptive range. However, the values for pregnancy do require revisiting (Millward 2012b).

The other purpose of requirement figures is as a diagnostic indicator of risk, in population groups rather than individuals. In this case indicators used to estimate prevalence of disease states, or deficit risk, are carefully chosen so as to strike an acceptable balance between false positives and false negatives. The main implication of adaptation for estimating risk of deficiency defined only as intakes < requirements, is a marked reduction in the prevalence of risk for most populations compared with that assessed according to the current model. As in the prescriptive context, this low risk of deficiency applies only to that of being unable to maintain N-balance after full adaptation with otherwise nutritionally adequate diets satisfying the energy demands. Whether such populations enjoy optimal protein-related health is a separate issue which can only be resolved with large scale interventions in populations which enjoy adequate household amenities. It has been suggested that maintenance

of N-balance can no longer be used as a surrogate of adequate protein-related health, but the current lack of quantifiable alternative indicators of 'protein sufficiency' is no excuse for ignoring the issue of adaptation.

9.7 **Protein quality evaluation**

Qualitative aspects of the metabolic demand

The dietary need for protein is to satisfy the metabolic demand for amino acids, classified in Table 9.5 in terms of their dispensability. Given the extensive amino acid interconversions between the dispensable amino acids, the dietary protein amino acid pattern need not exactly match the amino acid pattern of the demand, which in any case is poorly understood, as long as there is sufficient total amino acid N or sources of non-essential N such as ammonium compounds. However, there will be a minimum dietary requirement for the IAAs and those which become indispensable under specific physiological or pathological conditions (the conditionally indispensable amino acids). In fact, dietary amino acid mixtures rich in IAAs like egg protein are less effective in meeting the demand than mixtures with added non-essential N which maintain overall balance with lower levels of IAAs. This indicates that the demand includes an absolute need for non-essential N, which is less efficiently provided for by the oxidation of excess IAAs. In practice, discussion of protein quality is usually limited to the extent to which dietary protein provides enough of the IAAs shown in Table 9.5.

Protein quality in animals

Animal growth assays resulted in the widespread view of marked differences in the relative quality of plant compared with animal protein sources in human nutrition because of deficiencies of some IAAs. As shown in Table 9.3 there are important differences in the IAA content of the main classes of proteins. Cereals have low levels of lysine and in some cases tryptophan, and legumes have lower levels of the sulphur amino acids compared with animal source proteins. Thus, cereals and to a lesser extent legumes performed poorly in animal growth assays, but when combined, maximal growth occurred. This gave rise to the concept of 'complementation' in which the appropriate balance of IAAs is provided from combinations of plant proteins. However, it is now recognized that in human nutrition, with the rapid growth of the new-born infant slowing markedly after weaning, the nutritional demand for IAAs for tissue growth becomes a minor component of the total metabolic demand after the first year of life, as shown in Figure 9.13. Thus, animal growth assays of protein quality have little relevance for protein-quality evaluation in human nutrition.

Protein quality evaluation in human nutrition

Protein quality is influenced by *digestibility*, the amount of the protein which is absorbed from the intestine and *biological value*, the extent to which the amino acid pattern of the absorbed amino acid mixture matches that of the cellular demand.

Digestibility can vary through limitation by plant-cell walls and by anti-nutritional factors in plant foods. Values range from 60 to 80 per cent in legumes and cereals with tough cell walls such as millet and sorghum to 97 per cent for egg and dairy proteins. Anti-nutritional factors in legumes and seeds include amylase and trypsin inhibitors, tannins in most legumes, and cyanogens in lima beans.

Biological value (BV) varies according to the cross-match between the composition of the absorbed amino acid mixture and the pattern of the metabolic demand for maintenance and net protein deposition. During human growth and development, after the second year of life metabolic demands for growth are quite low, so that amino acid needs are mainly for maintenance. We know from extensive experimental animal work that the amino acid pattern for maintenance is quite different from that for growth (see Table 9.7), with a lower level of each IAA. N-balance studies in adults indicate very small differences when comparisons are made between single proteins, and no differences in the apparent MPR were observed between plant-based or animal-source diets in the meta-analysis of all N-balance studies in adults in the 2007

WHO report (FAO/WHO/UNU 2007). Because of these difficulties in assessing protein quality directly in human nutrition, the current approach is to predict quality from digestibility and amino acid scoring.

Predicting protein quality: PDCAAS and DIAAS methods

If the IAA pattern of the protein requirement is known (i.e., as mg IAA/g protein), then the measured BV of a dietary protein should be predictable as an amino acid score from its composition relative to that of the requirement pattern. If the digestibility and the amino acid score are both known, then overall protein quality in terms of digestibility and amino acid score can be predicted as:

Protein quality = digestibility × amino acid score

This was formalized by the FAO in 1991 (FAO/WHO 1991) as the Protein Digestibility Corrected Amino acid Score, PDCAAS.

Digestibility

Digestibility, the proportion of food protein which is absorbed, has been defined from measurements of the N content of the foods and N-loss in faeces, with 'true' digestibility adjusted for endogenous N loss: that is, faecal N loss on a protein-free diet. Concern has been expressed over correction for faecal as opposed to ileal protein digestibility in the calculation of PDCAAS. Ileal digestibility of dietary proteins is a specific measure of amino acid absorption in the small intestine (where

TABLE 9.7 Recommended amino acid scoring patterns for infants, children, and older children, adolescents, and adults

	His	Ile	Leu	Lys	SAA	AAA	Thr	Trp	Val
Tissue amino acid pattern (mg/g protein)[1]	27	35	75	73	35	73	42	12	49
Maintenance (adult) amino acid pattern (mg/g protein)[2]	15	30	59	45	22	38	23	6	39
Age group	**Scoring pattern mg/g protein requirement**								
Infant (birth to 6 months)[3]	21	55	96	69	33	94	44	17	55
Child (6 months to 3 years)[4]	20	32	66	57	27	52	31	8.5	43
Older child, adolescent, adult[5]	16	30	61	48	23	41	25	6.6	40

[1]Composition of mixed tissue proteins.
[2]Adult amino acid requirement pattern.
[3]Based on the gross amino acid content of human milk.
[4]The 6 month (0.5 year) requirement pattern.
[5]The 3–10 year requirement pattern.

most amino acid absorption occurs), as measured at the terminal ileum. In contrast, faecal digestibility indicates how much of the N in food has been absorbed by the organism. In fact, digestibility is more complex than usually assumed. On the one hand ileal digestibility is a measure of residual amino acids after absorption from both dietary and endogenous sources but ignores any nutritional value of nitrogen absorbed from the colon. Faecal digestibility is a measure of residual nitrogen after absorption from all sources, much of which is bacterial protein, to some extent a function of the bacterial biomass in the colon. This is related to dietary non-starch polysaccharide (NSP) intake, which supports bacterial growth after its fermentation, together with N deriving largely from urea salvage. Taken together this means that for human diets containing large amounts of non-digestible carbohydrate, faecal N may not be a reliable measure of digestibility. Thus, the concepts of both ileal digestibility and faecal digestibility are subject to important limitations.

The FAO has suggested that conceptually, ileal digestibility of individual amino acids was the appropriate measure from which to develop a digestibility-corrected amino acid score, proposing a new term the Digestible Indispensable Amino acid Score (DIAAS) (FAO 2013), expressed as a percentage value for the limiting amino acid:

DIAAS (%) = digestible dietary IAA in the dietary protein at the terminal ileum (mg/g protein);

digestible dietary IAA in the dietary protein at the terminal ileum (mg/g protein) and the denominator, which is

Dietary IAA in the reference amino acid scoring pattern (mg/g protein).

However, because ileal digestibility studies are difficult to conduct, only a very few human studies have been undertaken and the currently available data is insufficient to recommend adoption of this new approach. Thus, in the meantime, until sufficient data has accumulated, the FAO has suggested that faecal digestibility (in effect PDCAAS), should continue to be used (FAO/WHO 1991).

Amino acid scoring

To score a protein, the single limiting amino acid is identified from the ratios of each individual amino acid in the protein with that in a reference pattern. Values <1 indicate potential deficiency, with the limiting amino acid indicated by the lowest ratio. This is the extent of the limitation: the score. Thus, if the lysine score for wheat gluten is only 0.5, to achieve the required intake of lysine (and consequently all IAAs), the digestible intake of the gluten will have to be increased by 1/0.5 = 2 times the reference protein intake.

The reference pattern is that of an ideal protein which would provide requirement levels of all amino acids when fed at the protein requirement level. So that:

Reference IAA pattern (as mg/g protein) = amino acid requirement/kg/d

and the denominator, which is

protein requirement/kg/d.

Defining amino acid requirement levels and a reference protein scoring pattern

The definition of amino acid requirement values has been challenging and remains subject to considerable controversy (Millward 2012c). Current values for the adult maintenance requirement pattern shown in Table 9.7 are derived from both N-balance studies in adults and stable-isotope studies which include various types of amino acid oxidation balance studies with ^{13}C-labelled amino acids, none of which have been judged entirely satisfactory (FAO 2013). The requirement for the sulphur amino acids has been predicted from the obligatory oxidative loss (on the basis that the sulphur amino acids have the highest ratio of obligatory demand to concentration in tissue proteins). The 'best estimate' of different values is often chosen from a wide range, for example, for lysine, 30mg/kg/d chosen from a range of values from 12–45mg/kg/d. No suitable values exist for infants, children, or other population groups so that a factorial model has been developed. This is based on the amino acid requirements for maintenance, assumed to be the same for all ages, and for growth assumed to be the same as the composition of mixed tissue proteins. The final age-related pattern (Table 9.7), was calculated as a weighted mean of the growth and maintenance patterns. (See Weblink Section 9.8 for more details.)

9.8 Sources and general nutritional properties of plant proteins

Plant foods are chemically diverse, often containing secondary metabolites, which may be either beneficial phytoprotectants or anti-nutritional, adversely influencing digestibility, having allergenic properties, or generally undesirable such as oxalates.

Major current sources of plant source proteins (PSPs) in the human diet are cereals, especially wheat, rice, and maize, starchy roots such as cassava and potato, and legumes including peas and various beans, especially soya (although only 2–3 per cent of soya production is

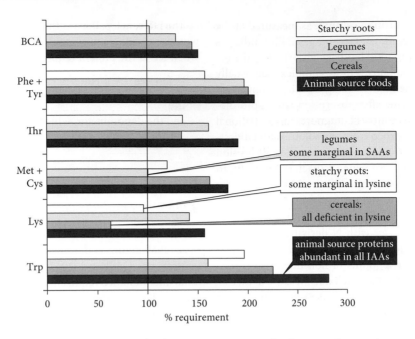

FIGURE 9.16 Amino acid profile of the main food protein sources as judged against the requirement pattern for older children, adolescents, and adults (see Table 9.7).

consumed as human food directly). Cereals represent globally the most important protein source, which is limited by low levels of Lys and, for maize, by Trp in the major protein fraction: that is, the prolamin storage protein.

Figure 9.16 shows the adequacy of the amino acid levels of the various PSPs as a percentage of that in the requirement pattern for older children and adults, for the main current food protein groups, starchy roots, legumes, cereals, and animal source proteins. Thus, some starchy root proteins like cassava and yam are marginally deficient in Lys, some legume proteins like pea are deficient in the SAA, while all cereals are deficient in Lys.

As to the significance of the low Lys, it is the case that there is considerable uncertainty in the current adult Lys value so that the deficiency may be more apparent than real. In the case of zein it is also low in Trp, one contributing factor to pellagra given the low bioavailability of niacin from maize (unless treated with lime), and the insufficient Trp limits its conversion to nicotinamide. Another important factor in the provision of Trp is its supply compared with that of the other large neutral amino acids (LNAAs). These are all transported into cells by the same LNAA transporter. When levels of these other LNAAs are high this will prevent the uptake of Trp into tissues. In fact, the ratio of Trp to LNAA is particularly low in maize and in sorghum, and both cereals have been implicated in pellagra.

In practice the protein quality and quantity in cereals can be quite variable. The relative amounts of storage as opposed to cytoplasmic proteins (as in the germ) influences the IAA profile so that that the profile of whole grain is better than that of gluten. Thus, the amino acid

quality worsens as the germ and bran are removed in flour production. Also, the relative amounts of storage and cytoplasmic protein can be altered with plant breeding, as with varieties of maize developed from opaque-2(o2), sugary-2(su2) hybrids with higher cytoplasmic protein:zein ratios. These are now in use under the name Quality Protein Maize (QPM). As shown in Table 9.8 these new maize strains have higher levels of all the IAA. Field trials in Ethiopia and elsewhere have shown better rates of weight gain and height growth in children where these hybrid strains are farmed. Protein sources currently under investigation as alternative human food protein include oilseeds such as rapeseed (canola), marine microalgae (e.g., chlorella), aquatic plants such as duckweed, and the cyanobacterium (blue-green algae) spirulina. The single-cell fungal mycoprotein (sold as 'Quorn') has been widely available, at least in the UK and Europe, for some time. In fact, all green leafy vegetables such as spinach are sources of high-quality protein (see Figure 9.17).

When the amino acid profile (biological value) of these various PSPs is judged against an animal source protein (ASP) such as beef, and the scoring pattern for schoolchildren, adolescents, and adults the most obvious features are that:

- all PSPs contain all IAA: often not appreciated;

- most PSPs contain more IAA than in the requirement pattern although relative proportions of some key amino acids like Lys, Trp and the SAAs vary;

- some (but by no means all) are indistinguishable from ASPs in terms of total amount and pattern of IAAs.

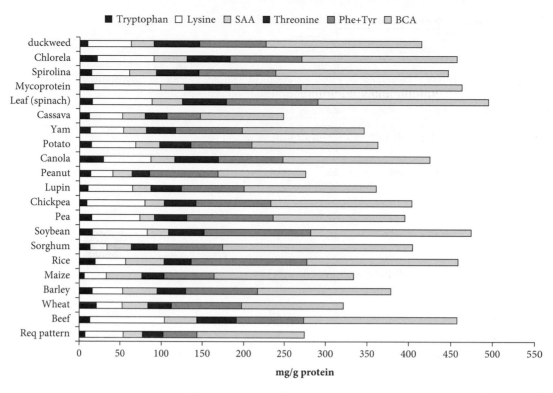

FIGURE 9.17 Amino acid profile of various plant source proteins (PSPs) is judged against beef and the requirement pattern for older children, adolescents, and adults (see Table 9.7)

It is clear that duckweed, chlorella, spirulina, mycoprotein, and leaf protein (of which spinach is typical of the food group), have similar levels of all the IAAs as meat and also have protein contents ranging from 25 to 50 per cent of the energy. Clearly with leaves in the human diet accounting for minor fractions of energy, leaves usually only represent a minor part of total protein. Nevertheless, as consumed in some countries such as Greece as Horta (steamed greens), at serving up to 500g, this would amount to 10g of very high-quality protein which would complement protein from bread to provide a balanced intake. The starchy roots exemplified by cassava, yam, and potato, are quite variable in terms of both amino acid profile and protein content. Cassava is quite poor in both respects. In contrast potato is a much better protein source with an IAA profile which exceeds the requirement for every IAA. With a P:E ratio of ≈11 per cent it has been the major source of protein in many traditional diets.

Oilseeds (canola) are used in the developed world mainly in animal feed, but have an IAA profile which exceeds the requirement for every IAA being especially rich in SAAs.

Legumes are traditionally viewed as protein-rich foods limited by their SAA but in the examples shown in Figure 9.17 (lupin, chickpea, pea, and soybean), only pea protein has less of the SAA compared with the reference pattern (about 70 per cent). They are all rich in Lys and

Trp especially soybean and pea and in many diets are consumed with cereals such as rice to complement the low Lys levels.

Taken together, these data show that in terms of their amino acid content there are many sources of PSPs which can be considered as alternatives to ASPs, and it is clear that currently available food sources make it relatively easy to construct nutritionally adequate vegetarian or vegan diets in terms of the quality of the protein content. Thus, their protein quality in terms of food sources may well only be limited by their digestibility.

Table 9.8 shows PDCAAS values of dietary protein sources, based on the scoring pattern for the older child, adolescents, and adults shown in Table 9.7 and their digestibility, and the PDCAAS-adjusted P:E ratios. The adjusted P:E ratio is the important measure: that is, available protein in foods determined by both protein content and quality.

ASPs generally perform well on both counts. Lys is the limiting amino acid for cereal proteins, yam, and cassava. Maize also contains less than the reference Trp level, but at 83 per cent of reference, compared with lysine at 60 per cent, Lys is limiting, and adjusting intake to supply Lys needs will supply more than enough Trp. The improved maize variety, QPM, has adequate Trp and 83 per cent of the reference lysine. Soya, in common with all legumes, has low levels of SAAs, but with this scoring pattern, is just sufficient.

TABLE 9.8 Protein and amino acid content and quality of animal and plant food sources

	P:E ratio	Lysine	Threonine	Sulphur AAs	Tryptophan	SCORE	Limiting AA	Digestibility	PDCAAS	Adjusted P:E ratios
Requirement	Pattern	48	25	23	6.6					
Beef	0.66	91	47	40	13	100		100	100	0.660
Egg	0.34	70	47	57	17	100		100	100	0.340
Cow's milk	0.19	78	44	33	14	100		100	100	0.194
Breast milk	0.060	69	44	33	17	100		100	100	0.060
Soya	0.388	65	38	25	13	100		90	90	0.349
Wheat	0.160	26	29	45	12	54	Lysine	95	51	0.082
Maize	0.130	29	36	29	5	60	Lysine	82	50	0.064
Improved maize	0.135	40	44	48	7	83	Lysine	80	67	0.090
Potatoes	0.100	54	38	29	14	100		82	82	0.082
Rice	0.072	36	37	40	11	75	Lysine	82	62	0.044
Yam	0.061	42	34	28	13	88	Lysine	80	70	0.043
Cassava	0.034	32	21	29	14	67	Lysine	80	53	0.018

Potatoes provide sufficient amounts of all amino acids and after correcting for digestibility the adjusted P:E ratio of potato at 8.2 per cent is higher than that of breast milk. However, even though the protein density of potato is adequate, the growth of a new-born infant could not be supported on mashed potato because its energy density is insufficient, with mashed potato too bulky to allow the new-born to eat enough to satisfy energy needs. Breast milk is high-fat and therefore energy-dense. For a young adult however, energy requirements per kg are only half that of infants so the potato could supply all of the energy needs. The high protein level in wheat together with its high digestibility means that it provides a much higher level of utilizable protein (PDCAAS-adjusted P:E ratio) than rice, yam, or cassava, even though wheat has the lowest Lys level of any staple at 54 per cent of the reference. Clearly cassava does badly mainly due to its low protein content so that only 1.8 per cent of its energy is utilizable protein.

9.9 Meeting future global protein requirements

Andy Salter

Whilst protein needs of human populations can be met from plant-derived foods, in practice for most populations, intakes represent a combination of animal and plant sources, although the relative proportions of each vary substantially on a continent basis (Figure 9.18). Availability, affordability, and cultural and personal preferences determine choice of protein sources, resulting in major population differences in both the amount and source of protein consumed, with Africa and Asia consuming mainly plant-sourced proteins (PSP)s, and the Americas, Europe, and Oceania mainly proteins from animal source foods (ASFs). However, dramatic changes in amount and type of protein can occur as countries' economic and social systems develop. For example, according to FAO food-balance tables, in 1987 the amount of protein available in China was approximately 62g/person/day, of which 11g were from ASFs. By 2017, this had increased to a total 101g of which 40g came from ASFs. Such shifts are frequently seen as countries become more economically stable and are perhaps driven by a perception that consumption of meat and dairy products is an aspiration rising from Western culture and associated with wealth. However, such shifts in the source of protein can have a major impact on health and, of increasing concern, on our environment. Diets rich in red and processed meat may comfortably meet protein requirements, but increase risk of chronic diseases such as cardiovascular disease and colorectal cancer. ASFs also tend to be energy dense, thus increasing the risk of obesity and related morbidities such as type 2 diabetes and non-alcoholic fatty liver disease, and, most importantly, are generally recognized as having detrimental environmental footprints.

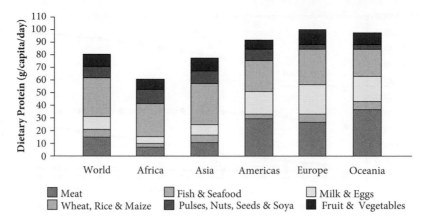

FIGURE 9.18 Sources of protein consumed in each continent of the world (g/capita/d).
Source: Reproduced from A M Salter & C Lopez-Viso (2021) Role of novel protein sources in sustainably meeting future global requirements. *Proc Nutr Soc* **80**, 186–194. Based on data reported by S H M Gorissen & O C Witard (2017) Characterising the muscle anabolic potential of dairy, meat and plant-based protein sources in older adults. *Proc Nutr Soc* **77**, 20–31. © The Authors 2021. Published by Cambridge University Press.

A large proportion of PSP crops is fed to farm animals (approximately 60 per cent, Figure 9.19), utilizing land and fresh water supplies which could be used directly for crops for human consumption. Figure 9.19 also indicates the relative inefficiency of the conversion of human-edible PSPs to edible ASFs. Effluent from animal agriculture is a major source of pollution of our waterways, and ruminant animals produce methane, a potent greenhouse gas. Taken together, the continuing increase in the global population, and the trend towards countries consuming more animal products as they become more economically stable, have led to increasing concerns over the impact of protein demand, particularly from animal products, on the environment.

Figure 9.20 shows an estimate of the increases in protein required to maintain the global adult population until the end of the century given UN-estimated population changes. These calculations indicate a 70 per cent increase in global protein requirements almost entirely associated with the growing population on the continent of Africa. In many high-income countries, requirements are predicted to remain stable but changing demographics mean that a much higher proportion of the protein required will be consumed by the elderly. This poses separate problems around providing adequate protein to maintain health in elderly persons, whose ability to consume protein-rich foods may be compromised by economic, physical, and mental changes frequently associated with aging.

The question of how to sustainably meet the global increase in protein requirement is of obvious concern given the uncertainty over the potential impact of

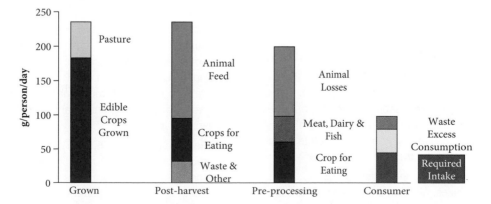

FIGURE 9.19 Global protein food chain, indicating the amount of protein (g/person/d), produced, harvested, and consumed, indicating losses from the human food chain.
Source: Reproduced from A M Salter & C Lopez-Viso (2021) Role of novel protein sources in sustainably meeting future global requirements. *Proc Nutr Soc* **80**, 186–194. Based on data presented by M Berners-Lee, C Kennelly, R Watson (2018) Current global food production is sufficient to meet human nutritional needs in 2050 provided there is radical societal adaptation. *Elem Sci Anth* **6**, 52. © The Authors 2021. Published by Cambridge University Press.

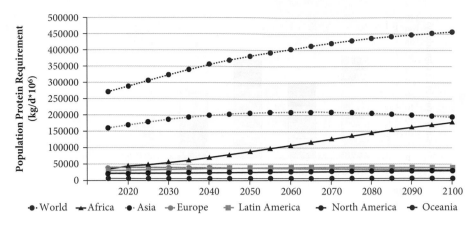

FIGURE 9.20 Estimate of global adult protein requirements until the end of the century. Based on an adult safe intake of 0.83g/kg/day and adjusted for regional mean body weights.

climate change on the ability to produce conventional crops, with major regional variations predicted. With much of the population of wealthier countries already consuming considerably more ASFs that is required to maintain good health, reducing intake in these regions could make a significant contribution, but this assumes the will to redistribute the excess protein to poorer populations. Similarly, reducing our overall dependence on ASFs and increasing the consumption of some of the more sustainable novel protein sources already discussed in Section 9.9 may also make a significant contribution. We could also make a shift from feeding animals human-edible crops, to more sustainable sources of protein such as insects (that could be fed on food waste) or bacteria (that can produce amino acids through fixation of nitrogen). Any such changes are going to require global cooperation and economic and political will beyond anything we have seen associated with global food systems to date.

KEY POINTS: PROTEIN REQUIREMENTS, QUALITY, AND SUSTAINABILITY

- Of the 21 amino acids within the proteome nine are nutritionally indispensable: six are dispensable, that is, capable of synthesis in the body in unlimited amounts, and six can become indispensable under conditions which limit their *de novo* synthesis.

- Individuals and populations adapt to the wide range of protein intakes. The minimum protein requirement (MPR), is the intake which balances all N-losses, that is, N-equilibrium in N-balance studies in adults, plus any special needs for growth, pregnancy, and lactation.

- The current value of the MPR for adults is the median of the wide range of values reported in a meta-analysis of N-balance studies, that is, the estimated average requirement (EAR), at 0.66g/kg/d. This is the intake meeting the obligatory demands, about 0.3g/kg/d, plus the average adaptive demand, assumed to be the maintenance requirement through all stages of the lifecycle. The safe individual-intake is 0.83g/kg/d, the EAR + 2sds of the inter-individual variation whilst the safe population intake is higher than the safe individual intake at ≈1.2g/kg/d, and is calculated as EAR + 2sds of individual protein intakes.

- Special needs are actual rates of protein deposition adjusted by an efficiency factor derived from the slopes of N-balance studies, which is likely to be overestimated and controversial, an important issue in pregnancy with requirements markedly overestimated and unsafe.

- Protein requirements are better explained by an adaptive metabolic demand model within which one component of the demand is adaptive amino acid oxidation, varying with the level of dietary protein intake, continuing throughout the diurnal cycle of feeding and fasting, and changing only slowly with variation in intake.

- The protein quality of human diets is predicted from protein or amino acid digestibility and amino acid score, calculated in comparison with an age-related reference amino acid pattern.

- Many plant-source proteins have amino acid scores comparable to animal-source proteins although all cereal proteins and some starchy root proteins are lysine limited while some legumes are sulphur amino acid limited.

- Global national protein consumption patterns reflect intake of animal source foods (ASFs). This intake increases with economic development with Europeans consuming 66 per cent more total protein than Africans, mainly due to nearly four times the consumption of ASFs. ASFs are energy-dense, their consumption, especially that of red and processed meat, increases the risk of obesity-related morbidities and other adverse health outcomes, and their production is very inefficient with a large adverse environmental footprint. The European/North American pattern of protein consumption is unsustainable globally and less desirable for human health compared with intakes of a higher proportion of plant-source proteins.

- Reducing global overall dependence on ASFs, developing more efficient and environmentally friendly animal-feeding systems, and increasing consumption of more sustainable and acceptable novel protein sources is the challenge for meeting protein needs with expanding populations in the developing countries.

 Be sure to test your understanding of this chapter by attempting multiple choice questions.

REFERENCES

Campbell W W, Johnson C A, McCabe G P et al. (2008) Dietary protein requirements of younger and older adults. *Am J Clin Nutr* **88**(5), 1322–1329.

FAO (2013) *Dietary Protein Quality Evaluation in Human Nutrition: Report of an FAO Expert Consultation.* 92. FAO Food and Nutrition Paper, Rome.

FAO (2014) *Research Approaches and Methods for Evaluating the Protein Quality of Human Foods: Report of a FAO Expert Working Group.* Rome.

FAO/WHO (1991) Protein quality evaluation: report of the Joint FAO/WHO Expert Consultation. FAO Food and Nutrition Paper 51. Rome;

FAO/WHO/UNU (2007) *Protein and Amino Acid Requirements in Human Nutrition: Report of a Joint FAO/WHO/UNU Expert Consultation.* WHO Tech R. 935. Geneva, WHO.

Millward D J (1995) A protein-stat mechanism for regulation of growth and maintenance of the lean body mass. *Nutr Res Rev* **8**(1), 93–120.

Millward D J (2003) An adaptive metabolic demand model for protein and amino acid requirements. *Br J Nutr* **90**(2), 249–260.

Millward D J (2012a) Nutrition and sarcopenia: Evidence for an interaction. *Proc Nutr Soc* **71**(4), 566–575.

Millward D J (2012b) Identifying recommended dietary allowances for protein and amino acids: A critique of the 2007 WHO/FAO/UNU report. *Br J Nutr* **108**(Suppl. 2), S3–S21.

Millward D J (2012c) Amino-acid scoring patterns for protein quality assessment. *Br J Nutr* **108**(Suppl. 2), 30–43.

Millward, D J (2021) Interactions between growth of muscle and stature: Mechanisms involved and their nutritional sensitivity to dietary protein: The protein-stat revisited. *Nutrients* **13**(3), 1–65. doi: 10.3390/nu13030729.

Rand W M, Pellett P L, Young V R (2003) Meta-analysis of nitrogen balance studies for estimating protein requirements in healthy adults. *Am J Clin Nutr* **77**(1), 109–127.

FURTHER READING

Darling A L, Millward D J, Lanham-New S A (2020) Dietary protein and bone health: Towards a synthesised view. *Proc Nutr Soc* **13**(1–8). doi: 10.1017/S0029665120007909

Haussinger D D, Gerok W, Sies H (1984) Hepatic role in pH regulation: Role of the intercellular glutamine cycle. *Trends Biochem Sci* **9**(7), 300–302. doi: https://doi.org/10.1016/0968-0004(84)90294-9

Millward D J (2008) Sufficient protein for our elders? *Amer J Clin Nutr* **88**(5), 1187–1188. PubMed PMID: 18996850.

Millward D J, Fereday A, Gibson *NR* et al. (2000) Human adult protein and amino-acid requirements: [^{13}C-1] Leucine balance evaluation of the efficiency of utilization and apparent requirements for wheat protein and lysine compared with milk protein in healthy adults. *Amer J Clin Nutr* **72**, 112–121. doi: 10.1093/ajcn.72.1.112

Millward D J & Garnett T (2010) Food and the planet: Nutritional dilemmas of greenhouse gas emission reductions through reduced intakes of meat and dairy foods. *Proc Nutr Soc* **69**(1), 103–118. doi: 10.1017/S0029665109991868.

Millward D J & Jackson A A (2004) Protein/energy ratios of current diets in developed and developing countries compared with a safe protein/energy ratio: Implications for recommended protein and amino-acid intakes. *Publ Hlth Nutr* **7**(3), 387–405. doi: 10.1079/PHN2003545

Millward D J & Jackson A A (2012) Protein requirements and the indicator amino-acid oxidation method. *Amer J Clin Nutr* **95**(6), 1498–1501.

Reeds P J (2000) Dispensable and indispensable amino-acids for humans. *J Nutr* **130**, 1835S–1840S. https://doi.org/10.1093/jn/130.7.1835S

Waterlow J C (1999) The mysteries of nitrogen balance. *Nutr Res Revs* **12**, 25–54. doi: 10.1079/095442299108728857

10 Alcohol metabolism: implications for nutrition and health

Vinood B. Patel and Victor R. Preedy

OBJECTIVES

By the end of this chapter, you should be able to:

- understand the varying intake of alcohol by different population and ethnic groups, and the contribution that alcohol makes to energy intake
- explain the main features, concepts, and consequences of alcohol metabolism
- understand how alcohol damages virtually all organs in the body including the liver
- describe the principal nutritional deficiencies in alcoholism
- identify causes associated with drinking in low socioeconomic status groups, and government policies to minimize alcohol consumption.

10.1 Introduction

The term alcohol is often interchanged with the primary alcohol, ethanol, and less commonly with ethyl alcohol. In the following text the word alcohol and ethanol will be used interchangeably. The consumption of alcoholic beverages is generally termed 'drinking' and dates back to over 9000 years ago when humans began fermenting alcoholic beverages. Today they are the most widely consumed beverages in the world and a leading cause of disability, morbidity, and mortality (WHO 2018). The oxidative metabolism of ethanol produces acetaldehyde and acetate, which are the current preferred names though there may be usage of systematic names, that is, for acetaldehyde and acetic acid these would be ethanal and ethanoate, respectively. However, the inadvertent consumption of certain alcohols such as methanol or ethylene glycol can produce toxic oxidative products, formaldehyde and oxalic acid, respectively.

Individuals will have preference for consuming different types of alcoholic beverages, for example wine, lager, ale, cider, spirits, or alcopops. However, some countries, regions within countries, or communities forbid the consumption of alcohol on religious, cultural, or moral grounds. Individuals may gain pleasure from the psychopharmacological effects of alcohol whereas others may react quite badly, with flushing, nausea and palpitations due to a genetic variation in alcohol- or acetaldehyde-metabolizing enzymes, producing high concentrations of acetaldehyde. Acute and chronic consumption of alcohol may cause malnutrition or act as a toxin and induce pathological changes in a variety of organ and tissues, such as the liver, brain, muscle, and the gastrointestinal tract. By contrast, a proportion of individuals consume moderate amounts of alcohol (one to two drinks/day), comprising up to 5 per cent of total dietary energy, and some data suggest that moderate alcohol consumption may be beneficial in reducing cardiovascular disease. However, some argue that its beneficial effect is controversial or outweighed by its detrimental effects. Recent guidelines suggest the cardioprotective effect is minimal or negligible (Department of Health 2015) and limited to women over the age of 55. Thus, it is important to take a balanced view of ethanol's effects.

Guidance on the Consumption of Alcohol by Children and Young People from the Chief Medical Officers of England, Wales, and Northern Ireland has suggested that children under 15 should not drink alcohol due to a range of damaging consequences. A common feature of excessive alcohol consumption is vomiting and coma with cognitive impairment as a result of long-term usage. Alcohol will lead to a lack of inhibitions, causing increased risk of drink driving accidents, crime, and risky sexual activity. Furthermore, women who are pregnant or about to become pregnant should avoid heavy alcohol consumption particularly in the first trimester as this can lead to neurological dysfunction such as that observed in foetal alcohol syndrome disorders and low birth weight. However, there are guidelines which suggest that pregnant women should avoid drinking altogether or not consume more than one or two units once or twice a week (Department of Health 2015). Drinking alcohol whilst breastfeeding should be avoided as breast milk will contain traces of alcohol and smell differently, thus affecting the baby's nutritional intake and/or feeding patterns.

The chemical nature of alcohol

In chemistry terms an alcohol is any organic compound with a functional hydroxyl group bonded to a carbon chain. As a consequence of its combined polar (OH group) and non-polar (C_2H_5 groups) properties, and because it is relatively uncharged, ethanol is miscible with water and can cross cell membranes by passive diffusion. It has the ability to dissolve lipids, such as in biological membranes and can act as a solvent for many organic compounds. Ethanol is produced from glucose via the fermentation of yeast to produce ethanol, carbon dioxide, and ATP. The source of carbohydrate dictates the type of alcoholic beverage. For example, beer is fermented from barley, wine from grapes, and cider from apples.

The immediate metabolite of ethanol oxidation, acetaldehyde (Figure 10.1), is a highly toxic and chemically reactive molecule that can bind irreversibly with proteins, DNA, RNA, and other molecules, forming complexes called adducts. Acetaldehyde is involved in liver disease pathology, where formation of acetaldehyde-protein adducts induces an immunological reaction. Readers are referred to the classic Novartis (formally CIBA) publication for additional reading (Novartis Foundation Symposium and Novartis 2007). Although dated it is comprehensive and has wide coverage discussing many scenarios of acetaldehyde toxicity. More recent reviews can be found in Patel (2016). Acetate, the product of acetaldehyde metabolism, is either oxidized peripherally to CO_2 in the Krebs (citric acid) cycle or used for synthesis of fatty acids and triglycerides. Acetate per se also has some biological activity, for example, it dilates resistance and capacitance blood vessels. It is also thought to affect mitochondrial fatty acid oxidation, reducing ATP concentrations.

Finally, in illicit or home brewed beverages and even in some commercially available beverages, there may be significant quantities of compounds that have putative toxic properties, such as congeners. These include diethylene glycol, acetaldehyde, acetone, methanol, and butanol.

The contribution that alcoholic beverages make to the energy intake of different population groups

Energy content of alcoholic beverages and the unit system

The chemical energy content of ethanol is 29.7 kJ (7.1 kcal) per g. In the UK, a unit of alcohol contains 10 mL of ethanol by volume and is equivalent to 8 g of ethanol (Table 10.1). However, there remains wide international variation in the amount of alcohol in a serving (from 7–14 g ethanol). Not all countries use the unit system (Table 10.1). The alcohol concentration of beverages can vary from 0.5 per cent (v/v) for low alcohol beers to 35-50 per cent (v/v) for distilled spirits such as vodka or whisky (Table 10.2). A unit of alcohol (10 mL or 8 g) of alcohol, is equal to a 125 mL glass of wine containing 8 per cent alcohol by volume or half a pint of 'ordinary' strength beer containing 3.5 per cent by volume. However, alcohol sold in UK pubs for most beers is around 4 per cent to 5 per cent (2.3 units and 3 units respectively, per pint), whereas a can of lager/beer/cider (440 mL) is 2 units. Wine is often sold as medium (175 mL) or large (250 mL) servings, containing around 13 per cent by volume (equating to around 2.3 and 3.3 units, respectively).

Recommended limits for alcohol consumption

Government guidelines by the UK Chief Medical Officers have recommended alcohol consumption of no more than 14 units/week for both men and women (Department of

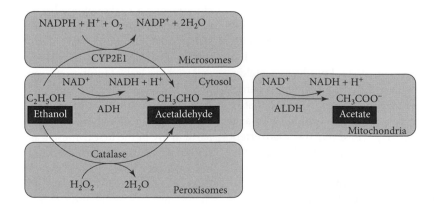

FIGURE 10.1 Oxidative pathways of alcohol metabolism.
Three major routes of ethanol oxidation depicting the conversion of alcohol to acetaldehyde and then acetate.

TABLE 10.1 The unit system

A. The unit system of alcohol consumption
One unit
Half a pint of beer at 3.5%
218 mL of beer at 4.5% (common alcohol concentration by volume)
One glass (125 ml) of wine at 8%
76 mL of wine at 13% (common alcohol concentration by volume)
One measure (50 ml) of fortified wine (sherry, port)
One measure (25 ml) of spirits (whisky, gin, vodka etc.)

B. Ethanol comprising one unit	
UK	8 g
Australia and New Zealand	10 g
USA	12 g
Japan	14 g

The unit system of alcohol ingestion is a convenient way of abstracting the amount of ethanol consumed by individuals and offers a suitable means to give practical guidance. The amount of alcohol in each unit will vary, for example depending on geographical location. Except for bars, the majority of UK bottled alcoholic beverages now display the total number of units, allowing consumers to monitor their unit intake.

Health 2015. Furthermore, the 14 units should be spread evenly over three days or more, and to include alcohol-free days for heavy drinkers. This advice is in contrast to previous maximal amounts recommended by the Royal College of Physicians, of 21 unit/week for men and 14 unit/week for women (Royal Colleges 1995). Previous governmental guidelines were based on maximum daily amounts, that is, no more than 3–4 and 2–3 units per day for men and women, respectively (Table 10.3).

The Health Survey for England reported that in 2018, 25.6 million people (57 per cent of people aged over 16) drank alcohol in the previous week of the survey. Taking the adult population as a whole, about 30 per cent of males and 15 per cent of females in the UK drink more than 14 unit per week, respectively. Around 24 per cent of men and 13 per cent of women in the UK are drinking harmful amounts of alcohol based on the full Alcohol Use Disorders Identification Test (AUDIT) measurement, where damage to health is likely. The National Health Service (NHS) estimates that around 9 per cent of men in the UK and 4 per cent of UK women show signs of alcohol dependence. However, there is considerable variation and usage of the terms used to describe alcohol consumption, for example, harmful, hazardous, alcohol use disorders, binge drinking, alcohol misuse, and so on. Many of these have precise definitions. The reader is advised to scrutinize precise definitions.

There are ethnic variations in the extent of alcohol consumption in the UK, with 22 per cent of Caucasian men drinking more than 14 units/week, compared to 5 per cent for Asian or 7 per cent for Black men. For women, the same ethnic patterns are seen as in men. There are also age-related changes in drinking patterns and this may also reflect sociological and demographic changes in the elderly

TABLE 10.2 Composition of alcoholic beverages

	Per 100 ml (all as g except energy)					
	Kcal	kJ	Alcohol	Protein	Fat	Carbohydrate
Alcohol free lager	7	31	Trace	0.4	Trace	1.5
Low alcohol lager	10	41	0.5	0.2	0	1.5
Lager	29	131	4.0	0.3	Trace	Trace
Special strength lager	59	244	6.9	0.3	Trace	2.4
Bitter	30	124	2.9	0.3	Trace	2.2
Cider (dry)	36	152	3.8	Trace	0	2.6
Wine (red, dry)	68	283	9.6	0.1	0	0.2
Wine (white, dry)	66	275	9.1	0.1	0	0.6
Wine (white, sweet)	94	394	10.2	0.2	0	5.9
Sherry (dry)	116	481	15.7	0.2	0	1.4
Spirits (various; 40% proof)	222	919	31.7	Trace	0	Trace

TABLE 10.2 *Continued*

	Per 100 ml (all as g)									
	Na	**K**	**Ca**	**Mg**	**P**	**Fe**	**Cu**	**Zn**	**Cl**	**Mn**
Alcohol free lager	2	44	3	7	19	Trace	Trace	Trace	Trace	0.01
Low alcohol lager	12	56	8	12	10	Trace	Trace	Trace	Trace	0.01
Lager	7	39	5	7	19	Trace	Trace	Trace	20	0.01
Special strength lager	7	39	5	7	19	Trace	Trace	Trace	20	0.01
Bitter	6	32	8	7	14	0.1	0.001	0.1	24	0.03
Cider (dry)	7	72	8	3	3	0.5	0.04	Trace	6	Trace
Wine (red, dry)	7	110	7	11	13	0.9	0.06	0.1	11	0.10
Wine (white, dry)	4	61	9	8	6	0.5	0.01	Trace	10	0.10
Wine (white, sweet)	13	110	14	11	13	0.6	0.05	Trace	7	0.10
Sherry (dry)	10	57	7	13	11	0.4	0.03	N	14	Trace
Spirits (various; 40% proof)	Trace	Trace	Trace	Trace	Trace	Trace	Trace	Trace	Trace	Trace

	Per 100 mL						
	Riboflavin	**Niacin**	**B6**	**B12**	**Folate**	**Pantothenate**	**Biotin**
	(mg)	**(mg)**	**(mg)**	**(µg)**	**(µg)**	**(µg)**	**(µg)**
Alcohol free lager	0.02	0.6	0.03	Trace	5	0.09	Trace
Low alcohol lager	0.02	0.5	0.03	Trace	6	0.07	Trace
Lager	0.04	0.7	0.06	Trace	12	0.03	1
Special strength lager	0.04	0.7	0.06	Trace	12	0.03	1
Bitter	0.03	0.2	0.07	Trace	5	0.05	1
Cider (dry)	Trace	0	0.01	Trace	N	0.04	1
Wine (red, dry)	0.02	0.1	0.03	Trace	1	0.04	2
Wine (white, dry)	0.01	0.1	0.02	Trace	Trace	0.03	N
Wine (white, sweet)	0.01	0.1	0.01	Trace	Trace	0.03	N
Sherry (dry)	0.01	0.1	0.01	Trace	Trace	Trace	N
Spirits (various; 40% proof)	0	0	0	0	0	0	0

Table 10.2 only gives an estimate of some of the compounds that will be present in alcoholic beverages. In addition, there will also be other compounds, which are not tabulated, such as fluoride, polyphenols, and other organic and non-organic compounds that impart characteristics of taste and smell. Data adapted from (Foods Standards Agency 2002).

population. It is reported that drinking more than 14 units a week is more common in the 55–64 age group for both sexes than any other age group. However, different patterns emerge if alcohol misuse is considered in terms of daily amounts. Recent trends have shown that binge drinking has decreased slightly in recent years (Statistics on Alcohol for England 2019). However, there are regional (North vs. South) and country variations (i.e., England vs. Scotland). Data obtained from surveys tend to underestimate alcohol consumption. As a result, seven-day drinking diaries are

being used to assimilate data by Health Survey England in conjunction with one-off surveys.

Drinking in the young and gender susceptibility

The results of a UK survey (Smoking, drinking, and drug use among young people in England 2018) has shown an overall decreasing trend in children aged 11–15 for

TABLE 10.3 Guidelines for maximal amounts of alcohol consumption

Guidelines by the Department of Health (UK) 2015
Men:
• Weekly: no more than 14 units/week
• Spread drinking of 14 units over 3 days
• Not advised: consistently drinking 4 or more units a day
Women:
• Protection: 1–2 units day, possibly protection against heart disease (past menopause)
• Weekly: no more than 14 units/week
• Not advised: consistently drinking 3 or more units a day
• Harmful: more than 1 or 2 units of alcohol, once or twice a week when pregnant or about to become pregnant. Safest to avoid drinking during pregnancy.

1 unit = 8 g ethanol (10 mL of pure ethanol)

TABLE 10.4 Consumption rates of different alcohol beverages in England 2017/18

	Consumption rates (units/week)	
	Men	Women
Spirits	1.8	1.6
Wine	4	5.4
Fortified wine	0.1	0.2
Normal strength beer/lager/cider	7.3	1.5
High strength beer & lager/cider	2.0	0.4
Alcopops	0.3	0.4
Total	**15.5**	**9.5**

Table 10.4 shows the variation in consumption of different alcohol beverages in the UK including low or no (zero) alcohol drinks. Variations in the consumption rates of different alcoholic drinks are often subject to socioeconomic and cultural factors. Note from 2008, consumption is calculated in units preventing direct comparison to previous data. Adapted from Health Survey for England (2019).

'drinking for the first time' (35 per cent in 2018, 39 per cent in 2013, compared to 61 per cent in 2003) and drinking in the last week (6 per cent in 2018, 9 per cent in 2013, compared to 25 per cent in 2003). However, about 70 per cent of 15-year-olds have reported drinking for the first time, compared to 14 per cent for 11-year-olds. The mean amount consumed by children aged 11–15 is approximately 10 units per week.

Drinking by schoolchildren and adolescents has at least six serious consequences: (a) alcohol poisoning and fatalities; (b) drinking in formative years will predict the extent of alcohol misuse or dependency later on; (c) drinking may be compounded by polydrug and other substance misuse including tobacco; (d) total lifetime intake of alcohol, rather than recent intakes, is a good predictor of alcohol-related harm; (e) tissues in the young are particularly sensitive to alcohol; (f) there is an association of underage or unsupervised dinking with poor academic performance and crime.

Men consume higher amounts of alcohol than women (Table 10.4) but women are more susceptible to alcohol-induced injury such as cardiomyopathy, skeletal muscle myopathy, brain damage, and liver disease. This may be related to lower clearance rates of alcohol in women during 'first pass metabolism' (the process where ethanol is metabolized by the gastrointestinal tract and liver first before it reaches the systemic circulation), as a consequence of either smaller liver size, differences in gastric alcohol metabolizing enzymes, endocrine factors, body fat composition, or even psycho-social factors in reporting alcohol consumption. Compared with men, women also have higher blood acetaldehyde concentrations following the same amount of alcohol per unit body weight. It has been estimated that whilst men will show an increased chance of developing liver disease at an intake rate of 40–60 g

ethanol/day, the threshold level for women is lower at 20 g/day. A comprehensive analysis of the vulnerability of women compared to men has been reviewed and readers are referred to this work (Fernandez-Sola et al. 2005).

Social impact of alcohol consumption

Generally speaking, people of low socioeconomic status are more than twice as likely to die from alcohol-attributable causes than those of high socioeconomic status. The impact of alcohol on mortality can be compounded by other risk factors such as smoking, poor health, and obesity. Furthermore, men are more likely to drink more than women and have a higher global burden of disease disability-adjusted life years (DALYs) than women. The consequences of a partner drinking more are several fold. Increased domestic/intimate violence towards partner, gambling money from the family budget, economic impact (loss of job). Parental drinking is also linked with alcohol use in young children and worse educational outcomes for children.

Alcohol has a large negative impact on mental health and well-being. Studies have shown alcohol use significantly increases the risk of suicide and depression (Preedy 2019; Patel & Preedy 2022). Consumption of high amounts of alcohol is also associated with increased aggression and is a leading cause of homicide. Alcohol consumption also increases the transmission of sexually transmitted diseases, such as HIV/AIDS due to increased risk of unprotected sex. Alcohol also depresses the immune system and in heavy alcohol consumers, this increases the risk of developing tuberculosis three-fold.

Policies to reduce alcohol consumption

On a yearly basis, globally, there are 3 million deaths (2.7 million men and 0.7 million women) attributable to alcohol (WHO 2018), and 52 million injury DALYs are attributable to alcohol. The major contributors include road injuries (373,000), self-harm (146,000), interpersonal violence (88,000), and falls (76,000). In the WHO European Region, 10 per cent of all deaths are attributable to alcohol. Current projections to 2025 indicate that total alcohol per capita consumption (15+ years) is expected to increase in approximately 50 per cent of WHO regions, with India showing the largest increase.

Alcohol policies include laws, rules, and regulations that aim to prevent or reduce alcohol-related mortality and DALYS (WHO 2018). These policies can exist at the global, national, or local regional level. Alcohol consumption-reducing strategies are multicomponent including a range of targeted approaches focusing on reducing alcohol availability, increasing the price of alcohol, marketing restrictions, minimum age purchase, and drink-driving surveillance and enforcement.

Drink-driving is a leading cause of road traffic accidents. In some countries, such as South Africa, this is up to 60 per cent. Globally, countries have varying degrees of blood alcohol concentration (BAC) limits for those who drink and drive, from zero tolerance (0.0 per cent) to 0.08 per cent, with most BAC limits either 0.05 per cent or 0.08 per cent. Many countries have no BAC limits. In December 2014 Scotland changed the drink-drive legislation from 0.08 per cent to 0.05 per cent. However, unexpectedly this did not result in a reduction in road traffic accidents when compared to England and Wales. The lack of effect was probably due to limited enforcement with random blood tests and public awareness campaigns. Other policies that have been suggested to reduce drink-related road accidents include sobriety check-points and introducing zero tolerance concentrations for young/novice drivers.

Increasing excise taxes on alcohol can have a direct impact on reducing alcohol consumption. This effect is greater for spirits than beer although less effective in high alcohol drinkers. Furthermore, in most countries the price of alcohol, in particular beer, has become more affordable. Despite this, some Eastern European countries have implemented strict policies. Russia, which at one stage had Europe's fourth highest consumption of alcohol (15.8 L/person) introduced a wide range of policies over a ten-year period from 2005, including changing the BAC drink-drive limit to 0.0 per cent, raising taxation on vodka, fines, and criminal prosecution for retailers selling alcohol to minors, and banning alcohol advertisement on the Internet. The net effect of these policies was a dramatic reduction in capita alcohol consumption by 3.5 L/year over the 2005–2015 period. Similarly, in 2017 Lithuania increased alcohol taxation by 29 per cent and 14 per cent on beer and vodka respectively, increased the minimum age of purchase from 18 to 20, and restricted the hours for selling alcohol after 8 pm as well as at sporting events. This led to a 4.8 per cent reduction in all-cause mortality in one year. Other countries have implemented targeted approaches: Thailand has banned alcohol sales within 300 metres of an educational setting, banned drinking in public parks, and set a BAC limit of 0.02 per cent for young drivers. Singapore has banned consumption of alcohol in all public places from 10.30 pm to 7.00 am and has tighter regulations in liquor control zones. South Africa has banned alcohol sales within 500 metres of places of worship, schools, and sporting facilities, and raised the minimum age to 21 for purchasing alcohol.

In May 2018 the Scottish government introduced the world's first minimum unit price for alcohol, which was set at 50 pence per unit, with the aim to target harmful drinkers and the low-cost high-strength alcohol market. This correlated with the price of cider increasing from £1.25/L to £3.80/L. Early follow-up studies have shown a clear benefit with cider sales deceasing by 20 per cent in 2020 and the volume of pure alcohol sold per person decreasing by approximately 4 per cent (from 7.4 to 7.1 litres). Further information on policies and modelling data to determine alcohol-attributable mortality and morbidity can be found at International Model of Alcohol Harms and Policies.

Severe acute respiratory syndrome coronavirus 2 and alcohol consumption

In 2019 an outbreak of a disease due to severe acute respiratory syndrome coronavirus 2 (SARS-CoV-2) was identified in China and the disease entity, Coronavirus disease 2019 (COVID-19), has now been declared a pandemic. The pandemic resulted in (at the time of writing) over 2 million deaths worldwide. The response of each country was different with regard to lockdowns, confinements, rules on travel, and social isolation as well as the provision of vaccines. This pandemic had an immediate effect on the consumption of alcohol-containing beverages and those seeking help and advice for their addictions. Some countries passed interim measures to ban the sale of alcohol. However, the episodic nature of the lockdowns and the degree to which guidelines were followed mean that current observation on the consumption of alcohol can only be treated as interim or preliminary. In the United Kingdom confinement was shown to shift the consumption of low-alcohol-containing beers to those of higher strength. Beer consumption fell by 40 per cent and this was offset by increased consumption of wines and spirits by 15 and 22 per cent, respectively. Overall, there was no overt increase in alcohol purchases when on-premises (bars and restaurants) and off-sales (shops, supermarkets) were taken together, that is, reduced purchases in bars and restaurants were offset by increased sales from shops and supermarkets and home deliveries. The caveat to these observations is that there will be changes in the pattern of alcohol consumption in subgroups as illustrated by Australian studies. These showed that during confinement there was a reduction in consumption of alcohol by young drinkers and a small increase in consumption by middle-aged women.

Energy and micronutrient content of alcoholic beverages

As mentioned earlier, one unit contains 8 grams of ethanol, which is equivalent to ten mL of ethanol and thus provides 234 kJ (56 kcal). This can underestimate the true energy content of alcoholic drinks since they also contain constituents, such as unfermented carbohydrates, amino acids, and fatty acids (see Table 10.2; Foods Standards Agency 2002) or when combined with 'mixers'(carbonated beverages) or fruit juices. Depending on the alcoholic beverage, the energy composition varies from about 126–921 kJ (30–220 kcal)/100 mL. Low- or zero-alcohol beverages will as expected have a lower energy content although this is compensated for by a higher carbohydrate content. Alcoholic beverages will also contain trace amounts of compounds that impart flavour or characteristics of taste and smell, for example, aliphatic carbonyls, other alcohols, monocarboxylic acids, sulphur-containing compounds, tannins, polyphenols, or minerals.

Ethanol's contribution to energy in the diet

In the UK the mean daily intake of alcohol in all men (19–64 yr, consumers and non-consumers) is 16.8 g (491 kJ or 118 kcal) (31.7 g for just consumers; 927 kJ or 222 kcal) and 8.1 g (237 kJ or 57 kcal) for all women (17.5 g for just consumers; 512 kJ or 123 kcal) (National Diet and Nutrition Survey 2018). Consideration must be given to the non-alcoholic energy contained within the beverages and possible under-reporting of alcohol consumption.

Most of the consumption of alcohol in the UK is in the form of beer (men) and wine (women) (Table 10.4). Overall (i.e., in alcohol consumers and non-consumers) the contribution of ethanol to total energy intake in the 19–64 yr age group is reported to be 5.2 per cent in men and 3.2 per cent in women (National Diet and Nutrition Survey 2018). In consumers, the corresponding contributions are 9.8 per cent and 6.9 per cent, respectively (National Diet and Nutrition Survey 2018).

The contribution made by ethanol-derived calories is significant in dependent alcoholics. In one study, patients attending an inner city Alcohol Misuse Clinic in the UK consumed on average 160 g ethanol/day; contributing about 60 per cent to dietary energy intake. Some studies have shown that alcohol may contribute to up to 85 per cent of dietary energy in some subgroups. However, alcohol consumption reporting is subject to errors. For example, underreporting of alcohol consumption is known to be commonly prevalent in all self-reporting methods. No food frequency questionnaires have been unequivocally validated in alcohol misusers. Typical patients with chronic liver disease may consume 160–250 g ethanol/day (1140–1770 kcal/day). This has nutritional consequences as ethanol may be perceived as being 'empty calories', that is, having negligible or minor quantities of micro- or macronutrients. High ethanol loads also impair the normal function of the liver and damage the intestinal tract (see Section 10.3).

There is now growing evidence that excessive alcohol intake increases the risk of type II diabetes. Consuming 6–8 units/day raises the risk by between 15 per cent and 75 per cent, with women at greater risk. However, some studies have reported a J-shaped relationship between alcohol and incidence of type II diabetes. The relationship between alcohol consumption and obesity is controversial and may relate to gender, genetic, and dietary factors as well as the concentrations of alcohol consumed. Obesity is not apparent in all alcoholics but in some subjects who consume moderate to high amounts of alcohol, obesity may increase. Some of this effect may be related to the effect of alcohol consumption on appetite. For example, in one study dietary intake following ingestion of 32 g of alcohol was 5786 kJ (1385 kcal) versus 4928 kJ (1179 kcal) when 8 g of alcohol was consumed.

Negative consequences of chronic alcohol ingestion

There are as many as 200 different alcohol-related disorders or conditions (Table 10.5), (Preedy & Watson 2005; WHO 2018; Patel & Preedy 2022) affecting the whole body. Many of the deleterious effects relate in some way to ethanol metabolism, altering cellular biochemistry either because of ethanol per se, or its immediate metabolite, acetaldehyde. Approximately 10–15 per cent of chronic alcohol misusers will have cirrhosis and 30 per cent will have gastrointestinal pathologies (Table 10.6). In terms of the gastrointestinal tract, all regions can be affected, from the mouth to the rectum. For example,

TABLE 10.5 Systems and tissues affected by alcohol misuse

[1] Hepato-Pancretobiliary

Hepatomegaly—fatty liver, alcoholic hepatitis, and fibrosis

Cirrhosis and hepatocellular carcinoma

Acute and chronic relapsing pancreatitis—malabsorptive syndrome

[2] Central, peripheral, and autonomic nervous systems

Acute intoxication

Progressive euphoria, incoordination, ataxia, stupor, coma, and death

Alcohol withdrawal symptoms including delirium tremens, morning nausea, retching and vomiting, nightmares and night terrors, blackouts, and withdrawal seizures

Nutritional deficiencies

   Wernicke–Korsakoff syndrome

   Pellagra

   Tobacco-alcohol amblyopia

Others

Cerebral dementia, cerebellar degeneration

Demyelinating syndromes—central pontine myelinolysis

Marchiafava–Bignami syndrome, associated with electrolyte disturbances

Foetal alcohol syndrome—full-blown syndrome, mental impairment, attention deficit and hyperkinetic disorders, specific learning difficulties

Peripheral nervous system

Sensory, motor, and mixed neuropathy

Autonomic neuropathy

[3] Musculoskeletal

Proximal metabolic myopathy, principally affecting Type II (white) fibres

Neuromyopathy secondary to motor nerve damage

Atrophy of smooth muscle of gastrointestinal tract, leading to motility disorders

Osteopenia—impaired bone formation, degradation, nutritional deficiencies (e.g., calcium, magnesium, phosphate, vitamin D)

Avascular necrosis (e.g., femoral head)

Fractures—malunion

[4] Genitourinary

IgA nephropathy

Renal tubular acidosis

Renal tract infections

Female and male hypogonadism, subfertility

Impotence

Spontaneous abortion

Foetal alcohol syndrome

[5] Cardiovascular

Cardiomyopathy, including dysrhythmias

Hypertension

Binge strokes

Cardiovascular disease (including stroke)

Myocardial infarction

[6] Dermatological

Skin stigmata of liver disease—rosacea, spider naevi, palmar erythema, finger clubbing

Skin infections—bacterial, fungal, and viral

Local cutaneous vascular effects

Psoriasis

Discoid eczema

Nutritional deficiencies (including pellagra)

[7] Respiratory

Chronic bronchitis

Respiratory tract malignancy

Asthma

Postoperative complications

[8] Oro-Gastrointestinal

Periodontal disease and caries

Oral infections, leukoplakia, and malignancy

Alcoholic gastritis and haemorrhage

Alcoholic enteropathy and malabsorption

Colonic malignancy

[9] Haematological

RBCs—macrocytosis, anaemia because of blood loss, folate deficiency and malabsorption, haemolysis (rarely)

WBCs—neutropenia, lymphopenia

Platelets—thrombocytopenia

Table 10.5 is designed to show that diseases associated with alcohol misuse are not confined to only the liver and brain. Virtually all tissues and organs systems can be adversely affected with only some being life threatening. Furthermore, not all individuals will develop a disease, possibly due to inherent protective, dietary, or genetic factors.
(Adapted from T J Peters & V R Preedy (1998) Metabolic consequences of alcohol ingestion. Novartis Foundation Symposium **216**, 19–24.

oral mucosal lesions have been shown to occur in as much as 28 per cent of chronic alcoholics. The relative risk of rectal cancers increases about four-fold in chronic alcohol misusers. Fatty liver will occur in 80 per cent of chronic alcoholics (but only 10–15 per cent will develop cirrhosis) and 50 per cent will have bone marrow changes (perturbing red blood cell morphology). Half of chronic alcoholics will have damaged skeletal tissue (osteoporosis, osteopenia, fractures including post-fracture malunion) whereas between 20–30 per cent will exhibit a spectrum of subclinical or clinical cardiac abnormalities (i.e., alcoholic cardiomyopathy) or other cardiovascular diseases including hypertension. A staggering 80 per cent of subjects will have skin lesions including those of vascular, fungal, bacterial, or viral origins and 40–60 per cent will have alcoholic myopathy. Abnormal gonadal function will occur in 50 per cent of male alcoholics.

TABLE 10.6 Prevalence of alcohol-induced pathologies in chronic alcohol abusers (A) and alcohol attributable fractions to selected global deaths/diseases/injuries (B)

A.	(%)
Skin disorders	80
Alcoholic myopathy	50
Bone disorders	50
Gonadal dysfunction	50
Gastroenterological disorders	30
Cirrhosis	15
Neuropathy	15
Cardiomyopathy	10
Brain disease (organic)	10
B.	**(%)**
Alcohol use disorders	100
Liver cirrhosis	48
Oral cavity and pharynx cancers	31
Pancreatitis	26
Poisoning	12
Falls	11
Liver cancer	10
Haemorrhagic stroke	9
Hypertension	7
Breast cancer	5
Ischaemic heart disease	3

A. The prevalence of alcohol-related disorders related to chronic alcohol-dependent subjects.

V R Preedy & R R Watson (2005) *Comprehensive Handbook of Alcohol Related Pathology*. Volumes 1–3. Academic Press, San Diego. WHO 2018 Global status report on alcohol and health 2018.

B. Another way of examining the effect of alcohol is to explain in terms of alcohol attributable fractions (WHO 2018).

Another way of examining the effect of alcohol is to explain in terms of alcohol attributable fractions, and this is shown in Table 10.6. As a rule of thumb, 50 per cent of chronic alcohol misusers will have one or more organ or tissue abnormalities. In England, in 2018 there were 7551 alcohol-related deaths, of which the majority were due to alcoholic liver disease (ALD) (ONS 2021). Globally, per annum approximately 3.0 million deaths (5.3 per cent of all deaths) are alcohol related (WHO 2018). There is, however, under-reporting of alcohol-related illnesses and conditions.

Very often, dependent drinkers smoke cigarettes or tobacco-related products, that is, they are addicted to nicotine and this has a greater effect on the development of disease than either addiction alone. This is particularly relevant with respect to cancers of the upper aerodigestive tract, and these synergistic effects of smoking and drinking have also been seen in the development of cirrhosis, possibly due to toxic metabolites of nicotine processed in the liver. The advent of smokeless cigarettes, such as e-cigarettes or vaping is a relatively new phenomena but there is little research on this in relation to alcohol consumption. However, one study showed a positive correlation between e-cigarette usage and the extent of alcohol consumption.

In Europe and the Americas, between 15–55 per cent of people attending hospital (as either inpatients or outpatients) or primary care centres are classified as dependent or hazardous alcohol abusers. However, fewer than 5 per cent of adults have such misuse or dependency recorded in their medical records. Prevalence rates of alcohol misuse will depend on geographical and socioeconomic factors. In London, UK, a third of all acute hospital admissions are alcohol related and the prevalence of alcohol misuse in in-patients in city hospitals may be as high as 50 per cent. In fracture clinics, 40–70 per cent of patients score positively for alcohol-related dependency or abuse syndromes. Overall, in 2018 there were over 1.1 million NHS admissions to Accident and Emergency (A&E) Departments due to alcohol consumption, placing a financial burden of £3.5 billion on the NHS. This compares to the overall cost of £21 billion to the UK economy as a consequence of alcohol misuse as it not only affects health but societal factors (police, judiciary, social departments, etc.).

Questionnaires to detect alcohol misuse and impact on health

There are several questionnaires designed to detect alcohol misuse. These questionnaires have been well validated and include The Alcohol Use Disorder Identification Test (AUDIT), Michigan Alcohol Screening Tool (MAST), Cut, Annoyed, Guilty, Eye-Opener (CAGE), Paddington Alcohol Test (PAT), Severity of Alcohol Dependence Questionnaire (SADQ), and others. Currently the gold standard is perceived to be the AUDIT questionnaire due to its wide applicability, translation into different languages, and international usage. In some circumstances these can be more useful than laboratory tests on serum, plasma, urine, or saliva, which only indicate recent alcohol consumption. However, these questionnaires do not give accurate information on the amount of alcohol consumed.

10.2 Alcohol metabolism

Many of the pathologies associated with excessive alcohol consumption are due to the damaging effects of acetaldehyde, and molecular and cellular metabolic changes (e.g., DNA methylation, redox state, anti-oxidant or endocrine status) associated with ethanol oxidation (See Figure 10.1 for a scheme of ethanol metabolism.) All biochemical pathways and cell structures have the potential to be targeted by ethanol or its related metabolites. Central to these effects is the liver, where 60–90 per cent of ethanol metabolism occurs. Up to 90 per cent of the substrates utilized in conventional metabolic pathways in liver may be displaced by ethanol oxidation. Ethanol ingestion can inhibit protein and fat oxidation in the body by approximately 40 and 75 per cent, respectively. The 2.5-fold increase in oxidation of carbohydrate after a glucose load is also abolished by ethanol. Oxidation of ethanol by gastric first pass metabolism will account for 5–25 per cent of ethanol oxidation due to the presence of ADH enzyme in the stomach, and 2–10 per cent of ingested ethanol will appear in the breath, sweat, or urine (that is, unmetabolized ethanol).

The metabolic fate of alcohol following digestion and absorption

Ethanol is rapidly absorbed, primarily in the upper gastrointestinal tract, and appears in the blood as quickly as 5 minutes after ingestion. Its distribution will approximate total body water. Its elimination thereafter will approximate to Michaelis-Menten kinetics, though zero-order elimination kinetics have also been described. Blood alcohol concentration depends on pathophysiological factors, such as absorption rate, first pass metabolism, the extent to which liver function has been altered, and blood flow. The rate at which alcohol is oxidized, or disappears from the blood, varies from 6 to 10 g per hour. This is reflected in plasma concentrations, which fall by 9–20 mg/100 mL/hour. In response to a moderate dose of alcohol of 0.6–0.9 g/kg body weight, the elimination rate from the blood is approximately 15 mg/100 mL blood/hour on an empty stomach though there is considerable variation. One study reported that alcohol elimination rates ranged from 14 to 31 mg/100 mL/hour.

Food in the stomach will delay the absorption of alcohol and blunt the peak blood alcohol concentration. The peak blood concentration is the point at which the rate of elimination equals the rate of absorption. Using a standard dose of ethanol/kg body weight, it has been shown that the peak is lower after a meal compared with an empty stomach. The time to metabolize the alcohol was two hours shorter in the fed state than the fasted state, indicative of a postabsorptive enhancement of ethanol oxidation which can be as much as 35–50 per cent.

The type of food taken with alcoholic beverages will also alter the peak ethanol concentration: after a standard dose of ethanol of 0.3 g/kg, meals rich in fat, carbohydrate, and protein result in peak ethanol concentrations of 16.6, 17.7, and 13.3 mg/100 mL respectively. Part of this variation may be due to increased portal blood flow in response to feeding which will essentially deliver more ethanol to the liver for oxidation.

Peak blood ethanol concentrations (mg/L) will depend on a combination of the concentration of ethanol in the beverage (g of pure alcohol per litre of beverage), the total amount of ethanol consumed in absolute terms (g of ethanol ingested), the fasting or fed state, and the other foods consumed with the alcoholic drink. Physiological and genetic factors will also play a role (see next section). One of the primary determinants of alcohol metabolism is the rate of gastric emptying. The small intestine is the main site of ethanol absorption.

First pass metabolism and the contribution of the stomach

First pass metabolism principally occurs in the liver (hepatic first pass metabolism), but a small proportion of alcohol is also metabolized by the stomach (gastric first pass metabolism). Stomach ADH (called sigma-ADH) is a different isoform from the enzyme in the liver. Physiological factors that influence gastric emptying will also influence the contribution of this pathway to ethanol elimination. In one study, where ethanol (0.3 g/kg body weight) was administered by different routes, it was calculated that the amount of ethanol absorbed (0.224 g/kg body weight) was 75 per cent of the administered dose: the difference being ascribed to first pass metabolism. The rate of gastric ethanol metabolism has been reported to be about 1.8 g of ethanol per hour. Reduced first pass metabolism and/or reduced gastric ADH will occur in *Helicobacter pylori* infection and during histamine H2-receptor antagonist therapy, this results in higher blood alcohol concentrations. There are also ethnic differences: those of East Asian origin have a lower stomach ADH/first pass metabolism compared with Caucasians. This will also result in a higher blood alcohol concentration. Chronic alcoholism reduces the capacity of this gastric route of ethanol oxidation due to the development of gastritis (which is an inflammation of the stomach).

Gender differences in alcohol metabolism

There are gender differences in the rate of ethanol elimination rates ascribed to first pass metabolism. The activity of gastric ADH in women is also lower than in men, though this is less apparent in women over 50 years old. Compared with men, women will have higher blood ethanol concentrations after an equivalent load. The lower first pass metabolism activities, lower blood volume, and higher body fat account for the higher blood ethanol concentrations in women, rather than differences in gastric emptying or rate of ethanol oxidation in the liver. It has however, been proposed that women and men have comparable peak blood alcohol concentrations when dosage is based on total body water.

The speed with which alcohol is distributed in body water

After ingestion, alcohol that is not immediately absorbed passes down the gastrointestinal tract. Very high ethanol concentrations occur in the small intestine compared with plasma. Effectively, there is a gradient down the gastrointestinal tract. For example, a dose of 0.8 g ethanol/kg body weight (equivalent to 56 g ethanol =7 unit = 3.5 pints of ordinary beer (3.5 per cent v/v), consumed by a 70 kg male) will result in blood ethanol levels of 100–200 mg/100 mL, 15–120 min after dosage. Maximum blood concentrations occur after about 30–90 min. Gastric concentrations of ethanol peak at 8 g/100 mL of luminal contents; jejunal concentrations are approximately 4 g/100 mL compared to approximately 0.15 g/100 mL in the ileum. Concentrations in the ileum reflect plasma concentrations, that is, from the vascular space. After about two hours ethanol concentrations in the stomach and jejunum will approximate concentrations in plasma. In the postabsorption phase, the distribution of alcohol in the body will reflect body water to the extent that, for a given dose of alcohol, blood concentrations will reflect lean body mass. The solubility of ethanol in bone and lipid is negligible. Whole blood concentrations (which includes plasma and cellular contents) of ethanol are about 10 per cent lower than plasma concentrations because red blood cells have less water than plasma.

Metabolism by alcohol and aldehyde dehydrogenases and other routes

Alcohol is oxidized to acetaldehyde by three major routes (Figure 10.1), namely:

(i) ADH (alcohol dehydrogenase; cytoplasm; (ii) MEOS (microsomal ethanol oxidizing system; endoplasmic reticulum), and (iii) catalase (peroxisomes). There are at least six classes of ADH, and oxidized substrates include steroids and some intermediates in the mevalonate

pathway as well as fatty acid ß-oxidation and retinoids (Lieber 2000).

Alcohol metabolism via ADH leads to excess production of the reducing equivalent NADH, so that the NADH/NAD$^+$ ratio increases, with a corresponding rise in the lactate/pyruvate ratio. The metabolism of acetaldehyde to acetate via aldehyde dehydrogenase (ALDH, principally in the mitochondria), also produces NADH, so exacerbating the elevated ratio. Changes in the cellular (via ADH) or mitochondrial (via ALDH) redox state may explain metabolic abnormalities in alcoholism such as: hyperlactacidemia, hyperuricemia, increased lipogenesis, decreased mitochondrial beta-oxidation of fatty acids, hypoglycaemia, reduced glycolysis, and disturbances in the tissue responsiveness to hormones. Other contributing abnormalities include lipid peroxidation, iron dysregulation, adduct formation, DNA damage, epigenetic modulations, altered gene expression, apoptosis, necrosis, perturbed proteolytic cascades, translational defects, hypoxia, Kupffer cell activation, altered antioxidant status, membrane changes, and alterations in cellular trafficking (Patel 2016). Extrahepatic tissues, for example, mouth, oesophagus, duodenum, jejunum, rectum, and muscle also contain ethanol metabolizing enzymes, leading to localized damage.

Ethanol oxidation via peroxisomal catalase is a minor pathway and requires the concomitant presence of a hydrogen peroxide (H$_2$O$_2$) generating system (See Figure 10.1). When there is an increase in H$_2$O$_2$ generation, for example, from the oxidation of long chain fatty acids in the peroxisomes, or increased mitochondrial hydrogen peroxide production, there may also be an increase in catalase-mediated ethanol oxidation.

The metabolite acetaldehyde is oxidized to acetate via NAD$^+$-dependent aldehyde dehydrogenase (ALDH). As with ADH, there are several classes of ALDH. To date nineteen members of the ALDH enzymes have been identified and are involved in the removal of exogenous and endogenous aldehydes (Patel 2016). Of these the mitochondrial ALDH2 is the important in terms of alcohol related pathology. The location of ALDHs in extrahepatic tissues such as heart may be protective whereas lower concentrations in brain may explain the vulnerability of CNS tissues in alcoholism (Kwo & Crabb 2002).

Acetaldehyde itself is a highly reactive toxic metabolite. As mentioned earlier, some acetaldehyde becomes bound to cellular constituents such as proteins, lipids, and nucleic acids generating harmful adducts. Adduct formation not only changes the biochemical characteristic of the target molecule but the new structure may also be recognized as 'foreign' (i.e., a neoantigen) thus initiating an immunological response (Novartis 2007).

Gene polymorphisms in ADH and ALDH enzymes may explain some of the pathologies of alcoholism, and why some individuals will develop certain diseases when others do not. About 50 per cent of East Asian origin populations (Taiwanese, Han Chinese, and Japanese) have a deficiency of ALDH2. After alcohol consumption this results in an elevation in acetaldehyde concentrations causing visible facial flushing (see Weblink Section 10.1 on facial flushing). The modified allele is designated ALDH2*2 (which has little or no metabolizing activity and is designated rs671 where rs is the reference SNP number) whilst the (normal) fully functional gene is ALDH2*1. If individuals with low ALDH activity continue to consume alcohol, then the high acetaldehyde concentrations will induce greater tissue damage. This has also been shown experimentally when agents such as cyanamide (an inhibitor of ALDH activity) can cause greater metabolic perturbations in alcohol-exposed tissues.

Whilst considerable work has been carried out into polymorphisms of the ALDH2 gene, most of its relevance pertains to those of East Asian origins rather than Caucasians. Nevertheless, work has been carried out on polymorphisms relating to ADH genes. These studies show that those with fast metabolizing polymorphisms (thus producing acetaldehyde concentrations much more quickly) are less likely to be hospitalized due to the effects of alcohol, drink less, and score lower on alcoholism screening tests.

Two minor but important non-oxidative pathways of ethanol metabolism result in the formation of phosphatidylethanol and fatty acid ethyl esters (FAEE). FAEE are formed from fatty acids and ethanol in reactions catalysed by either cytosolic or microsomal FAEE synthase. In the former reaction, the immediate precursor is fatty acid, whereas the microsomal pathway utilizes fatty acid CoA. The FAEE are broken down by a cytosolic hydrolase or may traverse the membrane into the intravascular space. Phosphatidylethanol is formed in a dose- and time-dependent manner when ethanol becomes the polar group of a phospholipid in a reaction catalysed by phospholipase D. It is found in the blood of alcoholics and due to its low metabolism, in organs exposed to ethanol, including liver, intestines, stomach, lung, spleen, and muscle. Phosphatidylethanol and FAEE are cytotoxic and may perturb protein synthesis and cell-signalling due to reduced phosphatidic acid production. FAEE have previously been used as a diagnostic biomarker of alcohol consumption.

Induction of microsomal cytochromes following repeated ingestion of alcohol

The MEOS is particularly important in heavy ethanol ingestion as it is an inducible pathway of ethanol metabolism. It is thus of particular significance in chronic ethanol misusers where the existing enzymes become saturated and unable to cope with the high ethanol load. The purified protein of MEOS is commonly referred to as cytochrome P450 2E1 (CYP2EI or 2EI) (although 1A2 and 3A4 are involved, see Zakhari 2006), and its induction is due to increases in mRNA concentrations and its rate of translation. Acute bouts of alcohol exposure can also lead to CYP2E1 induction. The MEOS system utilizes NADPH (Figure 10.1) and produces free radicals (hydroxyethyl, superoxide anion, and hydroxyl radicals), leading to increased cellular oxidative stress, particularly endoplasmic reticulum stress. The MEOS has a higher K_m for ethanol (8–10 mmol/L) compared with ADH (0.2 to 2.0 mmol/L).

The metabolic basis for 'fatty liver' of chronic alcohol ingestion

Alcoholic liver disease has three consecutive stages, namely fatty liver (steatosis), alcoholic hepatitis with fibrosis, and cirrhosis, though fatty liver may progress directly to cirrhosis (Patel 2016). The ability of the liver to develop steatosis in the presence of low fat diets has led to the hypothesis that the *de novo* synthesis of triacylglycerol may arise via increases in fatty acid synthesis in the liver. Fatty liver is clinically diagnosed when the lipid content of the liver is 5–10 per cent by weight. It occurs in about 80 per cent of chronic alcohol misusers and is usually asymptomatic but many pro-inflammatory pathways are initiated, and with continued alcohol consumption can lead to steatohepatitis. At this stage, patients are at significant risk and may be hospitalized. In many cases of acute alcoholic hepatitis, the mortality rate is up to 35 per cent, with a mortality rate at one month of 20 per cent. Fatty liver, however, is not itself fatal and occurs in a variety of other conditions such as hyperlipidaemia/obesity associated with insulin resistance. The biochemical features of alcoholic fatty liver are distinct from other non-alcohol fatty liver pathologies such as those due to diabetes, reflecting their different aetiologies. However, histologically alcoholic fatty liver disease is similar to diet induced non-alcoholic fatty liver disease.

Increased availability of free fatty acids in the liver present a greater biochemical 'target' for the free radicals generated as a consequence of alcohol metabolism. This leads to peroxidation of fatty acids within the liver, generating lipid peroxides, malondialdehyde, and 4-hydroxynonenal, which in turn can form aldehyde-protein adducts, that is, malondialdehyde-protein adducts and 4-hydroxynonenal-protein adducts. As with acetaldehyde-protein adducts, the lipid-derived protein adducts are immunogenic, promoting inflammation.

The lipid in affected liver is largely triacylglycerol, which may increase between 10–50 fold; there is also a less marked increase in esterified cholesterol. Various metabolic pathways are altered leading to the development of fatty liver. These include downregulation of peroxisome proliferator-activated receptor alpha, decreased AMP-activated protein kinase activity, leptin dysregulation, and these mechanism are covered more comprehensively in Patel (2016).

Lactic acidosis resulting from alcohol ingestion

The increased NADH/NAD$^+$ ratio following alcohol metabolism increases the lactate/pyruvate ratio leading to lactic acidosis in alcoholics, whereas poor nutrition/starvation, dehydration, depleted glycogen stores, and increased free fatty acids in the liver promotes the ketogenic pathway producing the predominant ketone body, ß-hydroxybutyrate. These effects can cause the blood pH to fall to 7.1, and hypoglycaemia may occur. In severe cases of ketoacidosis and hypoglycaemia, permanent brain damage and death may arise. However, the prognosis of alcoholic acidosis is generally good. These conditions may be exacerbated by thiamin deficiency and indeed thiamin deficiency per se may hasten acute episodes of lactic acidosis. The high concentration of lactic acid also impairs the kidney's ability to excrete uric acid and consequently blood uric acid concentrations rise (hyperuricemia), causing gout. Further information on the clinical effects of chronic and acute alcohol can be found in Weblink Section 10.1.

10.3 Toxic effects of chronic alcohol ingestion

Effects of alcohol on liver function

The pathological mechanisms leading to cirrhosis are complex and are still the subject of intensive research. Fatty changes, as described earlier, arise with micro- and macrovesicle fat droplets and is generally asymptomatic. This can be detected on ultrasound, CT, MRI, or fibroscan, and is associated with abnormal liver function tests (e.g., raised activities of aminotransferases in plasma), although these have low diagnostic sensitivity (50–70 per cent). Ethanol metabolism by both the MEOS and ADH pathways leads to excess free radical production in the cytosol and mitochondria, respectively. The major cellular antioxidant glutathione (a free radical scavenger) is also reduced in alcoholics, decreasing the cell's ability to dispose of free radicals. Mitochondrial damage occurs (reduced ATP production, release of cytochrome c). These changes eventually result in hepatocyte necrosis, and inflammation. Progression to

alcoholic hepatitis involves invasion of the liver by neutrophils. Gut derived bacterial endotoxin also stimulates Kupffer cells causing the release of pro-inflammatory cytokines. Giant mitochondria are visible and dense cytoplasmic lesions, known as Mallory bodies, are seen. Acetaldehyde contributes at this stage by stimulating stellate cells to produce collagen leading to fibrosis and lowers the cellular antioxidant (glutathione) concentrations. Alcoholic hepatitis can be asymptomatic but usually presents with abdominal pain, fever and jaundice, and in severe acute hepatitis, patients may have encephalopathy, ascites, and ankle oedema. Altered gut microbiome is now thought to be an important contributing factor to liver inflammation, promoting Kupffer cell activation and pro-inflammatory cytokine release. Continued alcohol consumption may lead to cirrhosis. At this stage increasing fibrocollagenous deposition occurs spreading throughout the hepatic architecture leading to scarring. There is ongoing necrosis with concurrent regeneration. This is classically said to be micronodular, but often a mixed pattern is present. The greater amount of fibrotic tissue deposited in the liver is correlated with the severity of cirrhosis. Alcoholics usually present with one of the complications of cirrhosis such as gastrointestinal haemorrhage (often due to bleeding from oesophageal varices), ascites due to low albumin synthesis, reduced clotting factor production leading to bleeding, encephalopathy, or renal failure. It is unclear why only a fraction of alcoholics develop cirrhosis. It has been suggested that there may be genetic factors, and that differences in immune response may play a role. Dietary factors may also contribute. For example, with inadequate intake of cysteine and glycine, glutathione production may be impaired. Poor intake of vitamins A, C, and E will also reduce the ability of the hepatocyte to cope with the oxidative stress imposed by alcoholism.

Effects of alcohol on skeletal muscle

Alcoholic myopathy is common, affecting 40–60 per cent all chronic alcohol abusers, and is a major cause of morbidity. It is characterized by muscle weakness, myalgia, muscle cramps, and loss of lean tissue; up to 30 per cent of muscle may be lost. Histological assessment correlates well with symptoms and shows selective atrophy of Type II muscle fibres. Reductions in muscle protein and RNA, with reduced rate of protein synthesis, also occur. Rates of protein degradation appear either unaltered, reduced, or increased depending on the degradation pathway investigated. Attention has focused on a role for free radicals in the pathogenesis of alcoholic myopathy. Cholesterol hydroperoxides are increased in alcohol-exposed muscle, implying membrane damage.

10.4 Alcohol and nutrition

Nutritional deficiencies are an important consideration in alcohol misusers, with the effect on nutrition generally linked to the type of alcohol consumer. Thus, it is important to distinguish between hazardous, harmful drinkers or dependant alcoholics, since this will correlate with the degree of nutritional damage. These aforementioned terms have been classified by the National Institute of Clinical Excellence (NICE) but in simple terms those described as 'hazardous' (heavy or binge) drinkers are at risk of physical and psychological harm but have no overt alcohol-related pathologies. Individuals categorized as 'harmful' have defined health problem or problems without demonstrable dependence but likely to develop dependence. Those who are 'addicted' or 'dependent' may have the same or worse pathologies than those described as harmful but at the same time exhibit a number of psychological or physical symptoms upon withdrawal of alcohol. Dependence may be categorized as mild or severe. Thus, in general the degree of nutritional impairment is: severe dependent > mild dependent > harmful > hazardous drinker.

Altered nutritional status is due to either inadequate dietary intake, sometimes as a result of alcohol displacing food in the diet, gastrointestinal damage affecting the absorption of nutrients, increased renal excretion, or damage within the hepatocyte itself. The types of foods purchased will also depend on the beverage preference. For example, in some communities the diets of non-drinkers are different from wine drinkers. Furthermore, there are differences between the diets of beer drinkers and wine drinkers. A landmark study which examined 3.5 million supermarket transactions showed that beer drinkers purchased more sugar, bread, eggs, tinned goods, chips, sausages, ketchup and mustard, butter or margarine, and soft drinks than wine drinkers. Beer drinkers also purchase less fruit and vegetables, low-fat cheeses, milk, and meat than wine drinkers.

The consequences of nutritional deficiencies are varied but can have significant effects on health. For example, circulating iron concentrations may be elevated in some alcohol misusers due to dysregulated intestinal iron absorption, causing increased hepatic tissue iron deposition which leads to liver injury from oxidative stress. There are a number of animal studies which show alcohol decreases hepcidin concentrations consistent with an increase in iron absorption. Hepatic stores of total retinoids (vitamin A) decrease in chronic alcohol misusers and correlate with severity of liver disease, whereas in very severe cases of alcoholism, classical symptoms of beri-beri and pellagra may arise, reflective of deficiencies of the vitamin thiamin and niacin, respectively (Watson & Preedy 2003).

There are no in-depth studies measuring micronutrient intake in alcohol misusers by comparison with the Lower Reference Nutrient Intake (LRNI). Of the few studies examining vitamin status in the UK, 95–100 per cent of alcohol misusers had intakes of folate and selenium below the UK Recommended Nutrient Intake (RNI), 50–85 per cent of all alcoholics had intakes of calcium, zinc, Vitamins A, B_1, B_2, B_6, and C lower than the RNI, and 45 per cent of subjects had intakes of magnesium and iron lower than the RNI. Clearly, more meaningful data are needed. However, an intake below the RNI itself does not imply malnutrition. Indeed, many alcoholics will have perfectly adequate nutrient status, such that studies on middle-class alcoholics, free from major organ disease, suggest that when malnutrition is present it is only mild to moderate. Alcohol will also adversely affect the metabolism of a number of nutrients including thiamin and it has been suggested that about half of alcoholics with liver disease will have thiamin deficiency. A UK study showed that 45 per cent of alcohol misusers without liver disease had either reduced activities of erythrocyte thiamin-dependent transketolase or a high activation ratio, both are suggestive of poor thiamin status. This is of concern as Wernicke's-encephalopathy/Wernicke-Korsakoff syndrome is a frequent manifestation of thiamin deficiency, particularly in alcohol misusers. Thiamin deficiency will arise from both inadequate intakes and alcohol-induced interference of active transport processes in the intestinal tract.

Acute or chronic alcohol impairs the absorption of galactose, glucose, other hexoses, amino acids, biotin, and vitamin C. There is no strong evidence that alcohol impairs the absorption of magnesium, riboflavin, or pyridoxine so these deficiencies will arise as a result of poor intakes and/or excess renal loss. Hepato-gastrointestinal damage of course may have an important role in impairing the absorption of some nutrients such as the fat-soluble vitamins, due to villous injury, bacterial overgrowth of the intestine, pancreatic damage, or cholestasis.

The muscle wasting that occurs in alcoholic myopathy arises directly as a consequence of alcohol or acetaldehyde on muscle, and is not associated with malnutrition per se. This suggests that there is a fundamental problem in assessing malnutrition in chronic alcoholics using anthropometric measures such as muscle or limb circumference due to the presence of alcoholic myopathy.

Some aspects of alcoholic liver disease can be reproduced in laboratory animals fed nutritionally complete diets with alcohol, thus excluding the direct consequence of malnutrition as a causative factor. However, the concomitant presence of alcoholism and malnutrition exacerbates organ damage and/or nutritional status. Due to the effects of alcohol and acetaldehyde on nutrient metabolism, the following nutrients have been studied in greater detail due to their direct impact on liver disease pathology.

Micronutrients

B vitamins

Vitamin B_{12}, also known as cobalamin, is a water-soluble vitamin responsible for haematopoiesis, DNA synthesis, and maintaining adequate nerve myelination. Severe deficiency may lead to pernicious anaemia. It is complexed to dietary animal protein and during digestion becomes bound to intrinsic factor, absorbed via the ileum, is distributed to tissues, and stored in the liver. In alcoholics the plasma concentration of vitamin B_{12} is often normal or elevated due to poor retention by peripheral tissues; the plasma concentration of holotranscobalamin is also elevated. However, liver concentrations are low due to reduced uptake, storage, or increased leakage. Thus, plasma concentration may not be a good indicator of vitamin B_{12} status in alcoholics and more useful information may be obtained from the plasma concentration of methylmalonic acid, which indicates functional B_{12} status.

Vitamin B_6 or the active form known as pyridoxal 5'-phosphate is required as a co-factor for transaminase activity. Since alanine aminotransferase activity is used as a marker to assess liver injury due to alcohol, a low concentration of vitamin B_6 can affect the interpretation. Folate deficiency is a frequent occurrence in alcoholics (80–95 per cent), resulting in megaloblastic anaemia (15 to 65 per cent of alcoholics). It stems from decreased gastrointestinal absorption due to reduced transport across basolateral membranes, decreased liver folate absorption, and increased renal excretion.

Deficiencies of vitamins involved in the methylation cycle may lead to an increased plasma concentration of homocysteine and lower concentration of s-adenosylmethionine (SAM) in alcoholics, the latter being an important methyl donor for histone and DNA methylation. SAM also plays a crucial role in maintaining mitochondrial function and is a precursor for glutathione synthesis, which is one of the main cellular antioxidants. Clinical studies have targeted SAM therapy in alcoholics, where a dose of 1 g/day for six months resulted in lower mortality rates but failed to improve on histological parameters.

Vitamin D

Vitamin D is a lipid soluble vitamin, largely synthesized in the skin via the action of UVB radiation. Vitamin D is transported to the liver and then to the kidneys where the active form 1, 25(OH)$_2$ vitamin D is produced. In alcohol consumers, plasma vitamin D concentration has been reported to be unchanged or lower than controls. One effect of alcohol could be malabsorption, since administration of vitamin D to alcoholics does not raise plasma 25(OH)D concentration. However, given that most of the vitamin D in the UK is produced by photo-conversion of 7-dehydrocholesterol in the skin rather than being acquired from the diet, intestinal malabsorption may not be a major issue. The main problem may relate to uptake of vitamin D precursors by the liver, aberrant conversion to 25(OH)D by CYP2R1, or poor release of this metabolite into the circulation. Alcohol also interferes with vitamin D precursor metabolism in the kidneys. Evidence from animal studies suggests that expression of the 24-hydroxylase enzyme (CYP24A1) is up-regulated by alcohol and this may lead to the production of the inactive metabolite, 24,25(OH)$_2$ vitamin D. Reduced sun exposure is another factor that needs to be considered as well, especially in older populations. The overall result of these perturbations may increase incidences of osteopenia and osteomalacia in alcoholics and therefore a greater risk of fractures. However, alcohol may also target bone tissue directly as shown by well-controlled alcohol dosing studies.

Alcohol and zinc

Zinc is one of the most abundant trace elements found in the body and can be obtained in the diet through the consumption of meat, cereals (e.g., 25–30 per cent of zinc in diet is found in cereals) and dairy products. Zinc is stored in the liver, muscle, bone, and kidneys and plays a crucial role in a range of cellular processes, through its action as a cofactor in zinc metalloenzymes and as a structural component in zinc finger transcription factors. In alcoholics, the concentration of circulating zinc correlates inversely with liver disease severity, with zinc concentration 50 per cent lower than normal healthy controls. The mechanism leading to low plasma zinc concentration can be attributed to low albumin concentration, since zinc is mainly bound to circulating albumin. At the cellular level, poor intestinal zinc absorption, altered hepatic metabolism, and increased renal excretion contribute to low plasma zinc concentration. Increased hepatic oxidative stress is also thought to cause zinc release from zinc proteins, leading to reduced liver zinc concentration. Current research has shown promising findings in animal models where zinc supplementation prevents biochemical and histological alterations in alcohol induced liver disease.

Alcohol and selenium

Selenium is found in a variety of foods (meat, fish, dairy products, cereals) but high doses, mainly as a dietary supplement, can be toxic. Selenium plays an essential role in the catalytic activity of selenoproteins, particularly the antioxidant glutathione peroxidase. In alcohol consumers, plasma selenium concentration is lower, possibly due to reduced intestinal absorption. The lower selenium concentration contributes to tissue pathology due to reduced glutathione peroxidase activity, leading to increased hepatic oxidative stress. Selenium supplementation in models of liver disease have shown protection against alcohol-induced oxidative injury (Patel 2016).

Alcohol and nutritional support

It is now widely recognized that the treatment of alcoholism should cover an assessment for malnutrition. The type of treatment will depend on the severity of the disease and any underlying nutritional abnormalities. Recent clinical trials have examined enteral and parenteral nutrition for the treatment of severe alcoholic hepatitis. Of the few random clinical trials undertaken the majority have shown a benefit to alcoholic liver disease (ALD) patients in terms of nutritional status and liver function. However, the long-term benefit remains unclear, due to small sample sizes. Parenteral nutrition, whilst more costly, also carries greater risk than enteral nutrition due to complications such as infection. There have been mixed responses in alcoholic hepatitis or alcoholic cirrhotic patients following parenteral nutrition, where nutritional status and survival rates have shown either an improvement or no change. It is likely the small sample size and heterogeneity of the sample population is partly responsible for these different outcomes.

10.5 Links between alcohol intake and risk of cardiovascular disease

A range of epidemiological studies have indicated that light to moderate amounts (1–3 unit per day) of alcohol is cardioprotective and reduces coronary heart disease, particularly in middle-aged men and post-menopausal women. There is a J or U-shaped mortality risk curve with increasing alcohol consumption. Here, a protective effect is observed at low concentrations of alcohol intake, around 20 g/day (approx. 1–2 units/day). The extent of this protection is variable and is attributed to increased HDL cholesterol concentrations, reducing circulating concentrations of fibrinogen, factor VII and plasminogen activator, inhibiting platelet aggregation and thus decreasing clot formation, and reduced LDL cholesterol oxidation in arterial walls. The reported cardioprotective effects of alcohol may be due to anti-oxidants or other substances in the beverages such as polyphenols in red wine (although it is now believed that all forms of alcohol can convey a cardioprotective effect). Indeed, large quantities of red wine containing catechins, quercetin, or resveratrol would need to be consumed to correlate with *in vitro* studies. However, recent UK guidelines suggest that the overall cardioprotective effects of alcohol is minimal when considered in light of the increased incidence of cancers, accidents, and other alcohol-related diseases.

There are other risk factors that are associated with alcohol consumption, such as smoking and obesity. Furthermore, there is a substantial body of evidence to support the notion that the total cumulative intake of ethanol (i.e., over a lifetime) will predict disease severity particularly of the heart, muscle, and liver. Clearly the best advice is for moderate alcohol consumption, combined with a healthier lifestyle of exercise and a well-balanced diet.

The risk–benefit of alcohol consumption can be seen in a J or U-shaped mortality curve. Once consumption goes beyond the threshold of 20 g/day and rises to approximately 70 g/day, no benefit is obtained, whilst consumption of greater than approximately 90 g/day is associated with an increased risk of coronary heart disease. The harmful effect of alcohol on cardiovascular mortality is distinct from the direct toxic effects on cardiac muscle, which leads to alcoholic cardiomyopathy. The main feature is a dilated left ventricle, causing reduced systolic contraction and lower cardiac output. The mechanisms are a reduction in cardiac contractile protein synthesis (particularly myosin heavy chain) and the toxic effects of acetaldehyde and fatty acid ethyl esters. Management of this disorder, without heart failure ensuing, can be obtained if alcohol abstinence/reduced alcohol intake is followed.

There is a relationship between alcohol consumption and blood pressure. The mechanism for hypertension that occurs after >2 drinks per day is possibly due to increased sympathetic overactivity that occurs from alcohol withdrawal after heavy drinking, also known as *'holiday heart syndrome'*. Heavy drinking is associated with an increased risk of stroke and arrhythmias. The precise relationship between ischaemic and haemorrhagic stroke and alcohol is less clear, but some studies suggest haemorrhagic stroke has a greater occurrence and the pattern is thought to follow a U or J-shaped relationship. As opposed to acute binge drinking, chronic heavy alcohol drinking is also associated with atrial fibrillation. This association with heavy alcohol drinking has been demonstrated in men, but there is evidence of an association with only moderate alcohol use in women (Klatsky 2015).

10.6 Links between alcohol intake and risk of cancers

Various research organizations have confirmed that alcohol poses a significant increase in risk of several types of cancer, including the mouth, pharynx, larynx, oesophagus, colon, breast, and liver. The International Agency for Research on Cancer has stated that alcohol is a carcinogen, with 3.5 per cent of all cancers attributed to chronic alcohol drinking. The carcinogenic properties of alcohol are considered to be due to the toxic effects of acetaldehyde, causing the formation of protein adducts, increased induction of cytochrome P450 2E1 leading to reactive oxygen species causing membrane peroxidation, altered histone acetylation/methylation and DNA methylation, and increased DNA adduct formation. The latter product is thought to display highly mutagenic properties and leads to a reduction in cell apoptosis. The World Cancer Research Fund suggests that one in five cases of breast cancer can be prevented by avoiding alcohol. Alcohol increases the concentrations of circulating oestrogen in women alcoholics, stimulates oestrogen receptor signalling in breast cancer cells, and nuclear transcription of oestrogen response genes. Studies suggest that the neurotoxic substance salsolinol derived from acetaldehyde and dopamine may be the agent responsible for these effects. Drinking alcohol >5 unit a day increases the association with hepatocellular carcinoma. Liver cancer usually arises from the development of cirrhosis, however the direct toxic effects of acetaldehyde following chronic alcohol consumption also needs to be recognized.

The risk of these cancers appears linear, with higher amounts of alcohol consumption associated with increased risk. There is no evidence of a 'safe threshold' or 'J shaped curve'. The form in which the alcohol is consumed has only a small impact, with beer and spirit drinkers having more cancers of the upper gastrointestinal tract than wine drinkers. As a rule, each unit consumed per day increases the risk of breast and prostate cancer by 5–10 per cent.

KEY POINTS

- Alcohol misuse is common: in the UK at least 25 per cent of the adult population drink more than recommended guidelines, with at least 2 million dependent on alcohol.

- The young (schoolchildren and adolescents) and women are particularly vulnerable or susceptible to the deleterious effects of alcohol and its metabolites.

- In alcohol misusers, the overall contribution of ethanol to total energy intake may rise to 60 per cent or higher.

- Alcohol absorption and metabolism is affected by a number of variables, including gastric alcohol-metabolizing enzymes, ethnicity, gender, presence of different foods, and body size.

- There are at least 200 different alcohol-related disorders or conditions.

- 50 per cent of chronic alcohol misusers will have one or more organ or tissue abnormalities.

- There are a number of routes of ethanol metabolism. The microsomal ethanol oxidizing system (MEOS) is particularly important in chronic alcoholism.

- The immediate metabolite of ethanol oxidation, acetaldehyde, is highly toxic.

- The effects of alcohol or acetaldehyde on the body are due to many processes, such as adduct formation, changes in protein, carbohydrate and lipid metabolism, membrane dysfunction, increased gut permeability, altered cytokines and impaired immunological status, perturbations in gene expression, enhanced apoptosis, reactive oxygen species/oxidative stress, and changes in intracellular signalling. Many of these will be exacerbated by malnutrition.

- About 50 per cent of alcoholics will have nutritional deficiencies and these can arise via a number of processes including poor dietary intakes, displacement of foods (empty calories theory), maldigestion, malabsorption, reduced liver uptake, and increased renal excretion.

ACKNOWLEDGEMENTS

With thanks to Professor Timothy J. Peters and Dr Ross Hunter for providing original material and Dr Paul Sharp for scientific input on micronutrients.

 Be sure to test your understanding of this chapter by attempting multiple choice questions.

REFERENCES

Department of Health (2015) *Alcohol Guidelines Review: Report from the Guidelines Development Group to the UK Chief Medical Officers*. Department of Health, London.

Fernandez-Sola J, Nicolas J M, Estruch R M, Urbano-Marquez A (2005) Gender differences in alcohol pathology. In: *Alcohol Related Pathology* Volume 1, 261–278. (Preedy V R & Watson R R eds). Amsterdam. Academic Press.

Foods Standards Agency (2002) *McCance and Widdowson's The Composition of Foods*. Royal Society of Chemistry, Cambridge.

Health Survey for England (2019) Health and Social Care Information Centre.

Klatsky A L (2015) Alcohol and cardiovascular diseases: Where do we stand today? *J Intern Med* **278**, 238–250.

Kwo P Y & Crabb D W (2002) Genetics of ethanol metabolism and alcoholic liver disease. In: *Ethanol and the Liver: Mechanisms and Management,* 95–129 (D I N Sherman, V R Preedy, R R Watson eds). Taylor & Francis, London.

Lieber C S (2000) Alcohol: Its metabolism and interaction with nutrients. *Ann Rev Nutr* **20**, 395–430.

National Diet and Nutrition Survey (2018) *Rolling Programme for 2014 to 2015 and 2015 to 2016*. Public Health England.

Novartis (2007) *Acetaldehyde-Related Pathology: Bridging the Trans-Disciplinary Divide (Novartis Foundation Symposia)* by Novartis Foundation. John Wiley, Chichester.

ONS (2021) *Alcohol-related Deaths in the United Kingdom, Registered in 2018*. Statistical Bulletin. Office for National Statistics, London.

Patel V B (2016) *Molecular Aspects of Alcohol and Nutrition*. Academic Press, Oxford.

Patel V B & Preedy V R (2022) *Handbook of Substance Misuse and Addictions: From Biology to Public Health*. Springer Nature, Amsterdam.

Peters T J & Preedy V R (1998) Metabolic consequences of alcohol ingestion. Novartis Foundation Symposium **216**, 19–24.

Preedy V R (2019) *Neuroscience of Alcohol: Mechanisms and Treatment*. Academic Press, Cambridge.

Preedy V R & Watson R R (2005) *Comprehensive Handbook of Alcohol Related Pathology*. Volumes 1–3. Academic Press, San Diego.

Royal Colleges (1995) Alcohol and the heart in perspective: Sensible limits reaffirmed. A Working Group of the Royal Colleges of Physicians, Psychiatrists, and General Practitioners. *J. Royal College of Physicians London* **29**, 266–271.

Smoking, Drinking and Drug Use among Young People in England (2018) The Health and Social Care Information Centre.

Watson R R & Preedy V R (2003) *Nutrition and Alcohol: Linking Nutrient Interactions and Dietary Intake*. CRC Press, London.

WHO (2018) *Global Status Report on Alcohol and Health 2018*. World Health Organization.

Zakhari S (2006) Overview: How is alcohol metabolized by the body? *Alcohol Res Health*, **29**, 245–254.

FURTHER READING

Awoliyi S, Ball D, Parkinson N et al. (2014) Alcohol misuse among university staff: A cross-sectional study. *PLoS One* 9:e98134. doi: 0.1371/journal.pone.0098134. eCollection.

Haber P S (2000) Metabolism of alcohol by the human stomach. *Alc Clin & Exp Res* **24**, 407–408.

Institute of Alcohol Studies (2008) Statistics on Alcohol: England.

Johansen D, Friis K, Skovenborg E et al. (2006) Food buying habits of people who buy wine or beer: Cross sectional study. *BMJ* **332**, 519–22.

Jones A W (2000) Aspects of *in-vivo* pharmacokinetics of ethanol. *Alc: Clin Exp Res* **24**, 400–402.

Mezey E (1985) Effect of ethanol on intestinal morphology, metabolism and function. In: *Alcohol Related Diseases in Gastroenterology*, 342–360 (H K Seitz & B Kommerell eds). Springer-Verlag, Berlin.

Roine R (2000) Interaction of prandial state and beverage concentration on alcohol absorption. *Alc: Clin Exp Res* **24**, 411–412.

Saunders J B & Devereaux B M (2002) Epidemiology and comparative incidence of alcohol-induced liver disease. In: *Ethanol and the Liver: Mechanisms and Management,* 389–410 (D I N Sherman, V R Preedy &d R R Watson eds). Taylor & Francis, London.

Statistics on Alcohol England (2019) Health and Social Care Information Centre.

World Cancer Research Fund International (2018) *Diet, Nutrition, Physical Activity and Cancer: A Global Perspective—The Third Expert Report*. World Cancer Research Fund International, London. Available at: https://www.wcrf.org/dietandcancer

PART 3

Micronutrient function and metabolism

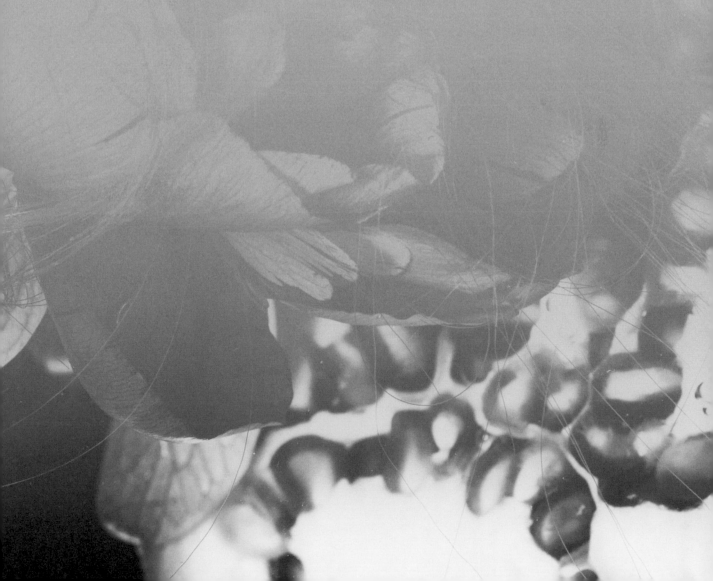

Water soluble vitamins

Helene McNulty and Kristina Pentieva

11.1 Introduction

Water soluble vitamins are essential organic compounds that are required by humans in small amounts from the diet. Water soluble vitamins act as coenzymes in: intermediary metabolism (reactions related to energy production and storage); glucose, amino acid, and fatty acid metabolism; transfer of single-carbon units; or have an essential role in maintaining antioxidant status.

In addition to the normal absorption of water soluble vitamins from foods that occurs in the small intestine, intestinal bacteria can synthesize relatively large amounts of some of these vitamins. Indeed, the bacterial synthesis of thiamin, riboflavin, nicotinic acid, biotin, and folate can all be increased when the diet contains relatively large amounts of non-starch polysaccharides (dietary fibre). Specific high affinity carrier-mediated transport systems for thiamin, riboflavin, nicotinic acid, biotin, and folate have been identified in the colon. In this way, it is likely that bacterial synthesis in the colon can make

a significant contribution to vitamin nutrition, but this remains to be more fully investigated.

11.2 Vitamin B_1: thiamin

Thiamin was the first member of the water soluble group of vitamins to be identified. Its discovery was associated with the disease beriberi, a condition described in Chinese medical texts as early as 2700 BC but only at the end of nineteenth century beriberi was linked to dietary inadequacy caused by the consumption of highly milled (polished) rice whereas brown rice containing the outer layers of the grain appeared to be protective. Decades later the beriberi preventative substance in the rice polishings was isolated and called *aneurine* and following the discovery of its chemical structure and synthesis, the name was changed to thiamin. While now largely eradicated, beriberi remains a problem in some parts of the world (e.g., Sub-Saharan Africa) among people whose diet is especially high in carbohydrates.

Forms of thiamin in foods and dietary sources

In foods of animal origin thiamin exists mostly as thiamin diphosphate (TDP, also known as thiamin pyrophosphate) whereas in plant and fortified foods it is predominantly in a non-phosphorylated form as free thiamin (Figure 11.1). The most concentrated food sources of thiamin are brewer's yeast, pork, the aleurone layer of cereal grains, most nuts, and legumes. Dietary survey data show that meat, bread, fortified breakfast cereals, and potatoes are the main contributors to dietary thiamin intake in adults (Table 11.1).

Cereal grains lose thiamin during refining (polished rice 80 μg/100 g) whereas parboiled rice retains most of the thiamin (190 μg/100g) since it migrates into the starchy endosperm during the procedure. In many countries, including the UK, there is a policy of thiamin enrichment of flour for bread to replace thiamin losses during the cereal-refining process. Some raw fish, shellfish,

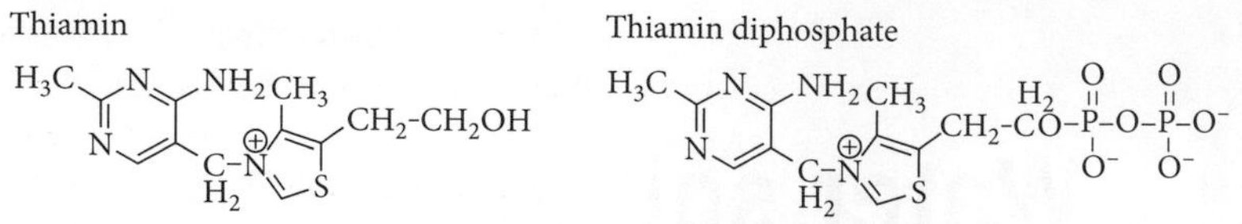

Thiamin

Thiamin diphosphate

FIGURE 11.1 Thiamin forms.

TABLE 11.1 Percentage (%) contribution of food groups to mean daily intakes of water soluble vitamins in adults 18–90 years (n 1500)*

Food Group	Water soluble vitamins						
	Thiamin	Riboflavin	Niacin	B_6	B_{12}	Folates	Vitamin C
Grains, rice, pasta, & savouries	3.1	1.9	3.0	1.5	2.6	2.0	0.8
Bread	16.9	5.3	12.5	5.4	0.4	14.4	0.1
Breakfast cereals	14.0	13.6	8.7	10.2	5.5	11.7	3.7
Biscuits, cakes, & pastries	2.1	1.5	1.5	0.8	1.0	1.2	0.3
Milk & yogurt	5.2	24.7	4.8	5.4	20.8	9.6	6.5
Creams, ice-creams, & chilled desserts	0.6	1.7	0.5	0.3	1.3	0.4	0.3
Cheeses	0.3	2.9	1.9	0.5	3.8	1.1	0.0
Butter, spreading fats, & oils	0.1	0.3	0.1	5.7	3.7	5.3	0.0
Eggs & egg dishes	0.8	3.2	1.5	0.8	5.7	2.0	0.1
Potatoes & potato dishes	11.3	1.3	3.7	14.6	0.2	9.6	15.6
Vegetables & vegetable dishes	7.0	2.5	3.0	5.4	0.4	10.4	22.3
Fruit & fruit juices	4.0	2.4	1.7	5.6	0.1	4.8	27.6
Fish & fish dishes	1.8	1.9	5.6	3.6	14.2	1.2	0.3
Meat & meat products	19.9	15.9	37.5	23.4	31.3	6.6	5.9
Beverages	0.2	9.2	6.3	6.2	1.3	10.1	2.6
Sugars, confectionery, preserves, & savoury snacks	1.8	2.4	1.5	1.9	1.0	1.3	2.3
Soups, sauces, & miscellaneous foods	2.8	1.6	1.1	1.3	1.3	1.8	2.5
Vitamin supplements	7.5	7.4	4.6	7.3	5.4	6.2	9.0
Nuts, seeds, herbs, & spices	0.6	0.2	0.7	0.4	0.0	0.3	0.0

*Data based on the Irish National Adult Nutrition Survey (NANS; 2008–2010). Summary Report. 2011 (www.iuna.net/surveyreports). The UK National Diet and Nutrition Survey (NDNS) of British adults shows very similar results for the contribution of food groups to mean daily intakes of water soluble vitamins.

and ferns contain anti-thiamin compounds that can destroy thiamin; the regular consumption of such foods may increase the risk of developing thiamin deficiency. The UV light, use of bicarbonate during cooking, and sulfite treatment of foods can cause major losses of the vitamin.

Absorption and transport of thiamin

Dietary thiamin phosphates are hydrolysed by intestinal phosphatases to free thiamin which is absorbed in the duodenum and proximal jejunum by a saturable active transport system at concentrations within the dietary range or by a process of passive diffusion at higher concentrations. The active transport system is saturated at relatively low concentrations, limiting the amount that can be absorbed to no more than about 2.5 mg (10 µmol) from a single dose. Microbiota in the large intestine can synthesize considerable amount of thiamin and studies have shown the existence of efficient carrier-mediated transporters in the colon suggesting that the bacterial source of thiamin may be used by the colonocytes. There is also active transport from the intestinal cells into the bloodstream. Much of the absorbed thiamin is phosphorylated in the liver, and both free thiamin and thiamin monophosphate circulate in plasma, bound to albumin. All tissues can take up both thiamin and thiamin monophosphate and are able to phosphorylate them to the active di- and triphosphates. A small amount of thiamin is excreted in the urine unchanged; the major excretory product is thiochrome. Sweat may contain up to 30–56 nmol/L of thiamin, and in hot conditions this may represent a significant loss of the vitamin.

Metabolic functions of thiamin

Thiamin in the form of TDP is involved in glucose and amino acid metabolism. It acts as a coenzyme for three multi-enzyme complexes: pyruvate dehydrogenase and 2-oxoglutarate dehydrogenase in central energy-yielding metabolic pathways and oxo-acid dehydrogenase responsible for the catabolism of the branched-chain amino acids (leucine, isoleucine, and valine). It is also a coenzyme for transketolase in the pentose phosphate pathway of carbohydrate metabolism. Thiamin also has a role in electrical conduction in nerve cells; TTP activates through phosphorylation the chloride channel in nerve membranes and has been suggested to have a direct role in the regulation of cholinergic neurotransmission.

Thiamin deficiency

A well-nourished human adult body contains only around 30 mg of thiamin and considering its fast turnover rate in the tissues (biological half-life of 10–20 days), the deficiency can develop rapidly during depletion. In thiamin deficiency there is an impaired provision of pyruvate to the citric acid cycle which especially under the conditions of relatively high carbohydrate diet, may result in increased plasma concentrations of lactate and pyruvate and consequently to life-threatening acidosis.

The main cause of thiamin deficiency is the low dietary intake of the vitamin in developing countries whereas in the Western countries predominantly chronic alcoholism is associated with this condition. At a high risk for thiamin deficiency are also individuals suffering from diabetes mellitus, inflammatory bowel disease, congestive cardiac failure, renal disease, cancer, HIV-AIDS, narcotic addicts, and those taking diuretics (Sechi & Serra 2007). Thiamin deficiency can lead to beriberi and Wernicke's encephalopathy with Korsakoff's psychosis. Beriberi presents in three clinical forms: dry beriberi is a sensory and motor peripheral neuropathy which usually occurs in older people and is associated with a more prolonged but less severe thiamin deficiency; wet beriberi involves congestive cardiac failure with oedema of the lower extremities and is linked to a high carbohydrate intake combined with high physical activity; acute pernicious (fulminating) beriberi (shoshin beriberi) mainly in infants with heart failure, increased pulse rate, and metabolic abnormalities but usually without peripheral neuritis. Wernicke's encephalopathy with Korsakoff's psychosis is a condition associated with chronic alcohol abuse and clinically presents with abnormal motor function, ataxia, amnesia, and confabulation. Results from post-mortem examination and brain imaging suggest that Wernicke's encephalopathy is significantly underdiagnosed.

Assessment of thiamin nutritional status

Erythrocyte transketolase activation coefficient (ETKac) is a functional biomarker of thiamin status which measures the ratio of the activity of thiamin-dependent enzyme transketolase before and after *in vitro* activation with prosthetic group TDP. ETKac is highly specific for thiamin and sensitive to depletion of the vitamin in the marginal-to-deficient range. Values of ETKac >1.25 are indicative of deficiency, and <1.15 are considered to reflect adequate thiamin status (Sauberlich 1999). Other biomarkers for assessment of thiamin status include urinary excretion of thiamin which reflects recent intake as well as thiamin diphosphate concentrations in erythrocytes (Table 11.2).

Please see Weblink Section 11.1 for information about the basis for setting thiamin requirements and dietary recommendations.

TABLE 11.2 Biomarkers of thiamin nutritional status[1]

	Adequate	Marginal	Deficient
Urinary excretion			
nmol/24h	≥375	150–375	<150
µg/24h	≥100	40–99	<40
µmol/mol creatinine	>28	12–28	<12
µg/g creatinine	≥66	27–65	<27
Urinary excretion over 4 h after a 19 nmol (5 mg) parenteral dose			
nmol	≥300	75–300	<75
µg	≥80	20–79	<20
Erythrocyte transketolase activation coefficient			
	<1.15	1.16–1.24	>1.25
Erythrocyte thiamin diphosphate			
nmol/L	>150	120–150	<120
µg/L	>64	50–64	<50

Source: Criteria based on H E Sauberlich (1999) *Laboratory Tests for the Assessment of Nutritional Status*. CRC Press, Boca Raton.

11.3 Riboflavin: vitamin B$_2$

Riboflavin was first isolated from milk whey as a yellowish compound and in the 1930s its structure and corresponding active coenzyme derivatives, flavin mononucleotide (FMN) and flavin adenine dinucleotide (FAD) were identified (Figure 11.2). Riboflavin coenzymes have a central role in oxidation-reduction reactions involved in energy metabolism and are essential for maintaining antioxidant status. Riboflavin is a regulator of cryptochromes, the blue-sensitive pigments in the eye that are responsible for sensitivity to day and setting circadian rhythms. The clinical signs of riboflavin deficiency are not specific and the condition is usually accompanied by other micronutrient deficiencies.

The predominant riboflavin forms in foods are FMN and FAD apart from milk and eggs, which contain relatively large amounts of free riboflavin. A small proportion of riboflavin in foods is covalently bound to enzymes and hence is not biologically available. Rich sources of riboflavin are foods of animal origin, particularly milk and dairy products, and to a lesser extent meat, eggs, and fish. Riboflavin enrichment of white flour is mandated in the USA and some other countries with the aim to restore the losses of the vitamin during milling and refining processes. Milk and dairy foods, meat, and fortified breakfast cereals are the main dietary contributors to riboflavin intake in adults (Table 11.1). Riboflavin and its coenzymes are heat stable but light sensitive and considerable losses of the vitamin occur when the foods are exposed to UV light.

Absorption and transport of riboflavin

Dietary FMN and FAD are released from proteins in food by the combined action of gastric acid and intestinal hydrolases. The released FMN and FAD are then hydrolysed in the intestinal lumen to yield free riboflavin, which is absorbed in the upper small intestine by a saturable carrier-mediated mechanism; the maximum amount of riboflavin that can be absorbed from one meal is about 27 mg. The absorption efficiency of dietary riboflavin is estimated to be 95 per cent. Alcohol consumption and the use of some drugs (i.e., chlorpromasine) can inhibit intestinal uptake of riboflavin. Normal intestinal microflora in the large intestine synthesizes riboflavin which is absorbed in the colon by a similar carrier-mediated process. Much of the absorbed riboflavin is phosphorylated in the intestinal mucosa and enters the bloodstream as riboflavin phosphate. Free riboflavin and FAD are the main transport forms in plasma, with a small amount of riboflavin phosphate. Most riboflavin in tissues is as coenzymes bound to flavoproteins (riboflavin-dependent enzymes); unbound coenzymes are rapidly hydrolysed to free riboflavin which leaves the cells. Thus, the intracellular phosphorylation of riboflavin to FMN and FAD is a form of metabolic storage of the vitamin, which is important for maintaining riboflavin homeostasis. When riboflavin is absorbed in excess of metabolic needs, little is stored in the body tissues and the excess is excreted, mainly through the urine. In animals, the maximum growth response is achieved with intakes that give about 75 per cent saturation of tissues, and the intake to achieve tissue saturation is that at which there is quantitative urinary excretion of the vitamin. Equally, there is very efficient conservation of tissue riboflavin in deficiency, with only a four-fold difference between saturation and the minimum tissue concentration seen in deficiency.

Metabolic functions of riboflavin

Riboflavin coenzymes function as electron carriers in a wide variety of oxidation and reduction reactions central to all metabolic processes, including the mitochondrial electron transport chain, energy production, oxidation of fatty acids, metabolism of drugs, and xenobiotics. As shown in the lower half of Figure 11.2, riboflavin can undergo either two single-electron reductions via the semiquinone flavin radical, or a two-electron reduction to the fully reduced form.

Riboflavin is also involved in the antioxidant defence system through FAD-dependent enzyme glutathione reductase, which maintains the optimal concentration of reduced glutathione, the most important antioxidant within the cells. In addition, FAD acts as a regulator of cryptochromes which are photosensitive flavoproteins in

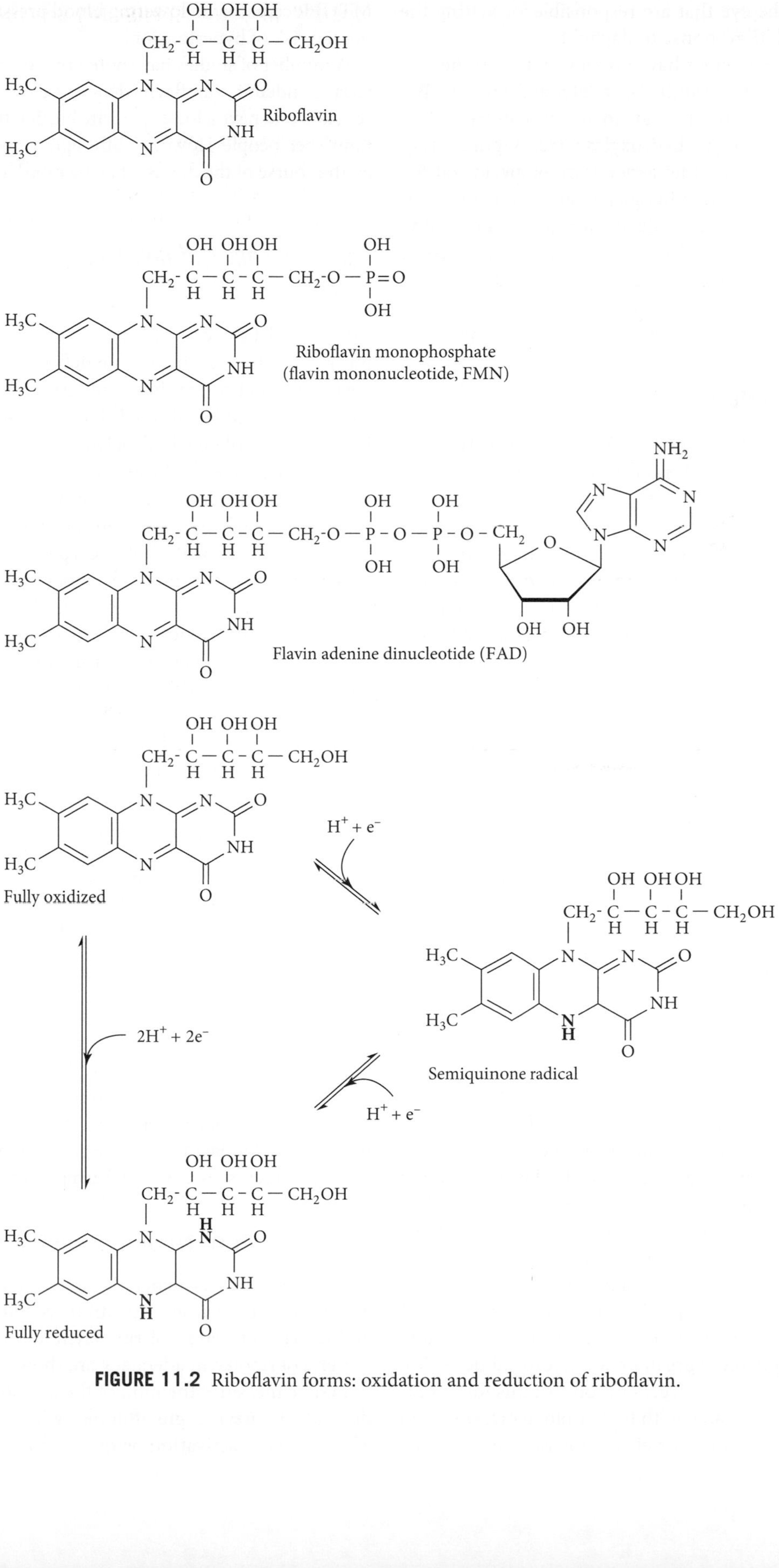

FIGURE 11.2 Riboflavin forms: oxidation and reduction of riboflavin.

the retina of the eye that are responsible for setting the circadian clock in response to daylight.

Riboflavin coenzymes have also a vital role in the metabolism of niacin, vitamin B_6, folate, and vitamin B_{12}. The conversion of tryptophan to niacin requires FAD-dependent kynurenine hydroxylase (see Figure 11.4); FMN is involved in the generation of pyridoxal-5′-phosphate which is the biologically active form of vitamin B_6 whereas FAD is a cofactor for methylenetetrahydrofolate reductase (MTHFR), the enzyme responsible for the conversion of 5,10-methylenetetrahydrofolate to 5-methyltetrahydrofolate which in turn is required for the remethylation of homocysteine to methionine.

Riboflavin deficiency

Riboflavin deficiency, also known as ariboflavinosis, clinically presents with lesions of the margin of the lips (cheilosis) and corners of the mouth (angular stomatitis), a painful desquamation of the tongue ('magenta tongue') and a seborrhoeic dermatitis, with filiform excrescences. Conjunctivitis with vascularization of the cornea and opacity of the lens (cataract) also may occur. Most of the clinical symptoms of ariboflavinosis are not specific or life-threatening and the condition is usually accompanied by other micronutrient deficiencies. Riboflavin deficiency is a common problem in low and middle income countries where the consumption of foods rich in riboflavin, especially milk and dairy products, is limited.

Although clinical riboflavin deficiency is rare in developed regions, evidence from population-based surveys shows that sub-optimal riboflavin status is much more widespread in the UK and other high income countries than previously recognized, especially among young women and during pregnancy (McNulty et al. 2019). Furthermore, riboflavin requirements may be higher in people homozygous for the common C677T polymorphism in the gene encoding the folate-metabolizing enzyme methylenetetrahydrofolate (MTHFR)—affecting 10 per cent of populations globally—as this genetic variant leads to an increased propensity for the riboflavin (FAD) cofactor to dissociate from the enzyme's active site, rendering it inactive, and in turn leading to impaired folate metabolism. Also, this polymorphism is linked with an increased risk of CVD via elevated homocysteine concentrations, however, deficient or low riboflavin status has a marked modulating effect on homocysteine and exacerbates this phenotype (McNulty et al. 2002), whilst supplementation with riboflavin corrects it (McNulty et al. 2006). Arguably of greater relevance to public health, however, is the increased recognition that this folate polymorphism is associated with higher blood pressure, and emerging evidence of a novel role for riboflavin (as the MTHFR cofactor) in lowering blood pressure and reducing the risk of hypertension.

A number of studies have noted that in areas where malaria is endemic, riboflavin-deficient people are relatively resistant and have a lower parasite burden than adequately nourished people. However, although parasitaemia is lower, the course of the disease may be more severe.

Secondary nutrient deficiencies in riboflavin deficiency

Riboflavin deficiency is associated with hypochromic anaemia as a result of secondary iron deficiency. There is an increased rate of iron loss from the gastrointestinal tract and some animal studies showed that the absorption of iron is impaired in riboflavin deficiency (Powers 2003). These effects appear to be modulated by changes in the morphology and cytokinetics of the small intestine, including an increased rate of enterocyte transit along the villi and fewer, but longer, villi with deeper crypts. Additionally, riboflavin deficiency may impair the mobilization of iron from intracellular stores, such as hepatic ferritin.

Riboflavin deficiency can lead to a functional vitamin B_6 deficiency, because pyridoxine oxidase is a flavoprotein that is very sensitive to riboflavin depletion. Impaired tryptophan metabolism in riboflavin deficiency can result in reduced synthesis of NAD, and may be a factor in the aetiology of pellagra.

Iatrogenic riboflavin deficiency

A variety of drugs, including phenothiazines, tricyclic antidepressants, antimalarials, and adriamycin are structural analogues of riboflavin and inhibit flavokinase, leading to functional deficiency because of impaired formation of the flavin coenzymes.

Neonatal hyperbilirubinaemia is usually treated by phototherapy. The peak wavelength for photolysis of bilirubin is the same as that for photolysis of riboflavin. Infants undergoing phototherapy show biochemical evidence of riboflavin depletion, but because photolysis products of riboflavin can cause damage to DNA, riboflavin supplements are avoided during phototherapy.

Assessment of riboflavin nutritional status

Riboflavin status is generally assessed by erythrocyte glutathione reductase activation coefficient (EGRac) and urinary excretion of riboflavin and its metabolites. Criteria of riboflavin adequacy are shown in Table 11.3.

EGRac measures the ratio of the activity of riboflavin-dependent enzyme glutathione reductase before and after *in vitro* activation with prosthetic group FAD;

TABLE 11.3 Biomarkers of riboflavin nutritional status

	Adequate	Marginal	Deficient
Urinary excretion[1]			
μg/g creatinine	≥80	27–80	<27
mol/mol creatinine	≥24	8–24	<8
μg/24h	≥120	40–119	<40
nmol/24h	≥300	100–299	<100
mg over 4 h after 5 mg dose	>1.4	1.0–1.4	<1.0
μmol over 4 h after 5 mg dose	>3.7	2.7–3.7	<2.7
Erythrocyte glutathione reductase activation coefficient[2]	≤1.26	1.27–1.39	≥1.4

[1] Criteria based on H E Sauberlich (1999). *Laboratory Tests for the Assessment of Nutritional Status.* CRC Press, Boca Raton.

[2] Criteria based on Wilson et al. (2013) *Hypertension* **61**, 1302–1308.

higher EGRac values are indicative of lower riboflavin status. EGRac is a functional biomarker and a preferable choice as it is sensitive to riboflavin intake, reflects tissue saturation, and indicates a long-term status.

Urinary excretion of riboflavin reflects dietary intakes under conditions of tissue saturation. Riboflavin excretion through urine is very low at dietary intakes below about 1.1 mg/day; thereafter there is a sharp increase with increasing intake indicating tissue saturation. Excretion is only correlated with intake in subjects who are maintaining nitrogen balance. In negative nitrogen balance there may be more urinary excretion than would be expected, as a result of net catabolism of tissue flavoproteins, and loss of their prosthetic groups. Higher intakes of protein than are required to maintain nitrogen balance do not affect biomarkers of riboflavin nutritional status.

Please see Weblink Section 11.1 for information about the basis for setting riboflavin requirements and dietary recommendations.

impaired niacin synthesis. Newer evidence links deficient riboflavin status with an increased risk of hypertension in people with a common polymorphism affecting folate recycling, whilst riboflavin supplementation can reduce blood pressure in these genetically at-risk individuals.

- Various drugs and also phototherapy for neonatal hyperbilirubinaemia can cause iatrogenic riboflavin deficiency.

- Riboflavin status is assessed by the erythrocyte glutathione reductase activation coefficient assay and urinary riboflavin excretion.

11.4 Niacin

Niacin was discovered as a chemical compound, nicotinic acid, produced by the oxidation of nicotine in 1867—long before the 1930s when its role as active vitamin was identified as well as its ability to cure pellagra in humans. The main metabolic role of niacin is as a precursor of the coenzymes nicotinamide adenine dinucleotide (NAD) and its phosphorylated form, nicotinamide adenine dinucleotide phosphate (NADP), which can be synthesized *in vivo* from the essential amino acid tryptophan. In developed countries, average intakes of protein provide more than enough tryptophan to meet requirements for NAD synthesis without any need for preformed niacin; it is only when tryptophan metabolism is disturbed, or intake of the amino acid is inadequate, that preformed niacin becomes a dietary essential. Thus, pellagra should be regarded as being due to a deficiency of both tryptophan and niacin.

KEY POINTS

- Riboflavin functions in its coenzyme forms, FAD and FMN, in a wide variety of reactions, including key reactions in energy metabolism and antioxidant status.

- Riboflavin deficiency is particularly common among populations in low- and middle-income countries, and there is accumulating evidence that sub-optimal riboflavin status is more widespread worldwide than previously recognized, including in high-income countries.

- Riboflavin deficiency can cause secondary deficiency of iron, functional deficiency of vitamin B_6 and

Forms of niacin in foods and dietary sources

The term niacin is the generic descriptor for three compounds that have the biological action of the vitamin: nicotinic acid, nicotinamide, and nicotinamide riboside (Figure 11.3). Tryptophan is not considered to be a niacin vitamer. Poultry, meat offal (especially liver and heart), fish, and yeast are good sources of preformed niacin; nuts and some fruits and vegetables provide significant amounts, as does coffee. Most of the niacin in cereals is in the form of nicotinic acid and it is biologically unavailable, since it is bound covalently to complex carbohydrates forming indigestible compound called niacytin. However, alkaline processing of grain foods results in hydrolysis of most of niacytin with release of highly bioavailable free niacin. Fortified breakfast cereals are also important source of niacin. In many countries in the world, including the UK, it is mandatory to enrich with niacin the flour in order to compensate the losses of the vitamin during grain processing.

The main dietary source of nicotinamide riboside is milk. Another source of niacin are protein rich foods which contain high amount of tryptophan that can be metabolized in the liver to NAD. Quantitatively, synthesis from tryptophan is more important than dietary preformed niacin. Total niacin intakes are calculated as mg niacin equivalents: the sum of preformed niacin plus 1/60 of tryptophan (the average equivalence of dietary tryptophan and niacin). The main contributors to dietary niacin intake in adults are meat and meat products (Table 11.1).

Absorption and transport of preformed niacin

Niacin is present in tissues, and therefore in foods, largely as the nicotinamide coenzymes; post-mortem hydrolysis is extremely rapid, so that much of the niacin of meat is free nicotinamide. Any remaining coenzymes are hydrolysed to nicotinamide in the intestine. A significant

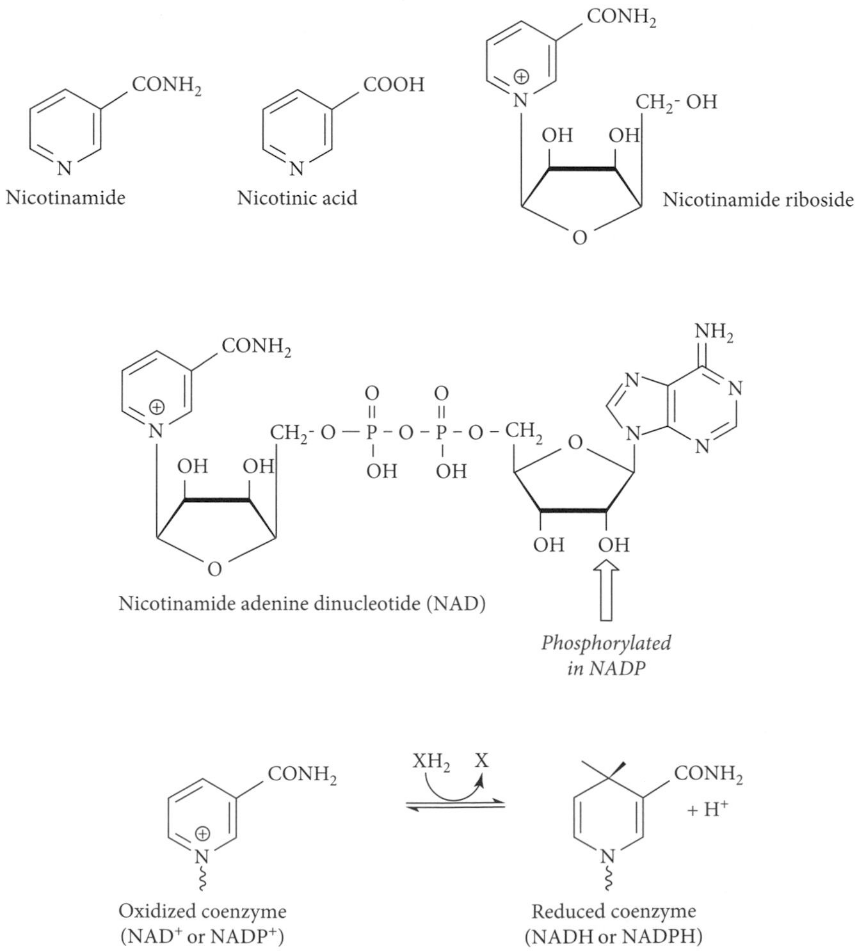

FIGURE 11.3 Niacin forms.

proportion of dietary nicotinamide may be deamidated to nicotinic acid by intestinal bacteria. Both nicotinic acid and nicotinamide are absorbed from the stomach and upper ileum. Nicotinamide riboside may be hydrolysed to nicotinamide in the intestinal tract, but there is some evidence that it is taken up by mucosal cells and may be available to tissues as the intact riboside. At low concentration, absorption occurs as saturable sodium-dependent facilitated process, but at higher concentrations, passive diffusion predominates. Nicotinamide generated by the microbiota in the large intestine is absorbed by a sodium-independent mechanism in the colon. The bioavailability of niacin from foods of animal origin is high (>70 per cent) but in some cereals, notably corn and millet, niacin is present as nyacytin and its bioavailability is less than 10 per cent unless hydrolysed by alkaline treatment during food preparation.

Niacin circulates in the plasma mainly as nicotinamide but nicotinic acid also can be found. In the tissues nicotinamide and nicotinic acid are converted to NAD and NADP, which is a form of metabolic trapping; however, there is a continuous turnover of the coenzymes and the tissue storage is very limited. The excess of niacin is metabolized in the liver to N^1-methylnicotinamide and its oxidation products, 2- and 4-methylpyridonecarboxamides which are excreted into the urine.

Synthesis of niacin from tryptophan

In the liver, NAD is synthesized from tryptophan by the kynurenine pathway as shown in Figure 11.4, and then hydrolysed to release nicotinamide, which is exported to other tissues. In nutritionally replete individuals on average 60 mg of tryptophan is converted to 1 mg niacin. This process is comparatively inefficient and it is prone to a high interindividual variability. It is controlled predominantly by the activities of the enzymes tryptophandioxygenase, kynureninehydroxylase, and kynureninase which in turn are dependent on the adequate supply of riboflavin, vitamin B_6 and iron and are affected by some hormones (glucagon, glucocorticoids, and oestrogen) as well as various drugs (i.e., izonazid). Diets with a high content of leucine also could depress the conversion of tryptophan to niacin (Bender 1983). It was reported that pregnant women and women on oral contraceptives have three-fold increased rate of conversion of tryptophan to ND. In addition, the production of NAD is minimal under the conditions of low tryptophan intakes as the available tryptophan is used primarily to maintain protein synthesis rather than to replenish niacin pools. Therefore, various nutrient deficiencies, use of drugs, metabolic disturbances, and physiological conditions could affect NAD synthesis with potential impact on niacin status.

Metabolic functions of niacin

Nicotinamide coenzymes NAD and NADP are involved in a wide variety of oxidation and reduction reactions in energy-yielding metabolism. In general, NAD^+ is involved as an electron acceptor in energy-yielding metabolism, and the resultant NADH is oxidized by the mitochondrial electron transport chain. The main involvement of NADPH is in reductive reactions of lipid and steroid biosynthesis; it participates also as co-dehydrogenase in oxidation of glucose 6-phospate to ribose 5-phosphate in the pentose phosphate shunt (Bender 2003).

NAD^+ is the source of ADP-ribose for the reversible modification of proteins. The transfer of ADP-ribose to the target proteins is catalysed by three classes of enzymes: 1) mono-ADP-ribosyltransferases which modify G-proteins that are involved in signal transduction; 2) poly-ADP-ribose polymerase (PARP) that function in DNA replication, DNA repair, and cell differentiation and PARP activity is strongly correlated with the lifespan of different species; 3) class of enzymes that are involved in the formation of cyclic ADP-ribose, which mobilizes calcium from intracellular stores in response to hormones.

The sirtuins are a family of enzymes that catalyse the deacetylation of lysine residues in histones and other proteins, and hence are involved in epigenetic regulation of gene expression. The reaction involves the hydrolysis of NAD^+ and the formation of O-acetyl ADP-ribose and nicotinamide. Sirtuin activity is limited by the availability of NAD^+, and hence depends on niacin nutritional status and the ratio of NAD^+:NADH, and hence the energy state of the cell. There is evidence from animal studies that over-expression of some sirtuins is associated with increased lifespan.

Niacin deficiency

Pellagra is the disease caused by the combined deficiency of niacin and tryptophan. It is characterized by a photosensitive dermatitis affecting regions of the skin that are exposed to sunlight. Advanced pellagra is also accompanied by a dementia or depressive psychosis, and there may be diarrhoea. Untreated pellagra is fatal. Pellagra was a major problem of public health for many years and continued to be a problem until the 1980s in some parts of the world. It is now rare, although outbreaks occur among refugees, and occasional cases are reported in alcoholics, and among people treated with isoniazid.

The synthesis of NAD from tryptophan (Figure 11.4) requires both riboflavin (for kynurenine hydroxylase) and vitamin B_6 (for kynureninase), and deficiency of either may lead to the development of secondary pellagra when intakes of tryptophan and preformed niacin are marginal. Similarly, an excessive intake of leucine, as occurs in parts

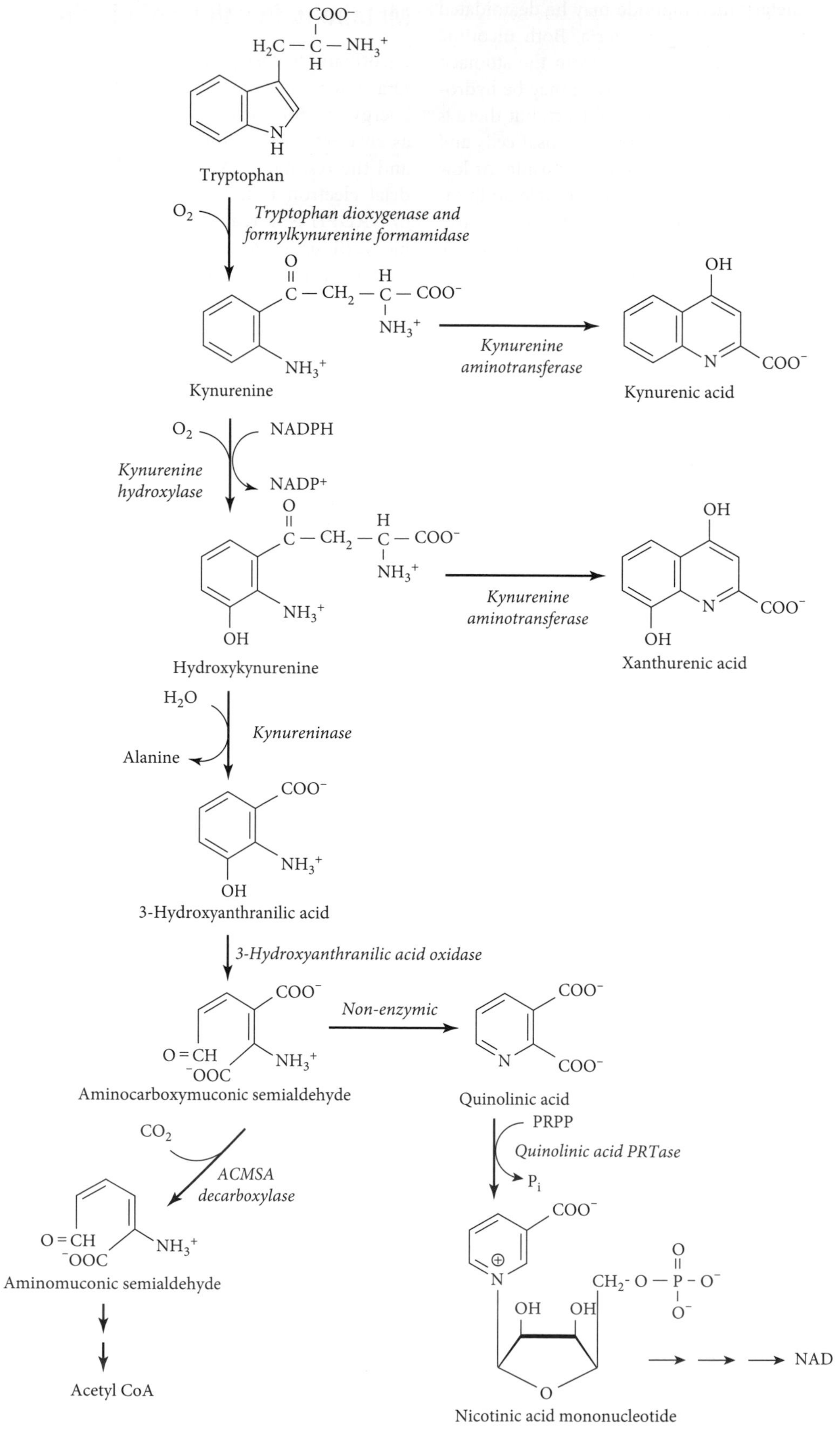

FIGURE 11.4 Kynurenine pathway.

TABLE 11.4 Biomarkers of niacin nutritional status[1]

	Elevated	Adequate	Marginal	Deficient
N^1-methylnicotinamide				
μmol/24 h	>48	17–47	5.8–17	<5.8
mg/g creatinine	>4.4	1.6–4.3	0.5–1.6	<0.5
mmol/mol creatinine	>4.0	1.3–3.9	0.4–1.3	<0.4
methylpyridonecarboxamide				
μmol/24 h	–	>18.9	6.4–18.9	<6.4
mg/g creatinine	–	>4.0	2.0–3.9	<2.0
mmol/mol creatinine	–	>4.4	0.44–4.3	<0.44
ratio, erythrocyte NAD: NADP				
	–	>1.0	–	<1.0

[1] Criteria based on: H E Sauberlich (1999) *Laboratory Tests for the Assessment of Nutritional Status*. CRC Press, Boca Raton.

of India where the dietary staple is jowar (*Sorghum vulgare*), inhibits tryptophan metabolism and may be a factor in the development of pellagra. Kynureninase is inhibited by oestrogen metabolites, also leading to reduced formation of NAD from tryptophan. This may explain the two-fold excess of women over men in most reports of pellagra during the early twentieth century.

Niacin excess

Very high intakes of nicotinic acid (3000 mg/day), usually used to treat hyperlipidaemia, can cause hepatotoxic effects with elevated concentration of liver enzymes. Lower doses of nicotinic acid are associated with a marked vasodilatation, flushing, burning and itching of the skin, and possibly also hypotension. In contrast large doses of nicotinamide (3000 mg/day) provided for up to three years appear to be well tolerated in several trials of patients with or at risk of developing type 1 diabetes. The European Food Safety Authority set an upper limit of intake of nicotinic acid at 10 mg/day, because of concerns about the potential hazard of vasodilatation among elderly people, and an upper limit of 900 mg/day for nicotinamide, since it does not cause any detectable side effects (EFSA 2006). The UK Expert Group on Vitamins and Minerals (2003) set an upper level of intakes for nicotinic acid at 17 mg/day and for nicotinamide at 500 mg/day.

Assessment of niacin status

The two methods of assessing niacin nutritional status are measurement of blood nicotinamide nucleotides and the urinary excretion of niacin metabolites; the latter approach is more commonly used, however neither

measurement is wholly satisfactory. Criteria for adequacy are shown in Table 11.4.

Please see Weblink Section 11.1 for information about the basis for setting niacin requirements and dietary recommendations.

KEY POINTS

- Niacin is not strictly a vitamin as it can be endogenously synthesized from tryptophan; niacin intake is calculated as mg niacin equivalents: preformed niacin + 1/60 tryptophan intake.

- Niacin functions as the nicotinamide ring of the coenzymes NAD and NADP in oxidation and reduction reactions; NAD is also involved in DNA repair and in the regulation of intracellular calcium concentrations in response to hormones.

- Niacin deficiency (pellagra) is rare with occasional outbreaks among refugees.

- Deficiency of riboflavin or vitamin B$_6$ as well as the usage of some drugs can impair tryptophan metabolism and may lead to the development of pellagra.

- Current methods of assessing niacin status are unsatisfactory.

11.5 Vitamin B$_6$

Vitamin B$_6$ has a central role in the metabolism of amino acids, is the cofactor for glycogen phosphorylase, and has a role in the modulation of steroid hormone action and regulation of gene expression. The vitamin is widely distributed in foods, and clinical deficiency is rare. However, marginal status, affecting amino acid metabolism and steroid hormone responsiveness, is relatively common. High doses of the vitamin, of the order of 100-times reference intakes, cause peripheral sensory neuropathy.

Forms of vitamin B$_6$ in foods and dietary sources

The generic descriptor vitamin B$_6$ includes six derivatives: the alcohol pyridoxine, the aldehyde pyridoxal, the amine pyridoxamine and their 5'-phosphates (Figure 11.5). All six vitamin B$_6$ derivatives are considered to have vitamin activity, as they can be converted in the body to the coenzymes pyridoxal 5'-phosphate (PLP) and pyridoxamine 5'-phosphate (PMP).

Vitamin B$_6$ is widely distributed in foods. Rich sources of vitamin B$_6$ are meat, fish, fortified breakfast cereals, potatoes and bananas, but also milk, nuts, beans and vegetables provide significant amounts. In foods of animal origin the main forms of vitamin B$_6$ are PLP and PMP which are considered highly bioavailable. However, in plant foods a significant amount of the vitamin is in glycosylated forms which have limited availability. Food processing and storage may affect the vitamin content of foods. A proportion of vitamin B$_6$ in foods may be biologically unavailable after heating, as a result of the formation of (phospho)pyridoxyllysine by reduction of the bond by which the vitamin is bound to the ε-amino groups of lysine in proteins. Pyridoxyllysine is a vitamin B$_6$ antimetabolite that also reduces the nutritional value of proteins in which lysine is the limiting amino acid.

Absorption and transport of vitamin B$_6$

The phosphorylated vitamin B$_6$ derivatives are hydrolysed by intestinal membrane bound alkaline phosphatase and absorbed by carrier-mediated diffusion in the jejunum. Intestinal microbiota can synthesize relatively large amounts of vitamin B$_6$, at least some of which is absorbed in the colon also by a carrier-mediated mechanism. There are no major differences in the

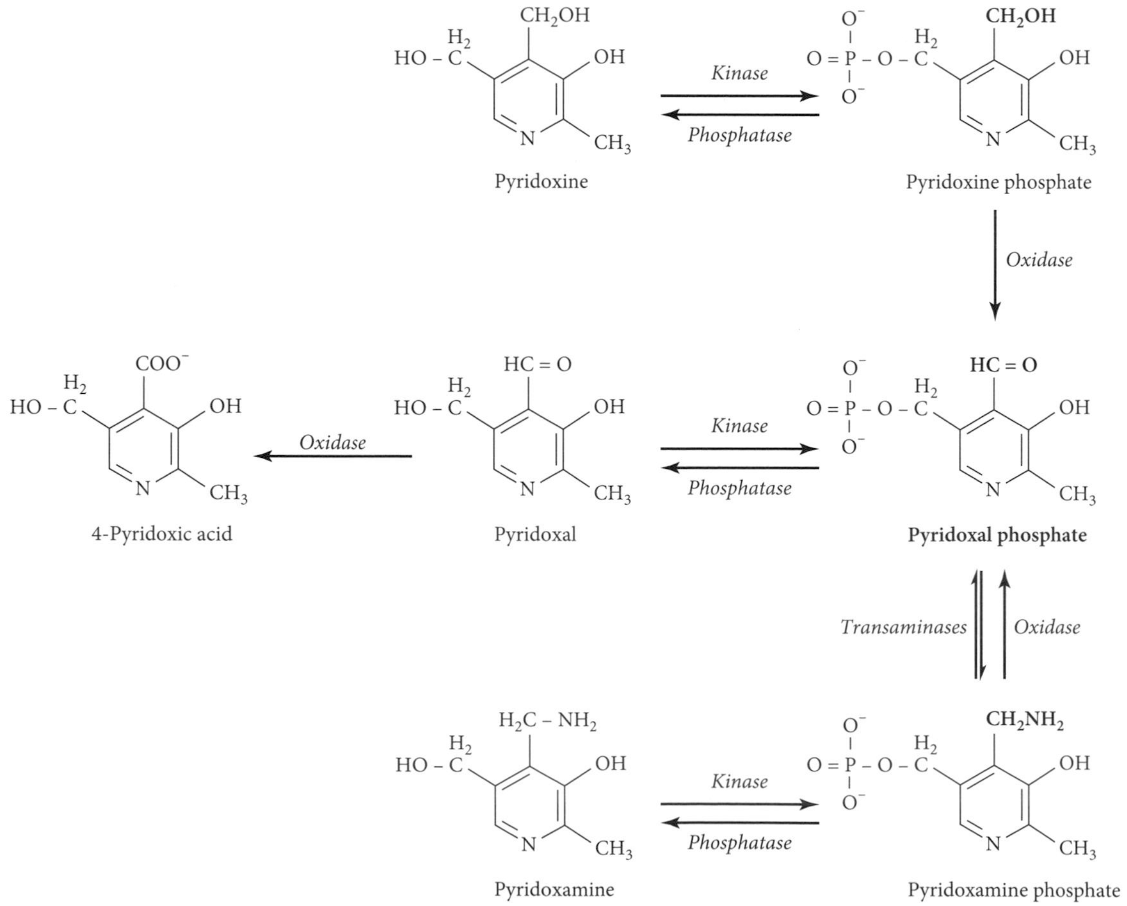

FIGURE 11.5 Interconversion of vitamin B$_6$ forms.

bioavailability of pyridoxine, pyridoxal, and pyridox-amine. The bioavailability of vitamin B$_6$ from supple-ments (usually in the form of pyridoxine hydrochloride) is almost complete and is estimated at 95 per cent. The bioavailability of vitamin B$_6$ from a mixed diet is around 75 per cent whereas only 50 per cent of glycosylated B$_6$ forms available in plant foods can be absorbed.

After intestinal absorption, B$_6$ derivatives are phos-phorylated within the intestinal mucosal cells which is a form of metabolic trapping of the vitamin but in or-der to cross the cellular membranes they undergo again through a process of dephosphorilation. The main forms of vitamin B$_6$ in the circulation are PLP and pyridoxal which are bound tightly to albumin. Vitamin B$_6$ de-rivatives are transferred via the portal circulation to the liver, where they are metabolized; vitamin B$_6$ in excess of requirements is catabolised to the inactive metabolite 4-pyridoxic acid, which is the main excretory product of vitamin B6 in the urine.

Some 80 per cent of the body's total vitamin B$_6$ is PLP in muscle, bound to glycogen phosphorylase. However, this PLP cannot be considered as a storage of vitamin B$_6$ as it is not released in B$_6$ deficiency, but rather during intensive exercise when muscle glycogen reserves are re-quired. Under these conditions PLP is available for redis-tribution to other tissues, and especially liver and kidney, to meet the body metabolic needs.

Metabolic functions of vitamin B$_6$

The metabolically active vitamin B$_6$ derivatives are PLP and PMP, which act as coenzymes in more than 100 re-actions of amino acid metabolism, one-carbon metabo-lism, glycogenolysis and gluconeogenesis, haem synthe-sis, niacin formation, and recycling of steroid hormone receptors from tight nuclear binding.

Vitamin B$_6$-dependent enzymes catalyse three main types of reactions involving amino acids: transamina-tion (permitting both utilization of amino acid carbon skeletons for gluconeogenesis or ketogenesis and also the synthesis of non-essential amino acids); decarboxylation to form a variety of biologically active amines, many of which are neurotransmitters; racemization leading to the formation of racemic mixtures of D- and L-amino acids, which have a role in signalling during brain development.

PLP is essential for the normal functioning of several enzymes in one-carbon metabolism which are respon-sible for the transfer of one-carbon units to folate de-rivatives used for the synthesis of purine and pyrimidine nucleotides and the remethylation of homocysteine to methionine. PLP is also a cofactor for enzymes involved in the transsulfuration pathway, where homocysteine is metabolized to cysteine and glutathione (the most im-portant antioxidant within the cells).

Steroid hormones act by binding to nuclear receptors that then bind to hormone response elements on DNA, regulating the transcription of specific genes. PLP acts to release the hormone-receptor complex from DNA bind-ing, so terminating the nuclear action of the hormone. In experimental animals, vitamin B$_6$ deficiency results in increased and prolonged nuclear uptake and retention of steroid hormones in target tissues, and enhanced end-organ responsiveness to low doses of hormones.

Vitamin B$_6$ deficiency and excess

Clinical deficiency of vitamin B$_6$ is extremely rare. Vulnerable population groups at risk of suboptimal vitamin B$_6$ status in developed countries are older people and pregnant women. Vitamin B$_6$ deficiency may result from a prolonged admin-istration of drugs that can form biologically inactive adducts with pyridoxal, such as penicillamine and isoniazid. The symptoms of vitamin B$_6$ deficiency include cheilosis, glossi-tis, fatigue, headaches, irritability, hypochromic microcytic anaemia, convulsive seizures, and abnormal electroen-cephalograms. Much of our knowledge of B$_6$ deficiency is derived from an outbreak in the early 1950s, which resulted from an infant milk preparation that had undergone severe heating in manufacture (Borschel 1995). The result was the formation of the antimetabolite pyridoxyllysine by reaction between pyridoxal phosphate and the ε-amino groups of lysine in proteins. Affected infants developed hypochromic microcytic anaemia, failure to thrive, hyperirritability, and convulsive seizures. They responded to the administration of vitamin B$_6$.

Vitamin B$_6$ toxicity

Supplements of 50–200 mg vitamin B$_6$/day are widely prescribed and self-prescribed for a variety of conditions, including premenstrual syndrome, depression, morning sickness in pregnancy, hypertension, and carpal tunnel syndrome, although there is little evidence of efficacy. There are promising results from trials of pyridoxamine to prevent non-enzymic glycation of proteins in diabetic patients and hence slow the development of the adverse effects of poor glycaemic control.

Large doses of pyridoxine (500–2000 mg/day) taken for several months or longer were associated with pro-gressive sensory ataxia presented by unstable gait, numb-ness of the feet and hands, impaired position sense and vibration sense in the distal limbs, and with less affected senses of touch, temperature, and pain. On withdrawal of the supplements there was substantial recovery of nerve function, although there was residual damage in some patients. However, all investigations in this area were not properly controlled. Animal studies have shown that vi-tamin B$_6$ in gross excess is neurotoxic, causing peripheral

TABLE 11.5 Biomarkers of vitamin B_6 nutritional status[1]

	Adequate status
plasma pyridoxal phosphate	>30 nmol (7.5 µg)/L
erythrocyte alanine aminotransferase activation coefficient	<1.25
erythrocyte aspartate aminotransferase activation coefficient	<1.80
urine 4–pyridoxic acid	>3.0 µmol/24h >1.3 mmol/mol creatinine
urine xanthurenic acid after 2 g tryptophan load	<65 µmol/24 h increase
urine cystathionine after 3 g methionine load	<350 µmol/24 h increase

[1] Criteria based on J E Leklem (1990) *J Nutr* **120,** Suppl 11, 1503–1507.

neuropathy. The European Food Safety Authority (EFSA 2006) set an upper level of 25 mg/day whereas the UK Expert Group on Vitamins and Minerals (2003) derived a safe upper level of vitamin B_6 intake at 10 mg/day.

Assessment of vitamin B_6 nutritional status

Vitamin B_6 status can be assessed by direct measurements of B_6 derivatives in blood and urine or by employing tests assessing B_6 biochemical function (activation of erythrocyte aminotransferases by PLP added *in vitro*; testing the ability of the body to metabolize a high dose of tryptophan or methionine). In general, studies have reported poor agreement between different vitamin B_6 biomarkers. Criteria of adequacy based on Leklem (1990) are shown in Table 11.5.

Plasma PLP concentration is the most widely used biomarker of vitamin B_6 status. It correlates well with vitamin B_6 intake and tissue PLP concentration. However, various physiological and lifestyle factors can affect plasma PLP values. There is a decline of plasma PLP with age and females have lower PLP concentrations than males; smoking and high alcohol consumption are associated with depressed PLP values (Bates et al. 1999). Low plasma PLP concentrations are common in chronic inflammatory conditions which is attributed to redistribution of PLP to the sites of inflammation rather than to a compromised vitamin B_6 status (Paul et al. 2013). To overcome these limitations, it has been recommended the assessment of PLP to be combined with other biomarkers of status.

The measurement of the activity of vitamin B_6-dependent enzymes, aspartate aminotransferase, and alanine aminotransferase in erythrocytes before and after *in vitro* activation with the prosthetic group PLP gives the opportunity to derive respective activation coefficients which are indicative for vitamin B_6 status; higher values of the activation coefficients relate to lower vitamin B_6 status.

The measurement of the urinary excretion of tryptophan metabolites after tryptophan load is a useful functional test for assessment of vitamin B_6 status, although it is rarely used currently. The oxidation of tryptophan (see Figure 11.4) includes the vitamin B_6-dependent enzyme kynureninase and in deficiency, after tryptophan load there is a considerable increase in the urinary excretion of xanthurenic and kynurenic acids. However, oestrogen metabolites inhibit kynureninase and give results that falsely suggest vitamin B_6 deficiency. The conversion of homocysteine (arising from the metabolism of methionine) to cysteine in the transsulphuration pathway (see Figure 11.7) includes two PLP dependent steps; in vitamin B_6 deficiency there is an increase in the urinary excretion of homocysteine after a loading dose of methionine. However, the metabolic fate of homocysteine is determined mainly by the need for cysteine and the rate at which it is remethylated to methionine, so increased excretion of homocysteine following a test dose of methionine cannot necessarily be regarded as evidence of vitamin B_6 deficiency.

Please see Weblink Section 11.1 for information about the basis for setting vitamin B_6 requirements and dietary recommendations.

KEY POINTS

- Vitamin B_6 functions as a coenzyme in a wide variety of reactions of amino acid metabolism, glycogen phosphorylase, and in the regulation of steroid hormone activity.

- Clinical deficiency of vitamin B_6 is very rare, but inadequate status is common in older adults and pregnant women.

- Vitamin B_6 status can be assessed by measuring plasma concentrations of PLP or other B6 derivatives, urinary excretion of 4-pyridoxic acid, erythrocyte aminotransferase activation coefficient, or measuring different metabolites after test doses of tryptophan or methionine.

- Very high doses of supplemental vitamin B_6 cause sensory nerve damage.

11.6 **Folates**

Folate is required for one-carbon metabolism and thus is essential for important biological pathways including DNA biosynthesis and repair, homocysteine metabolism, and methylation reactions. Maternal folate nutrition before and in early pregnancy is known to play a critical role in foetal development.

Forms in foods and dietary sources

Although the terms 'folic acid' and 'folate' are often used interchangeably, correctly, folic acid refers to the synthetic form of the vitamin, while folate refers to the natural forms as found in plant and animal tissues and thus in food sources. The richest sources of food folates are green leafy vegetables, asparagus, beans, legumes, liver, and yeast. Folic acid is found in the human diet only in fortified foods and supplements but is readily converted to the natural cofactor forms of folate after its ingestion. As shown in Table 11.1, the major dietary contributors to total folate intake are bread (14 per cent), breakfast cereals (12 per cent), vegetables (10 per cent), potatoes (10 per cent). Of note, however, all natural food sources of folate are poorly bioavailable compared with folic acid, therefore fortified foods provide the most important dietary source.

The structure of folate in its main cofactor forms is shown in Figure 11.6. All folate forms comprise three moieties: a pteridine, a p-aminobenzoic acid, and a glutamate residue. The parent compound, folic acid (pteroylglutamic acid), is completely oxidized and not found in nature. The natural folate forms are reduced molecules, with four additional hydrogen atoms to the pteridine (in the 5,6,7,8- positions), giving rise to the tetrahydro (THF) folate forms. Also, whereas folic acid is a monoglutamate, containing only one glutamic acid, most food folates exist as polyglutamate derivatives containing additional glutamate residues bound in peptide linkage to the gamma-carboxyl group. In addition, THF can carry

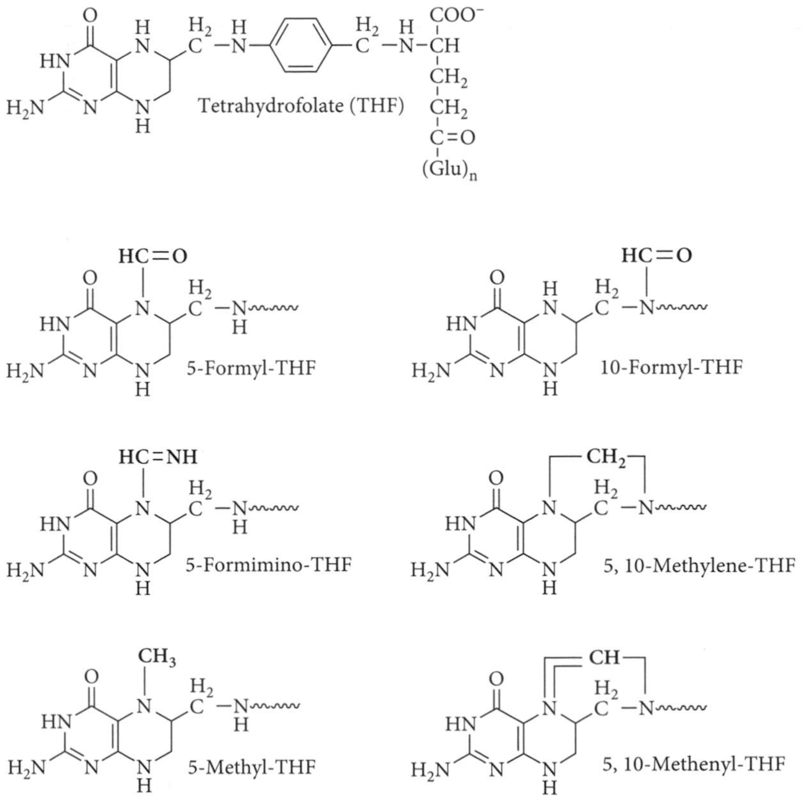

FIGURE 11.6 Folate forms: tetrahydrofolate and the one-carbon substituted folate derivatives.

one-carbon groups attached at the *N*-5 (methyl, formyl, or formimino), the *N*-10 (formyl) or bridging *N*-5 and *N*-10 (methylene or methenyl) positions of the pteridine ring, giving rise to a number of different cofactor forms of folate.

Absorption and transport of folate

The intestinal absorption of food folates is a two-step process. The first step involves the hydrolysis of folate polyglutamates to the corresponding monoglutamate derivatives, which occurs in the proximal part of the jejunum with the involvement of a brush-border conjugase enzyme and an optimal pH of 6.5. Once hydrolysed, the second step is the transport of monoglutamyl folates through the intestinal membranes into the enterocyte. Folates require transporters to cross cell membranes; these include the reduced folate carrier, the proton-coupled folate transporter (PCFT), and the folate receptor proteins, FRα and FRβ. Folic acid is a monoglutamate and thus does not require deconjugation before uptake by intestinal cells.

The bioavailability of naturally occurring food folates (i.e., proportion of ingested vitamin that is absorbed and available for metabolic processes) is limited and variable. As reduced folate forms, natural folates are inherently unstable outside living cells. Also, the ease with which folates are released from different food matrices and the removal of the polyglutamyl 'tail' (deconjugation) before uptake by intestinal cells can vary greatly. As a result, naturally occurring folates show incomplete bioavailability compared with folic acid at equivalent levels of intake. Apart from their limited bioavailability, natural folates in foods (particularly green vegetables) can be unstable during cooking, and this can substantially reduce the folate content of a food before it is ingested. In contrast, folic acid provides a highly stable and bioavailable vitamin form. The bioavailability of folic acid is assumed to be 100 per cent when ingested as a supplement, while folic acid in fortified food is estimated to have about 85 per cent the bioavailability of supplemental folic acid.

Metabolic functions of folate

Folates function as cofactors within one-carbon metabolism (See Figure 11.7). This involves the transfer and utilization of one-carbon units in a network of pathways required for essential processes, including DNA and RNA biosynthesis, amino acid metabolism, and methylation reactions. For folate to function effectively within this network, it interacts closely with vitamin B_{12}, vitamin B_6, and riboflavin.

Reduced folates enter the one-carbon cycle as THF which acquires a carbon unit from serine in a vitamin B_6-dependent reaction to form 5,10 methyleneTHF. This cofactor form, in turn, is either converted to 5 methyl-THF or serves as the one-carbon donor in the synthesis of nucleic acids, where it is required by thymidylate synthetase in the conversion of deoyxuridine (dUMP) to deoxythymidine (dTMP) for pyrimidine biosynthesis, or is converted to other folate cofactor forms required for purine biosynthesis. Methylenetetrahydrofolate reductase (MTHFR) is a riboflavin-dependent enzyme that catalyses the reduction of 5,10 methyleneTHF to 5 methylTHF. Once formed, 5 methylTHF is used by methionine synthase for the vitamin B_{12}-dependent conversion of homocysteine to methionine and the formation of THF. Methionine is activated by ATP to form S-adenosylmethionine, which then donates its methyl group to more than 100 methyltransferases for a wide range of substrates such as DNA, hormones, proteins, neurotransmitters, and membrane phospholipids.

Of note, given the metabolic interaction of folate with vitamins B_{12}, B_6, and riboflavin as outlined above, low status of one or more of these B vitamins, or polymorphisms in folate genes, can impair one-carbon metabolism, even if folate intakes are adequate.

Folate deficiency and excess

Clinical folate deficiency

The discovery of folate as an essential nutrient dates back to the 1930s, when a fatal anaemia of pregnancy was first described in India which was subsequently proven to be responsive to folate treatment. It is now known that clinical folate deficiency leads to megaloblastic anaemia, a condition is characterized by immature, enlarged blood cells (reflecting impaired DNA synthesis) which is reversible with folic acid treatment. Red blood cell and serum folate concentrations decline throughout normal pregnancy, but supplementation with folic acid prevents this decline and the occurrence of megaloblastic anaemia of pregnancy.

There are various causes of folate deficiency, arising from increased requirements, reduced availability, or both. Pregnancy is a time when folate requirement is greatly increased to sustain the demand for rapid cell replication and growth of foetal, placental, and maternal tissue. Certain gastrointestinal conditions, notably celiac disease, can also lead to deficient folate status through chronic malabsorption. Heavy alcohol consumption and certain commonly used drugs, such as phenytoin and primidone (anticonvulsants) and sulfasalazine (used in inflammatory bowel disease), also lead to folate deficiency.

It is important to appreciate that the absence of clinical folate deficiency does not necessarily mean that folate

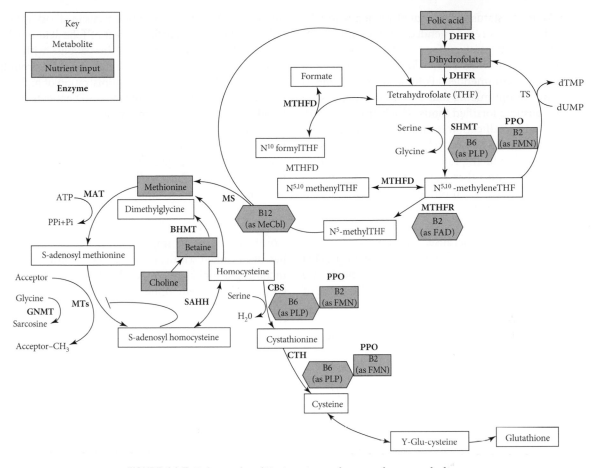

FIGURE 11.7 Folate, related B vitamins and one-carbon metabolism.

Folate, related B vitamins and one-carbon metabolism. B2, vitamin B_2 (riboflavin); B6, vitamin B_6; B12, vitamin B_{12}; BHMT, betaine-homocysteine methyltransferase; CBS, cystathionine β-synthase; CTH, cystathionine γ-lyase; DHFR, dihydrofolate reductase; dTMP, deoxythymidine monophosphate; dUMP, deoxyuridine monophosphate; FAD, flavin adenine dinucleotide; FMN, flavin mononucleotide; GNMT, glycine N-methyltransferase; MAT, methionine adenosyltransferase; MeCbl, methylcobalamin; MS, methionine synthase; MT, methyltransferases; MTHFR, methylenetetrahydrofolate reductase; MTHFD, methylenetetrahydrofolate dehydrogenase; PLP, pyridoxal 5'-phosphate; PPO, pyridoxine-phosphate oxidase; SAHH, S-adenosylhomocysteine hydrolase; SHMT, serine hydroxymethyltransferase; TS, thymidylate synthase.

status is optimal in terms of maintaining health. Thus in many developed countries folate deficiency (manifested as megaloblastic anaemia) may be relatively rare, but suboptimal folate status (detected by low red blood cell or serum folate concentrations) is commonly encountered, with adverse consequences for human health through the lifecycle, from pregnancy, through childhood, to preventing cardiovascular disease, certain cancers, and cognitive dysfunction in later life (Bailey et al. 2015).

Folate and neural tube defects (NTD)

Conclusive scientific evidence (i.e., from randomized controlled trials) has been available since 1991 that maternal folic acid supplementation in early pregnancy protects against NTDs, including spina bifida, anencephaly, and related birth defects. These major malformations occur as a result of failure of the neural tube to close properly in the first few weeks of pregnancy, leading to death of the foetus or newborn, or to various disabilities involving the spinal cord. The proven role of folate in preventing NTD has led to folic acid recommendations for women of reproductive age which are in place worldwide. Specifically, for the prevention of first occurrence of NTD (primary prevention), women are recommended to take 400 µg/day of folic acid from preconception until the end of the first trimester of pregnancy. Some women are considered to be at higher risk and thus are recommended to take higher folic acid doses (4 mg/day). This category includes women with a previous pregnancy affected by NTD and those taking certain anticonvulsant drugs.

There are, however, public health challenges related to implementing these recommendations in practice. Folic acid supplementation is a highly effective means to optimize folate status in individual women who take their supplements as recommended, but it is not an effective public health strategy for populations because in practice very few women start taking folic acid before conception as recommended. Folic acid-fortification is very effective in optimizing folate status in women who choose to consume fortified foods (Hopkins et al. 2015). When folic acid-fortification is undertaken on a mandatory basis, it has been shown to be effective in increasing folate status and reducing rates of NTD at a population level. In the UK and other European countries, policy to prevent NTD (i.e., based on advice to take folic acid supplements) has had little impact on preventing NTD, despite active health promotion campaigns over many years. This is primarily because, for many women, the early period when folic acid is protective against NTD may have passed before folic acid supplements are started. This has resulted in an unacceptably high rate of NTD in Europeans countries, recently estimated to be 1.6 times higher than in regions of the world with mandatory folic acid fortification policies in place (Khoshnood et al. 2015).

Folate and pregnancy outcomes beyond NTD

Apart from preventing NTD in early pregnancy, folate continues to play an important role during pregnancy. Deficient maternal folate status (and/or elevated plasma homocysteine) is associated with higher risk of a number of adverse pregnancy outcomes, including gestational hypertension, preeclampsia, placental abruption, pregnancy loss, low birth weight, and intrauterine growth restriction.

Maternal folate during pregnancy is also important for neurodevelopment and cognitive function in the child, especially in the first decade of life. The biological mechanism linking maternal folate with the offspring brain appears to involve folate-mediated epigenetic changes related to brain development and function. The most widely studied epigenetic mechanism for gene regulation is DNA methylation which depends upon the supply of methyl donors provided by folate and related B vitamins (Figure 11.7).

Folate and cardiovascular disease

The role of folate and the metabolically related B-vitamins (B_{12} and B_6) in cardiovascular disease (CVD) has been widely investigated, and there is good evidence linking poor folate status with an increased risk of CVD, and stroke in particular. Randomized controlled trials show that folic acid intervention can decrease the risk of stroke by as much as 18 per cent, and by over 25 per cent in people with no previous history of stroke. Although a

number of secondary prevention trials in at-risk patients failed to show any beneficial effect of folic acid in CVD, all such trials were performed in patients with well-established pathology. A reasonable conclusion from the available evidence is that the administration of high dose folic acid and related B vitamins to CVD patients is of no benefit in preventing another event.

Most research in this area has focused on plasma homocysteine as the relevant risk factor leading to CVD, but it is likely that folate (and metabolically related B vitamins) have protective roles in CVD that are independent of their homocysteine-lowering effects. Also, the C677T polymorphism in the gene encoding MTHFR, affecting 10 per cent of adults worldwide, leads to impaired folate metabolism and has been linked with several adverse health outcomes, including a higher risk of CVD by up to 40 per cent. Of note, there is emerging evidence suggesting that riboflavin—required in the form FAD as a cofactor for the folate metabolizing enzyme MTHFR—may be a key modifying factor linking this polymorphism with CVD via a novel and genotype-specific effect on blood pressure. Randomized trials published to date demonstrate lowering of systolic blood pressure, by between 6 and 13 mmHg, in response to low-dose riboflavin intervention in hypertensive patients with the variant *MTHFR* 677TT genotype.

Folate and cognitive dysfunction

Folate has a fundamental role in the nervous system throughout the lifecycle, from neural development in early life through to the maintenance of mental health and cognitive function in later life. A growing body of evidence indicates that lower status of folate and related B vitamins, and/or elevated homocysteine concentrations, are associated with a greater rate of cognitive decline in older age.

Folate and cancer

There is good epidemiological and animal evidence to link low folate status with an increased risk of cancer, while higher folate intake and status may play a protective role, particularly in the case of colorectal cancer. The most plausible biological mechanism explaining this relationship is that low folate leads to abnormal DNA synthesis and repair, and to reduced availability of S-adenosylmethionine for DNA methylation thus leading to aberrant gene expression. However, whereas higher folate status within the dietary range may be protective against some cancers, some remain concerned that exposure to excessively high folic acid intakes could increase the growth of pre-existing neoplasms.

Folate toxicity

No adverse effects are associated with the consumption of excess folate from food; concerns regarding safety are limited to the synthetic vitamin, folic acid. Concerns that folic acid could be harmful at high levels of exposure have delayed the implementation of effective public health policy on folic acid aimed at preventing NTD. Whilst over 80 countries worldwide to date (including the United States, Canada, and Australia) have passed regulations for the mandatory fortification of staple foods with folic acid, other countries (including almost all European countries) have delayed decisions to introduce mandatory fortification with folic acid on the basis of concerns regarding possible health risks.

Traditionally the health concern related to the potential risk that long-term exposure to high dose folic acid might 'mask' the anaemia of vitamin B_{12} deficiency in older people while allowing the associated irreversible neurological symptoms to progress. Although such cases have typically been seen at daily folic acid doses of 5000 μg (5 mg) and above, most authorities now recommend that all adults limit their intake of folic acid (supplements and fortification) to 1000 μg (1 mg) daily. Other evidence suggested that folic acid doses in excess of 1mg/d could potentially promote the growth of undiagnosed colorectal adenomas in those with pre-existing lesions. However, one meta-analysis of 50,000 individuals concluded that folic acid supplementation neither increased nor decreased site-specific cancer (Vollset et al. 2013). Moreover, in countries where mandatory fortification of flour with folic acid has been implemented, there has been a considerable reduction in the incidence of NTD, with no evidence of adverse effects on the population groups with high folic acid intakes.

An expert international panel tasked with reviewing all aspects of folate biology and biomarkers concluded that it was 'not aware of any toxic or abnormal effects of circulating folic acid' even from much higher exposures than those obtained through food fortification (Bailey et al. 2015).

Assessment of folate status

Folate is assessed by measuring concentrations in serum/plasma or in red blood cells, as determined by microbiological assay, or more typically in clinical laboratories, by protein binding assays. Serum folate is the earliest and most sensitive indicator of altered folate exposure and reflects recent dietary intake. Red cell folate parallels liver concentrations, accounting for about 50 per cent of total body folate, and thus reflects tissue folate stores. It responds slowly to changes in dietary folate intake but is a good indicator of longer term status in that it reflects

TABLE 11.6 Biomarkers of folate nutritional status in adults[1]

	Reference range		Deficiency	
	nmol/L	μg/L	nmol/L	μg/L
serum folate	≥10	≥4.4	<10	<4.4
red blood cell folate	≥340	≥150	<340	<150
plasma homocysteine[2]	5–15 μmol/L		>15 μmol/L	

[1] Criteria based on L B Bailey et al. (2015) *J Nutr* **147**, 1636S–1680S.
[2] Non-specific functional biomarker of folate status.

folate intake/status over the previous 3–4 months when circulating folate is incorporated into the maturing red cells. In addition, the measurement of plasma homocysteine concentration provides a functional indicator of folate status, on the basis that normal homocysteine metabolism requires an adequate supply of folate. With deficient or low folate status, plasma homocysteine is invariably found to be elevated. Sufficient folate status is indicated as a serum folate value ≥10, and a red blood cell folate concentration ≥340 nmol/L (Table 11.6). These cutoff values for folate insufficiency/deficiency were set on the basis of the metabolic indicator, plasma homocysteine (Bailey et al. 2015).

Please see Weblink Section 11.1 for information about the basis for setting folate requirements and dietary recommendations.

KEY POINTS

- Folates function as cofactors in one-carbon metabolism required for critical processes including DNA biosynthesis and methylation reactions. Vitamin B_{12}, vitamin B_6, and riboflavin interact closely with folate within the one-carbon network.

- Severe folate deficiency results in megaloblastic anaemia. Low maternal folate status in early pregnancy is associated with a higher risk of NTD, whilst maternal folic acid supplementation before and in early pregnancy protects against NTD.

- Low folate status leads to higher plasma homocysteine, which in turn is linked with an increased risk of adverse health outcomes including CVD and dementia in later life.

- Natural food folates can be unstable under normal conditions of cooking and have limited bioavailability compared with folic acid, the synthetic vitamin form. Fortified foods provide a highly bioavailable source of folate.

- Folate recommendations in many countries are now expressed as DFEs, devised to take into account the greater bioavailability of folic acid from fortified foods compared with naturally occurring food folates.

11.7 Vitamin B$_{12}$

Intestinal malabsorption, rather than inadequate dietary intake, can explain most cases of vitamin B$_{12}$ deficiency, at least in developed countries. Dietary deficiency of B$_{12}$ only occurs among strict vegans, but deficiency (or low status) as a result of impaired absorption is much more widespread throughout the world, especially among older people with atrophic gastritis. In low-income countries, however, vitamin B$_{12}$ deficiency is largely the result of low dietary intakes, typically in regions where vegan diets or limited animal foods are consumed, but gastrointestinal infections and infestations, along with host–microbiota interactions, may also be contributory factors (Green et al. 2017).

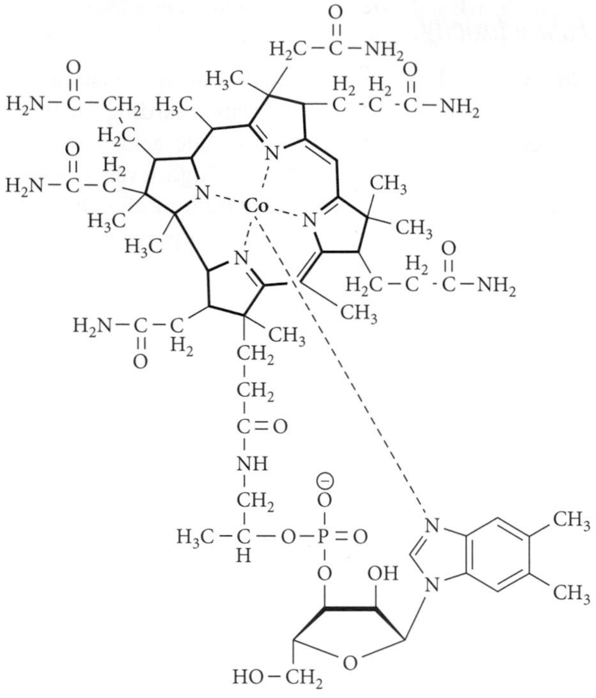

FIGURE 11.8 Vitamin B$_{12}$.

Forms of vitamin B$_{12}$ in foods and dietary sources

Vitamin B$_{12}$ is found only in foods of animal origin, especially meat, poultry, fish (including shellfish), and to a lesser extent, dairy products and eggs. There are no plant sources, therefore people consuming vegetarian, and particularly vegan, diets are most at risk dietary B$_{12}$ deficiency. See Table 11.1 for the typical food sources of vitamin B$_{12}$ in adults.

Vitamin B$_{12}$ has the largest and most complex chemical structure of all the vitamins and is unique in that it contains a central metal ion, cobalt. Thus, 'cobalamin' is the term used to refer to compounds having vitamin B$_{12}$ activity. The structure of vitamin B$_{12}$ is shown in Figure 11.8. The various B$_{12}$ forms have different groups chelated to the central cobalt atom: CN (cyanocobalamin); OH (hydroxocobalamin); CH$_3$ (methylcobalamin); 5'-deoxy-5'adenosine (adenosylcobalamin). The cofactor forms of vitamin B$_{12}$ used in the human body are methylcobalamin and 5-deoxyadenosylcobalamin; other B$_{12}$ forms ingested in foods are converted to these forms. Cyanocobalamin is more stable to light than the other cobalamins and therefore is the B$_{12}$ form of choice in most vitamin supplements and fortified foods; it is readily converted to 5-deoxyadenosylcobalamin and methylcobalamin once ingested.

Absorption and transport of vitamin B$_{12}$

The absorption of vitamin B$_{12}$ from food requires normal functioning of the stomach, pancreas, and small intestine. Stomach acid and enzymes free vitamin B$_{12}$ from food,

allowing it to bind to R-binder (haptocorrin). In the alkaline environment of the upper small intestine, haptocorrin is then digested by pancreatic enzymes, freeing vitamin B$_{12}$ to bind to intrinsic factor (IF), a protein secreted by the parietal cells in the stomach. The vitamin B$_{12}$-intrinsic factor complex is absorbed in the terminal ileum by receptor-mediated endocytosis in the presence of calcium, which is supplied by the pancreas. The IF-B$_{12}$ receptor consists of two components: cubilin and receptor-associated protein. Vitamin B$_{12}$ can also be absorbed by passive diffusion, but this process is very inefficient, only about 1 per cent absorption of the vitamin B$_{12}$ dose is absorbed passively. Inside the ileal enterocyte, B$_{12}$ is released by proteolysis and is bound to transcobalamin II for export from the enterocytes. Tissue uptake is by receptor-mediated endocytosis of holo-transcobalamin II, followed by proteolysis to release hydroxocobalamin, which may either undergo methylation to methylcobalamin in the cytosol, or enter the mitochondria to form adenosylcobalamin.

Another important component of vitamin B$_{12}$ absorption is the enterohepatic circulation. It is estimated that between 0.5 and 5.0 μg of B$_{12}$ is excreted in bile per day, and is readily reabsorbed across the ileal enterocyte, thus representing a mechanism for conserving vitamin B$_{12}$.

Metabolic functions of vitamin B$_{12}$

Vitamin B$_{12}$ is required as a cofactor for two mammalian enzymes: methionine synthase and methylmalonyl CoA mutase.

Vitamin B$_{12}$ in the form of methylcobalamin interacts closely with folate by acting as a cofactor for the folate-dependent enzyme, methionine synthase, which is required in the synthesis of methionine from homocysteine. Methionine, once formed, is activated by ATP to form S-adenosylmethionine, a methyl group donor used in many biological methylation reactions, including the methylation of a number of sites within DNA, RNA, and proteins (Allen et al. 2018). In the second reaction, vitamin B$_{12}$ in the form 5-deoxyadenosylcobalamin, is required as a cofactor methylmalonyl CoA mutase. This enzyme catalyses the conversion of methylmalonyl-CoA to succinyl-CoA, an intermediate step in the conversion of propionate to succinate. This conversion is an important step in the oxidation of odd-chain fatty acids.

Vitamin B$_{12}$ deficiency

Clinical deficiency

'Pernicious anaemia' is the clinical condition of severe deficiency of vitamin B$_{12}$, estimated to occur in about 2 per cent of individuals over 60 years of age. This arises from an autoimmune gastritis characterized by profound B$_{12}$ malabsorption owing to failure of intrinsic factor secretion. There are both haematological and neurological consequences.

Vitamin B$_{12}$ deficiency results in the development of megaloblastic anaemia which is identical to that seen in folate deficiency. The explanation why the clinical sign of megaloblastic anaemia is identical for a deficiency of either folate or vitamin B$_{12}$ is because DNA synthesis will be impaired in either case. In B$_{12}$ deficiency, folate recycling becomes impaired because of a decrease in the activity of the B$_{12}$-dependent enzyme methionine synthase, therefore 5-methyl THF cannot be converted to THF. Thus folate cofactors become 'trapped' in a form that cannot be used for DNA synthesis and, as a result, DNA synthesis becomes impaired.

Apart from causing anaemia, clinical B$_{12}$ deficiency affects the nervous system resulting in demyelination of peripheral and central neurons and neurological complications, including neuropathy and subacute combined degeneration of the spinal cord (Stabler 2013). The neurological damage in B$_{12}$ deficiency is caused by demyelination due to failure of methylation of arginine of myelin basic protein. The progression of neurological complications of B$_{12}$ deficiency is gradual and not reversible with treatment.

Less severe vitamin B$_{12}$ depletion

A less severe depletion of vitamin B$_{12}$ can arise from food-bound B$_{12}$ malabsorption as a result of atrophic gastritis leading to reduced gastric acid production

(hypochlorhydria). This in turn diminishes B$_{12}$ absorption from food because of the essential role of gastric acid in the release of B$_{12}$ from food proteins. Food-bound B$_{12}$ malabsorption commonly occurs in older adults and leads to sub-clinical deficiency, where there is metabolic evidence of deficiency but without the classical haematological or neurological deficiency signs.

There are number of adverse consequences of low, though not necessarily deficient, biomarker status of vitamin B$_{12}$, with evidence that low B$_{12}$ is associated with a higher risk of various diseases of ageing including cardiovascular disease, cognitive dysfunction, dementia, depression, and osteoporosis (Allen 2018). Many such adverse health outcomes are similar to those associated with low folate status (as discussed above in Section 11.6), perhaps not surprisingly, given the close metabolic interrelationship of B$_{12}$ with folate and their common role in sustaining methylation reactions.

Drug-induced vitamin B$_{12}$ deficiency

Nitrous oxide inhibits methionine synthetase, by oxidizing the cobalt of methylcobalamin. Patients can thus develop neurological signs after surgery when nitrous oxide is used as the anaesthetic agent. There are also historical reports of neurological damage due to B$_{12}$ depletion among dental surgeons and others occupationally exposed to nitrous oxide.

The use of proton-pump inhibitors (PPIs), gastric-acid suppression drugs (for treating gastroesophageal reflux disease and simple heartburn), leads to hypochlorhydria and thus the concern that these drugs could contribute to food-bound B$_{12}$ malabsorption, and ultimately, B$_{12}$ deficiency. The evidence to date is inconsistent, however, with some studies showing no association of PPIs with B$_{12}$ status, and others reporting an increased risk of B$_{12}$ deficiency with long-term PPI use. Metformin, the most commonly prescribed drug in type 2 diabetes, was reported many years ago to lead to vitamin B$_{12}$ malabsorption in up to 30 per cent of diabetic patients. More recently, large observational studies and systematic reviews have concluded that metformin use is indeed associated with significantly lower B$_{12}$ status, albeit the precise mechanism explaining this relationship is not well understood.

Assessment of vitamin B$_{12}$ status

Nutritional status of vitamin B$_{12}$ status is assessed using up to four biomarkers, both direct and functional biomarkers (Table 11.7). Serum total vitamin B$_{12}$ is measured using microbiological assay or automated competitive protein binding assays and this measurement has been the standard laboratory assay for many years, with deficiency typically identified as B$_{12}$ concentrations <148pmol/L

TABLE 11.7 Biomarkers of vitamin B_{12} nutritional status in adults[1]

	Reference range	Deficiency
serum total vitamin B12[2]	≥148 pmol/L	<148 pmol/L
	≥221 pmol/L	<221 pmol/L
serum holotranscobalamin[2]	≥35 pmol/L	<35 pmol/L
	≥45 pmol/L	<45 pmol/L
serum methylmalonic acid[2]	≤376 nmol/L	>376 nmol/L
	≤271 nmol/L	>271 nmol/L
	≤210 nmol/L	>210 nmol/L
plasma homocysteine[3]	5–15 µmol/L	>15 µmol/L

[1] Criteria based on L H Allen et al. (2018) *J Nutr* **148**, 1995–2027.
[2] Various international authorities have applied different cut-points to define vitamin B_{12} deficiency, and there is no universal agreement as to the cut-point for each biomarker which best reflects deficiency of B_{12}.
[3] Non-specific functional biomarker of vitamin B_{12} status

(Green et al. 2017). About 80 per cent of total vitamin B_{12} is metabolically inert, therefore the measurement of holotranscobalamin (holoTC), or 'active B_{12}', representing only the metabolically active fraction (20 per cent) of total B_{12} that is available for cellular processes, provides an alternative measure of B_{12} status. Although widely considered to be a robust biomarker of B_{12} status, serum HoloTC is not generally available in clinical settings.

Other measures of vitamin B_{12} status are the functional markers, methylmalonic acid (MMA) and homocysteine, that are found in elevated concentrations in B_{12} deficiency. With B_{12} depletion, the activity of the B_{12} dependent enzyme methylmalonyl CoA mutase is reduced, leading to an accumulation of the by-product MMA. Serum MMA is thus found to be elevated in the majority of patients with B_{12} deficiency, but there is no universal agreement as to the MMA cut-point above which deficiency is indicated. Also, with vitamin B_{12} depletion, the activity of the B_{12}-dependent enzyme methionine synthase is impaired, in turn leading to elevated homocystein, which provides another functional marker of B_{12} status. Plasma homocysteine is however influenced by other vitamins (particularly folate) and non-nutrient factors (including renal function), thus it is not a specific biomarker of B_{12}.

Although there are four biomarkers in use for assessing vitamin B_{12} status, there is a lack of agreement, both as regards the best biomarker and the cut-points to define deficiency (Hughes & McNulty 2018). Also, given the limitations of each of the direct (total B_{12} and holoTC) and functional (homocysteine and MMA) biomarkers, the use two or more B_{12} biomarkers is recommended to more accurately diagnose deficiency, particularly in older people who are most at risk.

Please see Weblink Section 11.1 for information about the basis for setting vitamin B_{12} requirements and dietary recommendations.

KEY POINTS

- Vitamin B_{12} plays an essential role in the synthesis of the citric acid cycle intermediate, succinyl-CoA, and in folate recycling within one-carbon metabolism vitamin B_{12} is important for methylation reactions and homocysteine metabolism.

- Severe deficiency of vitamin B_{12} termed 'pernicious anaemia' (as a result of intrinsic factor deficiency) causes megaloblastic anaemia and neuropathy.

- Low, though not necessarily deficient, status of vitamin B_{12} is associated with a higher risk of various diseases of ageing including cardiovascular disease, cognitive dysfunction, dementia, depression, and osteoporosis.

- Dietary deficiency of vitamin B_{12} occurs only in strict vegans as there are no plant sources of the vitamin. Food-bound B_{12} malabsorption as a result of atrophic gastritis, leading to subclinical B_{12} deficiency, is relatively common among older people.

- Vitamin B_{12} status can be assessed using up to four biomarkers, both direct (serum total vitamin B_{12} and serum holoTC) and functional biomarkers (serum MMA and plasma homocysteine), but there is a lack of agreement as regards the best biomarkers or cut-points to use for accurate diagnosis of B_{12} deficiency.

11.8 Vitamin C (ascorbic acid)

Vitamin C is a water soluble vitamin also known as L-ascorbic acid. Unlike most other animals, humans and other primates do not have the ability to synthesize vitamin C and must obtain it from the diet. Although it is synthesized as an intermediate in the gulonolactone pathway of glucose metabolism, in humans and other species for which it is a vitamin, one enzyme in this pathway, gulonolactone oxidase, is absent. Vitamin C is a potent antioxidant and an essential cofactor in numerous enzymatic reactions (e.g., in the biosynthesis of collagen, carnitine, and neuropeptides) and in the regulation of gene expression.

Forms in foods and dietary sources

Vitamin C (L-ascorbic acid) is an electron donor; it can sequentially donate two electrons (Figure 11.9). It can undergo oxidation to the monodehydroascorbate radical and dehydroascorbate, both of which have vitamin

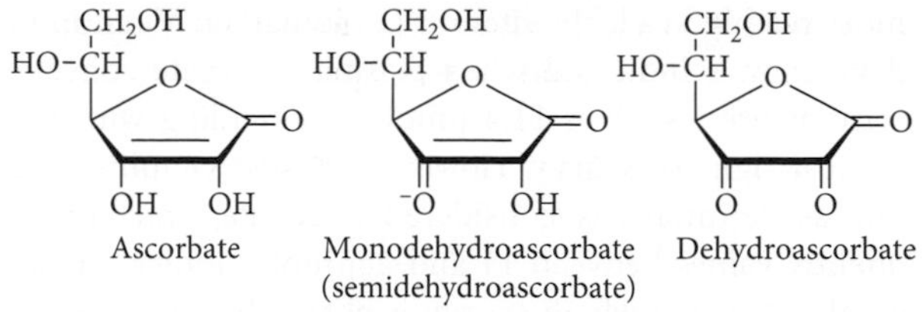

FIGURE 11.9 Vitamin C.

activity because they can be reduced to ascorbate. D-Iso-ascorbic acid (erythorbic acid) has some vitamin activity; it is not a naturally occurring compound, but is widely used, interchangeably with ascorbic acid, in cured meats and as an antioxidant in a variety of foods. Fruits and vegetables provide dietary vitamin C, with citrus fruits, kiwifruit, and blackcurrants providing particularly rich sources, and potatoes are an important source in many countries; see Table 11.1 for the main food sources of vitamin C in adults.

Absorption and transport of vitamin C

At intakes up to about 100 mg/day, some 80–95 per cent of dietary ascorbate is absorbed. With larger intakes, the absorption is lower (e.g., about 50 per cent of a 1.5 g dose is absorbed); unabsorbed ascorbate is a substrate for intestinal bacterial metabolism. Ascorbate is absorbed by active transport, while dehydroascorbate is absorbed by a carrier-mediated (equilibrium) transport, followed by reduction to ascorbate inside the intestinal epithelial cell. Ascorbate enters tissues by way of sodium-dependent transporters, while dehydroascorbate enters by way of the (insulin-dependent) glucose transporter, and is reduced to ascorbate intracellularly. Tissue uptake of dehydroascorbate is impaired in poorly controlled diabetes mellitus, and functional signs of deficiency may develop despite an adequate intake of vitamin C.

About 70 per cent of ascorbate in the blood circulation is in plasma and erythrocytes. The remainder is in white cells, which have a marked ability to concentrate ascorbate; mononuclear leukocytes achieve 80-fold, platelets 40-fold, and granulocytes 25-fold concentrations compared with plasma. There is no specific storage organ for ascorbate, the only tissues showing a significant concentration of the vitamin are the adrenal and pituitary glands. Although the concentration of ascorbate in muscle is relatively low, skeletal muscle contains much of the body's pool of 5–8.5 mmol (900–1500 mg) of ascorbate. The major fate of ascorbic acid is urinary excretion, either unchanged or as dehydroascorbate and diketogulonate. At plasma concentrations above about 85 μmol/l, the renal transport system is saturated, and ascorbate is excreted quantitatively with increasing intake.

Metabolic functions of vitamin C

Vitamin C is a potent reducing agent, meaning that it readily donates electrons to recipient molecules. Related to this oxidation-reduction (redox) role, two major functions of vitamin C are as a potent antioxidant and an essential cofactor in numerous enzymatic reactions (e.g., in the biosynthesis of collagen, carnitine, and neuropeptides) and in the regulation of gene expression.

Vitamin C is the primary water soluble antioxidant in plasma and tissues. Even in small amounts, vitamin C can protect critical molecules in the body, such as proteins, lipids, carbohydrates, and nucleic acids (DNA and RNA), from damage by free radicals and reactive oxygen species (ROS) that are generated during normal metabolism, by active immune cells, and through exposure to toxins and pollutants. Vitamin C also participates in redox recycling of other important antioxidants; for example, vitamin C can regenerate vitamin E from its oxidized form.

The role of vitamin C as a cofactor is also related to its redox potential. By maintaining enzyme-bound metals in their reduced forms, vitamin C is required by mixed-function oxidases in the biosynthesis or hydroxylation of several critical biomolecules, including noradrenaline, collagen and carnitine. These enzymes are either monooxygenases or dioxygenases. Moreover, several dioxygenases involved in the regulation of gene expression and the maintenance of genome integrity require vitamin C as a cofactor. These enzymes contribute to the epigenetic regulation of gene expression by catalysing reactions involved in the demethylation of DNA and histones.

Vitamin C has an important role in increasing the bioavailability of iron from foods by enhancing the intestinal absorption of non-haem iron (i.e., from non-meat sources). Finally, although *in vitro* studies show that vitamin C affects several components of the immune system, evidence from human studies is conflicting and therefore the effects of vitamin C in influencing immune function are unclear.

Vitamin C deficiency and excess

Severe vitamin C deficiency has been known for many centuries as the potentially fatal disease, scurvy. Although there is no specific site for vitamin C storage in the body, signs of deficiency do not develop until previously adequately nourished subjects have been deprived of the vitamin for 4–6 months, by which time plasma and tissue concentrations will have fallen considerably. The name scurvy for the vitamin C deficiency disease is derived from the Italian *scorbutico*, meaning an irritable person; deficiency is associated with listlessness and general malaise, and sometimes changes in personality and psychomotor performance. The behavioural effects are likely related to impaired synthesis of catecholamines as a result of reduced activity of dopamine β-monooxygenase.

Symptoms of scurvy include subcutaneous bleeding, poor wound healing, hair and tooth loss, and joint pain and swelling. Most of the signs of scurvy are due to impaired collagen synthesis. The earliest signs are skin changes, beginning with plugging of hair follicles by horny material, followed by petechial haemorrhage and increased fragility of blood capillaries leading to extravasation of red cells. Later there is haemorrhage of the gums and loss of dental cement. Wounds show only superficial healing in scurvy, with little or no formation of (collagen-rich) scar tissue, so that healing is delayed and wounds can readily be reopened. Anaemia is common in scurvy, and may be either macrocytic (indicative of folate deficiency), or hypochromic (indicative of iron deficiency). Folate deficiency may be epiphenomenal, since some of the major dietary sources of folate are the same as those of ascorbate. Iron deficiency in scurvy may well be secondary to reduced absorption of inorganic iron, and impaired mobilization of tissue iron reserves.

The tolerable upper intake level (UL) for vitamin C is set at 2 g/day, based on adverse gastrointestinal effects at higher doses. Because oxalate is a metabolite of vitamin C, there is some concern that high-dose supplemental vitamin C could increase the risk of calcium oxalate kidney stones, albeit the evidence from human studies is conflicting. Given the conflicting results, it seems prudent for individuals predisposed to oxalate kidney stone formation to avoid high-dose vitamin C supplements. Unabsorbed ascorbate in the intestinal lumen is a substrate for bacterial fermentation, which may explain the diarrhoea and intestinal discomfort reported in some studies with high doses of supplemental vitamin C.

Assessment of vitamin C status

Biomarkers of vitamin C nutritional status are shown in Table 11.8. Vitamin C status is based on plasma ascorbate measurements, or less commonly, on white cell (leukocyte) concentrations. Plasma ascorbate provides the most readily available vitamin C biomarker. Vitamin C deficiency is indicated when plasma ascorbate concentrations are less than 11.4 μmol/L, coinciding with the clinical signs of scurvy. However, plasma vitamin C as low as 28μmol/L is considered adequate, and consequently values between 11 and 28μmol/L indicate marginal deficiency where scurvy is absent but the risk for chronic diseases is elevated. Thus, marginal vitamin C status (sometimes referred to as hypovitaminosis C) is indicated by plasma ascorbate concentrations of 11.4–28.4 μmol/L. The highest plasma ascorbate values observed in human studies are 70–80μmol/L and concentrations plateau in that range, even with supplemental vitamin C intakes >100 mg/day, because of quantitative excretion of the vitamin as the renal threshold is exceeded (Padayatty & Levine 2016).

Different types of leukocytes have different capacities to accumulate ascorbate, so that a change in the proportion of granulocytes, platelets, and mononuclear leukocytes will result in a change in the total concentration of ascorbate/10^6 cells, although there may well be no change in status.

Please see Weblink Section 11.1 for information about the basis for setting riboflavin requirements and dietary recommendations.

Pharmacological uses of vitamin C

A significant number of people consume gram amounts of vitamin C supplements to treat both respiratory infection and inflammation. Supplements of vitamin C are widely consumed to protect against cancer, cardiovascular disease, and viral infections, although there is little evidence of efficacy. Results from the Physicians' Health Study in USA showed no benefits of vitamin C supplements. High doses of vitamin C are popularly recommended for the prevention and treatment of the common cold. The evidence from controlled trials is unconvincing, and a Cochrane review of the literature reported that high-dose supplementation of vitamin C does not reduce the incidence of the common cold in the general population.

TABLE 11.8 Ascorbate concentrations as biomarkers of vitamin C nutritional status[1]

		Deficient	Marginal	Adequate
plasma	μmol/L	<11.4	11.4–28.4	>28.4
	mg/100ml	<0.2	0.2–0.5	>0.5
leukocytes	pmol/10^6 cells	<1.1	1.1–2.8	>2.8
	μg/10^6 cells	<0.2	0.2–0.5	>0.5

[1] Criteria based on Schleicher et al. (2009) *Am J Clin Nutr* **90**, 1252–1263.

Ascorbate enhances the intestinal absorption of inorganic iron, and therefore it is frequently prescribed with iron supplements. Iron is absorbed as Fe^{2+} and not as Fe^{3+}; ascorbic acid in the intestinal lumen will both maintain iron in the reduced state and also chelate it. A dose of 25 mg of vitamin C, taken together with a semi-synthetic meal, increases the absorption of iron by about 65 per cent, while a 1 g dose gives a nine-fold increase.

KEY POINTS

- Vitamin C (L-ascorbic acid) is a potent antioxidant and an essential cofactor in numerous enzymatic reactions, including in collagen biosynthesis.
- Vitamin C also participates in the redox recycling of other antioxidants and is important in reducing the tocopheroxyl radical formed by oxidation of vitamin E in membranes and plasma lipoproteins.
- The signs of severe vitamin C deficiency (scurvy) are fatigue, mood and behavioural changes, subcutaneous bleeding and poor wound healing, along with skin lesions, hair and tooth loss, and joint pain and swelling.
- Vitamin C status is assessed by plasma and leukocyte ascorbate concentrations.
- Vitamin C enhances the absorption of inorganic iron very considerably. Foods containing vitamin C are best consumed in the same meal as iron-containing foods from plant sources in order to improve the absorption of iron in the gut.

11.9 Pantothenic acid

Pantothenic acid belongs to the group of B-vitamins and has a central role in virtually all aspects of metabolism as a functional moiety of coenzyme A (CoA) and as a prosthetic group of acyl carrier protein. Pantothenic acid is widely distributed in all foodstuffs where it is predominantly in the form of CoA; meat, milk, eggs, whole grain cereals, peanuts and almonds are considered good sources of the vitamin. It is absorbed in the small intestine through a saturable sodium-dependent carrier-mediated process with bioavailability estimated at around 50 per cent. Intestinal microbiota can generate pantothenic acid but the extent to which it is absorbed in the large intestine is unknown. Deficiency has not been unequivocally reported except in depletion studies, which have frequently also used the antagonist ω-methyl pantothenic acid. Deficiency symptoms include numbness of

feet and hands, sleep disturbances, and impaired motor coordination.

Prisoners of war in the Far East in the 1940s, who were severely malnourished, showed a new condition of paraesthesia and severe pain in the feet and toes, which was called the 'burning foot' syndrome or nutritional melalgia. The condition was cured by pantothenic acid supplementation. There seem to be no reports of neurological damage in deficient animals that would explain the 'burning foot' syndrome.

The naturally occurring vitamer is D-pantothenic acid, shown in Figure 11.10. Free pantothenic acid and its sodium salt are chemically unstable, and the usual pharmacological preparation is the calcium salt (calcium dipantothenate). The alcohol, pantothenol, is a synthetic compound that has biological activity because it is oxidized to pantothenic acid *in vivo*.

Pantothenic acid in the form of CoA is involved in a wide variety of metabolic reactions. CoA is essential for the energy-yielding metabolism in the mitochondria by providing substrates for the citric acid cycle. CoA-dependent catabolic processes such as glycolysis, β-oxidation of fatty acids, and oxidative degradation of amino acids generate acetyl-CoA which enters the citric acid cycle for oxidation and energy production. In addition, CoA is required for the synthesis of phospholipids, fatty acids, isoprenoid derivatives (cholesterol, steroid hormones, and vitamin D), and donates an acetyl group for the production of acetylcholine, the conversion of serotonin to melatonin and acetylation of sugars (N-acetylglucosamine). CoA is also involved in the acetylation of proteins, which is a way of altering protein conformation and function, hormone activation (e.g., adrenocorticotropin), and transcriptional regulation (histone acetylation). Furthermore, a number of cellular proteins are covalently modified with the

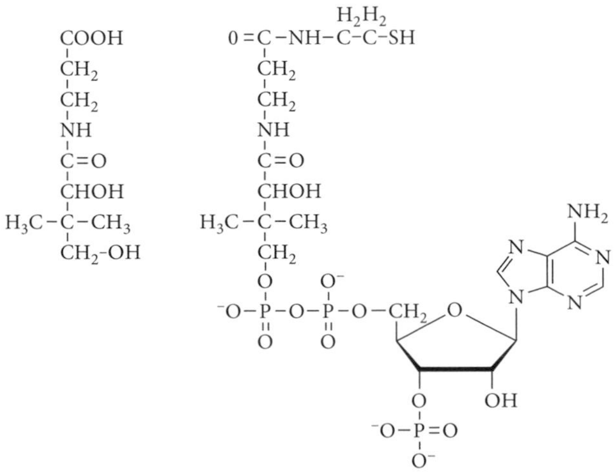

FIGURE 11.10 Pantothenic acid and CoA.

addition of lipid groups (mainly palmitoyl or myristoyl group), a process facilitated by CoA and known as protein acylation. This modification enables specific protein activities, such as the function of transmembrane receptors (e.g., insulin receptor), gap junction proteins, and signal transduction, whereas other acylated proteins, such as myelin (protein-lipid complex), are of structural importance.

Biomarkers for assessment of pantothenic acid nutritional status include measurement of vitamin concentration in whole blood, erythrocytes, and urine, however, criteria for their adequacy are not well developed. From the limited studies that have been performed it is not possible to establish requirements for pantothenic acid. Average intakes are around 4–6 mg/day; this is obviously adequate, since deficiency is unknown under normal conditions. Both EFSA and IOM set adequate intake for adults at 5 mg/day.

KEY POINTS

- Pantothenic acid is required for synthesis of coenzyme A and the acyl carrier protein for fatty acid synthesis.
- Pantothenic acid is involved in a wide variety of metabolic processes including energy generation, synthesis of cholesterol, vitamin D, and steroid hormones.
- Pantothenic acid deficiency is unknown except in depletion studies; the vitamin is widely distributed in foods.
- There is no evidence on which to base estimates of requirements.

11.10 Biotin (vitamin H)

Biotin was discovered as the protective or curative factor in 'egg white injury'—the disease caused in experimental animals by feeding diets containing large amounts of uncooked egg white. The glycoprotein avidin in egg white binds biotin, rendering it unavailable. Dietary deficiency of biotin with clinical signs is extremely rare. The vitamin is widely distributed in many foods, is synthesized by intestinal flora, and can be absorbed in the colon. However, there is increasing evidence that suboptimal biotin status may be relatively common. The few early reports of biotin deficiency all concerned people who consumed abnormally large amounts of uncooked eggs. More recently, signs of biotin deficiency (hair loss, scaly erythematous dermatitis, and ataxia) have been observed in patients receiving total parenteral nutrition for prolonged periods and without biotin supplementation. Up to half of women in the first trimester of pregnancy show metabolic abnormalities that respond to supplements of biotin, suggesting that marginal status may be widespread. Biotin deficiency in experimental animals is teratogenic, however, there are no data in humans for association between biotin deficiency in pregnancy and an increased incidence of foetal malformations.

The structure of biotin is shown in Figure 11.11; biologically active is the isomer D(+)-biotin. Food sources rich in biotin are liver, eggs, mushrooms, and some cheeses. It occurs in foods covalently bound to proteins by the formation of a peptide bond between the carboxyl group of the side-chain and the ε-amino group of a lysine residue, forming biocytin (biotinyl-lysine). Biocytin is hydrolysed by biotinidase in the pancreatic juice and intestinal mucosal secretions and it is absorbed through a saturable carrier-mediated process. Avidin in raw egg

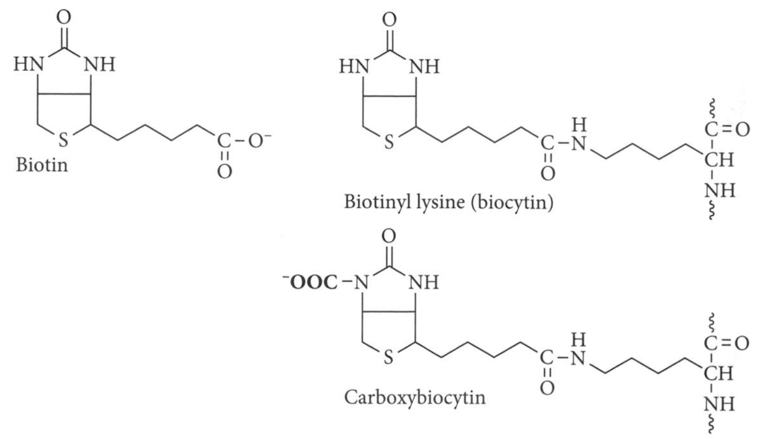

FIGURE 11.11 Biotin and biocytin.

white has a very high affinity for biotin and prevents its absorption in the small intestine.

Metabolically, biotin is of central importance in lipogenesis and gluconeogenesis, acting as the coenzyme for carboxylation reactions. There are four biotin-dependent carboxylases: acetyl-CoA carboxylase in fatty acid synthesis; pyruvate carboxylase for gluconeogenesis and replenishing citric acid cycle intermediates; propionyl-CoA carboxylase which takes part in the synthesis of methylmalonyl-CoA; and β-methylcrotonyl-CoA carboxylase which is involved in the degradation of the branch-chained amino acid leucine. The holocarboxylase synthetase that attaches biotin to the carboxylases also acts to assemble a multi-protein gene repression complex in chromatin, and to biotinylate histones and other proteins involved in gene expression and the cell cycle, as well as enolase, a key enzyme in gluconeogenesis, and a tumour suppression factor leading to reduced expression on some oncogenes. In addition, biotin induces the synthesis of a number of key enzymes of glycolysis and gluconeogenesis.

Specific biomarkers for assessment of biotin status are not well defined. Results of some small-scale intervention studies in adults have indicated that the decrease of urinary biotin excretion, the suppression of propionyl-CoA carboxylase in lymphocytes, as well as the increase of the ratios of acylcarnitines in urine (which reflect disturbances in biotin-dependent carboxylase activities) could be considered sensitive biomarkers for biotin depletion. However, the variability of these biomarkers in the general population is unknown and the criteria for their adequacy are not well developed.

There is little information concerning biotin requirements, and no evidence on which to base recommendations. Average intakes range between 15 and 70 µg/day. The European Food Safety Authority set an adequate intake for adults at 40 µg/day.

KEY POINTS

- Biotin functions as a coenzyme in carboxylation reactions, and has a role in regulating the cell cycle.
- Clinical deficiency of biotin occurs only in people consuming large amounts of uncooked egg white or receiving long-term total parenteral nutrition.
- Marginal biotin status is common in women in the first trimester of pregnancy.
- Biomarkers for assessment of biotin status are not well defined.
- There is no evidence on which to base estimates of requirements.

ACKNOWLEDGEMENTS

The authors wish to acknowledge the valuable contributions of their colleagues at Ulster University who assisted in preparing the tables for this chapter: Dr Maeve Kerr, Ms Michelle Clements (PhD Researcher), and Ms Oonagh Lyons (PhD Researcher).

 Be sure to test your understanding of this chapter by attempting multiple choice questions.

REFERENCES

Allen L H, Miller J W, De Groot L et al. (2018) Biomarkers of nutrition for development (BOND): Vitamin B-12 review. *J Nutr* **148**, 1995–2027.

Bailey L B, Stover P J, McNulty H et al. (2015) Biomarkers of nutrition for development: Folate review. *J Nutr* **147**, 1636S–1680S.

Bates C J, Pentieva K D, Prentice A (1999) An appraisal of vitamin B6 status indices and associated confounders, in young people aged 4–18 years and in people aged 65 years and over, in two national British surveys. *Publ Hlth Nutr* **2**, 529–535.

Bender D A (1983) Effects of a dietary excess of leucine on the metabolism of tryptophan in the rat: A mechanism for the pellagragenic action of leucine. *Brit J Nutr* **50**, 25–32.

Bender D A (2003) Niacin. In: *Nutritional Biochemistry of the Vitamins*, 2nd edn. Cambridge University Press, New York.

Borschel M W (1995) Vitamin B6 in infancy: Requirements and current feeding practices. In: *Vitamin B-6 Metabolism in Pregnancy, Lactation and Infancy*, 109–124 (D J Raiten ed.). CRC Press, Boca Raton.

EFSA Scientific Committee of Food (2006) *Tolerable Upper Intake Levels for Vitamins and Minerals*. European Food Safety Authority.

Expert Group on Vitamins and Minerals (2003) *Safe Upper Levels for Vitamins and Minerals*. Food Standards Agency, London.

Green R, Allen L H, Bjørke-Monsen A et al. (2017) Vitamin B12 deficiency. *Nat Rev Dis Primers* **3**, 17040.

Hopkins S M, Gibney M J, Nugent A P et al. (2015) Impact of voluntary fortification and supplement use on dietary intakes and biomarker status of folate and vitamin B12 in Irish adults. *Am J Clin Nutr* **101**, 1163–1172.

Hughes C F & McNulty H (2018) Assessing biomarker status of vitamin B12 in the laboratory: No simple solution. *Ann Clin Biochem* **55**, 188–189.

Khoshnood B, Loane M, de Walle H et al. (2015) Long term trends in prevalence of neural tube defects in Europe: Population based study. *BMJ* **351**, h5949.

Leklem J E (1990) Vitamin B-6: A status report. *J Nutr* **120**(Suppl 11), 1503–1507.

McNulty H, Dowey L C, Strain J J et al. (2006) Riboflavin lowers homocysteine in individuals homozygous for the MTHFR 677C->T polymorphism. *Circulation* **113**, 74–80.

McNulty H, McKinley M C, Wilson B et al. (2002) Impaired functioning of thermolabile methylenetetrahydrofolate reductase is dependent on riboflavin status: Implications for riboflavin requirements. *Amer J Clin Nutr* **76**, 436–441.

McNulty H, Ward M, Hoey L et al. (2019) Addressing optimal folate and related B-vitamin status through the lifecycle: Health impacts and challenges. *Proc Nutr Soc* **78**, 449–462.

Padayatty S J & Levine M (2016) Vitamin C: The known and the unknown and Goldilocks. *Oral Diseases* **22**, 463–493.

Paul L, Ueland P M, Selhub J (2013) Mechanistic perspective on the relationship between pyridoxal 5'-phosphate and inflammation. *Nutr Rev* **71**, 239–44.

Powers H J (2003) Riboflavin (vitamin B2) and health. *Amer J ClinNutr* **77**, 1352–1360.

Sauberlich H E (1999) *Laboratory Tests for the Assessment of Nutritional Status*. CRC Press, Boca Raton.

Schleicher R, Carroll M D, Ford E S, Lacher D A (2009) Serum vitamin C and the prevalence of vitamin C deficiency in the United States: 2003–2004 National Health and Nutrition Examination Survey (NHANES). *Am J Clin Nutr* **90**, 1252–1263.

Sechi G & Serra A (2007) Wernicke's encephalopathy: New clinical settings and recent advances in diagnosis and management. *Lancet Neurol*, **6**, 442–455.

Stabler S P (2013) Clinical practice: Vitamin B_{12} deficiency. *New Engl J Med* **368**, 149–160.

Vollset S E, Clarke R, Lewington S et al. (2013) Effects of folic acid supplementation on overall and site-specific cancer incidence during the randomised trials: Meta-analyses of data on 50,000 individuals. *Lancet* **381**, 1029–1036.

Wilson C P, McNulty H, Ward M et al. (2013) Blood pressure in treated hypertensive individuals with the MTHFR 677TT genotype is responsive to intervention with riboflavin: Findings of a targeted randomized trial. *Hypertension* **61**, 1302–1308.

12

Fat soluble vitamins

Mairead E. Kiely

OBJECTIVES

By the end of this chapter you should be able to:

- describe the dietary and other sources and biomarkers of nutritional status of each of the fat soluble vitamins
- identify risk factors and at-risk population groups for low nutrient status
- summarize the absorption, transport, and metabolism of each nutrient
- outline the main metabolic functions of each nutrient
- describe the main effects of fat soluble vitamin deficiency and links with health outcomes
- describe the metabolic interactions between fat soluble vitamins and other nutrients
- explain the effects of excessive intakes of each vitamin
- discuss and evaluate the scientific basis for dietary requirements for each vitamin.

12.1 Vitamin A: retinoids and carotenoids

Vitamin A deficiency is a serious public health nutrition problem worldwide and the most important cause of preventable blindness among children in regions at risk. In 2009, the WHO estimated that based on the currently accepted cut-off for low serum retinol concentrations (<0.7 mmol/L), 33 per cent of the preschool population and 15 per cent of pregnant women in populations at risk were deficient in vitamin A (Tanumihardjo et al. 2016). Due to the central role of vitamin A in immune function, even mild deficiency in children makes them susceptible to infections, including diarrhoea, respiratory infections, and measles. Vitamin A deficiency is most prevalent in low-income regions, particularly in Africa and Southeast Asia, and together with iron, vitamin A is the most high-risk nutrient in all settings.

Dietary sources

The term vitamin A includes both provitamin A carotenoids and retinol and its active metabolites. The term retinoid is used to include retinol and its derivatives and analogues, either naturally occurring or synthetic, with or without the biological activity of the vitamin. The main biologically active retinoids are shown in Figure 12.1 and there are others. Free retinol is chemically unstable and occurs mainly in foods as palmitate. Liver, full-fat dairy produce, fortified margarine, eggs, and oily fish are good sources of retinol.

Provitamin A carotenoids are carotenes that can be oxidized to retinaldehyde, which are nutritionally important as few foods are rich sources of retinol. The structures of α- and β-carotene are shown in Figure 12.1. Good dietary sources of carotenes are dark-green leafy vegetables and deeply coloured yellow and orange vegetables and fruit; sweet potatoes, carrots, pumpkins, sweet red peppers, mangoes, and melons. Fortified foods can be important dietary sources of preformed vitamin A. In low- to middle-income countries sugar, cereal flours, edible oils, margarine, and noodles can be fortified in a culturally acceptable manner that increases β-carotene intakes and retinol status (Tanumihardjo et al. 2016). Such fortification programmes are 2–4 times more cost-effective than supplement distribution.

International units and retinol equivalents

The total vitamin A content of foods is expressed as µg retinol equivalents (RE)—the sum of that provided by retinoids and carotenoids; 6 µg β-carotene is equivalent to 1 µg RE. Other provitamin A carotenoids yield a maximum of half the retinol of β-carotene, and 12 µg of these compounds = 1 µg RE. The Composition of Foods expresses food data for vitamin A as retinol, β-carotene

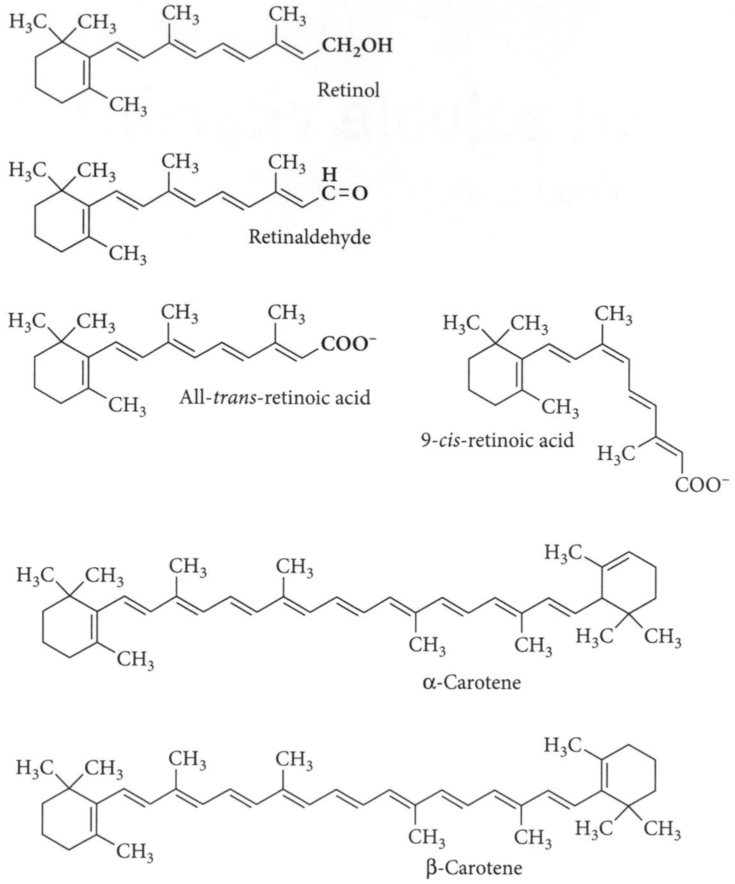

FIGURE 12.1 Vitamin A vitamers: retinoids and α- and β-carotene.

and RE (https://www.gov.uk/government/publications/composition-of-foods-integrated-dataset-cofid. The now obsolete international unit (IU) of vitamin A activity was based on a biological assay of the ability of the test compound to support growth in animals; 1 IU = 10.47 nmol of retinol = 0.3 μg free retinol or 0.344 μg retinyl acetate. Thus, 1 IU of vitamin A activity = 1.8 μg β-carotene or 3.6 μg of other provitamin A carotenoids. This is different in North America. In 2001, the Dietary Reference Values report introduced the term *retinol activity equivalent* (RAE) to take account of the incomplete absorption and metabolism of carotenoids; 1 RAE = 1 μg all-*trans*-retinol, 12 μg β-carotene, 24 μg α-carotene or β-cryptoxanthin. On this basis, 1 IU of vitamin A activity = 3.6 μg β-carotene or 7.2 μg of other provitamin A carotenoids.

Absorption and transport of vitamin A and carotenoids

Dietary retinyl esters are hydrolysed by pancreatic triacylglycerol lipase and intestinal brush-border phospholipase ensuring that 70–90 per cent of dietary retinol is absorbed. Uptake into enterocytes is by facilitated diffusion from lipid micelles, followed by binding to cellular retinol binding protein (RBP), then esterification to retinyl palmitate. Retinyl esters enter the lymphatic circulation and then the bloodstream in chylomicrons, together with dietary lipid and carotenoids.

Carotenoids are absorbed dissolved in lipid micelles. The absorption of carotene is between 5 and 60 per cent, depending on whether the food is cooked or raw and the amount of fat in the meal. As shown in Figure 12.2, β-carotene and other provitamin A carotenoids undergo oxidative cleavage to retinaldehyde in the intestinal mucosa. Retinaldehyde is reduced to retinol, then esterified and secreted in chylomicrons. A significant amount of carotene enters the circulation in chylomicrons without being oxidized to retinaldehyde and is secreted in very low-density lipoproteins (VLDL).

Central oxidative cleavage of β-carotene, as shown in Figure 12.2, gives rise to two molecules of retinaldehyde, which can either be reduced to retinol or oxidized to retinoic acid. However, the biological activity of β-carotene on a molar basis is considerably lower than that of retinol, not two-fold higher as might be expected. This may be

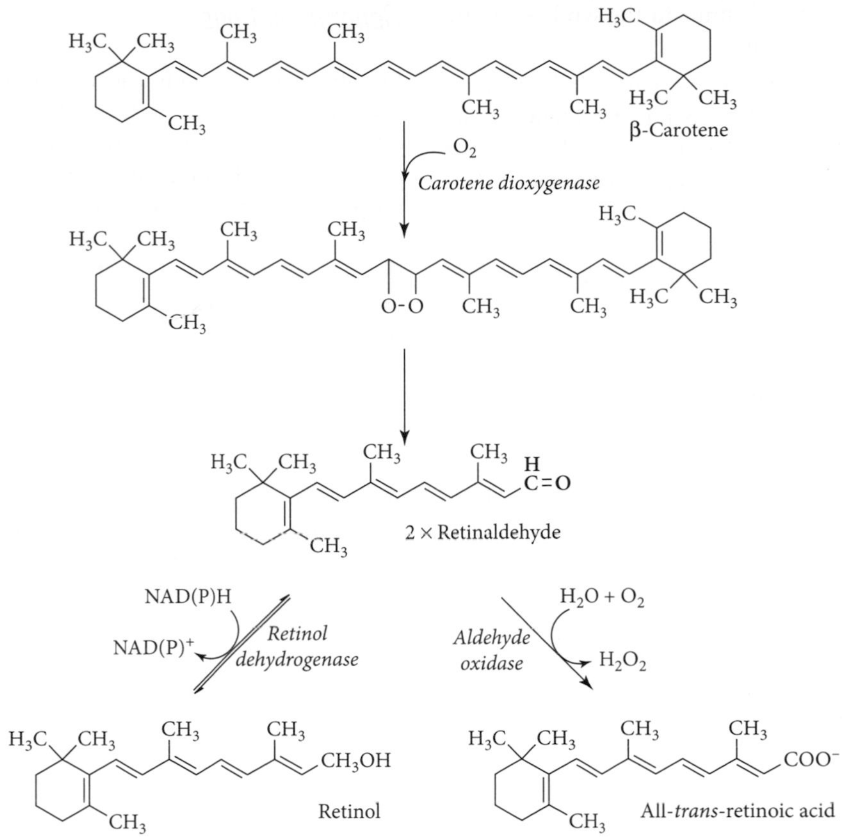

FIGURE 12.2 The reaction of carotene dioxygenase.

due to its limited absorption from the intestinal lumen; limited activity of carotene dioxygenase or excentric (asymmetric) cleavage. Excentric cleavage of β-carotene leads to formation of longer chain apo-carotenals (long-chain analogues of retinaldehyde), which can be oxidized to the corresponding acids, then undergo β-oxidation to yield all-*trans*-retinoic acid. However, as shown in Figure 12.2, the oxidation of retinaldehyde to all-*trans*-retinoic acid is irreversible, so while the products of excentric cleavage of β-carotene can yield retinoic acid, they cannot yield retinaldehyde, and therefore cannot support the physiological functions that require retinaldehyde.

Tissues can take up retinyl esters from chylomicrons, but most are left in the chylomicron remnants that are cleared by the liver. In the liver, retinyl esters are hydrolysed and free retinol is transferred to the rough endoplasmic reticulum, where it binds to apo-RBP (i.e., RBP without retinol). Holo-RBP (i.e., with retinol) is secreted as a 1:1 complex with the thyroid hormone binding protein, transthyretin, which protects it (as a small molecule) from kidney filtration and urinary loss. Impaired synthesis of RBP in protein-energy malnutrition can result in failure of retinol release from the liver, and functional vitamin A deficiency, even if liver reserves are adequate.

Binding to RBP serves to maintain vitamin A in aqueous solution, protect it against oxidation, and deliver it to target tissues. Moderate renal damage, or the increased permeability of the glomerulus in infection, may result in considerable loss of vitamin A. Cell surface receptors in target tissues take up retinol from the RBP–transthyretin complex and the apo-RBP is catabolised in the kidney.

Retinoic acid is the main metabolite of retinol and is a ligand for nuclear receptors involved in modulation of gene expression, rather than a catabolic product. It is formed in the liver and other tissues and transported bound to serum albumin. Retinoic acid cannot be reduced back to retinaldehyde or retinol. The main excretory product of both retinol and retinoic acid is retinoyl glucuronide, which is secreted in the bile. At high levels of intake, the capacity to metabolize retinol is saturated; excess retinol is toxic.

Metabolic functions of vitamin A

Vitamin A has four metabolic roles:

- as the prosthetic group of the visual pigments;
- as a nuclear modulator of gene expression;

- as a carrier of mannosyl units in the synthesis of hy-drophobic glycoproteins;
- in the retinoylation of proteins, particularly the regu-latory subunits of cAMP-dependent protein kinases, suggesting a role for retinoic acid in modulating hor-mones and neurotransmitters acting at the cell-surface.

Retinol and retinaldehyde in the visual cycle

The light-sensitive pigment in the photoreceptor cells of the retina is retinaldehyde bound to the protein opsin; excitation by a single photon results in the propagation of a nerve impulse. The pigment epithelium of the ret-ina isomerises all-*trans*-retinol to 11-*cis*-retinol, which may either be stored as esters or oxidized to 11-*cis*-retinaldehyde and transported to the photoreceptor cells bound to the inter-photoreceptor retinoid binding protein.

As shown in Figure 12.3, 11-*cis*-retinaldehyde binds to a lysine residue in opsin, forming rhodopsin (visual purple). Opsins are cell-type specific; they serve to shift the absorption of 11-*cis*-retinaldehyde from the ultra-violet into the visible range—either a relatively broad spectrum of sensitivity for vision in dim light (in the rods, with an absorbance peak at 500 nm) or more de-fined spectral peaks for differentiation of colours in stronger light (in the cones), with absorption maxima at 425 (blue), 530 (green), or 560 nm (red), depending on the cell type.

The absorption of light by rhodopsin results in a change in the configuration of the retinaldehyde from the 11-*cis* to the all-*trans* isomer, together with a con-formational change in opsin. This process, known as bleaching, as it causes the loss of the purple colour of rhodopsin, results in both the release of retinaldehyde from the protein and the initiation of a nerve impulse. Under conditions of low light intensity, the all-*trans*-retinaldehyde released from rhodopsin is reduced to all-*trans*-retinol, which is then transported to the pigment epithelium bound to the inter-photoreceptor retinoid binding protein. Under conditions of high light intensity, all-*trans*-retinaldehyde undergoes photo-isomerization to 11-*cis*-retinaldehyde and reduction to 11-*cis*-retinol in the photoreceptor cell. The rate-limiting step in ini-tiation of the visual cycle is the regeneration of 11-*cis*-retinaldehyde. Eyes contain only 0.01 per cent of the body's vitamin A, but they are so sensitive to vitamin A status that the speed of adjustment to dim light is di-rectly related to the amount of vitamin A available to regenerate rhodopsin. In vitamin A deficiency, when there is little stored 11-*cis*-retinaldehyde in the pigment epithelium, both the time taken to adapt to darkness and the ability to see in poor light are impaired; one vitamin A injection can relieve night blindness.

Genomic actions

Retinoic acid has both a general role in growth and a specific morphogenic role in development and tissue differentiation. These functions are the result of nucle-ar actions, modulating gene expression by activation of nuclear receptor proteins. Both deficiency and excess of retinoic acid cause severe developmental abnormali-ties. There are two families of nuclear retinoid receptors: the retinoic acid receptors (RAR), which bind all-*trans*-retinoic acid, and the retinoid X receptors (RXR), so called because their physiological ligand was unknown when they were first discovered. It is now known to be 9-*cis*-retinoic acid, which also binds to, and activates, the retinoic acid receptors. RXR forms active homodimers, and also forms heterodimers with the vitamin D recep-tor, the thyroid hormone receptor, and the peroxisome proliferation activated receptor (PPAR), involved in the expression of genes regulating lipid and carbohydrate metabolism, including enhancement of insulin signal-ling. In the presence of all-*trans*- or 9-*cis*-retinoic acid, the receptor heterodimers are transcriptional activators. However, the heterodimers will also bind to DNA in the absence of retinoic acid, in which case they act as repres-sors of gene expression.

Assessment of vitamin A status

Circulating retinol concentrations in the normal range are not related to habitual vitamin A intake and are not responsive to supplementation, as the relationship be-tween serum retinol and total body retinol content or liver retinol concentration is not linear. Vitamin A status can be measured using serum concentrations of retinol. Because RBP is released from liver only as the holo-protein, and apo-RBP is cleared from the circulation rap-idly after tissue uptake of retinol, the fasting plasma con-centration of retinol remains constant over a wide range of intakes. Serum retinol concentrations reflect liver reti-nol stores only when they are severely depleted (<20 µg retinol/g liver (<0.07 µmol/g)) or very high (>300 µg/g liver (1.05 µmol/g)). A serum concentration <0.7 µmol/L is considered deficient, while a serum concentration of >1.05 µmol/L is regarded as normal. Visual signs of vitamin A deficiency (night blindness) tend to appear at serum retinol concentrations <0.35 µmol/L. WHO criteria are that vitamin A deficiency is considered a pub-lic health problem when more than 20 per cent of the children aged 6–71 months have serum retinol levels <0.7 µmol/L.

The most sensitive assessment of vitamin A status is the relative dose response test (RDR)—a test of the abil-ity of a dose of vitamin A to raise the plasma concen-tration of retinol several hours later, after chylomicrons

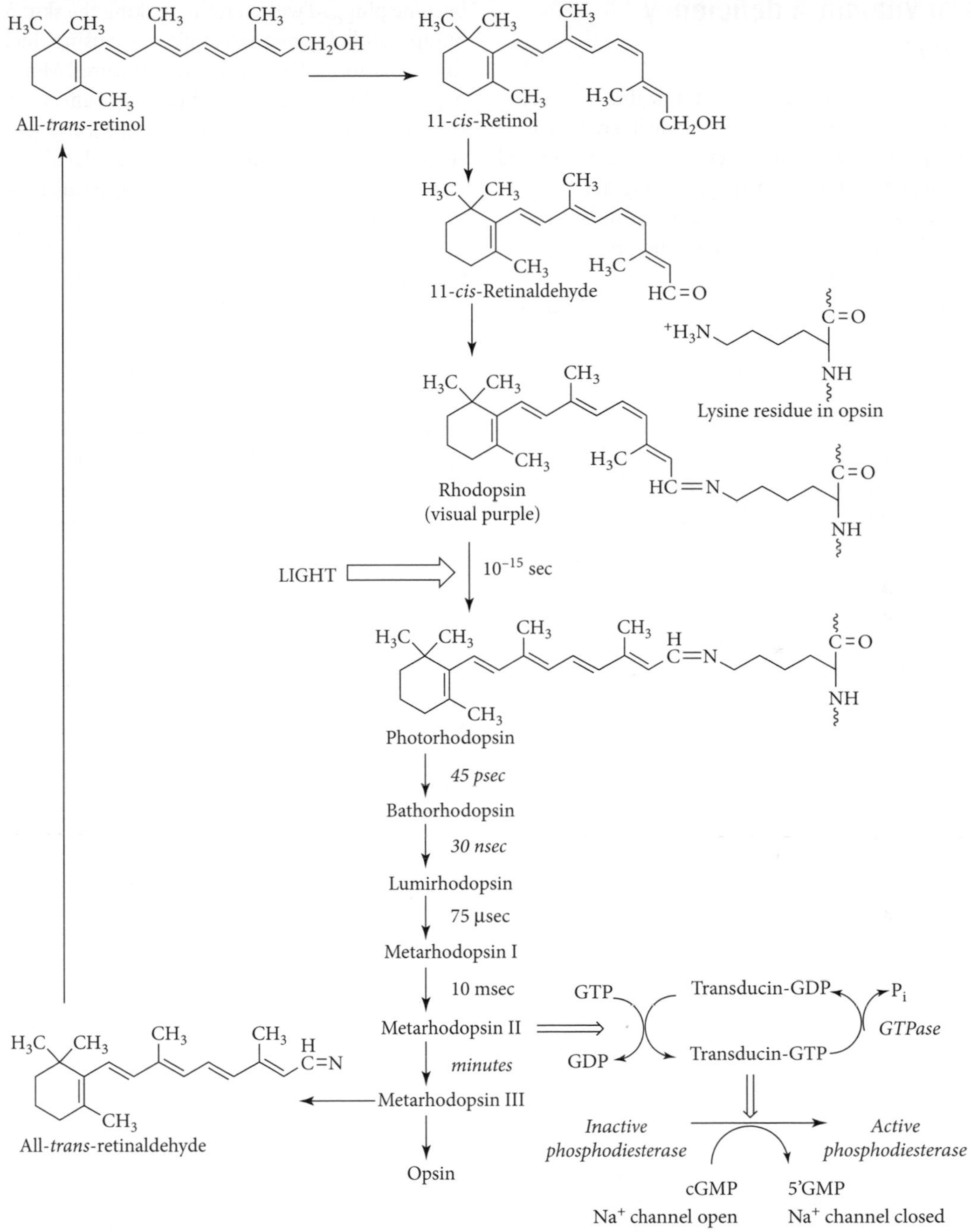

FIGURE 12.3 The role of vitamin A in the visual cycle.

have been cleared from the circulation. Marginal deficiency, which is associated with depletion of hepatic stores of retinyl ester, is identified by a rise of >15 per cent in the concentration of serum retinol after oral administration of a small dose of vitamin A; this is referred to as a positive RDR. An RDR >20 per cent indicates depletion of liver reserves of retinol to under 0.07 μmol/g.

Interpretation of plasma concentrations of retinol is confounded by the fact that both RBP and transthyretin are negative acute phase proteins, and their synthesis falls in response to infection. Similarly, both protein-energy malnutrition and zinc deficiency result in a low plasma concentration of retinol, despite possibly adequate liver reserves, as a result of impaired synthesis of RBP.

Effects of vitamin A deficiency and excess

Vitamin A deficiency tends to cluster within countries. Common features of areas of endemic deficiency include poverty, a high incidence of infectious diseases, limited infrastructure, and food insecurity, resulting in low availability and access to vitamin A-containing foods. In such settings, a vicious cycle often exists of vitamin A deficiency leading to increased susceptibility to and severity of infection, which, in turn, can reduce intake and accelerate body losses of vitamin A.

Vitamin A deficiency is the single most common preventable cause of blindness and is largely responsible for 0.5 million new cases of blindness annually on a global basis. Impaired dark adaptation, followed by loss of colour vision, especially reduced sensitivity to green light, can progress in more prolonged and severe deficiency leading to conjunctival xerosis—squamous metaplasia and keratinization of the epithelial cells of the conjunctiva with loss of goblet cells in the conjunctival mucosa, dryness, wrinkling, and thickening of the cornea (xerophthalmia). As the deficiency progresses, there is keratinization of the cornea, which is still reversible, although there may be residual corneal scarring. In advanced xerosis, yellow-grey foamy patches of keratinized cells and bacteria (Bitot's spots) may accumulate on the surface of the conjunctiva. Finally, there is ulceration of the cornea due to increased proteolytic action, causing irreversible blindness (see Figure 12.4).

As retinoic acid is involved in cell differentiation, other epithelial cells are also affected by vitamin A deficiency. Atrophy of the respiratory epithelium leads to an increased risk of infection. In the intestine, there is increased permeability to disaccharides, and a reduction in the number of goblet cells and mucus secretion. Differentiation of keratinocytes (immature epithelial cells) towards the production of mature epidermal cells, causes skin packing, making it hard and scaly. Hyperkeratosis (goose flesh) is an early symptom of vitamin A deficiency; hair follicles become plugged with keratin, making the skin rough and bumpy, and clogging sweat glands causing thickening of the palms and soles and affecting flexures. More seriously, hyperkeratosis causes loss of taste and affects the urinary tract, causing infection. Female genitals and seminal vesicles of testes can be affected, causing infertility.

Vitamin A has specific effects on immune cells; antigen-presenting cells (APCs) such as macrophages and dendritic cells can convert vitamin A to retinoic acid, which regulates differentiation of T-cells, promoting immune homeostasis, migration, and antigen-presenting capacity in response to infection. The increased susceptibility to infection and impairment of immune responses in vitamin A deficiency cause significant childhood mortality. An infection such as measles commonly triggers the development of xerophthalmia in children whose vitamin A status is marginal, because of loss of vitamin A as a result of infection. In addition to functional deficiency as a result of impaired synthesis of RBP and transthyretin in response to infection, there may be considerable urinary loss of vitamin A due to increased renal epithelial permeability and proteinuria.

In adults, excessive alcohol consumption or chronic use of barbiturates reduces liver reserves of vitamin A, both as a result of liver damage and also by induction of cytochrome P450 enzymes that oxidize retinol to retinoic acid. Vitamin A deficiency is also associated with reduced incorporation of iron into haemoglobin, causing functional iron deficiency anaemia, with increased deposition of iron in liver and spleen. Conversely, iron deficiency leads to reduced plasma retinol concentrations and increased liver stores of vitamin A that are not adequately mobilized.

Vitamin A supplementation and mortality

Since 2011, the WHO has recommended high-dose vitamin A supplementation (VAS) every 4–6 months for children aged between 6 and 60 months in at-risk countries. This policy is based on RCTs in the late 1980s and

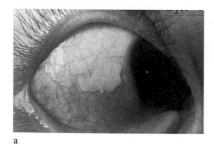
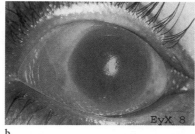
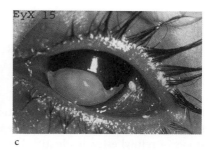

a   b   c

FIGURE 12.4A-C Advancing stages of Xerophthalmia.
(a) Xerophthalmia—Bitot's spot, an early sign of vitamin A deficiency. (b) Xerosis—more advanced xerophthalmia. The conjunctiva are dry, thick, and folded (see lateral fornix); the cornea is rough, pitted, keratinized. (c) Keratomalacia—late stages of vitamin A deficiency. The lower part of the cornea is necrotic, soft, white, and opaque.
Source: A-C Image © Teaching Aids at Low Cost.

early 1990s, six of which showed a marked effect on overall mortality, mostly due to reductions in incidence and death from diarrhoea, with additional benefits for reduction of blindness (Imdad et al. 2017). Vitamin A supplementation of infants below the age of six months is not yet recommended. Focus is on maternal supplementation for prevention of deficiency among newborn and young infants.

Toxicity of vitamin A

Vitamin A is both acutely and chronically toxic. Acutely, large doses of vitamin A (in excess of 300 mg in a single dose to adults) cause nausea, vomiting, and headache which disappear within a few days. After a very large dose there may also be itching and exfoliation of the skin, and extremely high doses can prove fatal. Single doses of 60 mg of retinol are given to children in developing countries as a prophylactic against vitamin A deficiency—an amount adequate to meet the child's needs for 4–6 months. About 1 per cent of children so treated show transient signs of toxicity, but this is considered to be acceptable in view of the considerable benefit of preventing xerophthalmia.

The chronic toxicity of vitamin A is a more general cause for concern; prolonged and regular intake of more than about 7500–9000 μg/day by adults (and significantly less for children) causes signs and symptoms of toxicity affecting the skin, central nervous system, liver, and bones. There is some evidence that lower habitual intakes, in excess of ~1500 μg/day, are associated with increased risk of osteoporosis and bone fracture. This may be the result of formation of RXR homodimers in the presence of large amounts of 9-cis-retinoic acid, leaving insufficient RXR to dimerize with, and activate, the vitamin D receptor, discussed later.

Vitamin A analogues (isotretinoin or 13-cis-retinoic acid (Accutane, Roaccutane) and etretinate) are highly teratogenic in humans. Distinctive features include malformations of the ear; heart defects; brain defects, including hydrocephalus, microcephaly, and cerebellar abnormalities; facial abnormalities (small mandible, cleft palate), and abnormalities of the thymus gland. Plasma concentrations of retinol associated with teratogenic effects are unlikely to be reached with intakes below 7500 μg/day. The lowest reported daily supplement associated with liver cirrhosis is 7500 μg/day taken for six years. As a cautionary measure, pregnant women are advised not to consume more than 3000 μg/day, which is the tolerable upper intake of preformed vitamin A for all adults. For infants and children aged <3 y, the upper level is 800 μg/d.

Carotenoids do not cause hypervitaminosis A, because of the limited oxidation to retinol. Accumulation of even abnormally large amounts of carotene seems to have no short-term adverse effects, although plasma, body fat, and skin can have a strong orange-yellow colour (hypercarotinaemia) following prolonged high intakes.

Carotenoids are antioxidants, trapping singlet oxygen generated by photochemical reactions or lipid peroxidation in membranes and there has been much speculation of a role for carotenoids in carcinogenesis—that deficiency may be a risk factor or that increased intake may be protective against cancer. Epidemiological and case-control studies show a negative association between β-carotene intake and a number of cancers. In an intervention trial in China, supplements of β-carotene, vitamin E, and selenium to a malnourished population led to a reduction in mortality from a variety of cancers. However, intervention trials in Finland (in which heavy smokers were given supplements of β-carotene and/or vitamin E) and the USA (involving smokers and people who had been exposed occupationally to asbestos dust, who received β-carotene and retinyl palmitate), both reported increased lung cancer mortality among those receiving carotene supplements. This is because carotene is an antioxidant under conditions of low oxygen concentration, as occurs in most tissues, where oxygen is bound to haemoglobin, myoglobin, and other proteins. However, under conditions of high oxygen concentration, as occurs in the lungs, carotene becomes a pro-oxidant, especially at high concentrations.

The basis for setting vitamin A requirements

The European Food Safety Authority (EFSA) revised its dietary recommendations for vitamin A (EFSA 2015a) using a factorial approach, considering the tight regulation of liver retinol stores, and the lack of a dose-response of serum concentrations to supplementation. Certain assumptions were made; a total body/liver retinol store ratio of 1.25 (i.e., 80 per cent of retinol body stores are in the liver); a liver/body weight ratio of 2.4 per cent; a fractional catabolic rate of retinol of 0.7 per cent per day of total body stores; an efficiency of storage in the whole body of ingested retinol of 50 per cent, and reference body weights for women and men in the EU of 58.5 and 68.1 kg, respectively (pooled from a set of relatively old data).

On this basis, targeting normal serum retinol concentrations, average requirements (AR) of 570 μg RE/day for men and 490 μg RE/day for women were derived. Reference Intakes of 750 μg RE/day for men and 650 μg RE/day for women were set, and scaled down for infants and children based on body weight (from 250 to 750 μg RE/day). Pregnant women are recommended to consume 700 μg RE/day, to allow foetal accretion of ~3600 μg retinol over the course of pregnancy and maternal tissue growth. An intake of 1300 µg RE/day was proposed for lactating women, allowing for an average amount of retinol secreted in breast milk of 424µg/day. In the UK, reference nutrient intakes (RNI) for vitamin A have been unchanged since 2001, at 700 µg RE/day for men and 600 µg RE/day for women.

KEY POINTS

- Vitamin A deficiency is a major public health problem worldwide, and the commonest preventable cause of blindness in regions at risk of vitamin A deficiency, particularly in Africa and Southeast Asia.

- Vitamin A is consumed as preformed retinol, or it may be synthesized from dietary carotenoids, which also act as antioxidants.

- The main functions of vitamin A are as the prosthetic group of the visual pigments, and as the ligand for a variety of nuclear receptors that regulate gene expression and tissue differentiation.

- Mild vitamin A deficiency severely impairs immune responses and leads to infection and diarrheal disease in children, a major cause of child mortality.

- Vitamin A is both acutely and chronically toxic in excess, and is also teratogenic.

12.2 Vitamin D

Vitamin D provokes intense interest and debate among the scientific and lay communities alike. Year on year, PubMed-listed journal publications with vitamin D in the title or abstract have increased, from ~1200 in 2000 to >5500 in 2020, while publications for vitamins A and E have remained steady at ~1200–1500 per annum during the same period. Vitamin D has been a nutrient of concern for many years due to its limited supply in the food system.

Vitamin D_3 (cholecalciferol) is obtained from the diet and supplements and can also be synthesized in the skin on exposure to sunshine. Derived from irradiation of ergosterol in plants, vitamin D_2 (ergocalciferol) is also present in the diet. The vitamin D metabolite, 25-hydroxyvitamin D (25(OH)D), which circulates in blood, and acts as the most commonly used biomarker of vitamin D exposure, is a lesser known dietary constituent. The chief challenge is that for those with few opportunities to receive sun exposure, or a reduced ability to synthesize vitamin D in the skin, diet is a critical source of vitamin D. However, naturally occurring sources of vitamin D, such as oily fish, are consumed irregularly and in low quantities, which means that habitual vitamin D intakes are usually low and the prevalence of deficiency is high.

Vitamin D is a prohormone; vitamin D metabolism generates calcitriol (1,25-dihydroxyvitamin D $(1,25(OH)_2D)$), which is involved in maintaining calcium homeostasis (i.e., keeping serum calcium within its physiologically correct range of 2.1–2.6 mmol/L), the regulation of cell proliferation and differentiation, modulation of the immune system and in the secretion of insulin, and thyroid and parathyroid hormones. This makes vitamin D interesting to endocrinologists and scientists of many disciplines, including nutrition.

Sources of vitamin D

Skin synthesis on exposure to sunlight

Vitamin D_3 synthesis begins when ultraviolet (UV) radiation of wavelengths between 280–315 nmol activates 7-dehydrocholesterol (7-DHC), an intermediate in the synthesis of cholesterol that accumulates in skin, but not other tissues (Webb 2006). As shown in Figure 12.5, cholecalciferol production is not an enzymatic process. After the initial photoisomerization step, pre-vitamin D_3 undergoes heat isomerization over several hours in the skin to form D_3. Pre-vitamin D_3 can also be isomerized into the inert isomers, lumisterol and tachysterol, or back to 7-DHC (Webb 2006). There is a limit to the amount of previtamin D_3 that will form in the skin. Prolonged sun exposure will only increase the inert isomers once this limit has been reached, which is well before a person has reached their minimal erythemal dose (MED), the amount of sun exposure that causes visible skin reddening. As vitamin D synthesis has already occurred prior to the development of sunburn, prolonged exposure produces no benefit. For this reason, brief, regular periods of sun exposure to larger amounts of skin are recommended for vitamin D synthesis without risk of sun damage.

Many factors interfere with skin synthesis of vitamin D. The solar zenith angle, which is the angle between the sun's rays and the vertical, is controlled by the season, latitude, and time of day. Therefore, when the sun is high in the sky (and your shadow is short), solar radiation has a short path through the atmosphere and the energy in a beam is concentrated. When the sun is low (and your shadow is long), the solar zenith angle is large and the radiation weakens as it travels through the atmosphere (Webb 2006). Above latitudes of about 40°, no cutaneous synthesis is possible for many months during winter, as illustrated in Figure 12.6. Other environmental factors, such as prevailing weather, cloud cover, and air pollution

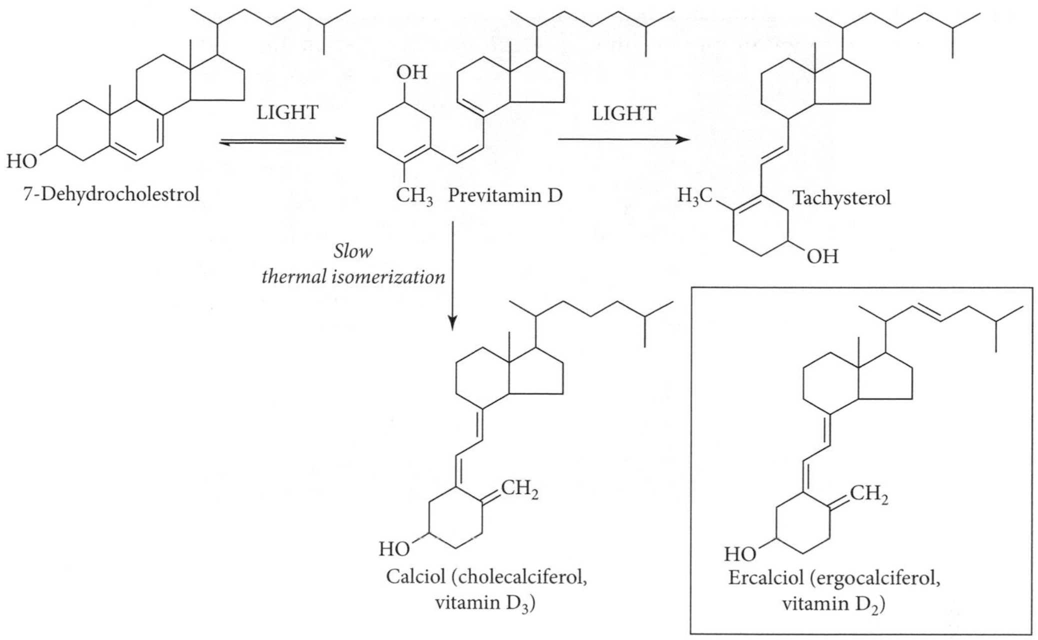

FIGURE 12.5 Photosynthesis of vitamin D in the skin. The box shows the structure of vitamin D_2.

influence availability of UVB of sufficient strength to make vitamin D. Personal factors relating to the efficiency of skin synthesis are skin pigmentation, because melanin absorbs UV radiation, preventing D_3 synthesis, and age, which decreases the amount of 7-DHC available. Other personal factors which limit sun exposure are clothing, sunscreen use, sun exposure preferences and modern life, urbanization, lack of recreational spaces, and long working hours.

Dietary intake

With restricted access to UVB, large sectors of the global population rely on vitamin D in the food supply to maintain adequate vitamin D status. Natural sources of cholecalciferol and 25(OH)D include oily fish, meat, and eggs and ergocalciferol is readily produced in mushrooms exposed to UVB. Liquid milk only contains appreciable amounts of vitamin D if it is fortified. Rich vitamin D food sources are consumed infrequently (e.g., oily fish), therefore those that have a low vitamin D content but are consumed frequently, such as beef and eggs, make an important contribution to vitamin D intakes (Kiely & Black 2012). Habitual vitamin D intakes frequently fall well below dietary recommendations and typical intakes in the EU are around 3–7.5 µg (120–300 IU, as 1 µg = 40 IU), depending on the country, with slightly higher values in the United States and Canada, particularly among supplement consumers. Low dietary intakes are correlated with vitamin D status in high latitude countries during wintertime, and there

is a recognition that diet is important for maintaining adequate vitamin D status. Challenges in assessing vitamin D intakes include the dietary assessment method selected, which needs to include sufficient numbers of days to capture consumption of rich sources, collection of supplement data and the completeness of the food composition data available, which have variable coverage of foods and vitamers. The UK Food Composition Tables includes data on vitamin D_3, D_2, and 25(OH)D, and are continually updated and are available at: https://www.gov.uk/government/publications/composition-of-foods-integrated-dataset-cofid.

Depending on legislation, some foods can be fortified with vitamin D_3 or D_2, including milk, spreads, yoghurt, cheese, juices, breads, and breakfast cereal. Bio-fortification or bio-addition of eggs, mushrooms, milk, and meats is also possible by fortifying the diet of laying hens or livestock with vitamin D or 25(OH)D or by UVB-irradiation of mushrooms or milk. There is compelling evidence that food fortification increases vitamin D intakes and serum 25(OH)D among children and adults with no adverse effects in modern times. While vitamin D_2 is effective at preventing vitamin D deficiency, it is not as efficient at increasing serum 25(OH)D concentrations as vitamin D_3. Some countries, notably Finland, have demonstrated the effectiveness of food fortification with vitamin D as a public health strategy; fortification of fluid milk and fat spreads increased serum 25(OH)D concentrations considerably between 2003 and 2011, effectively preventing vitamin D deficiency.

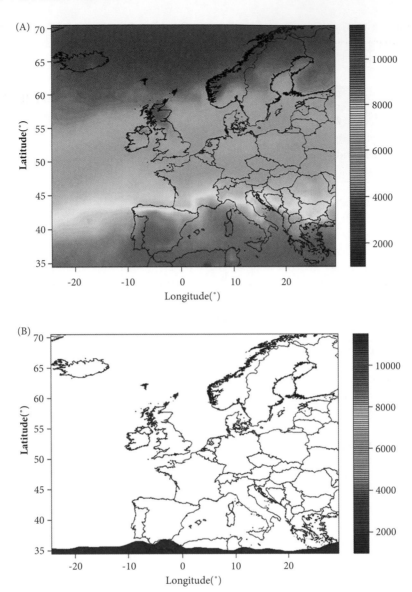

FIGURE 12.6 Mean monthly modelled UVB doses effective for pre-vitamin D$_3$ synthesis (Jm−2) across Europe for June (A) and December (B), based on average of data from years 2003–2012. Scale begins 1000 Jm−2.

Source: C M O'Neill, A Kazantzidis, M J Ryan et al. (2016) Seasonal changes in vitamin D-effective UVB availability in Europe and associations with population serum 25-Hydroxyvitamin D. *Nutrients* 30 Aug. **8**(9), 533.

Absorption, transport, and metabolism of vitamin D

Vitamin D synthesized in the skin enters the circulation bound to a vitamin D-binding protein (DBP). As a fat soluble vitamin in the gut, it is absorbed alongside fatty acids and phospholipids, transported through the lymph, and from there is incorporated into chylomicrons and taken up by the liver for further metabolism. Disorders that interfere with absorption of vitamin D, such as cystic fibrosis, inflammatory bowel disease, or gastrointestinal surgery can interfere with vitamin D absorption and contribute to deficiency.

The three main steps in vitamin D metabolism are 25-hydroxylation, 1α-hydroxylation, and 24-hydroxylation, performed by cytochrome P450 oxidases (CYPs) located either in the endoplasmic reticulum or the mitochondria (Bikle 2014). Hepatic 25-hydroxylase (CYP2R1) first converts vitamin D to 25-hydroxyvitamin D (25(OH)D). Circulating concentrations of 25(OH)D provide a biomarker of exposure indicating vitamin D status, with no distinction between vitamin D from the diet or skin synthesis. In the kidney primarily, 25(OH)D is converted by 1α-hydroxylase (CYP27B1) to its active form, calcitriol, or 1,25-dihydroxyvitamin

D (1,25(OH)$_2$D). One of the most important discoveries about the 1α-hydroxylase was that it is not located only in renal tissue, but is expressed in many cells of the body, including the epithelium of the skin, lungs, breast, intestine, and prostate; in cells of the immune system including T and B lymphocytes; in endocrine tissues including the thyroid and islet cells of the pancreas, and in osteoblasts and chondrocytes (Bikle 2014). This means that provided sufficient unbound (free) 25(OH)D is locally available, the machinery of cellular production of 1,25(OH)$_2$D is present, which has massive implications for non-skeletal effects of vitamin D. Further metabolism of 1,25(OH)$_2$D and 25(OH)D by 24-hydroxylase (CYP24A1) yields the degradation product calcitroic acid (Figure 12.7). The DBP, a member of the albumin family of binding proteins, transports vitamin D metabolites. Genetic differences in the structure of the DBP may influence vitamin D status.

Circulating 25(OH)D represents a reservoir in blood and there is some vitamin D in muscle. Although adipose tissue contains vitamin D, this is considered as sequestered, rather than stored, as it is not readily available. Obesity is associated with low vitamin D status and people with obesity have a lower dose-response both to vitamin D supplementation and to UV exposure compared with leaner individuals. Mendelian Randomization studies have shown that vitamin D deficiency is caused by obesity and not the other way around. Therefore, metabolic complications such as insulin resistance arising from adiposity are probably exacerbated by obesity-related vitamin D deficiency rather than caused by it.

Regulation of vitamin D metabolism

The classical physiological function of vitamin D is in the control of calcium homeostasis. Through the vitamin D receptor (VDR), present in most cells, 1,25(OH)2D enhances intestinal calcium absorption in the small intestine by stimulating the expression of the epithelial calcium channel and calbindin, the calcium binding protein, which regulates active transport through the cell. In the kidney, 1,25(OH)$_2$D increases tubular reabsorption of calcium, preventing calcium excretion. In osteoblasts, 1,25(OH)$_2$D increases the expression of the receptor activator of NFκB ligand (RANKL), which stimulates binding and maturation of preosteoclasts, and release of calcium and phosphorus from bone. In this way, serum calcium and phosphorous levels are maintained. Parathyroid hormone (PTH), secreted by chief cells of the parathyroid gland, stimulates production of 1,25(OH)$_2$D by increasing 1α-hydroxylase activity. Constant serum calcium concentrations are maintained through negative feedback; by controlling PTH and 24-hydroxylase expression, 1,25(OH)$_2$D effectively regulates its own production and

activity. In addition to serum calcium, phosphorus, fibroblast growth factor (FGF-23), and other factors can influence production of 1,25(OH)$_2$D.

Physiological functions of vitamin D

Genomic effects of 1,25(OH)$_2$D are through the VDR, a ligand-activated transcription factor that functions to control gene expression, which is present in most tissues. A large number of genes (~3 per cent of the human genome) are under the direct or indirect control of calcitriol, suggesting a broad spectrum of activities (Bouillon et al. 2019). Polymorphisms of the VDR (e.g., Taq1, Apa1, Bsm1) affecting the function of vitamin D are linked with musculoskeletal disorders, renal disease, and some immune conditions. The VDR recognizes a vitamin D response element, which enables the binding of a heterodimer comprised of a VDR molecule and a retinoid X receptor (RXR) molecule. This is the mechanism by which 1,25(OH)$_2$D regulates mineral metabolism in the intestinal and renal epithelium and in osteoblasts. Dimers formed with the unoccupied RXR receptor decrease gene expression; because of this, deficiency of vitamin A leads to impaired vitamin D function. Excess vitamin A can also impair vitamin D function as large amounts of 9-*cis*-retinoic acid leads to excess RXR homodimers, leaving insufficient RXR to form heterodimers with the VDR. Calcitriol also influences insulin secretion, the synthesis and secretion of parathyroid and thyroid hormones, and has a role in the regulation of cell proliferation and differentiation, the cell cycle, and apoptosis.

In addition to its nuclear actions, 1,25(OH)$_2$D has non-genomic functions, namely by regulating ion channel activity (especially chloride and calcium channels) (Figure 12.7). For example, in intestinal mucosal cells it recruits membrane calcium transport proteins from intracellular vesicles to the cell surface, resulting in a rapid increase in calcium absorption, before calbindin has been induced. It can also act via cell-surface receptors in many cells, leading to activation of protein kinase C and mitogen-activated protein kinases (MAP kinases), to inhibit cell proliferation and induce differentiation.

Much attention in the vitamin D field has shifted towards understanding its role in non-skeletal health outcomes, particularly in relation to immune function and non-communicable diseases as it impacts both innate and adaptive immune function. By the induction of human antimicrobial peptide genes (e.g., cathelicidin), *in vitro* studies have shown that 1,25(OH)$_2$D can profoundly boost the innate immune system to combat pathogenic infections. Therapeutic possibilities are also presented by the actions of 1,25(OH)$_2$D in T-cell proliferation and cytokine production and in the regulation of dentritic cells, influencing immune tolerance, with implications

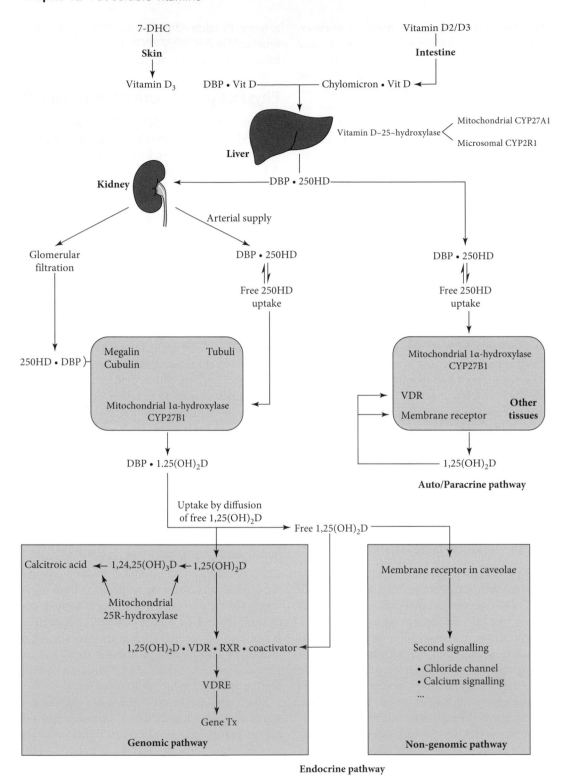

FIGURE 12.7 Metabolism and action of vitamin D and its metabolites, with special focus on renal and extrarenal production of 1,25(OH)2D and the genomic or nongenomic pathways of vitamin D action.

Source: R Bouillon, C Marcocci, G Carmeliet et al. (2019) Skeletal and extraskeletal actions of vitamin D: Current evidence and outstanding questions. *Endocr Rev.* **40**(4), 1109–1151.

for autoimmune disorders, such as multiple sclerosis, inflammatory diseases, and allergy.

Because of the role of vitamin D in cell differentiation, maternal vitamin D status in pregnancy is an important factor in foetal development. During early pregnancy, vitamin D metabolism changes; DBP concentrations increase and increases in 1-α-hydroxylase activity in the maternal kidney, placenta, and decidua lead to circulating 1,25(OH)$_2$D concentrations that are 2–3 times higher than usual by 12 weeks of gestation, with relatively stable serum 25(OH)D. These metabolic shifts appear to be unrelated to calcium metabolism. There are compelling data to support the possibility that the immunomodulatory effect of 1,25(OH)$_2$D is necessary for successful pregnancy implantation by adapting the immune response, creating an anti-inflammatory milieu and enabling foetal growth. Infants are born with circulating 25(OH)D concentrations reflective of maternal status, at about 60–80 per cent of maternal samples collected at delivery.

Vitamin D deficiency and health outcomes

Defined on the basis of circulating 25(OH)D concentrations, vitamin D deficiency is variably classified between <25 and <50 nmol/L (or 12–20 ng/mL), depending on the author or regulatory authority referenced. In the UK, and in this chapter, vitamin D deficiency is denoted by a 25(OH)D <25–30 nmol/L and low vitamin D status as <50 nmol/L. Estimates of the prevalence of vitamin D deficiency vary from 6–7 per cent in the United States and Canada to 13 per cent in Europe, 18 per cent in Africa, and 22 per cent in the UK. Persons with dark skin who are resident at high latitude are at much higher risk of vitamin D deficiency than their pale-skinned counterparts, with estimates of between 36 and 60 per cent in the UK. Pregnant women and newborn infants are at particularly high risk of vitamin D deficiency.

Rickets and osteomalacia

The growing years, when children and adolescents can become vulnerable to nutritional deficiency, are marked by rapid skeletal growth and development. Nutritional rickets results from a failure to mineralize newly formed bone. Epiphyseal cartilage continues to grow, but is not replaced by bone matrix and mineral. Therefore, nutritional rickets is a disorder of defective chondrocyte differentiation and mineralization of the growth plate and defective osteoid mineralization, caused by vitamin D deficiency and/or low calcium intake, leading to a prolonged state of hypocalcaemia (corrected serum total calcium <2.12 mmol/L) (Munns et al. 2016). The diagnosis of nutritional rickets is made on the basis of clinical

history, physical examination, and biochemical testing (PTH, serum calcium and phosphate, alkaline phosphatase, 25(OH)D), and is confirmed by radiographs. Osteomalacia is the defective remineralization of bone during normal bone turnover in adults, so that there is progressive bone loss, leading to pain, skeletal deformities, and muscle weakness (Figure 12.8). Morbidity associated with rickets and osteomalacia extend beyond bone pain and include hypocalcaemic seizures, failure to thrive, developmental delay, dental anomalies, and hypocalcaemic cardiomyopathy, which can be fatal, as documented in recent times in Britain.

Although disputed as the first description, as 'rickets' was listed in the Annual Bill of Mortality of the City of London in 1634, Francis Glisson published his treatise *De Rachitide* in 1650, which detailed the anatomical malformations and clinical symptoms of the condition, and noted that rickets afflicted the children of wealthy families more often than the poor (O'Riordan & Bijvoet 2014). This may have been due the prolonged use of wet nurses by the wealthy classes and was certainly due to inappropriate infant feeding practices. Reports of rickets as a leading cause of maternal mortality started to emerge, due to malformations of the pelvis (Figure 12.8). Increasing prevalence in the nineteenth century coincided with increasing urbanization and air pollution, and a lack of sunlight was identified as a major risk factor. In 1822, Sniadeki wrote that the direct action of the sun was 'the most efficient method for prevention and cure' and various authors reported on the effectiveness of fish liver oils in the treatment of rickets. Up to the early 1920s, ultraviolet light, sunlight, and cod liver oil were being used to treat rickets both in Europe and the United States. An 'accessory factor' proposed by Mellanby was later named

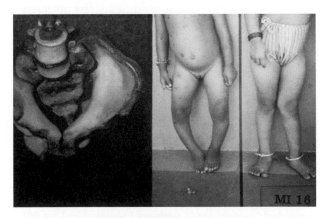

FIGURE 12.8 Nutritional rickets.
Both children have vitamin D deficiency (rickets). In one the knees are bent out (bowed), and in the other bent in (knock-kneed). The pelvis is deformed by rickets, with small pelvic outlet.
Image © Teaching Aids at Low Cost.

by McCollum as vitamin D. Hess and Steenbock discovered that UV irradiation of foods, such as milk and cereals, could increase their vitamin D content by activating DHC. Irradiated ergosterol was used for food fortification and the treatment of rickets, and by the 1930s fortification of milk and supplementation of infants was common practice, effectively eradicating rickets. However, quality control was inadequate, and excessive addition of vitamin D to milk resulted in hypercalcemia among young children in the UK, which effectively ceased the practice of vitamin D fortification.

This reversion and global migration from tropical to more temperate regions, has prompted the re-emergence of nutritional rickets as a significant, widespread, global health problem. It is entirely preventable by proper nutrition and adequate supplies of calcium and vitamin D during pregnancy and early life. As babies are born with circulating 25(OH)D concentrations that reflect maternal status, vitamin D deficiency during pregnancy increases the risk of nutritional rickets in infants. This is doubly important as the vitamin D content of breastmilk is low, so vitamin D status in new born infants is critical for the first several months. In infants and young children, a vitamin D intake of 10 μg/day from birth alongside provision of appropriate calcium-containing complementary foods at the correct time (around six months) prevents nutritional rickets (Munns et al. 2016).

Osteoporosis is the age-related loss of bone mineral and bone matrix, characterized by decreased bone density (mass/volume), leading to decreased mechanical strength, making the skeleton more likely to fracture. It affects up to 40 per cent of women and 12 per cent of men as they age, and is caused by genetic, personal, and lifestyle factors over the life-course. Osteoporosis carries enormous personal costs to individuals and to health services and there is a clear need to address nutritional risk factors, including vitamin D and calcium deficiencies, which lead to elevated PTH and demineralization of bone.

Although the presence of the VDR in muscle cells is still under debate, Vitamin D deficiency has adverse effects in both skeletal and cardiac muscle and has been linked with muscle weakness in both children and adults. Weakness of proximal muscles, indicated by delayed onset of walking in infants and difficulty climbing stairs in adolescents, is a clinical manifestation of vitamin D deficiency, which may coexist with skeletal features such as rickets. Muscle strength and function in the elderly is associated with an increased risk of falls, causing fractures. Randomized controlled trials (RCTs) have shown that vitamin D (especially in combination with calcium) decreases the incidence of hip fractures and other nonvertebral fractures, particularly in older frail adults.

Non-skeletal health outcomes

Many studies in different populations from around the world have described associations between low circulating levels of 25(OH)D and a large number of disorders, including cardiovascular disease (CVD), cancer, diabetes, obesity, infections, neurological and autoimmune diseases (Rejnmark et al. 2017). Problems with observational data, such as bias, confounding, and the possibility of reverse causation make it impossible to infer causation. A systematic review of 210 RCTs covering 54 earlier systematic reviews of non-skeletal effects of vitamin D supplementation reported that most trials produced null findings of vitamin D supplementation on cardiovascular disease (CVD), type 2 diabetes, weight loss, cancer, and depression. Three-quarters of studies on mortality as an outcome showed a benefit. As the vast majority of studies did not specify low vitamin D status as an inclusion criterion, the null hypothesis could not truly be rejected (Rejnmark et al. 2017).

The VITAL trial in the United States, which reported in 2019, was an RCT of vitamin D$_3$ (50 μg/day) with or without omega-3 fatty acids (1 g/day) or placebo for prevention of cancer and major CV events among almost 26,000 men > 50 y and women > 55 y. Following an average five-year follow-up period, vitamin D supplementation did not show benefit of reduced risk of disease. In New Zealand the VIDA study among >5000 participants aged between 50 and 84 years delivered a daily dose of around 75 μg/day of vitamin D$_3$ or placebo for more than three years. There were no effects on CVD, cancer, nonvertebral fractures, or falls but some benefits for bone mineral density, lung and arterial function, particularly if baseline 25(OH)D concentrations were low, were reported.

Links between vitamin D deficiency, lung function, and respiratory tract infections can be traced back to the time when patients suffering from tuberculosis (TB) were treated with sun exposure. In 2016, a systematic review of RCTs, with analysis of individual participant data, reported a modest protection of vitamin D supplementation, provided daily or weekly, against acute upper respiratory tract infections, particularly among subjects who were deficient at baseline. Since the start of 2020 and the COVID-19 pandemic, attention has focused on the links between vitamin D and immune resistance, particularly in relation to respiratory tract infections. At the time of writing (March 2021), four small vitamin D trials had been published among patients hospitalized with varying severity of COVID-19 symptoms. Studies on a role for vitamin D in prevention are underway. Currently, none of the data support taking vitamin D specifically to prevent infection with SARS-CoV-2. In the post-pandemic era, the global focus must be on prevention, including maintaining good nutritional status for immune function. This includes prevention of vitamin D deficiency.

Vitamin D excess and toxicity

Intoxication with vitamin D causes weakness, nausea, loss of appetite, headache, abdominal pains, cramp, and diarrhoea. More seriously, it causes hypercalcaemia (serum calcium of 2.75–4.5 mmol/L), which results in calcinosis (deposition of calcium in soft tissues including kidney, heart, lungs, and blood vessels), diffuse demineralization of bones and irreversible renal and cardiovascular toxicity. Hypercalcaemia can also lead to hypercalciuria, which may result in the precipitation of calcium phosphate in the renal tubules and the development of urinary calculi. The EFSA established a no observed adverse effect level (NOAEL) of 250 µg/day for 25(OH)D, and applying a generous safety margin, set a tolerable upper level (UL) of vitamin D intake of 100 µg/day for everyone over the age of ten years; 50 µg/day for 1–10 year olds; 35 µg/day for 6–12 months old and 25 µg/day for infants aged 1–6 months. These intake levels are impossible to achieve through diet only and relate primarily to supplementation. Long-term data on the safety of high doses of vitamin D are required to establish more secure ULs among adults. Recent adverse effects from high single-dose interventions, including increased fracture risk, mean that taking weekly or daily low-dose supplemental vitamin D is preferable to bolus doses.

Vitamin D requirements and dietary recommendations

Early estimates of vitamin D requirements were based on the amounts supplied in a spoonful of cod-liver oil, and required by house-bound elderly people to maintain serum concentrations of 25(OH)D at levels seen in younger people at the end of winter. For many years, no reference intakes were published for younger adults as it was assumed that sunlight exposure met nutritional requirements. RCTs carried out between 2008 and 2016 showed that vitamin D is required on an ongoing basis, because dermal vitamin D synthesis during the summer is insufficient to meet the requirements of people of all ages living at high latitudes all year round.

Setting nutritional recommendations for healthy populations is an iterative process based on the best available scientific evidence at that time. A transparent risk assessment process is the cornerstone of setting nutrient intake recommendations. The risk assessment specifies health outcomes linked with biomarkers of nutritional status and the levels of those biomarkers are titrated against the risk of adverse outcomes (e.g., serum concentration of 25(OH)D at which the risk of nutritional rickets escalates). Dietary intakes required to meet these biomarker thresholds are calculated using dose-response data from RCTs.

Vitamin D recommendations in North America (Institutes of Medicine 2011), Europe (EFSA 2016), and the UK (SACN 2016) are based on skeletal health outcomes, as evidence from RCTs in non-skeletal health outcomes was insufficient at the time. In the UK, a daily intake of 10 µg/day (8.5 µg in infants) was set to keep serum 25(OH)D >25 nmol/L in 97.5 per cent of the population. In the EU and North America, this value is 15–20 µg/day and the target serum 25(OH)D is 50 nmol/L. Some expert clinical guidelines lean towards higher intake recommendations of 25 µg/day, to ensure that patients meet the 50 nmol/L target. Vitamin D intake recommendations for pregnant and lactating women are the same as for non-pregnant adults. Infants should receive 8.5–10 µg/day from birth from supplements or formula.

Intakes of vitamin D are substantially below recommendations, and as vitamin D deficiency is widespread, it is timely to reflect on how governments can act to bridge the gap to prevent vitamin D deficiency. Properly regulated food fortification programmes Could ensure adequate vitamin D in the food supply for public health.

KEY POINTS

- Vitamin D_3 is synthesized in the skin on exposure to ultraviolet irradiation in sunshine and is also available in a few foods, such as oily fish, egg yolks, and meat. Vitamin D_2 in the diet is derived from irradiation of ergosterol in plants and fungi, e.g., mushrooms.

- People living at a latitude higher than about 40° require additional vitamin D from supplements or food fortification to maintain adequate vitamin D status as there is insufficient UVB for skin synthesis.

- The serum concentration of the vitamin D metabolite, 25-hydroxyvitamin D (25(OH)D), is the most commonly used biomarker of vitamin D exposure.

- Vitamin D has a major role in the maintenance of calcium homeostasis, and regulates cell proliferation and differentiation in many tissues, making it a key regulator of the immune, endocrine, skeletal, renal, cardiovascular, and neurological systems.

- Vitamin D deficiency and low calcium intakes lead to nutritional rickets in children and adolescents, and osteomalacia in adults, both of which are due to under-mineralization of bone.

12.3 **Vitamin E**

In 1922, Evans and Bishop observed that rats could not reproduce on a diet of rancid lard (i.e., devoid of vitamin E). Wheat germ oil provided the needed component, originally called 'factor X' and 'anti-sterility factor'. The wheatgerm oil was later purified and two compounds were extracted and named α- and β-tocopherol. Subsequently, γ- and δ-tocopherol, plus the four tocotrienols, were isolated from plant oils. As an effective lipid-soluble scavenger of lipid peroxyl radicals, vitamin E is the most important chain-breaking antioxidant in blood and biological membranes. Although present in low concentrations, it is extremely efficient in maintaining the integrity of membranes against lipid peroxidation, thereby protecting cells and lipoproteins.

From the time of its discovery, it took over 40 years before it was proved that vitamin E deficiency could cause disease in humans. Data came largely from clinical studies of haemolytic anaemia in preterm infants, a new human population at that time. Another 25 years passed before functions of vitamin E unrelated to its antioxidant activity were described. More recent studies have shown that vitamin E has roles in cell signalling, by inhibition or inactivation of protein kinase C, and in modulation of gene expression, inhibition of cell proliferation, and platelet aggregation (Traber, 2007). These effects are specific for α-tocopherol, and are independent of its antioxidant properties.

Forms in food and dietary sources

As shown in Figure 12.9, there are eight vitamers of vitamin E; the tocopherols have a saturated side-chain and the tocotrienols an unsaturated. The four forms of each (α, β, γ, and δ) differ in the methylation of the ring. Tocotrienols occur in foods as both the free alcohols and as esters; tocopherols occur naturally as the free alcohols; but acetate and succinate esters are used in pharmaceutical preparations because of their greater stability.

Table 12.1 shows the different biological activity of vitamin E vitamers. The (now obsolete) international unit (IU) of vitamin E potency was equated with the activity of 1 mg of (synthetic) DL-α-tocopherol acetate; on this basis D-α-tocopherol (*RRR*-α-tocopherol, the most potent vitamer) is 1.49 IU/mg. It is now usual to express the vitamin E content of foods in terms of mg equivalents of (*RRR*)- α-tocopherol (TE), based on their biological activities. For the major vitamers present in foods, total α-tocopherol equivalent is calculated as the sum of mg α-tocopherol + 0.4 × mg β-tocopherol + 0.3 × mg γ-tocopherol + 0.01 × mg δ-tocopherol + 0.3 × mg α-tocotrienol + 0.05 × mg β-tocotrienol + 0.01 × mg γ-tocotrienol.

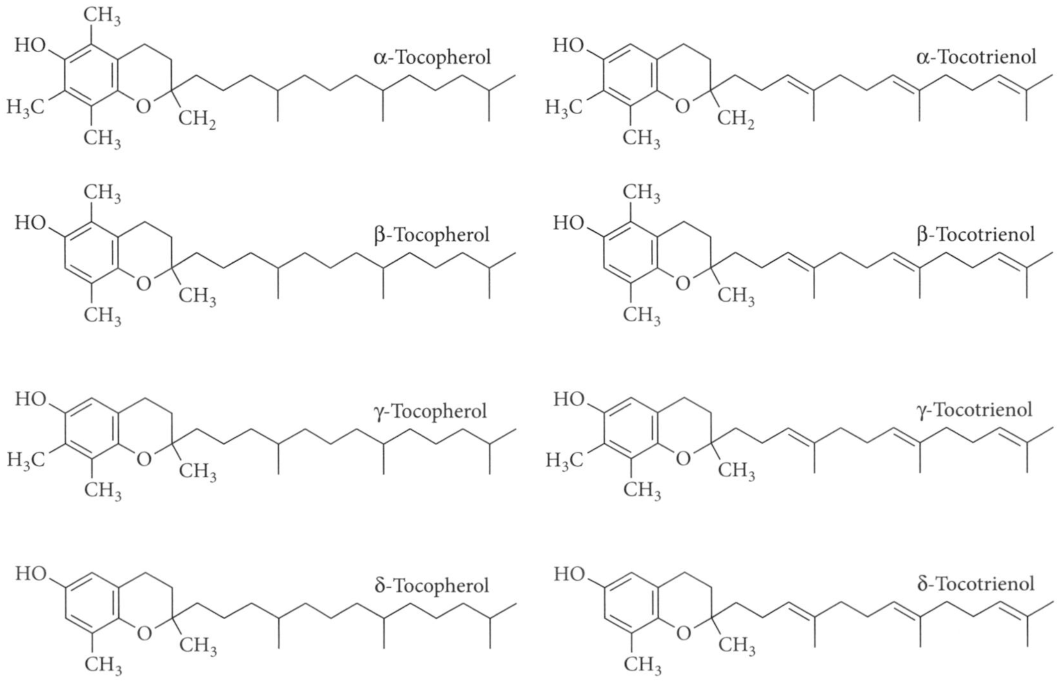

FIGURE 12.9 Vitamin E vitamers.

TABLE 12.1 Relative biological activity of the vitamin E vitamers

D-α-tocopherol (*RRR*)	1.49	1.0	1.0
D-β-tocopherol (*RRR*)	0.75	0.50	0.38
D-γ-tocopherol (*RRR*)	0.15	0.10	0.09
D-δ-tocopherol (*RRR*)	0.05	0.03	0.02
D-α-tocotrienol	0.75	0.50	0.12
D-β-tocotrienol	0.08	0.05	–
D-γ-tocotrienol	–	–	–
D-δ-tocotrienol	–	–	–
L-α-tocopherol (SRR)	0.46	0.31	0.11
RRS-α-tocopherol	1.34	0.90	–
SRS-α-tocopherol	0.55	0.37	–
RSS-α-tocopherol	1.09	0.73	–
SSR-α-tocopherol	0.31	0.21	–
RSR-α-tocopherol	0.85	0.57	–
SSS-α-tocopherol	1.10	0.60	–
RRR-α-tocopheryl acetate	1.36	0.91	–
RRR-α-tocopheryl acid succinate	1.21	0.81	–
all-*rac*-α-tocopherol	1.10	0.74	0.02
all-*rac*-α-tocopheryl acetate	1.00	0.67	–
all-*rac*-α-tocopheryl acid succinate	0.89	0.60	–

The naturally occurring compound is D-α-tocopherol, in which all three asymmetric centres have the *R*-configuration (2*R*, 4'*R*, 8'*R*, or all *R* (*RRR*) α tocopherol). Chemical synthesis yields a mixture of eight possible stereo-isomers which have different biological activity (all-*rac*-α-tocopherol). The synthetic all-*rac* mixture has 0.74 × the activity of *RRR*-α-tocopherol. In the USA and Canada, only the 2*R* isomers are considered to contribute to vitamin E intake, giving an equivalence of 0.45 IU/mg for synthetic all-*rac*-α-tocopherol.

Food sources of vitamin E are vegetable oils (particularly wheatgerrn oil, with 170 mg of α-tocopherol per 100g), nuts (almonds have 45 α-tocopherol per 100g), cereals (rye has 2.3 α-tocopherol per 100g), seeds, and green leafy vegetables. Corn, soybean, palm, and sesame oils are rich sources of γ-tocopherol. All unsaturated oils have added vitamin E as an antioxidant to protect the oil.

The dietary assessment of vitamin E is challenging, and intake estimates rely on the accuracy of recipe calculations including oil as well as the currency of the food composition data for major cooking oils. Average intakes are typically around 12 mg TE, with substantially higher intakes among users of nutritional supplements.

Absorption and transport of vitamin E

During digestion, vitamin E is extracted from its food matrix and absorbed in micelles with other dietary lipids. Esters are hydrolysed in the intestinal lumen by pancreatic esterase, and also by intracellular esterases in the mucosal cells. Absorption is mediated by cholesterol membrane transporters at the apical membrane of the enterocyte. There is a wide variation in absorption of vitamin E, due to various factors such as the food matrix, nutritional and health status, alcohol intake, or genetic variation in the expression of cholesterol membrane transporters, for example, scavenger receptor class B type 1 (SR-B1). In intestinal mucosal cells, all vitamers of vitamin E are incorporated into chylomicrons, most of which goes to the liver in chylomicron remnants.

In the liver, the α-tocopherol transfer protein (α-TTP) preferentially binds α-tocopherol rather than other tocopherols or tocotrienols and ensures its incorporation into nascent VLDL to be secreted by the liver into the circulation and distributed to body tissues. The other vitamers do not bind well to the α-TTP and are largely metabolized in the liver by cytochrome P450 linked ω-hydroxylase, followed by β-oxidation, conjugation, and excretion.

Metabolic functions of vitamin E

Vitamin E acts as a lipid-soluble antioxidant by intercepting peroxyl radicals, which are formed instantaneously when a lipid radical reacts with oxygen (Traber 2021). In this way, vitamin E is an important nutritional component of the antioxidant defence system, a complex network including endogenous and nutritional antioxidants, enzymes, and repair mechanisms, with mutual interactions and synergies among the various components.

During lipid peroxidation, a peroxyl radical (ROO⁻) oxidizes other organic compounds such as other PUFA to generate lipid hydroperoxides (ROOH) and fatty acid radicals (R⁻). Unrestrained lipid peroxidation can result in the generation of thousands of lipid peroxides and massive cellular damage. α-Tocopherol, located in or near the membrane, can react with peroxyl radicals before they attack, thus protecting PUFA. α-Tocopherol also provides a hydrogen for the reduction of the fatty acid radicals. These processes generate the tocopheroxyl radical, which is then reduced by other antioxidants, such as ascorbic acid and glutathione. By protecting PUFAs within membrane phospholipids, α-tocopherol preserves intracellular and cellular membrane integrity and stability, plays an important role in the stability of erythrocytes and the conductivity in central and peripheral nerves, and prevents haemolytic anaemia in preterm infants.

Non-antioxidant actions of vitamin E

α-Tocopherol, but not the other vitamers, modulates gene expression and transcription, including the scavenger receptor for oxidized LDL in macrophages and smooth muscle, either by binding directly to enzymes, competing with substrates, or changing their activity by a redox mechanism. It inhibits platelet aggregation and vascular smooth muscle proliferation. In monocytes, it reduces formation of reactive oxygen species, cell adhesion to the endothelium, and release of interleukins and tumour necrosis factor. As yet no response element for intracellular vitamin E binding protein has been identified on any of the proposed target genes. Dietary tocotrienols, but not tocopherols, have a cholesterol-lowering effect by reducing the activity of HMG CoA reductase.

In experimental animals, vitamin E deficiency depresses immune system function, with reduced mitogenesis of B- and T-lymphocytes, reduced phagocytosis and chemotaxis, and reduced production of antibodies and interleukin-2, suggesting a signalling role in the immune system.

Vitamin E deficiency and excess

Vitamin E deficiency symptoms include failure of placentation, neuromuscular impairment, haemolytic anaemia, retinopathy, reduced immunity, and inflammation (Traber 2021). Most of the data on vitamin E deficiency results from genetic abnormalities or occurs secondary to fat malabsorption syndromes, such as cystic fibrosis or chronic cholestatic hepatobiliary disease, and in two rare groups of patients with genetic diseases.

- Patients with congenital abetalipoproteinaemia, who are unable to synthesize VLDL, have undetectably low plasma levels of α-tocopherol and develop devastating ataxic neuropathy and pigmentary retinopathy.
- Patients who lack the hepatic TTP and suffer from AVED (ataxia with vitamin E deficiency) are unable to export α-tocopherol from the liver in VLDL.

In both groups of patients, the only source of vitamin E for peripheral tissues is recently ingested vitamin E in chylomicrons. Without supplements, they develop cerebellar ataxia, axonal degeneration of sensory neurons, skeletal myopathy, and pigmented retinopathy similar to those seen in experimental animals. In premature infants, whose reserves of the vitamin are inadequate, vitamin E deficiency causes haemolytic anaemia, caused by instability of the erythrocyte membrane. Vitamin E deficiency as a result of low dietary intakes is unusual.

Vitamin E has low toxicity, and many people take relatively large supplements for protection against atherosclerosis and coronary heart disease, arising from well-publicized observational studies that indicated beneficial effects of vitamin E for cardiovascular disease prevention. However, RCTs with vitamin E have provided little evidence that this is effective, and very high intakes of vitamin E may interfere with blood clotting by antagonizing vitamin K and potentiating anticoagulant therapy. It could be argued that trials of high-dose vitamin E are inappropriate and potentially risky; that vitamin supplementation does not treat non-communicable diseases, and that the life-long exposure to moderate daily intakes of this nutrient are key to its effectiveness in disease prevention.

No adverse effects from 540 mg α-TE/day as part of a dose-response study resulted in a tolerable upper intake level (UL) of 300 mg α-TE/day, ranging from 100–260 mg in children (EFSA 2015).

Assessment of vitamin E status

The most commonly used index of vitamin E status is the plasma concentration of α-tocopherol. The reference range is ~12–46 μmol/L. Plasma α-tocopherol concentrations <8 μmol/L are associated with neurologic disease and <12 μmol/L with increased erythrocyte fragility, on the basis of haemolysis induced *in vitro* by hydrogen peroxide. Vitamin E deficiency based on circulating α-tocopherol concentration <12 μmol/L in serum or plasma has been observed in Africa, Southeast Asia, and the west Pacific (Traber 2021). Epidemiological studies suggest that an optimum concentration for protection against cardiovascular disease and cancer is >30 μmol/L.

Vitamin E requirements and dietary recommendations

The North American dietary recommendations for vitamin E were based on the plasma concentration of α-tocopherol required to prevent *in vitro* haemolysis (12 μmol/L) and dose-response data to achieve this concentration. Therefore, the average requirement is 12 mg/day, and the RDA is 15 mg/day. In 2015, EFSA considered dietary reference values for vitamin E and set Adequate Intake (AI) values for all population groups (from 5–11 mg/day) based on observed intakes in healthy persons with no apparent deficiency (EFSA 2015b). The EFSA considered the existing data on biomarkers of vitamin E status inadequate and recommended research on pregnant women, children, and infants and the establishment of much more secure food composition data to estimate dietary intakes.

KEY POINTS
- Vitamin E is the major lipid-soluble antioxidant in cell membranes and plasma lipoproteins.
- Vitamin E deficiency only occurs in people with severe fat malabsorption, and rare patients with a genetic lack of β-lipoprotein or hepatic tocopherol transfer protein, who suffer devastating nerve damage.
- Premature infants have vitamin E deficiency and are susceptible to haemolytic anaemia.
- There is little evidence on which to estimate dietary requirements for vitamin E, but intakes of 12–15 mg/day protect membrane integrity.
- There is no consistent evidence that supplemental vitamin E is protective against non-communicable diseases.

12.4 Vitamin K

Vitamin K was discovered in the 1930s by a Danish researcher called Hendrik Dam who observed a haemorrhagic disease of chickens fed on solvent-extracted fat-free diets and cattle fed on silage made from spoiled sweet clover. The problem in the chickens was a lack of vitamin K in the diet, while in the cattle it was due to the presence of dicoumarol, an antimetabolite of vitamin K. It soon became apparent that an anti-haemorrhagic factor or 'Koagulation vitamin,' was required for blood clotting, but it was not until 1974 that its role as a coenzyme for the vitamin K-dependent carboxylase, required for the synthesis of prothrombin and many other functions, was revealed.

Vitamin K is a family of compounds with a common 2-methyl-1,4-naphthoquinone ring structure called menadione (vitamin K3) and an isoprenoid side chain at the 3-position (Figure 12.10). Phylloquinone (vitamin K1) is present in plant foods, especially green leafy vegetables, and menaquinones (vitamin K2; including MK-4 to MK-13) are in animal and fermented foods. Menaquinones, bar MK-4, are also produced by bacterial fermentation in the gut. Vitamin K is a rich area of research interest, with a quadrupling of publications in the field over the past 20 years.

Forms in foods and dietary sources

As shown in Figure 12.10, there are two naturally occurring vitamers: phylloquinone has a phytyl side-chain, while the menaquinones have a polyisoprenyl side-chain, with up to 15 isoprenyl units (most commonly 6–10), shown by menaquinone-n. The synthetic compounds menadione and menadiol are vitamin K3. Menadiol diacetate (acetomenaphthone) is used in pharmaceutical preparations, and two water-soluble derivatives, menadione sodium bisulphite and menadiol sodium phosphate, have been used for administration of vitamin K by injection and in patients with malabsorption syndromes which would impair the absorption of menadione, phylloquinone, and menaquinones, which are lipid soluble.

Phylloquinone is obtained in the diet from vegetables, especially green leafy vegetables such as broccoli, kale, collard greens, and lettuce; vegetable oils such as soybean and canola oil, and some fruits. Meat, dairy foods, and eggs contain menaquinones and fermented foods such as natto and other foods produced using bacterial fermentation. Phylloquinone in plant foods is bound to chloroplasts so it is less bioavailable than from oils or supplements. Consumption of cooked vegetables together with a little oil may enhance absorption (Shearer et al. 2012).

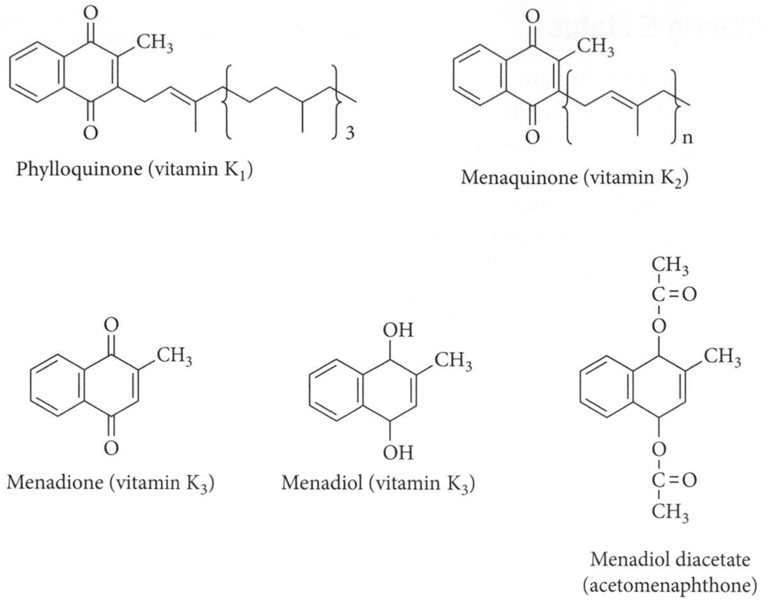

FIGURE 12.10 Vitamin K vitamers.

Absorption and transport of vitamin K

In the same way as the other fat soluble vitamins, phylloquinone is absorbed by enterocytes in the proximal small intestine in mixed micelles and incorporated into chylomicrons for transportation via the lymphatic system to the liver. Small quantities of phylloquinone are in circulation, carried mainly in lipoproteins. Human osteoblasts express several lipoprotein receptors to divest lipoprotein fractions of phylloquinone and enable synthesis of the vitamin K-dependent proteins in the bone matrix that regulate bone metabolism.

Dietary menaquinones have variable absorption; MK-7 is more efficiently and MK-4 and MK-9 are less efficiently absorbed than free phylloquinone (EFSA 2017). Menaquinones are synthesized by intestinal bacteria, but it is unclear how much they contribute to vitamin K nutrition, since they are extremely hydrophobic, and will only be absorbed from regions of the gastrointestinal tract where bile salts are present—mainly the terminal ileum. Therefore, the contribution of medium- and long-chain menaquinones produced by gut microbiota to vitamin K status is unclear.

The metabolic functions of vitamin K

The main metabolic function of vitamin K is as an essential cofactor in the post-translational modification (i.e., γ-glutamylcarboxylation) of glutamate residues (Glu) to γ-carboxyglutamate (Gla) (Willems et al. 2014). Early on, four vitamin K dependent proteins involved in blood coagulation, prothrombin (factor II) and factors VII, IX, and X, were discovered as a result of the haemorrhagic disease caused by deficiency. The function of γ-carboxyglutamate in these proteins is to chelate calcium, and induce a conformational change that permits binding of the proteins to membrane phospholipids. These, plus several other proteins, are members of the vitamin K dependent protein (VKDP) family.

Osteocalcin was the first extrahepatic VKDP identified and is involved in regulation of calcification of bone matrix (Willems et al. 2014). High circulating concentrations of under-carboxylated osteocalcin is an indicator of low vitamin K status, and is associated with low bone mineral density. Matrix Gla protein (MGP) purified from calcified cartilage is involved in the inhibition of inappropriate calcification of soft tissues. MGP, also present in vascular smooth muscle is of interest for cardiovascular health, as it may help to reduce vascular calcification, which contributes to endothelial dysfunction. Undercarboxylated MGP may therefore also be an indicator of low vitamin K status and a risk factor for vascular disease.

The more recently identified Gla-rich protein (GRP), also isolated from bone and cartilage, is associated with the upper zone of growth plate and cartilage matrix. Other proteins that undergo vitamin K-dependent carboxylation of Glu to Gla are nephrocalcin in the kidney cortex, hydroxyapatite and calcium oxalate in urinary stones, atherocalcin in atherosclerotic plaques, Gas6, which is involved in the regulation of differentiation and development in the nervous system, and proteins involved in cell signalling and apoptosis.

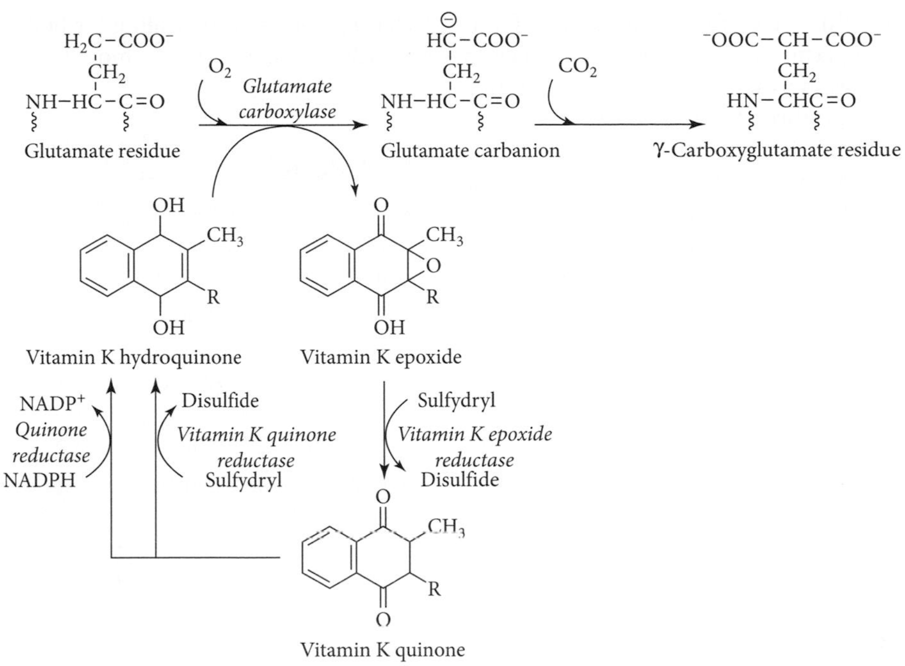

FIGURE 12.11 The reaction of the vitamin K dependent carboxylase.

Recycling and conserving in the vitamin K epoxide cycle

The vitamin K-epoxide cycle, which can be regarded as a local salvage pathway, is pivotal to both the function of vitamin K and to the conservation of its limited microsomal cellular stores (Shearer et al. 2012). Oxidation of vitamin K hydroquinone to the epoxide is linked to γ-carboxylation of the glutamate residue of vitamin K-dependent proteins (see Figure 12.11). Catalysed by vitamin K epoxide reductase, vitamin K epoxide is reduced to the quinone and subsequently reduced and recycled to the active hydroquinone substrate.

Vitamin K deficiency

Vitamin K deficiency results in prolonged prothrombin time (the time it takes for blood to clot), causing bleeding and haemorrhage because of impaired synthesis of the vitamin K dependent blood clotting proteins, although this is rare. Treatment with Warfarin or other anticoagulants can antagonize vitamin K activity and prothrombin. People with malabsorption may not absorb vitamin K efficiently and should receive supplements.

Synthesis of other VKDP is impaired in vitamin K deficiency. Under-carboxylated osteocalcin concentration is higher in people with low intakes of vitamin K who show no impairment of blood clotting. This may contribute to reduced bone mineralization and osteoporosis, but to date, vitamin K supplementation studies have been inconsistent for a role in osteoporosis prevention.

During pregnancy, only small quantities of phylloquinone cross the placenta from mother to foetus. Blood concentrations of phylloquinone in the full-term newborn are about half of that of the mothers and the phylloquinone concentration in cord blood is low (<0.1 nmol/L). Newborn infants also have low plasma levels of prothrombin and the other vitamin K dependent clotting factors (about 30–60 per cent of the adult concentrations, depending on gestational age). To a great extent this is the result of the relatively late development of liver glutamate carboxylase, but newborns are also short of vitamin K due to the placental barrier that limits foetal uptake of the vitamin. This is probably a way of regulating the activity of vitamin K dependent proteins in development and differentiation. Over the first six weeks of post-natal life, the plasma concentrations of clotting factors gradually rise to adult levels, but in the meantime, infants are at risk of potentially fatal intracranial haemorrhage, formerly called haemorrhagic disease of the newborn, now known as vitamin K deficiency bleeding in infancy. It is recommended to give all newborn infants prophylactic vitamin K1, either orally or by intramuscular injection.

Assessment of vitamin K status

Vitamin K status is not routinely assessed. Measurement of prothrombin time is the usual method of assessing vitamin K status, or monitoring the efficacy of anticoagulant therapy. This is the time taken for the formation of a fibrin clot in citrated plasma after the addition of calcium ions and thromboplastin to activate the extrinsic clotting system. The normal prothrombin time is 12–13 seconds; greater than 25 seconds is associated with severe bleeding. Measurement of fasting plasma phylloquinone is also possible, and there is an association between circulating phylloquinone and vitamin K1 intakes. Under-carboxylated osteocalcin and other under-carboxylated proteins in plasma are also a marker of vitamin K status, and responsive to supplementation. Urinary Gla excretion decreases during phylloquinone depletion and increases with repletion, although 24-hour urinary sampling is challenging. However, there is no biomarker for which a dose-response relationship has been established (EFSA 2017).

The basis for setting vitamin K requirements

Due to inadequate data to establish an average requirement in 2001, the Institute of Medicine established adequate intake values (AI) for vitamin K based on usual intakes in healthy populations, of 120 µg for men and 90 µg for women. The EFSA in 2017 also concluded that data were insufficient and set AI values based on body weight, considering a total body pool of phylloquinone of about 0.55 µg/kg body weight in healthy adults at steady state. On this basis, and in line with the UK DRV from 1991, the EFSA set an AI of l µg phylloquinone/kg body weight per day for all age and sex population groups. This translates to 70 µg/day for all adults, including pregnant and lactating women, with scaling for body weight down to 10 µg/day in infants of 7–11 months.

KEY POINTS

- Vitamin K is a family of compounds; phylloquinone (K1) and menaquinones (K2, including MK-4, 7, and 9) are nutritionally important.

- The main sources of vitamin K1 are green leafy vegetables and animal and fermented foods contain menaquinones. Although intestinal bacteria synthesize most menaquinones, it is not known to what extent this contributes to status.

- The metabolic function of vitamin K is as an essential cofactor for γ-glutamylcarboxylation of glutamate residues (Glu) in precursors of vitamin K dependent proteins involved in blood clotting and a range of other functions, including osteocalcin which regulates the calcification of bone matrix, and matrix Gla protein which inhibits inappropriate calcification of soft tissues.

- Clinically used anticoagulants for treatment of people at risk of thrombosis act as anti-metabolites of vitamin K.

- Due to low vitamin K status at birth, newborn infants are given prophylactic vitamin K to avoid intracranial bleeding.

 Be sure to test your understanding of this chapter by attempting multiple choice questions.

REFERENCES

Bikle D D (2014) Vitamin D metabolism, mechanism of action, and clinical applications. *Chem Biol* **21**(3), 319–329.

Bouillon R, Marcocci C, Carmeliet G et al. (2019) Skeletal and extraskeletal actions of vitamin D: Current evidence and outstanding questions. *Endocr Rev.* **40**(4), 1109–1151.

EFSA Panel on Dietetic Products, Nutrition and Allergies (2015a) Scientific opinion on dietary reference values for vitamin A. *EFSA Journal* **13**(3), 4028.

EFSA Panel on Dietetic Products, Nutrition and Allergies (2015b) Scientific opinion on dietary reference values for vitamin E. *EFSA Journal* **13**(7), 4149.

EFSA Panel on Dietetic Products, Nutrition and Allergies (2016) Scientific opinion on dietary reference values for vitamin D. *EFSA Journal* **14**(10), 4547.

EFSA Panel on Dietetic Products, Nutrition and Allergies (2017) Scientific opinion on dietary reference values for vitamin K. *EFSA Journal* **15**(5), 4780.

Imdad A, Mayo-Wilson E, Herzer K et al. (2017) Vitamin A supplementation for preventing morbidity and mortality in children from six months to five years of age. *Cochrane Database Syst. Rev.* **3**, Art. No.: CD008524.

Institute of Medicine (2011) *Dietary Reference Intakes for Calcium and Vitamin D.* National Academies Press, Washington, DC.

Kiely M & Black L J (2012) Dietary strategies to maintain adequacy of circulating 25-hydroxyvitamin D concentrations. *Scand J Clin Lab Invest Suppl* **243,** 14–23.

Munns C F, Shaw N, Kiely M et al. (2016) Global consensus recommendations on prevention and management of nutritional rickets. *J Clin Endocrinol Metab* **101**, 394–415.

O'Riordan J L & Bijvoet O L (2014) Rickets before the discovery of vitamin D. *Bonekey Rep.* **3,** 478.

Rejnmark L, Bislev L S, Cashman K D et al. (2017) Nonskeletal health effects of vitamin D supplementation: A systematic review on findings from meta-analyses summarizing trial data. *PLoS One.* **12(7)**, e0180512.

Scientific Advisory Committee on Nutrition (SACN) (2016) *Vitamin D and Health.* The Stationary Office, London.

Shearer M J, Fu X, Booth SL (2012) Vitamin K nutrition, metabolism, and requirements: Current concepts and future research. *Adv Nutr* **3**(2), 182–195.

Tanumihardjo S A, Russell R M, Stephensen C B et al. (2016) Biomarkers of nutrition for development (BOND)-vitamin A review. *J Nutr* **146**(9), 1816S–48S.

Traber M G (2021) Vitamin E. *Adv Nutr* **12**(3), 1047–1048.

Traber M G (2007) Vitamin E regulatory mechanisms. *Ann Rev Nutr* **27**, 347–362.

Webb A R (2006) Who, what, where and when-influences on cutaneous vitamin D synthesis. *Prog Biophys Mol Biol* **92**(1), 17–25.

Willems B A, Vermeer C, Reutelingsperger C P et al. (2014) The realm of vitamin K dependent proteins: Shifting from coagulation toward calcification. *Mol Nutr Food Res* **58**(8), 1620–1635.

FURTHER READING

Bender D A (2003) *Nutritional Biochemistry of the Vitamins,* 2nd edn. Cambridge University Press, New York.

Booth S L (2009) Roles for vitamin K beyond coagulation. *Ann Rev of Nutr* **29**, 89–110.

DeLucca H F (2008) Evolution of our understanding of vitamin D. *Nutr Rev* **66** (Suppl 2), s73–87.

Halliwell B (2012) Free radicals and antioxidants: Updating a personal view. *Nutr Rev* **70**, 257–265.

Noy N (2010) Between death and survival: Retinoic acid in the regulation of apoptosis. *Ann Rev Nutr* **30**, 201–217.

Pettifor J M (2004) Nutritional rickets: Deficiency of vitamin D, calcium, or both? *Am J Clin Nutr* **80** (Suppl), 1725S–1729S.

Saari J C (2012) Vitamin A metabolism in rod and cone visual cycles. *Ann Rev Nutr* **32**, 125–145.

Singh U, Devaraj S, Jialal I (2005) Vitamin E, oxidative stress, and inflammation. *Ann Rev Nutr* **25**, 151–174.

Zingg J M (2015) Vitamin E: A role in signal transduction. *Ann Rev Nutr* **35**, 135–173.

13 Minerals and trace elements

Ruan Elliott and Paul Sharp

OBJECTIVES

By the end of this chapter you should be able to:

- identify the most important food sources of minerals and trace elements
- understand the main features of their absorption, metabolism, and tissue distribution
- describe their important functions in the body
- appreciate the effects of deficiency and excess, and their importance in the UK and elsewhere
- discuss the basis for dietary recommendations.

13.1 Introduction

This chapter deals with the key minerals and trace elements essential to a number of important biochemical and physiological functions in the body. The focus is placed on dietary sources of the various minerals and their homeostatic regulation in the body (i.e., the balance between bioavailability and absorption versus regulation and excretion). This is followed by discussion of the major metabolic functions of the minerals and trace elements and the consequences of deficiency and excess of individual elements. Each section closes by discussing the current methodologies used to assess the body status of these minerals and trace elements and the basis of the current dietary recommendations for intakes of these micronutrients. Five minerals are dealt with in detail. Calcium and phosphorus are discussed together due to their close interrelationship in maintaining bone health. This is followed by a discussion of the essential roles of iron, zinc, iodine, and selenium in human metabolism. Further information on the essential nature of these micronutrients as well as their interactions with other nutrients will be provided elsewhere in this book.

13.2 Calcium and phosphorus

Calcium is the most abundant mineral in the body and the majority is contained within the adult skeleton in the form of hydroxyapatite, a complex crystalline form of calcium phosphate, $Ca_{10}(PO_4)_6(OH)_2$. Accordingly, phosphorus is also abundant in bone (approximately 85 per cent of total body phosphorus). More details of the roles of calcium and phosphorus in maintaining bone health are provided in Chapter 25.

Food sources

The most important dietary sources of calcium in the Western world are milk and other dairy products including yoghurt, cheese, and ice cream (Table 13.1). Phosphorus is also abundant in these products and its concentration is approximately equimolar with calcium. However, while the calcium content of these foods contributes 34 per cent of the UK adult daily dietary intake, dairy products contribute only 20–30 per cent of daily phosphorus intake. Cereals contribute a further 30 per cent of dietary calcium. Many cereal products in the UK are fortified with calcium (added as calcium carbonate). The Bread & Flour Regulations (1998) stipulate that the calcium content of all flours derived from wheat (with the exception of wholemeal flour) must be between 94 and 156 mg/100g. The Scientific Advisory Committee on Nutrition (SACN) reviewed the evidence for mandatory fortification and found that removal of the added calcium would have a significant negative impact on calcium intakes for several population groups in the UK, particularly older children and young adults.

The phosphate content of cereals and vegetables is relatively high due to the presence of phytic acid (inositol hexaphosphate) and these foods contribute a further 25–35 per cent of daily phosphorus intake. In addition, carbonated soft drinks, which are often rich in phosphates, contribute significantly to phosphorus intakes in some diets.

TABLE 13.1 Main dietary sources of minerals and trace elements in the UK diet (derived for the National Diet and Nutrition Survey: results for years 9 to 11 (2016 to 2017 and 2018 to 2019 data for adults aged 19–64)

	% contribution to total dietary intake				
	Calcium	**Iron**	**Zinc**	**Iodine**	**Selenium**
Cereals and cereal products	30	37	26	13	27
Milk and milk products	34	2	15	32	6
Eggs and egg dishes	3	5	4	9	10
Fat spreads	0	0	0	2	0
Meat and meat products	8	17	31	9	29
Fish and fish dishes	2	4	3	10	16
Vegetables and potatoes	8	16	11	4	5
Savoury snacks	1	0	1	0	0
Nuts and seeds	1	2	2	0	1
Fruit	2	4	1	2	0
Sugar, preserves, and confectionery	2	2	1	2	1
Non-alcoholic beverages	4	1	1	5	0
Alcoholic beverages	2	6	0	7	0
Miscellaneous	3	4	3	3	2
Average daily total intake	813 mg	10.1 mg	8.6 mg	154 µg	49 µg

The calcium content of meat and fish is low in comparison to dairy products and cereals but sardines, for example, are a rich source of both calcium and phosphorus because the bones are consumed. In addition to its abundance in bone, phosphate is also present at high levels in soft tissues such as muscle and therefore levels of phosphorus in meat and fish are much higher than those of calcium and contribute a further 25–35 per cent to daily phosphorus intake.

Whilst in most foods, calcium is largely present as simple organic and inorganic salts (some of which form larger complexes with other dietary components), dietary phosphorus occurs in several forms including inorganic phosphate, organic phosphoproteins, phosphorylated sugars, sugar alcohols (e.g., phytate) and phospholipids. The relative amounts of organic versus inorganic phosphate varies depending on the food source; for example, organic complexes prevail in meat, whereas 80 per cent of phosphorus in cereals and grains is phytic acid and 33 per cent of phosphorus in milk is present as inorganic salts.

Absorption and metabolism

Bioavailability

Absorption of calcium from a mixed diet is fairly constant (25–35 per cent). Bioavailability of calcium in milk and dairy products is improved by the presence of sugars (lactose) and casein phosphopeptides. The absorption from calcium salts used as supplements (such as lactate, carbonate, gluconate, and citrate) is similar to that from dairy products (30–40 per cent). However, other dietary acids, especially oxalate and phytate, inhibit calcium uptake. Oxalate (in many vegetables and fruit, including spinach, rhubarb, and strawberry) is the most potent inhibitor of calcium absorption and forms a highly insoluble complex, whereas phytate (found abundantly in cereals and cereal products) is less potent but is present at a much higher concentration in the intestinal lumen and is likely to be the major dietary inhibitor of calcium absorption—a dietary phytate to calcium molar ratio of 0.2 can increase the risk of calcium deficiency.

Phosphorus is absorbed very efficiently from the small intestine (approximately 60–70 per cent of dietary intake) and is always absorbed as inorganic phosphate—phosphate groups are liberated from organic compounds prior to absorption by alkaline phosphatase on the luminal membrane of intestinal enterocytes. This enzyme has little activity towards phytate, which is largely non-bioavailable in its natural form. However, yeast added to leavened bread contains significant levels of the enzyme phytase, which is lacking from human intestinal secretions. Phytase readily releases phosphate from phytate and about 50 per cent of phytate-derived phosphate in bread is thought to be absorbed.

Absorption

Intestinal calcium absorption occurs via both active transcellular and passive paracellular pathways. The transcellular route (Figure 13.1) is subject to tight homeostatic control by the calcium content of the diet and 1,25-dihydroxycholecalciferol (1,25(OH)$_2$D$_3$; vitamin D$_3$), whereas the passive paracellular route is not regulated and is non-saturable. The trigger for vitamin D$_3$ synthesis is the release of parathyroid hormone (PTH) in response to decreased plasma calcium. Elevated PTH does not regulate intestinal calcium absorption directly but acts indirectly by increasing the production of active vitamin D$_3$ in the kidney.

The mechanisms involved in phosphate absorption are less well defined (Figure 13.2). Phosphate absorption is regulated by long-term changes in dietary phosphorus content and body phosphorus status. In addition, there

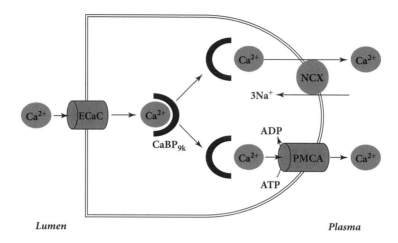

FIGURE 13.1 Mechanisms involved in intestinal calcium absorption.
Intestinal calcium absorption is subject to tight homeostatic control by the calcium content of the diet and 1,25-dihydroxycholecalciferol (1,25-(OH)$_2$D$_3$; vitamin D3). Uptake of calcium from the diet is mediated by epithelial Ca^{2+} channels (ECaC) located on the apical membrane of small intestinal enterocytes. Absorbed Ca^{2+} is picked up by the calcium binding protein calbindin D (CaBP$_{9k}$) and shuttled to the basolateral surface of the cell, where calcium is exported by the concerted action of a plasma membrane calcium ATPase (PMCA) and the sodium/calcium exchanger (NCX). Both the ECaC channels and calbindin are strongly upregulated by active vitamin D$_3$.

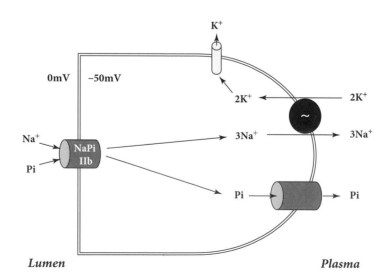

FIGURE 13.2 Mechanisms involved in intestinal phosphorus absorption.
Phosphorus is absorbed by intestinal enterocytes mainly as inorganic phosphate (Pi) via a sodium-dependent uptake process (NaPi IIb—the type IIb Na/Pi co-transporter). This transport pathway is sensitive to long-term changes in the phosphate content of the diet and vitamin D$_3$. Uptake is driven by the electrochemical gradient for sodium that is generated by the inside negative membrane potential and the Na/K ATPase, which maintains a low intracellular sodium concentration. Efflux of phosphate across the basolateral membrane is likely to utilize an as yet uncharacterized facilitated transport mechanism.

may be a regulatory role for vitamin D_3 in controlling dietary phosphate absorption.

Tissue distribution

The human body contains approximately 1.2 kg (30 moles) of calcium, 99 per cent of which resides in mineralized tissues such as the bones and teeth. The majority of calcium in these tissues is present as hydroxyapatite, the remainder is found in plasma (where plasma calcium is tightly regulated at 2.5 mmol/L), and extracellular fluid. Intracellular free calcium concentration is extremely low (approximately 100 nmol/L) but can rise dramatically following hormone- or neurotransmitter-induced release of calcium from the intracellular stores in the endoplasmic/sarcoplasmic reticulum, and influx from the extracellular fluid through plasma membrane calcium channels.

Body phosphorus is also largely associated with the bone; approximately 85 per cent of the ~600 g of phosphorus in the body is found in the skeleton. The remaining 15 per cent is found in the soft tissues and blood largely as phospholipids, phosphoproteins, and nucleic acids as well as inorganic phosphate.

Renal excretion and regulation

Approximately 50 per cent of plasma calcium is present in the ionized form (Ca^{2+}) and is freely filtered by the renal glomerulus. Most of this calcium is reabsorbed as the tubular fluid passes through the nephron and this process is tightly regulated by the action of parathyroid hormone (PTH) and vitamin D_3, which increase calcium reabsorption (Figure 13.3), and calcitonin, produced in the C-cells of the thyroid gland, which promotes calcium excretion. PTH is released in response

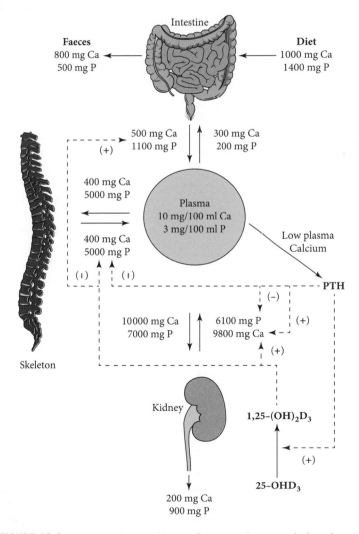

FIGURE 13.3 Homeostatic regulation of serum calcium and phosphorus.
PTH, parathyroid hormone; 1,25-$(OH)_2D_3$, 1,25-dihydroxycholecalciferol (vitamin D_3).

to a decrease in plasma calcium concentration and has several actions, including promoting the synthesis of $1,25(OH)_2D_3$ (the active form of vitamin D_3), which in turn stimulates intestinal calcium absorption. In addition, PTH (and vitamin D_3) promotes bone resorption and increases renal calcium reabsorption. Together, these coordinated actions serve to raise plasma calcium back to normal levels within minutes to hours. Thyroid synthesis of calcitonin is promoted in response to elevated plasma calcium levels and release of this hormone acts to antagonize the effects of PTH, inhibiting bone resorption and renal calcium reabsorption, to lower plasma calcium.

As a consequence of bone resorption induced by PTH and vitamin D_3, plasma phosphate concentration also rises. The excess phosphate, which could decrease ionized Ca^{2+} in the plasma, is eliminated via the kidney under the action of PTH, which inhibits its reabsorption in the renal tubules.

Functions

Bone mineralization

The major function of both calcium and phosphorus is in the formation of hydroxyapatite. Calcium is deposited in bone at a rate of 150 mg/day during adolescence as the skeleton develops. In the adult, there is a dynamic equilibrium between calcium and phosphate deposition and resorption (400 mg Ca and 500 mg P are exchanged between the bone and plasma every day) due to the constant remodelling associated with the maintenance of the healthy skeleton (see also Chapter 25).

Blood clotting

Integral to the formation of a blood clot is the presence of a series of irregular fibrils of the protein fibrin, formed by polymerization of a soluble precursor, fibrinogen. Deposition of fibrin is the endpoint of a complicated cascade of enzyme reactions involving at least ten other factors. This sequence of events can be initiated by blood coming into contact with a foreign surface such as collagen (the intrinsic clotting system) or following tissue damage (the extrinsic clotting system). Ca^{2+} (also known as factor IV in this context) is a key component for the activation of both intrinsic and extrinsic factor X activator complexes, the point at which the intrinsic and extrinsic coagulations systems converge. Calcium ions are further required at several subsequent stages of this cascade and a decrease in plasma calcium below 2.5 mmol/L is associated with a reduced ability to form blood clots.

Calcium and cell signalling

Intracellular free calcium concentrations are low (less than 100 nmol/L). Low cytosolic calcium is maintained because cell membranes are relatively impermeable to calcium. Any calcium that does enter the cell is rapidly removed from the cytosol through the action of a number of calcium transporters and channels present in the plasma membrane and in the membranes of intracellular organelles, in particular, the smooth endoplasmic/sarcoplasmic reticulum, which sequesters calcium and acts as an intracellular store. However, cytosolic calcium levels can increase dramatically following agonist-induced generation of the second messenger inositol-1,4,5-trisphosphate (IP_3), which binds to specific receptors on the smooth endoplasmic reticulum and leads to an increase in intracellular Ca^{2+} concentration by as much as 10-fold. The release of Ca^{2+} initiates further second messenger activity by binding to various target molecules, for example protein kinase C and calmodulin, which in turn trigger diverse cellular responses including cell division, cell motility, contraction, secretion, endocytosis, and fertility.

Phosphorus and energy metabolism

Phosphorus not incorporated into bone mineral is generally found in the soft tissues. Intracellular phosphorus (as the phosphate ion) participates in a number of processes associated with energy metabolism. Ultimately, energy produced from metabolism during oxidative phosphorylation in the mitochondria is stored as high-energy phosphate bonds in ATP (plus phosphocreatine in skeletal muscle).

A number of second messenger signalling cascades rely on phosphorylation or dephosphorylation as a mechanism for either activating or deactivating crucial enzymes. One of these second messengers is cyclic AMP, which is generated from ATP by the enzyme adenyl cyclase. Formation of cyclic AMP activates a specific protein kinase that modulates the activity of a number of proteins by adding phosphate groups. An example of one such target protein is glycogen phosphorylase, the rate-limiting enzyme in glycogenolysis.

Effect of deficiency and excess

There are obligatory daily losses of calcium in the urine and the faeces. If this calcium is not replaced by the dietary supply, plasma calcium is maintained by resorption of bone under the action of PTH. If a low intake of calcium persists, or if intestinal absorption is impaired over a prolonged period, bone resorption will cause a severe decrease in bone mass and this dietary imbalance, together with a number of other factors, contributes to an increased risk of osteoporosis (see Chapter 25). It is estimated that more than one-third of women and one-fifth of men will sustain an osteoporotic fracture during their lifetime.

Evidence for calcium toxicity is rare and adverse effects are limited to people taking high-level calcium supplements. However, the US Food and Nutrition Board has set a tolerable upper intake level of 2500 mg/day for adults (aged 19–50 years). On the basis of a number of *in vitro* studies, some concerns have been raised that high intakes of calcium might reduce the absorption of other essential trace elements, especially iron. However, a review of human studies found no evidence to suggest that chronic calcium supplementation alters body iron status in the long term (SACN 2010).

Since almost all food contains phosphorus, deficiency syndromes associated with inadequate phosphorus intake are extremely rare. When phosphorus intakes are low the body can adapt accordingly, increasing intestinal absorption and decreasing renal excretion. However, in some circumstances, for example in people with chronic gastrointestinal malabsorption syndromes or uncontrolled diabetes, there can be an imbalance in phosphorus homeostasis, which is regulated by demineralization of the bone to maintain plasma phosphate levels. In severe situations, this can result in rickets in children or osteomalacia in adults. Similarly, rare genetic defects in the renal reabsorption of phosphate can cause rickets.

Toxicity associated with high phosphorus intakes is rare and is likely to be a problem only when calcium intakes are low. Elevated phosphorus intakes may lead to an increase in plasma phosphate concentration, which is thought to be a risk factor for a reduced bone mass. When plasma phosphate increases it is buffered by calcium, resulting in a decrease in the plasma ionized calcium concentration. This decrease in ionized calcium is sensed by a calcium-sensing receptor on the chief cells of the parathyroid gland leading increased release of PTH in the serum. PTH is a major calcitropic hormone and promotes bone resorption to re-establish plasma calcium levels. To this end there has been some concern in recent years that increasing dietary phosphorus intakes due to the large consumption of processed foods, which are often rich in phosphates, could have a detrimental effect on bone health. However, as yet there are no strong epidemiological data to support this hypothesis.

Status assessment

Presently there are no reliable biochemical indicators accurately to assess calcium status. As we have seen above, plasma calcium is so tightly regulated that it bears little relationship to body calcium status. However, since 99 per cent of body calcium is retained within the skeleton, measures of bone mass such as the bone mineral content (of a specific region such as the femoral neck) and bone mineral density have proved to be useful indicators of body calcium status. The most common measure of phosphorus status is serum phosphate concentration.

The basis of dietary recommendations

The dietary reference values (DRVs) for calcium in the United Kingdom were determined by factorial analysis of the basal amounts of calcium required for bone growth and the maintenance of bone mineralization. In infants, calcium requirements (reference nutrient intake (RNI): 525 mg/day) are high and are met in part by the high bioavailability of calcium from breast milk and the high calcium concentration in infant formula (Abrams 2010). Calcium retention in children increases significantly between the ages of one and ten years as the skeleton develops. Absorption efficiency of calcium from a mixed diet in childhood is around 35 per cent, less than from breast milk, and accordingly, the RNI has been calculated at 350 mg/day at age two rising to 550 mg/day at age ten. Assuming retention increases in adolescence to 250 mg/day for girls and 300 mg/day for boys the RNIs increase accordingly to 800 mg/day and 1000 mg/day, respectively. In adults following cessation of growth, there is still a significant calcium requirement based on calcium losses of 150 mg/day in the urine and 10 mg/day in sweat, skin, and hair loss. Calcium absorption from an adult mixed diet is assumed to be 30 per cent, giving an RNI of 700 mg/day. The Dietary Reference Intakes in the USA are set at higher levels: 1300 mg/day for older children (aged 9–18 years); 1000 mg/day for adults (aged 19–50 years) and 1200 mg/day for older adults (51+ years old).

No recommendation has been made for increasing calcium intakes during pregnancy due to higher rates of absorption compared with non-pregnant women. However, in adolescent pregnancies where females are still growing it may be advisable to increase calcium intakes. During lactation, the mother requires increased calcium for milk production, most (if not all) of which is derived from the diet. Initial estimates suggested that an extra 550 mg calcium/day was required for this purpose although this has been questioned due to potential adaptations in maternal calcium metabolism during lactation.

In the UK and EU, phosphorus requirements are based on an equimolar intake; calcium and current recommendation is that the RNI for phosphorus should be equal to the calcium intake in mmol/day. Recently, the US Food and Nutrition Board set separate RDAs for phosphorus; for infants (0–6 months) 100 mg/day rising to 275 mg/day by 12 months of age; for children (1–3 years) 460 mg/day rising to 500 mg/day by age 8 years; for adolescents 1250 mg/day; for adults 700 mg/day. No further recommendations were made for pregnant or lactating adult women.

KEY POINTS

- Calcium and phosphorus both essential for bone mineralization and, as such, the majority of both within the human body is located in bone.
- Calcium also plays a critical role in blood clotting while phosphorous is required for energy metabolism and both are involved in cell signalling processes.
- Milk and dairy products and cereals and cereal products are major dietary sources of both calcium and phosphorus.
- Calcium and phosphorus homeostasis is regulated via the combined actions of calcitonin, PTH, and the biologically active form of vitamin D, 1,25(OH)$_2$D.
- Calcium deficiency over prolonged periods decreases bone mass and increases risk of osteoporosis. Phosphorus deficiency, which is very rare, can cause rickets in children or osteomalacia in adults.

13.3 Iron

Iron is an essential trace metal and plays numerous biochemical roles in the body, including oxygen binding in haemoglobin and acting as an important catalytic centre in many enzymes, for example the cytochromes. However, in excess, iron is extremely toxic to cells and tissues due to its ability to rapidly alter its oxidation state and generate oxygen radicals. Consequently, body iron levels must be tightly regulated to avoid pathologies associated with both iron deficiency and overload. The World Health Organization estimates suggest 1.6 billion people worldwide suffer from anaemia with approximately half of these being due to iron deficiency anaemia, while 1.5–2 billion people have iron deficiency, making this the most common nutritional deficiency syndrome. At the same time one person in 200 of northern European descent is genetically predisposed to the iron-loading disease haemochromatosis. The prevalence of these disorders highlights the importance of maintaining homeostatic control over iron nutrition.

Food sources

Dietary iron is found in two basic forms, either as haem or non-haem iron. Haem is found in meat and meat products that are rich in two major haem-containing proteins, haemoglobin and myoglobin. The most important dietary sources of haem iron are those that are eaten in significant quantities, though these are not necessarily the richest sources of haem. Between 25 per cent and 50 per cent of the total iron content of meat is haem; the

remainder is non-haem iron largely present in iron storage proteins such as ferritin. Therefore, haem iron accounts for approximately 5–10 per cent of the daily iron intake in industrialized countries, whereas in vegetarian diets and in developing countries the haem iron intake is negligible. In their recent review 'Iron and Health', SACN assessed the impact of reducing red meat consumption on iron intake. Red and processed meat accounts for approximately 10 per cent of iron intake in UK adults. It was estimated that reducing red and processed meat consumption to 70 g/day (in line with SACN recommendations) would have little impact on the number of adults consuming iron at levels below the lower reference nutrient intake (LRNI) (SACN 2010).

The main form of iron in all diets is non-haem iron, present in foods such as cereals, vegetables, pulses, and fruits in a number of compounds ranging from simple iron oxides and salts to more complex organic chelates (Table 13.1). Exogenous iron from the soil can be present in significant quantities on the surface of food.

Average dietary iron intakes in the UK are 10–11 mg/day. Cereals contribute between one-third and one-half of our daily iron intake, yet most of the naturally occurring iron in cereals is in the seed coat. However, since the 1950s in the UK, all wheat flours (other than wholemeal) have been fortified with iron by law so that they contain at least 1.65 mg iron/100 g flour. In reviewing the evidence for mandatory fortification, SACN found that removal of the added iron would have a significant negative impact on intakes for several population groups in the UK, particularly older children and young adults. In addition to flour, breakfast cereals and many infant foods are also commonly fortified with iron. However, there are concerns over whether the level of iron used to fortify these food products is appropriate, and whether the form used (usually elemental iron) is bioavailable.

Absorption and metabolism

The absorption of iron by duodenal enterocytes is influenced by a number of variables, especially dietary factors; for example, the iron content of foods, the type of iron present, and other dietary constituents that influence non-haem iron bioavailability. Absorption is also regulated in line with metabolic demands that reflect the amount of iron stored in the body, and the requirements for red blood cell production (Figure 13.4).

Bioavailability

Iron bioavailability varies between foods according to the forms of iron present, food processing and other components of the foods (see Weblink Tables 13.1 and 13.2). Haem iron is the most bioavailable form of iron.

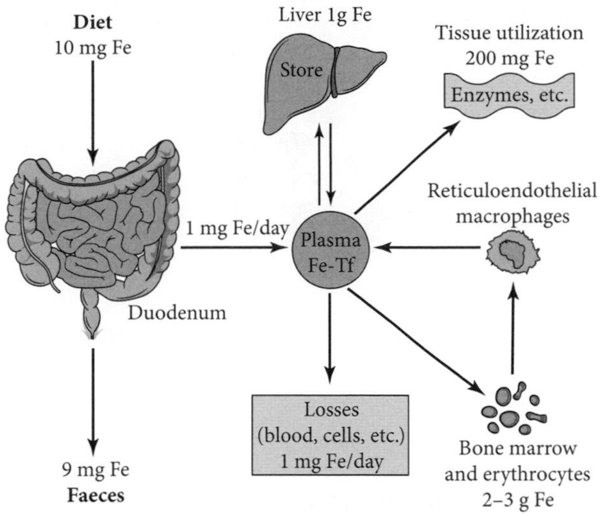

FIGURE 13.4 Body iron metabolism.
Seventy per cent of body iron resides at any one time
in the erythropoietic system: bone marrow, circulating
erythrocytes, and reticulo-endothelial macrophages; 25 per
cent is present in body stores in the liver. Approximately
1 mg Fe/day is absorbed from the diet to replace iron lost
through minor bleeding and cell shedding.

Although it accounts for only 5–10 per cent of dietary
iron in Western countries, absorption of iron from haem-
containing foods is some 20–30 per cent. Compared with
non-haem iron, haem absorption is less influenced by
the iron status of individuals. The calcium content of
the diet is thought to be the only other dietary factor to
influence haem iron absorption, though the mechanism
for this action is unknown. Food preparation alters haem
iron bioavailability; prolonged cooking of meat at high
temperatures is thought to degrade the porphyrin ring
allowing the iron centre to be removed and join the non-
haem iron pool.

Although non-haem iron is the most prevalent form
of dietary iron, only 1–10 per cent is absorbed due to
the profound influence of other dietary components.
The most potent enhancer is ascorbic acid (vitamin C),
which acts by reducing ferric iron to the more soluble and absorbable ferrous form. Other small organic
acids, such as citric acid, also promote the absorption
of non-haem iron, possibly by forming stable soluble
complexes with iron, thereby avoiding precipitation in
the gut lumen. Meat and fish, as well as being abundant
sources of haem and non-haem iron, also significantly
promote the absorption of non-haem iron. The nature of the so-called 'meat factor' is still unclear; protein (high levels of cysteine- and histidine-containing
peptides that could reduce ferric iron to ferrous);
carbohydrate (glycosaminoglycans), and phospholipid
(L-α-glycerophosphocholine) have all been shown to
increase non-haem iron bioavailability.

The best-known dietary inhibitors of non-haem
iron absorption are phytates, which are salts of inositol hexaphosphates found especially in cereal products.
Phenolic compounds found in all plant food sources
are also potent inhibitors of non-haem iron absorption.
Perhaps the best-known group of phenolic compounds
are the tannins found in abundance in tea and red wine.
Both the phytates and phenolic compounds are thought
to form chelates in the intestinal lumen rendering the
iron in a non-absorbable form.

In animals and plants, iron is stored intracellularly in
ferritin. Surprisingly little attention is given to the possibility that ferritin in foodstuffs may be an important dietary source of iron, although a number of recent studies have shown that humans absorb iron from ferritin in
foods. The mechanism of absorption is presently unclear;
some studies suggest ferritin can be absorbed intact via
endocytosis, while others have shown that iron is liberated from the ferritin core within the acidic milieu of the
stomach.

Intestinal absorption

For mechanisms of intestinal iron absorption see
Figure 13.5.

Transport and tissue distribution

Transferrin can bind two ferric iron molecules, delivering the iron to the sites of storage (mainly in the liver)
or utilization (e.g., the bone marrow) (Figure 13.4). Body
iron content is some 3–5 g (approximately 50 mg/kg body
weight). Of this, approximately 70 per cent is present in
the erythropoietic system, 20 per cent is stored as ferritin
and haemosiderin in the liver, 5 per cent is incorporated
into myoglobin in muscle, and 5 per cent bound or utilized by various enzymes. Clearly, erythrocyte production
and destruction accounts for the majority of metabolic
iron turnover in the body (some 20–30 mg/day). The
typical lifespan of a red blood cell is 120 days and after
this time senescent erythrocytes are engulfed by cells of
the reticuloendothelial system (a combination of splenic
macrophages and the Kupffer cells in the liver) and the
iron recovered from haemoglobin by the action of haem
oxygenase. This liberated iron is transported in the blood
bound to transferrin back to the bone marrow for new
red blood cell production or to the liver for storage.

Iron homeostasis

There are no known regulated excretory pathways for excess iron disposal from the body. Iron losses are therefore
restricted to that stored in cells shed from the lining of
the gastrointestinal and urinary tracts, skin and hair, and

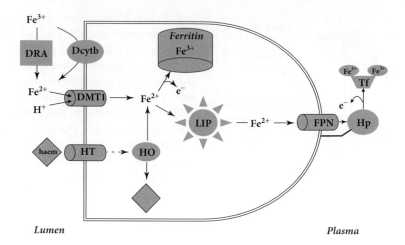

FIGURE 13.5 Mechanisms involved in intestinal iron absorption.

Haem and non-haem iron are absorbed in the duodenum through independent mechanisms (Figure 13.5). Haem is transported into enterocytes via a poorly characterized haem transporter (HT). Inside the enterocyte, the iron contained with the haem porphyrin ring is excised by the action of haem oxygenase (HO) and enters a common pool along with the non-haem iron. Dietary non-haem iron is largely present in its less soluble and non-absorbable ferric (Fe^{3+}) form. It is reduced to ferrous iron (Fe^{2+}) by the coordinated actions of dietary reducing agents (DRA), and the endogenous ferric reductase Dcytb (duodenal cytochrome b). Fe^{2+} is transported into the cell by the divalent metal transporter (DMT1). Uptake is driven by an inwardly directed H^+ gradient. Iron within enterocytes can be re-oxidized to Fe^{3+} and stored ferritin or it can enter a labile intracellular pool (LIP) and be processed for transport out of the cell via ferroportin (FPN), at which point it is immediately oxidized to Fe^{3+} by the basolateral membrane-bound ferrioxidase hephaestin (Hp) and loaded onto transferrin (Tf) for onward transport in the blood.

losses through bleeding. Basal losses of iron amount to approximately 1.0 mg/day in men and post-menopausal women and 1.3–1.4 mg/day in premenopausal women (due to menstrual blood loss). These losses must be replaced by dietary intake to maintain body iron levels. The body has three basic mechanisms for maintaining iron homeostasis: (1) continuous re-utilization of iron recovered from senescent red blood cells, (2) regulation of intestinal iron absorption, and (3) exploitation of an intracellular store (ferritin).

The main regulator of body iron homeostasis is hepcidin, a peptide synthesized in the liver. Hepcidin production is increased in response to elevated iron status and inflammatory stimuli (e.g., cytokines such as interleukin 6) and decreased when iron requirements are high. Hepcidin functions by inhibiting iron release from recycling macrophages, intestinal enterocytes, and the hepatic iron stores. This mechanism serves to control the flow of iron between the absorptive, storage, and recycling compartments and the bone marrow. In pathological situations, high levels of hepcidin lead iron to become trapped within the recycling macrophages resulting in inadequate erythropoiesis and ultimately the anaemia of chronic disease. Inflammation and high levels of hepcidin also contribute to the anaemia of critical illness observed in most patients admitted to intensive care units. However, in the absence of blood loss,

which is also a common contributing factor for these patients, anaemia of critical illness develops too rapidly to be explained simply by the effects of hepcidin on iron metabolism. The rapid onset appears to be due to changes in the turnover of red blood cells via a combination of reduced red cell production (erythropoiesis), programmed destruction of new red cells (neocytolysis), and premature death and degradation of mature red cells (eryptosis). While blood transfusions and iron infusions are sometimes used to as treatments for this type of anaemia, it is not entirely clear how beneficial these treatments are.

Functions

Iron-containing enzymes

Iron plays a major role in regulating energy production via oxidative phosphorylation. The cytochromes are haem-containing enzymes consisting of a globin chain plus a haem group containing one iron atom.

The cytochromes function as efficient electron carriers and play a crucial role in the mitochondrial oxidation of fuels and formation of ATP. Iron is also an important component of other enzymes involved in energy metabolism, for example, aconitase, succinate dehydrogenase, and NADH-dehydrogenase, in which iron is present as non-haem iron. Haem iron is an important component of the antioxidant enzymes catalase and peroxidase.

Pro-oxidant activity

Despite its essential role in metabolism, iron may also act as a pro-oxidant and is therefore potentially harmful. Under circumstances where iron accumulates to levels that exceed the total iron-binding capacity of proteins such as transferrin and ferritin, free iron promotes lipid peroxidation and tissue damage, raising the possibility that disturbances in iron metabolism play a pathogenic role in a number of diseases, including haemochromatosis (see below in section 'Haemochromatosis'), type 2 diabetes, rheumatoid arthritis, and non-alcoholic fatty liver disease. In these disorders, iron is deposited in tissues (liver, pancreas, synovium, etc.) and the underlying pathology is exacerbated by the pro-oxidant effects of the excess iron. Excess iron is also believed to be an aggravating factor in atherosclerotic cardiovascular disease. Although the precise mechanisms are complex and yet to be fully elucidated, accumulation of non-transferrin-bound iron both in the vasculature and in inflammatory macrophages in atherosclerotic plaques may promote vascular permeabilization, immune cell recruitment, inflammation, and oxidative stress (Vinchi 2021)

Immune function

There are complex interactions between iron metabolism, infection, and immune function. First, iron is essential for almost all organisms and it seems that a key part of the human host defence against infection is to remodel iron metabolism at both the tissue and subcellular level to deprive invading organisms of iron supply (so-called nutritional immunity). Infection and inflammation trigger a reduction in circulating iron levels by trapping iron in reticuloendothelial macrophages as a result of the action of hepcidin on ferroportin. Equally, intracellular iron may also be redistributed across different compartments to help combat intracellular pathogens. Conversely, pathogens have evolved a range of strategies to obtain iron from their host, emphasizing the importance of competition for this vital nutrient, Moreover, the remodelling of human iron metabolism in response to infection and inflammation is not without consequences for the host: it contributes to the anaemias of chronic disease and critical illness, as discussed above, and may increase risk of other pathologies.

Iron also contributes to the regulation of innate and adaptive immune function. Iron is an important growth promoter for several crucial immune responses including lymphocyte proliferation and generation of the bactericidal respiratory burst by neutrophils. Hence cell-mediated immunity may be compromised by iron deficiency. While the majority of this work has been carried out in animal models, there is evidence that iron status also affects the risk of infection in humans (reviewed in SACN 2010; Drakesmith & Prentice 2012).

Iron is also a growth factor for microorganisms. Most iron in the body is either sequestered in ferritin or protein-bound and therefore unavailable as a nutrient for microbial growth, but when there is a failure in body iron homeostasis, as in iron overload, an individual may be more at risk of infection. Iron supplementation of iron-deficient populations may also inadvertently increase the risk of infection, particularly in regions where malaria is endemic. This is perhaps best exemplified by the iron supplementation study in Pemba Island, in Tanzania's Zanzibar Archipelago, which was terminated due to increased incidence of adverse health events and rates of hospital admission (Sazawal et al. 2006). While, in general, the published data on iron supplementation and infection risk are inconclusive, there is some evidence for adverse effects in certain groups, for example, children at risk of diarrhoea. This area has been reviewed extensively (SACN 2010).

Effects of iron deficiency and overload

Anaemia

The most common cause of anaemia is nutritional iron deficiency. It is estimated that 1.6 billion of the world's population (largely in developing countries) are anaemic, with 50 per cent of those cases due to iron deficiency. Iron deficiency anaemia is characterized by the presence of small pale red blood cells containing reduced concentrations of haemoglobin (microcytic hypochromic anaemia).

The National Diet and Nutrition Survey indicates that approximately 49 per cent of adolescent females and 25 per cent of adult women in the UK have iron intakes below the LRNI (8.0 mg/day). Yet the prevalence of anaemia in these groups is lower, at approximately 7–9 per cent, and rates of iron deficiency (denoted by low serum ferritin) are approximately 15–17 per cent. This apparent discrepancy has led SACN to suggest the current dietary reference values for iron may be too high and should be reviewed (SACN 2010). Nonetheless, the prevalence of low iron in these groups is a cause for concern.

Whilst mild anaemia in many individuals is of little health consequence (due to a number of compensatory mechanisms such as increased cardiac output, diversion of blood flow to vital organs, and increased release of oxygen from haemoglobin), severe anaemia exceeds the body's ability to adapt, resulting in impaired oxygen delivery to the tissues. This in turn has deleterious effects on a number of important body functions.

Work performance

Work performance, particularly physical work capacity, is severely limited in anaemia. This is ascribed to two main mechanisms; decreased oxygen supply to tissues resulting from a deficit in haemoglobin; reduced activity of cytochromes in muscles impacting on the rate of oxidative phosphorylation and energy production. The relationship between iron deficiency and physical work capacity was the subject of a systematic review of the literature by Haas & Brownlie (2001). They found that aerobic and endurance capacities were both adversely affected by iron deficiency anaemia. Furthermore, activity levels in female tea pickers in Sri Lanka and in female cotton mill workers in China were decreased by anaemia in proportion to haemoglobin levels. Activity levels were increased following iron supplementation. This led additionally to an increase in work productivity.

Cognitive function

The relationship between iron deficiency and impaired performance in mental and motor tests in children is well established. Brain iron content increases throughout childhood and reaches its maximal levels in young adulthood between the ages of 20 and 30 years. Iron is particularly abundant in the basal ganglia, responsible for motor control, and hippocampus, which controls spacial awareness. Iron accumulation during infancy appears to be particularly important not only in determining both total brain iron content but also brain development; an early deficit in brain iron is not compensated for in later years. A number of studies have found that iron deficiency anaemia in early infancy is associated with poor psychomotor development and these effects can persist in the longer-term even if dietary intake levels are improved (Lozoff 2007; Benton 2008). Iron deficiency can also affect educational performance in older children; however, unlike deficiency in early infancy, there is evidence that supplementation can improve a number of markers including attention and concentration in class (Falkingham et al. 2010). In adolescents and adults, iron deficiency anaemia is associated with poor cognitive function, but again can be improved by supplementation (Murray-Kolb & Beard 2007).

Haemochromatosis

The majority of cases of primary iron overload is accounted for by the genetic disease hereditary haemochromatosis, an autosomal recessive disorder affecting mainly populations of Northern European descent. While several different forms of hereditary haemochromatosis exist, as a result of mutations in various genes involved in iron homeostasis, two risk variants have been mapped to the HLA-H (now called HFE) gene region of chromosome 6 that account for more than 80 per cent of the cases of hereditary haemochromatosis. The most common of these two variants has a carrier frequency of approximately one in ten and a homozygous frequency of one in 200 but not all those who are homozygous for the common variant will develop clinical iron overload.

Most cases of hereditary haemochromatosis are not diagnosed until patients present with clinical problems associated with organ failure (typically around 40–50 years of age). Simple and effective treatment of these patients can be achieved through the regular removal of excess body iron by phlebotomy. While it is not possible to treat haemochromatosis by dietary means, the Haemochromatosis Society have published dietary recommendations which include: avoiding iron supplements and breakfast cereals that are heavily fortified with iron, and high doses of vitamin C which would enhance uptake from the diet; restricting the consumption of red meat and offal that are rich in haem iron; consuming tea with meals to limit iron bioavailability.

Status assessment

Currently, there is no single test available to determine with complete accuracy perturbations in body iron status. Therefore, a wide variety of biochemical methods are employed to assess a number of key indices of iron metabolism (Table 13.2).

The basis of dietary recommendations

Daily basal iron losses occur as a consequence of desquamation of cells lining the gastrointestinal tract (0.14 mg/day) and urinary tract (0.1 mg/day); blood loss accounts for a further 0.38 mg/day and bile losses amount to 0.24 mg/day. Minor amounts are lost due to shedding of skin and hair. Thus, basal iron losses are estimated at 14 µg/kg body weight/day. In infants, children, and adolescents, in addition to basal losses, iron is also required for growth of the tissues and organs and for the expanding red blood cell mass. Within the first year of life the infant doubles its iron content and triples its body weight. Body iron content is again doubled between one and six years old. The growth spurt in adolescence also increases iron demand, as does the dramatic increase in haemoglobin concentration seen in males during puberty and the onset of menarche in females. It is estimated that women also lose an average of 0.7 mg/day through menstrual blood loss. However, menstrual blood loss is

TABLE 13.2 Common methods for assessing body iron status

Status indicator	Normal range	Additional information
Serum ferritin	30–300 µg/l	1 µg/l serum ferritin indicates 5–10 mg tissue stored ferritin; 12–15 µg/l indicates empty stores Acute phase protein—false high levels seen in infection and inflammation
Transferrin saturation	25–30%	Values below 16% are indicative of inadequate supply for erythropoiesis Values above 50–60% are generally indicative of haemochromatosis
erythrocyte protoporphyrin	<80 µmol/mol	Protoporphyrin is the final intermediate in haem synthesis Protoporphyrin haem levels rise when iron supply is limited Values >80 µmol/mol haem indicate iron deficiency anaemia Also increased by other diseases, resulting in increased erythroid turnover
Serum transferrin	2.8–8.5 mg/l	A measure of reticulocyte differentiation—shedding of TfR receptor into serum Detectable receptor levels increase in iron deficiency Also increased in all diseases increasing erythrocyte turnover Not affected by iron overload
Haemoglobin	120–180 g/l	Values below 110 g/l indicative of anaemia Not altered by iron overload Not altered in intermediate phases leading to anaemia
Mean cell volume	80–94 fl	Smaller erythrocyte volume in anaemia Cannot distinguish between iron deficiency anaemia and other anaemias

not normally distributed within the population but is skewed to the right, and consequently the estimated average requirement for females for iron is set at the 75th percentile.

Dietary intakes to satisfy the metabolic requirements and iron losses depend largely on the bioavailability of iron in the diet. In most industrialized countries typical diets will be rich in meat, poultry, and fish plus food containing high levels of ascorbic acid. Current UK guidelines assume that iron absorption from typical diets is 15 per cent and this has been used to calculate the current dietary reference values for iron. Assuming average endogenous iron losses in adults are 1.0 mg/day in men and 1.7 mg/day in women, the Estimated Average Requirements for iron are 6.7 and 11.4 mg/day in men and women, respectively; with the RNIs set at 8.7 mg/day for men and 14.8 mg/day for women.

The EAR for breast- or formula-fed infants aged 0–3 months is 1.3 mg/day, which trebles (in line with increased growth) over the next six months to 3.3 mg/day. There are no recommendations for increasing iron intake during pregnancy as the extra demand should be met by a combination of pre-existing body stores, lack of menstrual blood loss, and the increased intestinal absorptive capacity during the second and third trimesters. This recommendation was reviewed by SACN in 2010, which found that there is currently no evidence to support routine iron supplementation of pregnant women, but this should be kept under review. However, SACN supported the NICE guidelines that supplementation should be considered for women with Hb below 110 g/L in the first trimester and 105 g/L at 28 weeks pregnancy. Likewise, there are no recommendations for increasing iron intake during lactation, where iron losses (i.e., secreted in breast milk) are compensated for by the amenorrhoea associated with lactation.

KEY POINTS

- Iron serves a number of key functions in the human body, most notably in oxygen transport and energy production via oxidative phosphorylation.
- Iron is found in food in two forms: haem and non-haem iron. Haem is found in meats and meat products whereas plant-based foods contain essentially only non-haem iron. Haem iron is the more bioavailable form and typically only 1–10 per cent of non-haem iron is absorbed with various dietary components enhancing or inhibiting its absorption in the duodenum.
- While iron is essential, excess free iron is extremely toxic. Therefore, body iron levels must be tightly

regulated. Iron is very unusual in that whole body homeostasis is regulated exclusively at the level of absorption across the gut as there is no known regulatable excretory mechanism.

- The peptide hormone hepcidin, secreted by the liver, is responsible for this regulation.
- Iron deficiency is considered the most common nutritional deficiency syndrome, with an estimated 1.5–2 billion people globally who are iron deficient. Prevalence is particularly high in developing nations but is also a significant public health concern in developed nations.

13.4 Zinc

Human zinc deficiency was first noted in the 1960s in adolescents living in the Nile delta of Egypt and in rural Iran. Since these observations there has been a very significant increase in the understanding of human zinc metabolism. However, there are still significant nutritional questions to be addressed including the assessment of marginal zinc status.

Food sources

Daily zinc intake from an omnivorous diet is typically 10–12 mg/day. The zinc content of foods varies greatly (Table 13.1), with the highest levels found in meat, whole grains, and shellfish (particularly oysters). Animal sources provide the majority of dietary zinc in omnivorous diets (50–60 per cent of daily intake is from meat, fish, dairy, and eggs), mainly due to the high levels of zinc present in muscle (up to 50 mg/kg). In contrast, fat has very low zinc content (5 mg/kg). In many cultures, cereal products are the major dietary energy sources and therefore provide the majority of dietary zinc. In the UK cereal products account for approximately 25–35 per cent of zinc intake. It is important to note that the zinc content of cereals decreases with an increase in the level of refinement of flour.

Absorption and metabolism

Bioavailability and absorption

Zinc is variably absorbed from different food groups, with a range of foods and food components capable of enhancing or inhibiting absorption (see Weblink Table 13.3). Fractional absorption of around 30 per cent is typical of solid diets compared with 60–70 per cent from aqueous solutions. The major inhibitor of absorption is phytate, which is negatively charged at food pH and readily forms insoluble complexes with positively charged ions such as zinc, thereby limiting bioavailability. This inhibitory effect on absorption can be partly overcome by food preparation techniques; for example, addition of yeast during breadmaking increases phytase activity, reducing phytate levels. On the other hand, diets containing large quantities of unleavened bread are poor providers of zinc and are thought to be associated with the growth defects observed in adolescents living in the Nile delta of Egypt and rural Iran. There is also evidence that competition between zinc and other metals (e.g., copper, cadmium, and possibly iron) at the level of intestinal absorption may limit zinc bioavailability.

Animal protein is thought to act as an 'antiphytic' agent and enhances the bioavailability of zinc. It is thought that small peptides and amino acids released during digestion improve the solubility of zinc and protect against the formation of insoluble phytate complexes. A typical omnivorous diet should provide adequate zinc to maintain body homeostasis, since animal protein intake would be sufficient to outweigh the inhibitory effects of the phytate.

Endogenous zinc is also present in the small intestine as a consequence of pancreatic and biliary secretions (zinc is an essential cofactor for the carboxypeptidases involved in protein digestion). This source of zinc is also available for absorption and is taken up by the intestinal epithelial cells by the same route as dietary zinc. Zinc absorption is transporter-mediated and takes place in the small intestine, with the highest absorption rate in the jejunum (Figure 13.6).

Transport and tissue distribution

Zinc entering the blood from the intestinal enterocytes is transported in the portal circulation to the liver bound mainly to albumin (70 per cent) and α_2-macroglobulin (20–30 per cent). Total body zinc amounts to some 2–3 g in a typical adult and is primarily localized intracellularly. Only 0.1 per cent of total body zinc is found in the plasma whereas 60 per cent is found in skeletal muscle and a further 30 per cent is contained within bone (Figure 13.7).

Homeostatic regulation of body zinc content

There is no recognized storage pool of zinc in the body but a number of tissues, especially the liver, are highly active in redistributing zinc between body organs. A key feature of this regulation appears to be the intracellular zinc binding protein metallothionein, a cysteine-rich protein that is induced by high dietary zinc levels and is thought to act as an intracellular zinc buffer.

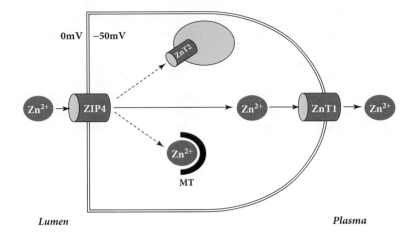

0mV −50mV

Lumen   Plasma

FIGURE 13.6 Mechanisms involved in the intestinal absorption of zinc.
Zinc absorption takes place in the small intestine with the highest absorption rate in the jejunum. Uptake is transporter mediated; ZIP4 is thought to provide the major uptake pathway for zinc from the diet. ZIP4 is mutated in the inborn error of zinc metabolism acrodermatitis enteropathica and is strongly regulated by dietary zinc status. Inside the cell, free zinc is bound by metallothionines (MT), small zinc-binding proteins that maintain a low cytosolic 'free' zinc concentration. In addition, zinc can be accumulated within cytoplasmic vesicles through ZnT2 where it can be incorporated into zinc-dependent proteins and enzymes. Zinc leaves enterocytes via ZnT1 and is transported in the portal circulation to the liver bound mainly to albumin (70 per cent) and α2-macroglobulin (20–30 per cent).

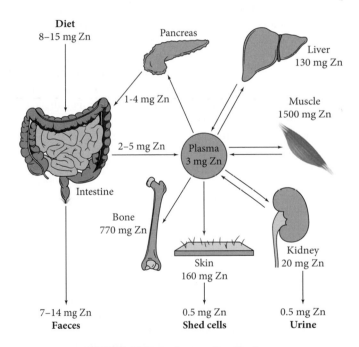

FIGURE 13.7 Body zinc distribution.
Dietary intake (8–15 mg/day) matches endogenous losses in the faeces, urine, and shed skin cells. Zinc is mainly an intracellular ion—hence low plasma but high tissue concentrations.

The small intestine ultimately controls body zinc content by regulating both the amount of dietary zinc absorbed and the quantity of endogenous zinc lost in the faeces mainly from the intestinal, biliary, and pancreatic secretions. Absorption (or reabsorption) of zinc is directly related to body zinc status. Zinc losses via the intestinal route include not only exogenous and endogenous zinc but also that lost as a consequence of shedding of cells lining the gastrointestinal tract. Total losses via this route normally range from 0.5 to 3 mg/day but can be greater in gastrointestinal disease.

In addition to intestinal losses, excretion of zinc via the kidney usually amounts to 0.5 mg/day in healthy adults. Most of the zinc filtered through the glomerulus is reabsorbed from the renal tubular fluid as it passes along the nephron. Urinary zinc excretion can be altered significantly in renal disease, insulin-dependent diabetes mellitus, alcoholism, and starvation.

Further zinc can be lost from the body via skin, hair, and sweat (about 0.5 mg/day). Also, zinc accumulates in the prostatic fluids and an ejaculation of semen contains up to 1 mg of zinc.

Functions

The major function of zinc in human metabolism is as a cofactor for over 100 known metalloproteins and enzymes; however, this may be an under-estimate as recent genome analysis suggests that 3–10 per cent of the human genome may encode zinc metalloproteins.

Several key enzymes involved in the synthesis of RNA and DNA are zinc dependent, including the DNA and RNA polymerases. In addition, zinc plays a key role in gene transcription as an essential structural component of the zinc finger motifs found in several nuclear hormone receptors (e.g., those for vitamin D, testosterone, and oestrogen) and transcription factors. The ability of zinc finger-containing transcription factors to bind DNA is highly zinc-dependent and may be lost in zinc deficiency.

In addition to zinc fingers, structural zinc centres are essential for several enzymes, including the antioxidant enzyme superoxide dismutase, which contains zinc and copper. Zinc exhibits further antioxidant actions, including induction of metallothionein synthesis, and the ability to bind to sulfydryl groups on various proteins, protecting them from oxidation.

Zinc has profound effects on carbohydrate metabolism. Zinc-deficient animals have significantly impaired glucose tolerance, due to a reduced ability to secrete insulin from the pancreas in response to an oral glucose load. In addition to co-release with insulin from the beta cell, zinc is thought to increase and stabilize insulin binding to its receptors on target tissues, prolonging its biological actions. Intriguingly, recent genome-wide analyses have linked a beta cell zinc transporter polymorphism with an increased risk of developing type 2 diabetes, further establishing the close relationship between zinc and insulin activity.

Zinc deficiency and excess

Zinc deficiency

A number of zinc deficiency states of varying severity occur, especially in low-income countries. Zinc deficiency, along with deficiency of vitamin A, is responsible for the largest disease burden attributable to micronutrient deficiencies globally in children under five years of age and is particularly prevalent in South Asia, Sub-Saharan Africa, and regions of Central and South America. In the UK, zinc intakes are sufficient to replace endogenous daily losses and therefore severe zinc deficiency does not occur.

The first observations of human zinc deficiency were made in the 1960s in Iran and Egypt (Prasad et al. 1963). It was suggested that zinc deficiency was the major contributing factor to a syndrome in which adolescents presented with dwarfism and hypogonadism. A number of similar cases were subsequently identified, mainly in Middle Eastern countries. The primary causes of zinc deficiency in these patients were two-fold: nutritional deficiency due to poor zinc bioavailability from a diet rich in unleavened bread; and geophagia (the practice of eating clay), which would further reduce zinc absorption.

Cases of severe zinc deficiency have been reported in industrialized countries and are characterized by dermatitis, diarrhoea, and impaired immunity leading to greater susceptibility to infections. This syndrome, called acrodermatitis enteropathica, does not have its origins in nutritional deficiency but rather develops due to an autosomal recessive inborn error of zinc metabolism leading to a decrease in the absorption of dietary zinc. Recently, an intestinal zinc transport protein (ZIP4) has been shown to be mutated and functionally impaired in a number of families affected with acrodermatitis enteropathica, demonstrating that this defective protein contributes to the aetiology of the disease.

Zinc deficiency inhibits growth and zinc supplementation studies have shown improved growth in infants and children (Krebs et al. 2014). The health problems associated with zinc deficiency are multifactorial and it has been estimated that these issues may contribute directly to 500,000 deaths annually in infants and children under five years of age.

Zinc is found associated with the skeleton where it is thought to play a central role in the turnover and metabolism of the connective tissues. Zinc deficiency has a negative influence on bone formation, which might result from imbalances in DNA synthesis and protein metabolism. Zinc is also an essential cofactor for a number of crucial enzymes including alkaline phosphatase, which plays a major role in bone mineralization, and collagenase, which is fundamental to the development and remodelling of bone structure.

Zinc excess

Whilst on the whole zinc is considered to have low toxicity, acute excessive intake (of the order of 2 g zinc) can cause symptoms such as nausea, abdominal pain, vomiting, diarrhoea, and fever. In practice, high zinc intake from diet is unlikely to present a problem as zinc in the body is tightly regulated with excess excreted in faeces and urine (Figure 13.7). There are some concerns that high doses of zinc (75–300 mg/day) over a prolonged period of time might adversely influence status of other important trace elements, in particular copper and iron, primarily by inhibiting their absorption. However, these findings are largely from *in vitro* and animal studies and there are too few human studies to substantiate this hypothesis.

Status assessment

The intracellular nature of zinc, its tissue distribution (and homeostatic redistribution), and the absence of a defined storage pool make the assessment of zinc status difficult. Several indices of zinc status are in use including zinc levels in plasma, erythrocytes, leukocytes, neutrophils, and hair as well as various physiological measurements, including the activity of a number of metalloenzymes, dark adaptation, and taste acuity, but all have limitations. However, recent studies suggest that plasma or serum zinc may still be a useful biomarker of zinc deficiency and respond to supplementation at the population level, but further research is required.

The basis of dietary recommendations

Endogenous losses of zinc are dependent on zinc intake and the homeostatic mechanisms controlling faecal loss. Zinc requirements are based on replenishing basal losses. A number of methods have been used to determine these, including metabolic balance studies, turnover time of radiolabelled endogenous zinc pools, and measurement of total endogenous zinc loss. In people eating a typical Western diet providing 10–12 mg zinc/day, endogenous losses amount to 2–3 mg/day and this amount needs to be replaced. The dietary reference values for the UK are based on these figures and assume that 30 per cent of dietary zinc is bioavailable for absorption. This translates into RNIs of 9.5 and 7.0 mg/day for men and women, respectively. For infants and children, factorial analysis has suggested that zinc losses are related to body size and that the zinc requirements for growth increase incrementally with age. Consequently, the RNI for infants has been set at 4.0 mg/day, rising to 9.0 mg/day in prepubescent children. Currently, the estimates for dietary zinc requirements for the elderly are the same as for the general adult population. However, there is evidence that zinc status is reduced in the elderly, possibly as a consequence of reduced absorptive efficiency, and this requires further research.

In the UK there is no recommendation for additional zinc in pregnancy as the additional requirements seem to be offset by adaptive responses. In lactation, additional zinc requirements have been calculated on the basis of a zinc secretion in milk of 2.13 mg/day, giving an increase in the RNI of 6.0 mg/day. Requirements fall after four months of lactation, as milk secretion decreases, to an additional 2.5 mg/day over the RNI.

KEY POINTS

- Zinc is associated with hundreds of different proteins in the human body where is serves either as a cofactor for enzymic activity or as a structural component.
- Zinc is found in most foods, but concentrations vary markedly with the highest levels in meat, whole grains, and shellfish.
- Absorption of dietary zinc is typically around 30 per cent and, like iron, other components of foods can inhibit (e.g., phytate) or enhance (e.g., animal protein) its absorption in the gut.
- Zinc deficiency in children inhibits growth. Deficiency can also cause dermatitis, diarrhoea, and compromised immune function, leading to greater susceptibility to infections.
- While serum zinc concentrations can be useful in detecting deficiency, a robust biomarker of zinc status that can detect moderate insufficiency remains elusive.

13.5 Iodine

Iodine is an essential micronutrient that forms a vital component of the thyroid hormones thyroxine (T_4) and triiodothyronine (T_3), which are crucial regulators of the metabolic rate and physical and mental development in humans. Iodine deficiency is prevalent in many areas in the world, where it constitutes a major problem of public health nutrition, but was thought to be relatively rare in the UK. However, a report of a cross-sectional study on UK schoolgirls published in *The Lancet* (Vanderpump et al. 2011) suggested that the UK was also iodine deficient. Based on current dietary intake data, at-risk groups include not only school-age children but also women of childbearing age and their babies. In response, SACN commissioned a review of iodine status in the UK (SACN 2014).

Dietary sources

Iodine is usually found in food as inorganic iodide or iodate. The iodine content of plants and cereals varies dramatically (from 10 µg/kg to 1 mg/kg) depending on the iodine content of the soil. Similarly, the iodine concentration in drinking water lies in the range 0.01–70 µg/l, depending on geographical location and therefore for many population groups does not constitute a major dietary source. The richest sources of iodine in the diet are marine fish, shellfish, and sea salt. Significant amounts of iodine are also provided by multivitamin and mineral supplements and by seaweed. In some regions of the world where iodine deficiency is a significant problem, extra iodine is provided in the diet in the form of iodide- or iodate-supplemented salt or bread.

For the UK population, cow's milk and dairy products represent the major sources of dietary iodine. The iodine content of cow's milk varies depending on the type of animal feed, iodine supplementation, and use of iodophor sanitizers during milking. The rapidly expanding consumer use of plant-based milk alternatives raises some concern in this context as, unless fortified, these products typically contain low levels of iodine (Bath et al. 2017). Consumption of plant-based milk alternative products is forecast to continue to grow in response to consumer concerns about sustainability of their diets. Most milk alternatives are already fortified with calcium, vitamins B_{12} and D, and routine addition of iodine as a fortification in these products would help address this concern.

Absorption and metabolism

Bioavailability

Iodine is rapidly and efficiently absorbed in the proximal small intestine as iodide. In addition, some organic iodine complexes, such as thyroid hormones added to animal feeds, can be absorbed intact. The remaining larger organic complexes are lost in the faeces. Other iodine-containing foods and iodates, which are often used for example as salt fortificants or as food additives, are readily broken down and reduced to iodide in the intestinal lumen.

Iodine content of foods can be significantly affected by the way that foods are cooked. For example, boiling foods reduces their iodine content by approximately half whereas frying decreases iodine by less than 20 per cent. Bioavailability and absorption is also influenced by other dietary components, especially by brassicas (e.g., cabbage, broccoli, etc.) that are rich in sulphur-containing glucosides (glucosinolates), which can liberate the 'goitrogens' thiocyanates and isothiocyanates that compete with iodide for absorption and tissue uptake. This is a particular problem is some developing countries, where goitrogens are abundant in staple crops such as cassava.

Transport and tissue distribution

Iodate is rapidly converted to iodide in the intestinal lumen and is absorbed by the intestinal sodium/iodide co-transporter. Unlike the majority of trace elements, iodide in the blood does not appear to be bound to plasma proteins and is available for uptake by all the tissues of the body. The majority of the circulating iodide (approximately 80 per cent) is rapidly taken up by the thyroid gland, but significant amounts are also accumulated by the salivary glands, choroid plexus (the area on the ventricles of the brain where cerebrospinal fluid is produced), and the lactating mammary gland. Uptake of iodide by all of these tissues employs a similar mechanism utilizing a sodium/iodide co-transporter, which is stimulated by the thyroid-stimulating hormone (TSH) released from the pituitary gland. Total body iodine levels are 15–20 mg in healthy adults. Once the iodine requirements for thyroid hormone production have been met, excess circulating iodide is removed from the blood by the kidney and excreted in the urine. Urinary iodide excretion is a good indicator of body iodine status.

Function: the thyroid hormones

Iodide is taken up from the circulation by the follicle cells of the thyroid and passes into the inner colloidal space where it is rapidly oxidized to iodine (I_2), by thyroid peroxidise (a haem-containing enzyme), and reacts with the tyrosine residues on thyroglobulin, a large glycoprotein, to produce monoiodotyrosine (MIT) and diiodotyrosine (DIT). These two precursors can condense to form the thyroid hormones T_3 and T_4. These remain bound to thyroglobulin and are stored within the colloid until the thyroid is stimulated by TSH, whereupon thyroglobulin is taken up by the follicle cells and acted on by lysosomal enzymes to liberate active T_3 and T_4, which are subsequently released into the circulation. T_4 has a relatively low biological activity and serves as a reservoir for the production of the more active T_3 following removal of a 5' iodine by selenium-dependent deiodinases present in the liver, kidney, muscle, and pituitary.

The major functions of the thyroid hormones are the maintenance of the metabolic rate, cellular metabolism, and growth. Their function is elicited by binding to nuclear receptors that in turn bind to DNA and regulate the transcription of several genes in target tissues, in particular the brain, heart, liver, and kidneys.

Effects of iodine deficiency and excess

Iodine deficiency causes a wide range of disorders collectively known as iodine deficiency disorders in which symptoms range from mild, such as goitre, to severe, including mental retardation or cretinism. The foetal brain is particularly susceptible to iodine deficiency if the mother is iodine deficient. Iodine deficiency in foetal life results in failure to myelinate the central nervous system, especially the cerebellum. The neurological damage caused by iodine deficiency in children can be irreversible and devastating, resulting in mental retardation, and hearing and speech defects.

Goitres are formed by enlargement of the thyroid gland. Iodine deficiency leads to a decrease in T_4 production, which in turn stimulates the pituitary gland to produce TSH. This increase in TSH production stimulates the thyroid follicles to enlarge and multiply, giving the characteristic goitre appearance. With very large goitres there can be additional problems such as blockage of the oesophagus and trachea as well as damage to the laryngeal nerves.

Most people are very tolerant to a wide range of iodine intakes, and no adverse effects of up to 2 mg iodine/day have been reported. These levels are unlikely to be achieved in the normal diet and are more likely to result from iodine contamination of food or water supply. Some individuals are sensitive to iodine and may experience mild skin irritations following higher than normal iodine intakes.

Status assessment

Assessment of iodine status in populations living in areas at risk of iodine deficiency is based on a number of methods. The size of the thyroid is measured by palpation or ultrasound, giving a measure of the number of goitres that can be classified in accordance with WHO/UNICEF/ICCIDD guidelines.

Urinary iodine is commonly used as an indicator of iodine intake and population status. Urinary iodine can be determined either as total urinary iodine excretion (UIE), based on 24-hour urine collections, or urinary iodine concentration (UIC), based on spot urine samples, with the latter being presented either as absolute concentration (µg/L) or corrected for creatinine concentration (µg/g creatinine). UIE is generally considered a good indicator of recent iodine intake, although homeostatic adaptions could lead to underestimates at very high and low intakes, and median UIE is used at a population level for assessing iodine status and the risk of iodine deficiency (values below 50 µg iodine/day indicate moderate to severe deficiency) (SACN 2014). UIC is more convenient to measure that UIE as spot urine samples can be used in place of 24-hour collections. If sufficient samples can be collected, the effects of confounders such as hydration, time of day, and day-to-day variation are evened out so that median UIC (µg/l) correlates well with median UIE. However, these confounders mean that UIC is certainly not suitable to assess iodine status for individuals.

Neonatal levels of TSH in the serum (normal range 0.5–5.5 mU/L) are used routinely in most developed countries to give an indication of congenital hypothyroidism.

Dietary recommendations

In adults, an iodine intake of 70 µg/day appears to be the minimum necessary to avoid the appearance of goitre and has therefore been set as the LRNI for the UK. The RNI has been set at 140 µg/day to allow for different dietary patterns. Recent estimates for iodine intake in the UK suggest means of 180 and 140 µg/day for men and women, respectively. Because of the possible risk of iodine toxicity, the Expert Group on Vitamins and Minerals (2003) in the UK recommended that iodine intake should not exceed 940 µg/day. Data from the National Diet and Nutrition Survey in the UK indicate that adolescent boys and girls (19–28 per cent with intakes below the LRNI) may be at risk of having low iodine status. This is a particular concern for women entering reproductive years as poor maternal iodine status during pregnancy can impair foetal development. Those excluding milk and fish from their diet may be at higher risk of having low iodine status.

UK DRVs do not currently advocate an increment in iodine for pregnant or lactating women. In contrast, the WHO recommends that pregnant and lactating women should have iodine intakes of 250 µg/day. The SACN Subgroup on Maternal and Child Nutrition considered there was insufficient evidence to recommend a revision to the DRVs for iodine for pregnant and lactating women.

KEY POINTS

- Iodine is essential as it forms a component of the thyroid hormones thyroxine (T_4) and triiodothyronine (T_3).
- The richest dietary sources of iodine are marine fish, shellfish, and sea salt. For the UK population, cow's milk and dairy products represent the major sources of dietary iodine.
- The bioavailability and absorption of dietary iodine is inhibited by goitrogens present in a range of foods such as cruciferous vegetables, soy, and cassava.
- Iodine deficiency is prevalent in many areas in the world. For many years iodine deficiency was thought to be relatively rare in the UK. However, more recent analyses which suggest that the UK also has iodine deficiency may be a significant public health concern, in particular, for at-risk groups such as school-

age children, but also women of childbearing age and their babies.

- Iodine deficiency causes disorders in which symptoms range from mild, such as goitre, to severe, including mental retardation or cretinism.

13.6 Selenium

Selenium is a metalloid mineral that exists in a number of oxidation states. It is found in all soils and rocks at varying concentrations, depending on geological factors. Plant selenium content is determined by the concentration in soils and this consequently enters the animal and human food chain. Hence the geographical variation in soil selenium content has a dramatic impact on dietary intake and selenium status in different populations globally. To circumvent these problems Finland, for example, has been adding selenium to its food chain by including selenate in fertilizers used to treat arable crops. This has been associated with a consequent rise in plasma selenium levels, close to values at which plasma glutathione peroxidase (GPx3; the main biomarker of selenium status) is optimally active.

Dietary sources

Selenium is found in a variety of foods, with cereals, meat (particularly offal), Brazil nuts, and fish providing particularly rich sources. However, many of these sources are consumed in small amounts and do not contribute significantly to selenium intake in the UK. The main sources of selenium in the UK diet are breads, cereals, fish, and meat. Until the mid-1980s, most wheat entering the UK food chain was imported from Canada. However, due to changes in the economic and political climate, together with the advent of modern food manufacturing technologies, UK and European wheat, which has lower selenium content, now predominates. Consequently, there has been a considerable reduction in selenium intakes and status in the UK over the last two decades. However, despite these changes, bread and cereal products still contribute 27 per cent of daily UK selenium intakes. Meat (29 per cent), fish (16 per cent), and eggs and dairy products (10 per cent) are the other major contributors (Table 13.1).

Selenium intakes are low in all population groups in the UK. National Diet and Nutrition Survey data report intakes for those aged 11 years and older are approximately 65–70 per cent of the RNI, with median intakes of 33–52 μg/day. The data also show that 24–59 per cent of these age groups have intakes below the LRNI.

Absorption and metabolism

Selenium is present in foods mainly in organic forms such as the amino acids selenocysteine (from animal products) and selenomethionine (mainly in cereals), but also exists in the environment as inorganic selenide, selenite, and selenate compounds. In garlic for example, up to 50 per cent of the selenium content is present as selenate. Selenite is usually only found in supplements and food fortificants.

The mechanisms involved in intestinal selenium absorption have not been fully elucidated; however, there is good evidence that selenomethionine and selenocysteine share common transport pathways with methionine and cysteine, respectively. Unlike many micronutrients, the absorption of selenium does not appear to be regulated by metabolic demand. The average absorption of selenium from food is approximately 80 per cent and there is evidence that organic forms (i.e., seleno-amino acids) are taken up more efficiently than inorganic compounds. In addition, there is evidence that the bioavailability of inorganic selenium compounds is decreased by heavy metals and by a high sulphur diet. In humans, all absorbed forms of selenium are metabolized to hydrogen selenide. This in turn is converted to selenophosphate, which is used to synthesize selenocysteine. Selenium in the form of selenocysteine is incorporated into a number of selenoproteins. Selenocysteine is the '21st amino acid' and has its own codon (UGA). In most translation events UGA acts as a stop codon for protein synthesis; however, in selenoproteins it permits the insertion of selenocysteine due to the presence of so-called SECIS (selenocysteine insertion sequence) elements in the 3' untranslated region of target mRNAs.

Function

Twenty-five human selenoproteins have been identified (see Weblink Table 13.4) and include iodothyronine deiodinase, responsible for the conversion of T_4 to T_3, thioredoxin reductase, which reduces nucleotides in DNA synthesis, and members of the glutathione peroxidase (GPx) family that are important antioxidant enzymes. Plasma GPx activity (GPx3) is used as a biomarker of selenium status.

Epidemiological evidence as well as data from animal studies point to a role for selenium in reducing cancer incidence. The Nutritional Prevention of Cancer trial (Clark et al. 1996), carried out in the USA, was the first human study to test the hypothesis that selenium supplementation could reduce cancer risk. Subjects with a history of non-melanoma skin cancer were given either selenium supplements (200 μg/day selenium-enriched yeast containing predominantly selenomethionine) or a placebo. Interestingly, while there was no direct effect of selenium supplementation on non-melanoma skin cancer there was a significant reduction in both total cancer incidence and cancer mortality in the supplemented group. In particular, cancers of the prostate, colon, and lung were greatly reduced in those in the supplemented group who had relatively low baseline selenium status, suggesting that the effects of selenium were tissue-specific and related to selenium status. The effect on prostate cancer incidence led to the much larger Selenium and Vitamin E Cancer Prevention Trial (SELECT) in 2001. However, in summer 2008 the trial was stopped, as there was no evidence of a beneficial effect of selenium supplementation on prostate cancer incidence (Lippman et al. 2009).

In the UK, SACN has reviewed the evidence for an association between selenium status and a range of health outcomes (SACN 2013). The report concluded that no adverse health consequences of dietary intakes at the levels typically seen in the UK or benefits of higher intakes have been convincingly demonstrated. However, a more recent meta-analysis identified significant inverse associations between selenium exposure and risk of various cancer types (Cai et al. 2016) and this remains an area of active research.

Effects of selenium deficiency and excess

The best-characterized selenium deficiency syndrome is Keshan disease, a cardiomyopathy that affects children and women of child-bearing years in rural China where soils are selenium deficient. Symptoms are further exacerbated by viral infection. Kashin-Beck disease, a form of osteoarthropathy observed in rural China, is also associated with severe selenium deficiency but other factors including low iodine status and food mycotoxins are also important in the aetiology of this disease.

Low selenium status has been associated with various other health problems including lower reproductive capacity, decreased thyroid function, and reduced antioxidant status. Pathologies associated with selenium deficiency are linked to decreased activity of a number of selenoenzymes.

Low selenium status is also an established risk factor for a number of viral infections, being associated with higher susceptibility to infections and more severe disease outcome. Recent preliminary studies have reported an inverse association between selenium status and risk of death due to COVID-19 infection (Moghaddam et al. 2020). However, more research is required to confirm any causal association and determine whether low selenium status is truly a risk factor for SARS-CoV2 infection

and COVID-19 mortality or whether inflammation due to the infection causes a decrease in selenium status with the magnitude of the decline being a surrogate for the severity of the disease.

While selenium is an essential micronutrient, and supplementation or fortification of foods may in many cases be advantageous, selenium is exceedingly toxic in excess. While selenosis is less common than selenium deficiency, a three- to four-fold change in selenium intake can be the difference between beneficial and harmful intakes. Symptoms of selenium excess (consumption of >800µg/day) include brittle hair and nails, skin lesions, and garlic odour on the breath due to expiration of the metabolite dimethyl selenide. The Expert Group on Vitamins and Minerals (2003) in the UK has set a safe upper limit for selenium intake at 450µg/day. In North America, the Institute of Medicine in 2000 set the upper level of tolerable intake at 400µg/day.

Status assessment

Selenium status can be assessed by measuring selenium levels in a number of tissues including plasma or serum. A serum selenium level of 100 µg/L is generally seen as indicative of nutritional adequacy. Current median plasma selenium levels, measured as part of the UK's National Diet and Nutrition survey are 72 µg/L in those aged 11–18 years and 86 µg/L in those aged 19–64 years. Selenium levels in red blood cells, hair, and toenails are seen as good indicators of long-term selenium intakes. The antioxidant activity of plasma glutathione peroxidase (GPx3) has emerged as a functional biomarker of selenium status but it is a poor indicator of status in individuals with moderate or high intakes. Current UK dietary recommendations are based on the calculated intake of selenium required to optimize the activity of the plasma GPx3.

Dietary recommendations

Keshan disease is not seen in populations where selenium intake is >20 µg/day. On this basis the World Health Organization has set the basal requirements (i.e., the amount required to prevent pathology) at 21 µg/day for males and 16 µg/day for females and its recommended intakes at 40 µg/day for males and 30 µg/day for females. The current UK RNI for selenium is 75 µg/day for men and 60 µg/day for women. UK selenium intakes decreased from the 1970s to the turn of the millennium (SACN 2013) and while the National Diet and Nutrition Survey suggests intakes have been stable over the last decade, intakes remain low (median intakes of 52 and 46 µg/day for men and women, respectively, aged 19–64 years). In contrast, selenium intakes in New Zealand have increased in recent years due to the import of selenium-rich wheat from Australia. Nutrient reference values in Australia and New Zealand were set in 2006 at 70 µg/day for men and 60 µg/day for women. In the USA, where median intakes are substantially higher than in the UK (154 µg/day for men and 98 µg/day for women), RDA for selenium is set at 55 µg/day with no gender difference.

KEY POINTS

- Selenium is required for the production of seleno-proteins, which are involved in antioxidant function, thyroid hormone production and regulation, DNA synthesis, and various other cellular processes.

- Selenium is found in a variety of foods, with cereals, meat, Brazil nuts, and fish providing particularly rich sources. The main contributors to selenium intake in the UK diet are breads, cereals, fish, and meat.

- A high proportion of dietary selenium is generally absorbed in the gut.

- The selenium content of plant foods is determined by the selenium content of the soil they grow in, which varies geographically having a profound effect on overall dietary intakes.

- Data from the National Diet and Nutrition Survey indicate that median selenium intakes are currently below the RNI for most groups of the population.

13.7 Copper

Copper is the third most abundant dietary trace metal and its major biological function is as a catalytic centre in numerous enzymes involved in redox reactions (see Weblink Table 13.5).

It is found at high levels in shellfish, liver, kidney, nuts, and whole grain cereals. The absorption efficiency ranges from 10–50 per cent and is variable depending on both requirements and bioavailability of the diet (see Weblink Table 13.6). The majority of absorption occurs in the duodenum via a carrier-mediated mechanism (Figure 13.8). Copper homeostasis is tightly maintained by regulating absorptive efficiency and biliary excretion.

Dietary-induced copper deficiency is relatively rare in human beings due to the plentiful and varied supply in the diet and the high efficiency of absorption. However, the assessment of marginal copper deficiency is problematic due to the lack of a sensitive and specific biomarker of copper status. Most reported incidences of deficiency are in association with prolonged diarrhoea and/or malnutrition, particularly in infants. Nutrient-nutrient interactions have also been linked with deficiency, with over-the-counter zinc supplements suppressing dietary copper absorption and ultimately resulting in deficiency.

Menke's disease, a rare congenital condition, involves failure of copper absorption and leads to severely

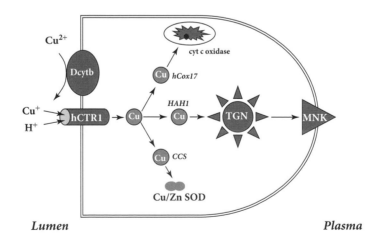

FIGURE 13.8 Mechanisms involved in the intestinal absorption of copper.
Dietary copper is largely present as Cu^{2+}; however, it is rapidly converted to Cu^+ in the reducing environment of the small intestinal lumen, together with endogenous copper reductase activity on the apical surface of enterocytes (most likely Dcytb also functions as a copper reductase). Cu(I) is transported into intestinal cells by hCTR1 (a proton-coupled transporter. Absorbed copper is immediately bound by a number of intracellular chaperones that direct copper to specific cellular sites; hCOX17 travels to the mitochondria, where its copper load is incorporated into cytochrome *c* oxidase; CCS delivers copper to the cytosolic Cu/Zn superoxide dismutase; HAH1 takes copper to the trans golgi network (TGN), where it is loaded onto the Menkes ATPase (MNK) or ATP7A protein. Copper-loaded MNK translocates to the basolateral membrane of the enterocyte, releasing its copper into the portal circulation where it is bound by histidine and albumin for delivery to the liver.

impaired mental development, inability to keratinize hair, and skeletal and vascular problems. These symptoms are associated with impairment of a number of copper-dependent enzymes. Dietary copper overloading is also rare due to the body's ability to excrete excess copper in the bile. These mechanisms fail in a second congenital disease, Wilson's disease, which leads to copper accumulation in the body, particularly in the liver and the basal nuclei of the brain, with consequent pathological damage.

KEY POINTS

- Copper is an essential cofactor in numerous enzymes.
- Rich dietary sources of copper include shellfish, liver, kidney, nuts, and wholegrain cereals.
- Absorption of dietary copper is normally in the range of 10–50 per cent and is dependent on the individual's status and dietary factors.
- Copper deficiency is rare in humans. However, little is known about the prevalence of marginal deficiency due to a lack of robust, sensitive biomarkers of status.
- The rare congenital disorders Menke's disease and Wilson's disease cause copper deficiency and overload, respectively.

13.8 Concluding comments

The detrimental health impacts of deficiencies in certain minerals and trace elements remain a matter of major global health concern. Equally, the true health impact of mineral and trace element insufficiencies or suboptimal status are not well defined and better biomarkers of status are urgently needed. Looking forward, dietary patterns in many countries need to change to address sustainability issues and this is likely to have profound effects on mineral and trace element intakes and bioavailability. Innovative solutions need to be found to resolve these challenges.

KEY POINTS

- Calcium and phosphorus are the major bone minerals; 99 per cent of total body calcium and 85 per cent of total body phosphorus are present in the bone and are in a dynamic equilibrium with the plasma that is controlled by the hormones PTH, $1,25\text{-}(OH)_2D_3$, and calcitonin.
- Non-haem iron absorption from the diet is strongly influenced by other dietary components that may act to promote (e.g., ascorbic acid) or inhibit (e.g., phytic acid) bioavailability. The amount of iron absorbed is usually sufficient to replace basal losses and satisfy metabolic demand for growth and red blood cell formation. However, if intake drops then there is a risk of developing iron deficiency anaemia, the most common nutritional deficiency disease in the world.
- Zinc is essential for a wide array of biochemical processes that play a central role in growth and development. Its major function is as a structural and/or catalytic centre in a number of metalloenzymes and proteins that rely on the presence of zinc for their activity.
- Iodine is an essential nutrient because it is a constituent of the thyroid hormones thyroxine and triiodothyronine. These hormones are required for normal growth and maintenance of the metabolic rate. If iodine intake is low, iodine deficiency diseases can become prevalent and symptoms can range from mild (e.g., goitre) to severe (e.g., cretinism).
- For these and many other key minerals and trace elements there is a requirement for better functional markers of body nutritional status. In particular, these markers where possible should relate to physiological or biochemical factors connected to the known essential function of these micronutrients and should be able to identify changes in body stores as well as dietary intakes.
- There is a requirement to fully evaluate specific health risks associated with marginal deficiencies of specific minerals and trace elements that have not developed into a full deficiency syndrome. Similarly, more information is required regarding the effects of excess intakes of many of these elements.

ACKNOWLEDGEMENTS

The authors would like to thank Sarah Bath, Susan Fairweather-Tait, Diane Ford, Linda Harvey, Rachel Hurst, Margaret Rayman, and Victor Preedy for their helpful comments and input.

Be sure to test your understanding of this chapter by attempting multiple choice questions. See the Further Reading list for additional material relevant to this chapter.

REFERENCES

Abrams S A (2010) Calcium absorption in infants and small children: Methods of determination and recent findings. *Nutrients* **2**, 474–480.

Bath S C, Hill S, Infante H G et al. (2017) Iodine concentration of milk-alternative drinks available in the UK in comparison with cows' milk. *Br J Nutr* **11**, 525–532.

Benton D (2008) Micronutrient status, cognition and behavioral problems in childhood. *Eur J Nutr* **47**(Suppl 3), 38–50.

Bread & Flour Regulations (1998) https://www.legislation.gov.uk/uksi/1998/141/contents/made

Cai X, Wang C, Yu W et al. 2016 Selenium exposure and cancer risk: An updated meta-analysis and meta-regression. *Sci Rep* **6**, 19213.

Clark L C, Combs G F Jr, Turnbull B W et al. (1996) Effects of selenium supplementation for cancer prevention in patients with carcinoma of the skin: A randomized controlled trial. Nutritional Prevention of Cancer Study Group. *JAMA* **276**, 1957–1963.

Drakesmith H & Prentice A M (2012) Hepcidin and the iron-infection axis. *Science* **338**, 768–772.

Falkingham M, Abdelhamid A, Curtis P et al. (2010) The effects of oral iron supplementation on cognition in older children and adults: A systematic review and meta-analysis. *Nutr J* **9**, 4.

Haas J D & Brownlie T 4th (2001) Iron deficiency and reduced work capacity: A critical review of the research to determine a causal relationship. *J Nutr* **13**, 676S–688S.

Krebs N F, Miller L V, Hambidge K M (2014) Zinc deficiency in infants and children: A review of its complex and synergistic interactions. *Paediatr Int Child Health* **34**, 279–288.

Lippman S M, Klein E A, Goodman P J et al. (2009) Effect of selenium and vitamin E on risk of prostate cancer and other cancers: The selenium and vitamin E cancer prevention trial (SELECT). *JAMA* **301**, 39–51.

Lozoff B (2007) Iron deficiency and child development. *Food Nutr Bull* **28**, S560–S571.

Moghaddam A, Heller R A, Sun Q et al. (2020) Selenium deficiency is associated with mortality risk from COVID-19. *Nutrients* **12**, 2098.

Murray-Kolb L E & Beard J L (2007) Iron treatment normalizes cognitive functioning in young women. *Am J Clin Nutr* **85**, 778–787.

Prasad A S, Mial A, Farid Z et al. (1963) Zinc metabolism in patients with the syndrome of iron deficiency anaemia, hypogonadism and dwarfism. *J Lab Clin Med* **61**, 537–549.

Sazawal S, Black R E, Ramsan M et al. (2006) Effects of routine prophylactic supplementation with iron and folic acid on admission to hospital and mortality in preschool children in a high malaria transmission setting: Community-based, randomised, placebo-controlled trial. *Lancet* **367**, 133–143.

Scientific Advisory Committee on Nutrition (2010) *Iron and Health Report.* https://www.gov.uk/government/uploads/system/uploads/attachment_data/file/339309/SACN_Iron_and_Health_Report.pdf (accessed June 2021).

Scientific Advisory Committee on Nutrition (2013) *Statement on Selenium and Health.* https://www.gov.uk/government/uploads/system/uploads/attachment_data/file/339431/SACN_Selenium_and_Health_2013.pdf (accessed June 2021).

Scientific Advisory Committee on Nutrition (2014) *Statement of Iodine and Health.* https://www.gov.uk/government/uploads/system/uploads/attachment_data/file/339439/SACN_Iodine_and_Health_2014.pdf (accessed June 2021)

Vanderpump M P J, Lazarus J H, Smyth P P et al. (2011) Iodine status of UK schoolgirls: A cross-sectional survey. *Lancet* **377**, 2007–12.

Vinchi F (2021) Non-transferrin-bound iron in the spotlight: Novel mechanistic insights into the vasculotoxic and atherosclerotic effect of iron. *Antioxid Redox Signal.* Published online 22 March 2021. doi: 10.1089/ars.2020.8167

FURTHER READING

Department of Health (1991) *Dietary Reference Values for Food Energy and Nutrients for the United Kingdom: Report on Health and Social Subjects 41.* HMSO, London.

Expert Group on Vitamins and Minerals (2003) *Safe Upper Levels for Vitamins and Minerals.* https://cot.food.gov.uk/sites/default/files/vitmin2003.pdf (accessed June 2021).

Haschka D, Hoffmann A, Weiss G (2020) Iron in immune cell function and host defense. *Sem Cell Dev Biol*. Published online 3 January 2021. doi: 10.1016/j.semcdb.2020.12.005

Public Health England (2020) *NDNS: Results from Years 9 to 11 (2016 to 2017 and 2018 to 2019).* https://www.gov.uk/government/statistics/ndns-results-from-years-9-to-11-2016-to-2017-and-2018-to-2019 (accessed June 2021).

Rayman M P (2000) The importance of selenium to human health. *Lancet* **356**, 233–241.

Scientific Advisory Committee on Nutrition (2012) *Nutritional Implications of Repealing the UK Bread and Flour Regulations.* https://assets.publishing.service.gov.uk/government/uploads/system/uploads/attachment_data/file/221137/sacn-uk-bread-flour-regulations-position-statement.pdf (accessed June 2021).

Theobald H E (2005) Dietary calcium and health. *Nutr Bull* **30**, 237–277.

Weiss G, Ganz T, Goodnough LT (2019) Anemia of inflammation. *Blood* **133**, 40–50.

WHO (1997) *The World Health Report 1997: Conquering Suffering, Enriching Humanity.* https://apps.who.int/iris/handle/10665/41900 (accessed June 2021).

Zhang J, Taylor E W, Bennett K (2020) Association between regional selenium status and reported outcome of COVID-19 cases in China. *Am J Clin Nutr* **111**, 1297–1299.

PART 4

Dietary requirements for specific groups

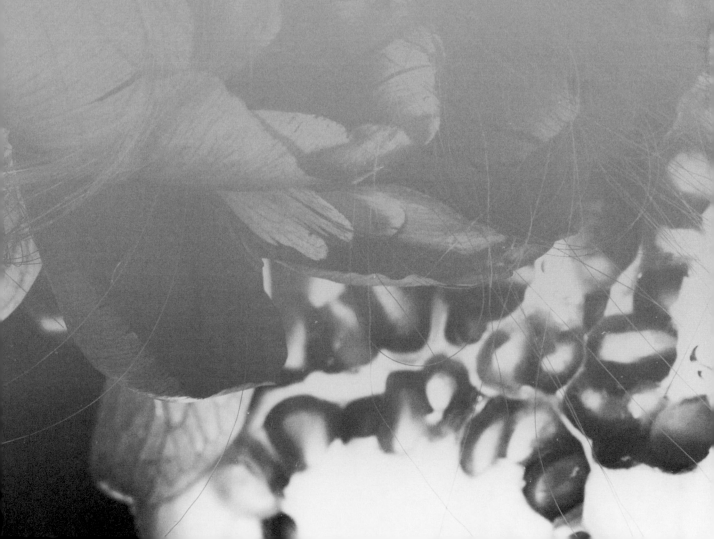

14

Infancy, childhood, and adolescence

Parul Christian and Charlotte Evans

OBJECTIVES

By the end of this chapter you should be able to:

- describe the changing characteristics of growth and maturation from birth through adolescence that alter nutrient requirements
- discuss the composition of human milk and compare with alternatives
- understand the importance of breastfeeding for the infant
- explain complementary feeding and associated issues
- explain the issues around childhood obesity
- understand nutritional considerations across childhood development stages.

14.1 Introduction

Growth, development, and functional maturation mark the postnatal period of infancy, childhood, and adolescence. Nutrition considerations are specific and strongly linked to chronologic age, although condition at birth is critical in setting the trajectory. Interactions between individuals' genetic endowments for growth and their nurturing environments, which begin *in utero*, determine body size and composition. Growth has specific nutritional needs but is not a steady process, proceeding rapidly in early life, slowing in middle childhood, and accelerating at puberty before linear growth ceases. Each stage of growth is strongly influenced by endocrine factors and a myriad of environmental and psychosocial factors depending on context. With increasing age also come physical and psychomotor maturation, which influence activity and body composition and, through feeding skills and food choices, dietary intakes.

The first thousand days of life, beginning at preconception through 24 months of age is considered a critical period of growth, with consequences for future health and function into adulthood. Since 1990, increasing prevalence of overweight and obesity in children and adolescents has been detected globally, but particularly in high-income, and more recently, middle-income countries. This has led to a dual burden of malnutrition in many countries with under-nutrition and obesity both being public health concerns, as well as specific nutrient deficiencies. These topics are discussed in more detail in this chapter.

14.2 Growth, development, and maturation

Physical growth, body composition, and maturation

Using the Intergrowth-21st birth weight standards for newly born infants, median birth weight at 40 weeks of gestation is 3.38 kg for boys and 3.26 kg for girls. By 4–6 months of age, the infant is typically expected to double, by one year triple, and by two years almost quadruple their weight. Median length of the reference standard is 49.9 cm for boys and 49.2 for girls at 40 weeks of gestational age at birth, and increases by 50 per cent by year one, 100 per cent by year four and triples by preadolescence. Growth velocity (rate of weight gain) declines dramatically over the course of the first two years of life, is stable during the period of childhood through the prepubertal period, until about 8–9 years of age when weight and height velocity begin to increase. The age at onset of puberty and the pubertal growth spurt and peak height velocity and when it occurs, differ between boys and girls and vary widely between individuals.

Each organ has a unique pattern of growth and maturation. At birth, brain weight is 25 per cent, and at five years 90 per cent, of expected adult brain weight. Seventy-five per cent of postnatal brain growth takes place in the first two years of life. By contrast, about 30 per cent of male adult body mass is acquired during adolescence.

TABLE 14.1 Body composition with age in childhood

Age	Mean weight (kg)	Whole body: water (% body weight)	Whole body: fat (% body weight)	FFM: water (% LBM)	FFM: protein (% LBM)
Birth	3.5	72	14	84	14
4 months	7	60	26	82	15
12 months	10	59	24	78	19
2 years	12	60	21	78	18
5 years	18	60	21	74	20
10 years	32	60	17	72	20
25-year-old men	70	60	12	72	21
25-year-old women	60	55	25	72	21

FFM, fat-free mass; LBM, lean body mass.

Source: E M E Poskitt (2003) Nutrition in childhood. In: *Nutrition in Early Life,* 291 (J B Morgan & J W T Dickerson eds). Copyright © 2003, John Wiley, Chichester. Reproduced with permission of John Wiley.

Table 14.1 shows age- and sex-related changes in body composition. After birth, total body water, especially the proportion of extracellular fluid component declines dramatically and percentage weight that is body fat (% BF) increases rapidly to a peak around six months old. Early infancy is followed by a period lasting until around five years old when the rate of fat deposition slows in comparison with the rate of lean tissue, leading to a drop in % BF. Typically, this is followed by a second phase of relatively rapid fat deposition, the adiposity rebound, which continues almost unabated in girls until growth ceases. In boys the adiposity rebound reverses with the rapid lean tissue deposition of late puberty. These periods of *physiological* change in fat/lean deposition tend to be obscured today by the prevalence of overweight and higher adiposity in a high proportion of young children and adolescents.

Of clinical relevance is the loss of weight in the first few days of life as extracellular fluid volumes fall in the changed hormonal environment of postnatal life. This weight loss is usually <10 per cent body weight and birth weight is generally regained by ten days of age. Where weight loss is >10 per cent of birthweight or where birthweight has not been regained by 14 days, clinical assessment of possible nutritional or health problems are indicated and nutritional intervention may be required.

The age of onset of the secondary sexual development characteristic of puberty is considered to occur between 8 and 13 years in girls and 9 and 13.5 years in boys, with a similar mean age (11.5 years) in both sexes. In girls, the first manifestations of puberty vary but the growth spurt always occurs early in the progression of puberty, with peak height velocity (PHV—the point of most rapid growth in height) on average 0.7 years after the onset of puberty and before menarche. In boys, testicular enlargement is the first observable sign of the onset of puberty. Growth acceleration in boys occurs relatively late in the pubertal process, with PHV occurring on average 1.5 years after the first signs of puberty and continuing longer than in girls. Peak rates of deposition of bone mineral occur 0.7 years after PHV in both sexes and peak bone mass (PBM) is achieved two years after cessation of growth (mean: girls 16 years; boys 18 years). Pubertal changes in body size and composition lead to greater differences in nutrient requirements between males and females than during childhood. Iron requirements are higher in girls, which is related to menstrual blood loss after onset of menarche. In adolescent boys increased lean body mass (LBM) leads to greater nutritional demands/kg body weight.

Endocrine control of somatotrophic growth

Hormonal regulation underlying the process of linear growth is specific during each life stage as described by Karlberg's biologically oriented mathematical model 'Infancy-Childhood-Puberty (ICP)' for growth. In addition, a complex interaction with a host of environmental and physiologic conditions exists. Infancy is an extended period of growth during which hormonal factors influencing growth, predominantly driven by IGF-2, may still persist. The childhood period is marked by the influence of the growth hormone (GH) -IGF-1 axis as well as GHRH (growth hormone releasing hormone) and somatostatin produced by the hypothalamus. On the other hand, the interaction between gonadal and adrenal steroid hormones along with growth hormone (GH) is essential for a normal adolescent growth spurt and maturation to occur (Christian & Smith 2018).

Neuromuscular development

Important developmental milestones occur with age which impact on children's independence and ability to feed themselves. These are presented in Table 14.2 and include ages when young children are expected to be able to feed themselves (to varying degrees) and when it is common for independence to increase to a level where it is common for demands to be irrational such as insisting on a bowl of a particular colour or only having ice cream for dinner. These milestones are of relevance for children's nutritional well-being and growth but children's individual food preferences are also important.

Food preferences develop throughout infancy and childhood. At the preschool age, children spend the majority of their time with family and carers and their food preferences develop from regular exposure to family foods. Optimal healthy food behaviours for this age group include eating with adults without television or other screens, being exposed to a variety of healthy foods such as fruits and vegetables, and learning how to eat with suitable implements such as knives and forks. Once at school and mixing with their peers, children take their cues for food preferences from their friends and peers as well as from their families. They can be influenced by the portrayal of foods on television and by other advertising pressures. Then, in adolescence, peer-approved fashions for certain foods and eating styles, as well as food advertising directed at adolescents, can lead to haphazard eating and bizarre diets with risk of compromising the good quality diets needed to meet the demands of growth and maturation. At home, the deliberate choice of unconventional foods and meal patterns may be used to express independence from the family. Those living away from home for the first time may have difficulty accessing foods and may lack the cooking skills required for a good diet. Lifestyles adopted in the adolescent years can continue into adult life.

Immunological development

Infants are born with an immature immune system that matures over the course of the first years of life. This delay in the development of the infant's immune system is offset by defence factors in human milk. One advantage of the delayed development is that more energy and nutrients can be diverted to the development of the central

TABLE 14.2 Age at average development of feeding/nutrition skills

Age of child	Feeding skills acquired
36 weeks gestation to birth	Integrated sucking and swallowing reflexes
Three months	Conveys bolus of food from front of mouth to back of mouth
Five months	Conveys objects placed in hand to mouth
	Drinks from hand-held cup with biting movements
Five and a-half months	Reaches out for objects and conveys them to mouth
Six and a-half months	Begins to make chewing movements
	Feeds self with biscuit, rusk, or other small item
	Transfers objects from one hand to the other
Seven months	Learns to shut mouth, shake head, and indicate 'No'
Nine months	Picks up raisin-sized object with thumb and forefinger
	Throws food to ground with great enthusiasm—and expects someone else to pick it up
Ten months	Holds beaker of liquid but drops it when finished
Twelve months	Tries to spoon feed but unable to stop rotation of spoon (and loss of food) before it reaches mouth
Fifteen months	Manipulates spoon and food on spoon to mouth
Eighteen months	Determined to be independent at mealtimes
Two years	Expresses own self and independence in—often irrational—food refusal; this spell may last some years
Five years	Eating in company with peers may lead to eating a greater variety of foods than previously accepted
	May also lead to strong preference for popular foods

Source: E M E Poskitt (2003) Nutrition in childhood. In: *Nutrition in Early Life*, 231 (J B Morgan & J W T Dickerson eds). Copyright © 2003, John Wiley, Chichester. Reproduced with permission of John Wiley.

nervous system and for growth. However, newborns are at an increased risk of bacterial infections especially related to the respiratory and gastrointestinal tracts. The non-specific immune response (innate immunity) comprising neutrophils and macrophages is limited in early life, developing gradually over time. Adaptive immunity (related to T and B cell responses) also takes longer to develop and respond due to the naïve status of the newborn.

Thus, in the early months of life human milk with its immune system plays a key role in providing protection against infection (see 'Development of gastrointestinal function' below). Human milk contains many immune factors and bioactives including maternal secretory immunoglobulin A (sIgA) which are transferred to the infant through breastfeeding and provide protection against pathogens that occur in the maternal environment. The immune factors in milk are protected from digestion in the gut. Breastfed infants have lower risk of common enteric infections, suggesting the presence of antimicrobial agents in human milk. Also, the lack of inflammatory reactions during the period of protection are attributed to anti-inflammatory agents in milk. Prolonged protection against certain diseases long after weaning has been shown, indicating the presence of immunoregulatory or immunomodulatory factors that modulate or regulate the development of the immune system in infants. The development of the immune system is closely tied to the development of infant gut function and maturity and existence of intestinal commensal and pathogenic bacteria (see Chapter 27).

Development of gastrointestinal function

The gastrointestinal tract is also immature and develops over the course of infancy and early childhood. Digestive function changes dramatically over the first year of life. Pancreatic lipase, amylase, and bile salt pool size are low in the newborn compared with older infants. Immaturity of gastrointestinal enzymatic function makes digestion and absorption less efficient with infant formula than with mother's milk in the first months of life. Fat absorption is lower in formula-fed infants than in breastfed infants. Lactase levels in the newborn are also quite low but increase as milk feeding begins. Digestion and absorption in breastfed infants are promoted by many specific components in human milk (Table 14.3). Many of these components act to promote growth of the intestinal epithelium and thus lengthening of intestinal villi and maturation of mucosal function.

Development of renal function

Young infants cannot dilute or concentrate their urine as much as older children and adults. This makes them particularly susceptible to fluid overload or to overload from food-derived non-metabolizable substances, which

TABLE 14.3 Some specific components of breast milk which facilitate nutrient absorption

Type of nutrient	Specific component of milk	Effects on gastrointestinal absorption
Carbohydrates Lactose		Digested to glucose and galactose which are readily absorbed
		Fermentation of any lactose in colon produces lactic acid and low pH which encourage growth of non-pathogenic colonic bacteria
		Facilitates absorption of calcium as soluble calcium lactate
Fats	Presence of bile salt stimulated lipase in breast milk	Helps digestion of fat in milk in young infants in whom pancreatic lipase activity is low
	Relatively small fat droplet size	Small droplets offer larger surface area for volume, encouraging enzymatic digestion
	Saturated fatty acids—palmitic acid	Position of palmitic and other saturated fatty acids in middle of triglyceride molecule facilitates fatty acid absorption as monoacylglycerol which is more readily absorbed than free palmitic acid.
		Good absorption of palmitic acid discourages precipitation of calcium as calcium palmitate
Nitrogen-containing compounds	Casein:whey ratio	More soluble whey proteins predominate
	Casein composition	Human milk casein micellar structure creates small, easily digested flocculates in stomach
	Urea	May be used as nitrogen source by colonic bacteria to combine with organic acids and form amino acids which can be absorbed
Micronutrients	Lactoferrin and other micronutrient binding compounds	Many specific binding compounds in breast milk facilitate absorption of iron, folic acid, vitamin B_{12}, zinc, and other micronutrients

are excreted via the kidney (the potential renal solute load: PRSL). The PRSL is expressed as milliosmoles per litre (mOsm/L) and indicates the total number of ionic or molecular particles in the fluid. For infant feeds it is calculated as: PRSL (mOsm/L) = Na + K + P + Cl + (protein, mg/175) when the dietary intakes of sodium (Na), potassium (K), phosphorus (P) and chloride (Cl) are expressed in mmol/L, and is regulated to ensure it is appropriate. The PRSL for breast milk is ≈ 93 mOsm/L; for infant formula ≈ 135 mOsm/L; and for cow's milk ≈ 308 mOsm/L. Cow's milk is not appropriate for infants due to the high PRSL.

14.3 Anthropometric assessment

Body weight and height are the most commonly used anthropometric assessments undertaken by clinicians and nutritionists as indicators of appropriate growth, and are used as proxy measures of nutritional status in childhood and adolescence (see Chapter 31). In 2006 the World Health Organization (WHO) produced new international growth reference standards using longitudinal measurements of healthy, breastfeeding children (exclusive in the first four months and continued through 12 months of age) from six different countries from birth to five years (https://www.who .int/publications/i/item/924154693X). These standards are used globally, to calculate anthropometric indices of weight-for-age, length/height-for-age, weight-for length z-scores (WAZ, L/HAZ, and WLZ) of physical growth using the median value for weight and height at a given age and sex. A Z-score below 2 (−2SD) is used as a cut-off to define underweight, stunting, and wasting to estimate the national, regional, and global burden of these conditions among under five-year olds.

On the other hand, body mass index (BMI—weight in kg/height in m^2) is also calculated using weight and height for children and adolescents to estimate the prevalence of overweight and obesity as described below. In the United Kingdom (UK) the Royal College of Paediatrics and Child Health (RCPCH) produced a suite of growth charts which incorporate the WHO growth data to four years and then use the UK 1990 data from 4–18 years (https://www.rcpch.ac.uk/resources/growth-charts).

Charts and resources (RCPCH)

Early years charts:

- 0–4 years
- neonatal and infant close monitoring (0–2 years)
- personal child health record, or 'red book' charts.

School-age charts:

- 2–18 years
- childhood and puberty close monitoring (2–20 years),
- Down syndrome (0–18 years)
- body mass index (BMI) (2–20 years)
- personal child health record, or 'red book' charts.

In 2012 the Scientific Advisory Committee on Nutrition (SACN) and the RCPCH agreed on a working definition of clinical overweight and underweight for the UK using the WHO/UK BMI charts (https://www.gov .uk/government/publications/sacn-statement-defining-child-underweight-overweight-and-obesity).

As defined by BMI:

- overweight is on or above the 91st centile
- obesity is on or above the 98th centile
- low BMI is on or under the 2nd centile
- very thin is on or below the 0.4th centile.

Using these charts normal growth can be seen as between the BMI 2nd centile and the 91st centile when plotted on weight and height growth charts. These new charts allow for monitoring of puberty and make it easy to predict expected adult height and to compare this to expected height based on parental height—a useful comparison if there a question about the 'normality' of a child's height.

UK 1990 growth charts display a range of centile curves from the 0.4 centile to the 99.6 centile. There is disjunction in the height for age curves at two years since supine length should be measured before this age and, with the exception of children who cannot stand, standing height is measured after this age. Standing height is always slightly less than supine length. When evaluating growth using these charts it is recommended that children with heights <0.4 centile should be assessed to exclude problems. A height >99.6 centile may indicate a growth problem but is much more likely to be a 'very tall normal'. Weight should be assessed in association with length/height.

Whilst single occasion growth assessments may be necessary, velocity of growth is usually more informative than size attained. Crossing the growth centiles downwards, even within the normal range, may indicate that assessment, or at least follow up, is advisable. Linear growth velocity is influenced by seasonality, with most rapid growth in spring/early summer, and ideally is recorded over a full year. In practice annual growth velocity may have to be inferred from increments recorded over shorter periods and extended to expected rates/year.

Undernutrition

For definitions of underweight, wasting, and stunting see Section 4.31 'Anthropometric assessment', and Chapter 31.

Overnutrition

In childhood, as in adult life, BMI is used to define overweight and obesity. In children, mean BMI varies non-linearly with age. The International Obesity Task Force (IOTF) (now called World Obesity Federation) defines childhood overweight and obesity as the BMI SD score or Z score (see Chapter 20) which, if maintained throughout childhood, would achieve the adult overweight and obesity BMI cut-off points of 25 and 30 kg/m^2 at 18 years. However, in the UK the 91st and 98th centiles are used as the clinical cut off points for overweight and obesity respectively, while the 85th and 95th centiles are usually considered the cut-off points for overweight and obesity in epidemiological studies (SIGN 115 2010, https://www.sign.ac.uk/assets/qrg115.pdf).

14.4 Nutrition in infancy

Breastfeeding prevalence and promotion

Recent reviews highlight the already widely recognized benefits for both mother and child from breastfeeding (Victora et al. 2016). Systematic reviews show evidence of protection against both short-term outcomes such as morbidity and mortality, and benefits to long-term outcomes such as higher intelligence, lower rates of overweight and diabetes, and reduced risk of breast and ovarian cancers. In the UK, government recommendations are based on the WHO recommendation that infants be exclusively breastfed until six months of age. Current UK Department of Health guidelines on infant feeding are:

- breast milk is the best form of nutrition for infants; it provides all the nutrients a baby needs (Table 14.4);
- exclusive breastfeeding is recommended until around the first six months of an infant's life;
- six months is the recommended age for the introduction of solid foods for both breast- and formula-fed infants;
- breastfeeding (and/or breast milk substitutes, if used) should continue beyond the first six months along with appropriate types and amounts of solid foods;
- mothers who are unable to, or choose not to, follow these recommendations should be supported to optimize their infants' nutrition.

Globally, the prevalence of exclusive breastfeeding for six months has increased but is estimated at only about 50 per cent, with introduction of other fluids and foods beginning earlier for the other half of the infants (Victora et al. 2016). The UK Infant Feeding Survey 2010 (https://digital.nhs.uk/data-and-information/publications/statistical/infant-feeding-survey/infant-feeding-survey-uk-2010) shows some promising improvements

TABLE 14.4 Advantages of human milk for young infants

Factor	Advantage
Colostrum	High in vitamin A, zinc, sIgA
Convenience	Ready to feed but convenience dependent on local acceptability of breastfeeding
Low cost	Mother may have stores of fat laid down in pregnancy to mobilize for provision of fat in breast milk. Mother does not have to eat expensive foods to produce milk
Clean	Human milk is not sterile but bacteria present are usually non-pathogenic and milk contains antibodies to bacteria in maternal gastrointestinal system
Composition	Appropriate amino acid profile; contains long-chain PUFA; high organic acid residues in infant large bowel may be converted to amino acids by colonic flora
Facilitated absorption of micronutrients	Binding proteins, such as lactoferrin, facilitate absorption of many micronutrients
Enzymes	Human milk contains enzymes the role of which is not understood in all cases. However, bile salt stimulated lipase may improve the efficiency of fat absorption in early infancy
Other non-nutritional factors in human milk	Human milk contains hormones, growth promoting factors, cytokines, and prostaglandins. The role of many of these is not clear but they may be relevant to the protective effects of breastfeeding against infection and possibly against non-communicable disease of later life
Anti-infective properties	These are varied: see Table 14.5

in the breastfeeding rates across the UK since 2005. At birth 81 per cent of mothers were breastfeeding, falling to 69 per cent by the end of week one. The rates for those being exclusively breastfed at six weeks differed across the UK with 24 per cent in England, 22 per cent in Scotland, 17 per cent in Wales, and 13 per cent in Northern Ireland exclusively breastfed. However, only around 1 per cent of infants were still receiving breast milk exclusively at six months in 2010. Breastfeeding rates continue to be higher for more educated women, older women, and black and minority ethnic women, highlighting the continuing problem of getting young and disadvantaged mothers to breastfeed in the first place and then to continue exclusive breastfeeding. Apart from the health and social advantages for mothers and children of breastfeeding in infancy, there are also potential economic advantages to communities from increasing breastfeeding rates (see Chapter 15).

Promoting successful lactation

Milk volume increases over the first month of lactation as infants develop appetite and grow. Early initiation (within one hour of birth) is also a WHO recommendation. Successful lactation is promoted by early initiation, frequent and night-time suckling especially in the first days of lactation; 'emptying' the breasts; and absence of other foods given to the baby. Once breastfeeding is established, it may be possible to reduce feed frequency to six to eight times a day, although infants vary widely in terms of their pattern of feeding. Night-time suckling, which particularly stimulates prolactin secretion, is important in maintaining high milk output.

Exclusive breastfeeding for six months may be recommended but this has to be viewed against the finding that most women in many contexts, especially in high-income countries, stop breastfeeding either exclusively or altogether long before that time. The reasons why mothers decide to feed infant formula from birth, or to switch from mother's milk to infant formula are complex. One of the main reasons women give for stopping exclusive breastfeeding is that they have 'insufficient' milk. The infants of these mothers are usually growing satisfactorily so they can be presumed to be receiving sufficient milk. Commonly, milk 'failure' results from inappropriate lactation practices, usually at initiation of breastfeeding, leading to lack of confidence by a mother in her ability to produce sufficient breast milk to satisfy her infant.

It is widely felt that more women could breastfeed successfully if they were given better advice and support in the perinatal period. UNICEF has developed standards as part of the baby-friendly hospital (BFH) Initiative (www.unicef.org.uk/babyfriendly/) aimed at hospital practices relating to breastfeeding. A hospital with a BFHI status incorporates several practices supporting breastfeeding in maternity and paediatric wards.

Mothers should be advised on infant feeding but their choices, once made, should be supported. Many need to or want to work outside their homes. Some formula feeding, whilst breastfeeding continues, may enable mothers to avoid abandoning breastfeeding altogether, although the introduction of other formula almost invariably leads to reduction in milk volume.

Human milk composition

The composition of human milk changes over the course of lactation. The first milk, colostrum, is low in volume and high in protein, especially sIGA, as well as vitamin A and zinc. Over the first few days of lactation as milk volume increases, milk composition modifies to 'transitional' and then 'mature' milk. Table 14.5 outlines the biochemical composition of colostrum and human milk, together with indications of the range of nutrients in modern infant formulas and neat cow's milk for comparison. Volume of milk produced and precise composition of breast milk vary between individual women. The fat content of human milk falls in concentration with duration of lactation. Secretory IgA levels decline gradually with time although lactoferrin levels remain high beyond six months' lactation. Milk composition also varies according to the time of day and the stage of feed, with fat levels being higher in the morning and towards the end of a feed. The fatty acid composition of human milk fat partly reflects the fatty acid composition of the mother's diet.

Carbohydrates

Lactose, the main carbohydrate (80 per cent) in milk, accounts for approximately 40 per cent of total milk energy. Its suitability as the carbohydrate in milk lies in its high solubility, promotion of protective intestinal flora, and facilitation of calcium absorption through the relative solubility of calcium lactate. Other carbohydrates in milk include monosaccharides, oligosaccharides, and protein-bound carbohydrates.

Proteins

Human milk protein is 30–40 per cent casein to 70–60 per cent whey. Whey proteins include lactalbumin, sIgA, lactoferrin, and lysozymes. Casein is a mixture of proteins associated with magnesium, phosphate, and citrate ions, bound with calcium as 'calcium caseinate complex'. Human milk casein forms smaller micelles with looser structure than the casein of cow's milk. The structure facilitates enzymic action. Precipitation of tough, undigested

TABLE 14.5 Comparative outline of energy, macronutrient, and selected micronutrients/100 ml of colostrum, mature human milk, infant formula, preterm formula, and cow's milk (information derived from various sources)

Nutrient	Colostrum	Mature human milk (whey dominated)	Infant formula	Preterm formula	Cow's milk
Energy (kcal) (kJ)	69 (290)	70 (295)	67 (280)	80 (335)	67 (280)
Protein (g)	2.0[a]	1.3	1.5	2.4	3.3
Fat (g)	2.6	4.2	3.6	4.4	3.8
Carbohydrate (g)	6.6	7.0	7.2	7.8	4.8
Calcium (mg)	28	35	46	100	115
Sodium (mg)	47	15	16	41	55
Zinc (mg)	0.6	0.3	0.6	0.07	0.4
Iron (mg)	0.1	0.1	0.8	0.9	0.05
Retinol (µg)	155	60	75	75	52
Vitamin D (µg)	N	0.04	1.0	5.0	0.03
Vitamin C (mg)	7	4	9	16	1

[a] Since much of this protein is sIgA it is not clear how much is digested and absorbed and how much remains in the gastrointestinal tract.
N: significant quantities but no reliable information.

casein curds in the stomach is less likely than with cow's milk or unmodified cow's milk formula. Heat treatment of cow's milk protein in the manufacture of infant formulas affects casein micellar structure and enhances digestibility.

The newborn liver has very small amounts of cystathionine β-synthase, an enzyme involved in the synthesis of cysteine from methionine. However, provided there is sufficient methionine in the diet, cysteine deficiency does not seem to arise. There is cystathionine synthase activity in organs other than the liver, which may account for this paradox.

About 25 per cent of total nitrogen in human milk is non-protein nitrogen, of which 50 per cent is urea, with small amounts of glucosamines, nucleotides, free amino acids, polyamines, and biologically active peptides. Taurine is present in unusually high amounts amongst the free amino acids. Levels of taurine are lower in cow's milk and infant formula is supplemented with taurine. Infants fed low-taurine diets conjugate less stable bile acids with glycine rather than taurine, although evidence of disadvantage for normal full-term infants does not exist. All amino acids are potentially essential in infancy if rapid protein synthesis (e.g., in catch-up growth) outstrips the synthesis of amino acids from precursors.

Fat

Although the quantities of fat in human and cow's milk are not very different, the component fatty acids differ. Human milk fat is higher in unsaturated fat, particularly the essential fatty acids, linoleic (18:2ω6) and α-linolenic (18:3ω3) acids as well the long-chain polyunsaturated fatty acids (LCPUFAs), viz. arachidonic (20:4ω6), eicosapentaenoic (20:5ω3), and docosahexaenoic (22:6ω3) acids. There is great interest in the role of LCPUFAs in neurological development since it was shown that the levels of LCPUFAs in the brains of infants who were breastfed were higher than in those fed unsupplemented cow's milk formula. LCPUFAs are seen as conditionally essential for fast growing premature infants who may have difficulty synthesizing these from precursors to meet the needs of the rapidly growing premature brain. The importance of these fatty acids in full-term infants has yet to be determined, but term infant formulas are commonly fortified with LCPUFAs.

The fats in human milk are more readily digested and absorbed than those of cow's milk since saturated fatty acids, especially palmitic, tend to be attached to the middle carbon of the glycerol molecule encouraging absorption bound to micelles as monoglycerides rather than free fatty acids. Fat absorption from breast milk is also facilitated by the presence of bile-salt-stimulated breast milk lipase, although this is probably not of great significance except in premature infants. Most infant formulas now contain fats derived predominantly from vegetable oils with different proportions of fatty acids from those found in human milk fat. The relative proportions of fatty acids in plant fats are largely determined by plant genetics. The relative proportions of fatty acids in the milk of mammals, human and otherwise, reflect in part dietary fatty acid content.

Human milk has a surprisingly high level of cholesterol. The explanation for this is not obvious. Human milk also contains relatively high levels of carnitine—an amino acid-derived compound which is involved in mitochondrial oxidation of fatty acids. Infants can synthesize carnitine, but premature infants and those undergoing very rapid catch-up growth may be unable to synthesize carnitine at a sufficiently rapid rate to meet demand. This may limit the rate of fatty acid oxidation.

Bioactives in human milk

Our knowledge of an existing immune system in human milk is limited, although there is increasing understanding of the existence and bioactivity of an array of different non-nutritional components including immunoglobulins, hormones, human milk oligosaccharides (HMOs), cytokines, proteins such as lactoferrin, peptides, cells, and mRNA. SIgA resists digestion, adheres to the intestinal mucosa, and can be detected in infants' stools. It prevents adherence of viruses and bacteria to mucosal cells (often a preliminary to pathogenic invasion), allowing destruction of pathogens by the phagocytic components of milk. SIgA specific to organisms affecting the mother–infant dyad appears rapidly in milk. Lymphocytes in the Peyer's patches in the small intestine are sensitized to organisms present in the maternal gastrointestinal tract but acquired from the infant during maternal contact. These lymphocytes, sensitized to produce relevant sIgA, seem specifically directed to the mammary glands and from there migrate into the milk. Human milk contains cells (macrophages, lymphocytes, neutrophils) and humoral components which protect infants against infection in the first months of life. Nutrient binding proteins in milk such as lactoferrin, which binds iron, facilitate absorption of some nutrients essential for microorganism growth, thus inhibiting pathogen multiplication in the gastrointestinal tract.

There are other factors in milk which discourage pathogen growth in the infant intestine. Colonization with *Lactobacillus bifidus* and *Bifidobacterium spp.* in the colon is promoted by the glycoprotein components of whey proteins and by *N*-acetyl-d-glucosamine containing HMOs (food for these commensal bacteria) in human milk but not present in cow's milk. *Lactobacillus* and *Bifidobacterium spp.* promote lactic and acetic acid production from metabolism of lactose reaching the large bowel, creating an environment of pH ≤5 which discourages growth of potential pathogens such as *E. coli* and *Shigella spp.* Other active protective compounds in human milk include anti-staphylococcal and anti-*Giardia* factors.

Enzymes and hormones

The roles of the many enzymes, other than human milk lipase, and of the hormones in human milk remain largely undetermined. One enzyme—glucuronidase—can cause minor problems in early infancy. Newborn infants are prone to jaundice due to poor hepatic capacity to form bilirubin glucuronide which is excreted via the bile. Jaundice can have pathological significance and is more likely not only with prematurity but with infection, dehydration, and undernutrition. Thus, any jaundice raises concerns. High levels of glucuronidase in the milk of some mothers deconjugate bilirubin glucuronide excreted in the bile, allowing resorption of bilirubin, increased bilirubin load on the liver, and 'breast milk jaundice'. This jaundice, usually developing at 7–10 days, is mild, benign, and occurs in otherwise healthy infants who are feeding and gaining weight well. It resolves gradually as liver function matures.

Transmission of infection via human milk

Mother's milk may transfer viral infection, most notably hepatitis B; cytomegalovirus (CMV); and human immunodeficiency virus (HIV), from mothers to infants. The risk from CMV in maternal milk is low since antibodies are also transferred unless mothers are acutely infected with CMV during lactation. Infants of mothers positive for hepatitis B surface antigen are at risk irrespective of the feeding method and should be actively immunized at birth. Risk of HIV infection increases with duration of breastfeeding such that 5 per cent infants are infected by six months and 15–20 per cent by 24 months of lactation. The risk is greatest when breastfeeding is not exclusive. Current UK advice for HIV-positive and high risk, but not serologically tested, women is to avoid breastfeeding. The risk to infants when mothers have consistently undetectable HIV viral load and are on HAART (highly active antiretroviral therapy) is likely to be low but remains to be empirically established. Thus, the advice to avoid breastfeeding should remain even for these mothers. The 2010 WHO guidelines in the context of HIV infection provide guidance that applies to many under-resourced settings and alongside the importance of antiretroviral therapy, local policies, and health services as well as the mother's choice (WHO/UNICEF 2016).

Vitamin supplementation in infancy and early childhood

Rapid growth rates in young infants demand relatively high levels of micronutrients. Infant formulas contain sufficient supplementary vitamins and iron for normal term infants until weaning at around six months old. The vitamin content of breast milk is affected by maternal nutritional status. Breastfed infants under six months do not need vitamin supplementation (other than the provision of vitamin K as discussed below) provided their mothers have adequate vitamin status during pregnancy and lactation. Low maternal levels of vitamins and/or low levels of maternal intake may be reflected in low levels of these nutrients in milk (Table 14.6). The vitamin most likely to be present in inadequate amounts in human milk is vitamin D, particularly for infants with mothers of Asian, African, and Afro-Caribbean or Middle Eastern origin. Northern climes also make children more at risk of vitamin D deficiency since the period of the year when sunlight of wavelength around 300nm, suitable for synthesis of vitamin D in the skin, is less. The UK Department of Health updated guidelines in 2016: https://www.gov.uk/government/publications/sacn-vitamin-d-and-health-report) that all

pregnant and breastfeeding women should take daily supplements containing 10μg of vitamin D and all infants and young children aged six months to five years should take daily supplements as vitamin drops to ensure the daily requirement of 7–8.5 μg vitamin D is met. Breastfed infants may need to receive drops containing vitamin D from one month of age if their mothers have not taken vitamin D supplements throughout pregnancy. As noted below, those infants who are fed infant formula fortified with vitamin D will not need vitamin drops until they daily intake falls below 500ml (see Chapter 12).

The UK Health Departments recommend a daily dose of vitamins A, C, and D for:

- breastfed infants from six months (or from one month if there is any doubt about the mother's vitamin status during pregnancy);
- formula-fed infants who are over six months and consuming amounts less than 500 ml per day;
- children under five years of age (https://www.nhs.uk/conditions/baby/weaning-and-feeding/vitamins-for-children).

'Healthy Start' is a UK-wide government scheme with a focus on the nutrient intakes of pregnant women, nursing mothers, and their children from low-income families. This scheme endorses the UK recommendations for supplementary vitamins A, C, and D for infants and children by offering vouchers which can be exchanged for milk, fresh and frozen fruit and vegetables, infant formula milk, as well as free adult and infant vitamins (www.healthystart.nhs.uk/). Daily supplements contain 7.5 μg vitamin D, 233 μg vitamin A, and 20 mg vitamin C.

TABLE 14.6 Nutritional problems associated with breastfeeding among undernourished mothers or those with low intake of nutrients

Specific factor	Examples of clinical problem
Low vitamin K content	Haemorrhagic disease of the newborn
Deficiencies secondary to maternal micronutrient deficiency:	
Vitamin B$_1$	Infantile beriberi
Vitamin B$_{12}$	Infantile B$_{12}$ deficiency
Vitamin D	Neonatal hypocalcaemia; neonatal rickets
High glucuronidase levels	'Breast milk jaundice'
Transmission of drugs, pesticides, dietary antigens, etc.	Many drugs are transmitted in breast milk in small quantities
	Infant may be affected my maternal lithium intake
	Risk of high level of dioxins in breastfed infants
	Dietary antigens can also cross in breast milk and occasionally in sufficient quantity to sensitize 'at-risk' atopic infants
Transmission of infection	Infant may acquire:
	HIV
	Hepatitis B and C viruses
	Cytomegalovirus
Transmission of antibodies	Neonatal hyperthyroidism
	Neonatal myasthenia gravis

Vitamin K and vitamin K deficiency bleeding (VKDB)

Plasma vitamin K levels are low in the newborn. At birth infants lack the colonic flora which synthesize vitamin K. The ability to absorb vitamin K from milk or formula varies, as does breast milk vitamin K (mean = 15 μg/l). Levels of vitamin K in cow's milk are higher than in human milk and modern infant formulas are supplemented with vitamin K. Thus, breastfed infants are at most risk of VKDB, which can cause minor bruising, blood loss from (for example) the umbilical stump, or major haemorrhage in the brain in later stages (> seven days after birth). Prophylaxis and treatment are vitamin K. All term newborn infants in UK should receive prophylactic oral (or parenteral if oral is not practical), vitamin K (1 mg) at birth to prevent early (first 24 hours) or classical (1–7 days age) VKDB. If the dose is vomited, it should be repeated. Infants who are then formula fed should have sufficient vitamin K from formula and do not need further supplementation with vitamin K. Recommendations for breastfed infants are to provide 1mg vitamin K orally, weekly or 0.25µg daily, preferably until three months of age. Premature infants, infants thought to be at high risk of bleeding postnatally, and infants with perinatal trauma, should receive 400 µg/kg, to a maximum of 1 milligram, vitamin K parenterally as soon as possible after birth. Refusal by parents to allow vitamin K supplementation, more likely when it is given parenterally, is associated with some cases of VKDB.

Donor human milk

When a mother's own milk is not available or sufficient, donor human milk acquired from established human milk banks may be the best alternative and is recommended by WHO and UNICEF in appropriate situations. The need for DHM is most evident for infants born extremely preterm in the neonatal intensive care units. DHM confers protection against necrotizing enterocolitis and other infections and is superior to preterm infant formula. Human milk banks (HMBs) collect milk from donors and should follow rigorous pasteurization processing to destroy bacterial and viral contamination. Use of donor human milk is limited based on the supply and it is important for mothers of preterm infants to receive intensive lactation counselling to establish breastfeeding once both the mother and infant are able.

Formula feeding

All infant formulas available in the market in high-income countries including the UK are highly modified from their base of cow or goat milk or soya protein. Many formulas are available and differ according to content, and whether they contain LCPUFAs, taurine, carnitine, or nucleotides. They also differ in presentation, such as traditional powder in tins with a measuring scoop, powder and scoop sachets, one-litre ready to feed (RTF) versions, and smaller 'one feed' RTF. Infant formulas are designed to provide appropriate nutrition for babies whose mothers are unable or choose not to breastfeed, or mothers who breastfeed in combination with formula. Essential nutrient components are in line with the compositional standards set out in the UK/EU legislation. Infant formula is suitable from birth until neat cow's milk is introduced, which should not be before 12 months.

Legal controls on the composition and use of formulas

Since the 1970s a series of UK government and international agency reports have made recommendations on the composition and promotion of infant formula. In 1981, the WHO adopted the International Code of Marketing of Breast Milk Substitutes promoting breastfeeding and the correct use of infant formulas. The implementation of this code is closely watched by UNICEF and other international non-governmental organizations. In 2006 the EU published a directive on the composition, labelling, and marketing of infant and follow-on formulas. Composition of infant formulas available in the UK are regulated by the Infant Formula and Follow-on Formula Regulations (Statutory Instruments 2007) and subsequent amendments. It should be noted that follow-on milks are not recommended by either the WHO or the UK Departments of Health.

Below is a list of types of infant formula currently available in the UK, from First Steps Nutrition Trust (www.firststepsnutrition.org/) which can be used from birth to one year of age.

- cow's milk based infant formula
- cow's milk-based infant formula for hungrier babies
- goat's milk based infant formula
- thickened (anti-reflux) infant formula
- soya protein-based infant formula
- lactose-free infant formula
- partially hydrolysed infant formula.

Currently there is only one completely animal-free soya-based infant formula available in the UK for mothers who wish to bring their baby up on a vegan diet.

Food allergy

Food allergy and intolerance and the maturation of the immune system in relation to dietary components are discussed in detail in Chapter 27. We briefly discuss the potential for infant diets to cause atopic conditions, particularly atopic eczema.

Greer et al. (2019) reviewed the evidence for associations between early nutritional interventions and the development of atopic disease. They concluded that among infants with a high risk of allergy there was no strong evidence to suggest that maternal dietary restrictions either during pregnancy or during lactation had any effect on the development of allergy. Exclusive breastfeeding for the first four months of life is associated with delay in onset of early childhood atopic dermatitis, cow milk protein allergy, and wheezing. Thus, the recommendation for early feeding of infants with a strong family history of atopic disease is the same as for other infants: exclusive breastfeeding for six months. For infants with a strong family history of atopic disease but who are not breastfed, there is no consistent evidence that hydrolysed milk protein infant formula may reduce the risk of developing infant and childhood atopic conditions. For those no longer breastfed after six months, again a hydrolysed cow milk protein formula should be suitable. To reduce the risk of childhood allergies such as coeliac disease and wheat allergy developing, avoidance of early (<four months) introduction of gluten is recommended, however there is no advantage in delaying introduction of complementary foods in potentially atopic infants beyond six months old. Introducing gluten before 12 months and while the infant is still breastfeeding may reduce the risk of allergy (ESPGHAN 2017).

Thus, the recommendations for infants with a strong family history of atopic disease would seem the same as those for other infants: exclusive breastfeeding for six months. If an infant formula is needed after six months' breastfeeding and the infant has developed evidence of cow milk protein allergy, a hydrolysed cow milk protein formula should be suitable. Infants at high risk of peanut or egg allergies are advised to seek medical advice and introduction of these foods is usually carried out gradually between six and 12 months (ESPGHAN 2017).

In the past, soy protein formulas were sometimes recommended for infants with evidence or risk of atopic conditions. The risk of allergy to soy protein is much the same as that for cow milk protein. Furthermore, there is now concern that soy protein formulas contain undesirably high levels of phytate, aluminium, and phyto-oestrogens. The effect of these compounds is not clear, but they would seem better avoided. The ESPGHAN Committee on Nutrition states that 'soy protein formulas have no role in the prevention of allergic diseases' (ESPGHAN 2017). If soy-based infant formulas are used after six months of age because of high risk of cow milk protein allergy or the expense of hydrolysed cow milk protein formulas, the infants should first be tested for allergy to soy protein.

14.5 Nutrition of low birthweight infants

Low birthweight (LBW, birthweight <2500g) may be due to being born preterm, that is, before the completion of 37 weeks' gestation, and/or small-for-gestational age (SGA), defined as below the tenth centile weight for gestational age. Nourishing LBW infants is more difficult when the problems of prematurity are added to those of small size. We only outline the issues here. Up-to-date neonatal textbooks provide greater detail.

Premature infants have proportionally less fat and higher total body water and extracellular fluid than term infants. Body volume to surface area, low glycogen stores, and immature, inadequate, adaptive physiology put LBW infants at above-average risk of hypothermia, hypoglycaemia, and infection. Management of feeding in LBW infants depends on gestational age, age since birth, birth weight, and whether there are associated medical or other problems. Sucking and swallowing are sufficiently coordinated for most infants over 35 weeks' gestation to be breast- or bottle-fed with care. Under about 36 weeks gestational age the sucking/swallowing reflexes are poorly coordinated with breathing. Gavage enteral feeding or even parenteral feeding are likely to be necessary until the swallowing processes are more mature. Further, independent of the poor coordination, apnoea and bradycardia may be associated with feeding, so small frequent feeds, hourly or even more frequently, are advisable. Feeding the preterm infant requires careful balance between meeting nutritional needs and avoiding the complications of large volumes of feed (Table 14.7).

Recent evidence would suggest that failure to meet macronutrient needs, predominantly for energy, very early in postnatal life increases the risk of retinopathy of prematurity and to later reduction in cognitive skills. Thus, whilst the emphasis in the recent past has been to initiate feeding—enteral or parenteral—very slowly and in small amounts because of the risks of high volume intakes, the emphasis has shifted toward maximizing the energy intake to the extent possible with careful monitoring. The emphasis is on initiating some enteral, or parenteral for those unable to tolerate enteral tube feeding, nutrition as soon after birth as the clinical state allows. For those who are parenterally fed, early trophic feeding, that is the provision of very small feeds of mother's own milk enterally as early as practical, may accelerate the rate at which infants later tolerate enteral feeds, thus promoting earlier discharge from the neonatal unit. For those who cannot cope with oral feeding, non-nutritive sucking on a pacifier (dummy) during or between tube feeds may accelerate the rate at which the infants learn to cope with oral feeds and seems to shorten the hospital stay.

Feeding regimens, enteral and parenteral, for LBW infants need to provide for very rapid growth, the nutrient costs of morbidity, and the problems of immature gastrointestinal, hepatic, and renal function. The milk of mothers delivering prematurely is higher in energy and protein than that of term mothers. Facilitated absorption of nutrients contributes to better tolerance of mother's own milk than of formula by preterm (PT) infants. These benefits and the immunological protection and possible maturing effect on the immature gastrointestinal tract of hormones and enzymes make own mother's milk the choice for introduction of enteral feeds in very LBW infants. However, the increased energy of PT breast milk may still be insufficient to support optimal growth in

TABLE 14.7 Problems developing in LBW infants which may relate to feeding and nutrition

Problem	Nutrition-related factor
Hypothermia and hypoglycaemia	Small glycogen content of liver; immature mechanisms for mobilizing low fat store of LBW infants; large surface area to volume encourages heat loss
Bradycardia and apnoea	Gastric overload with feed
Respiratory distress syndrome	Acutely, high fluid intakes and gastric distension can exacerbate cardiorespiratory problems, e.g., respiratory distress syndrome. Chronically, micronutrient deficiencies especially for antioxidants may contribute to development of chronic obstructive pulmonary dysplasia
Patent ductus arteriosus	Fluid overload may discourage closure of the ductus arteriosus and contribute to cardiac failure
Intestinal obstruction	Inspissated curd syndrome: obstruction with casein flocculates. Less likely in breast milk fed than cow's milk-based formula-fed infants. Necrotizing enterocolitis: intestinal stasis followed by eforation and erionitis; resulting from intestinal overload with distension of the bowel and ischaemia in the bowel wall
Eye disease of prematurity	Shortage of micronutrients and/or long chain PUFAs may contribute to eye damage: retrolental fibroplasia
Bone disease of prematurity	Insufficient intake of calcium and/or, phosphorus, vitamin D sodium, protein
Anaemia of prematurity	Rapid growth rate due to good nutrition outstrips capacity to form haemoglobin; deficiency of folic acid, riboflavin, copper; or haemolysis due to low vitamin E:essential fatty acid ratio
Reduced cognitive skills in later life	Insufficient intake of long-chain PUFAs to meet needs of rapidly developing brain; period of inadequate energy intake early in postnatal life

very PT infants. Expressing milk into containers results in cells and fat adhering to the containers and thus loss of some energy and protective effects. Cells are rapidly lost on storage although non-cellular immune factors such as sIgA and lactoferrin keep their activity much longer. The benefits of human milk are thus reduced when it cannot be fed at the breast especially if it is not 'own mother's milk'. Once enteral feeding is established, commercial or hospital laboratory developed 'human milk fortifiers' can be used to add extra nutrients to expressed breast milk fed to LBW infants if growth rates are faltering. There is no strong evidence that preterm infants benefit from energy- and protein-enriched formulas compared with normal infant formulas when discharged from neonatal units.

Nutritional requirements of LBW infants are estimated as those supporting growth equivalent to the expected growth and nutrient accretion of the post-conception-age-matched foetus. Such nutrient accretion is difficult to achieve in PT infants for minerals such as calcium, phosphorus, and zinc. For these reasons LBW infant formulas for use in neonatal units have increased minerals/unit volume as well as energy and protein (Table 14.5).

Supplements of vitamins A, C, and D and folic acid, enterally or parenterally depending on the method of feeding, should be given to LBW infants. Iron supplementation should be avoided before four weeks since it may enhance the production of free radicals but after this supplementation is advisable.

Nutrition, growth, and later disease

The work of David Barker and colleagues in the 1980s and 1990s led to research into the relation of foetal and early infant growth and nutrition with health and disease in adulthood (programming). Low birthweight, particularly with the SGA infant who shows rapid catch-up growth postnatally, especially after two years of age, is associated with increased prevalence of coronary heart disease and type 2 diabetes mellitus in adult life. The pathophysiological mechanisms for these and other foetal and infant programming events have been under intense investigation for decades although the actual causative mechanisms are still not well defined. The various possible explanations for programming (and varied mechanisms may be at work) include permanent alteration to an organ when there has been inadequacy of some (nutritional) factor at a critical stage of intrauterine life; epigenetic effects from intrauterine nutritional or other factors affecting methylation of DNA or the modification of histones resulting in gene transcription; and even permanent acceleration of the rate of cellular aging due to oxidative stress affecting, for example, telomere length.

Postnatally, parental attitudes and family circumstances influence infant and later childhood (and even adult) feeding and lifestyle practices profoundly, so the effects of prenatal or early postnatal suboptimal nutrition may be obscured in later life by the effects of the postnatal environment. Thus, it can be unclear whether associations, such as between infant feeding practices and later obesity, are the effects of true programming or the consequences of common environments and related lifestyle choices. It does appear however that the later expression of genes may be influenced by the events and environment *in utero* and early postnatal life. There is even some evidence from animal models that epigenetic effects could lead to transgenerational changes.

14.6 Complementary feeding

Definitions and the age of introduction of foods other than breast milk

Complementary feeding, previously known as weaning, has been defined by the WHO as 'the process starting when breast milk alone is no longer sufficient to meet the nutritional requirements of infants, and therefore other foods and liquids are needed, along with breast milk' (https://www.who.int/publications/i/item/924154614X). The term complementary feeding is used here to embrace the use of all foods and liquids other than breast milk or infant formula. The WHO recommends starting to introduce solid foods from six months and the complementary feeding phase usually refers to the period from 6–24 months.

Despite many countries recommending that the start of complementary feeding is from six months, many babies are introduced to complementary foods at a younger age. Earlier than recommended introduction of complementary feeds is associated with low maternal age, formula feeding, and maternal smoking. Breastfed infants are often switched to infant formula with the introduction of complementary foods. Cow's milk should not be introduced before 12 months due to the delay in development of renal function in infants, although small volumes may be added to complementary foods.

Digestion is not a problem for full-term infants introduced prematurely to complementary foods. For example, starchy foods (e.g., cereal) fed before pancreatic enzymes reach mature levels around four months of age, can be digested fully. Increased salivary amylase and intestinal glycoamylase activities compensate for low pancreatic amylase. However, there are nutrient intakes that are more likely to be inadequate during the first six months, such as iron. Iron requirements during complementary feeding can be met by consumption of iron-rich and iron-fortified foods and iron-fortified formula milk. The main recommendations for complementary feeding are presented in Table 14.8.

Suitable complementary foods

Complementary foods may be home prepared or commercially produced. Initially one small feed is introduced per day but feed frequency may increase quite quickly to two or three meals for infants aged six to eight months and then to three or four meals per day for infants aged

TABLE 14.8 Recommendations for complementary feeding*

1. Exclusive or full breastfeeding for about six months is a desirable goal.
2. The term 'complementary feeding' should include all solid foods and liquids other than breast milk, infant and follow-on formulas.
3. Recommendations for complementary feeding should embrace both breastfed and formula-fed infants.
4. Infants should be offered a variety of foods with different flavours and textures including bitter tasting green vegetables.
5. No sugar or salt should be added to complementary foods.
6. Avoidance or delayed introduction of potentially allergenic foods is unnecessary since this has not been convincingly shown to reduce allergies.
7. Iron requirements during complementary feeding should be met through consumption of iron-rich and iron-fortified foods.
8. Cow's milk should not be used as the main drink before 12 months. Small volumes may be added to complementary foods.
9. Early (<4 months) or late (>11 months) introduction of gluten should be avoided. Gradually introducing gluten while the infant is still breastfeeding may reduce the risk of coeliac disease.
10. Infants and young children should only receive a vegan diet with adequate medical or dietetic supervision.

* Summarized from ESPGHAN Committee on Nutrition (2017).

nine to 23 months. Offering a variety of foods from the onset of complementary feeding is recommended, particularly a range of fruits and vegetables including bitter tasting green vegetables. Later dietary preferences relate to early dietary experience and therefore feeding infants a diet of mainly sweet foods and lower intakes of fruits and vegetables could have disadvantages for later food choices which may subsequently lead to increased risk of tooth decay and excess energy intake. A review of European commercial baby foods commissioned by WHO Europe (Hutchinson et al. 2020), highlighted that a large proportion of baby foods are high in sugars, including savoury meals, with a third of the calories of all commercial infant foods coming from total sugars. Many infant foods contain more than 10 per cent free sugars with concentrated fruit juice being the most common form of sweetener. The authors recommended that added sugars and fruit be restricted in some food categories such as savoury meals and proposed a new nutrient profile module tool to help reduce inappropriate labelling and food promotion of infant foods.

Energy

Whilst breast milk (or formula) remains the main source of energy early in complementary feeding, cereal-based complementary foods, with energy density enhanced by additional fat sources, should be introduced early. Rice preparations are usually recommended since rice is gluten free. With wheat-based foods there is a slight risk of malabsorption from either temporary gluten intolerance following gastrointestinal infection, or permanent gluten intolerance in coeliac syndrome.

Fat

Fat in the diet can increase the energy density of foods, thus facilitating energy sufficiency and optimum infant growth with the relatively small volumes of food tolerated by infants' small gastric capacities. Fats also act as sources of fat-soluble vitamins, essential fatty acids, and exogenous cholesterol and enhance taste and food texture and thus palatability.

Protein

Dietary surveys consistently show mean protein intakes above recommended intakes in infancy in high-income countries. Provided breast milk, formula or, later, cow's milk intakes are around 500 ml/day, protein intakes are likely to be adequate even if complementary feeds are low in protein and amino acid variety (as may happen in diets with a single plant staple). However, very high protein diets of more than 15 per cent of energy intake, more common with large volumes of milk intake, increase the risk of subsequent overweight or obesity (ESPGHAN 2017).

Food consistency

Complementary feeding is a progressive process. Infants quickly learn to cope with solid and lumpy foods and ultimately foods which require chewing prior to swallowing. This progression is important. Prolonged partial breastfeeding without adequate addition of complementary foods, excessive juice or other sugary drinks intake, the offering of inadequate foods because of perceived food 'allergens', the inappropriate provision of a milk-free diet that is low in protein, can all lead to undernutrition. Prolonged bottle feeding, with few complementary foods beyond one year, can lead to failure to thrive (FTT) due to the low energy density of fluids proffered or to 'fussy' eaters due to failure to progress in variety of foods and food textures. Infants should be moved from fluids fed by bottle to predominantly fluids fed by cup over the second six months of life.

High solute load and low iron content make unmodified cow's milk unsuitable for early complementary feeding. Cow's milk protein intolerance (CMPI) can cause incipient or overt gastrointestinal blood loss. Current recommendations are that cow's milk should not be given as a drink to infants under one year. When it forms part of family recipes, small amounts may be safe before this age.

Vegan infants

Plant-based diets, including solely plant-based vegan diets, are becoming more common in many countries (see Chapter 17). Feeding infants according to a strict vegan diet (as opposed to eating occasional plant-based meals) is more challenging in terms of ensuring nutritional adequacy for infants and is not recommended without specialist medical and dietetic advice. Information provided by ESPGHAN (2017) concludes it is possible for a vegan diet to meet nutrient requirements with supplementation; however, the risks of poor infant health if not following specialist advice (for mothers and infants) are serious. ESPGHAN recommend that mothers supplement their diet through fortified foods or supplements with vitamins B_{12}, B_2, A, and D during breastfeeding and lactation. Care should be taken to ensure the infant is provided with enough energy, protein, vitamin B_{12} (0.4ug/day from birth, 0.5ug/day from six months), vitamin D, iron, zinc, folate, n-3 fatty acids (especially DHA), and calcium. A soya-based infant formula that is supplemented with micronutrients should be used as an alternative to breast milk as breast milk output declines soy products, beans and pulses, and tofu can be used for protein sources.

The nutritional status of the mother during pregnancy and lactation is important for infant health. Vegan mothers have higher levels of unsaturated fatty acids in their milk than omnivore mothers which is beneficial to the infant. However, in vegan mothers, low levels of milk vitamin B_{12} reflect low maternal intakes, and deficiency in infants may be exacerbated by breastfeeding even without evidence of maternal deficiency. Supervision of complementary feeding of vegan infants is strongly recommended (Baldassarre et al. 2020).

14.7 Nutrition in children and adolescents

Digestion and absorption in preschool children enable them to consume the same foods as adults but nutrient needs and feeding skills are different. Table 14.9 lists some of the issues to consider in child and adolescent nutrition. Children's small stomachs limit the amounts of food taken at any one meal. They do not (should not) consume food overnight. Young children should therefore be fed three meals a day and perhaps two or three small

TABLE 14.9 Modifications to quality of diet needed for child to progress from infant diet to that of adult

Age	Diet	Nutritional issue
Young infant	Wholly breastfed	Entirely liquid diet Quite low-energy food, 0.7 kcal/ml All essential nutrients in one food No fibre in diet 50% total energy from fat
Weaning diet	Milk: formula or breast plus some 'solid' foods such as purees and porridges	Diet low energy density, 1 kcal/g Very little or no fibre May be low in fat high in sugars
Young child	Mixed diet: may be quite limited in variety Little unprocessed meat Children often not keen to chew, and whole fruit largely absent and vegetables usually eaten only with reluctance	Varies according to food offered and children's pickiness Fat content may range from 25% to 40% total energy. Iron content may be low Fibre intake largely from breakfast cereals
Schoolchild	Diet may be influenced by school meal Packed lunches often brought from home Sugary drinks popular vegetables not popular	Improved standards of school meals have potential to reduce saturated fats and sugars Packed lunches often higher in saturated fats, sugars, and salt than school meals and lower in vegetables High contribution of sugars from drinks Fibre low and predominantly from breakfast cereals rather than fruits and vegetables
Adolescent	May be a balanced adult diet or a chaotic and irregular diet	Important that diet is encouraged to follow recommendations for adults whenever possible
Diets	Diet eaten away from home without supervision. Maybe excessive soft drink consumption and high fast food intake Other aspects of 'healthy living' important as well as diet	Diets high in saturated fat and sugars and low in vegetables

Source: E M E Poskitt (2003) Nutrition in childhood. In: *Nutrition in Early Life* .(J B Morgan & J W T Dickerson eds). Copyright © 2003, John Wiley, Chichester. Reproduced with permission of John Wiley.

between-meal nutritious snacks, the timing of which depends on mealtimes. Small eaters and 'fussy' eaters may have difficulty consuming sufficient energy to meet needs so recommendations for adults to consume <35 per cent dietary energy from fat do not apply to young children. The transition from >50 per cent dietary energy derived from fat provided by exclusive breastfeeding to <35 per cent energy derived from fat should be spread over the first five years of life. Similarly, adult recommendations for fibre intake should not apply in early childhood since high fibre content lowers food energy density and phytates reduce absorption of micronutrients.

In some countries, supplements are recommended for young children. For example, the UK government recommends supplementary vitamins A, D, and C to children under five years, and recognize the difficulties of achieving varied diets in young children. Nevertheless,

relatively few children, approximately 10 per cent according to figures from the UK National Diet and Nutrition Survey (NDNS), receive supplementary vitamins after infancy.

As children grow up they become more independent in their eating habits. Meals should be spread over the day and snacks regulated to facilitate recognition of hunger and satiety. Table 14.10 outlines the principles for good nutrition during different stages of childhood. However, the majority of children and adolescents in many countries do not consume optimal diets for health. Children, and adolescents in particular, typically consume diets containing higher than recommended levels of free sugars and lower than recommended levels of fruits and vegetables. Adolescents are also high consumers of fast food and sugary drinks which are associated with poor diet quality (Evans 2020).

TABLE 14.10 Outline of principles for good nutrition in childhood and adolescence

Age group	Nutritional principles
Children <3 y weaning diets	Should include three main meals and two smaller meals.
	Diets should be energy dense and varied to provide micronutrients:
	good sources of iron;
	vitamin C with each meal to encourage micronutrient absorption.
	Whole milk rather than skimmed or semi-skimmed at least until 2 years.
	High fibre diets should be avoided but fibre in diets gradually increased.
Preschool children 3–5 years	Three meals, two snacks but other food in between meals discouraged. Increasing foods with fibre content.
	Full fat milk is recommended until aged 5. However semi-skimmed milk is recommended for children who are overweight.
	Careful modification of diet if child progressively gaining weight and plotting on higher weight and BMI centiles.
	Encourage active lifestyle and discourage periods of inactivity in front of television, computers, and tablets.
	Discourage eating between meals and outside recognized snack periods.
	Fresh fruit as snacks. Water as beverage.
	Vegan diets need careful attention for energy adequacy and micronutrient bioavailability.
Schoolchildren 5–10 years	Control of high energy snack foods.
	Maintain active lifestyle both at home and in school—encourage walking—to school, etc.
	Encourage hobbies, interests, and activities other than television watching and computer games such as sports. Restrict screen time.
	Create nutritionally adequate and healthy school meals which are palatable and acceptable.
	Guard for dieting obsessions which might suggest early disordered eating.
Adolescents	Encourage adolescents to look after themselves and eat healthily despite peer pressures.
	Encourage activity. Discourage inactivity.
	Encourage high micronutrient intake.
	Approximately the equivalent of one pint of reduced-fat milk/day.
	Give support to those experiencing unhealthy weight gain.
	Recognize the stresses of adolescence and the need for adolescents to show independence and to develop their own lifestyles. Maintaining 'contact' with children at this critical point in life may ultimately be more important than an ideal dietary intake.

Source: E M E Poskitt (2003) Nutrition in childhood. In: *Nutrition in Early Life* .(J B Morgan & J W T Dickerson eds). Copyright © 2003, John Wiley, Chichester. Reproduced with permission of John Wiley.

Nutritional problems in children and adolescents

Childhood overweight and obesity

(See also Chapter 20.) Childhood overweight and obesity increased dramatically in many high-income countries from the early 1990s and is widely acknowledged as an epidemic worldwide. The UNICEF report on the State of the World's Children 2019 (UNICEF 2019) provides information from across the globe. Data from 41 Organization for Economic Cooperative Development (OECD and EU) countries indicate that the highest rates of childhood (ages 5 to 19 years) overweight and obesity are in the US, New Zealand, and Greece at more than 37 per cent. The prevalence of childhood overweight and obesity has increased by more than 20 per cent in nearly all of these countries and more than doubled in some, such as Turkey, Bulgaria, and Hungary. Childhood obesity is also becoming more of a problem in low- and lower middle-income countries where there has been a significant rise in childhood overweight and obesity over the last decade. While in higher-income countries the prevalence of obesity is highest in socioeconomically disadvantaged populations, the opposite is often seen in low-income countries.

Public Health England publishes annual figures for the National Child Measurement Programme (NCMP) in England; from four- to five-year-olds in school reception and from 10–11 year-olds in school year six. The figures from 2019/20 (PHE 2020a) indicate that in reception year four to five years) 23.0 per cent of children measured were overweight or obese with 9.9 per cent obese up from 22.6 per cent and 9.7 per cent respectively the previous year. In year six (10–11 years) 35.1 per cent were overweight or obese with 21.0 per cent obese up from 34.3 per cent and 20.2 per cent respectively for the previous year. Approximately 1 per cent of children are underweight (NHS digital 2020). There is a strong positive relationship between socioeconomic deprivation and obesity prevalence for children in England, with obesity prevalence being significantly higher in areas of deprivation. In 2019/20 children living in the least deprived households had rates of obesity of 6.0 per cent and 11.9 per cent at ages 5 and 11 years, while children living in the most deprived decile had rates of 13.3 per cent and 27.5 per cent respectively. The inequality gap has continually widened since national measurement records began in England in 2006 for both age groups. Continuing rises in levels of obesity at school entry are particularly disturbing since, in the past, the prevalence of overweight/obesity in the toddler age group was low and most children showed a physiological decline in percentage of body fat between one and five years.

Obesity is a matter of serious public health concern. In the short term, psychological distress and the physical handicap of gross size contribute to underachievement at school. Obesity in childhood and adolescence tracks into adulthood with obese children and adolescents being five times more likely to become obese adults than children of a healthy weight; increasing the risk of corresponding co-morbidities and medical complications. Particularly concerning, is the appearance of some complications previously associated with adult obesity being seen amongst adolescents. Type 2 diabetes mellitus in an obese child in teenager years is no longer a rarity.

Causes of childhood obesity

The vast majority of obese children have no recognizable underlying medical cause for their obesity. While there is an understanding that genetics may predispose people to gain weight, the modern obesogenic environment encourages the intake of energy dense, low nutrient food (and drink), and a decrease in physical activity levels. The main causes of childhood obesity are an excess of energy (calories) consumed and a lack of energy expended (WHO 2020). The WHO recognizes that there has been a global shift towards diets that are more energy dense with increasing intakes of foods high in fats and sugars, together with reductions in physical activity and increased sedentary behaviour. The increase in sedentary behaviour is due to both changes in recreation time such as increased screen time from television, computers, and smart phones as well changes in modes of transportation in an increasingly urbanized world. See Table 14.11 for a summary of the main factors contributing to childhood obesity. These are societal changes that are challenging and therefore need to be addressed on many different levels including individual, community, and national levels that tackle the food and activity environments.

Childhood overweight and obesity prevention is potentially an easier solution that targets the causes of obesity rather than solely focusing on obesity management. However, strategies thus far have not proved particularly effective. The WHO recommends the following: increased consumption of fruits and vegetables, legumes, wholegrains and nuts; reductions in fats, particularly saturated fats and sugars; and an increase in daily physical activity. Increased sedentary behaviour potentially increases the viewing of food and drink advertising, making adolescents particularly vulnerable.

Another sociological factor which may be contributing to the obesity epidemic is sleep duration. There is some evidence to suggest that the more hours children sleep, the less the prevalence of overweight/obesity. This may simply be that it is not possible to be sleeping and eating at the same time. However, both leptin and ghrelin, hormones associated with appetite suppression and appetite stimulus respectively, are affected by sleep with lack of sleep tending to raise circulating ghrelin and lower leptin levels so there could be physiological reasons for this finding. There are a growing number of studies of changing patterns of sleeping in children and it is likely that the availability of 24-hour Internet and screens in bedrooms could be contributing to less sleep for many children and adolescents.

There are very rare, single-gene defect, obesity syndromes associated with abnormalities of leptin metabolism. Such children usually show unrelenting increase in fatness from birth but dramatic and sustained fat loss with leptin therapy. Most childhood obesity 'secondary' to recognized conditions includes short stature as a feature, often with other clinical findings and/or psychodevelopmental problems. Prader–Willi syndrome is the most likely condition to present with obesity as the prime

TABLE 14.11 Environmental and lifestyle factors contributing to childhood obesity

Relevant issue	Contributing factors
Energy intake excessive for needs	Overall availability and accessibility of foods
	Variety of foods
	Variety of packaging/presentation of foods
	High energy density of many foods
	Prevalence of bought foods often leading to loss of organization to meals and snacks
	Decline in family meals possibly leading to poor control of intake
	Use of sweetened fruit juices and sugary drinks rather than water to drink
	Advertising and peer pressures
Energy expenditure inadequate to balance energy intake	Widespread use of cars even for short journeys
	Lack of public transport to meet public needs
	Risk on roads from traffic and undesirable characters so little opportunity for activity outside the home
	Decline in school sports programmes
	Widespread use of television, computer games, and smart phones as entertainment even for young children
	Overall sedentary lifestyles
Socioeconomic factors	Higher cost of 'healthy' diet high in fruits and vegetables
	Low self-esteem and mental health issues
	Environment of social deprivation
	Lack of authoritative parenting leading to erratic bedtimes and meal times
	Advertising and portrayal of food and eating in media
	Difficulties in accessing 'healthy' foods such as fruits, vegetables, and high fibre cereals in local shops in some areas

concern, although short stature and low IQ usually alert clinicians to the possibility of 'secondary' obesity.

Prevention of childhood overweight and obesity

In the past, management of obesity tended to focus on individuals. Public policies aiming to facilitate weight control were viewed as an unwelcome interference. However, there is now general recognition that preventing the development of childhood obesity is the main aim, and interventions need to focus, not just on individuals, but on creating a healthier food environment to ensure that healthy choices are accessible and affordable for all. Interventions at the national, community, and individual level to create sustainable lifestyle changes are explored here.

National policies targeting the food environment

National policies related to food target the whole population, albeit to varying degrees, and rely on voluntary or mandatory regulation of the food, hospitality, and/or retail industries. In 2017 the WHO published a list of 'Best Buys', the most cost-effective and other recommended interventions to reduce non-communicable diseases (NCDs), many of which relate to children (WHO 2011). Interventions to increase healthy diets (or reduce unhealthy diets) include eliminating trans fats and reducing sugars through reformulation of foods and taxation of sugary drinks, limiting portion and package sizes, social marketing to increase nutritious foods and reduce foods high in fats, sugars and/or salt; and limiting marketing of foods and beverages to children. Schools also provide excellent opportunities to improve the quality of food children eat during the school day; and many countries have standards in place for school meals that ensure fruits and vegetables and other nutritious foods are served daily.

Food and beverage companies have long been viewed as contributing to the obesogenic environment by manufacturing and heavily promoting products considered 'unhealthy' (that is, energy dense and micronutrient poor) to children. Advertising influences children's choices both in terms of categories and brands of foods chosen. There is now legislation in some countries to restrict food advertising in relation to television programmes predominantly directed at children. However, many organizations are advocating for more restrictions, particularly around digital marketing.

In a number of countries around the world further action has been implemented to reduce intakes of sugary drinks. The UK announced a sugary drinks industry levy (SDIL) in 2016 which was implemented two years later in 2018, giving beverage companies time to reformulate drinks to contain lower levels of sugar (PHE 2020b). Drinks containing less than 5g of free sugars per 100ml escaped the levy aimed at industry (not consumers). Evaluation to date (PHE 2020b) reports that many companies have reformulated and far more drinks available now contain less than 5g of sugars/100ml. Furthermore, data indicate that free sugars consumption has reduced in all age groups. Additional policies that have led to successful reformulation of foods include the reduction of artificial trans fats which has taken place in different countries either voluntarily (UK) or through mandatory regulation (US).

Community action

Regional authorities such as local councils in England also offer opportunities to improve the local food and physical activity environment to prevent childhood overweight and obesity as well as run programmes to manage childhood obesity. A number of programmes offer children of various ages interventions with help and lifestyle advice over a period of years, usually to young children and their mothers, or short courses involving not only dietary advice and exercise but support and advice over shopping, cooking, reducing sedentariness, changing behaviours, and tackling a wide variety of socioeconomic issues which contribute to obesity. Some communities are encouraging healthier food access from retailers and out-of-home food outlets; as well as more active transport such as promotion of cycling and walking to school. Policies for the prevention of childhood obesity may differ little from those recommended for treatment except they are aimed at whole community groups and have a more generic message.

Individual and family action

There are sustainable dietary and lifestyle practices which aim to balance energy intakes and expenditures in obese children. It is acceptable for young children who are still growing rapidly to have the individual target of keeping their weight static so they 'grow into their weight'. Extremely obese adolescents, already more than their expected adult weight and close to the end of linear growth, need different goals. They need to aim for some weight reduction through developing negative energy balance by increasing activity, which is not easy for the very obese, modifying energy intakes, and reducing sedentary behaviour. Quite small losses of body fat can improve morale, make physical activity easier, reduce orthopaedic discomfort, and decrease the prevalence of problems such as high blood pressure and insulin resistance, so older children who lose weight but do not achieve a healthy weight may nevertheless benefit.

Failure to thrive (FTT)

Failure to thrive (FTT) is failure to gain in weight and height at the expected rate, the expected rate being that indicated by the child's weight and/or height velocity compared with reference standards for age. Approximately 1 per cent of children are reported to be underweight in the National Measurement Programme in England at the

age of 4–5 years (NHS digital 2020). There are a number of possible causes for FTT although resolution through catch-up growth (CUG) can be very rapid if a treatable cause can be removed and there is provision of extra nutrients to support catch-up. Where resolution of the precipitating cause is impossible, as with cystic fibrosis and with some congenital heart disease with high cardiac output, increasing the energy and nutrient density of the diet so children ingest more energy and nutrients without increasing food volume can improve rates of growth for both height and weight, although it is not until weight is more or less expected for weight for height that linear growth accelerates and catches up.

Psychosocial deprivation (PSD) may be a common form of FTT although often unrecognized. Overt cases come from homes, where the nurturing environment is in some way deficient in affection and stimulus which enable normal growth. The explanation for the poor growth is multifactorial and is likely to be related to poverty and mental health issues of a parent or carer with inadequate food offered or ingested. Children with PSD may have low levels of growth hormone and ACTH secretion when in their deprived environments which are possible to normalize allowing catch-up growth upon improvement of the environment. However, this is often difficult due to the challenging socioeconomic circumstances and causes.

Iron deficiency

Healthy iron status in childhood is essential for growth, energy, and protein metabolism and distribution of oxygen around the body. Iron deficiency is one of the most common nutrient deficiencies in the world and exists with consumption of both traditional and modern diets. The most significant finding in iron deficiency is anaemia. Haemoglobin levels are physiologically lower in children (except the newborn) than in adults contributing to difficulty deciding whether there is functionally important anaemia. Depression, irritability, loss of appetite, apathy, evidence of impaired learning and cognition, and slowed growth rates have all been associated with iron deficiency anaemia (IDA). The extent to which these symptoms occur with iron deficiency in the absence of anaemia is not clear, but they are not dependent on the presence of anaemia to be manifestations of iron deficiency. Children with IDA are often tired and anorexic, look pale, and may be breathless. As IDA usually develops over months, physiological adaptations to falling haemoglobin may obscure symptoms until there is severe anaemia (e.g., Hb <40 g/L).

Most iron is transferred across the placenta in the last trimester so preterm neonates have less iron/kg body weight than term infants. In all infants, iron stores are used up in early growth as the blood volume increases with the increase in body size. By six months, and often earlier in preterm infants, iron stores are minimal. Timely complementary foods providing good sources of iron are necessary. Iron stores build up only after growth has slowed considerably around five years of age. Iron requirements increase in adolescence with increase in body size and the needs for growth and, in girls, the onset of menstruation. The UK NDNS of children found most children's iron intakes were adequate but nearly half of adolescent girls had intakes below the lower reference nutrient intake (LRNI) for iron. Amongst those with low iron status, non-meat-eaters and non-Caucasian adolescent girls had greatest risk of poor iron status. Table 14.12 outlines some of the clinical and dietary factors contributing to iron deficiency.

Dietary iron exists as haem in meat and non-haem in cereals, vegetables, eggs, nuts, and fish. The largest contributors in the UK diet are from iron-fortified cereals including breakfast cereals and bread, meat and meat products, and vegetables. Additional daily vitamin C in the form of fruits and vegetables and juice may help sustain iron in the more absorbable ferrous form. Treatment of iron deficiency with oral iron is effective.

Calcium and vitamin D

Adequate calcium and vitamin D are needed for optimal bone health. Calcium and vitamin D nutrition are inescapably linked. The active form of vitamin D, $1,25(OH)_2D$, is necessary for absorption of calcium, maintenance of circulating calcium levels, and bone mineralization (see Chapter 25). Children at most risk of deficiency are those growing most rapidly—infants, especially LBW, and adolescents. In vitamin D deficiency in children, calcification of cartilage at the growing end of the shaft of long bones fails and uncalcified cartilage and osteoid accumulate, causing swelling at the epiphyseal plate. Overall bone mineralization is poor. In extreme cases, soft long bones bend with weight bearing, leading to bowed arm bones in crawling children and bowed leg bones in those who are walking; known as rickets.

Although the total daily calcium increment is highest in adolescence, the proportion of calcium taken up compared with total body, or total bone mineral, weight is highest immediately after birth. In the term foetus, 300 mg of calcium are transferred across the placenta to the foetus each day. Calcium needs in adolescence are less clear, but recommendations range from 600 to 1200 mg calcium daily with recommendations higher in boys than girls due to boys' greater accretion of calcium in LBM at adolescence. Dietary phytates and substances competing with calcium binding sites may affect calcium requirements significantly in childhood (see Chapter 25).

The main sources of calcium in the UK are dairy products such as cow's milk and cheese. In countries where consumption of dairy products is less common, in China for example, the main sources are vegetables and legumes such as tofu, and calcium-fortified soya drinks. The main source of vitamin D both during and after infancy is from conversion of 7–dehydrocalciferol to cholecalciferol by ultraviolet light radiation of approximately 300 nm to the skin. This accounts for the seasonal changes in plasma 25(OH)D levels with peaks in August/September and troughs in February in the UK and similar northern latitudes. Most children with adequate summer sunshine exposure should have no need for dietary vitamin D. However, Black and Asian children who live in Northern Europe and have had little exposure to summer sunlight are at risk of low levels of circulating vitamin D

TABLE 14.12 Examples of causes of iron deficiency in infants and children

Basic cause	Clinical problem
Too little in: low iron intake	Prolonged breastfeeding
	Feeding with unmodified cow's milk in early life
	Complementary feeding diet low in iron
	Vegetarian diet with low haem iron content
	Poor appetite, vomiting
Too much out: abnormal iron losses	Overt bleeding; extensive bruising in haemorrhagic diathesis
	Insidious blood loss from gastrointestinal tract: reflux oesophagitis in infancy; cow's milk protein allergy; Crohn's disease, etc.
Failure to absorb: iron in diet is sufficient but it does not get into the body	Lack of vitamin C so iron in ferric form
	Non-haem iron sources
	Binding by high levels of phytates in diet
	Abnormalities of gastrointestinal tract, e.g., gluten-sensitive enteropathy
	Absorption reduced in febrile illnesses
Failure to utilize: iron not incorporated into blood cells	Fever
	Chronic disease
	Associated deficiencies of vitamin A, riboflavin, copper
	Some inborn errors of metabolism associated with hypochromic anaemia
Increased requirements: extra dietary iron needed	Catch-up growth in LBW infants
	Catch-up growth after severe malnutrition

TABLE 14.13 Examples of causes of vitamin D deficiency in infants and children

Basic cause	Clinical situation
Too little in: inadequate vitamin D synthesized or ingested	Lack of exposure to (in Europe) summer sunshine due to housebound existence; extensive covering by clothing (e.g., some Islamic girls); urban environment with nowhere to play outside
	Low intake of oily fish, liver, eggs, cream, margarine, and fortified breakfast cereals
	Failure to hydroxylate vitamin D to active forms in liver and kidney diseases
Too much out: excessive losses largely through biliary tract	Loss of inactive metabolites of vitamin D into bile secondary to drugs such as barbiturates and steroids
Failure to absorb: particularly fat malabsorption since vitamin D is fat soluble	Cystic fibrosis, biliary atresia, chronic hepatitis
	Possible binding in gut with high phytate diet leading to reduced enterohepatic circulation of vitamin D
Failure to utilize	Inborn errors of metabolism involving vitamin D
	Shortage of substrates for bone mineralization: calcium deficiency; phosphate deficiency in premature infants
Increased requirements	High levels of 1–25 OH vitamin D may be necessary to maintain adequate calcium absorption on low calcium diets or high phytate diets
	Catch-up growth
	Individuals with heavily pigmented skin in environments of lowish levels of sunshine wavelength 300 nm

metabolites. In children regularly requiring drugs such as anticonvulsants or cortisol, which stimulate metabolism of cholecalciferol to inactive metabolites excreted via the bile duct, there is increased susceptibility to vitamin D deficiency. Table 14.13 outlines some causes of vitamin D deficiency.

14.8 Conclusions

Growth, dependency, and development make infants and children susceptible to nutritional imbalance and deficiencies. In order to promote and encourage nutritious diets to children continuous action is needed at national, community, and individual levels. Children should not be exposed to food marketing or excessive amounts of foods high in saturated fats and sugars where they live, study, and play. It is wise for all health educators in contact with families of children to promote meals varied in content, texture, and taste for children. Family meal times without excessive pressure to eat or screen-based distractions are generally agreed to be the ideal. Snacks should be nutritious; and wholegrain cereals, fruits and vegetables should be seen as enjoyable components of the diet. These actions are proven to improve children's diets and health by reducing risk of both nutrient deficiencies and overweight/obesity. A sustainable and healthy lifestyle in childhood will ensure healthy children grow into healthy adults.

KEY POINTS

- There are sound and varied physiological reasons why human milk is most suitable for the young infant.

- The adaptations in modified formula milks accommodate the needs of young infants, even those with very low birthweight or with inborn errors of metabolism where mother's own milk may have limitations.

- The transition from breast or formula feeding to a solid diet is a time when young children are vulnerable to nutritional problems.

- Vitamin and mineral deficiencies present problems in child nutrition even in high-income countries.

- Overnutrition as overweight and obesity is a serious problem in public health and child nutrition. It presents challenges which involve not only diet but lifestyle change for prevention and cure.

- Adolescence is a period of both nutritional vulnerability and opportunity for nutritional intervention for a healthy adult lifestyle.

- Healthy childhood nutrition encompasses not only what is eaten but when and how it is consumed. It should be recognized as only one part of the nurturing environment which promotes healthy living as children grow and mature and can take on responsibility for developing their own personal lifestyles where healthy nutrition is a significant part of the balance between healthy eating and physical activity.

 Be sure to test your understanding of this chapter by attempting multiple choice questions.

REFERENCES

Baldassarre M E, Panza R, Farella I et al. (2020) Vegetarian and vegan weaning of the infant: How common and how evidence based? A population-based survey and narrative review. *Int J Environ Res and Publ Hlth* **17**(13), 4835 doi.org/10.3390/ijerph17134835

Christian P & Smith E R (2018) Adolescent undernutrition: Global burden, physiology, and nutritional risks. *Ann Nutr Metab.* **72**, 316–328

ESPGHAN (2017) *Complementary Feeding: A Position Paper by the nnEuropean Society for Paediatric Gastroenterology, Hepatology, and Nutrition (ESPGHAN) Committee on Nutrition. JPGN.* doi: 10.1097/MPG.0000000000001454

Evans, C E L (2020) Next steps for interventions targeting adolescent dietary behaviour. *Nutrients,* **12**(1), 190. doi.org/10.3390/nu12010190

Greer F R, Scott H, Sicherer A, Burks W (2019) The effects of early nutritional interventions on the development of atopic disease in infants and children: The role of maternal dietary restriction, breastfeeding, hydrolyzed formulas, and timing of introduction of allergenic complementary foods. *Pediatrics* **143**(4), e20190281. doi: https://doi.org/10.1542/peds.2019-0281

Hutchinson J, Rippin H, Threapleton D et al. (2020). High sugar content of European commercial baby foods and proposed updates to existing recommendations. *Matern & Chld Nutr.* **17**(1), e13020 https://doi.org/10.1111/mcn.13020

Karlberg J (1989) A biologically-oriented mathematical model (ICP) for human growth, *Acta Paediatrica* **78**(s350). https://doi.org/10.1111/j.1651-2227.1989.tb11199.xNHS Digital (2020) https://digital.nhs.uk/data-and-information/publications/statistical/national-child-measurement-programme/2019-20-school-year

Public Health England (PHE) (2020a) *Child Obesity: Patterns and Trends.* https://www.gov.uk/government/publications/child-obesity-patterns-and-trends (accessed February 2021).

Public Health England (PHE) (2020b) *Sugar Reduction: Report on Progress between 2015 and 2019.* https://www.gov.uk/government/publications/sugar-reduction-report-on-progress-between-2015-and-2019 (accessed February 2021).

Statutory Instruments 3521 (2007) *Infant Formula and Follow-on Formula Regulation.* http://www.opsi.gov.uk/si/si2007/uksi_20073521_en_1 (accessed February 2021).

UNICEF (2019) *The State of the World's Children 2019. Children, Food and Nutrition: Growing Well in a Changing World.* UNICEF, New York. ISBN: 978-92-806-5003-7. https://www.unicef.org/media/63016/file/SOWC-2019.pdf (accessed February 2021).

Victora C G, Bahl R, Barros A J D et al. (2016) Breastfeeding in the 21st century: Epidemiology, mechanisms and lifelong effect. *Lancet* **387**, 475–490

WHO (2011) *From Burden to 'Best Buys': Reducing the Economic Impact of Non-communicable Diseases in Low and Middle Income Countries.* https://apps.who.int/iris/bitstream/handle/10665/259232/WHO-NMH-NVI-17.9-eng.pdf (accessed February 2021).

WHO (2020) *Noncommunicable Diseases: Childhood Overweight and Obesity.* https://www.who.int/news-room/q-a-detail/noncommunicable-diseases-childhood-overweight-and-obesity (accessed February 2021).

WHO/UNICEF (2016) *Guideline: Updates on HIV and Infant Feeding: The Duration of Breastfeeding, and Support from Health Services to Improve Feeding Practices among Mothers Living with HIV.* World Health Organization, Geneva.

FURTHER READING

Poskitt E M E & Emunds L (2008) Management of childhood obesity. *Int J Hlth Care Qual Assur* **21**(4) https://doi.org/10.1108/ijhcqa.2008.06221dae.001

SACN (2018) *Feeding in the First Year of Life.* https://assets.publishing.service.gov.uk/government/uploads/system/uploads/attachment_data/file/725530/SACN_report_on_Feeding_in_the_First_Year_of_Life.pdf (accessed February 2021).

Stewart L & Thompson J (eds) (2015) *Early Years Nutrition and Healthy Weight.* Wiley-Blackwell, Chichester.

Waters E, Swinburn B, Seidell, J et al. (2010) *Preventing Childhood Obesity: Evidence Policy and Practice.* Wiley-Blackwell, Oxford.

15 Pregnancy and lactation

Victoria Hall Moran

15.1 Introduction

In the nine months between conception and birth, the foetus grows from a single cell to an organism capable of surviving outside the womb. Alongside this extraordinary process of growth and development are profound changes taking place in maternal anatomy and physiology, which include alterations to the cardiovascular, endocrine, and renal systems, together with marked changes in maternal body composition. Maternal diet and nutritional status need to be sufficient in pregnancy both to meet the additional nutrient demands that result from these changes, and those of the growing foetus, which is wholly dependent on an adequate nutrient supply to meet its needs. This dependence is demonstrated in the numerous experimental animal studies that show how manipulation of foetal nutrition can lead to impaired foetal growth and smaller size at birth. The importance of adequate maternal nutrition to support the process of reproduction has been widely recognized for many years, although the complexity of the supply line to the foetus (Figure 15.1) and the changing pattern of demand for individual nutrients across pregnancy can make it challenging to translate knowledge of nutrient needs into dietary recommendations in pregnancy.

Interest in maternal nutrition has increased following the findings of epidemiological studies conducted since the 1980s that have linked different patterns of prenatal growth to long-term health. For example, a consistent finding across a range of international studies is that lower birth weight is associated with a greater risk of development of non-communicable diseases (NCD), including cardiovascular disease and type 2 diabetes in

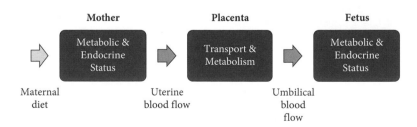

FIGURE 15.1 Factors along the foetal 'supply line' which can mediate the differences between maternal nutrition and foetal nutrition.
Redrawn with permission from J E Harding (2001) The nutritional basis of the origins of disease. *Int J Epidemiol* Feb; **30**(1),15–23. doi: 10.1093/ije/30.1.15. ©2001, Oxford University Press.

adult life. Importantly, this link cannot be explained by differences in adult lifestyle, such as patterns of diet and smoking. Growth impairment in foetal life, evident as smaller size at birth, appears to result in permanent 'programmed' changes in physiology and function, with the consequence that individuals who were smaller at birth are more vulnerable to develop these conditions later in life. The central role of nutrition as an influence on foetal growth led to the proposal that undernutrition in prenatal life is a cause of later cardiovascular disease and type 2 diabetes. We now recognize that both prenatal and early postnatal life are periods when lifelong physiology and function can be programmed, a concept that has been described as the developmental origins of health and disease (DOHaD). The link to NCDs, leading causes of mortality worldwide, has highlighted the need to take a lifecourse approach to understanding adult health and well-being, and this is embedded in international health policy (Figure 15.2).

Understanding the mechanistic basis of the observed links between early development and later health has become a theme of international research, with a view to formulation of future strategies to prevent cardiovascular disease and type 2 diabetes. There are a number of candidate mechanisms to explain how differences in early environment, such as differences in maternal nutritional status, can cause permanent changes in physiology and function. These include alterations to oocyte physiology and other effects via epigenetic mechanisms (Figure 15.3) (Hanson et al. 2015).

Epigenetic changes, such as DNA methylation, modify gene expression without changing the underlying DNA sequence (see Chapter 29). This enables one genotype to result in different phenotypes, depending on environmental cues, leading to lifelong differences in function. It also has the potential to influence the health of the next generation. A range of nutritional cues have been identified that point to roles of under- as well as over-nutrition, including nutrients involved in the methylation cycle (Hanson et al. 2015). As environmental factors that women may be exposed to before they know they are pregnant can cause permanent changes in the pattern and characteristics of development early in embryonic life, it highlights the significance of appropriate maternal nutrition both *before* as well as during pregnancy.

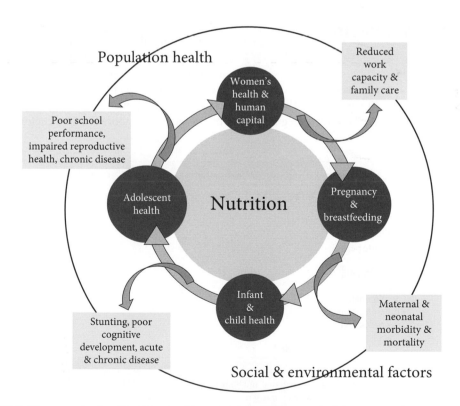

FIGURE 15.2 The central role of nutrition in determining health across the lifecourse and across generations. All stages in a woman's life are connected by the effects of good or poor nutrition both presently and at the previous life stages. Good nutrition in adolescence affects a woman's health and productivity, and her ability to go through pregnancy and breastfeeding successfully, which impacts infant and child health. Poor nutrition at any of these stages has negative consequences that disrupt the cycle and impact later life stages, including future generations.
Source: M A Hanson, A Bardsley, LM De-Regil et al. (2015) The International Federation of Gynecology and Obstetrics (FIGO) recommendations on adolescent, preconception, and maternal nutrition: 'Think Nutrition First'. *Int J Gynaecol Obstet* Oct; **131**(Suppl 4), S213–53. doi: 10.1016/S0020-7292(15)30034-5. With permission from Elsevier.

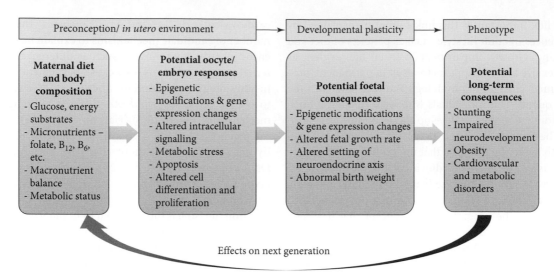

FIGURE 15.3 Effects of the preconception and *in utero* environment on offspring phenotype and future health. Maternal diet, body composition, and metabolic status provide cues to the oocyte, embryo, and foetus during critical periods of developmental plasticity. The availability and levels of essential micro- and macronutrients and energy substrates serve as cues that can trigger epigenetic and other responses affecting foetal development, which have lasting impact on offspring health and which can be passed in a similar manner to the next generation.
Source: M A Hanson, A Bardsley, LM De-Regil et al. (2015) The International Federation of Gynecology and Obstetrics (FIGO) recommendations on adolescent, preconception, and maternal nutrition: 'Think Nutrition First'. *Int J Gynaecol Obstet* Oct; **131** (Suppl 4), S213–53. doi: 10.1016/S0020-7292(15)30034-5. With permission from Elsevier.

15.2 **Pregnancy and birth outcomes**

The average duration of pregnancy is 266 days—38 weeks after conception, and 40 weeks from the date of the last menstrual period (LMP). Timing in pregnancy is described in terms of gestational age of the foetus, which is commonly divided into three trimesters (Figure 15.4). In relation to foetal development, two periods are defined: *embryonic* (0–8 weeks), during which implantation occurs, the placenta is established, and organogenesis begins, and *foetal* (nine weeks to term) that is characterized by growth and elaboration of structures. The estimated

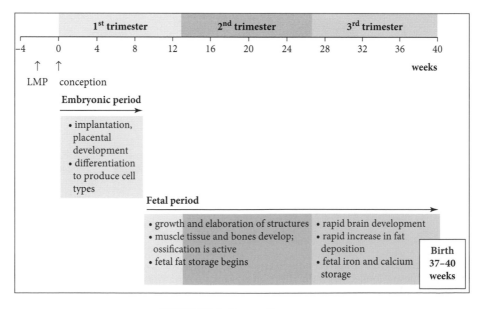

FIGURE 15.4 Stages of pregnancy
Redrawn from BNF (British Nutrition Foundation) J Buttriss, S Stanner et al. (2013) *Nutrition and Development: Short and Long Term Consequences for Health*. Wiley Blackwell. © 2013, John Wiley.

date of delivery (EDD) can be calculated from the LMP; if this is uncertain, foetal measurements obtained using ultrasound can be compared with standard growth curves in order to derive an EDD. The timing of birth is categorized as term or pre-term according to gestational age (see Box 15.1).

Pre-term births are associated with greater risk of complications in early postnatal life, and recognized as a key determinant of adverse infant outcomes, in terms of survival and quality of life. But the most frequently recorded pregnancy outcome is weight of the infant at birth. Low birth weight (LBW) is associated with poorer outcomes in infancy. However, as foetal growth is strongly linked to gestational age, LBW infants can include those born at lower gestational ages as well as term infants whose growth *in utero* has been impaired. Definition of foetal growth restriction therefore needs to distinguish infants who are smaller than expected for their gestational age (see Box 15.1) from those born early but at appropriate size.

15.3 Changes in maternal physiology

Pregnancy is a period of major changes in maternal physiology. Alterations to the maternal cardiovascular, endocrine, and renal systems take place alongside the remodelling and deposition of new tissue, and there are significant changes in maternal body composition.

Maternal changes in pregnancy can be characterized as follows:

- the changes precede foetal demands and are therefore anticipatory;
- the changes are in excess of possible foetal requirements;
- the maternal internal milieu is altered to favour placental exchange of nutrient substrates and foetal metabolic processes.

Blood volume expansion and haemodynamics

There is an increase in the circulating blood volume of approximately 1600 ml by the end of a pregnancy. Eighty per cent of this increase is required to perfuse the placental bed in the last few weeks of pregnancy. The increased blood volume consists of a rise in the plasma volume from 2500 ml in the non-pregnant state to 3800 ml at term, resulting in lower blood concentrations of some nutrients. Although there is an increase in red blood cell volume of 1400 ml to 1640 ml at term, and oxygen-carrying capacity is enhanced, the relatively greater increase in plasma volume leads to a fall in haemoglobin concentration from a non-pregnant average of 130 to 140 g/L to concentrations of 100 to 110 g/L in late pregnancy. This physiological change reduces blood viscosity and improves placental perfusion, and therefore oxygen and nutrient exchange at the placental bed. Heart rate increases from an average of 70 beats per minute before pregnancy to a rate of 80 to 85 beats per minute at full term. This is accompanied by an increase in the stroke volume, the amount of blood pumped by one cardiac cycle, from 65 ml in the non-pregnant state to 70 ml at term. The effect is to increase cardiac output, the amount of blood pumped per minute, from a non-pregnant average of 5000 ml to 6500 ml at term.

Respiratory system

Changes in the maternal respiratory system take place to meet an increasing requirement for oxygen as the pregnancy develops. The tidal volume, the amount of air inspired per breath, increases from a non-pregnant average of 400 ml to 550 ml at term; overall, oxygen consumption is increased by about 15 per cent. Chemoreceptors are reset to allow the mother to tolerate her PCO_2 at 4 kilopascal (kPa) rather than the non-pregnant average of 4.7 kPa. Carbon dioxide is excreted through the lungs, but foetal production of CO_2 must be excreted to the mother via the placenta before reaching her lungs. The physiological lowering of the homeostatic level of CO_2 (PCO_2) by about 15–20 per cent facilitates diffusion of CO_2 from foetus to mother, and is a vital physiological shift in maternal homeostasis. For the mother, the deeper breathing required to lower the PCO_2 produces the common symptom of breathlessness but her foetus benefits from an increased concentration gradient of CO_2 across the placenta and facilitated excretion of CO_2.

Renal system

Changes to the renal system occur early in pregnancy, modifying maternal fluid balance and increasing capacity for excretion of waste products. Renal plasma flow increases by about 30 per cent with a corresponding increase in glomerular filtration rate. Serum albumin falls, as a physiological change designed to reduce the plasma oncotic pressure, allowing increased renal glomerular filtration. An unwanted effect of the increased glomerular filtration is that significant glycosuria and aminoaciduria are common in normal pregnancy and can occasionally contribute to generalized nutrient depletion.

Gastrointestinal system

Nausea and vomiting are common problems in early pregnancy, affecting the majority of women. These symptoms may be related to the high circulating levels of human chorionic gonadotrophin (HCG), although observational studies have not shown association between HCG concentration in serum and symptoms in individual women. In about 1 per cent of cases, the vomiting is severe enough to require hospital admission and parenteral nutritional support; the condition is termed hyperemesis gravidarum. Although it is a major concern for many women that their nutrient intake may be compromised during the period of nausea and vomiting, there is no evidence of foetal harm, except in the most severe cases. In fact the evidence is that those women who suffer more from nausea and vomiting often have relatively higher birthweight infants.

There is a generalized relaxation of smooth muscle in the gastrointestinal tract from about ten weeks of gestation through to term. Relaxation of the cardiac sphincter in the stomach is thought to be the cause of the common symptom of acid reflux, which results in heartburn. Delayed gastric emptying can lead to a feeling of fullness after meals. There is an overall increase in transit time from stomach to caecum from around 52 hours in non-pregnant women, to 58 hours in pregnancy. Animal studies report an increase in villous density and absorptive area in the small bowel in pregnancy. Assuming such changes appear in human pregnancy, the combined effect would be to increase nutrient uptakes. Evidence for this is scant but calcium absorption has been shown to increase in early pregnancy, and iron absorption increases throughout pregnancy—up to 40 per cent higher towards the later stages. The reduced peristalsis in the large bowel is associated with increased water reabsorption and production of hard stools leading to the common symptom of constipation.

The placenta

Foetal growth and development are dependent on the integrity of the maternal–placental unit. The placenta is the interface across which nutrients, gases, and other compounds are transferred between mother and foetus; the placenta reaches a weight of around 1.5 kg by term. It is perfused on the foetal side by foetal blood and on the maternal side by maternal blood, exchange taking place across the cellular interface, which is the syncytiotrophoblast.

Major physiological changes take place in the terminal branches of the uterine artery at the placental site to allow expansion of the placental blood flow with advancing pregnancy. At term, the placental blood flow reaches 800 ml/min. The structural modifications of these terminal arteries follow invasion of the trophoblast into the uterine wall and the arterial walls in the first half of pregnancy. Placental transfer of nutrients takes place through four mechanisms: simple diffusion, facilitated diffusion, active transport, and pinocytosis. The placenta is also the source of several placental proteins with endocrine functions including HCG (the hormone detected in maternal blood or urine in pregnancy testing), human placental lactogen, and human placental growth hormone. The latter two have insulin counter-regulatory effects and therefore may modify the nutrient substrate mixture presented to the placenta by the mother. A number of other pregnancy-specific placental proteins have been identified, most of which appear to have a role in early pregnancy to facilitate uterine implantation.

Maternal homeostasis of foetal nutrient substrates

Maternal plasma concentrations of water-soluble and fat-soluble nutrients vary during pregnancy; some nutrients show a decrease in plasma concentration while others show an increase. Individual profiles vary widely, but in general, water-soluble nutrients and metabolites are present in lower concentrations in pregnant than in non-pregnant women, whereas fat-soluble nutrients and metabolites are present in similar or higher concentrations. Some of the decreases in nutrient concentrations observed early in gestation may be explained by haemodilution arising from the expansion of plasma volume. In addition, the hormonal changes that occur in pregnancy contribute to development of insulin resistance in later gestation. Insulin resistance is associated with an increased availability of all potential foetal substrates, particularly in the postprandial period when glucose, amino acid, and lipid levels are raised. Insulin resistance is a factor in enhanced lipolysis in the adipose tissue, which is seen during overnight fasting in pregnancy and generates glycerol and free fatty acids as potential foetal or maternal fuel.

15.4 Weight gain in pregnancy

Weight gain in pregnancy varies greatly, but most women gain between 10kg and 12.5kg over the course of their pregnancy. Maternal weight gain in pregnancy is accounted for by the weight of the foetus (representing about a third of the total weight gain), placenta and supporting tissues, increase in blood volume and extracellular fluid, and deposited fat. Definitions of optimal weight gain, as well as strategies to manage weight gain in pregnancy, are topics of current research. Both inadequate and excessive weight gains are associated with poorer maternal and infant health outcomes. Inadequate gestational weight gain is linked to lower birth weight and increased risk of SGA, and there is strong evidence for a U-shaped association between lower weight gain in pregnancy and preterm birth among normal weight and underweight women. Additionally, the rise in maternal obesity in recent decades has focused attention on the health consequences of excessive weight gain, both in the short and longer term. For example, there is a growing evidence base that shows greater adiposity and increased risk of insulin resistance and obesity among children born to mothers whose weight gain in pregnancy was excessive (Eitmann et al. 2020).

A comprehensive review of the influence of gestational weight gain on pregnancy outcome was published in 2009 which informed revision of the US Institute of Medicine's guidelines for weight gain during pregnancy. The IOM recommendations set ranges for 'adequate' or desirable weight gain, which are associated with optimal maternal and infant outcomes; these differ according to maternal body mass index (BMI) before pregnancy (see Box 15.2). There is widespread awareness of the IOM recommendations among policy makers, but currently there are no formal guidelines in the United Kingdom. Studies carried out in high-income countries, including the UK, consistently show that weight gain of a significant proportion of pregnant women fall outside the IOM recommended ranges of weight gain, and that excess gestational weight gain is common. It is important to recognize that weight gain recommendations based on data

from high-income countries may not apply universally; for example, lower ranges have been recommended for Asian women.

15.5 Obesity in pregnant and lactating women

It has been estimated that 40 per cent of adult women globally are overweight (BMI 25.0–29.9 kg/m^2) and 15 per cent are obese (BMI ≥30) (WHO 2020). In the UK the prevalence of maternal obesity has doubled over the last two decades and around one in five women attending antenatal care are now obese. Women who are obese are likely to be older in pregnancy, have a higher parity, and live in areas of high deprivation, compared with women who are not obese. Maternal obesity is associated with numerous adverse health outcomes, relating both to the immediate implications for pregnancy complications and longer-term health consequences in her offspring. Pregnant obese women are at greater risk of developing gestational diabetes, pre-eclampsia, gestational hypertension, depression, and are more likely to have an instrumental or caesarean birth and surgical site infection compared to non-obese women. Maternal obesity is also linked to increased risk of preterm birth, large-for-gestational-age babies, foetal defects, congenital anomalies, and perinatal death. There is increasing evidence that implicates maternal obesity as a major determinant of health during childhood and later adult life, including an increased risk of obesity, coronary heart disease, stroke, type 2 diabetes, and asthma. Maternal obesity may also lead to poorer cognitive performance and increased risk of neurodevelopmental disorders, including cerebral palsy. These outcomes may, in part, be mediated through changes in epigenetic processes, such as alterations in

---

BOX 15.2

Recommended ranges of weight gain in pregnancy (IOM 2009)

Prepregnancy BMI	Total weight gain range (kg)	Incremental weight gain during the 2nd and 3rd trimester; mean (range) (kg/week)
Underweight (<18.5 kg/m^2)	12.5–18	0.51 (0.44–0.58)
Normal weight (18.5–24.9 kg/m^2)	11.5–16	0.42 (0.35–0.50)
Overweight (25.0–29.9 kg/m^2)	7–11.5	0.28 (0.23–0.33)
Obese (≥30.0 kg/m^2)	5–9	0.22 (0.17–0.27)

both DNA methylation and in the gut microbiome. Despite the significant risks of maternal obesity, dieting whilst pregnant is currently not advised until its effectiveness and safety is further elucidated. Rather, current NICE guidelines recommend that women with BMI ≥30 should be helped to lose weight following their pregnancy, before becoming pregnant again.

There is accumulating evidence that obese women are less likely to breastfeed their babies than women who are not obese. Several systematic reviews have suggested that obese women are less likely to initiate breastfeeding, have a shortened duration of breastfeeding, have significantly delayed lactogenesis II (initiation of copious milk production), and lower milk transfer compared to women who are not obese. Obese women cite both physical and psychological barriers to the initiation and continuation of breastfeeding, including difficulties relating to larger breasts and positioning to breastfeed, perceived insufficient supply of breast milk, low confidence in their ability to breastfeed, negative body image, and embarrassment at breastfeeding in public. The evidence base on which interventions best support breastfeeding obese women is currently limited, but qualitative data suggests that positive support that is tailored to their specific needs has a beneficial effect (Chang et al. 2020).

15.6 Nutrition and health in the preconception period

During the embryonic period, when organs are being formed and the trajectory of foetal growth is established, the foetus is vulnerable to external factors in its environment. As such influences on the embryo occur early in pregnancy, often before women know they are pregnant, ensuring all women of childbearing age have a healthy body weight and adequate nutritional status, may be key to future reproductive success and ensuring the long-term health of their children. Interventions in pregnancy may be too late to be effective. For example, trials designed to avoid excessive gestational weight gain and improve perinatal outcomes have shown limited success, which may in part be due to their timing in pregnancy, as the effects of maternal metabolic condition on placental function and gene expression occur in the first trimester of pregnancy before most intervention trials are initiated. It is therefore a concern that so many women of childbearing age worldwide are overweight and obese and have habitual diets of poor quality that are energy-dense and micronutrient-poor. Although there is some evidence of changes in women's health behaviours, such as stopping smoking in pregnancy, dietary patterns appear to change little and there is little evidence of changes in health behaviours in anticipation of pregnancy. Establishing healthy dietary habits and body composition, particularly in adolescent girls and young women, are priorities to promote good periconceptional health and the health of the next generation.

15.7 Nutritional requirements during pregnancy and lactation

How nutrient requirements for pregnancy and lactation are calculated

Assessment of maternal nutritional requirements during pregnancy and lactation is complicated by the profound changes in maternal physiology during pregnancy that cause haemodilution, changes in the ratio of free to bound forms of nutrients, and alterations in nutrient turnover and homeostasis. For most nutrients, the factorial approach to setting requirements is used, and involves the addition of the estimated amount of the nutrient deposited in the mother and foetus to the requirements for non-pregnant women, together with an additional amount to allow for the inefficiency of utilization for tissue growth. For some nutrients such as for folate, where experimental data are available, recommendations can be based on the amount needed to maintain tissue levels and nutrient-dependent function.

Reference values for lactation are calculated using data relating to the quantity of milk produced during lactation, its energy and nutrient content, and the amount of maternal energy and nutrient reserves. The quantity and nutrient content of milk consumed by the infant who is growing well and maintaining appropriate biochemical indices of nutritional status are often used as proxies to assess maternal nutritional adequacy during lactation.

Nutrient requirements for pregnancy

The current UK recommendations for nutrient intakes in pregnancy (and lactation) are summarized in Table 15.1. The reference nutrient intake (RNI) represents an intake at which about 97 per cent of the population will meet their daily requirements. Considering the major physiological changes of pregnancy described above, maternal demand for all nutrients may be expected to increase. However, as pregnancy is associated with significant maternal metabolic adaptations, this helps to compensate for any extra requirements and enable optimal utilization of nutrients. Consequently, major changes in maternal intake are generally not warranted. The following sections describe the nutrients for which there are recommended

TABLE 15.1 RNI for women in the UK

Nutrient/day	Women (19–50 years)	Pregnancy	Lactation	
			(0–4 Months)	(4+ Months)
Energy (kcal)	1940	+200[a]		
Protein (g)	45	+6	+11	+8
Thiamin (mg)	0.8	+0.1[a]	+0.2	+0.2
Riboflavin (mg)	1.1	+0.3	+0.5	+0.5
Niacin (mg)	13	[b]	+0.2	+0.2
Vitamin B_6 (mg)	1.2	[b]	[b]	[b]
Vitamin B_{12} (µg)	1.5	[b]	+0.5	+0.5
Folate (µg)	200	+100	+60	+60
Vitamin C (mg)	40	+10	30	+30
Vitamin A (µg)	600	+100	+350	+350
Vitamin D (µg)	10	10	10	10
Calcium (mg)	700	[b]	+550	+550
Phosphorus (mg)	550	[b]	+440	+440
Magnesium (mg)	270	[b]	+50	+50
Sodium (mg)	1600	[b]	[b]	[b]
Potassium (mg)	3500	[b]	[b]	[b]
Chloride (mg)	2500	[b]	[b]	[b]
Iron (mg)	14.8	[b]	[b]	[b]
Zinc (mg)	7.0	[b]	+6	+2.5
Copper (mg)	1.2	[b]	+0.3	+0.3
Selenium (µg)	60	[b]	+15	+15
Iodine (µg)	140	[b]	[b]	[b]

[a] Third trimester only

[b] No increment

increased intakes, together with some minerals of key importance during pregnancy.

Energy

Pregnancy places increased demands for energy on the mother to support placental and foetal growth, growth of maternal tissues (uterus, breast, and adipose tissue), and deposition of fat reserves in preparation for lactation. Maintenance of the larger tissue mass and movement during pregnancy, particularly for weight-bearing activities after 25 weeks gestation also increases the demand for energy in pregnancy, although some of these costs may be offset by adaptive changes in activity and maternal metabolism. Energy requirements are influenced by a wide range of factors, such as rate of maternal weight gain, foetal growth rates, maternal lifestyle and activity levels, maternal body composition, and genetic factors. As such, estimates of the total energy cost of pregnancy vary widely and are subject to a high degree of inter-individual variation. Early work conducted by Hytten and Leitch estimated the total energy cost of pregnancy at 700,00 kcal (293 MJ) based on an increase in basal metabolism of 30,000 kcal (126 MJ), together with the additional requirement of 40,000 kcal (167 MJ) associated with increased body size, equivalent to an increment of 250 kcal/day (1.04 MJ/day).

Longitudinal studies of well-nourished women with access to ample food supplies, however, reveal only small increases in energy intake during pregnancy, and these are not always statistically significant or universal. Such studies suggest that pregnancy is associated with adaptive responses to conserve energy, which may involve adjustments in physical activity and an increase in work efficiency and an adaptation of the metabolic response in pregnancy.

There is large inter-individual variation in the metabolic response to pregnancy and, whilst BMR is increased in most pregnant women as would be expected with increasing tissue mass, BMR declines in early pregnancy in some women. This energy conserving response is most common in women who are undernourished with limited fat stores and relatively high physical activity, such as subsistence farmers in low-income countries. Women in high-income societies are more likely to conserve energy through a reduction in physical activity, and studies have generally found that such women walk less and sit more, but the changes are subtle. In the third trimester of pregnancy, in particular, women may spend less time performing strenuous activities.

Energy reference values for pregnancy in the UK are estimated using the factorial method, where increments in requirements are added to the mother's estimated average requirement (EAR) based on singleton pregnancies reaching term. In their review of dietary reference values for energy, the Scientific Advisory Committee on Nutrition (SACN 2011) in the UK concluded that, in the absence of new evidence, the Committee on Medical Aspects of Food Policy recommendation of an additional increment of 200 kcal/day (0.84 MJ/day) in the last trimester of pregnancy should be retained. This recommendation assumes an average pregnancy weight gain of 12.5 kg, an efficiency of energy utilization of 90 per cent and a mean birth weight of 3.4 kg and is equivalent to an extra total intake of approximately 18,000 kcal (75 MJ). It is noted that women who are underweight at the beginning of pregnancy, and women who do not reduce their activity, may have a higher EAR.

Current UK recommendations on energy intake in pregnancy differ from those set by the WHO and the USA. WHO recommendations, which are specific to women in societies with a high proportion of non-obese women, suggest an increase of 360 kcal/day (1.5 MJ/day) in the second trimester and 475 kcal/day (2.0 MJ/day) in the third trimester, based on an average gestational weight gain of 12 kg (FAO/WHO/UNU 2004). In the USA, the Reference Daily Intake (RDA) comprises incremental values of 340 kcal/day (1.43 MJ/day) in the second trimester and 452 kcal/day (1.9 MJ/day) in the third trimester.

Protein

Growth of the foetal, placental, and maternal tissues rely on an adequate supply of protein. Protein deposition in maternal and foetal tissues increases throughout pregnancy, with most occurring during the third trimester. Maternal nitrogen metabolism adapts from very early in pregnancy to increase nitrogen and protein deposition in the mother and foetus. Adaptations include lower urea production and excretion, lower plasma α-amino nitrogen, and a lower rate of branched chain amino acid transamination.

The UK RNI for protein is increased by 6g/day (to 51 g/day) throughout pregnancy. As intakes of protein in high-income countries range from 60 to 110 g/day, this increased demand will be met by most diets. However, maternal under-nutrition and malnutrition are major problems in the poorest developing countries and dietary protein requirements in these populations may not be met due to limited access to high quality foods, traditional food habits, food taboos and limited knowledge. A review of energy and protein intake in pregnancy found that, whilst high-protein supplementation may be harmful to the foetus, balanced energy and protein supplementation (where protein provides less than 25 per cent of the total energy intake) appears to increase birthweight and substantially reduce the risk of stillbirth and small for gestational age (Ota et al. 2015).

Fatty acids

Polyunsaturated fatty acids include the essential fatty acids linoleic acid (n-6 fatty acid) and alpha-linolenic acid (n-3 fatty acid) found mainly in seed oils. Longer chain more unsaturated derivatives (LCPUFA), which include eicosapentaenoic acid (EPA), arachidonic acid (AA) and docosahexaenoic acid (DHA), are the end products of the n-3 pathway but may not be synthesised in sufficient quantities and therefore must be obtained from dietary and/or supplement sources. N-3 LCPUFAs, obtained from fatty fish, such as salmon, tuna, mackerel, herring and sardines, and fish oil, are considered to possess protective properties for human health by impacting on immunological reactions. Changes in dietary patterns in industrialised countries worldwide have led to a shift in the dietary ratio of n-6/n-3 LCPUFA intake, favouring foods rich in n-6 LCPUFA sources. Current Western diets typically have a n-3:n-6 PUFA balance of between 1:10 and 1:30, far from the recommended ratio of 1:3 to 1:5. Diets with higher n-6 to n-3 ratios may contribute to the pathology of metabolic syndrome through inflammatory processes and other currently unrecognized mechanisms.

Foetal supply of fatty acids depends on maternal fatty acid status, placental function, and placental levels of fatty acid transporters and binding proteins. DHA and AA are preferentially transferred across the placenta, and many reports suggest that higher maternal DHA intake can increase infant DHA status Experimental research indicates that foetal exposure to dietary LCPUFAs can modulate metabolic pathways involved in gene expression, cell differentiation, and epigenetic modifications. Studies have demonstrated that omega-3 fatty acid supplementation lowers the incidence of preterm births, but its effect on reducing pregnancy complications are inconclusive (Godhamgaonkar et al. 2020).

The European Food Safety Authority (EFSA) advises that pregnant and lactating women should increase their DHA intake by 100-200 mg/day to compensate for oxidative losses of maternal dietary DHA and accumulation of DHA in body fat of the foetus/infant. It should be noted that high consumption of oily fish carries an increased risk of exposure to lipophilic industrial contaminants such as polychlorinated biphenyls and mercury, which can be harmful to foetal development. The Food Standards Agency (FSA) advises pregnant women to consume no more than two portions of oily fish per week, avoid eating shark, swordfish and marlin, and limit the amount of tuna consumed to no more than two steaks a week or four medium-sized cans a week. As consumption of *trans* unsaturated fatty acids appear to be associated with lower maternal and neonatal PUFA status, it may be beneficial to minimise the consumption of *trans* fatty acids during pregnancy, in line with general dietary guidance.

Thiamin

Thiamin, or vitamin B_1, plays an essential role in the utilization of energy, in carbohydrate metabolism and in the metabolic links between carbohydrate, protein and fat metabolism. The amount of thiamin needed is proportional to caloric intake (approximately 0.5 mg thiamin is required per 1000 kcal consumed). An increase in thiamin in the third trimester is recommended to accompany the suggested increase in energy intake. The prevalence of thiamin deficiency is relatively common in low-income populations, including many areas of rural Southeast Asia, where diets are high in refined or unfortified grains and low in animal source foods and legumes. In the UK there is little evidence of low thiamin status in the general population nor is there evidence of thiamin depletion arising as a result of pregnancy. Excessive vomiting in pregnancy, however, can cause thiamin depletion, in which case antenatal vitamins containing thiamin and other B vitamins may be beneficial.

Riboflavin

The RNI for riboflavin (vitamin B_2), a water-soluble vitamin that plays a central role as a coenzyme in energy-yielding metabolism, is increased in pregnancy. Riboflavin deficiency is commonly encountered in combination with deficit of other B vitamins in areas of poor overall nutrition, and deficiency is endemic in populations whose staple diet consists of rice and wheat, with low or no consumption of meat and dairy products. Poor riboflavin status is also common amongst some population groups in high-income countries, including the UK. For example NDNS data show that about 75 per cent of 11–18 year-old girls fall short of the biochemical threshold for adequate riboflavin status. To ensure adequate riboflavin supply, an increase in dairy products and/or meat may be necessary in some women.

Folate

Pregnancy places an increased demand on the supply of folate for the synthesis of DNA and for other one-carbon (1-C) transfer reactions, and as a consequence, pregnant women may be at increased risk of developing a folate deficiency. A number of pregnancy complications and poor birth outcomes, including placental abruption, preeclampsia, spontaneous abortion, stillbirth, foetal growth restriction, LBW, and prematurity, have been linked to impaired folate status and to hyperhomocysteinemia, and maternal folic acid supplementation has been associated with increased birthweight and decreased incidence of LBW and SGA.

An adequate supply of folate during the first four weeks of pregnancy is necessary for normal development of the foetal spine, brain, and skull and to protect against neural tube defects (NTDs), which include anencephaly, spina bifida, and encephalocele. NTDs are among the most common congenital malformations in neonates contributing to infant mortality and serious disability. The UK has the highest burden of NTDs in Europe, with an incidence rate ranging from 0.8 to 1.5 per 1000 births, depending on ethnic, geographic, and nutritional factors. The National Diet and Nutritional Survey (NDNS) in the UK indicated that among women of childbearing age, three-quarters had a red blood cell folate concentration lower than the threshold for optimal avoidance of folate-sensitive foetal neural tube defects (748nmol/L). Young women are particularly at risk, with significantly lower mean red blood cell folate levels in those aged 16 to 24 years compared to women aged 35 to 49 years. A landmark randomized controlled trial conducted by the Medical Research Council in UK (MRC Vitamin Study Research Group 1991), demonstrated that 50 per cent or more of NTDs could be prevented if women consumed a folic acid-containing supplement before and during the early weeks of pregnancy in addition to the folate in their diet. Based on this and similar research, it is recommended that all women planning a pregnancy should take a supplement containing 400 µg of folic acid per day prior to conception and until the 12th week of pregnancy, when the foetal spine is developing. These recommendations have been endorsed around the world.

Some 30 years after publication of the Medical Research Council study and years after various recommendations have been issued to promote folic acid supplementation to ensure adequate peri-conceptional folate concentrations for pregnant women, it is clear that Europe has failed to implement an effective policy for prevention of NTD by folic acid. A large study of 18 countries covering approximately 12.5 million births in Europe over the period 1991 to 2011, found that the overall prevalence of NTD remained comparable over this period. Similarly, the proportion of women taking prenatal folic acid supplements declined to 31 per cent in the same period. Those least likely to take folic acid supplements were women aged under 20 years and non-Caucasian women. One important limitation to the value of recommending folic acid supplementation prior to conception is that only about half of all pregnancies in the UK are planned.

Steps to achieve folate sufficiency have included the mandatory fortification of staple foods, usually grains, in more than 50 countries around the world, including the United States, Canada, Chile, Costa Rica, and New Zealand. In September 2021, the UK government announced that folic acid will be added to non-wholemeal wheat flour across the UK, preventing an estimated 200 neural tube defects per year (around 20 per cent of the annual UK total) (see Chapter 11.6).

Vitamin C

Vitamin C (ascorbic acid) is a water-soluble antioxidant and an important cofactor for enzymes involved in collagen and neurotransmitter synthesis. The vitamin also promotes iron absorption and release of iron from ferritin stores. The vitamin is actively transported across the placenta to the foetus which, in combination with haemodilution, results in a decline in maternal plasma vitamin C levels over the course of pregnancy. To compensate for this, UK dietary guidelines constitute an increased intake of 10mg/day during pregnancy. Vitamin C deficiency is rare in pregnancy but low status may be a feature of women who have generally poor quality diets. Vitamin C depletion in pregnancy has been associated with an increased risk of pregnancy complications, yet there is limited evidence to support the routine supplementation of vitamin C alone or in combination with other supplements for the prevention of foetal or neonatal death, poor foetal growth, preterm birth, or pre-eclampsia (Rumbold et al. 2015).

Vitamin A

Vitamin A is a fat-soluble vitamin that has antioxidant functions and plays an important role in supporting vision, protein synthesis, reproduction, growth, and development. Vitamin A is essential for the production of rhodopsin in the retina and deficiency of this vitamin is a major cause of preventable blindness in children in developing countries. Vitamin A deficiency is recognized as a major cause of illness in low-income countries and women of childbearing age are among the populations most at risk, with associated outcomes including foetal loss, LBW, preterm birth, and infant mortality. Adverse health consequences related to vitamin A consumption are also observed when intake is excessive; particularly in the form of supplements or from animal sources. The teratogenic effects of excessive vitamin A intake are evident in the developing embryo resulting in significant neurological and physiological birth defects. Current recommendations in the UK advise against the consumption of supplements containing vitamin A and of liver or liver products by pregnant women or women who might become pregnant. Dietary deficiency of vitamin A is rare in high-income countries and supplementation is rarely indicated.

Vitamin D

Vitamin D is a fat-soluble vitamin that plays an important role in bone metabolism and as well as a wide range of other functions including some anti-inflammatory and immune-modulating properties. Changes in vitamin D metabolism occur during pregnancy, leading to increased serum concentration of the biologically active form, 1,25-dihydroxy vitamin D_3 (1,25(OH)$_2$D), whilst serum 25-hydroxy vitamin D_3 (25(OH)D) concentration is unaffected. Without sufficient 1,25(OH)$_2$D, the intestine cannot absorb calcium and phosphate adequately, which leads to secondary hyperparathyroidism and a lack of new bone mineralization resulting in growth failure and rickets.

The prevalence of Vitamin D deficiency (which can be defined as serum 25(OH)D concentration <25nmol/L) varies by season and reports range from 9 per cent in the Americas, 13 per cent in Western Pacific, 23 per cent in Europe, to 79 per cent in Eastern Mediterranean. In the UK during the winter months of January–March, 29 per cent of adults, 19 per cent of children aged 4–10 years, and 37 per cent of children aged 11–18 years have a serum 25(OH)D concentration lower than 25nmol/L. The proportion of adults and children with serum 25(OH)D concentrations lower than 25nmol/L in the summer months decreases to 4 per cent and 6 per cent respectively (NDNS Rolling Programme 2019). Groups at particular risk of vitamin D deficiency in the UK include pregnant and breastfeeding women, particularly teenagers and young women; people who have no or limited exposure to the sun of sufficient intensity to induce cutaneous vitamin D synthesis, including people who cover their skin for cultural reasons, are confined indoors, or live in Scotland and Northern England; and individuals who have darker skin (e.g., people of African, African-Caribbean, or South Asian ethnic origin).

There are few controlled studies of vitamin D supplementation in pregnancy and as a consequence the optimal vitamin D regimen to prevent and treat vitamin D insufficiency *in utero* is not clear. In 2016 SACN published an evidence review of vitamin D and health and set a RNI of 10μg/day for adults and children of all ages, replacing previous advice that the RNI was required only for children under four years and older adults aged 65 years and over. Prior to this updated advice it was thought that cutaneous synthesis of vitamin D was sufficient for the majority of the population. A vitamin supplement containing vitamins C and D and folic acid is available free of charge in the UK through the Healthy Start Scheme, which provides help to low-income pregnant women or new mothers and their children under the age of four.

Calcium

There are major physiological changes in calcium metabolism during pregnancy to ensure an adequate foetal supply. In early pregnancy there is an increase in bone turnover with increased bone resorption liberating calcium to the plasma pool. There is also increased gut absorption of calcium and an increase of urinary calcium excretion in healthy pregnant women of about 100 per cent, which would suggest that the combination of the above factors is more than adequate to provide the calcium required for foetal skeleton and membranes. There is considerable remodelling of bone with an increase in bone mineral density in the long bones of the arms and legs and a reduction in bone mineral density in the pelvis and spine being a feature of late pregnancy. In the UK, no increase in calcium intake is recommended although problems may arise with exclusion diets in those who do not consume dairy products or in young mothers who are still growing themselves. In populations with low dietary calcium intake, such as those living in low-income countries, the WHO recommends daily calcium supplementation (1.5 g–2.0 g oral elemental calcium) for pregnant women to reduce the risk of pre-eclampsia (WHO 2018).

Iron

The combination of new tissue formation, haematopoiesis (blood formation) in the foetus and the mother, and typical blood losses at delivery, suggest a total iron requirement of approximately 800–900 mg over the course of pregnancy. The iron-sparing cessation of menstruation meets some of this need, but women still require an additional 1 mg/day in early pregnancy, rising to 6 mg/day in late gestation. This may justify iron supplementation in many women, were it not for evidence that small

bowel absorption of iron, along with other divalent cations, is greatly enhanced in pregnancy.

Iron deficiency is the most frequently occurring nutritional disorder in the world. Whilst it is particularly widespread in low-income countries, often exacerbated by infectious diseases, such as malaria, HIV/AIDS, hookworm infestation and schistosomiasis, it is the only nutrient deficiency which is also prevalent in high-income countries. Anaemia affects half a billion women of reproductive age worldwide. In 2011, 29 per cent (496 million) of non-pregnant women and 38 per cent (32.4 million) of pregnant women aged 15–49 years were anaemic (Stevens et al. 2013). Maternal iron deficiency anaemia (defined as haemoglobin (Hb) concentration <110g/L in first trimester, <105g/L in second and third trimesters, and <100g/L in postpartum period) is the most common nutrient deficiency in pregnancy and is associated with an increased risk of preterm delivery, LBW, peripartum haemorrhage, caesarean delivery, perinatal mortality, and low infant iron status. There is also increasing evidence that low pregnancy iron, particularly in the third trimester, may be associated with adverse neurodevelopment in offspring (Janbeck et al. 2019).

Whilst iron deficiency is thought to affect up to 23 per cent of pregnant women and 14 per cent of non-pregnant women in the UK, the National Institute of Clinical Excellence recommends against routine iron supplementation of pregnant women, as it can lead to constipation and other gastrointestinal symptoms and are expensive to administer on a population-wide scale. Guidelines published by the British Committee for Standards in Haematology recommend that non-anaemic women identified to be at increased risk of iron deficiency (such as those with previous anaemia, multiple pregnancy, consecutive pregnancies with less than a year's interval between, adolescents, and vegetarians) should have a serum ferritin checked early in pregnancy and be offered oral supplements if plasma ferritin concentration is <30 ng/ml.

It should be noted that a poor maternal diet resulting in anaemia is unlikely to occur in isolation and effects may not be correctable by a brief period of supplementation with iron. Recognizing the complexity of the aetiology of anaemia, the WHO recommends that public health strategies to prevent and control anaemia should include improvements in dietary diversity; food fortification with iron, folic acid and other micronutrients; distribution of iron-containing supplements; and control of infections and malaria.

Zinc

Zinc plays a critical role in a variety of complex processes during cell replication, maturation, and adhesion, such as DNA and RNA metabolism, signal recognition and transduction, gene expression, and hormone regulation. Consequently, an adequate supply of zinc is essential for the normal growth and development of the foetus. An estimated 17.3 per cent of the world's population is at risk of inadequate zinc intake. It has been suggested that a low serum zinc concentration may be associated with suboptimal outcomes of pregnancy, but studies have failed to find a consistent positive effect of zinc supplementation in pregnancy on reducing the risk of preterm births, stillbirths, low birthweight, or small for their gestational age when compared to no zinc supplementation or placebo (Carducci et al. 2021). It has been recommended that, rather than focusing on any specific nutrient supplementation in isolation, attention should be directed on addressing the underlying problem of poor nutrition in order to make any significant impact on morbidity and mortality. The WHO and UNICEF promote antenatal use of multiple micronutrient supplementation, including zinc, to all pregnant women where there are population level micronutrient deficiencies (WHO 2020), although it has been suggested that infant micronutrient supplementation (including zinc) may be more effective than maternal supplementation. Nevertheless, zinc deficiency remains an important global public health issue with its deficiency estimated to contribute to over half a million infant and young child deaths per year.

Iodine

Iodine is an essential element for the production of the thyroid hormones triiodothyronine and thyroxine, which both play a crucial role in brain development and neurological function during foetal and postnatal growth. There is clear evidence to show that insufficient intake of iodine during pregnancy and the immediate postpartum period increases the risk of neurologic and psychological deficits in children. In the children of women living in iodine-deficient areas where women have severe iodine deficiency, child IQ tend to be lower and attention deficit and hyperactivity disorders more prevalent compared to those in iodine-replete regions. For women with marginal iodine status, the demands of pregnancy and lactation can precipitate clinical and biochemical symptoms including increased thyroid volume and altered thyroid hormone levels.

Iodine deficiency affects more than 2.2 billion individuals (38 per cent of the world's population) and remains the leading cause of preventable intellectual impairment worldwide. Although iodine deficiency is more common in developing countries it is not confined to them and it has been reported that women of childbearing age and pregnant women in the UK are generally iodine insufficient. Although evidence is limited in the UK, children born to pregnant women with iodine insufficiency, as defined by World Health Organization thresholds of median urinary iodine concentration <150 µg/L, do not appear to have detrimental neurodevelopmental outcomes (Threapleton et al. 2021).

The evidence to support the widespread use of iodine supplementation during pregnancy remains inconclusive. Although women who take iodine supplements are less likely to develop hyperthyroidism than those who do not receive iodine, no consistent benefit is observed amongst outcomes such as reducing preterm birth, perinatal mortality, low birthweight, or neonatal hypothyroidism (Harding et al. 2017). Women who take iodine supplements are also more likely to experience nausea or vomiting during pregnancy.

Many experts condone preventive salt iodization as a cost-effective and sustainable solution to the problem. Since 1993 the World Health Organization (WHO) has conducted a global programme of salt iodization to boost dietary levels and prevent deficiency, focusing largely on low-income countries. Although some European countries, including Switzerland and Denmark, have signed up to the WHO programme it is currently not compulsory for manufacturers in the UK to add iodine to salt.

Nutrient requirements for lactation

The daily nutrient requirements during lactation are higher than requirements in pregnancy and RNIs are higher for a broad range of nutrients (see Table 15.1). With the exception of protein and energy, however, the exact nature of these requirements is not well understood and it is difficult to conclude whether breastfeeding women need to make significant changes to their diet. Factors such as the duration and intensity of lactation (whether the infant is breastfed exclusively or only partially) are likely to influence nutritional requirements as an exclusively breastfeeding woman will have greater energy and nutrient needs (with the exception of iron, attributed to the potential protective effect of lactational amenorrhoea) than a woman who is partially breastfeeding.

For most nutrients, any maternal inadequacy of their own nutritional status reduces the quantity rather than the quality of milk. Women can produce milk with adequate protein, carbohydrate, fat, folate, and most minerals even when their own supplies are limited. For these nutrients milk quality is maintained at the expense of the maternal nutrient stores. Many minerals are transferred into milk by active transfer rather than by passive diffusion and this process compensates for variations in maternal mineral status. In areas where micronutrient deficiency is prevalent, milk composition can be affected by marginal maternal deficiencies in fatty acids, iodine, and most vitamins.

Energy cost of lactation

The energy requirement of a breastfeeding woman is defined as 'the level of energy intake from food that will balance the energy expenditure needed to maintain a body weight and body composition, a level of physical activity and breast milk production that are consistent with optimal health for the woman and her child' (FAO/WHO/UNU 2004). This definition implies that the energy required to produce an appropriate volume of milk must be added to the woman's usual energy requirement and assumes that she resumes her usual level of physical activity soon after giving birth (FAO/WHO/UNU 2004).

The energy requirements for lactation are high. For exclusive breastfeeding, the energy cost of lactation is estimated at around 2.6 MJ/day. In well-nourished women, this may be subsidized by energy mobilization from tissues of approximately 0.7 MJ/day, resulting in a net increment of 1.9 MJ/day (454 kcal/day) above non-pregnant, non-lactating energy requirements. The energy needs are lower for a partially breastfeeding woman, depending on the extent to which non-breast milk foods are consumed by her infant. The mother must increase her energy intake in order to meet this additional demand. Although some studies suggest that metabolic adaptations and reduced physical activity act to conserve energy during lactation, it is generally believed that resting metabolic rate and thermogenesis are not significantly altered and the savings gained from a sedentary lifestyle are unlikely to have much impact on availability of energy for lactation.

UK dietary reference values for energy during lactation increase by 330 kcal per day in the first six months, that is, a daily energy intake of 2270 kcal. Thereafter, the energy intake required to support breastfeeding will be modified by changes in maternal body composition and the breast milk intake of the infant.

Protein

The protein content of milk varies with stage of lactation, with colostrum containing about 30g/L and mature milk comprising 8–9 g/L. The protein in breast milk is comprised of lactalbumin, casein, lactoferrin, and IgA. It is estimated that women require an additional 11 g protein/day over the first six months to meet this need. In high-income countries, most women consume protein well in excess of this amount and would not need to alter their intake when breastfeeding.

Fatty acids

The major long chain polyunsaturated fatty acid (LCPUFA) components of breast milk are docosahexaenoic acid (DHA), eicosapentaenoic acid (EPA), and alpha linoleic acid (ALA). DHA in particular plays a major role in infant neurological development. Research has indicated that the DHA content of breast milk may be linked to a child's IQ and immune function later in childhood. The transfer of LCPUFA from mother to her infant during lactation is dependent largely upon maternal status, but genetic influences are beginning to be recognized. For example, studies have provided evidence that genetic variation in fatty acid desaturases (FADS1 and FADS2) affect levels of saturated, monounsaturated, and (n-6) and (n-3) fatty acids in breast milk during lactation. Thus, genetic variation among women may influence maternal-to-infant transfer of fatty acids through breast milk and impact on the fatty acid nutrition of the breast-fed infant. Genetic variation in FADS has also been implicated in the transfer of DHA from maternal plasma to breast milk: plasma but not milk DHA of women carrying the minor allele did not respond to a fish oil supplement, whereas women with the major allele (homo- or heterozygous) showed the expected increase in DHA in plasma and milk.

Fish and seafood comprise the only food group that is a significant source of n-3 LCPUFA that also offers a range of nutrients that are frequently under-represented in habitual diets, including iodine, calcium, vitamin D, zinc, and iron. Breast milk ALA levels tend to be better conserved than DHA, but DHA responds sensitively to dietary DHA and higher breast milk levels have been found in women consuming diets with high levels of fish intake. LC-PUFA levels in breast milk may be low when maternal fish intake is low and/or ALA intake is low and linoleic acid (LA) intake is high, as diets with very high levels of LA dramatically reduce conversion of ALA to DHA. Women who do not eat seafood may wish to increase their intake of DHA through consumption of vegetable oils with higher ALA content (e.g., soybean or rapeseed oil).

B vitamins

B vitamins are a broad class of water-soluble vitamins that play important roles in cell metabolism. The recommended intake for several B vitamins is increased during lactation (i.e., thiamin, riboflavin, niacin, vitamin B_{12} and folate). Breast milk concentrations of thiamin, riboflavin, and vitamin B_{12} are strongly dependent on maternal dietary intake and status, and low maternal status can lead to the infant becoming deficient. Maternal supplementation of thiamin, riboflavin, and vitamin B_{12} has been shown to rapidly improve the breast milk concentration and status of the infant. Maternal intake of folate has little impact on breast milk concentrations or infant status and consequently there is no evidence that maternal supplementation with folate postnatally elicits beneficial effects in the infant.

Vitamin C

Vitamin C is an essential water-soluble antioxidant vitamin and in its role as a cofactor for mixed function oxidases, it participates in the synthesis of various macromolecules including collagen and carnitine. The breast milk of well-nourished mothers contains 30–80 mg/L vitamin C, and to compensate for the excretion of vitamin C into breast milk UK dietary guidelines recommend an increase intake of vitamin C by the mother during lactation. Whilst high dose vitamin C supplementation (>200 mg/day) has not been found to result in higher breast milk concentrations in well-nourished women, supplementation of women with low vitamin C status can double or triple breast milk concentrations, thereby improving the nutrient status of the breastfed infant. Although no cases of scurvy have been reported in breastfed infants, the potentially negative influences of low ascorbic acid intake in early life related to its antioxidant properties have yet to be fully evaluated.

Vitamin A

Vitamin A is needed for the growth and differentiation of cells and tissues and plays a key role in infant development. Vitamin A deficiency is a major public health problem in low-income countries and globally, it is estimated that 140–250 million children under the age of five years are affected. These children suffer a significant increased risk of illness and death from common childhood infections as measles and diarrhoea. Lack of vitamin A also causes severe visual impairment and blindness. Although high doses of vitamin A should not be consumed during pregnancy to avoid teratogenesis, high-dose vitamin A supplementation of breastfeeding women with low serum retinol levels has been shown to be an effective way of ensuring adequate supplies for the infant through colostrum and breast milk and preventing deficiency in the infant.

Vitamin D

Through its action in regulating calcium and phosphate absorption, vitamin D plays a vital role in the growth and development of bones. Infants in particular are at risk of vitamin D deficiency because of their relatively large vitamin D needs related to the high rate of skeletal growth. Breast milk contains relatively small amounts of vitamin D and may not be sufficient to prevent vitamin D deficiency in exclusively breastfed infants if sunlight exposure is limited. Babies born to mothers with low vitamin D status are at risk of developing rickets. Studies have indicated that supplementation of healthy breastfeeding mothers with doses of 100–160 μg/d vitamin D (4000–6400 IU/d) provide the breastfeeding infant with adequate levels of vitamin D despite both mother and infant being limited in sunlight exposure.

All adults and children in the UK are advised to take 10μg vitamin D daily. Healthcare professionals are urged to take particular care to ensure that those at greatest risk of vitamin D deficiency take supplements, including all pregnant and breastfeeding women, particularly teenagers and young women, infants, and children under four years, those who have limited skin exposure to sunlight, or who are of South Asian, African, or African Caribbean origin. All infants and young children aged six months to three years are advised to take a daily vitamin drops containing vitamin D. Infants who are breastfed may need to take these vitamin drops from one month of age if their mother did not take supplements throughout her pregnancy, but predominantly formula-fed babies will not need them as infant formula is fortified with vitamin D.

Calcium

During lactation, women typically lose 280–400 mg/day of calcium through breast milk (Kovacs 2011). To meet this increased demand the mother must mobilize calcium from her own skeletal reserves. Physiologic adaptations, such as upregulation of intestinal calcium absorption and bone resorption, provide much of the calcium in breast milk. Once breastfeeding has ceased, bone density is fully restored over the subsequent 6 to 12 months. Several investigations of the effect of maternal dietary calcium intake on breast milk levels indicate that they may be independent and calcium supplementation may have little effect on breast milk concentration.

Magnesium

Magnesium is a cofactor for over 300 enzymatic reactions, including those involved in DNA/RNA synthesis, protein synthesis, cell growth, and cellular energy production and storage. Maternal serum magnesium levels and the amount excreted into the breast milk remain relatively constant during lactation. Although the UK DRVs include an increment in magnesium during lactation, there is little evidence of a relationship between magnesium supplement intake and the concentration of magnesium in breast milk. In some countries, such the USA and Australia, no increment in magnesium above pre-pregnancy levels are recommended, as it is assumed that bone resorption and reduced urinary excretion throughout lactation compensate for the extra requirements for magnesium secretion in milk (IOM 1997).

Zinc

An adequate supply of zinc is essential for mammary gland function and for milk synthesis and secretion. The level of zinc in colostrum is 17 times higher than that in blood, but declines rapidly in the breast milk during the first three months postpartum. A number of studies of lactating women with marginal zinc status have revealed that homeostatic mechanisms can compensate for low maternal dietary zinc intakes. The proportion of dietary zinc absorbed in such women has been shown to increase by over 70 per cent compared to non-lactating women or pre-conception values. Current UK recommendations for zinc intake during lactation range from 9.5 to 13 mg per day depending upon months postpartum (see Table 15.1). Generally, these dietary recommendations should be met from usual habitual intakes.

Copper and Selenium

Copper and selenium are essential trace minerals and requirements of both are increased during lactation. Copper has a number of biochemical roles, most notably as a cofactor for enzymes that regulate various physiologic pathways such as energy production, iron metabolism, connective tissue maturation, and neurotransmission. Selenium is a component of selenoproteins that act as antioxidant enzymes and regulate thyroid hormone metabolism and immune function. The copper content of human milk is highest during early lactation and declines as lactation progresses, averaging approximately 250 μg/L in the first six months. Maternal intake of copper has little effect on breast milk concentrations or infant status. The selenium content of breast milk also decreases over the course of lactation and is influenced by maternal status.

15.8 Factors which may impact on the nutrition and nutritional status of the pregnant and breastfeeding mother

Veganism and vegetarianism

Vegetarian and vegan diets are becoming more widespread in the Western world, where these dietary patterns correlate with healthy lifestyles and higher incomes, unlike other settings in which 'traditional' vegan and vegetarian diets are often associated with lower energy intake and caloric restrictions as a result of lower incomes. Studies on vegan and vegetarian diet during pregnancy have not found evidence of an increased risk of severe, adverse pregnancy-related events, such as pre-eclampsia or major birth defects, providing the two main potential deficits (i.e., vitamin B_{12} and iron) were corrected, although the data relating to birthweight and duration of gestation are contrasting (Piccoli et al. 2015). In resource-poor settings however, several studies have shown that prenatal dietary supplementation (energy, protein, and micro elements such as iron and vitamins) of women with a vegetarian diet improves foetal outcomes, particularly birthweight.

Adolescence

Inadequate nutrient intake is a concern among a substantial proportion of adolescent girls. UK survey data reveals intakes below the lower reference nutrient intake (LRNI) for all minerals, including iron, where 49 per cent of girls ages 11–18 years fell below the LRNI (NDNS Rolling Programme 2019). Perhaps unsurprisingly, poor nutritional intake and status prevail during adolescent pregnancy and lactation. The increased nutritional demand to support pregnancy and lactation occurs at a time of rapid growth and development in adolescents, with almost 45 per cent of maximum skeletal mass and 15 per cent of adult height being gained during adolescence. Studies have suggested that adolescents who continue to grow in stature and accrue fat mass during pregnancy have infants who weigh less than those of non-growing adolescents, suggesting a competition for nutrients between the growing mothers and their infants.

Adolescent pregnancy is a global concern, particularly in areas of poverty and social disadvantage. Around 11 per cent of all births are to adolescent girls aged 15–19 years, and 95 per cent of these births occur in low- and middle-income countries. There are increased risks for both mother and newborn associated with early childbearing and pregnancy and childbirth complications are the second leading cause of death among 15 to 19 year-old girls. In undernourished populations, adolescents' linear growth has been shown to cease and lean body mass and percent body fat decline during pregnancy and lactation, indicating a depletion of maternal energy and nutrient reserves to meet the additional demands of pregnancy and lactation. Young maternal age has been associated with an increased risk of preterm delivery, low birth weight and neonatal mortality, and this relationship has been found to be independent of important known confounders such as low socioeconomic status, inadequate prenatal care, and inadequate weight gain during pregnancy. There is increasing agreement that nutrition support is necessary to improve the health of vulnerable women, as well as an escalating consensus that this process should begin in adolescence.

Smoking

Exposure to tobacco smoke has a negative effect on both male and female fertility and on pregnancy outcomes. Smoking during pregnancy increases the risk of serious complications such as miscarriage, stillbirth, ectopic pregnancy, and premature labour. Babies born to women who smoke during pregnancy are more likely to be born prematurely, to have low birth weights, and to have birth defects such as cleft lip. Environmental (second-hand) cigarette smoke exposure is associated with low birthweight and wheeze and respiratory infection in infants and children of non-smoking women. Smoking has been linked to lower blood concentrations of vitamin C, Vitamin D, and several B vitamins (B_{12}, B_6, riboflavin, and folate) and higher plasma concentrations of homocysteine. The negative influence of maternal smoking on breastfeeding duration has been well described in the literature, and has been shown even after adjusting for socioeconomic factors. Nicotine is readily transferred into human milk and can suppress milk production. There is evidence that smoking is associated with a reduction of daily milk output by approximately 250–300 ml and alters milk composition, such as fat concentration and iodine levels, due to the interference of thiocyanate in cigarette smoke and iodine transport mechanisms.

Alcohol consumption

A number of adverse outcomes are associated with the excess consumption of alcohol in pregnancy, such as foetal alcohol syndrome (FAS) and alcohol-related birth defects (ARBD), which manifest in restricted growth, facial abnormalities, and other physical defects, and learning and behavioural disorders in children. The impact of lower levels of consumption are less well understood. Studies have shown that the risks of low birth weight, preterm birth, and being small for gestational age may increase in mothers drinking above 1–2 units/day during pregnancy. In the UK the Chief Medical Officer recommends that the safest approach is to abstain from alcohol whilst pregnant or when planning a pregnancy.

The consumption of alcohol while breastfeeding is associated with decreased oxytocin release inhibiting the milk ejection reflex in a dose response manner. Alcohol is transferred readily into the milk, and alcohol levels in the milk correlate closely to maternal blood alcohol concentration. Whilst the effects of alcohol on the infant varies depending on the amount consumed, there is some evidence that even moderate alcohol consumption may be associated with reduced milk intake, reduced motor development, impaired sleep patterns, and increased risk of hypoglycaemia, although the literature is contradictory. The long-term effects of exposure to alcohol during lactation are unknown.

Caffeine consumption

Caffeine is a commonly consumed stimulant, found in beverages (tea, coffee, soft drinks), chocolate, and over-the-counter medicine. Caffeine is freely transported across the placental barrier and small amounts pass into milk during lactation. High maternal caffeine intakes (>300 mg or more than four cups of coffee a day) have been associated with greater risk of lower birth weight, even after adjustment for smoking and alcohol intake. Additional evidence is emerging of a possible association between high maternal caffeine intake and increased risk of miscarriage, but the evidence is mixed in relation to other outcomes. UK and international guidance is that women can safely consume up to 200mg of caffeine daily (equivalent to two cups of coffee or tea).

The impact of global events on maternal and child nutrition: a complex syndemic

Malnutrition in all its forms, including obesity, undernutrition, and other dietary risks, is the leading cause of poor health globally. Maternal undernutrition specifically affects a large proportion of women in many developing countries and is an important determinant of poor maternal, newborn, and child health outcomes. Major global events, such as climate change and the recent COVID-19 pandemic, will considerably compound these health challenges. These factors represent a global syndemic (a synergy of epidemics which interact with each other to produce complex sequelae that and share common underlying societal drivers) that affects most people in every country and region worldwide (Swinburn et al. 2019). Climate change directly exacerbates maternal malnutrition in a multitude of ways, such as adversely affecting crop yields and the average supply of essential nutrients; impacting on farmers' ability to reliably supply food to markets; and creating climate-induced migration that is accompanied by loss of livelihood. Similarly, the COVID-19 pandemic has affected the food, nutrition, and health security of vulnerable groups including pregnant and lactating women, impacting on social and health inequities. One mechanism by which this has happened is through a major decline in food security which exists when all people, at all times, have physical, social, and economic access to sufficient, safe, and nutritious food which meets their dietary needs and food preferences for an active and healthy life. Women and girls around the world often eat last and eat the least, and when food insecurity gets worse, women and girls are often affected the most. Global evidence shows that this direct increase in food insecurity is a serious public health concern, reversing years of development gain. Many have seen the recent pandemic as an opportunity to rethink the dysfunctional global food system upon which the vast majority of the world now depends and reconfigure the types of programmes, policies, and multilevel intersectoral coordination mechanisms that are needed to ensure food and nutrition security for all, including young children, adolescents, pregnant and lactating women (Pérez-Escamilla et al. 2020).

KEY POINTS

- Changes in maternal physiology during pregnancy favour foetal growth and development.
- Normal pregnancy involves maternal metabolic adaptations; in consequence significant changes in maternal nutrient intake are generally not necessary.
- Estimates of the energy cost of pregnancy vary widely; adaptive changes suggest that not all pregnant women require an increase in energy intake.
- Maternal peri-conceptional folate deficiency increases the risk of congenital abnormalities, and all women planning a pregnancy should take a supplement containing 400 μg of folic acid per day prior to conception and until the 12th week of pregnancy.
- Excessive maternal vitamin A intake increases the risk of congenital abnormalities.

- It is recommended that all adults and children take a vitamin D supplement of 10μg/day. Low-income pregnant women or new mothers and their children in the UK can receive Healthy Start vitamin supplements containing vitamins C and D and folic acid free of charge.
- Maternal iron-deficiency anaemia is the most common nutrient deficiency in pregnancy and is associated with an increased risk of poor pregnancy outcomes for mother and baby.
- Severe iodine deficiency is common in some regions of the world. Poor maternal iodine status can lead to neurological and psychological deficits in the offspring.
- For most nutrients, maternal homeostatic mechanisms ensure the quality of their milk, but mothers may need more calories to meet their nutritional needs while breastfeeding.
- Some factors can affect the nutrition and nutritional status of pregnant and lactating women (e.g., restrictive diets, young age, smoking, consuming alcohol) and particular care should be taken to avoid detrimental effects.

 Be sure to test your understanding of this chapter by attempting multiple choice questions.

REFERENCES

Carducci B, Keats E C, Bhutta Z A (2021) Zinc supplementation for improving pregnancy and infant outcome. *Cochrane Database Syst Rev* Mar 16 **3**(3), CD000230. doi: 10.1002/14651858.CD000230.pub6. PMID: 33724446; PMCID: PMC8094617.

Chang S, Glaria A A, Davie P et al. (2020) Breastfeeding experiences and support for women who are overweight or obese: A mixed-methods systematic review. *Matern Child Nutr* **16**(1), e12865.

Eitmann S, Németh D, Hegyi P (2020) Maternal overnutrition impairs offspring's insulin sensitivity: A systematic review and meta-analysis. *Matern Child Nutr* **16**(4), e13031.

FAO/WHO/UNU (2004) Human energy requirements: Report of a Joint FAO/WHO/UNU Expert Consultation. Rome, 17–24 October 2001.

Godhamgaonkar A A, Wadhwani N S, Joshi S R (2020) Exploring the role of LC-PUFA metabolism in pregnancy

complications. *Prostaglandins Leuko Essen Fatty Acids* **163**, 102203.

Hanson M A, Bardsley A, De-Regil L M et al. (2015) The International Federation of Gynecology and Obstetrics (FIGO) recommendations on adolescent, preconception, and maternal nutrition: 'Think Nutrition First'. *Int J Gynaecol Obstet* **131**(Suppl 4), S213–S253.

Harding K B, Peña-Rosas J P, Webster A C et al. (2017) Iodine supplementation for women during the preconception, pregnancy and postpartum period. *Cochrane Database Syst Rev* **3**(3), CD011761. doi: 10.1002/14651858.CD011761.pub2. PMID: 28260263; PMCID: PMC6464647.

IOM (Institute of Medicine) (1997) *Dietary Reference Intakes for Calcium, Phosphorus, Magnesium, Vitamin D, and Fluoride.* National Academies Press, Washington, DC.

IOM (Institute of Medicine) (2009) *Weight Gain During Pregnancy: Reexamining the Guidelines.* Institute of Medicine

(US) and National Research Council (US) and Committee to Reexamine IOM Pregnancy Weight Guidelines, Washington, DC.

Janbek J, Sarki M, Specht I O et al. (2019) A systematic literature review of the relation between iron status/anemia in pregnancy and offspring neurodevelopment. *Eur J Clin Nutr* **73**(12), 1561–1578.

Kovacs C S (2011) Calcium and bone metabolism disorders during pregnancy and lactation. *Endocrinol Metab Clin North Am* **40**(4), 795.

MRC Vitamin Study Research Group (1991) Prevention of neural tube defects: Results of the Medical Research Council Vitamin Study. *Lancet* **338**, 131–137.

National Diet and Nutrition Survey (2019) Years 1 to 9 of the rolling programme (2008/2009–2016/2017): Time trend and income analyses. National Diet and Nutrition Survey (publishing.service.gov.uk) (accessed 7 July 2021).

Ota E, Hori H, Mori R et al. (2015) Antenatal dietary education and supplementation to increase energy and protein intake. *Cochrane Database Syst Rev* **6**, Art. No.: CD000032. DOI: 10.1002/14651858.CD000032.pub3.

Pérez-Escamilla R, Cunningham K, Moran V H (2020) COVID-19 and maternal and child food and nutrition insecurity: A complex syndemic. *Matern Child Nutr*: e13036.

Piccoli G B, Clari R, Vigotti F N et al. (2015) Vegan–vegetarian diets in pregnancy: Danger or panacea? A systematic narrative review. *BJOG* **122**(5), 623–633.

Rumbold A, Ota E, Nagata C et al. (2015) Vitamin C supplementation in pregnancy. *Cochrane Database Syst Rev* **9**, Art. No.: CD004072. doi: 10.1002/14651858.CD004072.pub3.

SACN (Scientific Advisory Committee on Nutrition) (2011) *Dietary Reference Values for Energy.* The Stationery Office, London.

Stevens G, Finucane M, De-Regil L et al. (2013) Nutrition impact model study group (anaemia): Global, regional, and national trends in haemoglobin concentration and prevalence of total and severe anaemia in children and pregnant and non-pregnant women for 1995–2011: A systematic analysis of population-representative data. *Lancet Glob Health* **1**, e16–e25.

Swinburn B A, Kraak V I, Allender S et al. (2019) The global syndemic of obesity, undernutrition, and climate change: The Lancet Commission report. *Lancet* **393**(10173), 791–846.

Threapleton D E, Snart C J, Keeble C et al. (2021) Maternal iodine status in a multi-ethnic UK birth cohort: Associations with child cognitive and educational development. *Paediatr Perinat Epidemiol* **35**(2), 236–246.

WHO (2018) *WHO Recommendation: Calcium Supplementation During Pregnancy for Prevention of Pre-eclampsia and its Complications.* World Health Organization.

WHO (2020) *Obesity and Overweight.* World Health Organization, Geneva, 3 March. https://www.who.int/news-room/factsheets/detail/obesity-and-overweight

FURTHER READING

Bates B, Prentice A, Bates C et al. (2015) National diet and nutrition survey rolling programme (NDNS RP). Supplementary report: blood folate results for the UK as a whole, Scotland, Northern Ireland (years 1 to 4 combined) and Wales (years 2 to 5 combined). Public Health England, London. publishing.service.gov.uk

Buttriss J, Stanner S, Wyness L (2013) *Nutrition and Development: Short and Long Term Consequences for Health.* Wiley-Blackwell, London.

Fleming T P, Velazquez M A, Eckert J J (2015) Embryos, DOHaD and David Barker. *J Dev Orig Health Dis* **6**(5), 377–383.

Jiang H, Powers H J, Rossetto G S (2019) A systematic review of iodine deficiency among women in the UK. *Public Health Nutr* **22**(6), 1138–1147.

Maternal & Child Nutrition (Open access journal). https://onlinelibrary.wiley.com/journal/17408709

Moran V H, Lowe N M (eds) (2016) *Nutrition and the Developing Brain.* CRC Press, Boca Raton, FL.

NICE (2014) Maternal and child nutrition: Public health guidelines (PH11). https://www.nice.org.uk/Guidance/PH11

Report on Health and Social Subjects (1991) 41 Dietary Reference Values (DRVs) for Food Energy and Nutrients for the UK, Report of the Panel on DRVs of the Committee on Medical Aspects of Food Policy (COMA). The Stationery Office, London.

SACN (2011) *Dietary Reference Values for Energy.* The Stationery Office, London.

WHO (2020) *WHO Antenatal Care Recommendations for a Positive Pregnancy Experience. Nutritional Interventions Update: Multiple Micronutrient Supplements During Pregnancy (who.int).* World Health Organization, Geneva.

16 Ageing and older people

Elizabeth A. Williams

OBJECTIVES

By the end of this chapter you should be able to:

- describe the changes in physiological function during ageing, and the nutritional relevance of these changes
- explain how ageing influences energy and nutrient requirements
- discuss the limitations of methods used for the assessment of nutritional status in older people
- identify factors contributing to poor nutritional status in older people, and associated adverse health outcomes.

16.1 Introduction

The ageing population

The world's population is ageing. Global life expectancy has doubled since the mid-1800s and is now around 72 years. Life expectancy increased steadily over the nineteenth and twentieth centuries, driven largely by reductions in early- and mid-life mortality due to improved public health, advances in medicine, and a reduction in infections. More recent changes in life expectancy are a consequence of a decline in fertility rates coupled with better health in older age. The net effect is a dramatic increase in the proportion of the world's older adult population such that it is estimated that the number of adults aged 65 years and over will double from around 700 million to 1.5 billion people by 2050; representing around 16 per cent of the world's population, and will exceed 20 per cent by 2100 (Figure 16.1). Nearly every country in the world has seen this shift in population demographics, however regions of the world that have seen the fastest population ageing include low-middle income countries in Africa, Asia, and Latin America, and this trend is projected to continue. Additionally, in older age groups women outnumber men; in the 75–84 year age group it is estimated that there are only 72 men for every 100 women, reflecting the higher mortality rate for men over the lifecourse.

As the age profile of the world's population shifts to an older population it is more important than ever that consideration is given to those factors that determine healthy ageing. Ageing is associated with an increased risk of disability and disease. It is estimated that 23 per cent of the total global burden of disease occurs in people aged >60years. Cardiovascular diseases, cancers, chronic respiratory diseases, musculoskeletal diseases, and neurological and mental disorders make the main contribution to the disease burden in older people. However, there is a wide variation in health status of older people and life expectancy, both within and between countries. This may reflect differences in diagnosis of illness and quality of care but is also indicative of socioeconomic inequalities that include differences in lifestyle, including dietary behaviour. Many governments have set targets to improve healthy lifespan (i.e., compression of morbidity) and reduce inequalities in life expectancy. There is a well-recognized link between dietary behaviour and health, and even in the very old (>80years), diet quality is inversely associated with mortality risk.

Dietary habits and nutritional status in older people are determined by numerous factors, some acting throughout the lifecourse. In this chapter consideration will be given to physiological changes that occur during ageing that may impact on nutritional status in later life. Additionally, dietary intakes and nutritional status of older people will be discussed as well as the associations with health and illness.

The biology of ageing

The causes of ageing have been debated for many decades. Theories of ageing have broadly fallen into two categories: first, that we are programmed to age, and second,

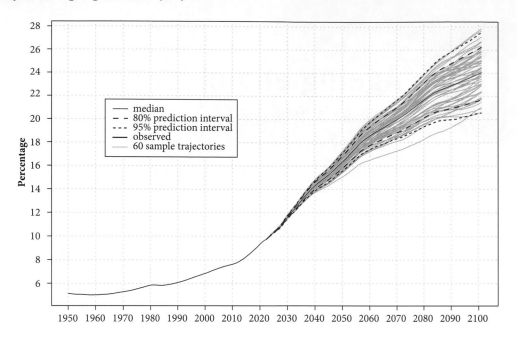

FIGURE 16.1 Percentage of the population aged 65 years or over

Source: © United Nations, DESA, Population Division. Licensed Under Creative Commons license CC BY 3.0 IGO. World Population Prospects 2022. https://population.un.org/wpp

that ageing in later life is due to the accumulation of damage at a cellular and molecular level. It is recognized that biological age (functional age) and chronological age (age since birth) are distinct, and with increasing life expectancy and the realization that the maximum lifespan is around 120 years, ambitions have turned to strategies to delay biological ageing. While genes are major determinants of biological age, lifestyle undoubtedly contributes to the ageing process. The role of diet in extending lifespan was first demonstrated in rodents in the classic calorie-restriction studies of McCay et al. in 1935. Since then, numerous studies have shown that dietary restriction in a variety of species can expand lifespan and delay biological ageing. Such studies are challenging to perform in humans, but there is some evidence from primates and humans that energy-restricted, micronutrient-replete diets have a beneficial effect on metabolism and molecular pathways associated with age-related diseases.

An important advance in biological ageing research has been the identification of nine cellular and molecular hallmarks of ageing (López-Otín et al. 2013). These hallmarks provide an important framework to investigate the mechanisms underlying the ageing process. The hallmarks are grouped into three categories: (i) the primary hallmarks that cause damage (genomic instability, epigenetic alterations, telomere attrition, and loss of proteostasis), (ii) those hallmarks that reflect compensation responses to the damage (mitochondrial dysfunction, deregulated nutrient sensing, and cellular senescence),

and (iii) the consequences that result in functional decline (stem cell exhaustion and altered intracellular communication). No single biomarker of ageing exists; however, several of these hallmarks of ageing are altered by diet and exercise.

16.2 Age-related changes relevant to nutrition

Ageing is associated with a progressive decline in physiological function and can result in loss of independence and poorer quality of life. Many of these changes have nutritional implications in terms of intake, requirements, and status. There are also accompanying pathological changes leading to an increased risk of infection, chronic disease, and death.

Body composition

Ageing is associated with a decline in lean body mass, coupled with an increase in fat mass (see Chapter 5). The decline in lean body mass occurs primarily in the muscle and is accompanied by infiltration of fat into the muscle and a redistribution of fat from the limbs to the central trunk of the body. This decline in muscle mass leads to impaired mobility, decreased muscle strength, and increased susceptibility to falls. The change in body composition coupled with a reduction in physical activity leads to reduced energy requirements, and energy intake often declines with age. Failure to match energy intake with the changes in energy requirements also puts older adults at risk of both positive and negative energy balance and consequently obesity and malnutrition. The prevalence of obesity increases with age, and then declines in the oldest age, as seen in the 2019 Health Survey for England (see Useful Websites Section 16.1). The survey reported that around a quarter of adults were obese, increasing to 36 per cent in adults aged 65–74 years and declining to 26 per cent in people aged ≥75years. The same survey reported 2 per cent of people aged 65–74 to be underweight. Other reports suggest that around 5 per cent of the older adult population are malnourished, whilst >33 per cent of adults aged >65 years have been found to be malnourished on admission to hospital.

Bone density

Peak bone mass is achieved around early adulthood, plateaus between the third and fourth decade of life and then progressively declines. There is an accelerated loss of bone mass in women around menopause associated with the decline in oestrogen, which leads to increased bone remodelling. Bone mass continues to decline postmenopause in women and in older men, but at a steady rate. The determinants of peak bone mass are multifactorial and include genetic, endocrine, nutritional factors, body size, and physical activity. Osteopenia and osteoporosis (bone loss) are more pronounced in women than men, due to a lower peak bone mass attained in early adulthood and the impact of menopause. Bone loss is associated with a high risk of fracture. Fracture risk in old age is in part determined by the dietary intake of calcium, vitamin D status, and lifestyle in childhood and early teens, and events during the menopause, highlighting the need for a lifecourse approach to tackling this problem in older age.

Hip fracture as a result of falls in older adults with osteoporosis and osteopenia is a significant health care burden. Over 70,000 hip fractures occur each year in the UK alone, with an estimated cost of £2 billion in treatment and management. Hip fractures are responsible for protracted hospital stays and are associated with a high risk of mortality, often as a result of exacerbation of comorbidities. Supplementation with vitamin D alone has yielded mixed results in terms of fracture prevention. The evidence for intervention with vitamin D and calcium in old age for the prevention of falls and fractures is encouraging, but the optimum doses of calcium and vitamin D needed for bone health in old age are yet to be defined (see Chapter 25).

Endocrine function

Ageing is accompanied by considerable changes in endocrine function, partly due to alterations in hormone secretion, changes in hormone receptor number, and alterations in hormone sensitivity. Important hormonal changes relevant to nutrition include a decline in growth hormone and insulin like-growth factor serum concentrations. A decline in pancreatic beta cell function with age leads to a reduction in insulin secretion, which contributes to the increased prevalence of type 2 diabetes. There are also significant changes in sex hormones with a reduction in testosterone in men and a halt to the cyclical production of oestrogen and progesterone in women at menopause. Such changes have been associated with an increase in intra-abdominal fat in both men and women, as well as a detrimental effect on bone health. Thyroid problems also have a higher prevalence in older adults, with a consequential impact on all tissues of the body and crucially on metabolic rate. Thyroid deficiency is thought to have a prevalence of around 6 per cent of women and 2 per cent of men over the age of 60 years, and is associated with decreased metabolic rate, weight gain, memory loss, reduced cognitive function, and fatigue. Subclinical hypothyroidism where levels of thyroxine are normal, but thyroid-stimulating hormone is elevated, is believed to affect around 15 per cent of women over the age of 60 years. Hyperthyroidism, which is present in around 3 per cent of people aged >85 years, is caused by elevated circulating thyroid hormone and leads to increased metabolic rate. A range of non-specific symptoms can occur with hyperthyroidism in older adults including weight loss, heart palpitations, depression, and tremor, and can make hyperthyroidism difficult to recognize.

Appetite

Ageing is associated with a decline in appetite known as the 'anorexia of ageing'. Appetite regulation is complex and is influenced by physiological, hedonistic, and external (societal and environmental) factors (Cox et al. 2020).

Physiological factors that are implicated in appetite regulation include energy expenditure, basal metabolic rate, body composition, gastrointestinal hormone secretion, motility, and the gut microbiome. Changes in gut hormone secretion and gastrointestinal motility have a direct effect on appetite and are well described in ageing. For example, there is an increased circulation of and sensitivity to the gut regulatory peptide hormone, cholecystokinin (CCK). CCK is secreted from the intestine in response to a meal; it slows gastric emptying and has a satiating effect on the individual. In addition to CCK, older adults also have higher concentrations of other anorectic hormones (leptin and insulin) in the fasted and postprandial states and report feeling less hungry when fasted compared with younger adults. Ghrelin is the only gut hormone known to increase appetite. There have been a small number of trials to investigate the impact of subcutaneous administration of ghrelin and the use of oral anamorelin, a ghrelin receptor agonist, on food intake and appetite regulation in frail older adults and cancer cachexia patients. These studies have demonstrated modest increases in appetite and food intake.

The hedonistic or pleasure system that control appetite and eating behaviour is complex and less well understood, although neuroimaging studies using functional MRI are helping to identify the reward centres of the brain that are responsive to foods. Taste, flavour, and sight of food are all important sensory cues to eat, and poor appetite is related to a loss of sensory perception. Taste and smell have been shown to deteriorate from age 60 onwards leading to many older adults reporting that foods taste bland. This can result in poor food practices such as the addition of table salt or sugar to foods to enhance the taste. Flavour enhancement of foods through the addition of seasonings and sauces (pepper, mustard, chilli) has been shown to increase food intake in hospitalized patients and in care home residents. Changes in orosensory perception in older age are thought to contribute to reduced sensory specific satiety in older adults which may contribute to a more monotonous diet.

The social and environmental circumstances in which food is consumed are of particular relevance for older adults who may be living alone or in a care setting. Strategies to improve the eating environment and make meal times more sociable, especially in care homes and hospital settings, have resulted in increased appetite and energy intake.

Gastrointestinal tract

(See also Chapter 4.) Changes that occur in the gastrointestinal tract with ageing may have a significant impact on nutrient intake and absorption and nutritional status. Loss of teeth and denture use can also have an impact on nutritional status as food choice can become limited (see Chapter 26).

There is evidence of malabsorption of certain nutrients with increasing age. Between 20 and 50 per cent of older adults are believed to suffer from an inflammation of the stomach lining known as atrophic gastritis, which causes a loss of gastric glandular cells. As a result, there is impaired secretion of hydrochloric acid, pepsin, and the glycoprotein intrinsic factor. Intrinsic factor is required for vitamin B_{12} absorption in the upper small intestine and so impaired intrinsic factor secretion leads to vitamin B_{12} deficiency. In addition, the reduced secretion of

hydrochloric acid (hypochlorhydria) can result in small bowel bacterial overgrowth, which can further compound the problem since the bacteria are also consumers of the available vitamin B_{12}. A common cause of atrophic gastritis is chronic infection with Helicobacter pylori, which can also contribute to the risk of gastric cancer and has an estimated prevalence of between 40–60 per cent in older adults living in developed countries.

There is some evidence that intestinal motility is altered in ageing. Gastric emptying is known to be delayed with increasing age and there are also reports of slower peristalsis and transit times, which contribute to feelings of fullness and a reduction in appetite. Gastrointestinal disorders that are common in older adults include constipation, diverticulitis, and colorectal cancer.

The gastrointestinal tract is home to trillions of bacteria, mostly found in the colon. These bacteria have long been known to play an important role in the colonic fermentation of carbohydrates, vitamin synthesis, and in gut health. More recently it has been recognized that changes in the composition of the gastrointestinal microflora of older adults and distinct gut microbiome patterns are associated with frailty, reduced cognitive performance, and long-term residential care. The impact of these changes has not been fully defined, although age-related gut microbiota dysbiosis is believed to alter metabolism, immune function and cognition, and to modulate the risk of a host of chronic diseases including obesity, diabetes, and cancer. The mechanisms of interaction between the gut microbiota, nutrition, and health is an area of intense research and is likely to inform novel healthy ageing strategies in the future (Vaiserman et al. 2017).

Immune function

(See also Chapter 27.) Ageing is associated with dysregulation in both innate and adaptive immune function leading to increased risk of and severity of infection, increased cancer incidence, and reduced antibody response to vaccination. The impact of ageing on the adaptive immune system is particularly pronounced. Age-associated atrophy of the thymus gland coupled with a lifetime exposure to pathogens results in a fall in the number of naïve T cells, leading to an overall change in the proportion of memory to effector T cells and a consequent decline in vaccine responsiveness. Within the innate immune system there is a reduction in macrophage function and an increase in the release of pro-inflammatory cytokines leading to a chronic inflammatory state reflected in raised circulating IL6. Natural killer (NK) cells increase in number with age, however, this is accompanied by an apparent decline in their function. Collectively, the changes to immune function that occur with age have been termed immunosenescence and the accompanying chronic low-grade inflammatory state is termed inflammaging.

The impact of immunosenecence and inflammaging on the vulnerability of older adults to infectious disease has been illustrated by the COVID-19 pandemic. Throughout the pandemic and in all countries, age has been the primary risk factor for disease severity and death from COVID-19. This risk has been further exacerbated in the older adult population due to the high prevalence of additional risk factors including obesity and type 2 diabetes.

Evidence suggests that poor nutritional status further exacerbates impaired immune function in older age. Protein energy malnutrition is associated with reduced lymphocyte proliferation, reduced antibody response to vaccination, and increased risk of infection. Deficits in micronutrients such as zinc, selenium, vitamins A, D, C, and E impair specific aspects of immune function, while dietary intake of omega-3 fatty acids helps modulate the inflammatory response (Calder et al. 2020). The role of nutrition in supporting the immune system and the impact of micronutrient deficiencies on risk and severity of infectious disease is yet to be fully defined, however there is little doubt that optimal nutrition status is important for the maintenance of immune function irrespective of age.

Cognitive function

Cognitive impairment and dementia increase with advancing age and as a consequence its global prevalence is predicted to triple by 2050. It is estimated that dementia affects one in 50 people aged 65–69 years and one in five people over the age of 85 years. Dementia is the umbrella term used to describe a range of symptoms, most notably impaired cognitive capacity and memory loss, which are associated with a number of brain conditions. The most common forms of dementia in older adults are Alzheimer's disease (AD) and vascular dementia (see also Chapter 24). Vascular dementia and AD often coexist and share similar risk factors. Whilst age is the biggest risk factor for dementia it is thought that one-third of a person's risk may be due to modifiable lifestyle factors including diet, exercise, and smoking.

There has been considerable interest in the potential of diet to reduce the risk of and delay the progression of dementia (see Useful Websites Section 16.1). Epidemiological studies have found associations between measures of cognitive function and dietary intake or nutrient status; however causality has been difficult to demonstrate and the results of randomized controlled trials (RCTs) have been mixed. Initial focus was on the role of single nutrients (B vitamins, antioxidants, polyphenols, long-chain PUFAs) and

specific foods (alcohol and coffee) to modify dementia risk. More recently the focus has turned to nutrient combinations and whole diet approaches (Nordic, Mediterranean, DASH diet).

Some epidemiological studies have shown associations between low B-group vitamin status, elevated plasma homocysteine, and cognitive decline. There is good evidence that elevated plasma homocysteine is an independent risk factor for AD and a number of intervention trials have investigated the potential for the homocysteine-lowering vitamins B_{12} and folate to influence cognitive function. However, a meta-analysis of 11 RCTs involving 22,000 people concluded that there was no significant effect of B vitamins on cognitive ageing (Clarke et al. 2014). Some promising results did emerge from the VITACOG study; a two-year randomized placebo control trial of B vitamins (B_6, B_{12} and folate) on memory and brain anatomy in people with mild cognitive impairment. This study showed reduced brain shrinkage and slower cognitive decline in those receiving treatment (de Jager et al. 2012).

Impaired redox status is a feature of AD; accumulation of products of oxidative damage to biomolecules is characteristic of the AD brain. For this reason, there has been interest in the possible protective role of dietary antioxidants on the development of and progression of AD. Particular attention has been given to vitamin C, which is found in high concentrations in the brain, and the fat-soluble antioxidant vitamin E. However, RCTs of these antioxidants have produced conflicting results. Recent Cochrane reviews have concluded that there is no good evidence that any vitamin or mineral supplement can maintain cognitive function in healthy middle- and older-age adults, or prevent or delay cognitive decline in people with mild cognitive impairment.

Epidemiological evidence suggests that dietary fat intake modulates the risk of cognitive decline, with fish oils (rich in omega-3 fatty acids) conferring protection, and diets high in saturated and trans-unsaturated fatty acids being associated with elevated risk of cognitive decline. However, Cochrane reviews of RCTs have found no effect of omega-3 fatty acid supplementation on cognitive function in cognitively healthy older people and no benefit for the treatment of dementia.

A few RCTs have investigated combinations of putatively beneficial nutrients, the most notable of which is the LipiDiDiet trial. This is a multinational (Northern European) randomized, placebo controlled intervention trial of a multinutrient combination (including omega-3 fatty acids, vitamins, and minerals) in people with early Alzheimer's disease. Initial analysis after 24 months did not find a significant benefit on the primary outcome measure of cognitive function but did show some benefits in terms of reduced cognition decline and reduced brain shrinkage. However, the clinical benefits were more apparent after 36 months intervention, suggesting that long-term interventions may be needed (Soininen et al. 2021).

Whilst it has been difficult to demonstrate with any certainty the benefits of individual foods and nutrients, evidence is accumulating that healthy dietary patterns such as the Mediterranean diet, the DASH diet, and the healthy Nordic diet may help protect against cognitive decline and AD, with higher adherence to these diets associated with a lower risk of cognitive decline and dementia (van den Brink et al. 2019). There is also growing interest and promising results from interventions that combine multidomain lifestyle intervention approaches that target several risk factors including diet and exercise to help preserve cognitive function.

Sarcopenia, sarcopenic obesity, and frailty

Sarcopenia is the term used to describe the age-related loss of muscle mass and low muscle function (strength or performance). It affects between 5–13 per cent of people aged 60–70 years, rising to between 11–50 per cent of people over the age of 80 years and can lead to loss of mobility, an increased risk of fractures and falls, impaired quality of life, and a higher risk of death. Factors that are thought to contribute to the development of sarcopenia include change in body composition with age, a sedentary lifestyle, prolonged periods of bed rest, poor nutritional status, hormonal status, inflammation, and chronic diseases such as cancer or heart failure. The most successful interventions to prevent sarcopenia have included an element of either strength or resistance training in combination with a protein supplement.

Sarcopenic obesity describes the phenotype that is characterized by muscle wasting of the limbs in the presence of obesity. The combination of obesity and sarcopenia in the same individuals increases the risk of disability and can be particularly difficult to recognize.

Physical frailty is common in the oldest old, that is, those >85 years, and is strongly associated with risk of death (Figure 16.2). One of the most commonly used definitions of frailty is that developed by Fried and colleagues that focuses on the physical aspects of frailty and defines frailty as the presence of three or more of the following characteristics—unintended weight loss, exhaustion, weakness, slow walking speed, and low physical activity (Fried et al. 2001). The other commonly used model is Rockwood's, that considers the accumulation of a number of health deficits generating a frailty index

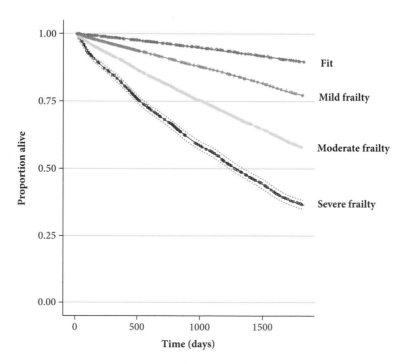

FIGURE 16.2 Five-year Kaplan Meier survival curve for the outcome of mortality for categories of fit, mild frailty, moderate frailty, and severe frailty (internal validation cohort).

Source: A Clegg, C Bates, J Young et al. (2016) Development and validation of an electronic frailty index using routine primary care electronic health record data. *Age Ageing* **45**, 353–360. ©2016. Published by Oxford University Press on behalf of the British Geriatrics Society.

score (Rockwood & Mitnitski 2007). In England, GP practices are now required to identify patients over the age of 65 with frailty, which has been facilitated by the development of tools such as the electronic frailty index that uses data held in primary care databases (Clegg et al. 2016). There is considerable overlap between frailty and sarcopenia, with older adults displaying elements of both syndromes.

Obesity

(See also Chapter 20.) There is an increasing prevalence of obesity in the older adult population, associated with a greater risk of a range of health problems and chronic diseases such as osteoarthritis, hypertension, diabetes, and cancer. Obese older adults have also been shown to have a poorer diet quality and are at greater risk of micronutrient deficiencies than older adults of healthy weight. However, the significance of overweight to mortality and morbidity in older adults is uncertain and there is some evidence that being overweight may confer protection in the adult population in general. A systematic review in 2013 concluded that whilst obesity (BMI>30) increased all-cause mortality rates compared with normal weight people, all-cause mortality was actually lowest in people in the overweight category with a BMI of 25–<30 (Flegal et al. 2013). More recent research has drawn similar conclusions that whilst underweight and obesity are undesirable, a BMI between 23–28 may be optimal for older adults. Other studies have concluded that maintaining a stable weight in older adult is beneficial and that unintentional weight loss is detrimental to health and is associated with higher mortality.

Multimorbidity

Multimorbidity is defined as the presence of two or more long-term health conditions and includes both physical and mental health conditions. The prevalence of multimorbidity increases with advancing age and is highest in those with lowest socioeconomic status.

Multimorbidity is the norm in older adults as shown in a landmark study using electronic patient records in Scotland (Barnett et al. 2012) and confirmed across many other developed countries. In a retrospective study of adult patient records in England the prevalence of multimorbidity was 27 per cent in the adult population as a whole, and in excess of 50 per cent in the 65–74 year age category, rising to >80 per cent in adults >85 years (Cassell et al. 2018). This study also demonstrated the socioeconomic gradient of multimorbidity as shown in Figure 16.3. Patients with multimorbidity have complex health needs and place a disproportionate demand on health services, accounting for >50 per cent of all GP consultations and >78 per cent of all prescriptions given to patients. This has implications for health service provision and has led to calls for restructuring of health services away from specialist treatment of single diseases to a more generalist approach.

Unsurprisingly, given that poor diet contributes to the development of many chronic diseases, it is also implicated in multimorbidity. A number of studies across the globe have considered the risk of multimorbidity and lifestyle factors, typically including smoking, alcohol consumption, fruit and vegetable consumption, physical activity, and obesity in their analysis. The results are largely consistent, with the greater the number of unhealthy lifestyle behaviours associated with increased risk of multimorbidity.

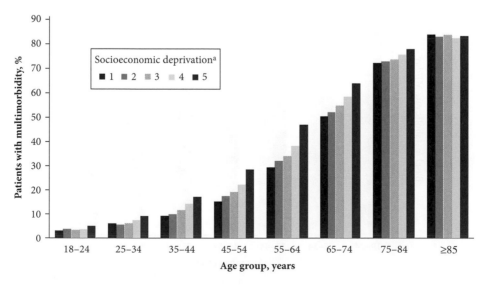

FIGURE 16.3 Prevalence of multimorbidity by age and socioeconomic status
[a] 1 is the quintile with the least socioeconomic deprivation, 5 is that with the greatest.
© British Journal of General Practice 2018. From A Cassell, D Edwards, A Harshfield et al. (2018) The epidemiology of multimorbidity in primary care: A retrospective cohort study. *Br J Gen Pract* **68**, e245–e251.

16.3 Changes in energy expenditure and in energy and protein requirements

Energy expenditure

(See also Chapter 6.) For the majority of older adults total energy expenditure decreases with age, due to reductions in basal metabolic rate and physical activity. The reduction in basal metabolic rate with increasing age (Figure 16.4) is mainly due to a change in body composition, principally a decline in lean body mass. There is considerable heterogeneity in the physical activity of the older adult population. However physical activity is markedly lower in those who are sick or infirm and it is recognized that physical activity declines with increasing age, being the lowest in the oldest old. As a consequence, the energy requirements for physical activity are lower in older adults than the adult population.

Energy requirements

Estimated average requirements (EAR) are the average requirements needed to maintain a healthy body weight in a healthy individual. EAR is calculated from the basal metabolic rate and physical activity level. For older adults the EAR is set lower than for young adults. Despite the reduction in energy requirements, micronutrient requirements are largely unchanged and consequently older adults need to consume a nutrient-dense diet. Daily energy requirements fall by approximately 100 kcals per decade; however, the presence of co-morbidities and infection may increase energy requirements or further reduce requirements because of immobility. Energy requirements may therefore deviate significantly from predictions due to underlying conditions.

Protein requirements

Estimating protein requirements in older age is complex, and there is a lack of consensus. Although requirements are set at the same level as for younger adults, on the basis of lower metabolic demand and a similar efficiency of utilization, requirements may be lower than for younger adults (see Chapter 9). It has been argued that older frail adults have a higher protein requirement due to the need to restore and maintain lean muscle mass. However, the counter argument is that elevated protein intake will result in an exacerbation of the age-related decline in renal function. There are also concerns that high protein intake will increase urinary calcium excretion and lead to bone loss, however the evidence to support this is mixed, with a number of studies suggesting a positive association between protein and bone health. Whilst the reference nutrient intake for protein in older adults has been set at the same level as for younger people (0.75g protein/kg/d), the optimum dietary protein intake is uncertain. The PROT-AGE study group and the European Society for Clinical Nutrition and Metabolism (ESPEN) have recommended intakes of 1–1.2g/kg/d for healthy older adults, up to 1.5g/kg/d for older people with acute or chronic disease and up

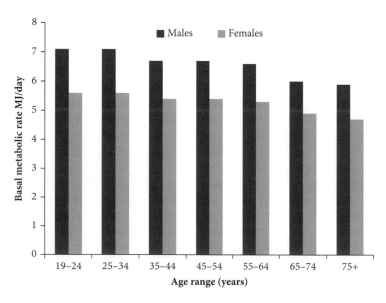

FIGURE 16.4 Changes in basal metabolic rate with age for men and women of average height for the age range and a BMI of 22.5kg/m^2

Source: Scientific Advisory Committee on Nutrition. Dietary reference values for energy (2011) (see Useful Websites list online).
© Crown copyright 2012. Scientific Advisory Committee on Nutrition. Dietary reference values for energy 2011.

to 2g/kg/d for malnourished older adults (Bauer et al. 2013; Deutz et al. 2014). There is a particular emphasis on the need for high quality protein in older adults, specifically animal protein, which contains a higher proportion of the branched chain essential amino acid, leucine, than plant protein. Leucine has a central role in the activation of the mTorc1 pathway that regulates muscle protein synthesis. Studies have shown that there is a threshold requirement of leucine to activate mTorc1. This threshold is higher in older adults due to a declining anabolic response to leucine with age. It is thought that consuming high quality protein coupled with physical activity will help prevent age-related sarcopenia. The timing of protein intake is also thought to be important, with protein intake evenly distributed throughout the day, resulting in greater net protein synthesis than if protein is consumed in a single meal.

16.4 Assessment of nutritional status in older adults

One of the primary purposes of nutritional assessment in older adults is for the detection of undernutrition, however there is no single recommended approach and nutritional assessment typically uses a combination of methods including the assessment of dietary intake, anthropometry, and biochemical analysis (see Chapters 30–32). Nutritional screening tools are recommended for the initial identification of individuals at risk of malnutrition and a more applied approach to the detection of nutritional problems is the assessment of physical function. The commonly used methods for nutritional assessment are summarized in Table 16.1.

TABLE 16.1 Commonly used methods for the assessment of nutritional status in older adults

Measures of nutritional status	Considerations and limitations in older adults
Dietary assessment: Weighed dietary intake Food frequency questionnaires Dietary history by interview Recall interviews (previous 24-hr) Computer/web-based dietary assessment	Weighed food diaries are particularly burdensome to the oldest old. Cognitive decline can make retrospective methods of dietary assessment unreliable. Diet stability in the older adults may improve recall. Multiple pass dietary recall may aid recall in older adults. Computer-based dietary assessment have been used with varying success in older adults.
Anthropometric measurements: Height, weight, and body mass index Skin fold thickness Mid-upper arm circumference (MUAC)	Alternative estimates of height (knee height, ulna length, or demi-span) are required for adults unable to stand. Presence of oedema will interfere with body weight measurement. BMI can underestimate body fatness in older adults due to change in body composition. May be difficult to obtain skinfold thickness at multiple sites in older adults with limited mobility.
Body composition Dual-X-ray absorptiometry (DXA) Bioelectrical impedance (BIA)	DXA has a low accuracy and precision for determining muscle mass. Presence of oedema or dehydration will distort the assessment of body composition using BIA.
Biochemical indicators Plasma vitamin B_{12} Serum ferritin Serum albumin	Many commonly used clinical biochemical indicators are affected by presence of infection, coexisting diseases, and multiple medications. Vitamin B_{12} is a concern in older adults, but plasma vitamin B_{12} levels are insensitive to functional B_{12} deficiency. Serum ferritin is elevated in the presence of an infection and inflammation. Serum albumin is affected by the presence of oedema, infection, and dehydration. It is widely used but should be interpreted with caution.
Nutrition screening tools History & physical examinations: Subjective global assessment (SGA) Malnutrition universal screening tool (MUST) Mini nutritional assessment (MNA)	Useful and recommended approaches for the identification of malnutrition and risk of malnutrition. MUST is the method most commonly used in UK hospitals. Presence or risk of malnutrition should prompt further nutritional assessment, monitoring, and intervention.

Assessment of dietary intake

Dietary assessment in older adults can present additional challenges over and above those encountered in the adult population. Age-related impairments such as reduced vision and hearing, and decline in mobility may hamper the acquisition of dietary intake data using methods such as weighed food diaries, 24-hour recall, and food frequency questionnaires. Cognitive decline may also make retrospective dietary assessment unreliable, and interviewer-led methods can be time consuming and burdensome to both the older adult and the interviewer.

In the past decade much effort has been put into the development of web-based and computer assisted 24-hour dietary recalls and food frequency questionnaire, and these technology-based approaches offer great potential to reduce the cost and improve the accuracy of dietary assessment. However, older adults can have difficulty with computerized methods designed for the general adult population, and consideration of computer and Internet access, computer literacy, and visual ability is needed in the design and application of new methods of dietary assessment.

Anthropometry and the assessment of body composition

(See also Chapter 31.) Anthropometry is the measurement of the size and dimensions of the human body and is a useful tool in the assessment of nutritional status. The anthropometric techniques commonly used to assess nutritional status in older adults are similar to those used for the adult population (weight, height, BMI, skinfold thickness, mid-arm muscle circumference); however, there are additional practicalities to be considered in this age group.

Weight is probably the most valuable indicator of nutritional status, and unintentional weight loss can also indicate the presence of an underlying disease. However, the presence of oedema in older adults can mask weight loss. Height and weight measures are also difficult to attain in older adults who are unable to stand. There are several methods of predicting height when a person is unable to stand; these include knee height, ulna length, and demi-span. All of these methods are highly correlated with adult height, which can be readily calculated using conversion charts. BMI is the most commonly used indirect indicator of body composition. However, BMI does not account for the age-related changes to body composition such as increased fat mass, and as a consequence BMI can underestimate body fatness in older adults.

Other useful anthropometric measurements in older adults are skinfold thickness and mid-upper arm circumference (MUAC), which are used for the indirect assessment of fat mass and fat-free mass respectively. It is recommended that skinfold thickness is measured at multiple body sites (biceps, triceps, subscapular, and supra-iliac) for assessment of total body fat; however, a single site, commonly the triceps, can be used in the old and infirm from whom multiple measurements may be difficult to obtain. Mid-arm circumference can be used as an indicator of malnutrition, with a circumference less than 22cm indicative of increased risk of malnutrition. MUAC coupled with triceps skinfold thickness can be used to derive mid-upper arm muscle circumference which is used as an estimate of muscle mass. One limitation of these methods is that neither skinfold thickness nor MUAC is sensitive to short-term changes in body fat or muscle mass.

Body composition can be assessed by a wide variety of techniques including MRI, CT scans, and doubly labelled water, but many of these are prohibitively expensive for use in routine clinical nutritional assessment. The more commonly used methods for assessment of body composition are dual-X-ray absorptiometry (DXA) and bioelectrical impedance (BIA). DXA was primarily developed for the assessment of bone density, but can also be used to assess fat mass, but has a low accuracy and precision for determining muscle mass. BIA estimates fat mass, fat-free mass, and total body water. This method is cheap, portable, and easy to use, but has a number of limitations including a lack of validation studies and reference values specifically for the elderly. A further limitation of BIA is that it is sensitive to changes in body water and consequentially the presence of oedema or dehydration can distort the measurement.

Biochemical indicators

Any of the biomarkers of nutritional status used for the population in general could be used for older people and the reader is directed to Chapter 32 for further information. There are a number of biomarkers of nutritional status that warrant special consideration when used in the older adult population. For many of the nutrients of concern in older adults no good biomarker exists and there are also several biomarkers that are affected by underlying disease conditions, inflammation, and/or medication use making their interpretation in older adults particularly challenging.

Vitamin B_{12} deficiency is common in older adults, but plasma B_{12} concentration, which is commonly used to determine vitamin B_{12} status, does not provide information about functional B_{12} deficiency. Where vitamin B_{12} deficiency is suspected the additional measurement of serum methylmalonic acid (MMA) and plasma holotranscobalamin is recommended.

Plasma ferritin concentration is considered to be the best indicator of iron stores in a healthy adult population. However ferritin is an acute phase protein and so plasma ferritin levels are raised in inflammatory states making it an unsuitable biomarker of iron deficiency in the elderly. An alternative marker that is increasingly used is the serum transferrin receptor, which is not affected by inflammation.

There is no reliable biochemical indicator of protein stores. Serum albumin is often used as a marker of protein status and thereby as an indicator of malnutrition. However serum albumin is affected by a wide variety of factors including dehydration, oedema, and infection, and so levels should be interpreted with caution in older adults. Furthermore, serum albumin concentration is insensitive to short-term changes in protein intake. In a patient population decreased serum albumin (hypoalbuminemia) is commonly used as a prognostic indicator of poor outcome, and in the older adult population in general it is associated with an increased risk of mortality and morbidity. However, serum albumin has been shown to be a poor predictor of the presence of malnutrition in a hospitalized setting. In a clinically stable population there is no evidence that serum albumin declines with age and so in these circumstances serum albumin may be a useful indicator of nutritional status.

Other biochemical markers of protein status include serum transferrin retinol binding protein. Protein loss can be measured using 24-hour urinary nitrogen and creatinine, but continence issues can make a complete collection difficult.

Nutrition screening tools

Nutrition screening tools are used in a variety of primary and secondary care settings to identify individuals who are malnourished or at risk of malnutrition and who would benefit from some form of nutritional intervention. Unlike laboratory assessment nutritional screening, tools are quick, simple, inexpensive, and easy to interpret. The NICE guidelines on malnutrition recommend that screening takes place on admission to hospital and weekly thereafter, on admission to care home, on initial registration with a GP, and whenever there is clinical concern. The identification of malnutrition or risk of malnutrition should then prompt further nutritional assessment, monitoring, and nutritional support by dietitians or nutritionally qualified medical staff.

The most widely used screening tool in the UK is the malnutrition universal screening tool (MUST) developed by the British Association for Parental and Enteral Nutrition (BAPEN) (see Useful Websites Section 16.2). MUST uses a combination of body mass index, unplanned weight loss, and presence of acute disease to identify presence and risk of malnutrition. The mini-nutritional assessment (MNA®) is another commonly used screening tool. The original MNA® was composed of 18 questions; the more recent short form is composed of six questions (MNA-SF®). Both tools have been validated, are easy to use, and have both paper-based and app versions. Another method still commonly used is that of subjective global assessment (SGA) which is based on the clinical history of the patient and on a physical examination.

Assessment of physical function

There are several methods for the objective assessment of physical function in older adults, and these are particularly useful for the identification of people at risk of sarcopenia and frailty.

Handgrip strength, measured with a handgrip dynamometer, is a commonly used method of assessing skeletal muscle function in older adults. Individuals are asked to perform a maximum handgrip contraction with the non-dominant hand 3–4 times. Maximum contraction is compared against age and gender-specific data. Handgrip strength declines with increasing age, presumably due to the age-related decline in muscle mass and strength. Handgrip strength has been shown to be a useful predictor of mortality. However, this method is not useful in older adults with arthritis.

Other useful functional assessment methods include the Short Physical Performance Battery which includes the assessment of walking speed, standing balance, and sit to stand.

16.5 Diet and nutritional status in older adults and health

What do older people eat?

Given the significant variation in the rates of physiological deterioration, and the burden of morbidity in older people, it should not be surprising that there is a considerable heterogeneity of dietary intake and nutritional status of older people. Furthermore, assessment of dietary intake and nutritional status can be more difficult in elderly populations than in younger adults and it is not always easy to disentangle effects of ageing per se from effects of inadequate dietary intake. The Euronut SENECA study of nutrition in the elderly in Europe, carried out in the 1990s among more than 2500 people between 70 and 75 years of age across Europe, reported considerable variability between and within countries for dietary intake, nutritional status, lifestyle, health, and physical and cognitive performance.

Useful data regarding dietary intake and nutritional status come from national or regional surveys, and systematic reviews of individual studies. The National Diet and Nutrition Survey (NDNS) (see Useful Websites Section 16.3) of the UK population reveals that, although dietary inadequacies do occur in older people, this group consumes diets that are generally closer to UK recommendations than younger adults. This has been observed in other populations, including the elderly Taiwanese, healthy Japanese elderly, and people aged 65 years and over in the USA.

Food intakes

Nutrient intakes in the UK population can be evaluated with reference to dietary reference values (DRVs). In the interpretation of food and nutrient intakes it is important to bear in mind that DRVs are generally determined using studies with younger adults and the optimum intakes for older adults are not well understood. In 2020 the UK Scientific Advisory Committee on Nutrition (SACN) published a position statement on nutrition for older adults living in the community (see Useful Websites Section 16.3). When dietary intake is considered in terms of food-based recommendations it is clear that in common with much of the UK population food intakes of older adults could be improved, with intakes of salt, free sugar, and saturated fat exceeding recommendations whilst intakes of fruit, vegetables, oily fish, and fibre fall short of recommendations. The consumption of red and processed meat exceeds recommendations in the 65–74 year-old age category whilst most older adults fail to achieve the UK recommendation of one portion of oily fish per week. Fruit and vegetable intake in the UK is low with a mean intake of 4.3 portions per day and only 32 per cent of 65–74 year-olds and 19 per cent of people aged 75+ achieve the recommended 5+ portions a day.

Similarly, data from the National Health and Nutrition Examination Survey (NHANES) in the USA show that less than one-third of people aged 60 years and above consume the recommended intakes of fruit and vegetables. NHANES and other studies point to dentition status as a determinant of fruit and vegetable consumption and diet quality generally. Compared with younger adults, older people are also found to have a higher mean consumption of 'table sugar, preserves, and sweet spreads'.

Energy and nutrients

Older people generally have lower food and energy intakes than younger adults, associated with lower BMR and energy requirements but a systematic review of energy and macronutrient intakes of older people shows that whilst mean energy intakes were lower than the chosen reference values, the majority of older people were in the normal body mass range or were overweight (ter Borg et al. 2015). Under-reporting of energy intake may explain this discrepancy. In the UK, men aged 65 years and older consume more than the recommended 35 per cent food energy from fat, in contrast with all other age and sex groups. Also, men and women aged 65 years and over consume more saturated fat as a percentage of total energy than all other age groups, above the recommended 10 per cent. Intakes of fat and added sugars have also been found to be above US recommended levels for about half of the elderly population in New York City (Deierlein et al. 2014), and this has been reported in other studies of older Americans. There is a marked variation in energy and macronutrient intakes of community-living older people but the overall picture is one of a higher than recommended contribution of fat and a lower than recommended contribution of protein to energy intake.

A relatively small proportion of older people is reported to be underweight (BMI <18.5), but weight loss, low BMI, and body weight are of concern because of the well-documented link with risk of protein-energy malnutrition, frailty, and hospitalization. NDNS reports a fall of approximately 10 per cent (men) and 8 per cent (women) in mean daily energy intakes in people from 19–64 years compared with those 65 years and older.

Dietary variety seems to be a determinant of diet quality in older people. Studies from the USA show that older people consume a wider variety of foods than younger adults but that those older people who consume a diet of low variety of energy dense foods and micronutrient-dense foods were more likely to have low intakes of energy and micronutrients and to have a low BMI.

The macronutrient composition of diets in the UK older adult population follows similar trends to younger adults. Whilst the mean energy intake is reported to be lower than estimated average requirements for both men and women, the percentage of energy derived from fat and saturated fat exceed requirements. Dietary fibre intake in UK older adults is low and falls far short of the 30g/d recommendation with mean fibre intake at just 18.7g/d in adults aged 65+ and only 6 per cent meeting recommendations, a factor that is likely to contribute to poor intestinal health and constipation on older adults.

A recent review of nutrient intakes across Europe using data from nationally representative surveys or large cohorts of adults aged ≥60 reveals a similar picture (Kehoe et al. 2019). The majority of 18 European countries studied reported mean dietary intakes in older adults that exceed the UK population recommendations for fat and saturated fat (33 per cent and 10 per cent of energy respectively). Mean intakes of salt are in excess of recommendations and fibre intakes are low.

Micronutrients

Lower food and energy intakes, together with an increased likelihood of conditions associated with impaired absorption and metabolism of micronutrients might be expected to lead to a higher prevalence of inadequate micronutrient status in the elderly than younger adults. However, the most recent data from the NDNS show that average intakes of most vitamins and minerals in people aged 65 years and older are above current recommended nutrient intakes (RNIs) for the UK. NDNS data from years 9–11 show mean intakes for the great majority of vitamins and minerals to be at or above the RNI in older people, suggesting a low risk of deficiency in this age group. Notable exceptions are vitamin D, selenium, and magnesium for which mean intakes are 32 per cent, 66 per cent, and 90 per cent of the RNI respectively in adults aged 65+. Selenium intakes in older people (>65 years) are lower than the lower reference nutrient intake (LRNI) in 47 per cent of this group and magnesium intakes are lower than the LNRI in 13 per cent. The LRNI is considered to meet the requirements of only 2.5 per cent of a population. The risk of low intakes increases in the oldest old, and in the UK, women over the age of 75 years are more likely to have low micronutrient intakes, particularly of vitamin A, riboflavin, selenium, iron, magnesium, and zinc.

Inadequate intakes of selenium and vitamin D were also reported in a systematic review and meta-analysis of 37 studies conducted in 20 Western countries examining micronutrient intakes and risk of deficiency in older people (ter Borg et al. 2015). Adequacy of micronutrient intake was assessed by comparison with Nordic EAR and prevalence of inadequate intake was calculated using the EAR cut-off method. For three of ten vitamins examined, more than 30 per cent of men and women were deemed to have inadequate intakes; these were thiamin, riboflavin, and vitamin D. Of ten minerals examined the percentage of the population at risk of inadequate intakes from food alone was 30 per cent or greater for calcium, magnesium, and selenium.

Data from the NHANES in the USA also show that an appreciable proportion of older adults in the USA consume less than the EAR for calcium (~13 per cent for men, 80 per cent from women), magnesium (~55 per cent), and vitamin D (close to 100 per cent). Low intakes of vitamin D, riboflavin, and B_{12} have been reported in older adult populations across Europe, and in Portugal and Poland more than 50 per cent of older adults have inadequate calcium intakes (Kehoe et al. 2019).

Whilst it is clear that nutrient intakes among older people differ according to geographical location and ethnic group there is agreement, from studies in different countries and using various criteria of adequacy, that older people are at risk of inadequate intakes of vitamin D, selenium, calcium, and magnesium.

Nutritional status and associated health outcomes

Given the influence that ageing can have on bioavailability of some nutrients, evaluation of the functional adequacy of diets in the elderly needs to include a consideration of biomarkers of nutritional status. Poor nutritional status, both under- and over-nutrition, has been associated with increased risk of adverse health outcomes, including cardiovascular disease, cognitive function, poor musculoskeletal health, and cancer. This section will consider evidence of poor nutritional status among older adults and associated adverse health outcomes.

Haemoglobin and iron status

Older adults have a higher prevalence of both low and high iron stores than younger people, and both are causes of concern. Anaemia, as defined by the WHO as a haemoglobin concentration <13g/dL for men and 12g/dL for women, has a global prevalence of around 17 per cent in the older adult population living in the community and is far greater (~47 per cent) in care home residents. However, using a combined index (haemoglobin and plasma ferritin), less than 5 per cent of community-living older people in the UK are iron deficient. Findings are similar in the USA although there is quite significant variability between ethnic groups. Plasma ferritin concentration tends to increase with age, but interpretation of plasma ferritin values in older people is complicated by chronic inflammation, which is common in old age.

Anaemia is a risk factor for a range of age-related conditions including frailty, dementia, falls, hospitalization, and mortality. The aetiology of anaemia in older-age adults is often difficult to determine and can be due to several factors, including inflammation, gastrointestinal problems, malignancy, infection with *Helicobacter pylori*, and medication; while iron deficiency is thought to account for only 12–25 per cent cases of anaemia.

Paradoxically, older adults are also at greater risk of high iron stores, particularly in women, thought to be due in part to the end of menstruation. Elevated iron stores in older people are concerning due to possible associations with increased risk of atherosclerosis, myocardial infarction, and diabetes; there may also be a link with levels of iron in the brain and risk of dementia, and for these reasons intakes in excess of requirements should be avoided.

Vitamin D and calcium status

(See also Chapter 12.) Many studies in Western countries report a high prevalence of poor vitamin D status among the elderly. Skin synthesis is the main source of vitamin D for most people, although diet does make a contribution and becomes particularly important when sunlight exposure is minimal. Biochemical thresholds for deficiency are variable, but are commonly <25nmol/L or <50nmol/L serum or plasma 25(OH)D. In the UK population, according to a threshold of 25nmolL plasma 25(OH)D, year round, about 24 per cent of women and 17 per cent of men aged 65 years and over living in the community have low vitamin D status. The percentage is higher in winter months. The prevalence of low vitamin D status is also higher among institutionalized elderly.

However, the scale of vitamin D deficiency among elderly people is a construct of the deficiency threshold used. The Institute of Medicine review of the dietary reference intakes of vitamin D and calcium proposed a threshold plasma 25(OH)D concentration of <30nmol/L to indicate increased risk of vitamin D deficiency and 50nmol/L to indicate the RNI. The ter Borg review of nutritional status in older people in Europe found a mean plasma 25(OH)D concentration of 56.2 (SD 14.0) nmol/L in men and 51.7 (SD 9.6) nmol/L in women. Habitual dietary intake of vitamin D in most adult populations is insufficient by itself to achieve proposed biochemical thresholds of adequacy. Previously the UK government advised that people over the age of 65 years should take a daily supplement of vitamin D. In 2016 this recommendation was revised, and current advice is that everyone over the age of four years should achieve a vitamin D intake of 10μg daily. It is recognized that in winter months this is likely to require the use of a supplement, however the most recent NDNS survey revealed that just 32 per cent of the older adult population consume a daily vitamin D supplement.

Inadequate vitamin D status has been linked with a number of adverse health outcomes but the evidence is strongest for musculoskeletal health and this has been the subject of several reviews over the last few years. Osteomalacia, which is the term describing poor bone mineralization in adults, can result from poor vitamin D status. Evidence also suggests that vitamin D supplements may reduce the risk of falls and increase muscle strength and function in older people, although evidence is mixed.

Folate and vitamin B$_{12}$ status

Red blood cell folate concentrations indicative of deficiency are reported in 6 per cent of free-living elderly and 16 per cent of older people living in care homes in the UK. This is in contrast to the USA, where, following mandatory fortification of flour with folic acid, the prevalence of folate deficiency is less than 1 per cent. Similarly, folate deficiency in the elderly is no longer a public health problem in other countries that have introduced mandatory fortification of flour with folic acid, including countries of Latin America and Canada.

There is uncertainty regarding the true prevalence of vitamin B$_{12}$ deficiency in older people, because of difficulties in interpreting plasma concentrations of vitamin B$_{12}$, which remains the most commonly used biomarker of vitamin B$_{12}$ status. Impaired absorption of vitamin B$_{12}$ is the most common cause of deficiency in older people. National surveys in the USA and UK report low plasma vitamin B$_{12}$ concentration in 4 per cent of elderly men and women. However, plasma vitamin B$_{12}$ concentration is an insensitive marker of functional vitamin B$_{12}$ deficiency; elevated serum methylmalonic acid (MMA) and plasma total homocysteine (tHcy) are biomarkers of poor vitamin B$_{12}$ status but elevated tHcy is not specific to vitamin B$_{12}$ and both are elevated when renal function is poor, which is more commonly seen in the elderly than younger people. NHANES reports elevated serum MMA and plasma tHcy in about 18 per cent of older people. There is current interest in the use of plasma holotranscobalamin as an alternative functional measure of vitamin B$_{12}$ status.

Some observational studies have shown an association between loss of cognitive function or dementia and biomarkers of status for folate, B$_{12}$ and other B-group vitamins in older people. However as described earlier in this chapter meta-analyses of randomized controlled trials have concluded there is insufficient evidence for a beneficial effect of homocysteine-lowering B vitamins on cognitive ageing. Similarly, whilst case-control and prospective cohort studies have generally shown a positive association between elevated plasma tHcy and risk of cardiovascular diseases, results of randomized controlled trials have generally failed to demonstrate a beneficial effect of homocysteine-lowering on cardiovascular outcomes.

Antioxidant nutrients

Antioxidant nutrients include vitamin C, carotenoids, and alpha tocopherol. There is little evidence that low intakes and/or biomarkers of poor status for these nutrients is prevalent among elderly people. However, in those people who do show evidence of poor status there may be an increased risk of mortality following stroke. Vitamin C deficiency can lead to scorbutic symptoms in some elderly people, including sublingual haemorrhage and poor wound healing, particularly among hospitalized elderly. Although there has been a great deal of interest in the possible role of antioxidant nutrients in lowering risk of cancers and CVD, results of randomized controlled trials have been inconsistent and interest in the possible protective value of antioxidant supplements against chronic disease has waned.

A study of elderly Italians (the InCHIANTI study) showed correlations between plasma concentrations of antioxidants and measures of physical performance and muscle strength, but there is little information from prospective studies to support these observations. There is some evidence to suggest that free radical damage may be important in other diseases of ageing, such Parkinson's disease and Alzheimer's disease, but the evidence lacks consistency. The use of vitamin E supplements may have modest cognitive benefits in older women. Supplementation with vitamin E has also been found to lead to significant delay in the progression of dementia. Generally though, the evidence is limited. A Cochrane review on the use of antioxidant supplements for the prevention of mortality found no evidence to support the use of antioxidants, with a suggestion that some antioxidant supplements may actually increase mortality.

Selenium and magnesium

Although data show a high prevalence of inadequate intakes of selenium and magnesium status, robust biomarkers are lacking for these nutrients. Selenium is an active component in various enzymes involved in immune function, oxidative stress, and muscle function. Selenium deficiency is associated with skeletal myopathy and cardiomyopathy, and low plasma selenium is reported to be independently associated with poor skeletal muscle strength in community-dwelling older adults. Further studies may clarify whether selenium supplementation can slow down the age-associated decline in muscle strength. Some studies suggest that selenium supplementation in people with marginal selenium status can improve immune function.

Stroke patients have been reported to exhibit deficits in magnesium in serum and cerebrospinal fluid. Acute magnesium or potassium deficiency can produce cerebrovascular spasm, and the lower the extracellular concentration of either magnesium or potassium the greater the magnitude of cerebral arterial contraction.

Sodium and potassium

Urinary sodium concentrations for people over the age of four years show that mean salt intakes are higher than recommended values across all age groups in the UK and are higher in the 65 year and over than for any other age group. Hypertension and stroke are both more common in older people than younger adults. Excess salt intake

causes hypertension not only through simple volume expansion but also through sodium-accelerated vascular smooth muscle cell proliferation, and it enhances thrombosis by the acceleration of platelet aggregation. An inverse association of potassium intake with stroke mortality has been reported, irrespective of hypertensive status. Clinical, experimental, and epidemiological evidence all suggest that a high dietary intake of potassium is associated with lower blood pressure.

Dehydration

Hydration status is rarely assessed in surveys. However, from the limited evidence it is clear that older adults are at particular risk of becoming dehydrated. Reasons for dehydration are multifactorial and include a decrease in sensitivity to thirst, changes in kidney function, and deliberate avoidance of fluids due to continence concerns. A variety of methods for assessment of hydration status have been used, making prevalence estimates difficult. However, it is estimated that around 20 per cent of care home residents in the UK and over a third of older adults admitted to hospital are dehydrated. Dehydration is linked with poor cognition, increased risk of falls, urinary tract infections, and increased mortality in hospitalized older adults.

16.6 Malnutrition and hidden hunger in older adults

Malnutrition is older adults can be difficult to recognize. According to the National Institute for Health and Care Excellence (NICE) a person is considered to be malnourished if they have:

- a BMI of less than 18.5kg/m^2;
- unintentional weight loss greater than 10 per cent in the past 3–6 months; or
- a BMI less that 20kg/m^2 and unintentional weight loss greater than 5 per cent in the past 3–6 months.

(See Useful Websites Section 16.4.) The issue of 'hidden hunger' is also a particular concern in older adults. Hidden hunger is a term used to describe individuals with adequate energy intake and normal/high body mass, but suboptimal intakes of nutrients resulting in adverse health outcomes. Hidden hunger in older adults is recognized as a societal concern in many Western societies.

Malnutrition is estimated to affect around 3 million people in the UK with 1.3 million older adults experiencing malnutrition, the majority of whom (93 per cent) live in the community. The annual health and social care cost of treating people with malnutrition in England alone is in the region of £19.6 billion per annum and in the USA the economic cost of malnutrition associated diseases in older adults is $51.3 billion per annum. The greatest prevalence of malnutrition is evident in hospitalized and institutionalized older adults. BAPEN has reported that over 30 per cent of people aged 65+ are malnourished or at risk of malnutrition on admission to hospital and a third of all care home residents are at risk of malnutrition. Malnutrition is associated with prolonged hospital stay, impaired wound healing, and increased mortality.

Reasons for malnutrition are multifactorial, but are often due to the catabolic effect of an underlying disease. For example, cachexia is a common feature of cancer and chronic obstructive pulmonary disease (COPD). People who have had a stroke or dementia often experience dysphagia (difficulty eating and swallowing) and are at particular risk of malnutrition. A reduced intake of food is one of the main causes of malnutrition in the elderly. This occurs due to a combination of physiological and psychosocial reasons including changes in taste, smell, and appetite coupled with social isolation, loneliness, bereavement, depression, and food insecurity. The many factors associated with poor nutritional dietary intake in older adults are summarized in Table 16.2.

The consequences of malnutrition are extensive. There is a loss of muscle and strength which impacts not only on a person's mobility but also on respiratory and cardiac muscle function, increasing the likelihood of chest infection and heart failure. Malnutrition also leads to impaired immune function, increased risk of infection, poor wound healing, and a loss of mucosal integrity leading to malabsorption, bacterial translocation, and increased risk of pressure sores. Malnutrition also has a psychological impact, causing apathy and depression, which often further exacerbate the situation. The association between malnutrition and mortality was assessed in 1203 free-living and special-housing residents between the ages of 60 and 96 in Sweden. Risk of malnutrition was found to be an independent predictor of mortality at seven-year follow-up (Naseer et al. 2016).

Nutritional support comprises oral nutritional support (food, supplement drinks, and sip feeds), enteral feeding (the delivery of nutrients to the gut), or parenteral nutrition (intravenous nutrition). Food intake can be enhanced by the use of oral nutritional supplements and is a relatively simple intervention in contrast to enteral or parenteral feeding that are only possible under close medical supervision. The use of oral nutritional supplements is associated with a range of clinical benefits including reduced infections, improved quality of life, fewer falls, and reduced hospital admissions in community-dwelling older adults requiring nutritional support; and reduced complications, reduced mortality, and reduced hospital stay when used in the hospital setting. However, where possible, nutrition support should take a 'food first' approach with a focus on encouragement of greater food intake.

TABLE 16.2 Factors associated with poor nutritional status in older people

Poor eyesight & hearing problems

Joint problems & hand tremors

Poor cognitive function

Nausea and vomiting

Poor appetite

Anorexia due to disease especially cancer

Medications

Lack of assistance at mealtimes

Poor dentition and chewing problems

Acute illness

Inability to shop/prepare food

Social isolation

Food insecurity

Loneliness, depression, and bereavement

KEY POINTS

- Ageing is associated with changes in body composition and deterioration in most physiological functions. There is a wide variation in the rate of loss of function.

- There is generally a fall in energy requirement and an associated decrease in energy intake in older age. For some older people energy intake falls below requirements, leading to weight loss and increasing the risk of frailty and hospitalization.

- Absorption and metabolism of some nutrients may be impaired in older people, increasing the risk of specific micronutrient deficiencies, including vitamin B_{12}.

- Older people are susceptible to inadequate intakes of vitamin D, selenium, calcium, and magnesium.

- Poor vitamin D status is common in older people, increasing the risk of osteoporosis, falls, and reduced muscle strength and function.

- There is increased prevalence of both low and high iron stores in older people, and both are detrimental to health.
- Adherence to a healthy or Mediterranean dietary pattern in older age is associated with reduced risk of a number of chronic diseases.

- NICE recommends the use of nutritional screening tools such as MUST for the detection of malnutrition.
- Older people are advised to consume a balanced, nutrient-dense diet and to remain as physically active as possible.

ACKNOWLEDGEMENT

The author sincerely acknowledges the contribution of Professor Hilary J. Powers who co-authored the previous edition of this chapter on which this chapter is based.

 Be sure to test your understanding of this chapter by attempting multiple choice questions. See the Useful Websites list for additional material relevant to this chapter.

REFERENCES

Barnett K, Mercer S W, Norbury M et al. (2012) Epidemiology of multimorbidity and implications for health care, research, and medical education: A cross-sectional study. *Lancet* **380**, 37–43.

Bauer J, Biolo G, Cederholm T et al. (2013) Evidence-based recommendations for optimal dietary protein intake in older people: A position paper from the PROT-AGE Study Group. *J Am Med Dir Assoc* **14**, 542–559.

ter Borg S, Verlaan S, Mijnarends D M et al. (2015) Macronutrient intake and inadequacies of community-dwelling older adults: A systematic review. *Ann Nutr Metab* **66**, 242–255.

Calder P C, Carr A C, Gombart A F et al. (2020) Optimal nutritional status for a well-functioning immune system is an important factor to protect against viral infections. *Nutrients* **12**, 1181.

Cassell A, Edwards D, Harshfield A et al. (2018) The epidemiology of multimorbidity in primary care: A retrospective cohort study. *Br J Gen Pract* **68**, e245–e251.

Clarke R, Bennett D, Parish S et al. (2014) Effects of homocysteine lowering with B vitamins on cognitive aging: Meta-analysis of 11 trials with cognitive data on 22,000 individuals. *Am J Clin Nutr* **100**, 657–666.

Clegg A, Bates C, Young J et al. (2016) Development and validation of an electronic frailty index using routine primary care electronic health record data. *Age Ageing* **45**, 353–360.

Cox N J, Morrison L, Ibrahim K et al. (2020) New horizons in appetite and the anorexia of ageing. *Age Ageing* **49**, 526–534.

Deierlein A L, Morland K B, Scanlin K et al. (2014) Diet quality of urban older adults age 60 to 99 years: The Cardiovascular Health of Seniors and Built Environment Study. *J Acad Nutr Diet* **114**, 279–287.

Deutz N E, Bauer J M, Barazzoni R et al. (2014) Protein intake and exercise for optimal muscle function with aging: Recommendations from the ESPEN Expert Group. *Clin Nutr* **33**, 929–936.

Flegal K M, Kit B K, Orpana H et al. (2013) Association of all-cause mortality with overweight and obesity using standard body mass index categories: A systematic review and meta-analysis. *JAMA* **309**, 71–82.

Fried L P, Tangen C M, Walston J et al. (2001) Frailty in older adults: Evidence for a phenotype. *J Gerontol A Biol Sci Med Sci* **56**, M146–156.

de Jager C A, Oulhaj A, Jacoby R et al. (2012) Cognitive and clinical outcomes of homocysteine-lowering B-vitamin treatment in mild cognitive impairment: A randomized controlled trial. *Int J Geriatr Psychiatry* **27**, 592–600.

Kehoe L, Walton J, Flynn A (2019) Nutritional challenges for older adults in Europe: Current status and future directions. *Proc Nutr Soc* **78**, 221–233.

López-Otín C, Blasco M A, Partridge L et al. (2013) The hallmarks of aging. *Cell* **153**, 1194–1217.

McCay C M, Crowell M F, Maynard L A (1935) The effect of retarded growth upon the length of life and upon the ultimate body size. *J. Nutr.* **10**, 63–79.

Naseer M, Forssell H, Fagerstrom C (2016) Malnutrition, functional ability and mortality among older people aged 60 years: A 7-year longitudinal study. *Eur J Clin Nutr* **70**, 399–404.

Rockwood K & Mitnitski A (2007) Frailty in relation to the accumulation of deficits. *J Gerontol A Biol Sci Med Sci* **62**, 722–727.

Scientific Advisory Committee on Nutrition (2011) *Dietary Reference Values for Energy.* The Stationery Office, London. https://www.gov.uk/government/publications/sacndietary-reference-values-for-energy

Soininen H, Solomon A, Visser P J et al. (2021) 36-month LipiDiDiet multinutrient clinical trial in prodromal Alzheimer's disease. *Alzheimers Dement* **17**, 29–40.

United Nations, Department of Economic and Social Affairs, Population Division (2022) World Population Prospects 2022. https://population.un.org/wpp/Graphs/Probabilistic/PopPerc/65plus/900

Vaiserman A M, Koliada A K, Marotta F (2017) Gut microbiota: A player in aging and a target for anti-aging intervention. *Ageing Res Rev* **35**, 36–45.

van den Brink A C, Brouwer-Brolsma E M, Berendsen A A M et al. (2019) The Mediterranean, dietary approaches to stop hypertension (DASH), and Mediterranean-DASH intervention for neurodegenerative delay (MIND) diets are associated with less cognitive decline and a lower risk of Alzheimer's disease: A review. *Adv Nutr* **10**, 1040–1065.

FURTHER READING

Bjelakovic G, Nikolova D, Gluud LL et al. (2012) Antioxidant supplements for prevention of mortality in healthy participants and patients with various diseases. *Cochrane Database Syst Rev*, CD007176. Available from: https://www.cochranelibrary.com/cdsr/doi/10.1002/14651858.CD007176.pub2/full

Edmonds C J, Foglia E, Booth P et al. (2021) Dehydration in older people: A systematic review of the effects of dehydration on health outcomes, healthcare costs and cognitive performance. *Arch Gerontol Geriatr* **95**, 104380.

Freisling H, Viallon V, Lennon H et al. (2020) Lifestyle factors and risk of multimorbidity of cancer and cardiometabolic diseases: A multinational cohort study. *BMC Med* **18**, 5.

Jiang M, Zou Y, Xin Q et al. (2019) Dose-response relationship between body mass index and risks of all-cause mortality and disability among the elderly: A systematic review and meta-analysis. *Clin Nutr* **38**, 1511–1523.

Johnson K O, Shannon O M, Matu J et al. (2020) Differences in circulating appetite-related hormone concentrations between younger and older adults: A systematic review and meta-analysis. *Aging Clin Exp Res* **32**, 1233–1244.

Jun S, Cowan A E, Bhadra A et al. (2020) Older adults with obesity have higher risks of some micronutrient inadequacies and lower overall dietary quality compared to peers with a healthy weight, National Health and Nutrition Examination Surveys (NHANES), 2011–2014. *Public Health Nutr* **23**, 2268–2279.

Most J, Tosti V, Redman L M et al. (2017) Calorie restriction in humans: An update. *Ageing Res Rev* **39**, 36–45.

Rebelo-Marques A, De Sousa Lages A, Andrade R et al. (2018) Aging hallmarks: The benefits of physical exercise. *Front Endocrinol (Lausanne)* **9**, 258.

Shannon O M, Ashor A W, Scialo F et al. (2021) Mediterranean diet and the hallmarks of ageing. *Eur J Clin Nutr*. **75**, 1176–1192.

Wawer A A, Jennings A, Fairweather-Tait S J (2018) Iron status in the elderly: A review of recent evidence. *Mech Ageing Dev* **175**, 55–73.

17 Vegan/vegetarian diets

Thomas A. B. Sanders

OBJECTIVES

By the end of this chapter, you should be able to:

- describe the different types of vegan/vegetarian diets
- understand the reasons why people follow vegan/vegetarian diets
- understand the relevance of plant-based diets in the context of climate change
- identify the nutrients most likely to be lacking from vegan/vegetarian diets
- understand how the health of vegans/vegetarians differs from that of omnivores.

17.1 Introduction

Vegetarianism can be defined as the voluntary abstinence from eating meat and fish. The commonplace use of the term vegetarian is to describe someone who does not eat animal flesh (meat, poultry, fish) but who includes eggs and dairy products in their diet. The terms *ovo-vegetarian, lacto-vegetarian,* and *ovo-lacto-vegetarian* describe people who consume eggs, milk, or both, respectively. The term *vegan* is used to describe people who consume no food of animal origin. Usually, the first stage in becoming a vegetarian is to give up consuming red meat; this is followed by the exclusion of poultry and fish. Many vegetarians aspire to becoming vegans. Veganism is a way of life that avoids the exploitation of animals. Besides avoiding food of animal origin, vegans commonly will not use products that have been derived from animals, such as leather, wool, and vaccines. Fruitarianism is an extreme form of veganism where dietary intake is restricted to raw fruits, nuts, and berries, which has resulted in severe malnutrition in children. Macrobiotic diets, which originate from the teachings of the George Ohsawa Macrobiotic Foundation, consist of relatively large amounts of brown rice, accompanied by smaller amounts of fruits, vegetables, and pulses; processed foods and *Solanaceae* species (tomatoes, aubergines, and potatoes) are avoided; meat and fish are permitted if they are hunted or wild. In practice, however, most macrobiotic diets are vegetarian and contain only small amounts, if any, of milk products. The terms pesco-vegetarian and demi-vegetarians are more modern terms that describe individuals who include fish and those that occasionally eat meat respectively. These various definitions all depend upon the exclusion of certain foods from the diet, whereas the nutritional quality of diet is dependent on foods consumed.

The risk of nutrient deficiency is greatest in childhood as requirements relative to body weight are greater, and children are unable to exert the same degree of control over what they eat as adults can. Indeed, there have been several reports of severe protein energy malnutrition as well as deficiencies of iron, vitamins B_{12} and D in infants and toddlers fed inappropriate vegetarian diets (Sanders 1999). Older children are less susceptible to the dietary strictures imposed by their parents because they can forage for food independently. However, children can be brought up healthily on vegetarian and vegan diets, providing sufficient care is taken to ensure the latter is sufficiently energy dense and is supplemented with micronutrients where appropriate.

People with eating disorders often report following a vegetarian/vegan diet but there is a lack of evidence to show that vegetarian/vegan diets increase the risk of developing an eating disorder such as anorexia nervosa. Nevertheless, the adoption of a vegetarian/vegan diet may be used as a device by people with eating disorders to restrict food energy intake. Vegetarian/vegan diets are also promoted by advocates of complementary and alternative medicine (e.g., naturopathic medicine). Lower intakes of animal fats and higher intakes of fruit, vegetables, wholegrains, nuts, and legumes and lower intakes of saturated fatty acids may confer lower risk of cardiovascular disease and cancer, but these benefits need to be offset by increased risks of iron and vitamin B_{12} deficiencies, and for vegans a low calcium intake, which increases the risk of bone fractures in later life. The current evidence shows that the overall health of vegetarians/vegans differs little from omnivores. Owing to the smaller number of subjects that have been studied for decades or longer, there is less certainty about the long-term health effects of vegan diets.

17.2 Reasons for following vegetarian diets

Abstinence from eating 'flesh' has been associated with asceticism throughout history. Several religions refer to unclean meats (e.g., Judaism, Islam, and Sikhism) and specify methods for the slaughter of animals, perhaps indicating early recognition that diseases could be transmitted from animals to humans. Some religions were more explicit and advocated avoiding meat. Vegetarianism is widely practised by believers of the Hindu and Buddhist faiths and is prescribed by the Seventh Day Adventist Church on health grounds. More restricted forms of vegetarianism are practised by the Jain sect, who will only eat food that grows above the ground, and Rastafarians advocate a vegan diet based mainly on fruit.

The term 'ethical vegetarian' is often used to describe those who do not belong to any specific religious group but choose to follow the diet because of their beliefs. The UK Vegetarian Society was founded in 1842. They believed that abstinence from eating 'flesh foods' would promote health and argued that killing animals for food was morally wrong. The vegetarian movement was associated with an emerging middle class who sought to improve themselves through education and temperance. It was also strongly linked to Quakerism and the Seventh Day Adventist Church. Indeed, the leading characters in the Food Reform Movement in the nineteenth century, such as Kellogg, Graham, Cadbury, and Elizabeth Fry, also espoused a vegetarian diet. The belief that avoiding meat is healthier is still widely held and has been bolstered by a series of food scares that involved intensive farming and food processing (e.g., salmonellosis in poultry, bovine spongiform encephalopathy in cattle, and contamination of foods of animal origin with man-made chemicals) as well as evidence from prospective cohort studies showing an association of red and processed meat consumption with increased risk of breast and colorectal cancer.

Concern about the impact of climate change has emerged as a major driver for a shift towards plant-based diets. Not only is less land needed to feed a family on a vegetarian diet than on one containing meat, but following a vegetarian diet reduces greenhouse gas emissions (carbon dioxide and methane). It has been estimated that the greenhouse gas emissions for vegan diets are only 2.89 kg CO_2 equivalents/head/day vs. 3.81 for vegetarians and 5.63 for moderate meat-eaters (Scarborough et al. 2014). The values are higher for vegetarians because of the amounts of methane produced by milk cows.

The increased popularity of vegetarian and vegan diets is apparent as many foods are labelled as suitable for vegetarians or vegans and airlines and restaurants now offer suitable options to cater for their changing customer base. Estimates suggest about 4 per cent of men and 7 per cent of women consider themselves to be vegetarians in the USA, whereas estimates in the UK range from 5–7 per cent of the population. The number of vegans has increased dramatically in the past few years particularly among younger adults and is the range 0.5–1 per cent of the UK population. However, most vegetarians live in India (~350 million) and China (~55 million).

Until recently, most of the dietary and nutritional studies of vegans and vegetarians were carried out on members of the Vegan and Vegetarian Societies or religious groups such as members of the Seventh Day Adventist Church. These participants are generally keen to demonstrate the healthiness of their diets. They may also differ in other aspects of lifestyle that have an impact on health such as not smoking and abstinence from alcohol (Seventh Day Adventists). Vegetarian/vegans may refuse to have immunization with vaccines produced in animals or containing animal products. Fewer studies have been conducted on the very much larger populations of vegetarians in India and China. Consequently, the conclusions drawn in this chapter apply mainly to the UK, North America, Western Europe, and Australasia.

17.3 Dietary adequacy

In order to ensure nutritional adequacy a wider variety of foods is needed on a vegetarian diet based on plant food compared with the diet of an omnivore. This is because the distribution of nutrients is more diverse in plants and the nutrient density (nutrients/kcal) is often lower than that of meat, fish, eggs, and milk. This is particularly relevant to poorer countries where food choice may be limited. Sir Robert McCarrison's research in India in the early part of the twentieth century was among the first to recognize overdependence on polished rice as a cause of malnutrition. There is also plenty of evidence to show that growth rates of infants reared on restricted vegetarian diets are retarded. In parts of India, where poverty and intestinal infestation are common, vegetarian children are more likely to be anaemic and shorter in stature compared with non-vegetarians. On the other hand, when people of Indian origin migrate to developed countries and still maintain their vegetarian dietary practices but

TABLE 17.1 Energy and proximate nutrient intakes in vegetarians, vegans, and omnivores

	Energy		% Energy		Fibre	
	(MJ/day)	Protein	Carbohydrate	Sugar	Fat	(g/day)
European men (UK)						
Omnivores	10.3	14.4	42	19	38	26
Vegetarians	9.4	12	48	21	37	34
Vegans	9.2	11.7	50	21	34	44
European women (UK)						
Omnivores	7.28	15.5	43	20	38	20
Vegetarians	7.67	12.6	46	20	37	33
Vegans	7.35	10.8	53	21	34	36
South Asian (UK)						
Men	9.3	12.6	47	–	38	23
Women	6.1	11.7	54	15	37	16

Source: Data derived weighed intakes from: A Draper, J Lewis, N Malhotra et al. (1993) The energy and nutrient intakes of different types of vegetarian: A case for supplements? **Br J Nutr** 69, 3–19; G J Miller, S Kotecha, W H Wilkinson et al. (1988). Dietary and other characteristics relevant for coronary heart disease in men of Indian, West Indian and European descent in London. *Atherosclerosis* 70, 63–72; S Reddy & T A Sanders (1992) Lipoprotein risk factors in vegetarian women of Indian descent are unrelated to dietary intake. *Atherosclerosis* **95**, 223–229.

consume more dairy products, the impact of the vegetarian diet on growth is more limited.

There are surprisingly few qualitative differences in the intake of proximate nutrients (protein, carbohydrates, and fats) between vegetarians/vegans compared with omnivores in developed countries (Table 17.1). Energy intakes appear to be similar, although reports, which have estimated energy intake by food frequency questionnaire, suggest that energy intakes are lower among vegans and vegetarians than omnivores. However, this is not the case for weighed food intake surveys. The likely reason for the discrepancy between the two methods is that the portion sizes of plant foods consumed by vegetarians and vegans are often larger than those consumed by omnivores, particularly in the case of cereals.

Some vegetarian diets are bulky and can restrict energy intakes of young children. For example, several studies have reported impaired growth in children under the age of five years fed vegan and macrobiotic diets associated with low energy intakes. Indeed, a fruitarian diet, which has the lowest energy density, has resulted in the death of infants in the UK.

Protein

Recommended intakes are easily met by vegetarian and vegan diets and protein quality will be high if the proteins are derived from a variety of dietary sources.

Protein intakes are slightly lower in vegetarians than in omnivores, typically supplying about 12 per cent of the energy intake as opposed to 15 per cent in omnivores (Table 17.1). However, these intakes support nitrogen balance. Although plant proteins have a lower biological value than meat, the protein quality of vegetarian diets differs little from that of diets containing meat, as the constituent amino acids in the different plant proteins mutually complement each other. Cereals are an important dietary source of protein, whose protein quality is limited by the lysine content, but they are relatively rich in sulphur amino acids especially cysteine and methionine. In contrast, the protein quality of pulses is limited by their sulphur amino acid content but they have a relatively high concentration of lysine. Consequently, when a cereal and pulse (e.g., baked beans on toast) are eaten together in a meal their amino acid profiles complement each other to provide high quality protein similar to that of meat. The proper cooking of leguminous seeds (beans, peas, and lentils) is important because legumes contain protease inhibitors and other naturally occurring toxins which are destroyed by heat treatment (a minimum of 100°C for ten minutes). Processed soya-containing foods are an important source of high quality protein in many vegetarian diets and soya milk is a useful and nutritious alternative to mammalian milks. Maize is a cereal that is deficient in tryptophan, but this is of no consequence if milk products, which are rich in tryptophan, form part of the diet.

Meat and fish, especially shellfish, are rich sources of taurine. Taurine is thought to be a conditionally essential nutrient in the newborn where the capacity to synthesize it from cysteine is limited. Taurine is used to form the bile salt taurocholic acid which aids fat digestion in infants because it has a lower critical micellar concentration than glycocholic acid. Lower rates of urinary excretion of taurine have been found in vegan women compared with meat-eaters and the concentration of taurine in human milk from vegans is lower, but there is no evidence that this is of any consequence to the health of the infant. High intakes of red meat may have adverse effects on health because they are associated with the fermentation of carnitine by gut microbiota to trimethylamine oxide (TMAO), which can be absorbed into blood. Elevated levels of TMAO have been found to be associated with increased risk of cardiovascular disease. It is perhaps not surprising that blood concentrations of TMAO are lower in vegetarians and vegans than meat-eaters.

Carbohydrates

The intake of complex carbohydrates is generally high in vegan/vegetarian diets owing to the higher consumption of wholegrain cereals compared with the diets of omnivores. Sugar intakes, on the other hand, are relatively similar. However, much of the sugar intake in vegan/vegetarian diets comes from consumption of fruit and fruit juice rather than from added sugar. Intakes of dietary fibre (non-starch polysaccharide) are between 50 per cent and 100 per cent higher than in subjects on an omnivorous diet, depending on the sources of dietary carbohydrate (i.e., wholegrain/wholemeal versus refined cereal): typical daily intakes of dietary fibre are in the order of 30 g in vegetarians and 40 g in vegans compared with about 18 g in omnivores. The higher fibre intake is associated with a faster faecal transit time and a larger faecal bulk (faecal weight is typically about 160 g/day in female vegetarians compared with 117 g in omnivores). There is some evidence to suggest that vegetarians have a lower prevalence of diverticular disease. Higher faecal bulk (186 g/day) has been reported in South Asian vegetarian women living in the UK, which is more strongly related to the amount of fermentable carbohydrate provided by foods such as legumes and onions reaching the colon rather than the total fibre intake. There are marked differences in the faecal microbiota between meat-eaters, vegetarians, and vegans. The vegan and vegetarian diets contain more fermentable carbohydrates that are fermented to short-chain fatty acids in the colon which have the effect of lowering faecal pH. The lower pH promotes the growth of different gut microbiota. This partly explains why fewer primary bile acids are fermented to secondary bile acids in vegetarians compared with meat-eaters.

Fats

The proportion of energy derived from fat is only slightly lower in vegetarians/vegans than in omnivores and is typically in the region of 30–37 per cent of the dietary energy compared with omnivores where intakes are typically 35–38 per cent. Saturated fatty acid intakes are slightly lower in vegetarians, being derived mainly from dairy fats, but markedly lower in vegans (around 6 per cent energy) in comparison with omnivores. Cholesterol is virtually absent from plant foods but instead they provide small amounts of phytosterols (<0.5 g/day) which inhibit the reabsorption of cholesterol from the intestinal tract. The intake of polyunsaturated fatty acids is usually greater in vegetarian and especially vegan diets because of their preference for nuts, oilseeds, and vegetable oils. The polyunsaturated fatty acid:saturated fatty acids ratio is much higher in vegans than in omnivores or vegetarians and this is reflected in their lower levels of total and low density lipoprotein (LDL) cholesterol concentrations (Bradbury et al. 2014; Sanders 2014) Intakes of linoleic acid (18:2 *n*-6) are typically 9 per cent energy in vegans compared to 5 per cent energy in omnivores with intermediate values in vegetarians. Intakes of α-linolenic acid (18:3 *n*-3) are more dependent on the types of oils used in food preparation but are often higher than in omnivores. Rapeseed and soya bean oil are rich sources of α-linolenic acid. The long-chain omega-3 fatty acids are lacking from vegan diets and in vegetarian diets depend upon intakes from eggs and dairy foods. Differences in the composition of dietary fat are well illustrated by the composition of breast milk in vegans, vegetarians, and omnivores (Table 17.2).

Nutrients usually provided by food of animal origin

Meat and fish provide several nutrients that are scarce or absent from common foods of plant origin and these include iodine, taurine, vitamin B_{12}, vitamin D, and long-chain omega-3 polyunsaturated fatty acids such as eicosapentaenoic (20:5 *n*-3, EPA) and docosahexaenoic acid (22:6 *n*-3, DHA). Most other vitamins are abundant in plant foods: thiamin and many other of the B-complex vitamins including folate are present in high amounts in unrefined cereals; while retinol is only present in eggs and milk fat; in vegetarian diets vitamin A is supplied by carotenoids. Carrots and orange sweet potatoes are good sources of β-carotene. Dark green vegetables, such as spinach, are rich sources of β-carotene, lutein, and folate. Fruit and vegetables are rich sources of potassium (K^+). Sodium (Na^+) in vegetarian diets is added in the form of salt. In the case of iodine and selenium the amount

TABLE 17.2 Polyunsaturated fatty acids (wt. % total fatty acids) in the breast milk of vegans, vegetarians, and omnivores

	Vegans	Vegetarians	Omnivores
Linoleic (18:2 *n-6*)	23.8	19.5	10.9
α-linolenic (18:3 *n-3*)	1.36	1.25	0.49
Dihomo-γ-linolenic (20:3 *n-6*)	0.44	0.42	0.4
Arachidonic (20:4 *n-6*)	0.32	0.38	0.35
Docosahexaenoic (22:6 *n-3*)	0.14	0.30	0.37

Source: T A Sanders, S Reddy (1992) The influence of a vegetarian diet on the fatty acid composition of human milk and the essential fatty acid status of the infant. *J Pediatrics* **120**, S71–S77, with permission from Elsevier.

present in plants is dependent on the soil concentration of the mineral. Sulphur-containing compounds such as glucosinylates present in many *Brassica* species can reduce iodine uptake by the thyroid gland. Several minerals such as iron, zinc, and calcium may be less well absorbed when derived from plant sources, particularly if the diet contains a lot of phytate which chelates these divalent cations.

Iron

Iron is found in a very wide variety of foods but its availability from foods of plant origin is variable. Eggs, dairy food, and rice are poor sources of iron. Good sources of iron for vegetarians include wheat, pulses, dark green vegetables (especially low oxalate varieties, such as kale), fortified cereals, dried fruit, and iron cooking equipment. A higher incidence of iron deficiency anaemia among women and infants has been found in vegetarians of South Asian origin in the UK and North America. Iron deficiency anaemia is also more prevalent among macrobiotic vegetarians in Holland whose diets contain large amounts of brown rice. It seems that risk of anaemia is greatest among those who rely on rice rather than wheat. Iron intakes appear to be relatively high in vegetarians and vegans whose staple food is wholemeal bread, although bioavailability is low. The UK Biobank Study (Tong, Key et al. 2019) found haemoglobin and red blood cell concentrations to be slightly lower compared with regular meat-eaters and that mild to moderate anaemia was more prevalent in vegetarians (both of British and Indian ethnic origin) especially in premenopausal women (Table 17.3).

Meat is an important source of iron in the diet as the haem form is particularly well absorbed compared with iron from plant sources. However, iron absorption from plant foods can be increased in the presence of vitamin C. Serum ferritin levels have been found in several studies to be strongly correlated with the intake of haem iron.

Indeed, serum ferritin concentrations are low (below 12 mg/L) in both white and Asian vegetarian women of childbearing age compared with women who eat meat. While vegetarians are more prone to developing iron deficiency anaemia than meat-eaters, the diet may be helpful for those who carry the gene predisposing to haemochromatosis (HFE) (which affects about 1 in 200 in people of Northern Europe extraction) because they are less likely to develop haemochromatosis.

Vitamin B₁₂

Vitamin B_{12} is made by bacteria particularly during fermentation in the gut and accumulates in animal tissues; it is usually absent from plant-based foods, with the exception of some algae (e.g., nori). Vitamin B_{12} produced by the gut microbiota in the human large bowel is not available for absorption in humans because the vitamin first needs to bind with intrinsic factor, which is secreted in the stomach, to facilitate its absorption in the terminal ileum. Intrinsic factor deficiency, antibodies to intrinsic factor, or other factors that impair the absorption of the vitamin are the most common causes of vitamin B_{12} deficiency and are unrelated to the dietary intake. However, a chronically low dietary intake can result in severe deficiency and even be fatal. It usually takes 3–4 years on a vegan diet to deplete liver stores of vitamin B_{12}. The signs and symptoms of vitamin B_{12} deficiency are megaloblastic anaemia and neuropathy. If the intake of folate in not high, patients usually present with the symptom of megaloblastic anaemia, which is characterized by extreme tiredness, lassitude, and shortness of breath. However, if the intake of folate is high the megaloblastic anaemia does not develop (referred to as masking) and the more insidious neuropathies may develop. White vegans and vegetarians tend to present with neurological signs of deficiency (paraesthesia and sub-acute combined degeneration of the spinal cord, and even optic nerve neuropathy) because of their high dietary intake of folate. British

TABLE 17.3 Mean haemoglobin, red blood cell counts, and percentage of participants defined as anaemic among individuals recruited into the UK Biobank Study (Tong, Key et al. 2019) according to ethnic origin, type of diet, gender, and menopausal status

		n	Haemoglobin g/L Mean (95% CI)	Red blood cell count (Mean (95% CI)	% Classified anaemic
Men	British omnivores	121,433	150.3 (150.2, 150.3)	4.74 (4.74,4.74)	2.9
	British vegetarians	2130	147.6 (147.1, 148.0)*	4.67 (4.66, 4.69)*	3.9
	British vegans	166	144.8 (132.2, 146.3)*	4.50 (4.45, 4.56)*	6.6
	Indian omnivores	2254	147.2 (146.7, 147.7)	4.92 (4.90, 4.94)	7.6
	Indian vegetarians	485	145.2 (144.1, 146.2)*	4.91 (4.87, 4.95)	12.9*
Premenopausal women	British omnivores	20,791	133.0 (132.9, 133.2)	4.28 (4.28, 4.29)	8.7
	British vegetarians	1638	130.6 (130.1, 131.1)*	4.22 (4.21, 4.24)*	12.8*
	British vegans	76	131.4 (129.1, 133.8)*	4.17 (4.09, 4.24)*	7.9
	Indian omnivores	565	127.8 (126.8, 128.7)	4.44 (4.41, 4.47)	20.5
	Indian vegetarians	229	125.1 (123.5, 126.6)*	4.42 (4.37, 4.47)	26.6
Postmenopausal women	British omnivores	67,145	136.6 (136.5, 136.6)	4.36 (4.35, 4.36)	3.4
	British vegetarians	2583	133.7 (133.4, 134.1)*	4.29 (4.28, 4.31)*	5.8*
	British vegans	149	134.7 (133.2, 136.1)	4.27 (4.21, 4.32)*	4.0
	Indian omnivores	981	130.6 (130.0, 131.2)	4.45 (4.42, 4.47)	13.3
	Indian vegetarians	614	127.9 (127.1, 128.7)*	4.42 (4.38, 4.45)	19.2*

* Denotes statistically significantly different $P<0.05$ from respective omnivore reference group.
Data taken from supplementary table 1-8 from T Y N Tong, T J Key, K Gaitskell et al. (2019) Hematological parameters and prevalence of anemia in white and British Indian vegetarians and nonvegetarians in the UK Biobank. *Am J Clin Nutr* **110**, 461–472.

vegetarians of Indian origin, whose intakes of folate are lower, are more likely to present with megaloblastic anaemia.

A cross-sectional study of the EPIC-Oxford cohort study (Gilsing et al. 2010) found 52 per cent of vegan and 7 per cent of vegetarians had serum cobalamin concentrations in the deficient range (<118 pmol/L) and a further 21 per cent and 7 per cent respectively had depleted concentrations (<150 pmol/L) (Table 17.4). The participants in the deficient range subjects showed definite metabolic

vitamin B_{12} deficiency (increased plasma methylmalonic acid, severely increased plasma homocysteine, and low holotranscobalamin concentrations). The mean plasma homocysteine concentration in B_{12}-deficient vegans was 26 mmol/L compared with 10 mmol/L in those with normal cobalamin concentrations. Elevated plasma homocysteine concentrations indicate a decreased capacity for the body to conduct methyl transfer reactions which are important in regulating epigenetic changes in DNA (see Chapter 29). The significance of the elevated plasma

TABLE 17.4 Serum cobalamin concentrations shown according to diet group from a cross-sectional analysis of the EPIC-Oxford cohort study

	Omnivores (n = 226)	Vegetarians (n = 231)	Vegans (n = 232)
Serum cobalamin (pmol/L)	281 (270–292)[a]	182 (175–189)[b]	122 (117–127)[c]
Deficient <118 pmol/L	1 (0)	16 (7)	121 (52)
Depleted 118–149 pmol/L	3 (1)	40 (17)	48 (21)
Sufficient >150 pmol/L	222 (98)	175 (76)	63 (27)

Mean values (95% CI) or n (%). Values not sharing the same superscript are statistically significantly different ($P<0.001$) using analysis of variance.
Source: Data from A M Gilsing, F L Crowe, Z Lloyd-Wright et al. (2010) Serum concentrations of vitamin B_{12} and folate in British male omnivores, vegetarians and vegans: Results from a cross-sectional analysis of the EPIC-Oxford cohort study. *Eur J Clin Nutr* **64**, 933–939. Copyright © 2010, Macmillan.

homocysteine in B_{12} deficiency is uncertain but it may increase the risk of neurological disorders, stroke, and possibly cancer. The UK Biobank Study (Tong, Key et al. 2019) found platelet counts to be lower in vegans, which can probably be attributed to their very low intake of vitamin B_{12}. Vitamin B_{12} deficiency resulting in severe and irreversible neuropathy can also arise in infants fed human milk deficient in the vitamin. The level of vitamin in human milk depends on the maternal blood levels of cobalamin. Useful amounts of vitamin B_{12} are found in milk (fresh, pasteurized, and UHT but not sterilized milk) and eggs. There is no good evidence to show that spirulina and fermented plant foods, such as kombucha or tempeh, are of any value in prevention of vitamin B_{12} deficiency. There is also a variety of useful dietary sources of vitamin B_{12} from fortified products that are acceptable to vegans such as soya milks, yeast extracts, some margarines, and breakfast cereals. Alternatively, a dietary supplement of the vitamin can be taken regularly. The importance of ensuring an adequate dietary intake of vitamin B_{12} cannot be overemphasized.

Vitamin D and calcium

Vitamin D is mainly provided by sunlight exposure but is required in the diet in Northern latitudes in winter. Vegetarian sources of the vitamin include ergosterol (D_2) made from the irradiation of fungi and yeasts, and cholecalciferol (D_3) made from the irradiation of sheep's wool. However, vitamin D_2 has a lower bioactivity than D_3 (about two-thirds). More recently cholecalciferol has been extracted from algae and this form appears acceptable to vegans. Serum 25-OH vitamin D concentrations were found to be lower in vegetarians and vegans compared with omnivores in the Oxford cohort of the EPIC study (Crowe et al. 2011) but there were strong seasonal variations showing the dominant effect of sunlight exposure on vitamin D status. A high prevalence of rickets was reported in Dutch children reared on macrobiotic vegetarian diets and in the 1970s it was prevalent among South Asians migrants to the UK from Africa. However, clinical vitamin D deficiency symptoms (rickets and osteomalacia) do not appear to be significant problems among the white vegetarian populations in the UK. However, the UK government recommends that all adults, especially the housebound and the elderly, take supplements of vitamin D during the winter months when *uv*-B exposure of the skin is minimal. Rickets among vegans and vegetarians appears to be more related to the bioavailability of calcium from the diet in the presence of low 25-OH-vitamin D concentrations. A high intake of phytate from cereals such as oats, brown rice, or unleavened bread appears to be an important precipitating factor causing clinical rickets especially when milk is absent from the diet. Vegetarians who are regular consumers of milk products are likely to have adequate intakes of calcium and are, therefore, less likely to suffer from vitamin D deficiency. Vegans often have calcium intakes well below the reference nutrient intake. Vegetarians appear to have a slightly increased risk (relative risk 1.25) of hip and leg fractures, but vegans have a much greater risk (relative 2.31) compared to regular meat-eaters (Tong et al. 2020). Consequently, vegans may well benefit from an increased intake of calcium.

Omega-3 fatty acids

DHA plays an important role in the development of the retina and the central nervous system. It can be synthesized from α-linolenic acid (18:3 *n*-3) or obtained preformed in the diet from food of animal origin, especially fish. Vegan diets are devoid of DHA but contain significant amounts of α-linolenic acid (Sanders 2014). However, the levels of DHA in blood, arterial, and breast milk lipids are approximately only one-third of those found in omnivores. Reddy et al. (1994) showed that cord arterial lipids from vegetarian pregnancies had a lower proportion of DHA and a correspondingly higher proportion of the n-6 isomer docosapentaenoic acid (22:5 n-6; DPA) compared with omnivores. This would imply that the ability to convert α-linolenic acid to DHA is not limited in vegetarians but is more influenced by substrate availability. In the European Prospective Investigation into Cancer (EPIC) study, the proportion of DHA in the plasma lipids of vegans and vegetarians was 59 per cent and 31 per cent lower respectively than that in omnivores. There was some evidence to suggest that the proportions of EPA and n-3 isomer DPA (22:5 n-3) were higher in subjects with a lower ratio of linoleic/linolenic acid in their plasma lipids. Short-term supplementation with linolenic acid increases EPA but not DHA in blood lipids, but relatively small amounts (200 mg) of preformed DHA led to a substantial increase. Thus, although there is a basal rate of DHA synthesis from α-linolenic acid in humans, the proportion in blood lipids is greatly augmented by small amounts of dietary DHA (Sanders 2014).

Supplements and foods containing DHA from algal source are available. While there are cogent reasons for believing that a supply of DHA is important for normal visual and neurological development in the neonate, as yet there is no clear indication that the consumption of DHA by vegans or vegetarians has any clear health benefits. Consequently, the requirements for omega-3 fatty acids are likely to be met by α-linolenic acid.

17.4 Pregnancy

Pregnancy and birthweights appear to be normal in white vegetarian women. However, the birthweight adjusted for gestational age, parity, and maternal height was 240g less and the duration of pregnancy was 5.6 day shorter in Asian vegetarians compared with matched white omnivore controls in the UK (Reddy et al. 1994). These findings are worrying as a low birthweight is associated with an increased risk of developing type 2 diabetes and other disorders in later life. Indeed, the incidence of type 2 diabetes is much higher in British vegetarians of South Asian origin compared with the general population in the UK.

17.5 Child growth and development

Widdowson and McCance in their classic experiment carried out at the end of World War II clearly demonstrated that children grow and develop quite normally on a diet consisting of plenty of bread and vegetables with minimal amounts of milk and meat. The growth rate of the white Seventh Day Adventist vegetarian population appears to be virtually indistinguishable from that of white omnivores except for a later age of menarche. Lower rates of growth, particularly in the first five years of life, have been reported in children reared on vegan and macrobiotic diets probably because of low energy intakes. A fruitarian diet has caused severe malnutrition in young children because it provides little sustenance except for sugar and water. Cereal gruels and brown rice are bulky, and toddlers may have difficulty meeting their energy needs if these foods account for a high proportion of the dietary intake. However, bread, nuts, and pulses are more energy dense. Despite these lower rates of growth in the first few years of life, catch-up growth occurs by the age of about ten years. Height is normal but there is still a tendency for these children to be lighter in weight for height than children on mixed diets. The significance of slightly slower rates of growth is debatable, but some vegan children showed impaired growth (Sanders 1999). It needs to be more widely recognized that severe nutritional deficiencies can occur in children reared on inappropriate vegetarian and vegan diets. It seems almost inevitable that more children will fall victims to their parental folly because of the increasing popularity of vegan diets, inaccurate information peddled on social media, and the trend amongst an opinionated minority to ignore conventional nutritional wisdom.

17.6 The health of vegetarians and vegans

The health of Western vegetarian groups has been extensively studied and generally appears to be good, providing the known dietary pitfalls—low energy density, inadequate intakes of iron, vitamins B_{12} and D—are avoided. Indeed, the diets followed are often more consistent with healthy eating guidelines than with the general population. Figure 17.1 shows a typical dietary pattern of British vegetarians/vegans. The increased availability of processed foods acceptable to vegans as well the fortification with micronutrients (iron, zinc, vitamin B_{12} and vitamin D) of some cereal products as well 'plant' milks have made it easier to follow a nutritionally balanced vegan diet. However, should a vegan opt not to consume fortified foods then dietary supplements of vitamin B_{12} are essential.

Assessing the health impact of vegetarianism is often confounded by other lifestyle factors such as socioeconomic status, non-smoking status, and health

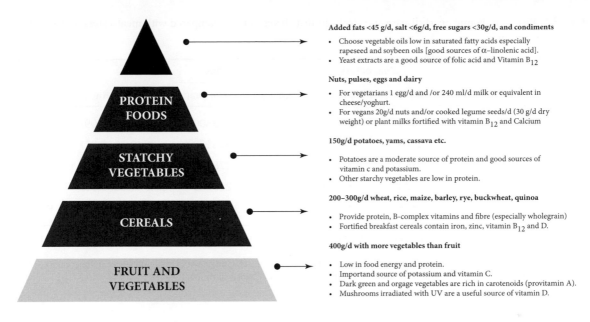

FIGURE 17.1 Healthy vegetarian/vegan diet food pyramid
Amounts based on 2000 kcal (8.4 MJ)/d without energy-containing beverages (other than milk or substitutes) and discretionary foods (e.g., crisps, confectionery, ice cream). Red denotes intakes need to be restricted, orange in moderation, green no restriction.

consciousness. One approach has been to compare robust biomarkers of cardiovascular disease (CVD) such as serum total cholesterol:high density lipoprotein cholesterol ratio (TC:HDL-C) and blood pressure. In the EPIC-Oxford Study, the mean TC:HDL-C ratio was 3.99 in vegans, 4.46 in vegetarians and 5.38 in moderate-meat-eaters (Bradbury et al. 2014). Blood pressure is also lower in vegetarians. A meta-analysis of randomized controlled trials (Gibbs et al. 2021) found that an ovo-lacto-vegetarian diet lowered systolic blood pressure by −5.47 mmHg (95 per cent CI −7.60, −3.34), a reduction that was comparable to that obtained with the dietary approaches to stop hypertension (DASH) diet, but a vegan diet only lowered blood pressure by a non-significant −1.30 mmHg (95 per cent CI −3.90, 1.29). However, salt intakes can be high on vegan diets especially as many processed foods are high in salt such as bread, meat substitutes, yeast extract, and soy sauce.

An almost universal finding has been that vegetarians and vegans are lighter in weight than their meat-eating counterparts. The Oxford cohort of the EPIC Study found a difference of 1 unit of BMI (Spencer et al. 2003). The lower BMI would be expected to be associated with a decreased risk of type 2 diabetes and gallstones. Indeed, the incidence of type 2 diabetes mellitus is lower in the predominantly white vegetarian populations that have been studied in the UK and USA. However, central obesity is common in South Asian vegetarians and they have correspondingly higher rates of type 2 diabetes. The EPIC-Oxford study showed that BMI tends to fall abruptly over

the age of sixty in vegetarians and especially vegans compared with meat-eaters, which suggests that elderly vegans may have difficulty maintaining muscle mass in old age (Spencer et al. 2003). As a low BMI is associated with increased mortality, particularly from respiratory illnesses, this is a concern. The age-related reduction of BMI in vegans is probably due to loss of muscle mass, which would tally with the lower concentrations of insulin-like growth factor 1, which is an anabolic hormone promoting growth in muscle mass, reported in vegans compared with omnivores.

Compared to the general population, most studies find vegetarians to have lower age-standardized mortality rates. However, these findings are confounded by factors other than diet that influence health, such as tobacco and alcohol use, socioeconomic status, physical activity, and educational attainment. For example, a comparison of members of the Seventh Day Adventist Church, the majority of whom are vegetarians, shows about half the rate of cancer compared with the average Californian population. However, similar observations have been made among other religious denominations that also proscribe smoking and alcohol use but who do not follow vegetarian diets, for example, in Mormons. Based on 18 years of follow-up, the EPIC-Oxford Study (Tong, Appleby et al. 2019) found the incidence of ischaemic heart disease to be lower in vegetarians compared with regular meat-eaters (Table 17.5). The absolute risk difference between vegetarians and meat-eaters per 1000 population over ten years predicted ten fewer cases of

TABLE 17.5 Risk of ischemic heart disease and stroke in British vegetarians compared with meat-eaters in the EPIC-Oxford cohort study

	Meat eaters		Vegetarians			
	Cases	Person years	Cases	Person years	Hazard ratio	Significance
Acute myocardial infarction	559	438001	145	278800	0.89 (0.73, 1.09)	P = 0.51
Ischaemic heart disease	2026	439125	496	276938	0.78 (0.70, 0.87)	P<0.001
Ischaemic stroke	340	438418	117	278383	1.12 (0.90, 1.41)	P = 0.59
Haemorrhagic stroke	173	438418	89	278383	1.43 (1.08, 1.90)	P = 0.04
Total stroke	678	438418	258	278383	1.20 (1.02, 1.40)	P = 0.06

Absolute risk differences (95%) per 1000 population over ten years are −10 (95% CI −13, −6.7) for ischaemic heart disease and +3.0 (95% CI 0.8, 5.4) for total stroke.

Source: Data from T Y N Tong, P N Appleby, K E Bradbury et al. (2019) Risks of ischaemic heart disease and stroke in meat eaters, fish eaters, and vegetarians over 18 years of follow-up: Results from the prospective EPIC-Oxford study. BMJ 366, l4897.

ischaemic heart disease but three extra cases of stroke, particularly haemorrhagic stroke. The excess incidence of stroke was unexpected but might be related to low vitamin B_{12} intake.

The absence of red and processed meats and the higher intake of dietary fibre in vegetarians compared with omnivores would be expected to reduce the risk of colorectal cancer. The US Seventh Day Adventist study found an overall lower rate of colorectal cancer in non-meat eaters (Orlich et al. 2015). However, the latter study included pesco-vegetarians who had a lower incidence of colorectal cancer than other vegetarian groups. This US cohort study also provides some evidence of a lower incidence of prostate cancer among white vegan men (Tantamango-Bartley et al. 2016). In the EPIC-Oxford cohort study, cancer incidence was lower in vegetarians than in the general population, but differences compared with meat-eating controls in the study were less after adjustment for tobacco and alcohol use (Key et al. 2014). The incidence of cancer of the stomach and lymphatic/haematopoietic system, especially myeloma, was significantly lower among the vegetarians (Table 17.6). However, the incidence of cancer of the brain, colon and rectum, breast, cervix, endometrium, ovary, and prostate and were remarkably similar. An analysis of causes of mortality up to the age of 90 in UK vegetarians and vegans compared to omnivores found few differences (Appleby et al. 2016) between groups—overall there was a slightly lower death rate from cancer but a slightly higher death rate from respiratory diseases, but overall mortality rates did not differ.

17.7 Conclusion

Vegetarian diets as consumed in developed countries differ little in terms of nutrient composition from those containing meat in developed countries. However, in developing countries, where the choice of foods is more restricted, nutritional deficiencies are more likely, particularly those of iron and vitamin B_{12}. In isolated areas where soil levels of selenium and iodine are low, selenium deficiency and goitre are also more likely to occur with plant-based diets. Thus plant-based diets are nutritionally adequate providing they are not restricted in variety or quality. Vegan diets need supplementation/fortification with vitamin B_{12} and probably calcium. Care needs to be taken to ensure that plant foods are adequately processed to inactivate anti-nutritive substances such as phytates, trypsin inhibitors, and cyanogenic glycosides. The higher intake of polyunsaturated fatty acids and lower intake of saturated fatty acids, lower BMI, and the higher intake of wholegrains, nuts, fruit, and vegetables probably explains the lower risk of ischaemic heart disease in vegetarians. In conclusion vegetarian/vegan diets can promote good health providing they are sensibly selected.

KEY POINTS
- Vegetarian diets are defined based on the foods excluded, whereas nutritional value depends on the foods included the diet.

TABLE 17.6 Relative risk of incident cancer among vegetarians and vegans compared to meat eaters in the EPIC-Oxford cohort study

Cancer site (ICD code)	Meat eaters (n = 32,491)		Vegetarians and vegans (n = 20,544)	
	No. of cancers	RR	No. of cancers	RR (95% CI)
Stomach (C16)	53	1.0	11	0.37 (0.19–0.69)*
Colorectum (C18–20)	382	1.0	154	1.03 (0.84–1.26)
Pancreas (C25)	80	1.0	22	0.73 (0.44–1.21)
Lung (C34)	166	1.0	58	1.16 (0.83–1.61)
Melanoma (C43)	191	1.0	71	0.79 (0.59–1.07)
Female breast (C50)	900	1.0	352	0.93 (0.82–1.07)
Endometrium (C54)	118	1.0	42	0.91 (0.62–1.33)
Ovary (C56)	148	1.0	56	0.86 (0.61–1.20)
Prostate (C61)	327	1.0	100	0.84 (0.66–1.07)
Kidney (C64)	57	1.0	21	0.90 (0.51–1.60)
Bladder (C67)	91	1.0	24	0.62 (0.38–0.99)
Brain (C71)	56	1.0	33	1.29 (0.78–2.13)
Lymphatic/hematopoietic tissue (C81–96)	284	1.0	79	0.64 (0.49–0.84)*
All sites (C00–97)	3275	1.0	1203	0.88 (0.82–0.95)*

* Denotes a statistically significantly difference from omnivores.
RR, relative risk with 95% confidence intervals adjusted for smoking habit, physical activity, and alcohol intake.
ICD, International Classification of Diseases, 10th revision.
Source: Data taken from T J Key, P N Appleby, F L Crowe et al. (2014) Cancer in British vegetarians: Updated analyses of 4998 incident cancers in a cohort of 32,491 meat eaters, 8612 fish eaters, 18,298 vegetarians, and 2246 vegans. *Am J Clin Nutr* **100**(Suppl 1), 378S–385S.

- People follow vegetarian diets for religious, ethical, and health reasons.
- Diets high in fruit and vegetables can be bulky and low in energy.
- Vitamin B_{12} is the nutrient most likely to be lacking in vegan and to a lesser extent in vegetarian diets. Deficiency causes megaloblastic anaemia, but this is masked by high intakes of folate so is more likely to present with neuropathy in vegetarians.
- Vegetarians are at increased risk of iron deficiency anaemia owing to the low bioavailability of iron from plant foods.
- Vegans show an increased risk of leg and hip fractures with increasing age.
- The lower BMI in vegetarians is associated with lower risk of type 2 diabetes but vegans may have difficulty maintaining weight in old age.

- Risk factors for cardiovascular disease are more favourable. The total cholesterol:HDL cholesterol is 0.33 and 0.80 lower in vegetarians and vegans compared with moderate meat-eaters and blood pressure is lower
- The incidence of ischaemic heart disease and cancer is lower in vegetarians than meat-eaters and risk of stroke is higher, overall life-expectancy is similar.
- Much larger long-term studies on vegans are needed to identify any subtle differences in their long-term effects on health.

 Be sure to test your understanding of this chapter by attempting multiple choice questions.

REFERENCES

Appleby P N, Crowe F L, Bradbury K E et al. (2016) Mortality in vegetarians and comparable non-vegetarians in the United Kingdom. *Am J Clin Nutr* **103**, 218–230.

Bradbury K E, Crowe F L, Appleby P N et al. (2014) Serum concentrations of cholesterol, apolipoprotein A-I and apolipoprotein B in a total of 1694 meat-eaters, fish-eaters, vegetarians and vegans. *Eur J Clin Nutr* **68**, 178–183.

Crowe F L, Steur M, Allen N E et al. (2011). Plasma concentrations of 25-dydroxyvitamin D in meat eaters, fish eaters, vegetarians and vegans: Results from the EPIC-Oxford Study. *Pub Hlth Nutr* **14**, 340–346.

Draper A, Lewis J, Malhotra N et al. (1993) The energy and nutrient intakes of different types of vegetarian: A case for supplements? *Br J Nutr* **69**, 3–19.

Gibbs J, Gaskin E, Ji C, Miller M A et al. (2021) The effect of plant-based dietary patterns on blood pressure: A systematic review and meta-analysis of controlled intervention trials. *J Hypertens* **39**, 23–37.

Gilsing A M, Crowe F L, Lloyd-Wright Z et al. (2010) Serum concentrations of vitamin B_{12} and folate in British male omnivores, vegetarians and vegans: Results from a cross-sectional analysis of the EPIC-Oxford cohort study. *Eur J Clin Nutr* **64**, 933–939.

Key T J, Appleby P N, Crowe F L et al. (2014) Cancer in British vegetarians: Updated analyses of 4998 incident cancers in a cohort of 32,491 meat eaters, 8612 fish eaters, 18,298 vegetarians, and 2246 vegans. *Am J Clin Nutr* **100**(Suppl 1), 378S–385S.

Miller G J, Kotecha S, Wilkinson W H et al. (1988). Dietary and other characteristics relevant for coronary heart disease in men of Indian, West Indian and European descent in London. *Atherosclerosis* **70**, 63–72.

Orlich M J, Singh P N, Sabate J et al. (2015) Vegetarian dietary patterns and the risk of colorectal cancers. *JAMA Intern Med*, **175**, 767–776.

Reddy S, Sanders T A (1992) Lipoprotein risk factors in vegetarian women of Indian descent are unrelated to dietary intake. *Atherosclerosis*, **95**, 223–229.

Reddy, S, Sanders, T A, Obeid, O (1994). The influence of maternal vegetarian diet on essential fatty acid status of the newborn. *Eur J Clin Nutr* **48**, 358–368.

Sanders, T A (1999) The nutritional adequacy of plant-based diets. *Proc Nutr Soc* **58**, 265–269. doi: 10.1017/s0029665199000361. PMID: 10466165

Sanders T A (2014) Plant compared with marine n-3 fatty acid effects on cardiovascular risk factors and outcomes: What is the verdict? *Am J Clin Nutr* **100**(Suppl 1), 453S–458S.

Sanders T A, Reddy S (1992) The influence of a vegetarian diet on the fatty acid composition of human milk and the essential fatty acid status of the infant. *J Pediatrics,* **120**, S71–S77.

Scarborough P, Appleby P N, Mizdrak A, et al. (2014) Dietary greenhouse gas emissions of meat-eaters, fish-eaters, vegetarians and vegans in the UK. *Clim Change,* **125**, 179–192.

Spencer E A, Appleby P N, Davey G K et al. (2003) Diet and body mass index in 38000 EPIC-Oxford meat-eaters, fish-eaters, vegetarians and vegans. *Int J Obes Relat Metab Disord* **27**, 728–734.

Tantamango-Bartley Y, Knutsen S F, Knutsen R et al. (2016) Are strict vegetarians protected against prostate cancer? *Am J Clin Nutr,* **103**, 153–160.

Tong T Y N, Appleby P N, Armstrong M E G et al. (2020) Vegetarian and vegan diets and risks of total and site-specific fractures: Results from the prospective EPIC-Oxford study. *BMC Medicine* **23**, 18353. https://doi.org/10.1186/s12916-020-01815-3.

Tong T Y N, Appleby P N, Bradbury K E et al. (2019) Risks of ischaemic heart disease and stroke in meat eaters, fish eaters, and vegetarians over 18 years of follow-up: Results from the prospective EPIC-Oxford study. *BMJ* **366**, l4897.

Tong T Y N, Key T J, Gaitskell K et al. (2019) Hematological parameters and prevalence of anemia in white and British Indian vegetarians and nonvegetarians in the UK Biobank. *Am J Clin Nutr* **110**, 461–472.

FURTHER READING

Appleby P N & Key T J (2015) The long-term health of vegetarians and vegans. *Proc Nutr Soc* **75**, 287–293.

Marriotti F (ed.) (2017). *Vegetarian and Plant-based Diets in Health and Disease Prevention.* Academic Press, London.

USEFUL WEBSITES

https://www.vegsoc.org/health (accessed March 2023)

https://www.nutrition.org.uk/healthyliving/helpingyoueatwell/veganandvegetarian.html (accessed March 2023 2021)

18

Dietary considerations for sport and exercise

Luc J. C. van Loon and Jorn Trommelen

OBJECTIVES

By the end of this chapter you should be able to:

- understand the basic physiology of muscle contraction
- appreciate the importance of usage of different fuels at different intensity exercise
- understand factors influencing fuel utilization in individuals under different circumstances
- understand the feeding strategies for optimum physical performance for different situations
- be aware of the evidence for the use of different ergogenic aids.

18.1 Introduction

Next to a certain genetic predisposition and participation in a regular and effective training regime, nutrition is a key factor in determining physical well-being and exercise performance capacity. Nutrition has become even more important for performance now that athletes have reached limits in training volume and intensity. This has led to a renewed interest among athletes, coaches, and exercise physiologists regarding the role of nutrition and the influence of gastrointestinal problems on physical performance and well-being (Maughan 2014). Clear-cut nutritional counselling, however, is often difficult for a number of reasons. First of all, there is no generally accepted nutritional recommendation for the recreational and/or elite athlete involved in heavy physical training. In contrast, most countries have recommended nutrient intakes (RNIs) or recommended daily allowances (RDAs) for nutrients for different age groups and genders. The RDAs for people involved in heavy physical work are generally derived from studies of people performing strenuous manual labour for their profession. Information about the daily requirements of the elite and/or recreational athlete, who will differ with respect to the relative work intensities, is lacking. Second, the elite

athlete's diet is often obscured by secrecy. Following the efforts to ban the use of performance-enhancing drugs, there has been an increasing interest in the potential for the use of specific food products and/or food ingredients to optimize exercise performance. The use of food in this respect is not new and dates back to the days when sports were first practised, around 500 BC in Greece.

Examination of the literature suggests that a well-balanced, healthy diet, compensating for the metabolic demands imposed by intense exercise training and competition, is all that athletes require to optimize their performance. However, the question remains as to what extent a well-balanced diet for the elite and/or recreational athlete differs from a balanced diet for the normal population. Based on an increasing number of scientific studies, certain dietary manipulations have been shown to improve physical performance, in particular endurance performance. Consequently, the development of specific sports supplements, and carbohydrate-rich sports drinks in particular, has shown a rapid increase over the last 30 years. More recently, there has been an increasing interest in functional food ingredients, an area in which sports nutrition has led in developing concepts and products. Unfortunately, in such a booming market the number of 'exotic' (sports) drinks and supplements, claiming to improve health and/or exercise performance without any reasonable scientific justification, is growing fast. In this chapter a general overview is given of the most important aspects of the athlete's nutrition, the practical use of nutritional supplements, and facts and fiction relating to the use of nutraceuticals as ergogenic aids.

18.2 Muscle structure and function

To perform physical exercise our bone structure needs to be moved by muscle force. To generate such force chemical energy has to be transformed into mechanical energy within the muscle. The basic structure of a muscle

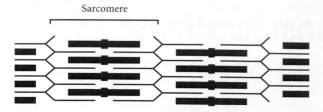

FIGURE 18.1 Schematic representation of the actin and myosin filaments in the sarcomere.
The sarcomere consists of thick myosin filaments between thin actin filaments. A muscle can shorten/lengthen when the myosin and actin filaments slide past each other. The latter is realized by the formation of cross-bridges between the actin and myosin filaments.

consists of an outer layer of connective tissue, which covers a number of small bundles each containing up to 150 individual muscle fibres (Weblink Section 18.1a). A muscle fibre represents the individual muscle cell and usually extends the entire length of a muscle. At both ends the muscle fibre fuses with a tendon, which is attached to the bone. The contents of a muscle fibre are enclosed within its plasma membrane (sarcolemma). Within the sarcolemma there are three main structures that play an important role in enabling muscle contraction. First of all, the muscle fibre contains numerous myofibrils, which represent the contractile elements of the muscle. These myofibrils are long strands of smaller subunits, the sarcomeres. Second, an extensive network of transverse tubules (T tubules) pass laterally among the myofibrils through the muscle fibre. This network allows nerve impulses to the sarcolemma to reach all myofibrils. In addition, another network of tubules, the sarcoplasmic reticulum (SR complex), runs in parallel with the myofibrils and serves as a storage site for calcium ions, which are essential to enable muscle contraction (see next section). The subunits of the myofibril, the sarcomeres, are the basic functional units of the muscle (Figure 18.1). The latter contain two types of small protein filaments, the actin and the myosin filaments, which are responsible for the shortening and lengthening of the muscle.

Muscle contraction

A single motor nerve innervates several muscle fibres, and together they are referred to as a single motor unit. Muscle contraction is preceded by a series of events (Weblink Section 18.1b). First, a motor nerve impulse is generated and conducted through the motor neuron towards its nerve endings. There, an electrical charge can be generated and conducted throughout the entire muscle fibre. The T tubules system allows the electrical charge to reach all myofibrils in the muscle fibre. This so-called action potential triggers the SR complex to release

its stored calcium ions. The calcium ions subsequently bind to the actin filaments that allow the myosin filaments to attach. When these actin–myosin cross-bridges are repeatedly formed and released, these filaments can slide past each other, causing the sarcomere to shorten/lengthen. As such, this sliding filament principle explains the shortening of the muscle during contraction. This process of muscle contraction requires energy, which is provided by adenosine triphosphate (ATP), the universal energy donor in the living cell. When the muscle is no longer stimulated, the flow of calcium ions is halted and the actin–myosin interaction inhibited. To enable muscle relaxation calcium ions need to be pumped back into the SR complex, which also requires ATP. Though this describes the process of a single contraction in an individual muscle fibre, it should be clear that an extremely complex, well-orchestrated series of muscle contractions within numerous muscle fibres from various muscle groups are needed to enable even the simplest of movements.

Muscle fibre types

Though all muscle fibres are generally of the same structure and function, certain differences between muscle fibres allow a classification into two main types (Weblink Section 18.1c). Skeletal muscle contains both type I and type II muscle fibres, also referred to as slow-twitch and fast-twitch fibres, respectively. This classification is based on the contractile and metabolic characteristics of the fibres. Type II fibre can be further classified into type IIa and IIb fibres. On average, most skeletal muscle contains about 50 per cent type I, 25 per cent type IIa, and 25 per cent type IIb fibres. However, there can be substantial differences in fibre type distribution between various muscle groups as well as between individuals. The main difference between the type I (slow-twitch) and type II (fast-twitch) muscle fibres is their contractile speed. The tension that can be generated is not much different between the individual type I or II muscle fibre. However, a type II motor unit can develop much greater strength than a type I motor unit. The latter is explained by the fact that a type II motor unit (recall that a motor unit is a motor nerve with the muscle fibres it innervates) contains a type II motor neuron which innervates between 300 and 800 different type II muscle fibres, whereas a type I motor neuron innervates only about 10 to 180 type I fibres. Besides the specific contractile differences, these muscle fibre types also show concomitant differences in their metabolic adaptation. The type I muscle fibres are relatively slow, more resistant to fatigue, and have a high aerobic capacity. The latter means that they are optimized to generate ATP by oxidative metabolic pathways (see Section 18.3). In contrast, the type II muscle fibres are

TABLE 18.1 Skeletal muscle fibre types

Muscle fibre	Type I	Type IIa	Type IIb
Contractile properties			
Contractile speed	Slow	Fast	Fast
Fatigue resistance	High	Moderate	Low
Motor unit strength	Low	High	High
Metabolic properties			
Oxidative capacity	High	Moderate	Low
Non-oxidative	Low	High	Highest capacity

Source: Adapted from J H Wilmore, D L Costill, W L Kenney (1994) *Physiology of Sport and Exercise*. Human Kinetics, Champaign, IL

fast, fatigue easily, and therefore need a metabolic system optimized to provide energy fast. This implies that they need to derive most of their ATP through non-oxidative (or anaerobic) metabolic pathways (see Section 18.3). An overview of the main differences between the type I, IIa and IIb fibres is provided in Table 18.1.

Clearly, type I muscle fibres are well suited for prolonged endurance exercise, such as marathon running, cycling, and so on. In contrast, the type II muscle fibres are predominantly activated during short-term high intensity activities and resistance exercise, such as sprinting, weight-lifting, and so on. Clearly, though the extent of type I and II motor unit recruitment depends on the physical activity, a combination of both is always apparent. Besides the sort (and intensity) of exercise, the duration of exercise can also result in changes in muscle fibre type recruitment. During prolonged endurance exercise the predominant use of type I motor units leads to the depletion of type I muscle fibre energy stores and subsequent fatigue. Consequently, in time more type II motor units need to be recruited to maintain exercise performance.

Fibre type composition has been reported to vary considerably between individuals, and is largely determined by genetic background. Elite athletes often have a muscle fibre type composition which more or less conforms to fibre type recruitment in their sport. For example, elite endurance runners and cross-country skiers often have a relatively high percentage of type I muscle fibres (up to 80 per cent) in their muscle tissue, whereas athletes such as world-class weight-lifters, body-builders, or sprinters tend to have relatively more type II muscle fibres. Clearly, genetic predisposition to develop a certain muscle fibre type composition (before birth and/or during the early years of childhood) can provide an advantage or disadvantage to excel in a certain sport. However, it should not be used as a predictive measure of athletic performance capacity, as muscle fibre type composition is merely one

of the numerous factors that attribute to performance capacity. There are some suggestions that prolonged exercise training can also modify muscle fibre type distribution. Whether this is true or not, all muscle fibres show a tremendous capacity to adapt to the demands imposed upon them by regular exercise training. As such, endurance training can substantially increase the aerobic capacity of any muscle fibre, whereas regular sprint and/or resistance training increases their anaerobic capacity.

18.3 Skeletal muscle substrate utilization

The immediate source of chemical energy required for skeletal muscle to contract is provided by the hydrolysis of ATP. Intracellular ATP stores are small (5.0–5.5 mmol/g wet muscle) and would be depleted within seconds of maximal contraction if not adequately replenished. In addition, close to maximal ATP levels are essential for normal (muscle) cell function. Hence, metabolic pathways for ATP resynthesis need to be activated directly in response to an increase in ATP demand. To maintain ATP levels and enable ongoing contractile activity, ATP synthesis rate needs to be tightly coupled to the rate of hydrolysis. This requires a highly efficient and responsive interaction between the various metabolic pathways responsible for ATP generation, especially during the transition from rest to exercise when the demand for ATP synthesis, at the muscular level, can increase more than 100-fold. Depending on the ATP demands, both substrate level phosphorylation and oxidative phosphorylation contribute to ATP synthesis. ATP synthesis from substrate level phosphorylation is required to sustain high intensity, dynamic exercise when the high ATP demands cannot be matched (entirely) by oxidative phosphorylation.

These ATP synthesis pathways include the phosphagen system and (anaerobic) glycolysis. The phosphagen system includes the breakdown of creatine phosphate, a high-energy compound stored in the muscle (25 mmol/g wet muscle), by creatine kinase, thereby providing energy for ATP synthesis. In addition, some ATP can be generated by the adenylate kinase reaction (2 ADP → ATP + AMP). For the cell to generate ATP by glycolysis, intracellular glucose or glycogen is converted to glucose-6-phosphate. Thereafter, in a series of enzymatic reactions, called glycolysis, glucose-6-phosphate is broken down to pyruvate, which under these conditions, is converted to lactate (see Chapter 7). Substrate level phosphorylation of ADP in these so-called anaerobic pathways allow high rates of ATP synthesis (5–6 times higher than the rate of ATP synthesis provided by oxidative phosphorylation) but can only be maintained for a relatively short period due to the depletion of creatine phosphate and the accumulation of by-products of the adenylate kinase reaction and anaerobic glycolysis, including adenosine diphosphate (ADP), adenosine monophosphate (AMP), inosine monophosphate (IMP), ammonia (NH_3), hydrogen ions (H^1), and inorganic phosphate (P). At rest and during exercise lasting more than ten minutes, the vast majority of ATP required for muscle contraction is generated through oxidative phosphorylation. The principal substrates that fuel this aerobic ATP synthesis are fat and carbohydrate. As the ability of skeletal muscle to synthesize ATP at rest and during continuous exercise depends on the availability of both fat and carbohydrate, these substrates ultimately need to be provided by dietary intake. Because ATP has to be generated continuously, a readily available pool of metabolic fuels is essential. Therefore, the human body contains a variety of storage sites for carbohydrate and fat.

Fuel storage

Carbohydrates are mainly stored as glycogen in skeletal muscle and in the liver. Carbohydrate stores are relatively small and normally range between 0.46 and 0.52 kg, corresponding to a total energy storage of 7.5–8.4 MJ (1,785–2,000 kcal). Fat is mainly stored as triacylglycerol (TAG) in subcutaneous and deep visceral adipose tissue. However, smaller quantities of TAG are present as lipid droplets inside muscle fibres (intramyocellular triacylglycerol or IMTG) (Hoppeler et al. 1985). In addition, some fat is present in the circulation as non-esterified or free fatty acids (FFA) bound to albumin, or as TAG incorporated in circulating lipoprotein particles (chylomicrons and very-low, low-, intermediate- and high-density lipoproteins or VLDL, LDL, IDL, and HDL respectively). In contrast to carbohydrate, fat stores are quite large and range between 9 and 15 kg in the average, non-obese male (body mass ~70 kg), corresponding to a total energy storage of 350–586 MJ (80,000–140,000 kcal). An overview of the energy stores is provided in Table 18.2. More than 97 per cent of our entire energy storage is covered by endogenous fat sources. This apparent preference for the storage of fat as a fuel source is quite practical as fat contains more than twice the amount of energy per unit of weight compared to carbohydrate. However, an important advantage of carbohydrate as a fuel source is that more ATP can be generated per unit of time compared to the oxidation of fat.

The rate at which ATP can be generated depends on the biochemical pathways that are followed as well as on the substrate (source) from which ATP is derived (Weblink Section 18.2). As mentioned, anaerobic ATP synthesis allows considerably higher rates of high-energy

TABLE 18.2 Fuel stores in an average person

Fuel source	In weight (g)	In energy (kJ)
Fat		
Plasma FFA	0.4	16
Plasma TG	4.0	156
IMTG	300	11 700
Adipose tissue	12 000	468 000
Carbohydrate		
Plasma glucose	5	90
Liver glycogen	100	1 800
Muscle glycogen	350	6 300

Based on estimates for a normal, non-obese person with a body mass of 70 kg. Fat provides 39 kJ/g and carbohydrate 18 kJ/g.

phosphate formation compared with the complete, oxygen-dependent, oxidation of carbohydrate and fat. In short, the maximal rate of ATP synthesis from fat can provide only enough energy to sustain exercise at 55–70 per cent of maximal oxygen uptake (VO_2max), depending on the training status. However, the rate of energy generated from carbohydrate (aerobic and anaerobic) can provide enough energy to sustain exercise up to 100 per cent of VO_2max. Consequently, during moderate intensity endurance type exercise, such as a marathon, ATP demands can be provided entirely by oxidative phosphorylation of both intra- and extracellular carbohydrate and fat. The relative contribution of fat and carbohydrate oxidation to total energy expenditure during exercise can vary enormously and strongly depends on exercise intensity, training status, and diet.

Exercise intensity and duration

Fat and carbohydrate are the main substrates that fuel aerobic ATP synthesis during prolonged exercise. Though both substrates will always contribute to total energy provision, their relative utilization has been shown to vary with the intensity and duration of exercise. Most conclusions about substrate utilization rates in relation to exercise intensity on a whole-body level have been derived from stable isotope studies (van Loon et al. 2001). Generally, the oxidation of blood plasma-derived FFA provides the majority of energy during low intensity exercise (~30 per cent of maximal oxygen uptake, or VO_2max), with little (or no) net utilization of intramuscular or lipoprotein derived TG and muscle glycogen and a relatively small amount of plasma glucose. During moderate intensity exercise (40–65 per cent VO_2max) fat oxidation, from an absolute point of view, reaches

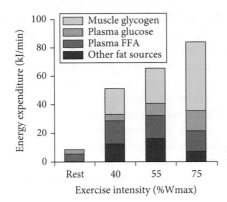

FIGURE 18.2 Substrate utilization during exercise of various intensities. Energy expenditure (expressed as kJ per minute) as a function of exercise intensity (expressed as percentage of maximal workload).
The relative contributions of plasma glucose, muscle glycogen, plasma FFA, and other fat sources (sum of intramuscular plus lipoprotein-derived TG) to energy expenditure are illustrated as described in the key.
Source: Adapted from L J van Loon, P L Greenhaff, D Constantin-Teodosiu et al. (2001) The effects of increasing exercise intensity on muscle fuel utilisation in humans. *Journal of Physiology* **536**, 295–304.

maximal rates, with the fat stores contributing about 50 per cent (40–60 per cent) of total energy expenditure (Figure 18.2). As plasma FFAs provide only about half of the energy derived from fat oxidation, intramuscular TAGs (IMTGs) are likely to form an important substrate source. However, the latter estimate also includes the use of lipoprotein-derived TAG, as the applied stable isotope techniques do not enable the distinction between TAG derived from the IMTG stores and TAG from circulating lipoproteins.

The majority of the carbohydrates oxidized during moderate intensity exercise are derived from muscle glycogen stores, with an estimated use of plasma glucose ranging between 15 per cent and 35 per cent of total endogenous carbohydrate utilization. As exercise intensity is further increased (up to 70–90 per cent VO_2max) fat oxidation decreases substantially, from both a relative as well as a quantitative point of view, accounted for by a decrease in the use of plasma FFA as well as intramuscular and/or lipoprotein-derived TAG sources. At the same time, endogenous carbohydrate oxidation rates increase exponentially (Figure 18.2), which is necessary to account for the high energy demands. Consequently, muscle glycogen becomes the most important substrate source during moderate to high intensity exercise.

Besides exercise intensity, exercise duration also plays an important role in substrate source utilization. As moderate intensity exercise is prolonged and muscle glycogen stores gradually decline, plasma FFA levels continue to increase with a concomitant increase in plasma FFA oxidation rates. Concomitantly, the use of endogenous carbohydrate stores (muscle and liver glycogen) and the use of other fat sources (IMTG and lipoprotein-derived TAG) decrease over time, compensated for by the increased plasma FFA utilization. As the endogenous carbohydrate storage capacity is quite limited, muscle and/or glycogen stores will become depleted during prolonged exhaustive endurance exercise. In this situation exercise intensity cannot be maintained since the ATP production will slow down due to the reliance on fat as a substrate. This phenomenon is often referred to as 'hitting the wall' and can occur as soon as 45 minutes after onset of intense exercise. The symptoms of fatigue during prolonged endurance exercise are strongly related to glycogen depletion in the exercising muscle, as shown in the early muscle biopsy studies by Bergström and Hultman (1966). Based on these physiological observations, it has become clear that the level of pre-exercise muscle glycogen concentration as well as carbohydrate ingestion during exercise are important to delay the onset of fatigue (see Section 18.4).

Training status

The relative contribution of fat and carbohydrate oxidation to total energy expenditure during exercise is also determined by training status. In their classic studies, Christensen and Hansen (1939) already observed that endurance training leads to an increased capacity to utilize fat as a substrate source and reduces the reliance on the limited endogenous carbohydrate stores during exercise. This training-induced change in substrate utilization has since been confirmed in numerous studies (Figure 18.3).

The contribution of fat as a fuel source during exercise has been shown to be increased in an endurance trained state, compared at the same absolute as well as the same relative (and therefore higher absolute) workload. The main metabolic adaptations to endurance training that are believed to be responsible for the increase in fat oxidative capacity include an increase in both the size and number of mitochondria and a concomitant upregulation of enzymes involved in the activation, mitochondrial transport, and oxidation of FA as well as enzymes involved in the TCA cycle and respiratory chain. In addition, endurance training has been shown to result in an increased capillary density of muscle tissue and increased FABP (FA-binding protein) and GLUT-4 (glucose transporter protein) content, suggesting an increased capacity to take up FFA and glucose from the circulation. Finally, endurance training has been shown to increase intramuscular triglyceride (IMTG) content in close proximity to mitochondria, suggesting increased availability for oxidation.

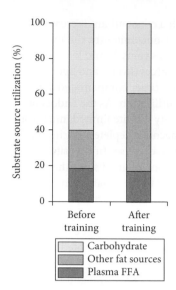

FIGURE 18.3 Substrate utilization and training status. Substrate utilization in the muscle as percentage of total energy expenditure at a given absolute intensity (64 per cent of pre-training VO_2max). The relative contributions of total endogenous carbohydrate, plasma FFA, and other fat sources (sum of intramuscular plus lipoprotein-derived TG) to energy expenditure are illustrated.
Source: Adapted from W H Martin 3rd, G P Dalsky, B F Hurley et al. (1993) Effect of endurance training on plasma free fatty acid turnover and oxidation during exercise. *American Journal of Physiology* **265**, E708–714.

Though the relative contribution of fat oxidation to energy expenditure increases substantially following endurance training, it is not quite clear which fat sources are utilized to a larger extent. Most studies do not report large differences in plasma FFA uptake in an endurance trained state. It has been suggested that the increase in fat oxidation observed in an endurance trained state is entirely accounted for by an increased contribution of IMTG utilization (Figure 18.3). In accordance with the increased reliance on IMTG as a substrate source in endurance athletes, studies have reported an increase in IMTG storage following endurance training. The regulation of IMTG metabolism also receives much interest in more clinical research as excess IMTG deposition in obese and/or type 2 diabetes patients has been linked to the development of defects in the insulin signalling cascade, causing skeletal muscle insulin resistance (Shulman 2000). Obviously, there are important differences between the increased IMTG accumulation in the obese and/or type 2 diabetes patient compared to the endurance athlete, as athletes generally show an improved insulin sensitivity compared to untrained individuals. The greater IMTG stores in the trained athlete should be considered as a preferred adaptive response to regular exercise training, and is by no means a representation of the consequences of excess fat intake and/or a sedentary lifestyle.

The observed decrease in the reliance on endogenous carbohydrate utilization in an endurance trained state has been shown to be accounted for by a decrease in both plasma glucose and muscle glycogen use during exercise (van Loon et al. 1999). The effects of endurance training on carbohydrate metabolism are not limited to changes in the reliance on the use of endogenous carbohydrate during exercise but also affect carbohydrate storage. Trained individuals have an increased capacity to store glycogen, which is attributed to an increase in insulin sensitivity, GLUT-4 content, and/or glycogen synthase activity. Clearly, endurance training increases performance capacity by reducing the reliance on the limited endogenous carbohydrate stores as a substrate, as well as by increasing the storage capacity of muscle glycogen.

18.4 Nutritional interventions to optimize performance

The first and clearest difference in nutritional needs between athletes and non-athletes is energy. The energy expenditure of a sedentary adult female/male amounts to approximately 8.5–12.0 MJ per day. Physical activity by means of training or competition will increase the daily energy expenditure by 2 to 4 MJ per hour of exercise, depending on physical fitness, duration, type and intensity of sport. For this reason, athletes must increase food consumption to meet their energy needs, according to the level of daily energy expenditure. This increased food intake should be well balanced with respect to an adequate macronutrient and micronutrient ingestion. The latter, however, is not always easy. Many athletic events are characterized by high exercise intensities. As a result, energy expenditure over a short period of time may be extremely high. For example, to run a marathon will use about 10–12 MJ. Depending on the time to finish, this may induce an energy expenditure of approximately 3.2 MJ/hour in a recreational athlete and 6.3 MJ/hour in an elite athlete who finishes in approximately 2 to 2.5 hours. A professional cycling race such as the Tour de France will cost an athlete about 27 MJ/day, a figure that will increase to approximately 40 MJ/day when cycling over mountain passes (Saris et al. 1989). Compensating for such high-energy expenditure by ingesting normal solid meals will pose a problem to any athlete involved in such competition, since the digestion and absorption processes will be impaired during intensive physical activity. These problems are not only restricted to competition days. During intensive training days, energy expenditure is also high. In such circumstances, athletes tend to ingest a large number of 'in between meals' providing up to 40 per cent of the total energy intake. These 'in between meals' are often energy-rich snacks, which are often low in protein and/or micronutrient content, such that the diet becomes unbalanced. Especially adapted nutritious foods/fluids which are easily digestible and rapidly absorbable may solve this problem. In addition to the energy needs and the limited capacity and time to digest and metabolize, there is the importance of muscle fuel selection. Metabolic capacity and power output depend on this selection. During prolonged moderate-to-high intensity exercise, the muscle cell depends mainly on carbohydrate as a substrate (see Section 18.3). Therefore, diet selection is not only a matter of energy but also a selection of the right substrate source.

In contrast with the efforts to ingest ample energy to account for the high-energy expenditure, some athletes (e.g., female runners, gymnasts, and ballet dancers) are known to reduce their energy intake, whereas their energy expenditure is high. Energy expenditure in the average sedentary subject ranges between 1.4 and 1.6 times basal metabolic rate (BMR), while reported energy intake in elite (female) gymnasts is usually below this level, despite the fact that they work out between three and four hours a day. This can probably be explained by two factors: underreporting and the urge to limit energy intake, the first being a result of the latter. Whether the intake data are reliable or not, some of these athletes limit their energy intake to reduce body mass and fat mass with major consequences for energy turnover and performance. Some of these athletes are also prone to developing eating disorders that will not only impair performance but can also strongly impair overall health.

Carbohydrate

In Section 18.3, under 'Exercise intensity and duration', it was shown that the muscle glycogen stores, from a quantitative point of view, form the most important substrate source during prolonged moderate-to-high intensity exercise. Combined with the fact that endogenous substrate stores are relatively small, it is not surprising that endurance performance capacity is often limited by the availability of endogenous carbohydrate. Therefore, athletes involved in moderate to high intensity exercise lasting more than 45–60 minutes should ensure that their muscle glycogen stores are optimized before the start of an important event. Much research has been performed to develop nutritional regimes to maximize pre-competition muscle glycogen stores. Bergström & Hultman (1966) introduced a dietary intervention which increased muscle glycogen concentration more than two-fold. This regime included an exhaustive cycling test followed by three days of a low carbohydrate diet, after which a second exhaustive exercise test was performed followed by three days of a high carbohydrate diet. Though muscle glycogen stores

increased more than two-fold following this classic glycogen loading regime, major disadvantages were reported when applied in practice. Side effects such as weight gain, gastrointestinal complaints, fatigue, irritability, and nervousness were reported. To reduce these side effects a modified glycogen supercompensation regime was introduced. This six-day glycogen loading regime included the tapering of training (reducing training volume, while maintaining training intensity) with an increase in the relative contribution of carbohydrate in the diet (from ~50 per cent up to 70 per cent of total energy intake). This modified regime results in slightly lower but comparable increases in muscle glycogen content without the apparent disadvantages of the classical regime. Carbohydrate loading and the tapering of training before competition as a means to maximize pre-competition muscle glycogen storage is nowadays widely practised by endurance athletes throughout the world.

Besides optimizing pre-competition muscle glycogen content, carbohydrates can be ingested during exercise to provide an additional exogenous carbohydrate source (Wagenmakers et al. 1993). Numerous studies have demonstrated that carbohydrate supplementation during prolonged endurance exercise increases exercise performance. The latter has been attributed to the maintenance of normal blood glucose concentrations and high carbohydrate oxidation rates throughout the latter stages of prolonged exercise. As an increase in plasma glucose availability leads to an increase in plasma glucose uptake and oxidation, it has often been suggested that carbohydrate supplementation also reduces the reliance on muscle glycogen. This glycogen sparing likely occurs during the early stages of an exercise bout and may be muscle fibre type specific.

Carbohydrate supplementation has been shown to be beneficial to performance in both endurance as well as non-endurance sports (such as soccer, tennis, and weight-lifting). Carbohydrates are generally provided during exercise in combination with water. A wide variety of studies have been performed to maximize exogenous glucose absorption and/or oxidation rates by varying the type, amount, and feeding schedule of carbohydrate-containing solutions (Cermak et al. 2013). Oxidation rates of glucose or glucose polymers ingested during exercise typically do not exceed 1.0–1.1 g per minute, because intestinal glucose uptake is limited. Therefore, for prolonged exercise lasting two to three hours, athletes are generally advised to ingest glucose (polymers) at a rate of up to 60–70 g per hour. Well-trained, elite endurance athletes competing longer than 2.5 hours may metabolize carbohydrates up to 90 g per hour, provided they consume a mixture of glucose (polymers) and fructose. As fructose is absorbed in the intestine via a different transporter from glucose, the combined ingestion of these so-called multiple transportable carbohydrates allows exogenous carbohydrate oxidation rates during exercise to reach ~1.5 g per min (Cermak et al. 2013).

In general, the ideal sports drink should contain such a carbohydrate content and composition that a total intake rate of 60–90 g carbohydrate per hour is practically achievable. On average the carbohydrate concentration in a sports drink should be around 6–10 per cent (certainly not exceeding 15 per cent) and osmolality should not be too high (~350 mosmol/kg), as this would result in impaired gastric emptying followed by gastrointestinal complaints. It is advisable to ingest 6–8 ml per kg body mass of such a sports drink during the warm-up period immediately before the start of training or competition to fill the stomach, which stimulates gastric emptying. Thereafter, during exercise, smaller amounts should be ingested (2–3 ml/kg) at 15–20 min intervals. The oxidation of exogenous carbohydrate can provide a substantial contribution to total energy expenditure (~20 per cent; see Figure 18.3). There are no differences in the oxidation rate of ingested carbohydrates (provided by a sports drink) during moderate intensity exercise between untrained men and trained endurance athletes, implying that both the recreational athlete as well as the elite athlete can profit from the use of carbohydrate-rich sports drinks during prolonged endurance exercise (van Loon et al. 1999).

The ergogenic benefits of carbohydrate ingestion during exercise may not be limited to prolonged endurance-type exercise. Improvements in performance have been reported during relatively short (<60 min), high-intensity (>75 per cent VO_2peak) exercise. This is surprising, as carbohydrate availability should not be a limiting factor during such conditions. Therefore, the ergogenic effect of carbohydrate ingestion has been proposed to also involve a signalling response from carbohydrate receptors in the oral cavity that supresses fatigue. Such a concept is supported by various studies that have now demonstrated that exercise performance can be enhanced by carbohydrate mouth rinsing (swirling a carbohydrate solution in the mouth for 5–10 seconds and spitting it out) (Carter et al. 2004). However, there are also ample studies that have failed to confirm an ergogenic effect of carbohydrate mouth rinsing. Therefore, it remains unclear if carbohydrate mouth rinsing is truly ergogenic and, if so, under what specific conditions.

Following prolonged endurance exercise, muscle glycogen stores are substantially reduced, therefore post-exercise restoration of muscle glycogen stores is considered an important factor in determining the time needed to recover. Under normal conditions, muscle glycogen is restored at a rate of only 3–7 per cent per hour, provided that an adequate amount of carbohydrate is ingested, and so approximately 24 hours are required for the muscle glycogen stores to be fully restored. However, in multi-day sports events (such as the Tour de France) available recovery time is far less than 24 hours. Muscle glycogen resynthesis rates are highest during the first few hours after exercise. Therefore, when trying to optimize performance capacity well within 24 hours after an exhaustive bout of exercise one should start to consume ample carbohydrate immediately after cessation of the first exercise bout. As solid food is often not too well tolerated immediately post-exercise, carbohydrate-rich (recovery) sports drinks are an effective alternative.

The combined ingestion of both glucose (polymers) with fructose or sucrose causes less gastrointestinal complaints and fructose ingestion has been shown to accelerate liver glycogen repletion. Therefore, both glucose (polymers) and fructose may be included in recovery drinks aiming to maximize endogenous glycogen repletion during the early stages of post-exercise recovery. Thereafter, moderate-to-high glycaemic index foods (bread, pasta, potatoes, etc.) should be consumed rather than low glycaemic index foods. Though carbohydrate intake rates of 0.8 g/kg/hour have been advised to optimize muscle glycogen synthesis rates, more recent studies have shown that about 1.2–1.5 g/kg/hour should be ingested, provided at short (30 min) intervals. The addition of an insulinotropic free amino acid/protein mixture to the carbohydrate-containing beverages may further accelerate muscle glycogen resynthesis rates when suboptimal amounts of carbohydrate are ingested.

Fat

The importance of optimizing muscle and liver glycogen storage and the use of carbohydrate supplementation during exercise to reduce fatigue and improve athletic performance has led to a quite detailed understanding of carbohydrate metabolism during exercise. Far less information is available on the role of fat as an endogenous and/or exogenous substrate during exercise. As discussed above in 'Training status', endurance training increases fat oxidative capacity, which reduces the reliance on the limited endogenous carbohydrate stores and subsequently improves endurance performance capacity. There is therefore considerable interest in potential (nutritional) interventions to increase fat oxidative capacity.

The effects of fat supplements and high fat diets on performance capacity have been the subject of concentrated research, particularly with respect to the use of medium-chain triacylglycerols (MCT). These C8 to C10 carbon triglycerides are more soluble in water and less likely to inhibit gastric emptying than long chain triglycerides (LCTGs). Furthermore, medium-chain fatty acid oxidation is not as limited by transport into the mitochondria as long-chain fatty acids are. Exercise studies with MCT supplementation have revealed that MCT is oxidized almost as rapidly as glucose and even better in combination with glucose. However, since MCT intake is limited due to gastrointestinal distress when ingesting larger amounts, their contribution to total energy expenditure is restricted to about 7 per cent. Also, most performance studies have not yet shown an additional effect to carbohydrate feeding alone.

As well as the use of such fat supplements during exercise, research has also focused on the effect of using long-term high fat diets on performance. The use of such diets provokes adaptive responses, leading to an increase in the capacity to oxidize fat as a fuel and to the sparing of endogenous carbohydrate stores during exercise. Though a few studies have reported an increase in time to exhaustion following the use of a high fat diet followed by a short-term carbohydrate-rich diet, most studies investigating the effects of high fat diets on exercise performance do not report an increase in exercise performance. In addition, from a health perspective, long-term high fat diets are not generally recommended and have been reported to result in a reduced capacity to maintain the desired exercise training intensity.

The recognition of the importance of carbohydrate has resulted in the recommendation of high carbohydrate diets in endurance athletes, and has led to a reduction in fat intake. In some cases this has been exaggerated, with fat intake being reduced to less than 20–25 per cent of total energy intake. There are data to suggest that this should not be recommended for athletes during recovery. Studies in trained athletes have shown that prolonged endurance exercise leads to a substantial net depletion of IMTG stores. Following post-exercise recovery the use of extremely low fat diets prevents the restoration of IMTG stores for up to 48 hours or more. In contrast, low and moderate fat-containing diets resulted in IMTG stores that did not return to pre-exercise levels until 24 hours of recovery. Clearly, more research needs to be performed to determine the importance of repleting IMTG stores for subsequent substrate use during exercise and their role in athletic performance.

Protein

For many years the effects of exercise on dietary protein requirements has been a controversial topic. Dietary reference values are intended to cover the needs of healthy individuals but not necessarily those who are physically active. Therefore reference values do not address the possible additional protein needs of physically active individuals. However, a considerable amount of experimental evidence indicates that regular exercise increases dietary protein needs (Hartman 2006). The latter is explained by the need for protein to repair exercise-induced muscle damage, as an additional substrate source during prolonged exercise, and to support training-induced muscle reconditioning with a concomitant increase in lean tissue mass.

The protein requirement of the individual athlete is likely to depend on the type, intensity, and duration of the exercise training regime. The majority of research has been focused on the impact of protein supplementation on adaptations to resistance-type exercise training. Current guidelines for athletes recommend a dietary protein intake between 1.6–2.2 g/kg body mass (Morton et al. 2018). Data from a nationwide survey in athletes indicated an average protein intake of ~1.5 g/kg body mass per day, which is likely to be a slight underestimation due to underreporting. In the extreme case of cyclists in the Tour de France, these elite athletes report an absolute protein intake of more than 3 g protein per kg per day, despite the fact that food selection was mainly focused on the consumption of carbohydrate-rich products. In short, where energy expenditure is matched by energy intake the suggested protein requirements are generally (more than) met through normal dietary intake in both strength as well as endurance athletes. However, restricted food intake, as reported in athletes trying to reduce body weight (especially female endurance runners, gymnasts, and weight-class athletes), is likely to result in inadequate protein intake.

A single bout of exercise increases both muscle protein synthesis as well as muscle protein breakdown rates, albeit the latter to a lesser extent, allowing protein balance to become more positive. Nonetheless, net muscle protein balance will remain negative in the absence of protein intake. Ingestion of protein and/or essential amino acids during post-exercise recovery inhibits protein breakdown and further stimulates muscle protein synthesis, leading to net muscle protein accretion (Tipton et al. 1999). Consequently, post-exercise protein ingestion has been suggested as an effective strategy to augment the skeletal muscle adaptive response to each successive exercise bout and so to facilitate skeletal muscle reconditioning. This has resulted in the widespread use of protein and/or amino acid supplements by strength as well as endurance athletes. The amount and source of dietary protein that is ingested and the timing of dietary protein consumption can be modulated to maximize post-exercise muscle protein synthesis rates and so optimize muscle reconditioning. Ingestion of whey protein seems most effective to rapidly increase muscle protein synthesis rates during the acute stages of post-exercise recovery. This can likely be attributed to its rapid digestion and absorption kinetics as well as its high leucine content. About 20 g of a high quality dietary protein should be provided during and/or immediately after cessation of exercise to maximize post-exercise muscle protein synthesis rates. Co-ingestion of (large) amounts of carbohydrate or additional crystalline leucine does not further augment post-exercise muscle protein synthesis rates when ample dietary protein is already provided. Protein ingestion prior to sleep has been shown to augment gains in muscle mass and strength during a prolonged resistance type exercise training programme (Snijders et al. 2015), and represents a commonly overlooked opportunity to consume a high-protein snack. A healthy diet with smart timing of the ingestion of dietary protein after exercise will augment the skeletal muscle adaptive response to exercise and likely improve exercise training efficiency (Cermak et al. 2012).

Fluids and electrolytes

Most of the energy required for intensive endurance exercise is produced by oxidative metabolism. Each litre of oxygen consumed will contribute to an energy production of approximately 20 kJ. However, 75 per cent of this energy is released as heat and only 25 per cent is used for mechanical work. In order to avoid hyperthermia, the produced heat must be transferred from the working muscles to the periphery, primarily by circulating blood. This heat is eliminated at the surface of the body by radiation, convection, and evaporation. Radiation and convection are the means by which dry heat is transferred to the immediate surroundings. Heat loss by convection can be increased substantially by wind or water. Evaporation of sweat is the most important way to eliminate heat when working in a warm environment because convection and radiation will be minimal under these circumstances. With intense physical activity sweat rates will then be maximized. At maximal sweat rates, a 70 kg male athlete may lose >30 ml/min or >1800 ml sweat/hour. Body size, training status, exercise intensity, and weather conditions all influence the sweating rate. One of the highest sweat rates ever reported amounted to 3.7 l/hour in a 67 kg marathon runner. Without appropriate rehydration, blood flow through the extremities and the skin will decrease. In addition, sweat response and heat flux will be diminished. This may cause hyperthermia with associated severe health risks when exercising in the heat. In any situation, dehydration substantially impairs exercise performance, with a ~10–15 per cent decrease in performance capacity with each degree increase in body temperature.

With sweat loss, electrolytes are also excreted. Because sweat contains fewer minerals than plasma, and water intake is inadequate to fully compensate the losses, the electrolyte concentration in blood normally increases as a result of intense endurance exercise. When water intake is equal to water loss by sweating, theoretically the plasma electrolyte levels, especially the major electrolytes sodium and chloride, should fall. However, due to the high plasma sodium content and the large extracellular space, this is not apparent until late in exercise, after significant amounts of water have been ingested. For example, if a 70 kg athlete (with 14 litres extracellular fluid) loses 6 litres of sweat (Na^+ concentration 30 mEq/l or 600 mg/l), then 180 mEq Na^+ would be lost. If the water loss occurs equally from extracellular fluid (containing 20–30 mEq Na^+) and intracellular fluid, then the extracellular Na^+ concentration would still be above 130 mEq/l. This may explain why drinking large amounts of plain water when competing in the heat has been shown to result in significant hyponatraemia in only about 10–20 per cent of ultra-endurance runners and triathletes. Therefore, in most exercise events lasting less than three or four hours, the major concern is that fluid and glucose be available to prevent exhaustion and heat stroke. In the case of maximal performance, the inadequate availability of carbohydrate and fluid may limit performance. In this respect both carbohydrate and sodium have beneficial characteristics stimulating water absorption. On the other hand, from a number of observational as well as experimental studies, it has become clear that there is a higher frequency of gastrointestinal distress symptoms when athletes are dehydrated, especially when fluid loss exceeds 4 per cent of body weight.

Restoration of fluid balance after exercise is an important part of recovery. Rehydration after exercise can be achieved effectively only if both electrolyte and water losses are replaced. The sodium content of most sports drinks lies within a range of 200 to 600 mg/l. These values are already at the lower level for an effective rehydration solution. An ideal post-exercise rehydration drink should contain around 1100 mg/l sodium to optimize water retention. In comparison, regular soft drinks contain virtually no sodium and are therefore less suitable as a rapid rehydration solution.

Micronutrients

Vitamins have attracted much attention in the world of sport because of their supposed capacity to enhance performance. Vitamin supplements are widely used by both professional and recreational athletes, often with extreme levels of intake exceeding 10 to 100 times the RDA/RNI. It is fair to assume that vitamin requirements are increased in the athlete involved in rigorous exercise training, due to the increased energy expenditure, sweat loss, core temperature, and so on. However, as the increased energy expenditure in the athlete is invariably compensated for by an increase in energy intake, assuming a healthy diet, increased vitamin intake fully or over-compensates for any increase in vitamin requirements. As such, there are no indications that long-term vitamin intake among athletes is in any way insufficient. However, athletes involved in regular intense exercise training who consume either an extremely low (<4.2 MJ/day) or high (>21 MJ/day) energetic diet for a prolonged period may be prone to vitamin and mineral deficiencies. The risk of developing any deficiencies on a very high energetic diet is explained by the fact that athletes requiring such an extreme high energy intake tend to consume energy-dense food and beverages (often during prolonged endurance exercise trials) which often tend to be of a lesser dietary quality. These athletes (such as cyclists during the Tour de France) as well as those consuming extremely low energy diets (such as female endurance runners, gymnasts, and ballet dancers) are prone to a marginal vitamin intake and can be at risk of developing vitamin deficiencies. Therefore, vitamin supplementation at moderate levels can contribute to achieving an adequate daily intake in these athletes. A vitamin deficiency may result in decreased performance and/or increased susceptibility to illness. However, vitamin supplementation at pharmacological levels has not been shown to improve performance.

In addition, vitamin E, C, and ß-carotene are important antioxidants. Exercise is associated with an increase in free radical production in skeletal muscle, with potential for tissue damage. It is suggested that especially in a state of overtraining the balance between oxidative stress and repair is disturbed, leading to increased lipid peroxidation and dysfunctional cell membranes. However, it has also been shown that athletes have increased levels of cellular antioxidant enzymes, such as superoxide dismutase (SOD), catalase, and glutathione peroxidase compared with untrained individuals. Most probably this is the result of a physiological adaptation to the higher oxidative stress and therefore it is questionable whether supplementation with antioxidants above the recommended intakes are of any benefit. So far, in this respect the available literature does not allow any claims. In fact, antioxidant supplementation may even attenuate the adaptive response to prolonged exercise training, as the oxidative stress induced by each training session may be one of the key signals required to allow muscle reconditioning.

With respect to the minerals, iron has probably attracted most attention regarding the advice on supplementation in athletes. Although other minerals, such as calcium, chromium, magnesium, zinc, copper, and selenium, are just as important, the status of these minerals is generally adequate in most athletes. Iron loss is increased in athletes, partly due to some iron loss in sweat, increased gastrointestinal and/or urinary blood loss, and foot strike haemolysis, as a consequence of the intensity and duration of the exercise. However, increased iron loss is usually well compensated for by the increased dietary intake associated with the higher energy expenditure. Though many athletes, in particular females, tend to be iron depleted, true iron deficiencies are rarely seen. Iron deficiency may be suspected in athletes who have a slightly lower than average haemoglobin (Hb), the so called athletes' anaemia (Hb <14 per cent in male and <12 per cent in female endurance athletes). This is usually a training-induced physiological condition, caused by an increase in plasma volume. Despite a slightly reduced haemoglobin, the total red cell mass is normal or greater than usual. Although athletes' anaemia may be associated with low serum ferritin, this physiological condition does not respond to iron supplementation. Therefore, iron supplement use is generally considered unnecessary and supplementation without medical indication is not advised. The latter is because excessive iron supplementation can cause gastrointestinal disturbances and/or even haemosiderosis. Iron supplements have not been shown to enhance performance capacity, except where iron deficiency anaemia pre-existed.

18.5 Ergogenic aids

Following the recognition of glycogen loading and carbohydrate supplementation as means to improve exercise performance, much attention has been directed towards the use of nutritional supplements to optimize sports performance. In response, various nutrition companies market specific sports nutrition products and sports nutrition has become a thriving branch of the food industry, and a wide variety of sports supplements is now available. These supplements range from the well-known sports drinks, high-energy bars and protein shakes to a multitude of supplements containing vitamins, minerals, free amino acids, for example. These supplements are, in some cases, useful to compensate (temporarily) for a less than adequate diet, to meet the special nutrient demands induced by intense exercise training, and/or to optimize exercise performance.

There is also interest in novel ergogenic supplements, often referred to as functional foods or nutraceuticals. From the term nutraceuticals it is already clear that these supplements should be categorized somewhere between nutritional supplements and pharmaceuticals. In most cases these nutraceuticals contain compounds that do occur in a normal diet but are present in amounts far above recommended levels, or the amounts typically provided by normal foods. Most of these supplements are proposed to increase performance capacity, often through a more pharmacological rather than a physiological effect. However, most of these claims rely on theoretical or anecdotal support rather than on documented results from scientific trials. It must be noted that such products are not registered in the same way as pharmaceuticals and so are not under appropriate legislation or obligatory safety requirements. At present these products can be advertised and marketed without the need for supporting evidence from clinical trials. Although the majority of the advertised nutraceuticals have not been clinically tested for their ergogenic potential nor for their safety, some nutraceuticals (e.g., caffeine, creatine, nitrate, and sodium bicarbonate) have received much attention from scientists because of their potential to affect muscle metabolism. Here, we will discuss the efficacy of a few currently popular ergogenic aids.

Caffeine

Caffeine is a well-known ergogenic aid, and was until January 2004 included on the banned substances list used by the IOC (International Olympic Committee), with an acceptance limit of 12 mg/ml in the urine. Several studies have provided evidence for the ergogenic properties of caffeine ingestion during prolonged endurance exercise tasks. These ergogenic effects have been observed even when caffeine was ingested in doses leading to urine concentrations well below IOC limits. An effective ingestion dose lies between 3 and 6 mg/kg body weight, comparable to the amount of caffeine in 2–6 cups of coffee. The most commonly used timing of caffeine supplementation is 60 minutes prior to the onset of exercise or during the latter stages of prolonged endurance type exercise. The performance-enhancing effect of caffeine is often attributed to its proposed stimulating effect on adipose tissue lipolysis and subsequent increase in fat oxidation rate. However, no direct *in vivo* evidence for this mechanism has yet been provided. A more plausible explanation would be the stimulating effect of caffeine on the central nervous system—the release and/or activity of adrenaline. Through this mechanism caffeine stimulates motivational aspects of behaviour, which probably explains the improved performance observed mainly during long-term endurance exercise tasks. Caffeine supplementation could be detrimental to performance capacity when an athlete is particularly sensitive to its diuretic effect. Other negative side effects associated with caffeine use include gastrointestinal distress, decreased motor control, shivering, headache, dizziness, and minor elevations in blood pressure and resting heart rate.

Creatine

As the amount of ATP in the muscle is limited, ATP stores can only provide enough energy to perform maximum exercise for several seconds. Some of the energy necessary to rephosphorylate ADP to ATP can be derived rapidly and without oxygen by the transfer of chemical energy from creatine phosphate. Creatine phosphate levels in the cell are more than three to five times greater than ATP, and creatine phosphate serves an important function as a high-energy phosphate buffer (see Section 18.3). In healthy individuals, the total endogenous creatine pool is approximately 120 g, of which 95 per cent is located in skeletal muscle. Each day 2 g is replenished by endogenous synthesis and dietary intake (meat and fish). Several studies have shown that oral creatine

(monohydrate) supplementation increases total creatine and creatine phosphate concentrations in human skeletal muscle. Ingestion of 20 g creatine per day for a period of 2–6 days can lead to a ~20 per cent increase in muscle creatine phosphate concentration. As an elevation in muscle creatine phosphate concentration increases the availability of energy for ATP rephosphorylation, it is not surprising that creatine supplementation has been shown to enhance performance in (repeated) high intensity, short-term exercise tasks, during which energy transfer is primarily derived from the ATP-creatine phosphate system. Consequently, creatine supplementation has become a common practice in both professional and amateur athletes (Terjung et al. 2000). Most research has focused on the ergogenic effects of (short-term) creatine supplementation on (repeated) high intensity exercise performance (such as sprinting, jumping, weight-lifting, volleyball, tennis, soccer, etc.). Several important possibilities concerning the long-term effects of creatine supplementation remain to be elucidated. Prolonged creatine use has been suggested to stimulate protein synthesis, especially when combined with exercise strength training. Others have speculated on the ergogenic possibilities of prolonged creatine use in endurance athletes as a means to increase oxidative capacity and increase fat oxidation. However, *in vivo* research has not shown any evidence for such an effect. In sports practice where body weight is an important factor in determining performance capacity of an athlete, it should be questioned whether the ergogenic metabolic effect of creatine use outweighs the concomitant increase in body weight secondary to water retention (1–2 kg) during the creatine loading phase. Though several concerns have been raised about possible side effects of creatine use, no evidence about associated health risks has been reported following creatine use.

Nitrate

Nitrate supplementation has been shown to improve exercise performance. This ergogenic effect is attributed to the conversion of nitrate into nitrite, which can subsequently be metabolized into nitric oxide. Nitric oxide has several functions that can potentially improve exercise performance, such as increasing muscle perfusion and improving oxygen efficiency. Liver and muscle tissue have demonstrated the capacity to convert nitrate into nitrite, and nitrite into nitric oxide. In addition, the mouth is an important site where nitrate is converted into nitrite by oral bacteria. This is illustrated by observations that the ingestion of antibacterial mouth wash attenuates the increase in plasma nitrite following the ingestion of nitrate.

Much early work demonstrated that nitrate supplementation can improve endurance-type exercise performance (Cermak et al. 2012), but nitrate supplementation has also been shown to improve performance during intermittent-type exercise tasks. The ergogenic effects of nitrate supplementation appear to be largely limited to amateur athletes, and may be less or even completely absent in professional athletes. Supplementation with natural nitrate sources such as concentrated red beetroot juice appears to be more effective than the ingestion of sodium nitrate. The minimal dose that appears to have an ergogenic effect is ~500 mg. This may be a single bolus prior to exercise, but daily supplementation for 3–6 days prior to competition has more consistently been shown to produce ergogenic effects.

Sodium bicarbonate

Sodium bicarbonate ($NaHCO_3$), also known as 'baking soda', can also be regarded as an ergogenic aid. Anaerobic glycolysis is the major source of energy supply during exercise of near maximal intensity lasting longer than 20–30 seconds (see Section 18.3). The total capacity of this energy system is limited by the progressive increase in the acidity of the intracellular environment by the accumulation of hydrogen ions. The latter inhibits muscle contraction by impairing the role of calcium in this process and by reducing the activity of several key glycolytic enzymes. This leads to muscular pain and fatigue and the inability to maintain exercise intensity for a more prolonged period. As bicarbonate represents the most important extracellular buffer for the accumulation of hydrogen ions, it has been suggested that an increase in plasma bicarbonate levels might delay the onset of muscular fatigue during prolonged anaerobic metabolism. Studies have indeed shown that bicarbonate loading in athletes (0.3 g $NaHCO_3$ per kg body weight taken 1–2 hours before exercise with 1 litre water) can improve performance in exercise tasks lasting between 0.5 and 6.0 minutes of near maximal performance. The use of bicarbonate does not seem to pose any major health risks, but gastrointestinal distress as well as hyperosmotic diarrhoea have often been reported following bicarbonate loading.

Ergogenic aids in sports practice

Though nutraceuticals such as caffeine, creatine, carnitine, and sodium bicarbonate have received much attention, scientific data on most ergogenic substances such as bee-pollen, royal jelly, hydroxycitrate, guarana, taurine, ginseng, yohimbe, colostrum, and various other exotic ingredients are either scarce or not available at all. Therefore, one should be aware of the nutritional quackery that offers athletes a wide variety of products and misleading information about preferred dosages and applications. Only a few nutraceuticals under specific circumstances and proper guidance have been shown to improve performance capacity. Whether the use of such

ergogenic aids should be encouraged in recreational and/or professional sports practice is highly questionable. More important is to realize that a healthy diet, compensating for the metabolic demands imposed by intense exercise training and competition, forms the basis of good nutritional practice and should be a first priority in any effort to optimize performance capacity.

KEY POINTS

- The main nutritional need of an athlete is to consume sufficient (healthy) foods to compensate for the increased energy expenditure.
- Muscle glycogen stores quantitatively form the most important energy source during moderate to high intensity exercise.
- Endogenous carbohydrate stores are relatively small and therefore often play a key role in limiting endurance performance capacity.
- Carbohydrate loading in the days preceding endurance competition represents an effective means to optimize pre-exercise glycogen contents.
- During prolonged endurance exercise (>45 minutes) carbohydrate ingestion can significantly improve performance capacity.
- Ingestion of low protein, carbohydrate-rich drinks and foods soon after cessation of exercise accelerates recovery.
- Protein requirements of athletes range between 1.6 and 2.2 g/kg bodyweight/day. These requirements are easily met by the increased food intake of most athletes.
- Fluids and electrolytes need to be supplied during prolonged exhaustive exercise to compensate for sweat loss.
- In general, vitamin and mineral needs in the athlete are fully compensated for by the increased dietary intake. Therefore, when a basic healthy diet is used, vitamin and mineral supplements are unnecessary. Vitamin and/or mineral supplementation does not improve performance capacity when no deficiencies are present.
- Only a few nutraceuticals (e.g., caffeine and creatine), under specific circumstances and proper guidance, have been shown to improve performance capacity. Caution is needed to assess the dubious nutritional merits of the wide range of sports supplements/nutraceuticals available to professional as well as recreational athletes.

 Be sure to test your understanding of this chapter by attempting multiple choice questions. See the Further Reading list for additional material relevant to this chapter.

REFERENCES

Bergström J & Hultman E (1966) Muscle glycogen synthesis after exercise: An enhancing factor localized to the muscle cells in man. *Nature* **210**, 309–310.

Carter J M, Jeukendrup A E, Jones D A (2004) The effect of carbohydrate mouth rinse on 1-h cycle time trial performance. *Med Sci Sports Exer* **36**, 2107–2111.

Cermak N M, Gibala M J, van Loon L J (2012) Nitrate supplementation's improvement of 10-km time-trial performance in trained cyclists. *Inter J Sport Nutr Exer Met* **22**, 64–71.

Cermak N M & van Loon L J (2013) The use of carbohydrates during exercise as an ergogenic aid. *Sports Med* **43**, 1139–1155.

Christensen E H & Hansen O (1939) V. Respiratorischer Quotient und O2-Aufnahme. *Skandinavisches Archiv Für Physiologie* **81**, 180–189.

Hoppeler H, Howald H, Conley K et al. (1985) Endurance training in humans: Aerobic capacity and structure of skeletal muscle. *J Appli Physiol (1985)* **59**, 320–327.

Martin W H, 3rd, Dalsky G P, Hurley B F et al. (1993) Effect of endurance training on plasma free fatty acid turnover and oxidation during exercise. *Amer J Physiol* **265**, E708–714.

Maughan R J (2014) *Sports Nutrition* Vol. 19: John Wiley, Chichester.

Morton R W, Murphy K T, McKellar S R et al. (2018) A systematic review, meta-analysis and meta-regression of the effect of protein supplementation on resistance training-induced gains in muscle mass and strength in healthy adults. *Brit J Sports Med* **52**, 376–384.

Hartman J W, Moore D R, Phillips S M (2006) Resistance training reduces whole-body protein turnover and improves net protein retention in untrained young males. *Appl Physiol Nutr Metab* **5**, 557–564.

Saris W, van Erp-Baart M, Brouns F et al. (1989) Study on food intake and energy expenditure during extreme sustained exercise: The Tour de France. *Int J Sports Med* **10**, S26–S31.

Shulman G I (2000) Cellular mechanisms of insulin resistance. *J Clin Inv* **106**, 171–176.

Snijders T, Res P T, Smeets J S et al. (2015) Protein ingestion before sleep increases muscle mass and strength gains during prolonged resistance-type exercise training in healthy young men. *J Nutr* **145**, 1178–1184.

Terjung R L, Clarkson P, Eichner E et al. (2000) American College of Sports Medicine roundtable. The physiological and health effects of oral creatine supplementation. *Med Sci Sports Exer* **32**, 706–717.

Tipton K D, Ferrando A A, Phillips S M et al. (1999) Postexercise net protein synthesis in human muscle from orally administered amino acids. *Amer J Physiol-Endro Met* **276**, E628–E634.

van Loon L J, Greenhaff P L, Constantin-Teodosiu D et al. (2001) The effects of increasing exercise intensity on muscle fuel utilisation in humans. *J Physiol* **536**, 295–304.

van Loon L J, Jeukendrup A E, Saris W H et al. (1999) Effect of training status on fuel selection during submaximal exercise with glucose ingestion. *J Appl Physiol* **87**, 1413–1420.

Wagenmakers A, Brouns F, Saris W et al. (1993) Oxidation rates of orally ingested carbohydrates during prolonged exercise in men. *J Appl Physiol* **75**, 2774–2780.

FURTHER READING

Burke L, Desbrow B, Spriet L (2013) *Caffeine for Sports Performance*. Human Kinetics, Champaign, IL.

Gleeson M, Jeukendrup A (2004) *SportNnutrition: An Introduction to Energy Production and Performance*. Human Kinetics, Champaign, IL.

Hultman E & Harris R (1988) Carbohydrate metabolism. In: *Principles of Exercise Biochemistry*, Vol. 27, 78–119 (Poortman J ed.). Karger, Basel. https://www.karger.com/Book/Toc/221519

Jeukendrup A E, Raben A, Gijsen A et al. (1999) Glucose kinetics during prolonged exercise in highly trained human subjects: Effect of glucose ingestion. *J Physiol* **515**, 579–589.

Rehrer N, Wagenmakers A, Beckers E et al. (1992) Gastric emptying, absorption, and carbohydrate oxidation during prolonged exercise. *J Appl Physiol* **72**, 468–475.

van Loon L J, Saris W H, Kruijshoop M et al. (2000) Maximizing postexercise muscle glycogen synthesis: Carbohydrate supplementation and the application of amino acid or protein hydrolysate mixtures. *Amer J Clin Nutr* **72**, 106–111.

Williams C & Devlin J T (1992) *Foods, Nutrition, and Sports Performance: An International Scientific Concensus, held 4–6 February, 1991 and organized by Mars, Incorporated with International Olympic Committee patronage*: Taylor & Francis, London.

Williams M H, Kreider R B, Branch J D (1999) *Creatine: The Power Supplement*: Human Kinetics, Champaign, IL.

Wilmore J H, Costill D L, Kenney W L (1994) *Physiology of Sport and Exercise*. Human Kinetics, Champaign, IL.

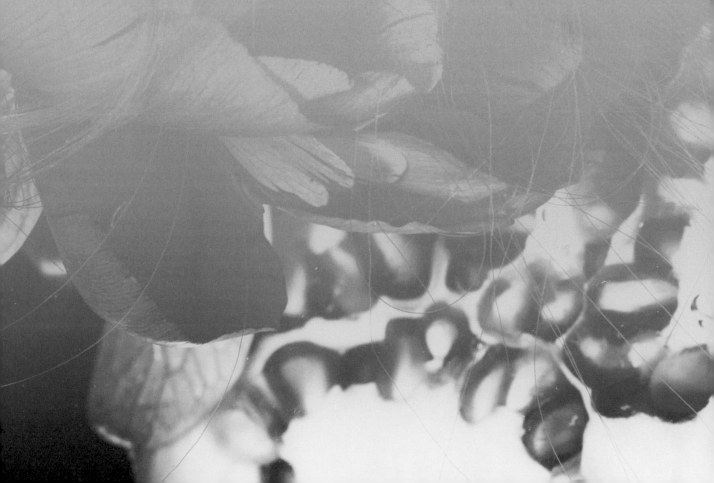

PART 5 Nutrition and disease

19 Nutrition and cardiovascular disease

Godfrey S. Getz and Catherine A. Reardon

OBJECTIVES

At the end of this chapter you should be able to:

- have an appreciation of the fact that dietary components can both increase and reduce the risk of cardiovascular disease
- have an understanding of lipoprotein metabolism and pathways involved in cholesterol and fatty acid absorption
- have an understanding of how diet influences plasma LDL-cholesterol levels
- have an understanding of how the gut microbiome contributes to increased risk of cardiovascular disease.

TABLE 19.1 Cardiovascular risk factors

Non-modifiable

- Male gender
- Age
- Ethnic background
- Genetic disorders of lipid metabolism

Potentially modifiable with changes in lifestyle or pharmaceuticals

- Cigarette smoking
- Obesity
- Diabetes
- High plasma LDL-cholesterol and triglyceride levels (VLDL)
- Low plasma HDL levels
- High plasma homocysteine levels
- Hypertension
- Physical inactivity
- Insufficient sleep
- Consumption of diets rich is saturated and trans unsaturated fatty acids, processed foods, refined sugars, sodium

19.1 Introduction

The role of nutrients in cardiovascular disease has been explored for many years. For the purposes of this chapter, the most common and pertinent cardiovascular diseases are myocardial infarction, stroke, and peripheral vascular disease, as well as hypertension, which serves as a major risk factor for myocardial infarction and stroke. Additional risk factors for these pathologies are obesity and diabetes. A major underlying basis for the genesis of myocardial infarction is atherosclerosis of the coronary arteries, which may also engender a milder degree of obstruction expressed as pain on effort or angina pectoris. Atherosclerosis in the carotid arteries contributes to stroke and atherosclerosis of the peripheral arteries gives rise to peripheral vascular disease, exemplified by intermittent leg pain. Atherosclerosis alone or complicated by supervening thrombosis gives rise to these clinical syndromes.

Many genetic and environmental factors increase the risk for developing atherosclerosis and cardiovascular diseases (Table 19.1). Among the environmental risk factors, a great variety of nutrients have been implicated in promoting atherogenesis. Dietary cholesterol and fatty acids of varying degrees of saturation have been widely implicated and explored in clinical studies as well as in studies in animal models. Other dietary components, including the amino acid homocysteine and various forms of carbohydrates, especially refined sugars, have also been demonstrated to influence atherogenesis. Vitamins and minerals also play a role as influential micronutrients that either promote or reduce cardiovascular disease. More recently the involvement of the intestinal microbiome in generating metabolites of nutrients that enhance their risk relationship has received attention. Independent of the atherosclerosis-based cardiovascular diseases are those resulting from hypertension, which may also be influenced by nutrient intake.

Atherosclerosis is a chronic inflammation of the large and medium sized arteries. Atherosclerotic lesions develop at select sites of the vascular system, largely determined by dynamics of the blood flow at those sites, with plaques developing at sites of disrupted blood flow. The arterial sites of these lesions that are of the greatest clinical relevance are the coronary, carotid, and femoral arteries. These lesions are characterized by the accumulation of monocyte/macrophage and smooth muscle cell-derived foam cells in the arterial wall intima. The lipid droplets of these cells are composed mainly of cholesteryl esters thought to be derived from native and modified LDL that enters the subendothelial space from the plasma and are ultimately taken up by the macrophages. Macrophages, as well as the endothelial cells overlying the sites of atherosclerotic lesions, secrete chemokines that attract further monocytes from the blood stream. Also present in the evolving atherosclerotic plaque or in the adventitia are cells of the adaptive immune system including B cells and a variety of proinflammatory T cells.

Macrophage-derived foam cells and other inflammatory immune cells secrete proinflammatory cytokines. As lesions advance their composition becomes more complex. Among these changes are some that are antiinflammatory. Beyond the cholesteryl esters of the lipid droplets, free cholesterol or cholesterol oxidation products may accumulate in lesional foam cells as a result of which they undergo apoptosis. If the phagocytic activity of neighbouring macrophages (i.e., efferocytes) is intact, the apoptotic cells will be effectively removed. If not, the apoptotic cells undergo secondary necrosis, which if substantial, forms the so-called necrotic core that promotes plaque instability and increased susceptibility to acute atherothrombosis.

Smooth muscle cells also migrate from the media to the intima where they express a synthetic, in contrast to a contractile, phenotype. These cells synthesize a variety of matrix proteins and proteoglycans that give rise to the fibrous cap, which protects the necrotic core from the pro-coagulant machinery of the blood. As the plaque evolves the constituent macrophages secrete a variety of lytic enzymes such as metalloproteinases, which can disrupt the protective fibrous cap and contribute to plaque rupture. Plaque rupture exposes the underlying procoagulant matrix and enables the attachment of platelets and the formation of a thrombus that often results in the acute clinical event, for example acute myocardial infarction (in coronary arteries) or stroke (carotid arteries). In summary, the mature atherosclerotic plaque is a very complex inflammatory milieu that develops over years or decades and is long clinically silent before it becomes a clinical event. This process is not irreversible, especially in the early stages when interventions could attenuate or slow down progression or even promote plaque regression. This somewhat simplified description highlights the participation of many cells of the innate and adaptive immune systems, and their derived biological mediators, in the evolution of vessel wall lesions.

19.2 Nutrients and atherosclerosis

LDL cholesterol

More than 100 years ago Windaus showed that the atherosclerotic lesions of the human aorta were highly enriched in cholesterol. The likely dietary origin of this cholesterol was shown by Anitschkow, who fed cholesterol to rabbits, which induced atherosclerotic lesion development (Goldstein & Brown 2015). The major lipoprotein carrying cholesterol into the artery wall is low-density lipoprotein (LDL). Some of the LDL entering the subendothelial space of the artery wall may be modified by local oxidation within the intima. The cells of the plaque, particularly macrophages and smooth muscle cells (especially in human plaques), take up native and modified LDL by macropinocytosis or receptor-mediated endocytosis respectively. Thus, the effect of the diet on circulating LDL levels is often taken as indicating alterations in the risk of developing atherosclerosis.

Cholesterol packaged in the LDL may be derived either from the diet (exogenous pathway) or from endogenous synthesis. These two sources are interrelated. Endogenous cholesterol synthesis is tightly regulated by the level of cellular cholesterol, which is influenced to a large extent by the level of cholesterol in the plasma (Table 19.2). How does diet influence plasma LDL-cholesterol levels?

As described in Chapter 8, dietary cholesterol is taken up into the enterocytes that line the lumen of the small intestine. Niemann-Pick C1-like 1 (NPC1L1) facilitates the trafficking of the absorbed cholesterol to the endoplasmic reticulum of the enterocyte, where much of the sterol is esterified by the enzyme acyl cholesteryl acyl transferase 2 (ACAT2). The cholesterol, along with triacylglycerides, are packaged in large lipid-rich lipoprotein particles called chylomicrons, with apoB48 as the major

TABLE 19.2 Factors influencing plasma cholesterol levels

- Cholesterol content of the diet
- Intestinal absorption of cholesterol
  - o Involves participation of bile acids, NPC1L1, ABCG5/G8, ACAT2, MTTP, apoB48
- Fat content of the diet
  - o Saturated/*trans* fatty acids increase, while polyunsaturated/monounsaturated fatty acids decrease
- Hepatic clearance of chylomicron remnants by LDL receptor related protein 1
- Hepatic clearance of LDL by the LDL receptor
- Endogenous cholesterol synthesis particularly by hepatocytes

apoprotein. The fate and metabolism of chylomicrons is described in Chapter 8.

The dietary cholesterol delivered to the liver may undergo a number of fates. It may be stored as cholesteryl esters; packaged in very low-density lipoproteins (VLDLs) along with triacylglycerides and secreted into the circulation; or it may be secreted into the bile either as cholesterol itself or as bile acids, which are metabolites of cholesterol.

An alternate route for the redistribution of intestinal absorbed cholesterol is via high-density lipoprotein (HDL). ABCA1 on the basolateral surface of the enterocyte promotes the generation of nascent HDL particles containing cholesterol, phospholipid, and apoA-I. These HDL particles are matured in the plasma. HDL receptors (e.g., scavenger receptor B1) on hepatocytes mediate the cellular uptake of HDL-associated cholesteryl esters, the so-called selective cholesteryl ester uptake pathway. The combined chylomicron and HDL pathways account for the transport of most of the dietary and bile derived cholesterol from the small intestine.

LDL is generated in the plasma from VLDL via the action of a variety of lipases and lipid transfer proteins. Plasma LDL homeostasis depends on its production from VLDL and its clearance from the plasma via the LDL receptor, especially the receptors on the surface of the hepatocyte. As plasma LDL levels increase more LDL finds its way into the artery wall, thus initiating the development of the atherosclerotic lesion. The continued elevation of plasma LDL contributes to the progression of atherosclerosis.

Entero-hepatic recycling of cholesterol

Dietary cholesterol is not the only source of the cholesterol in the lumen of the intestine. A good deal of the intestinally absorbed cholesterol recycles back to the intestinal lumen via an enterohepatic circulation pathway involving the liver, gall bladder, and the intestine. Hepatic cholesterol, derived either from the diet from the clearance of chylomicron remnants or from the clearance of plasma LDL and to some extent endogenously synthesized cholesterol, is exported from the liver into the bile canaliculi either as free cholesterol via an ABCG5/G8 dependent pathway or as bile salts/acids that are derived from hepatic cholesterol. The bile is stored in the gall bladder and is discharged upon feeding. The bile contains cholesterol, bile acids/salts, and phospholipid all of which facilitate lipid hydrolysis and absorption by intestinal enterocytes. The rate of formation of bile acids from cholesterol is regulated by the expression level of the rate-limiting enzyme cholesterol 7-α-hydroxylase. This enzyme is induced by a cholesterol load in the liver, although to a lesser extent for the human enzyme than for the murine enzyme. The expression of the enzyme is repressed by bile acids by complex transcriptional mechanisms involving hormones produced by the small intestine in response to high levels of bile acids (e.g., fibroblast growth hormone 15/19) and probably also by an indirect mechanism related to bile acid activation of the nuclear hormone receptor farnesyl X receptor (FXR). Destabilizing the mRNA may also contribute to the regulation of the level of this enzyme. This creates a self-regulating negative feedback on bile acid production. Bile acids also undergo enterohepatic cycling. The total bile pool is about 3 gm, which cycles between the liver, gall bladder, and intestinal lumen four to twelve times a day.

The role of cholesterol, and the LDL that carries it, in the genesis of atherosclerosis was established in two ways; from genetic diseases in humans affecting LDL receptor activity or levels (familial hypercholesterolemia and loss of function of proprotein convertase subtilisin/kexin type 9 (PCSK9)) and from dietary manipulation experiments in animal models. A very few well-performed clinical studies have also demonstrated the importance of LDL cholesterol in promoting atherosclerosis and cardiovascular diseases such as myocardial infarction and stroke.

Dietary cholesterol: evidence from animal models

Rabbits

It is fortuitous that Anitschkov chose the rabbit for his seminal experiments examining the effect of dietary cholesterol on atherosclerosis (Goldstein & Brown 2015; Steinberg 2013). While rabbits have low plasma cholesterol levels on low fat standard diet (<60 mg/dl, quite low by human standards), this species is quite susceptible to dietary cholesterol overload. In rabbits fed diets containing 6 per cent peanut oil and varying cholesterol levels (e.g., 0.05, 0.1, 0.15, 0.2, and 0.25 per cent of dietary mass) for up to 32 weeks, total plasma cholesterol increased progressively over time only in the animals fed the two highest levels of dietary cholesterol. As occurs in humans, there was considerable inter-individual variation in the plasma response to dietary cholesterol. In addition, atheromatous plaques were found in the aortic arch only in animals exposed to the diets with the two highest levels of cholesterol; at lower dietary cholesterol levels only early foam cell lesions were observed. In this study, coronary artery lesions were not responsive to the level of cholesterol in the diet. However, the co-existence of diabetes along with the cholesterol feeding results in coronary artery lesions even in the rabbit.

Mice

Currently, the most common animal model used for atherosclerosis studies is the mouse. The preference of the mouse relates to its size, the relative ease with which its genome can be modified, its fecundity and ease of cross breeding, and the fact that lesions develop in a reasonable time frame. However, unlike rabbits, wild type mice are relatively resistant to the development of atherosclerosis. Even in wild type animals fed a high fat and high cholesterol atherogenic diet, plasma cholesterol levels are only modestly elevated, probably because of the efficient clearance of VLDL and LDL and the mechanisms that down-regulate the endogenous synthesis of cholesterol. On the other hand, mice genetically modified to limit the clearance of cholesterol-containing apoB-lipoproteins (e.g., mice deficient in the LDL receptor or apoE, a major ligand for the LDL receptor) readily develop atherosclerosis with an atherogenic diet. ApoE deficient mice also develop lesions while being fed a standard low fat chow diet, though the atherogenic diet greatly augments lesion progression. The feeding of LDL receptor deficient mice a semi-synthetic low fat diet with varying levels of dietary cholesterol produced similar results with respect to plasma cholesterol and lesion development to those observed in rabbits (Teupser et al. 2003). Only at the highest dietary cholesterol levels did the mice develop reasonably sized lesions.

In both mice and pigs, the administration of adeno-associated virus expressing gain of function mutants of PCSK9 is sufficient to induce atherosclerosis in otherwise wild type mice fed an atherogenic diet. This is a valuable model for studying atherosclerosis since no prior genetic manipulation is required.

Humans

Conducting well-controlled dietary studies in human subjects is much more difficult than it is in animal models. The human diet is complex, and it is not easy to focus on a single nutritional variable. There are also nutrient interactions which make the categorization of the role of individual nutrient more ambiguous.

The absorption of cholesterol in humans is not as readily assessed as it is in experimental animals. The average daily intake of dietary cholesterol is about 350 mg. The fractional absorption of cholesterol is about 50 per cent, though the inter-individual variation is great, ranging between 20–80 per cent. The major sources of cholesterol in human diets are egg yolks, liver, fast foods, shellfish, red meat, and sausage. In the 2014 edition of the ACC/AHA Guidelines for Lifestyle Management to Reduce Cardiovascular Risk, dietary restriction of cholesterol is not included as a recommendation for lowering plasma LDL-cholesterol levels (Eckel et al. 2014). This is because

TABLE 19.3 Evidence that modulation of dietary components influences risk of cardiovascular diseases

	Supports	Does not support
Dietary cholesterol	Animal model dietary studies Genetic mutations	Human dietary studies (inconclusive)
Dietary fatty acids	Human dietary studies Animal model dietary studies	
Fat-soluble vitamins		Human clinical trials
Minerals (effect on hypertension)	Human clinical trials for potassium DASH diet	Human clinical trials for sodium (in context of healthy diet)
Gut microbiome Modifications of dietary components	Animal model dietary studies Human clinical studies	

the evidence to support the relationship between dietary cholesterol and LDL cholesterol levels in humans is inconclusive (Table 19.3). This probably reflects the complexity of the mechanisms regulating cholesterol homeostasis in the intestine affecting both absorption and export of cholesterol in lipoproteins, as well as the events controlling the clearance of intestinally derived lipoprotein particles containing dietary cholesterol by the liver and the intrahepatic responses to that cholesterol. On the other hand, the efficacy of ezetimibe, a pharmacological agent that specifically targets NPC1L1 to treat hypercholesterolemia, especially as a complement to statin, speaks to the importance of cholesterol absorption in contributing to the regulation of plasma cholesterol levels. With the current emphasis on individualized medicine, it is likely that future studies will throw light on these questions in individuals selectively exhibiting particular genetic patterns.

Given the uncertainty of the evidence about the role of dietary cholesterol in regulating plasma LDL, much of the focus in achieving ideal plasma LDL levels has been upon the nature and level of dietary fat, and of course upon the widely employed pharmacologic approaches. Statins are inhibitors of the rate-limiting enzyme of cholesterol biogenesis, HMGCoA reductase. They not only reduce cholesterol biosynthesis, but by reducing the endogenous regulatory cholesterol pool that influences the maturation of the transcription factor SREBP2, also upregulate the synthesis of the LDL receptor and hence the clearance of plasma LDL. As mentioned, ezetimibe, the inhibitor of NPC1L1 and cholesterol absorption, may have salutary effects.

Most exciting is the development of inhibitors of PCSK9, which also has the effect of enhancing LDL receptor mediated LDL clearance. Monoclonal antibodies to the enzyme significantly lower LDL levels when combined with statin therapy and reduce major adverse cardiovascular events.

Dietary fatty acids

Plasma cholesterol levels are not only determined by dietary intake but also by its biosynthesis in such tissues as the liver. One of the major regulators of cholesterol biosynthesis is the amount and nature of the fat in the diet. Cholesterol is presented to the absorbing enterocytes accompanied by a larger mass of free fatty acids and monoglycerides derived from the hydrolysis of dietary triglycerides. The fatty acids in the dietary triglycerides may be saturated, monounsaturated, or polyunsaturated. The polyunsaturated fatty acids belong to either the n-6 or the n-3 families. The major dietary sources of each of the fatty acids is distinct: saturated fatty acids are high in animal fat including dairy and meat, and certain tropical oils such as coconut oil; monounsaturated oils are found in Mediterranean oils like olive oil; polyunsaturated n-6 fatty acids are found in plant oils like corn and safflower oil; while the polyunsaturated n-3 fatty acids are mostly derived from plants and fish oils. Saturated fatty acids tend to occupy the 1 and 3 positions of the glycerol backbone of triglycerides, while unsaturated fatty acids preferentially occupy the 2 position.

Dietary fat provides about 35 per cent of food energy in the average Western diet. Saturated and monounsaturated fatty acids each furnish an average of 12–14 per cent, while polyunsaturated fatty acid accounts for about 6 per cent. Almost all of the unsaturated fatty acids have *cis* double bonds and where there is more than one double bond these are interrupted by a single methylene group. The metabolic fate of the various polyunsaturated fatty acids is distinct. Thus, linoleic acid (18:2 n-6 meaning it has two double bonds with the first double bond at the sixth carbon from the methyl end of the fatty acid) may be incorporated into complex lipids as such, or it may be elongated to arachidonic acid (20:4 n-6) before being incorporated into complex lipids (see Chapter 2). Additionally, arachidonic acid is the precursor of important signalling molecules, such as prostaglandins, isoprostanes, and leukotrienes, which regulate inflammation. The 18:2 n-6 fatty acids are present in the diet to the extent of 17 grams per day, while α-linolenic acids (18:3 n-3) are at much lower abundance (<2gm per day). The long chain C20 and C22 n-3 fatty acids may contain five or six double bonds respectively (eicosapentaenoic (C20:5 n-3) or docosahexaenoic acids (C22:6 n-3)). These fatty acids are found mostly in fish, especially deep-sea fish. There is also evidence that these fatty acids are the source of inflammation-resolving molecules called resolvins. The commercial hydrogenation for hardening of oils generates *trans* unsaturated fatty acids, which may also be found in meat and dairy products. In addition to *trans* fatty acids, conjugated linoleic acids (CLA) are found naturally in ruminant animal food products (e.g., beef, lamb, and dairy products). 18:2 *trans*-10, *cis*-12 and 18:2 *cis*-9, *trans*-11 isomers are the most abundant CLA. They are generated by bacterial linoleic acid isomerase mediated hydrogenation. Bacteria expressing the isomerase are present in the human intestine.

The intestinal absorption of fatty acids involves both passive diffusion and receptor-mediated pathways. CD36, which is present on the apical surface of enterocytes in the proximal small intestine, accounts for some of this absorption. But as CD36 has a high affinity for fatty acid and is readily saturated, it does not account for bulk absorption of fatty acids. Fatty acids are not only nutrients oxidized to provide energy but are also signalling agonists, influencing a variety of gastrointestinal hormones, such as cholecystokinin, glucagon like-peptide 1 and ghrelin, thus influencing stomach emptying, gall bladder contraction, intestinal motility, and neuroendocrine regulation of appetite by the brain.

The importance of the fatty acid composition of the diet for cardiovascular disease was highlighted by the seminal Seven Countries Studies of Ancel Keys. Keys and his colleagues showed that the extent of hypercholesterolemia and cardiovascular disease in the populations of these countries is related to the level of fat in their diets as well as the composition of this fat (Goldstein & Brown 2015). The lower the level of fat and the higher its unsaturated fat content, the lower is the extent of hypercholesterolemia and cardiovascular disease. For example, they studied several groups of Japanese men living in countries with different dietary habits and showed that the level of fat in the diet was linearly related to their serum cholesterol concentration. Since these studies, the dietary patterns in the countries that had diets rich in fruits and vegetables, olive oil, and low in animal products and refined sugar have shifted to diets rich in saturated fat, animal products, and processed foods, with accompanying increase in cardiovascular diseases.

Data from several human studies comparing the effect of recommended diets vs. control diets on the level of plasma LDL-cholesterol are compiled in the 2014 guidelines issued by the nutrition committee of the American Heart Association (Eckel et al. 2014). The recommended diets contained 5–6 per cent saturated fat and a total fat content of 20–27 per cent vs. 14–15 per cent saturated fat and 34–38 per cent total fat for the control diet. The substitution of the control diet by the recommended diet resulted in an 11–13 mg/dl reduction of LDL-cholesterol. For every 1 per cent of calories derived from saturated fatty acids that are replaced by carbohydrate, monounsaturated fatty acids (MUFA), or polyunsaturated fatty acids (PUFA), LDL-cholesterol was reduced by 1.2, 1.3, and 1.8 mg/dl respectively. The replacement of 1 per cent of energy derived from *trans* unsaturated fatty acids by equivalent energy derived from MUFAs or PUFAs, reduced LDL-cholesterol by 1.5 and 2.0 mg/dl, respectively. The nature of dietary carbohydrate in the context of a saturated fat diet influences the risk of myocardial infarction; with the risk being higher with high glycemic index carbohydrates.

In the Cardiovascular Health Study, total mortality and cardiovascular disease mortality were evaluated in association with plasma phospholipid saturated fatty acids of differing chain length at baseline (Fretts et al. 2016). Aggregate mortality was assessed over the next 16 years. Total mortality was higher with high plasma phospholipid palmitic acid (16:0), though cardiovascular disease mortality was not affected. Phospholipid stearic acid (18-0) levels had no influence on either mortality rate. On the other hand, both total mortality and cardiovascular disease mortality were reduced in association with increased levels of phospholipids with saturated fatty acids containing 22 or 24 carbon atoms. Similar mortality studies were performed on subjects based upon their baseline plasma phospholipid content of long chain (20–22 carbons) n-3 PUFA (eicosapentaenoic acid (EPA), DHA docosahexaenoic acid (DHA), and docosapentaenoic acid (DPA)) (Mozaffarian et al. 2013). Again, higher levels were associated with reduced total cardiovascular disease mortality. The risk ratio was calculated for the highest quintile/lowest quintile of these fatty acids, with the highest ratio subjects living 2.2 years longer than those with the lowest ratio. However, among older high-risk patients (>75 years) being treated with statins, the replacement of 18:2 n-6 fatty acids with the above n-3 long chain PUFA revealed no improvement in cardiovascular outcomes (Nicholls et al. 2020).

A very good review describes the pathophysiological effects of these fatty acids on pathways that impact on cardiovascular disease (Mozaffarian & Wu 2011). Fatty acids have been shown to affect plasma triglyceride levels, blood pressure, thrombosis, and arrhythmia. Many molecular mechanisms having been suggested, including a reduction in fatty acid synthesis and downregulation of the transcription factor SREBP1c. Meta-analysis of studies on cardiovascular disease risk with α-linolenic acid (18:3 n-3) consumption yielded inconsistent results (Pan et al. 2012). A novel mechanism has recently been reported (Grevengoed et al. 2021). They found that the biliary metabolite N-acyl taurine levels are relatively high in the bile in mice fed n-3 fatty acids. One derivative, C22:6 N-acyl taurine, inhibited luminal triacylglyceride hydrolysis and absorption in the intestine.

Other nutrients

To assemble as much available evidence as appears in published literature, Chareonrungrteangchai and colleagues have performed an umbrella review of several systematic reviews with meta-analyses of observational studies (SRMA) and randomized control trials (RCT) comparing diets of different composition (Chareonrungrueangchai et al. 2020) These include some of the nutrients discussed above. For example, the pooled risk ratios (RR) for total mortality for diets enriched in n-3 PUFA, legumes, nuts, chocolate, and vegetarian are reported as 0.89, 0.90, 0.68, 0.90, and 0.71 respectively. Feeding the Mediterranean diet had an RR of 0.55 for cardiovascular disease, 0.59 for cardiovascular mortality RR, and 0.64 for stroke. For cardiovascular disease mortality the RR in SRMA was 0.87 for vegetarian diets; legume 0.89; fish 0.82; high grain intake 0.68; and high chocolate 0.55. In RCT then RR was n-6 PUFA 0.95; n-6 PUFA 0.81, and low sodium 0.67. Other summaries of systematic reviews can be found in this review.

Essentially, all the above referenced studies have been performed in adults with or without clinical cardiovascular disease. However, it has become clear that subclinical early atherosclerotic lesions (determined at autopsy) are present in many children and adolescents. These studies indicated that the extent of these subclinical atherosclerotic lesions relates to lipid risk factors, and that the later development of clinical disease represents an expansion and progression of these early lesions. Thus, controlling lifestyle risks beginning in adolescence, including those attributable to diet, may lower the lifetime burden of atherosclerotic cardiovascular disease (Gidding et al. 2016). These strategies are in accord with the established low level of lifetime LDL cholesterol and reduced cardiovascular pathology associated with loss of function mutations of PCSK9, even when individuals have other cardiovascular risk factors.

Diabetes and obesity are additional risk factors for the development of atherosclerotic cardiovascular disease. Though detailed discussion of these entities is beyond the scope of this chapter, two recent reviews on these topics are included in the extended resources. (See Weblink Section 19.1.)

Water-soluble vitamins

The water-soluble B vitamins are essential for normal cellular functions and development (see Chapter 11). The B vitamins include vitamin B_1 (thiamin), B_2 (riboflavin), nicotinic acid, vitamin B_6 (pyridoxine), biotin, pantothenic acid, folate, and vitamin B_{12}. Deficiencies of riboflavin, folate, and B_{12} may be present in high-income countries and are overlooked. Some of these B vitamins have been shown to have cardiovascular effects.

Beriberi is the disease attributable to marked undernutrition due to a deficiency of thiamin in the diet. In the form of thiamin pyrophosphate it is a cofactor for some critical reactions of the citric acid cycle. Subjects with disorders of intestinal absorption and reduced intake of food, as in chronic alcoholism, may develop beriberi. The cardiac manifestations of this disease include elevated heart rate, endothelial dysfunction with vasodilation, and high output heart failure, dyspnea, and dilated cardiomyopathy.

Vitamin B_3 or nicotinic acid (niacin) has been shown to decrease LDL cholesterol, triglycerides levels, and lipoprotein (a). It also increases HDL cholesterol levels. Thus it has been widely used as a prescription medication to control hyperlipidemia. However, niacin treatment alone or in combination with statin therapy does not reduce the risk of major adverse cardiovascular events including myocardial infarction or stroke (HPS2-THRIVE Collaborative Group 2014).

Elevated serum levels of homocysteine are associated with increased vascular damage and risk of cardiovascular diseases, though some research raises questions about the role of homocysteine in vascular diseases. Both folic acid (vitamin B_9) and vitamin B_{12} participate in the remethylation of homocysteine to methionine and vitamin B_6 participates in the degradation of homocysteine, an important pathway regulating homocysteine levels. Supplementation with folate has been shown to decrease serum homocysteine levels and reduce the risk of stroke but has little impact on risk of myocardial infarction (Jenkins et al. 2021). Vitamin B_{12} and vitamin B_6 supplementation are not as effective. This one carbon cycle regulates methionine levels. Methionine is the primary methyl donor affecting many methylated substrates, including nucleotides and histones, which could have important effects on the epigenetic regulation of cardiovascular disease.

Fat-soluble vitamins

The fat-soluble vitamins are vitamins A, D, E, and K. Vitamin D is obtained from the diet but a more important source is dermal synthesis (see Chapter 12). While there is observational and epidemiological evidence that vitamin D deficiency may be associated with increased risk of cardiovascular disease and hypertension, clinical trials of vitamin D supplements have not noted a significant impact on these diseases. These trials were generally of poor quality, with small sample size and short duration. There is, however, evidence that vitamin D deficiency is associated with upregulation of the renin-angiotensin system and hence blood pressure. This vitamin D signalling system also negatively influences atherogenesis in mouse models. In preclinical studies, its deficiency may

be associated with dilated cardiomyopathy. Vitamin A encompasses retinoids and carotenoids, with β-carotene being the most prevalent carotenoid in the diet. Among its various functions, the carotenoids have antioxidant properties. A review of clinical studies of vitamin A and β-carotene supplementation on cardiovascular disease, involving at least nine years of follow up in the Women's Antioxidant Cardiovascular Study, show no significant impact on cardiovascular disease (Desai et al. 2014).

Vitamin E is present in plant oils. One of its major functional properties is that it is a powerful antioxidant. As oxidized LDL has been strongly implicated in atherogenesis, one might expect that supplementation with this vitamin might function to limit atherogenesis. However, a meta-analysis of vitamin E supplementation studies and randomized clinical trials has revealed either no measurable effect or only modest effects on human cardiovascular disease (Vivekananthan et al. 2003).

Vitamin K is mainly found in green vegetables with much lower levels in animal tissues. It is a cofactor for the γ-glutamyl carboxylase which carboxylates the glutamyl residues to form γ-carboxyglutamic acid (Gla) on the coagulation factors II, VII, IX, and X, as well as the anticoagulant proteins protein C, protein S, and protein Z. Proteins that have two Gla residues enhance the chelation of calcium. Newborns are often born deficient in vitamin K because of its poor passage across the placenta. Antagonists for this vitamin are widely used for the prevention of blood clotting in patients at risk for the development of coagulopathies. Other important Gla-containing proteins are osteocalcin of bone and the matrix glycoprotein of blood vessels. The matrix glycoprotein is thought to inhibit vessel calcification, so that treatment with vitamin K was thought to abrogate vessel calcification. However, recent reviews of randomized clinical trials do not bear out this expectation.

Other dietary factors may also impact cardiovascular diseases. For example, high fibre and polyphenol intake are associated with decreased risk for both coronary heart disease and stroke, while high fructose intake is associated with increased risk.

19.3 **The gut microbiome**

A great deal of recent investigation has focused on the role of the gut microbiome in modifying nutrients and energy homeostasis. The gut microbiome has a cellular mass that is at least equal to the mass of the cells in the host. The gut microbiome influences cardiovascular disease (Tang et al. 2019). It is important to recognize that the absorptive epithelium is in the small intestine, while the microbial organisms are concentrated in the colon. Furthermore, the composition of the microbiome is influenced by the diet and is not necessarily constant over time even in a single individual. Complex carbohydrates are metabolized to yield the short chain fatty acids acetate, propionate, and butyrate. These fatty acids can be oxidized to provide energy and can signal through G protein coupled receptors to influence blood pressure homeostasis. They may also have an impact on myocardial repair. Primary bile acids may also be transformed by the microbes to a multitude of secondary bile acids, which regulate energy metabolism, brown adipose tissue activation, and insulin sensitivity. The bacterial cell wall lipopolysaccharide and peptidoglycan promote vascular inflammation, insulin resistance, and enhance atherosclerosis in part by decreasing reverse cholesterol transport and increasing hyperlipidemia. These effects may be enhanced by gut leakiness that allows these bacterial products to gain access to the liver and blood vessels.

An important finding of the studies implicating microbiome in cardiovascular disease is that plasma levels of three metabolites of dietary phosphatidyl choline (lecithin) that are generated by intestinal enzymes and gut microbes, namely choline, trimethylamine (TMA), and betaine, a TMA containing species, are associated with increased cardiovascular risk. Trimethylamine is a gas that can be oxidized to trimethylamine oxide (TMAO) by oxidases in the liver. Plasma TMAO is a better predictor of adverse cardiovascular events in humans than traditional risk factors like LDL-cholesterol. The supplementation of diets with choline or trimethylamine increases atherosclerosis in mice. Further noteworthy is the presence of large amounts of lecithin in the bile, providing an endogenous source of choline for transformation to TMA. The acceleration of atherosclerosis may in part be due to a TMAO-mediated increase in macrophage cholesterol uptake and reduced macrophage cholesterol efflux. L-carnitine can also be metabolized to trimethylamine by gut microbes and, consistent with the role of TMAO, the feeding of carnitine to atherogenic mouse models also increases atherosclerosis. Red meat is a rich source of carnitine and this mechanism could contribute to the effect of red meat consumption on cardiovascular mortality.

Phenylacetylglutamine (PGA) is generated from a gut microbial metabolite of phenylalanine. PGA activates adrenergic receptors causing thrombosis and atherosclerosis.

An important strategy for proving the role of the microbiome is by faecal or colonic content transplant from mice in one metabolic state to recipients in another state. For example, feeding mice a diet enriched in saturated fats (e.g., lard) increases adipose tissue inflammation and changes the composition of the gut microbes, effects not seen with the feeding of diets enriched in polyunsaturated fats (e.g., fish oil) (Caesar et al. 2015). The transfer of colonic contents from fish oil fed mice to lard fed animals attenuated the effects of the saturated fat diet on adipose tissue inflammation. As stated above, short chain fatty acids regulate blood pressure homeostasis, mostly by lowering blood pressure. However recent studies indicate that microbiota derived metabolites beyond short chain fatty acids may impact blood pressure homeostasis.

19.4 Diet and hypertension

Hypertension is a multifactorial pathology which remains extremely prominent despite the availability of many forms of treatment, the so-called hypertension paradox. The pathogenesis of hypertension is affected by many environmental and host factors including the gut microbiome, which in part is regulated by the diet. Hypertension, along with other risk factors discussed in detail above, is a major risk factor for coronary artery disease and stroke. Elevated consumption of sodium and alcohol and increased body mass index are positively associated with blood pressure, while the consumption of potassium shows an inverse association (Chan et al. 2016). The recommended daily intake of sodium in adults is <2.3 g/day (about 1 teaspoon of table salt), which is much lower than the average intake of 3.4 g/day. The DASH (Dietary Approaches to Stop Hypertension) diet, which is low in sodium and high in potassium, has been fairly widely recommended as an environmental approach to limit elevations of blood pressure (Sacks et al. 2001). This diet is rich in fruit and vegetables, whole grains, fish, poultry, low fat dairy products, and nuts, and limited in consumption of alcohol, red meat, sweets, and sugar-rich beverages. Even in the context of the DASH diet, reducing the dietary content of sodium has beneficial effects on blood pressure. The variability in the response of individuals to sodium limitation is considerable and the regulation of blood pressure is so complex that it is not possible to readily make an a priori determination of an individual's sodium responsiveness.

As mentioned in Weblink Section 19.1, obesity is often accompanied by hypertension especially in those individuals with visceral adiposity. The regulation of blood pressure depends on the function of the renin-angiotensin aldosterone system, which is increased in obese individuals. Angiotensin II is produced from its precursor angiotensinogen by the action of the angiotensin converting enzyme 1 or 2 (ACE 1 or 2). Each of these components has been found in adipocytes including those of the perivascular adipose tissue (PVAT) associated with blood vessels. Angiotensin II contracts vascular smooth muscle cells while the further processing by ACE 2 results in products that are vasodilatory. The PVAT surrounding the abdominal aorta is protective, though with obesity this protection dissipates with the infiltration of inflammatory immune cells. The PVAT surrounding the thoracic aorta resembles brown adipose tissue, is more thermogenic, and is less inflammatory than the abdominal PVAT.

In summary, though there are certainly environmental contributions to the regulation of blood pressure, most subjects with hypertension are treated pharmacologically, though ideally in the context of a healthy, low sodium diet. Unfortunately, this approach is not uniformly completely effective. Physical exercise is an important part of the lifestyle guidelines for managing hypertension (Eckel et al. 2014).

19.5 **Conclusion**

Despite decades of study of the dietary regulation of the risk factors contributing to cardiovascular diseases, including hyperlipidemia, hypertension, and of course obesity and diabetes, new risk factors and environmental factors are still being uncovered. This is best illustrated by the role of the gut microbiome, which further study will surely reveal as making new contributions to these disorders. In these days of personalized medicine, individual variations in responsiveness to these factors will be clarified, presumably based on genetic and epigenetic variables.

KEY POINTS

- Lowering plasma LDL-cholesterol levels is important for reducing cardiovascular disease (CVD) risk but there is inadequate evidence to support restrictions on dietary cholesterol to achieve this.

- Saturated and *trans*-unsaturated fatty acids increase CVD risk; n-6 and n-3 polyunsaturated fatty acids decrease risk in part due to their effects on LDL-cholesterol levels.

- Attenuation of lifestyle risks beginning in adolescence, including those attributable to diet, may lower the lifetime burden of CVD.

- The modification of dietary components by the gut microbiome generates metabolites that increase risk for CVD.

- The DASH diet (rich in fruit and vegetables, whole grains, fish, poultry, low fat dairy products, and nuts, and low in red meat and sugar) especially in combination with low sodium intake limits elevations in blood pressure and stroke.

 Be sure to test your understanding of this chapter by attempting multiple choice questions. See the Further Reading list for additional material relevant to this chapter.

REFERENCES

Caesar R, Tremaroli V, Kovatcheva-Datchary P et al. (2015) Crosstalk between gut microbiota and dietary lipids aggravates WAT inflammation through TRL signaling. *Cell Metabol* **22**, 658–668.

Chan Q, Stamler J, Griep L M et al. (2016) An update on nutrients and blood pressure. *J Athero Thromb* **23**, 276–289.

Chareonrungrueangchai K, Wongkawinwoot K, Anothaisintawee T et al. (2020) Dietary factors and risks of cardiovascular diseases: An umbrella review. *Nutrients* **12**, 1088.

Desai C K, Huang J, Lokhandwala A et al. (2014) The role of vitamin supplementation in the prevention of cardiovascular disease events. *Clin Cardiol* **37**, 576–581.

Eckel R H, Jakicic J M, Ard J D et al. (2014) American College of Cardiology/American Heart Association Task Force on Practice Guidelines. 2013 AHA/ACC guideline on lifestyle management to reduce cardiovascular risk: A report of the American College of Cardiology/American Heart Association Task Force on Practice Guidelines. *Circulation* **129**(25 Suppl 2), S76–99.

Fretts A M, Mozaffarian D, Siscovick D S et al. (2016) Associations of plasma phospholipid SFAs with total and cause-specific mortality in older adults differ according to SFA chain length. *J Nutr* **146**, 298–305.

Gidding S S, Rana J S, Prendergast C et al. (2016) Pathobiological determinants of atherosclerosis in youth (PDAY) risk score in young adults predicts coronary artery and abdominal aorta calcium in middle age: The CARDIA Study. *Circulation* **133**, 139–146.

Goldstein J L & Brown M S (2015) A century of cholesterol and coronaries: From plaques to genes to statins. *Cell* **161**, 161–172.

Grevengoed T J, Trammell S A J, Svenningsen J S et al. (2021) An abundant biliary metabolite derived from dietary omega-3 polyunsaturated fatty acids regulates triglycerides. *J Clin Invest* **131**, e143861.

HPS2-THRIVE Collaborative Group; Landray M J, Haynes R, Hopewell J C et al. (2014) Effects of extended-release of niacin with laropiprant in high-risk patients. *N Engl J Med* **371**, 203–212.

Jenkins D J A, Spence J D, Giovannucci E L et al. (2021) Supplemental vitamins and minerals for cardiovascular disease prevention and treatment. *J Am Coll Cardiol* **77**, 423–436.

Mozaffarian D, Lemaitre R N, King I B et al. (2013) Plasma phospholipid long-chain ω-3 fatty acids and total and cause-specific mortality in older adults: A cohort study. *Ann Intern Med* **158**, 515–525.

Mozaffarian D & Wu J H (2011) Omega-3 fatty acids and cardiovascular disease: Effects on risk factors, molecular pathways, and clinical events. *J Am Coll Cardiol* **58**, 2047–2067.

Nicholls S J, Lincoff M, Garcia M et al. (2020) Effect on high-dose omega-3 fatty acids vs corn oil on major adverse cardiovascular events in patients at high cardiovascular risk: The STRENGTH randomized clinical trial. *JAMA* **324**, 2268–2280.

Pan A, Chen M, Chowdhury R et al. (2012) α-Linolenic acid and risk of cardiovascular disease: a systematic review and meta-analysis. *Am J Clin Nutr* **96**, 1262–1273.

Sacks F M, Svetkey L P, Vollmer W M et al. (2001) Effects of blood pressure of reduced dietary sodium and the dietary approaches to stop hypertension (DASH) diet. *N Engl J Med* **344**, 2–10.

Steinberg D (2013) In celebration of the 100th anniversary of the lipid hypothesis of atherosclerosis. *J Lipid Res* **54**, 2946–2949.

Tang W H W, Bäckhed F, Landmesser U et al. (2019) Intestinal microbiota in cardiovascular health and disease: JACC state of the art review. *J Am Coll Cardiol* **75**, 2089–2105.

Teupser D, Persky A D, Breslow J L (2003) Induction of atherosclerosis by low-fat, semisynthetic diets in LDL receptor-deficient C57BL/6 and FVB/NJ mice: Comparison of lesions of the aortic root, brachiocephalic artery, and whole aorta (en face measurement). *Arterioscler Thromb Vasc Biol* **23**, 1907–1913.

Vivekananthan D P, Penn M S, Sapp S K et al. (2003) Use of antioxidant vitamins for the prevention of cardiovascular disease: Meta-analysis of randomized trials. *Lancet* **361**, 2017–2023.

FURTHER READING

Cohen J C, Boerwinkle E, Mosley T H Jr et al. (2006) Sequence variations in PCSK9, low LDL, and protection against coronary heart disease. *N Engl J Med* **354**, 1264–1272.

den Hartigh L (2009) Conjugated linoleic acid effects on cancer, obesity, and atherosclerosis: A review of pre-clinical and human trails with current perspectives. *Nutrients* **11**, 370.

Glass C K & Witztum J L (2001) Atherosclerosis: The road ahead. *Cell* **104**, 503–516.

Hall J E, do Carmo J M, da Silva A A et al. (2015) Obesity-induced hypertension: Interaction of neurohumoral and renal mechanisms. *Circ Res* **116**, 991–1006.

Joseph J, Handy D E, Loscalzo J (2009) Quo Vadis: Whither homocysteine research? *Cardiovasc Toxicol* **9**, 53–63.

Kolodgie F D, Katocs A S Jr, Largis E E et al. (1996) Hypercholesterolemia in the rabbit induced by feeding graded amounts of low-level cholesterol: Methodological considerations regarding individual variability in response to dietary cholesterol and development of lesion type. *Arterioscler Thromb Vasc Biol* **16**, 1454–1464.

Wang D (2018) Dietary n-6 polyunsaturated fatty acids and cardiovascular disease: Epidemiologic evidence. *Prostaglandins Leuko Essential Fatty Acids* **135**, 5–9.

Witkowski M, Weeks T L, Hazen S L (2020) Gut microbiota and cardiovascular disease. *Circ Res* **127**, 553–570.

20 Obesity

Faidon Magkos and Arne Astrup

OBJECTIVES

By the end of this chapter you should be able to:

- know the definitions of overweight and obesity in adults
- describe the worldwide prevalence and secular trends in obesity
- summarize the physiological consequences of obesity and associated health risks
- discuss the most important causal factors in the pathogenesis of obesity
- evaluate different approaches for obesity prevention and treatment, including diet, exercise, drugs, and surgery.

20.1 Introduction

Obesity is defined as the excessive accumulation of fat in the body, to the extent it impairs normal physiological functions and increases risk of medical complications. Excess body fat is associated with the development of metabolic diseases such as type 2 diabetes, dyslipidaemia, and non-alcoholic fatty liver disease; cardiovascular disorders such as hypertension and coronary heart disease; gynaecological problems such as infertility and birth complications; certain types of cancer such as those of the endometrium, liver, kidney, pancreas, and colon; orthopaedic problems such as arthritis; pulmonary disorders such as obstructive sleep apnoea; and overall contributes to poor health.

20.2 Epidemiology of obesity

Definition of obesity

Obesity in adults is commonly defined, diagnosed, and classified by the body mass index (BMI), which is a simple index of body weight relative to height:

$$BMI = \frac{\text{weight (kg)}}{\text{height}^2 \text{ (m}^2)}$$

The World Health Organization (WHO) classifies BMI into underweight, normal weight, overweight, and obesity categories, which carry progressively increased risk to health, based primarily on North American mortality data (Table 20.1). BMI cut-offs are the same for both sexes and for all ages of adults, hence the BMI provides the most useful population-level measure of overweight and obesity. The recognition that Asian populations have excess body fat and greater disease risk at lower BMI values than Caucasian populations has led to additional BMI cut-off points for public health action, with overweight and obesity in Asians often being defined as a BMI value of 23.0–27.5 kg/m^2 and ≥27.5 kg/m^2, respectively.

The use of BMI is not without limitations. The relationship between BMI and per cent body fat is linear up to a BMI of approximately 40 kg/m^2 and levels off thereafter (i.e., plateaus), so that BMI values above 40 kg/m^2 do not reflect proportional increases in body fat. In addition, the same BMI may not correspond to the same degree of fatness in different individuals. For any given BMI value, women have more body fat than men, and within each sex, there is a large inter-individual variability in per cent body fat (by 15 per cent points or more). Furthermore, BMI cannot distinguish fat from lean tissues, so that increased BMI may misclassify individuals who have increased body weight because of increased muscle mass (e.g., many athletic populations).

Oftentimes, waist circumference is used as an index of abdominal fat deposition (i.e., central obesity) to inform increased disease risk for any given BMI category (Table 20.1). Still, this measurement does not differentiate between the abdominal subcutaneous and the intra-abdominal adipose tissue compartments. The latter is also referred to as visceral fat and is the depot more strongly associated with metabolic dysfunction and risk of disease. Another index sometimes used to evaluate fat distribution between the upper and the lower body is the waist-to-hip ratio. Besides these simple anthropometric measurements, there is a wide variety of methodologies available that can be used to estimate total body fat and

TABLE 20.1 Definition of obesity and disease risk

Classification	BMI (kg/m²)	Disease risk[a] relative to normal weight and waist circumference	
		Men <102 cm Women <88 cm	Men ≥102 cm Women ≥88 cm
Underweight	<18.5	Average[b]	
Normal weight	18.5–24.9	Average	Increased[c]
Overweight[d]	≥25 (25.0–29.9)	Increased	High
Obesity	≥30		
Class I	30.0–34.9	High	Very high
Class II	35.0–39.9	Very high	Very high
Class III[e]	≥40	Extremely high	Extremely high

The cut-off points for body mass index (BMI) and waist are based on White individuals. [a]Disease risk for type 2 diabetes, hypertension, and cardiovascular disease. [b]Risk for other clinical problems increases with underweight; [c]Increased waist circumference is a marker for increased risk even in persons with normal body weight; [d]The term 'overweight' refers to a BMI ≥25 kg/m², but most often is coined to specifically refer to the BMI range of 25–29.9 kg/m²; [e]Often referred to as severe, morbid, or extreme obesity.

regional fat deposition and distribution, typically with much greater precision, but also with considerably greater cost and technical difficulty. These include skinfold thickness measurements, hydrometry methods based on isotope dilution (water labelled with ²H or ¹⁸O), densitometric techniques based on water displacement (underwater weighing) or air displacement (BodPod), neutron activation analysis (whole-body ⁴⁰K counting), photon absorptiometry techniques such as dual energy X-ray absorptiometry (DXA), bioelectrical impedance analysis (BIA), ultrasound, computed tomography (CT), and magnetic resonance (MR) imaging and spectroscopy.

Prevalence and time trends of overweight and obesity

The global age-standardized mean BMI has increased from ~22 kg/m² in 1975 to ~24.5 kg/m² in 2016 among adult men and women (NCD Risk Factor Collaboration 2016). Accordingly, the prevalence of obesity has risen worldwide in almost every country in both sexes and across all age groups; still there are disparities in obesity rates based on geographical region, sex, age, race/ethnicity, urbanization, and socioeconomic status, which are likely of multifactorial origin. In particular, during the past four decades, the worldwide age-standardized prevalence of obesity has increased from 3.2 per cent to 10.8 per cent in adult men and from 6.4 per cent to 14.9 per cent in adult women. Obesity prevalence among adults has exceeded 30 per cent in men and women in high-income English-speaking countries and in several

countries in Polynesia and Micronesia, and in women in the Middle East and in southern and north Africa. At present, about 641 million adults (266 million men and 375 million women) worldwide are living with obesity, and many more have some degree of overweight. This alarming development have prompted obesity to be called a worldwide epidemic, or a pandemic.

20.3 Risks of obesity

Health impact

The increase in obesity rates has an important impact on the global incidence of a number of other diseases and conditions, including cardiovascular diseases, type 2 diabetes, certain types of cancer, osteoarthritis, obstructive sleep apnoea, and infertility, among others. The negative effects of obesity on multiple physiological functions result in dose-dependent increases in total and cause-specific mortality (most notably from type 2 diabetes and diseases of the liver, kidney, and the cardiovascular system) with increasing BMI within the range from 25 to 50 kg/m². The relationship between BMI and mortality is U- or J-shaped, with mortality rates generally being lowest for BMI values between 21 and 25 kg/m² and increasing linearly thereafter. Accordingly, life expectancy in men and women with overweight and obesity is reduced compared to those with normal body weight, by 1–2 years for those with BMI 25–30 kg/m², by 2–4 years for those with BMI 30–35 kg/m², by 5–6 years for those with BMI 35–40 kg/m², and by

8–10 years for those with BMI >40 kg/m^2. The years of life lost due to obesity are typically greater when obesity occurs at younger ages. Remarkably, the overall decrease in life expectancy due to obesity is similar to the decrease in life expectancy attributed to smoking. Obesity is one of the five leading risk factors for death globally, and in 2019 was responsible for more than 5 million deaths.

It is important to point out that many of the cardiometabolic complications of obesity can manifest even in the absence of increased BMI. Individuals with normal body weight but excessive accumulation of fat in a few key depots of the body, such as the visceral fat compartment (e.g., those with increased waist circumference) or the liver (e.g., those with hepatic steatosis), have increased risk for cardiometabolic disease. This implies that the overall health impact of excess fat deposition may be considerably larger than that reflected merely by high BMI values. On the other hand, there are also some individuals with obesity (about 7 to 13 per cent of all people with obesity) who are metabolically healthy, in that they do not present with any of the traditional obesity-related metabolic abnormalities (such as hyperglycaemia, dyslipidaemia, insulin resistance, etc.). However, this phenotype is most likely temporary and eventually, most of these individuals will convert to a metabolically unhealthy phenotype with increased risk to health.

Economic burden

Obesity has considerable economic consequences and has been estimated to account for 5–13 per cent of total healthcare expenditure in most countries during the years 2020–2050. The indirect financial burden is also important, as obesity in adults of working age may result in productivity losses through reduced workforce participation, increased work limitations, obesity-related sick leave, and work disability.

Social consequences

Individuals with increased body weight often experience heat intolerance and hyperhidrosis (increased sweating), intertrigo (inflammation of skin folds due to heat, moisture, friction, and lack of air circulation), difficulties with physical activities, and sexual problems of not only physical but also psychological nature. Additionally, individuals with overweight and obesity can suffer from low self-esteem and self-loathing, and also commonly have phobias about being in social or public situations due to previous experiences of weight bias and obesity stigma. Living with obesity has also been associated with lower academic performance, an overall shorter education, and reaching a lower socioeconomic status. These social and psychological problems often tend to be more pronounced than the somatic complications associated with obesity.

Medical complications of obesity

Type 2 diabetes

Overweight and obesity are major contributors to the development of type 2 diabetes in individuals who are genetically susceptible to the disease. Type 2 diabetes occurs 50–100 times more frequently in people with obesity than in those with normal body weight. Observational studies have demonstrated that overweight, obesity, physical inactivity, smoking, and dietary composition can account for almost all cases of type 2 diabetes. The genetic make-up determines if an individual living with obesity will develop type 2 diabetes or not and, in fact, several individuals with obesity who are not genetically predisposed to type 2 diabetes can maintain a normal glucose homeostasis throughout life. Type 2 diabetes is the most important medical consequence of obesity because it is common, has serious complications, is expensive to manage and difficult to treat, and reduces life expectancy by 8–10 years. Cardiovascular disease is a common complication of type 2 diabetes and a leading cause of death in this population (see also Chapter 21).

Type 2 diabetes takes several years to develop as a result of progressive metabolic defects that accompany obesity; impaired glucose tolerance and prediabetes are intermediate stages of disrupted glucose homeostasis before type 2 diabetes ensues. People with overweight and obesity who have impaired glucose tolerance or prediabetes have considerably increased risk for type 2 diabetes, but weight loss induced by lifestyle modification (as demonstrated in the US and Finnish Diabetes Prevention Trials and the Chinese Da Qing Diabetes Prevention Study) and pharmacologic interventions (as shown in the SCALE trial) decrease this risk by ~50 per cent (Magkos et al. 2020). Furthermore, many of the metabolic complications of type 2 diabetes are reversible with weight loss (induced by diet, drugs, or surgery), unless the diabetes has persisted for too long a time, with irreversible damage in the ability of pancreatic β-cells to secrete insulin. For example, the UK DiRECT (Diabetes Remission Clinical Trial) demonstrates that diet-induced moderate weight loss (5–9 per cent of baseline body weight at 1–2 years) leads to several-fold greater type 2 diabetes remission rates compared with a best-practice care control group (36–46 per cent vs. 3–4 per cent, respectively). Additional analyses suggest that the magnitude of weight loss per se rather the method (i.e., diet, exercise, drugs, or type of surgery) is the most important factor responsible for the prevention and treatment of type 2 diabetes (Magkos et al. 2020).

Hypertension

The risk of developing hypertension increases 5–6 times with obesity. Blood pressure is positively correlated with both the degree of obesity and waist circumference. Insulin resistance and compensatory hyperinsulinaemia, which are common obesity-related metabolic abnormalities, appear to be responsible for the increase in blood pressure. Insulin has an anti-natriuretic effect that results in an increase of both extracellular and intravascular volume. It is also possible that hyperinsulinaemia has a direct trophic effect on the smooth muscle cells of the arterioles, which can lead to a chronically hyperactive sympathetic nervous system. The hypertension associated with obesity is just as harmful as hypertension from other causes and should be controlled and treated with the same vigour as in patients without obesity. Even a small weight loss can result in a marked drop in blood pressure, and weight loss is a much more effective treatment than dietary salt restriction.

Cardiovascular disease

Obesity is associated with an increased risk of developing atherosclerotic cardiovascular disease, particularly heart failure and coronary heart disease, because of altered haemodynamics and heart structure. Also, central obesity (i.e., abdominal fat accumulation) is associated with a significantly increased risk for other types of cardiovascular disease. The risk of developing ischaemic heart disease and stroke is 2.5 and 6 times greater, respectively, among individuals with increased waist circumference compared to those with normal waist circumference, even when BMI is comparable. The risk of cardiac failure and atrial fibrillation is also markedly increased. The greater risk associated with central obesity is multifactorial, and includes an unfavourable blood lipid profile, hypertension, insulin resistance and type 2 diabetes, increased inflammatory markers such as C-reactive protein and fibrinogen, and reduced fibrinolytic activity. Even though weight loss improves many of the traditional cardiovascular risk factors, the cardiovascular outcomes of weight management remain controversial. Studies have suggested that weight gain, but also weight loss (particularly when it is rapid, unintentional, and experienced by individuals without obesity or older age), are linked to increased risk of ischaemic heart disease and/or stroke. However, among middle-aged patients with obesity and other comorbidities, intentional weight loss seems to be associated with reduced risk of cardiovascular events.

Arthritis

Osteoarthritis is a frequent complication of obesity, and obesity represents perhaps the strongest modifiable risk factor for developing osteoarthritis. Obesity-related osteoarthritis most commonly develops in the knees and ankles and has a significant negative impact on physical function and quality of life. A reduction in BMI by $2\,kg/m^2$ in women has been shown to reduce risk of developing osteoarthritis a decade later by ~50 per cent. Gout (arthritis urica) is also a frequent complication of obesity. Plasma urate is increased in the majority of patients with excess weight, but clinical symptoms are observed in only a minority.

Polycystic ovary syndrome

Polycystic ovary syndrome (PCOS) is a common endocrine disorder, affecting 6–8 per cent of women of reproductive age. PCOS is characterized by irregular or absent menses, impaired fertility, hirsutism, and acne. The pathogenic foundation of PCOS is insulin resistance, which causes an excess production of androgens by the ovaries. While there is clearly a genetic predisposition to PCOS, the presence of obesity worsens the degree of insulin resistance, and consequently increases the risk and worsens the symptomatology of PCOS. Weight management, including modest weight loss and maintenance of weight loss, or prevention of weight gain (also excess gestational weight gain), is a first-line treatment for women with PCOS. About 20–80 per cent of all women with PCOS have excess body weight (overweight or obesity), and weight loss has been shown to improve the clinical features of the disorder, including fertility rates. In fact, massive weight loss after bariatric surgery is associated with amelioration of insulin resistance and hyperandrogenism, restoration of ovulatory function, and subsequently promotes successful pregnancy with few maternal and neonatal complications.

Cancer

Obesity is associated with increased incidence of, and mortality from, several types of cancer. Observational cohort studies have found that obesity increases the risk of cancers of the endometrium, oesophagus, upper stomach, liver, kidney, thyroid, blood cells (multiple myeloma), brain (meningioma), pancreas, gallbladder, colon and rectum, ovaries and uterus, and breast (in postmenopausal women only). Subjects with obesity have an approximately 1.5–3.5 times greater risk of developing these cancers compared with normal-weight subjects, and about 5–45 per cent of the incidence of these cancers in both men and women can be attributed to overweight and obesity on a global scale, with considerable variability across sites. In addition to increased incidence, an increase in the BMI by 5 kg/m^2 increases cancer mortality by ~10 per cent. In men and women with a BMI ≥40 kg/m^2, the risk of dying from cancer compared to normal-weight individuals is increased by 50–60 per cent, and massive weight loss after bariatric surgery reduces cancer incidence and mortality, particularly among women. Even though data linking obesity to cancer comes from epidemiological studies that cannot establish cause-and-effect, evidence is consistent, and several mechanisms have been suggested to promote tumorigenesis in obesity. These include elevated levels of circulating insulin and insulin-growth factors (which promote cell proliferation), increased bioavailability of steroid hormones (particularly oestrogen), altered balance of adipokine secretion from adipose tissue (increased leptin which stimulates cell growth and decreased adiponectin with may have anti-proliferative effects), and chronic low-grade inflammation which can cause DNA damage.

Coronavirus disease 2019 (COVID-19)

Obesity increases the risk of severe disease and death due to COVID-19. A meta-analysis that pooled data from 399,000 patients found that people with obesity are 46 per cent more like to be infected by the SARS-CoV-2 virus, 113 per cent more likely to be hospitalized, 74 per cent more likely to be admitted to intensive care, and 48 per cent more likely to die, compared with individuals of normal body weight (Popkin et al. 2020). The reasons for these observations are not entirely clear but are likely of both mechanical and biological origin. Obesity is associated with compromised pulmonary function that makes ventilation difficult, including decreased expiratory reserve volume, functional capacity, and respiratory system compliance, as well as decreased diaphragmatic excursion because of excessive abdominal fat deposition. Furthermore, obesity is often accompanied by chronic low-grade inflammation, impaired immune function, and increased blood coagulation tendency, which may exacerbate morbidity in COVID-19 infections. For instance, there is evidence that the SARS-CoV-2 virus can directly infect adipocytes and elicit the inflammatory cytokine response that is the hallmark of severe COVID-19. Also, people with obesity are more likely than normal-weight people to have other comorbid conditions that are risk factors for severe disease or death due to COVID-19, including heart disease, lung disease, hypoventilation syndrome, obstructive sleep apnoea, asthmatic disease, hypertension, metabolic syndrome, and diabetes.

20.4 Causal factors

Obesity is a multifactorial disease, and each aetiological factor likely contributes to a different extent in different individuals. There is a complex array of biological, environmental, and societal factors that contribute to obesity (Figure 20.1) (Blüher 2019). Ultimately, however, the net result is the generation and maintenance of positive energy balance. Most cases of obesity are due to a combination of increased energy intake, owing to an inadequately functioning system of appetite regulation amidst an obesogenic environment, and decreased energy expenditure, owing to limited physical activity and increased sedentariness of daily life. From approximately 1910 to 1960, energy expenditure of the population decreased progressively because of technological advancements in the workplace and growing motorization. It is hypothesized that this reduction in energy expenditure was likely matched by a parallel reduction in energy intake which maintained energy balance, and thereby prevented big changes in body weight and the prevalence of obesity. In the years after World War II, however, an increase in the production and availability of energy-dense, carbohydrate-rich and fat-rich foods resulted in a 'flipping point' (around 1960–1970) which marked the beginning of a positive energy balance phase, in which the progressive increase in energy intake (that was of multifactorial origin; see section 'Environment and life-style factors') was disproportional to energy expenditure that either continued to decrease or remained constant at relatively low levels. This resulted in gradual weight gain and rising obesity prevalence rates.

Environment and lifestyle factors

Dietary energy content

Over the previous century—and particularly during the last 55 years or so—per capita calorie supply has been increasing consistently at a worldwide level (Our World in Data 2021). In the 1960s, the global average supply of dietary energy (that is, the available calories for consumers to eat) was 9.2 MJ (2200 kcal) per person per day. By 2013 this had increased to 11.7 MJ (2800 kcal) per person per day, representing a 27 per cent increase. To the extent calorie availability reflects calorie consumption, the number of people in the world consuming energy-richer diets has steadily increased since the 1960s, so that by 2017, more than half of the global population consumed high (10.0–11.7 MJ or 2400–2800 kcal per person per day) and very high (>11.7 MJ or >2800 kcal per person per day) calorie diets. In people with no predisposition to obesity, such diets can result in moderate increases in

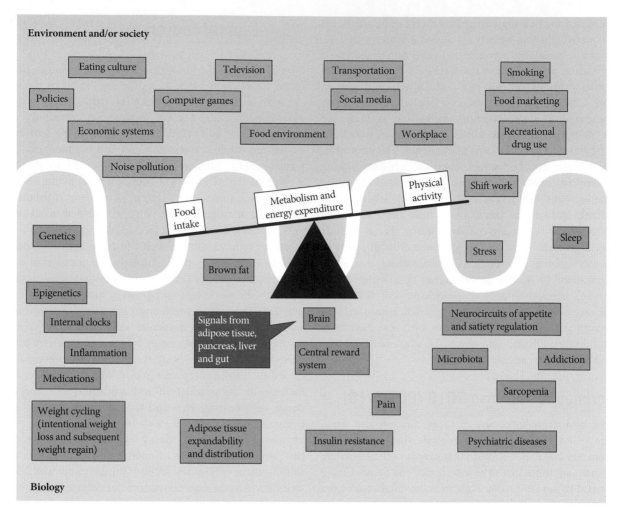

FIGURE 20.1 Complex biological, environmental, and societal factors contributing to positive energy balance and obesity. Individual physiology (e.g., genetic background) and behaviour (e.g., television viewing and transportation habits) are shaped by strong social and local environment factors and influence susceptibility to obesity. Weight gain is not caused by personal choice or by society but rather by the relationship between an individual and their environment.
Source: Reproduced with permission by Springer from M Blüher (2019) Obesity: Global epidemiology and pathogenesis. *Nat Rev Endocrinol* **15**(5), 288–298.

body weight, but weight gain can be considerable and eventually lead to obesity in predisposed individuals. Accordingly, dietary energy availability/consumption correlates directly with the prevalence of overweight and obesity across countries.

Energy density

The increase in dietary energy content over time may have been partly because of a change in the patterns of food consumption, with people being able to purchase and consume more food as a result of improvements in economy, but the most important contributing factor is likely a change in the nature of food supply itself. Advances in food science and technology have led to a near-complete transformation of the diet from a

traditional diet consisting mainly of home-cooked meals with minimally processed foods and food ingredients, to a modern diet with a substantial contribution of meals prepared outside home with convenient, highly processed, and energy-dense foods. Energy density is a relatively new concept that has been identified as an important factor in body weight control not only in adults but also in children and adolescents. Energy density is defined as the amount of energy per unit weight of a food or beverage (MJ per kg or kcal per gram). Foods with high water or fibre contents are generally low in energy density as water and fibre contribute considerable weight but little or no energy; these include, for example, foods that are naturally rich in water and/or fibre such as fruits, vegetables, and wholegrains, foods with a high water content such as soups and stews, foods like pasta and rice that absorb water during cooking, and foods like plain milk and yoghurt. On the other hand, foods with high fat or sugar contents are generally high in energy density; these include, for example, butter and cheese, confectionery, deep-fried foods, crackers and chips, sugary snacks and cereal, thick sauces, but also nuts and seeds. Over time, the development, marketing, and availability of foods with higher energy density, but also of foods with high energy content such as sugar-sweetened beverages, has dominated the food market and likely contributed to a passive increase in dietary calorie consumption. Still, a high dietary energy density is not always associated with weight gain; for example, low-carbohydrate diets are typically energy dense, but readily lead to weight loss in most patients with obesity, particularly those with type 2 diabetes.

Portion size

The rising prevalence of obesity has also been paralleled by increasing portion sizes in the marketplace. In the USA, for example, the serving size of soda has increased 6-fold since the 1950s, and the serving sizes of hamburger and French fries have both increased ~3-fold (Marteau et al. 2015). Similar trends have been documented in the UK and many European countries. Portion sizes are an important determinant of energy intake; the number of calories ingested by individuals at a meal is directly proportional to the serving size offered. To put this into perspective, simply having a sandwich for lunch every day provides approximately 30.6 MJ or 7300 kcal a year more today than it did in 1993, from the increase in the size of the bread slice alone. Currently, average serving sizes far exceed recommended portion sizes. This discrepancy results in confusion and impairs the ability of the public to estimate their calorie intake. In fact, research has indicated that most people are unable to accurately estimate portion sizes and, therefore, cannot accurately estimate their energy intake.

Physical activity and sedentariness

The level of physical activity has fallen dramatically during the second half of the previous century, with the replacement of manual labour with machines, and the increased use of every imaginable physical aid possible for housework, transport, and leisure pursuits. This trend has continued between 2001–2016 (Guthold et al. 2018) in high-income countries, with the prevalence of physical inactivity—defined as not meeting current recommendation for physical activity and health (i.e., engage in at least 150 min of moderate-intensity physical activity, or 75 min of vigorous-intensity physical activity per week, or any equivalent combination of the two)—increasing

from ~32 per cent to ~37 per cent. No changes were observed in low-income countries and a slight decrease was observed in middle-income countries, so that on a global scale, levels of physical inactivity did not change significantly between 2001–2016.

Currently, more than a quarter (~27.5 per cent) of all adults worldwide are insufficiently physically active, with women being less physically active compared with men; the corresponding sex-specific global prevalence of inactivity is ~32 per cent and ~25 per cent, respectively (Guthold et al. 2018). Also, a progressively more inactive lifestyle has a more detrimental effect on obesity in women than in men (note that the prevalence of obesity is also greater in women than in men; see section 'Prevalence and time trends of overweight and obesity'). Decreased regular physical activity and/or increased inactivity (sedentariness) increases risk for weight gain, because they favour induction of positive energy balance. On one hand, little time spent being physically active is directly linked with decreased total energy expenditure; on the other hand, more time spent being sedentary may also provide more opportunities for overeating. Regular physical activity increases energy expenditure and the capacity for fat oxidation but also improves appetite regulation, depending on the type, intensity, duration, and frequency of exercise. For example, the 30-minute daily physical activity recommendation by the American Heart Association and WHO is not enough to prevent weight gain and obesity in predisposed individuals; rather, 45–60 minutes of daily physical activity is required. High levels of habitual physical activity can prevent weight gain and obesity or induce weight loss, but they are not easily attained by the majority of individuals who struggle with excess weight.

Gut microbiome

In recent years, gut microbiota—the assortment of more than 10^{14} microorganisms that weigh approximately 1.5 kg and inhabit the gastrointestinal tract—has been implicated in the aetiology of obesity. There are reports demonstrating a reduction in gut microbiome diversity and microbial gene richness (count) in subjects with obesity compared to those with normal body weight, but there is still much debate on the exact signature of the obese microbiome. Gut microbiota mediate the interaction between the human host and the environment, by extracting energy from otherwise indigestible food components and producing metabolites and cytokines that can modulate host metabolism. For example, short-chain fatty acids can affect appetite regulation, and endotoxins (e.g., lipopolysaccharides) can aggravate low-grade inflammation. Several other pathways have been proposed as possible mediators of the link between the microbiome, metabolic health, and obesity. This includes gut barrier integrity, production of metabolites affecting satiety and insulin resistance, epigenetic factors, and metabolism of branched-chain amino acids and bile acids and subsequent changes in metabolic signalling and energy balance. There are also intervention studies demonstrating the direct role of diet in affecting the gut microbiome, but also the possible role of microbiome in differential responses to dietary treatment. However, definitive experimental evidence for a causal role of gut microbiota in human obesity is lacking.

Energy metabolism

Differences in daily energy expenditure between individuals result from differences in the amount of metabolically active tissues, the so-called lean body mass or fat-free mass (FFM), which are responsible for differences in basal (resting) metabolic rate; and differences in the level of habitual physical activity, which are responsible for differences in physical activity-induced thermogenesis. While absolute and proportional FFM (relative to fat mass) can by increased by various exercise training programmes, the resulting increase in resting metabolism because of greater FFM contributes little to total energy expenditure. Regular physical activity and the calories expended during exercise (i.e., physical activity-induced thermogenesis) is the only realistic way to increase total daily energy expenditure for most people. Energy metabolism is described in detail in Chapter 6.

With positive energy balance leading to weight gain, it is not only fat mass that increases. Weight gain from the normal body weight state to the obesity state consists of ~75 per cent fat (range: 50–90 per cent) and ~25 per cent FFM (range: 10–50 per cent); additional lean mass is needed to support the larger body size. This gain in FFM is composed mainly of increased organ and soft tissue mass (i.e., metabolically active tissues) and thus resting and total energy expenditure increase in parallel with increases in body weight and FFM. The basal metabolic rate (BMR) in MJ/day can be estimated as 1.55 + 0.09 × kg FFM (or in kcal/day as 370 + 21.5 × kg FFM). A person with an excess of 35 kg more than their ideal weight, of which about 12 kg is FFM, will have a BMR >1 MJ/day (or >240 kcal/day) higher than at their ideal body weight. In obesity, the level of physical activity is typically reduced, but because of the greater body weight, the energy expenditure in carrying out a given physical activity actually increases, as more mass needs to be shifted around. Accordingly, physical activity-induced thermogenesis typically does not decrease, and may even increase with obesity. Only when the actual levels of physical activity become very low, because of mobility problems or difficulty breathing or for other reasons, will obesity lead to a significant reduction in physical-activity induced thermogenesis.

Lipostatic regulation

Body fat mass is regulated through a lipostatic mechanism, which consists of a negative feedback between the fat depots and the brain (see also Chapter 6). The hormone leptin (from the Greek word 'leptos' meaning thin) is secreted by adipose tissue in amounts proportional to its size and conveys information about the size of body fat stores to appetite regulatory centres in the brain. Under normal circumstances, leptin signals to the central nervous system (CNS) that adequate energy stores are available (Figure 20.2) and consequently, other energy-consuming endocrinological effects necessary for normal growth and development can start or continue. Congenital leptin deficiency, caused by mutations in the leptin gene, leads to hyperphagia and morbid obesity in children as the brain does not receive the signal that adequate amounts of fat have been stored in the body. Leptin replacement therapy normalizes eating behaviour, metabolism, and body weight in these children. However, in the majority of individuals with excess body weight and fat (the so-called 'garden variety' obesity), the concentration of leptin in the peripheral circulation is about 10-fold higher than in normal weight individuals, suggesting the existence of leptin resistance. Leptin resistance may result from defects in the downstream signalling pathway (Figure 20.2) in a manner analogous to insulin resistance. It may also be that regulation by leptin is asymmetric, in that the absence or a very low leptin signal (as in congenital leptin deficiency) effectively trigger the CNS to engage mechanisms that preserve energy in the body (i.e., increase appetite and food intake) in times of famine/low energy supply, whereas the opposite cascade of events may not be operational in cases of abundance of energy (as in garden variety obesity).

The integration of both short-term and long-term regulation of energy balance takes place in the arcuate nuclei of the medial hypothalamus, where two different central neurons have opposing effects in a complex interplay involving several neurotransmitters and receptors (Figure 20.2). Low concentrations of leptin lead to activation of the neuropeptide Y/Agouti-related peptide (NPY/AgRP) neurons and inhibition of the pro-opiomelanocortin and cocaine- and amphetamine-regulated transcript (POMC/CART) neurons, leading to increased appetite, increased food intake, and weight gain. On the contrary, under normal physiological conditions and optimal homeostatic regulation of body weight, high concentrations of leptin lead to inhibition of the central NPY/AgRP neurons and activation of the POMC/CART neurons, and thus to reduced food intake and weight loss. The blood brain barrier is not freely permeable by leptin, and leptin transport into the brain is impaired in obesity, so that leptin concentrations in the

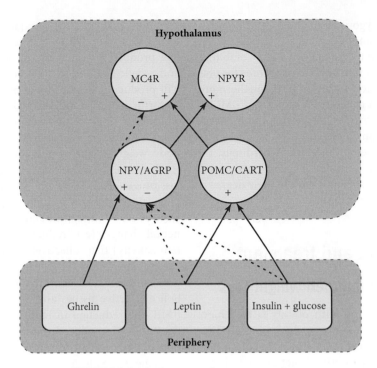

FIGURE 20.2 The leptin–melanocortin pathway

Activation of the orexigenic neurons produces the neurotransmitters neuropeptide Y (NPY) and agouti-related peptide (AgRP), which stimulate hunger and appetite. In contrast, activation of anorexigenic neurons, which express pro-opiomelanocortin (POMC) and cocaine- and amphetamine-regulated transcript (CART), increase release of alpha-melanocyte-stimulating hormone (α-MSH), which induces satiety and reduces appetite. NPY binds and activates the NPY receptor (NPYR) to stimulate appetite, while α-MSH binds to the melanocortin 4 receptor (MC4R) to suppress appetite. AgRP also binds to the MC4R but acts as an antagonist that inhibits binding of α-MSH to the receptor, and thereby leads to attenuated suppression of appetite.

cerebrospinal-fluid are only marginally greater and the ratio of brain-to-peripheral leptin levels is lower in people with obesity than those with normal body weight. This is likely responsible for the failure of central integration of energy balance in obesity.

Genetics

Gene abnormalities

Based on studies of families, twins, and adoptees, it has been established that heritable factors contribute to the development of obesity in a given environment. Typically, the genetic contribution to obesity is defined as polygenic, caused by genetic variance in several genes with a large environmental influence, or monogenic, caused by mutations in one gene with little environmental influence. Monogenic obesity is often subdivided into non-syndromic or syndromic obesity. Over 25 forms of syndromic obesity have been identified, the commonest one being the Prader–Willi syndrome (PWS). Non-syndromic monogenic obesity in humans is usually associated with genes expressed in brain areas related to appetite regulation. Congenital leptin deficiency is

characterized by severe hyperphagia, and individuals with these homozygous mutations in the leptin gene respond with considerable weight loss when treated with recombinant leptin injections. The most common form of monogenic obesity is caused by mutations in the melanocortin-4 receptor (MC4R), which occurs in 2–6 per cent of individuals with severe obesity (Figure 20.2).

Genetic predisposition

Genetic predisposition to obesity is generally based on polygenic obesity or common genetic variants. Genome-wide association studies (GWAS) have identified more than 300 single-nucleotide polymorphisms (SNPs) for various adiposity traits, grouped into seven categories: BMI (141 loci), body fat (15 loci), birthweight (8 loci), waist circumference or weight-to-hip ratio adjusted for BMI (97 loci), visceral adiposity (2 loci), waist circumference or weight-to-hip ratio not adjusted for BMI (26 loci), and extreme obesity (23 loci) (Goodarzi 2018). These variants have variable, and generally small, effect size. One of the variants with the largest effect size is located in the fat mass and obesity associated gene (FTO), which by itself is associated with 1.5 kg greater body weight in heterozygous form and 3 kg in homozygous form. This alone cannot explain the transition from the normal weight state to the obesity state. Combining many identified variants into genetic risk scores for obesity can account for up to 5 kg/m² in BMI. Furthermore, the association between genetic risk scores and BMI became stronger in recent times, suggesting the influence of genetic predisposition to obesity traits is augmented in modern obesogenic environments. Accordingly, the predisposition to obesity summed by genetic risk scores can help explain why individuals vary in their degree of obesity in the current environment.

20.5 Prevention and treatment

Principles of diet-induced weight loss

The goal for patients living with obesity is to achieve negative energy balance, mainly by decreasing food intake while still maintaining the satiating effect of the diet. As a general principle, the optimal diet to treat obesity should be safe, efficacious, nutritionally adequate, culturally acceptable, and economically affordable, and should ensure long-term compliance and maintenance of weight loss. Diet therapy consists of instructing patients how to modify their dietary intake to achieve a decrease in energy intake. Due to their larger body size, individuals with obesity have higher energy requirements for a given level of physical activity than their normal-weight counterparts (Figure 20.3). Reducing total energy intake will inevitably result in weight loss (Magkos 2020). During the early phase of weight loss, which lasts several weeks or months, body mass decreases rapidly but consists mainly of water, carbohydrate (glycogen), protein, and to a lesser extent fat; during the late phase, which extends for months or years, body mass decreases at a slow rate and consists mainly of fat (adipose tissue triglyceride). At the new body weight equilibrium, that is, when weight has again stabilized at a lower level, the weight loss consists of about 75 per cent fat mass and 25 per cent FFM. The relative contribution of fat and lean tissues to total weight loss remains constant across a wide range of weight losses and is not affected by the rate of weight loss (slower or faster, resulting from smaller or larger reductions in energy intake, respectively). The desirable rate of weight loss for most people with obesity is 0.5–1.0 kg per week, which reflects a sustainable rate of weight loss with lifestyle changes that can be permanent rather than temporary.

The overarching goal of reducing food intake can be achieved by setting an upper limit for energy intake. The larger the daily energy deficit the more rapid the weight loss. According to Wishnofsky's rule of thumb, stating that 7700 kcal (32.2 MJ) is equivalent to 1 kg of weight lost, a deficit of 300 to 500 kcal/day (1.3–2.1 MJ/day) will produce a weight loss of 300 to 500 grams/week, and a deficit of 500 to 1000 kcal/day (2.1–4.2 MJ/day) will produce a weight loss of 500 to 1000 grams/week. Unfortunately, diet-induced weight loss in real-life is nowhere near linear as this rule predicts, even when compliance to dietary calorie restriction is perfect; instead, body weight changes in a curvilinear fashion until it reaches a new steady-state (plateau). The rate of weight loss progressively slows as a result of a number of physiological adaptations, including hormonal changes (e.g., reduced triiodothyronine and leptin concentrations) and neural changes (e.g., reduced sympathetic nervous system activity) that collectively trigger reductions in total energy expenditure (Hall et al. 2012). As weight loss progresses, the loss of lean body mass consisting of metabolically active tissues and organs results in decreased BMR, the reduction in food intake is accompanied by decreased diet-induced thermogenesis, and the reduction of body weight results in decreased energy cost of movement (i.e., lower mass needs to be shifted around for a given pattern) and thus, decreased physical activity-induced thermogenesis. After modelling the physiological adaptions to weight loss, it has been estimated that every permanent decrease in energy intake by 22 kcal/day (~0.1 MJ/day) will lead to an eventual weight loss of 1 kg when the body weight reaches the new steady state. It will take nearly one year to achieve 50 per cent and three years to achieve 95 per cent of this weight loss. Gradually decreasing total energy expenditure as body weight is progressively lost effectively translates in a need to reduce energy intake even further to maintain a given energy deficit, which is not easy to accomplish.

Choosing the dietary energy deficit

The initial target of a weight loss programme should be to decrease body weight by 5–10 per cent from baseline. This amount of weight loss—even at the lower end of the range—has been demonstrated to result in significant improvements in body composition and fat distribution, metabolic function, and cardiometabolic risk factor profile (Magkos et al. 2016). Once this goal is achieved, a new target can be set. Patients will often want to lose more weight, but it should be remembered that even a 5 per cent weight loss is considered clinically significant, in that it improves cardiometabolic risk factors, decreases risk of co-morbidities, and decreases mortality associated with obesity-related complications. Several factors should be taken into consideration, for example, the degree of obesity of the individual, previous weight loss attempts, risk factors, co-morbidities, and personal and social capacity to undertake the necessary lifestyle changes.

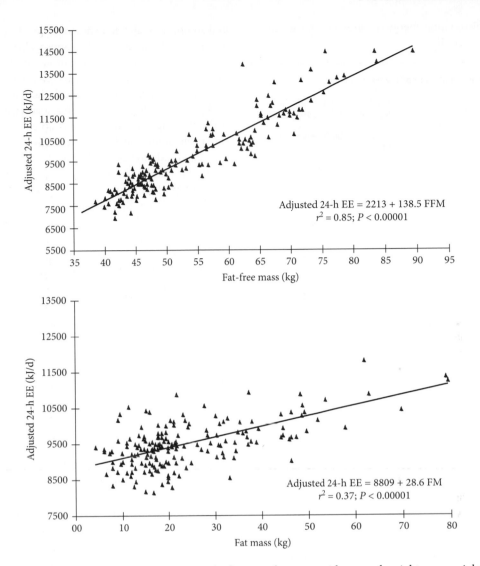

FIGURE 20.3 Energy expenditure (energy requirements) of men and women with normal weight, overweight, and obesity, relative to the amount of fat-free mass (FFM) (top) and fat mass (bottom). The 24-hour energy expenditure (EE) was measured by using whole-room indirect calorimetry inside respiration chambers and has been adjusted for differences in duration of exercise and spontaneous physical activity (top) and additionally for FFM (bottom).

Source: Reproduced with permission by Oxford University Press from B Klausen, S Toubro, A Astrup (1997) Age and sex effects on energy expenditure. *Am J Clin Nutr* **65**(4), 895–907.

To prescribe a diet with a given energy deficit, knowledge of the actual energy requirements for weight maintenance is required. It would seem natural to estimate patients' habitual energy intake from self-reported food consumption during a period of weight stability. However, these estimates are invalid due to systematic under-reporting of food consumption that can account for as much as 30–50 per cent of total energy intake. Energy requirements should therefore be assessed indirectly, by estimation of total daily energy expenditure. Whole-room indirect calorimetry (inside respiration chambers) or doubly labelled water (under free-living conditions) are methods that allow for measuring total

energy expenditure with high precision, but require specialized equipment and involve considerable cost, time (1–14 days), and technical difficulties. These techniques are seldom used in the context of the dietary treatment of obesity. Most commonly, resting metabolic rate (RMR) is measured or estimated and then multiplied by an activity factor to determine total energy expenditure. RMR can be measured within an hour by using bedside indirect calorimetry (canopy mode), or estimated by using prediction equations based on body weight, sex, and age (Table 20.2) or, for better accuracy, using equations that include body composition (FFM and fat mass) and even race/ethnicity among the predictors. Total energy

TABLE 20.2 Estimating energy requirements for weight maintenance in adults

BMR estimations based on body weight in women	
18–30 years	(0.062 × bodyweight in kg + 2.036) × 239 kcal/day
31–60 years	(0.034 × bodyweight in kg + 3.538) × 239 kcal/day
≥60 years	(0.038 × bodyweight in kg + 2.755) × 239 kcal/day
BMR estimations based on body weight in men	
18–30 years	(0.063 × bodyweight in kg + 2.896) × 239 kcal/day
31–60 years	(0.048 × bodyweight in kg + 3.653) × 239 kcal/day
≥60 years	(0.049 × bodyweight in kg + 2.459) × 239 kcal/day
Estimated total energy expenditure = BMR × PAL	
Activity level	PAL
Sedentary or light activity	1.5
Active or moderately active	1.7
Vigorous or vigorously active	2.0

The table is based on data obtained from the WHO/FAO (WHO/FAO/UNU 2004). Basal metabolic rate (BMR) is the minimal rate of energy expenditure and is measured under standard conditions of rest, fasting, immobility, and thermoneutrality. The Physical Activity Level (PAL) is an activity factor reflecting the overall lifestyle of an individual.

expenditure can then be calculated by multiplying RMR (in MJ/day or kcal/day) by the 'physical activity level' (PAL), which is a factor that reflects how active an individual's lifestyle is (Table 20.2). The energy content of the prescribed diet is then calculated as total energy expenditure (which equals energy requirement for weight maintenance) minus the desired energy deficit.

Clinical outcome and dietary adherence

Translating knowledge about the physiology of energy balance and weight loss into clinical practice requires a high degree of compliance on behalf of the patient, which can be difficult to obtain, and even more so to maintain. A variety of approaches, including diet, exercise, and pharmacotherapy, can be used to induce clinically significant weight loss in the short-term (on average, 5–9 per cent after six months), and among the various dietary treatments, low-calorie diets, high- and low-carbohydrate diets, high- and low-fat diets, high-protein diets, meal replacements, dietary regimens with and without behavioural and physical activity components,

and so on, have all been shown to be effective. However, a common characteristic of all approaches is the regain of lost weight over time, resulting in average weight losses of only 3–5 per cent after 2–4 years (Magkos 2020). Weight loss outcomes are better in clinical trials conducted in specialized clinics than in trials conducted by non-specialists without sufficient resources and access to allied health-care professionals such as dietitians and psychologists.

Dietary adherence is the cornerstone of successful treatment and has been shown repeatedly to be among the best predictors of long-term weight loss success, much better than diet per se or macronutrient composition; for example at two years in the POUNDS LOST (Preventing Overweight Using Novel Dietary Strategies) trial (Magkos 2020). At the same time, however, maintaining a high level of adherence is the most complicated part of the dietary treatment of obesity. To improve adherence, consideration should be given to the patient's food preferences, as well as to personal, educational, and social factors. Great efforts should be made to see the patient frequently and regularly, as increased patient support is positively associated with the success of any weight loss programme. Furthermore, long-term weight reduction is unlikely to succeed unless the patient acquires new eating and physical activity habits. These lifestyle changes can be facilitated by a combination of behaviour modification techniques which should comprise an integral part of the treatment programme.

Armamentarium of diet therapies for weight loss and maintenance

The dietary treatment of obesity is a rather 'new' development in experimental nutrition research, but diets for weight loss have been around for more than 2500 years. Therapeutic diets for obesity comprise several different weight-reduction regimens but can generally be classified into two distinct categories: hypocaloric (energy-restricted) diets and *ad libitum* diets. Low-caloric diets (LCD) usually provide 3.3–6.3 MJ/day (800–1500 kcal/day) and weight loss occurs independently of the diet composition. Very low caloric diets (VLCD) are modified fasts providing 0.8–3.3 MJ/day (200–800 kcal/day) that replace normal foods. *Ad libitum* diets do not restrict energy intake directly, but instead target a nutrient composition that will reduce food intake because of increased satiating effect.

Very low caloric diets (VLCD)

VLCDs (0.8–3.3 MJ or 200–800 kcal per day) aim to supply very little energy but all essential nutrients. Reducing the energy content of the diet requires an increased nutrient density. This can be difficult to obtain with natural foods if the diet is to be acceptable once the energy content of the diet becomes lower than 3.3 MJ/day (800 kcal/day). This has led to the commercial production of VLCDs supplemented with all nutrients in amounts that meet the Recommended Daily Allowance (RDA) criteria. Today, the 3.3 MJ/day (800 kcal/day) VLCD is the only version recognized as being both effective and safe. Patients using VLCDs have an increased risk of developing gallstones, and can also experience cold intolerance, dry skin, fatigue, dizziness, muscle cramps, headache, and hair loss. Therefore, use of such diets without medical supervision is not recommended.

Low caloric diets (LCD)

LCDs usually provide 3.3–6.3 MJ/day (800–1500 kcal/day) and normally consist of natural foods. The appropriate energy level chosen should be determined based on initial body weight, sex, and activity level. LCDs result in slower weight loss than VLCDs, but randomized trials demonstrate similar amounts of weight loss in the long term. Importantly, using LCDs rather than VLCDs to induce weight loss introduces healthy eating habits early in the weight management programme, giving a longer period in which to familiarize the patient with the dietary changes that are critical for long-term weight loss and maintenance. LCDs produce weight loss regardless of the duration of treatment and reduce body weight by an average of 8 per cent over three to 12 months.

Low-fat diets

Restricting intake of dietary fat remains the most effective means to reduce the energy density of the diet and hence the total energy intake. Accordingly, a meta-analysis of 37 clinical trials that randomized 57,000 participants to lower fat versus higher fat diets without the explicit intention to reduce body weight (i.e., *ad libitum* feeding), for a duration ranging from six months to more than eight years, found that dietary fat restriction induces a mean weight loss of 1.4 kg (Hooper et al. 2020). The amount of weight loss is influenced by the extent of reduction in dietary fat energy, the baseline dietary fat intake, and the pre-treatment BMI. On average, for every 1 per cent decrease in total energy as fat, weight decreases by 0.2 kg, whereas individuals with higher BMI and lower fat intake at baseline tend to lose more weight on low-fat diets.

Low-carbohydrate diets

Low-carbohydrate diets produce weight loss that is comparable to low-fat diets. In fact, an early meta-analysis reported that low-to-moderate carbohydrate diets (<45 per cent of total energy from carbohydrate) induce greater weight loss and improve metabolic risk factors to a greater extent than low-fat diets in the short-term (≤6 months) but not at later time points (12 months). This was confirmed by another meta-analysis that included 48 randomized trials with some 7300 individuals, in which no significant differences were found after six and 12 months between popular low-carbohydrate (e.g., Atkins, South Beach) and low-fat (e.g., Ornish) diets (Magkos 2020). Some differences among other popular dietary regimens were observed, but there were minimal and likely not clinically significant. Likewise, in the POUNDS LOST study, the largest and longest dietary intervention trial to date, no differences in weight loss over two years were found between subjects with overweight and obesity randomized to diets with carbohydrate contents ranging from 35 per cent to 65 per cent of the total energy (Magkos 2020).

High protein diets

Meta-analyses of replacing dietary carbohydrate with protein have shown that high-protein diets have more favourable effects on weight loss, body composition, and cardiovascular risk factors than low-fat diets, at least in the short-term (3–9 months). Differences between high-protein and low-fat dietary regimens attenuate with time, particularly when prescribed diets are energy-restricted, and become non-significant from one year into treatment onwards (Magkos 2020). More recently, a few large trials have evaluated the efficacy of dietary protein for weight maintenance after clinically significant weight loss (achieved by use of an LCD). The DiOGenes (Diet, Obesity, and Genes) study found that an *ad libitum* diet with high protein content (23 per cent of total energy intake) was more effective for maintaining weight loss than a diet with moderate protein content (17 per cent of total energy intake) for at least 6–12 months (Figure 20.4) (Larsen et al. 2010). The subsequent PREVIEW (Prevention of Diabetes Through Lifestyle Intervention and Population Studies in Europe and Around the World) project, however, which was ~3 times as large and ~6 times as long as the DiOGenes study, found that diets with high or moderate protein contents do not differ in their ability to maintain body weight over three years after clinically significantly weight loss (Figure 20.4) (Raben et al. 2021). Like for low-carbohydrate diets, the main challenge with high-protein diets is decreasing adherence over time that makes it difficult to distinguish between not following the diet prescription from decreasing efficacy of the diet per se. In a post-hoc analysis of data from the PREVIEW study, individuals who consumed ≥0.8 g/kg protein daily regained 1.5 per cent points less weight than those consuming <0.8 g/kg during

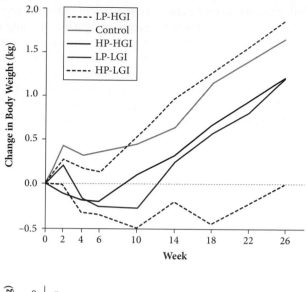

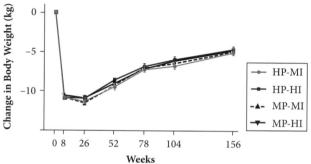

FIGURE 20.4 Weight maintenance with *ad libitum* diets differing in protein content in the DiOGenes (top) and PREVIEW (bottom) studies. In the DiOGenes trial, after a weight loss of ≥8 per cent induced by a low calorie diet, subjects were randomized to diets with high or low protein content (23 per cent and 17 per cent of total energy, respectively) and low or high in glycaemic index (2-by-2 factorial design), consumed *ad libitum* for six months, or a control diet which followed dietary guidelines in each participating country. Higher dietary protein content was associated with ~1 kg less weight regain over six months. In the PREVIEW trial, after a weight loss of ≥8 per cent induced by a low calorie diet, subjects were randomized to diets with high or moderate protein content (25 per cent and 15 per cent of total energy, respectively), consumed *ad libitum* for three years, in combination with moderate or high-intensity physical activity (2-by-2 factorial design). No differences in body weight were detected over three years.

Source: Reproduced with permission by Massachusetts Medical Society from T M Larsen, S M Dalskov, M van Baak et al. (2010) Diets with high or low protein content and glycemic index for weight-loss maintenance. *N Engl J Med* **363**(22), 2102–2113; and with permission by Wiley from A Raben, P S Vestentoft, J Brand-Miller et al. (2021) The PREVIEW intervention study: Results from a 3-year randomized 2 x 2 factorial multinational trial investigating the role of protein, glycemic index and physical activity for prevention of type 2 diabetes. *Diabetes Obes Metab* **23**(2), 324–337.

the three-year weight maintenance phase. At the very least, therefore, an adequate—but not necessarily high—protein intake facilitates maintenance of weight loss.

Low glycaemic index diets

Glycaemic index is defined as the ability of a standardized amount of bioavailable carbohydrate in a food to raise blood glucose concentration, relative to that of a reference food such as pure glucose or white bread (scale from 0 to 100). There is evidence from observational studies that foods with low glycaemic index, compared to foods with high glycaemic index, are associated not only with improved glucose metabolism but also with improved body weight outcomes (see also Chapter 7). However, well-controlled randomized clinical trials (e.g., GLYNDIET study) have failed to demonstrate differences in weight loss over six months between isocaloric energy-restricted diets with similar protein, fat, and carbohydrate contents but very different glycaemic index (low, 34; or high, 62) (Magkos 2020). Meta-analyses of energy-restricted and *ad libitum* dietary intervention trials find that diets with low glycaemic index or load (glycaemic load is the product of glycaemic index multiplied by the actual amount of the food carbohydrate consumed) have small and inconsistent effects on body weight compared to diets with high glycaemic index or load. An additional weight reduction by 0.6–1.1 kg is often found with low glycaemic index diets, but this does not always reach statistical significance. This beneficial effect is more robust (significant reduction by 1.8 kg) when the difference between comparison diets in glycaemic index exceeds 20 points (Zafar et al. 2019).

Other components of successful weight management

The real challenge in weight management is to maintain the reduced body weight and prevent subsequent relapse (Figure 20.4) (Magkos 2020). Successful weight loss maintenance, at least for six months or more, is associated with more initial weight loss, reaching a self-determined weight goal, having a physically active lifestyle and a regular meal pattern, and self-monitoring of eating and physical activity behaviours (Varkevisser et al. 2019). Internal motivation, social support, better coping strategies, and more psychological strength and stability (e.g., ability to handle stress) are also very important for successful weight management. In a systematic review of the efficacy of dietary treatment of obesity, including studies with long-term follow-up (≥3 years), success was defined as maintenance of all weight lost initially, or maintenance of at least 9 kg of weight loss. Initial weight loss ranged from 4 to 28 kg, and 15 per cent of the patients fulfilled one of the criteria for success; the success rate remained constant for up to 14 years of observation.

Diet combined with group therapy leads to better long-term success rates (27 per cent) than diet alone (15 per cent), or diet combined with behaviour modification and active follow-up; though active follow-up produces better weight loss maintenance than passive follow-up (19 per cent and 10 per cent, respectively). These findings stress once again the fact that long-term maintenance of weight loss is possible if efforts focus on changing multiple behaviours that aim at reducing energy intake and increasing energy expenditure, and continuously self-monitoring progress.

Exercise for weight loss and maintenance

Increased daily physical activity is an important component of successful weight control. Although it is not widely acknowledged, under carefully controlled laboratory conditions, increasing energy expenditure by exercise is as effective as decreasing energy intake by dietary restriction and results in similar and predictable decreases in body weight. In real life, however, exercise prescription is not as effective for weight loss as hypocaloric diet prescription. This is partly because of low adherence to the prescribed exercise but mainly because of multiple compensatory responses—both physiological and behavioural—that collectively trigger an increase in food intake and a decrease in other physical activities of daily life, which mitigate the exercise-induced increase in energy expenditure. Even in trials that devoted considerable resources to achieve near-perfect adherence to exercise (>90 per cent), but did not impose any sort of dietary control, actual weight loss was only 38–43 per cent of that predicted based on the energy expenditure of exercise, and clinically significant weight loss (≥5 per cent from baseline) was achieved by 46–62 per cent of the participants (Donnelly et al. 2013). These findings suggest that even in the best-case exercise scenario, individuals compensate on average for more than half of the energy expended during exercise, whereas one in two of them compensates to an extent that prevents successful weight loss.

Patients with overweight or obesity should gradually aim at increasing their physical activity, and as they lose weight, further increases in physical activity and exercise should be emphasized to help maintain weight. The impact of exercise on body weight control largely depends on the ability of patients to engage in adequate levels of physical activity. Although some weight regain is unavoidable, post hoc analyses from randomized clinical trials find a dose-dependent relationship between the amount of exercise and long-term weight maintenance, with optimal results obtained with ≥300 minutes or ≥2000 kcal of moderate or vigorous intensity exercise per week. Accordingly, the recommended level of exercise for weight control is at least 45–60 minutes of moderate-intensity physical activity on most days of the week. Besides effects on body weight, physical activity also improves many other health-related outcomes. For example, addition of an exercise component to diet restriction decreases the contribution of lean body mass to total weight loss (11–13 per cent) compared to diet only (24–28 per cent), so that ~0.5 kg of FFM is preserved across a wide range of weight losses. Exercise is also very potent (and more potent that diet per unit of weight loss) in decreasing intra-abdominal fat accumulation. These exercise-induced effects on body composition and body fat distribution are associated with improvements in many cardiometabolic risk factors, even in the absence of major reductions in body weight.

Pharmacotherapy for obesity

Currently, only a handful of drug treatments for obesity have been approved by the US Food and Drug Administration (FDA) and the European Medicines

TABLE 20.3 Available drugs for obesity treatment in the USA and EU

Generic name	Liraglutide	Semaglutide	Naltrexone/bupropion	Orlistat	Phentermine/Topiramate
Brand name	(Saxenda®)	(Wegovy®)	(Mysimba®, Contrave®)	(Xenical®, Alli®)	(Qsymia®)
EU status	Approved (2015)	Approved (2022)	Approved (2015)	Approved (2012)	Rejected
USA status	Approved (2014)	Approved (2021)	Approved (2014)	Approved (1999)	Approved (2012)
Mechanism of action	GLP-1 receptor agonist		μ-Opioid receptor antagonist/Noradrenaline and dopamine reuptake inhibitor	Pancreatic lipase inhibitor	Sympathomimetic, appetite suppressant (precise mechanism unknown)
Dose	3.0 mg injection (once daily)	2.4 mg injection (once weekly)	32 mg/360 mg (daily)	60–120 mg (3 times daily)	3.75/23 mg; 7.5/46 mg; 11.25/69 mg; 15/92 mg (once daily)
Main side-effects	Nausea, diarrhoea, vomiting, constipation		Nausea, vomiting, headache, dizziness	Steatorrhea, faecal urgency	Insomnia, dizziness, parasthesia
Indication	BMI ≥30 kg/m², or ≥27 kg/m² with other risk factors, i.e., type 2 diabetes, hypertension, or dyslipidaemia. In combination with dietary calorie restriction and increased physical activity.				

Agency (EMA) (Table 20.3) (Williams et al. 2020). All drugs are to be used in patients with a BMI ≥30 kg/m², or ≥27 kg/m² in association with other risk factors such as type 2 diabetes, hypertension, or dyslipidaemia, and always as adjuncts to conventional lifestyle treatment involving dietary calorie restriction and increased physical activity. They are all contraindicated in pregnancy and must be stopped prior to any conception attempts.

Orlistat is a specific inhibitor of pancreatic lipase, the enzyme secreted from the exocrine pancreas that is responsible for the enzymatic digestion of fat in the gastrointestinal lumen. Intake of the drug before meals reduces the absorption of dietary fat by 30 per cent and thereby results in negative energy balance via reduced calorie absorption. Orlistat leads to an extra weight loss of 2.9–3.4 kg/year over placebo. Treatment of individuals who have obesity and impaired glucose tolerance with orlistat in conjunction with lifestyle modification has been shown to decrease risk of progression to type 2 diabetes. Orlistat is often poorly tolerated due to common gastrointestinal side-effects including bloating, diarrhoea, faecal soiling, and flatulence. As there is a reduction in fat absorption, the absorption of fat-soluble vitamins (A, D, E, and K) and other nutrients (β-carotene, lycopene, flavonoids, etc.) may be impaired.

Liraglutide is a long-acting glucagon-like peptide-1 (GLP-1) analogue that binds to its receptor in the arcuate nucleus of the hypothalamus, thereby increasing satiety

and reducing food intake; it also delays gastric emptying. Liraglutide leads to an extra weight loss of 5.3–5.9 kg/year over placebo. The drug has also been shown to substantially reduce incidence of type 2 diabetes over a period of three years. Nausea, constipation, or diarrhoea can be experienced when liraglutide is initiated because it delays gastric emptying; however, these effects are transient, and symptoms quickly subside. Liraglutide is contraindicated in cases of personal or family history of medullary thyroid carcinoma or multiple endocrine neoplasia syndrome type 2. Semaglutide is a much more potent GLP-1 receptor agonist that produces a weight loss of about 12.5 kg over placebo with the same tolerability and safety profile as liraglutide. Importantly, structural modifications give semaglutide a more than 10-fold longer half-life than liraglutide (165 hours vs. 13–15 hours, respectively), thereby allowing dosing once per week as opposed to once per day. Semaglutide has also been shown to reduce cardiovascular events in patients with type 2 diabetes. There are several other available GLP-1 receptor agonists that have been licenced for type 2 diabetes (e.g., exenatide, lixisenatide, dulaglutide), as well as various combinations thereof (e.g., dual or triple agonists). Tirzepatide, for example, is a dual GLP-1/glucose-dependent insulinotropic polypeptide (GIP) receptor agonist that produces a weight loss of more than 20 kg over placebo; the drug received fast track designation by the FDA and will likely be approved for the treatment of obesity in 2023.

Naltrexone/bupropion is a combination therapy for obesity. Bupropion acts as a noradrenergic and dopaminergic reuptake inhibitor in the hypothalamus, and naltrexone is an opioid receptor antagonist. Together, they stimulate hypothalamic POMC neurons and mediate release of alpha-melanocyte-stimulating hormone (α-MSH) that reduces hunger and food intake (Figure 20.2). Naltrexone/bupropion therapy leads to an extra weight loss of 4.4 kg/year over placebo. Bupropion has been in use as an antidepressant for nearly two decades, and more recently as a smoking cessation aid in a sustained-release preparation. Side effects can include nausea, vomiting, dizziness, headache, and insomnia. Naltrexone/bupropion therapy is contraindicated for use with other opioid drugs or monoamine oxidase inhibitors, and also in patients with uncontrolled hypertension, seizure disorders, anorexia, or bulimia nervosa.

Phentermine/topiramate is a combination therapy for obesity approved for use only in the USA. Phentermine increases norepinephrine in the hypothalamus, thereby suppressing appetite, while topiramate modulates the gamma-aminobutyric acid (GABA) receptor that possibly also contributes to reduced appetite. Phentermine/topiramate therapy leads to an extra weight loss of 6.6–8.6 kg/year over placebo. Additionally, it improves several cardiometabolic parameters. Common side effects include paraesthesias, dysgeusia, dizziness, and dry mouth, whereas contraindications include pregnancy (risk of oral cleft), glaucoma (also a rare side effect), and hyperthyroidism.

There are several other emerging medications that hold promise for the future of obesity treatment. For example, inhibitors of the renal sodium-glucose co-transporter-2 (SGLT-2) which are currently licensed for the treatment of type 2 diabetes. These drugs (e.g., empagliflozin, ertugliflozin, dapagliflozin, canagliflozin) block the SGLT-2 transporter in the proximal tubule and thereby decrease glucose reabsorption. The result of augmented glycosuria is covert calorie loss in the urine (200–300 kcal/day) that promotes negative energy balance and weight loss.

Surgical treatment of obesity

Conventional weight management programmes consisting of diet, exercise, and pharmacological interventions fail in a substantial proportion of patients with severe obesity. Bariatric surgery is the treatment of choice for long-lasting weight loss in well-informed and well-motivated patients. Most surgical procedures for obesity have acceptable operative risks and are generally indicated for adults with a BMI ≥40 kg/m^2, or ≥35 kg/m^2 in conjunction with comorbidities such as type 2 diabetes, hypertension, or obstructive sleep apnoea, who have tried and failed to lose weight with intensive lifestyle and pharmacological interventions.

Currently, the commonest bariatric procedures worldwide are sleeve gastrectomy (~46 per cent of total), Roux-en-Y gastric bypass (~40 per cent of total), and adjustable gastric banding (~8 per cent of total). All these procedures lead to weight loss because of a significant reduction in food intake, albeit via different mechanisms. Gastric banding was once the most commonly performed bariatric surgery. It is a purely restrictive procedure that compartmentalizes the upper stomach by placing a tight, usually adjustable band around the entrance to the stomach. This mechanical restriction increases fullness while consuming significantly lower amounts of food. Advantages of gastric banding include the simplicity of the procedure, its easy revision or complete reversibility, and a low rate of perioperative complications. The efficacy of gastric banding is variable; a weight loss of ~20 per cent is typically achieved at one year after surgery. Gastric bypass employs a combined restrictive and malabsorptive approach. This procedure involves the creation of a small gastric pouch, divided and separated from the stomach remnant. The small intestine is then divided, creating a proximal biliopancreatic limb that transports the secretions from the gastric remnant, liver, and pancreas, and an alimentary (or Roux) limb, that is anastomosed to the new gastric pouch to drain consumed food. The distal end of the biliopancreatic limb is then anastomosed to the alimentary limb, creating a common channel where digestive enzymes mix with ingested food. As food is presented more rapidly to the small intestine, there is a robust release of gastrointestinal hormones such as GLP-1 and PYY, which promote satiety and facilitate a reduction in energy intake. Weight loss following gastric bypass is impressive and averages ~35 per cent at one year after surgery. In the landmark SOS study, maximum weight loss after both gastric banding and gastric bypass was achieved one year post-surgery, followed by variable weight regain in the subsequent years. At the 15-year follow-up, weight loss from baseline averaged 13 per cent in the gastric banding group and 27 per cent in the gastric bypass group, demonstrating excellent long-term weight loss maintenance (Adams et al. 2017). For gastric bypass in particular, and despite that weight loss varies widely between individual patients (just like with any weight loss intervention), >90 per cent of patients manage to maintain at least a 10 per cent weight loss from baseline at 12 years after surgery, ~70 per cent maintain at least a 20 per cent weight loss, and ~40 per cent maintain at least a 30 per cent weight loss; whereas, only 1 per cent of patients regain all post-surgical weight loss. Sleeve gastrectomy involves dividing the stomach along its vertical length to remove 75–85 per cent of its volume, creating a slender banana-shaped sleeve. Most of the cells producing the hunger hormone ghrelin are removed, thereby contributing to a decrease in hunger. In addition to being a restrictive procedure, the more rapid transit time of food into the intestine also results in increased secretion of several gastrointestinal hormones, such as GLP1 and PYY, which promote satiety. This procedure is fairly similar to gastric bypass in terms of long-term effects on body weight, cardiometabolic risk factors, and quality of life, and somewhat better in terms of postoperative complications.

The massive weight loss achieved by bariatric surgery has profound effects on cardiometabolic risk factors, disease outcomes, but also quality of life, rates of employment, and healthcare costs. For example, remission of type 2 diabetes occurs in ~57 per cent of patients after gastric banding and ~80 per cent of patients after gastric bypass, with corresponding per cent excess weight losses of ~46 per cent and ~60 per cent, respectively (Magkos et al. 2020). Importantly, bariatric surgery is associated with a reduction in total mortality by 30–40 per cent over 7–11 years of follow up. Although surgery brings about a variety of weight loss-independent metabolic changes, for example, in gut hormone responses, circulating bile acids and branched-chain amino acids, gut microbiota composition, and the temporal pattern of glucose and insulin responses to meal ingestion, most—if not all—of the improvements in body composition, body fat distribution, and 'hard' metabolic function endpoints (including insulin sensitivity in the liver and skeletal muscle, β-cell function and daily glucose and insulin profiles), are because of weight loss per se, as shown in studies in which patients achieved the same amount of weight loss with diet only or with surgery (Yoshino et al. 2020). There is little evidence for weight loss-independent, surgery-specific metabolic benefits, for example, a greater improvement in insulin sensitivity after weight loss induced by biliopancreatic diversion (a not so commonly performed procedure) than after matched weight loss induced by gastric bypass. Given that the effects of weight loss on most health outcomes are dose-dependent, it follows that much of the purported metabolic advantage of bariatric surgery over conventional dietary treatments is simply because of the far greater ability of bariatric surgery to reduce energy intake, and thereby induce greater weight loss and maintain it better.

KEY POINTS

- Overweight and obesity (increased weight relative to height, based on the BMI value) have become highly prevalent; worldwide, obesity affects 10–15 per cent of adults.

- Abdominal fat accumulation at any BMI (i.e., even among people with normal body weight) is associated with hypertension, dyslipidaemia, metabolic syndrome, type 2 diabetes, and cardiovascular disease.

- Weight loss in patients with obesity decreases the frequency and severity of many of the medical complications of obesity, with more weight loss likely having more pronounced effects.

- No single factor is responsible for obesity. Instead, there are many aetiological factors that synergize and each one likely contributes to a different extent in different individuals.

- Weight gain and obesity may be triggered in susceptible individuals in the modern obesogenic environment, because of increased availability of convenience foods with high energy density, increased portion sizes, and limited daily physical activity.

- Prevention and treatment of obesity require a reduction in calorie intake and an increase in physical activity energy expenditure, in conjunction with behaviour modification to facilitate lifelong changes in lifestyle.

- Hypocaloric diets consisting of normal foods or meal replacements are effective in inducing a weight loss of 5–10 per cent in most patients with obesity, but long-term weight maintenance is challenging.

- Regular exercise training is a useful adjunct to weight loss and weight maintenance, but a high level is usually required; lower levels still improve other health outcomes.

- Currently approved medications for weight loss, in conjunction with recommended changes in diet and physical activity, lead to an annual extra weight loss of about 3–13 kg over placebo.

- Bariatric surgery is the most effective therapeutic option for patients with severe or comorbid obesity who have repeatedly failed to lose weight with other methods. Surgery produces massive and sustainable weight loss.

Be sure to test your understanding of this chapter by attempting multiple choice questions.

REFERENCES

Adams T D, Davidson L E, Litwin S E et al. (2017) Weight and metabolic outcomes 12 years after gastric bypass. *N Engl J Med* **377**(12), 1143–1155. doi: 10.1056/NEJMoa1700459.

Blüher M (2019) Obesity: Global epidemiology and pathogenesis. *Nat Rev Endocrinol* **15**(5), 288–298. doi: 10.1038/s41574-019-0176-8.

Donnelly J E, Honas J J, Smith B K et al. (2013) Aerobic exercise alone results in clinically significant weight loss for men and women: Midwest Exercise Trial 2. *Obesity* **21**(3), e219–e228. doi: 10.1002/oby.20145.

Goodarzi M O (2018) Genetics of obesity: What genetic association studies have taught us about the biology of obesity and its complications. *Lancet Diabetes Endocrinol* **6**(3), 223–236. doi: 10.1016/S2213-8587(17)30200-0.

Guthold R, Stevens G A, Riley L M et al. (2018) Worldwide trends in insufficient physical activity from 2001 to 2016: A pooled analysis of 358 population-based surveys with 1·9 million participants. *Lancet Glob Health* **6**(10), e1077–e1086. doi: 10.1016/S2214-109X(18)30357-7.

Hall K D, Heymsfield S B, Kemnitz J W et al. (2012) Energy balance and its components: Implications for body weight regulation. *Am J Clin Nutr* **95**(4), 989–994. doi: 10.3945/ajcn.112.036350.

Hooper L, Abdelhamid A S, Jimoh O F et al. (2020) Effects of total fat intake on body fatness in adults. *Cochrane Database Syst Rev* **6**(6), CD013636. doi: 10.1002/14651858.CD013636.

Larsen T M, Dalskov S M, van Baak M et al. (2010) Diets with high or low protein content and glycemic index for weight-loss maintenance. *N Engl J Med* **363**(22), 2102–2113. doi: 10.1056/NEJMoa1007137.

Magkos F (2020) The role of dietary protein in obesity. *Rev Endocr Metab Disord* **21**(3), 329–340. doi: 10.1007/s11154-020-09576-3.

Magkos F, Fraterrigo G, Yoshino J et al. (2016) Effects of moderate and subsequent progressive weight loss on metabolic function and adipose tissue biology in humans with obesity. *Cell Metab* **23**(4), 591–601. doi: 10.1016/j.cmet.2016.02.005.

Magkos F, Hjorth M F, Astrup A (2020) Diet and exercise in the prevention and treatment of type 2 diabetes mellitus. *Nat Rev Endocrinol* **16**(10), 545–555. doi: 10.1038/s41574-020-0381-5.

Marteau T M, Hollands G J, Shemilt I et al. (2015) Downsizing: Policy options to reduce portion sizes to help tackle obesity. *BMJ* **351**, h5863. doi: 10.1136/bmj.h5863.

NCD Risk Factor Collaboration (2016) Trends in adult body-mass index in 200 countries from 1975 to 2014: A pooled analysis of 1698 population-based measurement studies with 19·2 million participants. *Lancet* **387**(10026), 1377–1396. doi: 10.1016/S0140-6736(16)30054-X.

Our World in Data (https://ourworldindata.org/) accessed 23 January 2021.

Popkin B M, Du S, Green W D et al. (2020) Individuals with obesity and COVID-19: A global perspective on the epidemiology and biological relationships. *Obes Rev* **21**(11), e13128. doi: 10.1111/obr.13128.

Raben A, Vestentoft P S, Brand-Miller J et al. (2021) The PREVIEW intervention study: Results from a 3-year randomized 2 x 2 factorial multinational trial investigating the role of protein, glycaemic index and physical activity for prevention of type 2 diabetes. *Diabetes Obes Metab* **23**(2), 324–337. doi: 10.1111/dom.14219.

Varkevisser R D M, van Stralen M M, Kroeze W et al. (2019) Determinants of weight loss maintenance: A systematic review. *Obes Rev* **20**(2), 171–211. doi: 10.1111/obr.12772.

Williams D M, Nawaz A, Evans M (2020) Drug therapy in obesity: A review of current and emerging treatments. *Diabetes Ther* **11**(6), 1199–1216. doi: 10.1007/s13300-020-00816-y.

Yoshino M, Kayser B D, Yoshino J et al. (2020) Effects of diet versus gastric bypass on metabolic function in diabetes. *N Engl J Med* **383**(8), 721–732. doi: 10.1056/NEJMoa2003697.

Zafar M I, Mills K E, Zheng J et al. (2019) Low glycaemic index diets as an intervention for obesity: A systematic review and meta-analysis. *Obes Rev* **20**(2), 290–315. doi: 10.1111/obr.12791.

FURTHER READING

Althoff T, Sosic R, Hicks J L et al. (2017) Large-scale physical activity data reveal worldwide activity inequality. *Nature* **547**(7663), 336–339. doi: 10.1038/nature23018.

Chao A M, Quigley K M, Wadden T A (2021) Dietary interventions for obesity: Clinical and mechanistic findings. *J Clin Invest* **131**(1), e140065. doi: 10.1172/JCI140065.

Johnston B C, Kanters S, Bandayrel K et al. (2014) Comparison of weight loss among named diet programs in overweight and obese adults: A meta-analysis. *JAMA* **312**(9), 923–933. doi: 10.1001/jama.2014.10397.

Lee C J, Sears C L, Maruthur N (2020) Gut microbiome and its role in obesity and insulin resistance. *Ann N Y Acad Sci* **1461**(1), 37–52. doi: 10.1111/nyas.14107.

NCD Risk Factor Collaboration (2017) Worldwide trends in body-mass index, underweight, overweight, and obesity from 1975 to 2016: A pooled analysis of 2416 population-based measurement studies in 128·9 million children, adolescents, and adults. *Lancet* **390**(10113), 2627–2642. doi: 10.1016/S0140-6736(17)32129–3.

Purcell K, Sumithran P, Prendergast L A et al. (2014) The effect of rate of weight loss on long-term weight management: A randomised controlled trial. *Lancet Diabetes Endocrinol* **2**(12), 954–962. doi: 10.1016/S2213-8587(14) 70200–1.

Sacks F M, Bray G A, Carey V J et al. (2009) Comparison of weight-loss diets with different compositions of fat, protein, and carbohydrates. *N Engl J Med* **360**(9), 859–873. doi: 10.1056/NEJMoa0804748.

21 Diabetes mellitus

Lutgarda Bozzetto, Brunella Capaldo, and Angela A. Rivellese

OBJECTIVES

By the end of this chapter you should be able to:

- understand the aetiology and pathophysiology of diabetes mellitus
- distinguish between type 1 and type 2 diabetes mellitus
- demonstrate an insight into the link between diabetes and obesity
- understand insulin resistance and how to reduce it essentially through lifestyle modifications
- understand nutritional implications of insulin therapy
- demonstrate a basic knowledge of how to prevent long-term complications of diabetes.

21.1 Introduction

Diabetes mellitus is a metabolic disorder of multiple aetiology characterized by chronic hyperglycaemia associated with impaired carbohydrate, fat, and protein metabolism. These abnormalities are the consequence of either inadequate insulin secretion or impaired insulin action, or both. Diabetes has been classified into four distinct types (Table 21.1): type 1, type 2, gestational diabetes mellitus, other specific types (American Diabetes Association (ADA) 2021). Type 1 diabetes (T1D) is characterized by a cell-mediated autoimmune destruction of pancreatic beta-cells that results in a partial or total inability to secrete insulin, and life-long need for insulin administration. Type 2 diabetes (T2D) is characterized by a progressive loss of insulin secretion on the background of insulin resistance. Individuals with this condition may not require insulin treatment either initially or ever, although it may be undertaken in some cases as the most appropriate blood glucose-lowering treatment. The specific aetiologies of this form of diabetes are yet to be determined, but it is known that most of these patients are obese or have increased body fat, predominantly in the abdominal region. Gestational diabetes is defined as any degree of glucose intolerance with onset or first recognition during pregnancy. Other specific types of diabetes include less common causes, for example, monogenic diabetes syndromes (such as neonatal diabetes and maturity-onset diabetes of the young [MODY]), diseases of the exocrine pancreas (such as cystic fibrosis), and drug- or chemical-induced diabetes (such as with glucocorticoid use, in the treatment of HIV/AIDS or after organ transplantation).

21.2 Epidemiology and diagnosis of diabetes mellitus

On the basis of the plasma glucose levels, diabetes can be diagnosed in four ways: (1) casual plasma glucose concentration >11.1 mmol/L (200 mg/dl), or (2) fasting plasma glucose (FPG) >7.0 mmol/L (126 mg/dl), or

TABLE 21.1 Aetiological classification of diabetes mellitus

1. Type 1 diabetes
   A. Immune-mediated
   B. Idiopathic
2. Type 2 diabetes
3. Gestational diabetes
4. Other specific types
   A. Genetic defects in beta-cell function (MODY)
   B. Genetic defects in insulin action
   C. Disease of the endocrine pancreas
   D. Endocrinopathies
   E. Drug or chemical induced
   F. Infections
   G. Uncommon forms of immune-mediated diabetes
   H. Other genetic syndromes associated with diabetes

(3) 2-hour plasma glucose >11.1 mmol/L (200 mg/dl) during an oral glucose tolerance test (OGTT) with 75 g of glucose, or (4) glycated haemoglobin (HbA1c) ≥6.5 per cent (48 mmol/mol). In the absence of specific symptoms of the disease each test must be confirmed on a subsequent occasion (with the same or a different test) to be diagnostic. The HbA1c test should be performed using a method that is certified by the National Glycohemoglobin Standardization Program (https://ngsp.org) and standardized or traceable to the Diabetes Control and Complications Trial (DCCT) reference assay. The HbA1c has several advantages compared with the FPG and OGTT: greater convenience (fasting not required), greater preanalytical stability, and less day-to-day variations. However, it has limited availability in certain regions of the developing world and a greater cost; in addition, it has a lower sensitivity (one-third fewer cases are diagnosed by this method) as compared with fasting plasma glucose, and is not well correlated with the average plasma glucose concentrations in certain individuals.

Some people show glucose levels that do not meet the criteria for diabetes but they are still too high to be considered normal. 'Prediabetes' is the term used for individuals with impaired fasting glucose (IFG) and/or impaired glucose tolerance (IGT) and indicates an increased risk for the future development of diabetes. IGT is diagnosed by the 2-hour plasma glucose after OGTT >140 and <200 mg/dl with fasting value <126 mg/dl, while IFG is defined by fasting plasma glucose >110 and <126 mg/dl. These two categories are strong risk factors not only for future diabetes but also for cardiovascular diseases. If untreated, approximately one-third of people with IGT develop T2D within 5–10 years, one-third remain stable, and one-third revert to normoglycaemia (Vaccaro et al. 1999). In individuals with IGT, the mortality due to cardiovascular and cerebrovascular disease is approximately twice that of people with normal glucose tolerance.

IGT and T2D are often associated with other metabolic disturbances and cardiovascular risk factors; this condition has been defined as the insulin resistance syndrome or metabolic syndrome. There is no internationally agreed definition of the metabolic syndrome, which is generally considered as an association of impaired glucose regulation (IGT or IFG) or T2D, raised arterial blood pressure, raised plasma triglycerides, low HDL (high density lipoproteins), and central obesity (see Chapter 20). A statement from the US National Cholesterol Education Program (NCEP) attempts to define diagnostic criteria for the metabolic syndrome based exclusively on these clinical parameters (Expert Panel on Detection, Evaluation,

and Treatment of High Blood Cholesterol in Adults 2001). Other abnormalities often associated with the metabolic syndrome are microalbuminuria, hyperuricaemia, non-alcoholic liver steatosis, and coagulation disorders. There is growing evidence pointing to insulin resistance as the common aetiological factor of this condition, considered to be associated with increased risk for cardiovascular disease (Sperling et al. 2015).

T2D accounts for almost 85–95 per cent of all cases of diabetes. Its estimated prevalence is 4–6 per cent of the population: two-thirds of these are diagnosed, while about one-third remains unrecognized. The prevalence is known to be much higher in older people and in some ethnic communities (up to 40 per cent of Pima Indians).

The global prevalence of T2D has almost quadrupled in recent years, from 135 million in 1995 to approximately 415 million in the year 2015. It is predicted to rise to around 642 million by 2040. The rise in prevalence is much greater in developing than in developed countries due to changes in lifestyle (high energy diets and reduced physical activity) and the rapid increase in overweight and obesity in developing countries (see also Chapters 20 and 33). Long-term complications of diabetes include retinopathy with potential loss of vision, nephropathy leading to renal failure, peripheral neuropathy with risk of foot ulcers, and autonomic neuropathy that contributes to erectile dysfunction and cardiac arrhythmia. However, most of the morbidity and mortality associated with diabetes is attributable to macrovascular complications such as myocardial infarction, heart failure, and acute stroke. Diabetes is associated with an age-adjusted cardiovascular mortality that is between two and four times that of the non-diabetic population, while life expectancy is reduced by five to ten years in middle-aged patients with T2D.

Several observational studies suggest that T2D diabetes is primarily a lifestyle disorder; the highest prevalence rates occur in developing countries and in populations undergoing 'Westernization' or modernization. Under such circumstances, it seems that genetic susceptibility interacts with environmental changes, such as sedentary lifestyle and overnutrition, leading to T2D. Populations with the highest recorded prevalence of T2D, such as Nauru or Pima Indians, share the common experience of change from a hunter-gatherer or agriculture-based lifestyle to one of sedentary living and a diet of energy-dense processed foods. A better understanding of the impact that lifestyle may have, not only on the risk of T2D but also on its key mechanisms, should help implement more effective preventive measures focused on more specific targets.

21.3 Aetiology and pathophysiology of diabetes mellitus

Type 1 diabetes

T1D is characterized by absolute insulin deficiency caused in most cases by immune-mediated destruction of the beta-cells (autoimmune type 1A). A minority of patients (10–20 per cent) with T1D, generally of African or Asian origin, have no evidence of autoimmunity although they are insulinopenic and ketosis-prone. This form of diabetes has been referred as type 1B. Insulin deficiency leads to multiple metabolic abnormalities resulting in hyperglycaemia and proneness to ketoacidosis, which require lifelong exogenous insulin replacement. The pathogenesis of T1D involves genetic, immunological, and environmental factors (Atkinson et al. 2014).

Genetic factors

T1D is a polygenic disorder with nearly 40 loci known to affect disease susceptibility. The major histocompatibility system (HLA), located on chromosome 6, provides ~30 per cent of genetic susceptibility. In particular, HLA class II (HLA-DR3, DQ2, and HLA-DR4, DQ8) genes show the strongest association with T1D, but also some non-HLA loci appear to be associated with the disease. Most of these loci (VNTR, PTPN22, CTLA4) affect the immune response and the maintenance of tolerance. Studies on families with multiple members affected by T1D have demonstrated that the risk of a sibling developing the disease is increased up to 27 times by the age of 16. However, between monozygotic twins, the concordance rate for the disease is 35–70 per cent (depending on the length of follow-up), indicating that beyond genetic factors, environmental influences, eventually occurring early in life are involved in disease development.

Immunological factors

The evidence that T1D is an autoimmune disease relies on lymphocytic infiltration of pancreatic islets, abnormalities of cell-mediated immune response, and circulating autoantibodies. Pancreatic tissues from individuals with recent onset T1D revealed the presence of macrophages, T- and B-lymphocytes, and other inflammatory cells giving a picture known as 'insulitis'. CD8 and CD4 T cells are the predominant population within the insulitis lesion. A key immunological feature of T1D is the presence of autoantibodies against β-cell antigens, that is, antibodies to islet cells (ICA), insulin (IAA), glutamic acid decarboxylase (GAD), carboxypeptidase H, and several other minor antigens. These autoantibodies can appear shortly after diagnosis or even months to a year before the clinical onset of the disease and tend to fall progressively thereafter. The early detection of these immunological markers can help identifying people with an increased risk to develop the disease, such as first-degree relatives of patients with T1D. In these subjects, the presence of multiple markers indicates a high risk (above 80 per cent) of developing the disease within the subsequent two years. Given the possibility to predict T1D, several clinical trials have been conducted to test different strategies to prevent or delay the disease (nicotinamide, subcutaneous or oral insulin, anti-CD3 antibodies, immunomodulators, etc.) but, disappointingly, none of these approaches has proved to be beneficial.

Environmental factors

Environmental factors could contribute to the pathogenesis of T1D through several mechanisms: (1) exerting a direct toxic effect on ß-cells, (2) triggering an autoimmune reaction against the ß-cell, (3) damaging ß-cells so as to increase their susceptibility to autoimmune destruction. Environmental factors include drugs or chemicals (e.g., alloxan, streptozotocin, pentamidine), viruses, and dietary factors. Among viruses potentially involved in human T1D, clinical evidence points to mumps, coxsackie B, cytomegalovirus, and rubella viruses as the most likely candidates. The viral aetiology is supported by seasonal variability in the incidence of the disease, with a peak in spring and autumn. In addition, clinical and epidemiological studies have shown a close relation between appearance of diabetes and preceding episodes of viral infections. The influence of dietary factors on the development of T1D is widely recognized. There is evidence of a close relationship between early exposure to cow's milk protein and incidence of T1D in childhood. In contrast, breastfeeding seems to offer some protection against the development of the disease. The hypothesis is that antibodies produced against bovine seroalbumin may cross-react with antigens of β-cells, triggering an autoimmune response. Early weaning and introduction of gluten have also been associated with the development of T1D; however, most of the available evidence is based on animal models. Recently, vitamin D deficiency has been speculated to play a role in the pathogenesis of T1D based on studies in animals showing that vitamin D protects β-cell function against interleukin-1b-induced inhibition. In contrast, human studies did not find any correlation between vitamin D status and increased risk of β-cell autoimmunity, leaving the question still unsettled.

In recent years, increased attention has focused on gut microbiota based on the observation that low gut microbiota diversity is associated with β-cell autoimmunity. Some studies have shown that the gut microflora of T1D patients is characterized by the dominance of bacteroides, a reduced bacterial diversity, and the lack of bacteria producing short-chain fatty acids. However, given the observational nature of the studies, no causal relationship can be established between intestinal microflora and T1D.

Clinical manifestations

Because subjects who develop T1D often have a rather abrupt onset of symptoms (polyuria, polydipsia (thirst), or even ketoacidosis), it was long assumed that β-cell damage occurs rapidly. In reality, the presence of islet cell autoantibodies occurs long before the clinical appearance of the disease at a time when there is no elevation in blood glucose and glucose tolerance is near normal. When fasting hyperglycaemia develops, at least 80–90 per cent of the functional capacity of β-cell is irreversibly lost. Insulin deficiency leads to multiple abnormalities of intermediary metabolism that culminate in hyperglycaemia and increased levels of ketone bodies with proneness to ketoacidosis (Figure 21.1). Hyperglycaemia results from both increased glucose production by the liver and reduced glucose utilization by peripheral tissue, mainly the skeletal muscle. Under conditions of insulin deficiency, an accelerated flux of gluconeogenic substrates (alanine, lactate, glycerol) takes place from peripheral tissues to the liver, which fuels gluconeogenesis. Glucose utilization decreases as a result of: (a) the lack of insulin stimulatory effect on glucose transport into muscle tissues; and (b) the increased availability of free fatty acids, which are known to inhibit glucose transport across the muscle membrane through operation of the glucose–fatty acid (Randle)

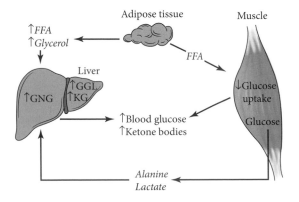

FIGURE 21.1 Consequences of insulin deficiency in type 1 diabetes mellitus.
GNG, gluconeogenesis; GGL, glycogenolysis; KG, ketogenesis.

cycle. In uncontrolled T1D, fatty acid mobilization from adipose tissue is markedly increased. Normally, in the liver fatty acids undergo beta-oxidation to acetyl CoA, which is totally oxidized in the Krebs cycle to water and carbon dioxide. When there is an excessive breakdown of fatty acids, as occurs in an insulin-deficient state, the capacity of the liver to oxidize all acetyl CoA is exceeded and two carbon fragments combine to form acetoacetate. Hepatic ketone body synthesis is further enhanced by the low insulin to glucagon ratio that critically regulates the activity of key enzymes of ketogenesis. All these metabolic abnormalities account for the classic symptoms and signs of the disease, such as glycosuria, polyuria, polydipsia, and weight loss. However, there is wide variability in clinical manifestations, with some individuals presenting acute signs of decompensation, and others being asymptomatic, thanks to good control with insulin therapy.

Type 2 diabetes

Although T2D has strong genetic components, modes of inheritance are largely unknown. One exception is the variant represented by MODY (maturity onset diabetes of the young) that conforms to autosomal dominant inheritance with high penetrance. The role of heredity in T2D is supported by familial aggregation, a concordance of 60–90 per cent for the disease in identical twins, and marked differences in its prevalence in different ethnic groups.

Both genetic and acquired factors contribute to T2D. Genetic factors somehow confer susceptibility to develop glucose intolerance, while non-genetic, environmental factors, mainly obesity and sedentary lifestyle, disrupt the fine balance between insulin secretion and insulin action. Of particular interest is also the evidence that dietary factors, especially a high intake of saturated fats, are associated with an increased risk of developing T2D, whereas a greater intake of polyunsaturated fatty acids is associated with a lower diabetes risk. Likewise, the quality of dietary carbohydrate affects diabetes risk. Update reviews and meta-analyses showed that diets with a higher glycaemic index are associated with higher risk of diabetes than are diets with a lower glycaemic index.

Patients with T2D have two major metabolic defects: (1) impaired insulin secretion, and (2) resistance to insulin action on target tissues, namely the liver, skeletal muscle, and adipose tissue.

Defect of insulin secretion

In T2D patients, fasting insulin levels have been reported as low, normal, or even elevated. This does not imply that insulin secretion is normal because, although fasting insulin is normal in absolute terms, it is inappropriately low for the ambient plasma glucose concentration.

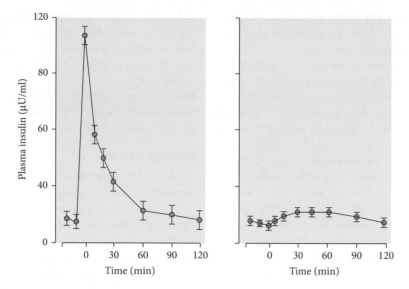

FIGURE 21.2 Insulin release elicited by an intravenous glucose load in normal subjects (left) and in patients with type 2 diabetes (right).

At matched plasma glucose concentration, normal subjects would have a much higher insulin concentration. The main abnormality of β-cell function in T2D is the loss of glucose-induced insulin secretion. In normal subjects a rapid rise in plasma glucose elicits a biphasic insulin release: a first (early) phase lasting 5–10 minutes and a second (late) phase persisting for the duration of hyperglycaemia (Figure 21.2). Although most of the insulin is secreted during the second phase insulin response, the first phase serves an important physiological function, that is, to stimulate glucose utilization by peripheral tissues and to inhibit glucose production by the liver in order to prevent an exaggerated increase in plasma glucose in the postprandial state. In patients with T2D, the first phase insulin response is characteristically lost while the late phase is preserved or only attenuated (Figure 21.2). The main features of the secretory defect are the following: (1) the defective insulin release is specific for glucose whereas insulin response to other secretagogues (arginine, isoproterenol, secretin) is substantially unaltered; (2) the potentiating effect of glucose on insulin response to secretagogues is reduced, confirming that in T2D the β-cell is selectively unresponsive to glucose; and (3) β-cell glucose unresponsiveness may be partially restored by correction of hyperglycaemia, suggesting that it may be, at least in part, an acquired defect caused by the toxic effect of high blood glucose on the beta-cells (glucose toxicity hypothesis). Other abnormalities of insulin secretion include disruption of the pulsatile insulin secretory pattern and an increased level of proinsulin, indicating an abnormal processing of insulin precursors within the β-cell.

Analyses of post-mortem pancreas specimens have revealed a 60 per cent lower β-cell volume in patients with T2D compared with weight-matched controls, due to an increased rate of β-cell apoptosis. In addition, some studies have shown the deposition in the islets of an insoluble fibril material, called amyloid, which is formed from a peptide designated 'islet amyloid polypeptide' (IAPP or amylin) that is co-secreted with insulin. Interestingly, studies in animals have shown a relation between dietary fat intake and production and/or secretion of amylin in the β-cells.

Another defect of insulin secretion in T2D is the reduction of the incretin effect. In normal subjects, after meal ingestion, the gastrointestinal tract produces various peptides, including the glucagon-like peptide-1 (GLP-1) and the glucose-dependent insulinotropic peptide (GIP)—collectively known as incretins—which act on β-cells to enhance the insulin release induced by hyperglycemia. In addition, GLP1-1 inhibits glucagon secretion, delays gastric emptying, and promotes weight loss by its appetite-suppressant effect. In patients with T2D, the incretin effect is markedly reduced since the GLP-1 response to nutrients is diminished. Moreover, there is a marked resistance to the stimulatory effect of GIP on insulin secretion. Although the mechanisms underlying the reduced incretin effect are not completely clear, the infusion of GLP-1 analogues in T2D patients is able to increase insulin release and to reduce glucose levels. Hence, a new incretin-based therapeutic strategy is now available for the management of T2D.

Insulin resistance

Insulin resistance is a state in which a given concentration of insulin produces a less than normal biological response. As known, insulin exerts its biological effects by initially binding to its specific cell-surface receptors. After this, a number of signals are generated that interact with a variety of effector units (enzymatic systems) leading to multiple metabolic effects. Insulin promotes the storage of nutrients by stimulating glycogen synthesis, protein synthesis, and lipogenesis and by inhibiting lipolysis, glycogen, and protein breakdown. In addition, insulin regulates water and electrolyte balance and stimulates cell growth and differentiation. This complex hormonal activity involves different intracellular pathways and mediators, which explains why some effects of insulin may be impaired (e.g., glucose metabolism) whereas others may not be (e.g., cellular growth). The cellular mechanisms underlying insulin resistance can involve insulin signalling, glucose transport, and metabolic pathways of intracellular glucose utilization (Cersosimo et al. 2015). However, it is still unclear what the primary defect is, which defect is genetically determined, and which is secondary to acquired factors, such as hyperglycaemia itself. With regard to glucose metabolism, it is known that insulin lowers blood glucose through two mechanisms: (1) by suppressing glucose production from the liver, and (2) by promoting the uptake of glucose by peripheral tissues, especially the skeletal muscle. Under conditions of insulin resistance, as occurs in T2D, the effect of insulin on the liver and peripheral tissues is impaired, thus producing the two major metabolic abnormalities observed in diabetic patients, that is, fasting and postprandial hyperglycaemia.

Liver

As a result of impaired insulin action on the liver, T2D patients tend to have abnormally increased glucose production in the post-absorptive state, which contributes to fasting hyperglycaemia. The excess of hepatic glucose production is almost entirely accounted for by an increased rate of gluconeogenesis, which is ~40 per cent higher in diabetic patients than in normal subjects. Such an increase is due to an increased supply of 3-carbon compounds (lactate, alanine, glycerol) from peripheral tissues to the liver as well as to a more efficient hepatic conversion of these substrates into glucose. Not only is hepatic glucose production increased in the basal state but it is less suppressed after glucose ingestion altering postprandial glucose homeostasis. In addition, the ability of the liver to take up and dispose of dietary glucose also seems to be reduced.

Skeletal muscle

Skeletal muscle is another important site of insulin resistance in patients with T2D. Indeed, the ability of insulin to stimulate glucose uptake by the skeletal muscle is reduced by 40–50 per cent in T2D patients compared with normal subjects. The impaired muscle glucose uptake

is due to a defect in a glucose transport step involving the translocation and/or activity of GLUT-4 or the glucose transporters located in skeletal muscle. In addition, the activity of two key enzymes that regulate non-oxidative (glycogen synthase) and oxidative (pyruvate dehydrogenase) glucose metabolism is reduced in diabetic patients. However, this defect is likely to be secondary to the reduced glucose transport.

Adipose tissue

Insulin profoundly influences adipocyte metabolism by stimulating glucose transport and triglyceride synthesis (lipogenesis), as well as inhibiting lipolysis. In T2D, particularly when associated with obesity, the ability of insulin to suppress lipolysis is markedly impaired. In addition, because glucose transport in the adipocyte is reduced, less glycerophosphate is formed through glycolysis and made available for triglyceride synthesis. The consequence is an increased flux of free fatty acids (FFA) from adipose tissue and a rise in their plasma concentration. FFA concentrations are elevated not only in the fasting state but they fail to be appropriately suppressed after meals. The result is a chronic elevation of FFA and triglyceride levels together with excessive deposition of fat in various tissues (liver, skeletal muscle) in patients with T2D, particularly when associated with obesity. In addition, the chronic elevation in plasma FFA levels has detrimental effects on both the β-cells and insulin action in peripheral tissues (lipotoxicity).

What are the defects causing β-cell dysfunction and insulin resistance?

β-cell dysfunction and insulin resistance are due to both genetic and acquired factors. An inherited defect in any part of the complex cascade that leads from glucose elevation to insulin processing and release could be a potential cause of β-cell dysfunction. Mutations in the glucokinase gene, in the β-cell glucose transporter (GLUT 2) gene, in mitochondrial DNA, and in the insulin gene itself have been identified in some diabetic patients. However, the importance of these alterations in the common form of T2D is questionable as these mutations have not been detected in populations with high prevalence of diabetes (e.g., Pima Indians). Other emerging factors associated with insulin secretory dysfunction include low birthweight (attributed to poor foetal development of the pancreas because of malnutrition) and deficiencies of some amino acids that seem to exert trophic effects on the β-cells.

With regard to insulin resistance, a number of candidate genes have been identified. Some mutations of the insulin receptor gene lead to altered insulin receptor biosynthesis, others impair the binding of insulin to its own receptor or insulin signalling. However, these mutations produce rare syndromes of severe insulin resistance but do not explain the insulin resistance associated with the common form of T2D. The current view is that insulin resistance is not due to a few 'major' genes but to a large number of 'polygenes', each with a relatively minor effect.

The role of acquired factors in deteriorating insulin secretion and insulin action is well expressed by the concept of 'glucose toxicity' and 'lipotoxicity'. According to the glucose toxicity hypothesis, a chronic increment in plasma glucose concentration leads to progressive impairment of insulin secretion and insulin sensitivity. This view is supported by animal studies showing that rats made diabetic by partial pancreatectomy develop hyperglycaemia and, concomitantly, a progressive impairment in insulin secretion and insulin action. Interestingly, when chronic hyperglycaemia is corrected by phlorizin, a substance that increases glucose urinary loss without any effect on the β-cell, both insulin secretion and insulin sensitivity substantially improve. Clinical studies in diabetic patients have shown that a tight glycaemic control, independent of the method by which it is achieved (diet, insulin, or hypoglycaemic agents), ameliorates considerably both insulin secretion and insulin action.

Like hyperglycaemia, the chronic exposure to elevated FFA levels may impair β-cell function and insulin sensitivity. This phenomenon has been referred to as 'lipotoxicity'. Once adipose tissue reaches its maximum capacity to expand, fat begin to accumulate in other tissues (liver, muscle, heart) causing apoptosis and inflammation. The damage produced by gluco-lipo-toxicity is responsible for the progression of the disease and the deterioration of glucose control over time.

Natural history of type 2 diabetes

Longitudinal studies have shown that the natural history of T2D progresses from the stage of normal glucose tolerance to impaired glucose tolerance to frank diabetes. The defects in both insulin secretion and insulin action occur at an early stage during the development of diabetes and progressively get worse because of the detrimental effects of environmental factors, such as obesity (particularly visceral fat accumulation), sedentary lifestyle, high fat intake, glucose toxicity, and lipotoxicity. To compensate for reduced peripheral insulin sensitivity, the β-cell usually increases its insulin secretion so as to prevent persistent hyperglycaemia. In this way, a near-normal blood glucose is maintained at the expense of increased insulin levels. However, when pancreatic β-cells are no longer able to compensate with an appropriate increase in insulin secretion (because of a genetic defect

or a progressive functional exhaustion), frank diabetes will develop (Weyer et al. 1999). Understanding the temporal sequence of changes in insulin secretion and insulin action as well as their interaction in the progression of T2D has important implications for the prevention of the disease and suggests that lifestyle modifications should preserve both insulin secretion and insulin sensitivity.

21.4 Vascular complications

Microangiopathy

Microangiopathy or microvascular complications of diabetes mellitus include retinopathy, nephropathy, and neuropathy, although the contribution of microangiopathy to neuropathy is still not completely understood. The incidence of retinopathy increases with duration of diabetes in type 1 diabetic patients: after 20 years of diabetes, over 95 per cent of patients have background retinopathy, although mostly without visual impairment. The risk of proliferative retinopathy, the main cause of blindness in type 1 diabetic patients, increases rapidly between 10 and 15 years after the onset of diabetes, then remains remarkably constant, reaching a cumulative risk of about 60 per cent after 40 years of diabetes. In T2D patients, retinopathy can be present at the onset of the disease (from 7 per cent to as many as 38 per cent) perhaps due to the fact that this condition sometimes remains unrecognized for years. It then increases, reaching, after 20 years, a prevalence of about 60 per cent for any kind of retinopathy and 20 per cent for proliferative retinopathy.

Diabetic nephropathy, characterized by the presence of albuminuria and/or a progressive decline in the glomerular filtration rate, affects 20–40 per cent of patients with T1D, while it is rarer in those with T2D. Still today, it represents, in industrialized countries, the leading cause of end-stage renal disease. The other chronic complication of diabetes is represented by diabetic neuropathy, affecting both the somatic and vegetative function of the peripheral neurological system. Diabetic microvascular complications result from the interaction of multiple genetic and metabolic factors, among which hyperglycaemia is almost certainly the most important one (Sheetz & King. 2002). In fact, all these complications are specific to diabetes and do not occur in the absence of long-lasting hyperglycaemia. The importance of hyperglycaemia in the determination and progression of microvascular complications has been proven beyond any doubt by the results of some intervention trials performed in the last few years in both type 1 and type 2 diabetic patients. These studies have clearly shown that improving blood glucose control significantly reduces the onset and progression of retinopathy, nephropathy, and neuropathy. Besides hyperglycaemia and genetic factors, which determine a variable degree of susceptibility and which have not been fully identified, other factors are important in determining microvascular complications, such as hypertension, smoking, and hyperlipidaemia. Hypertension has been shown to play a very important role, especially in the genesis of diabetic nephropathy. Intervention trials in type 2 diabetic patients have clearly shown that improving blood pressure control significantly reduces the onset and progression of both nephropathy and retinopathy.

How all these factors might lead to diabetic microangiopathy is still a matter of debate. Many hypotheses have been put forward and only partly proven. One possible pathway is illustrated in Figure 21.3. Hyperglycaemia, together with other factors, may lead, through different mechanisms, to abnormal endothelium function, haemodynamic effects, and changes in blood rheological characteristics. All these abnormalities may induce a thickening of basement membrane and an increased permeability with subsequent occlusive angiopathy, tissue hypoxia, and organ damage.

Macroangiopathy

Atherosclerotic arterial disease may manifest clinically as coronary heart disease, cerebrovascular disease, or peripheral vascular disease. In diabetic patients, atherosclerotic lesions are qualitatively similar to those present in the non-diabetic population, except that their progression appears to be accelerated. Both mortality and morbidity for all cardiovascular diseases are significantly increased in diabetic patients. In particular, both coronary heart disease and cerebrovascular disease mortality is increased from two- to four-fold in diabetic patients compared to the non-diabetic population. Moreover, about half of all lower limb amputations occur in diabetic patients.

The high prevalence of arterial disease in diabetes is explained partly by the increased frequency of the most important cardiovascular risk factors, and partly by other factors closely associated with diabetes (Table 21.2). Elevated LDL cholesterol remains one of the most important cardiovascular risk factors in diabetic patients. Intervention studies performed in diabetic patients with hypolipidaemic drugs, particularly statins, have clearly

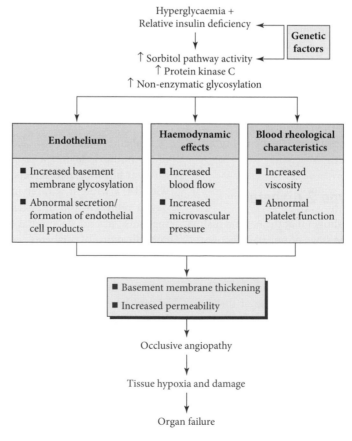

FIGURE 21.3 A possible pathogenic pathway of diabetic microangiopathy.
Source: Modified from J C Pickup and G Williams (1991) *Textbook of Diabetes*, Blackwell Scientific Publications, Hoboken, NJ.

TABLE 21.2 Main risk factors for cardiovascular disease in diabetic patients

General risk factors	Diabetes-related risk factors
Age	Hyperglycaemia
Smoking	Hyperinsulinaemia/insulin resistance
Dyslipidaemia	Microalbuminuria/proteinuria
Hypertension	
Hypercoagulability	
Obesity	
Physical inactivity	
Family history	

TABLE 21.3 Optimal goals for cardiovascular disease prevention in patients with diabetes

BMI <25 kg/m^2
- Waist circumference <94 cm for men and <80 cm for women
- Physical activity: 30 minutes of brisk walking every day (or any equivalent physical exercise)
- Abstinence from smoking
- Optimal blood glucose control (HbA$_{1c}$ <7%)
- LDL cholesterol <70mg/dl in high-risk patients or <55 mg/dl in very high-risk patients or in secondary prevention
- HDL cholesterol >40 mg/dl for men >50 mg/dl for women
- Plasma triglycerides <150 mg/dl
- Blood pressure <130/80 mmHg

shown that LDL cholesterol reduction significantly decreases cardiovascular events and mortality. Beside elevated LDL cholesterol, other lipid abnormalities, more typical of T2D, such as increased triacylglycerol, VLDL and IDL particles, and decreased HDL cholesterol, may have an important role in determining the high cardiovascular risk of diabetic patients.

With regard to the role of high blood pressure, the data are very consistent and similar to those for LDL cholesterol. There is strong evidence that reducing blood pressure significantly decreases cardiovascular events in diabetic patients. The data concerning the role of hyperglycaemia in explaining, at least in part, the excess cardiovascular risk of diabetic patients, are somewhat more controversial. Epidemiological data, as a whole, support the relationship between hyperglycaemia and cardiovascular disease, even if the association is not as strong as for microvascular complications. Moreover, intervention studies show that improvement in blood glucose control reduces the incidence of cardiovascular events in T2D patients, if the improvement is part of a multifactorial intervention that also includes targets for optimal plasma lipid and blood pressure levels (Gaede et al. 2003).

Insulin resistance and/or hyperinsulinaemia, typical features of T2D patients, are associated with an increased cardiovascular risk. It is not completely clear if this association is fully explained by the clustering of other cardiovascular risk factors with insulin resistance, such as lipid abnormalities, high blood pressure, coagulation abnormalities, high uric acid levels, low grade inflammation, and so on, or whether there is also a direct link between insulin resistance and/or hyperinsulinaemia and the atherosclerotic process. In any case, all these risk factors, together with others less well identified, may induce, through different mechanisms, endothelial injury. Different and repeated injuries increase the permeability of the endothelium to lipids as well as its adhesion to

monocytes and platelets. This in turn increases the procoagulant activities of the endothelium and increases the expression of vasoactive molecules, cytokines, and growth factors, which promote the migration of monocytes and proliferation of smooth muscle cells. In this way, the atherosclerotic process starts and will continue until the formation of the atherosclerotic plaque, with progressive narrowing of the arterial lumen and arterial stenosis or plaque rupture, thrombus formation, and acute ischaemia.

In conclusion, many factors may contribute to the excess cardiovascular risk typical of both T1D and T2D patients. Therefore, in order to reduce this risk, it is necessary to act not only on blood glucose control but also on plasma lipids, blood pressure, hypercoagulable state, obesity, physical activity, smoking habits, to try to reach the goals indicated in Table 21.3.

21.5 Management of diabetes mellitus

Diet

Nutritional management is the cornerstone of therapy for diabetes mellitus. In some patients with T2D it may be the only therapy required, in other types it allows a more accurate blood glucose control with lower doses of glucose lowering drugs or insulin. In any case, the aims of nutritional management must be not only the optimization of blood glucose control but also and, perhaps more importantly, the reduction of risk factors for cardiovascular diseases. Dietary recommendations for people with diabetes are very similar to those given to the general population for the promotion of good health. Of course, since diabetes mellitus is a chronic disease, diet therapy has to be considered as a lifelong practice and therefore it is essential that all nutritional programmes be adapted to the specific needs of the individual.

Total dietary energy

Most patients (60–70 per cent) with T2D are overweight or obese, and the frequency of overweight and obesity has increased over the last years also in T1D patients, especially those following intensive insulin therapy. All these patients should be encouraged to reduce their calorie intake and increase their energy expenditure. There is a very large body of evidence showing that body weight reduction, even if modest (5–10 per cent of basal body weight), is able to improve blood glucose control, reduce insulin resistance, and favourably affect the other cardiovascular risk factors, such as blood pressure and lipid abnormalities, often present in diabetic patients. Recently, it has been shown that a larger weight reduction (on average 15 kg) obtained by intensive dietary intervention (low-calorie diet of 800–900 kilocalorie) is even capable of determining a complete remission of the disease (about half of patients at one year and one-third at two years) in newly diagnosed T2D patients (Lean et al. 2018).

In addition to the treatment of the disease, weight reduction although modest (5–10 per cent of the initial body weight) is also one of the essential elements in the prevention of diabetes. In fact, different intervention studies, one in the USA (Diabetes Prevention Program Research Group 2002) one in Finland (Tuomilehto et al. 2001), and one in China (Da Qing study) have clearly shown that a moderate weight reduction (about 5 per cent of initial body weight), together with increased physical activity (at least ½ hour of brisk walking every day), and changes in the composition of the diet (reduction in saturated fat, increased consumption of dietary fibre), is able to prevent T2D, reducing its incidence in high risk individuals by 60 per cent. Moreover, the beneficial effects of this kind of intervention remain also many years after the conclusion of the study. From a practical point of view, weight reduction may be achieved by reducing the consumption of energy-dense foods, especially those rich in fat, to achieve a caloric deficit of 300–500 kcal/day. To be effective also in the long run this advice should be incorporated into structured, intensive lifestyle education programmes. The prescription of a very low energy diet (400–600 kcal) should be restricted to special cases (BMI >35 kg/m^2) and administered only in specialized centres.

Composition of the diet

The composition of the diet in terms of quantity/quality of nutrients, other bioactive components, such as dietary fibre and polyphenols and, in particular, in terms of different foods, is of fundamental importance in the dietary therapy of diabetes.

According to the recommendations for the nutritional management of patients with diabetes mellitus (Mann et al. 2004; ADA 2014), the most important aspect in relation to the composition of the diet is the consumption of saturated fat, <10 per cent of total energy or, even, <8 per cent for patients with high levels of LDL cholesterol. This strong recommendation is supported by the high rates of cardiovascular diseases in people with diabetes and by the fact that saturated fat intake has unfavourable effects on lipid metabolism (increase in LDL cholesterol), insulin resistance, and blood pressure. Together with saturated fat, cholesterol intake should also be reduced to less than 300 mg/day for all people with diabetes and less than 250 mg/day for those with raised LDL cholesterol. Taking into account the fact that protein intake should range between 10 and 20 per cent of total energy (0.8 g/kg body weight/day for those with established nephropathy), the remaining 80–90 per cent of the total energy should be provided by a combination of carbohydrates and unsaturated fat (especially monounsaturated fatty acids) according to clinical circumstances and local or individual preferences (Table 21.4).

As to carbohydrates, foods rich in dietary fibre (legumes, vegetables, fruit) and/or with low glycaemic index are recommended, while high glycaemic index foods must be restricted and replaced with unsaturated fat. Fibre-rich foods (legumes, vegetables, fruit, whole grain cereal-derived foods) have been shown to improve blood glucose control and lipid profile in both T1D and T2D patients (Riccardi et al. 2008). Moreover, a high consumption of vegetable fibre (25–29 g/day) as well as of whole grains is associated with a significant reduction in the risk of T2D, cardiovascular disease, and mortality (Reynolds et al. 2019). Therefore, an intake of dietary fibre of 15–20 g/1000 kcal, is recommended for diabetic patients as well as for the general population even though a lower amount may be also beneficial. This amount can be achieved with the consumption of five servings per day of vegetables and fruits, three servings per week of legumes, and two daily portions of whole grains.

Dietary fibre is only one of the factors able to modulate the glycaemic response to carbohydrate-rich foods, which can be influenced also by other factors such as the physical state of foods, type of starch, presence of antinutrients (substances that interfere with the absorption of nutrients, i.e., phytic acid, lectins, tannins). Due to the variety of factors that influence the impact of a meal on blood glucose, it is not possible to predict the postprandial glycaemic response of each food on the basis of its chemical characteristics. Therefore, it is necessary to examine the glycaemic response of foods *in vivo* and calculate their glycaemic index. This index is based on the

TABLE 21.4 Dietary recommendations for diabetes

Dietary fat	
Saturated fat + *trans*	<10% TE; <8% TE if LDL cholesterol high
Monounsaturated fat (*cis*)	10-20% TE
N-6 polyunsaturated fat	<10% TE
N-3 polyunsaturated fat	2–3 servings of fish/week and plant sources of n-3 fatty acids
Cholesterol	<300 mg/day
Dietary carbohydrates	
Carbohydrates	• 45%–60% TE according to metabolic characteristics of patients
Dietary fibre	• Ideally >40 g/day (or 20 g/1000 kcal), half of which soluble, but beneficial effects obtained also with lower and more feasible amounts • Cereal-based foods should, whenever possible, be wholegrain and high in fibre
Glycaemic index	• CHO-rich, low glycaemic index foods are suitable provided other components of the foods are appropriate
Dietary protein	
Protein	• 10–20% TE • For patients with T1DM and established nephropathy 0.8 g/kg normal body weight/day

* TE = Total Energy

Source: J Mann, K De Leeuw, Hermansen et al. on behalf of the DNSG of EASD (2004) Evidence-based nutritional approaches to the treatment and prevention of diabetes mellitus. *Nutr Met Cardiovasc Dis* **14**, 373–394.

increase in blood glucose concentrations (the incremental area under the curve of blood glucose concentrations) after the ingestion of a portion of a test food containing 50 g of carbohydrates, divided by the incremental blood glucose area achieved with the same amount (50 g) of carbohydrates present in an equivalent portion of a reference food (glucose or white bread). According to this index, carbohydrate foods can be divided into broad categories characterized by a high or low glycaemic index. Fibre-rich foods have a low glycaemic index, although some foods with low fibre content (pasta, parboiled rice) may also have a relatively low glycaemic index.

It is now well documented that a diet containing mainly low glycaemic index foods improves metabolic control in diabetic patients and may have favourable effects on other cardiovascular risk factors. Therefore, foods with a low glycaemic index (e.g., legumes, oats, pasta, parboiled rice, certain raw fruits) should replace, whenever possible, those with a high glycaemic index.

Beyond high-fibre, low-glycaemic diets, even low-carbohydrate diets (CHO <40 per cent total energy) may be useful for treating diabetes, according to some authors. In this regard, it is important to underline that results available so far show that this dietary approach can have favourable effects on glycaemic control in the short term (six months), but not in the long term (Sainsbury et al. 2018). Therefore, this kind of diet may be indicated only for short periods of time.

Moreover, the diet for diabetic patients, in accordance with general guidelines for the promotion of good health, should be moderately restricted in salt intake (<6 g/day) and alcohol intake (20–30 g per day, 250 ml of wine rich in polyphenols, less for those who are overweight, have high triglycerides, or high blood pressure).

Physical exercise

Aerobic physical exercise of moderate intensity but performed on a regular basis (daily or at least not less than four times/week) has been shown to improve blood glucose control, reduce insulin resistance, induce favourable effects on other cardiovascular risk factors, prevent the incidence of T2D and, finally, reduce cardiovascular and total mortality (Table 21.5).

Therefore, half an hour of brisk walking every day or, at least, four times/week is strongly recommended not only for diabetic patients but for all people. This kind of exercise may be carried out by any diabetic patient without particular precautions. Of course, diabetic patients may wish to perform heavier exercises (preferably

TABLE 21.5 Benefits of physical exercise

↓ blood glucose

↓ body weight

↑ HDL cholesterol

↓ blood pressure

↓ serum triglycerides

↓ fibrinogen

↓ platelet aggregation

↓ osteoporosis

↓ depression

↓ total and CVD mortality

TABLE 21.7 Guidelines for exercise in type 1 diabetes

- Monitor blood glucose before, during, and after exercise
- Avoid hypoglycaemia by:

  – taking 20–40 g extra carbohydrate before and hourly during exercise or reducing pre-injection insulin dosages by 30–50%

  – avoiding heavy exercise during peak insulin action

  – using non-exercising sites for insulin injection

- After prolonged exercise, monitor blood glucose and take extra carbohydrate to avoid delayed hypoglycaemia

aerobic); in these cases some precautions must be taken and some advices should be given (Table 21.6). In T2D patients, exercise increases peripheral glucose uptake but decreases insulin secretion; therefore, hypoglycaemia is rare and extra carbohydrate is generally not required. In T1D patients, glycaemic changes during exercise depend largely on blood insulin levels and, therefore, on insulin administration. Hyperinsulinaemia may cause hypoglycaemia, while hypoinsulinaemia, combined with counter-regulatory hormone excess, may lead to hyperglycaemia. The risk of hypoglycaemia may be reduced by consuming 20–40 g of extra carbohydrate before and hourly during exercise and/or by reducing pre-exercise insulin dosages (Table 21.7).

Glucose-lowering agents

When optimal blood glucose control (glycated haemoglobin, HbA1c <7.0 per cent, fasting and pre-meal blood glucose 80–120 mg/dl; 2-hour postprandial blood glucose 100–140 mg/dl) is not achieved in T2D patients by non-pharmacological approaches (nutritional management and physical exercise), glucose lowering drugs should be added, if specific contraindications are

TABLE 21.6 Recommendations for physical exercise in diabetic patients

Exercise characteristics

- Type: combined aerobic and resistance
- Intensity: 40–60% of VO_2max
- Duration: 30–60 minutes per session
- Frequency: at least 3 times/week, better daily
- Warm up before and cool off after each session
- Use of suitable shoes

not present. Drugs commonly used in the treatment of T2D are listed in Table 21.8. Beside metformin and sulphonylureas, the oldest and most known ones, the other drugs for the treatment of T2D patients are: (1) thiazolinediones—specifically pioglitazone—which acts on peripheral insulin action and is therefore useful instead of metformin or in addition to metformin and/or sulphonylureas, to further reduce insulin resistance and its associated metabolic abnormalities; (2) glinides (repaglinide, nateglinide), which increase insulin secretion, especially the first phase, with a very short time of action. These drugs may be useful in place of sulphonylureas when it is necessary to reduce postprandial blood glucose; (3) incretins: glucagon-like peptide-1 (GLP-1) analogues and inhibitors of dipeptidyl peptidase 4 (DPP-IV), the enzyme that rapidly decreases GLP1 levels. Both classes improve blood glucose control, potentiating glucose-stimulated insulin secretion. Therefore, they do not cause hypoglycaemia; the GLP1 analogues have also the added advantage of reducing body weight, reducing cardiovascular events, and slowing the progression of nephropathy; (4) Sodium-glucose co-transporters 2 (SLGT2) inhibitors act inhibiting the SLGT2 transporters located in the proximal nephron, thereby reducing glucose reabsorption and increasing urinary glucose excretion. Beside the improvement in blood glucose control, these agents induce modest weight loss and reduction in blood pressure. Similarly to GLP1 analogues, these agents reduce significantly total and cardiovascular mortality and delay the progression of nephropathy. The main side effects of SLGT2-inhibitors are genital mycotic infections, especially in women, and urinary tract infections as a consequence of glycosuria.

Following the results of the UK Prospective Diabetes Study (UKPDS), showing that metformin significantly reduces cardiovascular risk in overweight T2D patients, metformin is considered the first-choice drug in the treatment of T2D. Therapy with metformin is also useful in these patients because of the small weight loss that is generally associated with use of this drug (2–3 kg). The choice of medications added to metformin is based on

TABLE 21.8 Main glucose lowering drugs

Drug classes	Mechanisms of action	Disadvantages	Advantages
Biguanides	• Decrease in hepatic glucose production • Increase in peripheral insulin sensitivity	• Gastrointestinal disturbances • Lactic acidosis (rare)	• CVD
Sulphonylureas	• Increase in insulin secretion	• Hypoglycaemia • Weight increase	• Extensive experience • ↓ Microvascular risk (UKPDS)
Glinides	• Increase in first phase insulin secretion	• Hypoglycaemia (less than with sulphonylureas)	
Thiazolinediones	• Increase in peripheral insulin sensitivity	• Weight increase • Liquid retention • Heart failure • Bone fracture	• No hypoglycaemia • ↓ CVD events (Proactive)
α-Glucosidase inhibitors	Delay in carbohydrate absorption	• Gastrointestinal disturbances	
GLP-1 analogues	• Increase in insulin secretion (glucose-dependent) • Decrease in glucagon secretion (glucose-dependent) • Increase in satiety	• Gastrointestinal disturbances • Increase in heart rate • Injectable	• No hypoglycaemia • ↓ Weight • ↓ CVD events
DPP-IV inhibitors	• Increase in insulin secretion (glucose-dependent) • Decrease in glucagon secretion (glucose-dependent)	• Angioedema/urticarial and other immune-mediated dermatologic effects • Heart failure (?) • Acute pancreatitis (?)	• No hypoglycaemia • Neutral on body weight
SGLT-2 inhibitors	• Block glucose absorption by the kidney and ↑ Glycosuria	• Genital-urinary infections • Polyuria • Volume depletion	• No hypoglycaemia • ↓ Weight • ↓ Blood pressure • ↓ CVD events • ↓ CVD mortality

possible advantages/disadvantages of each of them and on the patient's clinical characteristics with particular regard to the presence of cardiovascular disease or renal disease; in these cases, GLP1 analogues and SGLT-2 inhibitors should be preferred.

A proposed flow-chart for the pharmacological treatment of T2D patients is shown in Figure 21.4. Metformin is indicated as first choice; if optimal blood glucose control is not achieved, other drugs (sulphonylureas/glinides or pioglitazone or incretins or SLGT2 inhibitors) should be added as dual or triple therapy; if the target values for blood glucose control are still not reached, bedtime insulin can be added or insulin therapy (3–4 administrations/day) can replace oral drugs. For each of these points a-glucosidase inhibitors, which slow down carbohydrate absorption, may be added. Insulin therapy may be also considered as earlier therapy especially in patients with very poor blood glucose control (i.e., HbA1c >9 per cent).

Insulin therapy

Insulin therapy is an essential lifesaving drug for T1D patients, and can be used also in the treatment of T2D patients when non-pharmacological therapy plus glucose lowering drugs are no longer able to achieve optimal blood glucose control. The types of insulin available today are generally based on human insulin, produced by DNA recombinant techniques, and, according to their time of action, may be divided into short-, intermediate- and long-acting insulin (see Table 21.9).

Beside human insulin, other types of modified insulin molecules have been introduced in clinical practice and are, now, the most utilized. These are the so–called insulin analogues, one type with a very short time of action (lispro, aspart, glulisine) and the other type with a very long time of action (glargine U-100, detemir, glargine U-300, degludec) (see Table 21.9). Short acting insulin analogues are very rapidly absorbed subcutaneously and therefore are

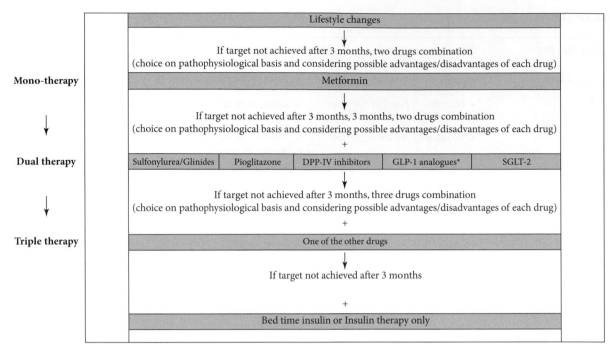

* Preferred for patients with ascertained cardiovascular disease
1) α glucosidase inhibition may be added to all the steps
2) Therapy may be also considered as earlier therapy in patients with very poor blood glucose control

FIGURE 21.4 Treatment of type 2 diabetes.

TABLE 21.9 Different types of insulin

Type	Time for onset of action	Time until peak action	Duration of action
Short acting insulins			
Regular human insulin	30–60 min	2–4 h	5–8 h
Rapid acting analogues			
Aspart	12–18 min	30–90 min	3–5 h
Glulisine	12–30 min	30–90 min	3–5 h
Lispro	15–30 min	30–90 min	3–5 h
Ultra–rapid acting insulin analogues			
Fast–aspart	5–10 min	20–30 min	3–5 h
Ultra–rapid lispro	5–10 min	20–30 min	3–5 h
Intermediate–acting analogues			
NPH	1–2 h	4–12 h	12–16 h
Lispro protamine	30–60 min	4–12 h	12–16 h
Long–acting analogues			
Detemir	1–2 h	6–8 h	Up to 24 h
Glargine U100	1–2 h	None	20–26 h
Glargine U300	1–2 h	None	Up to 36 h
Degludec	30–90 min	None	>42 h

particularly useful in controlling postprandial blood glucose. The long-acting insulin analogues produce quite stable basal insulin concentrations for up to 24 hours or longer.

Insulin therapy must be tailored to the individual patient and adjusted on the basis of blood glucose control. Therefore, there are no strict rules for its dosage and type of administration. However, some schemes of insulin therapy are more commonly utilized than others, the most common being that based on three injections of short-acting insulin before breakfast, lunch, and dinner plus an injection of long-acting insulin at bedtime (the so called basal-bolus scheme).

As an alternative to multiple daily injections (MDI), insulin can be administered by a wearable pump through continuous subcutaneous infusion (CSII). Most insulin pumps deliver rapid-acting insulin through a cannula, while a few are attached directly to the skin without tubing. Insulin is delivered as basal and bolus modality. Basal dose is supplied continuously at different programmable infusion rates (usually 3–5) over 24 hours to cover insulin needs between meals and at night. Boluses are injected before meals to cover food eaten or to correct a high blood glucose level. A bolus can be injected as a single shot or extended over time to cover late postprandial insulin demand as for meals rich in fat or protein.

To optimize and individualize insulin administration by MDI or CSII, blood glucose self-monitoring is essential. Since recently, continuous glucose monitoring (CGM) systems are available for clinical use. With CGM, glucose concentrations are continuously detected in interstitial fluid by a subcutaneous sensor and displayed on a receiver. This makes patients able to adapt insulin dosing or carbohydrate eating according to current glucose values and trends thus avoiding hypo- or hyperglycaemia. Sensor readings are also retrospectively available for diabetes care providers in dedicated clouds. They represent a unique opportunity to gain essential information for a successful control of postprandial glucose, such as the timing of insulin administration in relation to meals or the individual characteristics of postprandial glucose response, including the rate of gastric emptying. CGM may also highlight the effects of different nutritional composition of meals on glucose response. For these reasons, the use of CGM is increasing also in the field of nutritional research.

CGM can be combined with an insulin pump to automate insulin delivery according to blood glucose changes. Sensor augmented pump systems suspend insulin infusion when glucose is low or predicted to go low within the next 30 minutes. Hybrid closed loop systems are also available that not only suspend insulin infusion but also increase or decrease insulin delivery on the basis of sensor values. Within the hybrid system, pre-meal boluses are delivered by patients according to the carbohydrate content of food eaten.

Insulin therapy and nutritional considerations

Nutritional issues are essential in the proper management of insulin therapy. Pre-meal insulin doses must be adapted to the nutritional component of the meal and in particular to carbohydrate amount. According to individual preferences and social needing, patients may be given a fixed or flexible insulin regimen. People on fixed regimens receive alimentary plans with fixed amounts of carbohydrate. Patients on flexible regimens are trained to determine insulin doses according to their previous experience or by carbohydrate counting. Carbohydrate counting is an advanced skill according to which patients are trained to individuate foods containing carbohydrates and their portion size and calculate insulin dose based on their own insulin/carbohydrate ratio (I:CHO), which indicates how much carbohydrates are metabolized by one unit of insulin. This parameter varies between and within individuals and may change over time depending on insulin-sensitivity variations. Whatever insulin therapy utilized, doctors and patients must always try to reach and maintain optimal blood glucose control (HbA1c <7 per cent; fasting and pre-meal blood glucose 80–120 mg/dl; postprandial blood glucose 100–140 mg/dl).

KEY POINTS

- Diabetes mellitus is a major cause of morbidity and mortality. Its prevalence is increasing all over the world due to increased prevalence of obesity and low levels of physical exercise.

- Type 1 diabetes is characterized by severe insulinopenia and absolute need of exogenous insulin to prevent ketoacidosis and death.

- Type 2 diabetes is frequently associated with obesity. Pathogenic mechanisms are: a defect in beta-cell function and a decrease in insulin action at the level of peripheral tissues (insulin resistance).

- The incidence of diabetes can be reduced by about 60 per cent through lifestyle modifications, such as weight reduction, increased physical activity, and reduced saturated fat consumption.

- The main dietary recommendations for diabetic patients are: weight reduction of at least 5–7 per cent in overweight/obese patients; reduction of saturated fat (<10 per cent of the total caloric intake); increase in foods rich in fibre and with a low glycaemic index.

- Microangiopathic complications can be prevented or delayed by good glycaemic control.

- Cardiovascular complications can be reduced by about 50 per cent through multifactorial treatment of the major cardiovascular risk factors.

 Be sure to test your understanding of this chapter by attempting multiple choice questions.

REFERENCES

American Diabetes Association (ADA) (2021) Standards of medical care in diabetes. *Diabetes Care* **44**(Suppl 1), S14–S72.

Atkinson M A, Eisenbarth G S, Michels A W (2014) Type 1 diabetes. *Lancet* **383**, 69-82.

Cersosimo E, Triplitt C, Mandarino L J et al. (2015) Pathogenesis of type 2 diabetes mellitus. In: *Endotext*. MDText.com, Inc., South Darmouth (MA).

Diabetes Prevention Program Research Group (2002) Reduction in the incidence of type 2 diabetes with lifestyle intervention or metformin. *N Engl J Med* **346**, 393–403.

Gaede P, Vedel P, Larsen N et al. (2003) Multifactorial intervention and cardiovascular disease in patients with type 2 diabetes. *N Engl J Med* **348**, 383–393.

Lean M E, Leslie W S, Barnes A C et al. (2018) Primary care-led weight management for remission of type 2 diabetes (DiRECT): An open-label, cluster-randomised trial. *Lancet* **391,** 541–557.

Mann J, De Leeuw K, Hermansen B et al. on behalf of the DNSG of EASD (2004) Evidence-based nutritional approaches to the treatment and prevention of diabetes mellitus. *Nutr Met Cardiovasc Dis* **14**, 373–394.

Reynolds A, Mann J, Cummings J et al. (2019) Carbohydrate quality and human health: A series of systematic reviews and meta-analyses. *Lancet* **393**, 434–445.

Riccardi G, Rivellese A A, Giacco R (2008) Role of glycaemic index and glycaemic load in the healthy state, in prediabetes and in diabetes. *Am J Clin Nutr* **87**, S269–S274.

Sainsbury E, Kizirian N V, Partridge S R et al. (2018) Effect of dietary carbohydrate restriction on glycaemic control in adults with diabetes: A systematic review and meta-analysis. *Diabetes Res Clin Pract* **139**, 239–252.

Sheetz M J & King G L (2002) Molecular understanding of hyperglycemia's adverse effects for diabetic complications. *JAMA* **288**, 2579–2588.

Sperling L S, Mechanick J I, Neeland I J et al. (2015) The CardioMetabolic Health Alliance: Working toward a new care model for the metabolic syndrome. *J Am Coll Cardiol* **66**, 1050–1067.

Tuomilehto J, Lindstrom J, Eriksson J G et al. (2001) Prevention of type 2 diabetes by changes in lifestyle among subjects with impaired glucose tolerance. *N Engl J Med* **344**, 1343–1350.

Vaccaro O, Ruffa G, Imperatore G et al. (1999) Risk of diabetes in the new diagnostic category of impaired fasting glucose. *Diabetes Care* **22**, 1490–1493.

Weyer C, Bogardus C, Mott D M et al. (1999) The natural history of insulin secretory dysfunction and insulin resistance in the pathogenesis of type 2 diabetes mellitus. *J Clin Invest* **104**, 787–794.

FURTHER READING

Bhupathiraju S N, Tobias D K, Malik V S et al. (2014) Glycaemic index, glycaemic load, and risk of type 2 diabetes: Results from 3 large US cohorts and an updated meta-analysis. *Am J Clin Nutr* **100**, 218–232.

Colberg S R, Sigal R J, Fernhani B et al. (2010) Exercise and type 2 diabetes. The American College of Sports Medicine and the American Diabetes Association: Joint position statement. *Diabetes Care* **33**, e147–e167.

Li G, Zhang P, Wang J et al. (2014) Cardiovascular mortality, all-cause mortality, and diabetes incidence after lifestyle intervention for people with impaired glucose tolerance in the Da Qing Diabetes Prevention Study: A 23-year follow-up study. *Lancet Diabetes Endocrinol* **2**, 474–480.

Nauck M A, Vardarli I, Deacon C F et al. (2011) Secretion of glucagon like peptide-1 (GLP-1) in type 2 diabetes: What is up, what is down? *Diabetologia* **54**, 10–18.

Nolan C J, Damm P, Prentki M (2011) Type 2 diabetes across generations: From pathophysiology to prevention and management. *Lancet* **378**, 169–181.

Riccardi G, Rivellese A, Williams C (2003) The cardiovascular system in nutrition and metabolism. In: *Nutrition and metabolism*, 210–228 (Gibney M J, Macdonald I A, Roche HM eds). The Nutrition Society Textbook Series, Blackwell, Oxford

Tanasescu M, Leitzmann M F, Rimm E B et al. (2003) Physical activity in relation to cardiovascular disease and total mortality among men with type 2 diabetes. *Circulation* **107**, 2435–2439.

22

Cancers

Kathryn E. Bradbury, Aurora Perez-Cornago, and Tim J. Key

OBJECTIVES

By the end of this chapter you should be able to:

- describe the incidence of cancers worldwide and in the UK
- understand the pathophysiology of cancer
- appreciate the difficulties of linking dietary factors with cancer
- discuss the epidemiological data linking diet with cancer.

22.1 Introduction

Cancer is defined as a disease in which the normal control of cell division is lost, so that an individual cell multiplies inappropriately to form a tumour. The tumour may eventually spread through the body and overwhelm it, causing death.

Cancer has been recorded since the beginning of history, and was well known to the ancient Egyptians several thousand years ago. Until about a hundred years ago, however, cancer was a relatively minor cause of illness and death because it occurs mostly in old age, and few people formerly lived long enough to be at significant risk of developing cancer. Thus, concern about cancer as a major health problem has grown with the ageing of the population in high-income countries, where cancer now accounts for around a quarter of all deaths; cancer is also now a major health problem in low-income countries. An estimated 10 million people died from cancer in 2020 globally, and cancer is the first or second leading cause of premature death (i.e., at ages 30–69 years) in the majority of countries (Wild, Weiderpass & Stewart 2020).

Cancer can arise from the cells of different tissues and organs in the body; thus, there are many different types of cancer. The majority of human tumours are of epithelial origin (80–90 per cent), and cancers of connective tissues or haematopoietic cells are rarer. Cancer is caused both by external factors including tobacco, alcohol, ionizing radiation and ultraviolet light, and by internal factors such as inherited mutations, immune conditions, and hormones. The causes of cancer in different parts of the body vary—for example, tobacco causes cancer of the lungs and many other types of cancer, but does not appear to increase the risk for developing cancer of the endometrium. Dietary factors can also influence risk for developing cancer, but establishing the exact effects of diet on cancer risk has proved difficult and few dietary factors have been clearly shown to be important (World Cancer Research Fund 2018; Swerdlow & Peto 2020; Wild, Weiderpass & Stewart 2020).

22.2 Distribution of cancers throughout the world

Cancer rates worldwide

Worldwide, there were an estimated 19 million new cancer cases and 10 million deaths from cancer in the year 2020 (Ferlay et al. 2020). The most common cancers in terms of new cases were breast (2.3 million), lung (2.2 million), colorectal (1.9 million), prostate (1.4 million), and stomach (1.1 million).

The estimated numbers of new cancer cases in 2018 worldwide are shown in Figure 22.1 (top pie chart for men, bottom pie chart for women). Among men, there were 10.1 million new cases of cancer in 2020; lung cancer was the most common, followed by prostate and colorectum. Among women there were 9.2 million new cases of cancer in 2020; breast cancer was the most common, followed by colorectum and lung.

Cancer incidence varies between countries with a low and high human development index, but colorectal, breast, and prostate cancer are now common worldwide. In countries with low human development index the most

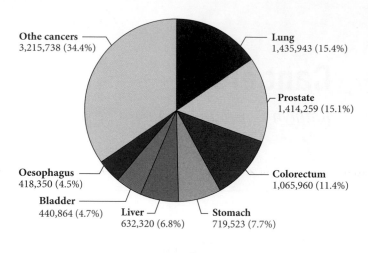

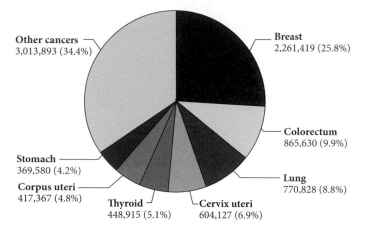

FIGURE 22.1 Number of new cases in 2020 worldwide for the most common cancers in men (top pie chart) and women (bottom pie chart).

common cancers in men are prostate, liver, and colorectum, and in women are breast, cervix, and colorectum. In countries with a very high human development index the most common cancers in men are prostate, lung, and colorectum, and in women breast, colorectum, and lung.

Table 22.1 shows the age-standardized incidence rates for common cancers in ten countries, selected to be representative of different parts of the world with good quality data on cancer incidence.

Stomach cancer rates are particularly high in Japan, China, and Russia.

Colorectal cancer rates are high in Japan, the UK, and New Zealand. Liver cancer rates are high in men in China and Japan. Lung cancer rates are high in men in all selected countries except India, and relatively high in women in the USA, the UK, New Zealand, and China. Breast cancer rates are very high in New Zealand, the USA, UK, Italy, and Argentina. Prostate cancer rates are very high in New Zealand, the UK, USA, and high in South Africa and Italy. Figure 22.2 shows wide variation in the incidence of stomach cancer worldwide, with very high rates in Russia, parts of Asia and South

America, and low rates overall in Australia and New Zealand, North America, and most of Europe. Figure 22.3 shows the incidence of colorectal cancer worldwide, again with large variation; there are very high rates in Australia and New Zealand, North America, and parts of Europe and low rates in Africa and India. As well as the large variation in current cancer rates between different countries around the world, time-trend studies show that cancer rates within populations can change substantially over decades, and studies of migrant populations have also shown gradual increases in cancer risk when people move from a low risk area to a high risk area. The large variations in rates between populations, and with time and migration, suggest that environmental factors such as diet are important in the aetiology of cancer.

Cancer rates in the UK

In 2017, the four most common cancers in the UK were cancers of the breast (55,000 cases per year), prostate (49,000 cases per year), lung (48,000 cases per year), and

TABLE 22.1 Age-standardized incidence rates, per 100,000 per year, for selected countries, 2020

Country	Stomach		Colorectal		Liver		Lung		Breast	Prostate
	men	women	men	women	men	women	men	women	women	men
Argentina	9.3	4.0	31.0	20.6	5.3	2.5	28.1	12.3	73.1	42.0
China	29.5	12.3	28.6	19.5	27.6	9.0	47.8	22.8	39.1	10.2
India	6.1	2.9	6.0	3.7	3.6	1.6	7.8	3.1	25.8	5.5
Italy	10.5	5.5	34.2	25.2	12.0	3.7	36.0	16.4	87.0	59.9
Japan	48.1	17.3	47.3	30.5	16.1	5.3	47.0	19.5	76.3	51.8
New Zealand	6.6	3.2	38.3	29.7	7.0	2.6	24.7	25.0	93.0	92.9
Russian Federation	21.5	8.5	34.4	23.9	6.6	2.7	49.1	7.8	54.9	43.7
South Africa	4.9	2.5	17.6	12.5	6.9	3.1	29.6	10.0	52.6	68.3
United Kingdom	5.4	2.7	40.0	29.0	7.1	3.7	35.2	29.9	87.7	77.9
USA	5.3	3.1	28.7	22.9	10.4	3.7	36.3	30.4	90.3	72.0

colorectum (42,000 cases per year), and these four types of cancer together accounted for over half of all new cases in that year (Cancer Research UK 2021).

Figure 22.4 shows the trends in incidence for four common cancers among men and women in England between 1971 and 2017. In men (Figure 22.4), the most common cancer is now prostate cancer; diagnosis of this cancer has shown a steep increase over the last twenty years, probably due to earlier and increased detection. Lung cancer was the most common cancer

among men for many years; incidence peaked in the 1970s and then declined steadily, principally due to earlier reductions in cigarette smoking, with a lag time of around 20 years. Colorectal cancer rates increased moderately during this period, while stomach cancer rates fell substantially.

In women (Figure 22.4), breast cancer is by far the most common cancer, and the incidence may still be increasing. The steep increase in incidence observed around 1990 was partly due to increased and earlier detection by

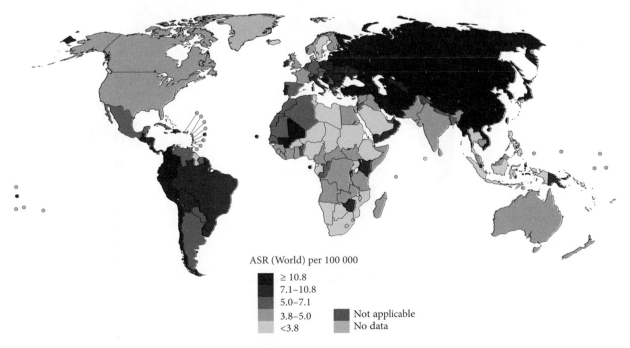

ASR (World) per 100 000

≥ 10.8
7.1–10.8
5.0–7.1
3.8–5.0
<3.8

Not applicable
No data

FIGURE 22.2 Worldwide variation in the incidence of stomach cancer.
Source: Data from Ferlay 2020.

Women

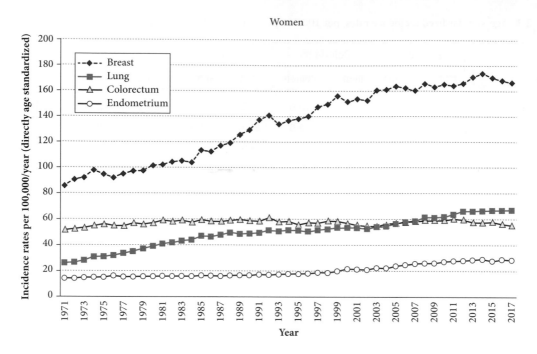

FIGURE 22.2 (*Continued*)

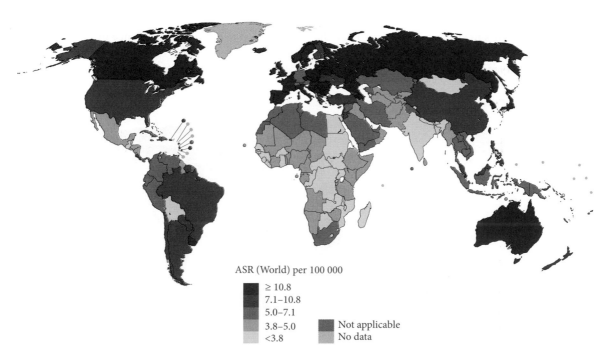

ASR (World) per 100 000

≥ 10.8
7.1–10.8
5.0–7.1
3.8–5.0   Not applicable
<3.8   No data

FIGURE 22.3 Worldwide variation in the incidence of colorectal cancer.
Source: Data from Ferlay 2020.

Men

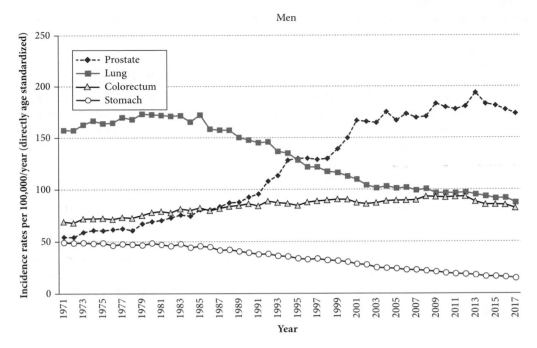

the national breast cancer screening programme. Rates of colorectal cancer in women have not changed dramatically in the last 40 years, whereas lung cancer has increased substantially due to earlier increases in cigarette smoking. Moreover, the incidence of endometrial cancer has increased moderately in the last 15 years.

22.3 Pathophysiology of cancer

Most cancers develop from a single cell that grows and divides more than it should, resulting in the formation of a tumour (growth). Cancers developing in most tissues take the form of a lump that grows, invades local non-cancerous tissue or adjacent tissues/organs, and may spread to other distant parts of the body through the bloodstream or lymphatic system. Cancers arising in the cells of the blood, such as leukaemia, do not usually form a lump because the cancer cells are floating freely throughout the bloodstream.

The change from a normal cell into a cancer, termed carcinogenesis, is a multistage process. Cancers represent a form of dedifferentiation that is associated with the loss of growth control and disturbances in the regulation of the cell cycle. Hanahan and Weinberg (2011) suggested that most cancers acquire the following capabilities: sustained proliferative signalling, evading growth suppressors, resisting cell death, enabling replicative immortality,

inducing angiogenesis, activating invasion and metastasis, reprogramming of energy metabolism, and evading immune destruction. The fundamental changes that determine carcinogenesis are mostly alterations (mutations) in the DNA; therefore, cancer can be viewed as a genetic disease at the level of somatic cells. Typically, the change from a normal cell to a cancer cell requires mutations in several different genes. Mutations in many different genes can result in cancer, but certain genes are frequently involved; in particular, cancer usually results from changes in the function of genes that control cell division (mitosis) and cell death (apoptosis). The key genes in carcinogenesis can be considered in two classes: oncogenes, genes that when over-activated lead to over-stimulation of cell growth and cell division, such as *Ras* and *Myc*; and tumour suppressor genes such as *p53*, which normally limit the rate of cell division but, if inactivated by a mutation, allow uncontrolled cell division. In addition to mutations, recent research has shown that epigenetic changes are also important in tumour development. Epigenetic mechanisms, for example DNA methylation, can regulate gene expression without changing the DNA sequence, and epigenetic changes can be influenced by environmental factors, including diet (Wild, Weiderpass & Stewart 2020).

The genetic mutations that lead to cancer development may be inherited (germline, i.e., the genome that a person has at birth) and/or arise during the lifetime of an individual (somatic) due to replication errors or the effects of agents such as ionizing radiation, chemical carcinogens, viruses, and endogenous damage (for example, as caused by oxidants). The development of cells into a new cancer is also strongly influenced by various endogenous and exogenous growth-promoting agents, especially hormones.

Chance also plays an important role in determining the occurrence of cancer in an individual. In simplified terms, if mutations in several key genes occur in several different cells then the behaviour of all these cells could remain normal, but if the same mutations all occur together within one cell, then this cell could give rise to a cancer. Chance, however, has little net effect on the incidence of cancer in populations (Swerdlow & Peto 2020).

The process of carcinogenesis in humans usually extends over many years or even decades, though this may vary by cancer site. Individuals who inherit a mutation in a key gene could be regarded as being born with the first step of carcinogenesis already present in all their cells. Mutations in cells can accumulate throughout life. Most deaths due to cancer are caused by the spread of the cancer from its site of origin into adjacent areas and to other parts of the body (metastasis).

22.4 Inherited genetic factors in carcinogenesis

At the cellular level cancer is a genetic disease, and genetic factors are involved in the determination of whether or not individuals develop cancer. For the common types of cancer, inherited highly penetrant genetic factors contribute to around 5–10 per cent of the cases of cancer in a population (Wild, Weiderpass & Stewart 2020). Inherited genetic factors (as opposed to mutations in genes that can occur during a person's lifetime) can be considered in two classes: high-risk mutations and low-risk genetic polymorphisms.

High-risk mutations

Inherited high-risk mutations confer a high risk for developing cancer, perhaps 10 to 50 times higher than the risk in individuals who do not have the mutation. The prevalence of these mutations, however, is low, generally around 1 in 1000 or less. As a consequence of the high risk conferred, these mutations cause clusters of cancers within families of closely related individuals that can be recognized by the medical profession and then studied by genetic epidemiologists. Well-known examples of genes which, when mutated, confer a high risk for common cancers are the mismatch repair genes *MLH1* and *MSH2*, which are associated with hereditary non-polyposis colorectal cancer; and *BRCA1* and *BRCA2* that increase the risks for breast, ovarian, and prostate cancer. At present,

there is no clear evidence that dietary factors can modulate the effects of genes such as these on cancer risk.

Low-risk polymorphisms

Low-risk polymorphisms are genetic variants that are termed polymorphisms (rather than mutations) because they occur at a prevalence of more than 1 per cent in a population. Such polymorphisms appear to confer a risk of cancer only moderately higher or lower (around 5 per cent to 50 per cent) than the risk in individuals with the 'wild type' allele (the most common phenotype in the population). Since the increase in risk is small to moderate, this class of genetic factor is less likely to cause obvious clustering of disease in families, and is more readily identified by very large population-based epidemiological studies that compare the prevalence of a particular allele in individuals affected by cancer with the prevalence in unaffected controls. In the last two decades, genome-wide association studies have identified a large number of new susceptibility alleles in the human genome, many of which are particular to specific types of cancer. It is possible that some of these polymorphisms may modify the impact of environmental factors such as diet on cancer risk, for instance by affecting the rate of detoxification or activation of mutagenic chemicals present in some foods. For example, a polymorphism in the gene coding for acetaldehyde dehydrogenase 2 reduces the activity of this enzyme, leading to higher circulating acetaldehyde and a higher risk of oesophageal cancer at the same level of alcohol intake (Wild, Weiderpass & Stewart 2020).

22.5 Non-dietary causes of cancer

Estimates of the proportions of cancer due to avoidable causes are shown in Figure 22.5 (adapted from Brown et al. 2018). These estimates are for the UK but would be broadly similar for most other high-income countries such as the USA. In lower income countries the proportion of cancer due to infective factors would be higher and therefore some of the other factors correspondingly lower, due particularly to higher rates of cancers of the liver (caused by hepatitis B virus and hepatitis C virus) and cervix (caused by human papillomavirus).

Tobacco

Worldwide, the most important preventable cause of cancer is tobacco, which causes more than 10 per cent of cancers in many high-income countries. Tobacco smoking causes cancers of the mouth, pharynx, oesophagus, stomach, colorectum, liver, pancreas, nasal cavities, larynx, lung, cervix, ovary, kidney, bladder, and myeloid leukaemia.

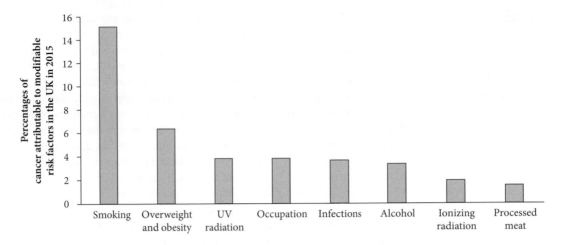

FIGURE 22.5 Percentages of cancer attributable to modifiable risk factors in the UK in 2015.
Source: Data from K F Brown, H Rumgay, C Dunlop et al. (2018) The fraction of cancer attributable to modifiable risk factors in England, Wales, Scotland, Northern Ireland, and the United Kingdom in 2015. *Br. J. Cancer* **118**, 1130–1141.

Infections

Infectious agents are responsible for about 13 per cent of cancers worldwide, with the proportion being higher in low-income countries. The most important numerically are cancers of the liver (hepatitis B virus and hepatitis C virus), cervix (human papillomavirus), and stomach (*Helicobacter pylori*). In some parts of the world parasites are important causes of cancer; for example, major causes of liver cancer (cholangiosarcoma) in China, Thailand, and some other parts of Asia are the liver flukes *Clonorchis sinensis* and *Opisthorchis viverrini*, and a major cause of bladder cancer in Egypt and East Africa is the trematode worm *Schistosoma haematobium*.

Hormonal and reproductive factors

Hormonal and reproductive factors are important determinants of three types of cancer in women, cancers of the breast, ovary, and endometrium (lining of the womb). Childbirth reduces the risk for all three of these cancers, probably through inducing terminal differentiation of the epithelial cells in the breasts, and stopping, for the duration of pregnancies, ovulation in the ovaries and cell division in the susceptible cells in the endometrium. In men, hormonal factors are also important in cancer of the prostate and may be important in the aetiology of cancer of the testis.

Ionizing radiation

The effects of ionizing radiation on cancer development depend on the amount of energy absorbed by the exposed organ or tissue. Ionizing radiation is estimated to cause about 5 per cent of cancers in high-income countries. In the UK, for example, approximately 3 per cent of lung cancers are due to naturally occurring radon gas inside buildings.

Other factors

Other factors including ultraviolet light, medical drugs, immunosuppression, occupational exposures, and pollution are each estimated to account for around 4 per cent or less of cancers (Swerdlow & Peto 2020).

22.6 Diet and cancer

Introduction

Research challenges

Research into the effects of diet on cancer is difficult. Different dietary factors may both increase or reduce the risk for developing cancer, and the size of the effect on risk may be small or moderate (usually less than two-fold), compared to the very large effects of some other agents (e.g., heavy smoking increases the risk for lung cancer by about 40-fold, and persistent HPV infection increases the risk for cervical cancer by about 100-fold). But probably the biggest difficulty in conducting research into diet and cancer is obtaining accurate estimates of long-term dietary intake, compounded by the interrelations of intakes of most foods and nutrients, the relatively narrow range in food choices within many populations, the biases in recall of food intake, and confounding by other lifestyle factors.

Types of evidence available

Much of the evidence suggesting the potential role of life-style and environmental factors in the determination of cancer risk comes from descriptive epidemiology, which looks at the variations of cancer rates with place and time (Swerdlow & Peto 2020). This research has shown that, for all the common cancers, rates vary widely (for many cancers by at least five-fold) between populations in different parts of the world, and that these variations are mostly not due to genetic make-up because rates change markedly when people migrate from one country to another, and also change with time within populations.

Many of the prominent hypotheses for effects of diet on cancer risk have been derived from so-called ecological studies, where population-level averages of food intakes in different populations are compared with cancer rates. For example, this type of study led to the hypotheses that high intakes of meat may increase the risk for colorectal cancer, and that high intakes of fat may increase the risk for breast cancer. Observations in ecological studies do not constitute evidence of a causal relationship, because the diets of the individuals who actually develop cancer are not measured, and many dietary and non-dietary factors vary between populations and could explain the variations in cancer rates.

Testing of hypotheses requires research on individuals, including observational studies that compare the diets of individuals who develop cancer with the diets of those who remain cancer-free, and randomized trials that test the effect of dietary supplements or dietary change on cancer incidence. More recently, so-called 'Mendelian randomization' studies have been developed to use genetic variants as indicators of exposures, including some dietary factors. Observational studies that attempt to retrospectively assess the diets of people who have already developed cancer can be relatively cheap and rapid to conduct, but the results are subject to biased recall of previous diet by cancer patients, and biased participation in the research particularly by health-conscious controls; thus small to moderate effects observed in this type of 'case-control' study should be interpreted cautiously. More reliable are prospective (or 'cohort' studies), in which diet is assessed among healthy (cancer-free) people who are then followed to relate diet at recruitment to subsequent cancer incidence; this type of study largely avoids the recall and selection biases of case-control studies but is slow and expensive, and the reliability of the results can be compromised by substantial errors in estimation of usual dietary intake, and by the difficulties of identifying and controlling for related dietary and non-dietary factors.

Biomarkers of nutritional intake such as blood concentrations of micronutrients and fatty acids can also be investigated in large prospective cohort studies. Some biomarkers can provide an estimate of relative dietary intake, avoiding the errors due to self-reporting, and can help elucidate biological mechanisms by which dietary factors may influence cancer risk. However, the stability of blood biomarkers within individuals over time is often unknown, and the relationship between dietary intake and the biomarker may not be clear. Furthermore, biomarkers of nutrient intake suffer some of the same problems as estimates of nutrient intakes, in that they are also subject to confounding by other dietary and non-dietary risk factors.

Where biological samples including genetic material are available it is also possible to investigate the relationship of genetic variants related to the intake, absorption, or metabolism of nutrients with disease risk. This approach has been developed as Mendelian randomization, whereby a genetic variant (or several variants) is used as an 'instrumental variable' which can, if certain assumptions are met, be used as an unbiased and unconfounded proxy measure of nutritional characteristics such as the concentration of a vitamin in the blood. Identifying genetic variants which predict dietary intake is more difficult, and few such variants have been identified so far.

Randomized controlled trials eliminate the biases that can affect observational studies, and positive results provide clear evidence that the intervention has caused the change in cancer risk. However, such trials have to be very large and therefore expensive, and can typically only test one or a few interventions over a few years; also, while it is relatively simple to test the effect of nutritional supplements, it is much harder to test the effects of changes in food intake because adherence to a prescribed diet is difficult to maintain long term and the participants cannot be blinded to the intervention. Thus, few results are available from trials of dietary factors and cancer risk, and negative results do not rule out the possibility that there could be an effect at a different dose or if the intervention had been implemented at a different age or had continued for longer.

Laboratory research also provides important information. Experimental data can demonstrate effects of dietary factors on cancer risk in animals or in cultivated human cells, and provide the basis for hypotheses of effects in humans and plausible mechanisms to explain associations observed.

Mechanisms for dietary factors predisposing to or protecting from cancer

Before discussing the evidence that links nutritional factors with the risk for individual cancer sites, we discuss in this section the types of mechanisms which underlie, or might underlie, these associations. Dietary factors may be associated with cancer development in various ways. At the cellular level, these may be via DNA damage and repair, cell proliferation, carcinogen metabolism, apoptosis, and cell differentiation. Hormonal regulation, inflammation, and immune responses can also be affected by dietary constituents. Moreover, dietary factors may also influence epigenetic changes, where gene expression is altered without changes to the DNA sequence.

Dietary mutagens

Diet may increase cancer risk by supplying mutagens that directly damage the host DNA, but there are few well-established instances of this mechanism. Examples are aflatoxin, Chinese-style salted fish, and aristolochic acid. Aflatoxin is a food contaminant produced by the fungus *Aspergillus flavus*; high levels of aflatoxin can occur in foods such as grains, oilseeds, nuts and dried fruit stored in hot and humid conditions, and, together with active hepatitis due to hepatitis B virus (HBV), or less commonly hepatitis C (HCV), can cause liver cancer (see next section). Chinese-style salted fish increases the risk for nasopharyngeal cancer (see next section), this food is usually allowed to partially decompose during processing and can contain high levels of nitrosamines. Aristolochic acid, which is present in Aristolochia plants (which are sometimes used in herbal medicine or can contaminate cereal crops), is considered a human carcinogen, increasing the risks of bladder cancer and upper urinary tract urothelial carcinoma (Wild, Weiderpass & Stewart 2020).

Other food components are also hypothesized to increase cancer risk due to their mutagenic effects, but evidence as to whether this does in fact occur in humans is not yet conclusive; for example, nitrites and their related compounds found in smoked, salted, and some processed meat products may be converted to carcinogenic *N*-nitroso compounds in the body, and this may explain the carcinogenic effects of processed meats. Meat and fish cooked at high temperatures can contain moderately high levels of heterocyclic amines and polycyclic aromatic hydrocarbons, which might increase risk in particular for colorectal cancer. Starchy foods like bread, breakfast cereals, and fried potatoes cooked at high temperatures contain moderate levels of acrylamide, which can be carcinogenic when administered at high doses to experimental animals, but epidemiological studies have not shown evidence of an effect of dietary acrylamide on cancer risk in humans, although exposure to acrylamide is not easy to measure.

Alcohol

Alcohol is one of the best-established and most important dietary risk factors for cancer. The metabolism of ethanol leads to the generation of the carcinogen acetaldehyde which can form DNA adducts and thus promote tumour growth. In addition, the effects of alcohol ingestion on metabolism in the liver may lead to impaired immunity, and the increase in breast cancer caused by alcohol intake may be due to effects on oestrogen metabolism, leading to increases in blood levels of oestrogens. Alcohol may also facilitate or enhance the role of other carcinogens via its ability to increase the absorption of carcinogens, or make mucosal cells more susceptible to chemical carcinogens.

Energy balance

The most striking finding that emerged from studies on diet and cancer in experimental animals conducted during the twentieth century was that major energy restriction (i.e., at least 30 per cent reduction) can substantially reduce the incidence of cancer. The implications of this observation for human populations are not clear, in particular because many of these experiments were carried out on weanling animals exposed to severe energy restriction that resulted in restricted growth.

Obesity in humans increases the risk of at least 13 types of cancer (adenocarcinoma of the oesophagus, gastric cardia, colorectum, liver, gallbladder, pancreas, breast (postmenopausal), endometrium, ovary, kidney, meningioma, thyroid, and multiple myeloma). The mechanisms by which obesity raises the risk for these cancers are not fully understood and may vary by cancer site. For cancers of the colorectum, pancreas, and kidney, the mechanisms may involve exposure to higher circulating insulin (e.g., Murphy et al. 2018). The increased risk for adenocarcinoma of the oesophagus may be due to increased reflux of gastric contents. For postmenopausal breast cancer and endometrial cancer the mechanism is probably increased production of oestrogens in the adipose tissue; postmenopausal women with obesity have much higher blood levels of endogenous (natural) oestrogens than women who have a normal weight, causing an increase in the risks for cancers of the breast and endometrium because oestrogens stimulate the growth and division of the cells in these tissues (Lauby-Secretan et al. 2016).

Dietary fibre

Diets high in fibre have generally been associated with a small reduction in the risk of colorectal cancer in both animal and human studies, although it is difficult to identify whether the 'dietary fibre' itself or other substances found in high fibre foods such as cereals, pulses, vegetables, and fruits are responsible for this apparent beneficial effect. Given the effects of fibre on the luminal characteristics and function of the bowel, several mechanisms have been proposed through which dietary fibre may reduce the risk for colorectal cancer. High intakes of fibre increase stool bulk, reduce transit time through the colon, and may thus minimize the absorption of carcinogens by the colonic mucosa. Dietary fibre may also reduce exposure to carcinogens through dilution of the gut contents and/or by binding the carcinogens for faecal excretion. Finally, fermentation of fibre in the large bowel produces short-chain fatty acids, such as butyrate, which may protect against colorectal cancer through the ability to promote differentiation, induce apoptosis, and inhibit the production of carcinogenic secondary bile acids by lowering the intraluminal pH.

It has also been suggested that dietary fibre may reduce the risk of hormone-dependent cancers by regulating sex hormone levels. A diet high in fibre can reduce the enterohepatic circulation of steroids and this may ultimately reduce serum hormone concentrations by causing increased faecal excretion; however, the importance of this potential mechanism is not established.

Vitamins and minerals

Consumption of adequate amounts of certain vitamins and minerals may help to reduce the risk for some types of cancer. Indeed, it has been suggested that the relatively high incidence of cancers of the oesophagus and stomach in some low and middle-income countries is partly due to micronutrient deficiencies. Micronutrient deficiencies may increase susceptibility in several ways; for example, deficiency may impair the integrity and repair of tissue structures. In addition, some micronutrients have antioxidant properties, and research in the 1990s explored the possibility that antioxidant nutrients might protect against cancer, but trials of antioxidant nutrients have so far been almost entirely negative. Overall, the evidence indicates that micronutrient deficiencies should be avoided (or corrected), but that consuming higher than the recommended requirements for vitamins and minerals is not beneficial and may have risks. Dietary folate and other B vitamins, as well as certain other dietary constituents, can also be also involved in epigenetic regulation (see section 'Epigenetic regulation' below).

Possible plant-derived anti-carcinogens

In addition to vitamins and minerals, many compounds in plants that are not nutrients are under investigation for possible anti-carcinogenic properties. Examples include: flavonoids such as quercetin found at high levels in apples, onions, and tea; carotenoids such as lycopene and lutein found in tomato products and green vegetables, respectively; isothiocyanates, predominantly

found in cruciferous vegetables; sulphur-containing compounds found predominantly in the *Allium* species garlic, onions, and leeks; and phytosterols, found in a wide variety of fruit and vegetables, which might have some anti-cancer properties via several pathways such as regulation of signal transduction pathways that regulate tumour growth and apoptosis. Phytoestrogens such as isoflavones derived from soybeans have received particular attention because of their possible anti-oestrogenic effects that could reduce risk for breast cancer in women and prostate cancer in men. However, research so far has not established a definite protective effect of any of these compounds for any type of cancer in humans.

Insulin-like growth factor-1

The cancer protective effect of energy restriction in animal models may be partly mediated by a reduction in circulating levels of the growth factor insulin-like growth factor-1 (IGF-1). High blood levels of IGF-1 are associated with an increased risk for some types of cancer in humans, such as colorectal, breast, and prostate cancer. Current data suggest that blood levels of IGF-1 in humans are positively related to intake of protein and especially protein from dairy products, although it is not clear if this apparent effect is big enough to result in material effects on cancer risk.

Epigenetic regulation

Change in gene expression without altering the DNA sequence is known as epigenetic regulation, the most well-known example being DNA methylation. In many genes, clusters of cytosine adjacent to guanine base (CpG) sequences are found in the promoter regions that normally act to increase expression of the gene. DNA methylation is the addition of a methyl group (CH3) to cytosine residues in these CpG sequences. When these sites are methylated in the promoter region, the transcription factors cannot bind to these sites and the gene is not expressed. During the differentiation of normal cells, DNA methylation can permanently switch off certain genes.

Dietary folate and other dietary constituents that can act as methyl group donors (e.g., choline, methionine, betaine) are essential for this epigenetic control (as well as for DNA synthesis). Dietary components which do not themselves provide methyl groups may also influence the methylation process. Other types of epigenetic processes include genomic imprinting, the deacetylation and methylation of histones involved in the mechanism that regulates promoter transcription, and chromatin modifications.

Inflammation

Inflammation is a generally protective physiological response to infection or trauma but in the long term inflammation can lead to DNA damage and promote the development of cancer by increasing cell proliferation, inhibiting apoptosis, and inducing the generation of new blood vessels. Dietary components can influence low-grade inflammatory processes. For example, heavy alcohol consumption can cause inflammation in the liver, which may alter metabolic processes, impair immune function, and increase the chances of developing liver cancer. There is also evidence that obesity is accompanied by generalized low-grade inflammation, which may in turn increase cancer risk at a number of sites in the body.

Overview of the associations of dietary factors with the risk for common types of cancer

Upper gastrointestinal tract

Cancers of the lip, oral cavity, pharynx, and oesophagus were estimated to account for 1,298,000 cases and 889,000 deaths worldwide in 2020. There are pockets of extremely high risk in some areas within countries, such as in Linxian in China and Golestan in Iran; in Linxian the incidence rates are more than 100-fold higher than in North America and Europe. In high income countries the main risk factors are alcohol and tobacco, and up to 75 per cent of these cancers are attributable to these two lifestyle factors. The mechanism of the effect of alcohol on these cancers is not fully understood, but may involve direct effects of salivary acetaldehyde derived from alcohol acting as a local carcinogen on the epithelium (Salasporo 2020). Infection with human papillomavirus is also associated with an increased risk of oral cancers. In addition, there is evidence that consuming drinks and foods at a very high temperature increases the risk for these cancers, and in 2016 IARC/WHO classified very hot drinks as probably carcinogenic to humans (Group 2A) (Loomis et al. 2016). Overweight/obesity is an established risk factor specifically for adenocarcinoma (but not squamous cell carcinoma) of the oesophagus. Early case-control studies suggested a protective role of fruit and vegetables, but the evidence from more recent prospective studies is weaker and might be explained by residual confounding due to smoking and alcohol.

In the very high risk areas of the world alcohol and tobacco use are not prevalent and do not explain the high risk. It has been proposed that a substantial proportion of cancers of the oral cavity, pharynx, and oesophagus in some low-income countries may be due to micronutrient deficiencies resulting from a restricted diet that is low in fruit and vegetables and animal products. The relative roles of various micronutrients are not yet clear, but deficiencies of riboflavin, folate, vitamin C, and zinc may be important. The results of trials in Linxian in China, aimed at reducing oesophageal cancer rates with micronutrient supplements, have shown no significant protective effects of a range of micronutrient supplements. The Golestan Cohort study found inverse associations between dietary intakes of fruit, vegetables, calcium and zinc and oesophageal cancer, but it is unclear whether these dietary factors explain the very high risk in this area.

One type of cancer within this group, nasopharyngeal cancer, is particularly common in Southeast Asia, and has been consistently associated with a high intake of Chinese-style salted fish, especially during early childhood, as well as with infection with the Epstein–Barr virus (Swerdlow & Peto 2020; Wild, Weiderpass & Stewart 2020). Chinese-style salted fish is a special product that is usually softened by partial decomposition before or during salting; some other types of preserved foods may also be associated with an increase in the risk for developing nasopharyngeal cancer.

Stomach cancer

Stomach cancer was estimated to account for 1,089,000 cases and 769,000 deaths worldwide in 2020. In 1975 stomach cancer was the most common cancer in the world, but mortality rates have been falling particularly in Western countries and stomach cancer is now relatively much more common in Asia than in Europe or North America. Infection with the bacterium *Helicobacter pylori* is an established risk factor for the development of stomach cancer, and a decline in the prevalence of infection with *Helicobacter pylori* in recent generations is an important factor in the decline in the incidence of stomach cancer. Dietary changes are also implicated in the recent decline in stomach cancer incidence. The introduction of refrigeration has led to a reduction in intakes of pickled and salt-preserved foods such as meat, fish, and vegetables, and has also facilitated year-round fruit and vegetable availability.

Substantial evidence suggests that a high total salt intake or a high intake of salted foods, such as salted processed meat and salt-preserved fish, increases the risk of stomach cancer. There are also prospective data showing an increased risk of stomach cancer with high intakes of pickled vegetables. These associations are thought to be causal, and may be due to salt itself or chemical carcinogens derived from nitrites; in some types of pickled vegetables mould or fungi develop during preservation and may also contribute to an increased risk of stomach cancer.

Early epidemiological studies suggested that risk of stomach cancer might be decreased by high intakes of fruit and vegetables, but more recent evidence from prospective studies does not clearly support a protective effect for fruit and vegetables. The findings of studies that have supplemented individuals with *Helicobacter pylori* with vitamin C, with or without standard antibiotic treatment have been mixed; some studies have shown vitamin C increased eradication rate but others have shown no effect (Mei & Hongbin 2018). There is some evidence, almost all from case-control studies, that dietary vitamin C is associated with decreased risk of stomach cancer. Prospective studies have also reported that high plasma levels of vitamin C are related to lower risk. In Linxian, China, combined supplementation with β-carotene, selenium, and alpha-tocopherol resulted in a significant reduction in stomach cancer mortality, but no significant benefit was obtained from vitamin C combined with molybdenum. In the few trials in patients with precancerous gastric lesions, some have shown greater regression with the use of supplements of vitamin C and/or β-carotene; differences in the trial duration, age of the participants, dose of vitamins and minerals, and sample sizes may explain the inconsistencies between trials.

Further prospective data are needed, in particular to examine whether some of the dietary associations may be partly confounded by *Helicobacter pylori* infection and whether dietary factors may modify the association of *Helicobacter pylori* with risk.

Colorectal cancer

Colorectal cancer is the fourth most common cancer in the world and was estimated to account for 1,932,000 cases and 935,000 deaths in 2020. Incidence rates are higher in high income than in low-income countries. Colorectal tumours may develop over a period of ten or twenty years. Most are thought to arise from a precursor lesion, the adenomatous polyp, and prevention or surgical removal of colorectal adenomas may decrease the occurrence of colorectal cancer. Diet-related factors may account for up to 80 per cent of the between-country differences in rates of colorectal cancer. Overweight/ obesity increases the risk of colorectal cancer; this has been shown in prospective cohort studies and recently confirmed in Mendelian randomization studies. Adult height, which is partly determined by the adequacy of nutrition in childhood and adolescence, is weakly positively associated with increased risk, and physical activity has been consistently associated with a reduced risk. Alcohol causes a moderate increase in risk. These factors together, however, do not explain the large variation between populations, and there is almost universal agreement that some aspects of a Western diet are a major determinant of risk. These may include the following dietary factors.

Meat

Early international studies showed a positive association between per capita consumption of red and processed meat and colorectal cancer mortality, and several biological mechanisms have been proposed through which meat may increase cancer risk. Nitrites and related compounds found in some processed meats may be converted to carcinogenic N-nitroso compounds in the colon, and mutagenic heterocyclic amines and polycyclic aromatic hydrocarbons can be formed during the cooking of meat at high temperatures. In addition, red meats are rich in haem iron, and high haem iron levels in the colon may increase the formation of N-nitroso compounds. In 2015 IARC/WHO classified processed meat as carcinogenic to humans (Group 1) and red meat as probably carcinogenic to humans (Group 2A) (Bouvard et al. 2015), and the latest World Cancer Research Fund report (2018) on diet and cancer concluded that the evidence that processed meat causes colorectal cancer is convincing and that the evidence for red meat is probable, while there is no evidence that poultry increases the risk of colorectal cancer. More research on the biological mechanisms of action will be informative.

Dairy products, calcium, and vitamin D

Dairy products, and milk intake specifically, have been consistently associated with a moderately lower risk of colorectal cancer in prospective cohort studies. Relatively high intakes of calcium from milk or other foods may reduce the risk for colorectal cancer, perhaps by forming complexes with secondary bile acids and haem in the intestinal lumen and thus inhibiting their damaging effects on the epithelium. Several observational studies have supported this hypothesis, and trials have suggested that supplemental calcium may have a modest protective effect on the recurrence of colorectal adenomas, but overall the evidence is not strong enough to draw a definite conclusion. There is also observational evidence that relatively high blood levels of vitamin D are associated with a relatively low risk for colorectal cancer, but this association might be confounded by other aspects of lifestyle such as time spent exercising outdoors, and evidence from Mendelian randomization studies does not support a causal relationship with vitamin D (Dimitrakopoulou et al. 2017).

Fibre

Burkitt suggested in the 1970s that the low rates of colorectal cancer in Africa could be due to the high consumption of dietary fibre, and there are several plausible mechanisms for a protective effect. Since Burkitt's hypothesis, evidence on the association of fibre with risk of colorectal cancer has become available from many prospective epidemiological studies conducted in a range of countries. Meta-analyses of these studies have shown that, on average, an increase in intake of total dietary fibre of 10 g per day is associated with a small reduction (approximately 7 per cent) in the risk for colorectal cancer (World Cancer Research Fund 2018). There have been a few randomized controlled trials of fibre supplementation; however, these trials were small and much larger trials would be needed to obtain reliable evidence on the effect of fibre supplementation on colorectal cancer risk (Yao et al. 2017).

Fat

As with meat, early international correlation studies showed a strong association between per capita consumption of fat and colorectal cancer mortality. One possible mechanism is that a high fat intake increases the level of potentially mutagenic secondary bile acids in the lumen of the large intestine. However, the Women's Health Initiative Dietary Modification Trial evaluated the effects of a low-fat eating pattern on risk of colorectal cancer in postmenopausal women and although the intervention group achieved a moderate reduction in their fat intake (as well as increasing their grain and fruit and vegetables intake), there was no evidence that the intervention reduced the risk of invasive colorectal cancer during the eight-year follow-up period. Taken together with the results of prospective observational studies, there is overall little evidence that relatively high intakes of fat increase the risk for colorectal cancer, at least in the middle-aged people studied in high-income countries.

Folate

In several observational studies, high dietary folate intake has been associated with reduced risk of colorectal cancer risk, and diminished folate status might contribute to carcinogenesis by altering gene expression or by increased DNA damage. However, it has also been suggested that, while a higher folate intake might be protective against initiation of colorectal neoplasia, it might also promote the growth of existing tumours. A pooled analysis of randomized controlled trials found no effect of folic acid supplementation on the risk of colorectal cancer, but had insufficient power to exclude a small effect.

A common polymorphism in a key gene involved in folate metabolism (methylene tetrahydrofolate reductase), that results in lower circulating folate concentrations, appears to be related to a slightly lower risk of colorectal cancer. However, the interpretation of these genetic data is not straightforward because the polymorphism results in various alterations to the fluxes of the folate metabolic pathway and other, connected metabolic pathways that may also influence cancer risk.

Liver cancer

Liver cancer was estimated to account for 906,000 cases and 830,000 deaths in 2020, worldwide. Approximately 80 per cent of cases of liver cancer occur in low-income countries, and liver cancer rates vary about 10-fold between countries, being much higher in Eastern and Southeast Asia than in Europe and North America. The major risk factor for hepatocellular carcinoma, the main type of liver cancer, is chronic infection with hepatitis B, and to a lesser extent, hepatitis C virus. Ingestion of foods contaminated with the mycotoxin aflatoxin, combined with active hepatitis virus infection, is an important risk factor among people in some low-income countries for example in Sub-Saharan Africa and southern China. Chronic, high alcohol consumption is established as the main diet-related risk factor for liver cancer in Western countries, probably via the development of cirrhosis and alcoholic hepatitis (Swerdlow & Peto 2020). Overweight/obesity also increases the risk of liver cancer. Several prospective studies have suggested that high coffee consumption is associated with a lower risk of liver cancer, but this needs further research and might be affected by reverse causation bias (that is, individuals with some physiological characteristics or early pathological changes which are associated with the development of liver cancer may experience symptoms which lead them to reduce their coffee intake).

Cancer of the pancreas

Cancer of the pancreas was estimated to account for 496,000 cases and 466,000 deaths in 2018, and is more common in high-income countries than in low-income countries. Overweight/obesity increases the risk, with an increase in risk of about 20 per cent for obesity. Diabetes is associated with an increased risk of pancreatic cancer. Some studies have suggested that risk may be increased by high intakes of meat or dairy products, or reduced by high intakes of fruits or vegetables, but data from prospective studies have been inconsistent and no associations with dietary intakes have been conclusively demonstrated.

Lung cancer

Lung cancer is the most common cancer in the world and was estimated to account for 2,207,000 cases and 1,796,000 deaths in 2020. Heavy smoking increases the risk by around 40-fold, and smoking causes over 80 per cent of lung cancers in high-income countries. The possibility that diet might also have an effect on lung cancer risk was raised in the 1970s following the observation that, after allowing for the effect of smoking, increased lung cancer risk was associated with a low dietary intake of vitamin A. Since then, numerous observational studies have found that lung cancer patients often report a lower intake of fruits, vegetables, and related nutrients (such as β-carotene) than controls. Several trials have tested the effects of supplements of β-carotene on lung cancer incidence. The hypothesis was that increased β-carotene intake would reduce the risk for lung cancer, but none of the trials found any evidence of benefit, and in fact two trials with a large number of cases of lung cancer, conducted among persistent smokers, found that the men who took the β-carotene supplements had a higher incidence of lung cancer than the men who did not.

The possible effect of diet on lung cancer risk remains somewhat controversial. Several recent observational studies have continued to observe an association of fruits and vegetables with reduced risk, but this association has been weak in prospective studies and the inverse association may only be present in current smokers. This apparent relationship is likely to be due to residual confounding by smoking, since smokers generally consume less fruit and vegetables than non-smokers, and smoking has such a large effect on lung cancer risk that such residual confounding could easily be large enough to cause the very small associations observed with fruit and vegetables. Overall, the evidence suggests that diet has at most a small effect on the risk of lung cancer, and in public health terms the overriding priority for reducing lung cancer rates is to reduce the prevalence of smoking.

Breast cancer

Breast cancer is the second most common cancer in the world and the most common cancer among women. Breast cancer was estimated to account for 2,261,000 cases and 685,000 deaths in women in 2020. Incidence rates are about three times higher in high-income countries than in low-income countries. Much of this international variation is due to differences in established reproductive risk factors such as age at menarche, parity and age at birth, and breastfeeding, but differences in dietary habits and physical activity may also contribute. In fact, age at menarche is partly determined by dietary factors, in that restricted dietary intake during childhood and adolescence can lead to delayed menarche. Adult height is positively associated with risk, and is partly determined by dietary factors during childhood

and adolescence in that restriction in food supply during growth can reduce adult height. Observational studies have consistently reported a moderately lower risk of breast cancer in women who report more physical activity, and this has been observed in both premenopausal and postmenopausal women. Oestradiol, and perhaps other hormones, play a key role in the aetiology of breast cancer. Recently, more data has become available to examine risk factors for different sub-types of breast cancer—primarily estrogen-receptor (ER)-positive (about 80 per cent of breast cancers) and ER-negative breast cancers, with the recognition that these different sub-types may have distinct risk factors. It is possible that any further dietary effects on risk for ER-positive breast cancer are mediated by hormonal mechanisms, whereas diet might affect the risk for ER-positive breast cancer by other, non-hormonal mechanisms.

Overweight/obesity

Obesity increases breast cancer risk in postmenopausal women by around 50 per cent, probably by increasing serum concentrations of free oestradiol. Postmenopausal women with obesity have circulating free oestradiol concentrations more than twice as high as those in lean postmenopausal women, because after the menopause the main source of circulating oestrogens is the conversion from androgens by the enzyme aromatase, particularly in the adipose tissue. Obesity does not increase risk among premenopausal women—in fact there is consistent and convincing evidence of an inverse association between body mass index (BMI) and premenopausal breast cancer, which is not well understood (Premenopausal Breast Cancer Collaborative Group 2018). However, obesity in premenopausal women is likely to lead to obesity throughout life and therefore to an eventual increase in breast cancer risk.

Alcohol

Apart from obesity, the only other well-established dietary risk factor for breast cancer is alcohol. There are a large amount of data from well-designed studies that consistently show a moderate increase in risk with increasing consumption, with about a 10 per cent increase in risk for an average of one alcoholic drink every day. The mechanism for this association is not known, but may involve increases in oestrogen levels.

Fat

Much research and controversy has surrounded the hypothesis that a high fat intake increases breast cancer risk. Although early case-control studies suggested that high fat intake may increase the risk of breast cancer, subsequent prospective observational studies and two large randomized controlled trials of a reduced fat intake do not support this hypothesis and imply that moderate variations in fat intake in adult women do not affect breast cancer risk.

Other dietary factors

The results of studies of other dietary factors including meat, dairy products, fruit and vegetables, fibre, and phytoestrogens with overall breast cancer are inconsistent and no convincing associations with these factors have been established. Some recent prospective studies have shown inverse associations between fruit, vegetables, and fibre intakes and ER-negative breast cancer, but further research is needed to confirm these associations and the possible underlying mechanisms.

Cancer of the cervix

Cancer of the cervix was estimated to account for 604,000 cases and 342,000 deaths in women in 2020. The highest rates are in Sub-Saharan Africa, Melanesia, and Central and South America. The major cause of cervical cancer is infection with certain subtypes of the human papillomavirus. Fruits, vegetables, and related nutrients such as carotenoids and folate have been inversely related with risk in a few observational studies; this could perhaps be because dietary factors may influence HPV persistence, but it seems more likely that these associations may be due to confounding by papillomavirus infections, smoking, and other factors, and overall there is little evidence that dietary factors influence the risk of this cancer.

Cancer of the endometrium

Endometrial cancer was estimated to account for 417,000 cases and 97,000 deaths in women in 2020, with the highest incidence rates occurring in Western countries. Risk is lower in women with high parity and with long-term use of combined oral contraceptives. Endometrial cancer risk is about three-fold higher in women with obesity than lean women. As with breast cancer, the effect of obesity in postmenopausal women on the risk for endometrial cancer is probably mediated by the increase in serum concentrations of oestradiol and the reduction in serum concentrations of sex hormone-binding globulin; in premenopausal women, the mechanism may involve the increase in anovulation and consequent increased exposure to oestradiol unopposed by progesterone. Adult height is also weakly positively associated with risk. Studies of dietary factors have provided some evidence that a high glycaemic load may be associated with a higher risk; a few studies have also suggested an inverse association of coffee drinking with risk, but overall findings are inconclusive.

Cancer of the ovary

Cancer of the ovary was estimated to account for 314,000 cases and 207,000 deaths in women in 2020, with the highest incidence rates occurring in Western countries. Risk is reduced by high parity and by long-term use of combined oral contraceptives. Adult height and obesity are both associated with a slightly higher risk, while obesity is associated with a slightly higher risk. Some studies have suggested that risk might be increased by high intakes of fat or dairy products and reduced by high intakes of vegetables, but the data are not consistent and overall are inconclusive.

Prostate cancer

Prostate cancer was estimated to account for 1,414,000 cases and 375,000 deaths in 2020. Prostate cancer incidence rates are strongly affected by diagnostic practices and therefore difficult to interpret, but mortality rates show that death from prostate cancer is about three times more common in North America and Europe than in Asia. Little is known about the aetiology of prostate cancer, and the only well-established risk factors are increasing age, family history, black ethnicity, and genetic factors. There is also evidence that relates obesity with a higher risk of aggressive forms of the disease, and height is positively associated with risk.

Hormones control the growth of the prostate, and interventions that lower androgen levels are moderately effective in treating this disease. Observational studies suggest that both free testosterone and IGF-1 may increase the risk of prostate cancer. Diet might affect prostate cancer risk by affecting IGF-1 levels, and data suggest that high intakes of protein, especially protein from dairy sources, may increase circulating levels of IGF-1; better understanding of these relationships is an important topic of current research.

Randomized controlled trials have shown that supplements of β-carotene, vitamin E, and selenium do not reduce overall risk of prostate cancer, and Mendelian randomization studies have provided no evidence for protective effects of selenium or vitamin D. Lycopene, primarily from tomatoes, has been associated with a reduced risk in some observational studies, but the data are not conclusive. Prospective studies in Asian men have reported that isoflavones, largely from soya foods, are associated with a lower risk of prostate cancer, but further research on this topic is needed.

Kidney cancer

Cancer of the kidney was estimated to account for 431,000 cases and 179,000 deaths in 2020. The range of geographic variation in incidence is moderate, with the highest incidence in high-income countries. Overweight/obesity is an established risk factor for cancer of the kidney, and may account for up to 30 per cent of kidney cancers in both men and women. Adult height is also positively associated with an increased risk of kidney cancer. Higher alcohol consumption appears to be associated with a lower risk of kidney cancer, but further research is needed to establish whether this is due to a real protective effect or whether it is due to confounding or reverse causation. Aristolochic acid from herbal medicines or crop contamination can cause higher risks of cancer of the renal pelvis (IARC Working Group on the Evaluation of Carcinogenic Risks to Humans 2012). Otherwise, there are only limited data on the possible role of diet in the aetiology of kidney cancer; some studies have observed an increase in risk with high intakes of meat and dairy products and a reduced risk with high intakes of vegetables, but the data are inconclusive.

Bladder cancer

Cancer of the urinary bladder was estimated to account for 573,000 cases and 213,000 deaths in 2020. The geographic variation in incidence between countries is about five-fold, with relatively high rates in high-income countries. Tobacco smoking increases the risk for bladder cancer, accounting for between one-third to two-thirds of all bladder cancers. Occupational risk factors, such as exposure to aromatic amines and polyaromatic hydrocarbons, also play a significant role. As explained above, aristolochic acid from herbal medicines or crop contamination can cause higher risks of bladder cancer, as can arsenic in contaminated water. There are no established dietary-related risk factors for bladder cancer.

Non-Hodgkin lymphoma

Non-Hodgkin lymphoma was estimated to account for 544,000 cases and 260,000 deaths in 2020. The range of geographic variation in incidence of total non-Hodgkin lymphoma is moderate, with the highest incidence in high-income countries. Obesity may increase the risk of some types of non-Hodgkin lymphoma, and adult height is also positively associated with risk. Studies of dietary factors have been inconclusive.

Other cancers

There are over 200 different types of cancer and for many of these, particularly the rare cancers, there has been little research on the possible role of diet. With continued follow-up in the many large existing cohort studies more data will become available and may show new associations of diet with cancer risk.

22.7 Recommendations for reducing cancer risk

Table 22.2 summarizes the associations of the common cancers with dietary factors.

The most important recommendation to reduce cancer risk is to not smoke. The main dietary-related recommendations to reduce cancer risk can be summarized as:

- maintain a healthy body weight
- limit alcoholic drinks
- avoid processed meat and limit red meat.

In addition, limit intake of salt-preserved foods, and eat a high-fibre diet, including fruits, vegetables, and wholegrains.

Furthermore, attention should be given to food safety by governments and food manufacturers and local suppliers to ensure the food and water supply is free from contamination with known carcinogens, for example aflatoxins found in mouldy foods such as cereals and nuts, and arsenic in artesian well water.

KEY POINTS

- The change from a normal cell to cancer requires mutations in several genes, and depends on endogenous processes, exogenous mutagens, and chance.
- Tobacco is the most important preventable cause of cancer worldwide.
- Diet-related factors, including obesity and alcohol, may account for around 20 per cent of cancers in developed countries.
- Obesity increases the risk for adenocarcinoma of the oesophagus, gastric cardia, colorectum, liver, gallbladder, pancreas, breast (postmenopausal), endometrium, ovary, kidney, meningioma, thyroid, and multiple myeloma.
- Alcohol causes cancers of the oral cavity, pharynx, larynx, oesophagus, colorectum, liver, and breast.
- High intakes of processed meat increase the risk for colorectal cancer, and high intakes of red meat probably increase the risk for this disease.
- Milk, calcium, and fibre may reduce the risk for colorectal cancer.
- Adequate intakes of fruit and vegetables may reduce the risk for several types of cancer of the gastrointestinal tract.

- In high-income regions, dietary intervention trials with micronutrient supplements have not shown benefit in reducing cancer risk, and high-dose micronutrient supplements may be harmful.
- For many cancers the importance, if any, of dietary factors is not clear.

 Be sure to test your understanding of this chapter by attempting multiple choice questions.

TABLE 22.2 Dietary risk factors, dietary protective factors, and other major environmental and lifestyle risk factors for common cancers*

Cancer	Dietary and diet-related risk factors	Dietary protective factors	Major non-dietary risk factors
Oral cavity, pharynx, and larynx	Alcohol		Smoking Human papillomavirus
Nasopharynx	Chinese-style salted fish		Smoking Epstein–Barr virus
Oesophagus	Alcohol (squamous cell carcinoma) Obesity (adenocarcinoma) Probably very hot drinks		Smoking
Stomach	Probably high intake of salt and salt-preserved foods		Infection by *Helicobacter pylori* Smoking
Colorectum	Alcohol Obesity Processed meat Probably red meat	Probably milk and calcium Probably dietary fibre Probably adequate fruits and vegetables	Sedentary lifestyle Smoking
Liver	Alcohol Obesity Foods contaminated with aflatoxin		Hepatitis viruses Smoking Liver flukes
Pancreas	Obesity		Smoking
Lung			Smoking
Breast	Alcohol Obesity (postmenopausal)	Obesity (premenopausal)	Reproductive and hormonal factors
Cervix			Human papillomavirus Smoking
Endometrium	Obesity		Reproductive and hormonal factors
Ovary	Obesity		Reproductive and hormonal factors
Prostate	Probably obesity for aggressive disease		None established
Kidney	Obesity Plants containing aristolochic acid (renal pelvis)		Smoking
Bladder	Plants containing aristolochic acid		Smoking Occupational risk factors
Non-Hodgkin lymphoma	Obesity		Viral infections Immunodeficiency

REFERENCES

Brown K F, Rumgay H, Dunlop C et al. (2018) The fraction of cancer attributable to modifiable risk factors in England, Wales, Scotland, Northern Ireland, and the United Kingdom in 2015. *Br J Cancer* **118**, 1130–1141.

Bouvard V, Loomis D, Guyton K Z et al. (2015) Carcinogenicity of consumption of red and processed meat. Lancet **16**, 1599–1600.

Cancer Research UK (2021) Cancer statistics for the UK. Available at: https://www.cancerresearchuk.org/health-professional/cancer-statistics-for-the-uk#heading-Zero (accessed 19 February 2021).

Dimitrakopoulou, Tsilidis K K, Haycock C et al. (2017) Circulating vitamin D concentration and risk of seven cancers: Mendelian randomisation study. *BMJ* **359**, j4761.

Ferlay J, Ervik M, Lam F, et al. (2020). *Global Cancer Observatory: Cancer Today.* International Agency for Research on Cancer, Lyon. Available at: https://gco.iarc.fr/today/about (accessed 18 February 2021).

Hanahan D & Weinberg R A (2011). Hallmarks of cancer: The next generation. *Cell* **144**, 646–674.

IARC Working Group on the Evaluation of Carcinogenic Risks to Humans (2012) Plants containing aristolochic acid. In: *Pharmaceuticals.* Volume 100 A. A review of human carcinogens.

Lauby-Secretan B, Scoccianti C, Loomis D et al. for the International Agency for Research on Cancer Handbook Working Group (2016) Body fatness and cancer: Viewpoint of the IARC Working Group. *N Engl J Med* **375**, 794–798.

Loomis D, Guyton K Z, Grosse Y et al. (2016) Carcinogenicity of drinking coffee, mate and very hot beverages. *Lancet Oncol* **17**, 877–878.

Mei H & Hongbin T (2018) Vitamin C and *Helicobacter pylori* infection: Current knowledge and future prospects. *Front Physiol* **9**, 1103.

Murphy N, Jenab M, Gunter M J (2018) Adiposity and gastrointestinal cancers: Epidemiology, mechanisms and future directions. *Nat Rev Gastroenterol Hepatol* **15**, 65970.

Office for National Statistics. Cancer Statistics: Registrations Series MB1 (2019) Available at: http://www.statistics.gov.uk/StatBase/Product.asp?vlnk=8843 (accessed 19 February 2021).

Premenopausal Breast Cancer Collaborative Group (2018) Association of Body Mass Index and age with subsequent breast cancer risk in premenopausal women. *JAMA Oncol* **4**, e181771.

Salapuro M (2020) Local actetaldehyde: Its key role in alcohol-related oropharyngeal cancer. *Visc Med* **36**, 167–173.

Swerdlow A & Peto R (2020) Epidemiology of cancer. In: *Oxford Textbook of Medicine* 6th edn. 1–86 (Firth J, Conlon C, Cox T eds). Oxford University Press, Oxford. https://academic.oup.com/book/41095/chapter-abstract/351063208?redirectedFrom=fulltext

World Cancer Research Fund International (2018) *Diet, Nutrition, Physical Activity and Cancer: A Global Perspective—The Third Expert Report.* London, UK: World Cancer Research Fund International. Available at: https://www.wcrf.org/dietandcancer (accessed 18 February 2021).

Wild C P, Weiderpass E, Stewart BW (eds) (2020) *World Cancer Report: Cancer Research for Cancer Prevention.* International Agency for Research on Cancer, Lyon. Available at: http://publications.iarc.fr/586 (accessed 19 February 2021).

Yao Y, Suo T, Andersson R et al. (2017) Dietary fibre for the prevention of recurrent colorectal adenomas and carcinomas. *Cochrane Database Syst Rev* CD003430.

FURTHER READING

Adami H, Hunter D J, Lagiou P, Mucci L (eds) (2018) *Textbook of Cancer Epidemiology.* (3rd edn). Oxford University Press, Oxford.

Key T J, Bradbury K E, Perez-Cornago A et al. (2020) Diet, nutrition, and cancer risk: What do we know and what is the way forward? *BMJ* **368**, m511.

Schatzkin A, Abnet C C, Cross A J et al. (2009) Mendelian randomization: How it can—and cannot—help confirm causal relations between nutrition and cancer. *Cancer Prev Res* **2**, 104–113.

23

Diseases of the gastrointestinal tract

John O. Hunter and Claire V. Oldale

OBJECTIVES

By the end of this chapter, you should be able to:

- describe the most common chronic diseases which affect the intestinal tract, their epidemiology and clinical presentation
- understand current views on the aetiology and what is understood of their pathophysiology and the role of the microbiome
- discuss the role of diet in their management.

23.1 Introduction

As the prime function of the gastrointestinal (GI) tract is the ingestion, digestion, and absorption of food, it is perhaps surprising that the medical profession has often underplayed the importance of diet in gut disorders. However, food can be a contributory causal factor of some diseases; nutritional deficiencies are frequently prominent, and diet is often the only practicable therapy.

In addition, the gut may be damaged by a wide range of external factors including bacteria and other microorganisms, trauma, radiation, and chemicals. Furthermore, although certain GI bacteria, such as *Salmonella* and *Campylobacter,* are well-known pathogens and produce disease by the release of exotoxins, a relatively new concept has now emerged; that an imbalanced microbiome may arise where no pathogens are present but where disturbances in microbial metabolism may cause disease. This has greatly increased the importance of nutrition in the management of GI disorders.

23.2 The colonic microbiome

The infant's gut is sterile at birth but quickly becomes colonized afterwards by an extremely complex population of microorganisms, known as the colonic microbiome.

It appears that immune tolerance in the early months of life allows the establishment of a gut flora which in health does not provoke an immune response. After three months or so immune tolerance ceases and thereafter all newly encountered microorganisms are treated as potential pathogens, thus rarely establishing themselves in the colon. This phenomenon, known as colonization resistance, keeps the microbiome remarkably stable.

The human colonic microbiome contains $10^{11\text{-}12}$ bacteria per gram of faeces, with over 500 different species and strains. The colon contains no oxygen, and most organisms are anaerobic. However, overgrowth of facultative anaerobes (aerobic bacteria which can survive in the colon) has been associated with disease. Nutrients for the microbiome stem from food residues entering the caecum, for digestion and absorption in the small bowel is incomplete; 5–25 g dietary fibre, 3–4 g fat, 12–14 g protein and as much as 20–40 g starch pass each day into the colon to be fermented by the microflora (Macfarlane & Macfarlane 2012). The composition of the microflora is influenced to some degree by diet, age, and geographical considerations. It may be significantly different between perfectly healthy individuals and this makes interpretation of the importance of changes in bacterial species alone tricky to evaluate.

Nevertheless, certain bacteria have been suggested to be beneficial. Indigestible sugars such as fructo-oligosaccharides modify bacterial composition of the dominant flora by increasing *Bifidobacteria spp.* In other studies, *Veillonella spp.* was found to be present in greater amounts in both the flora of racehorses after feeding with a supplement known to improve condition and also in that of elite distance athletes. Administration of *Veillonella* to mice increased their stamina, possibly via breakdown of L-lactic acid, thus reducing fatigue, converting it to propanoic acid, which could potentially serve as a further source of energy. Such studies suggest that these microorganisms may be beneficial to host health, but this has not been conclusively established.

Study of the microbiome has been hindered by the lack of reliable techniques to identify organisms present. Early studies depended on laboratory culture, which failed to identify fastidious bacteria needing unusual conditions for growth or easily overgrown by more vigorous organisms. The identification of bacteria in the microbiome is now usually performed by molecular approaches. The earliest involved analysis of 16S ribosomal rRNA. DNA itself can now also be sequenced for this purpose and may be extracted in order to gain insights into the functional significance of changes observed—a technique known as metagenomics. The data derived from these techniques are highly complex, requiring detailed computer analysis to reach reliable conclusions. Definite attribution of changes in the structure of the microbiome to specific foodstuffs or diseases frequently remains conjecture rather than established science (Magne at al. 2020).

Ascribing roles to the microbiome in the pathogenesis of disease becomes difficult when variations in the normal healthy state are not easily understood. A potentially more fruitful approach involves the study of metabolites (the metabolome) produced by the microbiome, separated by chromatography before identification by mass spectrometry. Principal Components Analysis (PCA) allows identification of those most significant. Such chemicals may be absorbed from the colon into the bloodstream and produce systemic inflammation, which may affect distant parts of the body.

Despite the stability of the microbiome in health, it can be damaged, particularly by infections and therapeutic courses of antibiotics, which may kill valuable bacteria. The faecal metabolome, for example has been shown to be abnormal in colon cancer, with concentrations of ammonia, acetaldehyde, and hydrogen sulphide, all toxic chemicals, significantly raised. Damage to the microbiome is the likely explanation for reports of irritable bowel syndrome (IBS) following events such as gastroenteritis or antibiotic therapy. The microbiome is different in IBS with overgrowth of facultative anaerobes. The excretion of hydrogen is greatly increased but falls dramatically when symptoms are relieved by diet. Other chemicals produced by the gut flora, including alcohols, aldehydes, ammonia, hydrogen sulfide, and ethyl esters of fatty acids have now been shown by metabolomic studies to be present in raised concentration in IBS. All are potentially toxic but clear when the conditions are put into remission by diet (Walton et al. 2013). Such chemicals may be absorbed from the colon into the bloodstream to affect distant organs.

Crohn's disease (see Section 23.7) also presents an abnormal microbiome which responds to diet. The actual bacteria responsible are yet to be definitively identified, but the microflora in Crohn's disease produce a range of toxic chemicals which are significantly reduced after successful dietary treatment. These include propanol and butanol, indoles, phenols, and the ethyl esters of propanoic and butanoic acids (Table 23.1) (Walton et al. 2016). Such results allow a confident conclusion that the microbiome is indeed relevant to disease pathogenesis. The role of diet in the management of both IBS and Crohn's disease is discussed later.

Abnormalities of the microbiome have been reported in a range of other diseases (Table 23.2) but so far there have been few reports of symptomatic improvement after treatment specifically to correct the microbiome, and their full significance therefore awaits further elucidation. Nevertheless, the potential importance of the gut microbiome to disease elsewhere in the body is shown by the demonstration that it influences the release of interferon, reducing viral infections, in lung stromal cells.

23.3 Manipulation of the colonic microbiome

For many years diet has been the best understood way of manipulating the microbiome for therapeutic purposes. However, alternative methods of treatment have been eagerly researched.

Probiotics and faecal transplantation

It has been suggested that restoring a 'normal' colonic microbiome by using high dose bacterial supplementation may be an effective way of treating colonic disorders. Probiotic bacteria are 'live micro-organisms which when administered in adequate amounts confer a health benefit on the host' (FAO/WHO 2016).

Owing to colonization resistance it is difficult to change the intestinal microflora and most probiotic bacteria, like pathogens, disappear rapidly from the GI tract when their administration ceases. Systematic reviews of the use of probiotic supplements in IBS management suggest insufficient evidence of efficacy to support overall recommendation for use in management, though some high dose, combination preparations show some beneficial effects on global symptoms and pain (Ford et al. 2018). A recent review by the American Gastroenterological Association recommended that probiotic use in GI disorders, such as *Cl difficile* infection and acute gastroenteritis, could be recommended only in the context of formal clinical trials (Su et al. 2020). Probiotics are unlikely to make symptoms worse, but practical guidelines suggest that when probiotic preparations are considered, they should be taken at the manufacturers recommended dose for a month before evaluation of symptom benefit.

Attempts have been made to improve the microbiome by administering a rich mixture of microorganisms in

TABLE 23.1 Changes in faecal chemicals before and after elemental feeding in patients with Crohn's disease

Compound	VOC concentration (ng/l), median (lower quartile, upper quartile)		
	Pretreatment	**Posttreatment**	**P-value**
Acetone	57 (38, 128)	80 (50, 104)	0.435
Propanoic acid	169 (0, 328)	12 (0, 84)	0.031*
Butanoic acid	1110 (316, 1596)	24 (0, 104)	0.001*
1-Propanol	229 (41, 892)	36 (0, 233)	0.025*
Propanoic acid, ethyl ester	19 (0, 117)	0 (0, 15)	0.008*
Butanoic acid, methyl ester	19 (7, 121)	0 (0, 1)	0.013*
Butanoic acid, ethyl ester	46 (4, 255)	0 (0, 15)	0.008*
p-Cresol	518 (118, 1160)	480 (144, 1051)	0.687
Indole	118 (54, 146)	20 (0, 128)	0.125
Dimethyl disulphide	83 (34, 683)	39 (0, 140)	0.113
1-Butanol	99 (57, 256)	58 (0, 199)	0.030*
Butanoic acid, 3-methyl	147 (48, 504)	0 (0, 45)	0.015*
Phenol	64 (16, 102)	24 (10, 177)	0.332

Abbreviations: CD, Crohn's disease; VOC, volatile organic compound. Concentrations of VOCs in faecal headspace from healthy controls and volunteers diagnosed with CD before and after treatment. Median and upper and lower quartile values are shown. Differences were examined by Wilcoxon's matched-pairs test, with $P<0.05$ being considered significant and indicated by an asterisk.
Source: Reproduced with author permission from C Walton et al. (2016) Enteral feeding reduces metabolic activity of the intestinal microbiome in Crohn's Disease: An observational study. *Eur. Jr. Clin. Nutr.* **70**, 1052–1056. Copyright © 2016, Macmillan, Springer Nature.

the form of faeces from healthy donors, administered via colonoscopy. This may prove to have value in *Cl. difficile infection,* when other treatments have failed.

Prebiotics and synbiotics

In an attempt to overcome immune rejection of foreign probiotic bacteria, attempts have been made to improve the metabolomic activity of the resident microbiome by administering prebiotics, which reach the colon undigested, and promote the growth of apparently beneficial organisms such as *Lactobacilli* and *Bifidobacteria.* These substances, predominantly fibres and sugars such as lactulose and oligosaccharides, may however also promote the growth of less helpful bacteria, producing wind and bloating, and there is little evidence to date that they have genuine therapeutic value. Synbiotics are mixtures of prebiotic chemicals and probiotic bacteria. Again, their benefits await confirmation.

Antibiotics

The links between food intolerance and the intestinal microbiome imply a potential therapeutic role for antibiotics, and this is indeed the case. The effectiveness of metronidazole on the microbiome, gas production, and symptoms in IBS was demonstrated by Dear et al. (2005), and antibiotics may also be effective in inducing remission in Crohn's disease (Levine et al. 2019). Rifaximin has also shown benefit in diarrhoea-predominant IBS (Ford et al. 2018), possibly due impairment of bacterial overgrowth in the small intestine (SIBO), but more likely a non-specific effect on the total microbiome. Antibiotic treatment, however, may cause side effects such as *C. difficile* overgrowth or the development of antibiotic resistance. Treatment can be expensive and benefits usually short-lived. For these reasons antibiotics have yet to achieve routine use for manipulation of the microbiome.

23.4 Dietary fibre

Low intakes of dietary fibre have clear associations with the development of several diseases. Meta-analyses reported in the SACN report on carbohydrates and health (SACN 2015) demonstrated that for every 7g/day increase in dietary fibre there were significant risk reductions for diabetes, cardiovascular disease, stroke, and

TABLE 23.2 Conditions in which an abnormality of the microbiome has been reported

Type of disease	Specific condition
GI disease	Irritable bowel syndrome (1)
	Inflammatory bowel disease (2)
Metabolic syndrome	Type II diabetes mellitus (3)
	Obesity (15)
	Non-alcoholic fatty liver disease (16)
Respiratory disease	Asthma (4)
Neurological disorders	Migraine (5)
	Parkinson's disease (6)
	Autism (7)
	Multiple sclerosis (8)
Autoimmune disorders (18)	Rheumatoid arthritis (9)
	Systemic lupus erythematosus (10)
	Immune thrombocytopaenic purpura (11)
	Ankylosing spondylitis (17)
Carcinoma	Colon (12)
	Breast (13)
	Pancreas (14)

For references associated with this table see the Chapter 23 Further Reading online.

colorectal and rectal cancers. Intakes of 25–29g fibre/day may offer greatest risk reduction and the national recommendation for total fibre intake in the UK is currently an average of 30g/day. In addition to its effects on disease prevention however, fibre exerts multiple effects within the bowel, and intakes form part of recommendations for a number of bowel conditions.

There are many types of dietary fibre, and the variation in molecular structure can substantially alter physicochemical properties and behaviour in the GI tract. These characteristics—solubility, viscosity, and fermentability—in turn exert differing functional, regulatory, and microbial effects, including impact on micronutrient bioavailability, gut transit time, and stool form (Gill et al. 2020). Certain fibres act as prebiotics and may enhance beneficial microbial numbers.

The solubility of a fibre source as a predictor of physiological or functional effect is now considered somewhat outmoded. Most fibrous foods contain fibre sources of varying solubility, which will simultaneously exert different physiological effects.

Viscosity (from gums, psyllium, B-glucans, and pectins) increases thickening of GI content and is thought to reduce gastric emptying rate and modulation of intestinal transit, increasing satiety through its reduction in postprandial glycaemic and lipaemic effects. Viscosity also exerts a cholesterol-lowering effect through binding of bile salts preventing reabsorption and can also increase stool bulk and reduce colonic transit.

The fermentability of fibre however, has the most direct impact on the colonic microbiome. Inulin type fructans, galacto-oligosaccharides, and resistant starches increase short-chain fatty acid (SCFA) concentrations in the colon following bacterial fermentation, which in turn affect colonic contractility and motility. These SCFAs may also mediate communication between the mucosal microbiota and immune system, with possible anti-inflammatory and immunomodulatory effects. SCFAs also play a role in the maintenance of intestinal barrier integrity and maintenance of GI homeostasis indirectly via reduction in colonic pH. It has been shown that a reduction in fermentable fibre intake reduces microbiome diversity, which may in turn reduce the structural mucosal integrity and that of the protective mucus layer.

The use of dietary fibres on the GI tract and their effect on the microbiome are now being more closely researched, however, specific recommendations for types and quantities of fibres based on their physicochemical and functional properties have been difficult to elucidate in IBS, inflammatory bowel disease (IBD), and diverticular disease and studies to date have not provided this level of detail.

TABLE 23.3 Causes of malabsorption

Mucosal	Coeliac disease	Giardiasis
	Intestinal infections	Whipple's disease
	Intestinal lymphangiectasia	Tuberculosis, tropical sprue
	Lymphoma	Hookworms and other nematodes
		HIV enteropathy
Intraluminal	Pancreatic insufficiency	Cystic fibrosis
	Bile-salt deficiency	Chronic pancreatitis
		Carcinoma of the pancreas
		Zollinger–Ellison syndrome
		Biliary obstruction
		Bile salt malabsorption
Structural	Post surgery	Gastric surgery
	Diverticula, fistulae, and stricture	Blind-loop syndrome
	Crohn's disease	Significant intestinal resections–'short-gut syndrome'
	Mesenteric arterial insufficiency	
	Radiation enteritis	

23.5 Malabsorption

Disease of the gastrointestinal tract may lead to malabsorption, either by failure of production of digestive enzymes, or by reduction of the mucosal surface area available for nutrient absorption by disease or surgery (Table 23.3).

The nutritional effects depend on the site of the gut involved. Deficiency of iron or vitamin B_{12} may be caused by gastro-duodenal disorders and lead to anaemia. Diseases of the small intestine and pancreas may cause classical symptoms of macronutrient malabsorption with large fatty offensive stools (steatorrhoea) and deficiencies. Protein-losing enteropathies such as intestinal lymphangiectasia, may cause oedema. Micronutrient losses may include calcium, magnesium, iron, zinc, copper, and other trace elements as well as folate. If the ileum is damaged B_{12} deficiency and bile-salt malabsorption may also occur. Severe and prolonged diarrhoea may lead to electrolyte imbalances, with sodium and potassium deficiency. Serum albumin concentrations may also be reduced secondary to leakage into the gut lumen. Unabsorbed fats may form insoluble soap complexes which bind unabsorbed calcium and fat-soluble vitamins.

Common symptoms of malabsorption are a failure of growth in children or weight loss in adults. Significant nutrient malabsorption may occur without obvious gastrointestinal dysfunction, presenting simply as an unexplained anaemia, reduced bone mineral density, or chronic fatigue in adults. Improvement of nutritional status may require adjunctive therapy in severe cases. In most, however, management of the underlying disorder and appropriate nutritional supplementation is often sufficient.

23.6 Coeliac disease

Epidemiology of coeliac disease

Coeliac disease is a chronic, autoimmune disease, precipitated by the consumption of the protein gluten in genetically predisposed individuals. Ingestion of gluten, found in wheat, rye, and barley, causes an inflammatory response, leading to villous atrophy of the small bowel and subsequent nutrient malabsorption.

Prevalence varies worldwide (from <0.25 per cent to >1 per cent), but a large-scale screening study in subjects from Finland, Italy, the UK, and Germany found a prevalence of coeliac disease of around 1 per cent, with a recent US study showing a prevalence of 0.71 per cent. It can present at any age, with 70 per cent of patients over the age of 20 at diagnosis. Female to male ratio ranges from 1.3:1 to 1.5:1. (Al-toma et al. 2019).

In addition to those with symptomatic disease there are patients in whom coeliac disease has developed but no symptoms are present (silent) and many more, carriers of HLA-DQ2.5 and DQ8 HLA-antigens, who have the genetic predisposition to develop coeliac disease but have yet to do so (latent). Thus, the true incidence may be greater than suspected—the so-called 'iceberg effect'.

The introduction of reliable serological markers such as IgA tissue transglutaminase (anti tTG2) and endomysial antibody supports accurate diagnosis and enables screening. Patients must be consuming a gluten-containing diet and total IgA levels must be tested concurrently. Checking coeliac serology is mandatory for patients presenting with other gastrointestinal and autoimmune disorders, and current guidance recommends investigation of serology in asymptomatic first-degree relatives. In those with a first-degree relative with coeliac disease the chances of developing the disorder rise to one in ten.

Pathology of coeliac disease

Small bowel biopsy in coeliac disease reveals characteristic changes. Intra-epithelial lymphocyte concentrations in the small bowel mucosa are increased, intestinal crypts are elongated and open out on to a flattened mucosal surface where there is villous atrophy to varying degree (Figures 23.1, 23.2). These structural changes decrease epithelial surface area available for digestion and absorption and can reduce digestive enzyme availability. The appearance of the mucosa in latent disease may appear superficially normal with residual villous structure; the absorptive surface however is less than the normal population. Multiple duodenal biopsies are indicated for clear diagnosis.

Clinical features

The classical features of symptomatic disease in childhood include weight loss or failure to thrive, anaemia, lassitude, and diarrhoea and may be accompanied by abdominal pain, steatorrhoea, blood loss, and dehydration.

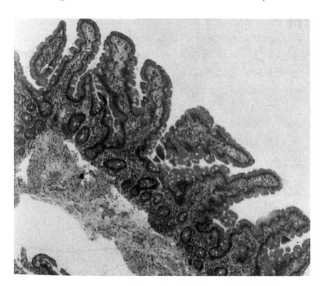

FIGURE 23.1 Normal jejunal mucosa.

FIGURE 23.2 A typical section of jejunal mucosa found in coeliac disease showing the flattened mucosal surface with villous atrophy (top arrow) and elongated crypts (bottom arrow).

Adult disease may also present with more subtle and variable symptoms such as anaemia, fatigue, mouth ulcers, bone thinning, neurological abnormalities such as ataxia and peripheral neuropathy, sub-fertility, and abnormal liver function. There is a cutaneous manifestation of the condition—dermatitis herpetiformis.

Latent and silent cases of coeliac disease show no symptoms whatsoever. These wide-ranging presentations can result in considerable delays in diagnosis, of up to 13 years from symptom onset. Until recently, definitive diagnosis demanded a duodenal biopsy confirming physical changes to bowel structure in addition to positive serology, but guidelines now suggest that children meeting specific serological and other criteria may be given a confirmed diagnosis without biopsy and it is possible that this may be extended to specific adult groups in the near future.

Aetiology of coeliac disease

The cause of coeliac disease was unknown until 1950 when the Dutch paediatrician Dicke reported significant improvement of symptoms in patients with the condition during war-time famine in the Netherlands, when supplies of wheat were scarce. He identified the toxic fraction in wheat as gluten. The major protein fractions of gluten are gliadin and glutenin and all forms of gliadin have been shown to be toxic to coeliac patients. It is believed that a small bowel mucosal enzyme, tissue transglutaminase (tTG), is important in modifying peptides derived from gliadin so that they become capable of forming autoantibodies.

Genetic factors are important in the pathogenesis of coeliac disease, and it has been shown that subjects with human leukocyte antigens (HLA) DQ2.5 and DQ8 are at much greater risk of developing the disease. The physical role of the HLA system is to present peptide fragments of antigens to T cells. A 33-amino acid peptide derived from gliadin has been shown to be remarkably stable and found in foods toxic to coeliac patients including wheat, barley, and rye but not in oats, rice, or maize. It is an excellent substrate for tTG and the deamidated peptide had a much higher affinity for DQ2. When incubated with T cells from coeliac patients it caused a marked increase in lymphocyte proliferation. This might lead to release of cytokines and mucosal damage. Thus, an autoantibody is formed by the action of tTG on gliadin, and coeliac disease is associated with other autoimmune disorders including type I diabetes, thyroid disease, and rheumatoid arthritis. Coeliac disease may be diagnosed by serological screening following the appearance of other autoimmune conditions.

Environmental factors also play a role. Exposure to gluten is essential, but other triggers for activation of the autoimmune response can occur at any age. Examples include GI infection, surgery, and pregnancy.

Nutritional deficiencies and dietary management

Damage to the gastrointestinal mucosa may result in reduced production of digestive enzymes and subsequent reduction in digestive capability. Intestinal villous atrophy may result in significant reduction of mucosal surface area, causing malabsorption of macro and micronutrients. Nutritional deficiencies will depend on the length of small intestine affected, and time from disease onset to diagnosis. Iron deficiency anaemia is common as the duodenum is the primary absorption site for iron. Calcium absorption maybe reduced by defective calcium transport mechanisms and low bone mineral density and osteoporosis are subsequently common chronic features of the disease. Low serum folate concentration is frequent in untreated disease but is not usually severe enough to be associated with megaloblastic anaemia. Vitamin B_{12} deficiency is less common but may develop secondary to folate deficiency or if the disease affects the distal ileum. Significant improvement in nutritional status typically follows a strict gluten-free diet and replacement therapies may not be required.

The treatment for coeliac disease requires lifelong exclusion of gluten from the diet, including the prolamins in wheat (gliadin), barley (hordein), rye (secalin), and traditionally also oats (avenin). However, avenin occurs in much lower concentrations than the prolamins present in wheat, barley, and rye, and pure, uncontaminated oats are no longer excluded on a gluten free diet. Standard oat products may be contaminated with gluten from other grains during harvesting, milling, and storage and should therefore be avoided.

Dietary sources of gluten can be obvious, but are also hidden in many manufactured products, which makes the diet challenging to follow. Ingredients lists on manufactured products must by law clearly highlight any allergen-containing ingredients, but some manufactured goods are also at risk of contamination with gluten during packaging or processing so may not be suitable. Gluten-free products must contain less than 20 parts per million of gluten to meet the international CODEX standard.

In addition to following a strict gluten free diet, recommendations of the British Society of Gastroenterology and Coeliac UK suggest that all patients with coeliac disease should be advised to consume 1000mg of calcium daily, using supplements if required, increasing to 1500mg if bone thinning is demonstrated (Al-toma et al. 2019).

In patients successfully avoiding gluten, transglutaminase antibodies become negative. Dietary education with regular input from a dietitian is essential to establish and maintain a diet that is gluten free and nutritionally complete. Guidelines recommend lifelong, annual monitoring of relevant blood parameters, clinical symptoms, and nutritional status. Complications of undiagnosed coeliac disease or failure to comply with dietary exclusion of gluten increases risk of intestinal lymphoma, osteoporosis, and nutritional deficiencies. In some patients, coeliac disease is associated with reduced splenic function and they should be immunized to reduce the consequent increase risk of infections such as pneumonia and meningitis.

Some patients fail to respond to the GF diet. Non-responsive, refractory coeliac disease may be considered when symptoms persist despite full adherence to a GF diet. It is important in these cases to rule out ongoing inadvertent exposure to gluten in the first instance, to review the diagnosis and investigate potential complications or coexisting conditions that may underlie persistent symptoms, such as IBS, lactose intolerance, bacterial overgrowth, microscopic colitis, or IBD. Following a diagnosis of refractory coeliac disease, patients require corticosteroid treatment to induce remission, but for those with with Type 2 refractory disease progression to lymphoma may occur and prognosis can be poor.

Non-coeliac gluten sensitivity (NCGS) is a relatively new clinical entity, where patients may report both GI and extra-intestinal symptoms relating to the ingestion of gluten specifically, but in the absence of diagnostic indices of coeliac disease (presence of serum antibodies and villous atrophy) or wheat allergy. There remains a degree of uncertainty surrounding the aetiology and pathophysiology of NCGS, and there have been no biomarkers identified to date, but patients' symptoms dramatically improve on a gluten free diet. It is not possible to currently differentiate these cases from those with food intolerant IBS, and there is considerable overlap in their management.

23.7 Inflammatory bowel disease

Epidemiology of inflammatory bowel disease

Inflammatory bowel disease refers to two similar but separate conditions, Crohn's disease (CD) and ulcerative colitis (UC). Both cause chronic and persistent gastrointestinal inflammation but whereas UC occurs only in the colon, CD may arise occur anywhere in the gut. The prevalence of both UC and CD are highest in Europe (505 and 322, per 100,000 per person years respectively).

IBD is now recognized to be the consequence of a complex interaction between genetic factors and the gut microbiome and this is discussed later. It was considered primarily to be a disease of Northern urbanized countries but is now known to be increasing throughout the world. A number of factors have been implicated. Of these by far the most important is smoking, but others include differences in breast feeding habits, risk of childhood infections, diet, vitamin D deficiency, use of medications such as antibiotics, oral contraceptives and non-steroidal anti-inflammatory agents, and psychological stress.

CD tends to occur earlier in life than UC. In childhood it is more common in boys but with the onset of puberty young women catch up. Peak incidence of CD is between the second and fourth decades of life. UC is similar.

Crohn's disease

Pathology

CD is a chronic inflammatory disorder of the GI tract, characterized by episodes of relapse and remission. It can affect any part of the gut, from mouth to anus, but most commonly affects the terminal ileum and colon. Segments of intestine affected by inflammation are often separated by apparently normal areas; known as 'skip' lesions, they may occur throughout the length of the bowel. Inflammation may extend through the layers of the gastrointestinal wall (transmural). Ulceration of the mucosal wall, and oedema and inflammation of the bowel in between, give the mucosal surface the cobblestone appearance that is typical of this disease. Fistulas, abnormal connections between two internal organs, or any part of the gastrointestinal tract and the skin, may arise in areas of the bowel that are severely affected. The bowel wall may thicken and the lumen narrow, predisposing to strictures and intestinal obstruction. There is an increased risk of small intestinal and colonic cancers.

Clinical features of Crohn's disease

Presentation varies according to the site and the extent of the disease. Symptoms associated with CD are abdominal pain, diarrhoea, weight loss, or in children failure to thrive, fever, and lethargy. Other organ systems may also be affected especially the eyes (iritis), the skin where painful nodules develop on the front of the legs (erythema nodosum) and joints, especially sacroiliac joints, knees, and ankles.

Biochemical characteristics include raised inflammatory markers such as C-reactive protein (CRP) or other acute phase proteins and a raised platelet count. Patients become anaemic and albumin concentrations in the blood are reduced. Increased migration of white cells through the bowel wall into the intestinal lumen occurs at sites of active disease. These white cells contain a protein known as calprotectin, which is resistant to degradation in the bowel lumen. Its concentration in the faeces provides an indication of the severity of inflammation present. White cells may be labelled with radioisotopes so that areas of active disease become visible on gamma camera scans (white-cell scanning). The final diagnosis is usually now confirmed by magnetic resonance imaging (MRI). This is widely used to follow changes in the progress of the disease although final confirmation of the diagnosis may require biopsies taken at endoscopy.

Aetiology of Crohn's disease

Both genetic and environmental factors are implicated in the development of CD. First-degree relatives of patients with CD are at increased risk and CD is known to be associated with specific ethnic groups. More recently, genome wide association studies have identified over 160 susceptibility loci/genes that are significantly associated with IBD The most significant environmental factor implicated in CD is smoking, which increases the risk of CD two- to five-fold; smoking also hampers response to treatment.

The role of diet has proved of considerable interest. Initially it was noticed that patients with CD ate considerably more refined sugar than healthy controls. Fat, dietary fibre, fruit and vegetable intake, margarine, and baker's yeast have also been implicated; however, data are difficult to interpret due to inconsistency in results and weakness of study design. The effect of disease state upon appetite, variations in food choice, and the reliability of dietary recall of foods eaten years previously hamper the validity of the research. Furthermore, intervention trials have shown that disease activity is not influenced by fibre or low sugar diets.

Fats have also been implicated in the pathogenesis and clinical course of CD. Epidemiological data show the incidence of CD in Japan has increased (Shoda et al. 1996), displaying a strong correlation with increased fat consumption, and an increased ratio of n-6 to n-3 fatty acids in the national diet, suggesting altered lipid metabolism as a factor. Replacement of n-6 arachidonic acid by other polyunsaturated fatty acids diverted from the n-3 pathway may reduce prostaglandin and thromboxane production suppressing inflammation.

The role of nutritional factors in the pathogenesis of CD is supported by studies on the importance of the faecal stream, containing gastrointestinal secretions, food residues, and bacteria. Diversion of the faecal stream away from the colon by performing an ileostomy allows CD in the lower bowel to heal. Colonoscopies were performed in patients given temporary ileostomies after gastrointestinal resections for CD to allow anastomotic healing. No evidence of CD was found in any patient. Six months after ileostomy reversal, all patients had endoscopic evidence of Crohn's recurrence. Evidence is now accumulating that key factors in the faecal stream affecting the development of CD are the microbiome and food residues. The importance of bacteria was suggested by effects of antibiotics in CD. Despite their demonstrated value in the treatment of CD, repeated antibiotic exposure, especially in childhood, can increase the risk of its development, as do attacks of bacterial gastroenteritis. No convincing specific bacterial pathogen has yet been demonstrated to cause CD. Suggestions that *Mycobacterium paratuberculosis* might cause and *Faecalobacterium prausnitzii* ameliorate CD have not been confirmed.

A possible mechanism for the link between microbiome, food, and inflammation in CD has been suggested by the demonstration in faeces and urine of chemicals produced by the gut flora. These include indole, ester, and alcohol derivatives of short-chain fatty acids. It is suggested these toxic substances stimulate a host immune response which leads to coating of over 80 per cent of organisms with immunoglobulin. The importance of food to this process was demonstrated by two weeks feeding with an elemental diet, which is virtually residue-free. Symptoms of CD improve and both toxic chemical production and immunoglobulin coating fall significantly. Furthermore, patients with CD in remission after two years primary dietary therapy show normal numbers (20 per cent) of bacteria coated with immunoglobulin (Walton et al. 2016).

The understanding that CD involves an immune attack on the intestinal microflora presents two main avenues for treatment. The immune response may be reduced by immunosuppressive drugs, or the activity of the microflora by antibiotics or diet. Many immunosuppressive agents are available for use in CD, as in UC. For many years the cornerstone of management has been corticosteroids, especially prednisolone, with immunomodulators such as azathioprine and methotrexate used for maintenance of remission. Latterly more sophisticated agents which block specific inflammatory cytokines such as TNF-alpha have been introduced. These include, among others, infliximab and adalimumab. Drugs releasing 5-aminosalicylic acid are often valuable in mild to moderate cases. Unfortunately, the success of immunosuppressive drugs tends to be variable with a significant risk of unpleasant side effects.

For these reasons, some patients are reluctant to consider immunosuppressive treatment. Antibiotics have been known for many years to produce worthwhile remissions but are still employed relatively sparingly. Faecal transplantation is being investigated. Modification of microbial activity by diet forms the basis of nutrition as a primary treatment for CD.

Nutritional deficiencies and dietary management

Dietary therapy in CD may be used to correct nutritional deficiencies or as primary treatment for the condition. Nutritional deficiency is a common complication of CD, affecting both macro- and micronutrients. Due to its relapsing and remitting nature, deficiency states may develop insidiously over a number of years remaining undetected until they are multiple and severe. Nutrient deficiency may arise by a number of different mechanisms, discussed below.

There is little evidence that deficiencies arise through inadequate dietary intake, though many patients develop anorexia or fear that eating will exacerbate symptoms. The pathology of the disease and associated medication do appear to limit absorption and handling of some nutrients. Changes in taste may be caused by deficiency of trace elements such as zinc, copper, and nickel or as a result of drug therapy. Strictures may cause abdominal pain and vomiting and may lead to malabsorption and intestinal losses.

Reduced intake, increased metabolism, and intestinal losses may cause protein energy malnutrition in the acute phase of the disease. Protein requirements in patients with IBD are therefore increased. Inflammation leads to increased production of cytokines, eicosanoids, catecholamines, and glucocorticoids which give rise to a catabolic response, protein breakdown, and negative nitrogen balance. Hypoalbuminaemia is common but is a result of the inflammatory response on acute phase plasma proteins apart from increased intestinal losses so is considered a marker of disease activity rather than a measure of nutritional status.

Serum concentrations of fat-soluble vitamins are affected by CD. Retinol may be decreased secondary to a reduced concentration of circulating retinol binding protein (which falls as part of the acute phase response) and may not reflect a deficiency state. Vitamin K produced endogenously by bacteria fermenting NSP in the colon may be reduced by antibiotic therapy or extensive large bowel resection.

Low serum levels of 25-hydroxycholecalciferol have been reported in 23–75 per cent of patients. Vitamin D metabolism and calcium homeostasis are closely linked (see Chapter 25) and many patients develop osteoporosis. Mechanisms include reduced dietary intake of calcium, malabsorption, and the direct effect of pro-inflammatory cytokines on bone metabolism. Bone density appears to be affected by steroid usage rather than disease activity, for measurements in patients whose CD was treated by diet were similar to control values and significantly higher than patients treated predominantly with corticosteroids (Dear et al. 2001). The importance of Vitamin D as a negative acute phase reactant is discussed elsewhere (Chapter 12).

Sulfasalazine competes with folate for absorption and cholestyramine binds many nutrients in the GI tract, especially fat-soluble vitamins. Iron deficiency anaemia is a frequent complication in CD and may develop secondary to gastrointestinal blood loss, reduced dietary intake, malabsorption, or small bowel resection. Oral iron supplementation is difficult as it frequently provokes symptoms. Serum magnesium levels are kept constant at the expense of body stores, whereas serum zinc may be reduced in the presence of inflammation reflecting reduced albumin concentrations, despite normal tissue levels. Selenium deficiency in CD has been reported. Proposed mechanisms include use of corticosteroids and bowel resection.

The role of diet as primary management

Diet was shown to produce remission in CD many years ago when patients awaiting surgery received total parenteral nutrition (TPN). Diet, however, is not widely used and remains controversial in many quarters.

TPN is invasive and expensive and the first stage of nutritional therapy involves enteral feeds which are equally effective. Elemental diet (ED) is a mixture of essential and non-essential amino-acids, glucose, lipids, vitamins, minerals, and trace elements. It is nutritionally complete but contains virtually no residue. Semi-elemental diets are polypeptide based and tend to contain a higher proportion of fat as medium chain triglycerides (MCTs), whereas polymeric diets contain whole proteins. The advantages of polymeric diets over ED are cost, greater energy density, palatability, and lower osmolality, which reduces the risk of highly concentrated feeds causing diarrhoea. Low long-chain triacylglycerol (TAG) content is crucial—feeds with less than 15 per cent of energy derived from long-chain TAG have been shown to be the most effective. ED will induce remission in 80–90 per cent of compliant cases.

Efficacy of enteral feeding is influenced by duration of treatment. Many paediatricians use enteral feeds (via naso-gastric tubes) for periods of up to three months and report that patients may thereafter return to a normal diet. Most adult patients prefer to take the feed orally and stop liquid diet as soon as symptomatic remission is achieved. This usually takes about two weeks with elemental diet but may be longer with polymeric presentations (up to six to eight weeks).

Remissions are disappointingly short if normal diet is resumed as soon as symptoms have cleared, but disagreement remains as to whether a second stage of dietary treatment is valuable to prevent relapse. It is possible to reintroduce normal food items one by one to see which induce symptoms (an elimination diet) but this takes time. As food reactions in CD may appear slowly and insidiously it is necessary to test each food for several days.

A validated approach to food reintroduction is provided by a low fat (<50g/day) fibre limited (<10g/day) exclusion diet (LOFFLEX). The core diet excludes foods known most likely to cause symptoms. If patients remain in remission on LOFFLEX, food reintroduction follows. Each food is tested for at least four days and those which trigger symptoms are excluded. In this way remission can be extended for considerable periods. In a series of 63 patients treated with elemental diet, 49 achieved remission and were offered LOFFLEX. Forty-one remained in remission until the end of the study six months later. Faecal calprotectin concentrations returned to normal, confirming mucosal healing (Hunter & Woolner 2018). Other maintenance diets have been suggested but remain to be clinically validated.

Results from LOFFLEX compare favourably with those achieved by pharmacological treatment. However, dietary therapy has other long-term benefits. In contrast to immunosuppressive treatment, the incidence of osteoporosis in patients treated by diet was no greater than in healthy age-matched controls and outcomes of pregnancy in women with CD treated by diet were no different from those in healthy women. Corticosteroids impair children's growth, but growth is normal when treated by diet. Dietary treatment for CD is inexpensive and harmless, and demand is likely to rise.

Ulcerative colitis (UC)

Pathology of ulcerative colitis

UC affects the mucosa of the colon starting from the anus and extending proximally. Mild cases may affect merely the rectum (proctitis) but commonly inflammation extends as far as the splenic flexure of the large intestine (left sided colitis) and severe disease may affect the whole colon (pan colitis).

The biochemical characteristics are similar to those of CD. However, if inflammation is limited to the rectum a rise in CRP is unusual. The colonic mucosa appears swollen and inflamed and in severe cases ulcers form, initially small and discrete but enlarging and coalescing until the colonic wall becomes denuded. The bowel may then dilate, a condition known as megacolon, which carries a high risk of perforation and is an emergency usually requiring surgery. Healing may result in scarring, stricture formation, or the development of inflammatory pseudopolyps from islands of persisting mucosa. There is a greatly increased risk of colon cancer in UC but this appears to be the consequence of prolonged chronic inflammation, as pseudopolyps are not pre-malignant.

Clinical features of ulcerative colitis

The characteristic symptoms of UC are diarrhoea and rectal bleeding. Blood loss may lead to anaemia and hypoalbuminaemia, which may be severe enough to cause peripheral oedema. Abdominal pain is common and is usually worse after meals. Anorexia leads to weight loss. Severe diarrhoea may cause loss of water and electrolytes leading to dehydration, hypomagnesaemia, and hypocalcaemia. As in CD, drugs used in the treatment of UC may also contribute to nutritional deficiency.

Aetiology of UC

The aetiology of UC is similar in many ways to that of CD. There are strong genetic influences, and the gastrointestinal flora is a major factor. No specific pathogen has been identified, but UC may follow attacks of gastroenteritis, and the gut flora in active disease is highly unstable, varying considerably over short periods of time. As in CD, the gut bacteria in UC are coated with immunoglobulin, but in contrast to CD, this cannot be reduced by diet. The microflora are believed to rely on colonic secretions, possibly mucus, as nutritional substrates. Unlike CD, smoking does not exacerbate UC, and indeed in distal cases may ameliorate it, perhaps by relieving constipation.

Colonocytes take up butyrate, which is derived from bacterial fermentation of fibre. It has been suggested that in UC this may be impaired by hydrogen sulphide and mercaptides which are present in higher concentration. Efforts to increase butyrate concentration in the colon, however, have been of limited success. There is no clear therapeutic value of butyrate enemas, nor the use of germinated barley foodstuff which contains substrates for SCFA production. Likewise, use of low sulphate diets to reduce sulphide formation remain unproven. *Psyllium* husk appears to reduce the risk of relapse by normalizing intestinal transit times rather than by increasing SCFA production.

The pharmacological treatment of UC is very similar to that described earlier for CD. In proctitis and left-sided colitis, corticosteroids and 5-ASA derivatives may be given by enema but are required systemically in severe disease. New biological treatments such as tofacitinib have recently been introduced.

Nutritional deficiencies and diet as treatment

There is little place for diet as primary therapy in UC. Patients may have food intolerances akin to those occurring in IBS but although avoiding the foods concerned may help relieve diarrhoea and pain, medication is still required to heal inflammation. Oral nutritional supplementation may be used to improve nutritional status but not as primary treatment. It is crucial to avoid constipation which may lead to relapses of proctitis and left-sided disease and in these cases non-fermentable bulking agents such as sterculia and linseed may be helpful.

23.8 Irritable bowel syndrome

Epidemiology of irritable bowel syndrome

The most prevalent disorder of the gastrointestinal tract is irritable bowel syndrome (IBS). It can account for up to 25 per cent of referrals to gastroenterologists and has a worldwide prevalence of 11.2 per cent. It is more likely to occur in women than men (2:1) and is not age dependent.

As yet there is no specific diagnostic test for IBS; it remains a diagnosis of exclusion, where no structural changes to the gut are seen on radiology or endoscopy, there is normal haematology and biochemistry, and stool cultures are negative. Faecal calprotectin is now recognized as a useful marker to exclude more serious pathology such as IBD, making luminal investigation unnecessary in younger patients.

As part of a comprehensive assessment, the Rome IV criteria for IBS demand a six-month history of abdominal pain and a change to stool form and consistency, which can inform diagnostic criteria for subtypes diarrhoea or constipation predominant (IBS-D and IBS-C respectively) or mixed bowel habit (IBS-M) (Lacy et al. 2016). Symptoms including bloating, flatulence, incomplete evacuation, urgency, and straining do not form part of formal diagnosis criteria but are often present.

Aetiology of irritable bowel syndrome

IBS is not a single entity, but a syndrome made up of several separate functional conditions, which produce abdominal pain with a change in bowel habit. It is a multifactorial condition, with genetic, environmental, and psychosocial factors, and complex interaction between the gut and brain. Pathophysiological mechanisms vary but are now known to include altered GI motility, visceral hypersensitivity, immune activation, alterations to intestinal permeability, and altered microbiome (Lacy et al. 2016).

About 20–25 per cent of IBS patients suffer from anxiety, which may lead to hyperventilation with aerophagia; these patients commonly present with pain, bloating, and flatulence, accompanied by other symptoms of anxiety. Chronic stress and depression may also lead to gastrointestinal symptoms.

IBS is characterized by dysmotility in the bowel, along with visceral hypersensitivity. An early study demonstrated that the pressure required to produce pain is significantly less in many with IBS when the rectum was distended mechanically with a balloon. Some cases are associated with constipation, but the majority (some 50 per cent of cases) are due to malfermentation of undigested food residues reaching an abnormal bacterial flora in the lower gut. This was first proposed by the Cambridge group (Hunter and colleagues) and is now supported by considerable evidence.

The link between food intolerance and IBS was first proposed by Jones et al. (1982). Fourteen out of 21 cases were shown to have symptoms reproducible by food ingestion with the link confirmed by double blind challenge using naso-gastric tubes. No evidence of IgE mediated food allergy was present in these patients. Food intolerance reactions in IBS disease differ from classical IgE reactions, often developing quite slowly after relatively large quantities of the food in question, and anaphylaxis is very uncommon.

It has been suggested that bacteria found in the small intestine may be responsible (small intestinal bacterial overgrowth or SIBO) but this claim is based on patterns of breath hydrogen produced by the non-digestible sugar lactulose. These are susceptible to changes in gastrointestinal transit, and in IBS jejunal aspirates for genuine small bowel overgrowth are almost invariably negative. Wherever the location of these bacteria however, therapeutic antibiotics are often effective (see Section 23.3), but their benefits are usually short lived and diet remains the cornerstone of treatment (Ford et al. 2018).

Dietary management of IBS

Dietary and lifestyle advice is considered the cornerstone of IBS management. Evidence-based guidelines produced by the British Dietetic Association (BDA) in 2016 provides a useful framework. A comprehensive assessment of dietary composition along with typical eating patterns and behaviours is fundamental to dietary management. First-line dietary intervention strategies may include manipulation (usually reduction) of dietary components that potentially cause difficulty, for example,

fibre and resistant starch; also fat, highly spiced foods, caffeine, alcohol, and chewing gum. The importance of a regular eating pattern without over-eating or rushing meals must also be taken into consideration. Symptoms frequently worsen after eating, and self-reported food intolerance with consequent dietary restriction can have a negative impact on quality of life. Given the importance placed by patients on food as a contributor to symptoms, self-management is often attempted. The BDA have produced clear, concise patient guidance for first-line manipulation which has been endorsed by the National Institute for Health and Care excellence.

Exclusion diets have been developed to manage IBS symptoms (Table 23.4), some with evidence of benefit, but data and rationale for food exclusions for others, however, may be limited. Any exclusion diet requires patience, prioritization, and time commitment; lifestyle implications may therefore mean these are an unrealistic option for some. It is important to acknowledge that given a significant central nervous system component to IBS symptoms in many patients, including anxiety, depression, and stress but also conditioning and expectation, any form of structured dietary manipulation where expert guidance and support are provided can result in symptom improvements in some patients.

There is no evidence to suggest that nutritional deficiencies occur directly as a result of untreated IBS symptoms, but risk of specific nutrient deficiencies may increase with food exclusions over prolonged periods, particularly where whole food groups are avoided (e.g., dairy foods).

The role of fibre in the diet is important in individuals with IBS (see Section 23.4). The use of a higher fibre diet in IBS is now limited to treatment of simple constipation. As fibre is a key substrate for fermentation, reducing fibre content of the diet to less than 10g per day may decrease gas production and consequent symptoms in patients presenting with diarrhoea, bloating, urgency, and pain. Wheat bran and other sources of insoluble fibre

TABLE 23.4 Exclusion diets used in the management of IBS

Paleolithic (stone-age) diet—no additives, grains, starches, lentils, dairy, sugars, preservatives

Specific carbohydrate diet—grain-free, sugar-free, starch-free, and unprocessed

Dairy free

Low fibre gluten free

Yeast and sugar free (the Candida diet)

Low fermentable oligosaccharides, disaccharides, monosaccharides and polyols (low FODMAP)

are therefore not recommended in patients with IBS, although there is a small amount of evidence to suggest linseed supplementation may result in improved stool output in those with constipation-predominant symptoms without the adverse side effects of excess fermentation. The combination of a poorly fermented fibre supplement such as methylcellulose, sterculia, or linseed with a low-fibre diet sounds paradoxical, but in practice can be highly effective.

Lactose is one of the most common food constituents regarded as a possible culprit in those with IBS, and lactose malabsorption may be present in up to half of patients, dependent on their ethnicity. If lactose is not digested in the small intestine, it will pass to the colon where it will be fermented. Hydrogen breath tests to identify lactose intolerance may not always be available or accurate, so dietary exclusion and re-challenge is an effective alternative. A dose of 12g lactose is recommended as a suitable challenge (equivalent to 125ml milk) with a food and symptom diary to assess response. Other components of milk may contribute to symptoms, and it is believed that dairy fat is of particular importance.

Gluten is also frequently suspected by patients as being implicated in their GI symptoms. However, it remains unclear to what extent it is the gluten component of gluten-containing grains that specifically contributes to symptoms, or whether other components within these foods (specifically certain starches or sugars) may be responsible. Several studies have investigated the efficacy of a gluten-free diet as a treatment for IBS, but with some conflicting results. Excluding wheat-containing foods not only reduces gluten consumption but also lowers intake of fructans (a fermentable starch), along with amylase-trypsin inhibitors, and all these components may produce GI symptoms.

TABLE 23.5 High FODMAP[1] foods (not an exhaustive list)

Fructose	Honey, agave nectar, Mango, high-fructose corn syrups, asparagus
Lactose	Milk, yogurt, cottage and cream cheeses, ice cream
Fructans	Wheat, rye, onion, garlic, leek, artichoke, inulin
Galacto-oligosaccharides	Cashew/pistachio nuts, lentils, beans, dried peas
Polyols	Apricots, avocado, sweetcorn, celery, cauliflower, sweet potato, sugar-free mints/chewing gum

[1] Fermentable oligosaccharides, disaccharides, monosaccharides, and polyols

There is growing evidence supporting the dietary restriction of short-chain fermentable carbohydrates to improve IBS symptoms; these include fructose, lactose, fructo- and galato-oligosaccharides, and polyols such as sorbitol and mannitol. Studies have shown beneficial effects of reducing these individually, particularly fructose, fructans, and polyols. However more recently work has focused on avoiding these collectively; this approach being known as the low FODMAP (fermentable oligo-, di- monosaccharides and polyols) diet (Table 23.5).

An early study demonstrated an improvement in all types of IBS symptoms in 74 per cent of IBS patients following a low FODMAP diet; subsequent studies have shown similar success rates (Figure 23.3). A recent systematic review and meta-analysis of seven studies (397 patients), comparing a low FODMAP diet to either habitual, or high FODMAP diet, demonstrated benefits to a low FODMAP diet compared to control (RR 0.69; 95 per cent

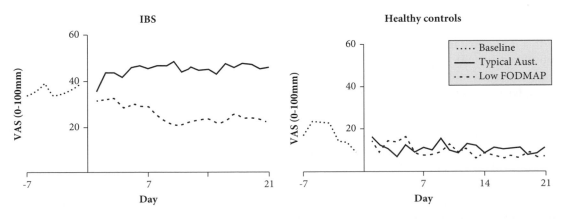

FIGURE 23.3 Visual acuity scores (VAS) for gastrointestinal symptoms in IBS and healthy controls on a typical Australian or low FODMAP diet

Source: Reproduced from E P Halmos, V A Power, S J Shepherd et al. (2014) A diet low in FODMAPs reduces symptoms of irritable bowel syndrome. *Gastroenterology* **146**(1), 67–75, with permission from Elsevier.

CI 0.54–0.88), but authors evaluated the evidence as low-quality, owing to small sample sizes, difficulty in blinding, and significant patient heterogeneity (Dionne et al. 2018). It is also important to note that these studies only evaluated the initial, short-term exclusion phase of the low FODMAP diet. Structured FODMAP reintroductions to improve dietary variety and to improve nutrition of both patient and the microbiome are considered an essential part to this approach. Changes to the colonic microbiome, with a reduction in beneficial *Bifidobacteria* spp. have been demonstrated after a short-term low FODMAP diet trial, so reintroduction of any well-tolerated FODMAP-containing foods is important. The use of specific probiotic supplementation alongside FODMAP exclusion may also prevent reductions in beneficial bacteria in those who are unable to reintroduce FODMAP-containing foods successfully. In practice, the low FODMAP diet is widely used but should be implemented with specialist dietetic guidance to ensure correct implementation, nutritional adequacy throughout, and interpretation of longer-term results for optimal success (McKenzie et al. 2016).

23.9 Intestinal failure

The European Society for Parenteral and Enteral Nutrition (ESPEN) has defined intestinal failure (IF) as 'the reduction of gut function below the minimum necessary for the absorption of macronutrients and/or water and electrolytes, such that intravenous supplementation is required to maintain health and/or growth' (Pironi 2016). The most recent recommendations include a functional and a pathophysiological classification for both acute and chronic IF and an additional clinical classification of chronic IF, based on energy and volume of supplementation requirements.

Functionally, IF is classified into three types characterized by severity, and the type and duration of nutritional support that is required in its management. Types 1 and 2 are considered acute IF and may resolve; type 3 on the other hand reflects a more chronic, irreversible picture where long-term parenteral nutrition (PN), plus fluid and electrolyte support is indicated.

- Type 1: short term (usually less than four weeks duration) and self-limiting, for example, postoperative ileus, critical illness, and intestinal obstruction. Short-term PN is often required until resolution.
- Type 2: may follow major GI surgery or trauma, abdominal sepsis, intestinal fistulae, or complex active Crohn's disease causing significant malabsorption. Duration is typically longer than one month, and

specialist multidisciplinary support is necessary to manage PN and fluid requirements.

- Type 3: chronic IF secondary to insufficient length of functioning bowel following major or multiple resections leading to a short bowel; possibly due to mesenteric infarct, volvulus, or severe radiation enteritis. Pan-intestinal dysmotility and pseudo-obstruction may also be causes of chronic IF. Long-term home parenteral nutrition (HPN) and other fluid, electrolyte, or metabolic support is indicated.

The pathophysiological classification of IF includes five major conditions: short bowel, intestinal fistula, intestinal dysmotility, mechanical obstruction, and extensive small bowel (SB) mucosal disease. In the case of short bowel, enterocutaneous fistula, or extensive SB disease, the primary mechanism of IF is malabsorption of ingested food. When intestinal dysmotility or intestinal mechanical obstruction occurs IF is primarily due to complete or partial restriction of oral/enteral nutrition due to feeding-related digestive symptoms.

A 'short bowel' is defined as a residual small intestinal length of <200cm, following bowel resection. The consequences of this will be dependent on the extent and sites of resection and the integrity and adaptation of the remaining bowel. It is recognized that minimum lengths of SB, specifically jejunum, are required to maintain adequate nutritional, electrolyte, or fluid status, without which parenteral support becomes necessary (Table 23.6). The presence or absence of a functioning colon will also affect absorptive capacity for fluid and electrolytes.

Intestinal adaptation can take place following resection, for up to two years following surgery. This affects the type of nutritional management required (i.e., oral, enteral, or parenteral) and this may change over time as adaptation occurs. Fluid and electrolyte management are key to reduce intestinal losses, reduce dehydration, and associated electrolyte imbalances. Oral fluid restrictions, with additional sodium and magnesium supplies,

TABLE 23.6 Relationship between length of remaining small intestine and supplementation requirements

Length of Bowel	Supplementation required
<100cm functioning jejunum	parenteral fluid and electrolyte
<75cm functioning jejunum	PN, fluid, and electrolyte
<50cm functioning jejunum with colon	PN, fluid, and electrolyte

are often indicated, particularly for those with a jejunostomy. Oral rehydration solutions with levels of sodium >90 mmol/l are beneficial in maintaining fluid balance. Pharmaceutical treatment to reduce GI secretions which may be helpful include H2 receptor antagonists, antidiarrhoeals, bile acid sequestrants, and somatostatin analogues such as octreotide.

Nutrient malabsorption is common. Additional calories and protein are indicated to compensate for this, and requirements of between 30–60kcal and 1.25–1.5g protein/kg/day are indicated. In patients with a jejuno-colic anastomosis, but not those with jejunostomy, provision of some fat in an oral or enteral diet as MCT may be beneficial to improve overall caloric and fat absorption. Micronutrient deficiencies are common in this group; regular monitoring and additional supplementation, sometimes above usual recommendations, is frequently required (Culkin et al. 2009).

23.10 Eosinophilic oesophagitis

Eosinophilic oesophagitis (EoE) is a chronic, non-IgE antigen mediated condition, characterized by eosinophilic infiltration of the oesophagus causing inflammation and fibro-stenosis. It occurs more often in male patients (3:1 male: female), the highest prevalence being in the third to fourth decades of life. The condition frequently presents as intermittent dysphagia. Symptoms range in nature and severity and differ between young children and adults. In young children symptoms may include food refusal, gastro-oesophageal reflux, vomiting, abdominal pain, and faltering growth. In older children and adults, predominant symptoms are reflux, solid food dysphagia, vomiting, abdominal and chest pain, over-chewing foods, and episodes of food impaction. These differences may be due to the progression of the disease from inflammation to fibrosis, causing stenosis Diagnosis requires endoscopy and oesophageal biopsy. The value of allergy testing to foods remains controversial, but expert assessment and appropriate management of co-existing IgE-mediated food allergies and co-morbid atopic conditions such as dermatitis, asthma, and allergic rhinitis/conjunctivitis is important, as these can exacerbate symptoms.

Medication and dietary management strategies are frequently used in EoE. A proportion of cases are responsive to proton-pump inhibitors (PPIs), which can provide an anti-inflammatory effect independent of acid suppression. Topical steroids such as swallowed aerosolized fluticasone or viscous budesonide are first-line pharmacologic therapies for treatment and are effective, though correct administration is crucial. Systemic corticosteroid therapy may help in difficult cases.

Dietary intervention comprising food elimination, followed by structured reintroduction, is effective for both children and adults. Different approaches to food elimination may be considered and it is important that any dietary approach undertaken is individualized. Symptom resolution may occur without histological remission, and repeated endoscopies required to determine the effectiveness of diet following a period of avoidance and reintroduction.

- As in Crohn's disease, elemental diet is effective in inducing remission, followed by structured reintroduction of individual foods to identify triggers. This can be successful but is expensive, challenging to implement, and may lead to social isolation.
- Empirical elimination diets—two (TFED) four (FFED) or six (SFED)food elimination diets: a step-up approach, from two (milk and wheat) to four (milk, wheat, egg, and soy), or six (as for four plus nuts and fish) food elimination as needed to induce remission has been shown to be effective in 43 per cent, 60 per cent, and 79 per cent of patients respectively, and may reduce the number of endoscopies and time needed to identify food triggers (Molina-Infanta et al. 2018)
- Targeted elimination diets involve food avoidance based on the history and results of allergy testing. This option is less frequently used, as serum food specific IgE, skin prick testing (SPT), atopy patch testing (APT), singly or in combination have poor positive predictive values for food triggers in adults.

23.11 Diverticular disease

Diverticular disease of the colon occurs in between 30 per cent and 50 per cent of adults in Western populations and is related to increasing age, obesity, and sometimes as a consequence of IBS. The formation of diverticula, which are permanent, follows increased pressure inside the colon. Small out-pushings of mucosa between the longitudinal muscle of the colon form pockets which may become infected (diverticulitis). Such inflammation may lead on to bleeding, abscesses, and strictures.

A healthy, balanced diet is recommended with good sources of fibre, such as methyl cellulose, along with the avoidance of constipation. Wheat bran is better avoided, but seeds, nuts, and fruit skins can be safely consumed. Reduced fibre intake may be needed during acute diverticulitis or its complications. Diverticular disease is not pre-malignant and surgery is usually only required for abscess or stricture.

KEY POINTS

- Chronic disorders of the intestine are associated with complex interactions between diet, the intestinal microflora, and intestinal mucosal immunity

- Diet may be of significant value in modifying intestinal damage and the clinical course of these diseases

- Manipulation of the gastrointestinal flora may provide a future means of treating these disorders.

 Be sure to test your understanding of this chapter by attempting multiple choice questions. See the Further Reading list for additional material relevant to this chapter.

REFERENCES

Al-toma A, Volta U, Auricchio R et al. (2019) European Society for the Study of Coeliac Disease (ESsCD) guideline for coeliac disease and other gluten related disorders. *UEG J* **7**(5), 583–613.

Culkin A, Gabe S M, Madden A M (2009) Improving clinical outcome in patients with intestinal failure using individualised nutritional advice. *J Hum Nutr Diet* **22**(4), 290–298.

Dear K L, Compston J E, Hunter J O (2001) Treatments for Crohn's disease that minimize steroid doses are associated with a reduced risk of osteoporosis. *Clin Nutr* **20**, 541–546.

Dear K L E, Elia M, Hunter J O (2005) Do interventions which reduce colonic bacterial fermentation improve symptoms of irritable bowel syndrome? *Dig. Dis. Sci.* **50**, 758–766

Dionne J, Ford A C, Yuan Y et al. (2018) A systematic review and meta-analysis evaluating the efficacy of a gluten-free diet and a low FODMAPs diet in treating symptoms of irritable bowel syndrome. *Am J Gastroenterol* **113**, 1290–1300.

Bajagai Y S & Klieve A (2016) Probiotics in animal nutrition: Production, impact and regulation, Paper no. 179. *FAO Anim. Prod. Health* (Makkar H P S ed.). FAO, Rome.

Ford A C, Harris L A, Lacy B E et al. (2018) Systematic review with meta-analysis: The efficacy of prebiotics, probiotics, synbiotics and antibiotics in irritable bowel syndrome. *Aliment Pharmacol Ther* **48**(10), 1044–1060.

Halmos E P, Power V A, Shepherd S J et al. (2014) A diet low in FODMAPs reduces symptoms of irritable bowel syndrome. *Gastroenterol* **146**(1), 67–75.

Hunter J O & Woolner J (2018) Validation of an algorithm for the dietary management of Crohn's disease. *Compl Nutr* **18**, 63–65.

Jones V A, Mclaughlin P, Shorthouse M et al. (1982) Food intolerance: A major factor in the pathogenesis of irritable bowel syndrome. *Lancet* **2**(8308), 1115–1117.

Lacy B E, Mearin F, Chang L et al. (2016) Bowel disorders. *Gastroenterol* **150**, 1393–1407.

Levine A, Kori M, Kierkus J et al. (2019) Azithromycin and metronidazole versus metronidazole-based therapy for the induction of remission in mild to moderate paediatric Crohn's disease: A randomised controlled trial. *Gut*, **68**, 239–247.

Magne F, Gotteland M, Gauthier L et al. (2020) The firmicutes: Bacteriodetes ratio: A relevant marker of gut dysbiosis in obese patients? *Nutrients* **12**, 1474. doi: 10.3390/nu12051474

Macfarlane G T & Macfarlane S (2012) Bacteria, colonic fermentation, and gastrointestinal health. *J AOAC Int.* **95**, 50–60.

McKenzie Y A, Bowyer R K, Leach H et al. (2016) British Dietetic Association systematic review and evidence-based practice guidelines for the dietary management of irritable bowel syndrome in adults (2016 update). *J Hum Nutr Diet* **29**, 549–575.

Molina-Infante J, Arias A, Alcedo J et al. (2018) Step-up empiric elimination diet for pediatric and adult eosinophilic esophagitis: The 2-4-6 study. *J Allergy Clin Immunol* **141**(4), 1365–1372.

Pironi L (2016) Definitions of intestinal failure and the short bowel syndrome *Best Pract Res Clin Gastroenterol* **30**, 173–185.

Shoda R, Matsueda K, Yamamoto S et al. (1996) Epidemiological analysis of Crohn's disease in Japan: Increased dietary intake of n-6 polyunsaturated fatty acids and animal protein relates to the increase incidence of Crohn's disease in Japan. *Am J Clin Nutr* **63**(5), 741–745.

Su G L, Ko C W, Bercik P et al. (2020) AGA Clinical practice guidelines on the role of probiotics in the management of gastrointestinal disorders. *Gastroenterology* **159**, 657–705.

Walton C, Fowler D P, Turner C et al. (2013) Analysis of volatile organic compounds of bacterial origin in chronic gastrointestinal diseases. *Inflamm Bowel Dis.* **19**, 2069–2078.

Walton C, Montoya M P B, Fowler D P et al. (2016) Enteral feeding reduces metabolic activity of the intestinal microbiome in Crohn's disease: An observational study. *Eur J Clin Nutr* **70**, 1052–1056.

FURTHER READING

Algera J, Colomier E, Simren M (2019) The dietary management of patients with irritable bowel syndrome: A narrative review of the existing evidence. *Nutrients* **11**, 2162.

Black C J & Ford A C (2020) Best management of irritable bowel syndrome. *Frontline Gastroenterol.* e-pub ahead of print. doi: 10–1136/flgastro-2019–101298.

Dahl C, Crichton M, Jenkins J et al. (2018) Evidence for dietary fibre modification in the recovery and prevention of reoccurrence of acute, uncomplicated diverticulitis: A systematic literature review. *Nutrients* **10**, 137.

Henderson C J, Abonia J P, King E C et al. (2012) Comparative dietary therapy effectiveness in remission of pediatric eosinophilic esophagitis. *J Allergy Clin Immunol* **129**(6), 1570–1578.

McKenzie Y A, Thompson J, Gulia P et al. (2016) (IBS Dietetic Guideline Review Group on behalf of Gastroenterology Specialist Group of the British Dietetic Association) British Dietetic Association systematic review of systematic reviews and evidence-based practice guidelines for the use of probiotics in the management of irritable bowel syndrome in adults (2016 update). *J Hum Nutr Diet* **29**(5), 576–592.

SACN (Scientific Advisory Group on Nutrition) (2015) Carbohydrates and Health Report. Available from: https://www.gov.uk/government/publications/sacn-carbohydrates-and-health-report (accessed 26 April 2016).

Steinbach E, Hernandez M, Dellon E S (2018).Eosinophilic esophagitis and the eosinophilic gastrointestinal diseases: Approach to diagnosis and management. *J Allergy Clin Immunol Pract* **6**(5), 1483–1495.

24 Nutrition and the nervous system

Saskia Osendarp, Domenico Sergi, and Lynda M. Williams

OBJECTIVES

By the end of this chapter you should be able to:

- know the mechanisms underlying nutrition and child development, as well as cognitive decline and Alzheimer's disease
- understand that a holistic dietary approach is preferential rather than a single nutrient approach in promoting child development and preventing cognitive decline
- be able to describe the dietary risk factors involved in poor child development and accelerated cognitive decline
- understand how these dietary risk factors promote child development and cognitive decline via different mechanisms.

24.1 General introduction

Nutrition can impact the nervous system and brain functioning in various ways. Like any other organ, the brain needs fuel and molecular building blocks to develop, mature, function, and prevent degeneration. Indirectly, nutrition impacts cognition through interactions between health status and the response of the environment: for instance, children who are undernourished are more apathetic and interact less with their care-givers and environment.

There has been a growing interest over the past few decades in how nutrition may affect the functioning of the brain throughout life and new brain imaging techniques have allowed researchers to gain more knowledge in the mechanisms on how specific nutrients affect specific areas of the brain. In this chapter we will describe separately the role of nutrition and environment during early life, when brain development is at its peak, and during aging when cognitive decline takes place.

However, new insights reveal that cognitive development and decline are to be seen as two sides of a continuous process throughout life. Early life nutrition may affect long-term cognitive and behavioural development throughout childhood into adult life, building a 'cognitive reserve'. This reserve affects cognitive functioning during adulthood and may eventually determine the rate of cognitive decline during aging, while nutrition continues to impact these processes.

24.2 Nutrition and cognitive development in early life and adolescence

Introduction

Child development is multidimensional, comprising cognitive, language, socioecoional, fine and gross motor domains. From conception to two years is particularly important because the brain is developing rapidly and is especially sensitive to risks and interventions. The early environment affects brain development and function and all future development builds on early development; for example, cognitive function on school enrolment determines to some extent future educational attainment. Poor nutritional status is usually associated with poor child development but it is also associated with many confounding variables including poverty, low levels of parental education, and inadequate home stimulation with little opportunity to learn. All these factors can themselves detrimentally affect child development, therefore associations between nutritional deficits with poor child development may not be causal and randomized supplementation trials are the best way of establishing a causal relation. In low and middle income countries (LMICS) poor nutrition is one of the main causes of poor development, and the most prevalent nutritional conditions that are associated with child development are poor linear growth or stunting (height for age ≤ −2 Standard Deviation (SD) of WHO references), micronutrient deficiencies, and intrauterine growth retardation (IUGR) (Walker et al. 2011).

The effect of undernutrition varies by severity, duration, and the age of the child. Some brain functions are developed during early life (Figure 24.1) and nutritional deficits can irreversibly impair development and later cognitive functioning. This has been shown for instance in children who had been anaemic under five years of age, and still performed less well at school at ten years of age even when the anaemia had been corrected, compared to children who were not anaemic during early childhood (Walker et al. 2011).

There are several possible mechanisms linking undernutrition to poor child development. Nutrition can have a direct effect on brain development and maturation via the modification of the brain's macrostructure (e.g., development of brain areas such as the hippocampus), microstructure (e.g., myelination of neurons), or functioning of neurotransmitters (e.g., dopamine levels or receptor numbers). Indirectly, nutrition can improve cognitive development via a general improvement of health (Figure 24.2). New evidence suggest that maternal nutrient restrictions and specific micronutrient (folate) deficiencies can affect neuro development and brain structure via lifelong changes in epigenetic regulation of gene expression at the level of DNA methylation. These effects occur even in the absence of low birth weight, and are still apparent during older age and associated with cognitive disturbances (Franke et al. 2020).

Undernourished children become apathetic, explore the environment less, are less active, less happy, and more fussy than well-nourished children. In response the mothers become less stimulating and tend to carry the child more. Similar behaviours have been found in severe acute malnutrition, stunted and iron deficient children, and their mothers. The more severely malnourished the child the greater the behaviour differences. The behaviours are known as 'functional isolation' and are thought to contribute to the child's poor development. In older children, improved health status will have a positive impact on school attendance, and also improve activity and social interactions with other children, leading to improved cognitive development and performance (Walker et al. 2011).

Maternal nutrition

Maternal stunting and low BMI are related to foetal growth restriction (Victora et al. 2021) and balanced protein/energy supplementation increases birth weight (Liberato et al. 2013). There are few studies linking maternal nutrition to child development. Infants born to mothers with low BMI (<18.5 kg/m^2) in Bangladesh had poorer development and behaviour at age seven months than those born to women with adequate BMIs, however the relation was no longer significant at 18 months (Liberato et al. 2013).

Macronutrient supplementation in pregnancy

Some early studies examined the effects of maternal protein/calorie supplementation on child development (Liberato et al. 2013). In Taiwan, supplementation in pregnancy and lactation benefitted the offspring's motor but not mental development at eight months of age, but there was no effect on IQ at five years. In Bogota, supplementation in the last trimester of pregnancy and given to the child for the first six months had transient benefits only. In Bangladesh, food supplements given early in pregnancy led to a small improvement in problem-solving at seven months compared with infants of mothers who received the supplement approximately four weeks later. The benefit occurred only in infants of mothers with a low BMI. The small difference in amount of supplement between the groups suggests that timing may be a critical factor, however, no benefits were apparent at 18 months (Liberato et al. 2013). In The Gambia, children whose mothers were supplemented for 20 weeks in pregnancy with balanced energy protein biscuits were no different in educational performance and cognitive ability during late adolescence from children whose mothers received a similar supplement during lactation. Overall, there is no evidence of sustained benefits from maternal protein calorie supplementation on child development, but there are very few studies and more research is needed.

Exposure to famine in utero

Natural experiments can be helpful and the Dutch famine provided an opportunity to analyse the effects of exposure to famine. Exposure *in utero* was associated with increased congenital neural defects and increased risk of schizophrenia, addiction to drugs or alcohol, antisocial behaviour, and major affective disorders in adulthood. An increased risk of schizophrenia was also found in two larger famines in different regions in China, suggesting that the findings are relevant to all populations and that some prenatal nutritional deficits can have latent effects that are not apparent at birth or early childhood. Recent evidence from the Dutch famine birth cohort study have indeed shown that an adverse prenatal environment, especially in the beginning of pregnancy, is associated with a range of adverse health effects including alterations of brain morphology that are even independent of birth weight and size at birth. These effects result in smaller brain volumes and accelerated cognitive ageing at later

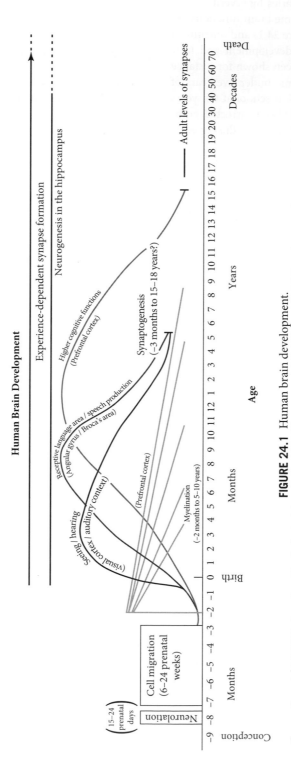

FIGURE 24.1 Human brain development.

Source: Copyright © 2001, American Psychological Association. R A Thompson and C A Nelson (2001) Developmental science and the media: Early brain development. *Am. Psychol.* **56**, 5–15. Animation: Gogtay et al. 2007.

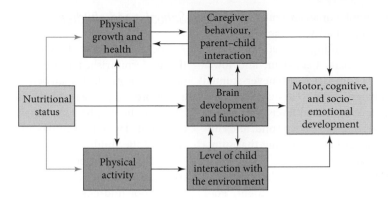

FIGURE 24.2 Conceptual framework on direct and indirect effects of nutritional status on child development.
Source: Reprinted from S P Walker et al. (2007) International Child Development Steering Group. Child development: Risk factors for adverse outcomes in developing countries. *Lancet* **369** (9556), 145–157, with permission from Elsevier.

age (mean age: 67.5 ± 0.9 years) in men, but not women (Franke et al. 2020).

Small for gestational age

Poor maternal nutrition before and during pregnancy is thought to be the main cause of foetal growth restriction in LMICs. Low birthweight, defined as weight that is less than 2500 g in a livebirth, is caused by two main underlying causes: being born too small (small-for-gestational age [SGA]), too soon (defined as preterm birth [gestational age <37 weeks]), or both. Annually, 23·3 million babies are estimated to be born SGA17 and 14·8 million are preterm (Victora et al. 2021). Systematic reviews show that being born SGA is associated with deficits in cognition, school achievement, and with behaviour problems in later childhood and adulthood. There is also some evidence of changes to brain structure in SGA children. However most studies come from developed countries, often with children identified prenatally and the aetiology may be different in these children. There are fewer cohort studies from LMICs and they often lack detailed data on gestational age. Cohorts from five LMICs found that birth weight was associated with years of schooling after controlling for postnatal growth. Similarly, in Thailand birth weight was related to IQ at nine years. Other studies looked at children who had low birth weight and were born at term. Most found deficits in mental and motor development in the first three years. Findings have been less consistent for older children, with no IQ deficits found at six years in Jamaica, at eight years in Brazil, or at 12 years in behaviour in South Africa (Walker et al. 2011). However, deficits were found in academic achievement in 15- to 16-year olds in Taiwan, and in Bangladesh low birth length was associated with low IQ at five years, controlling for gestational age and postnatal growth. One of the few studies based

in a low-income country that used the accepted definition of SGA was in Nepal (Christian et al. 2014) where SGA was associated with deficits in cognitive, executive, and motor function in seven- to nine-year old children. The evidence suggests that SGA is related to long-term deficits, but more adult follow-ups are needed.

Early childhood undernutrition

Many studies have shown that children with severe clinical malnutrition have long-term deficits in cognition, educational achievement, and behaviour. Nutritional rehabilitation alone is insufficient to improve their development after initial rehabilitation, but improving their subsequent environment by adoption or psychosocial intervention can make substantial improvements to cognition and school achievement.

Poor linear growth in early childhood affects many more children. It is partly caused by poor nutrition and is associated with poor child development. Length below -2 standard deviations of the WHO references is classified as moderate stunting which affects 149 million children under five years in low- and middle-income countries (LMICs) (Victora et al. 2021).

A meta-analysis of 68 observational studies in children under 12 years from LMICs identified ten cross-sectional studies adjusted for socioeconomic background. They found that one unit increase in height-for-age z-score (HAZ) was associated with improved cognitive ability: +0.24 SD in children under two years and +0.09 SDs in children over two years. In four prospective studies a unit increase in HAZ for children ≤2 years was associated with a +0.20 standard deviation increase in cognition at 5–11 years of age (Sudfeld et al. 2015). Motor development was also linked to early growth.

Five cohorts were followed to adulthood, and growth from birth to 24 months was associated with not completing secondary school. Jamaican adults who were stunted in the first two years had wide-ranging deficits in cognition, mathematics, reading, educational attainment, mental health and social behaviour, and lower wages (Walker et al. 2011) compared with never-stunted adults. The offspring of parents who were stunted in early childhood had poorer developmental levels, controlling for the parents' IQ, socioeconomic background, and children's height (Walker et al. 2011), suggesting that stunting may have an intergenerational effect.

The timing of poor growth appears to determine the presence and size of an effect on development. Poor growth in the first 24 months is the most likely to affect later cognition, executive function, and school attainment (Sudfeld et al. 2015) whereas poor growth after 24 months is less strongly associated with development, with little effect after three years.

Protein calorie supplementation in early childhood

Early macronutrient supplementation studies generally confirm the importance of the first 24 months (Liberato et al. 2013). Supplementation from the last trimester of pregnancy to two to three years of age in Guatemala and Colombia had concurrent mental and motor benefits. In Colombia benefits to reading readiness were found at seven years. Guatemala had the longest follow up, and exposure to supplementation from 0 to 36 months caused a benefit of 1.2 school grades in women only and an improvement of 0.25 SD in IQ and reading for both sexes and an increase in wages for males. The greatest impact came from exposure from birth to 24 months but there was no benefit from beginning supplementation after 36 months (Sudfeld et al. 2015).

Supplementation of children who were already undernourished had less consistent effects (Liberato et al. 2013). In Cali, Colombia, supplementation alone, beginning after age three years, had no cognitive benefits. In two Indonesian and one Jamaican study supplementation began before 24 months of age and concurrent benefits to child development were obtained in each. Two of the studies have reported long-term follow up. In Indonesia, a very small cognitive benefit was found at nine years of age, but only for children supplemented before 18 months. In Jamaica small cognitive benefits were found at ages seven to eight years, but were not sustained at 11–12 years. These findings suggest that sustained cognitive benefits only occur from providing adequate nutrition through the first two to three years and feeding malnourished children is unlikely to result in a catch-up in their long-term cognition and education without substantial improvements to their environment.

School-aged children

A Cochrane review of 18 school feeding studies found that undernourished or disadvantaged children in developing countries had small benefits in weight, attendance, and scores in mathematics. Younger children tended to benefit in height. Benefits in developed countries are less clear and depend on initial attendance and nutritional status of the children.

Short-term hunger may affect cognition and behaviour transiently. Studies have shown that when given breakfast schoolchildren's cognition is better by mid-morning compared with when they have no breakfast. Missing breakfast may have more negative effects on short-term cognitive performace in stunted or wasted children compared to better nourished children.

Micronutrient deficiencies and cognitive development

Micronutrient deficiencies are widely prevalent in many LMICs. Globally, an estimated two out of three women of reproductive age and one out of two children under five years of age are micronutrient deficient (Stevens et al. 2022): anemia affects 33 per cent of the world's population, and about half of the cases are due to iron deficiency. Zinc deficiency affects one in every two children in the few countries with information, and while successful universal salt iodization programmes have largely eliminated severe iodine deficiency, concern has shifted to 25 countries with mild to moderate deficiency that might lead to neurobehavioural effects (Victora et al. 2021). Micronutrients such as folate, iron, zinc, and iodine are involved in brain development, either directly via modification of brain structure and brain functions, or indirectly, and deficiencies are likely to impair cognitive, motor, and sociobehavioural functions (Prado & Dewey 2014).

Iron

Over the past three decades, a large effort has focused on understanding the relation between iron deficiency (ID) and development or behaviour in infants and young children. The relationship between iron and development may be either direct through an impact on brain function and structure or indirect through functional isolation. There is strong evidence that IDA (iron-defiency anaemia) is associated with poorer performance on (motor) development tests in infants and with lower scores on cognitive function tests and educational achievement tests in children. In a follow-up study from Costa Rica, children who had been iron deficient as an infant scored persistently more poorly on cognitive tests up to 19 years of age, and these effects were exaggerated in children from a high risk environment, suggesting an interaction between environment or social stimulation and ID (Osendarp et al. 2010) (Figure 24.3).

The iron content of the human brain increases continuously during development and up throughout the teenage period by active transferrin-receptor mediated transport of iron into the brain. ID impacts both neurogenesis as well as neurochemistry during brain development. Animal studies show structural impairments of the hippocampus in iron deficiency. ID also results in altered composition and amount of myelin in white matter. The importance of iron in neurochemistry is illustrated by the role of iron in the production of hormones from the monoaminergic pathways, particularly dopamine and norepinephrine (Osendarp et al. 2010).

Magnetic resonance imaging (MRI) has been used to map iron distribution in the brains of children and adolescents. The deposition of iron in the brain varies by region, and by age. Regions of the brain rich in iron in adulthood are not the regions that have a high iron content in early life and in addition, regions of the brain that are rich in iron are not necessarily the ones to be most affected by dietary ID. Potentially, the different regional needs for iron in the brain during different stages of neurodevelopment could impart a differential sensitivity of brain regions to nutritional deprivation of iron (Osendarp et al. 2010).

Pregnancy

Evidence from animal studies thus suggests a sensitive window of opportunity for reversing the detrimental effects of gestational iron deficiency on brain development, at least in rats and probably also in humans. However, relatively few studies have investigated the effect of maternal iron supplementation during pregnancy on subsequent cognitive development in the offspring and the results are inconclusive. In a study in Nepal, with a high prevalence of IDA, children whose mothers had received iron, folic acid, and vitamin A during pregnancy performed better on tests of non-verbal intelligence, executive function, and motor ability at seven to nine years of age, compared to children whose mothers had only received vitamin A during pregnancy. However, two other studies in China and Australia did not demonstrate effects of maternal iron supplementation on the Bayley Scores of Infant Development (BSID) at three, six, or 12 months of age, or on IQ at four years of age. Collectively, the evidence does not appear to show a clear benefit of antenatal iron supplementation on long-term child development (Larson & Yousafazi 2017).

Infancy

The results of intervention trials studying the impact of iron treatment on cognitive and motor development scores in ID infants have been inconsistent. Several trials and meta-analyses on trials gave oral iron supplements or placebo to children under two years of age. While benefits on social-emotional behaviour and motor development were observed, there was no evidence of benefit on cognitive development. There is also no evidence of longer-term effects of sometimes quite sustained courses of iron supplementation in young children on cognitive ability later in childhood (Larson & Yousafazi 2017). The lack of effect on cognitive development may be because such a benefit does not exist—perhaps because iron at these doses and

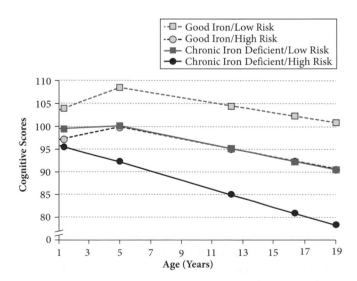

FIGURE 24.3 Long-term effects of iron deficiency (ID) and interaction with environment.
Source: T Shafir, R Angulo-Barroso, A Calatroni et al. (2006) Effects of iron deficiency in infancy on patterns of motor development over time. *Hum. Mov. Sc.* **25**(6), 821–838. Copyright © 2006 Elsevier B.V. All rights reserved.

via this route does not impact on brain development, or, as hypothesized by some authors, because the effects from ID on early brain development are irreversible.

The recognition of the role of iron in dopamine production allowed researchers to design more sensitive tests on the impact of iron deficiency on socioemotional development by focusing on behaviours that are known to be regulated by dopamine. Iron-sufficient infants showed best socioecoional development, as illustrated by their scores on shyness, orientation/engagement, and response to unfamiliar pictures, compared to iron-deficient anaemic infants who showed the worst scores and iron-deficient infants who had intermediate scores. These findings are consistent with an early disruption of the dopamine system, due to iron deficiency, and contribute to a growing understanding that altered affect and response to novelty are among the core deficits in early ID (Osendarp et al. 2010).

Older children

In contrast to trials undertaken in preschool children, treatment of anaemia with iron supplementation in older children demonstrates a benefit on global cognitive performance and IQ scores, particularly among anaemic children (Larson & Yousafazi 2017).

In summary, IDA during gestation and infancy is a strong risk factor for poor cognitive, motor, and socioemotional development of infants and children. Preventing iron deficiency through iron supplementation may have positive effects on motor development in infants and mental performance and socioemotional behaviour in older children.

Iodine

Iodine is an essential micronutrient that forms a vital component of the thyroid hormones thyroxine and triiodothyronine, which are crucial regulators of the metabolic rate and physical and mental development in humans. Thyroid hormones play a major role in the growth and development, function, and maintenance of the central and peripheral nervous system (Prado & Dewey 2014).

Pregnancy

Iodine deficiency during pregnancy leads to structural alterations in the brain of the foetus with long-term effects on mental development and performance. Severe iodine deficiency during pregnancy can cause cretinism, characterized by severe intellectual disability. Studies in pregnant women provided conclusive evidence that cretinism can be prevented by iodine supplementation before and during pregnancy. Moreover, the beneficial effects of iodine supplementation during pregnancy on cognitive development were still apparent in childhood

(Prado & Dewey 2014). In an iodine-deficient region in China, four- to seven-year old children whose mothers were given iodine during pregnancy performed better on a psychomotor test than those who were supplemented beginning at two years of age. Even mild iodine deficiency in the first trimester of pregnancy can negatively affect children's cognition eight years later. Among over 1000 eight-year old children in the UK, those whose mothers had been iodine deficient in the first trimester of pregnancy were more likely to have scores in the lowest quartile for verbal IQ and reading comprehension (Prado & Dewey 2014). Nevertheless, a recent meta-analysis suggest little to no benefit of iodine supplementation on motor and mental development in mild and moderate iodine deficient pregnant mothers, perhaps due to adaptations in thyroid function in contexts with low pre-pregnancy iodine intakes.

Infancy and childhood

Observational studies comparing populations from iodine-deficient areas with those in iodine-replete areas widely support the view that iodine deficiency reduces mental performance and leads to poorer achievement later in life. A meta-analysis of 18 studies completed in humans aged two months to 45 years indicated a general loss of 13.5 IQ points in chronically iodine deficient populations, compared to non-iodine deficient groups (Prado & Dewey 2014).

Eight randomized placebo-controlled intervention trials have addressed the question of reversibility of the consequences of iodine deficiency later in life. Of these studies, five found improvements in mental and cognitive performance in deficient schoolchildren following iodine supplementation. Three studies did not find an improvement on cognition, which may have been due in part to the fact that the iodine supplementation was not sufficient to improve the iodine status (Prado & Dewey 2014).

In conclusion, iodine is essential for brain development of the foetus and cognitive performance of infants and children. Studies have shown that iodine supplementation during pregnancy or childhood is effective for improving development and cognitive performance of children.

Zinc

Zinc is a key modulator of intracellular and intercellular neuronal signalling that is found in high levels in the brain, particularly the hippocampus, considered to be the area involved in learning and memory. In addition, zinc is essential for the activity of a large number of metalloenzymes, cellular functions including RNA and DNA synthesis, cellular growth, differentiation and metabolism.

The essential role of zinc in the central nervous systems is marked during brain growth, particularly between 24 and 40 weeks after conception, which is the period when the brain goes through extraordinary structural changes, and it is during this critical time that the brain is most sensitive to zinc deficiency. Its deficiency could interfere with various neurochemical processes regulated by zinc-dependent enzymes, including neurotransmission, the release of neurotransmitters, and subsequent neuropsychological behaviour. Research in animals showed that severe zinc deficiency, particularly during periods of rapid growth such as gestation or adolescence, is associated with alterations in brain development, increased emotional response to stress, reduced motor activity, and less accurate performance on measures of attention and short-term memory (Warthon-Medina et al. 2015).

Pregnancy

Despite evidence from animal studies for an important role of zinc during gestation when the brain develops rapidly, the evidence from human studies has not shown positive effects of zinc supplementation during pregnancy or infancy on child cognitive development. Randomized controlled studies of maternal zinc supplementation during pregnancy in the United States, Peru, Nepal, and Bangladesh have shown no effects or even negative effects of zinc supplementation during pregnancy on the motor and cognitive abilities of children between 13 months and nine years of age (Prado & Dewey 2014).

Infancy and childhood

The results of zinc supplementation during infancy or childhood on cognitive development are inconsistent. Of the nine studies providing zinc to infants beginning before the age of two years, only one study found a benefit on mental development but only in children who received psychosocial stimulation in addition to zinc (Warthon-Medina et al. 2015). One trial from Bangladesh resulted in a negative effect of zinc supplementation on mental development of infants. A 2009 meta-analysis of randomized controlled studies of zinc supplementation in infants did not find any evidence of benefits on BSID mental or motor skills but the authors concluded that the duration of supplementation in these studies may have been too short to permit detection of an effect (Prado & Dewey 2014). Of the nine studies that provided zinc for at least six months, four studies showed that zinc supplementation improved motor development and two other trials indicated that zinc supplementation in children under two years of age may increase activity levels (Prado & Dewey 2014). In older children two recent meta-analyses found no significant overall effect of zinc supplementation or zinc intakes on any indices of cognitive function:

intelligence, executive function, and motor skills, either in pre-school or school-aged children (Warthon-Medina et al. 2015).

In conclusion, zinc plays an important role as coenzyme in many physiological processes in the brain, and is therefore likely to have an impact on brain development and function. However, current evidence is inconclusive for effects of zinc on cognitive development and performance in infants and children.

Multiple micronutrients

Individuals who are deficient in one micronutrient are commonly at risk of deficiencies in others as well. Because micronutrient deficiencies often coexist and synergistic effects of micronutrients on physical functions may indirectly affect cognition, supplementing children with multiple micronutrients could likely have advantages over single micronutrient supplementation.

Pregnancy

Three randomized trials have reported positive effects of multiple micronutrient supplementation, compared to iron and folic acid alone during pregnancy on child development and motor development between the ages of six and 18 months, including motor development (Prado & Dewey 2014). A trial in Indonesia showed positive effects of maternal multiple micronutrient supplementation on motor and cognitive development at age 3.5 years in the children of undernourished and anaemic mothers. However, in Nepal, children whose mothers received iron and folate during pregnancy had better intelligence quotient (IQ), and better executive and motor functioning than the placebo group at seven to nine years of age, whereas children whose mothers had received multiple micronutrients scored higher only on a test of executive function, compared to mothers who had received vitamin A alone (Prado & Dewey 2014). The authors concluded that the smaller effect in the multiple micronutrient intervention group was possibly due to an inhibition of iron absorption by zinc.

Infancy

Studies of multiple micronutrient supplementation during infancy have shown some benefits immediately after the supplementation period on motor development or overall developmental quotient (Prado & Dewey 2014). In Mexico, infants between the ages of eight and 12 months who had received multiple micronutrient supplementation for four months were more active than those who had not received supplementation. However, none of these studies compared the multiple micronutrients with a control group receiving single micronutrients. A randomized trial in Bangladesh did not show an effect on mental or motor development in infants who received 16 micronutrients compared to infants who received one or two micronutrients (Prado & Dewey 2014), suggesting that providing multiple micronutrients may not be more beneficial than providing single micronutrients for infant development.

Older children

In 2001 Benton reviewed 13 studies that investigated the role of multiple micronutrients on cognition in children aged 6–16 years, of which most reported a positive effect of the micronutrient supplementation, mostly with nonverbal measures. The author postulated that performance on nonverbal tests results from basic biological functions, at least in part, and could be influenced by diet. In contrast, verbal intelligence comprises the acquired knowledge that was thought not to be affected by nutrition in the shorter term. Some limitations of the review were that no strict selection criteria were applied for inclusion or exclusion of studies and that the results of the studies were not pooled to quantify the effect of micronutrients on cognition. A more recent meta-analysis suggests that supplementation with multiple micronutrients in healthy school-age children may be associated with a small increase in fluid intelligence, whereas crystallized intelligence seems unaffected (Prado & Dewey 2014).

In conclusion, multiple micronutrient interventions can be considered an attractive intervention to tackle multiple deficiencies. A meta-analysis of nutrition interventions on mental development in children under two, suggested a trend toward greater benefit on mental development from interventions using multiple micronutrients as compared with single micronutrients. However, overall effect sizes were small and the authors concluded that there is a need for more research on nutrition interventions of adequate sample sizes for the outcome of interest, that is, mental development and for studies that combine nutrition interventions with other interventions known to affect cognitive development in children, such as social stimulation (Larson & Yousafzai 2017).

Essential fatty acids and cognitive development

Essential polyunsaturated fatty acids (PUFA) of the omega-3 (n-3) and the omega-6 (n-6) series are of critical importance during early life, and they are known to play an essential role in growth and development. The long-chain polyunsaturated fatty acids (LC-PUFA) eicosapentaenoic acid (EPA, n-3), docosahexaenic acid (DHA, n-3), and arachidonic acid (AA, n-6), can be formed from the precursors alpha-linolenic acid (n-3) and linoleic acid (n-6), respectively. However, the rates of

conversion of the precursor PUFA are low and are estimated to range only from 0.1 per cent to 10 per cent (1–3) (see Chapter 8). Moreover, recent evidence suggests that the conversion rates depend on common polymorphisms in the fatty acid desaturase (FADS) gene cluster, such that individuals with the less common genotypes have a very low ability to form EPA and DHA from its precursors, and may therefore benefit more from supplementation (Koletzko et al. 2014).

Approximately 25–30 per cent of the fatty acids in the human brain consist of polyunsaturated fatty acids (PUFA) of which the n-3 fatty acid docosahexaenoic acid (DHA) and the n-6 fatty acid arachidonic acid (AA) are major components. DHA and AA are rapidly incorporated into the nervous tissue of retina and brain during the brain's growth spurt, which mainly takes place from the last trimester of pregnancy up to two years of age (Koletzko et al 2014). Beyond development of the central nervous system, n-3 and n-6 fatty acids may influence brain function throughout life by modifications of neuronal membrane fluidity, membrane activity-bound enzymes, number and affinity of receptors, function of neuronal membrane ionic channels, and production of neurotransmitters and brain peptides.

Pregnancy and lactation

In several observational cohort studies, beneficial effects of n-3 or fish intake during pregnancy and/or lactation on developmental outcomes and cognition of the offspring up to 14 years of age have been reported, even after adjusting for potential confounding factors (Koletzko et al. 2014).

However, the findings from randomized controlled trials (RCTs) are conflicting. A meta-analysis of 11 RCTs involving 5277 participants, found no significant differences in standardized psychometric test scores for cognitive, language, or motor development in the offspring of mothers who had been supplemented with LC-PUFA (DHA and EPA) during pregnancy. There was a suggestion of higher cognitive development scores at later ages (2–5 years of age) in children of supplemented mothers.

Breastfeeding is an important source of essential fatty acids for infants, especially during the first six months of life, and supplementing lactating mothers with LC-PUFA may improve LC-PUFA intakes and health benefits in the infants. However, studies evaluating the effect of LC-PUFA supplementation during lactation on the cognitive and motor development of infants in later life reported inconsistent results. The inconsistency of findings from the randomized controlled trials providing LC-PUFA supplementation during pregnancy and/or lactation may have been due in part to the heterogeneity in outcomes measured and in methods of assessment used. Recent studies providing higher dosages of DHA and EPA, and assessing cognitive abilities in children at later ages (5 and 7.5 years of age) reported better performance on attention tests and information processing in children from previously supplemented mothers (Koletzko et al. 2014). In addition, there seems to be an interaction between breastfeeding and the fatty acid desaturases (FADS) genotype, such that breastfeeding is associated with higher IQ scores than bottle feeding in children with a FADS genotype linked to low endogenous LC-PUFA synthesis, but not in infants with a genotype supporting a more active LC-PUFA formation (Koletzko et al. 2014).

Infancy and older children

Supplementing formula for term infants with LC-PUFA, may be beneficial with regard to visual, motor, and cognitive outcomes in early childhood, and even more beneficial in later childhood. There seems to be a greater likelihood of benefit with higher dosages (DHA 0.32 per cent and above, AA 0.66 per cent and above) and longer duration (up to age one year) of a higher postnatal LC-PUFA supplementation (Koletzko et al. 2014). Further evidence for benefits of postnatal LC-PUFA intakes is derived from gene–nutrition interaction studies that report greater benefits of breastfeeding, which provides preformed LC-PUFA, in infants with genetically lower endogenous LC-PUFA synthesis. Supplementing infants from three or six until 18 months with fish-oil or LC-PUFA-rich complementary foods in The Gambia and Malawi, did not result in consistent benefits in their motor or mental development. However, the infants in these studies may not have been LC-PUFA-deficient, perhaps due to high maternal fish consumption and subsequent LC-PUFA levels in breast milk, possibly masking any effects of supplementary LC-PUFA (Prado & Dewey 2014).

Preterm infants, who are at risk of deficiency because fatty acids accumulate rapidly in the brain during the third trimester of pregnancy, may benefit more from LC-PUFA supplementation. There are consistent indications that LC-PUFA provided to preterm infants can have benefits for visual and cognitive outcomes, especially with higher DHA dosages. Certain subgroups of preterm infants, for example, with lower birthweight may achieve a greater benefit from preformed LC-PUFA (Koletzko et al. 2014).

In older children above two years of age, no benefits of additional LC-PUFA (DHA supplements or DHA enriched fish flour) were found on growth or cognition in studies from developing countries, or developed countries (Koletzko et al. 2014).

In conclusion, LC-PUFA interventions during pregnancy, lactation, and infancy may benefit later cognitive and visual performance in children, especially when

> ### KEY POINTS: NUTRITION AND COGNITIVE DEVELOPMENT
>
> - Child development is a multidimensional process and poor nutrition is a major risk factor for poor development.
> - Poor maternal nutrition leads to poor pregnancy outcomes, and impacts on longer-term child cognitive development; and protein-energy supplementation during pregnancy and through the first two to three years improves longer term cognition and education.
> - Micronutrients such as folate, iron, zinc, and iodine are involved in brain development, either directly via modification of brain structure and brain functions, or indirectly, and deficiencies are likely to impair cognitive, motor, and sociobehavioural functions.
> - Essential fatty acids are important building blocks of the brain, and EFA status during pregnancy, lactation, and infancy may impact later cognitive and visual performance in children.

higher dosages of DHA and EPA are provided and in populations that have genetically low endogenous LC-PUFA synthesis (Koletzko et al. 2014).

24.3 Nutrition and the nervous system in adulthood and the elderly

Introduction

Nutritional status is key in modulating cognition, with the evolution of the human brain made possible by changes in the availability and preparation of food, including the development of cooking. One of the most important advances was shore-based living and fishing which led to relatively higher intakes of the omega-3 fats, particularly DHA, which preceded the development of a higher brain-to body-weight ratio in humans.

However, in the last 50–60 years there has been a rapid change in dietary habits, with a decline in the intake of complex carbohydrates and fibre and an increase in the intake of processed food rich in saturated fats, refined grains, sugars, and advanced glycation end products (AGEs) typical of the Western diet. Selected studies on these dietary components are shown in Table 24.1, which includes relatively acute effects on cognition and the predisposition towards the development of age-related cognitive deficits. A pivotal factor in promoting these dietary changes is the accessibility of cheap highly palatable energy dense foods rich in fat and sugar.

TABLE 24.1 Selected human and animal studies investigating the effects of diet on cognitive function

Diet composition	Cognitive results	Postulated biological mechanism	Refs (see Chapter 24 Further Reading online)
High linoleic acid intake	Decreased cognitive function in humans	Oxidative stress	Kalmijn et al. 1997
Poor diet resulting in impaired glucose tolerance	Decreased cognitive function in humans	Disturbed glucose metabolism	Kalmijn 2000
Higher intakes of saturated fat and trans-fat	Decreased cognitive function in humans	Cholesterol levels atherogenic	Morris et al. 2004
Increased caloric and cholesterol intake	Decreased cognitive function in humans	not discussed	Zhang et al. 2006
High intake of omega-3 and DHA	Reduced risk of developing AD in humans	not discussed	Morris et al. 2003
High-fat diet (45% energy from fat)	Decreased cognitive function in rodents	Impaired insulin signalling	McNeilly et al. 2011
Lard-based diet (40% energy from fat)	Decreased cognitive function in rodents	not discussed	Greenwood, Winocur 1990
High-fat, high-glucose diet supplemented with corn syrup	Decreased cognitive function in rodents	Impaired insulin signalling and decreased BDNF levels	Stranahan et al. 2008
High-fat high-carbohydrate diet supplemented with vitamin E	Improved cognitive function in rodents	Decreased oxidative stress	Alzoubi et al. 2013
Lard and corn oil diet	Decreased cognitive function in rodents	Oxidative stress and decreased BDNF levels	Wu, Ying & Gomez-Pinilla 2004
High-saturated fats, trans fats and cholesterol	Decreased cognitive function in rodents	Neuroinflammation and decreased dendritic integrity in the hippocampus	Freeman et al. 2011
High-fat diet (60% energy from fat)	Rapid impairment of episodic memory in rodents	Increased adiposity and glucose intolerance	McLean et al. 2018

The impact of nutrients on the adult brain

The best studied nutrients with regard to their effect on memory and cognition are fats, particularly saturated fats, initially thought to act via increased metabolic and cardiovascular risk factors. As in brain development, omega-3 fats, DHA, and EPA remain key in adult brain function, with deficiency in these fatty acids leading to increased risk of cognitive defects. In support of this, DHA and EPA supplementation have been used to treat attention deficit hyperactivity disorder (ADHD), depression, and schizophrenia (Peet & Stoke 2005).

In contrast a high-intake of long-chain saturated fats is associated with cognitive and memory deficits. In rodents these cognitive deficits occur rapidly and are associated with diet rather than obesity per se. However, excess bodyweight is also associated with changes to brain structure and function with several cognitive deficits, including impaired episodic memory, seen in obese

adults. (Please see https://qbi.uq.edu.au/brain-basics/memory/types-memory for a very bref description of memory types.) In addition, a high intake of refined sugars is also associated both with the development of obesity and type 2 diabetes and, in the short term, with impaired memory in healthy adults and type 2 diabetics.

Studies on the impact of fruit and vegetables on cognitive function have mostly looked at the attenuation of age-related cognitive decline and failed to find any acute effects in young adults (Lamport et al. 2014).

Caloric restriction

Reduced caloric intake or caloric restriction (CR) results in increased longevity in yeasts and worms through to rodent and non-human primates by delaying age-related processes. CR has mostly been studied in rodents where it improves memory as well as increasing longevity. In older humans CR alone and in combination with exercise has been reported to improve executive function and memory, indicating a potential interaction between exercise and CR on cognition. Intermittent fasting is another form of CR which has been shown to improve mood.

Ketogenic diet

Ketogenic diets are high-fat, moderate-protein, very low carbohydrate diets which result in enhanced fat oxidation. This diet generates high levels of acetyl-CoA which instead of being metabolized via the citric acid cycle is channelled towards the production of ketone bodies: acetone, acetoacetate, and D-β-hydroxybutyrate in the liver. Once synthesized, ketone bodies are used as an energy source and as lipid precursors. Ketogenic diets are an effective treatment in refractory epilepsy and there is mounting evidence for their beneficial effects in neurodegenerative diseases such as Alzheimer's disease (AD) and in traumatic brain injury and stroke. One of the mechanisms by which ketogenic diets are thought to act is via improvements in mitochondrial function. Under normal circumstances a high-fat diet is considered as detrimental to cognitive function, but it is the metabolic route that fatty acids take which dictates its effect. In a ketogenic diet fats are channelled towards β-oxidation, whereas in a high-fat diet, which also contains carbohydrates, fats are not effectively oxidized but result in the synthesis of lipotoxic by-products which have negative effects on cognitive function and metabolic health. However, when the production of ketones overcomes the ability of tissues to oxidize them, they accumulate, causing a decrease in blood pH leading to ketoacidosis, making it important to monitor their plasma concentration. Studies also suggest that this dietary regimen deserves attention as an alternative nutritional therapy in neurodegenerative and cognitive diseases (Hallböök et al. 2012).

Amino acids and peptides

Tryptophan

Tryptophan is an essential amino acid mainly found in chicken, soya beans, cereals, tuna, nuts, and bananas. High-glycaemic index meals increase the availability of tryptophan. Tryptophan is a precursor of kynurenine and the neurotransmitter serotonin, which are both important players in mood and cognition with higher tryptophan levels related to improved cognition. The effects of tryptophan also arises from its role as a precursor of melatonin which in turn is important in sleep and the regulation of circadian rhythms.

Tyrosine

Another amino acid with potential effects on the brain is tyrosine. It has been shown to prevent cognitive impairment due to sleep deprivation and to aid acutely stressed volunteers by improving mood and cognitive function. The positive effect exerted by tyrosine on cognitive function may be attributable to its role as a precursor of the catecholamines: epinephrine, norepinephrine, and dopamine.

Carnosine

Carnosine is a dipeptide (β-alanyl-L-histidine) found mainly in meat and has been shown to have effects on mood and anxiety in animal models. In humans supplementation has been used to treat mental health problems and improve cognition in the elderly.

Nutrient-associated signals from the digestive system to the brain

Nutrition can influence the brain and cognition indirectly via hormonal signals from the gastrointestinal tract, adipose tissue, and the pancreas via the blood stream and vagal innervation. Vagal stimulation has been used to treat epilepsy and is associated with alleviating depression and improved memory performance. Gut hormones including glucagon-like peptide (GLP-1) from the intestine and ghrelin from the stomach have been shown to influence cognition and memory. Leptin from white adipose tissue, as well as its role in energy balance has a powerful and well-studied influence on memory with high densities of hippocampal leptin receptors. Insulin also has a positive impact on memory. Thus, the dysregulation of metabolism which occurs in obesity, and is marked by both leptin and insulin insensitivity, is strongly associated with the development of mental disorders.

Collectively, the ingestion of excess calories appears to have damaging effects on cognition while caloric restriction appears to be beneficial. These effects have been well documented in rodent models.

Impact of lifestyle on mood and cognition

Stress

Stressful events impact on the hypothalamic pituitary–adrenal axis, increasing the production of glucocorticoid hormones which act on the hippocampus, a brain region important in memory and cognition. An acute stress response helps the body to cope with a stressful event by inducing physiological changes. However, if stress becomes chronic it results in a series of adverse effects including mood alteration, anxiety, and cognitive dysfunction.

Exercise

Physical activity is preventive in cardiovascular disease, certain cancers, and obesity and is associated with reduced mental disorders such as AD, anxiety, and depression. Exercise has a positive effect on learning and memory, increasing the ability of the brain to adapt to environmental changes and respond to injury. Exercise has also been shown to counteract the negative effects of a high-fat diet in animal models with both endurance and resistance training resulting in similar outcomes on learning and spatial memory.

Sleep

Waste metabolites in the brain are removed via the glymphatic system, a name derived from glial and lymphatic as this process is dependent on glia and resembles the peripheral lymphatic system. The glymphatic system is the waste disposal system of the brain and is active during sleep. Inadequate sleep contributes to the neuropathology seen in AD, including the accumulation of Aβ plaques and tau tangles, changes in synaptic plasticity, and structure in brain regions associated with learning and memory. A decrease in the efficiency of the glymphatic system is seen in normal ageing. Sleep disruption is also associated with the dysregulation of body weight control, energy, and glucose homeostasis.

Alcohol consumption

Some studies suggest that low to moderate alcohol consumption may be associated with a reduction in cognitive decline, but evidence is mixed. There is also evidence that excessive alcohol consumption may be associated with an increased risk of alcohol-related dementia and alcohol-related brain damage. Alcohol abuse induces thiamine deficiency which is the main causes of Korsakoff's syndrome, which includes short-term memory loss, difficulties in acquiring new information, confabulation, and personality changes.

Caffeine

Caffeine is a methylated xanthine found in a variety of beverages, including tea and coffee. It is a mild stimulant of the central nervous system and produces its stimulatory effect by inhibiting the adenosine receptors A1 and A2A. Caffeine affects the levels of various neurotransmitters such as dopamine, norepinephrine, acetylcholine, gamma-aminobutyric acid, and glutamate. Caffeine enhances arousal and alertness and improves memory, mood, and vigilance with the lifelong consumption of coffee/caffeine linked to the prevention of cognitive decline. However, the evidence linking caffeine and cognitive performance is mixed and despite caffeine having been reported to prevent cognitive decline in healthy subjects some studies reported positive effects only in one sex and then in the oldest. Importantly, caffeine is only one of the active components in coffee, which contains a mixture of several bioactive compounds; these may have synergic effects with caffeine and be responsible for the positive effect of coffee on both motor and cognitive deficits seen in ageing rodents.

Nutrition and the nervous system: transitioning towards ageing

The increase in the ageing population, particularly in Europe and the USA has led to an increased burden of diseases associated with age-related cognitive decline (https://www.un.org/development/desa/pd/sites/www.un.org.development.desa.pd/files/undesa_pd-2020_world_population_ageing_highlights.pdf). Many of the risk factors associated with the development of accelerated brain ageing and dementia are linked to diet, with health professionals advising a balanced diet, a healthy weight, reducing cholesterol and blood pressure and adherence to guidelines for alcohol consumption (https://www.alzint.org/resource/nutrition-and-dementia/). Intrauterine and early life nutrition is key in the development of the nervous system and hence influences the development of cognitive decline at the other end of the life span. Early life nutrition may be influencing 'brain reserve' or the ability of the brain to maintain a relatively normal function despite future insults. This may help to explain the variability in susceptibility to accelerated cognitive ageing and dementia in the population.

A natural decline in certain aspects of memory and cognition is normal, with some degree of cognitive decline during ageing. This may include problems with memory, information processing speed, executive function, attention, and abstract reasoning. However, the incidence of more serious age-related cognitive decline including dementia is increasing. Sporadic, or late onset, AD is the most common form of dementia in the elderly, with less than 5 per cent of all cases due to early onset genetic causes. No effective therapy for AD is currently available with 81 million predicted to be affected by 2040 (Ferri et al. 2005).

Many of the diseases of ageing, including AD, are the consequences of chronic imbalances and dysregulation of important physiological pathways. Nutritional interventions to prevent or minimize this process must be pleiotropic and holistic to be effective. For example, the Western diet has numerous negative effects on cognition, while the Mediterranean diet has multiple positive effects, implicating many nutrients in exacerbating or ameliorating cognitive ageing.

Alzheimer's disease

The risk of developing AD increases with age with the risk doubling every five years after the age of 65. AD is characterized by neuronal loss, synaptic dysfunction, gliosis, insoluble aggregates of abnormally phosphorylated and ubiquitinated tau protein in the form of neurofibrillary tangles and amyloid–β protein fibrillary aggregates, and extra-cellular plaques (Querfurth & LaFerla 2010). Thus, there are two major hypotheses as to the causation of AD: either via the accumulation of hyperphosphorylated tau or amyloid–β aggregation, with AD therapies largely focused on trying to limit one or other of these pathologies.

The incidence of type 2 diabetes (see Chapter 21) also increases with age and it was assumed that this was coincident, but more recently a number of studies have shown that type 2 diabetes and AD are interrelated, with type 2 diabetes being a major risk factor for the development of AD. Longitudinal studies have shown that glucose intolerance and impaired insulin secretion are risk factors for dementia and that obesity, which strongly predisposes individuals to type 2 diabetes, is also linked to cognitive impairment and the development of AD. Weight loss after bariatric surgery improves cognition, an effect that appears to be linked to the rapid improvement in glucose tolerance. Also, caloric restriction reverses ageing induced changes and prevents amyloid-β and tau accumulation in animal models of AD and accelerated ageing (Mattson 2010).

These findings have led to a novel interpretation of the causes of AD which focuses on chronic peripheral hyperglycaemia coupled with the inability of the brain to respond to insulin, which has led to some researchers terming AD as *type 3 diabetes*. This has resulted in promising trials of drugs in AD which are normally used to treat diabetes, such as insulin and GLP-1 mimetics.

Vascular dementia

Vascular dysfunction in the brain can lead to neurodegeneration, mild cognitive impairment, through to more serious dementia and AD. The circulation in the brain is key in delivering nutrients to the brain and disposing of waste products. Any cessation in blood flow leads to neuronal damage within minutes. The sensitivity of the brain to altered blood flow has led to the two hit vascular hypotheses behind the development of AD. This incorporates the first hit of cerebrovascular damage which results in neuronal injury and degeneration initiating the accumulation of amyloid-β, which constitutes the second hit. Among the risk factors for vascular dementia are hypertension, type 2 diabetes, atherosclerosis, raised levels of homocysteine, and peripheral inflammation which includes several symptoms of the Metabolic syndrome discussed in detail below.

Metabolic syndrome and dementia

A hallmark of impaired metabolism is the metabolic syndrome which comprises a constellation of symptoms, defined by the International Diabetes Federation in 2006, including hypertension, dyslipidaemia, abdominal obesity, and impaired glucose tolerance and hyperglycaemia (https://www.idf.org/e-library/consensus-statements/60-idfconsensus-worldwide-definitionof-the-metabolic-syndrome.html). Amongst these symptoms it appears that hyperglycaemia is key in promoting cognitive decline. Indeed, midlife type 2 diabetes is associated with infarctions, reduced hippocampal and brain volume, and cognitive impairment. However, later life onset of these conditions has fewer effects on brain pathology and cognition. In cognitively healthy adults a high fasting blood glucose and greater glycaemic dietary load are related to poor perceptual speed as well as a greater rate of decline in general cognitive ability including verbal and spatial ability.

Dyslipidaemia

Another marker of the metabolic syndrome is dyslipidaemia, however, a direct relationship with cognitive decline is unclear. The brain is particularly rich in cholesterol and contains the highest concentration of any organ. Cholesterol is crucial for the functioning of the central nervous system and plays a pivotal role in synaptic plasticity, neurotransmitter release, and cellular trafficking. Variations in several genes involved in cholesterol metabolism and transport have been identified as AD risk factors. These are: apolipoprotein E (APOE), apolipoprotein J (APOJ CLU), ATP binding cassette sub-family A member 7 (ABCA7), and the sortilin related receptor (SORLI). Carriers of the APOE ε4 gene variant are at high risk of age-related cognitive decline. However, the relationship between dyslipidaemia and AD remains controversial with epidemiological studies showing conflicting evidence. Also, randomized clinical trials show divergent results on the effect of the cholesterol-lowering drugs, statins, on AD risk, with there being little exchange between circulating and brain cholesterol. Cell biology studies have shown that cholesterol is both harmful, as it interferes with β-amyloid production, and protective, with increased membrane cholesterol preserving membrane integrity when disrupted by β-amyloid toxicity. Cellular cholesterol also increases the generation of plasmin, a β-amyloid clearing enzyme. The overall conclusion is that there is support for a possible involvement of raised cholesterol levels in the development of AD and as a modifiable risk factor it is advised to keep cholesterol levels within the published guidelines, particularly if other risk factors are present.

Body weight and dementia

In contrast to the increased adiposity in mid-life acting as a risk factor in the development of AD, those suffering from dementia weigh less and have a lower BMI than cognitively normal individuals. Nonetheless, it is unclear whether protein energy malnutrition precedes or follows dementia. Weight loss in dementia due to inadequate dietary intake may be the result of dementia with symptoms including apraxia—no longer knowing how to eat; agnosia—no longer recognizing food as food; forgetting to eat; changes in olfaction and taste; difficulties with shopping and meal preparation, and finally depression and refusal to eat. There is also a natural decline of appetite drive in the elderly with a blunted response to fasting alongside an increase in the dietary requirements for protein due to a decrease in protein synthesis which also results in impaired muscle function. Hypermetabolism may be another factor linking weight loss and dementia.

Dietary influences on cognition in ageing

Diet types

The Mediterranean diet

A range of studies on the effects of single nutrients have struggled to confirm epidemiological data supporting a positive association between dietary components and cognitive decline, suggesting that there may be is a synergistic effect of different nutrients. Thus, dietary patterns rather than single nutrient supplementation appear to be key in cognitive ageing. Epidemiological evidence shows a positive association between cognitive function and diets characterized by a high intake of fruit, vegetables, nuts, cereals, legumes, fish, extra virgin olive oil, and a low intake of saturated fats and high-glycaemic index carbohydrates (Figure 24.4). This is best reflected in the Mediterranean diet, which is rich in antioxidants, is anti-inflammatory, and decreases the risk of metabolic syndrome, all of which are implicated in the pathogenesis of AD. Confirming these proposed positive effects, adherence to this dietary pattern has beneficial effects on cognitive function, with those with a mild cognitive impairment showing a lower risk of progression towards dementia when following a Mediterranean diet (Scarmeas et al. 2009).

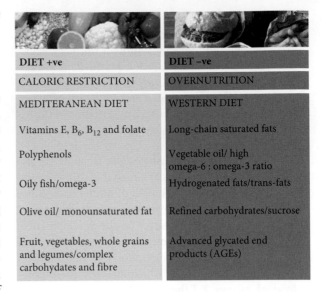

DIET +ve	DIET −ve
CALORIC RESTRICTION	OVERNUTRITION
MEDITERANEAN DIET	WESTERN DIET
Vitamins E, B_6, B_{12} and folate	Long-chain saturated fats
Polyphenols	Vegetable oil/ high omega-6 : omega-3 ratio
Oily fish/omega-3	Hydrogenated fats/trans-fats
Olive oil/ monounsaturated fat	Refined carbohydrates/sucrose
Fruit, vegetables, whole grains and legumes/complex carbohydates and fibre	Advanced glycated end products (AGEs)

FIGURE 24.4 Dietary patterns and components that are correlated with delayed (Diet +ve) and accelerated (Diet −ve) cognitive decline.

Dietary approaches to stop hypertension (DASH) and Mediterranean–DASH intervention for neurodegenerative delay (MIND) diets

Dietary approaches to stop hypertension (DASH) and Mediterranean–DASH intervention for neurodegenerative delay (MIND) diets are the most studied diets in relation to cognitive function. The DASH diet comprises fruit and vegetables, lean meat, whole grains, healthy fats, including omega-9 and omega-3 fatty acids, low-fat dairy products, and minimally processed fresh foods together with decreased sodium. The MIND diet combines the Mediterranean and DASH diets focusing on fresh fruit, vegetables and olive oil, berries, nuts, beans, green leafy vegetables, and wine in moderation. In parallel with the positive effect of the Mediterranean diet on cognition, adherence to either DASH or MIND is associated with a lower risk of AD and slower cognitive decline, with the strongest association for the MIND diet, further supporting the importance of dietary patterns, rather than nutrients in promoting cognitive health.

Dietary factors

Saturated fats

A high intake of saturated fat is associated not only with obesity and type 2 diabetes but also with cognitive impairment. The deleterious effects of high-saturated fat diets on memory and cognition have been demonstrated extensively in animal models. Importantly, in the elderly a high-saturated fat intake has been associated with a greater age-related decline in cognitive function and saturated fatty acids have been associated with the incidence of AD and Parkinson's disease, another neurodegenerative disease. It appears that the intake of fat 20–30 years before disease onset mainly contributes to disease risk. A summary of the human and animal studies linking a high-saturated fat diet with cognitive impairment is shown in Table 24.1. Thus, the consensus is that a high intake of saturated fats plays a major detrimental role in promoting the onset cognitive impairment via obesity and insulin insensitivity. Taken together with the protective effects of dietary restriction, which increases insulin sensitivity, this argues for role of insulin in the maintenance of cognitive ability.

Trans fats

The intake of hydrogenated, trans-unsaturated fats represents another risk factor for cognitive decline. One possible mechanisms is the ability of trans fats to raise LDL cholesterol levels, which in turn is associated with AD. However, the intake of processed food in which man-made trans fats are present is also associated with obesity, type 2 diabetes, and insulin insensitivity which are all risk factors for AD.

N-6 fatty acids

Western diets are characterized by a high intake of omega-6 fatty acids, notably linoleic acid, and a low intake of n-3 fatty acids, specifically, alpha linolenic acid, docosahexaenoic acid (DHA), and eicosapentaenoic acid (EPA), which is associated with the risk of developing dementia. The high intake of n-6 fatty acids in the Western diet is mainly from vegetable oil products, namely sunflower seed, soybean, corn, palm oil, and margarine.

A high dietary n-6:n-3 ratio leads to an increase in arachidonic acid, a substrate for cyclooxygenase and lipoxygenase enzymes, which promotes inflammation via the increased production of prostanoids while the levels of long-chain omega-3 fats which are neuroprotective are low. This both initiates and aggravates neuro-inflammation, typical of AD, with brain levels of arachidonic acid having been shown to be higher in AD. Also, the activity of the rate limiting enzymes involved in the metabolism of linoleic and linolenic acids decline with age, resulting in increased competition between the n-6 and n-3 fats. Indeed, a protective factor in the development of AD is the use of non-steroidal anti-inflammatory drugs (NSAIDS) which inhibit cyclooxygenase and the production of prostaglandins. However, once AD has developed, NSAID use has adverse effects on cognition. Decreasing the n-6:n-3 dietary ratio is a relevant dietary strategy that could sustain cognitive function and prevent the onset of dementia (https://www.fao.org/3/i1953e/i1953e00.pdf). Evidence thus suggests that a high n-6:n-3 ratio, typical in the Western diet, should be avoided in order to maintain optimal cognitive function.

N-3 fatty acids

A feature of the Western diet is an inadequate intake of n-3 fatty acids. DHA is more abundant in the brain than in any other tissue and dietary n-3 status is crucial for delaying the onset of age-related cognitive decline. The n-3 polyunsaturated fatty acids are components of the cell membrane and confer specific properties on the lipid bilayer depending on the length of the carbon chain and the degree of saturation, with DHA improving the flexibility of the lipid bilayer. DHA is also released from the plasma membrane, used to synthesize signalling molecules, notably the eicosanoids. The exchange of DHA between the brain and the plasma has been calculated as 4–5 mg/day. Thus, adequate dietary omega-3 is crucial and when dietary DHA intake is inadequate DHA sparing mechanisms are recruited, which lead to an increase in arachidonic acid and derivatives increasing inflammation in the brain. It has been shown, in animal models, that brain ageing is associated with a decrease in DHA and that this loss can be prevented by dietary supplementation.

Epidemiological studies have found a protective association between n-3 fatty acids, fish intake, and risk of AD, although findings are not consistent. Importantly, in randomized controlled trials the effects generated by n-3 fatty acid supplementation on cognition depend on the age and health status of the individuals tested, and while n-3 supplementation did not improve cognition in healthy elderly individuals, it did exert beneficial effects in people suffering from mild cognitive deficits, but not those affected by AD. However, it is also important to bear in mind that there are a number of factors which may confound the effects found in studies on n-3 supplementation and cognition: (1) the dose; (2) the type of n-3 PUFA used; (3) the duration of the intervention; (4) competition with the n-6 in the diet; (5) the n-3 fatty acid status of those tested.

In rodents, dietary supplementation with n-3 fatty acids can prevent neuro-inflammation, with the n-3:n-6 ratio being inversely associated with cognitive decline and hippocampal inflammation. Neurogenesis, which decreases with age, is also increased by n-3 fatty acid supplementation. Transgenic *Fat1* mice, which synthesize high levels of DHA, have increased neurogenesis and enhanced spatial learning abilities. Thus, despite some contradictory evidence in human supplementation studies, studies in rodents suggest that n-3 fatty acids play a role in preventing cognitive decline via defined mechanisms.

Oleic acid

Oleic acid is a monounsaturated fatty acids (MUFA) implicated in the prevention of the development of cognitive decline and dementia. Indeed, human studies showed that the consumption of extra-virgin olive oil, which is rich in oleic acid, increased verbal fluency and episodic memory and decreased the incidence of mild cognitive impairment. However, polyphenols and bioactives which are present in high concentrations in extra-virgin olive oil have both antioxidant and anti-inflammatory effects and thus have additional effects in mediating the benefits of extra-virgin olive oil on cognition. Nevertheless, oleic acid itself has been shown to be anti-inflammatory and may act synergistically with polyphenols.

Medium-chain fatty acids

Medium-chain fatty acids, despite being saturated, have a markedly different metabolism compared to long-chain saturated fats. Medium-chain fatty acids are oxidized more effectively, thus preventing the lipotoxic effects typical of an excess of long-chain saturated fatty acids. The consumption of medium-chain triglycerides exerts protective effects on cognition, confirmed in frail elderly individuals. The positive effects of medium-chain triglycerides on cognition may be via the modulation of lipid metabolism and promoting ketosis.

Carbohydrates

The intake of high-glycaemic index carbohydrates is associated with cognitive decline. In the elderly the risk of mild cognitive decline or dementia was highest in those who consumed the largest percentage of energy in the form of carbohydrates. A high sugar intake also plays a major role in the pathogenesis of impaired glucose tolerance and insulin insensitivity, which are associated with cognitive impairment. Not only do refined sugars and saturated fats in isolation contribute to impaired cognitive function, but their combination appears to be particularly detrimental to brain health, with the consumption of processed foods high in sugar and fat being associated with poor cognitive performance. These effects are seen in normal weight individuals with no history of diabetes or cardiovascular disease, suggesting that the diet itself is responsible for the cognitive impairment rather than obesity and associated conditions.

However, non-digestible carbohydrates including dietary fibre have positive effects on metabolic health, inflammation, and cognition. Non-digestible carbohydrates which modulate the gut microbiota are termed prebiotic and exert a beneficial effect on the central nervous system. Non-digestible carbohydrates escape digestion in the upper gut, reaching the colon undigested where they are broken down by the gut microbiota, leading to the synthesis of gas, short-chain fatty acids (SCFAs), and other metabolites. The effect of prebiotics on brain health may occur via SCFAs, modulation of the immune system of the host, and/or reduction of pathogenic bacteria. This latter mechanism is strengthened by the influence of the microbiota in the pathogenies of AD, via the activation of pro-inflammatory responses induced by LPS from gram-negative bacteria. An alternative pathway underlaying the effects of non-digestible carbohydrates may be via the release of the gut-derived peptides such as glucagon-like peptide-1 (GLP-1) and peptide YY (PYY).

Polyphenols and food bioactives

Polyphenols are secondary metabolites found in plants, particularly highly coloured fruit and vegetables such as blueberries, but are also found in tea, spices, herbs, and olive oil. Flavonoids are the most highly studied polyphenols for their effect on cognition. The influence of polyphenols on the brain depends on their bioavailability which in turn depends on their metabolism and their ability to cross the blood–brain barrier. Polyphenols are absorbed in the small intestine after deglycosylation. Once absorbed, polyphenols undergo biotransformation in the form of glucuronidation, methylation, thiol conjugation reactions, and sulfation. The gut microbiota also metabolizes polyphenols which in turn affects the composition of the gut microbiota. Finally, polyphenols or their metabolites need to cross the blood–brain barrier to influence cognition directly (Figure 24.5). However, the concentration of polyphenols in the brain has been shown to be relatively low at around 1 nmol/g tissue, thus, their action is unclear. However, the putative antioxidant effects exerted by polyphenols on the brain may be indirect and dependent on their action in the periphery or the regulation of the expression of genes with antioxidant function.

High levels of flavonoid intake have been found to decrease the risk of developing AD and to slow down cognitive decline in older subjects, improve language and verbal memory and cognitive performance during normal ageing. Specific polyphenols particularly catechins, theaflavins, and flavonols are associated with better episodic memory. The intake of fruit and vegetables, two good sources of polyphenols, lowers the risk of dementia but only in APOε4 allele non-carriers indicating that APOε4 allele can override the brain response to polyphenols. The consumption of grape or blueberry juice in older adults resulted in an improvement in memory performance. However, executive function was negatively associated with the intake of dihydrochalcones, catechins, proanthocyanidins, and

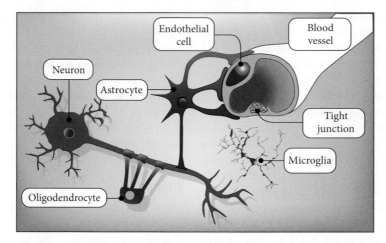

FIGURE 24.5 Major cell types in the brain in relation to the blood–brain barrier which is composed of specialized endothelial cells connected by tight junctions. Astrocyte end-feet also play a role in the formation of tight junctions.

flavonols and other studies have failed to detect an association between polyphenol intake and the risk of AD and dementia.

Cocoa and cocoa-derived products are good sources of flavonoids and the consumption of flavanol-rich cocoa and flavanol enriched drinks induce an increase in blood flow in the brain. Chocolate intake is associated with reduced cognitive decline, and a diet high in cocoa flavanol has been reported to enhance dentate gyrus function assessed by MRI.

A polyphenol that has provoked much interest is the stilbenid, resveratrol, which is a found in grapes, red wine, peanuts, blueberries, and the roots of Japanese knotweed. Resveratol is a potent antioxidant and anti-inflammatory which activates surtuin1 (SIRT1), an enzyme that deacetylates transcription factors and is a key component of memory and cognition suppressed in AD. However, the bioavailability of orally ingested resveratrol is low due to its rapid metabolism. In rodent models resveratol has been shown to reverse microglial activation, decrease the production of the proinflammatory cytokines IL-6, IL-1β, and TNFα, and reduce amyloid–β plaque formation. In humans there is evidence for the effect of resveratrol on the prevention of cognitive decline but bioavailability is a limiting factor. Besides improving glycaemic control and cardiovascular function, other mechanisms may be responsible for the effects of polyphenols on brain health and cognition, including neuronal protection against oxidative stress and inflammation. Apart from polyphenols, other food bioactives, such as phylloquinone, lutein, nitrate, folate, α-tocopherol, and kaempferol found in green leafy vegetables may slow down age-related cognitive decline, Gingko biloba extract has also been linked to brain health. Also, the dietary xanthophyls: lutein and zeaxanthin are highly concentrated in the macula lutea of the eye, giving it a typical yellow colour. Because

of this it is speculated that lutein and zeaxanthin play an important role in eye health and supplementation is recommended to maintain eye health during aging.

Vitamins

While ageing is generally characterized by a decrease in energy requirements, micronutrient requirements remain high. Despite the risk of nutritional deficiency when consuming a Western diet where the intake of fruit, vegetables, and whole grain products is low, older people in the UK are more likely than younger adults to achieve the recommended intakes of micronutrients through improved diet and supplement use. However, vitamin B_{12} deficiency is more common in the elderly than in younger age groups; this may be linked to the increased use of antacid medications and atrophic gastritis both of which interfere with vitamin B_{12} absorption.

Vitamins are key in cognitive function with low intakes of vitamin B_{12}, folate, riboflavin, and vitamin C all associated with poor memory and nonverbal abstract thinking tests in elderly subjects. The B vitamins are essential for energy production in the brain as cofactors in the glycolytic pathway and the citric acid cycle, and the production of adenosine triphosphate from glucose.

Folate, vitamins B_{12}, and B_6

The plasma concentration of homocysteine, regulated primarily by folate, vitamin $B_{12,}$ and vitamin B_6, is negatively linked to cognitive function. Increased risk of dementia is associated with high homocysteine and/or low vitamin B status. It has not been established that this association is causal and whether it is dependent on vitamin B status or not. Folate, and vitamins B_{12} and B_6 are important for one carbon metabolism and the generation of methionine from homocysteine. Methionine is converted into S-adenosylmethionine which is considered the universal methyl donor, important for DNA methylation, a key mechanism in epigenetic modification, and for the synthesis of neurotransmitters, myelin, and phospholipids. Methylation is also important in the regulation of amyloid-β levels as well as protein phosphatase 2A (PP2A) Bα subunit expression which is associated with tau phosphorylation.

Associations between low levels of circulating folate, vitamins B_{12} and B_6 with high levels of homocysteine linked to cognitive dysfunction, argue for increasing the intake of these vitamin (Smith et al. 2010). Indeed, supplementation with folic acid not only lowers plasma homocysteine but has also been shown to improve memory, information processing, and sensorimotor speed. B vitamin supplementation has also been shown to improve cognitive scores and slow down brain atrophy in grey matter regions specifically vulnerable to AD (Douaud et al. 2013). High levels of plasma homocysteine may be linked to impaired cognitive function via decreased cerebrovascular circulation, direct neurotoxicity of homocysteine, or the inhibition of methylation reactions and oxidative damage induced by homocysteine. However, these hypotheses rely on animal and cell studies, while in humans only small increases in circulating homocysteine are associated with raised risk of cognitive dysfunction and levels of homocysteine detected in the brain would not be neurotoxic. A more relevant mechanism is the necessity of vitamins B_6 and B_{12} and folate for one carbon metabolism and the importance of methylation reactions to brain function, through synthesis of neurotransmitters and key brain proteins.

Vitamins A, C, and E

Oxidative stress is involved in the pathogenesis of age-related cognitive decline and AD. Thus, the intake of the antioxidant vitamins A, E, and C might be expected to play a beneficial role in brain ageing. There is evidence for a protective role of vitamin E, with a higher intake associated with lower risk of AD and cognitive decline but findings are inconsistent. In prospective cohort studies focusing on vitamin A and C no associations with changes in cognitive functions have been found.

One of the difficulties of interpreting findings from studies of single antioxidant nutrients and cognitive function is that such studies do not take account of the considerable synergy between antioxidant nutrient function. Furthermore, they can rarely reproduce the complex profile of different antioxidants in the human diet.

Vitamin D

Vitamin D can be either synthesized during skin exposure to sunlight or consumed in the diet. Fatty fish is one of the few good dietary sources of vitamin D, but it also contains high levels of EPA and DHA which makes it difficult to distinguish the effect of vitamin D alone. Despite vitamin D deficiency prevalence being higher in the younger, vitamin D deficiency remains a problem in the elderly and patients affected by AD have lower circulating levels of vitamin D than healthy older people. Some studies have reported that higher vitamin D intake is associated with a decreased risk of dementia, but causality remains to be determined. The effect of vitamin D on brain function is an area of growing interest but the evidence base is currently limited (https://www.gov.uk/government/publications/sacn-vitamin-d-and-health-report).

Vitamin K

There are plausible mechanisms for a role for vitamin K in cognition. The brain contains several vitamin K-dependent proteins, one of which, growth arrest-specific 6 (Gas6), is expressed in the hippocampus of adult rats, and is particularly important in promoting the survival of both neurons and microglia. Also, vitamin K is involved in sphingolipid metabolism, which is important for memory formation. Cross-sectional studies have reported that a higher plasma concentration of vitamin K is associated with better performance in episodic memory tasks but generally evidence for a role for vitamin K in cognition is limited.

Trace elements

Zinc

Zinc is important in maintaining metabolic homeostasis in the elderly and is pivotal in protection against oxidative stress, being an integral part of the antioxidant enzyme superoxide dismutase. It is also important in proteins involved in DNA-damage repair such as p53. Thus, zinc deficiency might negatively affect antioxidant defence and compromise DNA repair and cause neuronal death. Zinc deficiency has been found in AD patients when compared to age-matched controls and AD patients have shown improvements with zinc supplementation. However, a more recent meta-analysis of studies linking zinc status to cognitive function, in normal aging, concluded that the findings of available studies to date were inconsistent (Warthon-Medina et al. 2015).

Selenium

Selenium is a trace element important in protection against oxidative stress, being a cofactor for the antioxidant enzyme glutathione peroxidase and a component of selenoprotein P. The importance of selenium in the brain is highlighted by the fact that in selenium deficiency the brain is the last organ to be depleted. Importantly, low levels of selenium have been associated with lower cognitive scores and increased cognitive decline.

Iron

Iron plays an important role in oxygen transport and storage, and any deficiency may lead to cerebral hypoxia and cognitive decline via decreased levels of oxygen reaching the brain. Indeed, studies have revealed that iron deficiency, independently of anaemia, increases the rate of cognitive decline. The ease with which iron deficiency can be treated emphasizes the need to monitor blood iron levels in the elderly.

Copper

Copper has an important role as a cofactor of many redox and antioxidant enzymes such as superoxide dismutase. However, it is toxic at high concentrations and causes oxidative cell damage and death. Impaired copper homeostasis has been reported to promote the formation of amyloid beta and neurofibrillary tangles with high copper levels associated with cognitive decline and AD.

Advanced glycation end oroducts (AGEs)

Advanced glycation end products (AGEs) result from the Maillard reaction; a chemical browning reaction between protein and sugar/fat when treated with dry heat and are present in high concentrations in processed food. Dietary AGEs have relatively recently been implicated in contributing to the development of dementia (Cai et al. 2014) with a time dependent decrease in insulin sensitivity and cognition shown in older individuals with high baseline AGEs. Dietary AGEs promote oxidative stress and have also been implicated in the development of type 2 diabetes and the ageing process in general via the chronic suppression of SIRT1. Conversely, studies restricting the intake of AGEs in the diets of diabetics have shown lower levels of inflammation and oxidative stress.

AGEs are also produced naturally in the body, as part of normal metabolism and the rate of formation is greatly enhanced in diabetes where circulating glucose is high. The Western diet not only contains large amounts of AGEs but is also high in fat and sugar, particularly fructose which increases AGE formation independently of glucose by increasing α-oxaldehyde levels. As part of normal ageing, irreversible AGE accumulation occurs in tissues containing long-lived proteins, such as collagen in the extracellular matrix, crystalins in the eye lens, and the basement membrane in the kidney. Thus, raised AGE levels in the body are a symptom of ageing but are also thought to be responsible for the aging process, particularly as their formation alters the structure and function of proteins, producing the hallmark features of ageing such as skin wrinkling and cataracts.

Plausible mechanisms underlying the effect of diet on cognitive ageing

(See Figure 24.6.)

Epigenetics

Nutrition can regulate gene expression via epigenetic mechanisms (see Chapter 29) including DNA methylation, histone modifications, and non-coding RNA. some of which have been shown to be involved in ageing and cognitive decline (Dauncey 2014). Many epigenetic changes can be reversed, pointing to the importance of dietary modification in the prevention and possible treatment of cognitive decline. Methyl donors such as folate, vitamins B_6 and B_{12}, choline and methionine are particularly important in this regard.

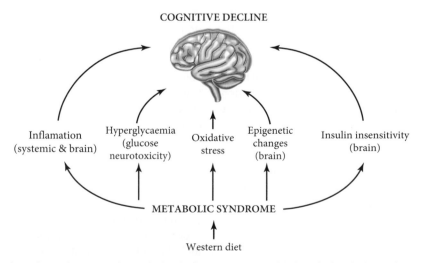

FIGURE 24.6 Pathophysiological states and metabolic dysfunction currently identified as linking the Western diet with the development of cognitive decline and Alzheimer's disease.

Inflammation

Systemic inflammation is a hallmark of obesity which is associated with deficits in memory and executive function. Studies in both humans and animal models show that the brain is particularly sensitive to inflammation. Regions involved in learning and memory, such as the cortex and the hippocampus, have high levels of cytokine receptors and the cytokines IL-1β and IL-6 have been shown to impair cognition and memory. Brain inflammation is a key feature in AD but whether inflammation precedes disease onset and is causative or not is questionable. Memory impairments and activation of microglia and astrocytes together with increased expression of pro-inflammatory mediators such as TNFα, IL-6, IL-1β in the hippocampus have been shown to be induced in rodents fed a high-fat diet, supporting a role for neuroinflammation in diet-induced neurodegenerative diseases and impairment of cognitive function.

Insulin insensitivity

Insulin insensitivity may predispose to cognitive impairment. In rodents, a high-fat diet not only promotes inflammation, but induces insulin insensitivity. Insulin crosses the blood–brain barrier (Figure 24.5) in a regulated and saturable fashion. Insulin receptors are widely expressed in the central nervous system particularly in the neurons of the temporal lobe of the cortex, hippocampus, hypothalamus, and amygdala with the temporal lobe and the hippocampus being the main sites of neurodegeneration in AD. Insulin is involved in learning and memory where it can improve working memory in both humans and animals, with intra-hippocampal insulin enhancing hippocampal-dependent spatial working memory. This role is highlighted by the fact that impaired insulin signalling correlates with decreased cognitive ability and AD and the fact that insulin signalling diminishes with AD progression. The important role that the loss of insulin signalling has in AD has led to AD being termed *type 3 diabetes*. Loss of central insulin signalling and hyperinsulinemia appear to increase amyloid-β oligomerization and promote tau hyperphosphorylation, providing a further relationship between impaired insulin signalling and neurodegenerative diseases.

Together with its role in regulation of metabolic pathways, insulin also acts as a vasodilator, a function which is impaired in individuals with insulin insensitivity. Cerebral hypoperfusion is associated with lower cognitive functions, dementia, and reduced brain volume.

Oxidative stress

In animal models chronic elevation of oxidative stress by diet accelerates cognitive decline. In rodents a high-fat diet induces oxidative stress, leads to a decrease in brain-derived neurotrophic factor (BDNF), and impaired cognitive performance, but supplementing the high-fat diet with vitamin E reverses these effects. The amount of fat in the diet appears to influence cognitive outcome as well the level of oxidative stress; a diet containing 40 per cent of energy from fat, mainly in the form of lard, was compared to one with 60 per cent fat. Only the 60 per cent diet induced oxidative damage in the hippocampus and impaired behavioural performance. However, the 60 per cent fat diet also produced a more marked increase in body weight and adiposity and consequently the metabolic dysfunctions and systemic inflammation that arise because of obesity may then mediate the increased hippocampal protein oxidation and cognitive decline.

Hyperglycemia and glucose neurotoxicity

The brain has a high requirement for glucose; 20 per cent of the body's glucose requirement is by the brain, even though the brain constitutes only 2 per cent of body weight. Neurons have the highest energy demand, requiring glucose for basal energy, the generation of action potentials and post-synaptic potentials, and the biosynthesis of neurotransmitters. Glucose is the preferred source of fuel for the brain, but it can be supplemented with lactate during periods of intense physical exercise and ketone bodies during starvation. Consequently, this makes the brain sensitive to any factors influencing its ability to take up and utilize glucose appropriately. The exact mechanisms of glucose uptake and utilization by the brain remain controversial. Glucose freely crosses the blood–brain barrier to access neurons mediated by the non-insulin dependent glucose transporter GLUT1, which is highly expressed in brain microvessels. Astrocytes were thought to play a key role in supplying energy to neurons via a process known as the astrocyte-neuron lactate shuttle (ANLS). In this process, the astrocyte is the primary target of blood-borne glucose to the brain with the astrocyte metabolizing glucose to provide lactate to the neuron. However, this mechanism has been challenged and by modelling the kinetics and cellular concentrations of neuronal and glial glucose and lactate transporters, it has been concluded that under normal conditions lactate shuttles from neurons to astrocytes, emphasizing the predominant role of glucose in neuronal metabolism.

The brain has not traditionally been considered as a target of damage associated with diabetic hyperglycaemia, but recent evidence has shown that the brain is sensitive to high levels of circulating glucose. As detailed above, neurons have different metabolic pathways to call upon to supplement glucose if they are faced with hypoglycaemia, including utilizing lactate and ketone bodies, pyruvate, glutamate, glutamine, and aspartate. However, the brain is not similarly protected against the effects of hyperglycaemia, which is usual in uncontrolled type 2 diabetes.

There is a strong association between the duration of diabetes and impaired cognitive performance suggesting that brain function may be impaired by prolonged elevations of blood glucose. Animal studies have shown that there is no protective effect of the blood–brain barrier against hyperglycaemia and that, as the brain tissue may be exposed to elevated levels of glucose, it is subject to the same long-term adverse effects of hyperglycaemia seen in peripheral tissues.

Excess glucose is neurotoxic to the brain via the polyol pathway, with hyperglycaemia increasing sorbitol and inositol levels in the cerebral cortex and hippocampus. This not only leads to an increase in sorbitol, but also results in a decrease in taurine which is key in maintaining tissue osmotic balance. The polyol pathway represents one of the first hypothesized links between hyperglycaemia and nerve dysfunction. The deleterious effects of activating this metabolic pathway are several. First, sorbitol is a strongly hydrophilic polyhydroxylated alcohol and as such does not diffuse readily through cell membranes but accumulates in the cell with effects on the cellular osmotic balance. Also, fructose produced by the polyol pathway can become phosphorylated to fructose-3-phosphate, which is broken down into 3-deoxyglucosone and methylglyoxal; both compounds are powerful glycating agents which contribute to the formation of damaging AGEs in the cell. Activation of the polyol pathway utilizes NADPH, which results in less of the cofactor being available for the synthesis of glutathione reductase, critical for the maintenance of the intracellular pool of reduced glutathione, potentially decreasing the ability of cells to respond to oxidative stress. In addition, the usage of NAD by sorbitol dehydrogenase leads to an increased ratio of $NADH/NAD^+$, which produces 'pseudohypoxia' linked to several metabolic and cell signalling changes. Thus, excess glucose, by activating the polyol pathway, can cause damaging changes to intracellular tonicity, generate AGE precursors, and subject cells to oxidative stress through both decreased antioxidant defence and increased generation of oxidant species, thereby initiating cellular damage via several different routes.

KEY POINTS: NUTRITION AND THE NERVOUS SYSTEM IN THE ELDERLY

- The Western diet is associated with increased risk of cognitive decline with ageing.
- The Mediterranean diet appears to be protective against age-related cognitive decline.
- Thus, a whole-diet approach is required to address cognitive ageing.
- Studies looking at single nutrients seldom show defined outcomes.
- Mechanistically, hyperglycaemia and insulin insensitivity appear to be major drivers of cognitive decline.

ACKNOWLEDGEMENTS

The authors acknowledge the contributions of Sally Grantham-McGregor to early versions of this chapter. Lynda Williams was funded by Rural and Environment Science and Analytical Services Division (RESAS) of the Scottish Government.

 Be sure to test your understanding of this chapter by attempting multiple choice questions. See the Further Reading list for additional material relevant to this chapter.

REFERENCES

(For additional references see Chapter 24 Further Reading online.)

Cai W, Uribarri J, Zhu L et al. (2014) Oral glycotoxins are a modifiable cause of dementia and the metabolic syndrome in mice and humans. *Proc Nat Acad of Sciences of the USA* **111**(13), 4940–4945.

Christian P, Murray-Kolb L E, Tielsch J M et al. (2014) Associations between preterm birth, small-for-gestational age, and neonatal morbidity and cognitive function ampng school-age children in Nepal. *BMC Pediatr.* **14**(58). doi: 10.1186/1471-2431-14-58. PMID: 24575933; PMCID: PMC3974060.

Dauncey M J (2014) Nutrition, the brain and cognitive decline: Insights from epigenetics. *Eur J Clin Nutr* **68**(11), 1179–1185. doi:10.1038/ejcn.2014.173.

Douaud G, Refsum H, de Jager C A et al. (2013). Preventing Alzheimer's disease-related gray matter atrophy by B-vitamin treatment. *Proc Nat Acad of Sciences of the USA* **110**(23), 9523–9528.

Ferri C P, Prince M, Brayne C et al. (2005) Global prevalence of dementia: A Delphi consensus study. *Lancet* **366**(9503), 2112–2117.

Franke K, Van den Bergh B R H, de Rooij S R et al. (2020). Effects of maternal stress and nutrient restriction during gestation on offspring neuroanatomy in humans. *Neurosci & Biobehav Reviews* **117**, 5–25, ISSN 0149-7634.

Gogtay N, Ordonez A, Herman D H et al. (2007) Dynamic mapping of cortical development before and after the onset of pediatric bipolar illness. *J Child Psychol Psychiatry* **48**(9), 852–862. doi: 10.1111/j.1469-7610.2007.01747.x. PMID: 17714370.

Hallböök T, Ji S, Maudsley S, & Martin B (2012) The effects of the ketogenic diet on behavior and cognition. *Epilepsy Res.* **100**(3), 304–309.

Koletzko B, Boey C C M, Campoy C et al. (2014) Current information and Asian perspectives on long-chain polyunsaturated fatty acids in pregnancy, lactation, and infancy: Systematic review and practice recommendations from an Early Nutrition Academy Workshop. *Ann Nutr Metab* **65**, 49–80. doi: 10.1159/000365767.

Lamport D J, Saunders C, Butler L T et al. (2014) Fruits, vegetables, 100% juices, and cognitive function. *Nutr Reviews* **72**(12), 774–789.

Larson L M & Yousafzai A K (2017) A meta-analysis of nutrition interventions on mental development of children under-two in low- and middle-income countries. *Matern Child Nutr.* **13**(1), e12229. doi: 10.1111/mcn.12229. Epub 2015 Nov 26. PMID: 26607403; PMCID: PMC6866072.

Liberato S, Singh G, Mulholland K (2013) Effects of protein energy supplementation during pregnancy on foetal growth: A review of the literature focusing on contextual factors. *Fd & Nutr Resrch* [S.l.], ISSN 1654-661X. Available at: <http://www.foodandnutritionresearch.net/index.php/fnr/article/view/20499>.

Mattson M P (2010) The impact of dietary energy intake on cognitive aging. *Front. Aging Neurosci.* **2**, 5. doi: 10.3389/neuro.24.005.2010

Osendarp S J, Murray-Kolb L E, Black M M (2010) Case study on iron in mental development: In memory of John Beard (1947–2009). *Nutr Reviews* **68**(Suppl 1), S48–S52. doi: 10.1111/j.1753-4887.2010.00331.x. Review. PubMed PMID: 20946368; PubMed Central PMCID: PMC3137944.

Peet M & Stokes C (2005) Omega-3 fatty acids in the treatment of psychiatric disorders. *Drugs* **65**(8), 1051–1059.

Prado E L & Dewey K G (2014) Nutrition and brain development in early life. *Nutr Reviews* **72**(4), 267–284. doi: 10.1111/nure.12102. Epub 2014 Mar 28. Review. PubMed PMID: 24684384

Querfurth H W & LaFerla FM (2010) Alzheimer's disease. *New Eng J of Med* **362**(4), 329–344.

Scarmeas N, Stern Y, Mayeux R et al. (2009) Mediterranean diet and mild cognitive impairment. *Archives of Neurol* **66** (2), 216–225.

Smith A D, Smith S M, de Jager C A et al. (2010). Homocysteine-lowering by B vitamins slows the rate of accelerated brain atrophy in mild cognitive impairment: A randomized controlled trial. *PloS One* **5**(9), e12244.

Stevens GA, Beal T, Mbuya MNN, Luo H, Neufeld LM; Global Micronutrient Deficiencies Research Group. Micronutrient deficiencies among preschool-aged children and women of reproductive age worldwide: a pooled analysis of individual-level data from population-representative surveys. Lancet Glob Health. 2022 Nov;10(11):e1590-e1599. doi:10.1016/S2214-109X(22)00367-9. PMID: 36240826.

Sudfeld C R, McCoy D C, Danaei G et al. (2015) Linear growth and child development in low- and middle-income countries: A meta-analysis. *Pediatrics* **135**(5), e1266–e1275.

Victora C G, Christian P, Vidaletti L P et al. (2021). Revisiting maternal and child undernutrition in low-income countries and middle-income countries: Variable progress towards an unfinished agenda. *Lancet* **397**, 1388–1399.

Walker S P, Wachs T D, Grantham-McGregor S et al. (2011) Inequality in early childhood: Risk and protective factors for early child development. *Lancet* **378**(9799), 1325–1338.

Warthon-Medina M, Moran V H, Stammers A L et al. (2015) Zinc intake, status and indices of cognitive function in adults and children: A systematic review and meta-analysis. *Eur J Clin Nutr* **69**(6), 649–661. doi: 10.1038/ejcn.2015.60. Epub 2015 Apr 29. Review. PubMed PMID: 25920424.

FURTHER READING

Black R E, Victora C G, Walker S P et al. (2013) Maternal and Child Nutrition Study Group: Maternal and child undernutrition and overweight in low-income and middle-income countries. *Lancet* **382**(9890), 427–451.

Brewer G J (2012) Copper excess, zinc deficiency, and cognition loss in Alzheimer's disease. *Biofactors* **38**(2), 107–113.

Cardoso B R, Roberts B R, Bush A I, et al. (2015) Selenium, selenoproteins and neurodegenerative diseases. *Metallomics* 7(8), 1213–1228.

Dineva M, Fishpool H, Rayman M P et al. 2020. Systematic review and meta-analysis of the effects of iodine supplementation on thyroid function and child neurodevelopment in mildly-to-moderately iodine-deficient pregnant women. **Am J Clin Nutr 112**(2), 389–412. doi: 10.1093/ajcn/nqaa071. PMID: 32320029.

Durga J, van Boxtel M P, Schouten E G, et al. (2007) Effect of 3-year folic acid supplementation on cognitive function in older adults in the FACIT trial: A randomised, double

blind, controlled trial. *Lancet* **369**(9557), 208–216. doi: 10.1016/S0140-6736(07)60109-3.

Grantham-McGregor S & Baker-Henningham H (2005) Review of evidence linking protein and energy to mental development. *Publ Hlth Nutr* **8**(7B), 1–12.

Larson L M, Phiri K S, Pasricha S. (2017). Iron and cognitive development: What is the evidence? *Ann Nutr Metab.* **71**(Suppl 3), 25–38. doi: 10.1159/000480742.

Murray E, Fernandes M, Fazel M et al. (2015) Differential effect of intrauterine growth restriction on childhood neurodevelopment: A systematic review. *BJOG.* **122**(8), 1062–1072.

Smith A D (2008) The worldwide challenge of the dementias: A role for B vitamins and homocysteine? *Food Nutr Bull* **29**(2 Suppl), S143–S172.

Walker S P, Wachs T D, Gardner J M et al. (2007) International Child Development Steering Group. Child development: Risk factors for adverse outcomes in developing countries. *Lancet* **369**(9556), 145–157. Review. PubMed PMID: 17223478.

25 Nutrition and the skeleton

Ian R. Reid

OBJECTIVES

By the end of this chapter you should be able to:

- understand bone remodelling
- understand the role of vitamin D and calcium nutrition in bone health
- appreciate the impact of body weight on bone mass
- understand the nature, pathogenesis, and management of osteoporosis, osteomalacia, and rickets
- understand what is known and not known regarding the role of other nutrients in bone health.

25.1 Introduction

The skeleton is the principal structural element of the body, necessary for all movement, for the protection of the internal organs, and for breathing. It plays a key role in the metabolism of the minerals calcium and phosphate, the former being critical in nerve and muscle function, and the latter in energy metabolism. Disease or dysfunction of the skeleton can result in pain, deformity, or fracture, which can impact on quality of life and on survival. As a result, maintenance of skeletal health across a wide variety of environments, lifestyles, and diets has had a high evolutionary priority.

For many years it has been widely believed that nutrition plays a key role in the maintenance of bone health. The demonstration, in the 1920s, that cod liver oil cured rickets suggested this, as did the later advocacy for calcium intake as an important determinant of bone mass, and of calcium supplements as key to the prevention and treatment of osteoporosis. This advocacy was substantially based on results from short-term calcium balance studies, since direct measurement of bone density was not available at that time. Curiously, it was already known in the 1960s and 1970s that fractures among the elderly were more common in high-calcium intake regions, such as Western Europe and North America, than they were in Africa and Asia. However, this was labelled as the 'calcium paradox' and set aside from threatening the calcium orthodoxy. In recent decades, it has become apparent that sunlight deprivation, rather than deficient diet, is the principal cause of vitamin D deficiency, and that calcium intake plays only a minor role in the genesis of osteoporosis or its management. However, public and professional interest in calcium and vitamin D metabolism and their impact on health remains high, so it is important that health professionals are well versed in this area. What has also emerged in recent decades is the key role of body weight as a determinant of bone density and of fracture risk, so nutrition other than calcium and vitamin D remains an important issue for bone health. A large number of other nutrients has been suggested to influence bone health, but critical evidence is lacking for many.

This chapter will review the structure of bone and the cellular events which provide remodelling and renewal of bone. It will then discuss the regulation of calcium metabolism and of the vitamin D endocrine system. A consideration of osteoporosis and osteomalacia, the two principal metabolic bone diseases, will then be undertaken, with a focus on the role of nutrition in each.

25.2 Bone structure and remodelling

Bone is a connective tissue consisting of two components: a protein scaffold (predominantly type I collagen) which determines a bone's architecture and provides tensile strength, reinforced by mineral plates which provide compressive strength (Figure 25.1). This mineral component of bone consists of hydroxyapatite crystals, which are formed from calcium and phosphate ions together

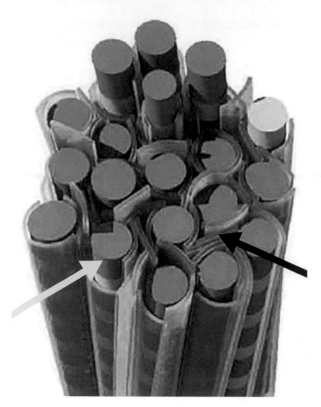

FIGURE 25.1 Model of ultrastructure of human cortical bone, based on studies with scanning transmission electron microscopy.

Collagen fibrils are shown as dark grey, as indicated by the white arrow. They are about 50 nm in diameter. Most of the mineral of human cortical bone is in the form of long, thin, polycrystalline plates, known as mineral lamellae (shown in lighter grey) which are curved to wrap closely around the collagen fibrils. In addition, stacks of closely packed uncurved lamellae occur between fibrils (red arrow).

Source: K Grandfield et al. (2018) Ultrastructure of bone: Hierarchical features from nanometer to micrometer scale revealed in focused ion beam sections in the TEM. *Calcif. Tissue Int.* **103**, 606–616, used with permission.

with smaller numbers of hydroxyl ions. The collagen scaffold of bone is produced by osteoblasts, and the mineral phase is subsequently laid down between the collagen fibrils, as long as calcium and phosphate concentrations in the extracellular fluid are normal and no mineralization inhibitors are present.

Regulation of mineralization is critically important, since mineral is an essential element of normal bone, but mineral deposition must be avoided in non-skeletal tissues where it will severely disrupt cell function and viability. Tight control of circulating levels of both calcium and phosphate is critical to maintaining this balance, as is the regulated activity of mineralization inhibitors such as pyrophosphate and mineralization-inhibiting proteins such as fetuin-A. Pyrophosphate is broken down by bone-specific alkaline phosphatase, thus allowing skeletal mineralization to proceed.

Like most other tissues, bone is in a constant state of renewal, with old tissue being removed and replaced. Bone removal is undertaken by osteoclasts which secrete both acid (to dissolve the mineral phase of bone) and proteolytic enzymes (to break down the collagen matrix). Bone is removed in small packets, which are then replaced by new bone produced by osteoblasts, as described above. The activities of both osteoblasts and osteoclasts are coordinated by osteocytes, cells of the osteoblast lineage which sit in lacunae scattered throughout bone. Osteocytes resemble neurones in that they have cell processes connecting adjacent osteocytes into a network which extends throughout bone and communicates with the cell surface. This three-dimensional osteocyte network extends throughout bone and senses skeletal loading. Bone turnover is regulated in response to this loading, as well as being influenced by circulating hormones and local cytokine production. Osteocytes regulate bone cell activity through the production of two proteins: RANKL, which stimulates osteoclast development, and sclerostin, which inhibits osteoblastic activity. Usually, about 20 per cent of the trabecular bone surface is undergoing remodelling at any time and, on average, the skeleton is replaced about every ten years as a result of this process.

Bone microarchitecture

Bundles of collagen fibres can be formed into an open meshwork of trabecular bone (also called spongy or cancellous bone), with the space between the trabeculae occupied by fat or by haematopoietic bone marrow. Alternatively, collagen fibres can be laid down without spaces to form denser cortical bone, also referred to as compact bone. This dense form of bone comprises about 80 per cent of skeletal mass. Figure 25.2 shows micrographs of both trabecular and cortical bone. Although trabecular bone accounts for only 20 per cent of total bone mass, it accounts for as much as 70 per cent of bone surface area and metabolic activity, which is mainly surfaced based. As a result, the trabecular envelope is much more dynamic, showing more rapid loss following menopause or glucocorticoid treatment, and more rapid increases in bone density following treatment of osteoporosis. Trabecular bone predominates in the ribs, ends of long bones and vertebral bodies, whereas cortical bone makes up the shafts of long bones.

25.3 Calcium metabolism

Calcium is present in biological systems as a positively charged ion. As well as being a major constituent of bone mineral (hydroxyapatite) it also plays critical roles in

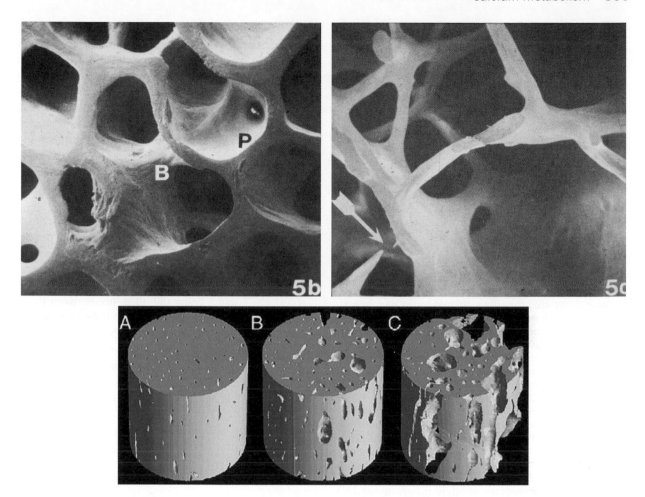

FIGURE 25.2 Trabecular and cortical bone, both normal and osteoporotic.
The upper panels show trabecular bone from the iliac crest in normal (left) and osteoporotic (right) individuals. The transition in architecture from exclusively plate-like structures to predominantly rod-like structures is noted and is the result of trabecular perforation by osteoclasts. Also note the osteoclast lacuna on the surface of the rod in the middle of the right-hand micrograph where perforation is imminent, and the wasting bone spicules in the lower left of the picture where a rod has previously been perforated. The lower image demonstrates the normal loss of cortical bone in women aged 20, 61, and 87 years, respectively.
Source: D W Dempster et al. (1986) A simple method for correlative light and scanning electron microscopy of human iliac crest bone biopsies: Qualitative observations in normal and osteoporotic subjects. *J Bone Miner Res* **1**, 15–21; D M Cooper et al. (2007) Age-dependent change in the 3D structure of cortical porosity at the human femoral midshaft. *Bone,* **40,** 957–965. Copyright © 1986 ASBMR. With permission from John Wiley.

neuromuscular and cardiac function, intracellular signalling, coagulation, and as a cofactor for some enzymes. Since these functions are critical to survival, extracellular calcium concentrations are tightly regulated so that stable levels are maintained in the face of widely varying intakes and losses (such as lactation). Sometimes the skeleton acts as a reservoir that can be drawn on to maintain this stability.

Total body calcium in the adult human is about 1 kg, 99 per cent in the skeleton in the form of hydroxyapatite. The remaining 1 per cent is outside of bone, mainly in the extracellular fluid (at a concentration of ~1.2 mmol/L). The ionized calcium equilibrates between plasma and the extracellular fluid, but in plasma about the same amount

of calcium again is bound to circulating albumin. The remaining 10 per cent of plasma calcium is complexed with anions, including phosphate and citrate, producing a total calcium concentration in plasma of 2.2–2.5 mmol/L. The ionized (or free) calcium is physiologically active and is under tight homeostatic control, as detailed in Figure 25.3.

The regulation of intestinal calcium absorption is a key part of this homeostatic control. Dietary calcium intake varies widely, from 100 mg/day in some parts of Africa up to more than 2000 mg/day in individuals with high intakes of dairy products. In Western countries, typical adult intakes are 500–1000 mg/day, and in Asian countries 200–500 mg/day, but increasing in recent decades as

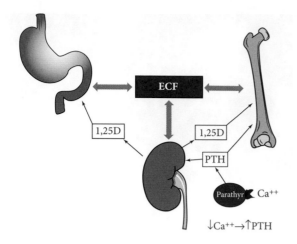

FIGURE 25.3 A simplified schema of calcium homeostasis. Three organs exchange calcium with the extracellular fluid (ECF): the gastrointestinal tract, bone, and the kidneys. These fluxes are primarily regulated by the parathyroid glands, which secrete parathyroid hormone (PTH) in response to reduced ambient levels of calcium. Calcium binding to calcium-sensing receptors on the surface of parathyroid chief cells reduces PTH secretion. PTH acts directly on bone to stimulate bone resorption, and on the kidneys, to promote renal reabsorption of calcium and the production of 1,25-dihydroxyvitamin D (1,25D). 1,25D primarily regulates calcium absorption in the proximal small intestine, but also stimulates bone resorption. Thus, PTH and 1,25D act in concert to maintain circulating calcium concentrations within a narrow physiological range. PTH also stimulates secretion of the phosphaturic hormone fibroblast growth factor-23 (FGF23) from osteoblasts and osteocytes, which in turn reduces 1,25D formation and PTH secretion, providing a negative feedback loop to this sequence of events.
Source: Copyright I R Reid, used with permission.

dairy products become more widely consumed. Between 10 per cent and 60 per cent of dietary calcium intake is absorbed, with a typical value of about 20 per cent in healthy adults taking a Western diet. At low calcium intakes, most absorption is in the upper small intestine via a cell-mediated pathway regulated by 1,25-dihydroxyvitamin D. Calcium enters enterocytes through an apical calcium channel, is transported across the cell by the protein calbindin-D9k, and is then extruded across the basolateral membrane by a calcium ATPase. When calcium intakes are high, this pathway is down-regulated and most calcium absorption takes place by passive diffusion between the enterocytic cells. Thus, calcium absorption efficiency is inversely related to calcium intake and its formal measurement of absorption is not usually helpful in clinical practice.

A tiny fraction of non-bone calcium is intracellular, being present at concentrations of about 100 nmol/L, almost 10,000-fold lower than in the extracellular fluid. The steep calcium gradient across cell membranes is maintained by ATP-dependent calcium pumps working in concert with calcium channels and sodium–calcium exchangers. The trans-membrane difference in calcium is critical to the electrical polarity of the membrane, on which the function of the nervous system and muscle depends. Calcium ions can also act as intracellular signals, with calcium release from intracellular stores or its entry through the cell membrane regulating cell functions, such as hormone release.

Dietary calcium requirement

Uncertainty remains regarding the optimal calcium intake for bone health. A detailed review of this has recently been published (Reid & Bristow 2020). Early studies used the calcium balance technique, which involves measuring calcium intake and total output (urine and faeces), the balance being the difference between these. More recently it has become possible to relate calcium intake to changes in regional or total bone mass over periods of years yielding a measurement of bone balance. This is more relevant than calcium balance when the task is to assess the impact of calcium intake on skeletal health and fracture risk.

Calcium balance studies from the 1930s to the 1950s concluded that calcium requirements could be met with daily intakes of a few hundred milligrams, and as low as 100 mg in men. Based on these findings, the World Health Organization concluded in 1962 that:

> Most apparently healthy people—throughout the world—develop and live satisfactorily on a dietary intake of calcium which lies between 300 mg and over 1000 mg a day. There is so far no convincing evidence that, in the absence of nutritional disorders and especially when the vitamin D status is adequate, an intake of calcium even below 300 mg or above 1000 mg a day is harmful. (Reid & Bristow 2020)

These conclusions were consistent with the observation that populations in Africa and Asia maintained good bone health and low fracture rates on intakes of about 300 mg/day. Calcium intakes of <200 mg/day can be associated with rickets developing in children even when vitamin D status is satisfactory.

In the 1970s it was demonstrated that calcium balance was directly related to calcium intake, and that postmenopausal women required an intake of almost 1500 mg/day to achieve zero balance. This analysis has proven to be mathematically flawed, since intake is used directly in the calculation of balance, so regressing one against the other will inevitably lead to the finding of a positive association. More recently, statistically appropriate methods have demonstrated 'that calcium balance was highly resistant to a change in calcium intake across a broad range of typical dietary calcium intakes (415–1740 mg/d)' and that calcium balance could be achieved with an average calcium intake of 741 mg/day (Hunt & Johnson 2007). A recent Chinese study found that zero calcium balance was achievable in adults aged <60 years with an intake of 300 mg/day.

The advent of bone densitometry has shown that there is little relationship between bone density and calcium intake in cross-sectional studies. Prospective bone density studies allow the estimation of bone balance over several years, and this is also unrelated to dietary calcium intake (Figure 25.4). The repeated demonstration of progressive bone density loss in older women with intakes up to 2 g/day directly contradicts the earlier balance studies which indicated that bone loss in older women could be completely prevented with adequate calcium intake, and this is consistent with the failure of calcium supplements to influence fracture rates (see section 'Calcium supplementation'). Observational studies also indicate that fracture incidence is not related to calcium intake (Bolland et al. 2015). Bone density studies, such as that in Figure 25.4, indicate that the calcium requirement in vitamin D-replete adults may be as little as 300 mg/day, consistent with the WHO conclusions from the 1960s. Aiming for greater than 500 mg/day represents a cautious interpretation of these data, making allowance for variation between individuals and for imprecision in assessing intake.

The absence of an international consensus regarding the optimal calcium intake of older adults is reflected in the range of current calcium recommendations: for example, 700 mg/d in the United Kingdom, 800 mg/d in China, 1000–1200 mg/in the United States, and 1300 mg/d in Australia and New Zealand.

Children might be more sensitive to adverse effects from very low calcium intakes, though observational studies are confounded by low protein intakes and low body weight. A study of calcium supplementation in Nigerian toddlers showed a 5 to 6 per cent increase in forearm bone density, but this benefit was lost within 12 months of cessation of the supplements. In Western populations, calcium intake does not significantly impact on bone mass accumulation or fracture in children.

Calcium supplementation

Many trials have assessed the effects of calcium supplements on bone turnover and density. Supplements result in an increase in serum calcium over the 4–8 hours after their ingestion. This results in partial suppression of parathyroid hormone, and bone resorption is also reduced. As a result, there is a benefit to bone density of about 1 per cent one year after the introduction of supplements. However, as shown in Figure 25.5, the gain is not progressive, so even after long-term use the difference between supplemented and non-supplemented individuals is still only about 1 per cent.

The major trials of calcium supplementation (i.e., recruiting >1000 participants) are set out in Table 25.1. The earliest of these (Chapuy 1992, 1994) showed clear prevention of hip and non-vertebral fractures, a finding which has not been reproduced in subsequent studies. This inconsistency has led to continuing controversy. There is similar inconsistency in meta-analyses, depending on whether these are restricted to community-dwelling

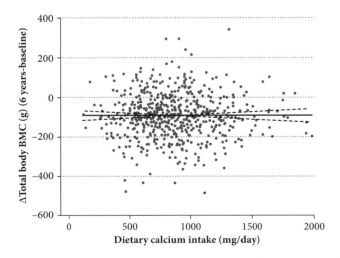

FIGURE 25.4 Absolute change (Δ) in total body bone mineral content (BMC) over six years in 698 osteopenic postmenopausal women not receiving bone-active medications, in relation to each woman's average calcium intake assessed at baseline, year three, and year six. The regression line (with 95 per cent CIs) for this relationship is shown (P = 0.99). *Source*: S M Bristow et al. (2019) Dietary calcium intake and bone loss over 6 years in osteopenic postmenopausal women. *J. Clin. Endocr.* **104**, 3576–3584, Copyright © 2019. Oxford University Press. Used with permission.

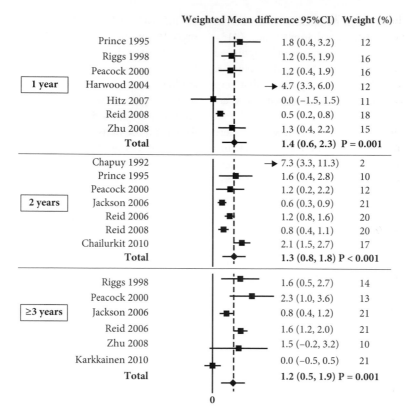

FIGURE 25.5 Random effects meta-analysis of effect of calcium supplements on percent change in total hip bone mineral density by duration of follow-up. There is a significant positive treatment effect at each time-point, but no evidence of greater effects with longer treatment periods.

Source: V Tai et al. (2015) Calcium intake and bone mineral density: Systematic review and meta-analysis. *BMJ* **351**, h4183. Copyright © 2015, BMJ, used with permission.

TABLE 25.1 Major trials of calcium ± vitamin D on fracture risk

Study	Setting	N	Age (years)	Calcium (mg/d)	Vitamin D (IU/d)	Duration (months)	Relative risk of fracture	
							Total	Hip
Chapuy 1992, 1994	Nursing home	3270	84 (6)	1200	800	36	0.83 (0.71, 0.97)	0.77 (0.62, 0.96)
RECORD 2005	Community	5292	77 (6)	1000	800	45	0.93 (0.82, 1.06)	1.10 (0.83, 1.47)
Porthouse 2005	Community	3314	77 (5)	1000	800	25	0.96 (0.70, 1.33)	0.71 (0.31, 1.64)
Women's Health Initiative 2006	Community	36282	62 (7)	1000	4000	84	0.97 (0.92, 1.03)	0.88 (0.72, 1.07)
Prince 2006	Community	1460	75 (3)	1200	0	60	0.87 (0.69, 1.10)	1.83 (0.68, 4.93)
Reid 2006	Community	1471	74 (4)	1000	0	60	0.92 (0.75, 1.14)	3.43 (1.27, 9.26)
Salovaara 2010	Community	3432	67 (2)	1000	800	36	0.83 (0.62, 1.11)	2.00 (0.37, 10.88)

Age data are mean (SD). Relative risks are given with 95 per cent confidence intervals. Where the confidence intervals do not include 1, data are bolded. Study participants were all female, except in the RECORD study, which was 15 per cent male.
Source: Data from Bolland et al. 2015. Copyright I R Reid, used with permission.

individuals, thus excluding the Chapuy study, or not. The different outcome in the Chapuy study is probably attributable to the presence of severe vitamin D deficiency in these women, and the fact that the intervention included vitamin D as well as calcium. Treatment of severe vitamin D deficiency has been shown to produce substantial increases in bone density, sufficient to reduce fracture risk. In contrast, calcium alone in non-deficient subjects produces only small changes in bone density, so it is not surprising that fracture incidence is not affected. As a result, recent meta-analyses published in major journals have found that calcium supplements do not impact on fracture incidence in community-dwelling adults (Bolland et al. 2015; Zhao et al. 2017), and this is reflected in position statements from the United Sates Preventive Services Task Force (Kahwati et al. 2018; Grossman et al. 2018) and the International Osteoporosis Foundation (Harvey et al. 2017).

Recently, an Australian study has reported reduction in fractures and falls in very elderly nursing home residents randomized to extra dairy products. Participants were vitamin D replete. The dairy intervention had a positive effect on body weight and circulating levels of insulin-like growth factor-1, which might be the origin of the anti-fracture efficacy since comparable doses of calcium alone do not produce this benefit.

The safety of calcium supplements has also become an issue of concern. Their gastrointestinal side-effects are well documented, and these have been shown in one trial to result in increased frequency of admissions to hospital. They increase the risk of kidney stones by about 20 per cent, and they may increase frequency of myocardial infarction by about the same amount, though this remains controversial. As a result, most authorities recommend that calcium is obtained from the diet rather than as supplements.

25.4 Vitamin D endocrine system

Vitamin D is not strictly speaking a vitamin, it is a prohormone, and can be formed by cutaneous synthesis. It is an organic molecule closely related to cholesterol which is formed in skin as a result of the direct action of ultraviolet light on 7-dehydrocholesterol. When sunlight exposure is sustained, other inactive compounds are produced, preventing the development of vitamin D intoxication. Melanin in skin absorbs ultraviolet light and so reduces vitamin D formation. In non-tropical areas, circulating 25-hydroxyvitamin D levels vary with the season, being lowest in late winter and peaking in late summer.

Animals produce vitamin D_3, also known as c(h)olecalciferol. In some plants, the closely related compound,

vitamin D_2 or ergolciferol, is produced by a similar process. This is widely used as a supplement. The terms vitamin D or calciferol refer to both vitamins D_3 and D_2. Naturally occurring dietary sources of vitamin D are limited to a few foods, including fish oils, egg yolk, butter, and liver. Vitamin D is fat soluble and absorbed in the upper small bowel. In the United States, fortification of food (particularly dairy products) with vitamin D is widespread, and use of vitamin D supplements is common in some countries.

Vitamin D and its metabolites circulate in plasma bound to a specific carrier protein, vitamin D binding protein. Medical conditions which change plasma protein concentrations (e.g., pregnancy, liver disease, malnutrition) change total levels of vitamin D and its metabolites without necessarily changing the free, physiologically active levels. Vitamin D itself is biologically inactive. It is converted to 25-hydroxyvitamin D in the liver, a largely unregulated process. 25-hydroxyvitamin D has a low affinity for the vitamin D receptor but is present in higher concentrations than other active metabolites so might contribute to regulation of target genes. The key step in vitamin D activation is the conversion of 25-hydroxyvitamin D to 1,25-dihydroxyvitamin D, which occurs predominantly in the kidney. This is a tightly regulated process, controlled by parathyroid hormone, calcitonin, 1,25-dihydroxyvitamin D itself, and phosphate, as depicted in Figure 25.6. Inactivation of both 25-hydroxyvitamin D and 1,25-dihydroxyvitamin D is mediated by 24-hydroxylation, a process principally regulated by local concentrations of 1,25-dihydroxyvitamin D. The effect of 1,25-dihydroxyvitamin D on catabolism of 25-hydroxyvitamin D explains the low levels of this metabolite in patients treated with calcitriol (i.e., 1,25-dihydroxyvitamin D) and in those with primary hyperparathyroidism.

The vitamin D endocrine system acts to maintain serum calcium concentrations to ensure that the roles of calcium in neuromuscular function, coagulation, and enzymatic action are maintained. It does this primarily through the stimulation of proximal intestinal calcium absorption by 1,25-dihydroxyvitamin D, and this metabolite also stimulates renal tubular reabsorption of calcium, the development and activity of bone resorbing cells (osteoclasts) and, in high concentrations, inhibition of bone mineralization. The latter two activities indicate that high concentrations of vitamin D metabolites can be deleterious to skeletal health, so it should not be regarded as a bone strengthening tonic.

Vitamin D receptors are expressed in many tissues, raising the possibility that it plays a key role in the regulation of skin, muscle, fat, immune function, neoplasia, and glucose metabolism. Vitamin D metabolites are used in the management of psoriasis, but do not have an established therapeutic role in other non-skeletal

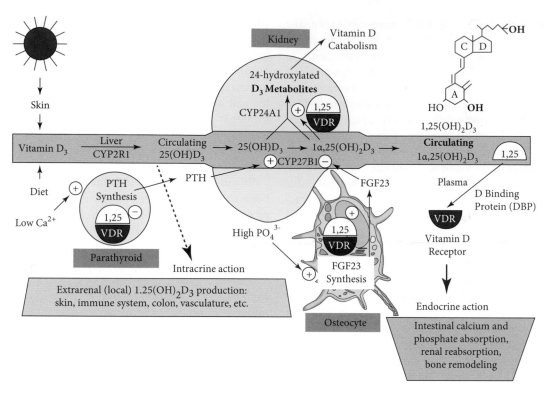

FIGURE 25.6 Regulation of vitamin D metabolism, and principal sites of its actions.
VDR = vitamin D receptor; $1,25 = 1\alpha,25(OH)_2D_3$ = 1,25-dihydroxyvitamin D. The enzymes involved in vitamin D metabolism are represented by CYP~, and stimulation of their activity is indicated by + and inhibition by –.
Source: Modified after M R Haussler et al. (2012) Molecular mechanisms of vitamin D action. *Calcif. Tissue Int.* **92**(2), 77–98
Copyright © 2012, Springer Science Business Media, LLC. Used with permission.

conditions at the present time. (See also Chapter 12, Section 12.2).

Defining Vitamin D deficiency

25-hydroxyvitamin D is the principal circulating vitamin D metabolite and is used to determine vitamin D status. Ingested and endogenous calciferols are converted to 25-hydroxyvitamin D with little regulation, so serum levels of this metabolite accurately reflect both excess and deficiency states. 25-hydroxyvitamin D concentrations are higher in those with greater sunshine exposure, which is influenced by dress, lifestyle, and latitude. Dark skin and obesity are associated with lower levels of 25-hydroxyvitamin D. The serum concentration of 25-hydroxyvitamin D is the preferred biomarker of vitamin D status, because it reflects exposure to both dietary intake and cutaneous synthesis, and it has a long half-life.

There is substantial disagreement as to what constitutes vitamin D deficiency, with 25-hydroxyvitamin D thresholds between 25 nmol/L and 75 nmol/L being in common use (divide by 2.5 to produce 25-hydroxyvitamin D levels in ng/mL). A lower limit of the reference range of 75 nmol/L is not based on the population distribution of 25-hydroxyvitamin D but rather on cross-sectional studies showing inverse associations between 25-hydroxyvitamin D concentrations and both mortality and morbidity. Association does not represent causation, and it is likely that these inverse associations arise from less time spent outdoors and lower levels of physical activity in those who are infirm, and from the confounding effects of obesity which reduces serum 25-hydroxyvitamin D levels and increases the risk of various diseases.

As vitamin D deficiency develops, the supply of 25-hydroxyvitamin D is no longer adequate to produce sufficient levels of 1,25-dihydroxyvitamin D, so parathyroid hormone levels increase leading to an increase in bone turnover and a fall in bone mass. Therefore, adequacy of 25-hydroxyvitamin D levels should be reflected by an absence of effect of vitamin D supplements on these endpoints. Such studies have demonstrated that vitamin D only suppresses parathyroid hormone when baseline serum 25-hydroxyvitamin D is <50 nmol/L. Recent post hoc analyses of trials assessing the effect of vitamin D on bone density demonstrated that positive effects were only observed in those participants whose baseline 25-hydroxyvitamin D was <25–30 nmol/L. This indicates that 25 nmol/L is a more realistic threshold for vitamin D deficiency in terms of bone health, and is consistent with the finding that patients with osteomalacia as a result of vitamin D deficiency have 25-hydroxyvitamin D levels well below this level. Away from the equator, there are seasonal variations in serum 25-hydroxyvitamin D of 20 nmol/L or more, so to ensure that the winter 25-hydroxyvitamin D nadir is >25 nmol/L, summer levels should be >45–50 nmol/L, giving some support for the use of that threshold. There is little experimental support for 75 nmol/L as the minimal safe level of 25-hydroxyvitamin D. If vitamin D is found to have effects on other biological endpoints, then the thresholds for those effects will not necessarily be the same as those found for bone.

Vitamin D supplementation

Global meta-analyses of the effects of vitamin D supplementation alone on bone density have been negative. However, studies which have assessed individual participants' bone density change in relation to their baseline levels of 25-hydroxyvitamin D, have found that when 25-hydroxyvitamin D is <25–30 nmol/L vitamin D supplementation results in significant increases in density (Reid et al. 2017). This is consistent with observations that patients with vitamin D-deficiency osteomalacia have substantial increases in bone density, often greater than 20 per cent, when treated with vitamin D. Individuals with baseline 25-hydroxyvitamin D >30 nmol/L have no bone density response to supplementation. This suggests that a 25-hydroxyvitamin D level of 25–30 nmol/L represents the lower bound of the range of vitamin D sufficiency, and individuals above this level are functionally replete.

Meta-analysis of trials comparing vitamin D supplementation alone with placebo show no effect on fracture (Bolland et al. 2018). As discussed above, the Chapuy trial in which most participants were vitamin D deficient, did find a reduction in hip and total fractures after supplementation with vitamin D plus calcium.

Vitamin D supplements in doses up to 2000 IU/day are considered to be safe. Above this level, some studies have found increases in falls and/or fractures, particularly when circulating levels of 25-hydroxyvitamin D are elevated above 120 nmol/L. Substantially higher vitamin D doses can result in hypercalcemia, which if undetected can lead to renal failure. Vitamin D has a long half-life so can be given infrequently in larger doses. However, large annual doses (500,000 IU) have been associated with increased falls and fractures. The range of beneficial and adverse effects of vitamin D are shown in Figure 25.7.

25.5 Body weight effects on bone

Body weight is related to bone density (positively) and fracture risk (inversely) throughout life (Johansson et al.

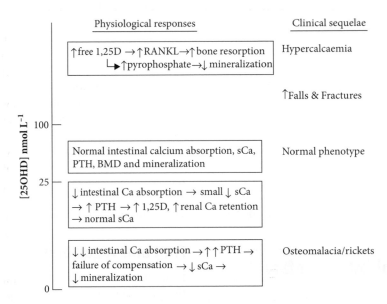

FIGURE 25.7 Physiological and clinical associations of high, normal, and low circulating concentrations of 25-hydroxyvitamin D (25OHD).
The concentrations of 25OHD at which the various changes occur vary between studies so indications on the figure are only approximate, and the y-axis is not linear. Secondary hyperparathyroidism occurs when 25OHD is <25–40 nmol/L, and clinical osteomalacia is usually only reported when 25OHD is <<25 nmol/L. The trials which have suggested that vitamin D supplements increase falls and fractures have achieved 25OHD concentrations >120 nmol/L. Hypercalcaemia is generally only found at very much higher 25OHD levels. sCa, serum calcium concentration.
Source: I R Reid (2018) High-dose vitamin D: Without benefit but not without risk. *J. Intern. Med.* **284**, 694–696, Copyright ©2018, John Wiley. Used with permission.

2014) and is a key element in most fracture risk calculators (e.g., FRAX). This is a dynamic relationship, with changes in weight directly impacting on bone mineral density (BMD). While the positive effect of body weight on bone density is consistent across many studies and sites of bone measurement, the relationship between body weight and fracture risk is more complex. High BMI is protective against most types of fractures, particularly hip fractures. Much of this effect is mediated by higher bone density. At the hip, however, fracture protection with high BMI is present even after correction for bone density, suggesting that padding of the greater trochanter may also play an important role. Some studies suggest that high BMI is positively associated with fractures of the upper arm and ankle, possibly indicating that these sites are vulnerable to the larger loads resulting from falls in those with higher BMI. In children, obesity is associated with increased risk of forearm fractures.

There is a complex literature which attempts to dissect the contributions of lean and adipose tissues to these weight effects on bone. Attempts have also been made to separate the effects of visceral fat from that of other adipose tissue. The difficulty is that all these tissue compartments are closely related to one another, so effects cannot be separated through simple multiple regression analysis because the tissue compartments are not truly independent variables. Probably both lean and fat masses impact on bone health, the former through mechanical effects of muscle on bone, and the latter through hormonal mediators such as oestrogen, insulin, leptin, amylin, and preptin.

Extreme weight loss associated with anorexia nervosa results in marked loss of bone mass as a result of nutrient deficiency and reduced levels of sex hormones, particularly oestrogen. Other endocrine effects of anorexia nervosa may contribute to bone loss, including reduced insulin-like growth factor-1, cortisol excess, and lower leptin levels. Patients with anorexia nervosa who do not recover their body weight and bone density have a substantially increased risk of fracture in later life. Calcium supplementation has little impact on bone health in anorexia.

25.6 Effects of other dietary components

While single factors such as calcium or vitamin D lend themselves to randomized controlled trials of supplementation, assessing the health effects of groups of nutrients or entire diets is much more complex. As a result, much of this literature is based on observational studies, with major risks of confounding from other lifestyle factors. Studies which have related bone health to broad

dietary patterns have not demonstrated consistent associations. Comparisons of bone density between those taking vegetarian diets (lacto-ovo and vegan) and meat consumers have generally shown that vegetarian populations have a slightly lower bone density, mostly accounted for by differences in body weight (Karavasiloglou et al. 2020). The EPIC-Oxford study followed up about 55,000 adults over >17 years and assessed the association of vegetarian diets with fracture incidence (Tong et al. 2020). Compared with meat eaters, the risk of hip fracture was higher in fish eaters (hazard ratio 1.26; 95 per cent CI 1.02–1.54), vegetarians (1.25; 1.04–1.50), and vegans (2.31; 1.66–3.22), after adjustment for BMI and lifestyle factors. Total fracture risk was also higher in vegans (1.43; 1.20–1.70). The increased risks were slightly attenuated but remained significant with additional adjustment for dietary calcium and/or total protein.

Protein

Since the bone matrix is mostly type I collagen, an adequate intake of amino acids is a prerequisite for normal bone formation. Bone growth is compromised if the protein supply is inadequate during childhood and adolescence, and increasing protein intake is likely to have contributed to progressive increases in stature among Western populations over the last 150 years. However, some studies have suggested that very high protein intakes might have a deleterious effect on bone by increasing urinary calcium loss. This might reflect mobilization of bone mineral to buffer the acid load resulting from protein catabolism, though it could also result from protein effects on the efficiency of dietary calcium absorption. While short-term protein feeding studies do demonstrate calciuric effects, there is little evidence from population studies that high protein intakes have a negative effect on bone health. On the contrary, provision of protein supplements to the frail elderly is associated with reduced fracture risk and improved recovery in those who have already fractured. These benefits might be mediated through effects on muscle as well as on bone.

Fruit and vegetables

Plant protein catabolism results in less acid production than animal protein, so it is proposed that fruit and vegetables might increase body pH and improve skeletal health. There is some observational evidence that a diet high in fruit and vegetables is associated with a reduced fracture risk (Brondani et al. 2019). However, a two-year randomized controlled trial of increased intake of fruit and vegetables in postmenopausal women found no benefit to bone density. Similar studies assessing effects on bone resorption markers have also been negative

(Brondani et al. 2019). Divergence of observational and intervention studies suggests residual confounding in the former. Fruit and vegetable intake is likely to be related to other dietary patterns and to lifestyle factors.

Some plants produce compounds which have weak oestrogen-like effects. These are known as phytoestrogens, and include isoflavones, which are found in foods such as soy. Higher bone density has been reported in East Asian populations with high soy intakes but similar associations have not been found in populations with lower intakes. Trials of supplements containing high doses of phytoestrogens have shown small beneficial effects on bone, but probably not of a magnitude that would impact on fracture risk.

Mineral and vitamin effects

Along with calcium, phosphate is the other key mineral constituent of bone, so the adequacy of its intake and the regulation of phosphate balance is critical to skeletal health. Normal concentrations of both calcium and phosphate in the extracellular fluid are necessary for bone mineralization to proceed normally. Supranormal levels of either mineral can result in ectopic mineralization of other tissues, with consequent damage to the affected tissue. Unlike calcium, about 20 per cent of total body phosphorus is outside the skeleton, since phosphorus is a key element in organic molecules involved in energy transfer via adenosine triphosphate (ATP), intracellular signalling via cyclic adenosine monophosphate (cAMP), in DNA and RNA, and in the phospholipids of cell membranes.

Phosphate is widespread in the diet so dietary phosphate deficiency is uncommon. Dairy, meat, grains, and legumes are rich in phosphate, and phosphate is a common food additive. In the United States, the recommended daily allowance of phosphorus (including organic and inorganic [phosphate] forms) is 700 mg/day, which is commonly exceeded by a factor of about two in Western diets.

Circulating phosphate levels fluctuate much more than do those of calcium, being influenced by time of day and by average phosphate intake. Renal tubular reabsorption of phosphate is a key determinant of circulating phosphate concentrations, and is regulated via both fibroblast growth factor-23 (FGF23, secreted from osteocytes) and parathyroid hormone (PTH). Intestinal phosphate absorption is stimulated by 1,25-dihydroxy-vitamin D, which is itself influenced by FGF23 and PTH (Peacock 2021). Hypophosphataemia is seen in those taking phosphate binders (e.g., some antacids), in alcoholism, and in states of FGF23 or PTH excess. Renal failure, or deficiencies of either FGF23 or PTH cause hyperphosphataemia.

The effect of high phosphate intake on the skeleton has been assessed by calcium balance studies. A meta-analysis of such studies concluded that high phosphate intake reduces urinary calcium excretion, but does not change calcium balance (Fenton et al. 2009). There is no evidence that phosphate intake significantly impacts on fracture risk.

There is little evidence that magnesium intake impacts on bone health in the general population. Rarely, severe magnesium deficiency can develop as a result of other illnesses or medications. This can impair parathyroid hormone secretion and action, resulting in hypocalcaemia. Magnesium is sometimes promoted as having a positive influence on bone and heart health, but there is little objective evidence to support this contention.

The handling of sodium and calcium are linked in the proximal renal tubule. Thus, sodium loading can increase urine calcium excretion but normal regulatory mechanisms caused this to be compensated for by increased absorption so sodium intake does not impact on bone density or fracture risk. It may, however, contribute to renal calculi in susceptible individuals.

Fluoride ion has been used as a treatment for osteoporosis, since it stimulates osteoblast activity and increases bone density. However, it interferes with normal bone mineralization and actually increases fracture rates rather than diminishing them. Studies of the effects of fluoridation of drinking water on fracture risk have produced inconsistent findings. Very high fluoride content occurs naturally in some places and produces skeletal fluorosis, characterized by bone overgrowth, bone pain, and fractures. Strontium supplementation has also been used to treat adult patients with osteoporosis. Strontium ions substitute for calcium in bone, increasing measured bone density and possibly decreasing incidence of vertebral fractures. This treatment has been abandoned in most countries because it also appears to increase risk of myocardial infarction. The doses used in osteoporosis treatment were far greater than typical dietary intakes. High doses of strontium in animals can cause osteomalacia. A high incidence of rickets has also been observed in geographical areas with very high content of strontium in the soil, but the importance of dietary intake as a determinant of skeletal health in most populations is unknown. *Aluminium* intoxication can also impaired mineralization and cause osteomalacia. This has been an issue in some renal dialysis centres but is seldom seen now since dialysate aluminium levels have been reduced.

It has become common to see individual vitamins or minerals suggested to be associated with osteoporosis. It is unlikely that most such associations are causative, since global food intake decreases as age and frailty advance, as does body weight, a further important determinant of skeletal health. Inter-correlations between intakes of a wide range of nutrients have the potential to make it challenging to determine which nutrient-bone associations are causative. While vitamin A intoxication causes hypercalcaemia, there is little evidence that this vitamin plays an important role in skeletal health. Observational studies have found an association between elevated plasma homocysteine and fracture risk. Folate, vitamins B_6 and B_{12} are involved in homocysteine metabolism, and supplementation with them lowers homocysteine levels. However, a large randomized trial of supplementation with these vitamins did not impact on fracture incidence (Stone et al. 2017). While vitamin C is necessary for collagen synthesis and scurvy is associated with bone pain and fractures, there is not consistent evidence that this vitamin influences bone health in the general population. Vitamin K facilitates carboxylation of the bone protein osteocalcin. Japanese investigators have produced some positive results from the use of vitamin K supplements, but meta-analyses are unconvincing for either bone density or fracture endpoints (Mott et al. 2020).

Beverages

High alcohol intakes are associated with an increased risk of fracture. This is partly associated with the increased risk of falling in those who become intoxicated, and to a direct toxic effect of alcohol on osteoblasts. High alcohol intake in men is associated with hypogonadism, which will lead to bone loss. Paradoxically, modest alcohol intakes have been associated with reduced risk of fractures in some studies, particularly in postmenopausal women. Intakes of one to two drinks daily appear to reduce hepatic metabolism of oestrogen, resulting in increased bone density and fewer fractures. In advanced old age, continued use of alcohol may be an indicator of continued social functioning and, therefore, a marker of well-being. This, and associations of alcohol use with other aspects of diet and lifestyle may confound some of these associations.

Caffeine has been suggested to impact on calcium and bone metabolism, so caffeinated drinks (including tea and coffee) might influence bone density and fracture risk. The evidence for this is not compelling. A recent meta-analysis suggested that high coffee intakes increased fracture risk in women but decreased it in men. In the Women's Health Initiative Observational Study, there was no association between caffeinated soft drink intake and incident hip fracture, though there was a small increase in hip fracture risk with non-caffeinated soft drinks. Despite the apparent fracture associations, soft drink intake had no impact on bone density. In teenagers, carbonated soft drinks have been associated with reduced bone density, but this could be confounded by associated changes in diet and lifestyle. The inconsistency of the findings suggests that if effects are present, they are minor.

25.7 Bone mass through the lifespan

By the end of the first trimester the human foetal skeleton is present as a cartilaginous template. Skeletal mineralization occurs principally in the third trimester when calcium uptake into bone exceeds 100 mg/day. Bone growth continues in the neonate, with recommended calcium intakes of 200–260 mg/day during the first year of life, usually met by breastfeeding. Recommended intakes in many Western countries climb to 500 mg/day for ages 1–3, and 700–1000 mg/day from childhood through to puberty. Such recommendations are guided by calcium balance and observational studies, and trial data are sparse in these age groups. Most Asian and African communities do not achieve these intakes, without clear evidence of adverse effects, as long as vitamin D status is satisfactory.

Bone growth accelerates during puberty, and more than 90 per cent of peak bone mass is achieved by age 18, with slower increase over the following few years. Peak bone mass is predominantly genetically determined, though intercurrent illness, medications (e.g., glucocorticoids, sex steroids), exercise, and malnutrition can all have effects.

Bone densities are stable from the mid-twenties until the decline in oestrogen levels in the few years before the menopause results in increased bone resorption and bone loss (Figure 25.8). Bone density decreases by several per cent per annum in the first five postmenopausal years, then at about 1–2 per cent/year through until the end of life. Densities also decrease in men, more slowly than in women, from about the age of 60 years. This is possibly the result of declining sex hormone levels at that time.

With ageing, individuals tend to maintain their ranking within the population distribution of bone density, so peak bone mass is thought to remain an important predictor of bone density in later life.

Pregnancy and lactation

Pregnancy and lactation increase calcium requirements. During pregnancy, the fractional absorption of calcium from the diet doubles, which meets the amount of calcium required by the foetus, of approximately 30 g. Lactation delivers about 200 mg/day of calcium to the neonate, much of this being provided by resorption of the maternal skeleton, this being effected by both osteoclasts and

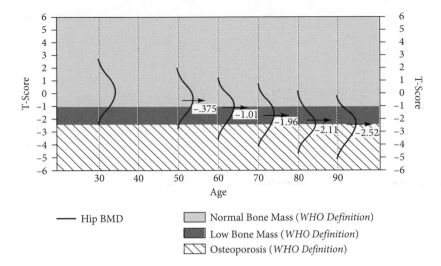

FIGURE 25.8 The distribution of hip bone mineral density and its decline with age in white women in North America.

Source: National Osteoporosis Foundation. Osteoporosis: Review of the Evidence for Prevention, Diagnosis and Treatment and Cost-Effective Analysis. Osteoporos Int 8 (Suppl 4), S7–S80 (1998), used with permission of SpringerNature.

osteocytes. Reductions in spine bone density of 5–10 per cent have been reported during six months of exclusive lactation, with lesser losses in the long bones. However, this loss is fully reversed at weaning, and parity and history of breastfeeding are not related to fracture risk later in life (Kovacs 2017).

25.8 Osteoporosis

Osteoporosis is widely defined as a condition 'characterized by low bone mass and microarchitectural deterioration of bone tissue, with a consequent increase in bone fragility and susceptibility to fracture'. The process of bone loss which results in osteoporosis is almost universal in women after menopause, and in men beyond their mid-sixties. Whether the degree of bone loss justifies the label of *osteoporosis* will depend on what operational definition is used, and there is wide variation in practice in this respect. A World Health Organization committee defined osteoporosis as a bone density more than 2.5 standard deviations below the mean value in the young normal population. However, this overlooks the fact that many other risk factors substantially impact on fracture risk, particularly age and fracture history. Health economics suggests that use of treatments for osteoporosis should be based on actual fracture risk rather than a single bone density threshold. Health economics also suggests that there should be different thresholds for medications according to their cost. Based on these considerations, it seems most appropriate to retain the qualitative definition of osteoporosis given above, but to determine pragmatic intervention thresholds according

to current circumstances. Since the process of bone loss is universal and both bone density and fracture risk are continuously distributed, it is not helpful to dichotomize individuals as either having or not having *osteoporosis*. The clinician's task is not to diagnose osteoporosis, but rather to assess fracture risk and to provide lifestyle and medication advice according to the level of that risk.

Fracture epidemiology

The fall in bone density after the menopause (Figure 25.8) is mirrored by a progressive rise in fracture numbers (Figure 25.9). As a consequence, about one in five white women suffer a hip fracture between the age of 50 years and death, with similar incidences for clinical vertebral fractures and wrist fractures. These 'classical' osteoporotic fractures only account for about half of the total number of fractures that occur in this age group. The remaining life-time risk of *any* fracture in 60-year-olds is >50 per cent for women and about 30 per cent in men. Fracture rates are highest in Europe and North America and lowest in Africa. Numbers of hip fractures are increasing in most countries, particularly in Asia. This results from progressive increases in age-adjusted incidence rates in non-Western countries, and from ongoing global increases in the population aged >70 years as a result of increasing longevity.

Hip fractures usually result from falls, particularly falls to the side. They are more common in winter. About 50 per cent of patients do not recover their former levels of mobility or independence after hip fracture. Vertebral fractures often occur after lifting rather than after falls, and only about a third of vertebral deformities come to

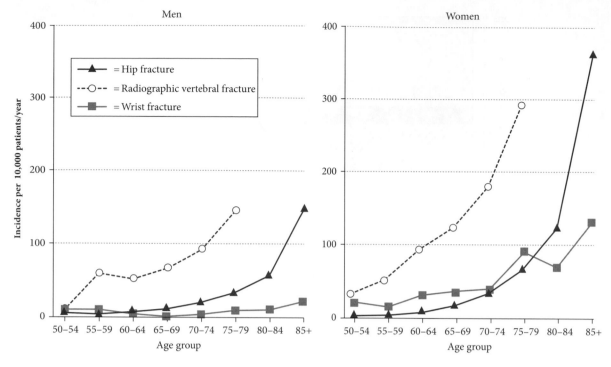

FIGURE 25.9 Fracture incidence by age and gender in England and Wales during the period 1988–1998.
Source: N C Harvey et al. (2019) *The Epidemiology of Osteoporotic Fractures: Primer on the Metabolic Bone Diseases and Disorders of Mineral Metabolism*, 9th edn. © 2019 American Society for Bone and Mineral Research, used with permission.

medical attention. Vertebral fractures result in height loss and kyphosis, with consequent chronic back pain. There is a 20 per cent excess mortality at one year after hip fracture, and a smaller but sustained increase in mortality after vertebral fracture.

Risk factors for fracture

Bone density is an important risk factor for fracture. At any age, bone density is determined by the amount of bone laid down during growth and the subsequent changes resulting from aging, drugs (e.g., glucocorticoids), or disease. Factors which maximize bone mass gain during childhood and adolescence (e.g., global nutritional status, physical activity, timing of puberty, genes), those that maintain it during adulthood (normal gonadal function, nutrition, exercise), and those that delay or slow its loss during menopause and old age (timing of menopause, fat and lean masses, physical activity, vitamin D status) will reduce fracture risk. Some of the common pathological and pharmacological contributors to osteoporosis are listed in Table 25.2.

Age increases fracture risk independently of its effect on bone density, possibly by way of effects on bone quality. The components of bone quality are continuing to be defined, but include the material properties of both the protein matrix and the mineral phase of bone, as well as bone architecture. The integrity of collagen cross-linking, the state of cross-link isomerization, and content of advanced glycation end-products (resulting from non-enzymatic glycation of collagen) impact on skeletal strength. Falling is an important risk for fracture. Fall risk is related to frailty, neuromuscular health, eyesight, medications, and safety of the home environment.

Patient assessment for osteoporosis frequently involves use of a fracture risk calculator, such as the FRAX (https://www.sheffield.ac.uk/FRAX/index.aspx) or Garvan (https://www.garvan.org.au/bone-fracture-risk) calculators, both available online. Both use age, sex, weight, bone density, and fracture history. Falls, ethnicity, smoking, alcohol, height, and glucocorticoid use are additional factors used by one or other calculators. Calcium intake and vitamin D status have not been shown to be useful fracture predictors.

Osteoporosis management

The prevention and treatment of osteoporosis involves correction of reversible risk factors (e.g., underweight, smoking, high alcohol intake), prevention of vitamin D deficiency by safe sunlight exposure (or use of supplements if this is not feasible), encouragement of safe exercise, falls prevention where appropriate, and the formal assessment of fracture risk with provision of treatments for fracture prevention in those with a fracture risk that

TABLE 25.2 Conditions and drugs associated with osteoporosis

Inflammatory disorders	*Low body weight*
Rheumatoid arthritis	Anorexia nervosa
Inflammatory bowel disease	HIV infection
Cystic fibrosis	*Immobilization*
Bone marrow disorders	Parkinson's disease
Multiple myeloma	Poliomyelitis
Mastocytosis	Cerebral palsy
Leukaemia	Paraplegia
Hypogonadism	*Defective synthesis of connective tissue*
Athletic amenorrhoea	
Haemochromatosis	Osteogenesis imperfecta
Turner's syndrome	Marfan's syndrome
Klinefelter's syndrome	Homocystinuria
Post-chemotherapy	*Miscellaneous*
Hypopituitarism	Chronic obstructive lung disease
Malabsorption	
Coeliac disease	Congestive heart failure
Gastrectomy	Pregnancy/lactation
Bariatric surgery	Ankylosing spondylitis
Liver disease	Hypercalciuric nephrolithiasis
Total parenteral nutrition	Depression
Endocrine disorders	*Drugs*
Cushing's syndrome	Glucocorticoids, alcohol,
Diabetes	medroxyprogesterone acetate
Thyrotoxicosis	anti-convulsants, methotrexate
?Hyperparathyroidism	heparin, cyclosporin, omeprazole
	aromatase inhibitors, glitazones
	androgen deprivation therapy

Copyright I R Reid, used with permission.

makes this cost-effective. The specific criteria for pharmaceutical intervention vary from country to country. A calculated ten-year hip fracture risk of 3 per cent is sometimes used, though for generic bisphosphonates treatment is cost-effective at a lower fracture risk, whereas for expensive new injectable medications a higher threshold is more appropriate. The advantage of the fracture calculators is that they can factor in multiple risk factors such as bone density, fracture history, age, weight, and family history. Some countries use specific individual criteria such as history of an osteoporotic fracture or a certain level of bone density as the criteria for intervention rather than their integration in a calculated fracture risk.

Anti-resorptive agents

Medications increase bone density and reduce fracture risk by either decreasing the activity of osteoclasts (and thus bone resorption), or by increasing osteoblast activity (and thus bone formation). Calcium supplements or oestrogen preparations were the original antiresorptive agents used, but this area is now dominated by the use of bisphosphonates. Bisphosphonates are small, phosphate-containing salts which bind avidly to the mineral surface of bone and inhibit the activity of osteoclasts. They remain in bone for years, being incorporated into bone mineral as new bone is laid down over bisphosphonate-covered surfaces. Bone resorption is reduced by >50 per cent and fracture rates reduced by 20–70 per cent, depending on the site of fracture and the agent used. Alendronate and risedronate are widely used oral agents, typically given as a weekly dose. They are poorly absorbed, so must be taken fasting, with water alone, at least half an hour before the first food of the day. Zoledronate is given as an intravenous infusion, typically at intervals of one to two years, though its duration of action may be >5 years.

More recently, denosumab has come into use. It is a monoclonal antibody directed against RANKL which is a key stimulator of osteoclast formation and activity. Denosumab is given as a six-monthly subcutaneous injection. It has efficacy comparable to that of the more potent bisphosphonates. However, unlike bisphosphonates, it has a rapid offset of effect after its discontinuation, so transition to another antiresorptive agent is necessary if it is to be discontinued.

Anabolic agents

Agents which increase bone formation have become available over the last two decades. Several analogues of parathyroid hormone or parathyroid hormone-related peptide are available, usually given as daily subcutaneous injections. Romosozumab is a recently released, monoclonal antibody which blocks the action of the osteoblast-inhibiting protein, sclerostin. As a result, bone formation is increased. Romosozumab also indirectly inhibits bone resorption. Romosozumab dramatically increases bone density and has been shown to have an anti-fracture efficacy greater than that of alendronate. It is given by monthly subcutaneous injections. These anabolic agents are only used for periods of one year (romosozumab) or 18–24 months (teriparatide and abaloparatide). Their effects do not persist following discontinuation, so patients must then be switched to an antiresorptive agent, either a bisphosphonate or denosumab.

Safety

Oral bisphosphonates cause upper gastrointestinal symptoms in about 20 per cent of patients. Intravenous bisphosphonates can cause a 'flu-like illness' after the first dose. This can comprise fever, musculoskeletal pain, or other symptoms of tissue inflammation. It occurs in about 30 per cent of bisphosphonate-naive subjects, and is not usually seen after second and subsequent doses. Denosumab is well-tolerated but does require rigid attention to the dosing schedule and a plan for crossover to another antiresorptive agent should discontinuation be necessary.

Bisphosphonates and denosumab have been associated with two very rare but distressing complications, osteonecrosis of the jaw and atypical femoral fractures. Osteonecrosis of the jaw is most commonly seen in patients with disseminated cancer receiving monthly treatment with either intravenous bisphosphonates or denosumab, who then undergo an invasive dental procedure such as a tooth extraction. In a few percent of cases, the extraction site fails to heal and becomes infected. These lesions can spread and be very difficult to control. The occurrence in osteoporosis patients is probably of the order of a few patients per 100,000 patient-years.

Atypical femoral fractures have been reported in most anti-resorptive agents, though the major evidence comes from the use of alendronate. They initially develop as stress fractures of the femoral shaft, usually progressing to a complete transverse fracture. In the early years of oral alendronate use, they occur in a few patients per 100,000 patient-years, but with >5 years use this rate can climb to >60 per 100,000 patient-years. The risk of these atypical fractures falls rapidly within months of drug cessation. As a result, the use of oral bisphosphonates is now usually punctuated by drug holidays of one to two years, taken after three to five years of treatment. In all patients with a fracture risk justifying drug intervention by conventional criteria, the number of other fractures prevented by antiresorptive drugs substantially exceeds the number of these rare femoral shaft fractures.

Teriparatide and abaloparatide have been associated with nausea, headache, and musculoskeletal pain, and the occasional development of hypercalcemia. Rat studies with both agents have shown an increased risk of osteosarcoma with long-term, high-dose treatment. However, substantial observational studies in humans have not shown any evidence of this effect.

25.9 Osteomalacia and rickets

Osteomalacia

Osteomalacia is a skeletal disorder resulting from defective mineralization of bone. As noted above, normal mineralization requires normal circulating concentrations of both calcium and phosphate, plus the absence of mineralization inhibitors. Vitamin D is critical to gastrointestinal absorption of both calcium and phosphate,

as well as playing a role in stimulating bone resorption. Because most vitamin D is produced in the skin as a result of sunlight exposure, vitamin D deficiency is predominantly a consequence of reduced time outdoors, habitual covering of the skin, living at high latitude, and pigmented skin (resulting in reduced efficiency of vitamin D production). Accordingly, vitamin D deficiency presents in veiled women, people with dark skin living at high latitude, and the frail elderly who seldom venture outside. It also occurs in the breastfed children of vitamin D-deficient mothers, particularly if the baby's sun exposure is low. Inherited or acquired abnormalities of vitamin D metabolism can also result in osteomalacia. There are a number of other causes of chronic hypophosphataemia, including dietary deficiency, malabsorption syndromes, alcoholism, tumours producing phosphaturic factors (such as FGF23), renal defects, genetic abnormalities in phosphate metabolism, and medications. In Africa, children with very low calcium intakes (<100 mg/day) have been shown to have rickets in the presence of normal vitamin D levels. The presence of calcium-binding factors in the diet, such as phytates, may contribute to this.

Pyrophosphate is an endogenous mineralization inhibitor which is important for the prevention of extra-skeletal calcification. It is broken down by alkaline phosphatase, so deficiency in this enzyme results in impaired mineralization. There are other exogenous factors that can bind to bone and inhibit mineralization, such as fluoride, strontium, and etidronate. Chronic acidosis also interferes with normal skeletal mineralization.

Under-mineralized bone has reduced strength, so is liable to deform and, sometimes, fracture. Bone pain and muscle weakness may be prominent clinical features of osteomalacia. In children, the absence of normal skeletal mineralization results in reduced growth and in developmental deformities, presenting the clinical picture of rickets.

Rickets

Rickets has been an important cause of childhood illness and deformity for many centuries. Following the Industrial Revolution, pollution and poor diet resulted in increased prevalence of the disease in England. Rickets continues to be an important disease in the developing world, and is still seen in developed countries particularly in children with dark skin living at high latitudes. In the breastfed infants of vitamin D-deficient mothers, rickets can develop within the first few months of life, though is more common between six and 18 months of age.

The clinical features of rickets were described by Whistler in 1645. Rachitic infants may have soft areas of the skull, known as craniotabes. In childhood, the classical presentation is with knock-knees or bowed legs, muscle weakness, and short stature. Widening of growth plates results in thickening of the wrists and ankles and enlargement of the costochondral junctions (sometimes referred to as a 'rachitic rosary'). Delay in motor developmental milestones and in eruption of teeth may occur. An increased risk of respiratory tract infections has been suggested but causation of this is disputed.

Bone is under-mineralized on X-rays, and the growth plates are widened and irregular. There may be pelvic deformities which can lead to narrowing of the birth canal causing difficulties later in life during childbirth. Serum alkaline phosphatase activity is increased, and hypophosphataemia or hypocalcaemia may be present, though normocalcaemia is common. Bone biopsy demonstrates excess osteoid and abnormal tetracycline labelling, both diagnostic features of impaired bone mineralization.

Following the First World War, there was an epidemic of rickets in Europe. Research at that time demonstrated that cod liver oil (which is rich in vitamin D) or sunshine exposure cured the condition, and these measures have remained the cornerstone of prevention and treatment since that time. Extra calcium is commonly provided when treating vitamin D deficiency rickets, since the remineralization of the skeleton increases calcium requirement. In the calcipenic rickets sometimes observed in Africa, where patients are already vitamin D replete, the provision of an adequate calcium supply is all that is required. Where rickets or osteomalacia result from other metabolic abnormalities (e.g., of phosphate or vitamin D metabolism) specific management of the underlying abnormality is required.

eral is laid down, providing compressive strength to the skeleton. The skeleton is continuously renewed throughout life.

- Bone density is directly related to body weight throughout life. Those with low body weight or who lose weight have an increased risk of most fracture types, particularly the hip.

- Most studies show no relationship between calcium intake and either bone density or fracture risk. At extremely low calcium intakes (i.e., <100 mg/day) bone mineralization is compromised and rickets can occur in children.

- Calcium supplements produce small, non-cumulative benefits to bone density but do not change fracture risk in community-dwelling adults.

- Vitamin D is principally produced in the skin as a result of sunlight exposure. Vitamin D deficiency is common in those who are seldom outdoors, who habitually cover their skin, or have dark skin.

- Vitamin D metabolites are key regulators of intestinal calcium and phosphate absorption. In severe vitamin D deficiency (i.e., 25-hydroxyvitamin D << 25 nmol/L), malabsorption of calcium and phosphate results in impaired bone mineralization.

- In severe vitamin D deficiency, supplements have a marked positive effect on bone density. In community populations with baseline 25-hydroxyvitamin D >30 nmol/L, consistent effects of supplements on bone density or fractures have not been demonstrated.

- Observational studies which correlate intake of individual nutrients with bone density are problematic to interpret because of inter-correlations between nutrients, and of nutrients with other lifestyle factors (such as exercise, smoking, and alcohol intake).

KEY POINTS

- Bone is a connective tissue, the essential structure of which is formed of type I collagen produced in osteoblasts. Between the collagen bundles, min-

 Be sure to test your understanding of this chapter by attempting multiple choice questions.

REFERENCES

Bolland M J, Grey A, Avenell A (2018) Effects of vitamin D supplementation on musculoskeletal health: A systematic review, meta-analysis, and trial sequential analysis. *Lancet Diabetes Endocrinol* **6**(11), 847–858.

Bolland M J, Leung W, Tai V et al. (2015) Calcium intake and risk of fracture: Systematic review. *BMJ* **351**, h4580.

Brondani J E, Comim F V, Flores L M et al. (2019) Fruit and vegetable intake and bones: A systematic review and meta-analysis. *PLoS ONE* **14**(5), e0217223.

Chapuy M C, Arlot M E, Delmas P D et al. (1994). Effect of calcium and cholecalciferol treatment for three years on hip fractures in elderly women. *BMJ* **308**, 1081–1082.

Chapuy M C, Arlot M E, Duboeuf F et al. (1992) Vitamin-D3 and calcium to prevent hip fractures in elderly women. *N Engl J Med* **327**(23), 1637–1642.

Fenton T R, Lyon A W, Eliasziw M et al. (2009) Phosphate decreases urine calcium and increases calcium balance: A meta-analysis of the osteoporosis acid-ash diet hypothesis. *Nutrit J* **8**, 41.

Grossman D C, Curry S J, Owens D K et al. (2018) Vitamin D, calcium, OR combined supplementation for the primary prevention of fractures in community-dwelling adults. US Preventive Services Task Force recommendation statement. *JAMA* **319**(15), 1592–1599.

Harvey N C, Biver E, Kaufman J M et al. (2017) The role of calcium supplementation in healthy musculoskeletal ageing: An expert consensus meeting of the European Society for Clinical and Economic Aspects of Osteoporosis, Osteoarthritis and Musculoskeletal Diseases (ESCEO) and the International Foundation for Osteoporosis (IOF). *Osteoporos Int* **28**(2), 447–462.

Hunt C D & Johnson L K (2007) Calcium requirements: New estimations for men and women by cross-sectional statistical analyses of calcium balance data from metabolic studies. *Am J Clin Nutr* **86**(4), 1054–1063.

Johansson H, Kanis J A, Odén A et al. (2014) A meta-analysis of the association of fracture risk and body mass index in women. *J Bone Miner Res* **29**(1), 223–233.

Kahwati L C, Weber R P, Pan H et al. (2018) Vitamin D, calcium, or combined supplementation for the primary prevention of fractures in community-dwelling adults: Evidence report and systematic review for the US Preventive Services Task Force. *JAMA* **319**(15), 1600–1612.

Karavasiloglou N, Selinger E, Gojda J et al. (2020) Differences in bone mineral density between adult vegetarians and nonvegetarians become marginal when accounting for differences in anthropometric factors. *J Nutrition* **150**(5), 1266–1271.

Kovacs C S (2017) The skeleton is a storehouse of mineral that is plundered during lactation and (fully?) replenished afterwards. *J Bone Miner Res* **32**, 676–680.

Mott A, Bradley T, Wright K et al. (2020) Correction to effect of vitamin K on bone mineral density and fractures in adults: An updated systematic review and meta-analysis of randomized controlled trials. *Osteoporos Int.* 31, 2269–2270. 10.1007/s00198-019-04949-0)

Peacock M (2021) Phosphate metabolism in health and disease. *Calcif Tissue Int* **108**(1), 3–15.

Reid IR, Bristow SM (2020). Calcium and Bone. *Handbook Exp Pharmacol*, **262**, 259-80.

Reid I R, Horne A M, Mihov B et al. (2017) Effect of monthly high-dose vitamin D on bone density in community-dwelling older adults: Substudy of a randomized controlled trial. *J Intern Med* **282**(5), 452–460.

Stone K L, Lui L Y, Christen W G et al. (2017). Effect of combination folic acid, vitamin B_6, and vitamin B_{12} supplementation on fracture risk in women: A randomized, controlled trial. *J Bone Miner Res* **32**(12), 2331–2338.

Tong T Y N, Appleby P N, Armstrong M E G et al. (2020). Vegetarian and vegan diets and risks of total and site-specific fractures: Results from the prospective EPIC-Oxford study. *BMC Medicine,* **18**(1), 353

Zhao J, Zeng X, Wang J et al. (2017). Association between calcium or vitamin D supplementation and fracture incidence in community-dwelling older adults: A systematic review and meta-analysis. *JAMA,* **318**(24), 2466–2482.

FURTHER READING

Bilezikian J P (ed.) (2019) *Primer on the Metabolic Bone Diseases and Disorders of Mineral Metabolism*, 9th edn. American Society for Bone and Mineral Research.

26 Dental disease

Paula Moynihan

OBJECTIVES

By the end of this chapter you should be able to:

- give definitions for dental caries, erosion, abrasion, and periodontal disease
- describe the indices used for the measurement of dental caries
- describe the epidemiology and trends of dental caries
- outline the structure of the tooth
- explain the mechanisms of the decay process
- summarize evidence for the role of dietary sugars and fluoride in the causation and prevention of decay
- compare the relative cariogenicity of various sugars and other carbohydrates
- summarize the interaction of protective effects of fluoride and destructive effects of sugars.

26.1 Introduction

Teeth are important in enhancing facial appearance and for integration into society, as well as being important for eating and speaking. Despite being associated with low mortality rate, dental diseases inflict considerable pain and anxiety and are costly to healthcare services. Economic analysis indicates that worldwide in 2015 the direct costs of dental diseases was US$356.8 and in EU Member States expenditure on dental diseases was 90 billion euros, being the third most costly non-communicable disease. Attitudes towards dental health have changed dramatically over the last few decades: in the 1950s a common 21st birthday present for young women was a set of dentures. Nowadays, people are retaining their teeth well into older age and it is a realistic expectancy that teeth are for life. However, despite these expectations, the Global Burden of Disease Survey reported untreated dental caries in the permanent dentition is

the world's most prevalent condition affecting 34.1 per cent of the population, peaking in 15–19-year olds, with no improvement since 1990. To address this in a cost-effective way dentistry needs to move away from a profession that is mainly concerned with the reactive treatment of dental diseases such as drilling and extraction of teeth, to a profession with a true preventive focus, an aspect in which diet plays an important role.

Human beings have two dentitions. The deciduous dentition begins to appear in the mouth at about six months of age, consists of 20 teeth and is shed by early adolescence: the permanent dentition consists of 32 teeth and supplements, and replaces the deciduous dentition between the ages of about six and 21 years.

Dental diseases include enamel developmental defects (e.g., hypoplasia and fluorosis), tooth wear (e.g., erosion, abrasion, and attrition) (see Weblink Section 26.1 on Erosion), periodontal disease (gum disease) (see Weblink Section 26.2 on periodontal disease), and dental caries (see Figure 26.1). Defects to enamel may occur while the teeth are forming. There are two main types of enamel developmental defect: fluorosis and hypoplasia. The former is caused by an excess ingestion of excess fluoride during tooth development and is considered further in Section 26.4. Enamel hypoplasia has many causes including infections, drug side effects, congenital defects and dental trauma; dietary deficiency is another cause, considered further in Section 26.4.

A summary of the most recent evidence for nutrition and periodontal disease is given in Weblink Section 26.2. However, the main focus of this chapter is on diet, nutrition, and dental caries.

26.2 Epidemiology of dental decay

Archaeological surveys have revealed that dental caries was rare until the nineteenth century when the prevalence and severity rose rapidly, 'poor teeth' being the most important cause of rejection of volunteers for service in

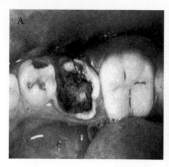
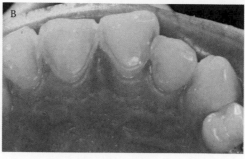

FIGURE 26.1 Caries (A) and erosion (B). Part B kindly supplied by Professor June Nunn.

the Boer War. The first surveys of the dental health of children in the UK were conducted between 1906 and 1908 and showed that 90 per cent of children were affected by decay with an average of four decayed teeth per child. From this time, the general trend shows that dental caries increased as consumption of sugars increased, and reached its peak in the late 1950s and 1960s; with interruptions to this trend occurring during the First and Second World Wars when the prevalence of dental caries fell by 40 per cent and 30 per cent respectively in the UK, due to a fall in sugars consumption. Surveys of adult and child dental health have been conducted in the UK since 1968. From 1985 standardized and coordinated surveys of child dental health have been conducted throughout the UK. In England, the National Dental Epidemiology Programme coordinated by Public Health England provides information on current disease levels and recent trends in dental health. Over the past four decades dental decay has declined considerably, largely due to the introduction of fluoride, especially in toothpaste. In 1983 the prevalence of dental caries (obvious decay experience) in 12-year olds in the UK was 81 per cent, and in 2013 this had reduced to 28 per cent for 12-year olds in England, Wales, and Northern Ireland. In 1983 the average number of teeth affected with dental caries per 12-year old child in the UK was 3.1 and in 2013 it was 0.8 for England, Wales, and Northern Ireland. For the primary dentition, despite year-on-year reductions in the prevalence of 5-year-olds with obvious caries experience from 30.9 per cent in 2008 to 24.8 per cent in 2015, there has been little improvement in recent years with 23.3 per cent affected in 2017 and 23.5 per cent in the most recent 2019 survey.

International figures: indices—DMFT, dmft

The severity of dental decay is measured using the dmft/ DMFT (primary dentition/permanent dentition) index, which is a count of the number of teeth that are decayed, missing, or filled. These indices are widely used throughout the world for monitoring levels of dental decay. The WHO criteria for assessing dental caries have been widely used in epidemiological surveys in which 'dental caries' is classified as when cavitation into the dentine of the tooth has occurred. Data on levels of dental caries from different regions of the world can be obtained from the WHO global oral health status report (WHO 2022) which provides data on the estimated prevalence of dental caries by WHO region and country, and the change in prevalence between 1990 and 2019. Despite improvement in dental health in many developed countries over the past 50 years, dental caries remains unacceptably high, affecting the majority of the world's population at age 12. Moreover, dental caries is a cumulative and progressive disease and increases with age: even in populations with low levels of dental caries at age 12, levels in adults are high and year-on-year increases in dental caries are observed for adult populations even where there is exposure to fluoride.

The global oral health status report estimated the prevalence of decay in the primary dentition of children globally to be 42.7 per cent, ranging from 38.3 per cent in high-income countries to 45.6 per cent in upper-middle income countries. In the UK in 2013 12 per cent of three-year-old children had dental caries, having one or more teeth affected. Data from the National Institute of Dental Health Research in the USA show that 27.9 per cent of 2–5-year-olds have decay experience and 20.5 per cent of 2–5 year-olds have untreated decay. New Zealand data from 2013 report 43 per cent of 5-year-olds have decay whereas in Sweden prevalence at age three is only 4 per cent. In low- and middle-income countries prevalence is generally higher, for example, in the Philippines, 88 per cent of 5-year-olds have decay, the majority untreated. The global oral health status report estimated the prevalence of dental caries of the permanent dentition of children aged five years and older is 28.7 per cent, ranging from 27.8 per cent in upper-middle income countries to 30.5 per cent in low-income countries.

Dental caries affects adults as well as children and the severity of the disease increases with age with almost all adults worldwide affected. The National Dental Epidemiology Programme for England, Oral Health Survey of Adults (aged 16+ years) Attending General Dental Practices in 2018 showed that although 82 per cent of participants had 21 or more natural teeth, 90 per cent had at least one filling and 27 per cent had untreated decay.

Geographic and social class differences

The prevalence of dental caries is very strongly related to social class. In the UK and many other countries there are marked differences in caries prevalence and severity between social classes and in some instances between geographical regions. Dental health shows a socioeconomic gradient with both prevalence and severity of dental caries being greater in deprived areas. The 2019 Oral Health Survey of 5-year-old children in England reported that 34.3 per cent of children from deprived areas versus 13.7 per cent of children in less deprived areas had caries experience. The survey showed higher levels of decay in the primary dentition in the North West of England (31.7 per cent, with an average of 1.2 affected teeth) compared with the South East (17.6 per cent, with an average of 0.8 affected teeth). In older children, most recent 2013 data show that the prevalence of decay in the permanent dentition for 12-year-olds is highest in Northern Ireland (57 per cent) compared with Wales (52 per cent) and England (32 per cent). Scotland was not included in this survey.

Dental health status may be assessed by looking at the proportion of a population that has no natural teeth, that is, are edentulous. In England, Wales, and Northern Ireland the proportion of adults that are edentulous has decreased dramatically since 1968 when 37 per cent of adults were edentulous to only 6 per cent in 2009. The National Dental Epidemiology Programme for England Oral Health Survey of Adults Attending General Dental Practices in 2018 showed 1.3 per cent of participating adults aged 65–74 years, and 4.4 per cent of those aged 74–85 years were edentulous. Adults from manual worker backgrounds are more likely to be edentulous: the 2009 Adult Dental Health Survey has shown that 98 per cent of adults from managerial and professional occupations were dentate compared with 90 per cent of those from lower socioeconomic occupation classifications. This survey showed that having 21+ teeth varied by age band, but there was little variation by ethnic group or by level of deprivation in area of residence.

26.3 The decay process

Today, the aetiology of dental caries is well established: dental caries is the localized loss of dental hard tissues as a result of acids produced by bacterial fermentation of sugars in the mouth. However, in the past, several theories for the causation of dental caries have been postulated, which are summarized below. It is now known that dental caries requires the presence of a cariogenic bacteria in the biofilm on the surface of the tooth and a sugars substrate to occur, but its development is influenced by the structure of the tooth, the salivary flow rate and composition, and the presence or absence of fluoride.

Theories of causation

Most research to support the theories of causation of dental caries favours the chemo-parasitic theory of W. D. Miller, although there were many earlier proposals (see Weblink Section 26.3 for extended information).

Ancient civilizations in China, Mesopotamia, and Greece believed that dental caries was caused by worms—the treatment for which was fumigation. In the 1920s and 1930s, dental caries was thought to be a deficiency disease and Lady May Mellanby (wife of Edward Mellanby who discovered vitamin D) believed that vitamin D deficiency was a major cause of dental caries due to its role in calcification. In the mid twentieth century some scientists thought that the protein matrix was the site of initiation of dental caries and devised the proteolytic theory of dental caries (Gottlieb's proteolytic theory). Later, in the 1960s, Jackson and co-workers proposed that dental caries was an autoimmune disease, another theory that never gained much credit. Of all the theories put forward, only the chemo-parasitic theory (also known as the 'acid theory') has persisted and is now widely accepted by dental scientists worldwide.

The structure of the tooth and the decalcification process

In order to describe the action of bacterial acids on the teeth an appreciation of the tooth structure is required. The teeth are composed of three mineralized tissues—enamel, dentine, and cementum (Figure 26.2). Dentine forms the bulk of the tooth and is mesodermal in origin. The dentine forms the roots of the tooth and is covered by a thin layer of cementum. The outer layer of the crown of the tooth consists of enamel, a hard substance that is ectodermal in origin. The teeth are supported by the alveolar bone of the maxilla or mandible, covered in epithelium. The epithelium around the necks of the teeth is called the gingivae (gums). The teeth are held in the alveolar bone by the periodontal ligament, allowing the teeth to move slightly. Enamel contains no cells, nerves, or blood vessels and is insensitive, but the dentine is very sensitive to many stimuli. The nerves and blood vessels

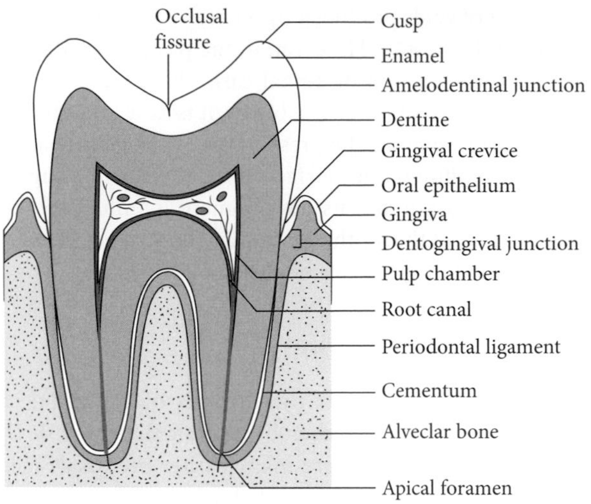
Occlusal fissure
Cusp
Enamel
Amelodentinal junction
Dentine
Gingival crevice
Oral epithelium
Gingiva
Dentogingival junction
Pulp chamber
Root canal
Periodontal ligament
Cementum
Alveclar bone
Apical foramen

FIGURE 26.2 Vertical section through a permanent molar tooth.
Source: Diagram drawn by D. S. Brown and reproduced with permission.

supplying the dentine come from the pulp that forms the soft centre of a tooth and is in turn supplied by nerves and blood vessels from the alveolar bone via the apical foramen of the tooth roots. Dental enamel consists of crystals of hydroxyapatite, a crystalline compound composed of calcium and phosphate arranged in a characteristic way in a thinly dispersed organic matrix.

The role of oral microorganisms in dental diseases and the role of dental plaque

An essential factor in the aetiology of dental caries is dental plaque, a white, slightly glutinous layer, which builds up on the surfaces of teeth when they are not cleaned. The plaque biofilm is composed of three main components: the pellicle, the plaque microbiome, and extracellular plaque matrix. The pellicle is the first layer to form on clean enamel and consists of salivary proteins including glycoproteins that are adsorbed onto the enamel surface. Microorganisms then become attached to the pellicle and multiply, forming colonies and developing to form a continuous layer increasing in depth. The microorganisms make up 70 per cent of plaque while the remaining 30 per cent consists of the plaque matrix. The polymers that make up much of the matrix are derived from dietary sucrose by enzymes secreted by the plaque microorganisms. Glucans (dextran and mutan) are formed mainly from dietary sucrose by glucosyltransferase of plaque streptococci (see Figure 26.3). The glucans help maintain plaque integrity due to their glutinous nature but also form an energy store from which bacteria feed during long breaks between the host's meals. Sucrose can also be converted to fructans (levans), which are rapidly metabolized by plaque enzymes. Dental plaque can usually be found in areas around the teeth that are least easily

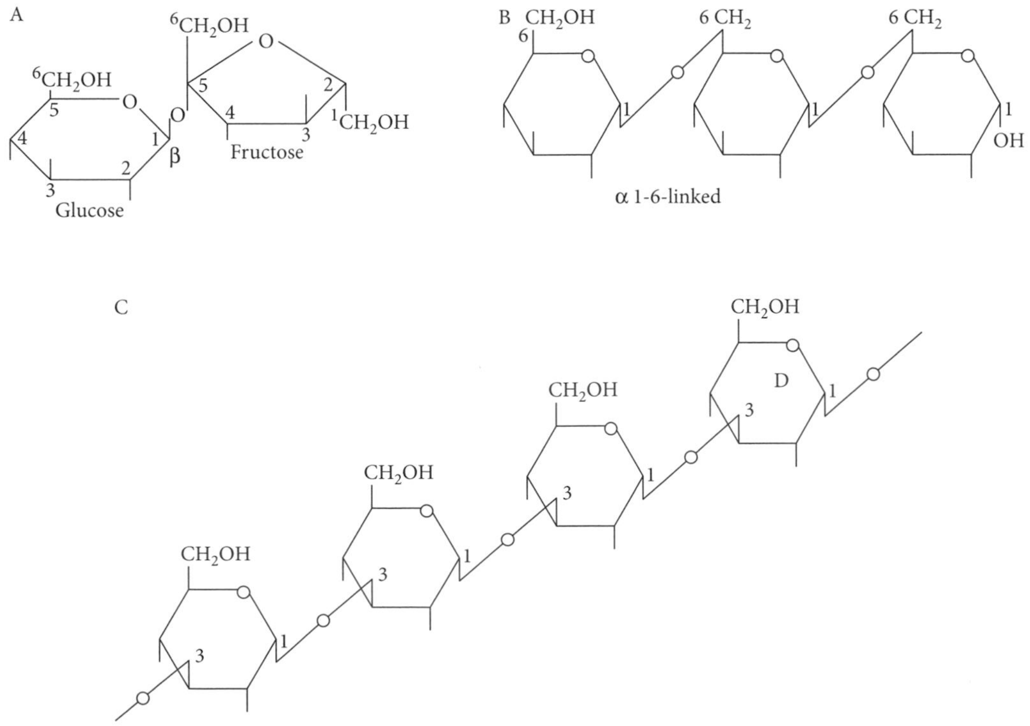
A
Glucose
Fructose
B
α 1-6-linked
C

FIGURE 26.3 Sucrose (A), dextran (B), and mutan (C).

cleaned, mainly in the pits and fissures of the occlusal (or chewing) surfaces, between adjacent teeth (proximal surfaces), or along the gingival margin of the tooth on the buccal (outer) or lingual (inner) surfaces of the teeth.

Dental caries arises as a result of an interaction between dietary sugars, plaque microbiome, and the host's tooth surfaces. Periodontal disease also occurs as a result of the presence of certain oral bacteria. It is therefore important to understand the composition of the oral microbiome and how this alters in health and disease. The main concentration of bacteria in the mouth is in the dental plaque, contributing to the majority of plaque volume. It is estimated that dental plaque contains over 500 different types of bacteria. The majority play no direct role in caries development but the type of microorganism present will influence the properties of the plaque.

Mutans streptococci are the main bacteria associated with dental caries and they play a crucial role in the initiation and progression of dental caries. They readily produce acids from dietary sugars and can synthesize extracellular plaque polysaccharides from sucrose, creating ideal conditions for bacteria such as lactobacilli, bifidobacteria, and some non-mutans streptococci, also associated with dental caries. To learn more about the cariogenic bacterial associated with dental disease visit our online resources at (Weblink Section 26.4). An account of the role of the oral microbiome in dental diseases is given by Kilian et al. (2016).

The caries process

Dietary sugars diffuse into the dental plaque where they are metabolized by plaque microorganisms to acid. Most of the acid produced is lactic, with some acetic, formic, and propionic acids also being produced. The acid produced reduces the pH of dental plaque, the mineral phase of enamel is dissolved by the plaque acids, and the caries process has begun. Enamel hydroxyapatite usually begins to dissolve around pH 5.5, which is sometimes referred to as the 'critical pH'. When the pH rises above this value, remineralization of enamel may occur. Saliva promotes remineralization as it contains bicarbonate which increases pH and encourages deposition of mineral in porous areas where demineralization of enamel or dentine has occurred. A demineralized lesion may therefore be remineralized; however, this is a slow process that competes with factors causing demineralization. If the pH in the mouth remains high enough for sufficient time, then complete remineralization may occur. However, if demineralization dominates, the enamel becomes more porous until finally the surface gives way and a cavity forms. The rate of demineralization is affected by the concentration of hydrogen ions (i.e., pH at the tooth surface) and the frequency with which the plaque pH falls below

the critical pH. Another relevant point is the amount of calcium and phosphate in plaque, since high levels of these minerals in plaque will help resist dissolution of the enamel. Overall, caries occurs when demineralization exceeds remineralization. The development of caries requires sugars and aciduric bacteria to occur, but is influenced by the composition of the tooth (the structure of the enamel can be altered by the diet while the teeth are forming), the quantity and composition of saliva (e.g., calcium and phosphate content and buffering power) and the time for which dietary sugars are available for fermentation.

The Stephan curve: the effects of different food combinations on plaque pH

Stephan, in the 1940s, pioneered work on the pH of dental plaque using microelectrodes. This work indicated that the resting pH of plaque was around 6.5 to 7 but, on exposure to sugars (glucose or sucrose), fell rapidly within a few minutes, to around pH 5. The rapid fall was followed by a slow recovery to baseline pH over the next 30–60 minutes. Stephan plotted pH against time and this time/pH graph is commonly referred to as a 'Stephan curve'. An example of a Stephan curve is shown in Figure 26.4. The Stephan curve has been commonly used to measure the acidogenic potential of a range of foods and this provides an indirect measure of the cariogenic potential of the food. However, it must be noted that measures of plaque pH alone must not be taken as a direct measure of cariogenic potential since these measurements take no account of protective factors in foods, the resistance of enamel, and salivary factors that influence the caries process. In order to determine the cariogenic potential of a food, data from plaque pH studies need to be interpreted alongside data from other types of study including animal studies and epidemiological surveys.

Consumption of different food combinations results in different patterns in plaque pH. This was clearly illustrated in a study that looked at the effect on plaque pH of consuming a sugary snack followed by either sweetened coffee or a 15 g piece of cheese (Figure 26.4). Consumption of cheese following a sugary snack almost abolishes the fall in plaque pH that usually results from sugars consumption. This effect of cheese is probably due to the stimulation of saliva by this highly flavoured food and its low sugars (lactose) content. Other foods that are good stimuli to salivary flow include peanuts and sugar-free chewing gum, and these also reduce the pH fall if consumed following a sugars-containing item. When sugars are consumed with other foods, the effect on pH is reduced, probably due to a diluting effect and the increased salivary flow due to mastication of other

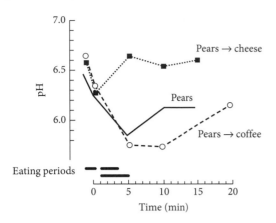

FIGURE 26.4 Stephan curves produced by eating either cheese or sugared coffee after tinned pears in syrup.
Source: A J Rugg-Gunn, W M Edgar, D A M. Geddes et al. (1975) The effect of different meal patterns upon plaque pH in human subjects. *Br. Dent.J.* **139**, 351–356. Reprinted by permission from Springer Nature.

foods. This is one reason why it is often recommended to consume sugary foods at mealtimes only (another reason is that limiting sugars to mealtimes reduces the frequency of consumption). A study that examined the effect of consuming breakfast items in a different order on plaque pH illustrates this diluting effect. The breakfast items were sugar-containing coffee, a boiled egg, and crispbread with butter. The smallest drop in pH was observed when all three items were consumed together and the largest drop in pH was observed when the sugared coffee was consumed alone. Therefore, consumption of one food may affect the acidogenicity of another.

Chewing sugar-free chewing gum following a sugars-containing snack has also been shown to increase the rate at which plaque pH returns to baseline. This has led to the advice to chew sugar-free gum following a sugary snack or meal.

26.4 Promoting and protective factors

There are many factors that have the potential to promote dental caries: poor oral hygiene (which contributes to markedly higher pathogenic bacteria in the oral microbiome) and intake of dietary sugars among them. Likewise, there are several factors known to be protective against dental caries, most notably fluoride exposure (both dietary and non-dietary) and restricted intake of sugars. However, there are also a number of other dietary factors that protect against dental caries, including milk, cheese, and xylitol, and knowledge of these factors assists in making dietary advice for dental health more positive.

Diet

Diet can affect the teeth while they are forming, before they erupt into the mouth (a pre-eruptive effect) and, once erupted, by a local direct effect. Much research was undertaken in the first half of the twentieth century on the pre-eruptive effect of diet on tooth structure. Deficiencies of vitamin D, hypocalcaemia, and protein–energy malnutrition (PEM) have been associated with enamel hypoplasia. This is an enamel developmental defect characterized by pits, fissures, or larger areas of missing enamel that become stained post-eruption and which render the tooth more susceptible to decay. A systematic review of the evidence for an impact of vitamin D status on risk of dental caries found evidence to suggest that, despite limitations of the quality of data, vitamin D supplementation may decrease incidence of dental caries (Hujoel 2013). A low serum concentration of 25-hydroxyvitamin D has also been associated with increased risk of dental caries. PEM and vitamin A deficiency can also cause salivary gland atrophy, reducing the quantity and affecting the composition of saliva, ultimately reducing the mouth's defence against plaque acids. However, in developing countries, in the absence of dietary sugars, undernutrition is not associated with dental caries. Undernutrition coupled with a high intake of sugars results in levels of caries greater than expected for the level of sugars intake. Despite considerable past interest in the pre-eruptive effect of diet on tooth decay, today, the post-eruptive local effect of diet in the mouth is considered to be much more important. There is a wealth of evidence for the association between diet and dental caries and most attention has focused on the important role of dietary sugars in the aetiology of caries.

The frequency of eating sugars, the amount eaten, and the cariogenicity of different dietary sugars have all to be considered. The cariogenicity of starches, the relative cariogenicity of naturally occurring and 'free' sugars ('added' sugars, plus the sugars in honey syrups and fruit juices and fruit juice concentrates), the possible effect of factors in foods that may protect against dental caries are important issues for health professionals and nutritional and dental scientists. The evidence relating diet to dental caries comes from a number of types of experiments, including human observational studies and human intervention studies, animal experiments, the aforementioned plaque pH studies and *in vitro* laboratory experiments. The strongest evidence comes from human epidemiological studies; however, it is important to consider collative evidence from all types of study in order to obtain an overall picture regarding cariogenicity of a product.

The role of dietary sugars

There is a wealth of epidemiological evidence showing an association between intake of dietary sugars and dental caries. In the past, when sugar (sucrose) intake and levels of dental caries have been compared on an inter-country basis using food balance data and WHO data on caries levels of different countries, a positive correlation has been found. Today, the relationship is not so evident in countries that have a high sugars intake since when all are exposed to a risk factor the relationship between the risk factor and the disease is less evidence. Moreover, modern diets contain an array of free sugars in addition to dietary sucrose.

Non-randomized intervention trials

There are no randomized controlled trials of the impact of sugars consumption on levels of dental caries. There have been two historical intervention studies both conducted in adults but neither of the studies was randomized. First, the Vipeholm study was conducted in a mental institution in Sweden shortly after the Second World War. The 964 patients (80 per cent of whom were male) were divided by wards into one control group and six test groups. Groups were given high sucrose intakes at meals only, or both at and between meals, in non-sticky (sucrose solution, chocolate) or sticky forms (caramels, toffees, sweet bread). The study was complicated but from the results it was concluded that: sugars consumption even at high levels is associated with only a small increase in caries increment if taken up to four times a day as part of meals; consumption of sugars between meals as well as at meals is associated with a marked increase in caries; and caries activity disappears on withdrawal of sugars from the diet. The highest caries increment was observed in the group that consumed 24 sticky toffees throughout the day. However, subtle differences between types of sugars were largely overridden by the effect of frequency. It would not be possible to repeat such a study today, as it would be unethical to prescribe high sugars diets knowing the association between sugars and dental caries.

The second human intervention study took place in Turku, Finland, in the 1970s. The aim of this two-year study was to investigate the effect on dental caries of nearly total substitution of sucrose in a normal diet with either fructose or xylitol. When only cavities were counted the results showed 56 per cent fewer cavities in the xylitol group than in the sucrose group but a similar number of cavities formed in the sucrose and fructose groups. The diet containing xylitol was therefore less cariogenic than the sucrose or fructose diets. After two years the mean DMFS increment was 7.2 in the sucrose group, 3.8 in the fructose group, and 0 in the xylitol group.

Observational studies

Cross-sectional epidemiological observation studies relating level of sugars consumption with dental caries experience have been popular as they are easy to do, but they can be misleading. This is because simultaneous measurements of diet and dental caries levels may not provide a true reflection of the role of diet in the development of the disease. (For fuller information on observational studies see Weblink Section 26.5.)

Other epidemiological studies have observed the level of dental caries in groups of people that habitually consume high or low amounts of sugars in their diet. Evidence for an association between sugars intake and dental caries also comes from population-based observational studies which show a marked increase in dental caries occurs in populations that have undergone the 'nutrition transition', that is, they have moved away from their traditional diets that were low in free sugars and adopted a Westernized diet high in free sugars. There are few epidemiological data relating dental caries to sugars consumption in populations consuming low levels of sugars. However, during the Second World War there was a reduction in sugar availability in many countries. Data exist from 11 European countries showing that a reduction in caries accompanied reduced sugar availability.

In the absence of randomized controlled trials on the impact of dietary sugars on caries development the next best available evidence probably comes from longitudinal cohort studies that enable sugars consumption to be related to development of dental caries over time. A systematic review of the evidence pertaining to amount of dietary sugars and risk of dental caries identified eight cohort studies (Moynihan & Kelly 2014). Five of the cohort studies identified in the systematic review had data that enabled the level of dental caries to be compared when the intake of free sugars was below compared with above 10 per cent of energy, and all showed lower dental caries when intake of free sugars was below 10 per cent energy. However, even at this level of sugars intake dental caries was not eliminated. Even low levels of dental caries in childhood are of significance since dental caries is a progressive and cumulative disease and therefore small reductions in dental caries in childhood are important to protect oral health throughout the lifecourse.

Based on systematic reviews of the evidence, WHO published a guideline 'Sugars Intake in Adults and Children' in which it made a strong recommendation that the intake of free sugars by individuals within populations is no more than 10 per cent energy (on an individual level) and a conditional recommendation that intakes are limited further to <5 per cent energy. In the UK, SACN (Scientific Advisory Committee on Nutrition) also systematically reviewed the evidence and recommended that the intake of free sugars on a population basis should be limited to <5 per cent energy intake to protect both oral and general health.

Frequency and amount of sugars

The results of experiments in rats have shown caries severity to increase with increasing sugars concentration of the diet when frequency of intake was similar, thus suggesting the amount of sugars eaten is related to the severity of caries. Evidence from studies in humans clearly indicates that the amount of sugars consumed and the frequency with which sugars are consumed are strongly associated (see Figure 26.5). Some studies in humans have shown that frequency of sugars intake is related to caries development, including the aforementioned Vipeholm study. However, very few studies have investigated the relationship between daily frequency of free sugars intake from all sources and development of dental caries. Many studies have used food frequency questionnaires to study the relationship between the intake of sugars-containing foods and drinks and caries, for example, the association between frequency of sweets or soft drinks and caries but these do not always report daily frequency, often asking how many times a week or month items are consumed. Only studies that measure both daily frequency and daily amount of sugars simultaneously can conclude which variable is more important, and these have shown that either amount of both are related to caries development. A recent systematic review of the epidemiological evidence pertaining to the frequency of intake of free sugars and caries risk

identified three relevant cohort studies (SACN 2015) none of which reported a significant association between the frequency of sugars consumption and caries risk, and overall it was concluded that the limited evidence showed did not show an association between frequency of consumption risk of caries. However, most studies have found that the two variables are highly correlated and so efforts to control one of these variables, should also control the other. Frequency is useful as a health education message when providing chairside advice as it can make advice to reduce sugars intake more tangible for the patient. However, when setting guidelines and policy it is more appropriate to set a goal in terms of amount as this will provide a tangible benchmark against which the intake of a population can be compared, and the success of health promotion strategies monitored. Moreover, reducing the frequency of free sugars intake alone is unlikely to have an impact on the occurrence of other non-communicable diseases associated with free sugars intake.

Starches and dental caries

Some postulate that all carbohydrate foods, sugars, and starches, should be considered cariogenic. This is because saliva contains amylase, which may hydrolyse dietary starch to form glucose, maltose, and maltotriose that can then be fermented by plaque bacteria to acids. However, starch is heterogeneous in nature; it varies in botanical origin, may be consumed in its raw form as in fruits and vegetables or it may be cooked, and varies widely in degree of refinement and processing, and all these factors must be considered when assessing the cariogenicity of starches. Current dietary guidelines promote the consumption of starch-rich foods, especially wholegrain varieties, such as bread, potatoes, pasta, rice, other cereal grains, and of fruit and vegetables. It is therefore important that the cariogenic potential of such foods is understood so that dietary advice for general health is not contraindicated in terms of dental health. A WHO Expert Consensus, published in 2003 concluded that there was little evidence from epidemiological, animal, and experimental studies of an association between intake of staple starchy foods and dental caries (WHO 2003). A more recent WHO-commissioned systematic review of the evidence pertaining to the effect of total dietary starch and of replacing rapidly digestible starches with slowly digestible starches on risk of dental caries, concluded that the best available evidence suggested no association between total starch intake and caries risk but that consumption of rapidly digestible starches (e.g., highly processed starchy foods) may significantly increase caries risk (Halvorsrud et al. 2019).

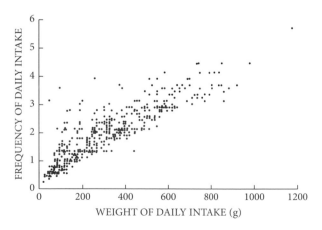

FIGURE 26.5 Plot of frequency of intake per day against weight consumed per day of confectionery, by 405 12-year-old English children. The correlation is +0.77.
Source: A J Rugg-Gunn (1993) Dental caries: The role of dietary sugars. In: *Nutrition and Dental Health*, 178 (Rugg-Gunn A J ed.). Oxford Medical Publications, Oxford. with kind permission of the author.

Fruit and dental caries

There is little evidence from epidemiological studies to show fresh whole fruit to be an important factor in the development of dental caries, and in fact apples have long been used as a symbol of oral health. Clinical trials of the effects of apple consumption on dental caries produced equivocal results; however, apples do contain polyphenols, which have an antibacterial nature. Despite this, animal studies have shown that fruit causes caries when consumed at very high frequencies, often in a pulped form (i.e., extrinsic), and some plaque pH studies have found fruit to be acidogenic, but less so than sucrose. It is important to consider, however, that plaque pH studies take no account of the protective factors found in fresh fruits, the fact that they provide a good stimulus to salivary flow, and any longer-term beneficial effects of fruit and vegetable consumption on the composition and characteristics of the oral microbiome. Based on the available evidence, it is unlikely that fruit consumption as part of a balanced diet increases risk of dental caries.

Many fruits are acidic, for example citric acid in citrus fruits, oxalic acid in rhubarb, tartaric acid in grapes, and malic acid in apples. This has led to concern that fruit consumption may contribute to erosive tooth wear. Fruit juices are more erosive than whole fruits, primarily because they are much more concentrated (1 x 200ml portion of orange juice contains the juice of approximately three oranges). In addition, consumption of whole fruit provides a good stimulus to salivary flow which neutralizes acid. Most data associating whole fruit consumption with dental erosion come from one-off case studies of unusual dietary habits. The WHO report 'Diet, nutrition and the prevention of chronic diseases' concluded that there was insufficient evidence to link whole fruit consumption to increased risk of dental erosion (WHO 2003). However, more recent data suggest consumption of fruit alone as opposed to with other foods may be a factor: one case-control study showed fruit consumption in-between meals, but not with other foods as meals, was associated with erosive tooth wear, whereas acidic drink consumption was strongly associated with erosion irrespective of timing and mode of consumption (O'Toole et al. 2017).

In conclusion, fruit juices contain free sugars and fruit acids along with fewer protective factors compared with whole fruits and they therefore pose a threat to dental health. By contrast, as habitually consumed, whole fruit does not pose significant risk to dental health especially if consumed as part of a meal. It could also be argued that replacing foods high in free sugars with fresh fruit would be likely to reduce dental decay.

Other carbohydrates and dental decay

Modern diets contain an increasing array of carbohydrates other than starches and sugars including maltodextrins, glucose syrups (collectively known as glucose polymers), and non-digestible oligosaccharides such as oligofructose and gluco-oligosaccharides. The latter are increasingly being used in foods as they are prebiotics, encouraging the growth of favourable colonic bacteria. Relatively much less is understood on the cariogenic potential of these carbohydrates. However, the limited data that are available suggest that they are not safe for teeth.

The role of fluoride in caries prevention

Fluoride increases the resistance of the teeth to decay primarily through an intra-oral topical effect but also as a lesser systemic effect on the teeth during the period of tooth development. Fluoride ingested during the period of the development of the enamel becomes incorporated into the enamel crystal structure and replaces the hydroxyl groups in hydroxyapatite to form fluoroapatite, which is more stable and resistant to demineralization. However, the main effect of fluoride is through its intra-oral effect: remineralization of enamel in the presence of fluoride results in the porous lesion being remineralized with fluoroapatite rather than hydroxyapatite. Furthermore, intra-oral fluoride inhibits bacterial sugars metabolism, which results in less acid production. The inverse relationship between fluoride in drinking water and dental caries is well established and benefits the teeth even when there is exposure to fluoride through other sources, for example, toothpaste. Overall exposure to fluoride reduces dental caries by approximately 50 per cent but does not eliminate dental caries.

Arguments for and against water fluoridation

The link between water fluoride content and dental caries prevention was first established in the USA in the early twentieth century and its effectiveness has now been demonstrated in over a hundred surveys in more than 20 countries including the UK. The first area in the UK to have an artificially fluoridated water supply was Birmingham in 1964, an area that consequently has seen a dramatic improvement in levels of dental caries. The reduction in dental treatment arising from fluoridation results in considerable savings to health costs, due to the fall in the number of extractions and general anaesthetics. There are examples where water fluoridation has been discontinued and subsequently levels of dental caries have increased, for example areas of Scotland including

Kilmarnock, Wick, and Stranraer where dental caries levels increased, despite the fact that fluoride toothpastes were available. A 25 per cent increase in dental caries was observed in some areas of Scotland over five years after removal of water fluoridation.

Fluoridation of drinking water can substantially decrease dental caries but an excess of fluoride during the development of the teeth may cause 'dental fluorosis', an enamel developmental defect that manifests as small white diffuse opacities with severe pitting and staining of enamel in more severe cases. For permanent teeth the period when there is greatest risk from excess fluoride is between two and six years. Severe fluorosis is rare in the UK and cases have usually been linked with excessive fluoride ingestion from eating toothpaste or misuse of fluoride supplements. Severe fluorosis is observed particularly in countries that have very high levels of fluoride in water supplies. Enamel fluorosis as well as skeletal fluorosis are found in large areas of India, Thailand, in the Rift Valley of East Africa, and in many Arab states. It is important to realize that fluoride is not the only cause of opacities in teeth.

The optimal level of fluoride in water is the level at which a substantial caries reduction is observed with a negligible prevalence of enamel fluorosis. In temperate climates including the UK the optimum concentration of fluoride is 1.0 mg/l, while in warmer climates it might be nearer 0.6 mg/l. Water fluoridation is endorsed by more than 150 science and health organizations including the International Dental Federation, the International Association for Dental Research, and the WHO. Despite this expert endorsement, there are small groups of people who strongly oppose water fluoridation on the grounds of perceived health risks and imposed treatment of the water supply. To learn more about this topic see Weblink Section 26.6.

In conclusion, fluoridation of the water supply is a caries-preventive measure with the potential to reach the sectors of the population that are at highest risk of caries. One could argue that in some areas, where caries levels are very low, water fluoridation might not be a cost-effective measure. However, even when caries levels are low in children, levels in adults are high and in some parts of the UK and other countries, including areas of social deprivation, prevalence of caries in childhood remains high and dental attendance, oral hygiene practice, and dietary habits are poor. For these reasons water fluoridation is a highly effective, economical public health measure.

Other sources of fluoride

Effective use of fluoride for caries prevention is strongly endorsed by the WHO. Systemically, exposure to fluoride can be gained through fluoridation of the water supply, salt fluoridation schemes, and through the provision of fluoridated milk to schools. Fluorides are also widely found in nature in addition to being naturally present in some water supplies. Fluoride is found in seafood (when bones are eaten), tea leaves, some beers, and in foods cooked in fluoridated water. Salt fluoridation has been successfully implemented in Switzerland since 1955 and fluoridated salt sits alongside non-fluoridated salt in supermarkets, allowing the consumer choice, which is politically advantageous. Fluoridated salt has also been effectively used in Jamaica, Costa Rica, and parts of France and Germany. In the UK school milk fluoridation schemes have run in Merseyside. Globally, school milk fluoridation schemes have also been introduced in the Russian Federation, Chile, and Thailand. School milk fluoridation provides a useful means of conveying the benefits of fluoride to children living in socially disadvantaged communities where the water supply is non-fluoridated. Evidence synthesis, albeit from observational data, has indicated a beneficial effect of drinking fluoridated milk on caries prevention in children.

The main effect of dietary fluoride is a local effect on the teeth whilst in the mouth, however, it also has a lesser systemic effect on the teeth after digestion and absorption. Fluoride in toothpaste and mouth rinse provides a mainly topical effect as these are not supposed to be swallowed.

Are sugars as strongly related to dental decay in the presence of fluoride?

Despite a marked effect of fluoride on caries prevalence, a relationship between sugars intake and caries still exists in the presence of fluoride. Longitudinal studies of the relationship between intake of dietary sugars and dental caries levels in children in the UK and USA have shown that the observed relationship between sugars intake and development of dental caries remains even after controlling for use of fluoride. A study conducted in northeast England reported on the decrease in dental caries levels during the Second World War in 12-year-old children from areas with naturally high and low water fluoride. Caries levels were lower in the high fluoride area in 1943 but, following the wartime sugar restriction, dental caries levels fell further by approximately 50 per cent, thus indicating that exposure to fluoride did not totally override the effect of sugar in the diet. Some argue that exposure to fluoride delays rather than prevents dental caries as even when levels of dental caries are low in children exposed to fluoride, levels by adulthood are high and an annual increase in dental caries of one carious surface per year has been reported for older adults residing in fluoridated USA (Sheiham & James 2014).

In a longitudinal study of over 1700 Finnish adults, the observed linear relationship between amount of sugar consumed and development of DMFT was weaker in those who brushed daily with fluoridated toothpaste compared with less than daily, thus indicating that fluoride reduced but did not eliminate the association between amount of dietary sugars and dental caries (Bernabé et al. 2016). It is likely that, in high-income countries where there is adequate exposure to fluoride, a further reduction in the prevalence and severity of dental caries will not be achieved without addressing the causes of social and environmental factors that drive a high sugars intake (Moynihan & Miller 2020).

Other dietary factors

Other minerals

Apart from fluoride there are other trace elements that influence dental caries although the influence is of relatively small importance. Dietary molybdenum, strontium, boron, and lithium are related to a lower caries experience in humans while higher selenium intakes are related to higher caries prevalence.

Other factors that protect against dental caries

The protective effect of some food components against dental caries has been recognized for decades. In the 1930s Osborn and Noriskin suggested that foods provided substances that protect against decay. Apart from the well-recognized role of fluoride, other dietary components such as phosphates, calcium, casein, and polyphenols may also have cariostatic properties. Milk, despite containing approximately 5 per cent sugar primarily as lactose, was one of the first foods to be described as cariostatic. Although lactose may be fermented to acid, it has been shown to be the least cariogenic of the common dietary sugars. Milk also contains high concentrations of calcium and phosphate and is also rich in casein. Many studies of several types have all indicated that milk is not cariogenic and may even protect against dental caries. The cariostatic nature of cheese is also well established with evidence from animal and human experimental and intervention studies demonstrating its cariostatic nature. Cheese is a strong gustatory stimulus to salivary flow, which conveys protection to the teeth. However, cheese has been shown to be cariostatic in animals that have had their salivary glands removed and so the cariostatic effect is not due to saliva alone. A high concentration of calcium and phosphate and the formation of casein phosphoproteins are thought to convey a strong anti-caries effect.

Inorganic phosphates protect against dental caries by increasing the availability of phosphate in plaque so that demineralization is resisted and remineralization encouraged. Organic phosphates protect mainly by binding to the tooth surface and reducing enamel dissolution. Phytates are the most effective of these compounds. Despite promising results from incubation and animal experiments for a cariostatic effect of phosphates, studies in humans showed them to be less effective, possibly due to the higher phosphate concentration of human compared with rat saliva. Phytates also reduce the absorption of some micronutrients, for example, iron and zinc, and are therefore unsuitable as caries-preventive food additives.

Recent interest is focusing on the cariostatic properties of polyphenolic compounds found in plant foods. Apples contain polyphenols yet clinical trials of the impact of apples on dental caries have shown equivocal results. Polyphenolic compounds in cranberries have antimicrobial effects and have been shown to reduce the adhesion of *Streptococcus mutans*. Both green and oolong teas contain polyphenolic compounds that have been shown to suppress *Streptococcus mutans* and reduce extracellular glucan formation. Other foods such as honey, chocolate, and liquorice all contain factors that protect against dental caries but the benefit of these factors is overridden by the negative effect of the high sugars content of these foods or in the case of liquorice the dark staining effects.

Sugar-free foods that stimulate salivary flow can be classed as caries protective and include sugar-free chewing gum. Plaque pH studies have shown that chewing sugar-free gum increases plaque pH (see Figure 26.6). Results of several clinical trials have shown that

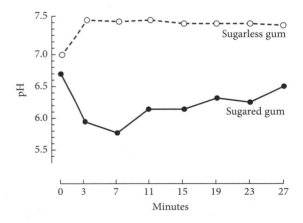

FIGURE 26.6 Measurements of plaque pH during chewing of sugared or sugarless (sorbitol-containing) chewing gum. *Source*: Reprinted by permission from Macmillan: *British Dental Journal*. A Rugg-Gunn, W G Edgar, G N Jenkins (1978) The effect of eating some British snacks upon the pH of human dental plaque. *Br. Dent. J.* **145**, 95–100. © 1978, Nature Publishing Group.

sugared gums are cariogenic when compared with sugar-free gum or no gum and sugar-free gums are caries-preventing compared with no gum. The most impressive results have been obtained with gums containing the non-sugar sweetener xylitol.

KEY POINTS

- The dental health of children in the UK has improved over the past four decades but the prevalence remains high, for example, almost a quarter of 5-year-old children have decay, with those from lower socioeconomic groups most affected.

- The prevalence of dental caries varies across the world with the highest prevalence in some low- and middle-income countries such as the Philippines, and the lowest levels in high-income countries including the UK and Sweden.

- The majority of adults attending dental practices have 21+ teeth and only 7.5 per cent of those aged 85+ attending the dentist were edentulous.

- Dental caries occurs due to demineralization of the dental mineralized tissues by acids derived from the bacterial metabolism of dietary sugars to acids.

- Acidogenic bacteria, for example, *Mutans streptococci*, have an important role in dental caries since these microorganisms can produce acid from sugars at a low pH. However, the overall composition of the oral microbiome may impact on the caries process.

- Epidemiological evidence shows that the amount of intake of sugars is related to dental caries. Data from experimental and animal studies suggests that frequency of intake of sugars is an important factor but there is insufficient epidemiological evidence to show a relationship between frequency of intake of sugars and risk of dental caries.

- Some foods protect against decay. These include milk, cheese, some plant foods, and foods that stimulate salivary flow.

- Fluoride, both dietary and non-dietary, is a highly effective caries-preventive measure but it does not eliminate dental caries. In populations exposed to fluoride, further reductions in caries levels will not occur without addressing the factors that drive a high sugars intake.

 Be sure to test your understanding of this chapter by attempting multiple choice questions. See the Further Reading list for additional material relevant to this chapter.

REFERENCES

Bernabé E, Vehkalahti M M, Sheiham A et al. (2016) The shape of the dose–response relationship between sugars and caries in adults. *J Dent Res* **95**(2), 167–172.

Halvorsrud K, Lewney J, Craig D et al. (2019). Effects of starch on oral health: Systematic review to inform WHO Guideline. *J Dent Res* **98**(1), 46–53.

Hujoel P P (2013) Vitamin D and dental caries in controlled clinical trials: Systematic review and meta-analysis. *Nutr Rev* **71**(2), 88–97.

Kilian M, Chapple I L C, Hannig J M et al. (2016). The oral microbiome: An update for oral healthcare professionals. *Brit Dent J* **221**, 657–666.

Moynihan P & Kelly S A M (2014) Effect on caries of restricting sugars intake: Systematic review to inform WHO guidelines. *J Dent Res* **93**, 8–18.

Moynihan P & Miller C (2020). Beyond the chair: Public health and governmental measures to tackle sugar. *J Dent Res* **99**(8), 871–876. doi:10.1177/0022034520919333.

O'Toole S, Bernabé E, Moazzez R et al. (2017). Timing of dietary acid intake and erosive tooth wear: A case-control study. *J Dent* **56**, 99–104. doi: 10.1016/j.jdent.2016.11.005. Epub 2016 Nov 14. PMID: 27856311.

Sheiham A & James W P (2014) A new understanding of the relationship between sugars, dental caries and fluoride use: Implications for limits on sugars consumption. *Publ Hlth Nutr*. **17**(10), 2176–2184. doi:10.1017/S136898001400113X

WHO (2003) *Diet, Nutrition and the Prevention of Chronic Diseases*. Technical Report Series 916. World Health Organization, Food and Agricultural Organization, Geneva.

WHO (2022) The WHO report: Global oral health status report: Towards universal health coverage for oral health by 2030. Available at: global-status-report-on-oral-health-2022 (accessed 30 April 2023).

FURTHER READING

Delivering Better Oral Health: An Evidence-based Toolkit for Prevention, 4th edn. (2023) https://www.gov.uk/government/publications/delivering-better-oral-health-an-evidence-based-toolkit-for-prevention (accessed 30 March 2023).

Fejerskov O, Nyvad B, Kidd E (2015) *Dental Caries: The Disease and its Clinical Management*, 3rd edn. Wiley Blackwell, Oxford.

Levine R S & Stillman-Lowe C R (2019) *The Scientific Basis of Oral Health Education,* 8th edn. Springer, Cham, Switzerland.

Najeeb S, Zafar M S, Khurshid Z et al. (2016) The role of nutrition in periodontal health: An update. *Nutrients* **8**(9), 530. doi:10.3390/nu8090530.

Peres M A, Macpherson L M D, Weyant R et al. (2019). Oral disease: A global public health challenge. *Lancet. Oral Hlth Sers.* **394**, 249–260.

Righolt A J, Jevdjevic M, Marcenes W et al. (2018). Global-, regional-, and country-level economic impacts of dental diseases in 2015. *J Dent Res* **97**, 501–507.

SACN (2015) *A Systematic Review of the Evidence: Carbohydrates and Oral Health*, available from: https://www.gov.uk/government/publications/sacn-carbohydrates-and-health-report (accessed 15/03/2031).

WHO (2015) *Guideline on Sugars Intake by Adults and Children.* WHO, Geneva.

WHO (2019). *Implementation Manual on Ending Childhood Caries.* WHO, Geneva.

Zohoori F V & Duckworth R M (eds) (2020) The impact of nutrition and diet on oral health. *Monogr Oral Sci.* **28**, 125–133. Karger, Basel. https://doi.org/10.1159/000455380

Immune regulation, food allergies, and food intolerance

Stephan Strobel and Carina Venter

27.1 Introduction

Following Jenner's seminal observation (1789) that exposure to cowpox could protect against smallpox ('vaccination'), it was realized that after successful recovery from a particular infectious disease, the same disease rarely occurred again. This altered adaptive reactivity is what we call specific immunity. Innate immunity is not dependent on prior antigen contact. Specific receptors provide interfaces between these integrated defence systems and the relationships between innate and adaptive immunity. This chapter discusses features of the immune system and explores the relationships between nutrition, immunodeficiencies, and interactions between micronutrient status and immunity. Finally, aspects of diagnosis and management of food allergy and food intolerance are discussed, with specific examples.

27.2 Structure and function of the immune system

Cells which participate in immune responses are collected in lymphoid organs such as the thymus, spleen, lymph nodes, and Peyer's patches. In this environment they can perform their functions very effectively and they disseminate immunity by migrating throughout the body. Cells of the immune system are dispersed between other tissue structures within the gut epithelium, lamina propria, and other mucosal sites (Brandtzaeg 2011) (see Figure 27.1).

Cells involved in immunity

Antigen-presenting cells are found mainly in the lymphoid organs and the skin. Their main role is to present antigen in a particular way to lymphocytes so as to start off antigen-specific immune responses. They include interdigitating cells in the thymus, Langerhans cells of the skin, interdigitating cells resident in T cell areas of lymph nodes, and as follicular dendritic cells in lymph nodes (see section further below on 'Oral tolerance').

T (thymus-dependent) lymphocytes perform important immunoregulatory functions via their secreted products and also act as effector cells, capable of killing other cells. Many immunological diseases, both immunodeficiency and abnormally enhanced reactivity, can ultimately be attributed to defects of T-cell regulatory function. In terms of protective immunity, T-cells are particularly important in defence against intracellular bacterial and protozoal pathogens, viruses, and fungi.

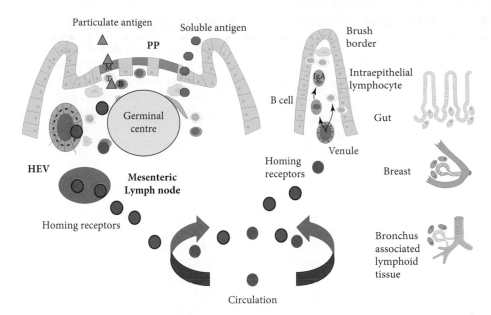

FIGURE 27.1 The gut-associated lymphoid tissue (GALT): schematic representation of the GALT and the recirculation and partial homing pathway of activated lymphocytes.
Once the T-lymphocytes in the Peyer's patches have become activated by gut-derived antigens they leave the mucosa facilitated by specific receptors and reach the systemic circulation via high endothelial venules, the mesenteric lymph nodes, and the thoracic duct. From here, some home back to the lamina propria, others (T-lymphocytes) may reach other mucosal sites and in this way distribute gut-derived 'information' throughout the mucosa-associated lymphoid tissues. HEV: high endothelial venules.

Natural killer (NK) cells can develop without thymic influence and provide an important link at the interface of innate and adaptive immunity. They are large granular lymphocytes which do not express markers of conventional T or B-cell lineage. These cells express low affinity Fc receptors for IgG (FcRg) and can kill target cells by using antibody-dependent and independent mechanisms. They may kill virus infected targets without prior sensitization through perforin, which is a cytolytic protein stored in secretory granules of NK cells. Many additional functions of NK cells have been described, including antigen-specific memory. They exhibit biological functions that are components of both innate and adaptive immunity, blurring the functional borders between these two arms of the immune response.

B-lymphocytes are independent of the thymus and in humans probably complete their early maturation within the bone marrow. When appropriately stimulated, B-lymphocytes undergo proliferation, maturation, and differentiation to form immunoglobulin-producing plasma cells. Eventually there are many identical daughter cells derived from a single B-cell, forming a clone. The enormous diversity of antibodies that an individual can produce is explained partly by rearrangements of nucleic acid within precursor B-cells and by random mutation.

Molecules of the immune system
Immunoglobulins

Immunoglobulin (Ig) molecules are the effector products of B-cells. Although they all have a broadly similar structure, minor differences within the main immunological classes (IgG, IgM, IgA, IgD, and IgE) are associated with a range of important biological properties. Molecules almost identical to secreted immunoglobulins are incorporated in the cell membranes of B-cells (surface Ig) expressing many related molecules concerned with antigen recognition and cell–cell communication.

In healthy adults, IgG accounts for more than 70 per cent of the immunoglobulins in normal serum. It is distributed equally between the blood and extracellular fluids. Infants receive IgG via placental transfer from around the 32nd week of gestation. Other immunoglobulin classes (IgA, IgM, IgE, IgD) are usually produced after birth and are not transmitted transplacentally. Increased IgM levels at birth can indicate an intrauterine infection.

IgA accounts for about 20 per cent of total serum immunoglobulins. Its function within the bloodstream and tissues is thought to be less important than its role as a secretory antibody. IgA is produced in two subclasses IgA_1 and IgA_2. IgA_2 is usually found in areas of greatest bacterial exposure (gastrointestinal tract) and is more resistant to bacterial proteases. In blood, IgA$_1$ is the predominant class. Major sites of IgA$_2$ synthesis are the laminae propriae underlying the respiratory tract, the gut, and other mucosae. Secretory IgA as a dimer linked by a joining 'J' chain confers immunity to infection by enteric bacterial and viral pathogens and may also be involved in the regulation of the commensal (= microbiome) gut flora. Oral immunization strategies are used to induce protective immunity to intestinal infections such as cholera, typhoid, and others.

IgE concentration in serum is generally low (>240 ng/ml). This is partly because it has a considerable affinity for cell surfaces and binds to mast cells, eosinophils, and basophils via low and high affinity receptors. Specific IgE antibodies can trigger immediate hypersensitivity reactions, which occur in atopic individuals, for instance in hay fever and anaphylactic food allergic reactions. The physiological function of IgE antibodies in humans is unclear but they appear to be important in defence against helminths.

Molecules on cell membrane

Cells exhibit characteristic cytological features. However, it must be appreciated that cells which appear morphologically identical may be functionally very different.

Identification of CD (cluster of differentiation) surface markers on lymphocytes is valuable in functional clinical diagnosis (e.g., of immunodeficiency syndromes) and in classification of malignancies. CD testing is usually done on a FACS (fluorescence activated cell sorter) with monoclonal antibodies. Over 400 distinct human CD markers have been described (http://www.hcdm.org).

Cytokines (interleukins)

Lymphocytes mediate their effects through direct cell-to-cell contact and through soluble factors such as cytokines or both. Over 30 of such cytokines have been described and various tests of antigen-specific cell-mediated immunity are available, based on the secretion of these factors by cells of the immune system (Satitsuksanoa et al. 2018) (see also Figure 27.2 and additional Weblink material).

Interleukin-1 (IL-1)

Many cell types in the presence of antigen or tissue injury secrete IL-1. Cells that produce IL-1 are also able to present antigen to T-cells: simultaneous stimulation of T-cells with processed antigen and IL-1 is required for initiation of a specific immune response. IL-1-producing cells include circulating monocytes, macrophages, fibroblasts, B-cells, and epithelial cells.

Interleukin-2 (IL-2)

IL-2 is secreted by activated T-cells and is responsible for amplification of the population of responding T-cells and for inducing the production of other cytokines from many cell types.

Tumour necrosis factor-alpha (TNF-α)

TNF-α is produced by mononuclear phagocytes *in vitro* when they are stimulated with bacterial endotoxin. TNF-α has effects on general cellular metabolism; it causes weight loss, fever, and acute phase reaction (associated

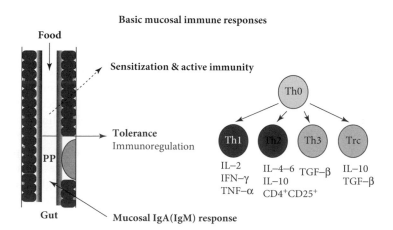

FIGURE 27.2 Schematic diagram of the basic principles of oral tolerance induction to protein antigen. Single or multiple feeds within the context of a normal microbiota will induce systemic tolerance and a mucosal immune response. This response may be favoured particularly if antigen gains access via the Peyer's patch. Under certain conditions, depending on the nature of the antigen and poorly understood host factors, systemic priming after oral antigen may also result. Activated Th and regulatory cells will secrete tolerogenic and/or inflammatory cytokines. PP: Peyer's patch; Trc: T-regulatory cell; Th1-3: Thelper cells.

with infections or neoplasia). It activates other mononuclear cells and granulocytes and increases their nonspecific killing capability. New members of this large family emerge regularly.

Interferon-gamma (IFN-γ)

IFN-γ is produced mainly by activated T-cells of the Th1 (T-helper cell) population following exposure to antigens or macrophages and natural killer (NK) cells. It increases class II expression on epithelial cells (among other cell and tissue types). This leads to increased antigen-presenting activity. It also activates macrophages for antitumor activity, in synergy with TNF-α.

Interleukin-4 (IL-4)

IL-4 is secreted by another functional subgroup of T-helper cells, the Th2 cells. This cytokine affects T- and B-lymphocytes, NK cells, macrophages, and others. IL-4 is particularly involved in cell growth and IgE isotype selection. Allergic patients tend to respond to allergens with IL-4 secretion whereas non-allergic patients predominantly secrete IFN-γ under these conditions.

Interleukin-10 (IL-10)

Interleukin-10 is produced primarily by lymphocytes and monocytes. It has many effects in immunoregulation and inflammation. It down-regulates the expression of Th1 cytokines, MHC class II antigens, and co-stimulatory molecules on macrophages and acts in a tolerance inducing fashion. It also enhances B cell survival, proliferation, and antibody production.

Interleukin-17 (IL-17)

Interleukin-17 is part of a large cytokine family and can regulate innate immunity and activities of phagocytes. It is secreted by a distinct Th-lymphocyte population (Th-17). Th 17 cells are characterized by their expression of IL-17A, IL-17F, IL-6, IL-8, TNF-α, IL-22, and IL-26, a mediator of inflammation. Cells that produce IL-17 are involved in adaptive immunity, promote autoimmunity, and have antiviral activity (Akdis et al. 2016). IL-17 may also inhibit the induction of tolerance to antigens.

Interleukin-22 (IL-22)

IL-22 promotes innate immunity to bacterial infection and is expressed by cells of the innate and adaptive immune responses in many biological disease models. IL-22 promotes pathological inflammation and tissue repair. It has regulatory functions of the innate immune system and is produced by CD3$^+$ (CD4$^-$ CD8$^-$) T cells that are detected in the skin epidermis, gut and lung epithelia, and also in blood.

Transforming growth factor beta (TGF-β)

TGF-β is a regulatory cytokine secreted by T-lymphocytes and affects B- and T-lymphocytes, macrophages, and other cells. It has anti-inflammatory, suppressive (IgG, IgM), and stimulatory (IgA) effects and is thought to play a major role in oral tolerance induction.

Cells and interleukins do not act in isolation and subsequent responses such as tissue damage/repair are the result of an immunological cascade.

The major histocompatibility complex

The MHC (major histocompatibility complex) gene cluster is located on chromosome 6. It contains genes coding for the 'human leukocyte antigen' (HLA) cell surface glycoproteins, which are found on many cells, not only leukocytes, and are involved in antigen presentation.

Although it was originally identified through its role in transplant rejection, it is recognized that proteins encoded on chromosome 6 are involved in many aspects of immunological recognition. These include interactions between different lymphoid cells as well as between lymphocytes and antigen-presenting cells.

Patterns of immune responses to fed antigen

The immune system of the gastrointestinal tract (GALT) has a number of different roles. When the route of entry of soluble or particulate antigen is through the intestinal epithelium or follicle-associated epithelium of Peyer's patches of immunologically normal and mature mammals, the general trend is towards suppression of immunity, in other words, specific suppression, *oral tolerance*. However, active immunization may also follow the feeding of antigen and this is typically in the form of harmless secretory IgA antibody. In some circumstances induction of a potentially adverse pathogenic immune reactions may occur after oral or mucosal antigen exposure. Thus, there may be either induction or suppression of a particular immune response, whether antibody- or T-cell-mediated, when antigen is encountered via the gut.

Regulation of gut immunity

T-lymphocytes are dispersed in the mucosa as well as in the organized lymphoid tissues of the GALT. Just as in the systemic immune system, gut T-lymphocytes have

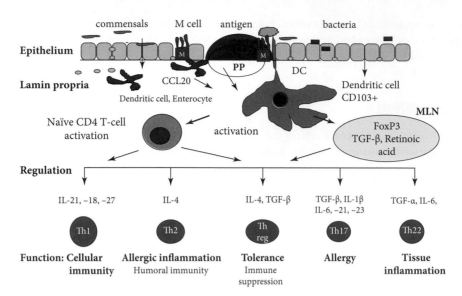

FIGURE 27.3 Schematic depiction of mucosal regulatory pathways.
Soluble and particulate antigens and bacteria are presented via the mucosal epithelium and underlying gastrointestinal lymphoid tissues (GALT). Depending on the nature and route of access to the underlying GALT complex, immunoregulatory processes are triggered. In the MLN CD103+ dendritic cells trigger a complex regulatory pathway via Foxp3, transforming growth factor-beta and retinoic acid. The functional regulatory outcomes are dependent on the activation of the cell types and cytokines depicted in the Regulation part of Figure 27.3. The outcomes may affect cellular immunity, allergic inflammation, tolerance (immune suppression), allergies, and tissue inflammation.
MLN: mesenteric lymph node.

two main functions: immunoregulatory and effector functions (see Figure 27.3).

Gut lymphoid cells generate protective immunity to infectious microorganisms and parasites. Immune responses may disrupt intestinal anatomy and function. This occurs when substantial tissue damage occurs as an unavoidable by-product of a protective immune response to an infectious agent. The classical example of this is tuberculosis or leprosy. Inappropriate immune responses, developing in response to harmless antigens such as foods, are well-recognized and can be responsible for the disease.

Oral tolerance (OT)

Oral tolerance describes a state of antigen-specific hyper-responsiveness or unresponsiveness after prior oral mucosal exposure. Breakdown of this homeostatic process is considered to be one cause of food hypersensitivity. The mechanisms by which tolerance is mediated include T-cell deletion, anergy (lack of immune response), and suppression. Cell-mediated delayed hypersensitivity reactions (Th1-type), which are implicated as pathogenetic mechanisms in the development of food-related gastrointestinal inflammation are particularly well suppressed. Regulatory events during the induction period are now better understood and involve a large number

of signalling events and secretion of transcription factors (Venter et al. 2020). The balance between tolerance (suppression) and sensitization (priming) is dependent on diverse factors including genetic background, nature of antigen, frequency and extent of administration, maturity of the immune system, and immunological status of the host. Dendritic cells of the lamina propria play a major regulatory role in the induction of tolerance to food antigens through presentation of antigen to regulatory T cells. Sampling of the antigen may be facilitated via DC cell sampling within the intestinal lumen.

Antigen administration during an ill-defined 'vulnerable' postnatal period is thought to have sensitizing and potentially tolerizing effects on (Kim et al. 2020). Larger antigen doses may cause T-cell deletion and anergy, whereas smaller doses lead to suppression through induction of IL-4/IL-10-secreting Th2 cells and cells secreting IL-10 and TGF-β. Food allergic diseases can be envisaged as being due to a breakdown in the usual physiological downregulation of immunity to dietary and other gut-derived antigens. It is still unresolved whether early postnatal antigen exposure or delayed introductions are likely to prevent or reduce allergic sensitization in the infant. Clinical studies which investigate the effects of early and later introduction of peanuts are summarized in guidelines and tend to favour early introduction of tree nut allergens combined with

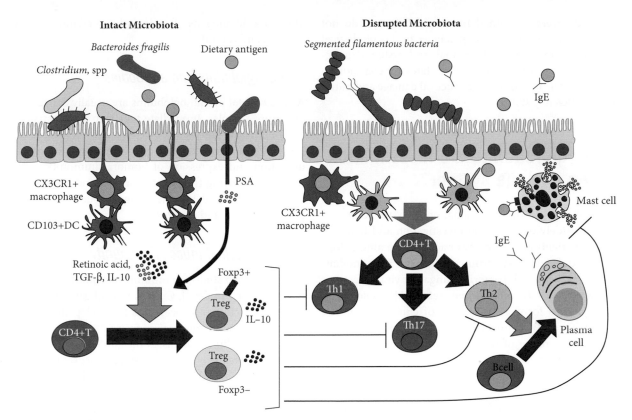

FIGURE 27.4 Interaction between gut microbiota and induction or disruption of oral tolerance.
Commensal bacteria stimulate epithelial cells, T cells, macrophages, and DCs and promote the differentiation of naive
CD4þ T cells into Tregs. Bacteroides fragilis-producing PSA promotes the production of Foxp3þpositive Tregs. Segmented
filamentous bacteria and other microbiota activate DCs and macrophages and induce the production of pathogen- or
antigen-specific inflammatory Th1 and Th17 cells via IL-1b, IL6, and IL23. Antigen-stimulated DCs promote Th2-cell
differentiation, followed by the induction of antigen-specific IgE-producing plasma cells. Dietary antigen IgE complex-
mediated degranulation of mast cells promotes IgE-mediated allergic inflammation. Healthy Treg induction is capable of
inhibiting the inflammatory process of Th1, Th2, and Th17 differentiation and mast cell activation.
DC, dendritic cell; Foxp3, forkhead box P3; PSA, polysaccharide A.
Source: D Tokuhara et al. (2019) A comprehensive understanding of the gut mucosal immune system in allergic inflammation.
Allergol. Int. **68**(1), 17–25, https://creativecommons.org/licenses/by-nc-nd/4.0

breast milk (Kopp et al 2022) (https://pubmed.ncbi.
nlm.nih.gov/35274076). Major confounding variables
in these studies are the genetic make-up of the population studied, general eating habits, maternal food intake
during and after pregnancy, duration of breastfeeding,
and mode of delivery (see Figure 27.4).

Allergic sensitization

Patients are clinically sensitized when they have generated high-enough specific IgE levels that permit elicita-
tion of an allergic reaction. A number of individuals may
have circulating antigen-specific IgE antibodies without
clinical symptoms on ingestion of a particular food. This
is one of the reasons why it is not possible to establish
clinical diagnoses on the level of antigen-specific IgE
alone. IgE antibody binds to high affinity receptors on
the surface of mast cells and basophils in such a manner

that contact between only a few membrane-associated
molecules and the inducing antigen may trigger the release of highly active mediators of inflammation, including histamine, proteolytic enzymes, leukotrienes, and
prostaglandins.

Aberrant immunity, malabsorption, and infection

Dietary antigens, eaten every day, reach the organized
lymphoid tissues in sufficient amounts to induce a variety of mostly harmless humoral and cellular immune responses. If a sensitization has occurred, the re-exposure
in a further meal may result in a local immune reaction
causing immediate or delayed reactions.

Severe immunodeficiency states, where infants lack all
or most important aspects of their immune system (e.g.,

X-linked severe combined immunodeficiency), do not cause primary nutritional deficiencies or morphological mucosal abnormalities. Intestinal and systemic infections resulting in failure-to-thrive states are common in these children and may cause mucosal damage and nutritional deficiency states.

27.3 Approach to the investigation of the immune system

Protocols for the clinical evaluation of systemic immunity are widely used for the investigation and management of patients with primary or acquired (malnutrition) immunodeficiency syndromes. Primary immunodeficiencies present with bacterial, fungal, or viral infections associated with the functional defect. Microbiological investigations form a major part of clinical diagnostic workup. Over 400 primary immunodeficiencies affecting an extensive range of deficiencies have been identified (Notarangelo et al. 2020).

Commonly used diagnostic procedures for immunodeficiency states

Responses to infant immunization and immunological memory, particularly with live vaccines (BCG), are valuable and can indicate normal or impaired cell-mediated immunity. Severe combined primary immunodeficiencies (SCID) usually present before 18 months of age when the humoral protection transferred during birth and breastfeeding has waned.

Blood examination

Examination of blood films and differential analysis and inspection of cell morphology are early investigations. The absolute lymphocyte count, which is normally low in peripheral blood, is an important finding. The differential diagnosis of an immunodeficiency needs to be confirmed in specialized laboratories performing genetic analysis and functional assays.

Lymphocytes and specific cell-mediated immunity

Supportive evidence for a specific cell-mediated abnormality (implying both normal afferent and efferent limbs) can be obtained by *in vivo* intra-dermal tests of delayed-type hypersensitivity, using a range of recall antigens to which the body will usually have been exposed.

These 'recall' antigens often include tuberculin, tetanus antigen, and candida.

Immunoglobulins and antibodies

Assays of total immunoglobulins are usually carried out on serum samples. Immediate type skin prick tests and serum analysis are used for *in vivo* and *in vitro* detection of IgE antibodies. More precise information on the induction and expression of humoral immunity is obtained by studying the primary and secondary immune responses to killed vaccine antigens, such as tetanus, pneumococcus, and measles.

Gastrointestinal mucosal immunity

It is difficult to study the immune system of the gastrointestinal tract in detail. Clinical tests of gastrointestinal immune function have been slow to develop and require functional assays, mucosal biopsies, and often short-term organ culture systems. Some guidelines for investigation of patients in whom intestinal mucosal immunodeficiency or hypersensitivity states may be present are given below.

Attention must also be paid to the roles of non-immunological *digestive* factors. These may act not only as alternative mechanisms of disease, mimicking immunological infectious disorders, but also as factors that will alter immunity in general such as malnutrition and micronutrient deficiencies.

Mucosal hypersensitivity reactions

Normally it appears that pathogenic antigen-specific immunity of T-cell origin does not develop in the intestine to enterically encountered antigens such as bacteria and foods. Three jejunal mucosal histopathological features of villous atrophy, crypt hyperplasia, and high intraepithelial lymphocyte count may reflect a cell-mediated inflammatory immune reaction within the mucosa (Nowak-Wegrzyn et al. 2017).

27.4 Nutrition and immunodeficiency disorders

Virtually any component of the immune system can be affected by nutritional status. The resulting immunodeficiency varies in severity from trivial to fatal. Immunodeficiency can also result from acquired diseases. This is well illustrated in severe forms of the acquired immunodeficiency syndrome (AIDS). Acquired immunodeficiency can be iatrogenic, for example, as a result of immunosuppressive treatment for autoimmune diseases. In addition to causing susceptibility to infection, immunodeficiency may be associated with abnormally regulated immune responses, as in allergy or autoimmunity. In general, an uncomplicated immunodeficiency itself has no effect on the digestive and absorptive capacity of the gut. Secondary effects often occur on the basis of diarrhoea and malabsorption when immunodeficiency is complicated by infections.

Primary immunodeficiencies

Abnormalities of polymorph function and deficiencies of complement components or antibodies may all result in susceptibility to bacterial infection. SCID can be caused by different gene defects. These are autosomal or X-linked and mostly lead to a deficiency affecting both antibody and cell mediated immunity. Affected infants are susceptible to commonly occurring, normally benign viral infections and may die from generalized chickenpox, measles, cytomegalovirus, or other viral infections.

Selective deficiency of the B-lymphocyte (antibody) system occurs in X-linked recessive hypo- or agammaglobulinemia. The lack of immunoglobulins is not absolute, but the patient fails to respond to antigenic stimuli and may fail to develop an immunological memory. Most patients with immunoglobulin deficiency suffer from 'acquired' or 'late-onset' hypogammaglobulinemia known as 'common variable immunodeficiency'. This is associated with an unusually high incidence of autoimmune disease.

Secondary immunodeficiencies

Secondary immunodeficiencies are not inherited. Immunoglobulin deficiency may result from abnormal losses of serum proteins into the gut, for example in intestinal lymphangiectasia. Drugs may also depress the immune system: cytotoxic drugs and those used for treatment of epilepsy such as phenytoin or drugs used in rheumatoid arthritis such as penicillamine may induce IgA deficiency or even hypogammaglobulinemia (low IgG).

Secondary cell mediated immune defects can occur in Hodgkin's disease or sarcoidosis and following infections such as leprosy, miliary tuberculosis, or measles. They may also result from loss of lymphocytes from the gut in protein-losing enteropathy or due to thoracic duct fistula, and can also be caused by treatment with cytotoxic drugs. Clearly, one of the most important causes of secondary immunodeficiencies, infection, and malnutrition relate to HIV infection and AIDS (Jin et al. 2021).

HIV infection affects the ability to absorb food and leads to increased nutrient losses. The subsequent malnutrition leads to increased severity of infection, delayed recovery from infections, and shortened life span. Even though HIV is transmitted via breast milk, great strides have been made worldwide in reducing the mother-to-child transmission through exclusive breastfeeding and ART (antiretroviral therapy) during pregnancy and the neonatal period. Neonatal HIV transmission rates have been reduced dramatically.

UNAIDS estimates that 38 million people now live with HIV/AIDS, 2 million acquire the infection each year, and 700,000 died of HIV/AIDS in 2019. Around 1.6 million children now live with HIV/AIDS and around 150,000 became newly infected in 2019 (UNAIDS Fact sheet 2020; see Useful Websites Section 27.2).

Pre-exposure prophylaxis (or PrEP) is a way for individuals who do not have HIV but who are at very high risk of HIV infection to reduce transmission by about 75 per cent on regular use. HIV affects acute phase proteins and micronutrient status. Reduced plasma concentrations of retinol, beta-carotene, folate, and iron occur in asymptomatic HIV positive subjects. Once opportunistic infections occur, systemic infection leading to release of cytokines results in anorexia; fungal infections of the mouth and throat lead to discomfort and pain in eating, particularly of solids; intestinal damage occurs, particularly secondary to cryptosporidium infection and persistent diarrhoea. Secondary infection with TB is common, leading to severe metabolic disturbances and malnutrition.

Growth failure, micronutrient malnutrition, and poor nutrition during the management of infection are frequently noted among children of parents with AIDS. Nutritional interventions through diet or particular feeds are an important route to maintaining 'health' in an HIV-positive subject who has progressed towards clinical AIDS. Ready-to-use therapeutic foods (RUTF) based on peanuts increase the fat free mass and reduce anaemia but do not improve zinc status.

SARS-CoV-2 (COVID-19) pandemic and inflammatory immune responses

The widespread outbreak of COVID-19, a newly identified zoonotic corona virus, has become one of the greatest medical and epidemiological challenges over the last decades. It has the propensity to affect different populations according to their age, pre-existing immunodeficiencies, and risk factors such as diabetes, obesity, cardiovascular systems, kidney disease, and ethnic background. A number of infections are silent and can transfer infections. This feature complicates the effectiveness of distancing and quarantine measures. A major feature is involvement of the lung, which leads to severe inflammation with much reduced oxygen intake often needing mechanical ventilation. Selenium status is likely to influence human response (Bermano et al. 2020) to the severe acute respiratory syndrome coronavirus 2 (SARS-CoV-2) infection, and Se status is one (of several) risk factors which may impact on the outcome of SARS-CoV-2 infection, particularly in populations where Se intake is sub-optimal or low. For more detailed, but rapidly changing therapeutic strategies such as dexamethasone administration and other treatment strategies please refer to updated reviews (Calder 2020).

27.5 Malnutrition and immune function

Hunger and malnutrition remain among the most devastating problems worldwide; obesity (BMI >35) and its associated co-morbidities, such as heart disease, hypertension, stroke, and maturity onset diabetes have reached alarming dimensions worldwide and are not limited to high-income countries.

Overnutrition

It has been predicted that by 2030 around 85 per cent adults of the USA will be overweight or obese. Predictions for the UK paint a similar, although less dramatic picture. Fifty per cent of adult men and 40 per cent of women and 25 per cent of children, dependent on educational attainment, are expected to be obese (BMI >35). Overweight affects 30–70 per cent and obesity affects 10–30 per cent of adults (WHO Europe 2020; see Useful Websites Section 27.3).

Obesity is correlated with increased circulating concentration of the adipose tissue hormone leptin, often associated with leptin resistance. Leptin is distributed to the brain and other tissues regulating immune function, normal fat loss, and appetite (see also Chapter 4). Individuals with obesity present with an increased secretion of the inflammatory cytokine TNF-α, reduced T-lymphocyte responses, insulin resistance, and an increased incidence of infections.

Obesity also affects the intestinal microbiota and is associated with the development of the autoinflammatory type 2 diabetes mellitus (Ampofo & Boateng 2020). Most associated risks and immunological changes are reversible with weight reduction and increased physical activity (see also Figure 27.5).

Undernutrition (malnutrition)

Severe malnutrition is the consequence of systemic deficiency of energy and nutrients over a prolonged period: the consequences of malnutrition include death, disability, and stunted physical and mental development (see Useful Websites Section 27.4).

Protein energy malnutrition (PEM) can occur in two clinical extremes—kwashiorkor and marasmus, with gradual transitions between these two extremes (see also Chapter 35). Marasmus usually occurs early in infancy

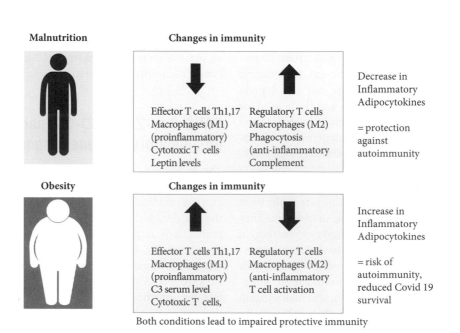

Both conditions lead to impaired protective immunity

FIGURE 27.5 Effect of malnutrition on immunity and host defence functions.
Both conditions may lead to impaired protective immunity, which is often reversible after correction of underlying nutritional abnormality.

TABLE 27.1 Effects of malnutrition on immunity and host defence functions

Deficiency state	Functions affected or enhanced	Infections
Protein energy malnutrition (acute)	Innate immunity: phagocytosis, macrophage activation, Adaptive immunity: T cell activation, memory, antibody production, cytokine secretion, leptin levels decreased	Opportunistic, viral, and helminth infections, including measles, tuberculosis, and influenza respiratory infections
Protein energy malnutrition (chronic)	Thymic and T cell development, innate immunity: complement, macrophage activation, adaptive immunity: immunoglobulins (IgG, IgA decreased) vaccine efficacy reduced, leptin levels decreased	Respiratory + skin + intestinal infections, other organisms: helminths, malaria, HIV, measles, influenza BCG, malaria, encapsulated organisms
Overnutrition	Innate immunity: leukocytes preactivated (interferon-γ), TNF-α increased), NK function + phagocytosis decreased Adaptive immunity: T cell activation reduced Leptin concentrations increased (often with leptin resistance combined)	Opportunistic and fungal infections

where children are wasted and grossly underweight but without oedema and skin changes. Patients with kwashiorkor, mostly in their second year of life or older, are growth-retarded, suffering from oedema, skin changes, abnormal hair, hepatomegaly, and apathy.

Undernutrition which leads to the impairment of the immune function can be due to insufficient intake of energy, of macronutrients, or of micronutrients. These deficiency states often occur in combination. PEM and micronutrient deficiencies related to iodine, vitamin A, iron, and zinc can present together. Proteins are essential for both cell replication and production of immunologically active molecules and receptors. It is therefore not surprising that nearly all forms of immunity are affected by PEM (see Table 27.1).

Clearly, the level and impact of undernutrition is most pronounced in low and middle-income countries. In high-income countries, although moderate deficiencies can be present in any age group, low-income is an important determinant of undernutrition. Additionally, undernutrition is prevalent among the institutionalized elderly, patients with eating disorders, substance users, individuals on unsupervised elimination diets, and in those recovering from major surgery (see also Figure 27.5).

Non-specific host defence and barrier functions

Skin, mucous membranes, and epithelial surfaces act as non-specific host defence mechanisms and are adversely affected in children with kwashiorkor. Skin lesions found in these compromised children enhance penetration and adhesion of infectious organisms. This leads to mucosal and gastrointestinal infections, further impairing the absorption of essential nutrients.

27.6 Influence of micronutrients on immune function

The developing child

There is very little reliable information on the effects of vitamins, saturated, and unsaturated fatty acids, trace minerals, and other food constituents on the infant's developing immune system. Many reports examine effects of corrections of moderate or severe deficiencies on immune responses in children (and animals) without reference to the maternal status. Very little is known of the effects of supplementation at the level of or above dietary reference values (DRVs) in a non-deficient population.

Vitamin A

Vitamin A and associated retinoids are essential for normal differentiation of epithelial tissues throughout the body. Vitamin A deficiency is the second most serious nutritional deficiency worldwide. It is associated with increased morbidity in children, particularly in those suffering from respiratory tract infections and measles. Vitamin A is important for both innate and adaptive immune responses, including cell-mediated immunity and antibody responses. Vitamin A deficiency leads to a reduced integrity of mucosal epithelia that leads to an increased susceptibility to ocular, respiratory, and gastrointestinal inflammatory and diarrhoeal diseases with subsequent malnutrition. Vitamin A deficiency has also been associated with diminished innate immune responses affecting phagocytic activity and NK cell function. Several studies have shown a marked decline in measles-related deaths in vitamin A deficient children orally supplemented with vitamin A. Vitamin A supplementation has also been shown to enhance serum antibody responses to common vaccines. Adverse effects of vitamin A deficiency are usually restored to normal after supplementation. Vitamin A supplementation in deficiency states generally reduces morbidity and mortality from infectious disease, especially in children. There is some evidence that vitamin A and its metabolite retinoic acid alter the balance of T-cell responses resulting in enhanced Th1 immunity. Vitamin A is important for the differentiation of regulatory T lymphocytes and suppressing Th17 differentiation. Vitamin A supplementation downregulates IFN-γ, TNF-α enhances IL4,5,10 secretion and responses to vaccines (Th2) response. Periodic vitamin A supplementation in HIV infection is associated with reduced mortality. There is no doubt that vitamin A supplementation in deficiency states is helpful.

Vitamin C

Vitamin C is a water-soluble antioxidant and appears to affect most aspects of the immune system. High concentrations are found in white blood cells. Reduced plasma levels are associated with reduced immune function, especially those affecting the bacterial-killing efficiency of white blood cells. Vitamin C regulates the immune system via antiviral and anti-oxidant properties. Reactive oxygen species (ROS), which can be scavenged by Vitamin C, play important roles in killing intracellular bacteria. Vitamin C has been shown in some studies to reduce inflammatory processes *in vivo* and *in vitro*, although the results are inconsistent.

Positive effects of vitamin C on the common cold have been disputed. Vitamin C levels in leukocytes fall rapidly during upper respiratory tract infections and return to normal during recovery suggesting that supplementation during this phase might be beneficial. Vitamin C supplementation (600mg/day) in ultra-marathon runners decreased the incidence of upper respiratory tract infections. A systematic Cochrane review concluded that Vitamin C supplementation may affect symptoms but does not reduce the incidence of common colds in the general population (Hemilä 2017).

Vitamin D

Vitamin D is a micronutrient for which there is an increasing body of evidence for wide ranging immune-enhancing and potentially harmful effects. The physiological function of vitamin D is to maintain serum calcium and phosphorus concentrations that maintain neuromuscular function, bone ossification, and immunity. Vitamin D enhances the efficiency of the small intestine to absorb dietary calcium and phosphorous. The active metabolite of vitamin D, 1,25 dihydroxycholecalciferol (1,25(OH)$_2$D$_3$), regulates the transcription of a large number of genes through binding to the vitamin D receptor (VDR) (Martineau et al. 2017). A deficiency of vitamin D, through dietary inadequacy or inadequate exposure to sunlight or both, may interfere with natural killer-cell function, resulting in increased susceptibility to infection.

1,25(OH)$_2$D modulates synthesis of interleukins and cytokines. It has been shown to downregulate the immunostimulatory IL12 while increasing secretion of the immunosuppressive IL10. Besides stimulating monocytes and macrophages, 1,25(OH)$_2$D$_3$ may act as an immunosuppressive agent by decreasing the rate of proliferation and the activity of both T- and B cells while inducing suppressor T cells. The VDR is expressed in a number of human tissues, including muscle tissue, liver, intestine, reproductive organs, and 1,25-(OH)$_2$D$_3$ exerts most of its actions after it has bound to its specific receptor, which is present on monocytes and activated lymphocytes. The hormone inhibits lymphocyte proliferation and immunoglobulin production in a dose-dependent fashion and may interfere with T-helper cell (Th) function. Vitamin D sufficiency may be an important protective factor for food allergy in the first year of life. Vitamin D insufficiency has been directly linked to adult acute respiratory distress syndrome (ARDS) highlighting the effects on the lung epithelium. Even though oral Vitamin D intake has health benefits, over-supplementation has been associated with hypercalcaemia, the hallmark of Vitamin D intoxication, which may occur if circulating 25(OH)D levels are constantly above 380–500 nmol/L. Serious harm has also been reported at lower levels of >125 nmol/L and the Institute of Medicine has set the upper limit for adults at 100µg/d.

Vitamin E

Vitamin E is a major lipid-soluble antioxidant. Deficiency states or supra-dietary intakes influence immune cell function. Vitamin E deficiency is uncommon; groups most likely to be deficient include premature infants, individuals on selective diets, alcohol and drug users, patients with gastrointestinal or hepatic disorders with malabsorption syndromes, and individuals with A-beta-lipoproteinemia. Vitamin E deficiency is associated with reduced antibody production and T-cell proliferation after mitogen stimulation.

Vitamin E supplements in the elderly ranging from 60mg to 800mg/day for 30 days showed significant improvements in cytokine production (IL-2), enhanced NK-cell function, delayed hypersensitivity responses, and enhanced hepatitis B vaccination antibodies. Supplementation of vitamin E in moderately deficient elderly resulted in increased resistance to infection.

Mineral and trace element deficiencies

Iron deficiency

Iron deficiency is the most common nutritional deficiency among children and women of childbearing age. Globally 2 billion people—over 30 per cent of the world's population—are anaemic, many due to iron deficiency. In resource-poor areas, infectious diseases frequently exacerbate this. In healthy organisms, iron is maintained at a stable concentration in the plasma, and it is stored in hepatocytes and splenic and hepatic macrophages at constant levels. Malaria, HIV/AIDS, hookworm infestation, schistosomiasis, and other infections such as tuberculosis are particularly important factors contributing to the high prevalence of anaemia in some areas. (http://www.who.int/nutrition/topics/ida/en/).

The relationship between iron status and immunity is complex. Iron is essential for electron transfer reactions, gene regulation, binding and transport of oxygen, and regulation of cell differentiation and cell growth. It is involved in the neutrophil killing of bacteria through generation of toxic hydroxyl radicals. A deficiency leads to impaired cellular immunity with a CD4/CD8 T-lymphocyte inversion. However, even though iron deficiency leads to impairment of the immune system, epidemiological evidence suggests that it is remarkably protective against high-grade parasitaemia and severe malaria. Some bacteria need iron for their multiplication and indiscriminate iron supplementation during this period could worsen the clinical situation.

It seems that T-lymphocyte numbers and function of neutrophils can be adversely affected by iron deficiency and restored to normal by iron supplementation.

Copper deficiency

Copper deficiency is rare in humans but can be encountered during extreme malnutrition. The most extreme case of human copper deficiency is Menkes disease, an X-linked recessive severe multisystemic immunodeficiency characterized by progressive neurodegeneration and marked connective tissue anomalies as well as typical sparse abnormal steely hair (kinky hair). Copper is a cofactor for several enzymes, including cytochrome C oxidase and superoxide dismutase. Copper is required for infant growth, host defence mechanisms, bone strength, red and white cell maturation, iron transport, cholesterol and glucose metabolism. Copper homeostasis is maintained over a wide range of intakes, mainly through changes in excretion. Deficiencies have been described in premature infants and in patients receiving total parenteral nutrition. Neutropenia was found to be the earliest and most common symptom of Cu deficiency in premature and low birthweight (LBW) infants. Children with a complete absence of the copper-carrying protein coeruloplasmin with subsequent severe copper deficiency suffer from impaired T-cell immunity, increased bacterial infections, and diarrhoea. A human study examining the effects of a low copper intake on immunity reported a reduced *in vitro* mitogen responsiveness of T-lymphocyte activation and circulating B-lymphocytes. Study participants exhibited an increased incidence of infections.

Zinc deficiency

Dietary zinc deficiencies are rare, although low plasma zinc levels can be found in individuals in resource-poor countries. Diarrhoea still causes 18 per cent of all deaths in children under five and accounts for nearly two million child deaths in low- and middle-income (LMIC) countries per year. Zinc is important for highly proliferating cells of the immune system and deficiency affects both innate and adaptive immunity. Zinc has antioxidant activity and is involved in the cytosolic defence against ROS generated by activated macrophages through its role as a cofactor for superoxide dismutase (SOD). Zinc supports Th1 responses and maintains skin and mucosal membrane integrity. Due to its major effects on the immune system and direct antiviral effects, zinc supplementation in children and adults with borderline zinc status can decrease lower respiratory and diarrheal illnesses (Barrea et al. 2020). Prophylactic zinc may reduce respiratory infections and improve growth in children with impaired nutritional status. While prophylactic zinc decreases mortality due to diarrhoea and pneumonia, it has not been shown to affect overall mortality. Care must be taken not to induce copper deficiency during zinc supplementation because of a competitive absorption of zinc and copper within enterocytes.

Infants with acrodermatitis enteropathica, a rare inborn error with a reduced ability to absorb zinc, show failure to thrive and show impaired lymphocyte proliferation and response to mitogens, decreased/inverted CD4/CD8 ratios, impaired NK activity, and cytotoxicity.

Selenium

Selenium is an essential trace element and is concentrated in tissues involved in the immune response, such as the lymph nodes, liver, and spleen. It is required for an optimal immune response and deficiency states affect innate and adaptive immunity. Selenium plays a key role in antioxidant function, through a cofactor role for glutathione peroxidase, which may be important for macrophage activation. Se is a cofactor of a number of transcription factors involved in cell-to-cell signalling.

Immune functions possibly affected by Se deficiencies are numerous and affect non-specific, humoral, and cell-mediated immune functions. Se supplementation, even in borderline deficiency states, has boosted cell-mediated immune responses, protects cells against oxidative damage, and downregulates inflammatory processes (Bermano et al. 2020). The beneficial effects of selenium supplementation are dependent on the underlying selenium status. Selenium supplementation may reduce cardiovascular disease as well as gastric and lung cancer in populations with low selenium levels.

The role of probiotics and effects

A number of important systematic reviews have been published examining aspects of prevention and treatment of food allergy (Fleischer et al. 2020). The World Allergy Organization report concludes that a probiotic may be used in pregnant women at high risk for having an allergic child. It may also be beneficial in women who breastfeed infants at high risk of developing allergic disease. Systematic reviews from the UK Food Standards Agency concludes that probiotic supplement during pregnancy, breastfeeding, and early life may prevent the development of eczema and sensitization to milk protein https://www.food.gov.uk/research/food-allergy-and-intolerance-research). (see also Figure 27.6).

Nutrigenomics and nutrigenetics

The developing field of nutrigenomics and nutrigenetics can be defined as the area of nutritional genomics, which first studies the role of specific genetic variants in the form of single nucleotide polymorphisms (SNPs) during the modulation of the response to dietary components, and second, the implications of such interaction and influence on health status and predisposition to nutrition-related diseases.

Allergic diseases, including food allergies, are complex and have a strong genetic component. The genotype, however, alone is unlikely to be the only driver of allergic disease, as the genotype has stayed relatively stable over recent decades during which allergic diseases have become a mounting epidemic. Epigenetic processes (DNA methylation and histone modifications) have the ability to regulate genetic and developmental processes in response to environmental signals, and these mechanisms may contribute to the development of food allergies. There is increasing evidence that the epigenetic mechanisms that regulate gene expression during immune differentiation are directly affected by dietary factors or indirectly through modifications in gut microbiota induced by different dietary habits (Acevedo et al. 2021).

The effect of diet on allergy outcomes Is therefore likely to be mediated by epigenetic changes including DNA methylation, rather than genetic changes. Children who develop allergic diseases have epigenetic signals affecting IgE production and T-cells already at birth.

Targeting these epigenetic markers with environmental factors such as dietary intake, may prevent aspects of the allergic march driven by Th2 immune processes.

Other examples demonstrate the possible effects of early peanut exposure in infancy and marked reduction of later development of peanut allergies. Furthermore, well-studied examples addressing the influence of nutrition on public health include the 'Mediterranean diet', which has demonstrated associated reduction of non-communicable diseases. Other examples demonstrate the possible effects of early peanut exposure in infancy and marked reduction of later development of peanut allergies (Agyemang & Sicherer 2019).

27.7 Food intolerance and food allergy

The increased public awareness of the relevance of diet to health and the absence of simple and reliable diagnostic tests have enhanced the idea that 'food allergy' is widespread (Sampath et al. 2021), broadly defined as adverse reaction to foods and food additives, may cause a wide range of distressing physical and psychological problems and chronic, disabling symptoms (see Further Reading Section 27.1).

Many claims for effective *in vitro* diagnostic tests for food sensitivity have been made; a reliable diagnosis of either food intolerance or allergy, however, relies on clinical methodology and supportive laboratory investigations. Recent advances in understanding the mechanisms of food intolerance help distinguish these from psychologically based aversion reactions to foods.

Scope and definitions

Adverse reactions to ingested food cause a wide variety of symptoms, syndromes, and diseases for which the general descriptive terms 'sensitivity', 'allergy', and 'intolerance' are often used. Except for immediate, IgE-mediated allergic reactions, these terms do not imply specific mechanisms for their pathogenesis and can be applied to a reaction with an unknown mechanism to a

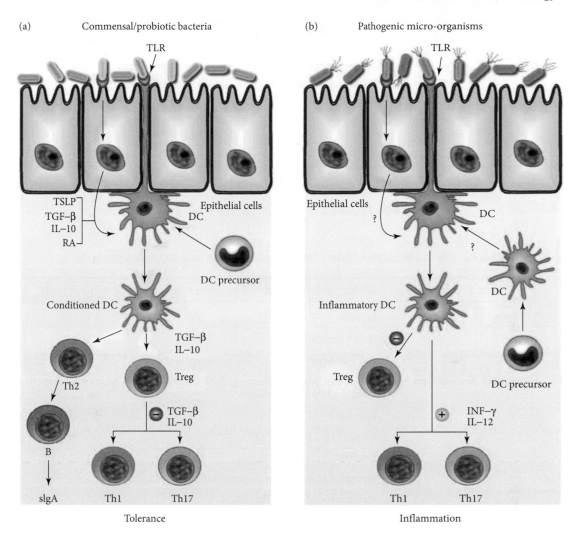

(a) Commensal/probiotic bacteria   (b) Pathogenic micro-organisms

Tolerance   Inflammation

FIGURE 27.6 Toll-like receptors, probiotics and immunotolerance.
Schematic view of the potential mechanism of action by which commensal bacteria and pathogenic bacteria interact with toll-like receptors (TLR) and elicit different immune responses. (a) Commensal and probiotic bacteria interact with intestinal epithelial-cell barrier and dendritic cells (DC) resident in the intestine. Some cytokines, including IL-10, transforming growth factor (TGF)-β, and thymic stromal lymphopoietin (TSLP) are expressed in intestinal epithelial cells, as a result of their interactions. Stimulation of cell TLR mediated by bacteria leads to up-regulation of TGF-β and IL-10, which in turn may limit the responsiveness of intestinal DCs resulting in the expansion and/or survival of T-cells with regulatory capacities, and limiting the ability of driving Th1, Th2, and Th17-cell responses. (b) Pathogenic bacteria have virulence factors that interact with intestinal epithelial-cell barrier and DCs resident in the intestine. Invasion of epithelium and direct interaction with DCs lead to activation of TLR and enhanced production of pro-inflammatory cytokines including interferon (INF)-γ and IL-12, which are capable of driving Th1, Th2, and Th17 response.
RA, retinoic acid; sIgA, secreted Ig A; Th, T helper cell; Treg, T regulatory cell.
Source: C Gómez-Lloriente et al. (2010) Role of toll-like receptors in the development of immunotolerance mediated by probiotics. *Proc Nutr Soc* **69,** 381–389, reproduced with kind permission.

clearly defined metabolic, pharmacological, or as yet undescribed immunopathological process (see Figure 27.7).

Adverse reactions not based on an immunological mechanism can be due to enzyme deficiencies, pharmacological effects (e.g., due to caffeine), non-immunological direct histamine containing or releasing effects (e.g., certain shellfish and cheeses), and direct irritation through gastric acid (oesophagitis) and spices.

Food intolerance and *food sensitivity* are often used for all reproducible, unpleasant (i.e., adverse) reactions to a specific food or food ingredient that are not psychologically based. The reaction may have a clearly defined metabolic, pharmacological, or immunopathological basis, or the mechanism may be unknown or disputed. The majority of adverse reactions to foods fall into this category. The provoking agent may be a single food or

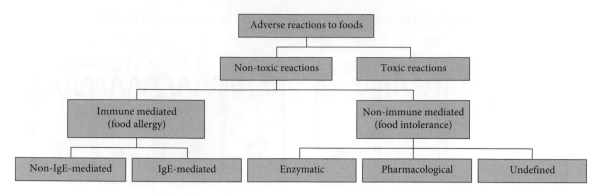

FIGURE 27.7 Allergic and non-allergic reactions to foods.

TABLE 27.2 Clinical features of adverse reactions attributed to food and food ingredients

System affected	Clinical features that could be caused by adverse reactions to foods
Skin	Urticaria
	Atopic dermatitis
	Angio-oedema
Gastrointestinal tract	Oral allergy syndrome (burning, itching of the lips and mouth and sometimes the larynx and pharynx)
	Pain, colic
	Nausea
	Vomiting
	Change in stool habit, e.g., looseness, frequency, blood, mucus
	Abdominal distension, flatulence
	Heartburn (gastro-oesophageal reflux)
	Failure to thrive
Respiratory tract	Asthma
	Rhinitis
Eyes	Watering eyes
	Conjunctivitis
	Peri-ocular pruritus
Cardiovascular system	Symptoms and signs of hypo- and hypertension
Blood	Symptoms and signs of haemolytic anaemia (rare)
Central nervous system	Headache
	Abnormal behaviour in children (including attention deficit hyperactivity disorder (ADHD))
	Fatigue
Generalized systemic	Anaphylaxis (circulatory collapse, wheeze, inability to swallow, and other symptoms)

ingredient, but sometimes—particularly in IgE-mediated food allergy—many different foods are involved.

IgE-mediated food allergic reactions are generally reproducible and there is evidence of an abnormal immunological reaction to the food and a plausible mechanism implicating immunological processes (Table 27.2).

Psychologically based food reactions (food aversions) comprise both psychological avoidance and psychological intolerance, where an unpleasant bodily reaction caused not directly by the food but by emotions associated with the food. Psychological intolerance normally does not occur when the offending food is given in a disguised form.

Diagnostic approaches

No single laboratory or *in vivo* test allows the definite clinical diagnosis of an immediate or delayed adverse reaction to food. A careful history and often a double-blind placebo-controlled food challenge are necessary to confirm or refute the diagnosis of a food allergic reaction (Table 27.3).

The development of component-resolved diagnostic tests may help to exactly identify which particular part of the antigen/allergen is the responsible trigger.

Allergic reactions to a given food may not appear regularly. They can be affected by the fat content of the meal, previous exercise, alcohol intake, medications, or other factors. See Table 27.4 for a summary of validated and non-validated tests and procedures for food allergies.

Elimination diet and challenge

Experienced clinicians, dieticians, and patients may have difficulties in elucidating the exact relationship between dietary constituents and the clinical reactions experienced. It has become established practice to use

TABLE 27.3 Diagnosis of food allergy and intolerance

History

Frequency, type, severity, seasonality of reactions, interval since food ingestion, coexisting intestinal disease

Clinical examination

Entity, degree, extension, overlap of symptoms

In vivo tests

Skin prick tests

Elimination diet

Open challenge

Double-blind placebo-controlled challenge

Gastrointestinal procedures

Intestinal permeability evaluation with and without challenge

Endoscopy before and after challenge

Biopsy

In vitro tests

Food-specific IgE antibodies

Cellular tests (lymphocyte/basophil proliferation to allergens)

TABLE 27.4 Clinical tests and procedures for food allergies

Validated	Unvalidated or bogus
Double-blind placebo-controlled	Food-specific IgG and IgG subclasses
Food challenge (DBPCFC)	Antibodies
Skin prick tests	Cytotoxicity test
Allergen specific IgE, (CRD)	Sublingual subcutaneous, and neutralization component resolved diagnosis
Endoscopy & histological examination	Immune complex measurements
	Electro-acupuncture
Intestinal permeability test	Vega testing
Respiratory function test	Applied kinesiology (DRIA) test
Allergen patch test	Hair analysis

as diagnostic criteria the objectively monitored effects of exclusion diets, blinded exposure, and provocation tests.

In young children with atopic eczema, for example, appropriate elimination and challenge protocols have shown that around 60 per cent respond positively to specific food challenges with dermal, gastrointestinal, and respiratory reactions. The most common foods associated with these reactions are cow's milk, egg, wheat, peanuts and tree nuts, soya, and less often, fish and shellfish. Detailed instructions for performing oral food challenges and for the selection of the most appropriate laboratory investigation or skin test have been published elsewhere.

Difficulties with placebo responses

Experience of elimination diet and open challenge protocols have revealed that patients' perceptions and doctors' diagnoses of food intolerance are invariably inaccurate. In patients with clear and convincing histories of adverse reactions, less than half of reactions can normally be confirmed on objective testing. For this reason, double-blind placebo-controlled food challenges (DBPCFCs) should be used and if negative, the food item should be introduced openly into the diet.

Coeliac disease, a special case of food intolerance

Coeliac disease is a particular form of food intolerance which is usually linked with the HLA DQ2 and DQ8 genetic phenotype and affects around 1.5 in 200 individuals. It is defined as lifelong gluten intolerance and responds normally to gluten withdrawal. The diagnosis is made with a small bowel biopsy with the help of specific serum antibodies such as tissue-transglutaminase. Coeliac disease is thought to be an autoimmune process triggered by a toxic gliadin fraction (see Chapter 23).

Skin prick (puncture) tests

In cases of suspected IgE-mediated reactions to food, a skin prick test may be performed. A small amount of an allergen in solution is placed on the skin and then introduced into the epidermis by gently puncturing the skin surface to facilitate allergen–IgE interaction. A positive reaction is manifested as the development of a weal, the diameter of which can be measured to grade the reaction. A positive test can confirm that the patient is atopic and can strengthen suspicions about probable precipitants. The diagnostic accuracy of a skin test varies according to the offending food. Negative reactions, particularly in atopic patients, have a high (95 per cent) accuracy of there *not* being an IgE mediated reaction. Positive tests have only a 50–60 per cent predictive accuracy.

ImmunoCAP and other test systems

The radioallergosorbent test (RAST) superseded by the quantitative ImmunoCAP test demonstrates food-specific serum IgE antibodies. The correlation of IgE measurements for the diagnosis of a particular allergy in individuals with eczema is limited and often indicate only a state of immunological sensitization without *clinical* reactivity. Overall, 10–15 per cent of individuals who exhibit skin test or ImmunoCAP sensitization may tolerate exposure to those tested allergens without an adverse clinical reaction.

Component resolved diagnostic systems

Component-resolved diagnostics (CRD) utilize purified native or recombinant allergens to detect IgE sensitivity to individual allergen molecules and have become of growing importance in clinical investigation of IgE-mediated allergies. Since these tests are still in the development phase, the availability of recombinant allergic components is limited although steadily increasing. Not all defined allergen components trigger clinical relevant reactions.

Atopy patch test (APT)

Atopic eczema in infancy and childhood is often caused or aggravated by common food allergens such as milk, egg, and wheat. During the APT, suspected allergens are applied to the patient's back under a dressing and allowed to remain in contact with the skin for 48 hours.

The area is then visually examined for reddening or evidence of infiltration. Prospective clinical studies in children have resulted in highly variable results and the test is unreliable.

Endoscopic studies and intestinal biopsy

This test involves swallowing a thin tube or an endoscope which is passed through the stomach into the small intestine where a small piece of the intestinal lining is removed by a cutting device. This procedure is often used in patients with a variety of slow onset gastrointestinal symptoms, such as frequent loose stools or features of unexplained iron deficiency, osteoporosis, weight loss, slower than expected gain in height, and other features of malnutrition. Intestinal biopsy is normally not indicated for the diagnosis of acute IgE-mediated gastrointestinal disease. In these conditions a biopsy may show degranulated IgE-positive mast cells or an eosinophilic infiltration on conventional histology.

Hydrogen breath test (lactase deficiency)

The hydrogen breath test measures the amount of hydrogen in a person's breath. Normally, very little hydrogen is detectable. However, undigested lactose in the colon is fermented by bacteria, and various gases, including methane and hydrogen, are produced. The hydrogen is absorbed from the intestines and exhaled. In the test, the patient drinks a lactose-loaded drink, and the breath is analysed at regular intervals. Raised levels of hydrogen in the breath indicate impaired digestion of lactose. This test is available for children and adults.

Lactose tolerance test (lactase deficiency)

Normally, when lactose reaches the digestive system, the lactase enzyme (beta-galactosidase) breaks it down into glucose and galactose. The liver then changes the galactose into glucose, which enters the bloodstream and raises the person's blood glucose level. If lactose is not or is incompletely broken down, the blood glucose level does not rise and a diagnosis of lactose intolerance is confirmed (see also under section 'Clinical features of lactose intolerance') (Fassio et al. 2018).

Non-validated tests for the diagnosis of food allergies

IgG antibodies

Currently, the determination of IgG antibodies to food has no predictive value for diagnosis and dietary management of patients with food allergic diseases. Measurement of food specific IgG$_4$ alone is not sufficient for a diagnosis of food allergy. Successful immunotherapy studies report an important increase in allergen-specific IgG4 levels which may be one of the important immunological mechanisms underlying induction of tolerance.

Symptoms, syndromes, and features of food-induced diseases

A wide range of factors determine the route to a state of allergic sensitization and allergic disease. The factors are often interlinked and modulated by the host's genetic disposition and health status (Figure 27.8).

Food allergies and intolerance in children

A wide range of conditions in childhood have been associated with food allergies and intolerance; these include eczema, wheeze, urticaria, mood changes, angio-oedema, epilepsy, failure to thrive, diarrhoea, vomiting, and gastrointestinal blood loss. There is some evidence that a small number of hyperactive children, often boys, respond to dietary measures with improvement of symptoms related to attention deficit hyperactivity disorder (ADHD) in a controlled trial.

Milk-induced colitis, mainly in breastfed children under two years of age, differs from ulcerative colitis in many clinical and pathological features. Small intestinal mucosal damage with malabsorption is best documented for cow's milk protein-induced enteropathy but can also occur with other food protein intolerance.

Food intolerance in adults

The incidence of adverse reactions to foods in adults is estimated to be 2–8 per cent of the population and estimated as 3–5 per cent in childhood. High estimates of 7.5 per cent in childhood and of 8 per cent in adults have

been proposed. This clearly highlights the scope of the problem and the importance of considering adverse reactions to foods in a sizeable number of patients. Table 27.5 shows the most common foods encountered as allergens.

Classic food allergic symptoms of asthma, urticaria, and anaphylaxis can be found in adults. Adults with food allergic reactions often have a history of adverse reactions to foods in childhood. Lower food allergy prevalence in adults suggests that there is some 'growing out of childhood allergies' often for milk and egg. *De novo* sensitization to foods in adults is less frequent. Adverse reactions to foods can affect all organ systems (Table 27.2) (see also Weblink Figure 27.1).

A number of patients report rather ill-defined and seasonal symptoms in relation to food intake. This could be due to a cross-reactivity between common inhalant allergens and food allergens on the basis of sharing common allergenic proteins (Table 27.6). This feature makes it often difficult to establish a sound diagnosis without resorting to elimination and challenge studies. The relationship between food intolerance and migraine is complex but its existence is well supported by clinical studies.

Migraine can be triggered by direct pharmacological actions on the vessel wall or by as yet unresolved immunological mechanisms. The threshold for the onset of a food-induced migraine is altered by many other factors such as fatigue, smoking, alcohol, and menstrual cycle.

There is some association between food allergy and *arthritis*, but there is little evidence that can withstand critical examination. *Psychiatric symptoms* such as irritability and depression may accompany other manifestations of food intolerance.

Other food-induced symptoms are *gastrointestinal* in origin. They include nausea, bloating, abdominal pain, constipation, and diarrhoea. These features are similar to those of the irritable bowel syndrome (IBS). The

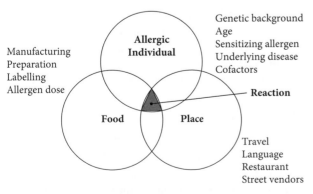

FIGURE 27.8 Factors involved in triggering allergic reactions.

TABLE 27.5 Most common foods encountered as allergens

Foods most commonly encountered as allergens	Others foods also known to trigger allergic reactions
Cow's milk	Fruits (kiwi, mango, banana)
Egg	Legumes (peas, lupine, lentils)
Soya	Seeds (sesame, sunflower)
Wheat	Spices (mustard, coriander)
Peanuts	Vegetables (celery, tomato)
Tree nuts	
Fish	
Shellfish	

TABLE 27.6 Cross reactivities between inhalant and food allergens

Inhalant allergen	Food allergen
Common	
Tree pollen	Apple, hazelnut, carrot, cherry, green kiwi, nectarine, peach, apricot, plum, celery, soya, fig
Less common	
Mugwort pollen	Spices, carrot, mango, celery, sunflower seeds
Celery–birch–mugwort–spice syndrome (mainly observed in birch pollen allergic patients with Bet v 1 sensitization)	
Natural latex	Avocado, banana, kiwi, tomato, chestnuts, peach, mango, papaya, cherry, celery
Rare	
Ficus benjamina	Figs, kiwi, banana, papaya pineapple, avocado
Bird feathers	Egg, poultry, offal
House dust mites	Crustaceans and molluscs

symptoms may arise either because of abnormal motility or because an individual has a reduced pain threshold to sensations associated with normal motility of the gut. Not surprisingly, these symptoms are often closely related to foods. Many people who avoid specific foods and have an unsubstantiated self-diagnosis of food allergy suffer from irritable bowel syndrome. Their often-self-imposed alterations in diet will influence gut motility, composition of bowel movements, and production of gas. In an introspective individual such physiological changes can reinforce the patient's concern. Recent studies explore the effects of anti-depressants and motility modulators in the treatment of irritable bowel syndrome.

Psychological food intolerance

There is no doubt that attitudes to food vary widely. Dieting, overeating, and food fads are extremely common. Alternatively, in a patient with an eating disorder such as anorexia nervosa, suggestions that the problem could be 'allergic' is seized upon to avoid the possible stigma of a psychiatric diagnosis. Perceived or imagined food intolerances can lead to child abuse and severe failures to thrive (induced illness) in children. Food fads and 'fashionable' dieting may lead to eating disorders in adolescents and adults.

Approaches to dietary treatment

Symptoms and clinical features may conform to well-recognized phenomena or to a disease associated with a specific food. This will be suspected from the history and confirmed by a small number of tests.

In other patients, food intolerance will form part of a wider differential diagnosis. When symptoms are mild, simple symptomatic treatment, such as non-sedating antihistamines, anti-diarrhoeals for occasional diarrhoeal bouts, or analgesics for headache, may be more appropriate initially than complex therapeutic diets. If food intolerance is suspected, a baseline elimination diet is introduced for some weeks; if symptoms and signs disappear, relevant foods are identified during a planned period of reintroduction and should be confirmed by placebo-controlled food challenges. If the challenge is negative, food needs to be openly introduced into the diet to confirm the challenge results.

Preventing food allergies

The food allergy and anaphylaxis guidelines of the European Academy of Allergy and Clinical Immunology do not recommend dietary restrictions for mothers during pregnancy and lactation. In particular circumstances, early allergen introduction in high-risk infants may be beneficial in reducing allergic sensitization through tolerance induction. Longer term results are outstanding and the effects of peanut exposure via aerosol or oral ingestion may affect the outcome. Exclusive breastfeeding is recommended for the first 4–6 months of life, which reduces the development of allergies (see also Weblink Tables 27.1, 27.2, 27.3).

Induction of allergen specific tolerance in food-allergic children

Clinical, 'rush' desensitization protocols, for example for antibiotic and hymenoptera allergies have been successfully used for a long time. Specific oral tolerance induction (S)OTI for IgE-mediated symptoms has been under development for use in milk, egg, and peanut- and gliadin-allergic children.

The underlying mechanisms relating to oral tolerance induction and desensitization procedures are still a matter of scientific investigation, and the longer-term outcomes of this treatment modality are unresolved. One of the major questions is whether the increase in tolerance after the treatment is related to oral 'tolerance' induction or whether it is based on immune suppression/desensitization. This distinction is important, since tolerance does not require the continuing allergen administration, whereas desensitization usually does. Performing desensitization protocols under routine hospital or ambulatory conditions may be hazardous with regard to potential serious side effect during allergen dose escalation. Overall advantages of SOTI (specific oral tolerance induction) could be an increase in the individual's threshold dose and thus a reduction of the risk of experiencing severe allergic reactions after accidental ingestion (Pepper et al. 2020).

Relevance of food allergy in diseases

Antibody responses to cow's milk proteins in humans

Milk-specific IgG antibodies are generally present in the serum of most children and in 10–20 per cent of adults. Patients with diffuse small bowel disease and enhanced intestinal permeability tend to have elevated titres of serum antibodies to many foods. IgE responses are of greater relevance than non-IgE responses in indicating the likely mechanism of food allergic diseases. IgE antibodies to food are general evidence of an atopic state (i.e., a predisposition to become allergic). If these food-specific IgE antibodies are particularly raised, they are often combined with a positive skin test. Adverse allergic (immediate anaphylactic) reactions to this food are likely and must be investigated and ruled out.

Managing the spectrum of food allergies and intolerances

It is important to define the underlying specific immunological mechanism causing adverse symptoms, especially in atopic patients who may suffer from symptoms which can be triggered by environmental allergens. Goldman et al. (1963) (Further Reading Section 27.2), proposed the diagnosis of cow's milk allergy in a patient if symptoms subsided after dietary elimination of milk and recurred within 48 hours after milk introduction. At times such a challenge can be too dangerous for highly allergic infants. Clinical improvement on a controlled elimination diet should be the first criterion for the routine diagnosis. If there is any doubt, a DBPCFC with the food in question needs to be performed. Eliminated foods need to be reintroduced into the children's diet at 6–12 months intervals to avoid unnecessary dietary restrictions.

Example: cow's milk allergy

The wide clinical spectrum of food (cow's milk protein) induced symptoms in childhood helps to illustrate these difficulties. Hill et al. 1995 (Further reading Section 27.3) assessed the relationship between clinical symptoms and serum antibody levels in children and found that the patients could be divided into three groups: first, those children who show immediate symptoms with small amounts of milk, evidenced by anaphylaxis, angiooedema, urticaria, and diarrhoea; second, those who develop symptoms often up to several hours after intake of moderate amounts of milk (approximately 200 ml) and in whom the skin test to the offending food was generally negative; and third, those mostly older children suffering from a poorly defined multisystem involvement, including, for example, skin, lung, gastrointestinal tract, central nervous system (migraine), who often required larger amounts more frequently and in whom symptoms could take over 24 hours to occur.

Atopic eczema

The incidence of atopic eczema is rising worldwide, and foods are among the many environmental factors that contribute to this distressing skin disease. A lack of bacterial stimulation of the gastrointestinal tract through excessive cleanliness ('hygiene hypothesis') has been put forward as an additional hypothesis and made plausible through a number of prospective studies in children living on farms (see Figure 27.9).

The strongest evidence of a role for food in the pathogenesis of atopic eczema comes from studies in which children with atopic eczema have responded well to exclusion diets in double-blind controlled crossover trials. Eczema can occur in exclusively breastfed infants. This can be linked to the transfer of food antigens from the mother's milk. Food intolerance and enhanced immune responsiveness to common foods are also features of atopic eczema in adults.

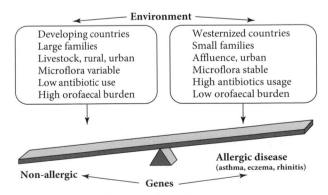

FIGURE 27.9 Balance of environmental and genetic factors which contribute to or suppress allergic sensitization: 'hygiene hypothesis'.
Epigenetic conditions, bacterial environment, host immune status and nutrition at particular times may tip the balance either way.

Asthma

Clinical observations suggest that milk intolerance is an occasional trigger of asthma. Asthma overall, is an important risk factor in patients suffering food allergic and in particular anaphylactic reactions. In about 5 per cent of patients with asthma, foods may cause worsening of their symptoms.

Cow's milk intolerance with malabsorption syndrome

The classic milk-induced malabsorption syndrome with an abnormal (flat) intestinal mucosa and failure to thrive has become less frequent. More often, infants and children suffer from vomiting, diarrhoea, eczema, urticaria and angio-oedema, recurrent otitis, and also constipation (see Table 27.2). In the classic form, gastrointestinal investigations show malabsorption, and jejunal biopsy reveals abnormalities of the jejunal mucosa, ranging from moderate villous atrophy to pathology indistinguishable from coeliac disease. In children with less severe symptoms, as is currently the more common disease presentation, the only evidence may be a mild infiltration with eosinophils or mast cells within an otherwise normal mucosa.

Cow's milk-sensitive colitis

An infant with food-sensitive colitis usually presents before the age of one year, with the alarming picture of loose stools that contain mucus and blood. An elimination diet leads to quick resolution of symptoms. This is mirrored by a rapid histological recovery on rectal mucosal biopsy. The pathology of the rectum differs from classic ulcerative colitis with preservation of mucosal crypts, without crypt abscess formation, and no depletion of goblet cell mucus. Additionally, there are substantial numbers of eosinophils and plasma cells in the infiltrate of the lamina propria. These infants respond well to elimination of cow's milk from their diet or from the mother's diet if they are still breastfed. Most children can tolerate cow's milk by the age of 2–3 years.

Principles of management

The treatment of cow's milk-induced allergic systems is the complete elimination of milk and of milk-derived products. *Extensively* hydrolysed infant formulas are tolerated in most infants who often suffer from the less severe form. These formulae are recommended for infants under six months of age. However, if symptoms do not subside within a week, amino acid-based formulas should be used. In all children who do not receive sufficient amounts of formula feeds, calcium supplementation must be given. Occasionally lactose intolerance overlaps with the syndrome of cow's milk protein-sensitive enteropathy. When there is extensive villous atrophy, loss of disaccharidase-containing mature enterocytes leads to a reduction in the disaccharidase activity of the small bowel mucosa. In this case, lactose intolerance is caused by cow's milk protein intolerance. Lactose intolerance due to primary lactase deficiency is extremely rare and lactose intolerance in a European population often indicates an underlying (mucosal) abnormality).

Non-allergic milk intolerance due to lactose

Lactose is the main disaccharide (beta-galactose-1,4-D-glucose) that is exclusively present in mammalian milk products. Its concentration is about 5 per cent and 7 per cent in cow's and human milk. It is a major source of energy in the young infant. Lactose is hydrolysed by the small intestinal lactase (ß-galactosidase), which is located in the brush-border of epithelial cells to create glucose and galactose after hydrolysis. Galactose is then rapidly converted into glucose in the liver.

The expression of lactase activity in the brush-border declines from the infant level to adult levels (a 10- to 20-fold reduction) between the ages of three and five years in 75 per cent of the world's population, while 25 per cent of the population appear to maintain infant levels of lactase in adulthood. Both males and females are equally affected. In Northern Europe individuals with lactase expression into adulthood are in the majority and are called 'lactase persisters'. In subjects of European descent, the genetic *LCT*-13910C>T variant is completely associated with the lactase-persistence phenotype. Low levels of this phenotype are reported for individuals of Central and South African descent.

An individual with lactase non-persistence expresses hypolactasia and will exhibit lactose maldigestion to some extent. Lactose maldigestion does not necessarily lead to lactose intolerance. Lactose maldigesters can, for example, demonstrate a positive breath hydrogen excretion test without displaying clinical symptoms of lactose intolerance.

Clinical features of lactose intolerance

The clinical effects of lactose ingestion are closely related to the ingested dose. There is a wide variation among individuals regarding the dose–response. The conventional lactose load, 50g, used in tolerance tests, equivalent of about three glasses of milk in one setting, produces usually symptoms in 80 per cent to 100 per cent of lactose-intolerant individuals. A challenge dose of 10–15g of lactose which is now often suggested, is equivalent to around ~250 ml of milk and will produce (usually slight) abdominal symptoms in 30–60 per cent of individuals. Distribution of total milk intake over the day will increase the total tolerance and may avoid the need for calcium supplementation. In some individuals clinical reactions have been described after intake of less than 1g of lactose, although most lactase-deficient individuals tolerate 7–10g of lactose without symptoms.

Abdominal pain in children

Recurrent abdominal pain in children is almost as common as irritable bowel syndrome. The post-weaning drop in intestinal lactase activity may occur as early as two years in some ethnic groups, or at five years in white-background children, so that schoolchildren occasionally may be intolerant of lactose. Studies of recurrent abdominal pain in children in the USA have shown clinical lactose intolerance in a substantial proportion, particularly in Black, Hispanic, and Asian children. Lactose intolerance associated with abdominal pain is particularly relevant in children of ethnic groups with a high prevalence of hypolactasia.

Diarrhoea after gastric surgery

Gastric surgery and surgery of the small intestine radically alter the physiology of the upper gastrointestinal tract. As noted above, the rate of gastric emptying may affect the tolerance to lactose in a susceptible individual. Lactose intake in a lactase non-persister after surgery may lead to bloating, faintness, discomfort known as 'dumping', and diarrhoea.

Secondary, reversible lactose intolerance

World Health Organization recommendations for management of children recovering from acute diarrhoea are that oral rehydration nutrition, including breastfeeding, should in general be introduced within 24 hours. Although many low lactose and modified milk preparations are available for nutrition of patients with acute and chronic diarrhoeas, these are likely to be more relevant in the management of immunologically based milk protein intolerance, or in chronic diarrhoeal disease, than during infectious acute gastroenteritis.

Treatment of lactose intolerance

As with other states of food intolerance, the reliable diagnosis of lactose intolerance relies on objective measurements of the clinical effects of the withdrawal and reintroduction of lactose. For a great number of individuals, especially infants and children, milk is such an important nutrient that before recommending a low lactose diet with the avoidance of milk, milk (i.e., lactose) intolerance should be formally confirmed. The only satisfactory treatment of lactose intolerance is a diet with low oral, tolerated, lactose content to avoid calcium and vitamin D deficiency states. Most lactose-intolerant patients can tolerate about 10g of fermented milk products such as cheese and yoghurt, in which the lactose content has been reduced. Lactose-reduced milk and milk products are widely available.

KEY POINTS

- Unbalanced or deficient nutrient intake may adversely affect the immune system in many different ways. Mild to serious deficiency diseases are its severe consequences. This also reduces the immune response to vaccinations in children and the elderly population. Deficiencies of minerals and vitamins are eminently of global importance. Both malnutrition, overweight and obese states adversely affect the immune system.

- The gastrointestinal tract and the associated lymphoid tissue are key players in the development oral tolerance. Its central role makes the gut lymphoid tissues probably the most important part of the body's immune system. This beneficial effect can be disrupted by intestinal infections including fungal colonization and their effects on the microbiome.

- A protective role of 'beneficial' intestinal bacterial colonization (probiotics) needs to be assessed for their adjuvant or clinical effect on an individual basis.

- Worldwide allergy prevalence has increased about to 8–10 per cent. A better immunological understanding of intestinal regulation has identified procedures which may be able to reduce the allergic immune response to highly allergic individuals (S(O) TI specific (oral) tolerance induction).

- International nutritional guidelines suggest that early feeding plans with careful introduction of allergenic foods during the first 4–6 months of life may be more likely to avoid sensitization rather causing it. This has been demonstrated for peanut, egg, and also gluten.

Be sure to test your understanding of this chapter by attempting multiple choice questions. See the Further Reading and Useful Websites lists for additional material relevant to this chapter.

REFERENCES

Acevedo N, Alhamwe B A, Caraballo L et al. (2021) Perinatal and early-life nutrition, epigenetics, and allergy. *Nutrients* **13**, 724. https://doi.org/10.3390/nu13030724.

Agyemang A & Sicherer S (2019) The importance of early peanut ingestion in the prevention of peanut allergy. *Expert Rev Clin Immunol* **15**, 487–495. https://doi.org/10.1080/1744666x.2019.1582331.

Akdis M, Aab A, Altunbulakli C et al. (2016) Interleukins (from IL-1 to IL-38), interferons, transforming growth factor β, and TNF-α: Receptors, functions, and roles in diseases. *J Allergy Clin Immun* **138**, 984–1010. https://doi.org/10.1016/j.jaci.2016.06.033.

Ampofo A G & Boateng E B (2020) Beyond 2020: Modelling obesity and diabetes prevalence. *Diabetes Res Clin Pr* **167**, 108362. https://doi.org/10.1016/j.diabres.2020.108362.

Barrea L, Muscogiuri G, Frias-Toral E et al. (2020) Nutrition and immune system: From the Mediterranean diet to dietary supplementary through the microbiota. *Crit Rev Food Sci* July 1–25. Ahead of print. https://doi.org/10.1080/10408398.2020.1792826.

Bermano G, Méplan C, Mercer D K et al. (2020) Selenium and viral infection: Are there lessons for COVID-19? *Brit J Nutr* **125**(6), 618–627. https://doi.org/10.1017/s0007114520003128.

Brandtzaeg P (2011) The gut as communicator between environment and host: Immunological consequences. *Europ J Pharm* **668**, S16–S32. https://doi.org/10.1016/j.ejphar.2011.07.006.

Calder P C (2020) Nutrition, immunity and COVID-19. *BMJ Nutrition Prev Health* **3**(1), 74–92. https://doi.org/10.1136/bmjnph-2020-000085.

Fassio F, Facioni M S, Guagnini F (2018) Lactose maldigestion, malabsorption, and intolerance: A comprehensive review with a focus on current management and future perspectives. *Nutrients* **10**(11), 1599. https://doi.org/10.3390/nu10111599.

Fleischer D M, Chan E S, Venter C et al. (2020) A consensus approach to the primary prevention of food allergy through nutrition: Guidance from the American Academy of Allergy, Asthma, and Immunology; American College of Allergy, Asthma, and Immunology; and the Canadian Society for Allergy and Clinical Immunology. *J Allergy Clin Immunol Pract* **9**, 22–43. e4. https://doi.org/10.1016/j.jaip.2020.11.002.

Goldman A S, Anderson D W Jr, Sellers W A et al. (1963) Milk allergy I: Oral challenge with milk and isolated milk proteins in children. *Pediatrics*, *32*, 425–443.

Hemilä H (2017) Vitamin C and infections. *Nutrients* **9**, 339. https://doi.org/10.3390/nu9040339.

Hill D J, Hudson I, Sheffield L J et al. (1995) A low allergen diet is a significant intervention in infantile colic: Results of a community-based study. *J Allerg Clin Immunol* **96**, 386–392.

Jin H, Biello K, Garofalo R et al. (2021) Examining the longitudinal predictive relationship between HIV treatment outcomes and pre-exposure prophylaxis use by serodiscordant male couples. *J Acquir Immune Defic Syndr* **86**, 38–45. https://doi.org/10.1097/qai.0000000000002522.

Kim E H, Jones S M, Burks A W et al. (2020) A 5-year summary of real-life dietary egg consumption after completion of a 4-year egg powder oral immunotherapy (eOIT) protocol. *J Allergy Clin Immunol* **145**, 1292–1295.e1. https://doi.org/10.1016/j.jaci.2019.11.045.

Martineau A R, Jolliffe D A, Hooper R L et al. (2017) Vitamin D supplementation to prevent acute respiratory tract infections: Systematic review and meta-analysis of individual participant data. *BMJ* **356**, i6583. https://doi.org/10.1136/bmj.i6583.

Notarangelo L D, Bacchetta R, Casanova J-L et al. (2020) Human inborn errors of immunity: An expanding universe. *Sci Immunol* **5**(49), eabb1662. https://doi.org/10.1126/sciimmunol.abb1662.

Nowak-Wegrzyn A, Szajewska H, Lack G (2017) Food allergy and the gut. *Nat Rev Gastroentero* **14**, 241–257. https://doi.org/10.1038/nrgastro.2016.187.

Pepper A N, Assa'ad A, Blaiss M et al. (2020) Consensus report from the food allergy research and education (FARE) 2019 oral immunotherapy for food allergy summit. *J Allergy Clin Immunol* **146**, 244–249. https://doi.org/10.1016/j.jaci.2020.05.027.

Sampath V, Abrams E M, Adlou B et al. (2021) Food allergy across the globe. *Allerg Clin Immunol* **148**, 1347–1364. doi: 10.1016/j.jaci.2021.10.018.

Satitsuksanoa P, Jansen K, Głobińska A et al. (2018) Regulatory immune mechanisms in tolerance to food allergy. *Front Immunol* **9**, 2939. https://doi.org/10.3389/fimmu.2018.02939.

Venter C, Eyerich S, Sarin T (2020) Nutrition and the immune system: A complicated tango. *Nutrients* **12**(3), 818. https://www.mdpi.com/2072-6643/12/3/818

WHO Europe (2020) http://www.who.int/nutrition/topics/ida/en/.

28

Eating disorders

Bruno Palazzo Nazar and Janet Treasure

OBJECTIVES

By the end of this chapter you should be able to:

- Learn how to identify eating disorder patients
- Differentiate between the main eating disorder syndromes
- Understand the recent changes in classification systems of eating disorders
- Reflect on the medical and psychiatric differential diagnosis of eating disorders
- Learn how to design a basic treatment plan for eating disorder patients.

28.1 Introduction

Definition

Eating disorders (ED) is a diagnostic grouping that encompasses disorders with severe disturbances in eating patterns (American Psychiatric Association 2013). Its main syndromes are anorexia nervosa (AN), bulimia nervosa (BN), binge eating disorder (BED), and avoidant/restrictive food intake disorder (ARFID). ED are complex syndromes impairing not only eating behaviours and the brain appetite regulatory system but also weight regulation and many other brain networks involved in body image perception, cognitive functioning, and emotional and social functioning and reward regulation (Treasure et al. 2020).

The Diagnostic and Statistical Manual, in its fifth edition (DSM-5) recognizes AN, BN, BED, and ARFID but also describes a category with atypical forms of these classic syndromes named 'other specified feeding and eating disorders' (OSFED). There was a particular concern among specialists to refine diagnostic criteria to decrease the use of catch-all categories like the DSM-4 'eating disorders' and 'not otherwise specified' (EDNOS). Developments have also been made for ICD-11(WHO

2020), where AN, BN, BED, ARFID, OSFED and 'feeding and eating disorders, unspecified' (UFED) are described. In this matter, there is a resemblance among the current ICD and DSM editions which will aid clinicians using either system to achieve a similar understanding of ED classification. Further, DSM-5 and ICD-11 are also included in the same chapter as classic eating disorders and two feeding disorders previously found in a separate chapter 'Feeding and eating disorders of infancy or early childhood'. Both DSM-5 and ICD-11 describe Pica disorder, but DSM-5 has 'rumination disorder', whereas ICD-11 has 'rumination–regurgitation disorder'.

The definitions encountered in the literature for this diagnostic group remain in a state of flux. Thus, the external borders of this diagnostic group are blurred, making room for other possible clinical entities such as orthorexia nervosa (an obsession with eating foods that one considers healthy), Muscle dysmorphia (a preoccupation with attaining an idealized muscular body), Purging disorder (recurring self-induced vomiting without other features of bulimia nervosa), and emetophobia (an intense fear of vomiting). This blurred area is composed mainly of OSFED and UFED which still need more research to define how they are best classified and conceptualized. For example, even though muscle dysmorphia (also know as vigorexia) was published as a specifier to body dismorphic disorder in DSM-5, there is enough research to argue maybe this syndrome could be understood as an eating disorder, under the unspecified eating disorders category.

It is important to differentiate eating disorders from disordered eating behaviours (DEB). The latter are not considered mental disorders and are frequently studied across epidemiological studies as a risk factor for ED, especially by nutrition specialists. Some examples of DEB might include very well studied behaviours (dietary restraint, fasting, over exercise, purging) but also some behaviours which still warrant further studies for their understanding (e.g., binge eating without subjective feelings of loss of control, binge eating without objective eating of large quantities of food, grazing). One step further to increase understanding of DEB, both in research and in clinical assessments, is to assess if they are associated with distress or impairment.

Epidemiology of eating disorders

AN epidemiology

Epidemiological global estimates of anorexia nervosa (AN) may differ according to study design. Prevalence refers to the number of cases of a given disorder in a population at a specific time, whereas point prevalence is the same concept but at a specific interval of time (e.g., one year), and incidence refers to the number of new cases arising in each determined period of time in a population group. Across the world AN prevalence usually varies from 0 up to 0.9 per cent in the general population; the point prevalence for AN in young women is 0.5 per cent. The lifetime prevalence ranges from 1.2 per cent to 2.2 per cent when AN is strictly defined.

The typical European incidence rate of AN is 8 per 100,000 of the population per year (Hoek 2006). Among all women, the AN incidence is 18.46 per 100,000 contrasted to 2.25 per 100,000 men. Interestingly, these numbers are higher in women aged 15–19, where the incidence rate is 270 per 100,000 using a strict AN definition and 490 per 100,000 using a broad AN definition. Of note, less than 20 per cent of ED cases present for treatment.

The incidence range is highest in women aged 15–19 years. Secular trends in AN show that these rates are stable for young women but there has been an almost three-fold increase in estimates of atypical AN in the past 40 years within women in their twenties and thirties. Instead of an actual rise in the number of ED sufferers, possible explanations for upward trend include a raise in awareness about ED, more available ED specialty services, and changes in the demographic structure of the population. Pediatric samples of AN tend to have a higher ratio of boys than adult samples. Among children, boys represent about 25 per cent of cases, whereas in adult samples they represent about 10 per cent.

BN epidemiology

This syndrome is usually reported to have a point prevalence varying from 0.9–1.5 per cent in females and 0.1–0.5 per cent in men. Using DSM-5 criteria, the lifetime prevalence is 2.3 per cent of the population, with studies reporting it no higher than 2.9 per cent. The atypical (or subclinical) forms of bulimia nervosa (BN) are frequent and its prevalence can be as high as 5.4 per cent of the population.

The incidence of BN has remained relatively stable in the last two decades following the early increase after it was defined in 1979, in the female 16–20 years-old group. Analysis of general practice databases in the UK demonstrated that the BN incidence was 20.7 in females and 11.8 in the general population. These numbers increase to 46.8 in the 15–19 years group but since the years 2000s they have become relatively stable.

BED epidemiology

Binge eating disorder (BED) is more common than AN or BN. The lifetime prevalence rates in the population are 3.5 per cent for women and 2 per cent for men (Hudson et al. 2007). In a large study in the USA, BED incidence rate per 1000 person-years was 6.6 in males and 10.1 in females, and in a study conducted in the UK it was 60/100,000.

Finally, it is important to highlight that ED can occur across the weight spectrum, in both genders and at any age. Also, there is wide diversity in socioeconomic status and ethnicity.

Etiology and risk factors

Multiple etiological factors have been identified as contributing to the development of an ED. Psychosocial factors are presented here. Others, including familial, genetic, and neurobiological are described in Weblink Section 28.1.

During the twentieth century ascetic ideals changed. It is noted that young girls are exposed to the ideal of thinness, whereas the 'muscular ideal' is valued for boys and overweight is stigmatized. Also, food is plentiful and is readily available and highly palatable through developments in food technology. Meals and eating are less socially constrained. These sociocultural factors have impact on the maintenance of all forms of eating disorders and predispose to the increased prevalence of loss of control and overeating. The mechanism appears to be through weight stigma and low mood associated with dieting behaviours.

The argument that all ED are culture-bound syndromes is weakened by the descriptions of AN individuals throughout history, going all the way back to the Middle Ages. Thus, AN could be regarded as a syndrome with a biological basis and a pathoplastic (i.e., changes in psychopathology and clinical presentation), sociocultural expression, which means that depending on the beauty or healthy ideal valued by society at a certain time, the pursuit of thinness would change to achieve that aim. For example, in the 1980s thinness was pursued as a beauty ideal, and more recently this pursuit has been associated with 'clean eating' and the pursuit of a healthier lifestyle.

On the other hand, disorders characterized by a loss of control and overeating appear to be more linked to cultural factors and these have a diagnostic overlap with obesity. From a clinical perspective, it might be of interest to understand how these etiological features present in each patient as this could help define treatment plans for individuals.

A classical study was conducted on Fiji island, where the prevalence of traits related to ED were measured one month and three years after Western television had been widely introduced. The level of ED traits increased from 12.7 to 29.2 per cent, and purging habits increased from 0 to 11 per cent (Becker et al. 2002). Longitudinal studies have highlighted the role of weight concerns, peer environment, and personality factors in the development of an ED. Approximately 10 per cent of women with elevated weight concerns go on to develop an ED (Solmi et al. 2021).

Family environment

Weight stigma within and outside of families as manifested by commenting/teasing about body shape and weight increases the risk, particularly of the binge eating forms of eating disorders. Similarly, stressful events and adversity also increase the risk of this form of eating disorder. In anorexia nervosa family dynamics can change after the development of an ED, as a means of coping with the problematic behaviours presented by the sufferer. These can serve to maintain the problem and include high levels of expressed emotion, a parenting style in which overprotective, controlling, hostile, and critical behaviours are prominent; collusion/accommodating (allowing ED behaviours to persist), and enabling (behaviours that cover up negative consequences of an ED); or denial of the disorder or its impact (Treasure & Nazar 2016).

The influence of peers has been demonstrated as a risk factor for ED symptoms. Although peer contact does not influence personality traits, it may influence the initiation of bulimic symptoms and modulate their frequency. In addition, peer pressure can manifest in the form of bullying, as well as teasing or other forms of verbal harassment, always targeting weight and eating behaviours (Eisenberg & Neumark-Sztainer 2008).

Activities and professions that are associated with, and value, a slim body and/or low weight tend to to be associated to higher rates of AN and BN. Elite athletes, especially those related to aesthetic sports (gymnastics, acrobatics, horse riding, or ballet dancers) are at risk. The term 'female athlete triad' was coined to characterize a common situation seen in this population of disordered eating, menstrual dysfunction, and low body mass (Tan et al. 2016). Similarly, fashion models are also a population of concern, and some countries have instituted policies regarding runway models having a minimum body mass index to be allowed to work. Another vulnerable group for the development of ED are college students, because of they represent the peak age of onset for eating disorders (18–25) and life changes/stress, promoted by the start of university life.

Life stressors can act as triggers for initiating the disorder. Also, internal (e.g., hormonal and physical changes during puberty) or external factors (e.g., family disruptions or peer-problems) may play a role in starting or maintaining the disorder.

28.2 Anorexia nervosa

Definition

Anorexia nervosa is considered a mental disorder that can be diagnosed by clinical interview of patients and family members. The clinical diagnosis of AN is based on Russell's triad (Russell 1973), consisting of:

1. Self-induced weight loss with starvation: There are different definitions available in order to determine low weight. DSM-5 states that being 15 per cent below a person's expected weight for their age, height, and sex would be the first criterion for a positive diagnosis (American Psychiatric Association 2013), whereas the ICD-11 describes the use of body mass index (BMI) of 18.5 kg/m^2 or less in adults or BMI-for-age under 5th percentile for children and adolescent as a cut-off (WHO 2020).

   DSM-5 has introduced BMI ranges to mark severity:
   - mild: BMI >17 kg/m^2
   - moderate: BMI 16–16.99 kg/m^2
   - severe: BMI 15–15.99 kg/m^2
   - extreme: BMI <15 kg/m^2

   However, ICD-11 describes a specifier 'with dangerously low body weight' for adults with a BMI under 14 kg/m^2 or children and adolescents with a BMI-for-age below 0.3rd percentile.

2. A specific form of psychopathology: Many mental symptoms expressing beliefs and cognitions concerning eating, weight, and/or body shape can be manifest prior to weight loss, but they might also occur as a consequence of malnourishment. Although many young women exhibit dissatisfaction with their body and weight, these ideas become extreme in AN. These over-valued perceptions of weight and body shape can become the core cognition of their mental functioning, dominating over other meaningful life domains.

3. Physical effects of malnutrition with all the stigmata related to the body response to lack of nutrients (e.g., hair loss, lanugo, dry skin) and an endocrine disturbance with abnormal hormonal function: This can be expressed in postmenarcheal females by amenorrhoea or loss of libido, and by sexual impotency in men.

Diagnosis

The diagnostic criteria in the systems of classification have changed over time. Patients that do not fulfill all criteria for AN on the initial assessment may require subsequent interviews with different sources of information.

Several strategies can be used by AN patients to lose weight. These behaviours can be compensatory to episodes of binge eating but also to trangression of diet rules. Purging behaviours include self-induced vomiting and laxative abuse. Other behaviours for weight loss include self-induced fasting, over-exercise, and skipping insulin injections in type 1 diabetes. Frequently, one strategy is used more frequently even if others are also present. However, this major strategy can also change over time. A person can use fasting in the early phases of the disease and then utilize purging as the AN progresses. Two subtypes of AN are described in the DSM-5: restricting type and binge-eating/purging type. The restricting subtype excludes those with regular binge eating or purging behaviour. A person with anorexia nervosa can alternate between restricting and bulimic subtypes at different times in their illness and can migrate into a different eating disorder such as bulimia nervosa (as explained in the 'Clinical features' section below). ICD-11 also offers three possible subtypes: restricting pattern, binge–purge pattern, or unspecified.

The criterion of amenorrhea has been omitted from DSM-5 and ICD-11 as it was demonstrated that AN patients with or without amenorrhea did not differ in clinical and prognostic outcomes. However, this sign is important as it can be an indirect marker of bone demineralization.

Apart from the aforementioned severity criteria, both DSM-5 and ICD-11 have course specifiers. In DSM-5, it should be specified if AN is in partial remission (either criterion B or C or both) is still present, or is in full remission. ICD-11 only has one course specifier, which is 'anorexia nervosa in recovery in normal body weight', for adults in recovery of AN with a BMI above 18.5kg/m^2.

Clinical features

Patients with AN present with a severe reduction in food intake with exclusion of high energy food items. Food choices become restricted, monotonous, and nutritionally unbalanced. During mealtimes, some behaviours can be observed: measuring and weighing portions, eating slowly, and in a ritualized manner such as starting with a particular item, careful cutting and counting chewing; hiding and disposing of food. Anything related to dieting and nutrition may be of interest to AN patients, who might consider themselves as 'experts' on nutrition. Some patients are also keen on cooking for others.

Since the term 'anorexia' means absence of hunger, it is potentially misleading as many patients do experience hunger. Also, patients report learning how to control or suppress normal internal signals related to feeding behaviour. It is possible that, when patients are able to overcome or ignore hunger and restrict intake they might experience this as rewarding. Consequently, starvation leads to adaptations that can disrupt normal digestion with a slow rate of gastric emptying, which can promote unpleasant physical feelings when normal food is ingested and may result in bloating. Fortunately, these adaptations are reversible with restitution of normal eating patterns and weight.

Hyperactivity and restlessness in AN may have different motivations. For some it is planned with a goal to burn calories or for high performance as an athlete. Also, in some cases a more automatic form of restlessness is observed. Possible explanations for this include a metabolic anomaly or anxiety. Laxative and diuretic misuse and self-induced vomiting may occur. Approximately a third of patients with AN report binge eating, although sometimes only the subjective component of loss of control over eating is present, lacking the objective consumption of large amounts of food.

Body image, a core element of some forms of eating disorders, probably includes a sensory perceptual, and an emotional and a cognitive component. These can present as overestimations of all or parts of body size, and the perception that body parts are too fat. However, not all patients present with this.

More general cognitive functioning is impaired, and patients or informants might complain of difficulties with concentration, attention, and the ability to use flexible and bigger-picture thinking styles. People with AN have difficulties reading intentions and emotions in others as well as signalling their own.

Comorbidity

Apart from the specific psychological symptoms related to weight and shape, other dimensions of mental functioning can be affected. Some symptoms arise as a complication of undernutrition and others may be comorbid conditions that predispose to vulnerability. A depressive syndrome is common, and patients present depressed mood, irritability, social withdrawal, and loss of sexual libido. Also, obsessive-compulsive symptoms may arise or become more pronounced. Anxiety disorders in childhood are an important risk factor for AN, and sufferers usually present social anxiety.

Markers of nutritional/medical status

The clinical workup of ED patients requires an assessment of nutrition and complications arising from weight loss practices. The main findings of AN on examination result from malnutrition. Some of the signs observed are low body temperature, bradycardia, postural hypotension, xerosis (dry skin), hair effluvium (hair loss), lanugo (growing of fine white hair over the body), nail changes, oral cheilitis, carotenoderma (orange pigmentation of the skin), and edema. A special concern with prepubertal patients is that growth can be stunted and there might be a failure in the development of secondary sexual characteristics (e.g., breast development).

Patients with AN might experience physical symptoms such as sleep disturbances, headache, fatigue, bloatedness, constipation, diffuse abdominal pain, and sensitivity to cold temperatures. It is not uncommon for patients to present with no physical complaints.

Laboratory findings may be in the normal range or with minimal changes in the initial stages of the disease although a wide range of possible abnormalities of all body systems can arise. Metabolic changes include hypercholesterolemia, hypoglycemia (frequently assymptomatic), and raised liver function test values. Endocrine changes can present as hypercortisolemia, high levels of cortisol, growth hormone and reverse T3, and low levels of testosterone, LH, FSH, and leptin. A mild normocytic and normochromic anaemia and a low white blood cell count are commonly seen.

Amenorrhea is a marker of the patient's nutritional status. If amenorrhoea is sustained, it can be associated with osteopenia. One possible outcome of this is osteoporosis and risk of bone fractures. Primary amenorrhoea, which starts in early teenage years and persists into young adulthood, may increase the risk since patients fail to form bone at a critical phase of development, Even though amenorrhea has been dropped as a diagnostic criterion in classification systems, it is important that all clinicians assess this during an initial interview and check if menses have returned as treatment progresses.

Differential diagnosis
Psychiatric disorders

Excluding the presence of coexisting depressive disorder can be difficult since many symptoms of depression are also consequences of semi-starvation. It is more straightforward to exclude the possibility of depression as the sole diagnosis if it is the case of a patient presenting weight loss, which is not self-induced, and the core psychopathology of anorexia nervosa is not seen. Psychotic syndromes may also be a possible differential diagnosis, as delusions relating to food, eating, and drinking can occur. One important differential diagnosis within the ED category is ARFID where patients can present food restriction and weight loss without any psychopathology of AN.

General medical conditions

Some of the presenting symptoms—loss of appetite, weight loss, amenorrhoea, unexplained vomiting—can be mimicked by a variety of medical conditions. Alternative diagnoses should be considered while recognizing the possibility of denial or deception. In this context, collateral informants can be helpful.

Although physical tests are not required for diagnostic purposes, all patients with anorexia nervosa should have a thorough physical examination and whatever investigations are indicated as a result. Bowel problems such as coeliac or inflammatory disorders, or endocrine disturbances such as Addison's disease are some of the commonest forms of differential diagnosis. Electrolytes and renal function should be checked in all those who frequently vomit or misuse significant quantities of laxatives or diuretics (see section 'Bulimia nervosa').

Course and outcome

Approximately 50–70 per cent of patients with anorexia nervosa recover, particularly if there is a short duration of untreated illness. For about a quarter of patients outcome is poor, with weight never reaching the healthy range. Some residual features are common, particularly a degree of overconcern with shape, weight, and eating, and mood problems. In cases where the ED persist, up to a quarter develop bulimia nervosa.

AN has the highest mortality across all mental disorders (Harris & Barraclough 1998). The standardized moratality rate (SMR) refers to how many individuals among the diseased group have died in comparison to what would be expected in the non-diseased population. Studies with 6–12 years follow up report a SMR of 9.6, while those with 20–40 years of follow up report a rate of 3.7. One in five patients with AN who die has committed suicide; the other deaths were due to medical complications of weight loss and malnourishment, especially cardiovascular malfunction.

Few consistent predictors of treatment outcome have been identified. Poorer prognosis is associated with extremely low weight, vomiting, failure of previous treatment and failure to respond to current treatment, disturbed family relationships, long duration of illness, late onset, and presence of severe psychiatric comorbidity. Some predictors of a good prognosis are a low weight loss and early onset. Usually, adolescents have better outcomes than adults, and younger adolescents better outcomes than older adolescents. It should be noted that some of these prognostic indicators have not been consistently replicated and may be more robust indicators of short-term rather than long-term outcomes.

Assessment

Patients with AN often come to treatment with their families or friends, who have persuaded them to seek help. Likewise, they may be reluctant and present with low motivation for treatment. Diagnosis is clinical, achieved through a careful history taking, focusing on the reasons why the low weight is being maintained. To help clarify the diagnosis, details on whether the patient is still making an effort to lose weight or if this is now more associated with the characteristic core psychopathology, is helpful. Again, equally important is to seek collateral reports from family and friends, as sometimes patients are unable to describe their behaviour accurately or truthfully, making assessment more difficult.

Treatment

The initial goals of treatment are to establish healthy eating habits and to restore normal weight. Later, underlying vulnerability factors that predispose to relapse or continued problems may need to be addressed. Multidisciplinary teams giving medical support alongside psychological and nutritional guidance are usually preferred.

No pharmacological agents have been found to be of unquestionable benefit to AN treatment. Clinicians may use benzodiazepines, antidepressants, and antipsychotics to facilitate treatment acceptance and to approach secondary symptoms.

Different psychological therapies have been demonstrated to be of use in all stages of AN. In adolescents with an early stage of illness, family-based therapy (FBT) in which the parents have an active role in promoting symptoms-change is recommended. In adult patients the therapist works to motivate the individual, perhaps with the support of others, to implement the behaviour-change strategies that are included in many psychotherapeutic models such as cognitive behavioural therapy for eating disorders (CBT-E) or the Maudsley model of treatment for adults with anorexia nervosa (MANTRA).

Different settings are available for patients according to medical risk and disease severity. The National Institute for Health and Care Excellence (NICE) 2017 recommend outpatient care as a first instance if risk is mild to moderate. If a patient is considered high risk, inpatient treatment should be considered. A number of factors need to be taken into account in the assessment of risk. These include: markers of medical instability, age and stage, medical or psychiatric comorbidity, and psychosocial context, especially family support (Treasure 2016).

Guidelines for the management of patients with severe medical morbidity treated in medical facilities can be found in the MARSIPAN (management of really sick patients with anorexia nervosa) (Robinson 2010). The MARSIPAN guidelines provide a framework for assessment, multidisciplinary guidance, and selection of treatments for patients with a BMI <15kg/m^2. It can guide clinicians how to assess risk through physical examination and explains when to consider different treatment interventions for severely ill patients. Also, guidelines for nutritional management are available.

Liaison with the psychiatric team and transfer to a specialist eating disorder unit (SEDU) when possible is recommended. In a proportion of cases it is necessary to use a mental health legislation specialist eating disorder unit to offer inpatients support with eating (in some cases nasogastric feeding), close monitoring in the initial phase when the risk of refeeding syndrome is high, or prevention of symptomatic ED behaviours (excessive water drinking or exercising). The multidisciplinary team consists of a medical doctor, psychologist, dietician, occupational therapist, and family therapist.

Refeeding syndrome refers to a metabolic complication of malnourished patients who are abruptly refed. It is caused by rapid shift in fluids and electrolytes, which can be potentially fatal. The hallmark biochemical feature is hypophosphataemia, but other biochemical changes can occur (hypokalemia, hypomagnesemia, abnormal glucose metabolism, thiamin deficiency, hypertriglyceridaemia) as well as cardiovascular and pulmonary disturbances.

There is current controversy about how to refeed severely ill AN patients. A balance must be found between feeding too quickly with the risk of refeeding syndrome or too slowly, in the so-called 'underfeeding syndrome', which may prolong the illness. Thiamine and other vitamin and mineral replacements are given initially with a soft diet. Within 48 hours, intake should be increased by 15–20kcal/kg/day, with a total amount of 3000 kcal per day as a later target. Careful monitoring is needed during this initial phase, with dietary supplements as needed.

A target rate of weight gain of 1 to 1.5 kg/week should be set for inpatient settings. During all this proccess, the patient needs to be encouraged to actively collaborate in the weight restoration. Dietetic input in devising an individualized nutrition plan is important. Patients start with an intake of 1000–1500 kcal/day and after two weeks might reach 2000 kcal/day. Intakes of up to 3500 kcal/day may be needed to achieve the desired rate of weight gain, creating a need to use supplements in addition to meals and snacks. The expected weight gain in outpatients is of 250–500g/week.

Novel treatment approaches for AN are discussed briefly in Weblink Section 28.2.

28.3 Bulimia nervosa

Definition

Bulimia nervosa was conceptualized as a distinct disorder from AN in the late 1970s, with its main features being described as: powerful and irresistible urges to overeat, use of devices to prevent weight gain (e.g., purging), and a morbid fear of becoming fat (Russell 1979).

One of the main features of BN is the occurrence of binge eating episodes. These refer to the loss of control over the type and amount of food eaten. The characterization of a binge eating episode requires the presence of both a subjective component, which is a strong irresistible impulse to eat, associated with an objective component of consumption of large amounts of food in a discrete period of time, for example, two hours.

Coupled with binge eating episodes are the compensatory behaviours, aimed at preventing weight gain. These can become a cycle of binge–compensation (e.g., binge–purge cycles). These strategies to counteract the effects of overeating can be harmful and disturb the normal physiological funtioning, promoting disturbances in hunger and satiety regulation which perpetuate the disorder.

A characteristic feature of BN is the morbid fear of becoming fat, and overvalued ideas about the role of weight and body shape. As a consequence, these distorted cognitions might influence other behaviours and social functioning. For example, a patient may avoid meeting friends out of fear of being perceived as fat, or have their self-esteem dampened by failing to lose a desired amount of weight.

Diagnosis

Bulimia nervosa is described within the main diagnostic classification systems. The criteria of DSM-5 states that binge–compensatory cycles occur at least once a week for three months, associated with an undue influence of body weight and shape on psychological funtioning which needs to be present for this diagnosis. The time criteria differ in ICD-11 as it requires one or more binge-compensatory episodes of at least one month to diagnose BN.

If a patient meets the criteria for AN, then it is not possible to diagnose BN (American Psychiatric Association 2013). ICD-11 specifies that if the patient is severely underweight and meet criteria, they should be classified as AN. There is a diagnostic hierarchy among ED syndromes, which means a patient with AN cannot receive a diagnosis of BN or BED. For example, a patient cannot be concurrently diagnosed with two ED syndromes, such as having comorbid BED with BN.

Two course specifiers are described in DSM-5: if in partial remission (when a patient previously met all criteria for BN but has recently presented with incomplete criteria for some time); or if in full remission (when a patient has previously met all criteria but no longer presents any criteria for a sustained period of time). It also has severity specifiers based on frequesncy of maladaptive compensatory behaviour:

- mild: compensatory behaviours occur on average 1–3 times/week
- moderate: on average 4–7 compensatory behaviours/ week
- severe: on average 8–13 compensatory behaviours/ week
- extreme: compensatory behaviours occur more than 14 times/week.

The ICD-11 does not specify a frequency criterion for the binge–compensatory cycles and only gives the aforementioned example of one or more cycles for at least a month. In addition to the compensatory methods described in DSM, ICD-10 described the possibility of diabetic patients skipping insulin doses, as a possible strategy for compensating the effects of overeating, but this is lacking in the current version. Of note, ICD-11 also allows for diagnosis with subjective binges only as it describes episodes of loss of control of overeating different from usual, instead of consumption of a large amount of food (also described as possible).

Clinical features

The background pattern of eating in BN is severe restriction and/or fasting often with a disrupted circadian pattern of feeding. Abnormal nutrient composition of meals may be one key component in the maintenance of BN. Frequently, BN patients diet excessively and restrict throughout the day. There is an association between energy deprivation and binge eating during a particular day. Studies using dietary recall estimate that BN patients consume around 3500–5000 kcal during a particular binge. Chronic malnutrition is another possible maintenance factor for the physiological abnormalities of BN. During non-binge days, there is overconsumption of proteins, while during binge days, there is overrepresentation of carbohydrates.

BN patients experience a number of medical symptoms including irregular menses, abdominal pain or distension, constipation, headache, fatigue. Possible signs noted on physical examination can include parotid gland enlargement, erosion of tooth enamel and periodontal disease (both secondary to self-induced vomiting), Russell's sign (small lacerations and callosities on the back of the hand, usually overlying the metacarpophalangeal and interphalangeal joints if the finger is used to induce vomiting).

Water–electrolyte imbalances occur, especially in purging patients. Dehydration can result in renal failure. Secondary hyperaldosteronism with substantial fluid retention can occur when diuretics and laxatives are discontinued. This rapid weight rebound leads to anxiety and return of the behaviours. Purging is associated with low levels of potassium with the increased risk of cardiac arrythmias. Self-induced vomiting can lead to metabolic alkalosis. On the other hand, laxative abuse can lead to metabolic acidosis.

Social cognition has been studied less in BN, but some social cognition deficits and interpersonal problems have been reported. Equally important, all areas of emotional functioning are dysfuntional in ED patients. Emotional functioning comprises emotional awareness and emotional regulation.

Psychiatric comorbidity is an important aspect of BN. It is common to find depressive and anxiety symptoms as well as impulsive behaviour in a subgroup of cases, with self-harm and substance misuse.

Assessment

Many patients do not seek treatment, as they are ashamed of their symptoms. Also, they value weight control and can present complaining of some features associated with the disorder rather than the disorder itself.

Sensitive questioning is needed to elicit the weight control methods. Deaths have been reported in some cases, as patients utilize weight loss products obtained from the Internet. As with AN, diagnosis is based on history-taking. Physical signs can include parotid enlargement, loss of dental enamel, and Russell's sign. Signs of dehydration and/or oedema may be present. Salt and water balance should be checked, as they can be abnormal. In extreme cases acute renal failure can occur.

Course and outcome

Bulimia nervosa often has a chronic course but with short periods of remission when circumstances change. Approximately 50–70 per cent of patients are still suffering from clinical eating disorder symptoms over a five-year follow up. Approximately 40 per cent of patients develop a secondary psychiatric diagnosis (e.g., depression) and they usually gain weight, on average 3.3kg, over this period. The mortality of BN patients is lower than that of AN, with a SMR of 1.93. Also, the rate of suicide is not as high as in AN but can account for up to 23 per cent of deaths.

Treatment

Nutritional management of BN aims at restitution of normal eating patterns with remission of binge/purge cycles. In the early phase of treatment, it is important to address what is a normal weight for that particular patient based on their physical constitution. Also, it should be explained that fluid retention and weight flunctuations might take place during treatment.

Cognitive behaviour therapy (CBT-BN) is effective in about 40 per cent of cases and this has been translated for self-management using books or the Internet to deliver the information and skills. This is particularly effective with a degree of guidance which requires less training and time than that required to implement individual CBT. This makes this a cost-effective solution which can be sustained over longer periods. Interpersonal psychotherapy (IPT) is also effective and is an alternative to CBT, although it takes longer (about 8–12 months) for its beneficial effects to be realized. Dialectic behavioural therapy, acceptance and commitment therapy, and mindfulness have also been tested with small but positive results.

Antidepressant drugs have some efficacy as treatments for BN are relatively straightforward to implement. However, the evidence suggests that few patients make a full and lasting response although they can help to supplement psychotherapy. Fluoxetine is approved by the Food and Drug Association in the USA (US FDA) for BN treatment. Topiramate has been shown to decrease binge eating, but side effects may limit its usefulness. Predictors of poor response to treatment include childhood obesity, low self-esteem, and personality disturbance.

28.4 Binge eating disorder

Definition

This disorder was first described by Alfred Stunkard, as one possible eating pattern seen in a subset of obese patients. They experienced eating binges at irregular intervals. During these binges, enormous amounts of food were consumed with an 'orgiastic quality', followed by 'severe discomfort and expressions of self-condemnation'

(Stunkard 1959), but without the use of compensatory behaviours as in BN. The tendency to overeat can occur early in life before the development of obesity.

Diagnosis

BED is described as a clinical entity only in the latest edition of DSM (American Psychiatric Association 2013) and ICD (WHO 2020). The binge eating episodes are similar to those observed in BN and are associated with cognitive and behavioural markers of loss of control overeating. These markers aid the clinician to assess loss of control as these patients usually have poor insight over their chaotic eating patterns. Three out of the five following behaviours are needed for a BED diagnosis in DSM-5:

- eating much more rapidly than normal
- eating until feeling uncomfortably full
- eating large amounts of food when not feeling physically hungry
- eating alone because of feeling embarrassed by how much one is eating
- feeling disgusted with oneself, depressed, or very guilty afterwards.

In ICD-11, in line with BN, the same frequency over time criteria (binges once a week for at least a month) is specified and no cognitive or behavioural markers of loss of control are required for diagnosis. As with BN, the diagnosis is also possible with the occurrence of subjective binges only.

As well as with other syndromes, DSM-5 also describes two course specifiers, if in partial or in full remission, with the same description as in BN. Severity specifiers were described also matching the frequency severity observed in BN. The difference is that instead of using compensatory behaviours (absent in BED), one should use binge eating frequency instead:

- mild: binge eating occurs on average 1–3 times/week
- moderate: average of 4–7 binge eating episodes/week
- severe: average of 8–13 binge eating episodes/week
- extreme: binge eating occurs on average more than 14 times/week.

Clinical features

There is a strong association between BED and obesity. It is possible for obese patients to develop BED and vice versa. In general, BED patients gain weight rapidly, have more weight oscillations than non-BED obese patients and reach higher weights. Also, BED patients frequently diet to lose weight. Eating binges are more frequent at the end of the day and may take place hidden from friends or family members. Medical complications of BED are exacerbated by obesity, which these patients have a 4.9 higher chance of developing.

Even if the eating binges are diagnosed the same way as in BN, they differ regarding content. Usually, subjects with BED consume less calories per binge episode than BN patients, with a caloric content of 1500–3000 kcal/binge. Moreover, binge episodes of BN patients usually are higher in carbohydrate and sugar content. BED patients tend to snack more than BN individuals but the frequency of atypical eating behaviours (nibbling, double-meals, nocturnal eating) does not differ, although they are higher than in non-ED controls.

Course and outcome

The BED remission in one year follow-up studies following RCTs yield rates of 25–80 per cent. These remission rates are significantly higher than expected for BN over this time frame. On a 12-year follow-up, the remission rate of BED was 67 per cent.

Moreover, the prevalence of morbid obesity is higher among the BED than BN group (Hudson et al. 2007). The mortality from BED has not been studied. A non-significant SMR of 2.69 was found with only two out of 68 BED patients dying over the course of a 12 year follow up. Also, in a five-year follow-up on a study about suicide in ED, no patient with BED had commited suicide.

Treatment

There may be three different treatment goals for BED: remission of binge eating episodes, improvement of associated comorbidity (e.g., depression), and weight loss. Many BED patients do not present for treatment but avail themselves of a variety of approaches to weight loss.

Many psychological interventions have been tested for BED. CBT, adapted as guided self-help (CBTgsh) is effective in the short-term in reducing binge eating. It is particularly useful as it can be used by non-specialist health providers who can optimize the effect if they can improve motivation and guidance. Moreover, CBTgsh is a cost-effective option. Individual or group formats of CBT have been tried. CBT can help approximately 70 per cent of patients to achieve binge eating remission with a longstanding effect after a four-year follow-up. Interpersonal psychotherapy (IPT) is another possible psychological treatment for BED, with comparable long-term outcomes to CBT. Also, IPT has been demonstrated to be more effective than other approaches for patients with chronic BED and low self-esteem.

A range of pharmacological agents has been tested. SSRI antidepressants are effective at reducing binge eating and mood problems although the weight reduction is modest. When compared to antidepressants the anticonvulsivant topiramate has a stronger effect on weight reduction but has problematic side effects, apart from its efficacy to binge eating. The antiobesity agent orlistat has been demonstrated to have a beneficial effect on binge eating. Lysdexamphetamine is the first agent to be FDA-approved for the treatment of moderate to severe cases of BED.

A proportion of BED patients fulfil the indications for bariatric surgery. Interestingly, this procedure may reduce binge eating as well as lead to weight loss. However, this is true for mild to moderate cases only. When binge eating does not respond to the surgery, it can be a factor participating on weight regain.

The other ED presentations are described in weblinks: avoidant-restrictive food intake disorder (ARFID) in Weblink Section 28.3; feeding disorders Weblink Section 28.4; other specified eating disorders (OSFED) in Weblink Section 28.5; and unspecified eating disorders (UFED) in Weblink Section 28.6.

KEY POINTS

- Eating disorders can present across a variety of patients throughout the weight span.
- The main current classification systems (DSM and ICD) reflect the same main syndromes for eating disorders (AN/BN/BED/ARFID/OSFED/UFED) and feeding disorders (PICA/rumination–regurgitation syndrome).
- It is important to have a high level of suspicion to establish the diagnosis of eating disorder in patients who still deny their symptoms, but for whom other conditions cannot explain their clinical trajectory.
- For all eating disorders, interventions based on psychological interventions and nutritional guidance are indicated. However, medication might play a role for BED and BN.
- All eating disorder treatment plans, regardless of setting, should aim at establishing a multidisciplinary approach.

 Be sure to test your understanding of this chapter by attempting multiple choice questions. See the Further Reading list for additional material relevant to this chapter.

REFERENCES

American Psychiatric Association (2013) *American Psychiatric Association: Diagnostic and Statistical Manual of Mental Disorders*, 5th edn, *DSM-5*. American Psychiatric Association, Arlington, VA.

Becker A E et al. (2002) Eating behaviours and attitudes following prolonged exposure to television among ethnic Fijian adolescent girls. *Brit J Psychiatr* **180**, 509–514.

Eisenberg M & Neumark-Sztainer D (2008) Peer harassment and disordered eating. *Int J Adolesc Med & Hlth* **20**(2), 155–164. doi: 10.1515/IJAMH.2008.20.2.155.

Harris E C & Barraclough B (1998) Excess mortality of mental disorder. *Brit J Psychiat* **173**, 11–53.

Hoek H W (2006) Incidence, prevalence and mortality of anorexia nervosa and other eating disorders. *Curr Opin in Psychiatry* **19**(4), 389–394. doi: 10.1097/01.yco.0000228759.95237.78.

Hudson J I et al. (2007) The prevalence and correlates of eating disorders in the National Comorbidity Survey Replication. *Biol Psychiatry* **61**(3), 348–358. doi: 10.1016/j.biopsych.2006.03.040.

Robinson P (2010) *MARSIPAN: Management of Really Sick Patients with Anorexia Nervosa*. The Royal College of Psychiatrists, London.

Russell G F M (1973) The management of anorexia nervosa. In: *Royal College of Physicians of Edinburgh, Symposium: Anorexia Nervosa and Obesity* (publication no. 42). T. and A. Constable, Edinburgh.

Russell G (1979) Bulimia nervosa: An ominous variant of anorexia nervosa. *Psychol Med* **9**(3), 429–448.

Solmi F et al. (2021) Changes in the prevalence and correlates of weight-control behaviors and weight perception in adolescents in the UK, 1986–2015. *JAMA Pediatrics* **175**(3), 267. doi: 10.1001/jamapediatrics.2020.4746.

Stunkard A J (1959) Eating patterns and obesity. *Psychiatr Quarterly* **33**, 284–295.

Tan J O A *i* (2016) Understanding eating disorders in elite gymnastics: Ethical and conceptual challenges. *Clinics in Sports Med* **35**(2), 275–292. doi: 10.1016/j.csm.2015.10.002.

Treasure J (2016) Applying evidence-based management to anorexia nervosa. *Postgrad Medical J* **92**(1091), 525–531. doi: 10.1136/postgradmedj-2015-133282.

Treasure J & Nazar B P (2016) Interventions for the carers of patients with eating disorders. *Curr Psychiat Reports* **18**(2), 16. doi: 10.1007/s11920-015-0652-3.

Treasure J *i*(2020) Cognitive interpersonal model for anorexia nervosa revisited: The perpetuating factors that contribute to the development of the severe and enduring illness. *J Clin Med* **9**(3), 630. doi: 10.3390/jcm9030630.

WHO (2020) *International Classification of Diseases and Related Health Problems (ICD-11)*. 11th edn. World Health Organization, Geneva, 57–60.

FURTHER READING

Dovey T M et al. (2008) Food neophobia and 'picky/fussy' eating in children: A review. *Appetite* **50**(2–3), 181–193. doi: 10.1016/j.appet.2007.09.009.

Guidelines for the Nutritional Management of Anorexia Nervosa (2005) *Royal College of Psychiatry Report*. Available at: http://www.rcpsych.ac.uk/docs/default-source/improving-care/better-mh-policy/college-reports/college-report-cr130.pdf?sfvrsn=c4aad5e3_2

Hilbert A & Brauhardt A (2014) Childhood loss of control eating over five-year follow-up. *Int J Eating Disord*, **47**(7), 758–761. doi: 10.1002/eat.22312.

Hilbert A, Hoek H W, Schmidt R (2017) 'Evidence-based clinical guidelines for eating disorders. *Curr Opin in Psychiat* **30**(6), 423–437. doi: 10.1097/YCO.0000000000000360.

Lindvall Dahlgren C, Wisting L, Rø, Ø. (2017) Feeding and eating disorders in the DSM-5 era: A systematic review of prevalence rates in non-clinical male and female samples. *J Eating Disords* **5**(1), 56. doi: 10.1186/s40337-017-0186-7.

McMaster Caitlin M et al. (2021) Dietetic intervention for adult outpatients with an eating disorder: A systematic review and assessment of evidence quality. *Nutr Rev* 79(8), 914–930. doi: 10.1093/nutrit/nuaa105.

Nazar B P et al. (2017) Early response to treatment in eating disorders: A systematic review and a diagnostic test accuracy meta-analysis. *Eur Eating Disord Rev* **25**(2), 67–79. doi: 10.1002/erv.249.

Ornstein R M et al. (2017) Treatment of avoidant/restrictive food intake disorder in a cohort of young patients in a partial hospitalization program for eating disorders. *Intl J Eating Disord* **50**(9), 1067–1074. doi: 10.1002/eat.22737.

Pennesi J-L & Wade T D (2016) A systematic review of the existing models of disordered eating: Do they inform the development of effective interventions? *ClinI Psychol Revi* **43**, 175–192. doi: 10.1016/j.cpr.2015.12.004.

Trindade A P et al. (2019) Eating disorder symptoms in Brazilian university students: A systematic review and meta-analysis. *Brazil J Psychiatr* **41**(2), 179–187. doi: 10.1590/1516-4446-2018-0014.

29 Diet and epigenetics

John C. Mathers

OBJECTIVES

By the end of this chapter you should be able to:

- understand the marks and molecules which constitute the epigenetic machinery and describe how this molecular machinery is used to regulate gene expression
- discuss the changes in epigenetic marks which occur during embryogenesis and across the life-course in humans
- use specific examples from studies in animals to illustrate how dietary and other factors influence epigenetic patterns and how this relates to phenotype
- discuss the effects of nutritional status, dietary patterns, and specific dietary factors on epigenetic marks and molecules in humans.

BOX 29.1
Definitions

What is epigenetics? Epigenetics is the area of science concerned with heritable changes in gene expression that do not involve changes to the underlying DNA sequence.

What is the epigenome? The epigenome is the constellation of marks and molecules (including DNA methylation, histone post-translational modifications, non-coding RNAs, and RNA post-translational modifications) responsible for regulating gene expression.

29.1 Introduction

During embryonic and foetal development, a single fertilized egg becomes a complex and wonderful human being. The information to direct this astonishing development resides within the human genome which contains approximately 3 billion base pairs (bp) and about 20,500 genes. Each of us has approximately 200 different cell types which make up our tissues and organs. These cell types arise from further differentiation of cells derived from the three embryonic germ layers, that is, the endoderm, mesoderm, and ectoderm. Although each of our nucleated cells contains exactly the same DNA sequence and the same genes, different cell types have distinctly different functions. In addition, the activity of individual cells is dynamic and responds to differences in physiological state, age, and environmental factors, including diet and nutritional status. For example, the cuboidal cells of the mammary gland are very metabolically active in secreting milk during lactation but are quiescent in the non-lactating woman. During fasting, liver cells hydrolyse glycogen and switch on gluconeogenesis to ensure that the body (and especially the brain) receives an adequate supply of glucose but, after a meal, adapt quickly to shut down gluconeogenesis and to store excess glucose as glycogen.

These inter- and intra-cell differences are mediated by characteristically different patterns of gene expression. Nearly half of all human genes are expressed in all tissues; these are the so-called house-keeping genes that are required by all cell types for the maintenance of basic cellular functions. However, about 12 per cent of genes are tissue specific; these are the genes which show at least five-fold higher expression in a specified tissue when compared with all other tissues. In addition, from moment to moment, each cell must regulate which genes are transcribed (switched on) or repressed (switched off), and also the intensity of expression of each individual gene to ensure that the cell has just the right amount of the corresponding protein. In diseases, patterns of gene expression (transcription) are dysregulated so that diseased tissues and organs are unable to carry out their normal functions. So, to maintain normal function and health, organisms have evolved complex, integrated, and multi-layered systems for regulating transcription in ways that are cell type and context specific. This regulation of transcription is achieved through the combination of how the genome is 'packaged' in the nucleus, epigenetic factors, and interactions with transcription factors. Transcription factors are proteins which bind to DNA in a sequence-specific manner and (usually) switch on expression of the corresponding gene. This chapter is concerned with epigenetics that includes aspects of the nuclear packaging of the genome.

29.2 The essentials of epigenetics

Epigenetic marks and molecules

Each human nucleated cell contains about 1.8 metres of DNA that is packaged into a nucleus with a typical diameter of only 6 μm. Importantly, this packaging must be organized so that the DNA is accessible for transcription, for DNA maintenance and repair and, where appropriate, for mitosis. The cell achieves this through 'smart packaging' of the DNA with other macromolecules including proteins and RNA—a multi-component consortium known as chromatin. Chromatin has a complex and dynamic organization. The first level of organization is the 'beads-on-a-string' type of structure in which 147 bp of DNA are wrapped around an octet of alkaline proteins (called histones) to form a nucleosome with a linker region of 50 bp of DNA separating one nucleosome from the next. The beads-on-a-string structure then coils into a 30 nm diameter helical structure. Chromatin can be considered to have two major domains: euchromatin consisting of relatively loosely packed chromatin in which genes are undergoing active transcription, and heterochromatin where the DNA is more tightly packed and which is transcriptionally silent (or inert).

The core histones H2A, H2B, H3, and H4 which make up the globular core of nucleosome are not just a convenient scaffold for DNA packaging. These histone proteins have 'tails' containing 20–35 mainly basic amino acids which protrude from the globular core. Importantly, these tails are post-translationally modified in multiple ways by covalent addition of other chemical groups. Such histone modifications (or decorations) include acetylation and methylation of lysine and arginine residues, phosphorylation of serines and threonines, ubiquitinylation and sumoylation of lysines, and ribosylation (the addition of one or more ADP-ribose moieties that occurs on several amino acid residues including arginine, serine, and cysteine). Some amino acid residues, for example, the lysine residue at position 4 on histone H3, that is, H3K4, can be covalently modified with up to three methyl groups. These post-translational

modifications (marks) are not fixed and appear to respond to the cellular environment and to cell status. Post-translational modifications of amino acids within the histone cores have also been described. Because of accessibility issues, these modifications may be added before the nucleosome is assembled and they appear to influence nucleosome dynamics. Although not yet well understood, it appears that the patterns of histone marks are a key component of the epigenetic machinery forming a 'histone code' which determines chromatin states and patterns of gene expression. In addition, recent discoveries have highlighted the role of H1 linker histones in regulating gene silencing by promoting both histone H3K9 methylation and localized chromatin compaction (Healton et al. 2020).

DNA itself is not a fixed, or static, molecule. Methyl groups are present at position C5 on some cytosine residues (referred to as 5mC) in CpG dinucleotides, that is, where the cytosine residue is followed by a guanine residue. In all nucleated cells, the pattern of DNA methylation is essentially bimodal. The large majority of methylated cytosine residues occur in transcriptionally silent regions of the genome. In contrast, most unmethylated CpGs are found in clusters known as CpG islands that are often upstream of genes which are actively transcribed. During DNA replication in humans and other mammals, the pattern of CpG methylation on the parental strand of DNA is copied on to the daughter strand by a specific enzyme called DNA methyltransferase 1 (DNMT1) see section 'The epigenetic machinery: writers, readers, and erasers' for more details. Patterns of DNA methylation are among the most extensively studied epigenetic marks, change throughout the lifecourse, are influenced by dietary and other environmental exposures, and are important regulators of gene expression.

Non-coding RNAs (ncRNA) are RNA species that are not translated into proteins. Some, such as ribosomal RNA and transfer RNA, have been known for decades and have well-described roles in translation. However, more recently, a richer repertoire of ncRNA has been described which includes a large number of small ncRNA, each containing approximately 22 nucleotides, known as microRNA (miRNA). The total number of miRNA encoded in the human genome is not known with certainty but is predicted to be around 2300 of which about 1115 are well described. MicroRNA regulate gene expression by base-pairing with complementary sequences in messenger RNA (mRNA) resulting in silencing of the corresponding mRNA. It has been argued that the co-option of RNA species, such as miRNA, to contribute to genome regulation allowed the development of much more complex multi-cellular organisms than would have been possible using the protein-dominated regulatory machinery which occurs in prokaryotes.

There are more than 170 modifications of RNA molecules that, together, are known as the epitranscriptome. This constellation of post-transcriptional modifications of RNA do not alter the RNA sequence per se. Although some of these modifications have been known for decades, it is only recently that their functional roles have become apparent. The majority of eukaryotic transfer RNAs (tRNAs) contain at least ten modifications per molecule and there are over 200 modifications on human ribosomal RNAs (rRNAs). The most widely studied modifications include the N6-methyladenosine (m6A) modification of messenger RNA (mRNA)—an essential regulator of mammalian gene expression—and 5-methylcytosine (m5C) and N1-methyladenosine (m1A) that have functional roles in ncRNA and also occur in mRNA. There are well-described patterns of modifications in both mRNA and tRNAs during development and in disease but the mechanisms through which the RNA modifications influence these processes remain to be discovered. Similarly, it is assumed that patterns of RNA modification respond to external stimuli (including, perhaps, nutrients) but how these stimuli regulate RNA modifications to influence transcription and translation is unknown.

Mitochondrial epigenetics

In addition to the DNA within the nucleus, human cells also contain DNA within the mitochondria—the cellular organelles that are best known for their role in ATP generation to provide the energy to power cell function. Mitochondrial DNA is derived entirely from the mother and consists of a circular genome containing 16,569 base pairs. Until recently, the mitochondrial contribution to epigenetic processes has been largely ignored, but it is now becoming evident that, just as for nuclear DNA, mitochondrial DNA contains 5mC. Importantly, patterns of mitochondrial DNA methylation appear to be influenced by environmental exposures, including dietary factors, and are associated with disease risk. For example, patterns of methylation at specific CpG sites with the mitochondrial gene predict risk of cardiovascular disease in adults with overweight and obesity (Corsi et al. 2020).

The epigenetic machinery: writers, readers, and erasers

During the S phase of the cell cycle, the DNA double helix is opened to produce a replication fork and each parental strand is copied to produce a complementary daughter strand. Initially, the newly synthesized daughter strand has no methylation marks but a specific enzyme, DNA methyl transferase 1 (DNMT1), is recruited to the replication fork and this enzyme copies the methylation

marks from the parental to the daughter strand. This process ensures that patterns of DNA methylation are retained during mitosis (underpinning the concept that epigenetic marks are heritable across cell generations) and DNMT1 is described as the maintenance DNMT.

During embryogenesis, other members of the DNMT family, notably DNMT3A and DNMT3B, are responsible for *de novo* methylation of the differentiating cell lineages. Unlike DNMT1, DNMT3A and DNMT3B have no preference for hemi-methylated DNA and, because there is no template for this *de novo* methylation, other mechanisms must be responsible for guiding the DNMTs to the appropriate CpG site. Although these mechanisms have not yet been elucidated, recent research has revealed sets of unique *de novo* DNA methylation target sites for both DNMT3A and DNMT3B during mammalian development. Although most *de novo* methylation is observed during embryogenesis, DNMT3A and DNMT3B (aided by another DNMT family member, DNMT3L) also catalyse *de novo* methylation in adult tissues including the germ cells and thymus. The methyl groups used by the DNMTs to methylate the 5' positions on cytosine residues are provided by the universal methyl donor *S*-adenosyl-methionine (SAM) which, in turn is produced in the one-carbon cycle from dietary compounds including folate, methionine, choline, and betaine.

Whilst the processes involved in methylating DNA have been known for many years, until recently, the mechanism responsible for removing methylation marks from DNA was a mystery. Of course, passive loss of methylation marks through failures by DNMT1 during DNA replication (in mitosis) might explain some epigenetic drift (see section 'Post-natal changes in epigenetic marks'). However, it was postulated that an active process of demethylation would be necessary to account for the rapid demethylation of the early embryo (discussed in 'Pre-natal changes in epigenetic marks') and other 'systematic' changes in DNA methylation patterns. It is now evident that a class of enzymes called ten-eleven translocation methylcytosine dioxygenases (TET) enzymes catalyse the progressive conversion of 5mC to 5-hydroxymethylcytosine (5hmC), then to 5-formyl cytosine (5fmC) and finally to 5-carboxycytosine (5caC). How 5caC is converted to cytosine (C) remains unclear (Figure 29.1). In addition, 5mC can be demethylated through an initial enzyme-catalysed step which converts the 5mC to a thymine (T) residue and then the base excision repair (BER) machinery 'repairs' the T and inserts a C at the appropriate place in the DNA sequence (Hackett & Surani 2013: Siesenberger et al. 2013).

As noted above, histone tails are subject to post-translational modifications with multiple small groups and molecules and, to date, more than 100 different histone modifications have been described. More than 50

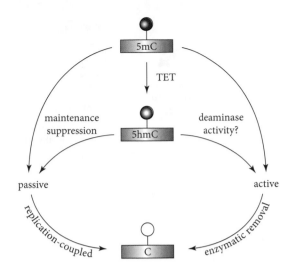

FIGURE 29.1 Active and passive mechanisms for removal of methyl marks from cytosine residues within DNA. 5mC, 5-methly cytosine; TET, ten-eleven translocation methylcytosine dioxygenase ; 5hmC, 5-hydroxymethyl cytosine; C, cytosine.

Source: J A Hackett & M A Surani (2013) DNA methylation dynamics during the mammalian life cycle. *Philos. Trans. Royal Soc. B*, **368**, 20110328. © 2012 The Authors. Published by the Royal Society.

one transcriptional event leads to the synthesis of multiple miRNA. Each miRNA is transcribed by RNA polymerase II and undergoes processing within the nucleus to produce a pre-miRNA hairpin before being exported to the cytoplasm for further processing. Eventually the mature miRNA, now about 22 nucleotides in length, is incorporated into the RNA-induced silencing complex (RISC) where the miRNA interacts with its mRNA target to silence transcription.

Each RNA modification appears to have its own 'readers', 'writers', and 'erasers' and this is true for the reversible mRNA modification, m⁶A. Methyltransferase-like 3 (METTL3) was the first demonstrated m⁶A writer, that is, an enzyme that catalysed the addition of a methyl group (donated by SAM) to the 6' position within adenine residues in RNA. Interestingly, the first demonstration of enzymatic demethylation of m⁶A involved the fat mass and obesity-associated (FTO) protein (Jia et al. 2011) which suggests that nutrition and/or obesity-related factors may influence the epitranscriptome.

29.3 Epigenetics across the lifecourse

Prenatal changes in epigenetic marks

Unlike the DNA sequence itself, it appears that patterns of DNA methylation in the offspring are not inherited from the patterns in the parental egg and sperm. In the early embryo, two major genome-wide DNA demethylation events coincide with significant development stages. The first occurs immediately following fertilization and affects the whole zygote. The paternal genome is demethylated, actively and rapidly, following fertilization. In contrast, there is passive demethylation of the maternal genome over a number of cell divisions because the maintenance DNA methyl transferase, DNMT1, is excluded from the nucleus. The second wave of demethylation occurs during the establishment of, and is restricted to, the primordial germ cells (PGCs) which are the direct progenitors of the sperm and oocytes in the offspring (Figure 29.2) (Siesenberger et al. 2013). These radical changes in epigenetic status allow the zygote to remove the epigenetic signature inherited from the parental gametes (in other words, to 'clear the slate') and, in so doing, to permit developmental totipotency. For reasons that are incompletely understood, imprinted genes do not undergo these demethylation events and retain the methylation patterns characteristic of their parent-of-origin. There are about 100 imprinted genes in the human genome which play important roles during development of the foetus and placenta, in nutrient exchanges

years ago, the first of these modifications, histone acetylation, was discovered and these early studies showed that hyper-acetylated histones were associated with active transcription. Post-translational modification of histones by acetyl groups adds negative charges to the positive lysine residues and, therefore, reduces interactions between DNA and the histones around which it is wrapped, so opening up the chromatin and allowing access by the transcriptional machinery. Histone acetylation also facilitates DNA replication and DNA repair. The covalent addition of acetyl groups to lysine (K) residues in histone tails is catalysed by a large family of histone acetyl transferases (HATs) and this is a very dynamic process with some acetyl groups being added and removed within minutes. There is a large family of histone deacetylases (HDACs) which catalyse the removal of acetyl groups and this family includes the sirtuins (or Sir2 proteins) which play roles in many aspects of health and disease, including ageing. Histone methyltransferases (HMTs) catalyse the transfer of up to three methyl groups to specific lysine and arginine residues on histones H3 and H4. As with DNA methylation, SAM is the methyl donor for these reactions. Two main families of histone demethylases, each with multiple members, catalyse the removal of methyl groups from histones.

MicroRNA are encoded at multiple locations across the human nuclear genome. Some miRNA genes occur within intronic regions of protein-coding genes whilst others are found in polycistronic clusters. For the latter,

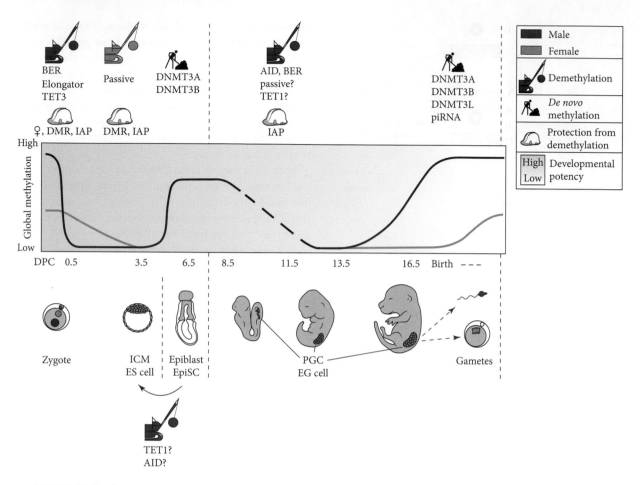

FIGURE 29.2 Changes in DNA methylation throughout the mammalian lifecycle showing stages at which genomic methylation shows major increases and decreases in males and females.
BER, base excision repair; TET, ten-eleven translocation methylcytosine dioxygenase; DNMT, DNA methyl transferase; PGCs, primordial germ cells; IAP, intracisternal A particle; AID, activation-induced deaminase; ICM, inner cell mass; ES cell, embryonic stem cell; EpiSC, epiblast-derived stem cell; DMR, differentially methylated region; DPC, days post coitum; piRNA, Piwi-interacting RNA.
Source: S Seisenberger et al. (2013) Reprogramming DNA methylation in the mammalian life cycle: Building and breaking epigenetic barriers. *Philos. Tran. Royal Soc. B*, **368**, 20110330. © 2012 The Authors. Published by the Royal Society.

between the mother and the foetus, in brain development and, later, in behaviour. Recent research suggests that a maternal protein called STELLA (also known as PGC7 or DPPA3) may protect these methylation sites by binding to chromatin-bearing H3K9 dimethylation (which marks certain paternal imprints) and preventing access by TET3 which would initiate active demethylation.

In the early stages of embryonic development, cells have low levels of DNA methylation and are epigenetically very similar to each other. However, by the blastocyst stage, different methylation patterns are evident in the outer trophectoderm cells compared with the inner cell mass. As embryogenesis progresses and the three embryonic germ layers, that is, the endoderm, mesoderm, and ectoderm appear, followed by organogenesis, the different cell lineages acquire characteristically different patterns of DNA methylation. This permits the remarkable phenotypic plasticity seen in the over 200 different cell types which arise in humans from a single fertilized egg.

Postnatal changes in epigenetic marks

In the early years of postnatal life, the vast majority of cells of a given type have similar epigenomes that are largely developmentally determined, that is, are dependent on information encoded within the genome. This is exemplified by comparisons of the epigenomes of cells from young monozygotic (MZ) twins, which show very similar patterns of DNA methylation and post-translational modifications of histones between members of each twin pair. However, during ageing, these epigenetic patterns become more diverse—a phenomenon described as epigenetic drift. The rate of accumulation of divergent DNA methylation patterns during ageing appears to be much greater than the accumulation of changes (mutations) in the genome itself. This is because the processes that police DNA methylation patterns are much less effective than the highly efficient and over-lapping systems of DNA repair which protect the genome from damage caused by endogenous and exogenous factors and by DNA copying errors in mitosis.

In parallel with these apparently random changes in epigenetic marks over time, recent evidence from genome-wide methylation studies indicates that there are a few hundred CpG sites which show remarkably consistent, and progressive, changes in methylation status during ageing. Some of these CpG sites gain methylation with time whereas at other sites, methylation status falls with age. These CpG sites have been described as an 'epigenetic clock' from which we can calculate DNA methylation age and it has been proposed that the latter may be a better indicator of biological age than is chronological age. The mechanism responsible for the 'epigenetic clock' is not known but it has been hypothesized that it measures the cumulative effect of an epigenetic maintenance system. The 'ticking' of the epigenetic clock may reflect a general progression of high- and low-methylated CpGs towards an intermediate level of methylation, that is, an age-dependent smoothening of the epigenetic landscape driven by an increase in entropy.

There is evidence of accelerated epigenetic clocks in a range of age-related conditions including Alzheimer's disease and Huntingdon's disease as well as in human genetic syndromes that accelerate biological aging, such as Werner's syndrome and Down syndrome. For example, diseased tissues such as tumours have a considerably 'older' DNA methylation age than have corresponding healthy tissues. In contrast, super-centenarians (humans who are at least 110 years old) and their offspring have decelerated epigenetic clocks. This is consistent with the concept that differences in capacity for somatic maintenance (including maintenance of the epigenome) explain, at least in part, inter-individual differences in the ageing trajectory and the risk of developing age-related frailty, disability, and disease.

There is growing evidence about the effects on the epigenetic clock of lifestyle factors (including diet, physical activity, and smoking) which influence human ageing. Much of this evidence is from observational studies showing that, for example, obesity may increase the epigenetic age of the human liver by 3.3 years for each ten unit increase in body mass index (BMI) (Horvath et al. 2014). More recently, analysis of cross-sectional data from the Women's Health Initiative (WHI) and the Invecchiare nel Chianti (InCHIANTI) study suggested that epigenetic age is reduced with increasing intake of fish and in those with moderate alcohol consumption and with higher concentrations of carotenoids in blood (a marker of fruit and vegetable consumption). Conversely, epigenetic age was accelerated by higher intake of poultry and in those with higher BMI. Of course, such cross-sectional analyses do not provide evidence of causality, for which intervention studies are required. Two recent intervention studies suggest that modification of dietary (and other lifestyle) factors may modulate epigenetic age. In a one-year pilot study, older residents in Poland who were randomized to a Mediterranean-like diet showed a trend towards epigenetic clock 'rejuvenation' although, curiously, no such improvement was seen in a group of Italian residents who underwent the same intervention (Gensous et al. 2020). Further, a small-scale, multi-component intervention targeting diet, use of specific dietary supplements, breathing exercises, stress management, and sleep advice for eight weeks resulted in lower epigenetic age at the end of intervention compared with baseline. These findings suggest that epigenetic age is plastic and can be modulated by dietary and other environmental factors. If confirmed in future studies, this would provide a rationale for the use of epigenetic age as a biomarker of how well individuals are ageing for use in epidemiological studies, interventions, and large-scale surveys.

29.4 Diet and epigenetics

Nutrition and epigenetics: establishing the connection in animal models

Cells, and indeed whole organisms (mammals and other animals), need to sense their environments (including nutritional status/exposure) on a continuous basis and to use that information to make decisions about gene expression. This provides a conceptual framework for the idea that epigenetics is a central process linking the genome and the environment with phenotype (Figure 29.3). Although there is now a wealth of empirical evidence from human studies (discussed in section 'Epigenetic actions of specific nutrients and other food components') which shows associations between nutrition and epigenetics, animal studies have been essential in establishing causal relationships between diet, epigenetic marks, and molecules and, in some cases, phenotype. In this respect, studies of diet during pregnancy in Agouti mice, of the feeding of honey bee larvae, and of the maternal behaviour of lactating rats have been particularly illuminating.

Mice

In a now classic study, Waterland and Jirtle supplemented the diet of pregnant viable yellow agouti (A^{vy}) mice with methyl donors in the form of folic acid, vitamin B_{12},

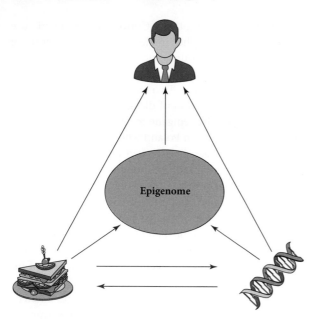

FIGURE 29.3 The epigenome is central to interactions between diet and the genome in determining phenotype.

Coat colour in agouti mice is associated with other important phenotypic characteristics. In particular, mice with lighter coloured (more yellow) coats are heavier than the leaner darker-coloured mice and are more likely to have obesity-associated diseases including diabetes and cancer. Interestingly, feeding dietary compounds which are not methyl donors to A^{vy} mice during pregnancy can also influence coat colour. For example, maternal dietary supplementation with genistein, the major isoflavone present in soyabeans and which is not a methyl donor, resulted in a higher proportion of offspring with pseudo agouti coats. Similar to observations in the methyl donor studies, supplementation with genistein resulted in increased methylation at six CpG sites within the A^{vy} IAP, the epigenetic changes were apparent in offspring tissues derived from all three germ layers and the changes persisted into adulthood.

These proof-of-principle studies suggest that the A^{vy} mouse model may be a useful 'epigenetic biosensor'. Feeding studies in A^{vy} dams might be used to screen nutritional and other factors with potential to alter epigenetic gene regulation and to investigate effects on adult phenotype.

Honey bees

In honey bees (*Apis mellifera*), queen bees are fertile, lay about 2000 eggs per day and up to 1 million eggs in lifetimes which can last for 3–4 years. In contrast, their genetically identical sisters, the worker bees, are infertile and normally live for only a few weeks. This dramatic difference in phenotype between the reproductive queens and sterile workers depends on the diet received by the larval insect. Bees which are fed royal jelly throughout the larval stage become queens whereas those female larvae destined to become workers are fed royal jelly for only three days and, after that time, are fed on beebread (a mixture of honey and pollen). Recent research has shown that biologically active phytochemicals, notably *p*-coumaric acid, present in beebread are responsible for 'chemical castration' of worker-destined larvae. In contrast, by feeding on royal jelly, (a glandular secretion produced by nurse bees, a relatively immature type of worker bee), larvae destined to become queens are protected from the inhibitory action of phytochemicals.

Establishment of the different castes (queens and workers) in female honey bees in response to nutrition during the larval phase of development is epigenetically mediated. This was demonstrated by silencing expression of the *de novo* DNA methyltransferase *Dnmt3* in newly hatched larvae and observing that the treated larvae emerged as queens with fully developed ovaries (Kucharski et al. 2008). Other experiments showed that *p*-coumaric acid upregulates *Dnmt3* expression. Silencing of *Dnmt3* was

choline, and betaine. They observed that the offspring of supplemented mothers were more likely to have darker coloured (pseudo agouti) coats than the offspring of unsupplemented mothers (which tended to be more yellow coloured) and that these phenotypic alterations appeared to be permanent. Importantly, Waterland and Jirtle went on to demonstrate that this phenotypic change in response to methyl donor supplementation in dams during pregnancy was due to altered epigenetic regulation of expression of the *agouti* gene. In unsupplemented A^{vy} mice, an IAP (an endogenous retroviral element called an intracisternal A particle) upstream of the *agouti* gene promotes constitutive transcription of the *agouti* gene. *Agouti* encodes a paracrine signalling molecule that causes follicular melanocytes to produce the yellow phaeomelanin pigment rather than the black eumelanin pigment so that the coat colour of the offspring of these mice is predominantly yellow. Maternal methyl donor supplementation increased CpG methylation of the IAP, effectively switching it off, resulting in reduced expression of the *agouti* gene and leading to the darker pseudo agouti fur on the offspring mice. Further, by assaying methylation in liver (derived from endoderm), kidney (from mesoderm), and brain (from ectoderm), these researchers showed that altered methylation at the A^{vy} locus in response to methyl donor supplementation was evident in tissues derived from all three germ layers of the early embryo. This demonstrates that nutritional modulation of A^{vy} methylation occurs early in embryogenesis, affects all tissues, and is maintained with high fidelity throughout development (Waterland & Jirtle 2003).

associated with reduced methylation of several CpG sites within the *Dcnt1* gene which encodes the protein dynactin. Dynactin is a complex protein that has roles in chromosome alignment and spindle organization during the cell cycle and may be important in regulating growth of the ovary in developing honey bee larvae. Recent findings have also suggested a role for miRNAs in determining honey bee phenotype by targeting non-methylated genes associated with neuronal development and complementing the regulatory effects of altered DNA methylation patterns (Ashby et al. 2016). However, as yet there is little information on effects of nutrition, for example, royal jelly or beebread, on miRNA expression in honey bees. Although derived from an insect rather than a mammal, these discoveries show that nutrition can influence epigenetic processes, specifically DNA methylation, leading to major changes in phenotype, including reproductive and behavioural status.

Rats

The way in which mother rats nurture their pups has long-term influences on the behaviour of the adult offspring. Dams that show high levels of licking/grooming (LG) behaviour and that adopt an arched-back nursing (ABN) posture, over the first week of lactation, produce offspring that are less fearful and that have lower hypothalamic–pituitary–adrenal (HPA) axis responses to stress in adulthood than do mothers which do not exhibit these 'good' nurturing behaviours. Cross-fostering the biological offspring of Low LG-ABN mothers to High LG-ABN dams produces offspring that resemble the normal offspring of High LG-ABN mothers. The reverse cross-fostering experiment produces offspring that resemble the normal offspring of Low LG-ABN mothers which is strong evidence that the maternal behaviours (rather than, for example, some genetic factor) are responsible causally for the altered offspring phenotype.

This maternal nurturing effect on offspring stress responsiveness is mediated epigenetically. Methylation across exon 1 of the *Gr* gene, which encodes the glucocorticoid receptor, is greater in the hippocampus in adult offspring of Low LG-ABN mothers, when compared with the offspring of High LG-ABN mothers. Indeed, this effect of maternal behaviour leaves a highly specific molecular fingerprint. The 5' CpG dinucleotide within the NGFI-A response element of the *Gr* gene is always methylated in the offspring of Low LG-ABN mothers but rarely so in the offspring of High LG-ABN dams. The cross-fostering regimes discussed above reverse the differences in the methylation of this 5' CpG dinucleotide and suggest a causal link between maternal behaviour and changes in DNA methylation. The absence of methylation at this maternal behaviour-sensitive epigenetic site

within the non-coding exon 1 promoter region, exon 1_7, permits NGFI-A binding and the recruitment of HATs (which acetylate the corresponding histones), and results in increased transcription of the *Gr* gene. These mechanisms may explain the long-term effects of maternal nurturing practices on *Gr* expression in the hippocampus and on HPA responses to stress in the adult offspring.

A recent study has investigated whether such processes occur in humans. At five months of age, DNA methylation of the glucocorticoid receptor (*GR*) gene and hypothalamic stress responses were compared in healthy infants who had been breastfed with healthy infants who had been formula-fed. Breastfeeding was associated with decreased methylation of CpG sites within the promoter of the *GR* gene that are involved in regulation of the hypothalamic–pituitary–adrenal and immune system responses. In addition, the breastfed infants showed decreased cortisol reactivity. These findings mirror those seen in the rat studies (discussed above) and may explain, in part, some of the positive effects observed in children who are breastfed. In addition, they support the concept that epigenetic mechanisms mediate the effects of early life experiences, including feeding, on cellular and whole-body defence systems that influence health across the life-source, including the ageing trajectory. For further discussion, see the section 'Early life nutrition, epigenetics, and health in later life' addressing the developmental origins of health and disease (DOHaD) hypothesis.

Epigenetics and obesity

As argued earlier in this chapter, cells and whole organisms (including humans), sense their environments on a continuous basis and use that information to direct gene expression so that the organism can mount an appropriate response. This environmental sensing is especially critical for supply of energy-yielding nutrients and is probably one of the earliest, and most evolutionarily conserved, survival and defence mechanisms. Since shortage of food, continuously or episodically, has been the norm for animals, including humans, during their evolution, there has been strong selective pressure for the development of effective sensing mechanisms for both current nutrient stores and exogenous sources of energy. The former is exemplified by changes in secretion of the hormones leptin and adiponectin in response to altered energy stores as triacylglycerols in adipose tissue which, in turn, modulate appetite and food intake. G protein–coupled receptors (GPRs) in the taste buds on the human tongue, and elsewhere in the oral cavity, allow us to detect the presence of sweet-tasting chemicals in foods which, in evolutionary terms, were signals for sugars. This ability to sense energy-yielding nutrients is not restricted to the mouth and, in the duodenum, so-called 'open' enteroendocrine cells sense sugars, D-amino acids, sweet proteins, and artificial sweeteners. We also have GPRs for fatty acids in the distal ileum and colon.

The evolutionary advantage of appropriate behavioural responses to such nutrient sensing provides support for the idea that obesity is a normal response to today's abnormal (in evolutionary terms) obesogenic environment. This argument discounts possible disadvantages of excess adiposity in respect of reproductive success which might be contingent on capacity for fight or flight. However, the huge inter-individual differences in adiposity which are evident in the face of apparently similar obesogenic environments suggests that either individuals sense those environments differently or that there are substantial differences in the way in which we respond. Since epigenetic processes are likely to be involved, what evidence is there for a link between epigenetics and obesity in humans? As noted above, obesity is associated with increased DNA methylation (epigenetic) age in the human liver, which may help explain why obesity accelerates the ageing process (Horvath et al. 2014).

Epigenome-wide association studies provide an important tool for identifying associations between demethylation and phenotypes such as adiposity. Such a study carried out on >10,000 adults identified differential methylation at 187 genetic loci that was associated with BMI. These genetic loci were associated with genes involved in lipid and lipoprotein metabolism, substrate transport, and inflammatory pathways. In addition, altered DNA methylation pattern predicted future development of type 2 diabetes. Evidence is accumulating that circulating microRNAs (miRNAs) act as a new class of endocrine factor.

Obesity is also accompanied by altered patterns of miRNA expression. Such miRNAs are released by several types of tissue, including adipose tissues, where they are involved in regulating differentiation of adipocytes. Recent findings show that such miRNAs are released from multiple tissues (including adipocytes), and are distributed in blood and that the profile of circulating miRNAs in obese individuals is different from that seen in lean individuals. These discoveries suggest that these miRNA may act as endocrine and paracrine messengers that facilitate crosstalk between metabolic organs (Ji & Guo 2019).

To be causally important in the aetiology of obesity, it is likely that epigenetic differences would occur either in the hypothalamus (causing dysregulation of appetite/reward systems) or in adipocytes (causing dysregulation of sensing of energy status). Leptin is encoded by the *LEP* (also called *OB*) gene and, in general, expression of *LEP* and leptin secretion from white adipose tissue increase as fat stores increase. The converse is seen with adiponectin,

which is encoded by the *ADIPOQ* genes. A recent investigation of severely obese adults (BMI 40–60 kg/m²) revealed that DNA methylation levels in the *ADIPOQ* gene were positively associated with BMI and waist circumference whereas methylation of *LEP* was negatively associated with BMI. In young children (17 months old), duration of breastfeeding (a behaviour linked with lower risk of adult obesity) was negatively associated with methylation of *LEP*.

Bariatric surgery approaches, such as gastric bypass, are being used increasingly to manage severe obesity (see Chapter 20). By restricting food intake and/or digestion and absorption of energy-yielding nutrients, bariatric surgery is highly effective in producing large, and sustained, weight loss. It is also a useful experimental model to investigate mechanisms involved in major changes in human adiposity. Using DNA extracted from adipose tissue biopsies, genome-wide methylation analysis identified hundreds of CpG sites across the genome that are differentially methylated after bariatric surgery (Boström et al. 2016). In most cases, the CpG sites were less methylated after surgically induced weight loss. Some of the affected CpG sites were in genes associated with obesity and related traits but, intriguingly, there was no change in methylation of *LEP*. This suggests that changes in epigenetic regulation of leptin synthesis in adipose tissue may not be causally important in the development of obesity, but it is also possible that bariatric surgery is unable to reverse all of the adverse effects of obesity on DNA methylation. Further, an epigenome-wide association analysis of DNA from whole blood from initially obese individuals (BMI = 30–73 kg/m²) revealed 4857 differentially methylated CpG sites at 12 months after bariatric surgery and a small deceleration in epigenetic age acceleration. This is in line with findings from a recent systematic review and meta-analysis that reported evidence of the reversibility of DNA methylation at specific loci in response to weight loss induced by bariatric surgery (ElGendy et al. 2020).

Regulation of energy balance by the hypothalamus involves a complex neural network in which specific orexigenic and anorexigenic neurotransmitters are released from specific neurons and these neurotransmitters modulate both food consumption and energy expenditure (see Chapter 6). There is some evidence in humans that methylation of the genes encoding neurotransmitters such as pro-opiomelanocortin (*POMC* gene) and the melanocortin 4 receptor (*MC4R* gene) is altered in obesity. However, much of this work has been done using blood cells, or other accessible sources of DNA, as a surrogate for the hypothalamus because of the ethical and practical difficulties of sampling the human brain during life.

Such observations provide some support for the idea that both sensing of energy status in adipose tissue and hypothalamic regulation of energy balance may be epigenetically dysregulated in obesity, but whether such effects are causes or consequences is not known.

Early life nutrition, epigenetics, and health in later life

The developmental origins of health and disease (DOHaD) hypothesis has revolutionized understanding of the antecedents of adult health and focused the attention of researchers, families, and policy makers on pregnancy and infanthood. This has led to the concept that an individual's health trajectory is established by exposures and events in the 1000-day window between conception and the child's second birthday. Among the most important of these exposures is nutrition—nutrition via the placenta *in utero* as well as infant feeding (breast or formula) and weaning practices. There is convincing epidemiological evidence, much derived from use of birth weight as a surrogate for nutrition *in utero*, that early life nutrition is a significant determinant of risk of multiple common complex diseases including cardiovascular disease, diabetes, and some cancers. Although studies in model organisms provide support for the hypothesis that these early life nutritional (and other) exposures are causal for later health, undertaking the types of intervention study which would be required to demonstrate causality in humans is challenging. As a step towards that goal, it would be very useful to know how different early life nutritional exposures are recorded and remembered over decades by the individual's cells and tissues. Epigenetic processes are a prominent candidate mechanism for such memorization (Goyal et al. 2019) since they provide a plausible mechanism for nutritional sensing (receiving information), for recording that exposure as changes in epigenetic marks, for remembering the exposure across cell generations (e.g., by copying epigenetic marks from the parental strand of DNA to the daughter strand during mitosis) and for revealing the phenotypic consequences of the exposure through altered gene expression. This has been described as the four Rs of (nutritional) epigenetics (Mathers 2008).

Experiments of nature

Disasters, such as famines, can provide valuable 'experiments of nature' in which to investigate the long-term consequences of major dietary changes early in life. The Dutch Hunger Winter which occurred over a period of about six months in 1944–1945 in the western part of The Netherlands resulted in very severe shortages of food for a well-defined time period. Although the famine was sufficiently severe to cause thousands of deaths, most women pregnant during the famine gave birth to live babies whose long-term health has been studied intensively. Sixty years after the Hunger Winter, there were significant differences in methylation of a number of genes, including the imprinted gene *IGF2*, in DNA from blood from those who had been exposed to famine *in utero* when compared with time- or sibling-controls without prenatal famine exposure (Figure 29.4). Importantly, the effects were most noticeable for those exposed to famine in early gestation rather than mid or late gestation (Tobi et al. 2014). Possible explanations for this finding are that epigenetic marks are more plastic to nutritional (and other) environmental exposures in early gestation, or that epigenetic changes which occur during embryogenesis and early foetal development are more long-lasting than epigenetic changes resulting from exposures later in pregnancy. The explanations are supported by the feeding studies in pregnant agouti (A^{vy}) mice discussed earlier in this chapter.

The Chinese Great Famine occurred between 1959 and 1961 and was so severe that >10 per cent of middle-aged and older Chinese adults had family member(s) who starved to death. Methylation of the *IGF2* gene was quantified in DNA from whole blood on adults (nearly 60 years after the famine) who had experienced famine as infants or foetuses and from those who were born after the famine. This revealed that exposure to severe famine was associated with higher *IGF2* methylation and that each unit increase in methylation was associated with 1.09-unit increase in total cholesterol concentration (Shen et al. 2019).

Seasonal fluctuations in dietary intake also provide an experiment of nature, albeit less dramatic than the Dutch Hunger Winter, in which to investigate how energy and nutrient availability during pregnancy affects the epigenome. In rural Gambia, West Africa, food availability is relatively scarce during the rainy season. Using blood samples, it has been observed that 9-year-old children, who had been conceived during that 'hungry' season of the year, had differently methylated metastable epialleles compared with control children conceived at other times of the year. Metastable epialleles (MEs) are alleles that are variably expressed in genetically identical individuals due to epigenetic modifications that were established during early development. The *agouti* locus, described earlier, is an example of a ME. There were positive associations between biomarkers of methyl donor supply in the Gambian mothers and methylation of MEs in their offspring which is reminiscent of the effects seen following B vitamin supplementation of A^{vy} mouse dams during pregnancy. One of the differentially methylated MEs was *VTRNA2-1*, which is a putative tumour suppressor and modulator of innate immunity. If such epigenetic changes result in changes in gene expression

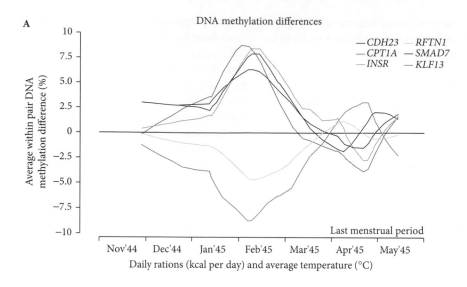

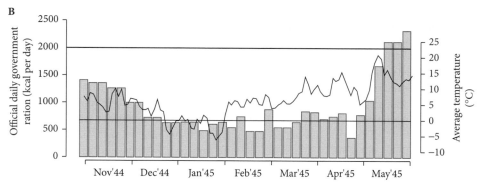

FIGURE 29.4 Effects of famine on methylation of selected genes in 60-year-olds who were *in utero* during the Dutch Hunger Winter.

(a) Differences in methylation of selected genes between exposed and non-exposed same sex siblings in relation to date of mother's last menstrual period; (b) Blue bars indicate official daily rations (in kcal/day) available during the Hunger Winter. The red line indicates the daily energy requirement of non-pregnant women, that is, 2000 kcal (8.4MJ) per day. The selected genes encode: *CDH23*, cadherin 23; *CPT1A*, carnitine palmitoyltransferase 1A; *INSR*, insulin receptor; *RFTN1*, Raftlin, Lipid Raft Linker 1; *SMAD7,* Mothers against decapentaplegic homolog 7; *KLF13,* Kruppel-like factor 13.

Source: E W Tobi et al. (2014) DNA methylation signatures link prenatal famine exposure to growth and metabolism. *Nat. Commun.* **5**, 5592. Published by Springer Nature.

and corresponding functional changes in the regulated pathways, this may provide a plausible mechanistic explanation for the greater risk of early death among adult offspring who were conceived during the 'hungry' season in rural Gambia.

Nutritional supplementation during pregnancy

Although there are many studies in animal models showing that manipulation of the maternal diet produces epigenetic changes (changes in both DNA methylation and in histone modifications) in multiple tissues of the offspring, there are few such studies in humans. In part, this is because of the ethical and practical difficulties of undertaking nutritional intervention studies in human pregnancies and also because blood, and less

commonly, buccal cells are the only tissue samples available from children. However, because women are recommended to take supplements providing 400 µg folic acid/d before, and for the first trimester of, pregnancy (to reduce the risk of offspring affected by neural tube defects), such peri-conceptual supplementation with folic acid provides a valuable model in humans in which to investigate effects of higher intakes of a methyl donor which could influence DNA methylation. Young children whose mothers had used folic acid supplementation in pregnancy had higher levels of methylation of the imprinted gene *IGF2*. Interestingly, *IGF2* was also one of the genes showing differential methylation in 60-year-olds who had been in early gestation when exposed to famine during the Dutch Hunger Winter and the Chinese Great Famine.

More broadly, there is growing evidence, largely from observational studies, that factors which influence one-carbon metabolism (and, potentially, methyl donor supply) in mothers during pregnancy are associated with changes in patterns of DNA methylation in their offspring. In addition to folate/folic acid, these factors include vitamin B_{12}, choline, and genetic variants such as *MTHFR* 677C>T (rs1801133) (see section 'Diet, genotype, epigenetics, and phenotype'), which affects folate status. These early observations suggest that both maternal and offspring genotype may modulate responses to maternal diet during pregnancy. Evidence for such diet:gene interactions across two generations highlights some of the complexities in undertaking studies in this area. Because they have identical genotypes, monozygotic (MZ) twins are often a useful model for separating effects of genetics from those of environmental exposures. However, this approach is unlikely to be suitable for studies of effects of maternal nutrition during pregnancy because, to a first approximation, both individuals in a MZ twin pair share the same maternal environment prenatally.

Assisted reproductive technology

In 1978, Louise Brown was the first baby to be born as a result of *in vitro* fertilization. Since then more than 8 million babies worldwide have been born following successful application of assisted reproductive technology (ART). Nearly 0.1 per cent of all children globally are conceived through ART although this figure is much higher in Western countries reaching 4–6 per cent in Belgium, Slovenia, and Denmark. Even without any further growth in use of ART (which is unlikely), it is predicted that by the end of this century there will be more than 150 million people alive who owe their existence to ART. ART often involves incubation of egg and sperm, the fertilized ovum and the developing embryo *in vitro* for 2–3 days before implantation of the embryo (now containing 8–16 cells) in the woman's uterus. This is a critical time for embryonic development, one in which there are radical changes in genome-wide patterns of epigenetic marks (Figure 29.2). Given that it is difficult to mimic *in vitro* the precise conditions that are found at the end of the fallopian tube where fertilization occurs naturally, it is possible that babies conceived by ART experience a different nutritional (and other chemicals) environment and that this exposure could influence their epigenome. Indeed, there is accumulating evidence of changes in methylation of imprinted and other genes in ART children, and it has been hypothesized that this is due to imbalances in components of one-carbon metabolism, notably folate and methionine.

What effects, if any, such epigenetic changes have on the immediate and longer-term health and well-being of ART children are poorly understood. There is evidence of higher proportions of low-birthweight babies and suggestions of higher prevalence of Angelman and Beckwith–Wiedemann syndromes (both of which are due to problems with imprinted genes) among those conceived by ART, but these observations need confirmation. Compared with naturally conceived offspring, children and adolescents born as a result of *in vitro* fertilization have greater risk of high blood pressure and other cardiovascular dysfunction. Some such offspring also appear to suffer metabolic consequences including increased fasting glucose and blood lipid concentrations. To date, there is no evidence that ART has adverse effects on cognitive outcomes or on the social, emotional, and motor functions of the resultant children.

The mechanisms through which periconceptional nutrition, including those associated with ART processes, may influence embryonic and foetal development and longer-term health of the offspring were reviewed recently (Fleming et al. 2018). For this purpose, the key events of the periconceptional period include the completion of meiotic maturation of oocytes, differentiation of spermatozoa, fertilization, and the resumption of mitotic cell cycles in the zygote through to implantation of the embryo on the wall of the uterus. As illustrated in Figure 29.5, maternal and paternal nutrition (including overnutrition associated with obesity) may influence embryonic health in multiple ways including through epigenetic and metabolic reprogramming. Studies in human and mouse embryos show that ART can modulate epigenetic profiles through the composition of, and duration of exposure to, the media use for culturing the fertilized eggs. This suggests that there are opportunities to optimize the nutritional conditions to which human embryos are exposed during ART. Investigating the effects of this nutritional environment (media composition) has the potential to improve the long-term health of the increasing number of babies whose conception is aided in this way. In addition, such experiments could advance our understanding of the intimate links between early life nutrition and epigenetics.

Dietary patterns and epigenetics

There is increasing evidence that whole diets, as distinct from intakes of specific foods or nutrients, are important for health and current research aims to find ways to describe, and to capture numerically, the complexity of whole diet in ways that are useful not only for research but also for large-scale surveys and public health interventions. Methods for studying and quantifying dietary patterns are relatively crude but, despite this limitation, it is apparent that certain dietary patterns are associated with better lifelong health. In general, these are

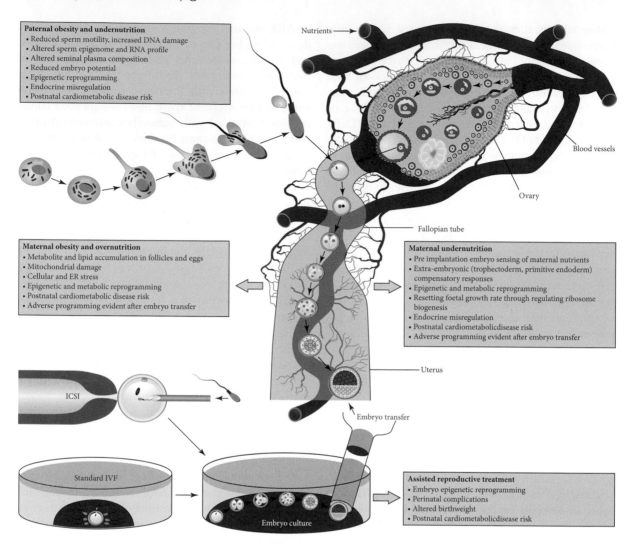

FIGURE 29.5 Overview of mechanisms through which maternal and paternal nutrition in the periconceptional period can influence embryonic development with implications for cardiometabolic health of the offspring.
These mechanisms include epigenetic and metabolic reprogramming. The nutritional composition of the media used for *in vitro* fertilization and embryo culture may have similar effects.
ER, endoplasmic reticulum; ICSI, intracytoplasmic sperm injection; IVF, in-vitro fertilization.
Source: Reprinted from *The Lancet*. T P Fleming et al. (2018) Origins of lifetime health around the time of conception: Causes and consequences. *Lancet* **391**,1842–1852.

plant-based diets containing a variety of foods with low intakes of animal products, other than fish, and are typified by the traditional Japanese and Mediterranean diets.

From a mechanistic perspective, such dietary patterns would be expected to influence patterns of gene expression and these transcriptional events might be mediated by epigenetic changes. Using buccal cells, researchers have observed higher expression of *MnSOD* and lower methylation of CpG sites at the *MnSOD* locus in vegetarian adults when compared with age-matched omnivores. *MnSOD* encodes a superoxide dismutase enzyme which is translocated to the mitochondria where it contributes to antioxidant defences, so the higher expression of *MnSOD* in the vegetarians might contribute

to their health. Among women in Southern Italy, adherence to a prudent dietary pattern (characterized by high intake of potatoes, cooked and raw vegetables, legumes, soup, and fish) was positively associated with *LINE-1* methylation—a marker of global DNA methylation. Global DNA methylation falls during ageing, and in common complex diseases such as cancer, so the higher global DNA methylation in these women may be a marker of better long-term health. However, whether change in global DNA methylation is a cause or a consequence of disease is not known. There is growing evidence that epigenetic mechanisms may mediate some of the beneficial effects of the Mediterranean dietary pattern across the lifecourse. For example, higher maternal adherence to a

Mediterranean diet in the periconceptional period is associated with favourable neurobehavioral outcomes and altered methylation of imprinted genes in early childhood. In prepubertal children who were encouraged to eat a more Mediterranean-like diet and to become more physically active, there was altered methylation of genes related to lipid metabolism and inflammation after 12 months intervention (Gallardo-Escribano et al. 2020). Finally, as discussed earlier in this chapter, slowing of the epigenetic clock in older Polish people who were randomized to a Mediterranean-like dietary intervention may indicate epigenetic mechanisms through which this dietary pattern contributes to healthier ageing (Gensous et al. 2020).

Dietary energy (caloric) restriction, epigenetics, and ageing

Restriction of dietary energy intake whilst maintaining adequate intake of nutrients (often called caloric restriction) results in increased longevity and lower risk of age-related disease in a wide range of animal models. Such energy restriction has multiple effects on cell and whole-body metabolism that reduce the accumulation of molecular damage which is a fundamental feature of the ageing process. Epigenetic dysregulation is a hallmark of ageing and it is now clear that dietary energy restriction produces long-lasting epigenetic effects that mediate expression of genes related to immuno-metabolic processes. Since raised systemic inflammation is one of the mechanisms that drive the accumulation of molecular damage, epigenetic reprogramming of such chronic low-grade inflammation by dietary energy restriction appears to contribute to healthier ageing and delays the onset of chronic age-related diseases (Hernández-Saavedra et al. 2019). Whether dietary energy restriction increases human longevity is not known and, for obvious ethical and practical reasons, it is unlikely that randomized controlled trials will be carried out to test that hypothesis. However, avoiding obesity—a consequence of dietary energy restriction—enhances healthy ageing, reduces the risk of age-related disease, and increases human longevity.

Epigenetic actions of specific nutrients and other food components

As discussed earlier, some nutrients, notably those involved in one-carbon metabolism, are important for epigenetics because they influence the provision of substrates (in this case, S-adenosyl methionine (SAM)) required by the epigenetic machinery. Food constituents can also influence epigenetic marks and molecules through their impact on the enzymes which catalyse the addition ('writers') or removal ('erasers') of epigenetic marks. Perhaps the best known of these diet-derived molecules is the short-chain carboxylic acid butyrate which, at physiological concentrations, is a potent histone deacetylase (HDAC) inhibitor. Through this action, butyrate opens up the chromatin structure, alters the expression of hundreds of genes, and has anti-neoplastic actions. Since butyrate is produced in substantial amounts by microbial fermentation of dietary fibre in the large bowel, it has been hypothesized that this may be one of the mechanisms through which dietary fibre promotes the health of this, and indeed, other organs. Butyrate is also present in milk and milk products including parmesan cheese.

Plant foods contain a wide array of secondary metabolites many of which have effects on the epigenome. The recent discoveries about the pathways used in DNA demethylation (Figure 29.1) have revealed a new role for vitamin C in providing reduced iron which is required for function of the TET enzymes that catalyse the conversion of 5mC to 5hmC. Other epigenetically active plant compounds include flavonoids such as genistein (discussed above in the section on the agouti mouse model), and multiple other compounds with diverse chemical structures including sulphoraphane, diallyl sulphide, and curcumin. Each of these compounds targets one or more components of the epigenetic machinery (Table 29.1). In addition, evidence is accumulating that several phytochemicals including curcumin, EGCG, and genistein modulate expression of miRNAs. For example, these polyphenols can reduce expression of miRNAs which promote tumorigenesis (onco-miRNAs or oncomirs) and/or activate expression of protective anticancer miRNAs (tumour suppressor miRNAs). It seems probable that these epigenetic actions of multiple secondary metabolites in a wide range of vegetables, fruits, cereals, nuts, and condiments contribute to the health-promoting effects of plant-based diets such as the Mediterranean diet.

29.5 Diet, genotype, epigenetics, and phenotype

Research using genome-wide association study (GWAS) approaches has revealed large numbers of genetic variants that are linked with important aspects of human phenotype such as height, adiposity, and risk of several common complex diseases, for example, cardiovascular disease, type 2 diabetes, and dementia. In many cases, the proportion of the variation in a particular phenotypic outcome, for example, BMI that can be explained by these

TABLE 29.1 Examples of epigenetically active nutrients and other compounds

Name of compound	Example food source	Class of compound	Epigenetic action
Alpha-tocopherol	Vegetable oils	Vitamin E	MicroRNA-mediated reduced DNMT3B activity
Anacardic acid	Cashew nuts	Phenolic lipid	HAT inhibitor
Ascorbic acid	Fruits and vegetables	Vitamin C	Required for DNA demethylation
Butyrate	Butter	Fatty acid	HDAC inhibitor
Caffeic acid	Coffee	Hydroxycinnamic acid	HDAC inhibitor
Curcumin	Tumeric	Polyphenol	HDAC inhibitor
Diallyl sulphide	Garlic	Organosulphur compound	HDAC inhibitor
Epigallocatechin-3-gallate (EGCG)	Green tea	Polyphenol	DNMT inhibitor
Genistein	Soy beans	Phytoestrogen	DNMT inhibitor / HAT activator
Quercetin	Onions	Flavonol	HAT inhibitor
Retinol	Liver	Vitamin A	
Selenium	Brazil nuts	Trace element	Reduces activity of DNMTs via regulation of one-carbon metabolism
Sulphoraphane	Broccoli	Isothiocyanate	HDAC inhibitor

DNMT, DNA methyl transferase; HDAC, histone deacetylase; HAT, histone acetyl transferase.

genetic factors is relatively small and this underscores the importance of lifestyle factors, notably diet, in determining health outcomes and other phenotypic characteristics. Indeed, evidence for the importance of diet:gene interactions in the aetiology of several phenotypes is accumulating. Given the role of the epigenome in mediating the effects of environmental exposures such as diet, it would not be surprising if variation in the genes encoding the epigenetic machinery, or involved in epigenetic mechanisms, influenced diet–epigenetic relationships.

A large number of genes involved in chromatin remodelling, DNA methylation, and microRNA biosynthesis are polymorphic, that is, they exist in two or more forms which are genetically determined (Table 29.2). In addition, mutations in such genes are associated with a range of genetic syndromes such as Rett syndrome (a rare neurological disorder that causes severe muscle movement disability) and which is associated with mutations in *MECP* leading to aberrant DNA methylation and increases expression of the retrotransposon *LINE 1*. MECP is an important 'reader' of DNA methylation. The MECP protein contains a methyl-CpG-binding domain that recognizes and binds to 5-mC regions and appears to be essential for nerve cell function.

Variants in genes encoding enzymes in the one carbon metabolism pathway, for example, *MTHFR* and *MTRR* affect the availability of methyl groups to form

SAM which is used to methylate both DNA and histones. Intakes of folate, riboflavin, vitamin B$_{12}$, and choline can modulate the effect of these genetic variants on methyl group availability. The phenotypic consequences of variants in the genes encoding epigenetic 'writers' such as DNMT and 'erasers' such as HDAC are emerging but, as yet, relatively little is known about how specific dietary factors can interact with those genetic variants to modulate phenotype.

29.6 Summary and conclusions

Epigenetic marks include the patterns of: (i) methyl (and hydroxymethyl) groups on cytosine residues within DNA, (ii) covalent additions of methyl, acetyl, phosphate, and other groups to histone tails within nucleosomes, and (iii) more than 170 modifications of RNA molecules that constitute the epitranscriptome. Together with a large number of miRNAs and other non-coding RNAs, these epigenetic marks and molecules constitute an integrated and multi-layered system that maintains DNA integrity and contributes to regulation of gene expression. During early embryogenesis there are widespread changes in epigenetic markings which contribute to cellular differentiation and ensure that each cell type, for example, an hepatocyte, a lymphocyte, or a neurone

TABLE 29.2 Examples of variants in genes that encode the epigenetic machinery or are involved in epigenetic mechanisms

Gene	Protein	Function	Common variant (rs number)
DNMT1	DNA methyl transferase 1	Maintenance DNA methyl transferase—copies methyl marks from parental to daughter strand during replication	53828G>A (rs2290684)
HDAC5	Histone deacetylase 5	Removes acetyl groups from histones	1698G>C (rs228769)
MTHFR	Methylenetetrahydrofolate reductase	Catalyses conversion of 5,10-methylenetetrahydrofolate to 5-methyltetrahydrofolate	*677C>T* (rs1801133)
MTRR	Methionine synthase reductase	Catalyses conversion of homocysteine to methionine	*66G>A* (rs1801394)

rs: Reference single nucleotide polymorphism.

expresses the constellation of genes which enables that cell to carry out its characteristic functions. Further, epigenetic marks change across the lifecourse with, in general, increasing diversity in epigenetic patterns.

Importantly, these epigenetic mechanisms provide a means for cells, and whole organisms, to respond appropriately to their environment, which includes nutrition. We sense our nutritional environments on a continuous basis and use that information to direct gene expression so that we mount an appropriate response. As yet, the ways in which dietary factors interact with the epigenome to influence our phenotype are not well understood. However, there are several good examples of specific nutrients and other food components which alter epigenetic marks and molecules. These include the B vitamin folate and other dietary methyl donors that contribute methyl groups to form SAM—the universal methyl donor—which is then used by the DNMT family of enzymes to methylate DNA and by histone methylases to methylate histones (the proteins around which DNA is wrapped in the nucleus). Other food-derived compounds influence the epigenetic machinery by inhibiting or activating the enzymes responsible for 'writing' or 'erasing' epigenetic marks. For example, vitamin C is essential for DNA demethylation through its role in providing reduced iron for TET enzyme activity. In addition, butyrate and curcumin are HDAC inhibitors, EGCG is a DNMT inhibitor, and genistein is an HAT activator.

The number of dietary components which have known, or potential, effects on the epigenetic machinery is huge and this produces both conceptual and practical challenges in designing studies aimed at understanding the epigenetic impact of whole diets or dietary patterns. However, it is increasingly evident that interactions between dietary choices and genotypes influence individual epigenomes and that these help to shape each person's phenotype and contribute to health and well-being across the lifecourse (Figure 29.3).

KEY POINTS

- The key epigenetic marks and molecules in humans include patterns of DNA methylation, post-translational modifications of the histones around which DNA is wrapped in nucleosomes, modifications of RNA molecules (the epitranscriptome), and small non-coding microRNA (miRNA).

- These epigenetic marks and molecules are important components of the molecular machinery that maintains DNA integrity and regulates gene expression.

- DNA methylation patterns alter radically during early embryogenesis and help to determine the characteristic transcriptional profiles of the 200 cell types found in humans.

- Nutritional exposures early in life can be 'recorded' as altered epigenetic patterns and influence phenotype later in life.

- A wide range of molecules including, for example, vitamin C, butyrate, sulphoraphane, diallyl sulphide, and curcumin alter epigenetic marks and may be important in explaining the health benefits of the foods from which they are derived.

 Be sure to test your understanding of this chapter by attempting multiple choice questions.

REFERENCES

Ashby R, Forêt S, Searle I et al. (2016) MicroRNAs in honey bee caste determination. *Scient Rep* **6**, 18794.

Boström A E, Mwinyi J, Voisin S et al, (2016) Longitudinal genome-wide methylation study of Roux-en-Y gastric bypass patients reveals novel CpG sites associated with essential hypertension. *BMC Med Gen* **9**, 20.

Corsi S, Iodice S, Vigna L et al. (2020) Platelet mitochondrial DNA methylation predicts future cardiovascular outcome in adults with overweight and obesity. *Clin Epig* **12**, 29.

ElGendy K, Malcomson F C, Bradburn D M et al. (2020) Effects of bariatric surgery on DNA methylation in adults: A systematic review and meta-analysis. *Surg Obes Rel Dis* **16**(1), 128–136.

Fleming T P, Watkins A J, Velazquez M A et al. (2018) Origins of lifetime health around the time of conception: Causes and consequences. *Lancet* **391**, 1842–1852.

Gallardo-Escribano C, Buonaiuto V, Ruiz-Moreno M I et al. (2020) Epigenetic approach in obesity: DNA methylation in a prepubertal population which underwent a lifestyle modification. *Clin Epig* **12**,144.

Gensous N, Garagini P, Santoro A et al. (2020) One-year Mediterranean diet promotes epigenetic rejuvenation with country- and sex-specific effects: A pilot study from the NU-AGE project. *Geroscience* **42**, 687–701.

Goyal D, Limesand S W, Goyal R (2019) Epigenetic responses and the developmental origins of health and disease. *J Endocr* **242**(1), T105–T119.

Hackett J A & Surani M A (2013) DNA methylation dynamics during the mammalian life cycle. *Philos Trans Roy Soc B* **368**, 20110328.

Healton S E, Pinto H D, Mishra L N et al. (2020) H1 linker histones silence repetitive elements by promoting both histone H3K9 methylation and chromatin compaction. *Proc Nat Acad Sci USA* **117**(25), 14251–14258.

Hernández-Saavedra, D, Moody L, Xu G B et al. (2019) Epigenetic regulation of metabolism and inflammation by calorie restriction. *Adv Nutr* **10**, 520–536.

Horvath S, Erhart W, Brosch M et al. (2014) Obesity accelerates epigenetic aging of human liver. *Proc Nat Acad Sci USA* **111**(43), 15538–15543.

Ji C & Guo X (2019) The clinical potential of circulating microRNAs in obesity. *Nature Rev Endocrin* **15**, 731–743.

Jia G, Fu Y, Zhao X et al. (2011) N6-methyladenosine in nuclear RNA is a major substrate of the obesity-associated FTO. *Nature Chem Biol* **7**(12), 885–887.

Kucharski R, Maleszka J, Foret S et al. (2008) Nutritional control of reproductive status in honeybees via DNA methylation. *Science* **319**(5871), 1827–1830.

Mathers J C (2008) Personalised nutrition. Epigenomics: a basis for understanding individual differences? *Proc Nutr Soc* **67**(4), 390–394.

Seisenberger S, Peat J R, Hore T A et al. (2013) Reprogramming DNA methylation in the mammalian life cycle: Building and breaking epigenetic barriers. *Philos Trans Roy Soc B* **368**, 20110330.

Shen L, Li C, Wang Z et al. (2019) Early-life exposure to severe famine is associated with higher methylation level in the IGF2 gene and higher total cholesterol in late adulthood: The Genomic Research of the Chinese Famine (GRECF) study. *Clin Epig* **11**, 88.

Tobi E W, Goema J J, Monajemi R et al. (2014) DNA methylation signatures link prenatal famine exposure to growth and metabolism. *Nature Commun* **5**, 5592.

Waterland R A & Jirtle R L (2003) Transposable elements: Targets for early nutritional effects on epigenetic gene regulation. *Mol Cell Biol* **23**(15), 5293–5300.

FURTHER READING

The following papers represent key conceptual advances in, or provide reviews of, important areas of the field relevant to nutrition and epigenetics.

Bernstein B E, Stamatoyannopoulos J A, Costello J F et al. (2020) The NIH roadmap epigenomics mapping consortium. *Nat. Biotechnol.* **28**, 1045–1048.

Fraga M F, Ballestar E, Paz M F et al. (2005) Epigenetic differences arise during the lifetime of monozygotic twins. *Proc Nat Acad Sci USA* **102**(30), 10604–10609.

Frye M, Harade B T, Behm M et al. (2018) RNA modifications modulate gene expression during development. *Science* **361**, 1346–1349.

Gilbert W V, Bell T A, Schaening C (2016) Messenger RNA modifications: Form, distribution, and function. *Science* **352**(6292), 1408–1412.

Horvath S (2013) DNA methylation age of human tissues and cell types. *Gen Biol* **14**(10), R115.

Link A, Balaguer F, Goel A (2010) Cancer chemoprevention by dietary polyphenols: promising role for epigenetics. *Biochem Pharmacol* **80**(12),1771–1792.

Strahl B D & Allis C D (2000) The language of covalent histone modifications. *Nature* **403**, 41–45.

Wahl S, Drong A, Lehne B et al. (2017) Epigenome-wide association study of body mass index, and the adverse outcomes of adiposity. *Nature* **541**, 81–86.

Weaver I C, Cervoni, N, Champagne F A et al. (2004) Epigenetic programming by maternal behavior. *Nature Neurosci* **7**(8), 847–854.

In addition, the following website is used by epigenetics researchers worldwide:

The National Institutes of Health (NIH) Roadmap Epigenomics Mapping Consortium has generated high-quality, genome-wide maps of several key histone modifications, chromatin accessibility, DNA methylation, and mRNA expression for hundreds of human cell types and tissues. The website provides access to both data and experimental tools and protocols.

The ENCODE (Encyclopedia of DNA Elements) Consortium is building a comprehensive parts list of the functional elements in the human genome which regulate gene expression.

https://www.encodeproject.org/

PART 6

Assessment of nutritional status

30 Anthropometry

Barry Bogin and Laura Medialdea Marcos

OBJECTIVES

By the end of this chapter you should be able to:

- To understand how various anthropometric measurements and the technology required to take these measurements are used to assess nutritional status.
- To appreciate how inaccuracies in their recording can be minimized.
- To explain the use of anthropometric references and standards to estimate levels of under- and over-nutrition.
- To understand that dietary intake and nutritional balance are key determinants of human physical growth and development but many additional factors, especially food insecurity and its social-economic-political-emotional (SEPE) causes and impacts, regulate human growth.

30.1 Introduction

Anthropometry is the scientific measurement and analysis of variation in the size and shape of the human body. Compared with the dietary, biochemical status, functional, and clinical status methods, anthropometry may provide a relatively quick and inexpensive means for the assessment of nutritional status. Anthropometric assessment most commonly involves the measurement of height, weight, fatness, and muscularity. It should also include the estimation of biological maturation, for example dental age. Sex and age are essential factors since these influence biology, behaviour towards an individual, and behaviour by that individual. To interpret their meaning, anthropometric measurements are usually related to references or standards, which reflect the growth of healthy and well-nourished individuals according to age and sex. The definition and rationale for the measured indicators and their references and standards are given below, in Section 30.2.

It is important to stress that divergence from a growth standard or reference may be due to nutritional experience or other factors. Starvation results in thinness and a deceleration, even cessation, of height growth. Overeating beyond energy requirements results in growth acceleration, including greater height in children and adolescents and greater fatness at all ages. Most of the nutritional impact on height growth is due to changes in maturational 'tempo' or speed—deceleration in case of undernutrition, and acceleration in case of nutrient excess. Most nutritionists are not sufficiently aware of the significance of 'tempo' when interpreting growth and focus only on size, also called growth amplitude, ascribing differences in size to genetic and environment determinants. All healthy infants and children have a tempo of growth, from slow to rapid. A slow tempo makes a person seem small, and a rapid tempo large, at the time of measurement, even though they may be quite average in size as adults. Further discussion of the importance of tempo in the anthropometric assessment of nutritional status, and some ways to measure it, is provided below.

Non-nutritional factors may cause deviations from the growth reference or standard. Those that regulate amplitude (amount of height, weight, etc.) and tempo (rate of change in body size) include infectious disease, endocrine system irregularities, lack of psychosocial stimulation, and repeated exposure to adverse childhood experiences (ACEs). Types of ACEs include physical or emotional abuse, neglect, persistent stress from insecurities related to food, housing, and education, and fear from racism (https://developingchild.harvard.edu/resources/stress-and-resilience-how-toxic-stress-affects-us-and-what-we-can-do-about-it/). These harmful impacts on physical growth, from nutrition to racism, are often consequences of poverty and the lack of social-economic-political-emotional (SEPE) resources needed for health. Anthropometry serves as a 'mirror of society', reflecting the quality of SEPE factors and, more generally, the material and moral conditions in which groups of people live (Tanner 1987; Bogin & Varela-Silva 2015; Bogin 2021b).

The World Health Organization (WHO) states that, 'Anthropometry is the single most universally applicable, inexpensive, and non-invasive method available to assess the size, proportions, and composition of the human body. Moreover, since growth in children and body dimensions at all ages reflect the overall health and welfare of individuals and populations, anthropometry may also be used to predict performance, health, and survival' (WHO Expert Committee 1995: 1–2). As such, anthropometry is a valuable tool for guiding public health policy and clinical decisions.

30.2 Anthropometric measurements: what and why

The WHO and other health authorities recommend anthropometry as the primary method of nutritional assessment because of the relatively low cost and portability of equipment needed, the relatively modest level of training required (for approved courses see Weblink Section 30.1 , and the high accuracy and precision relative to dietary methods. Examples of anthropometric equipment are shown in Figures 30.1 and 30.2. and for the measurement of standing height for people aged two years and older in Figure 30.3.

A wide range of anthropometric measurements can be made for ergonomic, anthropological, physiological, medical, and sports purposes, but a more limited list is appropriate to nutritional anthropometry (Table 30.1) for the assessment of undernutrition at various ages. For infants and younger children, the most useful measurements are height (or recumbent length for under two-year-olds), weight, arm circumference, head circumference, and skinfolds. For older children, adolescents, and adults, it is useful to also measure waist (abdominal) and hip (buttocks) circumference (Weblink Section 30.2 provides explanation of the measurements). These are useful in the assessment of overnutrition, which is usually done by analysing the waist-to-hip ratio (WHR, https://www.who.int/publications/i/item/9789241501491). Armspan, the distance between the furthest points of hands (usually the tips of the middle fingers) when the arms are outstretched to either side, can be used to estimate height if the person is unable to stand straight, as may occur in the elderly. Demispan, half the distance of armspan (or the distance between index–middle finger web and the sternal notch) may also be used.

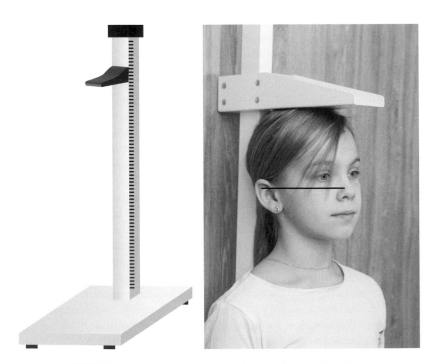

FIGURE 30.1 Anthropometric equipment for length/height.
Stadiometer for measuring stature. The subject stands on the base and the adjustable headpiece is brought into gentle contact with the crown of the head. The measurement is read via a counter incorporated into the headpiece. Note the head position, called the Frankfurt plane, in which the line is horizontal with respect to the floor and touches the lower border of the eye socket (orbit) and the upper border of the ear canal (auditory meatus).
Photo © Dmitry Naumov/Alamy Stock Photo.

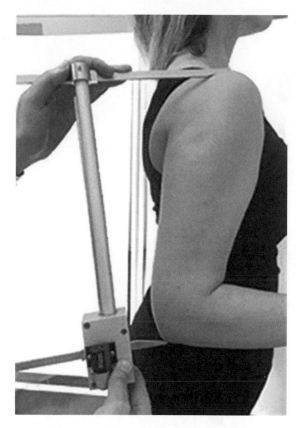

FIGURE 30.2 Anthropometric equipment for length/height.
Portable anthropometer for measuring any linear dimension, in this case humerus length. One end of the anthropometer is fixed and an adjustable piece slides along the calibrated column. The measurement is read from the window of the adjustable slide (circled).
Source: Reproduced from O Uzun, G Yegınoğlu, C E Öksüz et al. (2019). Estimation of stature from upper extremity: Anthropometric measurements. *J. Clin. Diagnostic Res.* **13**, AC09–AC15, with permission of the author.

Estimation of severe acute malnutrition using a smartphone app is described in Weblink Section 30.3.

Interpretation of anthropometric measurements

Height (or recumbent length) and weight are fundamental since they give the simplest measure of attained skeletal size (height or length), and of the total mass (weight). Height is a measure of accumulated size over time and barring severe trauma is not lost during the first three decades of life, even when people suffer prolonged starvation or illness, although the density of skeletal tissue in the bones may decline. Height declines after the age of 30 years by about 1.0 cm/decade, a trend noted throughout the world. In the elderly, postural changes due to osteoporosis, (micro) fractures of vertebra and other causes may accelerate the decline in height and some elderly people cannot stand erect or lie down in a fully stretched position. Total armspan or demispan has been shown to be a reliable and reproducible alternative measure of stature in the elderly and even in children as young as four years old (Banik et al. 2012).

Relatively low height-for-age may indicate a chronic deficiency in growth, due to prolonged illness, undernutrition, or exposure to ACEs and SEPE insecurities but may also indicate a slow tempo of growth (see Weblink Section 30.4).

Taking the growth tempo into account requires an estimation of biological maturation, most often using the formation of teeth or bones for people between the ages of about 2.0 to 18.0 years old. For infants <2 years there is insufficient dental and skeletal formation and by about age 18 years most of it is complete. Dental age is best assessed from age 5–18 years, as permanent tooth eruption takes place within this age range for most people across all maturational tempos. Biological maturation of the skeleton is not discussed here, as it requires the use of X-rays, usually of the hand and wrist, and this is impractical and often not ethically feasible for large-scale nutritional epidemiology (see Bogin 2021a for details on skeletal maturation). Estimating dental maturation is best done via X-rays of the jaws, to see the state of formation of teeth both erupted and unerupted. If X-rays are not feasible, the next best method is to count the number and type of erupted teeth. Tooth eruption takes place in an orderly and predictable sequence, with the first deciduous (baby) teeth usually being the central incisors, coming in at 6–8 months of age. The permanent teeth usually start coming in at about six years, beginning with the first molar (https://my.clevelandclinic.org/health/articles/11179-teeth-eruption-timetable). Early maturers (fast tempo) may erupt permanent teeth by five years of age. In principle, the more deciduous teeth an infant or child has, or the more permanent teeth a child, juvenile, or adolescent has, the more biologically mature. Undernutrition is known to delay the eruption of the teeth, that is, delay the tempo of maturation.

The use of biological age estimation as part of nutritional assessment is uncommon, even though important. Neither the WHO nor other major public health agencies mention biological maturation in the context of nutrition assessment. However, its use would be highly recommended, especially where registration of the legal age of individuals is not precise due to the lack of census records or other means of age verification.

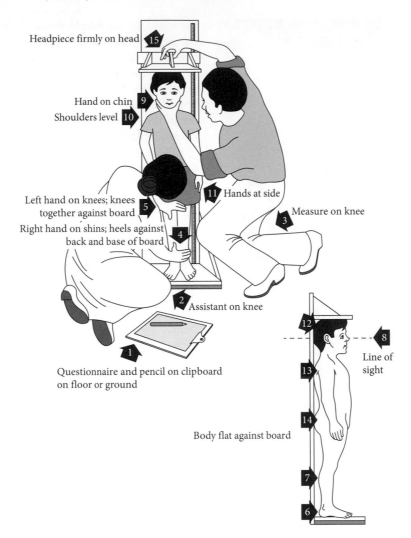

FIGURE 30.3 Measurement of standing height, from age two years and older.

TABLE 30.1 Recommended measurements for anthropometric nutritional assessment

Age group (years)	Practical field observations	More detailed observations
0–1	Weight, length	Head and arm circumference Triceps and subscapular skinfolds
1–5	Weight, length, height, arm circumference	Triceps and subscapular skinfolds
5–20	Weight, height, arm circumference	Triceps, subscapular, and medial calf skinfolds Calf circumference
>20	Weight and height	Arm and calf circumference Triceps, subscapular, and medial calf skinfolds Waist and hip circumferences (overnutrition only) Demispan (elderly subjects)

Statistical issues in the use of anthropometry for nutritional assessment

Any anthropometric variable has to be interpreted according to the age and sex of the person against the relevant standard or reference, and calculated as standard deviations or z-scores, usually denoted as zHT (z-score for height) or HAZ (height/length-for-age z-score), zWT (z-score for weight) or WAZ (weight-for-age z-score), and so on (see Figure 30.4 for explanation of standard deviations). Other acronyms for z-scores are also used and there is a confusing range of jargon and definitions (**see** Weblink Section 30.5). Use of z-scores allows the comparison of different anthropometric measurements, such as height, weight, or dental age which are not comparable in their original units (cm, kg, maturational years). When evaluating the effectiveness of a nutritional intervention, the change of z-score over time may be interpreted as success, for example, from −2.1 to −1.5, or failure (−1.5 to −2.1 z-score).

Anthropometric measurements can also be expressed as a percentile or centile of a normal distribution. Percentiles (centiles) and z-scores are compared, along with 'percentage of cases in 8 portions of the curve', 'cumulative percentages' and 'normal curve equivalents' in Figure 30.4 The WHO recommends the use of z-scores and the United States Centers for Disease Control, National Health, and Nutrition Examination Survey (CDC/NHANES) recommends the use of percentiles or z-scores. The anthropometric classifications proposed by WHO and the CDC are given in Table 30.2. There are biologically and clinically meaningful differences between the use of z-scores or percentiles, as is shown below (see Table 30.3).

In nutritional epidemiology the classification of stunting (zHT/HAZ < −2.0 or percentile/centile <5.0) is conventionally taken to mean inadequate nutritional status. A HAZ less than −2 standard deviations (< −2SD) of the median value of a growth standard or reference is defined as 'stunted'. There are recent criticisms of this interpretation and some of these are discussed below. Weblink Section 30.6 includes description of the origins of the term 'stunted' and references to how it has been used and misused in nutritional anthropometry. Nutritional anthropometry also uses WAZ or weight-for-height (WHZ). In contrast to height, measurements of weight reflect a more current indication of energy balance, as weight can be lost or gained relatively quickly. For infants and children under five-years old, a WAZ < −2SD is conventionally classified as 'underweight' and a WHZ < −2SD is conventionally classified as 'wasted'.

Arm circumference (i.e., mid-upper-arm circumference or MUAC, also circumferences of the calf or thigh) is used as a proxy for soft tissue mass and provides an indication of the total amount of bone, muscle, and fat at the mid-point of the upper arm. It is useful for assessing global nutritional status, especially the reserves of energy (adipose tissue) and protein (muscle tissue) of the body. In practice, measurements of MUAC may be both accurate and useful only in children between the ages of six months and 6.0 years due to the low association between nutritional status and changes in fat and muscle composition of the arm at younger and older ages. The cross-sectional area of bone in the upper arm (or leg) is assumed to be standard across populations, and unaffected by acute undernutrition. This is broadly acceptable, although it is well-known that the size and shape of bones are characteristics that vary due to age, sex, pregnancy

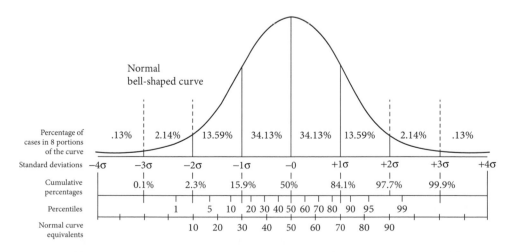

FIGURE 30.4 Percentile ranks (percentiles) compared to Normal curve equivalents.
Source: http://en.wikipedia.org/wiki/percentile_rank

TABLE 30.2 Anthropometric classifications proposed by the World Health Organization (WHO) and the Centers for Disease Control (CDC)

WHO	CDC
Use *Z* scores	Use percentiles
• Weight-for-height	• Weight-for-height
• Weight-for-age	• Weight-for-age
• Height-for-age	• Height-for-age
• BMI-for-age	• BMI-for-age
Weight-for-height	*Weight-for-height*
• Moderate acute undernutrition (wasting) < –3SD	• Acute undernutrition (wasting) ≤ percentile 3
• Severe acute undernutrition < –2SD	*Height-for-age*
Height-for-age	Chronic undernutrition (stunting) ≤ percentile 3
Chronic undernutrition (stunting) < –3SD	*BMI-for-age*
Severe chronic undernutrition > +2SD; ≤ +3SD	• Low weight < percentile 5
Weight-for-age	• Risk of overweight > percentile 85 and < percentile 95
• Overweight ≤ +3SD	• Overweight > percentile 95
• Obesity > +3SD	
BMI-for-age	
• Overweight > +2SD and < +3SD	
• Obesity > +3SD	

TABLE 30.3 Anthropometric classification for the assessment of physical growth and nutritional status

Indices-categories	Height-for-age	Weight-for-age	Weight-for-height	BMI-for-age	SumSF-for-age	UMA-for-age
I 0.0–5.0 percentile, Z < –1.650	Stunting	Underweight	Wasting	Light	Very slim	Wasting
II 5.1–15 percentile, –1.635 < Z ≤ –1,036	Below the mean	Below the mean	Below the mean	Below the mean	Below the mean	Below the mean
III 15.1–85 percentile, –1.036 < Z < +1.036	Average	Average	Average	Average	Average	Average
IV 85.1–95 percentile, +1.041 < Z < +1.645	Above the mean	Above the mean	Above the mean	Overweight	Above the mean	Above the mean
V 95.1–100 percentile, Z > +1.645	Tall	Overweight	Heavy for height	Obese	Obese	High muscular percentage

SumSF, sum of skinfolds; UMA, upper arm area.
* Unreliable values.
Source: Table created by Dr Maria Ines Varela-Silva, Loughborough University, adapted from Frisancho (2008), Centers for Disease Control, USA (http://www.cdc.gov/growthcharts), World Health Organization (http://www.who.org).

history, and physical fitness. These variations exist in similar ways within and between populations. Furthermore, skeletal growth is less plastic (affected by) nutritional circumstances, physical activity, ACEs, and SEPE insecurities than soft tissue mass. So long as circumference measurements are compared within groups similar for the sources of variation, the effect of bone size is negligible.

Estimation of body fat by skinfold assumes that subcutaneous fat is an indicator of total body fat. This is mostly true as many studies confirm a reasonably good correlation between skinfold thickness and more direct measures of fatness such as DEXA (dual energy X-ray absorptiometry) and air displacement plethysmography. But skinfolds only estimate fatness at the site of the measurement and various equations exist to convert skinfolds into an estimate of body fatness with variable reliability and accuracy, often affected by age, sex, ethnicity, total body size, and amount of muscle. DEXA and air displacement plethysmography can measure body fatness more directly—only dissection of a cadaver is direct—but require expensive machinery, located in a laboratory or hospital setting.

Waist and hip circumferences give a composite measure of fatness, both subcutaneous and visceral (see Weblink Section 30.2 and https://www.who.int/publications/i/item/9789241501491 which provides reference values). Lower limb skinfold and circumference measures are valuable, since it cannot be assumed that measures of the upper body are also representative of the lower body. Upper–lower body differences in fatness and muscularity are possible in individuals performing hard work on a regular basis. Adolescent and adult women and men, generally, have different distributions of upper–lower body fatness, with the 'female' patterns being more pear-shaped and the 'male' patterns being more apple-shaped. Many individual women and men will have reversed fatness patterns. Girl and boy infants, young children and older children not engaged in child labour tend to be more similar in body fatness and for these ages and cases upper body measures alone are useful in nutritional assessment.

In summary, for cost-effective population surveys the most common anthropometric measures of nutritional status in children are weight and height, either individually or combined, relative to growth standard or reference values. Table 30.4 summarizes the usefulness of WAZ, HAZ, and WHZ in nutritional assessment in children. An estimate of maturational tempo, such as dental age, would make the nutritional assessment more precise. Unfortunately, there are no growth references or standards that take into account both growth amplitude and tempo. Producing such a reference is a public health need.

The body mass index: pros and cons

Weight-for-height is also used to assess overnutrition in children and adults, and the body mass index (BMI = [weight(kg)/by height(m)^2] × 100) is now the most widely used measure for this purpose (Cole, Freeman & Preece 1995; Bogin & Varela-Silva 2012). For many people over age 20 years around the world BMI values of 18.5–24.9 are considered 'adequate', 'socially accepted', or 'healthy' weight-for-height (see

https://www.cdc.gov/healthyweight/assessing/bmi/adult_bmi/index.html). Adult undernutrition, called chronic energy deficiency (CED), is classified by the following BMI cut-offs: 17–18.5, grade I (moderate); 16–17, grade II (severe); below 16, grade III (severe acute) CED. BMI cut-offs of for overweight are 25–29.9 and values above 30 and 40 are used internationally to define moderate and severe obesity, respectively. Various nations and ethnic groups may have differently defined criteria. The World Health Organization, for example, published BMI cut-off recommendations for South Asians of >23.0 for overweight and 27.5, 32.5, and 37.5 for mild, moderate, and severe obesity respectively (WHO 2004).

In general, BMI is often assumed to be a measure of body fatness, but it is only a ratio of weight-for-height. It is essential to stress that the BMI does not measure body fat. BMI does reflect the diet and physical activity ecology of post-industrial societies; excessive energy intake and low levels of energy output and weight gain in this ecosystem is likely to be due to fat more than muscle. BMI is often an appropriate index of fatness in the general population, especially in the wealthier nations of North America, Western Europe, Australia, and Japan. However, in these nations, people in better than average physical condition, especially athletes with considerable muscle mass, will have a higher BMI but not excessive fat (Bogin & Varela-Silva 2012). Body shape and the proportion of leg length to total height also influence the BMI. In many poorer nations malnutrition and chronic disease can result in short stature due to relative stunting of the

TABLE 30.4 The usefulness of weight and height measures relative to reference data

	Weight-for-age	Height-for-age	Weight-for-height
Usefulness in populations where age is unknown or inaccurate	4	4	1
Usefulness in identifying wasted children	3	4	1
Usefulness in identifying stunted children	2	1	4
Sensitivity to weight change over a short time frame	2	4	1
Ease of accurate collection	2	3	2

Key: 1, excellent; 2, good; 3, moderate; 4, poor.
Source: After J Gorstein, K Sullivan, R Yip, et al. (1994). Issues in the assessment of nutritional status using anthropometry. *Bull World Health Organ* **72**, 273–283.

legs. Because the trunk of the body is more massive than the legs, such people will have a an adequate to high BMI but may have low fat and may even be undernourished.

For adults, and especially for infants and children, the BMI is a less sensitive marker of body fatness than are skinfold thicknesses. A major limitation with infants and children is that their body composition changes as they age, mainly due to the decrease in body water and increase in body fat (see Weblink Section 30.7). These changes in water, fat, and muscle make the use of the BMI almost useless for infants and children less than five years old. Body composition continues to change in older children and especially at the time of puberty and adolescence. These changes render the use of the BMI to estimate fatness to limited reliability for those aged 5–15 years old. A review of the history of use of the BMI and its many shortcomings may be found in Bogin and Varela-Silva (2012) and Bogin (2021a: 87–97) in the section titled 'The Body Mass Index: The good, the bad, and the horrid'.

Because of the limitations of the BMI, a variety of other anthropometric measurements have been proposed to estimate total fatness and the distribution of fat on the body. All of these have limitations for use with adults and none are justified for sub-adults. For adults, the measurement of waist and hip circumference and the calculation of the waist-to-hip ratio is one example. Another is the waist-to-height ratio. These focus on the central or abdominal distribution of body fatness because central fat is associated with high risks for diabetes, high blood pressure, and other metabolic disease. These types of measurements are criticized for being misleading in either a biological or a statistical sense (Burton 2020). An alternative to anthropometry are bioelectrical impedance devices (http://www.topendsports.com/testing/tests/BI.htm). However, these devices are not more accurate than skinfolds for the estimation of body fatness.

30.3 Growth standards and references

Between 2006 and 2009 the World Health Organization published new standards of growth in HAZ, WAZ, WHZ, head circumference-for-age, arm circumference-for-age, triceps skinfold-for-age, and other anthropometric dimensions for infants and children from birth to 5.0 years of age (https://www.who.int/publications/i/item/924154693X; https://www.who.int/publications/i/item/9789241547185 http://www.who.int/childgrowth/en/). The standards describe adequate infant and child growth under the WHO definition of 'optimal' environmental conditions. 'Optimal' conditions include several months of exclusive breastfeeding followed by

high-quality diet, good healthcare, mothers who did not smoke, and an emotionally secure home and family environment. In principle, the WHO growth standards may be applied to all under-5-year-olds everywhere, regardless of ethnicity, socioeconomic status, and type of feeding. The conventional interpretation of negative differences in growth of an individual or group from the standard is that such differences arise from problems of nutrition, health, and/or inappropriate psychosocial stimulation. This conventional interpretation is not always true and is explained in detail in Weblink Section 30.8.

In summary, using the standards requires knowing the age of the infant or child. If there is no official document, estimates can be made but if the age of a child cannot be estimated to within six months, then anthropometric estimation of nutritional status is not possible, as it may take the child outside the age range of the growth standards. The WHO standards differ from existing growth reference charts in several ways. First, standards describe 'how children should grow', in contrast to reference charts that describe how certain groups of children do grow. The standards are based on infants who were breastfed exclusively for the first six months after birth, making breastfeeding the biological 'norm'.

The WHO standards include growth indicators beyond height and weight, such as skinfold thicknesses, that are particularly useful for monitoring the increasing epidemic of childhood obesity. In addition, there are growth standards for head circumference and for motor development milestones that help to document adequacy of the brain and nervous system. The use of growth standards is valid only if infant and child populations under age 5-years-old grow similarly across the world's major regions when their needs for health and care are met, but not all populations grow in a similar fashion and the concept of a genetically determined 'growth potential' is incorrect (Bogin 2021a: 344–354). The WHO and many individual nations publish growth references for people over the age of 5.0 years. For recent British and US growth charts see http://www.rcpch.ac.uk/improving-child-health/public-health/uk-who-growth-charts/school-age-2-18-years/school-age-charts-an; http://www.cdc.gov/growthcharts/;. Other country-specific growth references avoid using the WHO standards. For example, Indonesia is characterized by short stature and when using WHO growth standards the prevalence of stunting reaches up to 43 per cent in several Indonesian districts. Yet many, if not most, of these children labelled as 'stunted' are healthy, vigorous, and performing well in school when assessed by paediatricians, human biologists, and teachers (Scheffler et al. 2020). The WHO standards appear to overestimate the burden of malnutrition in Indonesia resulting in a misapplication of resources to combat real malnutrition in the country.

Indonesian national growth reference charts (INGRC) were established in 2018 (Novina et al. 2020).

Uses, advantages, and limitations of anthropometry

Need for repeated measurement

Nutritional surveys do not normally follow up growth, but compare a single measurement with a growth standard or reference so that there is no way to infer the growth tempo and amplitude of the individual. A single measurement means that today the individual is so many SDs (standard deviations) under or over the standard curve, but it does not mean that this individual will continue to have this type of growth. Most healthy infants and children have periods of growth stasis or acceleration (mini growth spurts) during the year and consequent short-term changes in amplitude. Even with a HAZ of '0.00', indicating the 'perfect' median height-for-age and sex, does not always mean 'perfect' growth. The individual may have been much taller and has been decelerating in tempo due to malnutrition, exposure to toxic ACEs, or the loss of needed SEPE resources. Recovery from malnutrition is best confirmed by a rapid catch-up in the rate of growth from a lower to higher HAZ (Scheffler et al. 2021). It is recommended that at least two sets of anthropometric measurements be taken on the same individual, separated by several weeks. Then a longitudinal growth curve may be drawn and compared with growth velocity reference curves. Such 'rate-of-growth' references have existed since the 1950s (Cole 2012) but are not regularly used in nutritional epidemiology.

Accuracy of measurements

The major limitation of anthropometry is the extent to which measurement error can influence interpretation of nutritional status. Measurement error may arise from the use of uncalibrated or poorly calibrated equipment, or insufficient training of the anthropometrist. The relative simplicity of measurement can encourage investigators to skimp on training, but anthropometrists must achieve good levels of precision and accuracy (see Weblink Section 30.1) The need for repeated measurements has been explained above and can overcome some measurement error.

Use of arbitrary thresholds and ratios

As currently used, HAZ, WAZ, MUAC, and other anthropometric indicators for nutritional status depend on numerical thresholds, for example, HAZ < –2 SD being the threshold to define stunting. This is a purely mathematical threshold with little or no biological meaning (see Weblink Section 30.5 for additional explanation). An infant or child with a HAZ of –.99, –1.89 or –1.79 due to dietary deficiency, chronic infectious disease, or prolonged exposure to toxic ACEs suffers just as much as an officially 'stunted' youngster. The mathematical thresholds have some usefulness to target the worst-off people and allocate limited financial and human resources toward the alleviation of suffering. But public health workers, policy makers, and the public should not feel complacent about the suffering of people above the numerical thresholds.

The anthropometric indicators that we have mentioned throughout this chapter are likely to be influenced by factors such as age, sex, ethnicity, ACEs, and SEPE resources, among others However, the very definition of these indicators can lead to misinterpretation in some cases. For example, the WHZ indicator is a ratio between weight and height or length. In the presence of stunting, in which height or length is reduced compared to non-stunted individuals, the value of WHZ will be increased with respect to the same calculation in a non-stunted individual. Therefore, the result will be that the indicator overestimates WHZ of stunted individuals, masking the possible existence of acute malnutrition or risk of suffering from it. In a 2018 study by the NGO Action Against Hunger in Guatemala, with one of the highest stunting rates in the world, this error is described and it is explained how the misinterpretation of this indicator can lead to the non-identification of a high percentage of children suffering from acute malnutrition (https://www.accioncontraelhambre.org/sites/default/files/documents/estudio_relacion_entre_desnutricion_aguda_y_cronica696_0.pdf). To avoid this type of misinterpretation, it is always recommended to consider other anthropometric indicators such as zWT or WAZ, which are not influenced by height or length, and to consider measurements over time to analyse the trends in growth of infants and children.

30.4 Conclusion

Adequate weight-for-height does not mean that a person is well nourished and deviations from the desired value of a growth reference or standard does not mean that a person is malnourished. Many social-economic-political-emotional (SEPE) factors influence both the amount of growth and tempo of maturation. The limitations of nutritional anthropometry in the face of these issues were explained above in relation to the growth of schoolchildren in Indonesia, the misuse of BMI to estimate fatness, and in Weblink Section 30.6 entitled 'Protein energy malnutrition in historical perspective'. Despite these limitations, when applied correctly by well-trained personnel and with understanding of the ecology of the food, diet, and health of the society in which it is used, anthropometry remains of substantial value to assess nutritional status. Anthropometry is especially valuable when used as part of a comprehensive assessment of both physical growth and as an indicator of the quality of the material and moral conditions of society. To illustrate, Weblink Section 30.9 provides a case study of undernutrition in an infant girl named Precious, her mother, and her village. We discover that the undernutrition and poor health faced by Precious and her family come from the broader governmental and business structures and policies that lie 'upstream' from the world of Precious and her village (for more on upstream structural causes of malnutrition see Subramanian, Mejía-Guevara & Krishna 2016; Bogin 2021a: 438–439). The upstream societal structures and policies relate to agricultural supports, prices for farm products, food distribution, investments in water and sanitation infrastructure, health services, and quality education. Lack of these perpetuates poverty and deprives people of meaningful employment, dignity, and the hope that they can better themselves economically and socially. These are the major causes of the SEPE inequalities and insecurities in the village of Precious and her family and are the most immediate the causes of undernutrition. Anthropometric assessment of growth and nutritional status will remain necessary until those of us with privileged social status act in a concerted manner to improve material and moral conditions everywhere.

KEY POINTS

- Anthropometry is the most frequently used means of assessing the nutrition status of a population.
- Each age group and sex has its own most appropriate anthropometric measures.
- Estimates of both body size (growth amplitude) and biological maturation (growth tempo) are needed for meaningful assessment of nutritional status.
- The ease of collection of anthropometric measures should not mask the considerable potential for error in collection, but these errors are manageable if anticipated.
- The measurements are typically compared to measurements obtained for 'healthy' populations. These normative references or standards are continually being updated and reviewed and care must be taken to use the most appropriate one and to cite its use.
- Nutritional epidemiologists and public health workers need to be aware of the strengths and limitations of anthropometry. Body measurements such as height, weight, and fatness reflect the material and moral conditions of a society. These conditions include nutrition but also the many social-economic-political-emotional (SEPE) factors that are required for a healthy life.

ACKNOWLEDGEMENTS

The authors appreciate the assistance of PD Dr Christiane Scheffler, PhD, Institute of Biochemistry and Biology, Human Biology, University of Potsdam, and Professor Dr Michael Hermanussen, MD, University of Kiel, Aschauhof, 24340 Eckernförde-Altenhof, Germany for their review of the draft of this chapter and their many helpful comments.

 Be sure to test your understanding of this chapter by attempting multiple choice questions. See the Further Reading list for additional material relevant to this chapter.

REFERENCES

Banik S D, Azcorra H, Valentin G et al. (2012) Estimation of stature from upper arm length in children aged 4.0 to 6.92 y in Merida, Yucatan. *Indian J of Pediatrics* **79**(5), 640–646. doi: 10.1007/s12098-011-0580-0.

Bogin B (2021a) *Patterns of Human Growth*, 3rd edn. Cambridge University Press, Cambridge. doi: 10.1017/9781108379977.

Bogin B (2021b) Social-economic-political-emotional (SEPE) factors regulate human growth. *Human Biol and Publ Hlth.* 1, 1–20. https://doi.org/10.52905/hbph.v1.10.

Bogin B & Varela-Silva I (2012) The body mass index: The good, the bad, and the horrid. *Bull de la Société Suisse d'Anthropol* 18(2), 5–11.

Bogin B & Varela-Silva I (2015) The Maya project: A mirror for human growth in biocultural perspective. In: *Human Growth: The Mirror of the Society*, 3–23 (Sikdar M ed.). B.R. Publishing, Delhi.

Burton R (2020) The waist-hip ratio: A flawed index. *Ann Hum Biol* 47(7–8), 629–631. doi: 10.1080/03014460.2020.1820079.

Cole T J (2012) The development of growth references and growth charts. *Ann Hum Biol* 39(5), 382–394. doi: 10.3109/03014460.2012.694475.

Cole T J, Freeman J V, Preece M A (1995) Body mass index reference curves for the UK, 1990. *Arch Dis Child* 73(1), 25–9. doi: 10.1136/adc.73.1.25.

Frisancho A R (2008) *Anthropometric Standards: An Interactive Nutritional Reference of Body Size and Body Composition for Children and Adults.* University of Michigan Press, Ann Arbor, MI.

Novina N, Hermanussen M, Scheffler C et al. (2020) Indonesian national growth reference charts better reflect height and weight of children in West Java, Indonesia, than WHO child growth standards. *J. Clin. Res. Pediatr. Endocrinol* 12(4), 410–419. doi: 10.4274/jcrpe.galenos.2020.2020.0044.

Scheffler C, Hermanussen M, Bogin B et al. (2020) Stunting is not a synonym of malnutrition. *Eur J Clin Nutr* 74, 377–386 doi: 10.1038/s41430-019-0439-4.

Scheffler C, Hermanussen M, Soegianto S D P et al. (2021) Stunting as a synonym of social disadvantage and poor parental education. *Int J Environ Res.* 18(3), 1350. doi: 10.3390/ijerph18031350.

Subramanian S V, Mejía-Guevara I, Krishna A (2016) Rethinking policy perspectives on childhood stunting: Time to formulate a structural and multifactorial strategy. *Matern & Child Nutr,* 12, 219–236. doi: 10.1111/mcn.12254.

Tanner J M (1987) Growth as a mirror of the condition of society: Secular trends and class distinctions. *Acta Paediatr.Japonica* (overseas edn) 29(1), 96–103. doi: 10.1111/j.1442-200X.1987.tb00015.x.

WHO (2004) Expert consultation: Appropriate body-mass index for Asian populations and its implications for policy and intervention strategies. *Lancet,* 363(9403), 157–163. doi: 10.1016/S0140-6736(03)15268-3.

WHO Expert Committee (1995) *Physical Status: The Use and Interpretation of Anthropometry. WHO Technical Report Series 854.* Geneva. Available at: https://apps.who.int/iris/handle/10665/37003

FURTHER READING

Bogin B (2021a) *Patterns of Human Growth,* 3rd edn. Cambridge University Press, Cambridge.

Willett W (2012) *Nutritional Epidemiology,* 3rd edn. Oxford University Press, Oxford.

Dietary assessment

Bridget Holmes

OBJECTIVES

By the end of this chapter, the reader should be:

- able to understand the purpose of dietary assessment
- able to describe the various methods of dietary assessment used at national, household, and individual level
- aware of the strengths and weaknesses of the different approaches to dietary assessment at national, household, and individual level
- able to recognize the contexts in which it is appropriate to use specific methods
- aware of important factors in dietary assessment including portion assessment, use of technology, food composition tables, and assessment in special populations
- able to understand how to take into account the measurement errors associated with dietary assessment when interpreting results from dietary surveys.

31.1 Introduction

Dietary assessment involves the collection of information on the quantity, and usually frequency, of foods and drinks consumed over a specified time, and, using food composition tables, a calculation of energy and nutrient intakes. There is a wide variety of methods available to collect such information, each one with different strengths and weaknesses. Measurements can be performed at different levels (national, household, and individual level). The level of assessment and method must be selected with careful consideration, taking into account the exact purpose of the assessment.

The aim of this chapter is to explore the reasons for undertaking dietary assessment, to outline the different techniques that are available, to clarify which techniques are appropriate for specific purposes, and to consider the errors that arise when measuring diet and how to cope with them.

31.2 Objectives of dietary assessment

Before undertaking any dietary assessment, it is necessary to consider the exact purpose of the assessment, what is to be measured, in whom, over what time period, and how the measurements are to be collected. This will determine which technique is most appropriate for a given purpose, and avoid wasting resources using a technique that does not provide an appropriate measure or answer the research question(s).

What is the underlying purpose?

All dietary assessments aim to measure food consumption or to estimate the intake of nutrients or non-nutrients in individuals or groups (Cameron & van Staveren 1988). There will, however, be an *underlying purpose* which will dictate the level and nature of measurements to be made. For example, assessments may vary from very precise estimates of a single nutrient in metabolic balance studies, to broad estimates of the total quantity of food available for consumption for an entire country.

What is to be measured?

It is important to decide at the outset the level and need for accuracy and whether it is foods or nutrients (or both) that are to be assessed. Some studies may concentrate on patterns of food consumption and be less concerned with nutrient intake (e.g., in relation to dietary education and the promotion of healthy behaviours). Other studies may be focused primarily on nutrient intake; single nutrients, macronutrients, and/or micronutrients. Some studies may require a level of detail of nutrient intake that cannot be provided reliably by food composition tables (e.g., metabolic balance studies).

In a given culture, it is essential to know which substances constitute food or drink and which are taboo or unacceptable so that the assessment of diet is appropriately tailored. In France, for example, horse meat is sometimes consumed, whereas in England the consumption of horse meat is less acceptable. In some societies where food is gathered in the wild, nuts, flowers, berries, mushrooms, insects, and other foodstuffs may make important contributions to nutrient intake. Failure to identify all of the important food sources will lead inevitably to an underestimate of consumption and a mis-representation of intake. The same is true of beverages; it is vital to identify all fluid sources and take into account their composition so that energy and nutrient contributions are appropriately considered. It is also critical to understand for each culture how the food or drink may be consumed and what preparation or cooking takes place, as this will greatly impact the nutrient content of the item.

Whose diet is to be measured?

If diet is to be assessed for a country or region, and data are to be collected at an aggregate level, then the choice of method will be dictated largely by government decisions concerning information collected from growers, producers, importers, exporters, food processors and manufacturers, and those responsible for food storage. At the household and individual level, however, issues concerning literacy and level of education will be important in the choice of method. Age of subjects will also be important given that children under the age of 12 and some older people may have more trouble remembering accurately what, and how much, they ate and drank over the previous days or weeks. All methods should be adapted to the target population and be culturally sensitive.

When is diet to be assessed?

At the household and individual level, diet on weekdays and weekends may differ. There are seasonal variations in food availability in every country, often very marked in countries that have extreme wet and/or dry seasons or variations in employment and income (e.g., fruit picking in itinerant labourers). The duration over which diet is to be assessed also needs to be taken into account (e.g., single day, multiple days, week, month, year, etc.). The timing of administration of a dietary assessment (before or after a meal, for example) may influence the reported levels of consumption. A particularly difficult problem arises when the aim is to assess diet in the distant past, because evidence suggests it may be causally related to current disease state (e.g., the influence of past intakes of dietary calcium in relation to current risk of osteoporotic fracture).

How is diet to be assessed?

The choice of dietary assessment method will be largely influenced by the answers to the above questions—why? what? who? when? Of considerable importance is the level of resource available to undertake the dietary assessment and perform coding and analysis, in terms of financial resource and available subject, interviewer, and researcher time. Consideration also needs to be given to who will be doing the measuring and recording of diet, for example, subjects, nutritionists, clinicians, interviewers, as this may also influence method selection. The mode of assessment also needs to be considered, and data collection may be carried out using approaches including in person, by telephone, by mail, using on- or offline electronic tools or apps.

31.3 Methods of dietary assessment

There are five levels at which food availability and consumption can be measured (Figure 31.1). These range from national statistics through to individual consumption. Information from the finer levels of measurement (individual, household, or institution) can be built up to provide a picture of consumption at regional or national level. Further information on the Weblink includes detailed method definitions and strengths and limitations for household level (Weblink Section 31.1, Weblink Tables 31.2 and 31.3) and individual level techniques (Weblink Section 31.2), details on describing portions and amounts (Weblink Section 31.3), and details on how to describe methods when reporting findings from dietary investigations (Weblink Section 31.4 and Weblink Table 31.1).

Domestic food production (I)

Most governments require farmers and food producers to report how much food they produce. This is part of ongoing food surveillance which provides an overview of the adequacy of the national food supply. It also provides information about the levels of agricultural self-sufficiency (how much of a country's food supply is produced domestically). It is useful in planning food supplies and meeting dietary requirements.

This type of information is of limited value on its own, however. No country is entirely self-sufficient. Moreover,

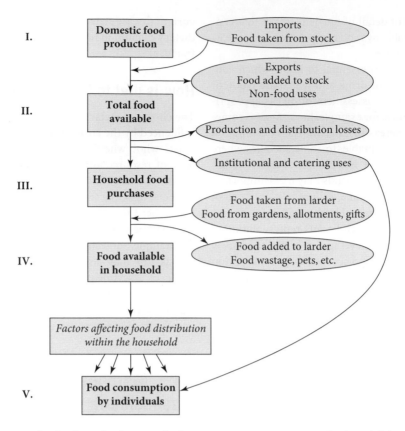

FIGURE 31.1 Points in the food supply chain at which it is convenient to measure food availability or consumption.

the quality of reporting varies between countries. High-income countries with large farms and highly mechanized means of production are likely to provide more complete and accurate information than countries in which a large part of the food production is based on subsistence agriculture with many smallholders and tenant farmers.

Total food available (II)

Of far greater usefulness in estimating a country's food supply and utilization are food balance data published on a regular basis by the Food and Agriculture Organization of the United Nations (FAO). Food balance sheets provide information on a country's food supply, during a specified reference period, through three components:

- Domestic food supply of food commodities in terms of production, imports, and stock changes.
- Domestic food utilization including livestock feed, seed, processing, losses during storage and transportation, export, and other uses. A distinction is made between the quantities 'utilized' and the food supplies available for human consumption.

- Per capita values for the supply of all food commodities (in kilograms per person per year), and the energy, protein, and fat content after application of appropriate food composition values.

Annual food balance sheets show the trends in the overall national food supply, disclose changes that may have taken place in the types of food consumed, that is, the pattern of the diet, and reveal the extent to which the food supply of the country is adequate in relation to nutritional requirements.

Food balance data were developed to capture food availability rather than actual intake. These data are widely used for making comparisons between countries to learn about the extent of hunger at the national level and global food security. They can be used in designing and targeting policies to reduce hunger (see Chapter 36) as well as understanding the relationships between diet and disease. They are also useful for assessing time trends in diet within countries, although they are not direct measurements of diet. Food balance data are available for analysis through FAO Corporate Statistical Database website, FAOSTAT. Food balance data are limited in three important ways. First, they provide only an estimate for the country as a whole and therefore there is no information on distribution of consumption subnationally within the country, for example by certain age or sex groups, rural/urban, socioeconomic status, and consequently there is no indication of any groups within the population that may be consuming amounts consistently above or below the average. Second, such data do not take into account home farming or production, hunting and gathering—which may be more common in lower income countries—or waste or losses of food from the system (e.g., domestic spoilage and plate waste, meals not eaten at home, or food given to domestic animals). This may introduce error and lead to an overestimate of apparent availability in the majority of cases. Third, the estimates give higher values than those derived from other types of surveys. In economically developed countries especially, the estimate of energy available for consumption may be as much as 25 per cent above the estimated requirement of the population. This suggests that there may be biases inherent in the reporting system which may lead to a distortion in estimated differences in consumption between countries. The agreement in ranking of countries based on a comparison of food balance data and other types of dietary survey, however, suggests that the data may be useful for certain epidemiological purposes. Compared to other dietary assessment methods at the household and individual level, food balance data are reported to be the least suitable for assessing individual-level intakes and dietary quality or for evaluating diet–disease burdens (Micha et al. 2018).

Past research has investigated how estimates published in food balance sheets from FAO compare with nationally representative, individual-based dietary surveys from the Global Dietary Database. FAO estimates were matched to survey data for 113 countries for 30 years of data (1980–2009) for fruit, vegetables, beans and legumes, nuts and seeds, whole grains, red and processed meats, fish and seafood, milk, and total energy. For all food groups and total energy, FAO estimates were found to substantially exceed or underestimate individual-based national surveys of individual intakes with significant variation depending on age, sex, region, and time. Calibration models were developed by the authors which can adjust the global, comprehensive and widely accessible FAO data, to facilitate a more accurate estimation of individual-level dietary intakes nationally and by age and sex (Del Gobbo et al. 2015). Other applications have used food balance data to provide information on the ability of a national food supply to ensure an adequate supply of micronutrients to its population, identify likely micronutrient deficiencies to facilitate fortification planning, and to estimate inadequate intakes of certain micronutrients (e.g., Wuehler et al. 2005; Wessells et al. 2012; Mark et al. 2016). Such applications, especially useful where national household or individual level surveys do not exist in a country, can

be useful to rank countries, and identify those at likely high risk of deficiencies relative to other countries. These countries can be then targeted for nationally representative dietary surveys and/or status assessments, as appropriate. Recently, FAO have committed to assign appropriate selected micronutrient values to all food balance data, extending the potential application for their use and validation with other methods.

Measurements at household level (III and IV)

There are a wide variety of household surveys, many of which measure food consumption, or 'apparent' consumption, where it is assumed that all food brought into the household is consumed by household members. There has been a dramatic increase in the number, quality, and availability of household surveys over the last thirty years, and by 2018 there were more than 845 surveys for at least 137 countries (FAO & The World Bank 2018). There are four main techniques used to assess food consumption at the household level: food accounts, inventories, household recall, and list-recall (Lagiou & Trichopoulou 2001). These are described in Weblink Section 31.1. The strengths and limitations of these methods are summarized in Weblink Tables 31.2 and 31.3.

One of the weaknesses of household surveys is the specificity of the food items, which is often lower than food composition table entries or individual level survey data. Household survey items represent broader food groups, rather than specific food items, and do not capture information on the level of food processing (e.g., fresh, dried) or level of preparation (e.g., raw, boiled, fried) of the items, important necessary detail to estimate the nutrient content of a food. Some research has inferred that, whilst there is little doubt that the household surveys are inferior to the 24-hour recall survey in terms of the precision of the food consumption data, data on food consumption obtained from household surveys can contribute to addressing the food consumption information gap, especially in low- and middle-income countries, where detailed individual level data is often unavailable or too expensive to collect, especially at a nationally representative level (Fiedler et al. 2013). Household surveys should be regarded as a complementary source of food and nutrition data, and not as a substitute for other sources of such information (Fiedler et al. 2012). In a relative ranking of four different dietary assessment methods, it was suggested household surveys can be used in combination with other methods including 24-hour recalls, to identify vehicles for food fortification and estimate coverage and impact due to their low cost and moderate validity (Coates et al. 2012).

Measurement of food consumption in individuals (V)

There are two main approaches to individual dietary assessment, *prospective* and *retrospective*. Prospective methods involve subjects collecting or recording their current diet, while retrospective methods require subjects to recall their recent or past diet. Both types of assessments have strengths and limitations, summarized in Table 31.1, and depend on the ability of the subject to provide accurate information. Exact definitions of techniques are given in Weblink Section 31.2.

Generally speaking, any measurement of diet will be biased in some way by the measurement process itself. There are therefore no entirely objective measures of an individual's food consumption or nutrient intake except in the controlled conditions of a metabolic unit. For example, the reported diet may not reflect the *actual* diet because subjects choose not to record or report certain items (especially 'unhealthy' foods like sweets and

TABLE 31.1 Strengths and limitations of measurements of individual food and beverage consumption

Method	Strengths	Limitations
Prospective methods		
General features	Current diet	Labour intensive
	Direct observation of what is eaten/drunk	Requires literacy and numeracy skills
	Duration of survey can be varied to meet requirements for precision of food consumption or nutrient intake	Subjects need to be well motivated and committed
		High level of respondent burden and commitment
	Do not rely on subject's memory	Usual consumption pattern may change due to inconvenience of recording, choice of foods which are easy to record, beliefs about which foods are 'healthy' or 'unhealthy'
		True consumption levels may be under-reported, particularly in overweight subjects
		Coding and data entry errors are common
Weighed diary	Precise portion size assessment	Food composition tables used to estimate nutrient intake
Unweighed diary	Fewer recording skills required compared with weighed diary	Loss of precision compared with weighed diary
	No scales needed	
Duplicate diet	Direct analysis of nutrient content of food (not dependent on food composition tables)	Very expensive
		Intense supervision needed
	Required in metabolic balance studies	Usual diet may not be consumed
Retrospective methods		
General features	Inexpensive	Biases caused by errors in memory, effect on usual consumption, perception and conceptualization of food portion sizes, presence of observer
	Quick	
	Can assess current or past diet	Daily variation in diet not usually assessed
	Lower respondent burden than required for prospective methods	Dependent on regular eating habits
		Food composition tables used to estimate nutrient intake
24-hour recall	Very quick	Prone to underestimate intake due to omissions
	Suitable for those with low literacy and numeracy skills	Interviewer burden is high
		Single observation provides poor measure
	Can be repeated to gain measure of daily variation and improve precision of individual intake	
Food frequency questionnaire	Suitable for large-scale surveys	Requires validation in relation to reference measure
	Can be posted	Literacy and numeracy skills needed if self-completed
	Short version can focus on specific foods or specific nutrients with few food sources	
Diet history	Assesses 'usual' diet	Over-reporting of foods believed to be 'healthy' (e.g., fruit)

alcoholic beverages for example). In addition, a record or recall of diet may not reflect the *usual* diet because, for example, a respondent may choose to report or consume more foods regarded as 'healthy' (such as fruits and vegetables) and fewer foods regarded as 'unhealthy' (like crisps or sweets). Subjects may also simplify their diet or eat less than usual in order to make the recording process easier. These errors are independent of those due to poor memory or inaccurate recording. The main consequence of these processes is an underestimation of dietary intake and a misrepresentation of healthy vs. unhealthy foods. Measurements of characteristics such as 'dietary restraint', 'social approval', and 'social desirability' help to identify subjects likely to misreport their food intake. Separating inaccurate reports of dietary intake as a result of under-recording and/or under-eating from those where subjects have consumed less than usual due to illness, dieting or fasting for example, can be facilitated by asking whether or not the subjects intake was typical or not and if not why. Such questions can also detect where subjects may have consumed more than usual, for example due to celebrations, special events, or religious festivals.

In order to quantify the intake of foods and drinks, an estimate of the portion weight of consumed items is required. Data entry errors made either at the recording stage or at the coding stage are common and the distortions that occur are often difficult to detect. The availability and use of good quality and up-to-date relevant country- or regional-food composition tables to convert food intake into nutrient intake is critical. Given the fact that the food supply is becoming more international, and more complex with an increasing number of foods, especially processed and pre-prepared foods and drinks appearing on the market, food composition tables must be updated on a regular basis in order to be representative of the foods available in a country. By combining dietary intake and food composition tables it is possible to determine whether or not the diet is nutritionally adequate. The use of food composition tables to estimate the energy and nutrient content of the diet will likely lead to further inaccuracies. The topic of measurement error is discussed in detail in Weblink Section 31.5.

Prospective methods: advantages and disadvantages

The main advantage of prospective methods is that they provide a direct measure of current diet. They can be carried out for varying lengths of time according to the level of accuracy of the estimate of food consumption or nutrient intake required. The main disadvantage of prospective methods is that they are labour-intensive for both the respondent and the interviewer. The respondent needs good literacy, language, and numeracy skills in order to provide an accurate record. This limits the usefulness of prospective methods in populations where literacy levels are low, unless trained interviewers are present (but the presence of an observer may result in a distortion of usual diet). Good respondent and interviewer motivation and a commitment to complete the record accurately and objectively are needed.

Retrospective methods: advantages and disadvantages

Retrospective methods require subjects to 'recall' their current or past diet. This may involve remembering the type and amount of all individual items consumed over a specified period of time (e.g., 24-hour recall), or creating a mental construct of 'usual' consumption involving recollection of both the frequency of consumption of specific foods or food groups and the amounts consumed. The main advantages of retrospective methods are that they are relatively quick to administer compared with prospective methods. They are also less expensive in terms of equipment and (except for repeat 24-hour recalls, see below) resources considering the time taken for interviewers to see subjects where data collection is in person. A further advantage of retrospective methods is that because there is a lower respondent burden than for prospective methods, the chances of obtaining a more representative sample of all consumers is increased. They can also be used to assess diet in the past, which may be relevant to epidemiological studies where the underlying causes of chronic diseases such as heart disease or cancer may lie in the past rather than the current diet.

The main disadvantage of retrospective methods relates to sources of bias. Errors in memory result in the omission of foods and drinks from the assessment. This may be a particular problem for some elderly subjects and for children under the age of 12 years. Subjects and interviewers must have good skills relating to the perception and conceptualisation of food portion size (the ability to develop an accurate mental construct of the amount of food consumed and to translate that construct into a description or selection of an appropriate food portion photograph which corresponds to the amount actually consumed). Amongst respondents, this is especially problematic in children under 12 years of age. In addition, the presence of an observer (interviewer) may cause subjects to over-emphasize what they perceive as the 'healthy' aspects of their diet and to minimize the 'unhealthy' aspects ('social desirability' and 'social approval' bias). Daily variation in diet is less readily assessed using retrospective methods (unless using repeated 24-hour recalls). Subjects who do not have regular eating habits will have difficulty describing the 'usual' frequency of consumption. And, as with most prospective methods, the use of food composition tables will introduce error into the estimates of energy and nutrient intake.

Prospective methods

- *Weighed diary or record.* The weighed diary, first described by Elsie Widdowson in 1936, was traditionally one of the most widely used techniques in dietary assessment. Subjects keep a record of all food and drink consumed. Each item is weighed prior to consumption using portable food weighing scales. Items leftover are also weighed. In practice, most weighed diaries include a proportion of items recorded in household measures. Despite the fact that this method is used less frequently today due to the burden associated with weighing items and the impact that recording has on intakes, it is still recognized to be one of the most accurate methods of assessing dietary intake. In low- or middle-income countries with low literacy levels the weighed record may be used in the presence of a trained interviewer. While such an assessment may provide very accurate estimates of intake, it can be intrusive and time consuming and may influence subject habits.

- *Household measures or unweighed diary or record.* This method is similar to the weighed diary, except that subjects record portion sizes in household measures (cups, glasses, bowls, spoonfuls, etc.) rather than weighing their food and drink on scales. Where country and population-relevant food photographs are available, they may be used to aid subjects with recording portion size. Records in household measures have the advantage of simplifying the recording process for subjects, and are less likely to impact intake compared with weighed records.

- *Duplicate diet method.* This technique requires subjects to weigh and record their food consumption at the time of eating. At the same time, subjects put aside an exact duplicate portion of each food consumed which is analysed chemically for energy and nutrient content. The main advantage of this method is that it is independent of errors associated with the use of food composition tables. It is best suited to metabolic balance studies in free-living populations.

- *Food checklist.* Respondents are provided each day with a pre-printed list of foods and drinks and asked to tick a box each time an item is consumed. A space is usually provided to record items consumed but not listed. Standard portion sizes may be indicated, or portion descriptions entered. This method is simple to use and well-liked by respondents, but the list of foods and drinks and the standard portions need to be relevant to the country and population being studied. Information collected is less detailed than with other

prospective methods and food consumption and nutrient intakes are less precise as a result.

Retrospective methods

- *24-hour recall.* The 24-hour recall (originally attributed to Wiehl in 1942) is now a widely used method. Subjects are asked to recall and describe every item of food and drink consumed over the previous 24 hours (usually midnight to midnight). The information is obtained through systematic repetition of open-ended questions. Amounts may be described in household measures or using food photographs. Trained interviewers may be used to carry out the data collection, especially in low- and middle-income countries. Interviewers must be thoroughly familiar with both the local diet and the food composition tables to be used, in order to probe subjects effectively and obtain adequate detail for subsequent coding of data and the transformation of the interview data into estimated nutrient intakes. In some settings, generally higher income countries, 24-hour recall data can be collected directly using specially designed, country and population relevant software (see section below 'Mode of administration and the use of technology'). In such settings, clear instructions should be integrated into the software, and alternative modes of administration should be made available where subjects are unable to complete the 24-hour recall without assistance. The 'multiple-pass' 24-hour recall is now in widespread use (consisting of several stages including an uninterrupted 'quick list' of items recalled; a detailed interview elaborating the quick list that determines detail and amounts; and a thorough review of the detailed interview). This multiple-pass method minimizes the opportunity for items to be forgotten. Regardless of the mode of administration, the 24-hour recall is relatively quick to administer compared with other individual level dietary assessment methods. However, it is prone to reporting errors and memory bias. As diets vary considerably from day to day, a single 24-hour recall is unable to provide information on day-to-day variation of food or nutrient intakes or to provide an accurate estimate of long term energy intake. For this reason, 24-hour recalls may be repeated for two or more days to improve the estimation of food and nutrient intakes. This approach is known as the repeated 24-hour recall.

- *Food frequency questionnaire.* Food frequency questionnaires (FFQ) are pre-printed lists of foods and drinks (or foods and drinks from given groups) on which subjects are asked to indicate the *typical* frequency of consumption over a specified time period in the past. Frequency responses are usually multiple choice and may range from several times per day to number of times per year, depending on the item and the time period that the FFQ covers. The number of foods on the list varies from a few questions on selected items (e.g., 20 items) up to a fully comprehensive list of items (e.g., 200 items) to assess total diet. Many FFQs inquire about frequency of consumption over the past year or previous six months, but other FFQs inquire about the past month or week. The questionnaire may be self- or interviewer-administered. FFQs may be qualitative containing no information on portion size, semi-or fully-quantitative with average portion specified, or subjects stating the average amount consumed on the days when the item is consumed using portion aids such as household measures or photographs. The use of average portion size information reduces the time required for administering and processing the FFQ, however there is a loss of precision at the individual level. Many FFQs are now either optically scannable or available digitally and this significantly speeds up the coding of information. Digital completion also reduces the likelihood of missed food items on the questionnaire as responses can be set to mandatory. An example of a segment of an FFQ is shown in Weblink Figure 31.1. The FFQ is an accepted method to measure typical diets and episodically consumed foods. The respondent burden is lower compared with other methods, however, errors may result from the inaccuracies in listing items, estimation of portion sizes, and frequency of consumption. For these reasons, only validated, country- and population-specific FFQs should be used. FFQs are widely used as the primary dietary assessment tool in epidemiological studies.

- *Diet history method.* The diet history was originally described by Burke in 1947 and although used less frequently today, is one of the oldest approaches for assessing diet. It is mostly used in clinical practice and is used to assess 'usual' diet over the recent past. Typically, a trained interviewer or dietician begins by carrying out a 24-hour recall which is further elaborated in a detailed interview lasting up to two hours. For each meal, subjects are asked to describe the range of foods that may typically be consumed, the frequency of their consumption, and typical amounts. Differences between weekdays and weekends and seasonal variations are clarified. The high costs and long duration of the interview limits the usefulness of this method in large epidemiological studies (Morán Fagúndez et al. 2015).

Brief methods of dietary assessment

Brief dietary assessment methods are useful when total diet does not need to be assessed. For example, simplified FFQs which contain far fewer food or drink items than would typically be included may be used in instances where a single nutrient or type of food or beverage is being estimated. Although such assessments can be made at a low cost with a low respondent and interviewer burden, they have several limitations including an inability to assess total diet and provide quantitatively precise information (Thompson & Subar 2013).

Dietary assessment toolkits

Given the multiple challenges that researchers face when diet needs to be accurately assessed, toolkits can aid users in the selection and implementation of the most appropriate dietary assessment method, or combination of methods, with the goal of collecting the highest quality dietary data possible, within local practical and financial restraints. Five toolkits are currently freely available online: The Diet, Anthropometry, and Physical Activity Measurement Toolkit (DAPA) (UK); The National Cancer Institute's Dietary Assessment Primer (USA); The Nutritools website (UK); the Australasian Child and Adolescent Obesity Research Network (ACAORN) method selector (Australia); the Danone Dietary Assessment Toolkit (DanoneDAT) (France), and aim to bring together information, including practical considerations, strengths and limitations of dietary assessment methods, guidance for method selection and study design, and recommendations for dietary data analysis in one place (Dao et al. 2019).

31.4 Considerations in dietary assessment

Portion size estimation

The assessment of portion size is one of the most important, but error prone, aspects of dietary assessment. In general, portion sizes of foods that are commonly bought and/or consumed in defined units (e.g., sliced bread, pieces of fruit, beverages in cans or bottles) may be more easily reported than irregularly shaped foods (e.g., meat, lettuce) or poured liquids (Thompson & Subar 2013). People are typically very poor at estimating weights of foods, unless weights are taken directly from packaging materials. Portion size aids are commonly used to help respondents estimate portion size. These may take the form of household measures, rulers, two-or three-dimensional pictures, or two-or three-dimensional food models. Country- and population-specific food photographs are time consuming and costly to produce, but offer a relatively straightforward and easy to use method to assess portion size. Digital food photographs can be integrated into software tools to assess dietary intakes to further facilitate assessment. Portion size aids designed for use with adults, based on adult portion sizes, require appropriate modification and validation in order to be appropriate for use with children.

Duration of the dietary assessment

In the past, for a given food or nutrient, the more days of information collected, the better the precision of the estimate. However, there is an increasing need to find a balance between achieving precise estimates of intake and maintaining recording enthusiasm of the subjects so as to minimize changes in patterns of usual consumption.

When records or recalls are used to assess diet, more than one day of dietary information is usually required. The number of days needed to measure dietary intake varies according to the research question, the population, the food or nutrient of interest, the level of variability within the population, the level of precision required, and the availability of resources to undertake the assessment and analysis. The recording is usually conducted over a period of one to four days, and up to seven days. One day is generally not considered to be suitable for individual assessment, because of the large intra-individual variability in daily food intake. However, due to the high costs associated with the collection of dietary intakes often only one day is collected, especially in low- and middle-income countries. Given that eating and drinking patterns vary between weekdays and weekend days and across seasons, assessments should be made with this in mind to provide representative coverage. While four to seven days increases the reliability of collected data, the subject burden is significantly increased and may lead to reduced compliance, greater drop-outs, and/or increased mis-reporting of dietary intakes (Thompson & Subar 2013). Performing analysis according to the reporting day may be used to reveal issues in reporting where several days of data are collected.

Mode of administration and the use of technology

The costs and burden associated with the collection of dietary information may be reduced by administering the instrument by telephone. Interviews by telephone can be significantly less expensive than face-to-face interviews and may also be less intrusive. Traditionally, 24-hour recalls have been performed by a face-to-face interview, but interviews being conducted by telephone have become increasingly common and are considered as a feasible alternative to face-to-face interviews. Portion size aids may be left with participants prior to the telephone interview. Difficulties reaching respondents at a convenient time to carry out the interview and repeated

interruptions during the interview may influence the level and accuracy of reporting.

The use of mobile telephones, digital cameras, and web-based systems are increasingly common in dietary assessment. Despite this, further research is still needed to investigate the validity of innovative dietary assessment technologies. Many new technology tools assessed in a recent evaluation showed close agreement to traditional methods of dietary intake, but gaps were wider when compared to more objective measures, including total energy expenditure from doubly-labelled water, although few tools have included such comparisons (Eldridge et al. 2019). Several computerized tools have been developed and made available, including FFQs, web-based 24-hour recalls, and food diaries. The use of technology makes traditional methods less arduous in terms of data collection and raises participation levels in populations less willing to participate in paper-based surveys, such as teenagers and adolescents. It also enables continued and remote access to participants, even during crisis situations such as the COVID-19 pandemic. Data can be transferred directly to the study centre and costs associated with coding are reduced significantly. Web-based assessments also allow personalized feedback to be provided immediately to subjects. Consideration needs to be given to the access to modern technology and the Internet and the necessary skills required for completing the method which makes the application of new technology inappropriate in some low- and middle-income countries, populations, and subgroups. Best-practice guidelines for reporting on new technologies for dietary assessment have also been proposed (Eldridge et al. 2019). Software developers are also working to create systems that operate on and offline which will facilitate use in low- and middle-income countries, and using-cloud based servers. Measurement errors associated with a finite list of foods and drinks, often with closed response categories, and the reduction in the amount of time spent by interviewers checking reported information with the subject should be considered.

Computer software is now available with the ability to analyse digital images of foods taken using smart phones and compare food intakes with standard portions. A calculation of energy and nutrient intakes is subsequently made. Despite advancements in technology over recent years, digital imaging is not sufficiently advanced to correctly identify and estimate portion sizes in 100 per cent of cases and certain foods (Martin et al. 2014). Certain foods that appear visually the same and impossible to distinguish using computer software may vary considerably in nutrient content, for example, standard and diet-carbonated beverages.

The use of computers and new technology has resulted in a broadening of the traditional definitions of dietary assessment methods. Technology has also facilitated the 'blending' of dietary assessment methods. The use of a blended approach has the potential to maximize the strengths of different methods, but also potentially means that limitations may be introduced. For example, a 24-hour recall performed using brief notes kept by subjects to aid as a memory prompt may help subjects to recall what they ate and drank, but may also impact the consumption on the recording day, something which is not usually an issue for the 24-hour recall method. Researchers should be cognizant of such impact and strive to maximize the strengths and minimize the limitations of such blended approaches. Factors to consider when selecting dietary software for data capture are provided in the section below.

Tables of food composition and dietary analysis software

The calculation of energy and nutrient content of foods and drinks consumed in dietary surveys is performed using tables of food composition. It is essential that food composition data is updated regularly to include new foods and drinks introduced on the market, and to update the nutritional value of foods and drinks that may have changed in composition over time, for example, as a results of changes in manufacturing processes or reformulations due to government health initiatives. Many countries have their own national tables of food composition, although they are of varying levels of quality and completeness. The UK food composition tables, McCance and Widdowson's 'The Composition of Foods', are updated on a regular basis and support the UK National Diet and Nutrition Survey rolling programme. The Composition of Foods Integrated Dataset (CoFID) integrates all available UK data as a single, consolidated and up-to-date dataset integrating new analyses and corrections on previously published data.

The International Network of Food Data Systems (INFOODS), established in 1984, is a worldwide network of food composition experts aiming to improve the quality, availability, reliability, and use of food composition data. FAO and INFOODS provide guidelines, standards, databases, capacity development tools, and technical assistance at country level to facilitate the generation, harmonization, and dissemination of quality food composition data. A detailed evaluation framework from FAO/INFOODS has been designed to assist food composition compilers in evaluating the quality of their published or unpublished tables, and assist users in evaluating the quality of published tables in a standardized manner. This framework will facilitate the understanding of the quality of food composition tables and identify where improvements can be made to improve quality.

Food composition tables form the basis of most dietary analysis software where nutrient analysis is possible. Dietary analysis software links the food or drink item consumed, with the appropriate code which in turn is linked to the nutrient composition. The software converts the amount reported to multiples of 100g. Software features differ depending on the intended use (for example, research or clinical practice). Traditionally, dietary analysis software was used only for analysis purposes, however, as the use of technology in dietary assessment increases for the purposes of data capture, the need for an all-encompassing software capable of both data capture and analysis has increased.

Key factors to consider when selecting dietary analysis software include:

- availability and cost of the software and the permitted number of users
- the possibility to tailor according to the survey design (number of days, method)
- quality and relevance (year, country, language) of the integrated food composition tables
- the extent and relevance of the food list, including date of last update
- the ability to search for items efficiently
- completeness of the database in order to gauge reliability (percentage of items with values assigned/missing)
- cost/potential to add (or remove) items, including recipes
- grouping of foods into categories, and possibility to adapt grouping
- the list of nutrients that can be reported
- types of data and visual output that can be generated by the software, including possibility to generate feedback for subjects
- the ability to carry out statistical testing.

Additional key factors to consider when selecting dietary software for data capture include:

- the dietary assessment method or combination of methods that can be used
- language, country, population group and appropriate validation
- researcher or subject interface, with ability to switch if needed
- ability to tailor software to ask specific research driven questions/food details
- ability to add (or remove) food items, including recipes

- integration of portion size assessments including food photographs
- ability to capture data online and offline, if needed
- availability and cost of the software, and future ownership of any developments.

31.5 **Dietary assessment in special populations**

The choice of dietary assessment method must be made considering not only the question to be answered but also the country and the population group in which the assessment is to take place. Assessing the intake of population subgroups, such as minority ethnic groups, low-income groups, pregnant women, or those at either end of life's spectrum (the youngest-young or the oldest-old) often presents specific problems in dietary assessment.

Infants and children. Capturing dietary information on very young infants presents numerous challenges, including those associated with difficulties measuring quantities of breastmilk consumed, collection of information on the precise type and preparation method of infant formula, together with difficulties recording quantity consumed (versus amount leftover, lost during feeding, or regurgitated). Diversification of the infant diet brings further challenges especially around determination of portions consumed and wasted. As children get older it is the variable nature of children's diets and rapidly changing eating habits that presents challenges for dietary assessment (Thompson & Subar 2013). Problems with recalling frequency as well as both the types and amounts of foods and drinks consumed and the ability to conceptualize portion size apply to younger children (Biró et al. 2002). Since younger children are less able to participate in dietary assessment, the ability of parents or caretakers to accurately report their children's food and drink intake is vital. Significant amounts of time spent in different locations, for example, school, crèche, childcare, or with different caretakers, may exacerbate the reporting of children's dietary intakes. While cognitive abilities are developed by adolescence, older children capable of reporting are often disinterested in providing complete and accurate reports (Livingstone & Robson 2000). Collecting information using technology may capture the interest of the adolescent age group.

Elderly. Dietary assessment in healthy older people without significant cognitive decline can be achieved, but methods that require recalling past diet, for example, 24-hour recall and FFQ methods are inappropriate if memory is impaired and sensory difficulties such as loss of hearing and vision and physical difficulties including being chair- or bed-ridden are present. In some cases the subject may have little involvement in food acquisition or preparation, thus limiting the subjects' ability to accurately name or describe the foods consumed (Adamson et al. 2009). Additionally, special diets are common in this group, for example, high fibre, soft foods due to difficulties in chewing, and methods may require adaptation as a result. Development and testing of methods are particularly important in this age group to ensure valid results (McNeill et al. 2009).

Pregnant and lactating women. Dietary assessment during pregnancy and lactation is extremely complex as a result of changes in energy and nutrient needs, appetite, and meal patterns. Social desirability bias may affect reporting in this group especially with regard to alcohol consumption, foods that should be avoided in pregnancy, and supplement intake. Specific information is needed on the stage of pregnancy in order to apply the appropriate dietary recommendations to investigate adequacy of the diet. The availability of validated dietary assessment methods in pregnant women is limited.

Ethnic populations. If the population to be assessed is composed primarily of, or contains, a significant proportion of minority ethnic groups for whom foods consumed and eating and cooking practices are not conventional, modifications to dietary assessment measures are required. Failure to appropriately adapt a dietary assessment method or tool will result in these groups being excluded from the assessment, or an incorrect capture of intakes in these groups. Assessing diets in these groups is difficult because any tool must capture the complexity of the diet, which may be a combination of ethnic foods and those commonly consumed by the autochthonous (native) population. Inclusion of ethnic foods and recipes in methods such as FFQs, or food lists used in technology-based 24-hour recalls, and inclusion of relevant and relatable portion measures should be considered. Quantification of specific portion sizes of traditional foods and dishes may be required, for example, when using food photographs it may be necessary to modify the range of portion sizes provided to incorporate smaller or larger portions. Food photographs should also consider culturally appropriate serving dishes and household utensils, with relevant fiscal markers to indicate a relatable portion reference. Food names should also be extended to include culturally specific names where possible. Interviewers should familiarize themselves with foods, drinks, recipes, differences in food names and portion sizes, along with methods of serving and eating/drinking in order to facilitate the collection of accurate dietary data. Cultural barriers may also need to be overcome in terms of recruitment, study participation, and method administration, along with population literacy and language levels. Food composition tables may also require updating to represent ethnic foods and recipes. Dietary acculturation is also an important factor to consider in survey design which has been linked to both positive and negative dietary choices (Ngo et al. 2009; Thompson & Subar 2013).

Low-income populations. Much of the evidence to date on dietary intake of low-income populations in high-income countries such as the UK is based on national surveys in which the number of low-income households is few, classification of households is based either on receipt of benefit or employment status, and cooperation rates in low-income households are low (Holmes et al. 2008). Few examples exist where low-income populations are studied in specific national surveys, and where they do exist (for example the UK low income diet and nutrition survey carried out in 2003–2005) they are not undertaken on a frequent basis or rolling programme. Studies undertaken in low-income groups often encompass several specific subgroups, including minority ethnic groups, the elderly, and families with large numbers of siblings. As a result, dietary assessment in such a diverse group may be problematic. Lower literacy, numeracy, and language skills exist alongside physical problems of record-keeping amongst the elderly and disabled. Drug and alcohol abuse create problems that impact not only on the quality of the data but also the safety and welfare of interviewers. Additionally, a higher likelihood of domestic chaos and stress factors arise among low-income households which mediate against accurate record-keeping and the ability to undertake dietary assessment (Holmes et al. 2008). Despite the importance of dietary data for a broad range of policy and research applications, there are also relatively few examples of nationally representative individual-level dietary surveys in low-income countries. Constraints to conducting large-scale dietary assessments include significant costs, time burden, technical complexity, and limited investment in dietary research infrastructure, including the necessary tools and databases required to collect individual-level dietary data in large surveys. To overcome these challenges the International Dietary Data Expansion (INDDEX) Project has developed INDDEX24, a digitalized mobile application integrated with a web database application, based on the 24-hour recall method, that aims to facilitate dietary data collection and processing, and decrease costs and effort related to implementation of food consumption surveys (Coates et al. 2017; Bell et al. 2020). Where dietary data does exist in low-income countries, whether from large or small sample dietary surveys, they are often not broadly accessible for use by researchers, policy makers, and other stakeholders. In response, FAO and the World Health Organization (WHO) jointly developed the FAO/WHO Global Individual Food consumption data Tool (FAO/WHO GIFT, available at: https://www.fao.org/gift-individual-food-consumption/en/), an open-access online data repository aimed at providing access to harmonized individual-level quantitative food consumption data, collected using 24-hour recalls or food records, especially from low- and middle-income countries (Leclercq et al. 2019). It is hoped that such initiatives will facilitate the collection and use of dietary data in low-income populations in the future.

31.6 Appropriate uses and validity of dietary survey methods

Appropriate uses of dietary survey methods

Appropriate uses of dietary assessment methods are summarized in Table 31.2. It is clear that certain techniques are limited to particular applications. For example, food balance sheets are appropriate for assessing both supply and availability at the country level, facilitating comparisons between countries and mapping trends in consumption within a country over time. Weighed records can be used to assess diet at the individual level and to build up pictures of regional or national consumption based on representative samples, but they may be too labour intensive for looking at distribution of food in families and cannot be used for metabolic studies. The footnotes provided in Table 31.2 clarify the specific problems, limitations, or advantages of using a particular technique in a particular setting.

The wide variety of techniques for assessing individuals' diets reflects both the difficulties and frustrations associated with attempts to measure diet without bias, and the importance attached to the need to obtain accurate measurements in many different circumstances. Improvements in dietary assessment techniques have been hampered by the lack of a readily measured absolute standard or reference of intake against which to assess the precision of other measures. The need to measure diet requires us to appreciate the many limitations, and to continue to develop and improve methods of dietary assessment pertinent to the population under investigation.

Validity and measurement error

Despite the fact that techniques of dietary assessment have changed little since the 1970s, there has been a growing awareness of the sources of bias in measurements of diet. Not only is there now an acceptance of the fact that all dietary assessment methods are influenced by the reporting process itself and no method is able to measure food intake without error, but also a recognition of the need to separate subjects with valid records from those who are believed to have inaccurate records. Dieticians, clinicians, epidemiologists, and researchers no longer accept that

TABLE 31.2 Appropriate uses of dietary survey methods (+++ very suitable, ++ moderately suitable, + limited application, −not suitable)

Level of dietary measurements	Dietary survey method							
	Food balance sheets	Household surveys	Surveys of individuals					
			Prospective			Retrospective		
			Weighed diary	Unweighted diarty	Duplicate diet[1]	24-hour recall	FFQ	Diet history
National	+++	+++	+[2]	+[2]	+[2]	++[2]	+[2]	+[2]
Regional	−	+++	+[2]	+[2]	+[2]	++[2]	+[2]	+[2]
Institution/group	−	+++	++[2]	++[2]	+[2]	++[2]	++[2]	++[2]
Household	−	++[3]	++[4]	++[4]	++[4]	++[4]	++[4]	++[4]
Individual	−	+[5]	+++[6]	+++[6]	+++[6]	+++[6]	+++[6]	+++[6]
Type of study								
Epidemiological	+++[7]	+++[7]	+++[8]	+++[8]	+[8]	+++[8]	+++[8]	++[8]
Clinical	−	−	+++	+++	++	+++[9]	+++	+++
Metabolic	−	−	+[10]	−	+++	−	−	−

[1] Includes other techniques of direct analysis (see text).

[2] Requires sample representative of population, institution, or group, or analysis weighted to reflect balance of subgroups.

[3] Requires store/larder inventory. Short-term measures (e.g., one week) may not reflect usual diet in individual households.

[4] The need for data from all household members may distort usual household food consumption patterns.

[5] Requires complex mathematical modelling of within-household food and nutrient distribution.

[6] Important to screen out individuals whose responses may not be valid (see text).

[7] Appropriate for ecological studies.

[8] See Margetts & Nelson (1997) for detailed discussion of use of dietary survey methods in epidemiological studies.

[9] Requires repeat 24-hour recalls for valid classification of subjects according to levels of intake (see text).

[10] Useful only if range of foods is of limited variation in composition, allowing reliable use of food composition tables.

their particular measure of diet is 'good enough'. Instead, they seek to determine the biases in their measurements in order to adjust for them when assessing the relationships between diet and health. Independent tests of validity are necessary to understand the relationship between what the method actually assesses and what it intends to measure. The potential impact of errors on findings can then be considered. The validation of dietary assessment methods usually involves a comparison of one method (test method) with another, which is considered more accurate (reference method). The main problems arising from inaccurate dietary measurements are:

- incorrect *positioning* of a country, household, or person in relation to the truth or some external reference measure (e.g., dietary reference values)
- incorrect *ranking* of countries, households or persons in relation to one another.

The first type of error can result in inappropriate investigations or actions being taken to remedy an apparent dietary deficit or excess that does not really exist. Alternatively, no action may be taken when some is needed (e.g., a true deficit is not detected because diet is overestimated). The second type of error tends to undermine the ability to assess relationships between diet and health (e.g., someone who properly belongs in the top quarter of the distribution of intake is classified in the bottom quarter, or vice versa). Again, this can lead either to inappropriate recommendations for improving health in the population or, more often, a failure to take action because the true relationship between diet and health is obscured by measurement error. Differences in modelled estimates of global dietary intakes were recently highlighted when data from two global initiatives were compared. Considerable differences for many countries were observed, far outweighing any reasonable differences that could be attributed to slightly differing years and age groups between the two estimates. Differences in estimates could suggest vastly different conclusions about diet quality, nutrient adequacy, and risk of non-communicable diseases, thus demonstrating that evidence on dietary intake is currently insufficient to produce robust, reliable, and replicable estimates of intake across countries globally (Beal et al. 2021).

Weblink Section 31.5 discusses measurement error and their effect and lists some of the sources of measurement error, their principal effects, some ways of taking errors into account in analysis, and ideas for dealing with them in practice. A summary is provided in Weblink Table 31.4.

KEY POINTS

- There is a wide variety of methods available for assessing diet. The method chosen should be validated and appropriate for the purpose, the population, and circumstances of the work being carried out.
- All methods for dietary assessment include errors. It is important:
  - not to take the measurements of food consumption and nutrient intake at face value;
  - to understand the likely sources of error; and
  - to appreciate how the errors may influence the interpretation of apparent associations (or apparent lack of associations) between diet and health.
- When reporting findings on dietary assessment, it is important to provide a full description of the sample characteristics (e.g., age, height, weight, body mass index, social class), number of subjects or households, dietary assessment method or methods, method or methods used for portion size assessment, number of days of measurement collected, mode of administration, food composition tables used, dietary data collection tool (software) for data capture, and, if relevant, dietary analysis software used. Detailed information should also be provided on method validation, and statistical approaches used in the analysis of the data.
- The use of technology is a key lever to reduce the burden associated with recording diet although further research on the validation of new approaches is still needed.
- Dietary assessment is a continually evolving science. Dietary assessment toolkits can facilitate the selection of an appropriate method for a particular setting, while recent initiatives in data collection and data dissemination demonstrate the increased interest and need for dietary data at the global level.

ACKNOWLEDGEMENTS

This chapter was originally authored by Dr Michael Nelson, Emeritus Reader in Public Health Nutrition, King's College London, and Director of Public Health Nutrition Research. It was updated in 2010, 2016, and 2023 by the current author.

Disclaimer: The views expressed in this publication are those of the author(s) and do not necessarily reflect the views or policies of the Food and Agriculture Organization of the United Nations.

 Be sure to test your understanding of this chapter by attempting multiple choice questions. See the Further Reading list for additional material relevant to this chapter.

REFERENCES

Adamson A J, Collerton J, Davies K et al. (2009) Nutrition in advanced age: Dietary assessment in the Newcastle 85+ study. *Eur J Clin Nutr* **63**, S6–S18.

Beal T, Herforth A, Sundberg S et al. (2021) Differences in modelled estimates of global dietary intake. *Lancet* **397**(10286), 1708–1709.

Bell W, Coates J, Some J et al. (2020) Validation and user experience study of INDDEX24, a novel global dietary assessment platform in Burkina Faso and Viet Nam. *Curr Dev Nutr* **4**(Suppl 2), 1161.

Biró G, Hulshof K F A M, Ovesen L et al. (2002) Selection of methodology to assess food intake. *Eur J Clin Nutr* **56**(Suppl 2), S25–S32.

Cameron M E & van Staveren W A (eds) (1988) *Manual on Methodology for Food Consumption Studies.* Oxford University Press, Oxford.

Coates J C, Colaiezzi B A, Bell W et al. (2017) Overcoming dietary assessment challenges in low-income countries: Technological solutions proposed by the International Dietary Data Expansion (INDDEX) Project. *Nutrients* **9**(3), 289.

Coates J, Colaiezzi B, Fiedler J L et al. (2012) A program needs-driven approach to selecting dietary assessment methods for decision-making in food fortification programs. *Food Nutr Bull* **33**(3), S146–S156.

Dao M C, Subar A, Warthon-Medina M et al. (2019) Dietary assessment toolkits: An overview. *Pub Hlth Nutr* **22**(3), 404–418.

Del Gobbo L C, Khatibzadeh S, Imamura F et al. (2015) Assessing global dietary habits: A comparison of national estimates from the FAO and the Global Dietary Database. *Am J Clin Nutr* **101**, 1038–1046.

Eldridge A L, Piernas C, Illner A K et al. (2019) Evaluation of new technology-based tools for dietary intake assessment: An ILSI Europe Dietary Intake and Exposure Task Force Evaluation. *Nutrients* **11**(1), 55.

Fiedler J L, Lividini K, Bermudez O I et al. (2012) Household consumption and expenditures surveys (HCES): A primer for food and nutrition analysts in low- and middle-income countries. *Food Nutr Bull* **33**(3), S170–S184.

Fiedler J L, Martin-Prével Y, Moursi M (2013) Relative costs of 24-hour recall and household consumption and expenditures surveys for nutrition analysis. *Food Nutr Bull* **34**(3), 318–330.

FAO & The World Bank (2018) *Food Data Collection in Household Consumption and Expenditure Surveys. Guidelines for Low- and Middle-Income Countries.* FAO, Rome.

Holmes B, Dick K, Nelson M (2008) A comparison of four dietary assessment methods in materially deprived households in England. *Public Health Nutrition.* **11**, 444–456.

Lagiou P, Trichopoulou A, DAFNE contributors (2001) DAta Food NEtworking: The DAFNE initiative: The methodology for assessing dietary patterns across Europe using household budget survey data. *Pub Hlth Nutr* **4**(5B), 1135–1141.

Leclercq C, Allemand P, Balcerzak A et al. (2019) FAO/WHO GIFT (Global Individual Food consumption data Tool): A global repository for harmonised individual quantitative food consumption studies. *Proc Nutr Soc* **78**(4), 484–495.

Livingstone M B E & Robson P (2000) Measurement of dietary intake in children. *Proc Nutr Soc* **59**, 279–293.

Mark H E, Houghton L A, Gibson R S et al. (2016) Estimating dietary micronutrient supply and the prevalence of inadequate intakes from national food balance sheets in the South Asia region. *Asia Pacific J Clin Nutr* **25**(2), 368–76.

Martin C K, Nicklas T, Gunturk B et al. (2014) Measuring food intake with digital photography. *J Hum Nutr Diet* **27**, S1:72–81.

McNeill G, Winter J, Jia X (2009). Diet and cognitive function in later life: A challenge for nutrition epidemiology. *Eur J Clin Nutr* **63**, S33–S37.

Micha R, Coates J, Leclercq C et al. (2018) Global dietary surveillance: Data gaps and challenges. *Food Nutr Bull* **39**(2), 175–205.

Morán Fagúndez L J, Rivera Torres A, González Sánchez M E et al. (2015) Diet history: Method and applications. *Nutr Hosp.* **31**(Suppl 3), 57–61.

Ngo J, Gurinovic M, Frost-Andersen L et al. (2009) How dietary intake methodology is adapted for use in European immigrant population groups: A review. *Br J Nutr* **101**, S86–S94.

Thompson F E & Subar A F (2013) Dietary assessment methodology. In: *Nutrition in the Prevention and Treatment of Disease,* edn,5–46 (Coulston A M, Boushey C J, Ferruzzi MG eds). Elsevier, Oxford.

Wessells K R, Singh G M, Brown K H (2012) Estimating the global prevalence of inadequate zinc intake from national food balance sheets: Effects of methodological assumptions. *PLoS One.* **11**, e50565. doi: 10.1371/journal.pone.005 0565.

Wuehler S E, Peerson J M, Brown K H (2005) Use of national food balance data to estimate the adequacy of zinc in national food supplies: methodology and regional estimates. *Publ Hlth Nutr.* **7**, 812–819.

FURTHER READING

Charrondiere U R, Stadlmayr B, Grande F et al. (2023) FAO/INFOODS Evaluation framework to assess the quality of published food composition tables and databases - User guide. FAO, Rome. Available at: https://doi.org/10.4060/cc5371en

de Quadros V P, Balcerzak A, Allemand P et al. (2022) Global trends in the availability of dietary data in low and middle-income countries. *Nutrients* **14**(14), 2987.

European Food Safety Authority (2009) General principles for the collection of national food consumption data in the view of a pan-European dietary survey. *EFSA Journal* **7**(12), 1435. [51 pp.]. doi:10.2903/j.efsa.2009.1435.

FAO (2018) *Dietary Assessment: A Resource Guide to Method Selection and Application in Low Resource Settings.* FAO, Rome. Available at: http://www.fao.org/3/i9940en/I9940EN.pdf

FAO (2001) *Food Balance Sheets: A Handbook.* FAO, Rome. Reprinted 2008. Available at: http://www.fao.org/3/x9892e/x9892e00.htm

FAO & Intake (2022) *Global Report on the State of Dietary Data.* FAO, Rome. Available at: https://doi.org/10.4060/cb8679en

Hedrick V E, Dietrich A M, Estabrooks P A, Savla J et al. (2012) Dietary biomarkers: Advances, limitations and future directions. *Nutr J* **11**, 109.

Margetts B M & Nelson M (1997) *Design Concepts in Nutritional Epidemiology,* 2nd edn. Oxford University Press, Oxford.

National Institutes of Health & Westat (2007) *National Children's Study Dietary Assessment Literature Review. Chapter 2: Pregnancy and Lactation.* National Institutes of Health, Applied Research Program and Westat, Rockville, MD. Available at: https://epi.grants.cancer.gov/past-initiatives/assess_wc/review/pdf/ncs_chapter2.pdf

32 Biochemical assessment

Hilary J. Powers

OBJECTIVES

By the end of this chapter you should be able:

- To introduce the application of biochemical assessment methods.
- To summarize the available choices of tissue and body fluid samples that may be collected for biochemical status assessment.
- To summarize current methods in use for biochemical assessment.
- To summarize the necessary precautions to be taken during sample selection, storage, and analysis, and to provide an outline of essential fieldwork and laboratory methodologies.
- To highlight some common problems of interpretation, and inter-relationships between the biochemical and other indices of nutritional status, for a composite and integrated picture of status and nutritional adequacy.

and of possible interfering and confounding factors. Therefore, it is essential to have access to suitable analytical equipment, suitable laboratory facilities, and relevant expertise, for sample collection, storage, sample analysis, and interpretation.

A biochemical marker may be selected for the purpose of estimating nutrient intake or assessing nutrient status more directly. In either case it may be used to make statements about nutrient adequacy in an individual or a population. The dietary intake of some nutrients is especially difficult to assess; in such cases a biochemical marker may be used to assess nutrient intake. Not every biochemical marker of nutrient status can provide useful information about intake. This will depend on factors intrinsic to the nutrient or the individual or group, and on the circumstances relating to details of sample collection and storage. Some of these factors will be considered later in the chapter and the potential value of biochemical markers of status as indicators of adequacy of intake will be discussed.

32.1 Introduction

Biochemical assessment ideally forms part of a coordinated set of nutritional investigations that may also include dietary assessment, anthropometry, physiological tests, and clinical investigations. A combination of these various modes of assessment can provide a detailed profile of an individual or a population group for use in nutrition-related health risk assessment and risk management.

The design of the biochemical aspect of a nutritional assessment depends on there being an available and suitable sample of body fluid or tissue for analysis, plus operationally feasible biochemical tests that are capable of measuring reliably the concentrations or functional adequacy of the nutrients of interest. The results must then be interpreted in the light of established normal ranges,

32.2 Biochemical markers of nutritional Intake and status

Biochemical markers have the following attributes:

- They can be considered as offering a more objective approach than estimates of dietary intake as they do not rely on the availability and reliability of information provided by individuals under study.
- Depending on the human material being used, and the particular biochemical marker, they can provide information about recent intake, intake over the previous few hours, or of much longer-term intakes, ranging from days to many months.
- They can often be measured with high specificity and accuracy.

- Some biochemical markers represent useful links between nutrient intake and risk of specific disease, and thereby constitute what are sometimes called 'intermediate disease endpoints'.

- They can be particularly useful when considering nutrient adequacy or as thresholds to aim for in population interventions.

- Many biochemical markers are nutrient-specific, and can be rapidly and predictably responsive to the correction of single nutrient deficiencies. Some, however, such as blood haemoglobin concentration, may reflect adequacy or deficiency of multiple nutrients, and should be interpreted accordingly.

- In a few cases, and particularly where intakes of the nutrients of interest cannot reliably be predicted from food tables, dietary nutrient intakes may be predicted more accurately from biochemical indices than from diet assessment. This is notably the case for urinary sodium, potassium, iodide, and fluoride, which are commonly used as markers for intakes of these elements, and urinary nitrogen, which is used as a marker for protein intakes.

Biomarkers of nutrient intake and status

It is useful to differentiate measurements of the concentration of a nutrient in serum, red blood cells, or urine from measurements which provide some insight into the functional impact of low or high intakes of a nutrient; the latter may be considered to be functional assays. For example, an increased concentration of methylmalonic acid in serum reflects a B_{12}-sensitive alteration in fatty acid metabolism; an elevated plasma homocysteine concentration reflects impaired remethylation in the methyl cycle and is responsive to intake of folate, vitamins B_2, B_6, and B_{12}. In some instances the biomarker may reflect both nutrient intake and altered risk of disease and may have value as an intermediate disease endpoint marker.

Protein and essential amino acids

Despite decades of interest in indicators of protein intake and status, and some early optimism that novel biomarkers would be identified, there are still no good validated biomarkers. The most commonly used index is serum or plasma albumin. This is lowered in conditions such as kwashiorkor, attributable to low protein intakes or poor protein quality. However, serum albumin may also be lowered by the acute phase reaction, and indeed, severely malnourished children commonly have infections that affect their acute phase status.

Serum albumin is traditionally measured by a dye-based assay, or by immunoassay. An alternative,

potentially more reliable index of inadequate protein supply is the plasma amino acid profile, since the essential amino acids, notably the branched-chain amino acids, are lowered when dietary protein is inadequate. Protein intake can be measured by 24-hour urinary nitrogen excretion, corrected for completeness of collection using PABA (p-amino benzoic acid). However, there are errors associated with other sources of urinary nitrogen, and the influence of different types of dietary protein (Bingham 2002).

Essential fatty acids

In the absence of a validated biomarker for total fat intake focus has rested instead on biomarkers of fatty acid intakes. By definition, essential fatty acids (EFAs) cannot be synthesized *de novo*, therefore there has been interest in profiling these fatty acids in serum or plasma as an indication of dietary intake (Arab 2003). However, interpretation is complicated by variability of lipoprotein profiles and by diurnal variation. Red blood cell membrane fatty acid profiles may also be used to determine EFA intake, and this approach has had considerable value in determining compliance in supplementation studies (Gadaria-Rathod et al. 2013). The fatty acid profile of adipose tissue biopsies has also proved informative about dietary intake (Hedrick et al. 2012).

Fat soluble vitamins

- Vitamins A and E are commonly measured in serum or plasma, together with carotenoids, by high performance liquid chromatography.

- Plasma retinol concentration falls when infection is present, also in protein energy malnutrition and zinc deficiency, so a low concentration must be interpreted with caution. A high level may be indicative either of adequate or high status.

- The major component of vitamin E in the plasma is alpha-tocopherol, and its concentration is considered to reflect vitamin E status. Plasma alpha-tocopherol concentration is usually expressed relative to plasma cholesterol or total lipids, since the vitamin E content of plasma is highly dependent on its lipid content. The concentration of a minor component of vitamin E in human tissue, gamma tocopherol, can also be measured in plasma but it is metabolized differently from alpha tocopherol and is not a reliable indicator of vitamin E intake.

- Vitamin D status is usually assessed by the concentration of 25-hydroxyvitamin D (25(OH)D) in serum or plasma (Prentice at al. 2008). The concentration reflects the combined exposure to dietary sources and cutaneous synthesis on exposure to sunlight. This compound is more useful than the active metabolite 1,25(OH)$_2$D as it has a longer half life; it is also present at a concentration about 1000 times higher than 1,25(OH)$_2$D. 25(OH)D is, however, a negative acute phase reactant and low plasma concentration may reflect an inflammatory state. Plasma or serum total 25(OH)D includes 25(OH)D$_2$ (from food and diet supplements) as well as 25(OH)D$_3$ (from the action of sunlight on the skin). Antibody-based methods provide information about total plasma or serum 25(OH)D whilst liquid chromatography methods have the additional benefit of being able to separate and quantify the two forms. Variation in methodologies has led to some difficulties of comparison between studies, and there is a move to international standardization of the measurement of 25(OH)D.

- Vitamin K status is assessed by measuring the rate of blood clotting, or more sensitively and specifically by PIVKA (protein induced by vitamin K absence or antagonism). The concentration of vitamin K in serum or plasma, or the degree of under-carboxylation of osteocalcin, a bone-related peptide, in plasma, may also be used to assess status. However, none of these approaches is wholly satisfactory (see Table 32.1).

Water soluble vitamins

Vitamin C status is usually measured by serum or plasma vitamin C concentrations, by HPLC (high performance liquid chromatography), and provides a reasonably good estimate of vitamin C intakes over the previous 1–2 weeks. The determination of vitamin C in lymphocytes, prepared from a 'buffy coat' layer from whole blood, also using HPLC, offers a method for determining longer term intakes and tissue concentration (Mitmesser et al. 2016).

- Of the B-vitamins, thiamin (B$_1$), riboflavin (B$_2$), and pyridoxine (B$_6$) status are commonly assessed by measuring the activation coefficients (that is, the ratios of *in vitro* enzyme activity with added cofactor, to that without the added cofactor), of specific erythrocyte enzymes that require one of these vitamins as part of their essential cofactors. These comprise: transketolase (requiring thiamin diphosphate) for thiamin status; glutathione reductase (requiring flavin adenine dinucleotide for riboflavin status), and certain erythrocyte amino acid aminotransferases (requiring pyridoxal phosphate) for vitamin B$_6$ status. These are considered functional assays. Direct measurement of the vitamin cofactors in plasma or red blood cells may

TABLE 32.1 Biochemical status indices for fat soluble vitamins and fatty acids

Vitamin/fatty acids	Most often used index	Other indices
Vitamin A and pro-vitamin A carotenoids	Plasma retinol and/or carotenoids *Plasma retinol is reliable when body stores are low, but less so when stores are adequate. Carotenoids are reliable over a wide range, but their absorption efficiency varies between individuals.*	Relative dose response (RDR) or modified RDR (MRDR) tests. Conjunctival impression cytology (with or without transfer) tests[a].
Vitamin D	Plasma 25(OH)D *Generally reliable, but one needs to be aware that this vitamin is generated from sunlight as well as occurring in the diet.*	Parathyroid hormone (PTH)[1] The ratio of 25(OH)D to 1,25(OH)$_2$D is being explored as an alternative to 25(OH)D alone.
Vitamin E	Plasma vitamin E:cholesterol ratio *Moderately reliable.*	Erythrocyte membrane vitamin E. Protection from hydrogen peroxide catalysed haemolysis.[1]
Vitamin K	Plasma vitamin K (phylloquinone) Under-carboxylation of osteocalcin, or PIVKA (protein induced by vitamin K absence or antagonism) *Fairly reliable.*	Potentially, the carboxylation of matrix gla protein.
Fatty acid profile	Plasma fatty acid profile *Reliable for recent intake.*	Erythrocyte membrane or fat biopsy fatty acid profile *Longer-term tissue-based indices.*

[a] Functional assays like these are often not entirely specific for just one nutrient but may be more relevant for overall health 'risk' as affected by several nutrients.

also provide information about dietary intake and tissue concentrations, but, for thiamin and riboflavin, dose-response data are lacking as are criteria for adequacy. For vitamin B_6, the direct measurement of plasma (or occasionally red cell) pyridoxal phosphate concentration is often the preferred index nowadays (Ueland et al. 2005; Taylor et al. 2020).

- Folate, vitamin B_{12}, biotin, and pantothenic acid status are commonly assessed by serum or plasma concentrations. Red cell folate is considered a better index of long-term folate intakes and body-stores than plasma or serum folate, but is subject to problems of poor inter-assay harmonization, in addition to being less straightforward, and is less routinely used. The differential measurement of a range of folate species in plasma or serum has, in recent years, been enhanced by the development of new assay methodologies based on liquid chromatography-tandem mass spectrometry.

- Vitamin B_{12} status is generally measured as the plasma or serum concentration, using an automated protein-binding immunoassay, but the predictive value for vitamin B_{12} deficiency is poor. The concentration of plasma or serum holotranscobalamin holds promise as a marker of vitamin B_{12} available to cells, and automated immunassays are available for this measurement (Valente et al. 2011). The concentration of methylmalonic acid (MMA) in plasma serum or urine increases if the availability of vitamin B_{12} for fatty acid metabolism is impaired and this measurement has value as a functional biomarker. There continues to be debate over appropriate cut-off points for each of these biomarkers of vitamin B_{12} status (Hill et al. 2013).

- Niacin status is usually assessed by the concentration of the primary urinary metabolite, N-methyl nicotinamide (NMN), or certain niacin-derived pyridones. Ideally, a 24-hour urine collection is used. Red blood cell niacin concentration is sensitive to niacin intake but criteria for adequacy are lacking (see Table 32.2).

Non-vitamin dietary organics

Compounds such as polyphenols, phytoestrogens, pterins, carnitine, choline, and flavonoids fall into the category of non-vitamin organics. Their physiological significance for human nutrition remains controversial and incompletely researched, and well-validated biomarkers of status are at an early stage of development. Carotenoids in serum or plasma can, however, readily be measured by high performance liquid chromatography in the same assay run for vitamins A and E. Alpha and beta-carotenes and beta-cryptoxanthin can all be converted in the body to vitamin A; in addition, all carotenoids are considered to have antioxidant properties.

Mineral nutrients

Macro-essential elements include sodium, potassium, calcium, phosphorus, magnesium, and chlorine (Na, K, Ca, P, Mg, and Cl). Some of these, (notably Na, K) can be monitored by 24-hour urine collections (Bingham 2002), or, less accurately, by spot urine samples, correcting for urine dilution using creatinine as the denominator. Others (e.g., Mg) are best studied in blood serum or plasma. Calcium and phosphorus status may be measured using health outcome indicators including bone-related enzyme activity and markers of bone turnover. Elements required in smaller amounts in the diet include iron, zinc, copper, selenium, chromium, manganese, and iodine (Fe, Zn, Cu, Se, Cr, Mn, and I).

- Iron status should be assessed using more than one marker, of which there are a number, and which can be categorized according to what aspect of iron adequacy they reflect. The most commonly used measure of iron status is haemoglobin (Hb) concentration, but this can be influenced by other factors including dietary intake of folate and vitamin B_2. Information about the size of red blood cells (mean cell volume, MCV) and their Hb content (mean cell Hb, MCH) can help determine whether iron deficiency is the cause of low Hb concentration. The serum concentration of iron and the transferrin receptor, serum transferrin saturation, and red cell zinc protopophyrin, are all used to assess availability of iron for tissues. Storage iron in tissues is usually assessed using serum ferritin concentration, which reflects hepatic ferritin iron. However serum ferritin is an acute phase protein and has limited value as a biomarker of iron status in the presence of infection. Serum hepcidin is showing some promise as a marker of iron status, being increased when iron stores are high and falling in response to iron deficiency.

- Zinc status can be assessed by plasma zinc levels (Gibson et al. 2008) but this index is strongly negatively affected by acute phase effects and by any insult that leads to a reduction in plasma protein concentration, notably albumin.

- Copper status can, albeit with caveats, be assessed by levels of plasma copper or of caeruloplasmin (the principal copper-carrying protein in plasma) or by the activity of the copper–zinc erythrocyte enzyme, superoxide dismutase; however there are no really robust status indices for this element.

- Selenium status can be assessed by serum, plasma, or red cell selenium levels, or by plasma selenoprotein P, or by one of several selenium-containing glutathione peroxidase enzymes in blood, usually assayed in whole blood or in washed red cells.

TABLE 32.2 Biochemical status indices for water soluble vitamins

Vitamin	Most often used index	Other indices
Vitamin B$_1$ (thiamin)	Erythrocyte transketolase AC[1] *Fairly reliable but not easy to measure*	Erythrocyte[2] or urinary[2] thiamin concentrations
Vitamin B$_2$ (riboflavin)	Erythrocyte glutathione reductase AC[1] *Reliable in principle; some doubt about suitable cut-off values for normality*	Erythrocyte[2] or urinary[2] riboflavin concentration
Vitamin B$_6$ (pyridoxine)	Serum[3] pyridoxal phosphate *Fairly reliable, but many different indices have been used for this vitamin*	Serum[3] vitamin B$_6$ profile[2] Erythrocyte aspartate or glutamate aminotransferase AC[2] (Urinary pyridoxic acid is an index of turnover)
Folate	Serum[3] folate	Erythrocyte folate[2] Serum or urine p-amino- or acetamido-benzoyl glutamate Plasma total homocysteine[4] Deoxyuridine suppression test[4] PMNL[5] nuclear lobe count[4]
Vitamin B$_{12}$	Serum[3] vitamin B$_{12}$; serum holotranscobalamin (a relatively new index) *Both are fairly reliable; the latter is more difficult to measure but is claimed by some to be more specific*	Serum[3] or urine methylmalonic acid Plasma homocysteine Transcobalamin II[2]
Niacin	Urinary N-methyl nicotinamide:2-pyridone ratio *Acceptable but not often measured*	Erythrocyte pyridine nucleotide concentration[2]
Biotin	Serum[3] biotin *Acceptable but not often measured*	Lymphocyte propionyl CoA carboxylase[2]
Pantothenate	Serum[3] pantothenate *Acceptable but not often measured*	
Vitamin C (ascorbate)	Serum[3] ascorbate or total vitamin C *Reasonably good*	Buffy coat vitamin C[2]

[1] AC, activation coefficient, which is the ratio of enzyme activity with added (vitamin-derived) cofactor to that without the added cofactor. The analogous tests of thiamin and vitamin B$_6$ status are based upon the same principle.

[2] Preferred by some laboratories, but reliability has been questioned.

[3] Serum is generally interchangeable with plasma for most of these assays, but haemolysis can catalyse degradation of oxidizable vitamins such as vitamins A and C and carotenoids.

[4] Functional assays that are not nutrient specific but may be relevant for overall health 'risk'.

[5] Polymorphonuclear leucocyte.

- Chromium and manganese have been measured in various blood fractions, but the interpretation of the results is difficult.
- Determination of iodine status generally relies on the measurement of urine iodine concentration (UIC), which is a good reflection of recent iodine intake. WHO advocate using a median urinary iodine concentration (mUIC) <100ug/L in children and adults and <150ug/L in pregnant women to indicate group or population deficiency. The use of UIC in individuals is less satisfactory mainly because of wide day-to-day variation and the need for multiple measurements over time (Bath & Rayman 2015). Serum thyroid stimulating hormone (TSH) and serum concentration of the thyroid hormones T3 and T4 are sometimes used as indicators of iodine status but they are relatively insensitive markers. Serum thyroglobulin concentration is proving useful as a sensitive marker of population iodine status (Pearce & Caldwell 2016).
- Elements with uncertain physiological roles, such as boron, silicon, vanadium, and lithium (B, Si, V, Li) have been studied to a minor degree in blood fractions, but little definitive status information is available (see Table 32.3).

TABLE 32.3 Biochemical status indices for minerals and trace elements

Mineral	Most frequently used index	Other indices
Sodium, potassium	Urinary excretion of Na, K *Both good measures of intake*	Plasma Na, K[1]
Calcium, phosphate	Plasma[2] ionized or albumin-adjusted calcium *The best index available for calcium, but not very reliable, except for extreme deficiency* *Plasma phosphate is a reasonable index of tissue phosphate status*	Bone-specific alkaline phosphatase[3]
Magnesium	Plasma Mg *Fair*	
Iron	Plasma Fe % saturation of transferrin Serum ferritin; serum soluble transferrin receptors; reticulocyte haemoglobin[5] *All are potentially useful indices, but iron status is complex, and difficult to measure unambiguously*	Erythrocyte concentrations of free protoporphyrin, blood haemoglobin, microcytosis. Hepcidin is currently being studied
Zinc	Plasma Zn *Fair—only really useful for detecting extreme deficiency*	Tissue zinc turnover rates Functional indices such as growth in young children
Copper	Plasma Cu or caeruloplasmin Superoxide dismutase-1 activity in erythrocytes *Fair, not very well-validated; the plasma indices are highly susceptible to acute phase effects*	
Selenium	Plasma Se *Reasonably good*	Erythrocyte Se Plasma or red cell glutathione peroxidase
Iodine	Urinary iodine (as a simple concentration or a ratio to creatinine) *Reasonable indicator of population intake* *Not very good for status of individuals, since it requires many repeat samples*	Goitre prevalence Serum or plasma levels of thyroid-stimulating hormone, thyroglobulin, or the tri-iodothyronine:thyroxine (T_3:T_4) ratio

[1] Hormonally controlled, but little affected by nutrition.

[2] Serum and plasma are often interchangeable, but some anticoagulants present in plasma samples may interfere with certain metal ion assays. Haemolysed samples must be avoided for most mineral assays.

[3] Non-specific functional marker. There are no widely used specific tests for Ca status.

Although it is difficult to measure fluoride status or nitrogen status by means of biochemical indices, fluoride *intake* can be monitored by its 24-hour urine excretion rates, and nitrogen (protein) *intake* can be approximately monitored by 24-hour nitrogen excretion rates.

Toxic elements

Some elements, including aluminium, mercury, lead, and cadmium (Al, Hg, Pb, Cd) have no known physiological functions or requirement by humans, but are of interest and concern because of their toxicity, if present as contaminants in food or from other environmental exposure. They can be measured in blood: Hg and Pb occur mainly in the erythrocytes; Al and Cd occur mainly in plasma.

The concepts 'optimum nutrition' and 'normal ranges' are discussed in Weblink Section 32.1.

32.3 Interpreting results of biomarker measurements

Inter- and intra-laboratory assay harmonization, and quality control of status assays

For some analytes, differences in the methods used across laboratories worldwide have posed significant problems for inter-study comparison and the definition of deficiency. Quality assurance schemes exist for some analytes, allowing laboratories to monitor performance and identify biases and variability. The UK external quality assurance scheme operates for clinical laboratories, as part of clinical governance. The Vitamin D external quality assurance scheme (DEQAS) is part of this scheme, and so is KEQAS (the vitamin K external quality assurance scheme). Some inter-laboratory 'round robin' sample-exchange programmes also exist, whose aim is to try to harmonize certain nutrient status assay methods and results, within and between different analytical laboratories (see also Weblink Section 32.2).

Validity, accuracy, and reproducibility

To be meaningful, biochemical assays must be valid, accurate, and reproducible. However, achieving this goal is challenging. First, the samples must be adequately collected, processed, and stored. For blood samples, the stability of the analyte and its distribution between the blood components must be considered. For example, in unseparated whole blood samples, homocysteine leaches rapidly out of the erythrocytes at room temperature, and the level in plasma then steadily rises, which can make the assay results impossible to interpret unless this artefact is effectively controlled. Vitamin C may deteriorate rapidly, especially in separated plasma or serum, or in red cell lysates, through oxidation. Red cell folate deteriorates if stored without a reducing agent (such as vitamin C). Carotenoids and some B-vitamins can be destroyed by light.

For trace elements, contamination is a major potential issue, since metal contaminants may be leached from commercially provided blood containers, stoppers, syringes, needles, or from skin surfaces. Urine samples may require an antimicrobial preservative if collected or stored at room temperature, for example, for 24 hours or longer, and they may need a marker for completeness of a 24-hour collection, such as urinary recovery from divided oral doses of para-amino-benzoic acid (PABA) (Bingham 2002). Hair and finger- or toe-nail samples must be thoroughly cleaned and must, if possible, represent a defined period of their growth. Saliva samples must be free of food contaminants. Blood samples that are used to provide plasma or white or red cells, must first be collected in an assay-compatible anticoagulant, then separated under controlled conditions of time and temperature. Separated samples must be stored frozen at minus 20°C or minus 80°C. Vitamin C needs the addition of an acidic, transition-metal-chelating preservative such as metaphosphoric acid, followed by storage at a very low temperature.

The chemical assay or measurement selected must be sufficiently specific (that is, capable of measuring just a single substance, or an appropriately interrelated group of substances), it must be accurate, precise that is, reproducible) and sensitive (that is, capable of measuring the very low concentrations that typically occur in body fluids). In more general terms, a wide variety of techniques often is available, and the challenge for the analyst is to optimize cost, accuracy, precision, and specificity. It may be desirable to measure, sequentially or simultaneously, more than one chemical form of a particular nutrient, if more than one form is present in the samples available, because each may be unique in its susceptibilities to interference, or in the nutritional information that it provides. Assays should be monitored by appropriate quality assurance procedures. These may include: (i) commercial control samples with assigned values (or acceptable ranges) of results for the analytes being measured; (ii) round robin inter-laboratory sample exchange or external quality assurance schemes, which provide consensus values from several laboratories; (iii) drift control samples monitoring inter-assay precision and enabling rejection of any assay runs that fall outside pre-determined performance limits. Drift is the tendency for an assay to alter its characteristics, such as its sensitivity, often imperceptibly but progressively, over a period of time, which can then give rise to inaccuracies or false conclusions. The necessary precautionary measures vary between analytes, and in the absence of external quality assurance schemes, it is important to try to cross-validate the preferred assays, for example, by comparing colorimetric or microbiological assays against 'gold-standard' high-performance liquid chromatography assays, and/or with mass-spectrometric analyses.

When reporting the results of an assay, it is very important to ensure that the units are appropriate, correct, and unambiguously stated. In the older literature, many concentration results were given in obscure, poorly defined, or non-standardized units. In the USA, analyte concentrations are frequently expressed as weight per volume, such as mg/litre, whereas in European countries molar or SI units (e.g., mmoles/litre) are generally preferred; this dichotomy can still sometimes cause confusion. The choice of a suitable comparison, that is, a 'normal' range of values is also essential. Different laboratories may use different assay procedures and even minor procedural differences can cause major discrepancies in the assay results. A 'normal' range may need to be laboratory-specific and frequently verified. Only where the assay has been well-harmonized between laboratories, can a broadly accepted population-comparison or normal range be used.

Another common pitfall is to assume that the only important influences on blood (especially plasma or serum) nutrient levels are the variations in nutrient intake and in tissue stores. A powerful influence is the common acute phase reaction, caused by infection and inflammation. The concentration of some nutrients in the circulation increases whilst others decrease during an acute phase reaction, however, they usually return to their original levels once the infection subsides. This is a very common cause of misinterpretation of biochemical status indices. Thus, a high serum ferritin concentration may imply iron sufficiency, but it may also arise through inflammation. The presence or absence of an acute phase reaction, and its severity, can be monitored by measuring an increase in acute phase protein concentrations in plasma; C-reactive protein (CRP) being the preferred index. Other acute phase proteins can provide more discriminatory information. Erythrocyte indices seem to be generally less vulnerable to this source of ambiguity than plasma or serum nutrient concentrations.

Temporal variations

Many blood and urine nutrients exhibit 24-hour and seasonal cycles of fluctuation, which may or may not reflect changes in nutrient intake. Thus, a sample taken at one time of day or during a particular season may yield a result that is quite different from that taken at a different time point. Intervention studies are especially prone to confounding by seasonal variations; hence the need for control groups that span exactly the same time interval. There may be secular changes over longer time periods, attributable to changing climatic conditions over several years, or changing lifestyles, food imports, and so on, which need to be monitored by periodic or regularly repeated surveys. The effects of public health interventions, such as the fortification of flour, or the removal of lead from petrol, may be overlaid by natural fluctuations or drift over time. Therefore, the reliable monitoring of intervention effectiveness must take such fluctuations and assay-drift-effects into account. This further underlines the need for robust quality control precautions and monitoring procedures.

Compartments and sample types

The fundamental aim of biochemical monitoring is to define whole-body adequacy for specific nutrients. However, the practicalities of assessment usually limit the researcher to readily accessible body fluids and surface tissues, the concentration in which may be modulated by blood homeostasis or selective and controlled nutrient or metabolite excretion. Certain mineral elements, notably calcium, copper, and zinc (Bates et al. 1997; Gibson et al. 2008) lack satisfactory biochemical indices, partly because the most vulnerable tissues and compartments cannot easily be monitored. For most vitamins, on the other hand, there are blood or urine indices that

adequately reflect body stores and/or intakes. For the alkali metals sodium and potassium and for iodine and fluorine, the amount excreted in the urine, preferably in a timed (e.g., 24-hour) urine sample is a fairly good measure of recent intake, but for many other nutrients, urine has no value as a source of status or intake information. Other potential sources of information for some nutrients include breast milk (for certain nutrients, in the special case of lactating women), or hair, nails, cheek cell scrapings, white blood cells, or saliva (see Weblink Section 32.3 for further information on compartments and samples).

32.4 Factors affecting the relationship between nutritional status indices and nutrient intakes

Many unexpected pitfalls have been encountered when attempting to predict nutrient intakes from biochemical status indices. They have included:

1. Intakes and status indices having different time-courses from each other, and also differing responses to the effects of various confounding factors. To monitor a relationship between a nutrient intake and the corresponding biochemical status index reliably, it is advisable to estimate the diet immediately before the collection of body fluids or tissue samples for status analysis. The optimum diet estimation period may be a few hours or several weeks or months, depending on the rate of turnover of the compartment and the analyte. In practice, the maximum period that is feasible for detailed diet-monitoring of individuals is about seven days, although for population studies or surveys, longer data-collection periods may be achievable by sequential or periodic monitoring of different groups of individuals.

2. Other confounding factors affecting the relationships between nutrient intake and status index values include:
   - Variable chemical forms and stabilities of nutrients in food, and of analytes in status samples.
   - Variable absorption efficiency and bioavailability of nutrients in foods and diets; also confounding effects of nutrient interactions. For instance, iron absorption is enhanced by vitamin C and by protein, but is negatively influenced by phytate, tannins, or high intakes of certain divalent metal ions such as zinc or copper, which compete with iron for intestinal absorption.

- Variable rates of nutrient turnover, for example, catabolic breakdown of many vitamins; variable nutrient distribution between body compartments; variable excretion rates and pathways. For instance, vitamin C and folate turnover rates are both increased in smokers; vitamin A excretion is increased during infections.

- Effect of homeostatic mechanisms, especially in the blood and tissues. For instance, plasma calcium is continuously maintained within very narrow limits in plasma by hormonally controlled mechanisms which extract it from, or deposit it in, the calcified tissues such as bone. As a result, plasma total calcium concentrations cannot effectively reflect overall body calcium status. Plasma retinol is likewise maintained within narrow limits in people who are adequately supplied with vitamin A, mainly by homeostatic control of plasma retinol-binding protein concentrations which in turn impose an upper limit on plasma retinol concentrations. Plasma sodium and potassium levels are homeostatically controlled by the adrenal mineralocorticoid hormones.

- Variable tissue growth rates at sites such as nails and hair, which affects the interpretation of hair or nail concentrations, for example, of trace elements.

- Problems of sample contamination, especially for trace element work.

- Effects of age, gender, ethnic group, genetic polymorphisms, and so on, on status index values and their normal ranges.

- Acute phase effects on biochemical status indices.

Markers of nutrient intake

See also Weblink Table 32.1.

For just a few nutrients there can be a relatively reliable prediction of intakes from measurements of key indices in urine: for example, of nitrogen intake from nitrogen (metabolite) excretion in timed urine collections; of Na, K, I, F intakes from urinary Na, K, I, F excretion rates over a 24-hour collection period (Bates et al. 1997).

For a further set of nutrients the biochemical status markers in blood or urine predict only approximate intake categories (e.g., high, medium, or low) but cannot provide a precise quantitative estimate of intakes. These include plasma vitamin C, the plasma carotenoids, plasma, or red blood cell levels of B-vitamins. Some markers are capable of predicting intake categories only if other supply sources are minimal; thus, for instance, plasma 25(OH) D levels can sometimes predict vitamin D intake, but only if irradiation sources such as sunlight do not make a major contribution. Some blood-based indices can distinguish between inadequate and adequate intakes, but not between adequate and high intakes: these include plasma retinol and plasma zinc. Others can distinguish between adequate and high intakes, but not between deficient and marginally adequate intakes; these include urinary thiamin, riboflavin, and vitamin C, because urine levels represent overflow once tissue demands are satisfied. One practical question, namely, whether the measurement of specific nutrient indices such as circulating vitamin C and/or carotenoid concentrations can be used as a surrogate measure of specific food-type intake, for example, of fruit and vegetables continues to be of considerable interest, since these blood biomarkers can often be measured more cheaply and accurately in populations rather than intake estimates.

Certain moderately useful status markers are essentially useless as indicators of intake, except under very strictly defined circumstances and with additional markers or precautions; these include plasma levels of many mineral elements, for example, plasma zinc or copper. Plasma iron, magnesium, and selenium levels can all provide both useful status information and a limited degree of prediction of nutrient intakes (Gibson et al. 2008. Iron absorption, however, is strongly modulated and limited, both by dietary components that limit its bioavailability, and by feedback mechanisms within the gut, which reduce absorption rates when body store levels are adequate or high.

32.5 **Emerging biomarkers**

Advances in the use of -omics over recent years have opened up the possibilities of applying such techniques to the development of new biomarkers of nutrient intake or dietary pattern. Particular progress has been made in the application of targeted or untargeted metabolomics approaches, using a number of analytical platforms including GC-MS (gas chromatography–mass spectrometry) and LC-MS (liquid chromatography–mass spectrometry). Untargeted metabolomics can be useful for hypothesis-generating. It has been used to compare metabolomic profiles between groups with different dietary patterns, for example, and can generate candidate metabolite markers of a particular dietary pattern. The matrices of interest include human breast milk, urine, serum, plasma, and red blood cells.

The metabolic fingerprint of a particular matrix may be predictive of disease and/or reflective of dietary patterns. For example, there is evidence to suggest that components of the urinary metabolome may reflect intake of a particular food groups, such as meat and meat products; a serum or plasma metabolome may be useful in differentiating a high glycaemic diet from a low glycemic diet (Pico et al. 2019; Collins et al. 2019; Maruvada et al. 2020). Metabolomics approaches can generate a large number of candidate dietary biomarkers but considerable effort is required to validate these putative biomarkers, ideally in large cohorts, and much more work needs to be done in this area.

In conclusion, the available biochemical markers of nutrient intakes and of tissue status are potentially powerful tools but are often subject to misuse and misinterpretation. They can provide valuable objective information to help define key relationships between dietary nutrient intake on the one hand, and diet-related disease (or impaired physiology) on the other, but to do so they require careful choice, measurement, and interpretation. For further information on analytical techniques for blood vitamin and mineral status see Weblink Section 32.4.

KEY POINTS

- Biochemical indices of nutrient status may, in certain cases, reflect recent intakes, or they may reflect body stores, or a combination of these. If carefully chosen, performed, and interpreted, they can fulfil the need for objective assessments of nutrient adequacy, at the whole- body level.

- Confounding factors such as homeostatic mechanisms, acute phase effects, a complex distribution of nutrients between body compartments that changes over time, can complicate their interpretation, unless these modulating factors are properly understood and allowed for.

- A lack of proper precautions during sample collection and index measurement can result in incorrect conclusions.

- Despite these caveats, carefully selected and properly validated biochemical indices provide a powerful and valuable adjunct to the other types of evidence of nutritional adequacy or inadequacy which is available both for individuals and populations. They form an essential component of nutritional research, and of public health surveillance programmes.

ACKNOWLEDGEMENTS

This chapter was updated and developed from material written by the late Chris Bates.

 Be sure to test your understanding of this chapter by attempting multiple choice questions. See the Further Reading list for additional material relevant to this chapter.

REFERENCES

Arab L (2003) Biomarkers of fat and fatty acid intakes. *J Nutr,* **133**, 925S–932S.

Bates C J, Thurnham D I, Bingham S A et al. (1997) Biochemical markers of nutrient intake. In: *Design Concepts in Nutritional Epidemiology*, 2nd edn, 170–240 (Margetts B M & Nelson M eds). Oxford University Press, Oxford.

Bath S C & Rayman M (2015) A review of the iodine status of UK pregnant women and its implications for the offspring. *Environ Geochem Hlth,* **37**, 619–629.

Bingham S A (2002) Biomarkers in nutritional epidemiology. *Public Health Nutr,* **5**, 821–827.

Collins C, McNamara A E, Brennan L (2019) Role of metabolomics in identification of biomarkers related to food intake. *Proc Nutr Soc.* **78**, 189–196.

Hedrick V E, Dietrich A M, Estabrooks P A et al. (2012) Dietary biomarkers: Advances, limitations and future directions. *Brit J Nutr,* **11**, 109–123.

Gadaria-Rathod N, Dentone P G, Peskin E et al. (2013) Red blood fatty acid analysis for determining compliance with Omega3 supplements in dry eye disease trials. *J Ocular Pharmacol Therap,* **29**, 837–841.

Gibson R S, Hess S Y, Holz C et al. (2008) Indicators of zinc status at the population level. *Brit J Nutr* **99**, S14–S23.

Hill M H H, Flatley J E, Barker M E et al. (2013) A vitamin B_{12} supplement of 500ug/d for eight weeks does not normalize urinary methylmalonic acid or other biomarkers of vitamin B_{12} status. *J Nutr,* **143**, 142–147.

Maruvada P, Lampe J W, Wishart D S et al. (2020) Perspective: Dietary biomarkers of intake and exposure-exploration with omics approaches. *Advan Nutr,* **11**, 200–215.

Mitmesser S H, Ye Q, Evans M et al. (2016) Determination of plasma and leukocyte vitamin C concentrations in a randomized, double-blind, placebo-controlled trial with

Ester-C(®). *Springerplus* **5**, 1161. doi: 10.1186/s40064-016-2605-7. eCollection 2016

Pearce E N & Caldwell K (2016) Urinary iodine, thyroid function, and thyroglobulin as biomarkers of iodine status. *Amer J Clin Nutr,* **104**, 898S–901S.

Pico C, Serra F, Rodriguez A M et al. (2019) Biomarkers of nutrition and health: New tools for new approaches. *Nutrients,* **11**, 1092. doi:10.3390/nu11051092

Prentice A, Goldberg G R, Schoenmakers I (2008) Vitamin D across the life cycle: Physiology, and biomarkers. *Amer J Clin Nutr,* **88**, 500S–506S.

Taylor A J, Talwar D, Lee S J et al. (2020) Comparison of thiamine diphosphate high-performance liquid chromatography and erythrocyte transketolase assays for evaluating thiamine status in malaria patients without beriberi. *Amer J Trop Med Hyg,* **103**, 2600–2604.

Ueland P-M, Ulvik A, Rios-Avila L et al. (2015) Direct and functional markers of vitamin B_6 status. *Ann Rev Nutr,* **35**, 33–70.

Valente E, Scott J M, Ueland P-M et al. (2011) Diagnostic accuracy of holotranscobalamin, methylmalonic acid, serum cobalamin, and other indicators of tissue vitamin B_{12} status in the elderly. *Clin Chem* **57**, 856–863.

FURTHER READING

Erdman J W, Macdonald I A, Zeisel S H (eds) (2012) *Present Knowledge in Nutrition,* 10th edn. Wiley-Blackwell Oxford, 1328.

Gibson R S (1990) *Principles of Nutritional Assessment.* Oxford University Press, Oxford. 691.

Sauberlich H E (1999) *Laboratory Tests for the Assessment of Nutritional Status.* CRC Press, Boca Raton. 486.

Scientific Advisory Committee on Nutrition (SACN) (2010) *Iron and Health.* The Stationery Office, London.

Zempleni J, Suttie J W, Gregory J F (eds) (2013) *Handbook of Vitamins,* 5th edn. CRC Press, Boca Raton. 605.

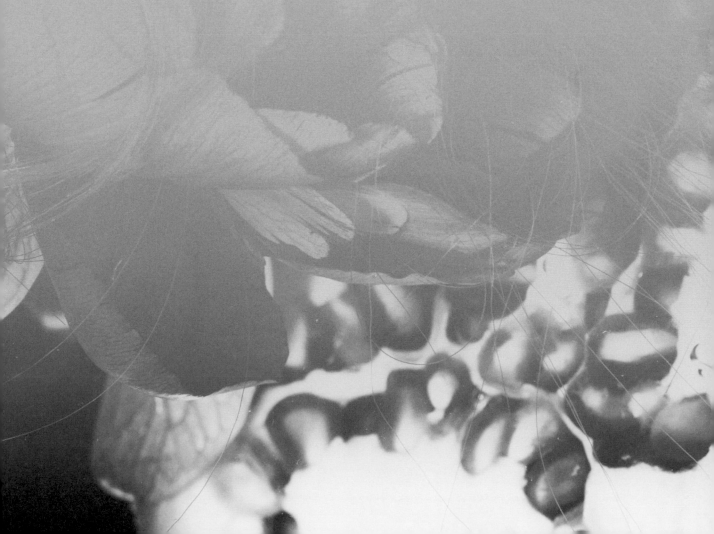

PART 7

Public health nutrition

33 The science of epidemiology

Annhild Mosdøl and Eric Brunner

OBJECTIVES

By the end of this chapter you should be able to:

- define nutritional epidemiology
- summarize its aims and limitations
- outline the concepts of population, exposure, outcome, and epidemiological effect
- define validity and repeatability
- interpret results of a study taking into account possible effects of bias, chance, and confounding, and the question of generalizability
- summarize criteria for assessing evidence of causality
- explain the main features of different epidemiological study designs, including their strengths and weaknesses.

33.1 Introduction: what is epidemiology?

The COVID-19 pandemic has raised the common understanding of both epidemics and epidemiological concepts. Epidemiology is the study of the distribution and causes of disease in populations. Communicable-disease epidemiology provides tools to monitor, track, and respond to threats like the Corona virus (SARS-CoV-2) and the infection caused. Diet-related health questions are situated mainly in the non-communicable or chronic disease domain of epidemiology. It addresses the following type of questions: Why is breast cancer more common in the USA than in Japan? Can diets rich in fruit and vegetables reduce blood pressure? Do dietary differences explain social gradients in cardiovascular disease mortality?

The word *epidemiology* comes from Greek and is made up of the words *epi*—upon, *dēmos*—the population, and *-logy*—the study of. Epidemiology provides methodological tools for investigations of health and disease at the *population level*. It can be defined as: (1) studies of the distribution of diseases and health-related conditions; (2) investigation into their causes (aetiology); (3) analysis of how these diseases best can be prevented and controlled in a population. Nutritional epidemiology is one speciality of epidemiology applied to nutrition-related concerns including the role of nutrition in the causes and prevention of ill health (Willett 2012). Throughout the previous chapters, there have been numerous examples of studies spanning from the molecular to the population level. Different types of study are complementary to understand the nature of health and disease.

Knowledge of epidemiology is useful for all health practitioners. Most papers published in medical or nutritional journals use at least basic epidemiological terminology. To interpret what you read, and to understand the evidence, whether it is to update yourself on nutritional advances or to identify the best evidence-based treatments in a clinical setting, you will need epidemiology. In public health and health promotion, epidemiology is important both to provide knowledge for action and to evaluate the effect of interventions.

This chapter aims to be a first introduction to epidemiology to give the reader skills that enable them to critically read and interpret epidemiological studies. The first three sections describe the basic concepts of epidemiology (Section 31.2), important aspects to consider when interpreting results (Section 31.3), and the most common study types (Section 31.4) with examples from nutritional epidemiology. Section 31.5 discusses some central issues that are particularly relevant in nutritional epidemiology.

33.2 Important concepts in epidemiology

Epidemiology provides a framework for designing our own studies and for evaluating existing evidence. The process starts by posing a good question. A good epidemiological question can be described in the same way as a hypothesis—a statement in a form that allows it to be tested and proven false. A question such as 'Is vitamin C good for health?' is too unspecific to be addressed scientifically. On the other hand 'Will taking 100 mg of vitamin C daily for five years reduce the risk of myocardial infarction?' is a question that can be answered. To formulate clear research questions you need to specify who you are interested in studying, *the population*, what you are interested in studying the effects of, *an exposure*, and what *outcome* you are interested in finding out about. The relation between the exposure and the outcome is evaluated by *measures of association and effect*. These are the essential elements of all epidemiological studies, no matter the study design used.

The population at risk

In 2019 there were estimated to be 17.2 million overweight pre-school children in Asia while there were 3.9 million in Latin America and the Caribbean (UNICEF, WHO & IBRD 2020). In which region was overweight the more common? The limited information given above is not sufficient to answer this question since we lack information about the total number of children—in other words *the denominator*. Since the total population of 0–5 year olds was larger in Asia (358 million) the proportion of overweight children was lower, 4.8 per cent compared to 7.5 per cent, compared to Latin America and Caribbean with 52 million 0–5-year-olds. To describe the magnitude of disease at a group level it is necessary to take account of the size of the relevant source population. Many of the measures in epidemiology are proportions, ratios, and rates where the defined study population is the necessary denominator.

The population of interest for a specific question is called the *population at risk*. It can for instance be defined as people living in world regions, countries, or districts or by characteristics such as age and gender. Sometimes, the population of interest is very narrow, such as the children of mothers who used vitamin A supplements during pregnancy. The phenomenon or event under study defines these boundaries. In a study of dietary effects on prostate cancer, women are irrelevant and should not be included in the study population. Thus, the population at risk is the group of people to whom an event can occur. The population at risk is also sometimes called the *target population*.

The concept of *risk* is fundamental in epidemiology. We are all accustomed to and use the word risk in everyday life settings. Most people associate the word with some form of danger or vulnerability. We have an intuitive feeling of some risks being either larger or smaller than other risks. However, in epidemiology, risk refers to a *probability* that a defined event is going to happen in a given population during a time period. If we have a situation where an event can occur or not in a group of people, the risk of it happening will vary between 0 (it will never occur) and 1 (it will always occur). Probability is a concept related to the likely distribution of alternative events or outcomes. For example, the proportion of boys among all babies born is generally about 52 per cent. We can say that the risk of having a boy is 0.52.

The outcome

In epidemiological studies, the term *outcome* refers to the disease or health-related variable being studied. Although the word presumably derives from clinical studies and implies some kind of follow-up, it is used in all kinds of epidemiological study designs to label the health-related phenomenon being studied. *Endpoint* is another word used to describe the study outcome of interest. Death and disease occurrence are referred to as 'hard' endpoints or outcomes because there is great certainty about the health status of the individual in question. A 'soft' outcome is one where there is some degree of uncertainty, such as a self-reported doctor diagnosis or self-rated health, when subjectivity comes into play. Disease and mortality outcomes may appear too infrequently within the limited time frame of a study. Instead, the investigator may decide to opt for a 'surrogate' outcome. A study of dietary prevention may need to continue for years if the investigator measures coronary events as the primary outcome, but only for months if blood cholesterol lowering is used as a surrogate measure of effect.

Outcomes are typically measured either as binary (consisting of two categories) or continuous variables, although other types of outcome variables may be studied. A common binary outcome measure is mortality, when the population is divided into the dead and the alive at the end of a study. Likewise, disease status can be categorized as cases or non-cases, controls or referents. The outcome variable may also measure a continuum of severity rather than an all-or-none phenomenon. Examples of continuous outcome measures are plasma homocysteine concentration, systolic blood pressure, body weight, or intake of specific nutrients. These measures may be presented and analysed as continuous measures, but it is often practical to categorize them into a binary form based on a defined cut-off level. Body mass index (BMI) is a continuous variable, but can be analysed as the proportion of participants with BMI over 30 kg/m^2. Although such cut-off points can be useful, it is important to remember that in general they are arbitrary.

When a binary outcome is being studied, there are three important types of measurement: incidence, prevalence, and mortality rates. *Incidence* is defined as the number of new cases appearing in the study population during a specified period of time. If the study population is roughly constant:

$$Incidence = \frac{Number\ of\ new\ events\ in\ a\ defined\ period}{Population\ at\ risk\ at\ the\ beginning\ of\ the\ period}$$

In many cases, the assumption of a constant study population holds true, although birth, death, loss to follow up or migration may be sufficiently large to cause difficulties. This estimate is typically used in routine health statistics. If a population of 100,000 people saw 190 new cancer cases over a period of one year, the cancer incidence would be 0.0019 per person year. By convention this figure is often presented per 1000 person years in order to get a number closer to one, here giving 1.9 cancer cases per 1000 person years.

The most accurate way of measuring incidence is to divide the number of new cases by the *person-time at risk*. Person-time is calculated by summing up the time each individual is at risk during the measurement period. This measure takes into account that when a person becomes a case that person is no longer part of the population at risk. To calculate person-time at risk you need information about the date of each event when the individual ceases to contribute to the observation time.

$$Incidence\ rate = \frac{Number\ of\ new\ cases}{Total\ person\text{-}time\ at\ risk}$$

Let's consider how person-time at risk is observed in a cohort study. The British Whitehall II study of civil servants recruited 10,308 participants between 1985 and 1988. No new participants enter the study after 1988. By 2020, approximately 1000 participants had died after follow-up time of 35 years. Roughly, the total observation time is 10,000 persons x 35 years = 350,000 person years. The incidence rate is 1000 deaths/350,000 person-years or 2.9 deaths per 1000 person-years.

The *prevalence* is the proportion of a population that meets the definition of a case at a given point in time, for instance the percent of a population who has type 2 diabetes mellitus at a given date. A prevalence figure is not a rate as it has no dimension of time. Prevalence may be used to estimate the disease burden in a population, for instance to plan health services.

Incidence and prevalence describe different aspects of the disease burden in a population and their relationship depends on the nature of the disease and the effectiveness of treatment. For a diseased person there are three prospects: continue to be ill, recover, or die. The risk of death among the diseased is described by the *case-fatality rate*, the incidence of death among the cases. When the duration of a disease is short, its prevalence in the population will be lower than if the duration of the disease is long given the same incidence rate. It follows from this fact that efforts to improve the survival with a disease without curing it will increase its prevalence (and potentially the burden of care). For instance, treatment of diabetes

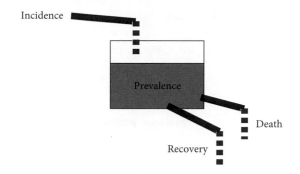

FIGURE 33.1 The relationship between incidence, the prevalence pool, and duration of the disease.

patients has improved substantially, but the disease is rarely cured and patients may require medical care for the rest of their lives. Figure 33.1 illustrates how the prevalence pool is related to incidence, recovery, and death.

Mortality rates are a type of incidence rate—the incidence of death. Crude mortality rates are calculated from the observed number of deaths and the size of the population of interest. Mortality rates for specific diseases are calculated using deaths due to the specific cause in the numerator.

$$Crude\ mortality\ rate = \frac{Number\ of\ cause\text{-}specific\ deaths\ in\ a\ year}{Mid\text{-}year\ population}$$

Crude mortality rates may hide useful information needed to understand the health and disease patterns of different populations. Crude cancer death rates are lower in Western Africa than in Australia. It may look as though the West African population experiences fewer cancer deaths and is less exposed to risk factors for cancer. However, the difference in mortality rates is partly due to the different age distribution in the two regions. Cancer rates are higher in older people. A population with a high proportion of older persons will have higher crude cancer mortality rates compared to a young population given the same age-specific cancer death rates. Breaking down mortality rates for specific groups, most commonly by gender and age, will contribute to a better of understanding of death rates over time and between populations. Crude and adjusted rates may be applied to measures of disease incidence in the same way as measures of death. Box 33.1 provides definitions of the terms used in this section.

The exposure

An *exposure* is any factor suspected to modify the risk of the outcome of interest. Exposures include a broad range of factors, from microbiological agents to all kinds of physical, social, or psychological variables. Studies of the health consequences of exposure can be either based on observations of what happens naturally in a population,

> **BOX 33.1**
> Definitions of incidence, prevalence, and mortality rates
>
> $$\text{Incidence rate} = \frac{\text{Number of new cases}}{\text{Total person-time at risk}}$$
>
> $$\text{Prevalence proportion} = \frac{\text{Number of existing cases at one time point or period}}{\text{Population at same place and time}}$$
>
> $$\text{Crude death rate} = \frac{\text{Number of deaths (defined place and time)}}{\text{Mid period population (same place and time)}}$$
>
> $$\text{Age specific death rate} = \frac{\text{Number of deaths in a particular age group (defined place and time)}}{\text{Mid period population (age group, same place and time)}}$$
>
> $$\text{Cause-specific death rate} = \frac{\text{Number of deaths due to a particular cause (defined place and time)}}{\text{Mid period population (same place and time)}}$$

or when participants are assigned to a certain exposure, referred to as an *intervention*, in an experimental study. Common exposures in nutritional epidemiology include intake of nutrients, foods, and food patterns, as discussed further in Section 5.3. You may have noticed that some of the examples of exposure variables could also be outcome variables. In a study of stroke, blood pressure may be the exposure variable of interest. In another setting, blood pressure can be the outcome of interest and in a study of how it is affected by exposure to high vegetable intakes. A factor or variable is designated as exposure or outcome depending on the objective of the study and the relationship being examined.

As with outcome variables, exposure variables may be presented in a binary form by dividing the population into the exposed and the unexposed. Exposure variables on continuous scales can be used as such or be categorized into two or more groups with graded levels of exposure using relevant cut-points. For instance vegetable intakes measured on a continuous scale in g/day can be categorized into three exposure levels defined as intakes <200 g/day, 200-400 g/day and >400 g/day. The way the exposure is measured depends on the question being asked. If the outcome is a food-borne disease, it may be sufficient to know whether the study participants have eaten shellfish from a certain shop last week or not. When studying thyroid cancer, the average shellfish consumption over time may be more relevant.

In experimental settings, the exposure of interest is an *intervention*. Usually this is hypothesized to be a protective factor, such as a treatment or a drug, as it would be unethical to expose participants to suspected harmful factors. Exposures can also be called risk factors. When established, a risk factor is a characteristic that is

established or suspected to increase the probability of a particular disease or malign condition to develop.

Measures of association and effect

In general, an *association* refers to the statistical link between two variables or factors. Several measures are used to summarize the association between an exposure and an outcome. The choice of suitable measures for a given study will depend on whether the exposure and outcome variables are in binary or continuous format (Weblink Box 33.1).

Binary exposure: binary outcome

Risk ratio, also called relative risk (RR), is the measure of effect most commonly used by epidemiologists when both the exposure and the outcome are binary variables. You calculate the risk ratio by dividing the risk, or incidence, of the outcome in the exposed group by the risk of the outcome in the unexposed group. For instance, if 40 persons in a group of 1000 exposed to a diet high in processed meat develop colon cancer, the risk of colon cancer is 40/1000 in this group. If only 20 persons develop colon cancer in a group of 1000 persons with low intake of processed meat (here considered the unexposed), the risk ratio of colon cancer between the groups is (40/1000)/(20/1000) = 2. A risk ratio of one indicates that there is no difference in incidence between the exposed and the unexposed. When the risk ratio is higher than one, as in the example above, the defined exposure poses a hazard; while a risk ratio lower than one means that the defined exposure is protective. Another commonly used measure is the attributable risk (AR) which is the incidence among the exposed minus the incidence among the non-exposed. Attributable risk estimates the size of the excess risk due to a particular exposure or risk factor.

The *odds ratio* (OR) is frequently used to describe an association and is an alternative to risk ratio outlined above. The term *odds* refers to a measure of likelihood based on a ratio rather than a proportion. The odds of throwing the number three with a dice is one to five, meaning that you are five times more likely to get one of the other five numbers. In comparison, the *probability* that you will get number three is one in six (1/6 or 16.7 per cent chance). In a case-control study (see Section 31.4), the odds ratio is calculated by dividing the odds of exposure among the cases by the odds of exposure in the non-cases. Consider the relationship between obesity (the exposure) and gestational diabetes (the outcome). If 50 of 100 women with gestational diabetes were obese compared with 20 of 100 non-diabetics, the odds of obesity among the diabetics are 50/50 (odds = 1), and the odds of obesity among the non-diabetics are 20/80 (odds = 0.25). In our example the odds ratio is (50/50) / (20/80) = 1/0.25 = 4. The odds of being obese among the women with gestational diabetes are four times greater than among the healthy women. An odds ratio of one indicates that there is no difference in the odds of being exposed among the cases as compared to the controls.

Table 33.1 shows a standard 'two by two' table for studies where both the exposure and the outcome are in a binary form and give the formulas to calculate risk ratio, attributable risk, and odds ratio. The study participants will fall into four groups (denoted cells a, b, c, and d) depending on whether they are either exposed or non-exposed combined with their status as a case or a control (alternatively referent or non-case). When the risk ratio and odds ratio can be calculated on the same set of data, these will often be similar. How closely they match

depends on the ratio of cases compared to the controls (see Weblink Box 33.1).

Binary exposure: continuous outcome

If the exposure variable is binary and the outcome is a continuous variable, the effect due to the exposure can be expressed as the mean difference in outcome between the two groups. An example is the mean difference in serum homocysteine levels at the start and the end of the study among participants given a supplement containing a vitamin B complex compared to the participants given placebo tablets.

Continuous exposure: continuous outcome

When both the exposure and the outcome variables are continuous, the observation pair for each individual in the sample can be plotted against each other. Statistical methods such as correlation and regression analyses are used to describe such relationships, typically as the average difference in outcome variable per unit difference in the exposure variable.

Continuous exposure: binary outcome

The combination of continuous exposure and binary outcome is seen frequently in nutritional epidemiology. The exposure is often a measure of nutrient intake, nutritional status, or biological variable such as circulating serum homocysteine level, while the outcome can be development of a disease such as myocardial infarction. The statistical technique used in these situations is logistic regression and the results are presented as odds ratios. In this case, the odds ratio is obtained using a different technique than the 'two by two' table described in Table 33.1, but is interpreted in the same way.

TABLE 33.1 Risk ratio, attributable risk and odds ratio: quantifying the health effect of a risk factor

	Cases	Non-cases	Total
Exposed	a	b	a + b
Unexposed	c	d	c + d
Total	a + c	b + d	

Risk ratio, $RR = \dfrac{risk_{exposed}}{risk_{unexposed}} = \dfrac{a/(a+b)}{c/(c+d)}$

Attributable risk, $AR = Risk_{exposed} - Risk_{unexposed} = [a/(a+b)] - [c/(c+d)]$

Odds ratio, $OR = (a/c)/(b/d) = ad/bc$

Quality of the measurements

Poorly measured variables, either exposure or outcome, can lead to misleading results and incorrect conclusions. The word *validity* is used in different situations, usually to describe whether aspects of a study can be considered to represent the truth. The validity of a measurement is an expression of the degree to which it assesses the aspect it is intended to measure (Porta 2016). The waist-to-hip ratio is commonly used to measure central obesity, but gives a rather poor and indirect estimate of intra-abdominal fat. A dual-energy X-ray absorption (DEXA) scan would give a more valid and direct estimate of the true fat mass.

The reproducibility of a measurement is another quality marker. It indicates whether a method is capable of producing the same result when used repeatedly under the same conditions. Repeatability is both a feature of the variable itself, for instance natural biological variation in the variable measured, and the systematic or technical error of its measurement (how well the measurement is performed). Measurement of adult height is highly reproducible. We would expect to find the same value when measured repeatedly over some weeks and it is relatively easy to minimize errors in measurement. Blood glucose, on the other hand, varies naturally over the course of a day. To get a valid estimate of a person's blood glucose levels it is important both to decide the most relevant aspect of glucose metabolism and to standardize the measurement procedures. By deciding to measure fasting blood glucose, variation due to fluctuations after meals is minimized. Repeat measurements of a variable may provide better estimates of the true average and will also quantify its reproducibility.

Figure 33.2 illustrates the difference between the validity and reproducibility. In all four cases we tried to capture the true value with six measurements on a continuous scale. When all six measurements are positioned closely to each other, the result is highly reproducible (A&B). If they are situated over a wider range of values, reproducibility is low (C&D). If the measurements are systematically distanced on one side from the true value, the validity is poor (B&D). This is *information bias* (see Section 31.3). Measurements symmetrically around the true value indicate high validity (A&C). Picture A is our goal—high validity and reproducibility as the measurements are positioned closely and the average is close to the truth.

Any measurement will to some extent deviate from the 'true' value. The most valid measurement methods are often relatively burdensome and expensive. Investigators

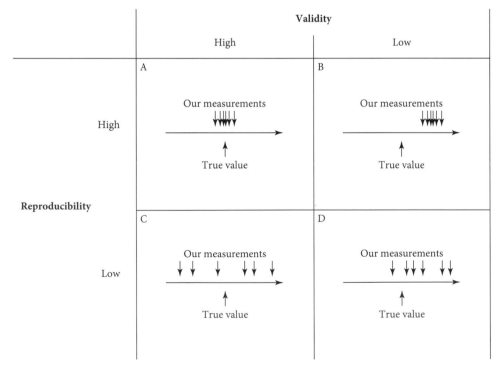

FIGURE 33.2 Illustration of the terms validity and reproducibility, showing four examples of six repeat measurements aiming to capture the same true value.

may choose to use a less valid method for measuring the variables of interest, but it is important to assess the size of unavoidable errors so that they may be taken into account when analysing and discussing the results. To establish the validity of a measurement we need to compare it with an absolute or gold standard method. The gold standard is usually the best existing or available test for the measure of interest. The more practical or cheaper measurement method is here called the test method while the gold standard is called the reference method.

If we have exposure or outcome variables in a binary form, the validity of a measurement can be determined by classifying the subjects both by the test method and by the reference method. By comparing the results from the two methods, four measures can be derived: sensitivity, specificity, positive predictive value, and negative predictive value. These measures are defined in Weblink Section 33.1 and Weblink Box 33.2. For variables in a continuous form, the validity is estimated by comparing the agreement or degree of association between the test method and the standard reference method. Such data are often presented as correlations or by cross-tabulation of findings according to quartiles of the distribution for each of the two measurement methods. This gives a picture of how well the survey test method is able to rank study participants compared to the standard reference method.

33.3 Interpretation of results

The survey is completed and the measures collected, but how should the results be interpreted? And how do you interpret studies reported in scientific journals? The epidemiologist is primarily interested in assessing and evaluating associations or effects, which address questions of disease causality, prevention, or treatment. In general, an observed association may be the product of four distinct influences. These are bias, chance, confounding, or cause. Before we even start to consider the importance of the results in question, the first three influences—bias, chance, and confounding—should be considered carefully.

Bias

Bias refers to measurement errors or misclassification that lead to results that consistently deviate from the truth. Many sources of bias have been identified, but they fit into two main categories: selection bias and information bias. When study participants have been selected to the study in a way that distorts the results, we have *selection bias*. This can arise in many ways. In a clinical setting, the terminally ill patients may be more willing to try out a novel treatment with unknown risks attached

to it than those with good prospects. If the new treatment performs badly it may be difficult to know whether the treatment really was poor or the patients would have died anyway. Important questions to ask when assessing a study sample can be: In what way were participants selected for investigation, for example, were they selected or were they volunteers, and how representative were they of the target population with regard to the study question?

When people are invited to join a study, some will not take part. The size of this sample relative to the original group of invitees is described as the *response rate*. Participants who volunteer to take part in studies are likely to differ from the general population, for instance by being more highly educated or more health conscious. The prevalence of the disease being studied can differ among responders compared to the population from which they came. Faced with such studies you might ask: What was the response rate and are the responders and non-responders likely to have differed in important ways? It is easier said than done to make sure that study groups are representative of the population at risk, but not all deviations pose a problem: It depends on the study objective. If a study sample is not fully representative of the study population, the results are likely to misrepresent the true disease burden. Still, the relationship between an exposure and an outcome may be correctly estimated even when there is considerable selection bias.

If the quality of information we collect deviates systematically from the truth we have *information bias*. Such bias may differ between groups. For example, dietary assessment studies repeatedly show that obese subjects tend to underestimate their energy intake more than non-obese subjects do (Maurer et al. 2006). Thus, it is necessary to be cautious about self-reports of diet in a study of the causes of obesity. Information bias can also arise when asking people about past exposures. A mother that has recently given birth to a child with neural tube defects may ask herself what could have caused it. She may recall supplement use and diet more meticulously than a mother with a healthy baby may. This type of information bias is called recall bias. As with selection bias, information bias is potentially important if its magnitude differs in relation to the presence or absence of the study outcome.

Bias is caused by imperfections in design or implementation of a study. It is difficult or even impossible to remove its effects at a later stage. Almost all studies are somewhat biased. This does not mean that they are scientifically unacceptable or that their results should be disregarded. If a study has been analysed and interpreted carefully, the investigators will have addressed this question themselves. There is no simple formula for measuring biases and quantifying their effects. Each study must be judged by assessing the probable impact of biases on the study question and then to allow for them when drawing conclusions.

The role of chance

When do we know that a numerical difference is meaningful? If a group of children that had been breastfed for at least four months had an average primary school reading test score of 67 while children that never had been given breast milk had score of 65, are they really different? We need to enter the territory of statistics and uncertainty. Chance can be thought of as variation due to sampling variation, biological differences, and other unknown factors. Statistical methods provide measures to quantify the element of chance and account for the uncertainty in the results of our study. It is important to recognize that a study utilizes a sample, but our primary interest lies in what the study sample tells us about the population from which it is drawn.

Hypothesis testing is used to take account of the effects of chance and is central to many areas of science. If we estimate the mean level of a parameter in one sample, say mean fasting blood glucose, we would get a slightly different result by chance if the measurement was repeated in another sample from the same population. The procedure called hypothesis testing starts by defining a *null hypothesis*, which usually is a statement saying that there is no real difference (only differences by chance) between the two groups we are studying. The *p-value* gives the probability of getting the observed or a more extreme result (that is, even further away from the null hypothesis) if the null hypothesis is true, or rephrased: What is the likelihood that we observe this difference between two groups if in reality they are similar? Usually a p-value less than 5 per cent ($p < 0.05$) is considered to give a statistically significant result leading to rejection of the null hypothesis.

Significance testing assesses whether a result, such as a difference between groups, is likely to be due to chance or some real effect. It cannot prove that it is one or another. Note that failing to reject the null hypothesis is not the same as saying that it is true. With a p-value of 0.05, the investigator runs a risk of falsely rejecting the null hypothesis on average 5 per cent of the time. Two possible errors can be made when using the p-value to make a decision: To conclude that a difference exists when it does not, is called a *type I error* (false positive or alpha error). Conversely, concluding that a difference does not exist when in fact it does by failing to reject a false null hypothesis is called a *type II error* (false negative or beta error).

Confidence intervals provide a measure of the uncertainty in data by giving a lower and upper confidence limit

of the true value in question. The intervals have a given probability, usually 95 per cent, sometimes 99 per cent, that the true value in the underlying population is included. We want confidence intervals to be narrow because this means that the estimate is likely to be more precise than with a wide confidence interval. The confidence interval and the p-value use the same equations to estimate the distribution of the likely true value. If the value corresponding to the null hypothesis falls between the lower and upper confidence limits of the probable true value, the null hypothesis is not rejected. For instance, the null hypothesis for the difference in disease risk between the exposed and non-exposed groups is that risk ratio of one. If the observed risk ratio is 1.05 and the 95 per cent confidence interval is 0.96–1.14 then the null hypothesis is not rejected (or put another way, it is accepted). The 95 per cent confidence interval indicates also that a new study investigating the same question could find a risk ratio anywhere between 0.96 and 1.14. Since confidence intervals combine information about chance effects and precision, it is preferred against hypothesis testing using the p-value alone.

Confounding

A *confounder* is any factor that can cause or prevent the outcome of interest and at the same time is associated with the exposure. This double association of a second exposure, that is, association both with the outcome and the exposure of interest, distorts the observed association between the exposure of interest and the outcome. Confounding is therefore about the confusion that may arise because two or more potentially causal influences are mixed up with one another.

Let us assume we have recorded the drinking habits in a study sample. During the observation period, the heavy consumers of alcohol have a higher rate of lung cancer than the participants drinking little or no alcohol (the observed association). Can the relationship between lung cancer and drinking habits be due to another risk factor for this type of cancer that is also associated with alcohol intake? If smokers tend to drink more than non-smokers there is an association between alcohol consumption and smoking habits. Further, smoking is an important risk factor for lung cancer, so the observed association between alcohol consumption and lung cancer is confounded by smoking habits. A confounding factor may produce a spurious effect as above, or can hide a real effect.

Confounding may be regarded as a type of bias, but contrary to other biases confounding can often be adjusted for in different ways. In the example above, the association between alcohol consumption and lung cancer can be analysed separately in smokers and non-smokers in a *stratified analysis*. Age is a common confounder in many studies. If we want to compare two groups with different age distributions, the risk estimates may be *standardized* to a common age distribution. Confounding can also be controlled for in statistical models, which simulate the stratified analysis described above. All these modifications are only possible, though, if the relevant confounding factor has been measured. If confounding is considered at the design stage, it can be controlled by *exclusion of sub-groups* that may cause interpretation problems. In experimental settings, random allocation of participants (*randomization*) to two treatments minimizes confounding, as any confounding factors will tend to be equally distributed in the two groups by chance.

Causality

Smoking as a risk factor for lung cancer has one of the largest risk ratios demonstrated for any non-communicable disease. Still, we have all heard about old men smoking 40 cigarettes a day living into their nineties. How can we regard smoking as a cause of lung cancer when many smokers never get the disease? Epidemiology is the science that takes a health-related hypothesis into the real world and attaches a probability to it. When considered together with other types of evidence, such as that derived from laboratory-based studies, epidemiology is a powerful tool for testing whether a proposed disease mechanism is important for population health.

What guidelines should be used for judging the evidence? Causation can rarely be proven without doubt, but if the possible presence of chance, bias, or confounding has been assessed carefully, the question of causality can be approached in different ways. Sir Austin Bradford Hill suggested a list of aspects to consider when judging whether an association between an exposure and an outcome could be causal (Hill 1974). However these are not absolute criteria for a causal relationship. The nine points are often referred to as the Bradford Hill criteria:

1. *Strength of the association:* A strong association, as measured by relative risks, increases the likelihood that it reflects a causal relationship. Confounding may also create strong associations, but if an association is to be completely explained by confounding, the confounder must carry an even higher risk for the disease under study. Weak associations may also be causal and important for public health if the exposure is common.

2. *Consistency:* Tabloid newspapers may make headlines of one study showing that eating carrots increases the risk of stroke, but the scientific community will judge such findings against the results of other studies. When a body of studies consistently point in the same direction over time and across populations, the evidence is strengthened. As studies giving null results often fail to be published, the available studies may not represent the full span of evidence concerning a given risk factor–disease relationship. This is called publication bias.

3. *Specificity:* The concept of specificity is that one particular exposure leads to a specific disease and that one disease always is triggered by the same cause. This criterion is relevant in some situations. Severe vitamin C deficiency, for instance, will manifest itself as scurvy. Increasingly, though, we discover that many diseases can have multiple causes and that some exposures may contribute to the development of more than one condition.

4. *Temporal sequence of cause and effect*: If a statistically significant association is found between two variables, but the presumed cause occurs after the effect rather than before it, the association cannot be causal. Logically, the temporal sequence of events, with exposure occurring before outcome, is an absolute criterion for causality. For example, a cross-sectional study showing that obese people eat less than thin people, tells us as little about dieting to lose weight as about the causes of obesity.

5. *Biological gradient:* A dose–response relationship where the risk increases progressively with higher exposure is generally strong support for a causal interpretation. Yet, a dose–response curve may have many different shapes. We do not always know where on this curve the observed range of exposure is situated.

6. *Biological plausibility and coherence:* Biological plausibility and coherence with existing knowledge will strengthen the confidence that an association is causal. Still, many epidemiological studies have pointed to possible risk factors for disease that could not be explained by biological knowledge at the time they were published. Such findings often spark off further studies to examine possible mechanisms in depth.

7. *Experiment (reversibility):* When an epidemiological association can be reversed, it is strong confirmation that the link most likely is causal. Reversion appears when reduction in a factor shown to increase the disease risk gives a corresponding drop in disease risk. An example is patients that manage to lower their salt intake and then experience reduced blood pressure.

The concept of a *cause* is interesting and worth thinking about. Many non-communicable diseases involve several causal factors. If a factor triggers a disease every time it is present, it is called a *sufficient cause*. Apart from some genetic abnormalities, there are very few conditions where this holds true. Even in the case of infections such as food poisoning, symptom severity will depend on the vulnerability of the person exposed. The bacteria are on the other hand a *necessary cause* of the food poisoning. If a necessary cause is absent, the disease cannot occur. We have already defined a *risk factor* as a characteristic that is known to increase the probability that a particular disease or malign condition will develop. Risk factors are neither sufficient nor necessary causes of disease. Smoking is a good example, as many smokers never get lung cancer while a smaller number of non-smokers do. Smoking is nevertheless the most important cause of lung cancer. Nutritional exposures are very often risk factors—they increase the likelihood of diseases to occur but are neither sufficient nor necessary causes.

Only under certain circumstances, such as serious injuries, will a causal factor trigger a disease or a health outcome directly. For most health outcomes, there will be *causal pathways* between the exposure and the outcome. Fruit and vegetable intake is thought to exert part of its protective effect on development of cancers by increasing the circulating antioxidant level, which might protect DNA from oxidative damage and prevent cells from becoming malignant (WCRF/AICR 2018). For someone interested in the biochemical aspects of nutrition, the molecular actions of nutrients are the most interesting aspects of this causal pathway. For those giving specific dietary advice to individuals, the dietary aspects of preventing cancer are most relevant. Among public health nutritionists, there may be particular interest in the social, cultural, and psychological determinants of food habits. All of these aspects contribute to a fuller understanding of the incidence of the disease at the population level.

33.4 Epidemiological study designs

Some elements of planning and designing studies

When someone starts planning an epidemiological study, the most important task is to formulate the essential questions the study seeks to answer. These questions will guide the decisions that need to be taken regarding the study sample, how exposure and outcome are measured, and the preferred study design. It is at this stage that potential biases best can be controlled or minimized. If a relevant variable is missed during the implementation, for instance a confounder, it can rarely be made up for after the data collection is completed.

Who should be included in the study sample and how many participants are needed? These two questions are very important. The goal is usually to generate results that have a wider application beyond the study participants, in other words, to get results that are *generalizable* to the whole target population or the population at risk. If we could include everyone from the target population in our study, the question of generalizability would not be an issue. This is rarely feasible or desirable. The generalizability of a study is closely related to the *external validity*. Validity has earlier been defined as the degree to which a measurement assesses the aspect it is intended to measure (Section 31.2 on quality of measurements). External validity concerns the extent to which the study captures accurately the phenomenon, such as a diet–disease relationship, as it is in the target population.

Often the study sample is formed in a two-stage process. First, the researchers select a subset of the target population on defined criteria to form the study population. It can be through random selection from the general population or by choosing an accessible group that can be followed up over time. The final study sample is the subset of the study population that chooses to participate in the study. This proportion of the original sample is the response rate (see Section 31.3 on bias). It is usually desirable to get a sample that is representative of the target population in terms of age, sex, and other demographic factors. Characteristics of external validity will differ with the study question. For instance, in a study exploring causal relationships it is important that the exposure–outcome relationship found in a study sample represents the association in the target population. If prevalence is important, the level of risk factors or outcomes of interest should be captured with minimal selection and information bias.

The other crucial task when selecting a study population is to determine the necessary number of participants. It is vital that the sample size is big enough to detect as statistically significant any true differences between the study groups. Performing studies that are too small to answer the main objective is a waste of resources and can be considered unethical. It is possible to calculate the necessary sample size first based on an expected prevalence or perhaps the size of an expected effect, such as weight loss in a trial of calorie restriction and second, on a specific statistical significance level for the primary research hypothesis. Such *power calculations* are described briefly in the Weblink Section 33.2.

The best study design in a particular situation depends on the question being asked. Figure 33.3 gives a schematic overview of the main study designs, and the following two sections will go through their features, strengths, and weaknesses (see Weblink Section 33.3 for examples). Study designs differ according to whether the researchers observe (*observational* studies) or attempt to influence the exposure (*experimental* studies).

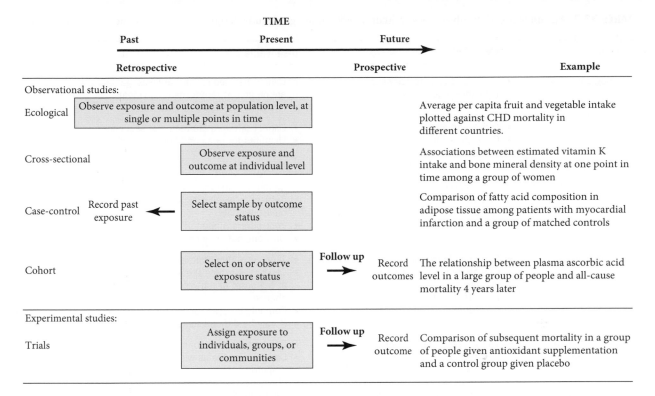

FIGURE 33.3 A schematic overview of epidemiological study designs and their main characteristics. See Weblink Section 33.3 Examples of epidemiological designs.

Studies differ according to the time order of collecting information on exposure and outcome. They are called *retrospective* when information is gathered about exposures in the past. They are *prospective* when exposure status is measured at the beginning of the study and the sample is followed over time to gather information on outcomes. The preferred design depends on the study question itself, characteristics of the main exposure and outcome variables, how strong existing evidence is concerning the research question, and the resources available. Data collection is expensive. There may be datasets from studies already completed that can be analysed to answer the question at much lower cost than by carrying out a new study. The following study designs are among the most common (Table 33.2).

Observational study designs
Ecological studies

Many of the hypotheses regarding diet–disease relationships originated from simple comparison of cause-specific death rates in different countries matched with information about exposures available at the population level. An example is death rates due to cardiovascular disease and average consumption of fruit and vegetables (see Weblink Figure 33.1). Such studies can be inexpensive

and quick, as this type of information is freely available from the websites of international organizations such as the World Health Organization and Food and Agriculture Organization of the United Nations. Time trend studies comparing levels of exposure and outcome over time in the same population, say per cent energy from fat by year and corresponding mortality rates due to cardiovascular diseases, are also useful. Another classic design is to compare disease rates of migrants with the disease rates in their native and destination country (*migrant studies*). Ecological studies match information about exposures and outcomes at a group level: They are prone to the *ecological fallacy*, which is a type of confounding. An association observed at a group level may not reflect a causal association at the individual level.

Cross-sectional studies

In cross-sectional studies, outcome and exposure status is determined simultaneously. Such studies are also called prevalence studies as they measure the proportion of people with a condition at given point in time. This is useful information for planning health services. Cross-sectional studies are poorly suited to study rare diseases or diseases of short duration, though, as only few people in the population will have the relevant condition at any given time. Data from cross-sectional studies also form

TABLE 33.2 Advantages and disadvantages of different epidemiological study designs

Studies	Advantages	Disadvantages
Ecological studies	• Quick and easy to carry out • Hypothesis generating	• Cannot be used for causal investigations since the data are on groups rather than individuals • Not good for hypothesis testing
Cross-sectional studies	• Quick and easy to carry out • Gives prevalence of condition in population • Hypothesis generating	• Cannot differentiate temporal sequence • Unsuited for rare conditions or conditions of short duration • Not good for hypothesis testing
Case-control studies	• Quick and easy to carry out • Can study many risk factors • Require relatively few participants • Suited for studies of rare diseases	• Cannot provide measures of risk, only odds ratios • Subject to recall bias, difficult to validate the measurements • Poor differentiation of temporal sequences • Only one outcome can be studied
Cohort studies	• Can provide measures of risk • Can study many exposures and outcomes • Permits quality control of the exposure measurements	• Time consuming and costly, many participants required • Can only examine the exposures that were measured at the onset of the study, relevant factors to study may change • Can only be used for common outcomes • It may be difficult to keep track of the study sample
Randomized controlled trials	• The gold standard for evaluating treatments or interventions • Trials are experimental studies, providing stronger evidence of causation than observational studies • Allow strong control of the parameters to minimize bias and confounding	• Can be time consuming and costly • Can have problems with participants not complying or dropping out of the study • The generalizability may be limited • Unethical if exposure is suspected to be a health hazard, then observational studies are the source of evidence

weak evidence to explore aetiological associations. Since both exposure and outcome are measured at the same time, it is impossible to know if one appeared before the other. You may, for instance, observe that people with high blood cholesterol eat more low fat foods than the general population. In this chicken-and-egg situation, we do not know what proportion of the dietary difference was due to awareness of increased coronary risk, neither can we conclude that low fat foods raise blood cholesterol.

Case-control studies

In a case-control study, patients that have developed a specific disease or condition (the cases) are compared with disease-free individuals (the controls) that are representative of the population the cases come from (see Weblink Figure 33.2). The extent and frequency of past exposures are compared between the cases and the controls to determine if these are different. This type of study is relatively quick to conduct and is well suited to identifying risk factors for rare as well as common diseases. An example is assessment of wheat consumption before their first birthday among children that have developed coeliac disease compared to healthy children of the same age. Case-control studies allow examination of

several exposures as suspected causes at the same time, and for adjustment for possible confounders. Odds ratios are used to present the results. The disadvantage of retrospective case-control studies is that the exposure measurements may rely on recall and thereby introduce recall bias. Biological measurements may be 'objective' but they can also be biased, as levels may change after disease onset.

Cohort studies

In cohort studies, exposure status is measured in a group of people that is followed over time, recording which individuals develop the outcome of interest (see Weblink Figure 33.3). If the exposure is measured well before disease develops, information bias and behaviour changes due to emerging symptoms will be minimized. The effect measure in cohort studies is usually risk ratio, comparing the incidence of disease among those exposed to some factor compared with the unexposed cohort members.

Several large cohort studies are in progress to investigate dietary effects on cancer risk, including the European prospective investigation into cancer and nutrition (EPIC) multicentre study, the nurses' health study in the USA and the Whitehall II study in the UK. These studies have followed thousands of people over several years before there were enough cases to analyse. This long follow-up time is a disadvantage of cohort studies, because the cost of running a study for many years is high. Meanwhile, many participants drop out, exposure levels change, and researchers may wish other risk factors had been measured at study baseline.

Experimental study designs
Randomized controlled trials

Trials are experimental studies that assess the effects of an intervention or treatment by comparing it with an established treatment or *placebo* (inactive treatment). In these studies, the researchers assign either the intervention or the control treatment to study participants. To be a randomized controlled trial, the participants must be randomly allocated to the groups. Ideally, neither the participants nor the investigator should know who is in which group, meaning the study is double blinded. If the randomization procedure is truly unbiased, it will minimize confounding since participant characteristics are evenly distributed into the two groups. Trials provide strong evidence that any observed effect or difference between the study groups at the end of the experiment is due to the intervention (the exposure), not the result of any difference between the groups at baseline. Some studies randomize groups of people, for instance schools or hospitals, to the intervention and control condition rather than randomizing individuals. This design is called a cluster-randomized controlled trial.

The intervention in a randomized controlled trial can be a medical intervention, a drug, a nutrient supplement, a food or an eating pattern. A trial can have one of two aims: To evaluate either *efficacy* or *effectiveness*. This essential distinction should always be explicit. Efficacy refers to the benefit of an intervention under ideal circumstances (can it work?). Effectiveness, on the other hand, refers to the degree of benefit under real-life, routine circumstances (does it work?).

There are several difficulties with trials. First, trials depend on motivated participants who consent to active involvement in an experiment. Such participants may not be typical of the population of interest in a variety of ways. Second, in most cases, a trial would be unethical and therefore not feasible if the treatment involves a potential hazard to the participants. Third, trials are often expensive and may require a long follow-up period, for example, if chronic disease prevention is the objective. Fourth, although researchers assign and hope to control exposure levels, compliance is a challenge: to what extent do participants receive the assigned treatment or follow the instructions they have been given? Fifth, blinding can be difficult or impossible. In particular, when the intervention involves a food, an eating pattern, or some other behaviour, it cannot easily be camouflaged.

Community intervention studies

Community intervention studies involve interventions designed to promote health at group, community, or population level. Interventions may take a number of forms, using mass media, supermarkets, workplaces, and other settings believed to influence population behaviour. Preferably, such intervention studies involve a control group, but uncontrolled before-and-after designs are often found in the literature. Whereas a randomized controlled trial follows a fixed group of individuals over time, a community intervention study may use one or more surveys, not necessarily with the same respondents, in order to estimate the intervention effect in the community.

Systematic reviews

Sir Austin Bradford Hill suggested that consistency in results over time and across populations was an important criterion when judging whether an association between an exposure and an outcome could be causal (Hill 1974). Systematic reviews can summarize the results of studies with the same or similar research questions (see Weblink Figure 33.4). In a systematic review, explicit and reproducible methods are used to search for and select primary studies, to assemble the results and assess the validity of the findings. Instead of disregarding small studies that are underpowered on their own, results can be combined to increase the effective sample size and the precision of the overall effect estimate. This summarizing technique using statistics to assemble a pooled estimate of an effect is called a meta-analysis (Weblink Section 33.4). Well-conducted meta-analyses provide precise estimates of the effects of interest. It may also be possible to examine heterogeneity in results obtained from individual studies. However, results from systematic reviews are potentially subject to publication bias, although it can sometimes be possible to estimate its likely magnitude by analysing the pattern of results.

The validity of the findings from a systematic review will depend on the quality of the underlying results from the primary studies relating to the presence of bias, chance, and confounding. The GRADE approach (grading of recommendations assessment, development, and evaluation) is a systematic and explicit approach used to consider how near the estimated effect from a systematic review is to the true effect for a given outcome (Guyatt et al. 2011).

The Cochrane Collaboration (http://www.cochrane.org/) was founded in 1993 with an aim 'to help people make well-informed decisions about health care by preparing, maintaining and ensuring accessibility to systematic reviews'. The collaboration has been instrumental in developing systematic review methodology (Higgins et al. 2020). A number of other groups and organizations have been formed to support evidence-based healthcare decisions and recommendations through systematic reviews of the evidence. Initially, systematic review methodology predominantly focused on how to combine the results from clinical studies in the form of randomized controlled trials. Increasingly, systematic review methods are used to address other types of research questions and study designs.

33.5 Nutritional epidemiology

Epidemiology has expanded from its early focus on infectious disease to wide-ranging research concerning the aetiology of chronic diseases. Diet is a relevant exposure for many chronic diseases. The methodological challenges of nutritional epidemiology are active research topics. This section presents some of the important issues.

Dietary factors as exposure variables

What is a healthy diet? Before you can answer this question, you must consider what 'diet' is. Dietary intake is not one exposure variable, but a wide range of components or aspects. Important exposure measurements can be the intake of energy, nutrients, single food items, or food groups. In addition to nutrients, foods contain a large number of non-nutrient bioactive compounds that may affect well-being and risk of disease. Freshness and the bioavailability of nutrients may be additional dimensions to consider. Research into the glycaemic (blood sugar raising) effect of carbohydrates has broadened our understanding of how foods with similar nutrient content can produce very different physiological responses. Thus, dietary exposures can be determined by the amount of food, energy, or nutrients entering the mouth, and also by the quantity and rate of uptake of a food component by the body.

A dietary component can have only weak effects alone, but be influential when combined or in synergy with other dietary factors. Researchers may therefore be interested in the effect of food combinations or food patterns. In nutritional epidemiology, two main approaches are analyses of a priori and a posteriori dietary patterns. The first approach, a priori food patterns, is to define dietary patterns according to assumed health benefits. Examples are the Mediterranean diet, the New Nordic Diet or Healthy Eating index. A priori food patterns can be used to develop experimental diets in randomized

controlled trials, but are also relevant in observational studies where each individual's dietary intake is scored according to its resemblance with the dietary pattern. A posteriori dietary patterns are patterns of food choices found in dietary data sets using statistical methods such as principal components or cluster analyses. A posteriori food patterns have been identified in British adults (Pryer et al. 2000). One out of four patterns identified was characterized by high intakes of white bread, butter, margarine, sugar/confectionery, and tea, while another pattern had high intakes of whole grain bread, fish/shellfish, and fruits/nuts.

Measurement of dietary exposure

In general, a person's diet varies considerably from day to day, and dietary habits may change over the years. Day-to-day variations in intake mean it is important to consider carefully the most appropriate exposure frame given the study objective and corresponding need to measure relevant aspects of the diet. For instance, if we want to study whether dietary factors modify the risk of chronic diseases, the average diet over time is probably the most important aspect to capture. Dietary assessment methods are covered in Chapter 31. Here it is important to note that each method has its strengths and weaknesses from an epidemiological perspective.

Section 31.2 (quality of measurements) defined the validity of measurements as the degree to which a measurement assesses the aspect it is intended to measure, and the reproducibility as the degree to which a measurement produces the same result when used repeatedly. Most dietary assessment methods rely on some kind of self-report from the study participants. It is well documented that most dietary data are subject to considerable misreporting, particularly as underestimation of food intake overall (total energy intake) and as selective underreporting or over-reporting of some types of foods.

One of the central purposes of epidemiology is to examine relationships between an exposure and an outcome. The most important aspect is often the individual's exposure ranked relative to other members of the population. If all study participants underestimated their food intake to the same extent, it could still be determined who had low, medium, and high intakes of a nutrient. However, such uniform patterns of bias are not observed in practice, though, as patterns of misreporting are associated with psychosocial and behavioural characteristics of study participants (Maurer et al. 2006). Self-report dietary data are therefore less accurate and precise than the ideal for epidemiological studies. Nutritional epidemiology continues to face the challenge of large measurement errors, partly random and partly systematic. Much research effort in nutritional epidemiology has been devoted to developing dietary assessment methods suitable for studies involving thousands of people. Important aspects of the development process include documenting the validity and reproducibility of these methods, investigating the error structure of dietary data, and exploring ways to handle these errors in statistical analysis.

Measurement error is a problem because the observed effect of diet on disease risk can deviate considerably from the true effect. Epidemiologists find that if the exposure of interest is measured only with non-differential measurement error, implying that the errors are similar among the exposed and non-exposed, the effect estimate will be *biased towards null*. This means that the effect estimate for the diet–disease association will be underestimated. In some situations, the association may be missed altogether. If the errors are systematically different between study groups, the non-uniform or differential measurement errors can lead to over- or underestimates of a nutrient-health outcome effect.

Biomarkers of dietary intake are biochemical analytes in body fluids and tissues selected as predictors for levels of nutrient intake and tissue status, and thereby avoid the problem of self-reported dietary intake (see section 31.4). We can assess some, but not all aspects of the diet using biomarkers. Due to the substantial cost of collection and laboratory analysis, biomarkers may not be a practical alternative to dietary assessment in large epidemiological studies. It is important to remember that although biomarkers can provide objective estimates of dietary intake, observed diet–disease relationships based on biomarkers may still be subject to confounding. This is discussed further in the next section.

Confounding of diet–disease relationships

Rigorous understanding of nutritional epidemiology requires that one get to grips with the issue of confounding. A confounder is any factor that can cause or prevent the outcome and at the same time is associated with the exposure in a way that distorts the observed exposure–outcome association. A prominent feature of dietary variables is that they are strongly correlated with each other. For instance, persons with a high intake of vitamin C tend to have high intakes of fibre, β-carotene, and other antioxidants because of the combinations of nutrients in the same foods. Statistical analysis of dietary patterns as described earlier points to covariance in food choices, such that certain sets of foods tend to be avoided or eaten often. For instance, some adults have a tendency to have high intakes of both white bread, butter, confectionery, and fast food. In this example, food intakes are positively correlated with each other, but they can also be inversely correlated. For example, people who often eat fish tend to have low intakes of meat and vice versa.

Such associations can contribute to confounding of an exposure–outcome association. Observational studies find that high fish consumption is associated with lower incidence of colorectal cancer but is considered to be confounded by meat intake (Knuppel et al. 2020). Dietary habits are also associated with a range of non-dietary health behaviours affecting health. Eating a healthy diet may for instance be associated with being physically active and a non-smoker. Sociodemographic variables like education, work, and income levels may confound diet–disease relationships too.

It may be difficult to estimate a pure treatment effect in studies involving food because humans need a rather constant intake of energy and fluids. Any induced changes to the sources of these two variables are likely to be compensated for by changes in other sources. Possible confounding by total energy intake is an important aspect to consider in all studies of diet–disease relationships. This can be illustrated by comparing two people with the same fibre intake of 23 g/day. If one person is a small woman with an energy intake of 7 MJ while the other is a large physically active man with an energy intake of 14 MJ, the former has a fibre-rich diet while the other has a fibre-poor diet. Thus, dietary data will usually be energy adjusted for instance by expressing the intake per 10 MJ, by a statistical adjustment called the residuals method (Willett 2012) or by introducing an energy term in the analysis.

In epidemiology, it is common to try to isolate the effect of a single exposure by holding others constant through statistical modelling techniques. This corresponds to the method of standardization as described in Section 31.3 on confounding. However, when variables are strongly correlated with the factor being studied, adjusting for the influence of other factors in statistical models may be difficult and potentially misleading. Even in trials where the researchers produce all the food the participants should eat, it is difficult to manipulate only one dietary factor at the time. For instance, when you want to change the glycaemic index of a meal, you will most likely change the type and amount of fibre, protein, or fat content as well. In summary, it is very difficult to isolate the effect of a single nutrient on disease risk from the effect of other, correlated aspects of diet and lifestyle in observational studies.

Several aspects of confounding are illustrated by the case of β-carotene and risk of cardiovascular mortality (see Figure 33.4). In the 1990s, several observational studies showed that people eating more fruit and vegetables rich in β-carotene and having higher circulating serum β-carotene levels have lower rates of cardiovascular disease (risk ratios below the value one). However, when β-carotene supplementation was tested in large randomized controlled trials, there was no beneficial effect (risk ratios around or above one) (Egger et al. 1998).

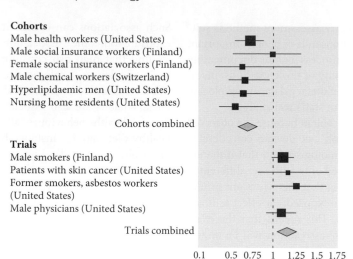

FIGURE 33.4 Results from a systematic review on the effects of serum β-carotene (exposure) and rates of cardiovascular diseases (outcome). The black squares show the relative risks (risk ratios) of individual studies with 95 per cent confidence intervals as horizontal lines.

Three of several possible explanations for this finding are related to confounding: (1) the observed relationship between β-carotene and cardiovascular mortality was confounded by other nutrients or bioactive components in fruit and vegetables; (2) β-carotene was a marker of a 'healthy' dietary pattern, and thus confounded by the combined the effect of several dietary practices; (3) the effect was due to healthy lifestyle or higher socioeconomic position among those with high levels of circulating β-carotene, confounding the observed relationship.

Future perspectives in nutritional epidemiology

Some of the most important public health problems today are coronary heart disease, diabetes, and cancers (Kivimäki et al. 2015). These disease groups have multiple and complex causal pathways where dietary factors are part of a larger picture. Dietary determinants of disease often fall into the risk factor category: they increase the probability that a particular disease or malign condition will develop, but are not classified as sufficient or necessary causes. There are numerous studies of the role of diet in the aetiology in these diseases, but it takes considerable research effort to synthesize the evidence regarding a diet–disease relationship. The World Cancer Research Fund and American Institute for Cancer Research have published extensive reviews of the evidence regarding how food, nutrition, and physical activity influence the risk of cancers (WCRF/AICR 2018). All these reviews are based on explicit criteria for how to collect, discuss, and judge the evidence as presented in the report's method section. Understanding of diet–disease relationships may change as evidence accumulates and systematic review methods become more sophisticated. Translating this

knowledge into dietary change at population level is a related and important research topic (Rees et al. 2013).

For many dietary variables, all individuals in a population are exposed and the range of exposure in that population may be rather narrow. This can pose a problem when we want to study how some dietary component affects risk of disease. In a population where everyone eats exactly the same amount of fruit and vegetables, we would not be able to show any positive or negative effects on the risk of cancer—no matter how strong the real effect is. It is more difficult to detect an association if the range of exposure in the study population is narrow compared to when it is wide. Some of the large cancer cohort studies, for instance the EPIC study, is designed to overcome this problem by studying diet–disease relationships across populations with diverse, and thus larger variation in diets.

Many of the diseases we would like to study, particularly cancer, have a long lag between the start of the relevant exposure and the manifest stages of the disease. This is the *latency period*. It takes many years for a healthy cell to undergo malignant transformation, and for that cancerous cell to develop into a detectable tumour. The relevant dietary exposure affecting the cancer risk may be 10 to 60 years back in time (WCRF/AICR 2018). Exposures could also exert their effect during critical periods in life. Researchers need to have a theory of relevant time dimensions to handle these challenges properly.

Developments in genetics have added a new research dimension in epidemiology that is relevant in the nutritional context. Individual differences in biological responses to dietary exposure are seen as a result of common genetic variation. This results in gene–environment interaction or, when the diet is involved, gene–nutrient interaction. Knowledge of how common gene variants influence nutrient status provides a novel method by which nutrient–disease effects can be investigated (Davey Smith 2011; Swerdlow et al. 2016). Since genetic variation is distributed randomly between individuals in a suitably stratified population, it is possible to produce findings from observational studies that are free of the major problem of confounding.

It is likely that diet has an important influence on many if not all common non-communicable diseases. Narrow ranges of exposure, possible synergies among nutrients, and small individual nutritional effects each present a challenge in studies of diet–disease relationships. Nevertheless, public health nutrition based on good science has a large contribution to make. Since everyone eats, a small change in diet at the population level may have a large positive effect on disease rates at a population level. The potential for dietary prevention of disease has yet to be fully understood.

KEY POINTS

- An outcome is the disease or health-related variable being studied, while an exposure is any factor that may influence the risk of the outcome.
- Measures of association and effect are the quantified relationship between exposure and outcome.
- An observed exposure–outcome association can be due to one of four factors: bias, chance, confounding, or a true causal relationship.

- Causality can rarely be proven on the basis of observational evidence alone, but if bias, chance, and confounding have been considered and ruled out, and the Bradford Hill criteria are broadly met, then causal inference may be appropriate.
- The primary question a study seeks to answer guides the decisions regarding the nature and size of the study sample, how exposure and outcome are measured, and the preferred study design.
- Observational study designs (ecological studies, cross-sectional studies, case-control studies, and cohort studies) permit examination of the world as it is, including its hazards.
- Experimental studies (trials) are powerful tools for testing health-related hypotheses (usually treatment or prevention).
- Systematic reviews are means of summarizing the existing evidence regarding a defined research question.
- The effects of diet can be studied at a number of levels: nutrients, non-nutrients, foods, and food patterns. Great care is needed in dietary assessment and in the capture of the relevant period of exposure.
- Dietary variables are strongly interrelated. It may be difficult to single out the specific dietary component which has the health effect of interest. This is the challenge of nutritional epidemiology.

 Be sure to test your understanding of this chapter by attempting multiple choice questions. See the Further Reading list for additional material relevant to this chapter.

REFERENCES

Davey Smith G (2011) Use of genetic markers and gene–diet interactions for interrogating population-level causal influences of diet on health. *Genes Nutr* **6**(1), 27–43. https://pubmed.ncbi.nlm.nih.gov/21437028/

Egger M, Schneider M, Smith G D (1998) Meta-analysis: Spurious precision? Meta-analysis of observational studies. *BMJ* **316**, 140–144.

Guyatt G, Oxman A D, Akl E A et al. (2011) GRADE guidelines: 1. Introduction—GRADE evidence profiles and summary of findings tables. *J Clin Epidemiol* **64**(4), 383–394.

Higgins J P T, Thomas J, Chandler J et al. (eds) (2020) *Cochrane Handbook for Systematic Reviews of Interventions Version 6.1* (updated September 2020). The Cochrane

Collaboration. Available from www.training.cochrane.org/handbook.

Hill A B (1974) *Principles of Medical Statistics*, 9th edn. *Lancet*, London.

Kivimäki M, Vineis P, Brunner E J (2015) How can we reduce the global burden of disease? *Lancet* **386**(10010), 2235–2237.

Knuppel A, Papier K, Fensom et al. (2020) Meat intake and cancer risk: Prospective analyses in UK Biobank. *Intern J Epidemiol* **49**(5), 1540–1552.

Maurer J, Taren D L, Teixeira P J et al. (2006) The psychosocial and behavioral characteristics related to energy misreporting. *Nutr Reviews* **64**(2 Pt 1), 53–66.

Porta M (2016) *A Dictionary of Epidemiology,* 6th edn. Oxford University Press, New York.

Pryer J A, Nichols R, Elliott P et al. (2000) Dietary patterns among a national random sample of British adults. *J Epidemiol and Community Hlth* **55**, 29–37.

Rees K, Dyakova M, Wilson N et al. (2013) Dietary advice for reducing cardiovascular risk. *Cochrane Database Syst. Rev.* **6**(12), CD002128.

Swerdlow D I, Kuchenbaecker K B, Shah S et al. (2016) Selecting instruments for Mendelian randomization in the wake of genome-wide association studies. *Int J Epidemiol.* https://pubmed.ncbi.nlm.nih.gov/27342221

United Nations Children's Fund (UNICEF), World Health Organization, International Bank for Reconstruction and Development/The World Bank (2020). *Levels and Trends in Child Malnutrition: Key Findings of the 2020 Edition of the Joint Child Malnutrition Estimates.* World Health Organization, Geneva.

WCRF/AICR Expert Panel (2018) *Food, Nutrition, Physical Activity and Cancer: A Global Perspective.* The Third Expert Report. WCRF/AICR, London.

Willett W (2012) *Nutritional Epidemiology,* 3rd edn. Oxford University Press, New York.

FURTHER READING

Introduction to epidemiology

Bhopal R (2016) *Concepts of Epidemiology,* 3rd edn. Oxford University Press, London.

Nutritional epidemiology

Willett W (2012) *Nutritional Epidemiology,* 3rd edn. Oxford University Press, New York.

34 Sustainable and inclusive global food systems

Marc J. Cohen

OBJECTIVES

By the end of this chapter, you should be able to:

- compare world food supply with population needs
- discuss the socioeconomic, agroecological, and health constraints on future food production
- explain the factors impeding access to food
- discuss the causes and impacts of increased food prices in the late 2000s
- demonstrate the multiple functions of agricultural and rural development in achieving equitable and sustainable food security
- elucidate the role of various approaches to agricultural research, including agroecology and biotechnology, in achieving food security
- make clear the relationship between sustainable natural resource management and food security
- define the terms food security, food insecurity, food supply, food demand, hunger, famine, Green Revolution, and entitlements.

34.1 Introduction

At the 1996 World Food Summit (WFS), high-level representatives of 182 nations and the European Union, including over 100 heads of state and government, agreed to the goal of reducing the number of undernourished people to half of the 1990 level by no later than 2015. They characterized the persistence of hunger as 'unacceptable' (FAO 2015). More recently, in 2015, the United Nations 2030 Agenda for Sustainable Development included 'zero hunger' as Sustainable Development Goal (SDG) Number 2.

A quarter century on from WFS, hunger is actually on the rise. Although the number of food-insecure people in the world declined for several decades prior to the 2010s, the hunger population initially remained flat thereafter. Then, between 2014 and 2019, the number of food-insecure people increased by 60 million to 690 million (FAO et al. 2020), leaving one of every 11 people on the planet hungry. Looking at food insecurity from a somewhat different perspective, 2 billion people in the world did not have regular access to safe, nutritious, and sufficient food in 2019, and 3 billion could not afford a healthy diet (FAO et al. 2020). With less than ten years to go until the SDGs' deadline of 2030, achieving zero hunger may seem extremely challenging at best, and a pipedream at worst.

Absolute food shortages are much less of a concern today than in the past, and food production and stocks remain more than adequate to meet everyone's minimum calorie requirements. Rather, *access* to food is the critical problem. Compounding low incomes and lack of productive resources among food-insecure populations is the more recent issue of the use of food and feed crops to produce energy; this contributed to food-price spikes in 2008 and 2010–2011. Also, climate change is likely to have a significant impact on food security and nutrition in the coming decades.

Despite overall supply-side adequacy, many analysts continue to worry that human population growth will outpace the earth's ability to produce adequate food supplies. This concern, voiced by Thomas R. Malthus over two centuries ago, remains a preoccupation, notwithstanding the successful role of science, technology, and public policy in addressing this dilemma. Growing demand for animal products in the developing world (OECD/FAO 2020) further fuels concern about overburdening the earth's 'carrying capacity'.

Two decades into the twenty-first century, the paradoxical nature of contemporary food insecurity is all too clear: the overwhelming majority of the world's hungry people live in rural areas and depend on food production and other agricultural activities for their livelihood. Growing integration of global markets, driven in part by advances in transportation and communications, holds great promise for better matching food supplies and needs. But rich-country farm subsidies and trade barriers make it difficult for poor farmers in developing countries to reap potential gains.

This chapter examines whether technological and institutional innovation will continue to allow food production to keep pace with population growth and rising food demand, as well as the environmental costs that this will entail. It also focuses on access to food and the bearing that globalization has on food security. The chapter concludes by looking at the policy actions needed to achieve food security in developing countries—particularly the importance of broad-based agricultural and rural development—while assuring sustainable management of the natural resource base upon which food production depends (see Weblink Section 34.1 for a review of general publications related to this chapter).

34.2 Definitions of 'food security' and 'hunger'

According to the Food and Agriculture Organization of the United Nations (FAO), '*Food security* exists when all people, at all times, have physical and economic access to sufficient, safe, and nutritious food to meet their dietary needs and food preferences for an active and healthy life' (FAO 2015). This definition is in keeping with the principle that everyone has a right to adequate food, to be free from hunger, and to enjoy general human dignity, enshrined in the International Bill of Human Rights (Weblink Section 34.2 (a) on the right to food). *Food insecurity*, then, is the absence of food security. *Hunger* means a state, lasting for at least one year, of inability to acquire enough food, defined as a level of food intake insufficient to meet dietary energy requirements (FAO 2015). *Hidden hunger* is a term sometimes used for micronutrient malnutrition.

The availability of adequate food is a necessary condition for achieving food security, but it is not sufficient. Of equal importance are *access to food, appropriate utilization of food*, and the *stability* of both availability and access (FAO 2015). Thus, even when food supplies are satisfactory, food insecurity may persist because people lack access, whether by means of production, purchase, public social protection programmes, private charity, or some combination of these, to available food. In addition, people may fail to consume sufficient quantities of food or a balanced diet even when supplies are ample (Weblink Sections 34.2 (b) on food insecurity in Sub-Saharan Africa). Sudden shocks (such as an economic, climatic, or health crisis) or cyclical events (such as seasonal food insecurity) may cause a loss of access to food.

34.3 World food supply

Past trends and current supply situation

Scholars and policy makers alike have long worried about how to balance food supplies with the demands of a rapidly growing population. Malthus claimed that population increases by a geometric ratio, whereas food production only increases by an arithmetic ratio, and thereby exerts a natural 'check' on population growth. Malthusian worries may have reached their apogee in the late 1960s and early 1970s. Then, analysts wrote off much of Asia as 'a hopeless basket case'. The threat of famine gripped West and Central Africa, Ethiopia, and Bangladesh, and it seemed as though Malthus's 'check' had kicked in with a vengeance. A spate of popular and academic tomes appeared in the industrialized countries on such topics as 'lifeboat ethics' and 'triage'.

The Green Revolution

In fact, between 1961 and 2002, per capita food production and agricultural productivity rose, while real prices for key food commodities fell steadily (World Bank 2007). By 2002, developing-country cereal harvests, at 1.2 billion tonnes, were triple those of 40 years earlier, while the population was a little over twice as large. The major factor controverting Malthusian predictions was a rapid increase in the output of cereals, the main source of calories in developing countries, as farmers in Asia and Latin America widely adopted high-yielding varieties, and governments, especially in Asia, implemented policies that supported agricultural development. As discussed below, food prices have shown greater volatility and remain above the levels of the early 2000s as of December 2020. Nevertheless, as Figure 34.1 indicates, average dietary supply adequacy (i.e., calorie availability) rose from 113 per cent of minimum needs in 2001 to 119 per cent in 2018. In Sub-Saharan Africa, the rise was from 92 per cent to 97 per cent, and in South Asia, from 105 per cent to 110 per cent. It is important to note that because these figures are based on averages, they say nothing about how the available food is actually distributed, nor do they indicate anything with regard to micronutrients.

Yield gains—attributable to a great extent to farmers' adoption of high-yielding cereal crop varieties bred at international agricultural research centres and adapted to local conditions at national agricultural research institutions—accounted for much of the increase in cereal output and calorie availability over the past six decades. Area expansion played a much less important role. Planting of these varieties coincided with expansion

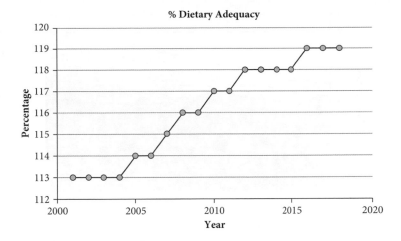

FIGURE 34.1 Global average dietary supply adequacy, 2000–2019.
Source: FAO FAOSTAT data (http://www.fao.org/faostat/en/#home).

of irrigated area and fertilizer use; this phenomenon is characterized as the *Green Revolution*. Yields increased broadly in the developing world, especially in Asia, including South Asia, with Sub-Saharan Africa lagging behind (World Bank 2007) (Weblink Section 34.3 (a) on the Green Revolution).

This evidence tends to support the views of anti-Malthusians, who argue that population growth spurs technological and institutional innovations to address the problems of resource scarcity (e.g., Pingali 2012).

Socioeconomic and environmental impact of the Green Revolution

Initial studies of the socioeconomic impact of the Green Revolution suggested that large-scale farmers were the main adopters, while small farmers faced access constraints, and benefited less because larger aggregate harvests meant lower prices. Also, some analysts contended that the technology encouraged mechanization, resulting in rural job losses. But this view is contested in later studies that argue that small farmers adopted high-yielding seeds after witnessing larger farmers' success, and that production gains generally created more non-farm rural employment opportunities than were lost. In Asia, new breeding efforts developed seeds more suitable to the needs of poorer farmers (e.g., with less need for purchased inputs such as fertilizers), and policies sought to improve provision to small farmers of services such as technical advice, credit, marketing, and access to inputs. Cereal prices did decline substantially, benefiting non-farm poor consumers in rural areas and cities alike, as well as poor farmers who were net purchasers of food (as most poor farmers are in many developing countries). Even farmers who produced more than they consumed gained, as technological advances

reduced unit costs of production and hence increased profits (Pingali 2012).

The Green Revolution had environmental benefits as well. Increased yields on existing farmland alleviated the need to clear new land in order to boost production. This is estimated to have preserved over 300 million hectares (equivalent to more than the combined total farmland of the USA, Canada, and Brazil) of forests and grasslands, including considerable wildlife habitat, thereby conserving biodiversity and limiting atmospheric releases of carbon that can cause global warming.

At the same time, widespread planting of high-yielding varieties of cereal crops in some instances has contributed to environmental problems, such as increased soil salinity and lowered water tables in irrigated areas; water, air, and soil degradation resulting from excessive agricultural input use, and human health problems due to heavy pesticide use. Harm to non-target species from pesticides offset some of the biodiversity gains (Pingali 2012; World Bank 2007).

Recent cereals supply trends

Between 2000 and 2018, production of primary crops increased by nearly 50 per cent, with total output exceeding 9 billion tonnes. Throughout this period, cereals (maize, rice, and wheat) accounted for about one-third of total production (Figure 34.2).

Supply trend of non-cereal crops

There is some evidence that increased cereal output has come at the expense of pulses and vegetables that could improve dietary quality, especially among low-income people. In Bangladesh, real cereal prices declined by 40 per cent between 1975 and 2000, thanks to widespread

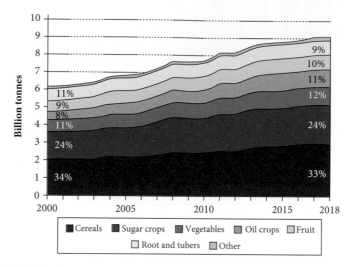

FIGURE 34.2 World crop production by commodity group, 2000–2018.
Note: Percentages on the figure indicate the shares in the total.
Source: FAOSTAT.

adoption of Green Revolution varieties, but prices for lentils (a protein- and micronutrient-rich legume), other vegetables, and animal products increased by 25–50 per cent. That said, as Figure 34.2 indicates, the share of vegetables in total world primary crop production did increase slightly between 2000 and 2018, from 11 per cent to 12 per cent.

Related to this, there was concern in the 1960s and 1970s that protein deficiency was the main nutrition problem in developing countries. This led to a major plant breeding effort aimed at developing 'protein quality' maize in the early 1970s. However, by mid-decade, nutritionists came to agree that insufficient calorie intake was the cause of the apparent 'protein gap' rather than deficiency of protein per se (Weblink Section 34.3 (b) on the 'protein gap'). Breeding thereafter focused on increasing the calorie supply by increasing the quantity of cereals available rather than their quality. Since the late 1990s, breeders have renewed their focus on nutritional quality with efforts to develop micronutrient-dense staples.

While cereals and livestock products are the main dietary staples, many low-income people in developing countries rely on root and tuber crops such as cassava, yams, potatoes, and sweet potatoes, either as their main staple or as a supplementary source of calories, vitamins, and protein. In Sub-Saharan Africa, root and tuber consumption accounts for 20 per cent of caloric intake, and cassava is the staple of 200 million people, second only to maize as the leading source of calories. These crops, particularly cassava, were once viewed as 'crisis foods' by poor consumers, who switched to consumption of these cheaper foods when cereal prices rose prohibitively. Poor farmers value cassava roots because they can be stored

in the ground for up to 18 months, and as the crop is vegetatively propagated, there is no need for purchased inputs. Its price tends to be rather stable. In recent years, many governments have come to see cassava as a strategic food-security crop, as well as an increasingly important source of feed and biofuel. Output is currently growing at 3 per cent per year, or well above population growth (OECD/FAO 2020).

Food loss and waste

FAO has estimated that up to one-third of global food supplies are lost or wasted post-harvest and all along the food chain, including in retail establishments, food-related businesses, and households. This stems from discarding still safe and edible food on aesthetic grounds or because it has reached its 'sell by' date. About 14 per cent of losses occur prior to the retail level. Reducing food loss and waste is incorporated into the SDGs (in indicator 12.3) and would contribute to more sustainable livelihoods and natural resource management, as well as reduced greenhouse gas emissions.

Future supply outlook

Cereals output is expected to increase by 14 per cent over 2020–2029. Virtually all of the gains are likely to come from higher yields, rather than expansion of area planted. Production gains are projected to exceed growth in demand, with falling real prices. Aggregate global food supplies should remain adequate to meet minimum needs, with normal stocks. Root and tuber production is projected to grow by 18 per cent during the same period (OECD/FAO 2020).

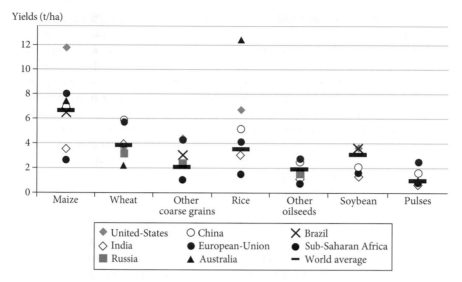

FIGURE 34.3 Projected crop yields for selected countries and regions in 2029.
Source: OECD (Organization for Economic Co-operation and Development)/FAO (2020) *OECD-FAO Agricultural Outlook 2020–2029*. FAO/Paris: OECD, Rome. https://doi.org/10.1787/1112c23b-en (accessed February 2021).

Crop yields are expected to grow substantially in Sub-Saharan Africa (16 per cent) and the Near East and North Africa (12 per cent) during 2020–2029. In North America, Europe, and Central Asia, where yields are much higher, growth rates will slow (OECD/FAO 2020) (Figure 34.3).

Population growth

Despite expansion of global food supplies, fears of an imbalance between population growth and food production endure. World population is expected to increase by 26 per cent between 2019 and 2050, from 7.7 billion to 9.7 billion. Over half of the added population will live in Sub-Saharan Africa, where the population was growing at growing at 2.7 per cent per year in 2019, compared to a global growth rate of 1.1 per cent (UNDESA 2019). (Weblink Section 34.3 (c) population growth.)

Subsistence and commercial agricultural systems

Farming systems in developing countries, and especially those in which low-income and small-scale farmers are engaged, can be conceptualized as lying on a continuum between pure subsistence agriculture, in which farm families produce exclusively for their own consumption, and pure commercial farming, in which producers sell 100 per cent of output. In practice, most small-scale farms in developing countries undertake a mix of self-provisioning and production for the market.

To the extent that farmers produce for the market and specialize, they take on additional risks. Small farmers may address some of the problems they face through collective action, for example, by forming associations and cooperatives to gain access to inputs, markets, or services. Commercial production frequently means gains in income, which in turn may lead to improvements in household nutrition. In contrast, subsistence or near-subsistence farm households tend to be extremely poor, and family members frequently work on other farms or at non-farm jobs in order to earn additional income. Nevertheless, the capacity to engage in self-provisioning offers smallholder households a crucial coping strategy in the face of shocks. Realities on the ground in developing countries involve a vast variety of locations on the above subsistence to commercialization continuum and an equal variety of outcomes in terms of profitability and food and nutrition security.

Constraints to production

Small farmers in developing countries face many problems in their struggle for sustainable livelihoods. These include socioeconomic, political, agroecological, climate change, and health constraints.

Socioeconomic and political factors

Among the socioeconomic and political limitations on food production are public policies and investments that are biased against poor farmers and consumers, women, and less-favoured areas; inadequate infrastructure; inequitable access to land and other critical resources; poorly functioning and poorly integrated markets: and lack of access to credit and technical assistance.

Whether or not people own assets, such as land or livestock, makes a crucial difference as to whether they will be poor and food insecure, as well as whether they will be able to produce food for themselves or for sale. Rural land ownership is extremely unequal in Latin America, despite extensive land reform efforts, and in some countries in Sub-Saharan Africa, such as Kenya, South Africa, Zimbabwe, and Malawi. Worldwide, landless rural people who depend on wage income tend to be among those most likely to be poor, although in Sub-Saharan Africa, smallholder farmers are just as likely to be poor. When low-income people do own land, it is generally of poor quality, with less certain access to water and less secure rights. When poor people have secure rights to assets and resources such as land, water, credit, information, and technology, they are more likely to invest in land management and sustainable natural resource use. Secure ownership or use rights over land mean that poor people have more stable incomes (either in the form of food directly consumed or the cash from selling their produce), and farm output often rises once secure property rights are in place.

Gender bias

Gender discrimination significantly constrains food production. Women play a central role as producers of food, managers of natural resources, income earners, and caretakers of household food security and nutrition. Yet female farmers frequently have less access to agricultural advisory services, education, and credit, and to labour, fertilizer, and other inputs than male farmers do, and often face restrictions on their right to own or control land (Table 34.1). In both Africa and Latin America, extension services and technical assistance focus primarily on male farmers (Botreau & Cohen 2020). In Burkina Faso, men and children provide more labour to farm plots controlled by men than to women's plots, while women primarily contribute their own labour on plots they control. Women's plots have 20–30 per cent lower yields.

Research confirms that total household agricultural output would increase if there were more equitable allocation of labour and inputs. If women farmers had the same access to productive resources as male cultivators, they could increase yields on their farms by up to 30 per cent—reducing the number of hungry people in the world by up to 17 per cent. When female African farmers obtain the same levels of education, experience, access to services such as extension and farm inputs that currently benefit male producers, they increase their yields for maize, beans, and cowpeas (all crops consumed by poor people) by 22 per cent over current levels (Botreau & Cohen 2020; Quisumbing 2003).

In addition, the social status and degree of empowerment of women has a direct bearing on the sustainability of natural resource management. To the extent that rural women enjoy ownership or control over land, the more likely they are to undertake natural resource-conserving measures. Poor rural women also often expend much time and effort in fetching fuel, usually in the form of wood, for cooking, lighting, and agricultural processing. Affordable technologies that reduce the time and effort women need to spend searching for fuel allow them to spend more time on other tasks, for example, child care, which is a key input into good child nutrition. At the same time, the development of affordable alternatives to wood as an energy source can reduce deforestation. In a

TABLE 34.1 Share of household agricultural land area held by women, men and jointly (%)

Country (date)	Definition of ownership	Women	Men	Joint
Ethiopia (2011–2012)	Registered	15	45	39
Malawi (2010–2011)	Owned	40	42	18
Niger (2011)	Owned	9	62	29
Nigeria (2010)	Right to sell (use as collateral)	4	87	9
Tanzania (2010–2011)	Owned	16	44	39
Uganda (2009–2010)	Owned	18	34	48
Bangladesh (2011–2012)	Documented	10	88	2
Timor-Leste (2007)	Land manager	12	88	n.a.
Tajikistan (2007)	Owner	14	86	n.a.
Viet Nam (2004)	Owner	72	15	13

Note: n.a. = not applicable.
Source: Doss & Quisumbing (2021).

similar 'win–win' approach, integrated pest (IPM) management technologies can help protect natural resources while also reducing the time women and children must spend weeding crops (Quisumbing 2003).

Lack of voice and policy bias

Low-income and food-insecure people frequently face an array of policy biases and lack political voice and organizations that are accountable to them and capable of articulating their interests to policy makers and other power holders. As a result, policies tend to benefit people who are already well off, and policy makers tend to give low priority to the needs of poor and hungry people or programmes that would benefit them. World Bank interviews of low-income people in developing and transition countries found that they regard their situation as one where freedom and the power to control one's life are lacking. A low-income Jamaican woman compared poverty to 'living in jail, living in bondage, waiting to be free'. Strong organizations that are representative and democratically controlled can ensure that poor people can engage in active citizenship.

Agroecological constraints

In many developing countries, poverty, low agricultural productivity, and environmental degradation interact in a vicious downward spiral. This is especially pronounced in resource-poor areas that are experiencing high rates of population growth and are home to many millions of food-insecure people, particularly in South Asia and Sub-Saharan Africa, as well as on the hillsides of Central America and Southeast Asia. Agricultural growth, poverty alleviation, and environmental sustainability are not necessarily complementary. Poor farmers throughout the developing world often face low soil fertility and lack of access to plant nutrients to enhance or maintain the soil, along with variable weather and acid, salinated, and waterlogged soils that contribute to low yields, production risks, and natural resource degradation (Ruben et al. 2006).

Less-favoured areas may be 'less-favoured' by nature or by humans. They include lands that have low agricultural potential because of limited and uncertain rainfall, poor soils, steep slopes, or other biophysical constraints, as well as areas that may have high agricultural potential but have limited access to infrastructure and markets, low population density, or other socioeconomic constraints. These areas are home to about 40 per cent of the world's chronically poor people. Low agricultural productivity and land degradation are severe and deforestation, overgrazing, and soil erosion and soil nutrient depletion are widespread (Ruben et al. 2006).

Some natural resource degradation in agricultural areas has been caused by the misuse of modern farming inputs (especially pesticides, fertilizers, and irrigation water in high-potential areas). But a great deal of environmental degradation is concentrated in resource-poor areas that have not adopted modern technology.

Globally, degradation between 1945 and 1990 caused cumulative crop productivity losses of 5 per cent, with mean reductions for Sub-Saharan Africa of 6.2 per cent (World Bank 2007). Since 1990, 25 per cent of cropland and 30 per cent of land used for grazing livestock has degraded (Place et al. 2021). Losses to pests reduce potential farm output value by 50 per cent. In developing countries, losses greatly exceed agricultural aid received.

Better management of farm inputs can reduce their negative environmental effects without a loss in farm yields. This includes integrated pest management, judicious use of synthetic pesticides and pest-resistant plant varieties, improved application of fertilizers (both mineral and organic) and water, and no- or low-tillage crop management.

Unless properly managed, fresh water may well emerge as the key constraint to global food production, particularly in Central and Western Asia, North Africa, and much of Sub-Saharan Africa, where population growth is expected to continue to be high and exploitable per capita water resources are quite low. While water supplies are adequate in the aggregate to meet demand for the foreseeable future, water is poorly distributed across countries, within countries, between seasons, and among multiple uses. Agriculture accounts for 70 per cent of global water consumption, and irrigated agriculture produces 40 per cent of the world's food on less than one-third of the cropland. However, two-thirds of the world's population lives with water scarcity for at least one month per year (Ringler et al. 2021).

Because of excessive withdrawals, a number of large rivers no longer reach the sea during a part of the year. Overuse of groundwater leads to falling water tables and saline intrusion, making further cultivation of large swaths of land impossible. Poorly managed use of agricultural chemicals pollutes water, posing a serious public health threat (Ringler et al. 2021).

The International Food Policy Research Institute (IFPRI) projects that industrial and household water demand will grow faster than agricultural demand between 2020 and 2050, but agriculture will remain the primary water user throughout that period. Even a moderate expansion in irrigated agriculture during that period can offset declining yields and rising food prices due to climate change (Ringler et al. 2021) (Weblink Section 34.3 (d) agroecological constraints).

Climate change

There is now scientific agreement that climate change and variability will have substantial and long-term negative effects on food security and nutrition, particularly because of more frequent and intense droughts and floods. These will reduce food production and consumption, as well as dietary diversity, and increase the incidence of diarrhoeal and other infectious diseases (Thompson & Cohen 2012; OECD/FAO 2020). Projections by IFPRI indicate that climate change will cause yield declines of 5–7 per cent for cereals, soybeans, and sugar crops between 2015 and 2050, leading to real price increases of 10–15 per cent. This will leave 500 million people at risk of hunger, including 30 per cent more vulnerable people in Sub-Saharan Africa than there would be in the absences of climate change (Rosegrant et al. 2021).

Deforestation and agricultural activities in developing countries account for over 20 per cent of the greenhouse gas emissions that can cause climate change. Climate change is likely to exacerbate declining reliability of irrigation water supplies and competition for water, particularly in the world's driest areas, where 1 billion people live in extreme poverty. Agricultural output in developing countries is expected to decline by 10–20 per cent by 2080 as a result (World Bank 2007; Thompson & Cohen 2012). In 2019, according to the International Organization for Migration, disasters from natural hazards (mainly heavy storms) displaced nearly 25 million people, putting them at risk of food and water insecurity, ill health, malnutrition, and increased likelihood of conflict.

Appropriate agricultural practices can mitigate climate change by increasing soil carbon sinks and reducing greenhouse gas emissions, at low cost. These practices include reduced deforestation, more sustainable forest management, and adoption of agroforestry, which integrates tree and crop cultivation (Thompson & Cohen 2012) (Weblink Section 34.3 (e) on the impacts of climate change).

Health constraints

Infectious disease has a significant bearing on food production, distribution, and consumption. Initial analysis of the impact of the coronavirus pandemic that emerged in 2019 (COVID-19) on food security indicates that it has not disrupted food production. However, widespread lockdowns associated with efforts to contain the disease have affected employment opportunities and the movement of food from farm to market to fork, and may also ultimately affect farmers' access to seeds, tools, fertilizer, and pesticides. For consumers, the pandemic has had severe impacts on access to food, especially for poor people, due to loss of income and employment. According to the FAO, depending on the depth of global economic contraction brought on by COVID-19, the disease may have rendered as many as 132 million people hungry in 2020 (FAO et al. 2020). The global fight against COVID-19 risks taking time, attention, and resources from efforts to combat other deadly diseases that have a bearing on food security.

Nearly 26 million people in Africa live with HIV/AIDS (more than two of every three people living with HIV/AIDS worldwide), with 1.1 million new infections and 470,000 deaths in 2018 (WHO-Africa 2021). HIV/AIDS limits labour participation in agriculture (by those afflicted and those affected, i.e., caring for the afflicted), and leads to a decline in the transfer of farmer knowledge across generations, weaker collective action, weaker property rights, a declining asset base, breakdown of social bonds, loss of livestock, and reliance on crops that are easier to produce but less nutritious and economically valuable.

A majority of the people affected by HIV/AIDS work in agriculture. As labour becomes depleted due to disease, new cultivation technologies and crop varieties need to be developed that do not rely so much on labour, yet allow crops to remain drought-resistant and nutritious. Innovations such as farmer field schools can facilitate the transfer of community-specific and organization-specific knowledge within and between generations. Making institutions, including agricultural research centres, more client-focused can help natural resource management remain effective in the presence of weakened social capital (i.e., the norms and networks that allow collective action, especially at the community level) and property rights. For example, where there are large numbers of women widowed by AIDS, gender-equitable land ownership rights are ever more important.

Because malaria often strikes during harvest time, it also threatens food output. According to the World Health Organization, malaria affected 229 million people and killed 409,000 in 2019, with 94 per cent of the cases and deaths in Africa. Agricultural practices also pose threats to health. Poor irrigation practices increase the spread of malaria-carrying mosquitoes. Excessive and poorly managed pesticide spraying leads to poisoning that affects nearly 400 million farmers annually (44 per cent of the world's agriculturalists), with 11,000 fatalities per year. Large-scale livestock operations raise the risk that animal diseases, such as avian influenza, will spread to humans (World Bank 2007).

It is critical to explore integrated efforts to address development problems across sectors. For example, the use of drip irrigation can make agricultural water use more efficient while denying habitat to malaria-carrying mosquitoes.

34.4 **World food demand and access to food**

Demand for food

Food demand derives primarily from income growth, population growth (see 'Population growth', above), and urbanization. According to the World Bank (2021), the economic effects of the coronavirus pandemic drove an economic contraction of nearly 1 per cent in low-income countries in 2020, and will mean limited economic growth in 2021–2022. The pandemic erased gains against poverty in many developing countries and emerging markets, with poverty returning to the levels of 2017.

In 2018, 55 per cent of the world's population lived in urban areas, compared to 30 per cent in 1950. By 2050, the figure is expected to reach 68 per cent. China, India, and Nigeria alone will account for 35 per cent of the growth in the world's urban population (UNDESA 2019).

When people move to cities, their lifestyles become more sedentary, and women experience higher constraints on their time due to work and care responsibilities. As a result, urban dwellers tend to shift consumption to foods that require less preparation time (e.g., from sorghum, millet, maize, and root crops to rice and wheat), and to more meat, milk, fruit, vegetables, and processed foods.

Cereals are the most consumed agricultural product. Demand for animal feed in both developed and developing countries will drive overall growth in demand for maize and other coarse grains between 2020 and 2029, as demand for cereal-based biofuels stagnates in the developed world. Projections indicate that feed demand will increase by 18 per cent during the decade. Meat production will be 12 per cent higher in 2029 as compared to 2017–2019, with poultry demand in developing countries driving the increased output (OECD/FAO 2020).

Food-price volatility

Between 1961 and 2002, real world food prices declined steadily, reflecting increased cereals output that resulted from adoption of Green Revolution technologies. There was a sharp price drop between 1975 and 1985. Beginning in 2002, real food prices began to rise, and by the mid-2008 had climbed 64 per cent. In nominal terms, wheat and maize prices were triple the level of early 2003, while the price of rice ballooned five-fold. Milk prices also tripled, while beef and poultry prices doubled (Figure 34.4). A number of structural and conjunctural forces aligned to drive food prices up, including rising energy prices, subsidized biofuel production (Weblink Section 34.4b), income and population growth, globalization, urbanization, land and water constraints, underinvestment in rural infrastructure and agricultural innovation, lack of access to inputs, and weather disruptions. National trade policies aimed at easing the effects of the price increases, such as export bans and import subsidies, contributed to volatility on international food markets. Depreciation of the US dollar after 2005 also pushed the dollar prices of commodities upwards, and contributed to increased speculation as a hedge against further dollar depreciation (Botreau & Cohen 2020).

Increased food prices benefit households that are net producers of food, but have an adverse impact on households that are net buyers. Most urban households (regardless of income level) are net buyers, as are many poor rural households in developing countries (Table 34.2 shows figures at the time of the 2008 food-price spike). When food prices rise (or incomes fall), families will try to maintain staple consumption, often by purchasing cheaper staples and eliminating animal source foods,

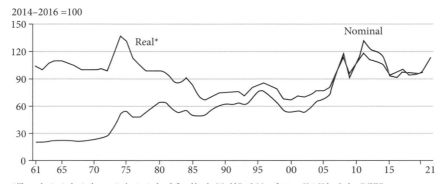

FAO Food Price Index in nominal and real terms

*The real price index is the nominal price index deflated by the World Bank Manufactures Unit Value Index (MUV)

FIGURE 34.4 Global food prices, 1961–2021.
Source: FAO Food Price Index (http://www.fao.org/worldfoodsituation/foodpricesindex/en/), January 2021.

TABLE 34.2 Net buyers of staple foods (% of reference population)

Country	All households	Urban poor	Rural poor
Bangladesh	77	96	83
Guatemala	91	98	82
Malawi	93	99	95
Nicaragua	90	94	73
Pakistan	84	96	83
Tajikistan	91	97	77
Viet Nam	46	100	41

Source: FAO, *The State of Food Insecurity in the World 2008,* Rome: FAO, accessible at ftp://ftp.fao.org/docrep/fao/011/i0291e/i0291e00.pdf

fruits, and vegetables from the food budget (Botreau & Cohen 2020). As a result, a food price increase of 50 per cent in Bangladesh will lead to a 30 per cent rise in iron deficiency among women and children, with negative effects on maternal mortality and children's physical and cognitive development and their productivity and earnings as adults (World Bank 2007).

After peaking in mid-2008, cereal prices declined 30–40 per cent in the third quarter of the year, due to the worldwide recession, good weather, and farmers' favourable production responses to higher prices in many countries. Food prices spiked again in 2010–2011, but declined steadily over the next five years, as farmers again responded to higher prices with greater production. Since 2016, prices have shown a high degree of volatility, and turned sharply upwards during the last quarter of 2020 and early 2021 (Figure 34.4). Many analysts believe that food-price volatility has become 'the new normal' in an

era of global climate change and high weather variability. Real prices have remained substantially above those of 2002–2006 in the years since, leading most observers to conclude that prices will not return to those levels, due to continued strong demand for cereals for food and feed, as well as to structural land and water constraints.

Policies that enhance the productivity of smallholder farmers will help them to respond to higher prices with increased output, thereby eventually benefiting consumers with increased supplies and reduced prices. Private sector actors in food-processing and retail industries have an important role to play in ensuring supply response (Weblink Section 34.4 (a) rising food prices).

Current state of food insecurity and future outlook

According to FAO, as of 2019, there were 690 million food-insecure people in the world (Table 34.3). Food insecurity is global and can be found even in the richest countries. However, the problems are more severe and affect a far greater proportion of the population in developing countries. The number of hungry people in the world has increased each year since 2014, with an additional 60 million people living in food insecurity by 2019. The population living in hunger in Sub-Saharan Africa rose by 16 per cent, to 235 million people, between 2015 and 2019. South Asia and Sub-Saharan Africa are home to more than 70 per cent of all food-insecure people, and form hunger's centre of gravity. Although the proportion of the world's people living in food insecurity fell from 12.6 per cent in 2005 to 8.9 per cent in 2015, it essentially remained at that level through 2019 (FAO et al. 2020). Real people with names and faces stand behind these hunger numbers (Box 34.1).

TABLE 34.3 Food insecurity by region, 2005, 2018, and 2030 (projected) (millions of people and percentage of population)

Region	2005		2018		2030	
	Number	%	Number	%	Number	%
World	825.6	12.6	678.1	8.9	841.4	9.8
Sub-Saharan Africa	174.3	23.9	221.8	21.4	411.8	29.4
West Asia and North Africa	42.6	10.9	45.4	8.9	63.5	10.4
Latin America and the Caribbean	48.6	8.7	46.6	7.3	66.9	9.5
East Asia	118.6	7.6	no data	<2.5	no data	<2.5
South East Asia	97.4	17.3	64.2	9.8	63	8.7
South Asia	328	20.6	261	13.8	203.6	9.5

Source: FAO, IFAD, UNICEF, WFP & WHO (2020) *The State of Food Security and Nutrition in the World 2020: Transforming Food Systems for Affordable Healthy Diets*. FAO, Rome.

BOX 34.1
Profiles of hunger

Kone Figue is a mother of six in Ponoundogou in northern Côte d'Ivoire. She and her husband farm six hectares of government-owned land. They grow cotton and groundnuts, mostly for sale, and maize (to eat and feed to their two cows) on about half the land, along with some yams and cassava. They plant the rest of the land to rice, their family's main staple food.

But the rice crop does not stretch out to provide a whole year's worth of meals, so the family ends up consuming the yams and cassava, even though they much prefer rice (preferably spiced up with some of the groundnuts).

Kone weeds and harvests the rice by hand, with the aid of simple tools like a hoe and a sickle. Her husband clears the land and sows the seeds. It is backbreaking work. Kone cannot afford to buy fertilizer or chemical weed killers, and even with manure from her cows, her yields are meagre.

They [the children] sometimes just get sick for no reason. Sometimes it's because of lack of food. We are poor. We have no money to buy or to feed ourselves. Now, everything is so expensive that we can only buy pasta, salt, and oil. Some days we have nothing to eat but chichita [a drink] because there is no money.
—A mother, Voluntad de Dios, Ecuador

I am very worried about my children's future. Now we have 2.5 acres of land. I have three sons. When they divide the land, each one of them will get only 0.8 acre, which is of no use. I don't know how they will survive.
—Fulmani Mandi, Jharkand, India

Before the coronavirus outbreak, Bone Kortie, 43, sold milk in Paynesville, a suburb of Monrovia, Liberia. Kortie was so well known for her goods that she earned the nickname 'Cold Milk' from her regulars. When COVID-19 hit, however, her business dried up. 'People are afraid to buy the milk', she says. She was losing so much money that she decided to stop selling it. This left Kortie and her eight children with just one meal a day—and on particularly bleak days, no meals. Those days were 'no food days', she says.

Even before COVID-19 hit, four out of five Syrians lived below the poverty line, one of the consequences of an almost decade-long war. Now, the coronavirus pandemic has amplified the crisis.

Marwan, a farmer in Eastern Ghouta, Syria, commented, 'Two months ago, we started to feel the impact of the coronavirus crisis. Our income was dwindling, and food prices continued to skyrocket. What we earned from last season's harvest couldn't cover my family's basic expenses. . . . Purchasing new seeds, after prices have increased dramatically, was out of the question, and so, for us, preparing for next season's harvest was out of reach.'

Access to food

The growing number of food-insecure people in the world may seem paradoxical in light of adequate food availability (see 'Past trends and current supply situation', above). However, current food availability is sufficient to provide everyone with their minimum calorie requirements *if the food were distributed according to need*; food is *not* so distributed.

People obtain access to the food that is available through *entitlements*, that is, the amount of food or other necessities that they can command based on their income and assets, given the legal, political, social, economic, and cultural context in which they live. Thus, a farmer may produce food, but may have to deliver some of her harvest to a landlord before she can consume what remains. Likewise, a wage earner's income permits her to command a certain amount of food. Government programmes, private charity, and gifts are other forms of entitlement (Weblink Section 34.4 (d) on social protection programmes).

Poverty drives food insecurity

Food insecurity persists primarily because of poverty. Low-income people cannot afford to buy all of the food they need, even though poor households typically spend 50–70 per cent of their income on food. In addition, poor people frequently lack access to land and other productive resources, and so cannot produce food for themselves. Female-headed households are disproportionately poor, food-insecure, and without land or other assets (Botreau & Cohen 2020). According to the World Bank, as of 2014, 774 million people, or 10.6 per cent of the world's population, lived on the equivalent of less than US$1.90 per day, in a state of extreme poverty, and could not meet their needs for food and the other necessities of life on a sustainable basis (Table 34.4). The Bank estimates that the economic disruption that resulted from the coronavirus pandemic pushed an additional 100 million people into extreme poverty in 2020, with over the half the increase in South Asia and one-third in Sub-Saharan Africa.

TABLE 34.4 People living on less than $1.90 per day (2011 PPP), 1990 and 2014 (millions of people and percentage of population)

Region	1990		2014	
	No.	**%**	**No.**	**%**
World	1912	36	774	11
Sub-Saharan Africa	284	56	408	42
East Asia and Pacific	977	61	53	3
South Asia	552	49	262	15
Latin America and the Caribbean	66	15	25	4
Middle East & N. Africa	15	7	10	3
Europe and Central Asia	31	9	2	

Source: World Bank Povcalnet Database (https://pip.worldbank.org/poverty-calculator) January 2021.

Despite rapid urbanization in developing countries, the overwhelming majority of the world's extremely poor people (80 per cent) remain in rural areas (De la Ocampo et al. 2018). However, as the world is becoming more urban, poverty and food insecurity are increasing in the cities. Poor urban dwellers are more dependent on money income, often have fewer opportunities to grow their own food, and frequently require access to childcare in order to pursue income-earning opportunities. The needed resources to address urban food insecurity may not be land so much as economic opportunities, such as secure employment at a wage adequate to meet basic needs, or the chance to own a business, as well as access to social safety-net programmes (Weblink Sections 34.4 (c) poverty drives food insecurity and 34.4 (d) social protection programmes).

Food emergencies

Disasters from natural hazards, violent conflicts, economic collapse, political crises, or some combination of these factors create food emergencies. Whilst these emergencies are often transitory in duration, they may have long-lasting impacts on the affected people. Research in Africa and Asia indicates that drought can lead to child malnutrition that in turn causes poor school performance and reduced lifetime earnings. Protracted conflicts cause ongoing displacement, disruption to livelihoods, and severe food insecurity. For example, in November 2020, the FAO-hosted Integrated Food Security Phase Classification system reported that more than 13 million people in Yemen faced acute food insecurity due to the ongoing violence there.

That same month, the UN World Food Programme estimated that climate-related and conflict shocks combined with the economic losses due to COVID-19 had left 271 million people acutely food insecure (WFP 2020). Of particular concern are refugees and internally displaced persons, who often live in camps and depend on humanitarian aid for survival. At the end of 2019, the number of forcibly displaced people reached 79.5 million, a figure not seen since World War II.

Even after conflict ends, the costly burden of reconstruction may leave many people food insecure for years. Landmines continue to maim and keep land out of production long after fighting ends.

Not only does violent conflict cause hunger, but hunger can also contribute to conflict, especially when resources are scarce and perceptions of economic injustice are widespread, as in Rwanda in 1994 or Central America in the 1970s and 1980s. The high food prices of recent years have led to widespread protests, particularly in urban areas of both the developing and developed countries, and sometimes to violence.

Where armed conflicts and civil strife occur, governments and the international community must give priority to conflict resolution and prevention. It is essential to expand and strengthen early warning systems and timely response mechanisms for food and political crises, to include conflict prevention in food security and development efforts, and to link food security and long-term sustainable development to humanitarian assistance programmes. Savings from conflict avoidance should be calculated as returns to aid and development spending. Humanitarian assistance must include agricultural and rural development components that lead to secure livelihoods and build sustainable social and agricultural systems. Development programmes should be implemented so as to avoid competition and foster cooperation among groups or communities, especially in conflict-prone areas.

Famine is a catastrophic disruption of the social, economic, and institutional systems that provide for food production, distribution, and consumption. Contemporary famines stem less from crop failure than from the political and financial failures of governments to prevent famine and respond effectively. The emergence of 'new-variant famines', in which HIV/AIDS interacts with violence, disasters from natural hazards, and/or political–economic failure, means that the margin for coping and recovery has narrowed greatly among vulnerable people, especially in Sub-Saharan Africa.

34.5 Globalization, international trade, and food security

Global agricultural exports have more than tripled in value and more than doubled in volume since 1995, exceeding US$1.8 trillion in 2018. The volume of global cereal trade grew nearly five-fold between 1961 and 2019, led by developing-country imports of wheat for food and maize for feed. Developed countries (mainly the USA, European Union, and Australia) heavily dominate world agricultural export sales, along with a handful of developing countries (mainly Argentina and Brazil, with India, Pakistan, Thailand, and Vietnam lagging far behind). Global cereal exports are projected to rise to 517 million metric tons in 2029, compared to 300 million metric tons two decades earlier (OECD/FAO 2020).

Globalization—that is, the growing integration of global markets for goods, services, and capital resulting in part from technological developments in transportation, information, and communications—offers significant new opportunities for economic growth in most developing countries, but it also carries significant risks: the inability of many developing-country industries to compete in the short term; the potential destabilizing effects of uncontrolled short-term capital flows; and increased exposure to price risk and worsening inequality as many poor people and marginalized regions may get left behind. Managing risks while exploiting growth opportunities will be a key challenge for developing countries in the years ahead.

The 1994 Uruguay Round of trade negotiations under the aegis of the General Agreement on Tariffs and Trade (GATT) resulted in creation of the World Trade Organization (WTO) and the inclusion of agriculture into rule-governed trade for the first time (Weblink Section 34.5 (a) on GATT and WTO). In response to the WTO Agreement on Agriculture and structural adjustment programmes enacted in the 1980s and 1990s with the strong encouragement of aid donors, many developing countries have liberalized food and agricultural trade (Weblink Section 34.5 (b) on structural adjustment). The developed countries have not reciprocated, instead maintaining barriers to high-value imports from developing countries such as beef, sugar, groundnuts, dairy products, and processed goods. Losses due to these trade barriers are not offset by developed countries' preferential trade schemes for specified quantities of certain developing-country exports.

Wealthy and upper-middle-income countries transferred more than US$500 billion per year (more than 3.5 times the annual development assistance provided to developing countries) to their own farmers between 2017 and 2019. Depending on how these farm payments are structured, they can have depressing effects on world prices, making developed-country exports cheaper than domestic produce and export crops in many developing countries. For example, European Union (EU) sugar subsidies under the Common Agricultural Programme (CAP) secured a 40 per cent world market share, at the expense of such non-subsidizing developing-country

exporters as Malawi, Thailand, and Zambia (Weblink Section 34.5(c) on the Common Agricultural Policy (CAP)). The EU has made some reforms to this system, but still makes substantial annual payments to its sugar farmers. Under pressure from the United States, Haiti virtually eliminated its tariff on imported rice in the mid-1990s. As a result, sales of subsidized US rice severely undercut local producers, accelerated migration to the cities as agricultural livelihoods collapsed, and increased Haitians' exposure to volatile global rice prices. In the early 2000s, subsidies to less than 25,000 US cotton farmers exceeded the gross national incomes of some of the poor West African cotton exporting countries that depend on cotton revenues for livelihoods and government budgets. The subsidies helped the United States to capture 40 per cent of the global cotton market, and the WTO declared them to be in violation of trade rules.

In 2007, the World Bank estimated that developed-country trade barriers and subsidies deprived developing countries of $17 billion per year in export earnings, a sum four times greater than annual aid to agriculture and rural development. But many developing countries still lack the administrative, technical, and infrastructural capacity to meet quality requirements in export markets and take full advantage of existing global trade rules or influence the creation of new ones (World Bank 2007).

More open global agricultural markets would increase developing countries' share of agricultural exports, and would be especially beneficial to countries in Latin America and Sub-Saharan Africa. However, agricultural trade liberalization would also increase the cost of food imports, leaving net-food-importing countries worse off. Low-income net purchasers of food would be particularly adversely affected. It is therefore important to have safety-net programmes available to address the impact of anticipated trade liberalization on these vulnerable groups (World Bank 2007).

In the Doha Round of international agricultural trade negotiations, begun in 2001 under WTO auspices, and still stalemated two decades later, developing-country delegates have proposed a 'development box' and a 'food security box'. Basically, these would allow developing countries to protect and subsidize efforts to achieve food security and agricultural and rural development. Developing countries have also sought increased access to developed-country markets. The Group of 20 (G20), composed of developing-country agricultural exporters and led by Brazil, China, India, and South Africa, has pressed developed-country exporters to reform their trade policies, but disagreements between the United States and European Union have stalled the negotiations. According to a study by IFPRI, a successful conclusion to the long-running WTO negotiations could lead to increased food production, better trade-related infrastructure, and reduction in tariffs, trade-distorting domestic farm subsidies, and export subsidies, thereby increasing global consumer welfare substantially (Bouet & Laborde 2017).

Coalitions with certain groups of higher income countries may help developing countries to improve their bargaining position in pursuing better access to industrialized countries' markets and other measures to ensure that trade supports food security. Without appropriate domestic economic and agricultural policies, however, developing countries in general and poor people in particular will not fully capture potential benefits from trade liberalization. The distribution of benefits will be determined largely by the distribution of productive assets.

Another aspect of food globalization is the expansion of supermarkets in developing countries. Issues related to this development include whether poor farmers will be able to meet quality standards and whether large-scale food marketing will meet the needs of poor consumers in terms of both affordability and accessibility.

34.6 Broad-based agricultural and rural development is key to food security

Ironically, although food insecurity results more from problems of access to food than from lack of food availability, broad-based agricultural and rural development must be at the centre of any strategy to achieve food security in the developing world. In order for development to be broad-based, economic growth in rural areas generated by farm and non-farm activities is necessary, but not sufficient. Policies must ensure that smallholder farmers and other rural poor people, including women and men alike, have political voice and access to the economic opportunities resulting from that growth.

Broad-based agricultural and rural development is essential to food security because the substantial majority of poor and food-insecure people remain rural, and agriculture and associated rural activities will remain their main sources of income (De la Ocampo et al. 2018). Low agricultural productivity in developing countries results in high unit costs of food, poverty, food insecurity, poor nutrition, low farmer and farm worker incomes, little demand for goods and services produced by poor non-agricultural rural households, and urban unemployment and underemployment. Research has shown that for every new GB pound of farm income earned in low-income developing countries, income in the economy as a whole rises by up to £2.60, as growing farm demand generates employment, income, and growth economy-wide. As agricultural production increases, it generates demand for inputs and implements. The need to process food and

agricultural raw materials also stimulates rural non-farm activities, which offer employment, management, and entrepreneurial opportunities for rural poor people, including women (World Bank 2007).

The development of well-functioning and well-integrated markets for agricultural inputs, commodities, and processed goods, especially in rural areas, will contribute enormously to poverty alleviation, food security, and the overall quality of life in developing countries (World Bank 2007). A *market* can be any context in which the sale and purchase of goods and services takes place. There need be no physical entity corresponding to the market; it might consist of a global telecommunications network on which company shares are traded. A *market economy* is an economic system in which decisions about the allocation of resources and production are made on the basis of prices generated by voluntary exchanges among producers, consumers, workers, and owners of the factors of production (i.e., land, labour, and capital).

Economists have long argued that market performance improves and marketing costs fall when the government no longer monopolizes trade and a competitive private sector emerges. Yet even as the government reduces its role, competent public administration will remain essential to promote and protect human rights, enforce contracts, maintain grading and quality control standards, regulate market conduct and investment, maintain public safety and health, create infrastructure (roads, storage facilities, and water works), provide agricultural research and technical advisory services, implement credible and sustainable macroeconomic policies, and provide a favourable environment for savings and investment and accurate and transparent incentives for consumers and producers alike. In other words, the public sector must continue to ensure the provision of *public goods.*

Markets alone cannot guarantee equity. Key public policies and investments must ensure that:

- poor farmers have access to yield increasing crop varieties,—including drought- and salt-tolerant and pest-resistant varieties—improved livestock, and other yield-increasing and environment-friendly technology

- poor farmers likewise have access to productive resources, including land, water, tools, fertilizers, and pest management

- smallholders have opportunities to participate in production of export crops, as this will have spillover effects on input use and food crop productivity, increase access to markets, and have a beneficial impact on income and food security

- institutional barriers to the creation and expansion of small-scale rural credit and savings institutions are removed, and credit is accessible to small-scale farmers, traders, transporters, and processing enterprises (Weblink Section 34.6 (a) on credit and other financial services)

- primary education, health care, clean water, safe sanitation, and good nutrition are available for all.

To succeed, agricultural and rural development programmes must be implemented within an appropriate policy context. This includes good governance—the rule of law, transparency, sound public administration, democratic and inclusive decision-making, and respect for human rights. Democratic governments are more likely to be responsive to the needs of all their citizens and to make food security a high priority. To ensure responsive policies, poor people need accountable organizations that articulate their interests. Farmer associations and cooperatives can help ensure that small farmers have access to inputs, credit, markets, and opportunities to engage in more diversified, higher-value crop production. In addition, trade, macroeconomic, and sectoral policies must not discriminate against agriculture, poor people, women, or less-favoured areas (World Bank 2007).

Boosting public investment

Developing-country governments underinvest in agriculture and rural development, despite their critical role in poverty alleviation and economic growth. Many policy makers believe that agriculture is a declining sector, and have put resources instead into industry and urban development, which tend to have more politically potent constituencies. Long-term declines in real food prices up until the early 2000s contributed to a sense of complacency about agriculture among both donor and developing-country governments. Also, in the 1980s and 1990s, there was a tendency in development circles to stress natural resource management, gender equality, and democratization without linking programmes to the agricultural context that remains central to the livelihoods of most poor people.

In 2017, according to FAO data, African governments devoted an average of 2.3 per cent of their budgets to agriculture, down from 3.7 per cent in 2001. For the Asia-Pacific region, the figure was 3 per cent in 2017. World Bank data indicate that African governments spent an average of 4.5 per cent on the military that year. In the 2003 Maputo Declaration, African heads of state had pledged to raise agriculture's budget share to 10 per cent, a pledge renewed in the 2014 Malabo Declaration (Weblink Section 34.6 (b) on the Comprehensive African Agricultural Development Programme).

Aid to agriculture

Official development assistance (ODA) donors have also underinvested in agriculture and rural development. In 2017, according to the FAO, donors provided $11 billion to the sector, accounting for just 4.4 per cent of ODA disbursements. In 1979, the figures were $16 billion and 20 per cent, respectively. In 2017, Africa received 41 per cent of ODA to agriculture. In addition to the important question of aid quantity, donors need to rethink their four-decade emphasis on reducing governments' economic role, which has contributed to developing-country underinvestment in agriculture.

Agricultural research: a global public good

Public agricultural research—that is, agricultural research that is publicly funded and generally carried out by national government agencies or international organizations—played a crucial role in the success of the Green Revolution. More recently, it is the public sector that has carried out virtually all of the research on so-called 'orphan crops', that is, crops widely consumed by poor people but for which markets are poorly developed and offer little profit potential, for example, varieties of beans such as cowpeas and coarse grains such as millet. The private sector, in contrast, focuses on agricultural research for which there is a market and profit potential, for example, hybrid maize, soybeans, and fruits and vegetables that are traded internationally. The private sector is unlikely to undertake much research needed by smallholder farmers in low-income developing countries because expected profits from disseminating the fruits of this research are unlikely to cover investment costs.

Despite the importance of public investment in pro-poor agricultural research, low-income-country governments and aid donors have provided only modest resources in recent years. Between 2000 and 2016, public agricultural research spending in low- and middle-income countries more than doubled. However, in Sub-Saharan Africa, the increase was only 44 per cent (2.8 per cent annually) (Table 34.5). In 2016, low-income countries spent 0.34 per cent of their agricultural gross domestic product on agricultural research, well under the UN target of 1 per cent. The donor-supported international agricultural research centres of the Consultative Group on International Agricultural Research focus mainly on research relevant to low-income-country agriculture. Their budgets declined more than 20 per cent in real terms between 2014 and 2017 (Beintema et al. 2020).

Pro-poor agricultural research must join all appropriate scientific tools and methods—including agroecology,

TABLE 34.5 Public agricultural research spending, 1981, 2000, and 2016 (by region and income class)

Region/country (number of countries per region)	2011 purchasing power parity dollars (billion)			Share (%)		
	1981	2000	2016	1981	2000	2016
Low- and middle-income countries (128)	8.3	12.9	28.2	39	42	60
Sub-Saharan Africa (44)	1.3	1.6	2.3	6	5	5
India	0.5	1.6	4.0	2	5	9
China	0.2	1.0	7.7	1	3	16
Other Asia-Pacific (23)	1.8	2.6	3.7	8	9	8
Latin America-Caribbean (24)	2.8	3.1	4.7	13	10	10
West Asia-North Africa (22)	1.3	2.3	4.5	6	7	10
High-income countries (51)	12.8	18.0	18.6	61	58	40
World (179)	21.1	30.9	46.8	100	100	100

Source: Beintema N, Nin Pratt A, Stads G-J (2020) *ASTI Global Update*. International Food Policy Research Institute (IFPRI), Washington, DC.

conventional plant breeding, and genetic engineering—with better utilization of indigenous knowledge. It is important that poor farmers have access to insights into agricultural development from the full range of approaches to tackling their problems (Weblink Section 34.6 (c) agricultural research as a global public good; and Weblink Section 34.6 (d), a case study on international agricultural research).

Agroecology

Although high-yielding Green Revolution technologies have been responsible for enormous productivity increases among small-scale farmers in Asia, many farmers in the region's less-favoured areas have been bypassed. The desire to find ways of assisting the developing world's poorest farmers, combined with concerns about excessive dependence on external inputs such as fertilizers, pesticides, and irrigation water embodied in the first generation of Green Revolution technologies, has stimulated interest in alternative or complementary approaches, including *agroecology*. This aims to reduce the amount of purchased external inputs that farmers have to use. Instead, it relies heavily on available farm labour and organic material available on the farm, as well as on improved knowledge and farm management.

One of the great strengths of this approach is that it promotes sustainable management of natural resources and active participation by farmers in identifying problems as well as designing and implementing appropriate solutions at the farm and community levels. Such participatory technology development can be extremely effective in appropriate solutions to production problems.

The market for organic produce is expected to reach $220 billion by 2024, with sales mainly in North America,

Europe, China, and Japan. If developing-country smallholders using agroecological methods can meet organic certification standards, they may be able to benefit from this rapidly growing market (World Bank 2007) (Weblink Section 34.6 (e), what is agroecology?).

The potential of modern agricultural biotechnology for food security

(This section draws on ideas in Pinstrup-Andersen & Cohen (2001) and World Bank (2007)). (Weblink Section 34.6 (f) on biotechnology.)

It is possible that the introduction of modern agricultural biotechnology into developing countries can contribute to increased productivity, lower unit costs and prices for food, preservation of forests and fragile land, poverty reduction, and improved nutrition. This depends on whether the technology is accessible to and adopted by poor people, on the economic and social policy environment, and on the nature of the intellectual property rights arrangements governing the technology. By raising productivity in food production, agricultural biotechnology could reduce the need to cultivate new lands and help conserve biodiversity.

Modern agricultural biotechnology offers many potential benefits to smallholder farmers in developing countries: productivity gains, resistance to pests and diseases without high-cost purchased inputs, heightened crop tolerance to adverse weather and soil conditions, and enhanced durability of products during harvesting or shipping. Biotechnology research could aid the development of drought-tolerant crops, to the benefit of small farmers and poor consumers. The development of cereal plants capable of capturing nitrogen from the air could contribute greatly to plant nutrition, helping poor farmers who often cannot afford commercial fertilizers, and who experience low yields as a result. Biotechnology may offer cost-effective solutions to micronutrient deficiencies, such as provitamin A-rich rice, and be used to develop protein-enhanced crops or edible vaccines. Cotton farmers in China who have adopted insect-resistant cotton have reduced their use of highly toxic insecticides.

Some forms of modern agricultural biotechnology are uncontroversial, such as the use of genetic markers in plant breeding and tissue culture. The latter approach was used to develop 'new rice for Africa', a variety that combines African resistance to harsh weather with broad Asian leaves that prevent weeds from absorbing sunlight and reduce the time that poor female farmers in West Africa must spend hand-weeding their crops.

However, transgenic technology, also known as genetic modification, has proved highly controversial. This involves the transfer of genes between species, and even across the boundaries of the animal, plant, and microorganism kingdoms.

Except for limited work on rice and cassava, little research on genetically modified (GM) food crops has focused on the productivity and nutrition of poor people. In 2018, according to the International Service for the Acquisition of Agribiotech Applications, commercial farms in the USA, Canada, Argentina, and Brazil accounted for 85 per cent of GM crop plantings, with the USA alone accounting for 39 per cent, and with most of the hectarage devoted to maize, cotton, soybean, and canola that tolerate herbicides and/or are resistant to pests. To date, private firms have carried out most of the research on GM crops, so it is hardly surprising that they have focused on large-scale, market-oriented producers. Moreover, these companies have subjected their research processes and products to intellectual property rights protection, which may create barriers to their use in pro-poor public sector research efforts, although there are some examples of public–private partnerships that have overcome these barriers. Additional public and philanthropic resources are needed in support of the appropriate research in developing countries, which also require appropriate institutions and policies to manage public health and environmental risks (Weblink Section 34.6 (g) on the potential of biotechnology for food security).

34.7 Policies for sustainable management of natural resources

A high degree of complementarity amongst agricultural development, poverty reduction, and environmental sustainability is more likely when agricultural development is broad-based and inclusive of small- and medium-sized farms, market-oriented, participatory and decentralized, and driven by technological change that enhances productivity without degrading natural resources. It is particularly important to 'get the incentives right', as subsidy policies may encourage unsustainable practices. Reforms may be politically difficult, for example, replacing generalized fertilizer subsidies with subsidies targeted to low-income farmers. Paying farmers to protect forests and watersheds can improve carbon sequestration (thereby mitigating climate change), biodiversity conservation, and maintenance of flows for drinking and irrigation water. Policies aimed at achieving sustainable agricultural development must take into account the role of property rights and collective action in natural resource management. Many natural resource management technologies and practices take years to give full returns, for example, terracing hillsides to prevent degradation. Without secure rights to resources, farmers lack incentives to adopt these approaches. Some technologies need to be adopted over a wide area to be effective, for example, integrated pest management (IPM), so adopting farmers must cooperate with their neighbours in collective action.

Dietary patterns likewise have a bearing on achieving win–win–win outcomes for agricultural development, poverty reduction, and environmental protection. See Weblink Section 34.7 (a) on sustainable healthy diets for further discussion.

Promoting sustainable development in less-favoured areas

Although productivity is lower in less-favoured areas, these zones usually have comparative advantage in some agricultural production or non-farm activities if investment in infrastructure and institutions is adequate. Active engagement of communities, including women, is essential for sustainable management of both natural resources and conflicts over their use. Research has found that public investment in less-favoured areas of China and India results in high returns that sometimes exceed those to investment in favoured areas in terms of both economic growth and poverty reduction. Investments in agricultural research, education, roads, and irrigation have greater incremental impact in less-favoured areas in these two countries, in part because of neglect of opportunities for investment in these areas. In Sub-Saharan Africa, where overall agricultural public investment is low, additional investment is needed in both high-potential and less-favoured areas (Ruben et al. 2006; World Bank 2007).

Soil fertility management

Low soil fertility and lack of access to reasonably priced fertilizers, along with failure to replenish soil nutrients, must be rectified through efficient and timely use of organic and mineral fertilizers and improved soil management. Mineral fertilizer use should be reduced where heavy application causes environmental harm. Fertilizer subsidies that encourage excessive use should be removed, but subsidies may remain necessary for less-favoured areas where current use is low and soil fertility is being diminished (World Bank 2007).

Integrated pest management

Until the 1990s, developing-country governments and aid donors alike encouraged use of synthetic pesticides. Now, consensus is emerging on IPM. It has a variety of definitions, but it is generally understood as a flexible approach to pest management that draws upon a range of methods to produce a result that combines the greatest value to the farmer with environmentally acceptable and sustainable outcomes. The techniques used may include traditional crop management—crop rotation, intercropping, mulching, tillage, and the like. IPM may also use pest-resistant crop varieties (developed through conventional breeding or genetic engineering), biological control agents, biopesticides, and, as a last resort, judicious use of synthetic pesticides. The options used by the farmer depend on the local context—agroecological needs, and availability and affordability of the various alternatives (World Bank 2007) (Weblink Section 34.7 (b) for an IPM case study).

Water policy reform

Comprehensive water policy reform is needed to help save water, improve use efficiency, and boost crop output per unit. Such reforms will be difficult, due to widespread practices and cultural norms that treat water as a free good and vested interests benefiting from current arrangements. Reforms might include secure and tradable rights for users, and subsidies targeted to poor water users in place of general subsidies. Devolving irrigation infrastructure and management to user associations, combined with secure access to water, will provide incentives for efficient use. Appropriate technology is needed to support conservation incentives. It can be difficult in practice to negotiate and enforce agreements allocating rights over groundwater or establishing the rights of upstream and downstream users, but agreements that all users perceive as fair have a greater chance of succeeding. It is essential to assure that indigenous people, pastoralists, and smallholder farmers, including women, have access to water (World Bank 2007).

34.8 Conclusion

Implementing the policy changes outlined in this chapter will be expensive, and will require difficult political choices. But the task is far from impossible. A 2020 analysis has estimated the cost of ending hunger by 2030 while doubling the incomes of smallholder farmers and meeting the commitments of the Paris climate accord. This would require a doubling of current levels of global aid to agriculture, rural development, and nutrition to $26 billion per year over 2020–2030, combined with additional domestic revenue mobilization of $19 billion per year in low- and middle-income countries (Laborde et al. 2020). The $33 billion in needed additional annual spending amounts to less than 0.3 per cent of current global gross domestic product (GDP), and compares to US$1.9 trillion in global military spending in 2019 alone (according to the Stockholm International Peace Research Institute). Moreover, the cost of *not* making the investments is also substantial: the FAO estimates that economic productivity losses due to undernutrition amount to US$1.4–2.1 trillion per year, or 1–1.5 per cent of global GDP. Accelerated progress toward sustainable food security will depend upon the willingness of developing- and developed-country governments, international aid agencies, non-governmental organizations, business and industry, and individuals to back their anti-hunger rhetoric with action, resources, and changes in behaviour and institutions. The research community has a moral obligation to monitor the presence or absence of such changes.

KEY POINTS

- Technological and institutional innovation has permitted food production to more than keep pace with population growth.

- Food security requires more than just adequate food availability; it also is a matter of access to the food that is available, its appropriate utilization, and the stability of availability and access.

- Small-scale farmers face many constraints, including lack of access to productive resources, natural resource degradation, and health crises.

- Climate change will have severe impacts on agriculture and food security, especially in Sub-Saharan Africa; the most vulnerable people will suffer earliest and the most.

- Rising hunger since 2014 makes it extremely difficult to achieve the Sustainable Development Goal of zero hunger by 2030 with business as usual.

- Discrimination, political disempowerment, violent conflict, and disasters from natural hazards contribute to food insecurity.

- Food insecurity is concentrated in the rural areas of South Asia and Sub-Saharan Africa, but urban food insecurity is increasing along with rapid growth of urban areas.

- The development of well-integrated markets is necessary but not sufficient for food security; public policies are also essential to ensure that rural poor people, including women as well as men, have access to resources and services.
- At present, developing-country governments and aid donors substantially underinvest in agricultural and rural development, including pro-poor agricultural research.

- Efforts to achieve food security must take sustainable natural resource management, including the role of property rights and collective action, into account.

Be sure to test your understanding of this chapter by attempting multiple choice questions. See the Further Reading and Useful Websites lists for additional material relevant to this chapter.

REFERENCES

(For additional references and Further Reading see Chapter 34 Further Reading online.)

Beintema N, Nin Pratt A, Stads G-J (2020) *ASTI Global Update*. International Food Policy Research Institute (IFPRI), Washington, DC. https://ebrary.ifpri.org/utils/getfile/collection/p15738coll2/id/134029/filename/134242.pdf (accessed February 2021).

Botreau H & Cohen M J (2020) Gender inequality and food insecurity: A dozen years after the food price crisis, rural women still bear the brunt of poverty and hunger. In: *Advances in Food Security and Sustainability,* Volume 5, 53–117 (Cohen M J ed.). Academic Press, Cambridge, MA.

Bouet A & Laborde Debucquet D (eds) (2017) *Agriculture, Development, and the Global Trading System: 2000–2015*. International Food Policy Research Institute (IFPRI), Washington, DC.

De la Ocampo P A, Villani C, Davis B et al. (2018) *Ending Extreme Poverty in Rural Areas: Sustaining Livelihoods to Leave No One Behind*. Food and Agriculture Organization of the United Nations (FAO), Rome. http://www.fao.org/3/CA1908EN/ca1908en.pdf (accessed February 2021).

Doss C & Quisumbing A (2021) Gender, household behavior, and rural development. In: *Agricultural Development: New Perspectives in a Changing World,* 503–528 (Keijiro O & Shenggen F eds). International Food Policy Research Institute, Washington DC. https://doi.org/10.2499/9780896293830

FAO (2015) *The State of Food Insecurity in the World 2015*. FAO, Rome. http://www.fao.org/3/a-i4646e.pdf (accessed February 2021).

FAO, IFAD (International Fund for Agricultural Development), UNICEF (UN Children's Fund), WFP (World Food Programme) & WHO (World Health Organization) (2020) *The State of Food Security and Nutrition in the World 2020: Transforming Food Systems for Affordable Healthy Diets*.

FAO, Rome. https://doi.org/10.4060/ca9692en (accessed February 2021).

Laborde D, Parent M, Smaller C (2020) *Ending Hunger, Increasing Incomes and Protecting the Climate: What Would it Cost Donors?* Cornell University, Ithaca, NY/ IFPRI, Washington, DC/ International Institute for Sustainable Development, Winnipeg. https://www.iisd.org/system/files/publications/ending-hunger-what-would-it-cost.pdf (accessed February 2021).

OECD (Organization for Economic Co-operation and Development)/FAO (2020) *OECD–FAO Agricultural Outlook 2020–2029*. FAO, Rome/ OECD, Paris. https://doi.org/10.1787/1112c23b-en (accessed February 2021).

Pingali P L (2012) Green revolution: Impacts, limits, and the path ahead. *PNAS* **109**, 12302–12308.

Pinstrup-Andersen P & Cohen M J (2001) Modern agricultural biotechnology and developing-country food security. In: *Genetically Modified Organisms in Agriculture: Economics and Politics*, 179–189 (Nelson G C ed.). Academic Press, London.

Place F, Meinzen-Dick, R, Ghebru H (2021) Natural resource management and resource rights for agricultural development. In: *Agricultural Development: New Perspectives in a Changing World*, 595–625 (Otsuka K & Fan S eds). IFPRI, Washington, DC.

Quisumbing A R (ed.) (2003) *Household Decisions, Gender and Development: A Synthesis of Recent Research*. Johns Hopkins University Press, Baltimore, MD & London.

Ringler C, Perez C, Xie H (2021) The role of water in supporting food security: Where we are and where we need to go? In: *Agricultural Development: New Perspectives in a Changing World*, 661–680 (Otsuka K & Fan S eds). IFPRI, Washington, DC.

Rosegrant M W, Wiebe K, Sulser TB et al. (2021) Climate change and agricultural development. In: *Agricultural*

Development: New Perspectives in a Changing World, 629–660 (Otsuka K & Fan S eds). IFPRI, Washington, DC.

Ruben R, Kuiper M H, Pender J (2006) Searching development strategies for less-favoured areas. *NJAS—Wageningen J of Life Sci* **53**, 319–342.

Thompson B & Cohen M J (eds) (2012) *The Impact of Climate Change and Bioenergy on Nutrition*. Springer for FAO, Dordrecht, The Netherlands.

UNDESA (UN Department of Economic and Social Affairs), Population Division (2019) *World Population Prospects: The 2019 Revision*. UN, New York. https://population.un.org/wpp/Publications/ (accessed February 2021).

WFP (2020) *WFP Global Update on COVID-19: November 2020, Growing Needs, Response to Date and What's to Come in 2021*. WFP, Rome. https://www.wfp.org/emergencies/covid-19-pandemic (accessed February 2021).

WHO-Africa (WHO Regional Office for Africa) (2021) *HIV in the WHO African Region*. WHO-Africa, Brazzaville. https://www.afro.who.int/sites/default/files/health_topics_infographics/WHO_INFOgraphics_HIV.pdf (accessed February 2021).

World Bank (2007) *World Development Report 2008: Agriculture for Development*. The World Bank, Washington, DC.

World Bank (2021) *Global Economic Prospects*. The World Bank, Washington, DC.

Malnutrition: global trends

Catherine Geissler and Anna Lartey

OBJECTIVES

By the end of this chapter you should be able to:

- explain the meaning of public health nutrition
- describe the main types of malnutrition world-wide, their prevalence, trends, causes, and consequences
- analyse causes by level, such as immediate, underlying, and root
- describe the deficiencies that may occur in particular risk groups.

35.1 General introduction

Global nutrition here refers to public health nutrition worldwide. Public Health Nutrition concerns the nutritional health of population groups rather than individuals. It involves the assessment of the extent and distribution of types of malnutrition and their causes, the policies and interventions of nutrition-related organizations, and the effectiveness of these actions. Another term that is sometimes used is Public Nutrition, as the causes of, and interventions for poor nutrition should be more widely based than through the health sector only.

Many of the components of public health nutrition are addressed in previous chapters: methods of assessment of nutritional status (Chapters 30–32); types of malnutrition including micronutrient deficiency diseases (Chapters 11–13) and the non-communicable diseases (NCDs) (Chapters 19–22). The effects of nutrition on specific organs, including the gastrointestinal tract (Chapter 23), cardiovascular system (Chapter 19), adipose tissue (Chapter 20), nervous system (Chapter 24), skeleton (Chapter 25), and teeth (Chapter 26); and the distribution of nutritional problems throughout the lifecycle from conception to old age (Chapters 14–16) are also relevant to this chapter. Methods of assessing the extent and distribution of malnutrition and its relation to the consumption of particular foods and nutrients and to other environmental factors

are reviewed in the chapter describing the science of epidemiology (Chapter 33). Nutritional epidemiology can be applied to understand causes of malnutrition, with other information coming from our knowledge of the availability, function, and requirements for individual nutrients and other food components, as well as underlying factors such as food supply and access, food safety, patterns of food consumption, and the important relationship between nutrition and physical activity.

The terms used to describe nutritional status and the economic situation of countries have changed over the years as the concepts have developed. For example, the term 'overnutrition' in contrast to 'undernutrition' is incorrect as it implies excess intakes of all nutrients which is not the case in people who are overweight or obese. The word 'hunger' is often used to denote undernutrition but this term is avoided in this chapter as it denotes the subjective feeling of lack of food rather than objective measures of inadequate energy intake. The terms 'developed' and 'developing' countries are also inaccurate, as are other commonly used terms such as 'industrialized and non-industrialized'. The currently accepted terms are high-income countries (HIC), low- and middle-income countries (LMIC), and low-income countries (LIC). These terms will therefore be used throughout this chapter. In this chapter the term 'malnutrition' encompasses both 'undernutrition' or 'undernourishment' (due to some combination of deprivation of food, care, and health) and 'overweight and obesity' (due to a combination of the excess consumption of some macronutrient diet components, and/or too little physical activity).

The extent of various types of malnutrition differs between low-, middle-income, and high-income countries. For example, general undernutrition and specific vitamin and mineral deficiencies are more prevalent in LMICs, whereas chronic diseases, otherwise called non-communicable diseases (NCDs), related to unhealthy diets such as obesity, coronary heart disease (CHD), and diabetes, have been more prevalent in HICs. However, pockets of undernutrition exist in HICs (see 'Specific groups' Section 35.4) and NCDs are expanding rapidly in the LMICs.

Undernutrition and overweight and obesity co-exist to the greatest extent in countries undergoing rapid socio-economic and demographic transitions such as Brazil and China. This co-existence is termed the 'nutrition transition' or the 'double burden'. The challenge of balancing interventions to address undernutrition and overweight and obesity is particularly great for these countries.

The main focus of this chapter is an overview of the extent, distribution, and causes of malnutrition using the food–care–health model of the UNICEF conceptual framework. Causes of malnutrition are examined at the immediate, underlying, and root or basic levels.

35.2 Extent and causes of malnutrition

Introduction

Descriptions of the extent and causes of, and responses to malnutrition, are usefully classified by stage in the lifecycle. Typically, discussion of causes and hence interventions begin with the infant. This is because dramatic growth failure most typically occurs between the ages of 12 to 18 months. When growth does fail, normally measured as stunting (see Chapter 31), there is a low probability of meaningful catch-up growth. The impairments in cognitive function that can accompany stunting are also largely irreversible. Hence early infant malnutrition has consequences for later life and potentially for the next generation. Malnourished babies become malnourished adolescents who are less able to learn in school and adults who are less productive in the labour market. Malnourished female babies will be more likely to be malnourished girls and women who give birth to malnourished babies. Malnutrition is therefore intergenerational. Discussion of causes and interventions should begin at preconception; the 'Barker Hypothesis' or 'foetal programming hypothesis' (see Chapter 14) proposes that poor maternal diet at critical periods of development in the womb can trigger an adaptive redistribution of foetal resources (including low birth weight). Such adaptations affect foetal structure and metabolism in ways that predispose the individual to chronic diseases later in life. Whenever possible, this section reviews the extent and causes of malnutrition by stages in the lifecycle.

Assessment of extent and worldwide distribution of malnutrition

Collection of reliable data on the extent and distribution of malnutrition is beset with difficulties, particularly for LMICs. First, different UN agencies have

different regional grouping labels, and in some cases the same regional label includes different countries. Moreover, the data are not collected in a comparable manner across countries or are just not collected at all. Anthropometric data for infants from the World Health Organization (WHO) tend to have the highest level of quality control and are hence the most useful for between-country comparison. Food deprivation data from the Food and Agriculture Organization (FAO), called 'Undernourishment' or 'Hunger', give the appearance of allowing comparisons across countries and time, but it is well known that comparability largely rests on the validity of some questionable statistical assumptions about the within-country distribution of food consumption. The data on adult anthropometry and micronutrient deficiency are also very patchy. Nevertheless, a rough approximation of the extent, pattern, and trends of the malnutrition problem can be estimated using existing data. Previously, the international organizations concentrated on documenting only nutritional status in LICs. However, since about 2000 they have included the epidemiology of obesity and diet-related degenerative diseases, the NCDs.

Undernourishment

The prevalence of undernourishment (POU) is a key indicator for measuring hunger. The Food and Agriculture Organization of the United Nations (FAO) reports annually on hunger prevalence for all countries. POU provides an estimate of the proportion of the world's population that does not have enough food to eat and thus are unable to meet their energy needs for an active and healthy life (FAO 2020). POU is calculated using national data from food balance sheets, specifically national food supplies, food consumption, and food energy needs of the population. POU is one of the Sustainable Development Goal 2 indicators to monitor hunger reduction globally. The State of Food Security and Nutrition in the World Report 2021 (FAO 2021) estimates that in 2020, 768 million people (9.9 per cent of the world's population) were undernourished. Before COVID-19, over a ten-year period the POU has been increasing steadily, with about 10 million additional people per year and nearly 60 million additional undernourished people added in five years. COVID-19 has exacerbated the situation; there has been an increase of 118 million more hungry people in 2020 alone compared with 2019 (Figure 35.1). Due to long-term COVID-19 effects, it is estimated that around 660 million people will still be undernourished in 2030 compared to the 30 million previously estimated without COVID-19 effect (FAO 2021).

Since 2017, FAO started reporting on a new indicator called the Food Insecurity Experience Scale (FIES) which enables various degrees of food insecurity

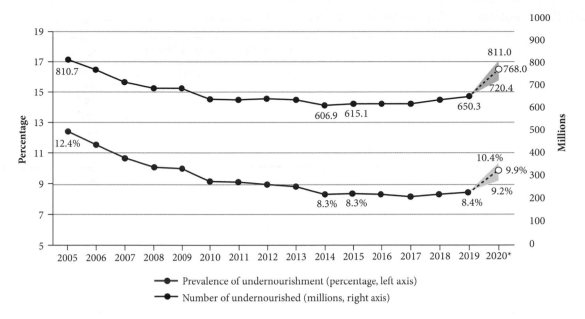

FIGURE 35.1 Prevalence and number of people undernourished in the world. Between 720 and 811 million people in the world were undernourished in 2020—as many as 161 million more than in 2019.
Source: FAO, IFAD, UNICEF, WFP, and WHO 2021. *The State of Food Security and Nutrition in the World*.

categorization: moderate and severe. FIES is a direct measure of people's access to food. Compared to POU, which relies on national level data on food availability and consumption, FIES uses data collected from individuals regarding their experience and the degree of the experience of food insecurity (SOFI 2017). Moderately food-insecure individuals may have been compelled to reduce the quantity of food they would normally consume sometime during the year, while severe food-insecure individuals would have gone for entire days without any food due to food not being affordable or available. The SOFI 2020 report estimates 690 million people experienced severe food insecurity while over 2 billion people globally did not have access to safe, nutritious, and sufficient food (were food insecure) in 2019 (FAO 2020). However, in 2020 due to the COVID-19 effect, 928 million people were severely food insecure, while the number that experienced moderate or severe food insecurity was 2.37 million, an increase of about 320 million in 2020 alone (FAO 2021) (Table 35.1). There are gender dimensions to food insecurity, with more women than men in every region and globally being affected by both moderate and severe food insecurity. COVID-19 further widened the gender gap in the prevalence of moderate or severe food insecurity in 2020. The prevalence of moderate or severe food insecurity was 10 per cent higher among women than men in 2020, compared to 6 per cent difference in 2019 (FAO 2021).

The Africa region showed the highest prevalence in undernourishment of 18.0 per cent in 2019 but increased to 21.0 per cent in 2020 mainly due to the COVID-19 effect (Table 35.1). The prevalence of undernourishment

in Africa was significantly higher than the global prevalence of 8.4 per cent in 2019 and 9.9 per cent in 2020 (FAO 2021). Although Asia is the region with the highest number of undernourished people, it is predicted that Africa will overtake Asia in having the highest number of undernourished persons by 2030 if the current trend continues.

No doubt, the COVID-19 pandemic has worsened the hunger situation due to its economic and social impacts, affecting people's livelihood and pushing millions into poverty and food insecurity. There are concerns that COVID-19 would erode gains made in reducing undernourishment and food insecurity, even in regions that have shown some progress, and could hold back meeting the achievement of the Sustainable Development targets, 2.1 on hunger reduction and food insecurity (Table 35.2).

Stunting

In 2012, the World Health Assembly endorsed a set of global nutrition targets to be achieved by 2025 and called for policies and actions to reduce the number of stunted children under five years by 40 per cent and to reduce and maintain childhood wasting to less than 5 per cent by 2025. Stunting (low length or height for age) is defined as length or height below two standard deviations of a reference healthy population of the same age (Table 35.3). Stunting, an indicator of poor growth, starts *in utero*, when poor maternal nutrition during pregnancy does not support proper foetal growth and development. After birth, inadequate child nutrition and frequent infections are common causes of child stunting. Stunting is

TABLE 35.1 Prevalence of undernourishment (POU) in the world, 2005–2020

	Prevalence of undernourishment (%)							
	2005	2010	2015	2016	2017	2018	2019	2020
World	12.4	9.2	8.3	8.3	8.1	8.3	8.4	9.9
Africa	21.3	18.0	16.9	17.5	17.1	17.8	18.0	21.0
Northern Africa	8.5	7.3	6.1	6.2	6.5	6.4	6.4	7.1
Sub-Saharan Africa	24.6	20.6	19.4	20.1	19.5	20.4	20.6	24.1
Eastern Africa	33.0	28.4	24.8	25.6	24.9	25.9	25.6	28.1
Middle Africa	36.8	28.9	28.7	29.6	28.4	29.4	30.3	31.8
Southern Africa	5.0	6.2	7.5	7.9	7.3	7.6	7.6	10.1
Western	14.2	11.3	11.5	11.9	11.8	12.5	12.9	18.7
Asia	13.9	9.5	8.3	8.0	7.8	7.8	7.9	9.0
Central Asia	10.6	4.4	2.9	3.2	3.2	3.1	3.0	3.4
Eastern Asia	6.8	<2.5	<2.5	<2.5	<2.5	<2.5	<2.5	<2.5
South-eastern Asia	17.3	11.6	8.3	7.8	7.4	6.9	7.0	7.3
Southern Asia	20.5	15.6	14.1	13.2	13.0	13.1	13.3	15.8
Western Asia	9.0	9.1	14.3	15.0	14.5	14.4	14.4	15.1
Western Asia and North Africa	8.8	8.2	10.5	10.9	10.7	10.6	10.7	11.3
Latin America and the Caribbean	9.3	6.9	5.8	6.8	6.6	6.8	7.1	9.1
Caribbean	19.2	15.9	15.2	15.4	15.3	16.1	15.8	16.1
Latin America	8.6	6.2	5.1	6.2	6.0	6.1	6.5	8.6
Central America	8.0	7.4	7.5	8.1	7.9	8.0	8.1	10.6
South America	8.8	5.7	4.2	5.4	5.2	5.4	5.8	7.8
Oceania	6.9	5.3	6.1	6.2	6.3	6.2	6.2	6.2
North America and Eurpoe	<2.5	<2.5	<2.5	<2.5	<2.5	<2.5	<2.5	<2.5

Source: *The State of Food Security and Nutrition in the World 2021: Transforming Food Systems for Food Security, Improved Nutrition, and Affordable Healthy Diets for All* (Internet). FAO, Rome. (Cited 30 July 2021). Available from: https://doi.org/10.4060/cb4474en

quite widespread in low- and middle-income countries. Global estimates indicate that in 2020, there were 149.2 million stunted children under five years. These figures do not reflect the effect of COVID-19. COVID-19's negative impact on household incomes, food insecurity, and closure of essential nutrition services would be expected to increase the numbers of stunted children. Although progress has been made globally in reducing stunting prevalence from 33.1 per cent in 2000 to 22.0 per cent in 2020 (FAO 2021), the numbers are still unacceptably high. There are wide variations in stunting reduction rates. Between 2000 and 2020 the following reductions were reported for Africa (45.5 per cent to 30.7 per cent), Asia (37.9 per cent to 21.8 per cent), Latin America and the Caribbean (LAC, 18.0 per cent to 11.3 per cent) (FAO 2021). The rate of reduction in many regions is too slow

to meet the achievement of the global stunting target set by the WHA. In terms of the numbers affected by stunting, it is decreasing in all regions except Africa where the numbers have increased from 54.4 million in 2000 to 61.4 million in 2020, due to rapid growth in the under-five population. The reduction in stunting numbers between 2000 and 2020 are as follows: for Asia from 135.9 million to 79.0 million, and for LAC from 10.2 million to 5.8 million (FAO 2021).

Wasting

Wasting, defined as low weight-for-length/height (too thin), is an acute form of malnutrition. The condition is widespread among children under five years in low-income countries. Children who present severe

TABLE 35.2 Sustainable Development goal 2: goal, targets, and indicators for hunger and improved nutrition

Goal 2	End hunger, achieve food security and improved nutrition and promote sustainable agriculture
Target 2.1	By 2030, end hunger and ensure access by all people, in particular the poor and people in vulnerable situations, including infants, to safe, nutritious, and sufficient food all year round
Indicators	• 2.1.1 Prevalence of undernourishment • 2.1.2 Prevalence of moderate or severe food insecurity in the population, based on the food insecurity experience scale (FIES)
Target 2.2	By 2030, end all forms of malnutrition, including achieving, by 2025, the internationally agreed targets on stunting and wasting in children under five years of age, and address the nutritional needs of adolescent girls, pregnant, and lactating women and older persons
Indicators	• 2.2.1 Prevalence of stunting (height for age < –2 standard deviation from the median of the World Health Organization (WHO) child growth standards) among children under five years of age. • 2.2.2 Prevalence of malnutrition (weight for height > +2 or < –2 standard deviation from the median of the WHO child growth standards) among children under five years of age, by type (wasting and overweight) • 2.2.3 Prevalence of anaemia in women aged 15 to 49 years, by pregnancy status (percentage)

Source: Global indicator framework for the Sustainable Development Goals and targets of the 2030 Agenda for Sustainable Development. https://unstats.un.org/sdgs/indicators/Global%20Indicator%20Framework%20after%202021%20refinement_Eng.pdf ©(2020) United Nations. Reprinted with the permission of the United Nations.

TABLE 35.3 Global nutrition targets endorsed by the World Health Assembly and their extension to 2030

	2025 target	2030 target
Stunting (SDG)	40 per cent reduction in the number of children under five who are stunted	50 per cent reduction in the number of children under five who are stunted
Anaemia (SDG)	50 per cent reduction in anaemia in women of reproduction age	50 per cent reduction in anaemia in women of reproduction age
Low birthweight	30 per cent reduction in low birthweight	30 per cent reduction in low birthweight
Childhood overweight (SDG)	No increase in childhood overweight	No increase in childhood overweight
Breastfeeding	Increase the rate of exclusive breastfeeding in the first six months up to at least 50 per cent	Increase the rate of exclusive breastfeeding in the first six months up to at least 70 per cent
Wasting (SDG)	Reduce and maintain childhood wasting to less than 5 per cent	Reduce and maintain childhood wasting to less than 3 per cent

Source: WHO & UNICEF (2017) The extension of the 2025 maternal, infant, and young child nutrition targets to 2030. Discussion paper. WHO, Geneva (also reproduced in State of Food Security and Nutrition 2021 available at https://doi.org/10.4060/cb4474en).

or moderate wasting are at an increased risk of death. According to the joint estimates release by UNICEF, WHO, and the World Bank (FAO 2021), in 2020 globally 6.7 per cent of children (45 million) under five years were wasted. Asia contributed 70 per cent (31.9 million) and Africa 27 per cent (12.1million) (FAO 2021).

The world is still behind in meeting the global nutrition target for wasting of 3 per cent by 2030. This calls for focused action, especially in high-burdened regions. Stunting and wasting can occur concurrently in the same child, increasing the risk of mortality. The *Lancet* series on maternal and child nutrition of 2021 reports a reduction in concurrent stunting and wasting from 7 per cent in 2000 to 4.7 per cent in 2015 (Victora et al. 2021). New evidence shows that the incidence of stunting and wasting are highest during the first six months of life.

Features of wasting and stunting

Underweight (weight for age) can be due to wasting and/or stunting. The main cause of wasting is negative energy balance, when available metabolizable energy is less than energy expenditure, due to insufficient intake of food, often of low energy density, exacerbated by malabsorption of nutrients, infection-related anorexia, and increased energy costs of infection or fever. 'Environmental enteropathy' is common in LICs and results in reduced absorptive surface in the intestine and transfer of intestinal bacteria into the blood and increased inflammation. This leads to both the reduced energy intake and increased energy expenditure that is compensated by energy stored in adipose and lean tissue, usually muscle, resulting in wasting.

As food insecurity affects children first, many classifications of malnutrition were devised in relation to children, initially focused on growth. Growth failure is distinguished between ponderal (weight) and linear growth, that is, thinness and shortness or stunting, measured as weight-for-length/height, or linear growth: length-for-age until the age of two years, and height-for-age thereafter.

Previously-used descriptive terms for childhood malnutrition such as chronic energy deficiency or protein energy malnutrition, imply simple causes and so are now rarely used. Currently preferred terms are severe acute malnutrition (SAM) and moderate acute malnutrition (MAM) which are defined by anthropometry without reference to cause (see Chapter 31).

SAM, which has clear physical signs and high fatality is easily recognized but borderline or moderate acute malnutrition, MAM, is so common that it often escapes notice. A stunted underweight child, living among similarly sized age-mates, can appear to be thriving. Biological variation in height and body build is well accepted and differences in height and weight in children from different countries and races were previously thought to result mainly from genetic variation. However, the World Health Organization child growth standards (2006) show that children from healthy, adequately resourced environments around the world grow similarly at least to age five years (www.who.int/tools/child-growth-standards).

Stunting has been considered as an adaptation to long-standing underfeeding, in that a smaller child requires less food. Clearly there are limits to such adaptation beyond which health suffers, with increased morbidity and mortality among stunted children. Height often falters in response to repeated infection as well as nutrient depletion. Therefore, variation in height, especially during the early years of life, is due more to nutritional and socio-economic deprivation than to ethnic variability.

Among adults, body mass index (BMI, defined as weight in kg/height in meters²) is used to define undernutrition as well as overweight and obesity. There are concerns that BMI cut-offs may not be appropriate for all genetic groups. To date this has focused mainly on the high BMI cut-offs of South Asians, but there are also concerns as to whether low BMI cut-offs are appropriate.

The risk of inadequate nutrition begins in foetal life, especially during the last trimester of pregnancy when the foetus undergoes most weight gain. Risk factors in the first few months of life include early non-exclusive breastfeeding, replacing breast milk with low nutrient and energy-dense foods (see Chapters 14 and 15). Breast milk substitutes may also be unsuitable because of a high renal solute load, as with fresh cow's milk, or low energy density as with diluted cow's milk or incorrectly reconstituted formula. Complementary foods provided in later infancy may have low energy and nutrient density. Traditional weaning porridge often has much lower energy density than breast milk and poor protein quality due to low levels of certain essential amino acids, inadequate vitamin A, and poorly available iron and zinc. SAM is commonly seen in children during the second year of life, and kwashiorkor (see section 'Kwashiorkor' below) in somewhat older children is triggered by infection, particularly by measles and diarrhoea accompanied by poor feeding. HIV infection increases the risk of SAM which can be associated with malabsorption and secondary infection linked to HIV immune failure. In highly HIV-endemic regions of southern Africa, HIV-infected children comprise a large proportion of children presenting to hospitals with SAM. Therefore, these children should be tested for HIV to allow antiretroviral therapy which greatly improves their chance of recovery. Stunting is also often linked to HIV infection. HIV has an important cross-generational effect with high rates of orphans, who risk malnutrition and early death, even when seronegative.

Multiple factors cause nutritional stunting. Linear growth is determined by nutritional factors that act at the growth plate of long bones, together with growth hormone and insulin-like growth factor (IGF-1). Certain amino acids, specifically leucine, and several vitamins and minerals such as calcium, zinc, copper, vitamin D, and possibly vitamin A influence linear growth. Single nutrient supplementation studies are confounded by interaction between micronutrients and possible concurrent deficiencies. Diet 'quality', implying animal source foods such as dairy, meat and eggs, fruits and green vegetables, may have a greater benefit for growth than any specific nutrient. Stunting and underweight have also been reported in young European children on alternative, macrobiotic, diets. In the acute phase response to infection various cytokines produced may slow bone growth. Cohort studies have demonstrated growth faltering due to frequent or chronic infections including HIV, or to worm infestation. Catch-up in linear growth during nutritional rehabilitation rarely starts until weight recovery is well under way or may not occur at all.

The climate of many tropical countries is characterized by a long dry and a short rainy season, when most farm work is done. Stored staples may not last until the next harvest, and so the time of planting is also the 'hungry season'. Even when people consider themselves as 'food secure', the marked seasonality of fruit and vegetables limits continued access to micronutrients so that seasonal variation in prevalence of SAM and MAM is common. Even short periods of undernutrition, especially if they coincide with key stages of child development, can be detrimental to health and development. Community nutrition support for young children is important, particularly those with MAM who might progress to SAM, thus the need for lipid-based nutritional supplements (LNS) or other food-based interventions during the hungry season (see Chapter 36).

Severe acute malnutrition (kwashiorkor)

Oedema in SAM (kwashiorkor) is incompletely understood and likely due to several factors; (see Chapter 31 Weblink Section 31.6) for further details about research into this. Nutritional oedema may be triggered by infection. Reactive oxygen species are generated during the cellular response to infection. Their adverse effect on metabolism is normally held in check by several antioxidant mechanisms, many of which require nutrients such as vitamins A, C, and E, zinc, selenium, and copper. Iron, specifically free iron, may act as a pro-oxidant. In malnourished people, oxidative stress to structural lipids may cause cell membranes to become permeable to sodium and potassium, which leak with water into the extracellular space, causing oedema. An imbalance between oxidative stress triggered by infection and protective antioxidants may therefore play a critical role in predisposing the undernourished child to kwashiorkor. Other suggested mechanisms include the presence of aflatoxin (Soriano et al. 2020), an altered profile of intestinal bacteria (Smith et al. 2013) and increased transfer of bacteria or their products across the intestinal wall into the circulation, leading to increased systemic inflammation.

Effects of malnutrition on short- and long-term health and growth

The effect of malnutrition on health depends on the timing and duration of nutritional stress. When low birthweight is due to intrauterine growth retardation, the immediate effects are a high neonatal death rate due to hypoglycaemia, hypothermia, and infection. Malnutrition increases a child's vulnerability to other illnesses, especially infection. Even though malnutrition is rarely recorded as a cause of death, it contributes to the case fatality of many illnesses. General malnutrition often coexists with micronutrient deficiencies so the contributions of individual nutrient deficiencies to mortality are difficult to estimate; however, nutritional deficiencies during the foetal period and early life are estimated to contribute to about 45 per cent of under-five-year-old mortality (Black et al. 2013).

SAM, especially if accompanied by oedema, has a high case fatality, around 20 per cent in hospital. It is difficult to distinguish the long-term effects of an episode of SAM from those of persistent socioeconomic deprivation. Prolonged stunting has an adverse effect on cognitive development. The survivor of early malnutrition may recover completely, but remain stunted in the longer term, or have a delayed adolescent growth spurt. Delayed growth and final short stature in women contribute to delivery of low birthweight infants, obstruction in labour, and thereby to high maternal mortality. If early deprivation is followed by nutritional excess in adult life, the survivor is at increased risk of chronic diseases such as diabetes and heart disease. Adverse foetal programming plays a part in the diseases associated with the nutritional transition that is evident in a number of LICs, characterized by rising intakes of fat, sugar, and meat and by reduction in physical activity (see Chapter 20). Secular trends towards an increase in BMI, and increased prevalence of obesity, already evident in urban populations, are now affecting the rural poor in rapidly developing regions.

Childhood overweight and obesity

Although undernutrition remains a major challenge in children, other forms of malnutrition—overweight and obesity, are increasing rapidly. Globally 5.7 per cent (38.9 million children) were affected by overweight in 2020 (UNICEF/WHO/World Bank Group 2021), and of this 24 per cent and 48 per cent respectively were in Africa (10.6 million) and Asia (18.7 million). Latin America and the Caribbean (LAC) has 3.9 million and Oceania 0.1 million overweight children. In terms of trends between the period 2000 to 2020, only Africa showed a reduction in the prevalence of overweight children from 6.2 per cent to 5.3 per cent, but all other regions showed an increase: Asia 4.5 per cent to 5.2 per cent; LAC 6.8 per cent to 7.5 per cent; Oceania 5.2 per cent to 8.0 per cent (UNICEF/WHO/World Bank Group 2021). What makes childhood overweight and obesity concerning is the fact that it can persist into adulthood. An obese child has a higher risk of becoming an obese adult.

Adult overweight and obesity

Global adult obesity, which was 11.4 per cent in 2010, is expected to reach 16.1 per cent by 2025 (World Obesity Atlas 2022). Increased prevalence of obesity has been observed in all regions of the world (SOFI 2020). Over two billion people were either overweight or obese in 2015 and of these more than 600 million (equivalent to 13 per cent of the world's adult population) were obese (Swinburn et al. 2019). In terms of regional distribution, North America and Europe (26.9 per cent) Oceania (23.6 per cent), and Latin America and Caribbean (24.2 per cent) presented high prevalence of adult obesity in 2016 (Table 35.4). The prevalence of adult obesity in Africa (12.8 per cent) and Asia (7.3 per cent) remains relatively

TABLE 35.4 Progress in child stunting and overweight, and adult obesity: most regions are making some progress, but not enough to achieve global targets; adult obesity is worsening in all subregions

	Child stunting (%)			Child overweight (%)			Adult obesity (%)		
	2012	2019	2030	2012	2019	2030	2012	2016	2025
World	24.8	21.3	~	5.3	5.6	×	11.8	13.1	×
Africa	**32.3**	**29.1**	~	**4.8**	**4.7**	~	**11.5**	**12.8**	×
Northern Africa	19.8	17.6	~	10.1	11.3	×	23.0	25.2	×
Sub-Saharan Africa	34.5	31.1	~	3.8	3.6	~	8.0	9.2	×
Eastern Africa	38.5	34.5	~	4.0	3.7	~	5.3	6.4	×
Middle Africa	34.4	31.5	~	4.8	5.1	×	6.7	7.9	×
Southern Africa	30.4	29.0	~	11.7	12.7	×	25.0	27.1	×
Western	30.6	27.7	~	2.3	1.9	✓	7.4	8.9	×
Asia	**27.0**	**21.8**	~	**4.4**	**4.8**		**6.1**	**7.3**	×
Central Asia	14.9	9.9	✓	7.3	6.2	~	15.6	17.7	×
Eastern Asia	7.9	4.5	✓	6.4	6.3	~	4.9	6.0	×
Southeastern Asia	29.4	24.7		5.5	7.5	×	5.4	6.7	×
Southern Asia	38.0	31.7	~	2.5	2.5	✓	4.5	5.4	×
Western Asia	15.9	12.7	~	7.7	8.4	×	27.2	29.8	×
Western Asia and North Africa	17.8	15.2	~	8.9	9.9	×	25.3	27.2	×
Latin America and the Caribbean	**11.4**	**9.0**	~	**7.2**	**7.5**	×	**22.2**	**24.2**	×
Caribbean	10.3	8.1	✓	6.2	7.0	×	22.0	24.7	×
Central America	16.0	12.6	~	6.5	6.9	×	25.1	27.3	×
South America	9.2	7.3		7.6	7.9		21.1	23.0	×
Oceania	**37.9**	**38.4**	~	**7.3**	**9.4**	×	**21.3**	**23.6**	×
Australia and New Zealand	n.a	n.a		16.2	20.7	×	27.0	29.3	×
North America and Europe	**n.a.**	**n.a.**		**n.a.**	**n.a.**		**25.0**	**26.9**	×
Northern America	2.7	2.6	✓	8.0	8.9	×	32.9	35.5	×

✓ On-track; ~ off-track with some progress; × off-track with no progress. Empty cells indicate no data.

Source: FAO, IFAD, UNICEF, WFP, and WHO (2020) *The State of Food Security and Nutrition in the World 2020: Transforming Food Systems for Affordable Healthy Diets*. FAO, Rome. https://doi.org/10.4060/ca9692en

low on average, but there are wide variations in the prevalence within each region. For example, the prevalence of obesity in South Africa is 27 per cent, a figure similar to that of North America and Europe. Globally, no country is on track to halt the rising obesity situation. Obesity has long-term health effects, being a major risk factor for non-communicable diseases such as cardiovascular disease, diabetes. and certain cancers.

The rising prevalence of obesity globally has cast the focus on food systems as a major driver. Food systems are failing on several fronts (Béné et al.2019) in not addressing the food security situation. Even though enough food is produced globally to feed everyone, there are still 690 million people who are undernourished (FAO 2020). The COVID-19 effect has increased the number to 811 million globally. Food systems are failing to deliver the nutrient-rich food ingredients such as fruits and vegetables, animal source foods, and fibre-rich grains affordably. The result is that healthy diets remain unaffordable for more than three billion of the world's population (FAO 2020), with Asia (1.85 billion) and Africa (1.0 billion) being most affected. With urbanization and rising incomes, the quest for convenience has led to substantial shifts in diets with the increased consumption of highly processed foods that are high in calories, sugar, fat, and salt but of low nutrient densities (Reardon et al. 2021). Poor diets are among the leading risk factors contributing to the global burden of disease. Eleven million adult deaths in 2017 were attributed to dietary risk factors (Reardon et al. 2021). Poor diet and undernutrition in early childhood is now known to be associated with increased risk of NCD in later life.

Micronutrients

Micronutrient undernutrition data are also far from perfect, with patchy data on the prevalence of the three major worldwide micronutrient deficiencies—iodine, iron, and vitamin A (see also Chapters 12, 13, 29). In this chapter, data on anaemia is from the WHO Global Health Observatory dated 2019. The WHO vitamin and mineral nutrition information system (VIMNS) has not updated vitamin A, urinary iodine, or goitres estimates, but will update estimates of iodine in the near future. Data presented here is from the 2021 iodine scorecard (https://www.ign.org/cm_data/IGN_Global_Scorecard_2021_7_May_2021.pdf). In relation to vitamin D, WHO reviewed the literature on rickets in 2019 (https://www.who.int/publications/i/item/9789241516587) and a summary is presented here.

Iron deficiency

The WHO estimates that iron deficiency is the most common and widespread nutritional disorder in the world (https://www.who.int/data/gho/data/themes/topics/anaemia_in_women_and_children, affecting more people than any other condition, constituting a public health condition of epidemic proportions. It affects a large number of children and women in LMICs, and is the only nutrient deficiency which is also recognized as being significantly prevalent in high-income countries (see Weblink Section 35.1).

For iron deficiency, anaemia prevalence is used as a proxy. In HICs this is likely to be a close approximation, but in the LMICs anaemia can occur due to factors other than iron deficiency, including deficiencies of other micronutrients such as folate, vitamins B$_{12}$ and riboflavin, and diseases such as malaria, other parasitic infections including hookworm infestation and schistosomiasis, current infectious diseases including tuberculosis, and other pathologies. Anaemia is a serious problem throughout the lifecycle, both in high-income and low-income countries. It is a particular problem for pregnant women, who have particularly high haematinic needs. Although ubiquitous in many low-income countries, the impact of iron deficiency and anaemia is hidden in the statistics of overall death rates, maternal haemorrhage, reduced school performance, and lowered productivity. The health consequences erode the development potential of individuals, societies, and national economies. The major health consequences include poor pregnancy outcome, impaired physical and cognitive development, increased risk of morbidity in children, and reduced work productivity in adults. Anaemia contributes to 20 per cent of all maternal deaths.

The WHO global database on anaemia is the only source of anaemia estimates at country, regional, and global level, based on the blood concentration of haemoglobin. The data coverage is best for preschool-age children, pregnant women, and non-pregnant women (70 per cent and more). Coverage for other population groups is much lower (30–40 per cent for school-age children, men, and the elderly). The prevalence of anaemia in 2019 in various age groups of the world population is shown in Table 35.5. The highest global prevalence is in preschool-age children (6–59 months) affecting 269 million children, followed slightly lower by pregnant women (15–49 years) affecting 32 million, then non-pregnant women, with 538 million anaemic. Africa and Southeast Asia are the WHO regions with the highest prevalence in each of these groups and Europe, Western Pacific, and the Americas the lowest. Since 2000 the global prevalence of anaemia in children under five has slowly decreased from 48.0 per cent, but from 2010 it has been stagnant, at least until the global COVID-19 pandemic. Since 2000 the global prevalence of anaemia in women of reproductive age has also been stagnant while the prevalence in pregnant women has decreased slightly up to the start of the 2019 pandemic.

Causes of iron deficiency

As with other nutrients, iron status depends on the balance between intake and losses. Diets low in iron are especially common where the main dietary source is non-haem iron in plant foods, which is absorbed less well than haem iron. Avoidance of meat by adults is more

TABLE 35.5 Prevalence of anaemia in children (6–59 months), pregnant women (15–49 years) and non-pregnant women

	Preschool children	Pregnant women	Non-pregnant women
	%	%	%
Global	38.8	36.5	29.6
Africa	60.2	45.8	39.8
SE Asia	49.0	47.8	46.5
E. Med	42.7	36.8	34.8
Americas	16.5	18.9	17.8
Europe	20.3	23.4	18.6
W. Pacific	19.4	21.3	16.2

Source: WHO global database on anaemia 2019.

commonly due to poverty than culture or religion, flesh foods being eaten on special occasions. In meat-free diets, cereals, especially millet, are a major iron source with additional contributions from dark green vegetables and pulses. However, such plant-based diets often have low iron bioavailability due to high phytate and little haem, and population differences in iron status are largely explained by differences in the dietary bioavailability of iron. Diets have been classified by the WHO according to bioavailability of iron, with nominal values of 5 per cent, 10 per cent, and 15 per cent for low, medium, and high bioavailability respectively, according to their content of iron absorption inhibitors, such as phytate, calcium, tannin and oxalates, and enhancers, such as ascorbic acid, other fruit acids, and a factor in flesh foods. Vegetarian and vegan diets are associated with iron deficiency in both adults and children (see Chapter 17) and dieting to lose weight is a risk factor in adolescent girls. Even when the adult diet is high in flesh foods and enhancers of iron absorption, complementary foods fed to young children may be deficient in iron unless cereals are fortified. Conditions such as coeliac disease or environmental enteropathy, where the gut absorptive area is reduced, and chronic inflammatory bowel disease are also risk factors.

Blood loss, especially menstrual loss in healthy women, is an important cause of iron deficiency, explaining variation in status and iron requirements within populations. In the humid tropics, persistent low-grade blood loss due to intestinal parasites, such as hookworm and whipworm, also contributes to iron deficiency, as does blood loss due to urinary or intestinal schistosomiasis. IDA (iron deficiency anaemia) due to insidious blood loss may also be the mode of presentation of intestinal tumours.

The major iron store in the newborn is in haemoglobin itself, and infants are born with a high haemoglobin concentration. Low birthweight infants, especially if they are preterm, have lower blood volume and smaller iron stores than infants born at normal weight. Catch-up growth during the early months of suckling on a diet that is naturally low in iron leads to an increased risk of iron deficiency in infancy. Although breast milk is low in iron, its iron is highly bioavailable. The risk of iron deficiency is increased if the infant is introduced early to unmodified cow's milk (which may cause microscopic intestinal blood loss), or to unfortified complementary foods. Young children with behavioural feeding problems are also at risk of iron deficiency.

For other nutrient deficiency anaemias see Weblink Section 35.2.

Iodine deficiency

Iodine deficiency is the most prevalent, yet easily preventable, cause of brain damage in the world. Serious iodine deficiency during pregnancy can result in stillbirth, spontaneous abortion, and congenital abnormalities such as cretinism, which is a serious and irreversible form of mental retardation that affects people living in iodine-deficient areas, especially in Africa and Asia. However, the less visible but more pervasive mental impairment that reduces general intellectual capacity is of far greater significance. Iodine deficiency is one of the main causes of impaired cognitive development in children.

One of the most visible manifestations of iodine deficiency is goitre (see Chapter 13) and this is used as one indicator of deficiency, although the currently preferred method of detection is urinary iodine excretion as a measure of iodine intake (see Weblink Section 35.3). Iodine deficiency was still very prevalent in many areas of the world, apart from the Americas, including Europe, Eastern Mediterranean regions, and parts of Africa in 2007 (Table 35.6). The number of countries where iodine deficiency is a public health problem has been greatly reduced since around 1990 through salt iodinization programmes. Data indicate progress since 2007 when the WHO reported that the total number of countries with insufficient intakes was 47 (de Benoist et al. 2008). The number of countries with iodine deficiency as a public health problem then decreased from 110 to 54 between 1993 and 2003, to 47 in 2007, 30 in 2013, and to 19 in 2017 (Gizak et al. 2017) In the last couple of decades there has therefore been substantial progress towards elimination of iodine deficiency although we have not yet met the 1990 UN world summit for children goal of eliminating iodine deficiency worldwide. However nine countries have excessive intakes, with the emerging risk of iodine-induced hyperthyroidism in susceptible groups following the introduction of iodized salt (Figure 35.2 global scorecard of iodine nutrition in 2021).

Causes of iodine deficiency

With the exception of seafood, the iodine content of all foods, animal or vegetable, depends on that of the soils in which they grow. Iodine deficiency occurs when people subsist on food grown on iodine-deficient soils and have little access to food from elsewhere. There is an iodine cycle in the environment in which iodine present in soil and sea water as iodide is oxidized to elemental iodine and then evaporates. Iodine is also leached out of soil by heavy rain and flooding, and river water is often relatively high in iodine in the tropical rainy season. Iodine reaches the sea, increasing its iodine concentration and that of marine plants and fish. Recycling of volatile iodine in rainwater is insufficient to replenish iodine-poor soils, which results in geographical variability of iodine in soils. Poverty and isolation are characteristic of areas of classical IDD (iodine deficiency disorder). Up to 100 million people living in the highlands bordering the Rift

TABLE 35.6 Number and prevalence (%) of insufficient iodine intake (UI *<100µg/L) in school-age children (6–12 years) and general population (all age groups) by WHO regions in 2007

Region	Region	School-age children		General population	
	Number (millions)	Prevalence %	Number (millions)	Prevalence %	Number (millions)
Africa	13	40.8	57.7	41.5	312.9
Americas	3	10.6	11.6	11.0	98.6
SE Asia	0	30.3	73.1	30.0	503.6
Europe	19	52.4	38.7	52.0	459.7
E Mediterranean	7	48.8	43.3	47.2	259.3
W Pacific	5	22.7	41.6	21.2	374.7
TOTAL	47	31.5	266.0	30.6	2,008.8

UI = International Unit.

Source: Adapted from Benoist B de, McLean E, Andersson M et al. (2008) Iodine deficiency in 2007: Global progress since 2003. *Food Nutr. Bull.* **29**(3), 195–202.

Valley stretching from Ethiopia via Kenya, Tanzania, Uganda, and the Democratic Republic of the Congo (DRC) south to Malawi are at risk of IDD. In such areas, notably the DRC, the problem is aggravated by consumption of the tuber cassava, which contains a toxin, linamarin, a cyanogenic glucoside found in the leaves and roots of cassava. This goitrogen is hydrolysed to release cyanide, which, after further metabolism to thiocyanate, inhibits uptake of iodine by the thyroid gland, which hypertrophies further, increasing the size of the goitre. Other goitrogens occur in cabbage, bamboo shoots, and bacteria in contaminated water.

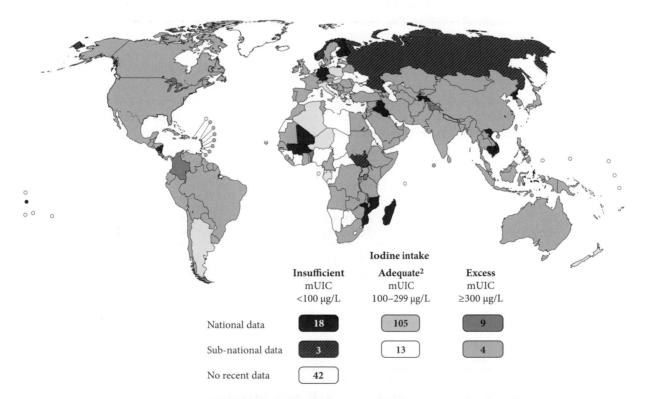

FIGURE 35.2 Global score card map 2021

Notes: In population monitoring of iodine status using urinary iodine concentration, school-age children serve as a proxy for the general population.

Adequate iodine intake corresponds to the median urinary iodine concentration values in the range of 100–299 µg/L, and includes categories previously referred to as 'adequate (100–199 µg/L) and 'more than adequate' (200–299 µg/L).

Vitamin A deficiency (VAD)

Vitamin A deficiency is the leading cause of preventable blindness in children and increases the risk of disease and death from common childhood infections such as diarrhoeal disease and measles (see Chapter 12). In pregnant women, vitamin A deficiency occurs especially during the last trimester when demand by both the unborn child and the mother is highest. In high-risk areas it causes a high prevalence of night blindness during this period and may increase the risk of maternal mortality. Vitamin A deficiency is a public health problem in more than half of all countries, especially in Africa and Southeast Asia, especially in young children and pregnant women in low-income countries.

Severe vitamin A deficiency manifests itself clinically as night blindness and corneal xerosis (see Chapter 12). Subclinical vitamin A deficiency in preschool children is defined as serum retinol levels <0.7μmol/L. Models have been constructed to determine the relationship between the two indicators so that the more frequently recorded clinical signs can roughly be projected to the subclinical level. While clinical signs of vitamin A deficiency are relatively rare, subclinical deficiency in preschool children is estimated to be widespread in LICs, and this increases the likelihood of morbidity and mortality. In pregnant women the estimated prevalence is lower than in preschool children. It is especially low in the Americas, but particularly high in the Western Pacific. Despite a marked increase in data submitted to the WHO, there are still many countries that lack national prevalence data. The WHO estimated in 2009 that vitamin A deficiency affects 250 million preschool children, and in vitamin A-deficient areas a substantial proportion of pregnant women. Around 250,000 to 500,000 vitamin A-deficient children become blind every year, half of them dying within 12 months of losing their sight.

- Globally, night blindness is estimated to affect 5.2 million preschool-age children (0.9 per cent of the population at risk of VAD) and 9.8 million pregnant women (7.8 per cent of the population at risk of VAD).

- Low serum retinol concentration (<0.70 μmol/l) globally affects an estimated 190 million preschool children (33.3 per cent of the preschool-age population) and 19.1 million pregnant women (15.3 per cent of pregnant women in populations at risk of VAD).

- The WHO regions most affected by vitamin A deficiency for preschool children and pregnant women respectively were: Southeast Asia (49.9, 17.3 per cent) and Africa (44.4, 13.5 per cent), followed by the Eastern Mediterranean (20.4, 16.1 per cent), Europe (19.7, 11.6 per cent), the Americas (15.6, 2.0 per cent), and the Western Pacific (12, 21.5 per cent).

More recent data (Stevens et al. 2015) indicates a decrease in the prevalence of vitamin A deficiency between 1991 and 2013 in all low- and middle-income countries and UN regions except Sub-Saharan Africa where deficiency has increased (Table 35.7). The number and percentage of child deaths attributable to vitamin A deficiency by region is shown in Table 35.8.

Causes of vitamin A deficiency

Deficiency of vitamin A is likely when a low intake is further reduced by disruption of metabolic pathways due to disease or other concomitant deficiency, and where requirements are enhanced by infection. Poor people in many tropical countries have limited access to vitamin A

TABLE 35.7 Estimated prevalence of vitamin A deficiency* by region in 1991, 2000, 2013

WHO region	Prevalence in 1991 (%)	Prevalence in 2000 (%)	Prevalence in 2013 (%)	PP of decreases (%)
Central Asia, Middle East, and North Africa	19	14	11	0·76
East and Southeast Asia and Oceania	42	21	6	0·99
Latin America and the Caribbean	21	15	11	0·89
South Asia	47	46	44	0·54
Sub-Saharan Africa	45	49	48	0·45
All low-income and middle-income countries	39	33	29	0·81

* Defined as a serum retinol concentration lower than 0·70 μmol/L.
A PP (posterior probability) greater than 0·50 indicates a decrease and a PP smaller than 0·50 indicates an increase in deficiency.

Source: Reprinted from Stevens G A, Bennett J E, Hennocq Q et al. (2015) Trends and mortality effects of vitamin A deficiency in children in 138 low-income and middle-income countries between 1991 and 2013: A pooled analysis of population-based surveys. *Lancet Glob. Health* **3**, e528–e536, with permission of Elsevier.

TABLE 35.8 Number in thousands and (proportion %) of child deaths attributable to vitamin A deficiency by region in 2000 and 2013

	Diarrhoea		Measles		All causes	
	2000	2013	2000	2013	2000	2013
C. Asia, Middle East, & N. Africa	3.7(8.2)	1.2(6.5)	0.4(5.0)	0.0(4.3)	4.1(0.9)	1.2(0.4)
E & SE Asia & Oceania	11.5(10.5)	1.4(4.1)	2.3(7.2)	0.2(2.0)	13.9(1.2)	1.6(0.3)
L.America & Caribbean	2.5(7.6)	0.5(5.9)	0(NA*)	0(NA*)	2.5(0.6)	0.5(0.3)
South Asia	90.9(18.0)	34.8(17.0)	12.3(11.2)	5.1(11.1)	103.2(3.0)	39.9(2.0)
Sub-Saharan Africa	102.8(19.0)	56.6(18.4)	39.7(11.5)	5.9(13.0)	142.5(3.4)	62.5(2.0)
All LMICs	211.4(17.2)	94.5(16.5)	54.8(11.0)	11.2(11.0)	266.2(2.7)	105.7(1.7)

NA = not available. * A population attributable fraction was not calculated because no measles deaths occurred in the L. America and Caribbean region during this time period.

Source: Reprinted from Stevens G A, Bennett J E, Hennocq Q et al. (2015) Trends and mortality effects of vitamin A deficiency in children in 138 low-income and middle-income countries between 1991 and 2013: A pooled analysis of population-based surveys. *Lancet Glob. Health* **3**, e528–e536, with permission of Elsevier.

as retinol available in liver, fish-liver oils, egg yolks, and dairy products, and rely on its precursor, β-carotene, which is present in orange fruits, dark green leaves, and some roots. Plant-derived β-carotene provides about three-quarters of dietary vitamin A, and the supply is markedly reduced in the long dry seasons.

Imbalance between supply and requirement may also influence the seasonality of vitamin A deficiency. This can occur if the season of maximal availability of β-carotene does not coincide with the post-harvest season of plentiful energy and protein intake, which facilitates the childhood growth spurt. The consumption of red palm oil, with its exceptionally high β-carotene content, was until recently limited to West Africa but is now more widespread. Mangoes, orange-fleshed sweet potatoes, and some indigenous fruits have a high content of β-carotene and have been shown to be effective at improving vitamin A status.

Vitamin A deficiency is largely responsible for the global annual toll of new cases of preventable blindness. The cumulative effects of retinol deficiency on the epithelia of the surface of the eye explain the progressive signs of xerophthalmia from dryness of the conjunctiva, Bitot's spots, thickening of the cornea, ulceration, and scarring. Dryness of the eye, from which the term *xerophthalmia* is derived, is due to failure of tear production in the lachrymal ducts, exacerbated by reduced mucus production due to fewer goblet cells in the conjunctiva (see Chapter 12, Figure 12.4). Night blindness, an early sign due to failure of conversion of retinol to retinal in the retina, is only evident in children old enough to toddle about, although night blindness in pregnant women is a useful indicator of population vitamin A status. Xerophthalmia is more commonly seen in Asia than in Africa.

Vitamin D deficiency

Prevalence

Few countries collect national statistics on rickets so it is difficult to estimate the global prevalence. Nevertheless, physicians in many high-income countries are concerned about apparent increases in the prevalence of rickets, particularly among children whose families originate from Asia, Africa, or the Middle East. In Africa, Asia, and the Middle East calcium deficiency is a contributing factor to rickets which can occur even where sun exposure and vitamin D status are adequate (see Chapter 12, Figure 12.8).

There has recently been increased attention to measuring subclinical vitamin D deficiency, usually assessed as plasma levels of 25-hydroxyvitamin D (25OHD), rather than just the clinical deficiency, rickets. The prevalence of low plasma 25OHD appears to be increasing and this does not seem to be entirely due to increased frequency of measurement or the ongoing debate as to what constitutes an adequate plasma 25OHD concentration. Because different plasma 25OHD cut-offs for adequacy are used by different researchers, global prevalence of low 25OHD is hard to estimate.

Vitamin D deficiency is of concern in many regions of the world and has been the subject of several reviews over recent years (see Chapter 12). The WHO notes that there are few population representative data available on the prevalence of rickets and no national registrations of rickets cases, but that rickets practically disappeared from high-income countries when vitamin D food fortification programmes started; however, some reports have shown a re-emergence in the past 10–20 years, related to less sun exposure because of the use of sun-creams, skin coverage to avoid skin cancer, and less time spent outdoors. Increased migration across the world and rural to urban migration increases the risk of deficiency especially in those with darker skins living in temperate climates. A prevalence in children greater than 1 per cent has been suggested to warrant a public health response, but this is challenging without a clear definition and aetiology.

SACN (2016) observes that at latitudes below 37°N, UVB radiation is sufficient for year-round vitamin D synthesis but at higher latitude, vitamin D is not synthesized during the winter months. In the UK, sunlight-induced vitamin D synthesis is only effective between late March/early April and September and not from October onwards throughout the winter months. Mean plasma 25(OH)D concentrations across age groups in the UK range between 40 and 70 nmol/L but lower for institutionalized adults (around 30 nmol/L). The proportion of the population (by age) with a plasma 25(OH)D concentration <25 nmol/L was:

2–8 per cent in young children(5m–3y);

12–16 per cent in 4–10 year olds;

20–24 per cent in teenagers (11–18 years);

22–24 per cent in adults (19–64 years);

17–24 per cent in older adults ≥65y and above;

and nearly 40 per cent of institutionalized adults.

For all age groups in the UK, mean plasma 25(OH)D concentration was lowest in winter and highest in summer. Around 30–40 per cent of the population had a plasma 25(OH)D concentration <25 nmol/L in winter compared to 2–13 per cent in the summer. A large proportion of some population groups did not achieve a plasma/serum 25(OH)D concentration ≥25 nmol/L in summer (17 per cent of adults in Scotland; 16 per cent of adults in London; 53 per cent of women of South Asian ethnic origin in Southern England; and 29 per cent of pregnant women in NW London) (SACN 2016).

Causes of vitamin D deficiency

The metabolism of vitamin D and calcium are closely linked and clinical deficiency symptoms such as rickets are often a result of a combination of the two deficiencies. This is supported by evidence showing that supplementation with both calcium and vitamin D together may be more effective in treating the deficiency disease than supplementation of either nutrient alone.

The main source of vitamin D is synthesis from 7-dehydrocholesterol in the skin when acted on by UVB rays of sunlight. At latitudes far from the equator there is generally a seasonal variation in prevalence of low plasma 25OHD which corresponds to the seasonal changes in amount of sunlight. Several factors contribute to the increased prevalence of deficient plasma levels of 25OHD (see Chapter 12). There is a consistent inverse association of body mass index with plasma 25OHD which may partly result from sequestration of vitamin D in the large body fat pool but may also occur if obese people spend less time in outdoor activity. Dark skin pigments decrease penetration of the UVB rays to the site of vitamin D synthesis so people of African or South Asian origin living in northern Europe and North America are particularly vulnerable to these seasonal differences in vitamin D synthesis from sunlight. Nevertheless, the lower 25OHD levels seen in people of African origin often do not translate into lower bone mineral density suggesting other factors, for example genetic, are involved.

Few foods contain much vitamin D; examples are fish-liver oils and fortified foods such as dairy products in some countries. Small fish for which bones are eaten, and dairy products, are also good sources of calcium.

Other micronutrient deficiencies

Other fairly common micronutrient deficiencies—folate, B_{12}, copper, riboflavin, zinc, niacin, thiamine, and vitamin C—are discussed in Weblink Section 35.4 and the clinical signs of deficiencies are shown in Weblink Figures 35.1–35.12.

35.3 Classification of causes of malnutrition

Introduction

The causes of malnutrition are represented in the United Nations Children's Emergency Fund (UNICEF) conceptual diagram (Figure 35.3) for child undernutrition but similar analysis can be applied to all age groups. The diagram indicates that the causes lie at many levels (immediate, underlying, and basic) and in many sectors (agriculture, health, water, education, and employment to name a few). It is important to understand the causes of malnutrition as this can guide the nature of interventions needed to correct and policies to prevent malnutrition. For example, if food availability at the household level is adequate, but infant girls are not receiving enough food, then there are social inequalities or lack of knowledge that need to be recognized and addressed. If infants are receiving sufficient food in terms of energy but have many micronutrient deficiencies then diet quality is a key constraint. If infants are having frequent bouts of diarrhoea due to poor water quality, then even a good diet

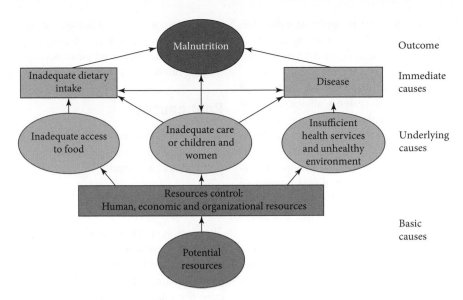

FIGURE 35.3 UNICEF conceptual framework for causes of malnutrition

will not prevent malnutrition. Finally, if the quality and quantity of food intake is sufficient, as is the health and sanitation environment, but the child is failing to thrive, it might be due to poor interaction between infant and parent or caretaker (e.g., psychosocial stimulation). Similar logic can be applied to other age or social groups. The first examples described in the following paragraphs focus on undernutrition, followed by examples related to overweight and obesity.

In reference to undernutrition, if people do not eat enough food they are likely to be deficient in many nutrients and susceptible to specific deficiency diseases. On the other hand, in reference to overweight and obesity, people may consume excess energy but, depending on the quality of their diet, they may have a poor balance of macronutrients and also be deficient in certain micronutrients, eventually leading to the chronic diseases, the NCDs, sometimes misleadingly called the 'diseases of affluence'.

Immediate causes

Undernutrition

The most immediate causes of malnutrition in young children (and also adults) are poor diet and infection. If the child is not able to ingest enough food, and of good quality, this can lead to undernutrition. Infection and an inadequate diet reinforce each other. Infection reduces the intake of nutrients by diminishing appetite, inhibiting nutrient absorption, and increasing nutrient requirements for combating infection, while poor diet can reduce the effectiveness of the immune system (see Chapter 27).

Overweight and Obesity

Immediate causes of overweight and obesity are excess intake of energy, through bottle feeding in infants, regular consumption of high fat, energy-dense foods and beverages, and high frequency of snacking coupled with reduced physical activity.

Underlying causes

Undernutrition

Inadequate dietary intake and high disease burden are influenced by underlying factors which can be grouped as household food insecurity, poor care for mothers and children, and the inadequate provision of health services and an unhealthy environment. Food and nutrition security exists when all people, at all times, have physical, social, and economic access to sufficient, safe, and nutritious food that meet their dietary needs and food preferences for an active and healthy life.

A food-insecure household cannot get reliable access to food of adequate quantity and quality, consistent with good health, either from their own production, or in the market. Household food insecurity has many interrelated causes: unemployment, low-incomes, high food prices, poor crop yields, low rates of exchange between food and non-food, and a lack of access to assets, including land, water, agricultural extension, and credit. Extreme crop failure, especially if repeated over several years, sometimes exacerbated by civil conflict and health outbreaks such as AIDS and COVID-19, manifests itself as widespread food deprivation and increased mortality. If a household is food insecure, it is likely that infants and other vulnerable groups within the household have an inadequate diet. The 2019 coronavirus pandemic has had a profound influence on health, mortality, reduced economic activity, unemployment, and hence poverty (see Sections 35.2, 35.4).

Poor health services, unclean drinking water, and non-existent hygiene disposal systems all increase the likelihood of infection, particularly diarrhoea, which is the third leading cause of disability-adjusted life years (DALYs) lost in children under the age of four, accounting for 17 per cent of DALYs in this age group, as estimated by Murray et al. (2015) who had first developed the concept. DALY is a measure of the overall disease burden, expressed as the number of years lost due to ill-health, disability, or early death.

Care is defined as the activities and practices of care givers including care for women (e.g., time for resting and appropriate food during pregnancy), breastfeeding and feeding of young children, psychosocial stimulation of infants, food preparation and storage practices, hygiene practices and care for children during illness, including diagnosis and health seeking behaviour. Care for mothers and infants is essential if food at the household level is to be given to infants in the right form at the right time in the right quantities. Care is also essential if hygiene practices are performed that minimize exposure to the hazards of the health environment and minimize the need for health services. The provision of care requires accurate information about the best practices, and the time and authority to implement them. Emphasis is on young children and women as they are the most vulnerable groups of the community but of course older children and men can also be affected.

Overweight and obesity

Underlying causes of overweight and obesity are mainly the converse, including overall food security, availability of plentiful, varied, and relatively inexpensive food products (although not always the most healthy), food advertising, a sedentary lifestyle at work and leisure due to labour-saving devices, transport, computers, work pressures, and increased perceptions of risk for children playing outside and walking or cycling to school.

Root causes

The basic or root causes of all forms of malnutrition, both undernutrition, overweight, and obesity are essentially political and economic.

Undernutrition

There are very few instances of high levels of undernutrition, as opposed to overweight and obesity, above GPD per capita levels of $4000 (Vollmer et al. 2014). Income growth is important for reducing undernutrition, however, the relationship is not as tight as one might think. Figure 35.4 uses survey data to model the effects of rapid income growth over a sustained time period for 12 countries. Overall, the study finds that a 10 per cent increase in income produces a 5 per cent reduction in the rate of undernutrition. Only in three countries does a rapid income growth rate alone achieve the millennium development goal of halving the 1990 rate of undernutrition by 2015. So income growth is extremely helpful in decreasing undernutrition rates, especially when income distribution is more equal, but it is by no means sufficient, and given the wide range of levels of undernutrition at a given GDP per capita, some may say it is not necessary.

Other factors beyond income growth are obviously essential for good nutrition at all ages, such as good levels of education, social equity, and enlightened government behaviour. Good governance—by which we mean institutions that give voice to all parts of society, respect for the civil, political, economic, social, and cultural rights of its citizens, and appropriate levels of investment in public goods such as safety, research, roads, health, and education—is more likely to produce these conditions. Good

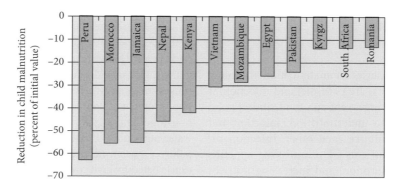

FIGURE 35.4 Percentage decline in malnutrition due to 2.5 per cent annual growth in per capita income, 1990s to 2015.
Source: Haddad L et al. (2003) Reducing child malnutrition: How far does income growth take us? © World Bank, Washington, DC:.

education levels mean that individuals know how to access, assess, and use information that is helpful to the attainment of good nutrition status (such as the right types of foods to consume, what to do in the case of diarrhoea, and the optimal duration of exclusive breastfeeding).

The status of women relative to men is a dimension of society that is crucial to the nutritional status of women and of infants. Empirical work using nationally representative data from 36 countries (Smith et al. 2003) showed that the poor status of women relative to men in South Asia plays an important role in explaining what has become known as the 'Asian enigma'. Yet, even if the status of women relative to men in South Asia does not improve, the numbers of stunting in that region is at least declining, which is not the case in Sub-Saharan Africa where the numbers of stunting are on the increase. The special case of Sub-Saharan Africa has been mentioned frequently. Some of the factors behind the tragic deterioration in nutrition status in Sub-Saharan Africa are summarized in Box 35.1.

Overweight and obesity

The basic or root causes of overweight and obesity are also essentially political and economic, and they can be very wide-ranging. Factors such as employment, income, work

BOX 35.1
Factors behind nutritional decline in Sub-Saharan Africa

- Closed markets in the developed world (especially the EU) for their exports
- HIV/AIDS—generating a health and development crisis and undermining the ability to respond to the crisis
- Wars and refugee movements
- Military expenditure
- Drought
- Crop and livestock disease
- Many ecosystems making technology diffusion and adaptation difficult
- Under-investment in agricultural research
- No economy serving as regional driver (Nigeria and South Africa's under-whelming economic performance)
- Low levels of human capital in terms of literacy
- Unfair terms of trade for natural resources on world markets
- Rapid population growth and urbanization not supported with commensurate infrastructure and market institutions.

pressures with associated lack of available 'free' time for exercise and food preparation, psychological pressures from overwork or lack of work, and other causes leading to comfort eating, food prices, taxation policies, advertising policies, environmental factors such as lack of facilities for transport including cycling, and for sport, can all contribute to overweight and obesity (see Chapter 20).

35.4 Specific groups at risk of malnutrition

Refugees

Conflict and climatic disaster are the main triggers for large-scale population movements. People migrate within their own or to another country, fleeing from danger or seeking food and security. The United Nations High Commission for Refugees (UNHCR) estimates that there are 26 million refugees across the world and that over half are children. They are defined and protected in international law and must not be expelled or returned to situations where their life and freedom are at risk. The total number of displaced people worldwide is 79.5 million. This includes not only refugees but also asylum seekers and people displaced within their own countries. (UNHCR.org accessed 11/2/21). In many cases refugees and displaced people find shelter in emergency camps. Famines brought about by conflicts or poor weather conditions among others and resulting in poor agricultural yields have been the cause of much displacement. Famines have been a problem in the past over centuries but for several decades have not been a major nutritional concern until recent political conflicts, the COVID-19 pandemic, and climate change are threatening famine in Madagascar, Afghanistan, and Ethiopia. Weblink Section 35.5 provides further information about famine.

Patients with prolonged stay in hospitals and care homes

Undernutrition in patients is associated with prolonged stay in hospital. Undernutrition may be estimated by anthropometry or by a nutrition risk assessment 'tool' or indicator. Such 'tools' provide a means of identifying at admission those at increased risk of malnutrition, for whom nutritional intervention is advisable (see Weblink Section 35.6).

Conditions such as mental illness and neurological incapacity, for example, coma or stroke interfere with eating; others, such as illness in the elderly, in the context of a poorly staffed ward, are also associated with reduced intake. The body's response to conditions such as infection, surgery, or burns, and cardio-respiratory disease also has a 'nutritional cost'. The risk of nutritional deterioration is increased by existing malnutrition, inability to eat, underfeeding, or an increased protein–energy requirement for infection or healing. Estimates of energy and protein requirements during illness were reduced when research indicated that the high requirements of the acute phase do not apply throughout convalescence. Overfeeding, especially during total parenteral nutrition (TPN) is linked to the development of hepatic steatosis, an early sign of TPN-associated liver disease. Various preventive strategies are now advised to avoid nutrient deficits or excesses, to provide pro-active antibiotic treatment, and to modulate the immune response by addition of nutrients such as n-3 fatty acids or arginine.

Patients with renal failure are at risk of negative energy and nitrogen balance and micronutrient deficiency due to uraemic anorexia, adaptive changes in protein turnover, and loss of micronutrients during dialysis. Folate deficiency may present as failure of anaemia to respond to treatment with erythropoietin. Micronutrient supplementation, especially with the B vitamins, is recommended. Surgical patients are also at increased risk of deficiency when, for example, nutrient absorption has been compromised by gastric bypass or gut resection. Supplementation with B vitamins and some minerals may be indicated. Malnutrition may slow down the metabolism of drugs, and also affect the response to treatment of infection.

Prevention of malnutrition in the community and in care homes is one of the elements of primary care for the elderly and other vulnerable groups. Strategies to improve the nutrition of vulnerable patients include improving the choice and quality of meals and identifying those who need help with feeding.

Alcoholics

Alcoholism (see Chapter 10) is a global problem, the nutritional consequences of which are worse when superimposed on an already impoverished diet. A heavy drinker consumes many 'empty calories' from alcohol with lower consumption of healthy foods since meals may be nutritionally inadequate or omitted due to forgetfulness or inebriation. Signs of malnutrition may become evident. Alcohol-related liver disease includes steatosis, hepatitis, and cirrhosis, all of which may be associated with primary or secondary malnutrition.

Micronutrient deficiency is common among alcoholics, partly explained by dietary lack but also due to increase excretion. Biochemical and clinical deficiency of the B vitamins may be accompanied by deficiency of magnesium, zinc, and calcium, the latter partly due to

excessive urinary loss and partly to vitamin D deficiency. Alcohol can interfere with nutrient metabolism which is particularly evident for thiamin. Chronic thiamin deficiency may result in polyneuropathy or the complexities of the Wernicke–Korsakoff syndrome (WKS). WKS may present acutely with signs mimicking those of acute alcohol toxicity such as unusual eye movements, an ataxic gait, and finally stupor and coma. Memory tends to fail more gradually. Treatment with thiamin is effective if given in time, otherwise the changes are irreversible.

Low socioeconomic groups in high-income countries

The demography of middle-income and affluent countries is changing as a result of immigration and increased survival into old age, both of which may affect rates of poverty and malnutrition. The groups most vulnerable to poverty are single-parent families, older people, refugees, and ethnic minorities. Isolation, depression, and memory deficits associated with dementia or alcoholism may exacerbate the situation. Ethnic minorities, especially when isolated by language and educational disadvantage, are particularly prone to poverty and undernutrition. Unemployed urban dwellers are at increased risk of poverty and malnutrition. Immigrants, especially recent arrivals unfamiliar with the 'bureaucratic systems' in the host country, are at particular risk of food insecurity and actual hunger.

There is often a divergence between macro- and micronutrient status such that deficiency of iron and other micronutrients coexists with obesity in both children and adults. This paradoxical combination of 'excess and deficiency' prompted a provisional extension of the definition of 'food insecurity' to include 'lack of access to balanced diet'. Low income in the UK is a risk factor for unbalanced dietary intakes that are high in energy and deficient in micronutrients. Failure to access healthy food is not the inevitable sequel of 'poverty and ignorance'. Among the poor of wealthy countries, decisions about where to shop and what to buy are often determined by distance, or problems with urban transport. Practical difficulties with food storage compound the problems of homeless people. Multicomponent strategies are needed to deal with the complex nutrition problems associated with poverty (see Chapter 36).

Vegetarians and vegans

Vegetarian diets (see Chapter 17) cover a wide range including vegan and macrobiotic diets and so cannot be considered as a single group. In general the diet of vegetarians may be less restricted than is commonly thought.

Plant-based diets are nutritionally adequate providing they are not restricted in variety or quality and vegetarians' higher intake of fruit and vegetables results in a lower risk of ischaemic heart disease. However, deficiencies can occur as diets high in fruit and vegetables may be bulky and low in energy and therefore inadequate in young children. Vitamin B_{12} is the nutrient most likely to be lacking in vegan and to a lesser extent in vegetarian diets, leading to elevated plasma homocysteine concentrations and increased risk of neurological disorders. Vegetarians are at increased risk of iron-deficiency anaemia as iron from plant foods has low bioavailability. In vegan diets long-chain n-3 fatty acids are absent but intakes of linoleic acid are high, which may inhibit the synthesis of long-chain polyunsaturated fatty acids needed in foetal brain development. Vegans may have difficulty maintaining weight in old age.

Pregnancy and early childhood

Poor nutrition linked to lack of education has cross-generational effects in all societies. Early onset of child-bearing and short birth interval are associated with low maternal weight gain and low birthweight. The immediate risks of low birthweight are reduced by effective neonatal care, but the sequelae of adverse foetal programming are not. If such infants are then 'overfed' in childhood, they are particularly prone to obesity and NCDs as adults. Adolescent mothers with poor diets are at particular risk due to the nutritional demands of their own growth and that of their infant. The risks of micronutrient deficiency during periods of rapid early child growth are greater among the poor and among those eating diets with low micronutrient bioavailability such as vegetarians.

Older people

The nutritional requirements of older people (see Chapter 16) reflect reduced energy expenditure, but sustained protein requirements (see Chapter 9). Micronutrient deficiency may be due to dietary inadequacy, physical problems with eating, or malabsorption, sometimes associated with disease or bacterial overgrowth of the gut. Since oxidative stress is responsible for some of the effects of aging, antioxidant micronutrient status in the elderly has implications for health. Less skin exposure to sunlight, plus less effective skin metabolism of precursors of vitamin D, increases the risk of deficiency of this nutrient. The diseases associated with ageing such as coronary artery disease, Alzheimer's disease, poor cognitive functioning, and cancers all have nutritional components. Non-nutritional factors including physical activity and social integration also play a part in the nutritional health of older people. Joint or movement disorders such as tremor or ataxia reduce mobility, thereby contributing to obesity, which, when combined with sarcopenia and consequent muscle weakness, results in further reduction in activity. By contrast, physical fitness is associated with good nutritional health, although it may not be clear which factor initiates the virtuous cycle.

35.5 Progress against the global nutrition targets

Since 2010 the global community has been working towards a common goal to address the fragmentation in the nutrition front. To achieve this, efforts are underway to harmonize nutrition targets that all countries can adopt and adapt. In 2012, the World Health Assembly endorsed the six global nutrition targets to be achieved by 2025 (Table 35.3). These targets were adjusted to align with the SDG (sustainable development goals) timeline of 2030. For stunting, a 50 per cent reduction in the number of children less than five years; for anaemia a 50 per cent reduction in women of reproductive age; for low birth-weight 30 per cent reduction; childhood overweight to be reduced and maintained to less than 3 per cent; exclusive breastfeeding for the first six months up to at least 70 per cent; wasting to be reduced and maintained at less than 3 per cent by 2030. The sustainable development goal 2 also sets targets for the elimination of hunger and childhood stunting, wasting, and overweight by 2030 (Table 35.2).

In 2013, the WHA endorsed the NCD global action plan to monitor the progress in non-communicable disease prevention and control, with the overall aim of reducing premature mortality due to NCDs by 25 per cent by 2025. The 2020 global nutrition report provides an independent assessment of the state of nutrition by tracking global and country progress on these targets. The GNR2020 analysis showed that 88 countries are off track and not likely to meet all the six global nutrition targets by 2030. Fifty countries were on track for only one target, 35 countries for two targets, 13 were on track for three targets, and only eight were on track for achieving four targets by 2030. At the current pace, most countries will not be able to meet the nutrition targets by the proposed 2030 timeline.

Overweight and obesity in children and adults present a special concern, as these conditions are increasing in all regions of the world. It has been indicated that without a reversal of the trend it is not likely the world will achieve global obesity targets, but even more concerning is the fact that the prevalence of obesity estimated at 16.1 per cent by 2025 could overtake underweight in women (World Obesity Atlas 2022).

35.6 COVID-19: impact on nutrition

The COVID-19 pandemic has challenged the world's systems in ways never imagined. It started as a health crisis in which severe outcomes were seen more in persons with underlying health conditions such as obesity, diabetes, and other NCDs. COVID-19 mitigation measures put in place caused serious disruptions on many fronts. On the economic front, massive loss of jobs and livelihoods have pushed millions into poverty. The World Bank estimates that COVID-19 will result in an additional 88 million to 115 million people moving into extreme poverty by 2022. COVID-19-induced job losses are estimated to be about 345 million in 2020 alone. Prior to COVID-19, concerns were raised about the failures in the global food systems regarding an inability to reduce food insecurity, deliver on healthy diets, reduce inequities within the food systems, and its environmental impact with respect to greenhouse gas emissions. Food systems are depleting natural resources—land, water, energy—and are accelerating agro-biodiversity loss.

The advent of COVID-19 amplified the deficiencies in the food systems. COVID-19-induced poverty is expected to further increase the number who cannot afford a healthy diet. The United Nations estimates that before COVID-19, three million people could not afford a healthy diet (SOFI 2020). Disruptions in social protection safety nets, for example school closures, left 370 million children without the much-needed school meals (WFP 2020). COVID-19 disruptions in essential nutrition services for women and children in low- and middle-income countries have aggravated an already dire situation. Modelling estimates of the impact of COVID-19-related shocks from economic, health, and food systems disruptions on maternal and child nutrition indicate that by 2022, the COVID-19 impact will likely result in an additional 9.3 million wasted under five-year-old children, 2.6 million stunted children, 168,000 child deaths, and 2.1 million cases of maternal anaemia. Future productivity losses due to additional stunting and child mortality could be USD\$29.7 billion (Osendarp et al. 2021).

The extent of the full impact of COVID-19-related disruptions on nutritional status is yet to be realized. COVID-19 is compelling the global community to build forward better by transforming food systems to become equitable, resilient, and to deliver on healthy diets that are sustainable for humans and for the planet. Moving in this direction, the UN Secretary-General convened the food systems summit in September 2021 to facilitate the transformation of the food systems and the achievement of the SDGs. The overall goal of the summit is to launch bold transformative actions that will speed up the achievement of the 17 SDGs through five clear objectives: (i) ensuring access to safe and nutritious food for all; (ii) shifting to sustainable consumption patterns; (iii) boosting nature-positive production at sufficient scale; (iv) advancing equitable livelihoods and value distribution; (v) building resilience to vulnerable shock and stress.

In summary, the COVID-19 pandemic has led to disruptions in all systems, be it economic, social, health, and food, all of which have direct impacts on nutritional status. As disruptive as COVID-19 has been, it presents an opportunity to build sustainability and resilience into health and global systems. COVID-19 shows the interconnectivity of all these systems.

KEY POINTS

- Globalization, urbanization, aging, and global warming represent some of the challenges to the achievement of good nutrition status.

- The rising prevalence of overweight and obesity and related chronic diseases is an additional concern in poor countries in transition.

- Malnutrition occurs in hospitals in both rich and poor countries and coexists with poverty, even in high-income countries. Among refugees, immigrants, the unemployed, the homeless, and older people, malnutrition may be exacerbated by alcoholism.

- The causes of undernutrition and overweight and obesity are classified at three levels—immediate, underlying, and root.

- The immediate causes of undernutrition relate to poor food intake and infection; the immediate causes of overweight and obesity relate to excess energy intake and/or inadequate energy expenditure.

- The underlying causes of undernutrition relate to household food insecurity, poor support for caring practices, and a weak health and sanitation environment; the underlying causes of overweight and obesity relate to household food security, a plentiful, relatively inexpensive and varied supply of energy-dense food, and low levels of activity.

- The root causes of both are economic and political in nature.

- The status of women is key in promoting good nutrition.

- Globally, progress in reducing all forms of malnutrition has been slow. Special attention will need to be given to regions such as Asia and Sub-Saharan Africa, that carry a high burden of malnutrition.

- Coronavirus (COVID-19) is the most recent additional challenge to health and nutrition.

ACKNOWLEDGEMENTS

The authors acknowledge the contribution of Suzanne Filteau and Chaza Azik in sections of this chapter with material adapted and updated from the previous edition of the book.

Be sure to test your understanding of this chapter by attempting multiple choice questions. See the Further Reading and Useful Websites lists for additional material relevant to this chapter.

REFERENCES

Béné C, Oosterveer P, Lamotte L et al. (2019) When food sysems meet sustainability: Current narratives and implications for actions. *World Dev.* **113**, 116–130.

de Benoist B, McLean E, Andersson M et al. (2008) Iodine deficiency in 2007: Global progress since 2003. *Food Nutr Bull* **29**(3), 195–202.

Black R E, Victora C G, Walker S P et al. (2013) Maternal and child undernutrition and overweight in low-income and middle-income countries. *Lancet* **382**, 427–451.

FAO, UNICEF, WFP, and WHO (2020) The state of food security and nutrition in the world. In: *Transforming Food Systems for Affordable Healthy Diets* (Internet). FAO, Rome (accessed 17 June 2021). Available from: https://doi.org/10.4060/ca9692en.

FAO, UNICEF, WFP, and WHO (2021) The state of food security and nutrition in the world. In: *Transforming Food Systems for Food Security, Improved Nutrition and Affordable Healthy Diets for All.* (Internet). FAO, Rome (accessed 30 July 2021). Available from: https://doi.org/10.4060/cb4474en.

Gizak M, Gorstein J, Andersson M (2017) Epidemiology of iodine deficiency. In: *Iodine Deficiency Disorders and their Elimination*, 29–43 (Pearce E N ed.) Springer, Cham. doi: 10.1007/978-3-319-49505-7_3

Murray C J L, Barber R M, Foreman K J et al. (2015) Global, regional and national disability-adjusted life years (DALYs) for 306 diseases and injuries and healthy life expectancy (HALE) for 188 countries 1990–2013. *Lancet* **386**, 2145–2192.

Osendarp S, Akuoku J K, Black R E et al. (2021) The COVID-19 crisis will exacerbate maternal and child undernutrition and child mortality in low- and middle-income countries *Nature Food* **2**, 476–484. https://www.worldobesity.org/resources/resource-library/world-obesity-atlas-2022

Reardon T, Tschirley D, Liverpool-Tasie L S O et al. (2021) The processed food revolution in African food systems and the double burden of malnutrition. *Glob Food Sec* **28**, 100466.

SACN Scientific Advisory Committee on Nutrition (2016) *Vitamin D and Health*. The Stationery Office, London.

Smith L C, Ramskrishnan A, Ndiaye A et al. (2003) The importance of women's status for child nutrition in developing countries: IFPRI research reports 131. IFPRI, Washington DC.

Smith, M I, Yatsunenko T, Manary M et al. (2013). Gut microbiomes of Malawian twin pairs discordant for kwashiorkor. *Science* **339**, 548–554.

SOFI (2017 and 2020) *The State of Food Security in the World*. FAO, Rome.

Soriano J M, Rubini A, Morales-Suarez-Varela M et al. (2020) Aflatoxins in organs and biological samples from children affected by kwashiorkor and marasmic-kwashiorkor: A scoping review. *Toxicol* **85**, 174–183.

Stevens G A, Bennet J E, Hennocq Q et al. (2015) Trends and mortality effects of vitamin A deficiency in children in 138 low-income and middle-income countries between 1991 and 2013: A pooled analysis of population based surveys. *Lancet Globl Hlth* **3**, e528–e536.

Swinburn B A, Kraak V I, Allender S et al. (2019) The global syndemic of obesity, undernutrition, and climate change: The *Lancet* commission report. *Lancet* **393**, 791–846.

UNICEF United Nations Children's Fund, World Health Organization, International Bank for Reconstruction and Development/The World Bank (2021) *Levels and Trends in Child Malnutrition: Key Findings of the 2021 Edition of the Joint Child Malnutrition Estimates*. World Health Organization, Geneva.

Victora C G, Christian P, Vidaletti L P et al. (2021) Revisiting maternal and child undernutrition in low-income and middle-income countries: Variable progress towards an unfinished agenda. *Lancet* **397**, 1388–1399.

Vollmer S, Harttgen K, Subramanyam M A et al. (2014) Association between economic growth and early childhood undernutrition: Evidence from 121 demographic and health surveys from 36 low-income and middle-income counties. *Global Hlth* **2**(4), e225–e234.

WFP World Food Porgramme (2020) *State of school feeding worldwide 2020* (Internet).

WFP (accessed 22 June 2021) Rome. Available from: https://www.wfp.org/publications/state-school-feeding-worldwide-2020

World Obesity Atlas (2002) https://www.worldobesity.org/resources/resource-library/world-obesity-atlas-2022

FURTHER READING

Global Panel on Agriculture and Food Systems (2020) Future food systems: For people, our planet and prosperity (Internet) [cited 22 June 2021]. Available from: http://www.glopan.org/foresight2/

Headey D, Heidkamp R, Osendarp S et al. (2020) Impacts of COVID-19 on childhood malnutrition and nutrition-related mortality. *Lancet* **396**, 519–521.

NCD Risk Factor Collaboration (2016) Trends in adult body-mass index in 200 countries from 1975 to 2014: A pooled analysis of 1698 population-based measurement studies with 19.2 million participants. *Lancet* **387**, 1377–1396.

Pearce E N, Andersson M, Zimmerman M B (2013) Global iodine nutrition: Where do we stand in 2013? *Thyroid* **23**(5), 523–528. doi:10.1089/thy.2013.0128

UN Food Systems Summit (2021) UN food systems action tracks (Internet) (cited 22 June 2021). Available from: https://www.un.org/en/food-systems-summit/action-tracks.

World Bank Group (2020) Poverty and shared prosperity: Reversal of fortune (Internet) (cited 22 June 2021). Available from: https://www.alnap.org/help-library/poverty-and-shared-prosperity-2020-reversals-of-fortune.

World Health Organization (2006) Child growth standards. Available: www.who.int/childgrowth/en/

36 Global nutrition policies and interventions

Catherine Geissler and Anna Lartey

OBJECTIVES

By the end of this chapter, you should be able to:

- list the roles of the governments, international agencies, and food industries in the nutrition of populations
- describe the types of direct (nutrition-specific) and indirect (nutrition-sensitive) measures to improve nutritional status of populations
- explain the evolution of emphasis in types of policy interventions used over the last half century to reduce malnutrition
- describe the current challenges for food and nutrition in high-income countries and low- to middle-income countries
- explain the reasons for the recent unprecedented momentum for nutrition.

36.1 General introduction

The main focus of this chapter is an overview of the efforts to alleviate and prevent all forms of malnutrition in both high-income countries (HICs) and low- and middle-income countries (LMICs). International agencies, governments, and the private sector all have important roles in determining and managing nutrition of populations. With increasing urbanization, the proportion of the population that is self-sufficient in food production is correspondingly decreased. The nutritional status of populations becomes increasingly dependent on the food industry which itself has become increasingly globalized. By definition, the private sector is motivated by profit as well as service to the community and is the target of much recent criticism especially in relation to the obesity epidemic (Monteiro et al. 2013).

Governments have an important role to play in the promotion of good nutrition and the elimination of malnutrition. Their roles may include: the collection of monitoring information on nutritional status in the country through surveys; consultations on specific policies; providing guidelines on good nutrition; providing information about the nutritional value of products, for example the saturated fat content of a particular food; regulating claims that can be made by the food industry about the health benefits of food products; mandatory fortification of certain foods that are of particular importance for equity; and taxation or subsidies on selected foods. Another rationale for government involvement is based on a human rights perspective—democratically elected governments have obligations with regard to many human rights, including the right to adequate nutrition, and should be directly accountable to the populations they serve.

International agencies perform an important global role in collecting information on a wide range of aspects of nutrition and food, publishing research reports and epidemiological information, providing advice, setting standards, giving guidance to countries, and providing practical support in healthcare and agriculture as well as feeding in emergencies. Currently used direct (nutrition-specific) and indirect (nutrition-sensitive) nutrition interventions and policies are described, identifying their strengths and weaknesses. Direct nutrition interventions have a stated objective of improving nutrition, and indirect interventions or policies may affect nutrition status in important ways but do not necessarily have this as their primary goal. The chapter also traces the evolution of views on food and nutrition interventions from the 1950s to the present day, including new challenges facing food and nutrition security for today and in the future.

36.2 Types of interventions

Much is known about how to combat the different forms of malnutrition. This is especially true for undernutrition but less so for overweight and obesity which is a more recent worldwide phenomenon with a shorter history of interventions on which to draw conclusions. However, a clear distinction between undernutrition and obesity and its related diseases becomes more blurred as more is understood about the effects of foetal programming, indicating that foetal and infant adaptation to nutritional stress increases susceptibility later in life to the chronic diseases associated with obesity. Therefore, many interventions apply at both ends of the malnutrition spectrum. The section below describes interventions for addressing all forms of malnutrition, which are often classified into two categories: Direct interventions (nutrition-specific) and indirect interventions (nutrition-sensitive).

Table 36.1 summarizes direct and indirect interventions, and ways in which indirect interventions can be strengthened for nutrition. This has relevance for both high-income countries (HICs) and low- and middle-income countries (LMICs).

(For further examples of such interventions in the UK see Weblink Sections 36.1 and 36.2 and Chapters 19–22.)

Nutrition-specific interventions

The evidence for nutrition-specific interventions, such as micronutrient supplementation, are based on efficacy trials that tell us what works under controlled conditions, and effectiveness trials and cost-effectiveness estimates which tell us what works under real-life conditions.

The UNICEF conceptual framework on the causes of undernutrition identifies three levels of causality: (i) immediate or proximal causes (poor diets, infections, and poor care); (ii) underlying or intermediate causes (household food security, health water, sanitation, maternal care; (iii) basic or distal causes (political and economic systems at national level). Over the years, most of the interventions for addressing maternal and child malnutrition have focused on the immediate causes which involve direct interventions mainly delivered in health systems and for which evidence regarding efficacy using randomized controlled trials is readily available. The *Lancet* maternal and child nutrition series of 2013 (Ruel et al. 2013) defined these direct nutrition-specific interventions and programmes as those 'that address the immediate determinants of foetal and child nutrition and development'. These interventions when administered within the 1000-day window are efficacious in reducing child stunting. The concept of the first 1000 days refers to the period from conception through the age of two years. This period is very crucial for the growth and development of the foetus and child and its long-term health outcomes. There is also increased evidence for the use of multiple micronutrient supplements on positive outcomes during pregnancy and this has been recommended for scaling up (Heidkamp et al. 2021).

In general, the direct interventions focus on (a) improving breastfeeding rates up to six months of age, (b) improving the quality and quantity of foods that complement breastfeeding beyond six months of age, (c) improving the quality and quantity of food consumed by adolescent girls and pregnant and lactating women, (d) supplementing diets of all individuals with micronutrient capsules and/or fortifying foods with added micronutrients during processing, (e) improving the quality of nutrition information provided to parents and caretakers, and (f) increasing the diversity of diets consumed by all individuals via home-based production of small livestock, fruit, and vegetables.

Although nutrition-specific interventions can reduce the undernutrition burden, the pace of reduction is not fast enough to achieve the sustainable development goal 2 and other global nutrition targets for stunting (see Chapter 35). The majority of nutrition-specific interventions are delivered through the health system, a delivery channel which tends to have low coverage. There is the need to scale up all available potential interventions, be it nutrition-specific or nutrition-sensitive to reach the global goals.

An extensive literature review on the efficacy of direct interventions targeting the main undernutrition outcomes in developing countries, that is, low birthweight at term (or more precisely intrauterine growth retardation (IUGR)), stunting, and the three main micronutrient deficiencies—iron, iodine, and vitamin A was published by Allen and Gillespie in 2001. The main conclusions are summarized in Weblink Section 36.3 and the full document is available online (https://www.adb.org/publications/what-works-review-efficacy-and-effectiveness-nutrition-interventions). Examples of specific interventions for the prevention and treatment of vitamin A deficiency, iodine deficiency, iron deficiency, and vitamin D deficiency in poor countries are given in Weblink Section 36.4. Examples of interventions for refugees and emergencies are in Weblink Section 36.5.

There is less information on the efficacy and effectiveness of approaches to overweight, obesity, and diet-related NCDs. What is known about approaches to preventing or reducing the predisposing factors of intrauterine growth retardation and stunting that also affect subsequent adult conditions is addressed in Chapters 19, 20, and 21 and treatment of obesity is discussed in Chapter 20. Although no such overall review of the effectiveness of various interventions in reducing the chronic nutrition related diseases, as distinct from undernutrition, has yet been carried out there are now some useful databases of interventions. For example, the World Cancer Research Fund (WCRF) has produced their 'Nourishing' framework (www.wcrf.org/int/policy/nourishing-framework) intended to promote healthy diets and reduce obesity. The interventions or policy actions in the domains of food environment, food system, and behaviour change are grouped into ten categories such as nutrition labelling standards and restricting food marketing, and for each

TABLE 36.1 Direct (nutrition-specific) and indirect (nutrition-sensitive) actions to reduce malnutrition including undernutrition and overnutrition

Objective	Direct interventions	Indirect interventions
Improving pregnancy outcome	• Target supplements to undernourished women: preconception weight <40–45 kg, or low attained weight during pregnancy; low BMI or height are less useful indicators. • Trimester 3 most effective to improve birthweight but intervene as soon as possible, and for as long as possible. • Provide energy: or encourage consumption of more of habitual diet (if protein intake is adequate). • Improve dietary quality and provide multiple micronutrients. • Provide iodine in areas with endemic deficiency. • Other risk factors for low birthweight are young maternal age at conception, so target interventions at those still growing.	• Improve the status of women to lower age at first marriage • Microcredit, targeted to women • More emphasis on education of girls • Improved maternity benefits
Improving child growth	• Improve breastfeeding with exclusive breastfeeding for 4–6 months. • Continue breastfeeding during complementary feeding. • Introduce national and international guidelines on complementary feeding; when, what/dietary quality, how much, micronutrients? • Increase energy intake (improves weight, not length) by increasing energy density most often (via reductions in water content of food). • Increased protein intake (usually limited benefit). • Add animal foods (dried skim milk improved growth in 12/15 trials, but fewer impacts of fish and meat) • Introduce micronutrient fortification of cereal staples (important) • Use of multiple micronutrient supplementation (promising)	• Agricultural research to be more focused on diet quality and nutrition outcomes • Agricultural production systems more in tune with child-care needs • Improved water, sanitation and health service delivery (better quality, better targeted)
Preventing and treating anaemia	*Pregnancy* • Fe supplements increase maternal haemoglobin and iron status and increase infant Fe status for 6 months after birth • No conclusions as to benefits for maternal and infant health & function • Daily (as opposed to weekly) supplements curing pregnancy are more effective *Infancy* • Supplement all LBW infants with Fe from 2 months. • Other need for Fe supplementation is uncertain (Cut-offs? Morbidity? Benefits for function?) *Children* Daily or weekly Fe supplements give improved mental and motor function *Adults* • Fe supplements lead to positive work performance even for iron deficiency/mild anemia, and tasks with moderate effort • Increased ascorbic acid from local foods not effective • Fe fortification of wheat (Venezuela), salt (+iodine in India), dry milk (Chile) effective • NaFeEDTA (an iron fortificant) shows good potential and increased Fe status when added to salt, soy sauce etc. • Multiple micronutrients supplements/fortificants may be more effective • Plant breeding for iron-dense cereals shows some promise, but awaiting efficacy and effectiveness trials • Food-based solutions cannot rely on plant sources—animal sources are critical	• Agricultural research to be more focused on diet quality and nutrition outcomes • Improved status of women for improved intrahousehold food distribution • Improved legislation for fortification • Improved technology for fortification

TABLE 36.1 *Continued*

Objective	Direct interventions	Indirect interventions
Preventing and treating iodine deficiency	• Salt iodization is crucial. • Prevent cretinism by iodine to the mother during first trimester but no later than second trimester. • Supplementation late in pregnancy may improve infant function. • Not clear that iodine supplementation in deficient children improves cognition or growth. • Iodized oil to 6 week old infants reduces mortality in first 2 months by 72%	• Improved legislation for fortification
Preventing and treating vitamin A deficiency	*Pregnancy* • low-dose vitamin A or betacarotene supplements in pregnancy decrease maternal mortality by 40%; also increases hemoglobin *Infants and Children* • High dose maternal supplementation at birth followed by breastfeeding leads to a 64% reduction in mortality under 12 months, 23% reduction in mortality 6-60 age group and major reduction (40%) in HIV mortality • Can also increase growth of malnourished children • Urgent need to (a) accelerate food fortification and (b) improve the availability of vitamin A rich foods • Continue genetically modified approaches, but with appropriate safety standards	• Improved legislation for fortification • Agricultural research to be more focused on diet quality and nutrition outcomes • Improved status of women for improved intra-household food distribution
Preventing diet-related chronic disease	*Mass media* may be able to play a role in nutrition education in developed countries but not enough experience of what will work eg in Asia. *Dietary Guidelines:* used in high-income countries to explicitly shift the diet towards healthy components, but difficulties when large pockets of undernutrition co-exist. *Food processing* modifications, (e.g., changes resulting in differing fat absorption); shifts in breeding, feeding, and market trim practices in the livestock sector can contribute to lower levels of fat in meat over time. *School-based efforts:* School-based initiatives offer important possibilities for improving diet and activity patterns; however few initiatives have made a marked improvement in this area and surprisingly few have been carefully evaluated. Others e.g., workplace interventions, social marketing campaigns, etc	• Food price policy to encourage the consumption of healthier foods • Regulation relative to nutrient content of the diet, food labelling, dietary claims, standards for institutional meals, food advertising and other marketing • Environmental policies to facilitate increased activity • Education re levels of exercise

there are examples of policy actions, countries that have implemented them, and what the actions involve.

WCRF also launched in 2020 a database of policy actions that have been implemented around the world, targeted at getting people, particularly youth and adolescents, to be more active. This database is called 'Moving', part of the EU-funded project that aims to tackle childhood obesity (https://www.wcrf.org/policy/policy-databases/nourishing-framework/). WHO has an eLibrary of evidence for nutrition actions (eLENA) (https://sdghelpdesk.unescap.org/e-library/e-library-evidence-nutrition-actions-elena) which provides evidence informed guidance for nutrition interventions that are relevant to both malnutrition and obesity and related NCDs. The eLENA project was first initiated in 2009 and gained momentum the following year when the World Health Assembly requested WHO 'to strengthen the evidence base on effective and safe nutrition actions to counteract the public health effects of the double burden of malnutrition'. Most of the cited interventions can be classified as direct but there is the occasional indirect intervention such as conditional cash transfers.

WHO also set up the global database on the implementation of nutrition action (GINA) (https://extranet.who.int/nutrition/gina/en/home) in 2012 to provide information on the implementation of nutrition policies and interventions. Information is collected from a variety of sources and users are invited to submit their data directly and to share information on how programmes are implemented, including country adaptations and lessons learned. This information is classified by country for those countries that have submitted information.

The Cochrane Collaboration was established in 2004 to provide a database of systematic reviews for the use of healthcare providers and policy makers to make well-informed decisions. This includes reviews on interventions relevant to nutrition including obesity.

Other reviews analyse the cost-effectiveness of various interventions such as reducing salt intake, promoting fruit and vegetable consumption, and for cardiovascular disease prevention. For an overview of this approach to cost-effectiveness see Cobiac et al. (2013). Interventions in the form of clinical trials are described for cardiovascular diseases in Chapter 19, prevention and treatment measures for obesity in Chapter 20, the management and prevention of diabetes in Chapter 21, and of cancer in Chapter 22.

Nutrition-sensitive interventions

Nutrition-sensitive interventions and programmes are those that address the underlying determinants of nutrition, in particular foetal and child nutrition and development, such as agriculture, food security, social safety nets, women's empowerment, education, water, and sanitation (Ruel et al. 2013). Nutrition-specific interventions alone are not enough to address the malnutrition challenge. There is the need to address the underlying causes of malnutrition (see Weblink Section 36.6). In the last five years, there have been calls for food systems transformation (Willett et al. 2019) to enable healthy diets. Food systems are defined as the sum of actors and interactions along the food chain—from input supply and production of crops, livestock, fish, and other agricultural commodities to transportation, processing, retailing, wholesaling, and preparation of foods, and to consumption and disposal. Scrutiny of food systems has brought into sharp focus the contribution of nutrition-sensitive programmes, especially agriculture and its role in improving nutrition.

Agriculture has the potential to influence nutrition outcomes through several pathways: (i) improvement in household food security; (ii) direct impact on diet quality; (iii) improving household income; (iv) women's empowerment (Ruel et al. 2013). A comprehensive review of the evidence for agriculture's impact on nutrition outcomes concludes that 'nutrition-sensitive agriculture can improve a variety of outcomes for mothers and children, especially when these programs include nutrition and health, behaviour change communication and carefully designed intervention to empower women' (Ruel et al. 2013). Other nutrition-sensitive agriculture programmes such as promotion of nutrition-enhanced biofortified crops and homestead gardens have resulted in improved micronutrient status (Haas et al. 2016; Michaux et al. 2019). Social protection programmes can be made nutrition-sensitive through targeting to reach vulnerable groups, for example, programmes to empower women, and conditional food and cash transfers linked to access to healthy foods can be nutrition-sensitive (Fenn et al. 2017). Water, sanitation, and hygiene (WASH) programmes that improve household access to safe water, and improved human waste management programmes have led to improved nutrition (Bekele et al. 2020).

Nutrition-sensitive interventions also focus on (a) lowering the price of foods to consumers via subsidies or vouchers, (b) improving access to income (more work, higher wages, higher productivity) and in more affluent countries, social security such as pensions and unemployment benefits, (c) improving the ability to borrow and save to smooth consumption from one period to the next, (d) improving the ability of women to make decisions that improve nutrition, (e) improving education and the ability to acquire and use information, (f) improving access to water, sanitation, and preventative and curative health services, (g) raising the productivity and nutrient content of crops and livestock, and (h) controlling the marketing of unhealthy food, especially to children. Overall, multisectoral actions that go beyond direct health systems interventions to include actions that address the underlying determinants of malnutrition are critical for making a bigger impact on reducing malnutrition (Heidkamp et al. 2020).

36.3 The evolution of policies and programmes

Introduction

This section outlines the changes internationally in food and nutrition concepts, causes, interventions, and policies from the 1950s to the present day, and highlights some current issues. The types of interventions that have been most strongly promoted by governments and international agencies have followed changes in perceptions of what were the most important problems, and their causes. In general, the perceptions around nutrition interventions and policies have moved from a 'food production-only' perspective at the national level to rather sophisticated perspectives on food as a human right.

The 1950s

The 1950s focused on producing enough food at the national level to potentially feed all people. Nutrition was considered to be primarily an issue of insufficient food quantity. The focus was on expanding agricultural output by bringing more land into cultivation. As the availability of good quality land for new production became limited, the returns from land expansion fell.

Applied nutrition programmes

There was increased recognition that an integrated approach to improving nutrition was required and 'applied nutrition programmes' were promoted by the international agencies. These were broad-based community development programmes designed to improve the quantity and quality of local food production and use, and to improve income. Components included protection against infections, nutrition education, horticulture, animal production, fishponds, food storage, pest control, cottage industry cooperatives, fortification, and improved transport. Some of these, such as implemented in South Korea, were later evaluated and found to be successful but because of their limited geographic outreach had little global impact.

The 1960s

In the 1960s the focus, while still on food production, switched to yield, or output per hectare. The 'Green Revolution', led by the newly formed Consultative Group on International Agricultural Research (CGIAR) and based on the research of Norman Borlaug, was an effort to develop high-yielding varieties of crops that would help to meet the increasing demand fuelled in large part by rapid population growth in the 1960s.

Green Revolution

The 'Green Revolution' succeeded in preventing large-scale famine—yields increased sharply, especially for the farmers that could afford the complementary inputs for the new varieties. Consumers benefited from lower cereal prices, allowing them to diversify diets into foods richer in proteins and nutrients. However, the new techniques were not particularly environmentally friendly, relying on pesticides, herbicides, inorganic fertilizers, irrigation and dams (see Chapter 34).

The 1970s

In the 1970s, the idea of broad-based agricultural growth as a driver of economic growth took hold more generally.

Income from agriculture

Agriculture was recognized to be not just about producing food, but about producing income, both for farmers and for the rural non-farm entrepreneurs from whom farmers purchased goods. The importance of income (or economic access) for reducing malnutrition was promoted by Amartya Sen's analysis of famines as failures of entitlement. There was also an increased recognition of the role of education, especially women's education, and nutrition-specific education, in promoting nutrition status.

National nutrition planning

The concept of national nutrition policies/planning also developed. It had been assumed in the development literature that as countries became more affluent, there would be a 'trickle-down effect' into improved nutrition. In the 1970s Alan Berg in *The Nutrition Factor* (Berg 1973) developed the idea that specifically targeting nutrition would improve human resources by reducing mortality, morbidity, the use of health services, educational loss, and so on, and so be a lever to economic development. An intersectoral planning approach was promoted and adopted by the development agencies. The World Food Conference in Rome in 1974, following an international food crisis in previous years, was instrumental in stimulating widespread analyses of the world food supply. Resolutions from the conference included the recommendation that countries should have a national nutrition policy, and many countries were provided with technical support by FAO and USAID to do so.

Nutrition policies in high-income countries

The 1970s also saw an increased interest in nutrition policy for developed countries, with the growing realization that existing legislation, which was based on food purity and the prevention of adulteration as well as control of deficiencies, was not adequate to deal with the changing nature of nutrition problems towards chronic nutrition-related diseases. For further information about the evolution of nutrition policy in high-income countries with emphasis on the UK see Weblink Sections 36.1 and 36.2. Several government advisory, professional, and consumer bodies in the USA, UK, and other countries have recommended appropriate dietary goals with the common theme of reducing fat, sugar, and salt intake and increasing the intake of dietary fibre, fruit, and vegetables.

Barriers to policy

In contrast to policies related to undernutrition that promote increased consumption of nutrients, the advice to reduce the intake of certain nutrients appeared to be a threat to some sectors of the food industry. There was considerable opposition to the advice and arguments about the validity of the evidence on which the recommendations were based. Additional constraints to updating food and nutrition policy included legislation designed to prevent adulteration and maintain quality as previously perceived, such as minimum fat levels in milk and premiums on animals with high fat content. However, the proposals were gradually accepted and incorporated into government policies, while industry recognized new opportunities in the production of high fibre, low fat, low salt and sugar food products. Norway was the first developed country to have an integrated food and nutrition policy as promoted in the World Food Conference in 1974. The UK and other European countries subsequently developed food and nutrition policies within their health and agriculture sectors (see the 1990s below).

The 1980s

Role of women

The 1980s saw greater recognition of the role that women play in child nutrition through their economic as well as their reproductive roles. The income that women control empowers them to make decisions that benefit their own health and the health of children.

Diet quality

The decade also saw the recognition of the role of diet quality in the promotion of nutrition status. The understanding emerged that micronutrients play a more general role in child survival, beyond deficiency diseases, and led to the promotion in LMICs of small-scale home gardening, capsule distribution, and fortification programmes that remain popular today. Such programmes had been in use for several decades in HICs, such as providing land for kitchen garden allotments and the provision of supplements to children and mothers during the world wars, and the fortification of white flour with vitamins and minerals from that period. Compulsory fortification of margarine with vitamins A and D has existed in the UK since 1967 and margarine regulations came into force in 1971.

Food insecurity

In the 1980s the concept of food insecurity was formalized as being not just about lack of access to food today, but also about the risk of losing access to food tomorrow. The risk of losing access is debilitating and leads to actions to cope with that risk, but these have very high costs such as the short-term necessity to use up natural resources, the need to rely on a diversity of income sources, and uneconomic storage of food at the household level.

Food–care–health

Finally, in the 1980s the importance of the nutrition triumvirate food–care–health, the concept devised by UNICEF, came to be accepted. Interactions became clearer: poor quality drinking water could undermine household food security; without adequate care for mothers, children could not be breastfed; without time to seek health support, preventive, and often curative, healthcare would not be accessed.

The 1990s

Nutrition goals

In the USA efforts to establish a scientific basis for human nutrition began with Wilbur Olin Atwater, who published the first dietary recommendations for Americans in 1894.

In 1977, the US Senate Select Committee on Nutrition and Human Needs, chaired by Senator George McGovern, published the dietary goals for the United States, recommending increasing complex carbohydrates and naturally occurring sugars from 28 per cent to 48 per cent of energy intake, reducing refined and processed sugars to about 10 per cent of energy intake, reducing fat from 40 per cent to 30 per cent of energy intake, reducing eating saturated fat to 10 per cent of energy intake, reducing cholesterol consumption to 300 milligrams daily, and reducing salt intake to 5 grams daily. A second version of the report was published in 1980, with less stringent changes from the standard American diet. The dietary guidelines for Americans have been published every five years.

Many other countries have developed nutrition goals and guidelines and reference values, based on recommended levels of specific nutrients or food groups, in many cases accompanied by diagrams in the shape of a plate, pyramid, or pagoda to demonstrate the recommended relative quantities (Figure 36.1).

In the UK explicit nutritional goals were set for the first time in the 1992 government health policy 'The health of the nation' which focused on five key areas for action: coronary heart disease and stroke, cancers, mental illness, HIV/AIDS and sexual health, and accidents, the first two of which are diet-related. The diet and nutrition targets for 2005 from the 1990 baseline were: to reduce the average per cent food energy from saturated fats by at least 35 per cent (to no more than 11 per cent food energy); to reduce the average per cent food energy from total fat by at least 12 per cent (to no more than about 35 per cent food energy); to reduce the per cent of men and women aged 16–64 who are obese by at least 25 per cent and

35 per cent respectively (to no more than 6 per cent of men and 8 per cent of women) and: to reduce the per cent of men drinking more than 21 units of alcohol per week and women drinking more than 14 units per week by 30 per cent (to 18 per cent of men and 7 per cent of women).

By concentrating on these targets it was expected that the associated dietary changes and reduction in obesity would have beneficial consequences on such diseases as cancer, osteoarthritis, and diabetes. A nutrition task force was set up to oversee implementation of action, promote coordination and cooperation between interested parties, and establish mechanisms for monitoring and evaluating progress. For further information about the evolution of nutrition policy in high-income countries with emphasis on the UK see Weblink Sections 36.1 and 36.2.

Food safety

Food safety became a major concern in the 1990s, particularly in HICs. European consumers in particular, having lost faith in the science establishment

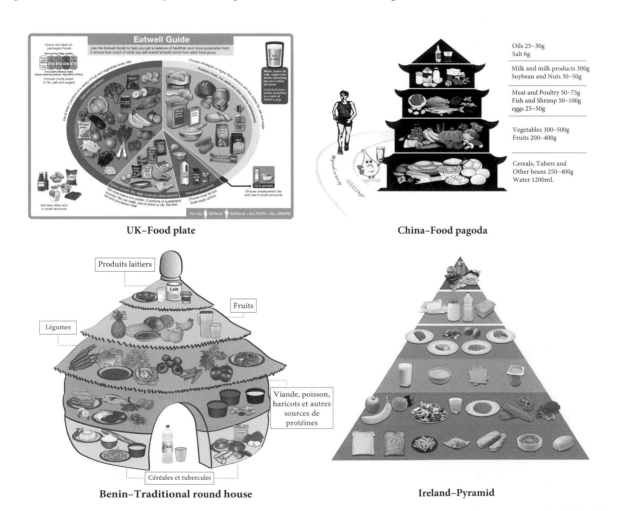

FIGURE 36.1 Visual food guides from different countries.

due to its initial assurances about the lack of danger to human health of BSE ('mad cow disease'), became extremely cautious about the safety of the food they purchased. And they were generally able and willing to pay a premium for assurances about the food production and processes used. In the UK this distrust also led to a revision of government structure via the Food Standards Act 1999, so that agricultural and consumer food interests which had been combined within the Ministry of Agriculture Food and Fisheries (MAFF) were separated into the Food Standards Agency (FSA) (www.food.gov.uk) to champion consumer interests, and the Department for the Environment Food and Rural Affairs (DEFRA) to look after agriculture (https://www.gov.uk/government/organisations/department-for-environment-food-rural-affairs).

A body similar to the FSA was later established within the European Union, the European Food Safety Authority (EFSA) in 2002 (www.efsa.europa.eu/en). In the United States food and nutrition are regulated by the Food and Drug Administration (www.fda.gov). The capacity required to adhere to the new food safety expectations made it harder for developing country exporters to gain market share in the developed world, harming their own food security through constraints to export opportunities. Conflated with these new food safety expectations have been concerns about genetically modified (GM) foods (see Chapter 34 and in 'The 2000s' below).

Food poverty

Also in the 1990s, awareness increased that pockets of food poverty still existed in HICs, following the increased economic inequality that developed in the 1980s in the UK and other countries, partly due to government cutbacks in welfare programmes. This led to reports and actions to relieve the constraints of the poor (Riches 1997) and echoed a similar period in the United States in the 1960s about 'hunger in America' which resulted in new welfare programmes such as WIC (Women, infants, and children: food and nutrition service, USA) (see Weblink Section 36.2).

Right to food

The concept of the right to food came of age from the 1990s. It was first formalized in the 1948 Universal Declaration of Human Rights, and then confirmed in milestones such as the 1966 Covenants on Civil and Political Rights and on Economic, Social, and Cultural Rights, the 1989 Convention on the Rights of the Child, and the 1996 World Food Summit hosted by FAO. The concept is that governments have a duty to respect, protect, facilitate and fulfil if necessary, the rights of the

> **BOX 36.1**
> ## Supreme Court of India's ruling on right to food
>
> On 23 July 2001, the court said: 'In our opinion, what is of utmost importance is to see that food is provided to the aged, infirm, disabled, destitute women, destitute men who are in danger of starvation, pregnant and lactating women and destitute children, especially in cases where they or members of their family do not have sufficient funds to provide food for them. In case of famine, there may be shortage of food, but here the situation is that amongst plenty there is scarcity. Plenty of food is available, but distribution of the same amongst the very poor and the destitute is scarce and non-existent leading to malnourishment, starvation and other related problems.'
>
> On 28 November 2001, the court issued directions to eight of the major schemes, calling on them to identify the needy and to provide them with grain and other services by early 2002. For example, for the targeted public distribution scheme directions were that 'The States are directed to complete the identification of BPL (below poverty level) families, issuing of cards, and commencement of distribution of 25 kgs grain per family per month latest by 1st January, 2002.'
>
> PUCL Bulletin, July 2001. Supreme Court of India. Record of proceedings. Writ petition (civil) no. 196 of 2001.

individual to secure adequate food and nutrition. In a legal sense the right to food can be used by one branch of a country's government to compel other branches to promote food and nutrition security, as happened in India in 2003 (see Box 36.1) or it can be used by the UN to 'name and shame' countries that do not give food and nutrition issues prominence. By identifying roles and responsibilities it clarifies for different actors their ability to make demands and for others to fulfil their obligations.

The 2000s

In the current millennium there has been unprecedented interest in nutrition globally and nationally as being fundamental to health and to economic development. The rapidly changing structure of society has implications for nutrition and health, and these are some of the issues outlined in this section, followed by examples of action plans in Europe, the UK, and internationally.

Obesity and NCDs

The main nutrition issue since the millennium in high-income countries is the continuing increase in the prevalence of obesity, in both adults and children, with concomitant increases in some NCDs. But this is now an international concern and WHO published in 2013 its global action plan for the prevention and control of non-communicable diseases 2013–2020 (WHO 2013–2020). The main focus is on four NCDs: cardiovascular disease, cancer, chronic respiratory disease, and diabetes. It pays little attention, however, to dementia, the most common form of which is Alzheimer's disease, which has become an increasing concern. From 2009, World Alzheimer's reports have documented various aspects of the disease, and the 2014 report included nutrition as an aspect of lifestyle and concluded that dementia needs to be included in WHO and national NCD planning. The 2021 report concluded that 'governments must prepare for a tsunami of demand for healthcare services as a result of global ageing populations, improved diagnostics, including biomarkers, and emerging pharmacological treatments'. In 2013 the G8 held a dementia summit in London which was effective in galvanizing the international community to agree to a significant increase in research and funding. In 2014 Public Health England included dementia in its seven priorities of their policy paper 'From evidence to action' based on the evidence that Alzheimer's disease had risen in the rank order of causes of premature mortality from 24th to 10th place between 1990 and 2010 (Figure 36.2).

In the UK from the early 2000s the problem of obesity hit the political agenda strongly, resulting in much political and public debate, including the role of government, industry, parents, and the individual, with conflicting views about the nanny state and the need for government to take more vigorous action. However, despite these policies and varied interventions, obesity continues to grow in the UK as in many other countries. Increasing research also refines the association between various food factors with aspects of health and so determines policy. For example, the USA has already undertaken folate fortification of flour and the UK did likewise in 2021.

Environment

Environmental concerns related to food and nutrition include food safety and genetically modified (GM) foods which continue to be important issues in both HICs and LICs (see Chapter 34). The growth of the organic food market has also received attention, although a study commissioned by the UK Food Standards Agency showed that organic foods had no superior nutritional benefits (Dangour et al. 2009). The issues of pesticides and fertilizers were not included in the study.

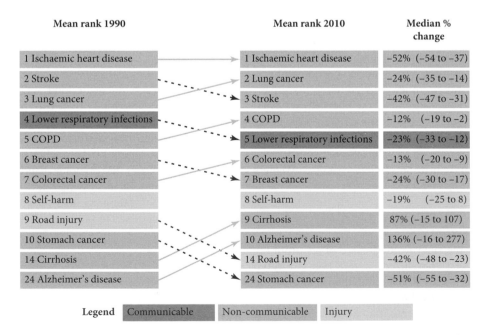

FIGURE 36.2 Causes of premature mortality in the UK: rank order 1990 and 2010. From 1990 to 2010, the years of life lost to ischaemic heart disease, stroke, and lung cancer reduced by 52 per cent, 42 per cent, and 24 per cent respectively, but these remain the top three causes of premature mortatility in the UK.
Source: Public Health England 'From evidence to action' 2014.

Genetically modified (GM) foods

To date, GM foods have realized benefits largely for producers in terms of higher productivity and lower costs despite no obvious benefits to consumers other than perhaps lower prices. Consumers in the USA have been consuming GM soybean products from most of the 1990s. But European consumers and many in the USA are concerned that the food safety and environmental safety issues related to GM foods have not been adequately researched.

Difficult policy issues have been raised: will US food aid shipments containing some GM foods not be admitted into hunger-wracked areas for fear of introducing GM seeds into the developing countries agricultural systems that then prevent them from exporting to Europe and other countries? The mandatory labelling of foods as 'containing GM organisms' was put forward as one solution to such issues, allowing consumers to make informed choices, but this has been opposed by GM producers who claim that it would be too expensive to keep the GM and non-GM crops separate throughout the food distribution chain. GMs are also involved in nutrition from the health side (vaccines and other drugs), but here the issue is not so much safety as access—the drug companies are resisting differential pricing regimes for rich and poor countries, making the argument that such pricing schemes will diminish their incentive to innovate. There are some new institutional arrangements, for example, public–private interactions such as GAVI on vaccines (https://www.gavi.org/our-alliance/about) and GAIN on food fortification (https://www.gainhealth.org/impact/programmes) that are seeking to create incentives for the private sector to develop drugs and fortified foods for the benefit of the poor.

Global warming

The major food-related environmental concerns are the effects of global warming and the ecological sustainability of the food supply, including the trade of produce that could be supplied locally. FAO estimates that climate change undermines current efforts to address undernutrition and hits the poorest hardest, especially women and children, through three main causal pathways: impacts on household access to sufficient, safe, and adequate food; impacts on care and feeding practices; and impacts on environmental health and access to health services. In the most affected areas these impacts can lead to transitory or permanent migration, which often leaves female-headed households behind. Some forecasts anticipate 24 million additional malnourished children by 2050—almost half of them in Sub-Saharan Africa. Poor health and undernutrition in turn further undermine people's resilience to climatic shocks and their ability to adapt.

Food prices and inequalities

Sharp rises in food prices that occur sporadically are a clear threat to food security and nutrition (Shrimpton et al. 2009). In many countries, including the UK, inequalities in health and food poverty are increasing and over recent years many people in high-income countries have had to resort to food banks for emergency food (Riches & Silvasti 2014).

Globalization

Globalization, namely the rapidly increasing levels of global food trade, and movements of finance, labour, and information present opportunities and risks for improved nutrition. Included here are regional trade associations such as the EU of which its common agricultural policy has an impact on nutrition in Europe through agricultural subsidies on beef production and sugar beet but little support for fruit and vegetables (Walls et al. 2016). Governments need to invest in infrastructure and market institutions that allow them to take advantage of new opportunities. In the case of the UK, having voted in June 2016 to leave the EU, these institutions must be reorganized. If they have not already established welfare programmes, governments also have to design and implement safety-net programmes that protect and compensate those who stand to lose from more open markets.

Urbanization

Urbanization is progressing rapidly throughout the world. Not only are people shifting to urban areas, but so too is the concentration of undernutrition and overweight and obesity. The poor in urban areas are equally at risk of undernutrition as the poor in rural areas, perhaps more so since they have little access to the means to produce their own food. Interventions that work in rural areas cannot be assumed to work as well in urban areas without redesign. This is because the main differences in urban areas—a reliance on food purchase, the large numbers of mothers working away from home, and water and air pollution from waste and fuels make nutrition more susceptible to loss of regular employment, food price fluctuations, the quantity and quality of child-care for low-income working mothers, and illnesses caused from environmental contaminants.

Demography

The demographic profile of the developing world is changing rapidly. The aging of the populations of all countries, both HICs and LMICs, is a result of lower infant mortality rates and lower birth rates. The shift in the demographic profile puts new pressure on the ability to finance undernutrition efforts adequately through public finance, due to the new demands for spending on obesity-related issues such as diabetes and heart disease. The presence of older family members, if they live close by, can however help parents to cope with the multiple demands of work and child rearing.

Life expectancy

In contrast to most of the world, where life expectancy has until recently been on the rise, in one area, Sub-Saharan Africa, it is on the decline, along with most indicators of malnutrition. AIDS has loomed large over Africa, China, India, Cambodia, and Russia, killing adults of working age and very young infants. Since 2019 the pandemic of coronavirus COVID-19 has had a further negative impact on life expectancy even in HICs. Adult mortality, the chronic illness that precedes it, and the caring for the adult with AIDS all serve to undermine the ability of workers to earn, farmers to produce food and families to purchase it, mothers to care for children, and for the provision of public goods such as safe water and healthcare. Good maternal nutrition status is also thought to minimize mother-to-child transmission via birthing, although breastfeeding does increase the risk if the mother is HIV-positive.

Diet transition

The diet transition in the developing world seems to be accelerating—for both rich and poor, rural and urban (see Chapter 1). It is a transition towards the coexistence of undernutrition and overweight and obesity. It is driven by changing preferences fuelled by some of the factors discussed above (in the context of undernutrition) such as growing incomes, changing relative prices, urbanization, aging, changing food choice options from changes in food technology and changes in the food distribution systems, and by a legacy of low birthweights from the previous generation.

What can food and nutrition policy do to make this transition healthier? Some options such as food price policies, food regulations, and dietary guidelines are outlined in Weblink Section 36.7. We should note that these options have had mixed success in the industrialized countries and the policy trade-offs in the LMICs are even more complicated. For example, price policy efforts to overcome obesity by making some foods with high saturated fat content more expensive might well undermine efforts to overcome undernutrition because those same foods may be rich in micronutrients. The effectiveness of public health anti-smoking policy is often referred to as a model, but it should not be leaned on too heavily as food

is essential, tobacco is not. There are plenty of areas in which additional technical research is needed to assess competing risks and to help develop policy options for the issue of the diet transition.

Technology

Rapidly advancing electronic technology provides new opportunities relevant to nutrition. One example is in educational support for capacity building in nutrition, where resources are limited, through online courses (Geissler 2015). Other examples include the use of web-based methods and smartphones to improve the quality of dietary assessment, an essential component of public health nutrition (www.nutritools.org).

36.4 Unprecedented momentum for nutrition

Periodically, global bodies set goals on food and nutrition to create a united front and momentum towards their achievement. At the beginning of the millennium there were the millennium development goals (MDGs) which had a strong focus on low- and middle-income countries. In 2015 the UN General Assembly adopted the sustainable development goals (SDGs) described as universal goals that apply to all nations, irrespective of the socioeconomic level. The SDGs are owned and led by countries. Countries have the freedom to set national targets and work towards their achievement. While the MDGs came with eight goals and 21 targets, the SDG has 17 ambitious goals and 162 targets (Figure 36.3), and comes with the solemn pledge of leaving no-one behind. The SDG give visibility to nutrition by incorporating it in SDG 2—to end hunger, achieve food security and improved nutrition, and promote sustainable agriculture.

Although malnutrition has been one of the major global challenges facing humanity, it was not given the same level of attention as other health challenges that emerged later such as HIV, which has its own global funding facility as indicated on its website (https://www.theglobalfund.org/en/hivaids). The neglect of nutrition became obvious, when the mid-term review of the MDGs showed that countries not likely to meet the MDGs were those with high burdens of malnutrition, especially stunting. The *Lancet* maternal and child undernutrition series of 2008 was a wakeup call that brought to the fore the high mortality and disease burden from malnutrition, and called for urgent attention to the situation (Black et al. 2008). The *Lancet* maternal and child nutrition series of 2013 provided evidence that impactful interventions exist that can be scaled up.

FIGURE 36.3 Sustainable development goals.
The sustainable development goals (SDGs) were set up in 2015 by the United Nations General Assembly, intended to be achieved by 2030, to succeed the millennium development goals which ended in 2015.
https://sdgs.un.org/goals

However, linking these nutrition-specific interventions with interventions that address the underlying causes of malnutrition will greatly accelerate progress (Bhutta et al. 2013), but to deliver results and impact, leadership for nutrition at all levels is fundamental (Gillespie et al. 2013). The complexity of the causes of malnutrition require multisectoral collaborations.

By 2021, much had changed for nutrition by the time the *Lancet* maternal and child undernutrition series were released. The global processes that contributed to moving nutrition to the global agenda are presented in subsequent sections.

Lancet maternal and child undernutrition series 2021

The *Lancet* maternal and child undernutrition (MCN) series 2021 was released when the world was engulfed with the COVID-19 pandemic and was at the same time racing towards the SDG timeline of 2030. Trends in global stunting prevalence were decreasing, although the numbers were still unacceptably high for Asia and Sub-Saharan Africa. The 'standing together for nutrition' consortium analysis showed that COVID-19 could erode the little progress made on child nutrition—stunting, wasting, and mortality especially in low- and middle-income countries (Osendarp 2021).

A key recommendation from the *Lancet* maternal and child undernutrition series 2021 was that lasting improvement cannot be sustained without concerted actions across all sectors including health, social protection, and food sectors (Heidkamp et al. 2021). Although the *Lancet* MCN series have traditionally focused on undernutrition, by 2021 the need for strategies to address malnutrition in all its forms, including adolescents and school-age children, have grabbed attention (Keats et al. 2021).

The scaling up nutrition movement (SUN)

The SUN movement came into being in 2010 to address some of the shortcomings identified in the *Lancet* MCN series of 2008. SUN was presented as a country-led, country-driven movement. SUN countries were encouraged to scale up the ten evidence-based cost-effective interventions of the *Lancet* 2008. SUN was explicit in highlighting that nutrition is too important to be left to one sector alone, often confined to health. The constitution of SUN reflects the need to bring all sectors on board, thus the movement's constituent networks comprise donors, UN agencies, civil society, businesses and country focal points and a secretariat that coordinates the global processes. The structure and focus of the SUN movement has changed over the years from focusing on undernutrition only, especially stunting, to now embracing a food-systems approach to nutrition by 2020.

SUN is in its third five-year strategic plan. Its composition, which still has the networks, also now includes an executive committee responsible for providing strategic guidance and the SUN lead group made up of eminent personalities that provide overall leadership to the Movement. By 2023, SUN country membership stands at 62 with four Indian states. The SUN strategy 3.0 (2021–2025) (SUN Movement 2020b) appears to be a gamechanger in moving from focusing on undernutrition only to now addressing all forms of malnutrition. SUN strategy 3.0 aims to achieve greater impact on nutrition through food systems, health systems, and social protection systems. SUN movement's strategic review of 2020 (SUN Movement 2020a) credits the movement with 'raising the visibility of nutrition within global and national agendas'.

Nutrition-for-growth summits

The first nutrition-for-growth summit was held in London, UK in June 2013. World leaders signed a global compact promising to prevent at least 20 million children from being stunted, save at least 1.7 million lives by 2020, and ensure that at least 500 million pregnant women and children under two years are reached with effective nutrition interventions. Donors secured over $4 billion USD to tackle undernutrition and $19 billion USD for nutrition-sensitive programmes by 2020.

The nutrition-for-growth summits are strategically linked to the Olympics to highlight the critical role of nutrition for performance. A nutrition-for-growth event was held in Milan, Italy in 2017, to prepare for the main summit to be held in Tokyo, Japan in 2020. However, COVID-19 disruptions in 2020 led to the postponement of the summit to December 2021. The Tokyo nutrition-for-growth summit is considered a watershed moment as it marks the end date for the commitments made in the 2013 summit and starts the final decade of action for the delivery of the SDG. The Tokyo nutrition-for-growth summit was a unique opportunity for securing new commitments for nutrition.

Three thematic priority areas of focus were: (i) integrating nutrition into universal health coverage; (ii) building food systems that promote safe, sustainable, and healthy diets that support people and planetary health; (iii) effectively addressing malnutrition in fragile and conflict affected contexts, supporting resiliency. The outcome of the nutrition-for-growth summit will be a compact document on specific, measurable, achievable, relevant, and time-bound (SMART) nutrition commitments across the thematic areas.

The second international conference on nutrition (ICN2)

The second international conference on nutrition (ICN2) was jointly organized by the Food and Agriculture Organization of the United Nations and the World Health Organization in November 2014. Among others, the purpose was to assess progress made since the 1992 international conference on nutrition, and to bring sectors together in aligning sectoral policies to improve nutrition.

The outcomes of ICN2: were (i) the Rome declaration on nutrition in which FAO and WHO members committed to ten actions to eradicate hunger and prevent all forms of malnutrition; (ii) a framework for action—this is a set of voluntary policy options and strategies that governments may adopt, as needed, in their national policies to achieve improved nutrition (FAO-WHO 2014). ICN2 is remembered as the event that put food systems for nutrition on the global agenda. It also shifted the global narrative from focusing on undernutrition to addressing all forms of malnutrition. FAO and WHO members recommended to the UN General Assembly to consider securing a decade of action on nutrition from 2016–2025, and this was approved in April 2016.

Decade of action on nutrition (2016–2025)

In April 2016 the UN General Assembly proclaimed the period 2016–2025 as the decade of action on nutrition (https://www.un.org/nutrition). The FAO and WHO were charged to lead the implementation of the nutrition decade in collaboration with other relevant UN agencies and in consultation with other international and regional organizations. The declaration of the nutrition decade was quite significant as for the first time, nutrition issues will be discussed at the highest decision-making body of the United Nations every two years when the UN secretary-general's report on the implementation of the decade is tabled.

The work programme for the nutrition decade identified six cross-cutting action areas for focus: (i) sustainable, resilient food systems for healthy diets; (ii) aligned health systems providing universal coverage of essential nutrition actions; (iii) social protection and nutrition education; (iv) trade and investment for improved nutrition; (v) safe and supportive environments for nutrition at all ages; (vi) strengthened governance and accountability for nutrition (FAO-WHO 2017).

Sustainable development goals (2015–2030)

The declaration of the sustainable development goals 2030 is another pivotal moment for nutrition. The nutrition community was quick to acknowledge that without addressing nutrition, the SDG cannot be achieved. The then UN system standing committee on nutrition analysed the contribution of nutrition to the SDG and was clear that at least 12 out of the 17 SDG targets are intrinsically linked to nutrition (UNSCN 2017).

The period 2020–2030 has been declared the decade of action for the achievement of the SDG. The nutrition community is riding on this to highlight nutrition. The year 2020, which was taken as the year of action for nutrition, was disrupted by the emergence of COVID-19 and rescheduled the year of action to 2021 for hosting major nutrition events, such as the UN food systems summit and the nutrition-for-growth summits held in 2021.

UN food systems summit

The rising prevalence of overweight and obesity was the impetus that cast the focus on food systems. Not only are food systems not delivering on safe and nutritious foods, an underlying cause of undernutrition, but is also leading to the consumption of high calorie, highly processed, nutrient poor foods—and thus promoting overweight and obesity. Existing food systems are not sustainable for the planet, contributing about 30 per cent of the global greenhouse gas emissions.

There has been a global call for food system transformation to save human health and the planet (Willett et al. 2019). The UN secretary-general's food systems summit was held in September 2021. COVID-19 has highlighted the deficiencies in the global food systems and presents the opportunity to build back better.

The vision is to use the UN food systems summit to launch bold, gamechanging actions and solutions to deliver progress on all the 17 SDGs. To support this vision the summit will pursue the following five specific objectives: (i) ensure access to safe and nutritious food for all; (ii) shifting to sustainable consumption patterns by promoting and creating demand for healthy and sustainable diets; (iii) boosting nature-positive production at scale, advancing equitable livelihoods, and value distribution; (v) building resilience to vulnerabilities shocks and stress.

It is expected that the food systems summit will culminate in significant actions and commitments that will enable the achievements of the SDG.

Global nutrition report (GNR)

The nutrition-for-growth event in 2013 led to several outcomes among which are the creation of the global nutrition report (GNR) and the global panel on agriculture and food systems for nutrition. The GNR, which is released annually, tracks the commitments made at the nutrition-for-growth summits. The country profiles of the GNR track progress towards global nutrition targets. Using the latest data on diet, the burden of malnutrition, nutrition strategies, financing and social determinants of nutrition, the report gives an independent but comprehensive assessment of the global nutrition situation. The website (https://globalnutritionreport.org) provides a rich source of information on global and country level progress on nutrition.

Global panel on agriculture and food systems for nutrition

The global panel was established in 2013 as an outcome of the nutrition-for-growth summit in UK. The global panel comprises an independent group of eminent internationally recognized leaders committed to improving nutrition. The global panel organizes high-level roundtable meetings through which national and global policy makers are brought together to discuss different aspects of food systems and its impact on nutrition. The global panel is known for its high quality, evidenced-based technical and policy briefs, as well as its foresight report on the future of food systems with focus on LMICs (Global Panel 2020). The global panel uses its high-level standing to influence policy makers to prioritize nutrition actions within their food systems.

UN nutrition

UN nutrition came into being in February 2020 and was formally endorsed by the United Nations economic and social council (ECOSOC) in July 2020. UN nutrition is the result of the merger between the UN systems standing committee on nutrition (UNSCN) and the UN network for the scaling up nutrition (SUN) movement. This merger responds to the UN reforms to address concerns about fragmentation within the international nutrition community.

UN nutrition coordinates UN agencies on issues at the global and country levels (see Weblink Section 36.8). Its core functions include: (i) adopting coherent and unified advocacy and policy actions for nutrition; (ii) ensuring UN alignment behind the SDG and other global goals and national plans; (iii) promoting and supporting knowledge-sharing across the UN and beyond; (iv) disseminating and translation global guidance developed by UN agencies into country-level action (UN Nutrition 2020). UN nutrition also serves as the UN Network for the SUN movement.

36.5 Conclusion

Nutrition has received unprecedented attention in the last ten years and has moved nutrition onto the global agenda. Many organizations that hitherto did not have nutrition in their strategies and policies now do. The narrative around nutrition is changing in many respects: (i) moving from a focus on undernutrition to addressing all forms of malnutrition, because with the rising prevalence of overweight and obesity globally it is now recognized that malnutrition is a global issue that affects all countries; (ii) poor diet is a key driver of the current nutrition situation, thus the focus has turned to food systems transformation to deliver healthy and sustainable diets; (iii) the focus on food systems has expanded the narrative to make food systems not only healthy but also sustainable.

The EAT-*Lancet* commission report on healthy diets from sustainable food systems (Willett et al. 2019) has been the impetus for linking food systems to human health and planetary health. The commission called for building sustainability into the way food is produced, consumed, and disposed of. The COVID-19 pandemic has wrought disruptions in health, food, economic, and social systems, and has exposed huge inequities in these food systems. To build back better, the call is to make food systems healthy, resilient, equitable, and sustainable (IFPRI 2021).

Global leaders have made pledges and commitments at the food systems summit. The world is looking for these gamechanging, high ambition, actionable solutions that will enhance the achievement of the SDG by 2030.

KEY POINTS

- Direct or nutrition-specific interventions are those involving the provision of food or nutrients to individuals at all stages of the lifecycle and have been shown to be efficacious, effective, and cost-effective in reducing undernutrition.

- Indirect or nutrition-sensitive interventions do not have improvements in nutrition as an explicit objective, but can have powerful indirect and supportive effects on nutrition status.

- These relate to agriculture, income generation and maintenance, the status of women, education, water access (quality and quantity), sanitation and health services (preventive and curative), and environmental measures to moderate activity.

- Several databases have been developed for interventions in both undernutrition and overweight and obesity: for example, the WCRF 'nourishing' framework, the WHO eLENA, the WHO GINA, and the Cochrane database of systematic reviews.

- Public policy for malnutrition reduction has moved from a focus on food production, through to food access to diet diversity, to the food–care–health model.

- Globalization, urbanization, global warming, and rapid population growth represent important current challenges to the achievement of good nutrition status.

- The increase in obesity and related chronic diseases is an additional concern in LMICs with the double burden of malnutrition.

- The right to food presents opportunities to put pressure on governments to address malnutrition

- The international governance of nutrition must continue to be strengthened to accelerate nutritional improvements, taking advantage of the current unprecedented momentum for nutrition.

- The UN food systems summit provides the opportunity for countries to come up with gamechanging solutions to transform their food systems to deliver on healthy diets for all.

 Be sure to test your understanding of this chapter by attempting multiple choice questions. See the Further Reading and Useful Websites lists for additional material relevant to this chapter.

REFERENCES

Bekele T, Rahman B, Rawstorne P (2020) The effect of access to water, sanitation and handwashing facilities on child growth indicators: Evidence from the Ethiopia Demographic and Health Survey 2016. *PLoS One* **15**, e0239313.

Cobiac L J, Veerman L, Vost T (2013) The role of cost-effectiveness analysis in developing nutrition policy. *Ann Review Nutr* **33**, 373–393.

Dangour A D, Dodhia S K, Hayter A et al. (2009) Nutritional quality of organic foods: A systematic review. *AJCN* **90**(3), 680–685. doi:10.3945/ajcn.2009.28041

FAO-WHO (2014) *Second International Conference on Nutrition: Report of the Joint FAO/WHO Secretariat on the Conference* (Internet). FAO, Rome (accessed 3 July 2021). Available from: http://www.fao.org/3/a-i4436e.pdf.

FAO-WHO (2017) *UN Decade of Action on Nutrition 2016–2025 Work Programme.* (Internet). FAO, WHO Rome (accessed 3 July 2021). Available from: https://www.un.org/nutrition/sites/www.un.org.nutrition/files/general/pdf/work_programme_nutrition_decade.pdf

Fenn B, Colbourn T, Dolan C et al. (2017) Impact evaluation of different cash-based intervention modalities on child and maternal nutritional status in Sindh Province, Pakistan, at 6 mo and at 1 y: A cluster randomised controlled trial. *PLoS Med* **14,** e1002305.

Global Panel (2020) *Future food systems: For people, our planet and prosperity 2020.* (accessed 23 October 2021). Available from: https://www.glopan.org/wp-content/uploads/2020/09/Foresight-2.0_Future-Food-Systems_For-people-our-planet-and-prosperity.pdf

Haas J D, Luna S V, Lung'aho M G et al. (2016) Consuming iron biofortified beans increases iron status in Rwandan women after 128 days in a randomized controlled feeding trial. *J Nutr* **146,** 1586–1592.

Heidkamp R A, Piwoz E, Gillespie S, et al. (2021) Mobilising evidence, data, and resources to achieve global maternal and child undernutrition targets and the sustainable development goals: An agenda for action. *Lancet* online 7 March. https://doi.org/10.1016/S0140-6736(21)00568-7

IFPRI International Food Policy Research Institute (2021) *Global Food Policy Report: Transforming Food Systems after COVID–19.* International Food Policy Research Institute, Washington, DC. https://doi.org/10.2499/9780896293991.

Keats E C, Das J K, Salam R A et al. (2021) Effective interventions to address maternal and child malnutrition: An update of the evidence. *Lancet Child Adolesc Health* **5,** 367–384.

Monteiro C A, Moubarac J C, Cannon G et al. (2013) Ultra-processed products are becoming dominant in the global food system. *Obesity* **14**(Suppl 2), 21–28. doi: 10.1111/obr.12107

Michaux K D, Hou K, Karakochuk C D et al. (2019) Effect of enhanced homestead food production on anaemia among Cambodian women and children: A cluster randomized controlled trial. *Matern Child Nutr* **15**(Suppl 3), e12757.

Osendarp S, Akuoku J, Black R et al. (2021) The potential impacts of the COVID-19 crisis on maternal and child undernutrition in low and middle income countries. Research-square.com

Riches G (ed.) (1997) *First World Hunger: Food Security and Welfare Politics.* Macmillan, Basingstoke.

SUN Movement (2020a) *Strategic Review of the Scaling Up Nutrition (SUN) Movement, 2019–2020 Report* (Internet) Geneva (accessed 3 July 2021). Available from: https://scalingupnutrition.org/wp-content/uploads/2020/04/SUN-Strategic-Review-Final-Report_ENG.pdf.

SUN Movement (2020b) *SUN Strategy 3.0 (2021–2025)* Geneva (accessed 3 July 2021). Available from: https://scalingupnutrition.org/wp-content/uploads/2020/07/SUN-Strategy3_draft_MAIN-DOCUMENT_ENG.pdf.

UNSCN (2017) *Global Governance for Nutrition and the Role of UNSCN* (Internet). FAO, Rome (accessed 31 July 2021). Available from: https://www.unscn.org/en/resource-center/UNSCN-Publications?idnews=1653.

UN Nutrition (2020) *UN Nutrition* (Internet) FAO, Rome (accessed 2021 June 29). Available from: https://www.unnutrition.org/.

Walls H L, Cornelsen L, Lock K et al. (2016) How much priority is given to nutrition and health in the EU Common Agricultural Policy? *Food Policy* **59,** 12–23.

WHO Global Action Plan for the Prevention and Control of NCDs (2013–2020) https://www.who.int/publications/i/item/9789241506236

Willett W, Rockstrom J, Loken B et al. (2019) Food in the anthropocene: the EAT-*Lancet* commission on healthy diets from sustainable food systems. *Lancet* **393,** 447–492.

FURTHER READING

Allen L H and Gillespie S R (2001) *What Works? A Review of the Efficacy and Effectiveness of Nutrition Interventions.* ACC/SCN Nutrition Policy Paper No. 19. Asian Development Bank. Nutrition and Development Series No. 5. Manila. (www.adb.org/publications/what-works-review-efficacy-and-effectiveness-nutrition-interventions).

Berg A (1973) *The Nutrition Factor in National Development.* The Brookings Institute, Washington DC.

Black R E, Allen L H, Bhutta Z A et al. (2008) Maternal and child undernutrition: Global and regional exposures and health consequences. *Lancet* **371,** 243–260.

Bhutta Z A, Das J K, Rizvi A et al. (2013) Evidence-based interventions for improvement of maternal and child nutrition: What can be done and at what cost? *Lancet* **382,** 454–477.

Geissler C (2015) Capacity building in public health nutrition. *Proc Nutr Soc* **74,** 430–436. doi:10.1017/S0029665114001736

Gillespie S, Haddad L, Mannar V et al. (2013) The politics of reducing malnutrition: Building commitment and accelerating progress. *Lancet* **382,** 552–569.

Heidkamp R A, Piwoz E, Gillespie S et al. (2021) Mobilising evidence, data, and resources to achieve global maternal and child undernutrition targets and the sustainable development goals: An agenda for action. *Lancet* **397,** 1400–1418.

Riches G & Silvasti T (eds) (2014) *First World Hunger Revisited: Food Charity or the Right to Food?* Palgrave Macmillan, Basingstoke.

Ruel M T, Alderman H, and the Maternal and Child Nutrition Study Group (2013) Nutrition-sensitive interventions and programmes: How can they help to accelerate progress in improving maternal and child nutrition? *Lancet* **382,** 536–551.

Shrimpton R, Prudhon C, Engesveen K (2009) The impact of high food process on maternal and child nutrition. *SCN News* **37,** 60–68.

Glossary

abrasion Tooth wear caused by brushing

acceptable daily intake (ADI) The amount of a food additive that could be taken daily for an entire lifespan without appreciable risk. Determined by measuring the highest dose of the substance that has no effect on experimental animals, and then dividing by a safety factor of 100

acetal Product of addition of alcohol to aldehyde

acetomenaphthone Synthetic compound with vitamin K activity; vitamin K_3, or menaquinone-0

achlorhydria Deficiency of hydrochloric acid in gastric digestive juice

acid Chemically, compounds that dissociate (ionize) in water to give rise to hydrogen ions (H^+); they taste sour

acid foods, basic foods These terms refer to the residue of the metabolism of foods. The mineral salts of sodium, potassium, magnesium, and calcium are base-forming, while phosphorus, sulphur, and chlorine are acid-forming. Which of these predominates in foods determines whether the residue is acidic or basic (alkaline); meat, cheese, eggs, and cereals leave an acidic residue, while milk, vegetables, and some fruits leave a basic residue

acidogenicity Ability to produce acid through bacterial metabolism

acidosis An increase in the acidity of blood plasma to below the normal range of pH 7.3–7.45, resulting from a loss of the buffering capacity of the plasma, alteration in the excretion of carbon dioxide, excessive loss of base from the body, or metabolic overproduction of acids

acrodermatitis enteropathica Severe functional zinc deficiency due to failure to secrete an as yet unidentified compound in pancreatic juice that is required for zinc absorption

acrodynia Dermatitis seen in animals deficient in vitamin B_6. There is no evidence for a similar dermatitis in deficient human beings

acrylamide A chemical that can be generated when the amino acid asparagine is heated above 100°C in the presence of sugars

active transport Energy-requiring transport of solutes across cell membranes against the prevailing concentration gradient

acute phase proteins A variety of serum proteins synthesized in increased (or sometimes decreased) amounts in response to trauma and infection, thus confounding their use as indices of nutritional status

acute phase reaction Increase or decrease in the concentration of certain proteins, or other substances including micronutrients, in blood serum following infections or tissue-inflammatory reactions. This reaction represents the body's normal response in counteracting the deleterious effects of a noxious stimulus, but the changes in distributions of proteins and nutrients between body compartments can confound attempts to measure nutrient status by blood assays in those situations where an acute phase reaction is present. Its presence can be monitored, for example by measuring C-reactive protein (CRP) or α_1-antichymotrypsin (ACT) (both of which are increased by infections or inflammation) in serum or plasma

additive Any compound not commonly regarded or used as a food which is added to foods as an aid in manufacturing or processing, or to improve the keeping properties, flavour, colour, texture, appearance, or stability of the food, or as a convenience to the consumer. The term excludes vitamins, minerals, and other nutrients added to enrich or restore nutritional value. Herbs, spices, hops, salt, yeast or protein hydrolysates, air, and water are usually excluded from this definition

adenine A nucleotide, one of the purine bases of the nucleic acids (DNA and RNA). The compound formed between adenine and ribose is the nucleoside adenosine, which can form four phosphorylated derivatives important in metabolism: adenosine monophosphate (AMP), (also known as adenylic acid); adenosine diphosphate (ADP); adenosine triphosphate (ATP); cyclic adenosine monophosphate (cAMP)

adenocarcinoma Cancer of the glandular epithelium

adenosine diphosphate (ADP) See adenosine triphosphate; energy metabolism

adenosine triphosphate (ATP) The coenzyme that acts as an intermediate between energy-yielding (catabolic) metabolism (the oxidation of metabolic fuels) and energy expenditure as physical work and in synthetic (anabolic) reactions. See energy metabolism

adequate intake Where there is inadequate scientific evidence to establish requirements and reference intakes for a nutrient for which deficiency is rarely, if ever, seen, the observed levels of intake are assumed to be greater than the requirements and thus provide an estimate of intakes that are (more than) adequate to meet needs

adipocyte A fat-containing cell in adipose tissue

adiponectin Hormone secreted by adipose tissue that seems to be involved in energy homeostasis; it enhances insulin sensitivity and glucose tolerance, as well as oxidation of fatty acids in muscle

adipose tissue Body fat storage tissue, distributed under the skin, around body organs, and in body cavities. Composed of cells that synthesize and store fat, releasing it for metabolism in fasting. Also known as white adipose tissue, to distinguish it from the metabolically more active brown adipose tissue which is involved in heat production to maintain body temperature. The energy yield of adipose tissue is 34–38 MJ/g (8000–9000 kcal/kg)

adiposis Presence of an abnormally large accumulation of fat in the body—also known as liposis

adipsia Absence of thirst

adulteration The addition of substances to foods, etc., in order to increase the bulk and reduce the cost, with intent to defraud the purchaser

aerobic (1) Aerobic micro-organisms (aerobes) are those that require oxygen for growth; obligate aerobes cannot survive in the absence of oxygen. The opposite are anaerobic organisms, which do not require oxygen for growth; obligate anaerobes cannot survive in the presence of oxygen. (2) Aerobic exercise is physical activity that requires an increase in heart rate and respiration to meet the increased demand of muscle for oxygen, in contrast with maximum exertion or sprinting, when muscle can metabolize anaerobically

afflatoxins A family of toxins produced by certain fungi, particularly Aspergillus species, which infect crops for human consumption such as maize (corn), wheat, peanuts, and tree nuts

agalactia Failure of the mother to secrete enough milk to feed a suckling infant

ageusia Loss or impairment of the sense of taste

agnosia Inability to interpret sensory information a result of brain damage

agricultural biotechnology The application of molecular biology to agriculture including, but not limited to, transgenic techniques in which scientists develop plant and animal varieties that contain genes from other species

AIDS Acquired immune deficiency syndrome; *see* HIV

alactasia Partial or complete deficiency of the enzyme lactase in the small intestine, resulting in an inability to digest the sugar lactose in milk, and hence intolerance of milk

alanine A non-essential amino acid

albumin (albumen) A group of relatively small water-soluble proteins, e.g., ovalbumin in egg white, lactalbumin in milk. Plasma or serum albumin is one of the major blood proteins, which transports certain metabolites including non-esterified fatty acids in the bloodstream. Serum albumin concentration is sometimes measured as an index of protein–energy malnutrition. Often used as a non-specific term for proteins (e.g., albuminuria is the excretion of proteins in the urine)

alcohol Chemically, alcohols are compounds with the general formula $CnH_2n_{-1}OH$. The alcohol in alcoholic beverages is ethyl alcohol (ethanol, C_2H_5OH)

alcohol units For convenience in calculating intakes of alcohol, a unit of alcohol is defined as 8 g (10 ml) of absolute alcohol

aldosterone A steroid hormone secreted by the adrenal cortex; controls the excretion of salts and water by the kidneys

alkali (or base) A compound that takes up hydrogen ions and so raises the pH of a solution

alkaline tide Small increase in blood pH after a meal as a result of the secretion of gastric acid

alkaloids Naturally occurring organic bases which have pharmacological actions. Many are found in plant foods, including potatoes and tomatoes (the *Solanum* alkaloids), or as the products of fungal action (e.g., ergot), although they also occur in animal foods (e.g., tetrodotoxin in puffer fish, tetramine in shellfish)

alkalosis *See* acidosis

allele One of two or more alternative forms of a gene located at the corresponding site on homologous chromosomes

allergen A chemical compound, commonly a protein, which causes the production of antibodies, and hence an allergic reaction

allergy Adverse reaction to foods caused by the production of antibodies

allotriophagy An unnatural desire for abnormal foods; also known as cissa, cittosis, and pica

alpha helix (α-helix) Common secondary structure in proteins

alpha linkage (α-linkage) Bond formed by ring closure of a sugar with the hydroxyl group to the right of the chain in the Fischer projection formula (q.v.)

alveolar bone Spongy part of jawbone that supports the teeth

amenorrhoea Cessation of menstruation, normally occurring between the ages of 40 and 55 (the menopause), but sometimes at an early age, especially as a result of severe undernutrition (as in anorexia nervosa) when body weight falls below about 45 kg

Ames test An *in vitro* test for the ability of chemicals, including potential food additives, to cause mutation in bacteria (the mutagenic potential). Commonly used as a preliminary screening method to detect substances likely to be carcinogenic

amines Formed by the decarboxylation of amino acids. Three are potentially important in foods—phenylethylamine (formed from phenylalanine), tyramine (from tyrosine), and tryptamine (from tryptophan)—because they stimulate the sympathetic nervous system and can cause increased blood pressure. In sensitive people they are one of the possible dietary causes of migraine

amino acid profile The amino acid composition of a protein

amino acids The basic units from which proteins are made. Chemically, they are compounds with an amino group ($-NH_3^+$) and a carboxyl group ($-COO^-$) attached to the same carbon atom

aminoaciduria Excretion of abnormal amounts of one or more amino acids in the urine, usually as a result of a genetic disease

aminogram A diagrammatic representation of the amino acid composition of a protein. A plasma aminogram is the composition of the free amino acid pool in blood plasma

aminopeptidase An enzyme secreted in the pancreatic juice which removes amino acids sequentially from the free amino terminal of a peptide or protein (i.e., the end that has a free amino group exposed). Since it works at the end of the peptide chain, it is an exopeptidase

aminotransferase Any enzyme that catalyses the reaction of transamination

amygdala Part of the limbic system in the brain involved in memory consolidation

amylases Enzymes that hydrolyse starch. Alpha-amylase (dextrinogenic amylase or diastase) acts to produce small dextrin fragments from starch, while β-amylase (maltogenic amylase) liberates maltose, some free glucose, and isomaltose from the branch points in amylopectin Salivary and pancreatic amylases are α-amylases

amylodyspepsia An inability to digest starch

β-amyloid Amino acid peptide sections that come from a protein in the fatty membrane surrounding nerve cells, and clump into plaques that accumulate all over the brains of Alzheimer's disease patients or in small clumps that may block cell-to-cell signalling at synapses

amylopectin The branched-chain form of starch, with branches formed by α1–6 bonds. About 75–80 per cent of most starches; the remainder is amylose

amylose The straight-chain form of starch, with only α1–4 bonds. About 20–25 per cent of most starches; the remainder is amylopectin

anabolic hormones Natural or synthetic hormones that stimulate growth and the development of muscle tissue

anabolism The process of building up or synthesizing

anaemia A shortage of red blood cells, leading to pallor and shortness of breath, especially on exertion. Most commonly due to a dietary deficiency of iron or excessive blood loss. Other dietary deficiencies can also result in anaemia, including deficiencies of vitamin B_{12} or folic acid (megaloblastic anaemia), vitamin E (haemolytic anaemia), and rarely vitamin C or vitamin B_6

anaemia, haemolytic Anaemia caused by premature and excessive destruction of red blood cells; not normally due to nutritional deficiency, but can occur as a result of vitamin E deficiency in premature infants

anaemia, megaloblastic Release into the circulation of immature precursors of red blood cells because of deficiency of either folate or vitamin B_{12}

anaemia, pernicious Anaemia due to deficiency of vitamin B_{12}, most commonly as a result of failure to absorb the vitamin from the diet. There is release of immature precursors of red blood cells into the circulation (megaloblastic anaemia) and progressive damage to the spinal cord (subacute combined degeneration), which is not reversed on restoring the vitamin

anaphylaxis A severe and potentially life-threatening reaction to a trigger such as an allergy

anaerobes Micro-organisms that grow in the absence of oxygen. Obligate anaerobes cannot survive in the presence of oxygen; facultative anaerobes grow in the presence or absence of oxygen

anaerobic threshold The level of exercise at which the rate of oxygen uptake into muscle becomes limiting and there is anaerobic metabolism to yield lactate

anaphylaxis/anaphylactic shock Severe and rapid allergic reaction that can cause death

aneuploidy An abnormal number of chromosomes, usually associated with miscarriage or developmental abnormalities, such as Down syndrome where there are three copies of chromosome 21 instead of two

aneurysm Local dilatation (swelling and weakening) of the wall of a blood vessel, usually the result of atherosclerosis and hypertension; especially serious when it occurs in the aorta, when rupture may prove fatal

angina (angina pectoris) Paroxysmal thoracic pain and choking sensation, especially during exercise or stress, due to partial blockage of a coronary artery (blood vessel supplying the heart) as a result of atherosclerosis

angio-oedema Presence of fluid in subcutaneous tissues or submucosa, particularly of the face, eyes, lips, and sometimes tongue and throat; may occur during an anaphylactic reaction

angiotensin-converting enzyme (ACE) Enzyme in the blood vessels of the lungs which activates angiotensin. Many of the drugs for treatment of hypertension are ACE inhibitors

angular stomatitis A characteristic cracking and fissuring of the skin at the angles of the mouth; a symptom of vitamin B_2 deficiency, but also seen in other conditions

anion A negatively charged ion

anomers Isomers of a sugar differing only in configuration at the hemiacetal carbon atom

anorectic drugs (anorexigenic drugs) Drugs that depress the appetite, used as an aid to weight reduction. Apart from sibutramine (Reductil), most have been withdrawn from use; diethylpropion and mazindol are available but are not recommended

anorexia Lack of appetite

anorexia nervosa A psychological disturbance resulting in a refusal to eat, possibly with restriction to a very limited range of foods, often accompanied by a rigid programme of vigorous physical exercise to the point of exhaustion. The result is a very considerable loss of weight, with tissue atrophy and a fall in basal metabolic rate. It is especially prevalent among adolescent girls; when body weight falls below about 45 kg there is a cessation of menstruation

anosmia Lack or impairment of the sense of smell

anovulation Failure to ovulate spontaneously

antacids Bases that neutralize acids, generally used to counteract excessive gastric acidity and to treat indigestion: sodium bicarbonate, aluminium hydroxide, magnesium carbonate, and magnesium hydroxide

anthocyanins Violet, red, and blue water-soluble colours extracted from flowers, fruits, and leaves

anthropometry Body measurements used as an index of physiological development and nutritional status; a non-invasive way of assessing body composition. Weight for age provides information about the overall nutritional status of children; weight for height is used to detect acute malnutrition (wasting); height for age is used to detect chronic malnutrition (stunting). Mid-upper arm circumference provides an index of muscle wastage in undernutrition. Skinfold thickness is related to the amount of subcutaneous fat as an index of over- or undernutrition

antibody Immunoglobulin that specifically counteracts an antigen or allergen (*see* allergen)

antidiarrhoeal Drug used to treat diarrhoea by absorbing water from the intestine, altering intestinal motility, or adsorbing toxins

antidiuretic Drug used to reduce the excretion of urine and so conserve fluid in the body

antiemetic Drug used to prevent or alleviate nausea and vomiting

antienzymes Substances that inhibit the action of enzymes. Many inhibit digestive enzymes and are present in raw legumes. Most are proteins, and therefore are inactivated by heat

antigen Any compound that is foreign to the body (e.g., bacterial, food, or pollen proteins or complex carbohydrates) which, when introduced into the circulation, stimulates the formation of an antibody

antihistamine Drug that antagonizes the actions of histamine: those that block histamine H_1 receptors are used to treat allergic reactions; those that block H_2 receptors are used to treat peptic ulcers

antihypertensive Drug, diet, or other treatment used to treat hypertension (high blood pressure)

antilipidaemic Drug, diet, or other treatment used to treat hyperlipidaemia by lowering blood lipids

antimetabolite Compound that inhibits a normal metabolic process, acting as an analogue of a normal metabolite. Some are useful in chemotherapy of cancer; others are naturally occurring toxins in foods, frequently causing vitamin-deficiency diseases by inhibiting the normal metabolism of the vitamin

antimotility agents Drugs used to reduce gastrointestinal motility and hence reduce the discomfort associated with diarrhoea

antimutagen Compound acting on cells and tissues to decrease initiation of mutation by a mutagen

antioxidant Substance that retards the oxidative rancidity of fats in stored foods. Many fats, especially vegetable oils, contain naturally occurring antioxidants, including vitamin E, which protect them against rancidity for some time

antioxidant nutrients Highly reactive oxygen radicals are formed during normal metabolism and in response to infection and some chemicals. They cause damage to fatty acids in cell membranes, and the products of this damage can then cause damage to proteins and DNA. A number of different mechanisms are involved in protection against, or repair after, oxygen radical damage, including a number of nutrients, especially vitamin E, carotene, vitamin C, and selenium. Collectively, these are known as antioxidant nutrients

anti-rachitic Preventing or curing rickets

anti-sialagogues Substances that reduce the flow of saliva

antivitamins Substances that interfere with the normal metabolism or function of vitamins, or destroy them

apastia Refusal to take food as an expression of a psychiatric disturbance

aphagosis Inability to eat

apocarotenal Aldehydes formed by oxidation of carotenes, other than retinaldehyde

apoenzyme The protein part of an enzyme which requires a coenzyme for activity, and is therefore inactive if the coenzyme is absent

apolipoprotein The protein of plasma lipoproteins without the associated lipid

aposia Absence of sensation of thirst

apositia Aversion to food

apraxia Inability to perform tasks as a result of brain damage

arachidonic acid A long-chain polyunsaturated fatty acid (C20:4 ω6)

arginine A basic amino acid; not a dietary essential for adults, but infants may not be able to synthesize enough to meet the high demands of growth

ariboflavinosis Deficiency of riboflavin (vitamin B$_2$)

arm–chest–hip index (ACH index) A method of assessing nutritional status by measuring the arm circumference, chest diameter, and hip width

aromatic ring Stable ring structure with π electrons delocalized around the ring as in benzene

arterial restenosis Rate of re-occlusion (narrowing) of arteries after artificial (i.e., mechanical) removal of the accumulated plaque coatings. Used as a measure of susceptibility to atherosclerotic disease. It can be followed non-invasively, e.g., by ultrasound measurements

arteriosclerosis Thickening and calcification of the arterial walls, leading to loss of elasticity. Occurs with ageing and especially in hypertension

arthritis Painful, swollen, and/or inflamed joints

ascites Abnormal accumulation of fluid in the peritoneal cavity, occurring as a complication of cirrhosis of the liver, congestive heart failure, cancer, and infectious diseases

ascorbic acid Vitamin C, chemically L-xyloascorbic acid, to distinguish it from the isomer D-araboascorbic acid (isoascorbic acid or erythorbic acid) which has only slight vitamin C activity

ash The residue left behind after all organic matter has been burnt off; a measure of the total content of mineral salts in a food

asparagine A non-essential amino acid; the β-amide of aspartic acid

aspartic acid (aspartate) A non-essential amino acid

asthma Chronic inflammatory disease of the airways which renders them prone to narrow too much. The symptoms include paroxysmal coughing, wheezing, tightness, and breathlessness. Asthma may be caused by an allergic response or may be induced by non-immunological mechanisms

astringency The action of unripe fruits and other foods to cause contraction of the epithelial tissues of the tongue; believed to result from destruction of the lubricant properties of saliva by precipitation by tannins

astrocyte Glial cell of the central nervous system that is shaped like a star

atheroma Fatty deposit composed of lipids, complex carbohydrates, and fibrous tissue which forms on the inner wall of blood vessels in atherosclerosis

atherosclerosis Degenerative disease in which there is accumulation of lipids, together with complex carbohydrates and fibrous tissue (atheroma), on the inner wall of arteries. Leads to narrowing of the lumen of the arteries

atrophy Wasting of normally developed tissue or muscle as a result of disuse, ageing, or undernutrition

attrition Tooth wear caused by grinding

auxotrophe Mutant strain of a micro-organism which requires one or more nutrients for growth that are not required by the parent organism. Commonly used for microbiological assay of vitamins, amino acids, etc.

availability Bioavailability or biological availability. In some foodstuffs, nutrients that can be demonstrated to be present chemically may not be fully available when they are eaten because the nutrients are chemically bound in a form that is not susceptible to enzymic digestion

avitaminosis The absence of a vitamin; may be used specifically (e.g., avitaminosis A) or generally to mean a vitamin-deficiency disease

axial position Substituent in a six-membered non-aromatic ring is above or below the average plane of the ring

bacteria Unicellular micro-organisms, ranging from 0.5 to 5 μm in size. They may be classified on the basis of their shape: spherical (coccus), rod-like (bacillus), spiral (spirillum), comma-shaped (vibrio), corkscrew-shaped (spirochaete), or filamentous. Other classifications are based on whether or not they are stained by Gram's stain, aerobic or anaerobic, autotrophic or heterotrophic

bacteriophages Viruses that attack bacteria, commonly known as phages

basal metabolic rate (BMR) The energy cost of maintaining the metabolic integrity of the body, nerve and muscle tone, respiration, and circulation. For children it also includes the energy cost of growth. Experimentally, BMR is measured as the heat output from the body, or the rate of oxygen consumption, under strictly standardized conditions: 12–14 hours after the last meal, completely at rest (but not asleep), and at an environmental temperature of 26–30°C to ensure thermal neutrality

bdelygmia An extreme loathing of food

behenic acid Very chain-saturated fatty acid (C22:0)

beriberi Result of severe and prolonged deficiency of vitamin B_1, especially where the diet is high in carbohydrate and poor in vitamin B_1

beta-linkage (β-linkage) Bond formed by ring closure of a sugar with the hydroxyl group to the left of the chain in the Fischer projection formula (q.v.)

beta-pleat (β pleat) Common secondary structure in proteins

beta-sheet flattened form Common secondary structure in proteins

bifidogenic Promoting the growth of (beneficial) bifidobacteria in the intestinal tract

bifidus factor A carbohydrate in human milk which stimulates the growth of *Lactobacillus bifidus* in the intestine. In turn, this organism lowers the pH of the intestinal contents and suppresses the growth of pathogenic bacteria

bile Alkaline fluid produced by the liver and stored in the gall bladder before secretion into the small intestine (duodenum) via the bile duct. It contains bile salts, bile pigments (bilirubin and biliverdin), phospholipids, and cholesterol

bile salts (bile acids) Salts of cholic and deoxycholic acid and their glycine and taurine conjugates secreted in the bile; they enhance the digestion of fats by emulsifying them

bilirubin, biliverdin Bile pigments, formed by the degradation of haemoglobin

binge–purge syndrome Feature of the eating disorder bulimia nervosa, characterized by the ingestion of excessive amounts of food and the excessive use of laxatives

bioactives Non-essential dietary compounds that have an effect on a living organism, tissue, or cell

bioassay Biological assay; measurement of biologically active compounds (e.g., vitamins and essential amino acids) by their ability to support growth of micro-organisms or animals

bioavailability Fraction of an ingested nutrient that is absorbed and used for a defined function in the body (*see* availability)

biocytin The main form of the vitamin biotin in most foods, bound to the amino acid lysine

bioelectrical impedance (BEI) Method of measuring the proportion of fat in the body by the difference in the resistance to the passage of an electric current between fat and lean tissue

bioflavonoids *See* flavonoids

biofortification Food fortification achieved by plant breeding or genetic modification to give a higher content of nutrients

bioinformatics Application of computational techniques to extract meaning from complex biological data

biological value (BV) Proportion of absorbed nitrogen that is retained for maintenance and/or growth

biotin A vitamin, sometimes known as vitamin H, required for the synthesis of fatty acids and glucose, among other reactions, and the control of gene expression and cell division

birth canal The bony birth canal comprises the passage through the pelvic bones; there is a soft tissue component to the birth canal in the form of the vagina and its supporting tissues

bisfuran polycyclic compounds Compounds with several rings including more than two five-membered oxygen-containing rings

Bitot's spots Irregularly shaped foam-like plaques on the conjunctiva of the eye, characteristic of vitamin A deficiency, but not considered to be a diagnostic sign without other evidence of deficiency

biuret test Chemical test for proteins based on the formation of a violet colour when copper sulfate in alkaline solution reacts with a peptide bond

black tongue disease Sign of niacin deficiency in dogs; the canine equivalent of pellagra

bland diet A diet that is non-irritating, does not overstimulate the digestive tract, and is soothing to the intestines; generally avoiding alcohol, strong tea or coffee, pickles, and spices

blood plasma Liquid component of blood, accounting for about half its total volume; a solution of nutrients and various proteins. When blood has clotted, the resultant fluid is known as serum

blood sugar Glucose; normal concentration is about 5 mmol/L (90 mg/L), and is maintained in the fasting state by mobilization of tissue reserves of glycogen and synthesis from amino acids. Only in prolonged starvation does it fall below about 3.5 mmol/L (60 mg/L). If it falls to 2 mmol/L (35 mg/L), there is loss of consciousness (hypoglycaemic coma)

B-lymphocytes Bursa-equivalent lymphocytes. After maturation into plasma cells they produce antibodies (immunoglobulins) during humoral responses in immunological reactions. They were first discovered in the bursa of Fabricius in the chicken—hence the name

body density Body fat has a density of 0.90 g/cm^3, while the density of fat free body mass is 1.10 g/cm^3. Determination of density by weighing in air and in water, or by measuring body volume and weight, permits calculation of the proportions of fat and lean body tissue

body mass index (BMI) An index of fatness and obesity. The weight (in kilograms) divided by the square of height (in metres). The acceptable (desirable) range is 18–25. Above 25 is overweight, and above 30 is obesity. Also called Quetelet's index

borborygmos (plural borborygmi); audible abdominal sound produced by excessive intestinal motility

borderline substances Foods that may have characteristics of medication in certain circumstances, and which may then be prescribed under the National Health Service in the UK

botulism A rare form of food poisoning caused by the extremely potent neurotoxins produced by *Clostridium botulinum*

bovine somatotrophin (BST) The natural growth hormone of cattle; biosynthetic BST is used in some dairy herds to increase milk production (approved for use in the USA in 1993; prohibited in the EU)

bovine spongiform encephalopathy (BSE) A degenerative brain disease in cattle, transmitted by feeding slaughter-house waste from infected animals. Commonly known as 'mad cow disease'. The infective agent is a prion; it can be transmitted to humans, causing early-onset variant Creutzfeldt–Jakob disease

bradycardia An unusually slow heartbeat (less than 60 beats/min)

bradyphagia Eating very slowly

bromatology The science of foods

bronze diabetes *See* haemochromatosis

brown adipose tissue (brown fat) Metabolically highly active adipose tissue, which is involved in heat production to maintain body temperature, as opposed to white adipose tissue, which is storage fat and has a low rate of metabolic activity

Brunner's glands Mucus-secreting glands in the duodenum

buffers Salts of weak acids and bases that resist a change in the pH when acid or alkali is added

bulimia nervosa An eating disorder, characterized by powerful and intractable urges to overeat, followed by self-induced vomiting and the excessive use of purgatives

butyric acid A short-chain saturated fatty acid (C4:0)

cachexia The condition of extreme emaciation and wasting seen in patients with advanced cancer and AIDS. Due partly to an inadequate intake of food and mainly to the effects of the disease in increasing metabolic rate (hypermetabolism) and the breakdown of tissue protein

cadaverine Low molecular weight polyamine with biological activity, the decarboxylation product of lysine

caecum The first part of the large intestine, separated from the small intestine by the ileocolic sphincter

calcidiol The 25-hydroxy-derivative of vitamin D, also known as 25-hydroxycholecalciferol; the main storage and circulating form of the vitamin in the body

calciferol Used at one time as a name for ercalciol (ergocalciferol or vitamin D$_2$), made by the ultra-violet irradiation of ergosterol. Also used as a general term to include both vitamers of vitamin D (vitamins D$_2$ and D$_3$)

calcinosis Abnormal deposition of calcium salts in tissues. May be due to excessive intake of vitamin D

calciol The official name for cholecalciferol, the naturally occurring form of vitamin D (vitamin D$_3$)

calcitriol The 1,25-dihydroxy-derivative of vitamin D, also known as 1,25-dihydroxycholecalciferol; the active metabolite of the vitamin

calorie A unit of energy used to express the energy yield of foods and energy expenditure by the body; the amount of heat required to raise the temperature of 1 g of water through 1°C (from 14.5 to 15.5°C). Nutritionally the kilocalorie (1000 calories) is used (the amount of heat required to raise the temperature of 1 kg of water through 1°C), and is abbreviated as either kcal or Cal to avoid confusion with the cal. The calorie is not an SI (Système Internationale) unit, and correctly the joule is used as the unit of energy, although kcal are widely used (1 kcal = 4.18 kJ; 1 kJ = 0.24 kcal)

calorimeter (bomb calorimeter) An instrument for measuring the amount of oxidizable energy in a substance by burning it in oxygen and measuring the heat produced

calorimetry The measurement of energy expenditure by the body. Direct calorimetry is the measurement of heat output from the body as an index of energy expenditure, and hence energy requirements

canbra oil (canola oil) Oil extracted from selected strains of rapeseed containing not more than 2 per cent erucic acid

cancer A wide variety of diseases characterized by uncontrolled growth of tissue

canola A variety of rape which is low in glucosinolates. Canola oil (canbra oil) contains less than 2 per cent erucic acid

canthaxanthin A red carotenoid pigment which is not a precursor of vitamin A

capric acid A medium-chain fatty acid (C10:0)

caproic acid A short-chain fatty acid (C6:0)

caprylic acid A medium-chain fatty acid (C8:0)

carbohydrate The major food source of metabolic energy: the sugars and starches. Chemically they are composed of carbon, hydrogen, and oxygen in the ratio $Cn:H_2n:On$

carbohydrate by difference Historically it was difficult to determine the various carbohydrates present in foods, and an approximation was often made by subtracting the measured protein, fat, ash, and water from the total weight

carboxypeptidase An enzyme secreted in the pancreatic juice which hydrolyses amino acids from the carboxyl terminal of proteins

carcinogen A substance that can induce cancer

cardiomyopathy Any chronic disorder affecting the muscle of the heart; may be associated with alcoholism and vitamin B_1 deficiency

caries Dental decay caused by attack on the tooth enamel by acids produced by bacteria that are normally present in the mouth

cariogenic Causing tooth decay (caries) by stimulating the growth of acid-forming bacteria on the teeth; sucrose and other fermentable carbohydrates

cariogenicity Ability to cause caries

cariostatic Preventing tooth decay

carnitine A derivative of the amino acid lysine; required for the transport of fatty acids into mitochondria for oxidation

carnosine A dipeptide, β-alanyl-histidine, found in the muscle of most animals. Its function is not known

carotenes The red and orange pigments of plants; all are antioxidant nutrients. Three are important as precursors of vitamin A: α-, β-, and γ-carotene

carotenoids A general term for the wide variety of red and yellow compounds chemically related to carotene that are found in plant foods

carotinaemia Presence of abnormally large amounts of carotene in blood plasma. Also known as xanthaemia

casein About 75 per cent of the proteins of milk are classified as caseins; a group of 12–15 different proteins

catabolism Those pathways of metabolism concerned with the breakdown and oxidation of fuels and hence provision of metabolic energy. People who are undernourished or suffering from cachexia are sometimes said to be in a catabolic state, in that they are catabolizing their body tissues without replacing them

catalase An enzyme that splits hydrogen peroxide to yield oxygen and water; an important part of the body's antioxidant defences

catalyst An agent that participates in a chemical reaction, speeding the rate, but itself remains unchanged

cathepsins (kathepsins) A group of intracellular enzymes that hydrolyse proteins. They are involved in the normal turnover of tissue protein, and the softening of meat when game is hung

cation A positively charged ion

CD4+cells Helper T-cells, part of the immune system, recognized by the presence of cluster of differentiation antigen 4 (CD4) on their exterior cell surfaces

CD8+cells Suppressor T-cells recognized by the presence of cluster of differentiation antigen 8 (CD8) on their exterior cell surfaces

cellobiose A disaccharide of glucose-linked β-1,4, which is not hydrolysed by mammalian digestive enzymes; a product of the hydrolysis of cellulose

cellulase An enzyme that hydrolyses cellulose to glucose and cellobiose. It is present in the digestive juices of some wood-boring insects and in various micro-organisms, but not in mammals

cellulose A polysaccharide of glucose-linked α-1,4, which is not hydrolysed by mammalian digestive enzymes; the main component of plant cell walls

cereal Any grain or edible seed of the grass family which can be used as food (e.g., wheat, rice, oats, barley, rye, maize, millet). Collectively known as corn in the UK, although in the USA corn is specifically maize

chalasia Abnormal relaxation of the cardiac sphincter muscle of the stomach so that gastric contents reflux into the oesophagus, leading to regurgitation

cheilosis Cracking of the edges of the lips, one of the signs of vitamin B_2 (riboflavin) deficiency

chelating agents Chemicals that combine with metal ions and remove them from their sphere of action, also called sequestrants (e.g., citrates, tartrates, phosphates, and EDTA)

chemoreceptors Specialized receptors mainly found in the carotid body, within the carotid arteries of the neck, which are sensitive to changes in the blood concentration of oxygen (PO2), carbon dioxide (PCO2), and acidity (pH)

chemotaxis Movement of phagocytes towards invading bacteria, cell debris, or foreign particles

chief cells Cells in the stomach that secrete pepsinogen, the precursor of the digestive enzyme pepsin (*see* pepsin)

Chinese restaurant syndrome Flushing, palpitations, and numbness associated at one time with the consumption of monosodium glutamate, and then with histamine, but the actual cause of these symptoms after eating various foods is not known

chitin The organic matrix of the exoskeleton of insects and crustaceans; also present in small amounts in mushrooms. It is an insoluble and indigestible non-starch polysaccharide, composed of *N*-acetylglucosamine. Partial deacetylation results in the formation of chitosans

chitosan Modified chitin; marketed as a fat binder to reduce fat absorption and aid weight reduction, with little evidence of efficacy

chlorpropamide Drug used in the treatment of diabetes; stimulates secretion of insulin. An oral hypoglycaemic agent

cholagogue Substance that stimulates the secretion of bile from the gall bladder into the duodenum

cholecalciferol Vitamin D

cholecystokinin Hormone that stimulates gall bladder and pancreatic secretion

cholestasis Failure of normal amounts of bile to reach the intestine, resulting in obstructive jaundice. May be caused by bile stones or liver disease

cholesterol Principal sterol in animal tissues, an essential component of cell membranes, and the precursor of the steroid hormones. Not a dietary essential, since it is synthesized in the body

cholestyramine Drug used to treat hyperlipidaemia by complexing bile salts in the intestinal lumen and increasing their excretion, thus increasing their synthesis from cholesterol

choline A derivative of the amino acid serine; an important component of cell membranes, as the phospholipid phosphatidylcholine (lecithin). It is synthesized in the body and is a ubiquitous component of cell membranes; therefore it occurs in all foods, so that dietary deficiency is unknown

cholinergic agonist Drugs that stimulate receptors activated by acetylcholine

chromosome Collection of genes packaged with histone proteins found in the nucleus of cells; humans have 23 pairs (*see* histones)

chronic energy deficiency A term recently introduced to describe adult malnutrition. Commonly defined by wasting or a low body mass index (*see also* protein-energy malnutrition (PEM))

chylomicrons Plasma lipoproteins containing newly absorbed fat, assembled in the small intestinal mucosa and secreted into the lymphatic system, circulating in the lymph and bloodstream as a source of fat for tissues. The remnants are cleared by the liver

chyme The partly digested mass of food in the stomach

chymotrypsin An enzyme involved in the digestion of proteins; secreted as the inactive precursor chymotrypsinogen in the pancreatic juice

cis Stereochemistry at a carbon–carbon double bond with both substituents on the same side of the bond

cissa An unnatural desire for foods; alternative terms are cittosis, allotriophagy, and pica

citrulline An amino acid formed as a metabolic intermediate, but not involved in proteins and of no nutritional importance

cittosis An unnatural desire for foods

coagulation A process involving the denaturation of proteins, so that they become insoluble; it may be effected by heat, strong acids and alkalis, metals, and other chemicals. The final stage in blood clotting is the precipitation of insoluble fibrin, formed from the soluble plasma protein fibrinogen

cobalamin Vitamin B_{12}

cocarcinogen A substance which, alone, does not cause the induction of cancer, but potentiates the action of a carcinogen

Codex Alimentarius Originally Codex Alimentarius Europaeus; since 1961 part of the United Nations FAO/WHO Commission on Food Standards to simplify and integrate food standards for adoption internationally (http://www.codexalimentarius.org)

coding sequence (cds) Region of an mRNA transcript that encodes for protein synthesis (*see* RNA)

codon Triplet sequence of the bases adenine, cytosine, guanine, thymine (A, C, G, T) which encodes a specific amino acid used during translation of mRNAs

coeliac disease Intolerance of the proteins of wheat, rye, and barley; specifically, the gliadin fraction of the protein gluten. The villi of the small intestine are severely affected and absorption of food is poor. Stools are bulky and fermenting from unabsorbed carbohydrate, and contain a large amount of unabsorbed fat (steatorrhoea) (*see* gliadin)

coenzymes Organic compounds required for the activity of some enzymes; most are derived from vitamins

cognition Mental process of acquiring knowledge and understanding through thought, experience, and the senses

cognitive Relating to cognition

cohort study Systematic follow-up of a group of people for a defined period of time or until a specified event; also known as longitudinal or prospective study

colectomy Surgical removal of all or part of the colon to treat cancer or severe ulcerative colitis

colitis Inflammation of the large intestine, with pain, diarrhoea, and weight loss; there may be ulceration of the large intestine (ulcerative colitis)

collagen Insoluble protein in connective tissue, bones, tendons, and skin of animals and fish; converted into the soluble protein gelatine by moist heat

colloid Particles (the disperse phase) suspended in a second medium (the dispersion medium); can be solid, liquid, or gas suspended in a solid, liquid, or gas

colon Also known as the large intestine or bowel; it terminates at the rectum, where faeces are compacted and stored before voiding

colostomy Surgical creation of an artificial conduit (a stoma) on the abdominal wall for voiding of intestinal contents following surgical removal of much of the colon and/or rectum

colostrum The milk produced during the first few days after parturition; it is a valuable source of antibodies for the newborn infant. Animal colostrum is sometimes known as beestings

commercial farming Agricultural production primarily or exclusively for the market, in contrast with subsistence farming

Committee on Medical Aspects of Food Policy (COMA) Previous name for the permanent Advisory Committee to the UK Department of Health, now called the Scientific Advisory Committee on Nutrition (SACN)

complementary DNA (cDNA) A complete or partial copy of a mature mRNA transcript which does not contain intronic sequences (*see* DNA, RNA, intron)

complementation This term is used with respect to proteins when a relative deficiency of an amino acid in one protein is compensated by a relative surplus from another protein consumed at the same time. The protein quality is higher than the average of the separate values

confabulation A memory disturbance, defined as the production of fabricated, distorted, or misinterpreted memories about oneself or the world, without the conscious intention to deceive

conjugated linolenic acid An isomer of linolenic acid in which the double bonds are conjugated, rather than methylene-interrupted

constipation Difficulty in passing stools or infrequent passage of hard stools

contaminants Undesirable compounds found in foods, residues of agricultural chemicals (pesticides, fungicides, herbicides, fertilizers, etc.), through the manufacturing process, or as a result of pollution

convenience foods Processed foods in which a considerable amount of the preparation has already been carried out by the manufacturer

coprolith Mass of hard faeces in the colon or rectum due to chronic constipation

coprophagy Eating of faeces. Since B vitamins are synthesized by intestinal bacteria, animals that eat their faeces can make use of these vitamins, which are not absorbed from the large intestine, the site of bacterial action

corn (*see* cereal)

corrinoids (corrins) The basic chemical structure of vitamin B_{12} is the corrin ring; compounds with this structure are corrinoids, whether or not they have vitamin activity

creatine A derivative of the amino acids glycine and arginine, important in muscle as a store of phosphate for resynthesis of ATP during muscle contraction and work (*see* ATP)

creatinine The anhydride of creatine, formed non-enzymatically; urinary excretion is relatively constant from day to day and reflects mainly the amount of muscle tissue in the body, so the amounts of various components of urine are often expressed relative to creatinine

cretinism Underactivity of the thyroid gland (hypothyroidism) in children, resulting in poor growth, severe mental retardation, and deafness

cristal height A measure of leg length taken from the floor to the summit of the iliac crest. As a proportion of height, it increases with age in children, and a reduced rate of increase indicates undernutrition. *See also* anthropometry

critical pH The pH below which tooth enamel demineralizes

Crohn's disease Chronic inflammatory disease of the bowel, of unknown origin; also known as regional enteritis, since only some regions of the gut are affected

cryptoxanthin Yellow hydroxylated carotenoid found in a few foods; a vitamin A precursor

crystallized intelligence The ability to use learned skills, knowledge, and experience

cultivar Horticultural term for a cultivated variety of plant that is distinct, and is uniform and stable in its characteristics when propagated

cyanocobalamin One of the vitamers of vitamin B_{12}

cyanogen(et)ic glycosides Organic compounds of cyanide found in a variety of plants; chemically cyanhydrins are linked by glycoside linkage to one or more sugars. Toxic through liberation of cyanide when the plants are cut or chewed

cysteine A non-essential amino acid, but nutritionally important since it spares the essential amino acid methionine. In addition to its role in protein synthesis, cysteine is important as the precursor of taurine, in the formation of coenzyme A from the vitamin pantothenic acid, and in the formation of the tripeptide glutathione

cystic fibrosis A genetic disease due to a failure of the normal transport of chloride ions across cell membranes. This results in abnormally viscous mucus, affecting especially the lungs and secretion of pancreatic juice, and hence impairing digestion

cystine The dimer of cysteine produced when its sulfydryl group (−SH) is oxidized, forming a disulfide (S–S−) bridge (*see* cysteine)

cytochrome P450 A family of cytochromes which are involved in the detoxication system of the body (phase I metabolism). They act on a wide variety of (potentially toxic) compounds, both endogenous metabolites and foreign compounds (xenobiotics), rendering them more water-soluble and more readily conjugated for excretion in the urine

cytochromes Haem-containing proteins. Some cytochromes react with oxygen directly; others are intermediates in the oxidation of reduced coenzymes. Unlike haemoglobin, the iron in the haem of cytochromes undergoes oxidation and reduction

cytokines Small protein molecules that are the main communication between immune system cells and other tissues. Secreted by immune cells. Have both stimulating and suppressing action on lymphocytes and immune response. Include TNF-α, IL-1, and IL-6, which are pro-inflammatory cytokines produced especially by macrophages very early following infection

cytosine One of the pyrimidine bases of nucleic acids (*see* pyrimidines)

dark adaptation The time taken to adapt to seeing in dim light; an index of vitamin A status, as adaptation is slower and less complete in vitamin A deficiency

deciduous teeth The first set of 20 teeth that appear during infancy and are lost during childhood and early adolescence as the adult (permanent) teeth erupt. Also known as milk teeth or first teeth

dehydroascorbic acid Oxidized vitamin C, which is readily reduced back to the active form in the body and therefore has vitamin activity

dehydrocholesterol The precursor for the synthesis of vitamin D in the skin

delayed hypersensitivity An exaggerated immune response that is delayed for a day or more. It is mediated by the response of T-cells to a foreign antigen or allergen

dementia Chronic or persistent disorder of the mental processes caused by brain disease or injury; marked by memory disorders, personality changes, and impaired reasoning

dentate With natural teeth

dentine Mineralized tissue beneath tooth enamel

deoxyribonucleic acid (DNA) The genetic material in the nuclei of all cells. A linear polymer composed of four kinds of deoxyribose nucleotide—adenine, cytosine, guanine, and thymidine (A, C, G, and T)—linked by phosphodiester bonds which is the carrier of genetic information. In its native state DNA is a double helix

Department of the Environment, Food and Rural Affairs (Defra) Website http://www.defra.gov.uk

dermatitis A lesion or inflammation of the skin; many nutritional deficiency diseases include more or less specific skin lesions, but most cases of dermatitis are not associated with nutritional deficiency and do not respond to nutritional supplements

designer foods Alternative name for functional foods

desmosines The compounds that form the cross-linkage between chains of the connective tissue protein elastin

desmutagen Compound acting directly on a mutagen to decrease its mutagenicity

dextrins A mixture of soluble compounds formed by the partial breakdown of starch by heat, acid, or amylases

dextrose Alternative name for glucose. Commercially the term 'glucose' is often used to mean corn syrup (a mixture of glucose with other sugars and dextrins), and pure glucose is called dextrose

diabetes insipidus A metabolic disorder, characterized by extreme thirst, excessive consumption of liquids, and excessive urination, due to failure of secretion of the antidiuretic hormone

diabetes mellitus A metabolic disorder involving impaired metabolism of glucose due to either failure of secretion of the hormone insulin (insulin-dependent or type 1 diabetes) or impaired responses of tissues to insulin (non-insulin-dependent or type 2 diabetes)

diarrhoea Frequent passage of loose watery stools, commonly the result of intestinal infection; rarely as a result of adverse reaction to foods or disaccharide intolerance

dietary fibre Material mostly derived from plant cell walls which is not digested by human digestive enzymes; a large proportion consists of non-starch polysaccharides

dietary folate equivalents (DFE) Method of calculating folic acid intake taking into account the lower availability of mixed folates in food compared with synthetic tetrahydrofolate used in food enrichment and supplements: 1 µg DFE = 1 µg food folate or 0.6 µg synthetic folate; total DFE = µg food folate + 1.7 × µg synthetic folate

dietary reference intakes (DRIs) US term for dietary reference values. In addition to average requirement and RDA, include tolerable upper levels (ULs) of intake from supplements (*see* dietary reference values)

dietary reference values (DRVs) A set of standards of the amounts of each nutrient needed to maintain good health. People differ in the daily amounts of nutrients they need; for most nutrients the measured average requirement plus 20 per cent (statistically two standard deviations) takes care of the needs of nearly everyone. In the UK this is termed reference nutrient intake (RNI); elsewhere it is known as recommended daily allowance or intake (RDA or RDI), population reference intake (PRI), or dietary reference intake (DRI). This figure is used to calculate the needs of large groups of people in institutional or community planning. Obviously, some people require less than the average (up to 20 per cent or two standard deviations less). This lower level is termed the lower reference nutrient intake (LRNI) (also known as the minimum safe intake (MSI) or lower threshold intake (LTI)). This is an intake at or below which it is unlikely that normal health could be maintained. If the diet of an individual indicates an intake of any nutrient at or below LRNI, then detailed investigation of their nutritional status would be recommended. For energy intake only a single dietary reference value is used—the average requirement—because there is potential harm (of obesity) from ingesting too much

dietetic foods Foods prepared to meet the particular nutritional needs of people whose assimilation and metabolism of foods are modified, or for whom a particular effect is obtained by a controlled intake of foods or individual nutrients

dietetics The study or prescription of diets under special circumstances (e.g., metabolic or other illness) and for special physiological needs such as pregnancy, growth, or weight reduction

diet-induced thermogenesis The increase in heat production by the body after eating due to both the metabolic energy cost of digestion and the energy cost of forming tissue reserves of fat, glycogen, and protein. It is approximately 10 per cent of the energy intake but varies with composition of the diet

dietitian (UK), dietician (US) One who applies the principles of nutrition to the feeding of individuals and groups, plans menus and special diets, supervises the preparation and serving of meals, and instructs in the principles of nutrition as applied to selection of foods

differentiation The process of development of new characters in cells or tissues

digestibility The proportion of a foodstuff absorbed from the digestive tract into the bloodstream, normally 90–95 per cent. It is measured as the difference between intake and faecal output, with allowance being made for that part of the faeces that is not derived from undigested food residues (shed cells of the intestinal tract, bacteria, residues of digestive juices)

diglycerides (diacylglycerols) Glycerol esterified with two fatty acids; intermediates in the digestion of triacylglycerols, and used as emulsifying agents in food manufacture (*see* glycerol)

3-diol (3-MCPD) A food-processing contaminant formed as a by-product during the processing of edible oils and fats

dipsesis (dipsosis) Extreme thirst, a craving for abnormal kinds of drink

dipsetic Tending to produce thirst

dipsogen Thirst-provoking agent

dipsomania Morbid craving for alcoholic drinks

disaccharidases Enzymes that hydrolyse disaccharides to their constituent monosaccharides in the intestinal mucosa: sucrase (also known as invertase) acts on sucrose and isomaltose, lactase on lactose, maltase on maltose, and trehalase on trehalose

disaccharides Sugars composed of two monosaccharide units; the nutritionally important disaccharides are sucrose, lactose, and maltose. *See* carbohydrate

diuresis Increased formation and excretion of urine; it occurs in diseases such as diabetes, and also in response to diuretics

diuretics Substances that increase the production and excretion of urine

diverticular disease Diverticulosis is the presence of pouch-like hernias (diverticula) through the muscle layer of the colon; it is associated with a low intake of dietary fibre and high intestinal pressure due to straining during defecation. Faecal matter can be trapped in these diverticula, causing them to become inflamed, causing pain and diarrhoea—the condition of diverticulitis

docosahexaenoic acid (DHA) A long-chain polyunsaturated fatty acid (C22:6 ω3)

docosanoids Long-chain polyunsaturated fatty acids with 22 carbon atoms

docosapentaenoic acid A long-chain polyunsaturated fatty acid (C22:5 ω3 or ω6)

Douglas bag An inflatable bag for collecting exhaled air to measure the consumption of oxygen and production of carbon dioxide, for the measurement of energy expenditure by indirect calorimetry

duodenum First part of the small intestine, between the stomach and the jejunum; the major site of digestion

dysgeusia Distortion of the sense of taste—a common side effect of some drugs

dyspepsia Pain or discomfort associated with eating. Dyspepsia may be a symptom of gastritis, peptic ulcer, gall bladder disease, hiatus hernia, etc. If there is no structural change in the intestinal tract, it is called 'functional dyspepsia'

dysphagia Difficulty in swallowing; commonly associated with disorders of the oesophagus

dysphoria Unpleasant mood characterized by an exaggerated feeling of depression and unrest

ectoderm the outermost layer of cells or tissue of an embryo in early development, or the parts derived from this, which include the epidermis and nerve tissue (*see also* endoderm, mesoderm)

ectomorph Body type characterized by slight bone structure and muscle mass (*see also* endomorph, mesomorph)

edentulous Without natural teeth

eicosanoids Compounds formed in the body from long-chain polyunsaturated fatty acids (eicosenoic acids, mainly arachidonic acid) formed by cyclo-oxygenase or lipoxygenase, including the prostaglandins, prostacyclins, thromboxanes, and leukotrienes, all of which act as local hormones and are involved in inflammation, platelet aggregation, and a variety of other functions

eicosapentaenoic acid (EPA) A long-chain polyunsaturated fatty acid (C20:5 ω3)

eicosenoic acids Long-chain polyunsaturated fatty acids with 20 carbon atoms

electrolytes Chemically, salts that dissociate in solution and will carry an electric current; clinically, used to mean the mineral salts of blood plasma and other body fluids, especially sodium and potassium

emaciation Extreme thinness and wasting, caused by disease or undernutrition

embolism Blockage of a blood vessel caused by a foreign object (embolus) in the circulation, such as a quantity of air or gas, a piece of tissue or tumour, a blood clot (thrombus), or fatty tissue derived from atheroma

emetic Substance that causes vomiting

emulsifying agents Substances that are soluble in both fat and water, and enable fat to be uniformly dispersed in water as an emulsion

emulsion An intimate mixture of two immiscible liquids (e.g., oil and water), one being dispersed in the other in the form of fine droplets

enamel Mineralized tissue that forms hard outer surface of tooth

endocrine glands Ductless glands that produce and secrete hormones. Some respond directly to chemical changes in the bloodstream; others are controlled by hormones secreted by the pituitary gland under the control of the hypothalamus

endoderm the innermost layer of cells or tissues of an embryo in early development, or the parts derived from this, which include the lining of the gut and associated structures (*see also* ectoderm, mesoderm)

endomorph In relation to body build, means short and stocky (*see also* ectomorph; mesomorph)

endopeptidases Enzymes that hydrolyse proteins (i.e., proteinases or peptidases) by cleaving peptide bonds within the protein chain, as opposed to exopeptidases that remove amino acids from the ends of the chain

endoplasmic (sarcoplasmic) reticulum Enclosed membranous system with the cytoplasm of cells. It has a number of functions which include acting as an intracellular store for calcium

endothelium Layer of cells lining blood vessels

endotoxins Toxins produced by bacteria as an integral part of the cell, so they cannot be separated by filtration

energy The ability to do work. The SI (Système Internationale) unit of energy is the joule, and nutritionally relevant amounts of energy are kilojoules (kJ, 1000 J) and megajoules (MJ, 1,000,000 J)

energy metabolism The various reactions involved in the oxidation of metabolic fuels (mainly carbohydrates, fats, and proteins) to provide energy (linked to the formation of ATP (adenosine triphosphate) from ADP (adenosine diphosphate) and phosphate ions)

enmeshment A condition where two or more people weave their lives and identities around one another so tightly that it is difficult for any one of them to function independently

enrichment The addition of nutrients to foods. Although often used interchangeably, the term fortification is used for legally imposed additions, and the term enrichment means the addition of nutrients beyond the levels originally present

enteral nutrition Tube feeding with a liquid diet directly into the stomach or small intestine

enteritis Inflammation of the mucosal lining of the small intestine, usually resulting from infection. Regional enteritis is an alternative name for Crohn's disease

enterocolitis Inflammation of the mucosal lining of the small and large intestine, usually resulting from infection

enterocytes The layer of epithelial cells that line the small intestine and are responsible for the absorption of nutrients from the diet

enterogastrone Hormone secreted by the small intestine which inhibits the activity of the stomach. Its secretion is stimulated by fat; hence, fat in the diet inhibits gastric activity

enterohepatic circulation Reabsorption from the small intestine of many of the compounds secreted in bile

enterokinase Obsolete name for enteropeptidase

enteropathy Any disease or disorder of the intestinal tract

enteropeptidase An enzyme secreted by the small intestinal mucosa which activates trypsinogen (from the pancreatic juice) to the active proteolytic enzyme

enterotoxin Substances toxic to the cells of the intestinal mucosa, normally produced by bacteria

enteroviruses Viruses that multiply mainly in the intestinal tract

entitlements The amount of food or other necessities that people can command based on their income and assets, given the legal, political, economic, and cultural context in which they live

E-numbers Within the EU food additives may be listed on labels either by name or by their number (E-number) in the EU list of permitted additives

enzyme A protein that catalyses a metabolic reaction, thus increasing its rate. Enzymes are specific for both the compounds acted on (the substrates) and the reactions carried out

enzyme activation assays Used to assess nutritional status with respect to vitamins B_1, B_2, and B_6. A sample of red blood cells in a test tube is tested for activity of the relevant enzyme before and after adding extra vitamin; enhancement of the enzyme activity beyond a standard level serves as an index of deficiency

enzyme induction Synthesis of new enzyme protein in response to some stimulus, normally a hormone, but sometimes a metabolic intermediate or other compound

enzyme-linked immunosorbent assay (ELISA) Extremely sensitive and specific analytical technique using antibodies linked to an enzyme system to amplify sensitivity

enzyme repression Reduction in synthesis of enzyme protein in response to some stimulus such as a hormone or the presence of large amounts of the end-product of its activity

eosinophil A white blood cell with granules that can be stained by eosin dyes. Eosinophils participate in allergic and hypersensitivity reactions

eosinophilia myalgia syndrome (EMS) A syndrome characterized by debilitating muscle pain and high eosinophil count

epialleles A group of genes that differ in the extent of methylation but are otherwise identical

epigenetics The area of science concerned with heritable changes in gene expression that do not involve changes to the underlying DNA sequence

epigenome the constellation of marks and molecules (including DNA methylation, histone post-translational modifications, and non-coding RNAs) responsible for regulating gene expression

episodic memory A person's memory of specific autobiographical events; the collection of past personal experiences that occurred at a particular time and place

equatorial position Bond to a substituent in a six-membered non-aromatic ring is parallel to one of the bonds in the ring

ergocalciferol Vitamin D_2

ergosterol A sterol isolated from yeast. When treated with ultraviolet light, it is converted to ergocalciferol; the main industrial source of vitamin D

erosion Tooth wear caused by dietary, digestive, or environmental acids

erucic acid A toxic mono-unsaturated fatty acid (C22:1 ω9) found in rape seed (*Brassica napus*) and mustard seed (*Brassica junca* and *Brassica nigra*) oils. Low erucic acid varieties of rape seed (canola) have been developed for food use

eructation The act of bringing up air from the stomach, with a characteristic sound. Also known as belching

erythorbic acid The D-isomer of ascorbic acid; used in foods as an antioxidant but has little vitamin activity

erythrocyte Red blood cell

erythrocyte glutathione reductase activation coefficient (EGRAC) An index of riboflavin status

erythropoiesis The formation and development of the red blood cells in the bone marrow

erythropoietin Hormone that controls erythropoiesis (red blood cell production) in the bone marrow

essential amino acid index (EAA index) An index of protein quality

essential fatty acids (EFAs) Polyunsaturated fatty acids of the ω6 (linoleic acid) and n-3 (linolenic acid) series, which are essential dietary components because they cannot be synthesized in the (human) body. They are essential components of cell membranes; they are also precursors of prostaglandins, prostacyclins, and related hormones and signalling molecules

ester Compound formed by condensation between an acid and an alcohol (e.g., ethyl alcohol and acetic acid yield the ester ethyl acetate). Fats are esters of the alcohol glycerol and long-chain fatty acids. Many esters are used as synthetic flavours

esterases Enzymes that hydrolyse esters (i.e., cleave the ester linkage to form free acid and alcohol). Those that hydrolyse the ester linkages of fats are generally known as lipases, and those that hydrolyse phospholipids as phospholipases; *see* ester

estimated average requirement (EAR) The mean requirement of a group of individuals for a nutrient

ethanol Systematic chemical name for ethyl alcohol

ethanolamine One of the water-soluble bases of phospholipids. Also called 2-aminoethanol

ethylene diamine tetra-acetic acid (EDTA) A compound that forms stable chemical complexes with metal ions (i.e., a chelating agent). Also called versene, sequestrol, and sequestrene

European Food Safety Authority (EFSA) website http://www.efsa.ei.int

European Society for Parenteral and Enteral Nutrition (ESPEN) website http://www.espen.org

eutrophia Normal nutrition

exchange list List of portions of foods in which energy, fat, carbohydrate, and/or protein content are equivalent, thus simplifying meal and diet planning for people with special needs

exclusion diet A limited diet excluding foods known possibly to cause food intolerance, to which foods are added in turn to test for intolerance (allergy)

exon Segment of a gene that is represented in the mature mRNA product. Contains the code for producing the gene's protein. Each exon codes for a specific portion of the complete protein, separated by introns that have no apparent function (*see* intron)

exopeptidases Proteolytic enzymes that hydrolyse the peptide bonds of the terminal amino acids of proteins or peptides (*see* peptides)

exothermic Chemical reactions that proceed with the output of heat

exotoxins Toxic substances produced by bacteria which diffuse out of the cells and stimulate the production of antibodies that specifically neutralize them (antitoxins)

extremophiles Micro-organisms that can grow under extreme conditions of heat (thermophiles and extreme thermophiles, some of which live in hot springs at 100°C) or cold (psychrophiles), in high concentrations of salt (halophiles), high pressure, or extremes of acid or alkali

extrinsic factor Protein secreted in the stomach that is required for the absorption of vitamin B_{12}

F$_2$-isoprostanes Oxidation products of lipids which are considered to be useful indices of oxidative damage (e.g., by oxygen free radicals)

faeces Body waste, composed of undigested food residues, remains of digestive secretions that have not been reabsorbed, bacteria from the intestinal tract, cells, cell debris and mucus from the intestinal lining, and substances excreted into the intestinal tract (mainly in the bile)

famine A catastrophic disruption of the social, economic, and institutional systems that provide for food production, distribution, and consumption

fasting Going without food. The metabolic fasting state begins some 4 hours after a meal, when the digestion and absorption of food are complete and body reserves of fat and glycogen begin to be mobilized

fat Chemically, fats (or lipids) are substances that are insoluble in water but soluble in organic solvents such as ether, chloroform, and benzene, and are actual or potential esters of fatty acids. The term includes triacylglycerols (triglycerides), phospholipids, waxes, and sterols. In more general use the term 'fats' refers to the neutral fats, which are triacylglycerols, mixed esters of fatty acids with glycerol. *See* fatty acids, glycerol

fat, neutral Fats that are esters of fatty acids with glycerol, triacylglycerols

fat-replacers Substances that provide a creamy fat-like texture used to replace or partly replace the fat in a food. Made from a variety of substances with lower energy content

fatty acids Organic acids consisting of carbon chains with a carboxyl group at the end. The nutritionally important fatty acids have an even number of carbon atoms. In addition to their accepted names, fatty acids can be named by a shorthand giving the number of carbon atoms in the molecule (e.g., C18), then a colon and the number of double bonds (e.g., C18:2), followed by the position of the first double bond from the methyl end of the molecule as n- or ω (e.g., C18:2 n-6, or C18:2 ω6)

fecundability *See* fertility

Fenton-type reactions The formation of hydroxyl radical (OH–) from hydrogen peroxide reacting with iron (II) or copper (I) ions in a process first observed by Fenton in 1894

ferritin An iron-carrier protein found in large amounts in liver and spleen, and also at low concentrations in serum, which can be measured as an index of iron stores in the body. However, it is also sensitive to the acute phase reaction. Its concentration in serum increases markedly during infection or inflammation; *see* acute phase reaction

fertility The ability to reproduce, which differs from fecundability, which is the ability of a woman to become pregnant

fibre, crude The term given to the indigestible part of foods, defined as the residue left after successive extraction under closely specified conditions with petroleum ether, 1.25 per cent sulphuric acid, and 1.25 per cent sodium hydroxide, minus ash. No real relation to dietary fibre

fibre, dietary *See* dietary fibre

fibrin The blood protein formed from fibrinogen which is responsible for the clotting of blood

Fischer projection formula Simple method of representing the configuration of stereogenic centres of chemical structures as the intersection of vertical and horizontal lines

flatulence (flatus) Production of gas (hydrogen, carbon dioxide, and methane) by bacteria in the large intestine. May be caused by a variety of foods that are incompletely digested in the small intestine

flavin The group of compounds containing the iso-alloxazine ring structure, as in riboflavin (vitamin B$_2$), and hence a general term for riboflavin derivatives

flavin adenine dinucleotide (FAD) One of the coenzymes formed from vitamin B$_2$ (riboflavin)

flavin mononucleotide (FMN) One of the coenzymes derived from vitamin B$_2$ (chemically, riboflavin phosphate)

flavonoids Widely found plant pigments; glycosides of flavones; the sugar moiety may be either rhamnose or rhamnoglucose. Some have pharmacological actions, but they are not known to be dietary essentials although they make a contribution to the total antioxidant intake, and some are phytoestrogens

flavoproteins Enzymes that contain the vitamin riboflavin, or a derivative such as flavin adenine dinucleotide or riboflavin phosphate, as the prosthetic group. Mainly involved in oxidation reactions in metabolism

fluoroapatite Crystal incorporating fluoride that forms tooth enamel

fluorosis Damage to teeth (white to brown mottling of the enamel) and bones caused by an excessive intake of fluoride

foam cells Macrophages that have accumulated very large amounts of cholesterol as a result of uptake of (chemically modified) low density lipoprotein. They infiltrate arterial walls and lead to the development of fatty streaks and eventually atherosclerosis

folinic acid The 5-formyl derivative of the vitamin folic acid; more stable to oxidation than folic acid itself, and commonly used in pharmaceutical preparations. The synthetic (racemic) compound is known as leucovorin

food, foodstuffs Any solid or liquid material consumed by a living organism to supply energy, build and replace tissue, or participate in such reactions. Defined by the FAO/WHO Codex Alimentarius Commission as a substance, whether processed, semi-processed, or raw, which is intended for human consumption, and includes drink, chewing gum, and any substance that has been used in the manufacture, preparation, or treatment of food, but does not include cosmetics, tobacco, or substances used only as drugs. Defined in EU directives as products intended for human consumption in an unprocessed, processed, or mixed state, with the exception of tobacco products, cosmetics, and pharmaceuticals

Food and Agriculture Organization of the United Nations (FAO) Founded in 1943; headquarters in Rome. Its goal is to achieve freedom from hunger worldwide. According to its constitution the specific objectives are 'raising the levels of nutrition and standards of living … and securing improvements in the efficiency of production and distribution of all food and agricultural products'; website http://www .fao.org

Food and Drug Administration (FDA) US government regulatory agency; website http://www.fda.gov; website for FDA consumer magazine http://www.fda .gov/fdac

Food and Nutrition Information Center (FNIC) Located at the National Agricultural Library, part of the US Department of Agriculture; website http://www.nal .usda.gov/fnic

food chain The chain between green plants (the primary producers of food energy) through a sequence of organisms in which each eats the one below it in the chain and is eaten in turn by the one above. Also used for the chain of events from the original source of a foodstuff (from the sea, the soil, or the wild) through all the stages of handling until it reaches the table

food exchange *See* exchange list

food insecurity The absence of food security

food poisoning May be due to (1) contamination with harmful bacteria or other micro-organisms; (2) toxic chemicals; (3) adverse reactions to certain proteins or other natural constituents of foods; (4) chemical contamination. The commonest bacterial contamination is due to species of *Salmonella*, *Staphylococcus*, *Campylobacter*, *Listeria*, *Bacillus cereus*, and *Clostridium welchii*. Very rarely, food poisoning is due to *Clostridium botulinum* (botulism)

food pyramid A way of showing a healthy diet graphically by grouping foods and the amounts of each group that should be eaten each day based on nutritional recommendations

food science The study of the basic chemical, physical, biochemical, and biophysical properties of foods and their constituents, and of changes that these may undergo during handling, preservation, processing, storage, distribution, and preparation for consumption

food security When all people, at all times, have physical and economic access to sufficient safe and nutritious food to meet their dietary needs and food preferences for an active and healthy life. Food security requires more than just adequate food availability; it also is a matter of access to the food that is available and appropriate utilization

Food Standards Agency (FSA) Permanent advisory body to the UK Parliament through Health Ministers, established in 2000 to protect the public's health and consumer interests in relation to food; website http:// www.food.gov.uk

food technology The application of science and technology to the treatment, processing, preservation, and distribution of foods. Hence the term food technologist

Foods for Specified Health Use (FOSHU, Japanese) Processed foods containing ingredients that aid specific bodily functions, as well as being nutritious (functional foods)

formiminoglutamic acid (FIGLU) test A test for folate nutritional status, based on excretion of formiminoglutamic acid (FIGLU), a metabolite of the amino acid histidine which is normally metabolized by a folic acid dependent enzyme

formula diet Diet composed of simple substances that require little digestion, are readily absorbed, and leave a minimum residue in the intestine: glucose, amino acids or peptides, and mono- and diacylglycerols, rather than starch, proteins, and fats

fortification The deliberate addition of specific nutrients to foods as a means of providing the population with an increased level of intake

fractional test meal A method of examining the secretion of gastric juices; the stomach contents are sampled at intervals via a stomach tube after a test meal of gruel

free radicals Highly reactive chemical species with one or more unpaired electrons

free sugars Mono- and disaccharides added to food, plus sugars in fruit juices, honey, and syrup

freeze-drying Also known as lyophilization. A method of drying in which the material is frozen and subjected to a high vacuum. The ice sublimes off as water vapour without melting. Freeze-dried food is very porous, since it occupies the same volume as the original and so rehydrates rapidly. There is less loss of flavour and texture than with most other methods of drying. Controlled heat may be applied to the process without melting the frozen material; this is accelerated freeze-drying

fructo-oligosaccharides Oligosaccharides consisting of fructose

fructosan A general name for polysaccharides of fructose, such as inulin. Not digested, and hence a part of dietary fibre or non-starch polysaccharides

fructose Also known as fruit sugar or laevulose. A six-carbon monosaccharide sugar (hexose) differing from glucose in containing a ketone group (on carbon-2) instead of an aldehyde group (on carbon-1)

fruit The fleshy seed-bearing part of plants (including tomato and cucumber, which are usually called vegetables)

fruitarian A person who eats only fruits, nuts, and seeds; an extreme form of vegetarianism

functional foods Foods eaten for specified health purposes because of their (rich) content of one or more nutrients or non-nutrient substances which may confer health benefits

galactans Polysaccharides composed of galactose derivatives

galactose A six-carbon monosaccharide differing from glucose only in the position of the hydroxyl group on carbon-4

gall bladder The organ situated in the liver which stores the bile formed in the liver before its secretion into the small intestine

gallstones (cholelithiasis) Concretions composed of cholesterol, bile pigments, and calcium salts formed in the bile duct of the gall bladder when the bile becomes supersaturated

gamma carboxyglutamate (γ-carboxyglutamate) A derivative of the amino acid glutamate found in prothrombin and other enzymes involved in blood clotting, and the proteins osteocalcin and matrix GLA protein (MGP) in bone, where it has a function in ensuring the correct crystallization of bone mineral. Its synthesis requires vitamin K

gastric inhibitory peptide (GIP) A hormone secreted by the mucosa of the duodenum and jejunum in response to absorbed fat and carbohydrate, which stimulates the pancreas to secrete insulin. Also known as glucose-dependent insulinotropic polypeptide

gastrin Polypeptide hormone secreted by the stomach in response to food, which stimulates gastric and pancreatic secretion

gastritis Inflammation of the mucosal lining of the stomach; may result from infection or excessive alcohol consumption. Atrophic gastritis is the progressive loss of gastric secretion with increasing age

gastroenteritis Inflammation of the mucosal lining of the stomach and/or small or large intestine, normally resulting from infection

gastroenterology The study and treatment of diseases of the gastrointestinal tract

gastroplasty Surgical alteration of the shape and capacity of the stomach, without removing any part. Has been used as a treatment for severe obesity

gastrostomy feeding Feeding a liquid diet directly into the stomach through a tube that has been surgically introduced through the abdominal wall. *See also* enteral nutrition, nasogastric tube

gavage The process of feeding liquids by tube directly into the stomach

gene Physical and functional unit of heredity. It is the entire DNA sequence necessary for the synthesis of a functional polypeptide or RNA molecule; *see* DNA, RNA

gene expression Overall process by which information encoded by a gene is converted to an observable phenotype, usually in the form of a protein; *see* phenotype

gene transcription Conversion of genomic DNA into mRNA; *see* DNA, mRNA

gene translation The synthesis of protein molecules using the triplet code of mRNA; *see* mRNA

generally regarded as safe (GRAS) Designation given to food additives when further evidence is required before the substance can be classified more precisely (US usage)

generic descriptor The name used to cover the different chemical forms of a vitamin that have the same biological activity

genetic polymorphisms Changes (i.e., variability) in one or more base pairs in the DNA gene sequence encoding a specific protein, resulting in the substitution of different amino acids in that protein which may subtly alter its function. Such polymorphisms in human populations are of increasing research interest because in some cases they may alter requirements for certain nutrients

genome The complete genetic information of an organism; *see* proteome, metabolome

genomics The study of the structure and function of genes; *see* proteomics, metabolomics

genotype The entire genetic constitution of an individual cell or organism

ghrelin A peptide hormone secreted by cells in the gastrointestinal tract that both stimulates the secretion of growth hormone and regulates feeding behaviour and energy balance by acting on the hypothalamus

giardiasis Intestinal inflammation and diarrhoea caused by infection with the protozoan parasite *Giardia lamblia*

gingivitis Inflammation of the gums (gingiva)

gliadin One of the proteins that make up wheat gluten. Allergy to, or intolerance of, gliadin is coeliac disease

glial cells Non-neuronal cells that surround neurons and provide support for and insulation between them

globins Proteins that are rich in the amino acid histidine (and hence basic), relatively deficient in isoleucine, and contain average amounts of arginine and tryptophan. Often found as the protein part of conjugated proteins such as haemoglobin

globulins Globular (as opposed to fibrous) proteins that are relatively insoluble in water, but soluble in dilute salt solutions

glomerular filtration rate The rate at which fluid is filtered through the kidney at the renal glomerulus, usually expressed in ml/min. It increases in normal pregnancy and is decreased in renal failure

glossitis Inflammation of the tongue; may be one of the signs of riboflavin deficiency

glucagon A hormone secreted by the α-cells of the pancreas which causes an increase in blood glucose by increasing the breakdown of liver glycogen and stimulating gluconeogenesis

glucans Soluble but undigested complex carbohydrates; found particularly in oats, barley, and rye

glucocorticoids Steroid hormones secreted by the adrenal cortex which regulate carbohydrate metabolism

glucomannan A polysaccharide consisting of glucose and mannose

gluconeogenesis Synthesis of glucose from non-carbohydrate precursors, such as glycerol, lactate, and a variety of amino acids

gluco-oligosaccharide A non-digestible oligosaccharide of glucose containing alpha-1,2 and alpha-1,6 glycosidic links; *see* oligosaccharides

glucosan A general term for polysaccharides of glucose, such as starch, cellulose, and glycogen; *see* polysaccharides

glucose A six-carbon monosaccharide sugar (hexose), with the chemical formula $C_6H_{12}O_6$, occurring free in plant and animal tissues and formed by the hydrolysis of starch and glycogen. Also known as dextrose, grape sugar, and blood sugar

glucose polymers Oligosaccharides of glucose linked with alpha-1,4 and alpha-1,6 glycosidic links

glucose syrup A type of glucose polymer

glucose tolerance The ability of the body to deal with a relatively large dose of glucose is used to diagnose diabetes mellitus

glucose tolerance factor (GTF) An organic complex containing chromium

glucosides Complexes of substances with glucose. The general name for such complexes with other sugars is glycosides

glucosinolates Substances occurring widely in plants of the genus *Brassica* (e.g., broccoli, Brussel sprouts, cabbage); broken down by the enzyme myrosinase to yield, among other products, the mustard oils that are responsible for the pungent flavour (especially in mustard and horseradish). There is evidence that the various glucosinolates in vegetables may have useful anti-cancer activity, since they increase the rate at which a variety of potentially toxic and carcinogenic compounds are conjugated and excreted

glucosuria (glycosuria) Appearance of glucose in the urine, as in diabetes and after the administration of drugs that lower the renal threshold

glucuronides A variety of compounds that are metabolized by conjugation with glucuronic acid to yield water-soluble derivatives for excretion from the body

glutamate Salts of glutamic acid

glutamic acid A non-essential amino acid

glutamine A non-essential amino acid, the amide of glutamic acid

glutathione A tripeptide (γ-glutamyl-cysteinyl-glycine) which is involved in oxidation–reduction reactions, the conjugation of foreign substances for excretion, and the transport of some amino acids into cells; *see* peptides

glutathione peroxidases A group of selenium-containing enzymes that protect tissues from oxidative damage by removing peroxides resulting from free-radical action, linked to oxidation of reduced glutathione; part of the body's antioxidant protection; often used as a biochemical index of selenium status

glutathione reductase Enzyme in red blood cells for which flavin adenine dinucleotide (FAD), derived from vitamin B_2, is the essential cofactor. It converts oxidized to reduced glutathione, with reduced nicotinamide dinucleotide phosphate (NADPH) as co-substrate. Activation of this enzyme *in vitro* from red cell extracts by added FAD provides a means of assessing vitamin B_2 nutritional status, sometimes known as the erythrocyte glutathione reductase activation coefficient (EGRAC) test

gluten The protein complex in wheat, and to a lesser extent rye, which gives dough the viscid property that holds gas when it rises. There is none in oats, barley, or maize. It is a mixture of two proteins, gliadin and glutelin. Allergy to, or intolerance of, gliadin gluten is coeliac disease

gluten-free foods Formulated without any wheat or rye protein (although the starch may be used) for people suffering from coeliac disease; *see* coeliac disease

gluten-sensitive enteropathy Coeliac disease

glycaemic index The ability of a carbohydrate to increase blood glucose compared with an equivalent amount of glucose. Glycaemic load is the product of multiplying the amount of carbohydrate in the food by the glycaemic index

glycation A non-enzymic reaction between glucose and amino groups in proteins resulting in formation of a glycoprotein; the basis of many of the adverse effects of poor glycaemic control in diabetes

glycerides Esters of glycerol with fatty acids

glycerol A trihydric alcohol ($CH_2OH–CHOH–CH_2OH$), also known as glycerine. Simple or neutral fats are esters of glycerol with three molecules of fatty acid, i.e., triacylglycerols, sometimes known as triglycerides

glycine A non-essential amino acid, chemically the simplest of the amino acids, it is amino-acetic acid, CH_2NH_2COOH

glycoalkaloids A group of nitrogen-containing compounds which are naturally produced by Solanaceae plant family, which includes potatoes, tomatoes and aubergines

glycocholic acid One of the bile acids

glycogen The storage carbohydrate in the liver and muscles, a branched polymer of glucose units

glycogenolysis The breakdown of glycogen to glucose for use as a metabolic fuel and to maintain the normal blood concentration of glucose in the fasting state. Stimulated by the hormones glucagon and adrenaline (epinephrine)

glycolysis The first sequence of reactions in glucose metabolism, leading to the formation of two molecules of pyruvic acid from each glucose molecule

glycoproteins Proteins conjugated with carbohydrates such as uronic acids, polymerized glucosamine-mannose etc., including mucins and mucoids

glycosides Compounds of a sugar attached to another molecule. When the sugar is glucose, they are called glucosides

glycosidic Ether-type bond formed by the hydroxyl group of a sugar displacing the hydroxyl group of a second sugar or other molecule

goitre Enlargement of the thyroid gland, seen as a swelling in the neck; may be hypothyroid, with low production of hormones, euthyroid (normal levels of the hormones), or hyperthyroid

goitrogens Substances found in foods (especially *Brassica* spp. but also including groundnuts, cassava, and soya bean) which interfere with the synthesis of thyroid hormones (glucosinolates) or the uptake of iodide into the thyroid gland (thiocyanates), and hence can cause goitre, especially when the dietary intake of iodide is marginal

Gomez classification One of the earliest systems for classifying protein–energy malnutrition in children, based on percentage of expected weight for age: over 90 per cent is normal, 76–90 per cent is mild (first-degree) malnutrition, 61–75 per cent is moderate (second-degree) malnutrition and less than 60 per cent is severe (third-degree) malnutrition

gout Painful disease caused by accumulation of crystals of uric acid in the synovial fluid of joints; may be due to excessive synthesis and metabolism of purines, which are metabolized to uric acid, or to impaired excretion of uric acid

Gram-negative, Gram-positive A method of classifying bacteria depending on whether or not they retain crystal-violet dye (Gram stain) after staining and decolorizing with alcohol

Green Revolution A process in which cereal crop yields increased as a result of farmer's adopting high-yielding varieties bred at international agricultural research centres and adapted to local conditions at national agricultural research institutions. Planting of these varieties has coincided with expansion of irrigated area and fertilizer use

growth faltering A term indicating that a child's linear or ponderal growth is falling away from the reference trend. This implies that weight or height has been measured at intervals over a period of time

growth hormone Somatotrophin, a peptide hormone secreted by the pituitary gland that promotes growth of bone and soft tissues. It also reduces the utilization of glucose and increases breakdown of fats to fatty acids; because of this it has been promoted as an aid to weight reduction, with little evidence of efficacy. Sometimes abbreviated to hGH (human growth hormone); growth hormone from other mammals differs in structure and activity

guanine One of the purines; *see* purines

gum Substances that can disperse in water to form a viscous mucilaginous mass. They may be extracted from seeds, plant sap, and seaweeds, or they may be made from starch or cellulose. Most (apart from dextrins) are not digested and have no food value, although they contribute to the intake of non-starch polysaccharides

haem The iron-containing pigment which forms the oxygen-binding site of haemoglobin and myoglobin. It is also part of a variety of other proteins, collectively known as haem proteins, including the cytochromes; *see* haemoglobin

haemagglutinins *See* lectins

haematemesis Vomiting bright red blood, due to bleeding in the upper gastrointestinal tract

haematinic General term for those nutrients, including iron, folic acid, and vitamin B_{12}, required for the formation and development of blood cells in bone marrow (the process of haematopoiesis), deficiency of which may result in anaemia

haemochromatosis Iron overload; excessive absorption, and storage of iron in the body, commonly the result of a genetic defect. In most cases it is caused by a recessive gene, i.e., it can only be passed on if both parents are carriers of the gene predisposing to the disorder. Around one in seven people in northern Europe are carriers of the recessive gene. Homozygotes are susceptible to iron toxicity from high absorption of dietary iron, which can lead to tissue damage (including liver cancer, heart disease, and diabetes) and bronze coloration of the skin. Sometimes called bronze diabetes. The disorder is usually treated by regular venesection, a procedure similar to blood donation, where around 500 ml of blood is removed

haemoglobin The red haem-containing protein in red blood cells which is responsible for the transport of oxygen and carbon dioxide in the bloodstream

haemorrhoids (or piles) Varicosity in the lower rectum or anus due to congestion of the veins; caused or exacerbated by a low fibre diet and consequent straining to defecate

haemosiderin An iron-storage protein

halal Food conforming to the Islamic (Muslim) dietary laws. Meat from permitted animals (in general grazing animals with cloven hooves, thus excluding pig meat) and birds (excluding birds of prey). The animals are killed under religious supervision by cutting the throat to allow removal of all blood from the carcass, without prior stunning. Food that is not halal is haram

halophiles (halophilic bacteria) Bacteria and other micro-organisms able to grow in high concentrations of salt

haplotype Collection of single-nucleotide polymorphs (SNPs) which are inherited as a group; *see* SNP

haram Food forbidden by Islamic law (opposite of halal)

Harvard standard Tables of height and weight for age used as reference values for the assessment of growth and nutritional status in children, based on data collected in the USA in the 1930s. Now largely replaced by the NCHS (US National Center for Health Statistics) standards

hazard analysis critical control process (HACCP) The identification of critical process points that must be controlled to produce safe foods

health foods Substances whose consumption is advocated by various reform movements, including vegetable foods, whole grain cereals, food processed without chemical additives, food grown on organic compost, supplements such as bees' royal jelly, lecithin, seaweed, etc., and pills and potions. Numerous health claims are made, but there is rarely evidence to support them

heartburn A burning sensation in the chest, usually caused by reflux (regurgitation) of acid digestive juices from the stomach into the oesophagus. A common form of indigestion, treated by antacids

heavy metals Naturally-occurring metallic elements that have a high density compared with water. They include essential nutrients (such as iron, selenium and zinc), as well as systemic toxicants such as mercury and lead

heat of combustion Energy released by complete combustion, as, for example, in the bomb calorimeter. Values can be used to predict energy physiologically available from foods only if an allowance is made for material not completely oxidized in the body

hedonic scale Term used in tasting panels where the judges indicate the extent of their like or dislike of the food

Hegsted score Method of expressing the lipid content of a diet, calculated as $2.16 \times$ % energy from saturated fat $- 1.65 \times$ % energy from polyunsaturated fat $- 0.0677 \times$ mg cholesterol. *See also* Keys score

Helicobacter pylori Bacterium commonly infecting the gastric mucosa in patients with ulcers; the underlying cause of ulcers, and implicated in the development of gastric cancer

helminths and nematodes Helminth is a general term meaning worm. The helminths are invertebrate parasites characterized by elongated, flat or round bodies, that develop through egg, larval (juvenile), and adult stages. The flatworms include flukes and tapeworms. Roundworms are nematodes. These groups are subdivided according to the host organ in which they reside, e.g., lung flukes, extraintestinal tapeworms, and intestinal roundworms

hemicelluloses Complex carbohydrates included as dietary fibre, composed of polyuronic acids combined with xylose, glucose, mannose, and arabinose. Found together with cellulose and lignin in plant cell walls; most gums and mucilages are hemicelluloses; *see* dietary fibre, cellulose, lignin

hepatitis Inflammatory liver disease, characterized by jaundice, abdominal pain, and anorexia. May be due to bacterial or viral infection, alcohol abuse, or various toxins

hepatomegaly Enlargement of the liver as a result of congestion (e.g., in heart failure), inflammation, or fatty infiltration (as in kwashiorkor)

hesperidin A flavonoid found in the pith of unripe citrus fruits; a complex of glucose and rhamnose with the flavonone hesperin; *see* flavonoids

hexoses Six-carbon monosaccharides such as glucose or fructose

hiatus hernia Protrusion of a part of the stomach upwards through the diaphragm. The condition occurs in about 40 per cent of the population; most people suffer no ill effects, but in a small number there is reflux of stomach contents into the oesophagus, causing heartburn

hippocampus The elongated ridges on the floor of each lateral ventricle of the brain, thought to be the centre of emotion, memory, and the autonomic nervous system

histamine The amine formed by decarboxylation of the amino acid histidine. Excessive release of histamine from mast cells is responsible for many of the symptoms of allergic reactions. It also stimulates secretion of gastric acid, and administration of histamine provides a test for achlorhydria; *see* achlorhydria

histidine An essential amino acid with a basic side chain

histones Proteins rich in arginine and lysine that occur mainly in the cell nucleus and are concerned with the regulation of DNA; *see* DNA

holoenzyme An enzyme protein together with its coenzyme or prosthetic group

homeostasis (homeostatic) The control of concentration of key components (in blood etc.) to ensure constancy and physiological normalization of their concentrations

homocysteine A sulphur amino acid formed as an intermediate in the metabolism of methionine; it is demethylated methionine. Normally present at only low concentrations (e.g., less than $10 \mu M$ in serum or plasma). High blood concentrations of homocysteine (occurring as a result of poor folic acid, vitamin B_6, and B_{12} status, and in certain other dietary and medical situations) have been implicated in the development of atherosclerosis, heart disease, and stroke

homogenization Emulsions usually consist of a suspension of globules of varying size; homogenization reduces these globules to a smaller and more uniform size

hormones Compounds produced in the body in endocrine glands and released into the bloodstream, where they act as chemical messengers to affect other tissues and organs

Human Genome Project International collaboration of laboratories to decipher the DNA sequence of the entire human genome

human immunodeficiency virus (HIV) The virus which causes AIDS (acquired immune deficiency syndrome). In many developing countries this virus is acquired by heterosexual intercourse, and may also be acquired 'vertically', either at the time of birth or in breast milk (around 25–30 per cent of all vertical transmission is by breastfeeding mothers). The level of access to specific treatment is still much lower in poor countries

hunger A condition in which people lack the basic food intake to provide them with the energy and nutrients for fully productive and active lives; it is an outcome of food insecurity

hybridization Formation of duplexes by RNA and/or DNA sequences (can be DNA–DNA, RNA–RNA, or DNA–RNA) *see* DNA, RNA

hydrogenated oils Liquid oils hardened by hydrogenation

hydrogenation Conversion of liquid oils to semi-hard fats by the addition of hydrogen to the unsaturated double bonds

hydrolyse (hydrolysis) To split a complex compound into its constituent parts by the action of water, either enzymically or catalysed by the addition of acid or alkali

hydroxyapatite The calcium phosphate complex which is the main mineral of bones and teeth; the crystalline forms make up dental enamel

hydroxylysine Amino acid found only in connective tissue proteins (collagen and elastin); incorporated into the protein as lysine and then hydroxylated in a vitamin C dependent reaction

hydroxymethylglutaryl CoA (HMG CoA) reductase inhibitors Drugs that inhibit the enzyme HMG CoA reductase, the controlling enzyme of cholesterol synthesis, used in the treatment of hypercholesterolaemia

hydroxyproline Amino acid found mainly in connective tissue proteins (collagen and elastin); incorporated into the protein as proline and then hydroxylated in a vitamin C dependent reaction. Peptides of hydroxyproline are excreted in the urine and the output is increased when collagen turnover is high, as in rapid growth or resorption of tissue

hydroxyproline index Urinary excretion of hydroxyproline is reduced in children suffering protein–energy malnutrition. The index is the ratio of urinary hydroxyproline to creatinine per kilogram of body weight, and is low in malnourished children

25-hydroxyvitamin D The most abundant vitamin D derivative, found mainly in blood serum (plasma), which is usually measured to define vitamin D status. It is the precursor of the hormone, 1,25-dihydroxyvitamin D, which is formed in the kidney by an enzyme which is homeostatically controlled so as to liberate the appropriate amount of the hormone required for the correct control of calcium transport

hyper- Prefix meaning above the normal range, or abnormally high

hyperalimentation Provision of unusually large amounts of energy, either intravenously (parenteral nutrition) or by nasogastric or gastrostomy tube (enteral nutrition)

hypercalcaemia, idiopathic Elevated plasma concentrations of calcium believed to be due to hypersensitivity of some children to vitamin D toxicity. There is excessive absorption of calcium, with loss of appetite, vomiting, constipation, flabby muscles, and deposition of calcium in the soft tissues and kidneys. It can be fatal in infants

hyperchlorhydria Excess secretion of hydrochloric acid in the stomach due to secretion of a greater volume of gastric juice rather than a higher concentration

hypercholesterolaemia Abnormally high blood concentrations of cholesterol. Generally considered to be a sign of high risk for atherosclerosis and ischaemic heart disease

hyperglycaemia Elevated plasma concentration of glucose caused by a failure of the normal hormonal mechanisms of blood glucose control

hyperinsulinism Excessive secretion of insulin, resulting in hypoglycaemia

hyperkalaemia Excessively high blood concentration of potassium

hyperkinetic syndrome (hyperkinesis) Mental disorder of children, characterized by excessive activity and impaired attention and learning ability. Has been attributed to adverse reactions to food additives, but there is little evidence for this

hyperlipidaemia (hyperlipoproteinaemia) A variety of conditions in which there are increased concentrations of lipids in plasma: phospholipids, triacylglycerols, free and esterified cholesterol, or unesterified fatty acids

hyperphosphataemia Excessively high blood concentration of phosphate

hypersalivation Excessive flow of saliva

hypersensitivity Heightened responsiveness induced by allergic sensitization. There are several types of response including that associated with allergy

hypertension High blood pressure; a risk factor for ischaemic heart disease, stroke, and kidney disease. May be due to increased sensitivity to sodium

hypertonic A solution more concentrated than the body fluids; *see* isotonic, hypotonic

hypervitaminosis Toxicity due to excessively high intakes of vitamins

hypo- Prefix meaning below the normal range, or abnormally low

hypocalcaemia Low blood calcium, leading to vomiting and uncontrollable twitching of muscles if severe; may be due to underactivity of the parathyroid gland, kidney failure, or vitamin D deficiency

hypochlorhydria Partial deficiency of hydrochloric acid secretion in the gastric juice

hypogeusia Diminished sense of taste; an early sign of marginal zinc deficiency, and potentially useful as an index of zinc status

hypoglycaemia Abnormally low concentration of plasma glucose; may result in loss of consciousness—hypoglycaemic coma

hypoglycaemic agents Drugs used to lower blood glucose concentrations in diabetes mellitus

hypokalaemia Abnormally low plasma potassium

hypophosphataemia Abnormally low blood concentration of phosphate

hypoplasia (of enamel) poorly developed, under-mineralized enamel

hypoproteinaemia Abnormally low plasma protein concentration

hypothermia Low body temperature (normal is around 37°C)

hypothyroidism Underactivity of the thyroid gland, leading to reduced secretion of thyroid hormones and a reduction in basal metabolic rate; commonly associated with goitre due to iodine deficiency

hypotonic A solution more dilute than the body fluids

hypovitaminosis Vitamin deficiency

iatrogenic A condition caused by medical intervention or drug treatment; iatrogenic nutrient deficiency is due to drug–nutrient interactions

idiosyncrasy Unusual and unexpected sensitivity or reaction to a drug or food

ileitis Inflammation of the ileum

ileostomy Surgical formation of an opening of the ileum on the abdominal wall; performed to treat severe ulcerative colitis

ileum Last portion of the small intestine, between the jejunum and the colon (large intestine)

ileus Obstruction of the intestines

immunoglobulin A member of a family of proteins from which antibodies are derived. There are five main classes in humans: IgM, IgG, IgA, IgD, and IgE

in vitro Literally 'in glass'; used to indicate an observation made experimentally in the test tube, as distinct from the natural living conditions (*in vivo*)

in vivo In the living state, as distinct from *in vitro*

inanition Exhaustion and wasting due to complete lack or non-assimilation of food; a state of starvation

index of nutritional quality (INQ) An attempt to provide an overall figure for the nutrient content of a food or a diet. It is the ratio between the percentage of the reference intake of each nutrient and the percentage of the average requirement for energy provided by the food

indigestion Discomfort and distension of the stomach after a meal (also known as dyspepsia), including heartburn

infarction Death of an area of tissue because its blood supply has been stopped

ingredient Any substance used in the manufacture or preparation of a foodstuff and still present in the finished product, even if in an altered form. Contaminants and adulterants are not considered to be ingredients

inorganic Materials of mineral, as distinct from animal or vegetable, origin. Apart from carbonates and cyanides, inorganic chemicals are those that contain no carbon. *See also* organic

inosine monophosphate (IMP) One of the purine nucleotides; *see* purines, nucleotide

inositol A carbohydrate derivative, a constituent of phospholipids (phosphatidyl inositols) involved in membrane structure and as part of the signalling mechanism for hormones which act at the cell surface

insulin Polypeptide hormone that regulates carbohydrate metabolism

insulin resistance Changes in the biological activity of insulin-sensitive peripheral tissues, which result in reduced disposal of nutrients such as glucose from the plasma for any given concentration of insulin

insulinaemic index The rise in blood insulin elicited by a test dose of a carbohydrate food compared with that after an equivalent dose of glucose

interferon One of a family of naturally occurring proteins produced by the cells of the immune system, attacking viruses, bacteria, tumours, and other foreign substances

integrated pest management (IPM) A flexible approach to pest management which draws on a range of methods to produce a result that combines the greatest value to the farmer with environmentally acceptable and sustainable outcomes

interleukins (ILs) Soluble polypeptide mediators produced by activated lymphocytes and other cells during immune and inflammatory response, such as IL1, IL6 (*see* cytokines)

International Network of Food Data Systems (INFOODS) Created to develop standards and guidelines for collection of food composition data, and standardized terminology and nomenclature; website http://www.fao.org/infoods

International Union of Food Science and Technology (IUFoST) website http://www.iufost.org

International Union of Nutritional Sciences (IUNS) website http://www.iuns.org

international units (IU) Used as a measure of comparative potency of natural substances, such as vitamins, before they were obtained in a sufficiently pure form to measure by weight

intervention study Comparison of an outcome (e.g., morbidity or mortality) between two groups of people deliberately subjected to different dietary or drug regimes

intestinal flora Bacteria and other micro-organisms that are normally present in the gastrointestinal tract

intestinal juice Also called succus entericus. Digestive juice secreted by the intestinal glands lining the small intestine, containing a variety of enzymes

intestinal polyposis Appearance of polyps (growths) on the surface of the intestine, mainly in the rectum and large intestine, which may in some cases be precursors of cancerous growths. Their measurement may be potentially useful as a functional index of pro-carcinogenic versus anti-carcinogenic influences, some of which may be nutrient-sensitive

intestine The gastrointestinal tract; more specifically the part after the stomach, i.e., the small intestine (duodenum, jejunum, and ileum), where the greater part of digestion and absorption takes place, and the large intestine

intrinsic factor A protein secreted in the gastric juice which is required for the absorption of vitamin B_{12}; impaired secretion results in pernicious anaemia

intron Segment of a gene that is transcribed but then removed from the primary transcript by splicing and so is not present in the mature mRNA product. Non-coding sequence of DNA between exons; *see* exon

inulin Soluble but undigested fructose polymer found in root vegetables. Also called dahlin and alant starch (although it is a non-starch polysaccharide)

inversion Applied to sucrose, means its hydrolysis to glucose and fructose (invert sugar)

invert sugar The mixture of glucose and fructose produced by hydrolysis of sucrose, 1.3 times sweeter than sucrose. So called because the optical activity is reversed in the process

invertase Enzyme that hydrolyses sucrose to glucose and fructose (invert sugar); also called sucrase and saccharase

iodine number (iodine value) Carbon–carbon double bonds in unsaturated compounds can react with iodine; this provides a means of determining the degree of unsaturation of a fat or other compound by the uptake of iodine

iodized salt Usually 1 part of iodate in 25,000–50,000 parts of salt, as a means of ensuring adequate iodine intake in regions where deficiency is a problem

ion An atom or molecule that has lost or gained one or more electrons, and thus has an electric charge.

Positively charged ions are known as cations because they migrate towards the cathode (negative pole) in solution, while negatively charged ions (anions) migrate towards the positive pole (anode)

ion-exchange resin An organic compound that will adsorb ions under some conditions and release them under others

ionization The process whereby the positive and negative ions of a salt or other compound separate when dissolved in water. The degree of ionization of an acid or alkali determines its strength (*see* pH)

ionizing radiation Electromagnetic radiation that ionizes the air or water through which it passes, e.g., X-rays and gamma-rays. Used for the sterilization of food, etc., by irradiation

irradiation A method of sterilizing and disinfecting foods using ionizing radiation (X-rays or gamma-rays) to kill micro-organisms and insects. Also used to inhibit sprouting of potatoes

irritable bowel syndrome (IBS) Also known as spastic colon or mucous colitis. Abnormally increased motility of the large and small intestines, leading to pain and alternating diarrhoea and constipation; often precipitated by emotional stress

ischaemia Inadequate blood supply to a tissue

ischaemic heart disease or coronary heart disease Group of syndromes arising from failure of the coronary arteries to supply sufficient blood to heart muscles; associated with atherosclerosis of coronary arteries

islets of Langerhans The endocrine parts of the pancreas; glucagon is secreted by the α-cells and insulin by the β-cells

isoenzymes Enzymes that have the same catalytic activity but different structures, properties, and/or tissue distribution

isoleucine An essential amino acid, rarely limiting in food; one of the branched-chain amino acids

isomers Molecules that have the same empirical formula but differ in position of substituents or functional groups

isotonic Solutions with the same osmotic pressure; often refers to a solution with the same osmotic pressure as body fluids. Hypertonic and hypotonic refer to solutions that are more and less concentrated

isotopes Forms of elements with the same chemical properties, differing in atomic mass because of differing numbers of neutrons in the nucleus. Radioactive isotopes are unstable, and decay to stable elements, emitting radiation in the process. Stable isotopes can be detected only by their different atomic mass. Since they emit no radiation, they are safe for use in labelled compounds given to humans

isozymes *See* isoenzymes

jejuno-ileostomy Surgical procedure in which the terminal jejunum or proximal ileum is removed or bypassed; formerly used as a treatment for severe obesity

jejunum Part of the small intestine, between the duodenum and the ileum

joule The SI (Système Internationale) unit of energy; used to express energy content of foods

keratomalacia Progressive softening and ulceration of the cornea due to vitamin A deficiency. Blindness is usually inevitable unless the deficiency is corrected at an early stage

Keshan disease A disease occurring in parts of China where selenium deficiency is believed to be a problem. The cardiomyopathy of the disease is believed to be of viral origin

ketoacidosis High concentrations of ketone bodies in the blood

ketogenic diet A diet poor in carbohydrate (20–30 g) and rich in fat; causes accumulation of ketone bodies in tissue

ketone bodies Acetoacetate, β-hydroxybutyrate, and acetone. Acetoacetate and acetone are chemically ketones; although β-hydroxybutyrate is not, it is included in the term ketone bodies because of its metabolic relationship with acetoacetate

ketones Chemical compounds containing a carbonyl group (C–O), with two alkyl groups attached to the same carbon; the simplest ketone is acetone (dimethylketone: $(CH_3)_2–C–O$)

ketonuria Excretion of ketone bodies in the urine

ketosis High concentrations of ketone bodies in the blood

Keys score Method of expressing the lipid content of a diet, calculated as $1.35 \times (2 \times$ % energy from saturated fat – % energy from polyunsaturated fat) $+ 1.5 \times \sqrt{}$(mg cholesterol/1000 kcal)

kGy Kilogray; a unit of radiation intensity

kilo As a prefix for units of measurement, one thousand times (i.e., 10^3); symbol k

Kjeldahl determination Widely used method of determining total nitrogen in a substance by digesting with sulphuric acid and a catalyst; the nitrogen is reduced to ammonia which is then measured. In foodstuffs most of the nitrogen is protein, and the term crude protein is the total 'Kjeldahl nitrogen' multiplied by a factor of 6.25 (since most proteins contain 16 per cent nitrogen)

koilonychia Development of (brittle) concave fingernails, commonly associated with iron-deficiency anaemia

Korsakoff's psychosis Failure of recent memory, although events from the past are recalled, with confabulation; associated with vitamin B_1 deficiency, especially in alcoholics

kosher The selection and preparation of foods in accordance with traditional Jewish ritual and dietary laws. Foods that are not kosher are treif. The only kosher flesh foods are from animals that chew the cud and have cloven hoofs, such as cattle, sheep, goats, and deer; the hindquarters must not be eaten. The only fish permitted are those with fins and scales; birds of prey and scavengers are not kosher. Moreover, the animals must be slaughtered according to ritual, without stunning, before the meat can be considered kosher

Krebs cycle Or citric acid cycle; a central pathway for the metabolism of fats, carbohydrates, and amino acids

kwashiorkor (from the Ga language of West Africa) A disease which occurs frequently in young (weanling) children in some developing countries where weaning foods are of poor quality. It is characterized by oedema (swelling due to extracellular fluid accumulation), failure to thrive, abnormal hair appearance (dyspigmentation), often enlarged liver, and increased mortality risk. Associated especially with poor diets that are low in protein and other nutrients, and also with frequent infections; *see* protein–energy malnutrition

lactase The enzyme that breaks down lactose (milk sugar) to galactose–glucose in the small intestine

lactation The process of synthesizing and secreting milk from the breasts

lactic acid The acid produced by the anaerobic fermentation of carbohydrates. Originally discovered in sour milk, it is responsible for the flavour of fermented milk and for the precipitation of the casein curd in cottage cheese

lacto-ovo-vegetarian One whose diet excludes meat and fish but permits milk and eggs

lactose The carbohydrate of milk, sometimes called milk sugar; a disaccharide of glucose and galactose

lactulose A disaccharide of galactose and fructose which does not occur naturally but is formed in heated or stored milk by isomerization of lactose. Not hydrolysed by human digestive enzymes, but fermented by intestinal bacteria to form lactic and pyruvic acids. Thought to promote the growth of

Lactobacillus bifidus and so is added to some infant formulas. In large amounts it is a laxative

laxative Or aperient, a substance that helps the expulsion of food residues from the body. If strongly laxative it is termed purgative or cathartic. Dietary fibre and cellulose function because they retain water and add bulk to the contents of the intestine; Epsom salts (magnesium sulfate) also retains water; castor oil and drugs such as aloes, senna, cascara, and phenolphthalein irritate the intestinal mucosa. Undigested carbohydrates, such as lactulose and sugar alcohols, are also laxatives

lecithin Chemically lecithin is phosphatidyl choline; a phospholipid containing choline. Commercial lecithin is a mixture of phospholipids in which phosphatidyl choline predominates; *see* choline, phospholipids

lectins Proteins from legumes and other sources which bind to the carbohydrates found at cell surfaces. Therefore, they cause red blood cells to agglutinate *in vitro*, hence the old names haemagglutinins and phytoagglutinins

legumes Members of the family Leguminosae, consumed as dry mature seeds (grain legumes or pulses) or as immature green seeds in the pod. Legumes include the groundnut (*Arachis hypogaea*), and the soya bean (*Glycine max*), grown for their oil and protein, the yam bean (*Pachyrrhizus erosus*) and the African yam bean (*Sphenostylis stenocarpa*), grown for their edible tubers as well as seeds

leptin Hormone secreted by adipose tissue that acts to regulate long-term appetite and energy expenditure by signalling the state of body fat reserves

less-favoured areas Lands that have low agricultural potential because of limited and uncertain rainfall, poor soils, steep slopes, or other biophysical constraints, as well as lands that may have high agricultural potential but have limited access to infrastructure and markets, low population density, or other socioeconomic constraints

lethal dose 50% (LD50) An index of toxicity; the amount of the substance that kills 50 per cent of the test population of experimental animals when administered as a single dose

leucine An essential amino acid, rarely limiting in foods; one of the branched-chain amino acids

leucocytes White blood cells, normally 5000–9000/mm^3; include polymorphonuclear neutrophils, lymphocytes, monocytes, polymorphonuclear eosinophils, and polymorphonuclear basophils. A 'white cell count' determines the total; a 'differential cell count' estimates the numbers of each type. Fever, haemorrhage, and violent exercise cause an increase (leucocytosis); starvation and debilitating conditions a decrease (leucopenia)

leucocytosis Increase in the number of leucocytes in the blood

leucopenia Decrease in the number of leucocytes in the blood

leucovorin The synthetic (racemic) 5-formyl derivative of folic acid; more stable to oxidation than folic acid itself, and commonly used in pharmaceutical preparations. Also known as folinic acid

levans Polymers of fructose (the principal one is inulin) that occur in tubers and some grasses

Lieberkühn, crypts of Glands lining the small intestine which secrete the intestinal juice

lignans Naturally occurring compounds in various foods that have both oestrogenic and anti-oestrogenic activity (phytoestrogens)

lignin (lignocellulose) Indigestible part of the cell wall of plants (a polymer of aromatic alcohols). It is included in measurement of dietary fibre, but not of non-starch polysaccharides

limosis Abnormal hunger or excessive desire for food

linoleic acid An essential polyunsaturated fatty acid (C18:2 ω6)

α-linolenic acid An essential polyunsaturated fatty acid (C18:3 ω3)

γ-linolenic acid A non-essential polyunsaturated fatty acid (C18:3 ω6)

lipase Enzyme that hydrolyses fats to glycerol and fatty acids

lipectomy Surgical removal of subcutaneous fat

lipids A general term for fats and oils (chemically triacylglycerols), waxes, phospholipids, steroids, and terpenes. Their common property is insolubility in water and solubility in hydrocarbons, chloroform, and alcohols

lipids, plasma Triacylglycerols, free and esterified cholesterol, and phospholipids; present in lipoproteins in blood plasma. Chylomicrons consist mainly of triacylglycerols and protein; they are the form in which lipids absorbed in the small intestine enter the bloodstream. Very low density lipoproteins (VLDLs) are assembled in the liver and exported to other tissues, where they provide a source of lipids. Lipid-depleted VLDL becomes low density lipoprotein (LDL) in the circulation; it is rich in cholesterol and is normally cleared by the liver. High density lipoprotein (HDL) contains cholesterol from LDL and tissues that is returned to the liver. *See also* hypercholesterolaemia; hyperlipidaemia

lipodystrophy Abnormal pattern of subcutaneous fat deposits

lipoedema Condition in which fat deposits accumulate in the lower extremities, from hips to ankles, with tenderness of the affected parts

lipofuscin A group of pigments that accumulate in several body tissues, particularly the myocardium, and are associated with the ageing process

lipoic acid Chemically, dithio-octanoic acid, a coenzyme (together with vitamin B_1) in the metabolism of pyruvate and in the citric acid cycle. Although it is an essential growth factor for various micro-organisms, there is no evidence that it is a human dietary requirement

lipolysis Hydrolysis of fats to glycerol and fatty acids

lipopolysaccharide (LPS) Bacterial-derived antigenic material which promotes an immune response in animals and humans

liposuction Procedure for removal of subcutaneous adipose tissue in obese people using a suction pump device

lipotropes (lipotrophic factors) Compounds such as choline, betaine, and methionine that act as methyl donors; deficiency may result in fatty infiltration of the liver

Listeria A genus of bacteria commonly found in soil; the commonest is *Listeria monocytogenes*. They can cause food poisoning (listeriosis)

low birthweight (LBW) Used as shorthand to describe babies born at weight less than 2.5 kg. Average birthweight is close to 3.5 kg (WHO reference mean). LBW can result from delivery before term (pre-term) or from intra-uterine growth retardation (IUGR) due to many causes including foetal undernutrition

lower reference nutrient intake (LRNI) Set two standard deviations below the EAR for a nutrient. Intakes of nutrients below this point will almost certainly be inadequate for most individuals; *see* reference nutrient intake

lutein A hydroxylated carotenoid (xanthophyll); not vitamin A active

luxus consumption An outdated term for diet-induced thermogenesis

lycopene Red carotenoid, not vitamin A active

lymph The fluid between blood and the tissues in which oxygen and nutrients are transported to the tissues, and waste products back to the blood

lymphatics Vessels through which the lymph flows, draining from the tissues and entering the bloodstream at the thoracic duct

lysine An essential amino acid of special nutritional importance, since it is the limiting amino acid in many cereals

lysozyme An enzyme present in tears and in body secretions and fluids that helps in the destruction of bacterial cell walls

macrocytes Large immature precursors of red blood cells found in the circulation in pernicious anaemia and in folic acid deficiency, due to impairment of the normal maturation of red cells; hence macrocytic anaemia

mad cow disease Bovine spongiform encephalopathy

Maillard reaction Non-enzymic reaction between lysine in proteins and sugars on heating or prolonged storage

malnutrition Disturbance of form or function arising from deficiency or excess of one or more nutrients

malondialdehyde An oxidation (degradation) product of unsaturated fatty acids, often used as an index of pro-oxidant action and of potential damage to lipids by pro-oxidant species such as oxygen free radicals

maltase Enzyme that hydrolyses maltose to yield two molecules of glucose; present in the brush-border of the intestinal mucosal cells

maltodextrin A polymer of glucose made by acid hydrolysis of starch

maltose Malt sugar, or maltobiose; a disaccharide consisting of two glucose units linked $\alpha 1$–4

mannosans Polysaccharides containing mannose

mannose A six-carbon (hexose) sugar found in small amounts in legumes, manna, and some gums. Also called seminose and carubinose

marasmic kwashiorkor The most severe form of protein–energy malnutrition in children, with weight for height less than 60 per cent of that expected, and with oedema and other symptoms of kwashiorkor

marasmus An old term still in common use in Anglophone developing countries. The adjective (marasmic) described abnormally small and thin infants. As noun and adjective the term was later used by nutritionists to define a weight less than 60 per cent of the reference mean weight for age. This definition is still used in resource-poor areas where stature is not measured. Now the term protein–energy malnutrition (PEM) is more commonly used

market Any context in which the sale and purchase of goods and services takes place. There need be no physical entity corresponding to the market; it might consist of a global telecommunications network on which company shares are traded

market economy An economic system in which decisions about the allocation of resources and production are made on the basis of prices generated by voluntary exchanges among producers, consumers, workers, and owners of the factors of production (i.e., land, labour, and capital)

mast cells Cells found predominantly in connective tissue, although a specialized population of mast cells is found in mucosal sites (e.g., the gut). Following degranulation, mast cells release preformed and newly synthesized mediators of inflammation, including histamine

mastication Chewing, grinding, and tearing foods with the teeth while it is mixed with saliva

maternal death Death of a woman whilst pregnant or within 42 days of delivery

MaxEPA Trade name for a standardized mixture of long-chain marine fatty acids, eicosapentaenoic acid (EPA, C20:5 ω3), and docosohexaenoic (DHA, C22:6 ω3) acids

medium-chain triacylglycerols Triacylglycerols containing medium-chain (C: 10–12) fatty acids, used in treatment of malabsorption; they are absorbed more rapidly than conventional fats, and the products of their digestion are transported to the liver, rather than in chylomicrons

megavitamin therapy Treatment of diseases with very high doses of vitamins, many times the reference intakes; little or no evidence for its efficacy. Vitamins A, D, and B_6 are known to be toxic at high levels of intake

melanocortin A group of peptide hormones including adrenocorticotropic hormone (ACTH) and melanocyte-stimulating hormone (MSH); derived from proopiomelanocortin (POMC) in the pituitary gland

menadione, menaphthone, menaphtholdiacetate Synthetic compounds with vitamin K activity; vitamin K_3, sometimes known as menaquinone-0

menaquinones Bacterial metabolites with vitamin K activity; vitamin K_2

menarche The initiation of menstruation in adolescent girls, normally occurring between the ages of 11 and 15. The age at menarche has become younger in Western countries, possibly associated with a better general standard of nutrition, and is later in less developed countries

menhaden Oily fish (*Brevoortia patronus, B.tyrannus*) from the Gulf of Mexico and Atlantic seaboard of the USA, a rich source of fish oils

mesoderm Middle layer of cells or tissues of an embryo, or the parts derived from this (e.g., cartilage, muscles, and bone) (*see also* ectoderm; endoderm)

mesomorph Description given to a well-covered individual with well-developed muscles. *See also* ectomorph; endomorph

mesophiles Pathogenic micro-organisms that grow best at temperatures between 25 and 40°C; usually will not grow below 5°C

metabolic equivalent (MET) Unit of measurement of heat production by the body; 1 MET = 50 kcal/h/m² body surface area

metabolic water Produced in the body by the oxidation of foods: 100 g of fat produces 107.1 g of water, 100 g of starch produces 55.1 g of water, and 100 g of protein produces 41.3 g of water

metabolic weight Energy expenditure and basal metabolic rate depend on the amount of metabolically active tissue in the body, rather than total body weight; (body weight)$^{0.75}$ is generally used to calculate the weight of active tissue

metabolism The processes of interconversion of chemical compounds in the body. Anabolism is the process of forming larger and more complex compounds, commonly linked to the utilization of metabolic energy. Catabolism is the process of breaking down larger molecules to smaller ones, commonly oxidation reactions linked to release of energy. There is an approximately 30 per cent variation in the underlying metabolic rate (basal metabolic rate) between different individuals, determined in part by the activity of the thyroid gland

metabolome The complement of metabolic reactions and metabolic products of an organism

metabolomics The study of the metabolome

metalloenzyme An enzyme having a metal (e.g., zinc or copper) as its prosthetic group

metalloproteins Proteins containing a metal

metaphosphoric acid A form of phosphoric acid which is used as a preservative for vitamin C (ascorbic acid) because its addition to biological fluids, such as serum or urine, lowers the pH and chelates (i.e., inactivates) the pro-oxidant metal ions such as ferrous and cupric ions

methaemoglobin Oxidized form of haemoglobin (unlike oxyhaemoglobin, which is a loose and reversible combination with oxygen) which cannot transport oxygen to the tissues. Present in small quantities in normal blood, increased after certain drugs and after smoking; found rarely as a congenital abnormality (methaemoglobinaemia). It can be formed in the blood of babies after consumption of the small amounts of nitrate found naturally in vegetables and in some drinking water, since the lack of acidity in the stomach permits reduction of nitrate to nitrite

methionine An essential amino acid, one of the three containing sulphur; cystine and cysteine are the other two. Cystine and cysteine are not essential, but can only be made from methionine, and therefore the requirement for methionine is lower if there is an adequate intake of cystine and cysteine

3-methyl-histidine Derivative of the amino acid histidine, found mainly in the contractile proteins of muscle (myosin and actin)

methylmalonic acid An intermediate in the metabolic turnover of succinic acid; this substance typically accumulates in conditions of vitamin B_{12} deficiency, and can be measured in serum or urine, as a functional index of vitamin B_{12} status

micelles Emulsified droplets of partially hydrolysed lipids, small enough to be absorbed across the intestinal mucosa

micro Prefix for units of measurement, one millionth part (i.e., 10^{-6}); symbol μ (or sometimes mc)

microbiological assay Method of measuring compounds such as vitamins and amino acids using micro-organisms. The principle is that the organism is inoculated into a medium containing all the growth factors needed except the one under examination; the rate of growth is then proportional to the amount of this nutrient added in the test substance

microbiome A community of microorganisms co-existing in a particular habitat, such as the colon

microglia A type of glial cell located throughout the brain and spinal cord, accounting for 10–15 per cent of all brain cells and acting as the first and main form of active immune defence in the central nervous system; the resident macrophage cells

micronutrients Vitamins and minerals, which are needed in very small amounts (micrograms or milligrams per day), as distinct from fats, carbohydrates, and proteins, which are macronutrients since they are needed in considerably greater amounts

mid-upper arm circumference (MUAC) A rapid way of assessing nutritional status, especially applicable to children; *see* anthropometry

migraine Type of headache, characterized by usually being unilateral and/or accompanied by visual disturbance and nausea

milli Prefix for units of measurement, one thousandth part (i.e., 10^{-3}); symbol m

mineral salts The inorganic salts, including sodium, potassium, calcium, chloride, phosphate, sulfate, etc. So called because they are (or originally were) obtained by mining

mineralocorticoids The steroid hormones secreted by the adrenal cortex which control the excretion of salt and water; *see* steroids

minerals, trace Those mineral salts present in the body, and required in the diet, in small amounts (parts per million)

minerals, ultra-trace Those mineral salts present in the body, and required in the diet, in extremely small amounts (parts per thousand million or less); known to be dietary essentials, although rarely if ever a cause for concern since the amounts required are small and they are widely distributed in foods and water

Ministry of Agriculture, Fisheries and Food (MAFF) Former UK Ministry now replaced by Defra and the FSA

miscarriage Spontaneous loss of a pregnancy before 24 weeks of gestation

mitochondrion (plural mitochondria) Subcellular organelles in all cells apart from red blood cells in which the major oxidative reactions of metabolism occur, linked to the formation of ATP from ADP; *see* ADP, ATP

monoamine oxidase Enzyme that oxidizes amines; inhibitors have been used clinically as antidepressant drugs, and consumption of amine-rich foods such as cheese may cause a hypertensive crisis in people taking the inhibitors; *see* amines

monoaminergic pathways Networks of neurotransmitters, such as norepinephrine, serotonin, and dopamine, that are involved in the regulation of arousal, memory, and emotion

monosaccharides Group name of the simplest sugars, including those composed of three (trioses), four (tetroses), five (pentoses), six (hexoses), and seven (heptoses) carbon atoms. The units from which disaccharides, oligosaccharides, and polysaccharides are formed

monosodium glutamate (MSG) The sodium salt of glutamic acid, used to enhance the flavour of savoury dishes and often added to canned meat and soups

mRNA An RNA copy of genomic DNA containing genes for translation into protein; *see* RNA, DNA, genome

mucilages Soluble but undigested complexes of the sugars arabinose and xylose found in some seeds and seaweeds

mucin Viscous mucoprotein secreted in the saliva and throughout the intestinal tract; the main constituent of mucus

mucopolysaccharides Polysaccharides containing an amino sugar and uronic acid; constituent of the mucoproteins of cartilage, tendons, connective tissue, cornea, heparin, and blood group substances; *see* polysaccharides

mucoproteins Glycoproteins containing a sugar, usually chondroitin sulfate, combined with amino acids or peptides; occur in mucin. *See* glycoprotein

mucosa Moist tissue lining, for example, the mouth (buccal mucosa), stomach (gastric mucosa), intestines, and respiratory tract

mucous colitis Irritable bowel syndrome (q.v.)

mucus Secretion of mucous glands, containing mucin; protects epithelia

mutagen Compound that causes mutations and may be carcinogenic; *see* mutation

mutation a permanent structural alteration in DNA

mycoprotein Name given to mould mycelium used as a food ingredient

mycotoxins Toxins produced by fungi (moulds), especially *Aspergillus flavus* under tropical conditions, and *Penicillium* and *Fusarium* species under temperate conditions

myoglobin Haemoprotein mainly found in muscle where it serves as an intracellular storage site for oxygen

myristic acid A medium-chain saturated fatty acid (C14:0)

myxoedema Low metabolic rate as a result of hypothyroidism; commonly the result of iodine deficiency

nano Prefix for units of measurement, one thousand-millionth part (i.e., 10^{-9}), symbol n

naphthoquinone The chemical ring structure of vitamin K; the various vitamers of vitamin K can be referred to as substituted naphthoquinones

nasogastric tube Fine plastic tube inserted through the nose and thence into the stomach for enteral nutrition

National Center for Health Statistics (NCHS) standards Tables of height and weight for age used as reference values for the assessment of growth and nutritional status of children, based on data collected by the US National Center for Health Statistics in the 1970s. The most comprehensive set of such data, and used in most countries of the world

National Health and Nutrition Examination Survey (NHANES) Conducted by the US National Center for Health Statistics (NCHS), Centers for Disease Control and Prevention, designed to collect information about the health and diet of people in the USA

natural killer cells (NK cells) Specialized T-cells with the continuous task of identifying and eliminating cells recognized as being foreign or non-self

nature-identical Term applied to food additives that are synthesized in the laboratory and are identical to those that occur in nature

neonatal Within the first 28 days of life

nephropathy Diabetic complication that involves the kidney and may lead to chronic renal failure and dialysis

net dietary protein–energy ratio (NDpE) A way of expressing the protein content of a diet or food taking into account both the amount of protein (relative to total energy intake) and the protein quality. It is protein–energy multiplied by net protein utilization divided by total energy. If energy is expressed in kcalories and the result is expressed as a percentage, this is net dietary protein calories per cent, NDpCal%

net protein ratio/retention (NPR) Weight gain of a test animal plus weight loss of a control animal fed a non-protein diet per gram of protein consumed by the test animal

net protein utilization (NPU) The proportion of nitrogen intake that is retained, i.e., the product of biological value (q.v) and digestibility (q.v)

net protein value (NPV) A way of expressing the amount and quality of the protein in a food; the product of net protein utilization and protein content per cent

neural tube defect Congenital malformations of the brain (anencephaly) or spinal cord (spina bifida) caused by failure of closure of the neural tube in early embryonic development

neuropathy Diabetic complication that involves peripheral and autonomic nervous system

neuropeptide Y A peptide neurotransmitter believed to be important in the control of appetite and feeding behaviour, especially in response to leptin

neuropsychological the structure and function of the brain as they relate to specific psychological processes and behaviours

niacin The generic descriptor for two compounds that have the biological activity of the vitamin: nicotinic acid and its amide, nicotinamide. In the USA niacin is used specifically to mean nicotinic acid, and niacinamide for nicotinamide

niacinamide US name for nicotinamide, the amide form of the vitamin niacin

niacinogens Name given to protein–niacin complexes found in cereals; *see* niacytin

niacytin The bound forms of the vitamin niacin, found in cereals

nicotinamide One of the vitamers of niacin

nicotinamide adenine dinucleotide and its phosphate (NAD, NADP) The coenzymes derived from niacin. Involved as hydrogen acceptors in a wide variety of oxidation and reduction reactions; *see* nicotinamide

nicotinic acid One of the vitamers of niacin

night blindness Nyctalopia. Inability to see in dim light as a result of vitamin A deficiency

ninhydrin test For the amino group of amino acids. Pink, purple, or blue colour is developed on heating an amino acid or peptide with ninhydrin

nitrogen conversion factor Factor by which nitrogen content of a foodstuff is multiplied to determine the protein content; it depends on the amino acid composition of the protein. For wheat and most cereals it is 5.8; rice, 5.95; soya, 5.7; most legumes and nuts, 5.3; milk, 6.38; other foods, 6.25. In mixtures of proteins, as in dishes and diets, the factor of 6.25 is used. 'Crude protein' is defined as $N \times 6.25$

***N*-nitroso compounds** A group of chemicals that occur ubiquitously. They are formed in the environment and can be absorbed from food, water, air, and industrial and consumer products, can be formed within the body from precursors in food, water, and air, can be inhaled from tobacco smoke, and are naturally occurring

No Adverse Effect Level (NOAEL) With respect to food additives, equivalent to No Effect Level

No Effect Level (NOEL) With respect to food additives, the maximum dose of an additive that has no detectable adverse effects

non-digestible oligosaccharide An oligosaccharide that is not digested (or is minimally digested) in the upper gastrointestinal tract

non-esterified fatty acids (NEFAs) Free fatty acids in the blood, as opposed to triacylglycerols

non-starch polysaccharides (NSPs) Those polysaccharides, other than starches, found in foods. They are the major part of dietary fibre and can be measured more precisely than total dietary fibre; they include cellulose, pectins, glucans, gums, mucilages, inulin, and chitin (and exclude lignin); *see* dietary fibre

nor- Chemical prefix to the name of a compound, indicating: (1) one methyl (CH_3) group has been replaced by hydrogen (e.g., noradrenaline can be considered to be a demethylated derivative of adrenaline); (2) an analogue of a compound containing one less methylene (CH_2) groups than the parent compound; (3) an isomer with an unbranched side chain (e.g., norleucine, norvaline)

noradrenaline (norepinephrine) Hormone secreted by the adrenal medulla together with adrenaline (epinephrine); also a neurotransmitter. Physiological effects similar to those of adrenaline

norepinephrine *See* noradrenaline

norovirus Highly contagious viruses that can be transmitted from an infected person, contaminated food or water, or by touching contaminated surfaces. Infection causes gastroenteritis with diarrhea, vomiting, and stomach pain. Symptoms develop 12 to 48 hours after exposure and recovery within one to three days. Norovirus results in about 685 million cases of disease and 200,000 deaths globally a year. It is common both in the developed and developing world

northern blotting Widely used technique for detecting mRNAs by hybridization with specific probes following transfer of RNA onto a solid support, such as a nylon membrane; *see* mRNA, RNA, hybridization

Norwalk-like virus Viral infection similar to that first reported in Norwalk, USA, which causes an intestinal illness that occurs in outbreaks

nucleic acids Polymers of purine and pyrimidine sugar phosphates; two main classes: ribonucleic acid (RNA) and deoxyribonucleic acid (DNA); *see* ribonucleic acid, deoxyribonucleic acid

nucleoproteins The complex of proteins and nucleic acids found in the cell nucleus

nucleosides Compounds of purine or pyrimidine bases with a sugar, most commonly ribose. For example, adenine plus ribose forms adenosine. With the addition of phosphate a nucleotide is formed; *see* purines, pyrimidines

nucleotides Compounds of purine or pyrimidine base with a sugar phosphate

nutraceuticals Term for compounds in foods that are not nutrients but have (potential) beneficial effects

nutrient density A way of expressing the nutrient content of a food or diet relative to the energy yield (i.e., per 1000 kcal or per MJ) rather than per unit weight

nutrients Essential dietary factors such as vitamins, minerals, amino acids, and fatty acids. Metabolic fuels (sources of energy) are not termed nutrients, so a commonly used phrase is 'energy and nutrients'

nutrification The addition of nutrients to foods at such a level as to make a major contribution to the diet

nutrition The process by which living organisms take in and use food for the maintenance of life, growth, and the functioning of organs and tissues; the branch of science that studies these processes

Nutrition Labeling and Education Act 1990 NLEA, the basis of current US food labelling

nutrition surveillance Monitoring the state of health, nutrition, eating behaviour, and nutrition knowledge of the population for the purpose of planning and evaluating nutrition policy. In developing countries in particular, monitoring may include factors that may give early warning of nutritional emergencies

nutritional genomics (nutrigenomics) The field encompassing the interactions between nutrients and the genome and gene products, the function of gene products, and the identification and understanding of the genetic basis for individual and population differences in the response to diet

nutritionist One who applies the science of nutrition to the promotion of health and control of disease, instructs auxiliary medical personnel, and participates in surveys

nyctalopia *See* night blindness

nycthemeral Relating to a physiological time unit, 24 hours made up of one day and one night

obesity Excessive accumulation of body fat. A body mass index above 30 is considered to be obese (and above 40 grossly obese)

obstipation Extreme and persistent constipation caused by obstruction of the intestinal tract

obstructed labour Failure of descent of the foetal presenting part (usually the head) because of disproportion between the size of the head, which may be unusually large, and the size of the bony birth canal, which may be unusually small. An important cause of maternal and/or foetal death if Caesarean section is not available to the labouring woman

oedema Excess retention of fluid in the body; may be caused by cardiac, renal, or hepatic failure or by starvation (famine oedema)

oestrogens Steroid hormones principally secreted by the ovaries, which maintain female characteristics

Office of Dietary Supplements (ODS) Office of the US National Institutes of Health; website http://dietary-supplements.info.nih.gov

oleic acid A mono-unsaturated fatty acid (C18:1 ω9)

oligoallergenic diet Comprised of very few foods or an elemental diet used to diagnose whether particular symptoms are the result of allergic response to food

oligodipsia Reduced sense of thirst

oligofructose (or fructo-oligosaccharide) Polymer of fructose made from sucrose or inulin, a non-digestible oligosaccharide

oligopeptides Polymers of four or more amino acids; more than about 20–50 are termed polypeptides, and more than about 100 are considered to be proteins

oligosaccharides Carbohydrates composed of 3–10 monosaccharide units (with more than 10 units they are termed polysaccharides). Those composed of fructose, galactose, or isomaltose have prebiotic action and encourage the growth of beneficial intestinal bacteria

omophagia Eating of raw or uncooked food

oncotic pressure That part of plasma osmotic pressure exerted by proteins

opsomania Craving for special food

oral tolerance A state of antigen-specific hyperresponsiveness or unresponsiveness after prior mucoal exposure

orexigenic Stimulating appetite

orexins Two small peptide hormones produced by nerve cells in the lateral hypothalamus, believed to be involved in stimulation of feeding

organic Chemically, a substance containing carbon in the molecule (with the exception of carbonates and cyanide). Substances of animal and vegetable origin are organic; minerals are inorganic. The term 'organic foods' refers to 'organically grown foods', meaning plants grown without the use of (synthetic) pesticides, fungicides, or inorganic fertilizers, and prepared without the use of preservatives

organoleptic Sensory properties, i.e., those that can be detected by the sense organs. For foods, it is used particularly of the combination of taste, texture, and astringency (perceived in the mouth) and aroma (perceived in the nose)

ornithine An amino acid that occurs as a metabolic intermediate in the synthesis of urea but is not involved in protein synthesis, so is of no nutritional importance

ornithine–arginine cycle The metabolic pathway for the synthesis of urea

orotic acid An intermediate in the biosynthesis of pyrimidines; a growth factor for some micro-organisms and at one time called vitamin B_{13}. There is no evidence that it is a human dietary requirement; *see* pyrimidines

osmophiles Micro-organisms that can flourish under conditions of high osmotic pressure (e.g., in jams, honey, brine, pickles), especially yeasts (also called xerophilic yeasts)

osmosis The passage of water through a semi-permeable membrane, from a region of low concentration of solutes to one of higher concentration. Reverse osmosis is the passage of water from a more concentrated to a less concentrated solution through a semi-permeable membrane by the application of pressure

osmotic pressure The pressure required to prevent the passage of water through a semi-permeable membrane from a region of low concentration of solutes to one of higher concentration, by osmosis

osteocalcin A calcium-binding protein in bone, essential for the normal mineralization of bone. Its synthesis requires vitamin K, and is controlled by vitamin D

osteomalacia The adult equivalent of rickets; a bone disorder due to deficiency of vitamin D, leading to inadequate absorption of calcium and loss of calcium from the bones

osteopenia Decreased calcification or density of bone

osteoporosis Degeneration of the bones with advancing age due to loss of bone mineral and protein as a result of decreased secretion of hormones (oestrogens in women and testosterone in men)

oxidases (oxygenases) Enzymes that oxidize compounds by removing hydrogen and reacting directly with oxygen to form water or hydrogen peroxide

oxidation The chemical process of removing electrons from an element or compound, frequently together with the removal of hydrogen ions (H^+). The reverse process, the addition of electrons or hydrogen, is reduction

oxycalorimeter Instrument for measuring the oxygen consumed and carbon dioxide produced when a food is burned, as distinct from a calorimeter, which measures the heat produced

oxyntic cells Secretory cells in the stomach that produce the hydrochloric acid and intrinsic factor of the gastric juice; also known as parietal cells

P450 enzymes/proteins Cytochrome P450 proteins are mainly drug-metabolizing enzymes, but are also important for metabolizing some endogenously derived compounds such as cholesterol, prostacyclin, thromboxane A_2, vitamins A and D, etc. The P450 proteins are categorized into families and subfamilies. In humans there are 18 families and 43 subfamilies. Most drugs are metabolized by three families: CYP1, CYP2, and CYP3. The subfamily CYP3A is the most important and abundant protein, known to metabolize at least 120 drugs

palatinose Isomaltulose, a disaccharide of glucose and fructose, an isomer of sucrose

palmitic acid A saturated fatty acid (C16:0)

palmitoleic acid A mono-unsaturated fatty acid (C16: 1 ω9)

pancreas A gland in the abdomen with two functions: the endocrine pancreas (the islets of Langerhans) secretes the hormones insulin and glucagon; the exocrine pancreas secretes pancreatic juice

pancreatic juice The alkaline digestive juice produced by the pancreas and secreted into the duodenum

pancreatin Preparation made from the pancreas of animals containing the enzymes of pancreatic juice. Used to replace pancreatic enzymes in cystic fibrosis as an aid to digestion

pangamic acid Chemically the *N*-di-isopropyl derivative of glucuronic acid. Claimed to be an antioxidant, and to speed recovery from fatigue. Sometimes called vitamin B_{15}, but there is no evidence that it is a dietary essential, nor that it has any metabolic function

panthenol The biologically active alcohol of pantothenic acid

pantothenic acid A vitamin of the B complex with no numerical designation

para-aminobenzoic acid (PABA) Essential growth factor for micro-organisms. It forms part of the molecule of folic acid and therefore is required for the synthesis of this vitamin. Mammals cannot synthesize folic acid, and PABA has no other known function; there is no evidence that it is a human dietary requirement. Not normally present in human diets, but can be used to validate 24-hour urine collections because an oral dose, given at each of three meal-times, is rapidly and quantitatively excreted in the urine

paracellular Movement of solute between cells

parageusia Abnormality of the sense of taste

parathormone Commonly used as an abbreviation for the parathyroid hormone; correctly a trade name for a pharmaceutical preparation of the hormone

parathyroid hormone The hormone secreted by the parathyroid glands in response to a fall in plasma calcium; it acts on the kidney to increase the formation of the active metabolite of vitamin D (calcitriol)

parenteral nutrition Slow infusion of a solution of nutrients into the veins through a catheter. This may be partial, to supplement food and nutrient intake, or total (TPN, total parenteral nutrition), providing the sole source of energy and nutrients for patients with major intestinal problems

pareve (parve) Jewish term for dishes containing neither milk nor meat. Orthodox Jewish law prohibits mixing of milk and meat foods or the consumption of milk products for 3 hours after a meat meal

parietal cells *See* oxyntic cells

Parkinson's disease a progressive disease of the nervous system marked by tremor, muscular rigidity, and slow imprecise movement, chiefly affecting middle-aged and elderly people; associated with degeneration of the basal ganglia of the brain and a deficiency of the neurotransmitter dopamine

PARNUTS EU term for foods prepared for particular nutritional purposes (intended for people with disturbed metabolism, or in special physiological conditions, or for young children). Also called dietetic foods

parosmia Any disorder of the sense of smell

passive transport Movement of solutes across cell membranes from an area of high concentration to an area of lower concentration

pathogens Disease-causing bacteria, as distinct from those that are harmless

pectin Plant tissues contain hemicelluloses (polymers of galacturonic acid) known as protopectins, which cement the cell walls together. As fruit ripens, there is maximum protopectin present; thereafter it breaks down to pectin, pectinic acid, and finally pectic acid, and the fruit softens as the adhesive between the cells breaks down

pellagra Disease due to deficiency of the vitamin niacin and the amino acid tryptophan

pentosans Polysaccharides of five-carbon sugars (pentoses)

pentoses Monosaccharide sugars with five carbon atoms. The most important is ribose

pentosuria The excretion of pentose sugars in the urine. Idiopathic pentosuria is an inherited metabolic disorder almost wholly restricted to Ashkenazi (North European) Jews, which has no adverse effects. Consumption of fruits rich in pentoses (e.g., pears) can also lead to (temporary) pentosuria

pepsin An enzyme in the gastric juice which hydrolyses proteins to give smaller polypeptides; an endopeptidase

peptidases Enzymes that hydrolyse proteins, and therefore are important in protein digestion

peptide linkage CONH linkage formed by reaction of an amine group of one amino acid with the carboxylic acid group of a second amino acid

peptides Compounds formed when amino acids are linked together through the −CO−NH−(peptide) linkage. Two amino acids so linked form a dipeptide, three a tripeptide, etc.

pesco-vegetarian Vegetarian who will eat fish, but not meat

petechiae (petechial haemorrhages) Small pinpoint bleeding under the skin; one of the signs of scurvy

pH Potential hydrogen, a measure of acidity or alkalinity. Defined as the negative logarithm of the hydrogen ion concentration. The scale runs from 0, which is very strongly acid, to 14, which is very strongly alkaline. Pure water is pH 7, which is neutral; below 7 is acid, above is alkaline

phagocyte A blood cell that ingests and destroys foreign particles, bacteria, and cell debris

phagomania Morbid obsession with food; also known as sitomania

phagophobia Fear of food; also known as sitophobia

pharmacokinetics A branch of pharmacology concerned with the movement of drugs in the body from the time of administration to excretion

pharmafoods Alternative name for functional foods

phase I metabolism reactions The first phase of metabolism of foreign compounds (xenobiotics), involving metabolic activation. These reactions occur mainly in the liver, but also in the small intestine and lungs, and comprise the microsomal or mixed function oxidase system, NADPH-dependent enzymes, and cytochrome P450 proteins. Generally regarded as detoxication reactions, but may in fact convert inactive precursors into metabolically active compounds and be involved in activation of precursors to carcinogens

phase II metabolism reactions The second phase of the metabolism of foreign compounds, in which the activated derivatives formed in phase I metabolism are conjugated with amino acids, glucuronic acid, or glutathione to yield water-soluble derivatives that are excreted in urine or bile

phenolic hydroxyl group Hydroxyl group attached to an aromatic ring

phenols 'Alcohol-like' compounds that have the hydroxyl group bound to a benzene ring

phenotype The observable characteristics of a cell or organism as distinct from its genotype

phenylalanine An essential amino acid; in addition to its role in protein synthesis, it is the metabolic precursor of tyrosine (and hence noradrenaline (norepinephrine), adrenaline (epinephrine), and the thyroid hormones)

phlebotomy (venesection) Removal of blood. This serves as a simple method for reducing body iron levels in people with haemochromatosis

phosphates Salts of phosphoric acid; the form in which the element phosphorus is normally present in foods and body tissues

phosphatidic acid Glycerol esterified to two molecules of fatty acid, with the third hydroxyl group esterified to phosphate; chemically diacylglycerol phosphate; intermediate in the metabolism of phospholipids

phospholipids Glycerol esterified to two molecules of fatty acid, one of which is commonly polyunsaturated. The third hydroxyl group is esterified to phosphate and one of a number of water-soluble compounds, including serine (phosphatidylserine), ethanolamine (phosphatidylethanolamine), choline (phosphatidylcholine, also known as lecithin), and inositol (phosphatidylinositol)

phosphoproteins Proteins containing phosphate, other than as nucleic acids (nucleoproteins) or phospholipids (lipoproteins), e.g., casein from milk, ovovitellin from egg yolk

phrynoderma Blocked pores or 'toad-skin' (follicular hyperkeratosis of the skin) often encountered in malnourished people. Originally thought to be due to vitamin A deficiency but possibly due to other deficiencies; also occurs in adequately nourished people

phylloquinone Vitamin K

physical activity level (PAL) Total energy cost of physical activity throughout the day, expressed as a ratio of basal metabolic rate. Calculated from the physical activity ratio for each activity, multiplied by the time spent in that activity

physical activity ratio (PAR) Energy cost of physical activity expressed as a ratio of basal metabolic rate

phytase An enzyme (a phosphatase) that hydrolyses phytate to inositol and phosphate

phytate (phytic acid) Inositol hexaphosphate, present in cereals (particularly in the bran), dried legumes, and some nuts as both water-soluble salts (sodium and potassium) and insoluble salts of calcium and magnesium. Contributes significantly to the daily intake of phosphorus but is also a major inhibitor of the absorption of iron and zinc

phytate inositol polyphosphate A plant acid which binds divalent metal ions such as ferrous iron, zinc, etc., and makes these ions less bioavailable for intestinal absorption from the food

phytoalexins Substances, often harmful to humans, which increase in plant tissues when they are stressed, as by physical damage, exposure to ultraviolet light, etc.

phytoestrogens (phyto-oestrogens) Compounds in plant foods, especially soya bean, that have both oestrogenic and anti-oestrogenic action

phytohaemagglutinin A lectin and mitogen; capable of promoting a rapid proliferation of immune cells

phytoprotectants Various compounds in plant foods that have protective effects against diseases but are not dietary essentials and hence are not classified as nutrients

phytosterol General name given to sterols occurring in plants, the chief of which is sitosterol

phytotoxin Any poisonous substance produced by a plant

phytylmenaquinone Vitamin K

pica An unnatural desire for foods; alternative words are cissa, cittosis, and allotriophagy. Also a perverted appetite (eating of earth, sand, clay, paper, etc.)

pico Prefix for units of measurement, one million-millionth part (i.e., 10^{-12}); symbol p

plaque (1) Dental plaque is a layer of bacteria in an organic matrix on the surface of teeth, especially around the neck of each tooth; may lead to development of gingivitis, periodontal disease, and caries. (2) Atherosclerotic plaque is the development of fatty streaks in the walls of blood vessels

pleiotropic Gene producing more than one effect

poliomyelitis Poliomyelitis, commonly called polio, is a highly infectious disease, caused by the poliomyelitis virus mostly contracted by children. Most infections produce no symptoms, but 5-10 % infected with polio have flu-like symptoms. In 1 in 200 cases, the virus destroys parts of the nervous system, causing permanent paralysis in the legs or arms. In 1955, a vaccine was introduced on a wide scale. Polio was eliminated from the Western Hemisphere by 1994, and has greatly decreased worldwide but children with permanent paralysis due to this virus exist in some Asian countries

polycystic ovarian syndrome (PCOS) Commonly recognized cause of anovulatory infertility which is associated with multiple small ovarian cysts, high androgen levels, and insulin resistance

polydipsia Abnormally intense thirst; a typical symptom of diabetes

polymer large molecule made up of a chain of smaller molecules

polymerase chain reaction (PCR) A technique for amplifying defined regions of DNA; *see* DNA

polymorphonuclear leucocyte The most numerous of the white blood cells (also known as neutrophil)

polyols Sugar alcohols

polypeptides *See* peptides

polyphagia Excessive or continuous eating

polyphenols Common name for several families of complex organic molecules. While many of these molecules are thought to have important functions they also act as major inhibitors of iron absorption; *see* phenols

polysaccharides Complex carbohydrates formed by the condensation of large numbers of monosaccharide units (e.g., starch, glycogen, cellulose, dextrins, inulin). On hydrolysis, the simple sugar is liberated

polyunsaturated fatty acids Long-chain fatty acids containing two or more double bonds, separated by methylene bridges: $-CH_2-CH=CH-CH_2-CH=H-CH_2-$

ponderal index An index of fatness, used as a measure of obesity: the cube root of body weight divided by height. Confusingly, the ponderal index is higher for thin people, and lower for fat people

ponderocrescive Foods tending to increase weight; easily gaining weight, the opposite of ponderoperditive

ponderoperditive Stimulating weight loss

postprandial Occurring after a meal

postprandial lipaemia The gradual increase in the concentration of blood triacylglycerol following consumption of a meal containing fat

post-translational modification Alterations to the nascent protein produced by translation

prebiotics Non-digestible oligosaccharides that support the growth of colonies of potentially beneficial bacteria in the colon

precursors, enzyme Inactive forms of enzymes, activated after secretion; also called pro-enzymes or zymogens

pre-eclampsia A complication of pregnancy when high levels of blood pressure are combined with heavy proteinuria; if untreated can lead to maternal and/or foetal death

prenylated Molecule with isoprene (2-methylbutadiene) substituent

prion Infective agent(s) responsible for Creutzfeldt–Jakob disease, kuru, and possibly other degenerative diseases of the brain in humans, scrapie in sheep, and bovine spongiform encephalopathy (BSE). They are simple proteins and, unlike viruses, do not contain any nucleic acid. Transmission occurs by ingestion of infected tissue

probiotics Preparations of live micro-organisms added to food (or used as animal feed), claimed to be beneficial to health by restoring microbial balance in the intestine. The organisms commonly involved are lactobacilli, bifidobacteria, streptococci, and some yeasts and moulds, alone or as mixtures

procarcinogen A compound that is not itself carcinogenic, but undergoes metabolic activation in the body to yield a carcinogen

products of conception The foetus, placenta, amniotic fluid, and foetal membranes

pro-enzymes Inactive precursors of enzymes, activated after secretion; also called zymogens

progoitrins Substances found in plant foods which are precursors of goitrogens

proline A non-essential amino acid

promoter DNA sequence within a gene which controls the start of transcription for that gene

pro-opiomelanocortin A precursor of melanocortin

prosthetic group Non-protein part of an enzyme molecule; either a coenzyme or a metal ion. Essential for catalytic activity. The enzyme protein without its prosthetic group is the apoenzyme and is catalytically inactive. With the prosthetic group, it is known as the holoenzyme

proteases Enzymes that hydrolyse proteins

protein All living tissues contain proteins; they are polymers of amino acids, joined by peptide bonds. There are 21 main amino acids in proteins, and any one protein may contain several hundred to over a thousand amino acids, so an enormous variety of different proteins occur in nature. Generally, a polymer of relatively few amino acids is referred to as a peptide (e.g., di-, tri-, and tetrapeptides); oligopeptides contain up to about 50 amino acids; larger molecules are polypeptides or proteins

protein-bound iodine The thyroid hormones, tri-iodothyronine and thyroxine, are transported in the bloodstream bound to proteins; measurement of protein-bound iodine, as opposed to total plasma iodine, was used as an index of thyroid gland activity before more specific methods of measuring the hormones were developed

protein, crude Total nitrogen multiplied by the nitrogen conversion factor = 6.25

protein efficiency ratio (PER) Weight gain per weight of protein eaten

protein–energy malnutrition (PEM) A spectrum of disorders, especially in children, due to inadequate feeding. Used to describe children who are wasted or underweight due to insufficient intake of macronutrients. In fact, PEM is commonly associated with multiple micronutrient deficiencies. Marasmus is severe wasting and can also occur in adults; it is the result of a food intake inadequate to meet energy expenditure. Kwashiorkor affects only young children and includes severe oedema, fatty infiltration of the liver, and a sooty dermatitis; it is likely that deficiency of antioxidant nutrients and the stress of infection may be involved. Emaciation, similar to that seen in marasmus, occurs in patients with advanced cancer and AIDS; in this case it is known as cachexia; *see* chronic energy deficiency

protein–energy ratio The protein content of a food or diet expressed as the proportion of the total energy provided by protein (17 kJ/g, 4 kcal/g). The average requirement for protein is about 7 per cent of total energy intake; average Western diets provide about 14 per cent

protein hydrolysate Mixture of amino acids and polypeptides prepared by hydrolysis of proteins with acid, alkali, or proteases; used in enteral and parenteral nutrition and in supplements

protein induced by vitamin K absence or antagonism (PIVKA) Under-carboxylated prothrombin, liberated when vitamin K supplies are suboptimal, which is potentially less functionally efficient than the fully carboxylated form in supporting normal blood clotting rates; hence the defect in blood clotting that arises in severe vitamin K deficiency. Assay of PIVKA is a more sensitive index in mild vitamin K deficiency than clotting times, and therefore it has been used as a vitamin K status index

protein quality A measure of the usefulness of a dietary protein for growth and maintenance of tissues and, in animals, production of meat, eggs, wool, and milk. It is only important if the total intake of protein barely meets the requirement. The quality of individual proteins is unimportant in mixed diets, because of complementation between different proteins

protein retention efficiency (PRE) The net protein retention converted to a % scale by multiplying by 16, then becoming numerically the same as NPU

protein score A measure of protein quality based on chemical analysis

proteinases Enzymes that hydrolyse proteins

proteolysis The hydrolysis of proteins to their constituent amino acids, catalysed by alkali, acid, or enzymes

proteome The protein complement of a cell, tissue, or organism translated from its genomic DNA sequence

proteomics The study of the proteome

prothrombin Protein in plasma involved in coagulation of blood. The prothrombin time is an index of the coagulability of blood (and hence of vitamin K nutritional status) based on the time taken for a citrated sample of blood to clot when calcium ions and thromboplastin are added

protoporphyrin Haem molecule minus iron (i.e., the organic ring structure without the central metal ion), the accumulation of which, in red cells, indicates either iron deficiency or other situations of impaired iron incorporation into haem (e.g., that caused by lead poisoning). Its quantification in red blood cells can be used as a functional index of iron status

provitamin A substance that is converted into a vitamin, such as 7-dehydrocholesterol which is converted into vitamin D, or those carotenes that can be converted to vitamin A

proximate analysis Analysis of foods and feedstuffs for nitrogen (for protein), ether extract (for fat), crude fibre, and ash (mineral salts), together with soluble carbohydrate calculated by subtracting these values from the total (carbohydrate by difference). Also known as Weende analysis, after the Weende Experimental Station in Germany which, in 1865, outlined the methods of analysis to be used

P/S ratio The ratio of polyunsaturated to saturated fatty acids. In Western diets the ratio is about 0.6; it is suggested that increasing it to near 1.0 would reduce the risk of atherosclerosis and coronary heart disease

psychomotor Relationship between mental and muscular activity such as movement and coordination

psychrophilic organisms Bacteria and fungi that tolerate low temperatures. Their preferred temperature range is 15–20°C, but they will grow at or below 0°C; the temperature must be reduced to about −10°C before growth stops, but the organisms are not killed and will regrow when the temperature rises

ptyalin Obsolete name for salivary amylase

ptyalism Excessive flow of saliva

pulses Name given to the dried seeds (matured on the plant) of legumes such as peas, beans, and lentils

purines Nitrogenous compounds (bases) that occur in nucleic acids (adenine and guanine) and their precursors and metabolites; inosine, caffeine, and theobromine are also purines

putrescine Low molecular weight amine with biological activity

pyridoxal phosphate 5-Phosphate of the aldehyde form of pyridoxine (vitamin B_6); this is the major form of the vitamin in blood and tissues, and is commonly measured as an index of vitamin B_6 status

pyrimidines Nitrogenous compounds (bases) that occur in nucleic acids: cytosine, thymidine, and uracil

pyruvate Salts of pyruvic acid

pyruvic acid An intermediate in the metabolism of carbohydrates, formed by the anaerobic metabolism of glucose

QUAC stick Quaker arm circumference measuring stick: a stick used to measure height which also shows the 80th and 85th centiles of expected mid-upper arm circumference. Developed by a Quaker Service Team in Nigeria in the 1960s as a rapid and simple tool for assessment of nutritional status

quantitative ingredients declaration (QUID) Obligatory on food labels in the EU since February 2000; previously legislation only required declaration of ingredients in descending order of quantity, not specific declaration of the amount of each ingredient present

Quetelet's index *See* body mass index

reciprocal ponderal index An index of adiposity; height divided by cube root of weight

reducing sugars Sugars that are chemically reducing agents, including glucose, fructose, lactose, and the pentoses, but not sucrose

reduction The opposite of oxidation; chemical reactions resulting in a gain of electrons, or hydrogen, or the loss of oxygen

reference dose (RfD) Maximum acceptable oral dose of a toxic substance, such as a pesticide, as set by the US Environmental Protection Agency (EPA)

reference intakes (of nutrients) Amounts of nutrients greater than the requirements of almost all members of the population, determined on the basis of the average requirement plus twice the standard deviation, to allow for individual variation in requirements and thus cover the theoretical needs of 97.5 per cent of the population

reference man, woman An arbitrary physiological standard; defined as a person aged 25, weighing 65 kg, living in a temperate zone with a mean annual temperature of 10°C. Reference man performs medium work, with an average daily energy requirement of 13.5 MJ (3200 kcal). Reference woman is engaged in general household duties or light industry, with an average daily requirement of 9.7 MJ (2300 kcal)

reference nutrient intake (RNI) Defined by COMA (Committee on Medical Aspects of Food Policy for the UK Department of Health), most recently in 1991, as being the amount of each nutrient that is sufficient to meet the needs of the majority (mean + two standard deviations) of healthy people in a defined population, or subgroup of it. Approximately equivalent (in concept, but not necessarily in magnitude) to the US or WHO RDAs (recommended dietary amounts)

reference standards (international reference standard)/growth standards These refer to databases recording the linear and ponderal growth of healthy children. They include anthropometric data collected on suitably large samples and analysed with precise specifications to provide a useful basis for reference

relative protein value (RPV) A measure of protein quality

renal plasma flow Rate of passage of plasma through the kidneys, directly related to glomerular filtration rate

renal threshold Concentration of a compound in the blood above which it is not reabsorbed by the kidney, and so is excreted in the urine

respiratory quotient (RQ) Ratio of the volume of carbon dioxide produced when a substance is oxidized to the volume of oxygen used. The oxidation of carbohydrate results in an RQ of 1.0; of fat, 0.7; and of protein, 0.8

respirometer *See* spirometer

restoration The addition of nutrients to replace those lost in processing, as in milling of cereals. *See also* fortification

resveratrol (3,4′,5-trihydroxystilbene) A fat-soluble antioxidant and anti-inflammatory polyphenolic compound, produced by certain plants in response to stress and found in grapes (skins), red wine, grape juice, peanuts, cocoa, blueberries, bilberries, and cranberries

reticulocyte Immature precursor of the red blood cell in which the remains of the nucleus are visible as a reticulum. Very few are seen in normal blood as they are retained in the marrow until mature, but on remission of anaemia, when there is a high rate of production, reticulocytes appear in the bloodstream (reticulocytosis)

retinal (retinaldehyde), retinene, retinoic acid, retinol Vitamers of vitamin A

retinal maculopathy Deterioration of the macula (central region of the retina) that occurs progressively in older people, thus irreversibly impairing vision. Thought to be exacerbated by pro-oxidant action such as by oxygen free radicals

retinoids Compounds chemically related to, or derived from, vitamin A, which display some of the biological activities of the vitamin, but have lower toxicity; they are used for treatment of severe skin disorders and some cancers

retinol binding protein (RBP) A plasma protein which specifically binds retinol (the alcohol form of vitamin A) and prevents it from being excreted. In vitamin A sufficiency, all the RBP is bound to retinol in a 1:1 complex, but in vitamin A deficiency the protein may become partially desaturated so that the ratio of protein to retinol then exceeds 1:1, and this can be measured as an index of vitamin A status

retinopathy Diabetic complication that involves the retina and may lead to blindness

reverse transcriptase Viral enzyme that catalyses the conversion of mRNA into cDNA; *see* mRNA, cDNA

rexinoids Compounds chemically related to vitamin A that bind to the retinoid X receptor but not the retinoic acid receptor

rhamnose A methylated pentose (five-carbon) sugar

rhodopsin The pigment in the rod cells of the retina of the eye, also known as visual purple, consisting of the protein opsin and retinaldehyde, which is responsible for the visual process. In cone cells of the retina the equivalent protein is iodopsin

riboflavin Vitamin B_2

ribonucleic acid (RNA) A linear single-stranded polymer composed of four types of ribose nucleotide—adenine, cytosine, guanine, thymine (A, C, G, U)—linked by phosphodiester bonds and formed by the transcription of DNA. The three types of cellular RNA—rRNA, tRNA, and mRNA—play different roles in protein synthesis

ribose A pentose (five-carbon) sugar which occurs as an intermediate in the metabolism of glucose; especially important in the nucleic acids and various coenzymes

rickets Malformation of the bones in growing children due to deficiency of vitamin D, leading to poor absorption of calcium. In adults the equivalent is osteomalacia. Vitamin D resistant rickets does not respond to normal amounts of the vitamin but requires massive doses. Usually, a result of a congenital defect in the vitamin D receptor, or metabolism of the vitamin; it can also be due to poisoning with strontium

risk factor A factor that can be measured to indicate the statistical or epidemiological probability of an adverse condition, effect, or disease. Does not imply that it is a causative factor, nor that reversing the risk factor will reduce the hazard

salatrims Poorly absorbed fats, used as fat replacers; triacylglycerols containing short- and long-chain fatty acids

saliva Secretion of the salivary glands in the mouth: 1–1.5 L secreted daily. A dilute solution of the protein mucin (which lubricates food) and the enzyme amylase (which hydrolyses starch), with small quantities of urea, potassium thiocyanate, sodium chloride, and bicarbonate

salivary glands Three pairs of glands in the mouth, which secrete saliva: the parotid, submandibular, and submaxillary glands

salt Usually refers to sodium chloride, common salt, or table salt (chemically, any product of reaction between an acid and an alkali is a salt). The main sources are either from mines in areas where there are rich deposits of crystalline salt, or deposits left by the evaporation of seawater in shallow pans (known as sea salt)

sarcopenia Age-related loss of skeletal muscle mass and strength

satiety The sensation of fullness after a meal

Schilling test A test of vitamin B_{12} absorption and status

Scientific Advisory Committee on Nutrition (SACN) provides expert advice to the UK Food Standards Agency and Department of Health

scurvy Deficiency of vitamin C leading to impaired collagen synthesis, causing capillary fragility, poor wound-healing, and bone changes

sedoheptulose (sedoheptose) A seven-carbon sugar which is an intermediate in glucose metabolism by the pentose phosphate pathway

serine A non-essential amino acid

serum Clear liquid left after the protein has been clotted; the serum from milk, occasionally referred to as lacto-serum, is whey. Blood serum is blood plasma without the fibrinogen. When blood clots, the fibrinogen is converted to fibrin, which is deposited in strands that trap the red cells and form the clot. The clear liquid that is exuded is the serum

Shigella Bacteria that grow readily in foods, especially milk, and cause bacterial dysentery

sialagogue Substance that stimulates the flow of saliva

sialorrhoea Excessive flow of saliva

siderosis Accumulation of the iron-storage protein haemosiderin in liver, spleen, and bone marrow in cases of excessive red cell destruction and on diets exceptionally rich in iron

single-cell protein Collective term used for biomass of bacteria, algae, and yeast, and also (incorrectly) moulds, of potential use as animal or human food

single-nucleotide polymorphism (SNP) Changes of individual bases in DNA sequences of the same species which account for population variance

sitapophasis Refusal to eat as an expression of mental disorder

sitology Science of food

sitomania Mania for eating, morbid obsession with food; also known as phagomania

sitophobia Fear of food; also known as phagophobia

sitosterol The main sterol found in vegetable oils

skinfold thickness Index of subcutaneous fat and hence body fat content. Usually measured at four sites: biceps (midpoint of front upper arm), triceps (midpoint of back upper arm), subscapular (directly below point of shoulder blade at angle of 45°), supra-iliac (directly above iliac crest in mid-axillary line); *see* anthropometry

socially acceptable monitoring instrument (SAMI) A small heart-rate counting apparatus used to estimate energy expenditure

sol A colloidal solution, i.e., a suspension of particles intermediate in size between ordinary molecules (as in a solution) and coarse particles (as in a suspension). A jelly-like sol is a gel

solanine Heat-stable toxic compound (a glycoside of the alkaloid solanidine) found in small amounts in potatoes, and in larger and sometimes toxic amounts in sprouted potatoes and when they become green through exposure to light. Causes gastrointestinal disturbances and neurological disorders

somatotrophin Growth hormone

sorbestrin Sorbitol ester of fatty acids; developed as a fat replacer

sorbitol (glycitol, glucitol) A six-carbon sugar alcohol found in some fruits and manufactured from glucose. Although it is metabolized in the body, it is only slowly absorbed from the intestine and is tolerated by diabetics

Southern blotting Technique for detecting genomic DNA sequences by hybridization following transfer of DNA onto a solid support, such as a nylon membrane. Named after its inventor, Professor Edwin Southern; *see* DNA, hybridization

Spanish toxic oil syndrome Disease that occurred in Spain during 1981–1982, with 450 deaths and many people chronically disabled due to consumption of oil containing aniline-denatured industrial rapeseed oil, sold as olive oil. The precise cause is unknown

specific dynamic action (SDA) Archaic term for diet-induced thermogenesis

sphygmomanometer Instrument for measuring blood pressure

spirometer (respirometer) Apparatus used to measure the amount of oxygen consumed (and in some instances carbon dioxide produced) from which to calculate the energy expended (indirect calorimetry)

sprue Disease in which the villi of the small intestine are atrophied and food is incompletely absorbed, followed consequently by undernutrition and weight loss

squamous cell carcinoma Cancer of the flattened (squamous) epithelium

stable isotopes Atoms with differing numbers of neutrons (and hence differing atomic weights) which are not radioactive, i.e., are not unstable. Molecules that are 'labelled' with relatively rare stable isotopes can be used as metabolic markers for nutrients as they enter the bloodstream and the body stores, and undergo turnover in the body. Measurement is by mass spectrometry

stachyose Tetrasaccharide sugar composed of two units of galactose and one each of fructose and glucose. Not hydrolysed in the human digestive tract but fermented by intestinal bacteria

stanols Analogues of cholesterol that inhibit its absorption from the intestinal tract

staple food The principal food (e.g., wheat, rice, maize, etc.) which provides the main energy source for communities

starch Polysaccharide, a polymer of glucose units; the form in which carbohydrate is stored in the plant. It does not occur in animal tissue

starch blockers Compounds that inhibit amylase and so reduce the digestion of starch. Used as a slimming aid, with little evidence of efficacy

State Registered Dietitian (SRD) legal qualification to practise as a dietitian in the UK

statins A family of related drugs used to treat hypercholesterolaemia. They act by inhibiting hydroxymethylglutaryl CoA reductase (HMG CoA reductase), the first and rate-limiting enzyme of cholesterol synthesis

stearic acid Saturated fatty acid (C18:0)

steatopygia Accumulation of larger amounts of fat in the buttocks

steatorrhoea Excretion of faeces containing a large amount of fat and generally foul-smelling

steatosis Fatty infiltration of the liver; occurs in protein–energy malnutrition and alcoholism

stercobilin One of the brown pigments of the faeces; formed from the bile pigments, which, in turn, are formed as breakdown products of haemoglobin

stercolith Stone formed of dried compressed faeces

stereoformula/stereochemical A two-dimensional representation of a three-dimensional molecular structure that indicates the spatial arrangement of bonds

stereo-isomerism Occurs when compounds have the same molecular and structural formula, but with the atoms arranged differently in space. There are two subdivisions: optical and geometrical isomerism

steroids Chemically, compounds that contain the cyclopenteno-phenanthrene ring system. All the biologically important steroids are derived metabolically from cholesterol; they include the sex hormones (androgens, oestrogens, and progesterone) and the hormones of the adrenal cortex

stilbenoid Hydroxylated derivatives of stilbene, belonging to the family of phenylpropanoids, produced by plants and bacteria, of which resveratrol is a biologically important example

sterols Alcohols derived from the steroids, including cholesterol, ergosterol in yeast, sitosterol and stigmasterol in plants, and coprosterol in faeces

stiparogenic Foods that tend to cause constipation

stiparolytic Foods that tend to prevent or relieve constipation

strain Horticultural term for seed-raised plants exhibiting certain desirable characteristics but are not stable or predictable enough when propagated to be a cultivar

striatum A part of the brain below the cortex that plays an important part in the reward system

stroke Also known as cerebrovascular accident (CVA); damage to brain tissue by hypoxia due to blockage of a blood vessel as a result of thrombosis, atherosclerosis, or haemorrhage

structural congenital malformations Abnormalities of the foetus or infant where developmental abnormalities have resulted in malformations of organs of the body

stunting Reduction in the linear growth of children, leading to lower than expected height for age, generally resulting in lifelong short stature. A common effect of protein–energy malnutrition, and associated especially with inadequate protein intake

subcostal angle Angle in degrees subtended by the costal margins in the midline of the body

subsistence farming Agricultural production exclusively to meet the needs of the farm household, in contrast with commercial farming

substantial equivalence Term used to denote oil, starch, etc., from genetically modified crops which does not contain protein or DNA, and cannot be distinguished from the same product from the unmodified crop

substrate The substance on which an enzyme acts. The medium on which micro-organisms grow

sucrase The enzyme that hydrolyses sucrose to yield glucose and fructose (invert sugar). Also known as invertase or saccharase

sucrose Cane or beet sugar. A disaccharide composed of glucose and fructose

sugar alcohols Also called polyols, chemical derivatives of sugars that differ from the parent compounds in having an alcohol group (CH_2OH) instead of the aldehyde group (CHO); thus mannitol from mannose, xylitol from xylose, lacticol from lactulose (also sorbitol, isomalt, and hydrogenated glucose syrup). Several occur naturally in fruits, vegetables, and cereals. They range in sweetness from equal to less than half of sucrose. They provide bulk in foods such as confectionery (in contrast to intense sweeteners), and so are called bulk sweeteners. They are slowly and incompletely metabolized, are tolerated by diabetics, and provide less energy than sugars. They are less cariogenic than sucrose

superoxide dismutases Enzymes which remove potentially harmful (pro-oxidant) superoxide from the body; they require zinc and copper, or manganese, for their enzymatic activity, and their assay can, for instance, be used to measure copper status

surfactants Surface active agents; compounds that have an affinity for fats (hydrophobic) and water (hydrophilic) and so act as emulsifiers (e.g., soaps and detergents). Used as wetting agents to assist the reconstitution of powders, including dried foods, to clean and peel fruits and vegetables, and in baked goods and comminuted meat products

sweeteners Four groups of compounds are used to sweeten foods: (1) the sugars, of which the commonest is sucrose; (2) bulk sweeteners, including sugar alcohols; (3) synthetic non-nutritive sweeteners (intense sweeteners), which are many times sweeter than sucrose; (4) various other chemicals such as glycerol and glycine (70 per cent as sweet as sucrose), and certain peptides

synbiotics Combination nutritional supplements composed of probiotics and prebiotics; *see* prebiotics, probiotics

T_3 Tri-iodothyronine, one of the thyroid hormones

T_4 Thyroxine (tetra-iodothyronine), one of the thyroid hormones

tachycardia Rapid heartbeat, as occurs after exercise; may also occur, without undue exertion, as a result of anxiety, and in anaemia and vitamin B_1 deficiency

tachyphagia Rapid eating

tagatose Isomer of fructose (D-lyxo-2-hexulose) obtained by hydrolysis of plant gums and used as a bulk sweetener; 14 times as sweet as sucrose. Not metabolized to any significant extent, so does not raise blood sugar and has zero energy yield

Tanner standard Tables of height and weight for age used as reference values for the assessment of growth and nutritional status in children based on data collected in England in the 1960s. Now largely replaced by the NCHS (US National Center for Health Statistics) standards. *See also* anthropometry

tannins Polyphenol plant constituents which bind divalent metal ions such as ferrous iron, zinc, etc., and make these ions less bioavailable for intestinal absorption from the food. They have an astringent effect in the mouth, precipitate proteins, and are used to clarify beer and wines. Also called tannic acid and gallotannin

Taq DNA polymerase A thermophilic bacterial enzyme which catalyses the synthesis of double-stranded DNA; utilized in PCR; *see* DNA, PCR

taste The tongue can distinguish five separate tastes: sweet, salt, sour (or acid), bitter, and savoury (sometimes called umami, from the Japanese word for a savoury flavour), due to stimulation of the taste buds. The overall taste or flavour of foods is due to these tastes, together with astringency in the mouth, texture, and aroma

tau protein Normally bound to microtubules, structures that form the skeleton of the brain cells and the 'tracks' for the transport of signals throughout the cells; in Alzheimer's disease the tracks become disrupted as tau detaches and forms tangles

taurine A derivative of the amino acid cysteine (aminoethane sulfonic acid). Known to be a dietary essential for cats (deficient kittens are blind) and possibly essential for humans since the capacity for synthesis is limited, although deficiency has never been observed. In addition to its role in the eye and nervous system, it is important for conjugation of the bile salts

T-cells Thymus-derived lymphocytes or thymocytes. Comprise Th1 and Th2 lymphocytes and natural killer (NK) cells. Th1 lymphocytes are important in cell-mediated immunity and IL-2 and IFN-γ production. Th2 lymphocytes are associated with the promotion of antibody-mediated immunity and production of cytokines IL-4, IL-6, IL-10, and IL-13

teratogen Substance that deforms the foetus in the womb and so induces birth defects

teratogenesis Process by which a harmful stimulus (e.g., drugs) can cause structural congenital malformations; almost always active during the period of embryogenesis

tetany Oversensitivity of motor nerves to stimuli; particularly affects face, hands, and feet. Caused by reduction in the level of ionized calcium in the bloodstream and can accompany severe rickets

tetraenoic acid Fatty acid with four double bonds (e.g., arachidonic acid)

thermic effect of food (TEF) Alternative term for diet-induced thermogenesis

thermoduric Bacteria that are heat-resistant, but not thermophiles; they survive pasteurization. Usually not pathogens but indicative of insanitary conditions

thermogenesis Increased heat production by the body, either to maintain body temperature (by shivering or non-shivering thermogenesis) or in response to intake of food and stimulants such as coffee, nicotine, and certain drugs. *See* diet-induced thermogenesis, brown adipose tissue

thermophiles (thermophilic bacteria) Bacteria that prefer temperatures above 55°C and can tolerate temperatures up to 75–80°C

thiamin Vitamin B_1

threonine An essential amino acid

thrombin Plasma protein involved in the coagulation of blood

thrombokinase (or thromboplastin) An enzyme liberated from damaged tissue and blood platelets; it converts prothrombin to thrombin in the coagulation of blood

thrombosis Formation of blood clots in blood vessels

thymidine, thymine A pyrimidine; *see* nucleic acids

thyroglobulin The protein in the thyroid gland which is the precursor of the synthesis of the thyroid hormones. The thyroid-stimulating hormone of the pituitary gland stimulates hydrolysis of thyroglobulin and secretion of the hormones into the bloodstream

thyrotoxicosis Overactivity of the thyroid gland, leading to excessive secretion of thyroid hormones and resulting in increased basal metabolic rate. Hyperthyroid subjects are lean and have tense nervous activity. Iodine-induced thyrotoxicosis affects mostly older people who have lived for a long time in iodine-deficient areas, have had a long-standing goitre, and then have been given extra iodine. Also known as Jod–Basedow or Basedow's disease

thyroxine One of the thyroid hormones

T-lymphocytes Thymus-dependent lymphocytes which, amongst other functions, help B-lymphocytes during immunological responses and provide protection from intracellular microbial infection

tocol, tocopherol, tocotrienol Vitamers of vitamin E

tocopherols The chemical descriptor for the most important series of compounds that have vitamin E activity. There are a range of tocopherols, distinguished by different Greek letter prefixes, which differ in their food origins and in their biological activities in the body

tolerable upper intake level (UL) of a nutrient Maximum intake (from supplements and enriched foods) that is unlikely to pose a risk of adverse effects on health

treif Foods that do not conform to Jewish dietary laws; the opposite of kosher

trans Stereochemistry at a carbon–carbon double bond with the two substituents on opposite sides of the bond; *See cis*

transaminase Any enzyme that catalyses the reaction of transamination

transamination The transfer of the amino group ($-NH_2$) from an amino acid to an acceptor keto- (or oxo-) acid

transcellular Movement of solutes through cells. This may involve passive and/or active transport

transcription factors Proteins which influence the activity of genes through modulation of their promoters

transcriptome The total mRNA complement of cells, tissues, or organisms transcribed from the genome; *see* mRNA, genome

transcriptomics The study of the transcriptome

transferrin receptor A tissue protein, also found in blood serum, which has a specific recognition-affinity for the iron-transporting protein transferrin. It increases in concentration in conditions of chronic iron deficiency, and therefore is used as an iron status indicator. Unlike most other iron status indices, it is not confounded by the presence of an acute phase reaction (q.v.)

transketolase Enzyme which interconverts certain sugar phosphates, and which requires thiamin diphosphate (TPP) as its essential cofactor. Assay of this enzyme in red cell extracts is commonly used as a biochemical index of thiamin status, i.e., by measuring the ratio of the enzyme activity both with and without added TPP

trehalose Mushroom sugar, also called mycose, a disaccharide of glucose

triacylglycerols Sometimes called triglycerides; lipids consisting of glycerol esterified to three fatty acids (chemically acyl groups). The major component of dietary and tissue fat. Also known as saponifiable fats, since on reaction with sodium hydroxide they yield glycerol and the sodium salts (or soaps) of the fatty acids

Trichinella spiralis Trichinella spiralis is a parasite of carnivores. Infections are acquired by eating uncooked muscle containing encysted larvae from infected animals, usually pigs. The larvae are digested from the cyst, pass to the small intestine and burrow beneath the epithelium where they develop into adults. Adult worms live in the intestine for up to two months and the female produces larvae, which migrate to skeletal muscle where they invade, develop and encyst. The larvae can remain alive for many years in the cell

triglycerides *See* triacylglycerols

tri-iodothyronine One of the thyroid hormones

trimester The 40 weeks of pregnancy are divided conventionally into three trimesters: 0–13 weeks, 14–26 weeks, and 27 weeks until delivery

Trolox Trade name for a water-soluble vitamin E analogue

trypsin A proteolytic enzyme of the pancreatic juice; an endopeptidase

trypsinogen The inactive precursor of trypsin, secreted in the pancreatic juice

tryptophan An essential amino acid; the precursor of the neurotransmitter 5-hydroxytryptamine (serotonin) and of niacin

tumour necrosis factor (TNF-α) *See* cytokines

tyramine The amine formed by decarboxylation of the amino acid tyrosine

tyrosine A non-essential amino acid, formed in the body from phenylalanine; the precursor of the synthesis of melanin (the black and brown pigment of skin and hair), and of adrenaline (epinephrine) and noradrenaline (norepinephrine)

ubiquinone Co-enzyme in the respiratory (electron transport) chain in mitochondria, also known as co-enzyme Q or mitoquinone

ubiquitinated A protein that has been modified by the attachment of ubiquitin molecules

umami Name given to the special taste of monosodium glutamate, protein, certain amino acids, and the ribonucleotides (inosinate and guanylate). The Japanese name for a savoury flavour, now considered one of the five basic senses of taste

under-five A shorthand term for children under 5 years of age; a period of rapid growth and relatively high nutritional requirements. Mortality in children at this age is a commonly used public health indicator. Clinics targeted at this age group are called under-five clinics. They traditionally combine nutrition and growth monitoring, and immunization and simple curative treatment

United Nations Children's Fund (UNICEF) originally the United Nations International Children's Emergency Fund; website http://www.unicef.org

United Nations University (UNU) website http://www.unu.edu

unsaturation Introduction of double bonds

untranslated region (UTR) Areas of the mRNA transcript that do not encode for protein synthesis but may contain features that control the regulation of gene expression. These are found both proximal (5′) and distal (3′) to the coding sequence; see mRNA

urea The end-product of nitrogen metabolism, excreted in the urine

uric acid The end-product of purine metabolism

urticaria An itchy rash which results from inflammation and leakage of fluid from the blood into the superficial layers of the skin in response to various mediators. Synonyms are 'hives' or 'nettle rash'

US Department of Agriculture (USDA) Created as an independent department in 1862; website http://www.usda.gov

US Recommended Daily Allowances (USRDA) Reference intakes used for nutritional labelling of foods in the USA

van der Waals forces Interaction through space of two non-polar groups

vascular dementia A form of dementia caused by reduced flow of blood to the brain

vegans Those who consume no foods of animal origin (vegetarians often consume milk and/or eggs)

verbal intelligence The ability to analyse information and solve problems using language-based reasoning, involving reading or listening to words, conversing, writing, or thinking

verbascose A non-digestible tetrasaccharide, galactose–galactose–glucose–fructose, found in legumes; fermented by intestinal bacteria, causes flatulence

villi, intestinal Small finger-like processes covering the surface of the small intestine in large numbers (20–40/mm^2), projecting some 0.5–1 mm into the lumen. They provide a large surface area (about 300 m^2) for absorption in the small intestine

vitafoods Foods designed to meet the needs of health-conscious consumers which enhance physical or mental quality of life and may increase health status

vitamers Chemical compounds structurally related to a vitamin, and converted to the same overall active metabolites in the body. Thus they possess the same biological activity

vitamin Thirteen organic substances that are essential in very small amounts in food

vitamin Q Ubiquinone

vitaminoids Name given to compounds with 'vitamin-like' activity; considered by some to be vitamins or to partially replace vitamins. Include flavonoids, inositol, carnitine, choline, lipoic acid, and the essential fatty acids. With the exception of the essential fatty acids, there is little evidence that any of them is a dietary essential

waist:hip ratio Simple method for describing the distribution of subcutaneous and intra-abdominal adipose tissue

wasting/wasted An old term meaning abnormally thin after weight loss. The term has been used more recently by nutritionists to define weight for height significantly less than the reference range

water activity, a_w The ratio of the vapour pressure of water in a food to the saturated vapour pressure of water at the same temperature

Waterlow classification A system for classifying protein–energy malnutrition in children based on wasting (the percentage of expected weight for height) and the degree of stunting (the percentage of expected height for age). See also Wellcome classification

weaning foods Foods specially formulated for infants aged between 3 and 9 months for the transition between breast- or bottle-feeding and normal intake of solid foods

Weende analysis Proximate analysis; see proximate analysis

Weight-control Information Network (WIN) of the National Institute of Diabetes and Digestive and Kidney Diseases; websites http://www.niddk;nih.gov/health/nutrit/nutrit.htm

Wellcome classification A system for classifying protein–energy malnutrition in children based on percentage of expected weight for age and the presence or absence of oedema. Between 60 per cent and 80 per cent of expected weight is underweight in the absence of oedema, and kwashiorkor if oedema is present; under 60 per cent of expected weight is marasmus in the absence of oedema, and marasmic kwashiorkor if oedema is present

Wernicke–Korsakoff syndrome The result of damage to the brain as a result of vitamin B$_1$ deficiency, commonly associated with alcohol abuse. Affected subjects show clear signs of neurological damage (Wernicke's encephalopathy) with psychiatric changes

(Korsakoff's psychosis) characterized by loss of recent memory and confabulation (the invention of fabulous stories). *See also* beriberi

wholefoods Foods that have been minimally refined or processed, and are eaten in their natural state. In general nothing is removed from, or added to, the foodstuffs in preparation. Wholegrain cereal products are made by milling the complete grain

World Cancer Research Fund (WCRF) Website http://www.wcrf.org

World Food Programme (WFP) Part of the Food and Agriculture Organization of the United Nations; intended to give international aid in the form of food from countries with a surplus; website http://www.wfp.org

World Health Organization (WHO) Headquarters in Geneva; website http://www.who.in

xanthelasma Yellow fatty plaques on the eyelids due to hypercholesterolaemia

xanthophylls Yellow-orange hydroxylated carotene derivatives

xanthosis Yellowing of the skin associated with high blood concentrations of carotene

xenobiotic Substances foreign to the body, including drugs and some food additives

xerophthalmia Advanced vitamin A deficiency in which the epithelium of the cornea and conjunctiva of the eye deteriorate because of impairment of the tear glands, resulting in dryness and then ulceration, leading to blindness

xylitol A five-carbon sugar; said to have an effect in suppressing the growth of some of the bacteria associated with dental caries

Xylitol sugar-alcohol (polyol)

xylose Pentose (five-carbon) sugar found in plant tissues as complex polysaccharide; 40 per cent as sweet as sucrose. Also known as wood sugar

zymogens The inactive form in which some enzymes, especially the protein digestive enzymes, are secreted, being activated after secretion. Also called pro-enzymes, or enzyme precursors

Appendix
Dietary reference values

United Kingdom

Dietary reference values (DRVs) were revised in 1991 (Department of Health 1991). The Scientific Advisory Commission on Nutrition (SACN) published revised DRVs for energy in 2011, for carbohydrates in 2015, and for vitamin D in 2016. The term 'dietary reference value' applies to a range of intakes based on an assessment of the distribution of requirements for each nutrient. DRVs apply to groups of healthy people and are not appropriate for those with disease or metabolic abnormalities. The DRVs for one nutrient presuppose that requirements for energy and all other nutrients are met. For most nutrients, three values are given.

EAR Estimated average requirement, which assumes normal distribution of variability.

LRNI Lower reference nutrient intake, 2 SD below EAR.

RNI Reference nutrient intake, 2 SD above EAR. Where only one value is given in summary tables this is the value chosen.

Other values used are as follows.

Safe intakes Some nutrients are known to be important but there are insufficient data on human requirements to set any DRVs. A safe intake is judged to be a level or range of intakes above which there is no risk of deficiency and below a level where there is a risk of undesirable effects.

Population average This is used for specifying carbohydrate (fibre) and fat needs.

United States of America

The Recommended Dietary Allowances (RDAs) set in 1998 have now been replaced by the Dietary Reference Intakes (DRIs). These comprise four reference values.

EAR Estimated average requirement

RDA Recommended dietary allowance

AI Adequate intake

UL Tolerable upper intake level

These reference values have been revised and produced by the Food and Nutrition Board, Institute of Medicine (IOM). The revised values have been put together over the period 1997–2014.

Europe

The European Food Safety Authority (EFSA) Panel on Dietetic Products, Nutrition, and Allergies has undertaken a thorough review of dietary reference values for Europe since values were published in 1993. The revision was completed in 2019 and the revised values are summarized at https://www.efsa.europa.eu/en/topics/topic/drv. The tabulated values are for PRIs where available, otherwise AIs are given. A very useful resource is the DRV finder, an online tool giving access to all EFSA-revised DRVs (https://efsa.gitlab.io/multimedia/drvs/index.htm).

LTI Lowest threshold intake

AR Average requirement

AI Adequate intake

PRI Population reference intake: mean requirement +2 SD.

Acceptable range Range of safe values given where insufficient information is available to be more specific.

Food and Agriculture Organization (FAO) and World Health Organization (WHO)

The Joint Food and Agriculture Organization of the United Nations/World Health Organization of the United Nations (FAO/WHO) expert consultation on human vitamin and mineral requirements met in 1998. The resulting recommendations were published in 2001, and these are included here. Energy requirements were published in the same year and are reported here. Protein and amino acids and energy requirements were updated in 2007; a reference is given to the resulting report.

Units

Units vary for different nutrients.

Energy (kcal/day, kJ/day, or MJ/day)

All energy values are based on the Schofield equations (WHO/FAO/UNU 1985) and so should be similar for each source. Any variation occurs because the equations are based on weight and activity within broad age bands.

Carbohydrate and fat

These are expressed as a percentage of total energy intake, including 5 per cent alcohol, or as a percentage of food energy, excluding alcohol.

Protein (g/day or g/kg/day)
Most nutrients (g/day, mg/day, or μg/day)

Some exceptions are:

niacin: mg/1000 kcal or mg/MJ

vitamin B_6: mg/g protein

vitamin E: mg/g polyunsaturated fatty acids.

Iron, Zinc

Requirements depend on the bioavailability from the diet, which may be low, moderate, or high (see Chapter 13). The UK and USA values assume Western diets of high availability.

Age bands

The different national and international sources of data have used different age bands in some instances. This is particularly evident for some nutrients in the updated EFSA dataset. Where necessary these have been adjusted in the tables to correspond as closely as possible to the most frequently used age bands.

REFERENCES

Department of Health (1991) *Report on Health and Social Subjects 41. Dietary Reference Values for Food Energy and Nutrients for the United Kingdom. Committee on Medical Aspects of Food Policy.* HMSO, London.

EFSA Dietary Reference Values. https://www.efsa.europa.eu/en/topics/topic/drv.

EFSA DRV Finder https://efsa.gitlab.io/multimedia/drvs/index.htm

FAO/WHO (2001) *Human Vitamin and Mineral Requirements.* Report of a joint FAO/WHO expert consultation. FAO, Bangkok, Thailand. Rome.

Food and Nutrition Board, Institute of Medicine (1997) *Dietary Reference Intakes for Calcium, Phosphorus, Magnesium, Vitamin D and Fluoride.* National Academies Press, Washington, DC.

Food and Nutrition Board, Institute of Medicine (1998) *Dietary Reference Intakes for Thiamin, Riboflavin, Niacin, Vitamin B6, Folate, Vitamin B12, Pantothenic Acid, Biotin and Choline.* National Academies Press, Washington, DC.

Food and Nutrition Board, Institute of Medicine (2000) *Dietary Reference Intakes for Vitamin C, Vitamin E, Selenium and Carotenoids.* National Academies Press, Washington, DC.

Food and Nutrition Board, Institute of Medicine (2001) *Dietary Reference Intakes for Vitamin A, Vitamin K, Arsenic, Boron, Chromium, Copper, Iodine, Iron, Manganese, Molybdenum, Nickel, Silicon, Vanadium and Zinc.* National Academies Press, Washington, DC.

Food and Nutrition Board, Institute of Medicine (2005) *Dietary Reference Intakes for Energy, Carbohydrate, Fiber, Fat, Fatty Acids, Cholesterol, Protein and Amino Acids (Macronutrients).* National Academies Press, Washington, DC.

Food and Nutrition Board, Institute of Medicine (2011) *Dietary Reference Intakes for Calcium and Vitamin D.* The National Academies Press, Washington, DC.

Henry C J K (2005) Basal metabolic rate studies in humans: Measurement and development of new equations. *Publ Hlth Nutr* **8**(7A), 1133–1152.

Mann J, Cummings J H, Englyst H N et al. (2007) FAO/WHO scientific update on carbohydrates in human nutrition: conclusions. *Eur J Clin Nutr* **61**(Suppl 1), S132–S137.

SACN (2011) Dietary reference values for energy. Available at: https://www.gov.uk/government/publications/sacn-dietary-reference-values-for-energy.

SACN (2015) *Carbohydrates and health report.* Available at: https://www.gov.uk/government/publications/sacn-carbohydrates-and-health-report.

SACN (2016) Vitamin D and health. Available at: https://www.gov.uk/government/publications/sacn-vitamin-d-and-health-report.

WHO/FAO/UNU (1985) *Energy and Protein Requirements.* WHO, Technical Report Series, Geneva, 724.

WHO/FAO/UNU (2007) *Protein and Amino Acid Requirements in Human Nutrition: Report of a Joint WHO/FAO/UNU Expert Consultation.* WHO Technical Report Series No. 935. WHO, Geneva.

Tables of dietary reference values

TABLE A1a Estimated average requirement (EAR) values for energy, for infants 1–12 months, UK

Age range (months)	EAR[1]					
	Breastfed		Breast milk substitute fed		Mixed feeding or unknown[2]	
	MJ/kg/day (kcal/kg/day)	MJ/day (kcal/day)	MJ/kg/day (kcal/kg/day)	MJ/day (kcal/day)	MJ/kg/day (kcal/kg/day)	MJ/day (kcal/day)
Boys						
1–2	0.4 (96)	2.2 (526)	0.5 (120)	2.5 (598)	0.5 (120)	2.4 (574)
3–4	0.4 (96)	2.4 (574)	0.4 (96)	2.6 (622)	0.4 (96)	2.5 (598)
5–6	0.3 (72)	2.5 (598)	0.4 (96)	2.7 (646)	0.3 (72)	2.6 (622)
7–12	0.3 (72)	2.9 (694)	0.3 (72)	3.1 (742)	0.3 (72)	3.0 (718)
Girls						
1–2	0.4 (96)	2.0 (478)	0.5 (120)	2.3 (550)	0.5 (120)	2.1 (502)
3–4	0.4 (96)	2.2 (526)	0.4 (96)	2.5 (598)	0.4 (96)	2.3 (550)
5–6	0.3 (72)	2.3 (550)	0.4 (96)	2.6 (622)	0.3 (72)	2.4 (574)
7–12	0.3 (72)	2.7 (646)	0.3 (72)	2.8 (670)	0.3 (72)	2.7 (646)

[1] Calculated as TEE + energy deposition (kJ/day).
[2] These figures should be applied for infants when there is mixed feeding and the proportions of breast milk and breast milk substitute are not known.
Note: EFSA Scientific Opinion on energy requirements for Europe can be found in: *EFSA J.* 2013, **11**(1), 3005.

TABLE A1b Population estimated average requirement (EAR) values for energy, for children aged 1–18 years, UK[1]

Age (years)	EAR MJ/day (kcal/day)		
	PAL[2]	Boys	Girls
1	1.40	3.2 (765)	3.0 (717)
2	1.40	4.2 (1004)	3.9 (932)
3	1.40	4.9 (1171)	4.5 (1076)
4	1.58	5.8 (1386)	5.4 (1291)
5	1.58	6.2 (1482)	5.7 (1362)
6	1.58	6.6 (1577)	6.2 (1482)
7	1.58	6.9 (1649)	6.4 (1530)
8	1.58	7.3 (1745)	6.8 (1625)
9	1.58	7.7 (1840)	7.2 (1721)
10	1.75	8.5 (2032)	8.1 (1936)
11	1.75	8.9 (2127)	8.5 (2032)
12	1.75	9.4 (2247)	8.8 (2103)
13	1.75	10.1 (2414)	9.3 (2223)
14	1.75	11.0 (2629)	9.8 (2342)
15	1.75	11.8 (2820)	10.0 (2390)
16	1.75	12.4 (2964)	10.1 (2414)
17	1.75	12.9 (3083)	10.3 (2462)
18	1.75	13.2 (3155)	10.3 (2462)

[1] Calculated from BMR × PAL. BMR values are calculated from the Henry equations (Henry 2005) using weights and heights indicated by the 50th centile of the UK–WHO Growth Standards (ages 1–4 years) (https://www.rcpch.ac.uk/resources/uk-who-growth-charts-0-4-years), and the UK 1990 reference for children and adolescents.

[2] PAL, physical activity level.

Note: EFSA Scientific Opinion on energy requirements for Europe can be found in: *EFSA J.* 2013, **11**(1), 3005.

TABLE A1c Population estimated average requirement (EAR) values for energy, for adults, UK

Age range (years)	Men		Women	
	Height[1] (cm)	EAR[2] MJ/day (kcal/day)	Height[1] (cm)	EAR[2] MJ/day (kcal/day)
19–24	178	11.6 (2772)	163	9.1 (2175)
25–34	178	11.5 (2749)	163	9.1 (2175)
35–44	176	11.0 (2629)	163	8.8 (2103)
45–54	175	10.8 (2581)	162	8.8 (2103)
55–64	174	10.8 (2581)	161	8.7 (2079)
65–74	173	9.8 (2342)	159	8.0 (1912)
75+	170	9.6 (2294)	155	7.7 (1840)
All adults	175	10.9 (2605)	162	8.7 (2079)

[1] Values for illustration derived from mean heights in 2009 for England (Health Survey for England 2009) (http://content.digital.nhs.uk/pubs/hse09report).

[2] Median PAL = 1.63.

Note: EFSA Scientific Opinion on energy requirements for Europe can be found in: *EFSA J.* 2013, **11**(1), 3005.

SACN (2011)

TABLE A1d Dietary reference values for energy for males, USA

Age	AEA*[1]	
	(MJ/day)	(kcal/day)
0–3 months	2.7	650
4–6 months	2.7	650
7–9 months	3.5	850
10–12 months	3.5	850
1–3 years	5.4	1300
4–6 years	7.5	1800
7–10 years	8.3	2000
11–14 years	10.4	2500
15–18 years	12.5	3000
19–50 years	12.1	2900
51–59 years	19.6	2300
60–64 years	9.6	2300
65–74 years	9.6	2300
75+	9.6	2300

*A more recent consideration of energy requirements has led to these being expressed according to age, gender, BMI, and physical activity level (Food and Nutrition Board, Institute of Medicine 2005).

[1] AEA, average energy allowance.

FAO/WHO recommendations can be found in FAO/WHO/UNU (2001) *Human Energy Requirements. FAO Technical Report Series 1.* Report of a Joint FAO/WHO/UNU Expert Consultation (2001).

TABLE A1e Dietary reference values for energy for females, USA

Age	AEA*[1]	
	(MJ/day)	(kcal/day)
0–3 months	2.7	650
4–6 months	2.7	650
7–9 months	3.5	850
10–12 months	3.5	850
1–3 years	5.4	1300
4–6 years	7.5	1800
7–10 years	8.3	2000
11–14 years	9.2	2200
15–18 years	9.2	2200
19–50 years	9.2	2200
51–59 years	9.2	2200
60–64 years	9.2	2200
65–74 years	9.2	2200
75+	9.2	2200
Pregnancy	+1.2	+300
Lactation	+2.1	+500

* A more recent consideration of energy requirements has led to these being expressed according to age, gender, BMI, and physical activity level (Food and Nutrition Board, Institute of Medicine 2005).

[1] AEA, average energy allowance.

TABLE A2 Dietary reference values for protein (g/day)

Age	UK		USA RDA/AI
	EAR[1]	RNI[1]	
0–3 months	–	12.5	1.52[2]
4–6 months	10.6	12.7	1.52[2]
7–9 months	11.0	13.7	11.0
10–12 months	11.2	14.9	11.0
1–3 years	11.7	14.5	13
4–6 years	14.8	19.7	19
7–10 years	22.8	28.3	19
Males			
11–14 years	33.8	42.1	34
15–18 years	46.1	55.2	52
19–50 years	44.4	55.5	56
50+ years	42.6	53.3	56

TABLE A2 *Continued*

Age	UK		USA RDA/AI
	EAR[1]	RNI[1]	
Females			
11–14 years	33.1	41.2	34
15–18 years	37.1	45.0	46
19–50+ years	36.0	45.0	46
50+ years	37.2	46.5	46
Pregnancy		+6	71
Lactation		+11	71

[1] Based on egg and milk protein; assume complete digestibility.
[2] AI, g/kg/day.
Note: WHO recommendations can be found in FAO/WHO/UNU (2007) *Protein and Amino Acid Requirements in Human Nutrition. WHO Technical Report Series 935.*
EFSA Scientific Opinion on DRVs for protein can be found in: *EFSA J.* 2012, **10**(2) 2557.

TABLE A3 Dietary reference values for fat and carbohydrate for adults as a percentage of daily total energy intake, and for dietary fibre (g/day)

	UK	USA[1]	WHO		Europe RI[3]
	Population average		Lower[2]	Upper	
Total fat	Not more than 35	20–35	15	30	20–35
of which saturated fat	Not more than 11				
Total carbohydrate	50	45–65	50	75	45–60
of which free sugars	Not more than 5		0	10	–
Dietary fibre	30[4]		–	–	25[5]
Males		30–38[5]			
Females		21–25[5]			

[1] Acceptable distribution range.
[2] Population nutrient goal.
[3] RI, Reference intake.
[4] As defined by SACN (2015).
[5] AI.
Mann et al. (2007)

TABLE A4 Dietary reference values for vitamin A (μg retinol equivalent/day)

Age	UK			USA RDA/AI	FAO/WHO Safe Intake	Europe PRI
	LRNI[1]	EAR	RNI			
0–12 months	150	250	350	500[1]	375	250
1–3 years	200	300	400	300[1]	400	250
4–6 years	200	300	400	400	450	300
7–10 years	250	350	500	400	500	400

TABLE A4 *Continued*

Age	UK			USA RDA/AI	FAO/WHO Safe Intake	Europe PRI
	LRNI[1]	EAR	RNI			
Males						
11–15 years	250	400	600	600	600	600
15–50+ years	300	500	700	900	600	750
Females						
11–50+ years	250	400	600	600–700	600	650
Pregnancy			+100	770	800	700
Lactation			+350	1300	850	1300

[1] AI.

TABLE A5 Dietary reference values for vitamin D (µg/day)

Age	UK RNI*	USA RDA/AI*	FAO/WHO RNI	Europe AI
Males and females				
0–6 months	8.5–10[1]	10[2]	5	–
7 months–3 years	10[1]	15[3]	5	15
4–6 years	10	15	5	15
7–10 years	10	15	5	15
11–24 years	10	15	5	15
25–50+ years	10	15	5	15
65+	10	15[4]	15	15
Pregnancy and lactation	10	15	5	15

* Assuming minimal exposure to sunlight.
[1] Safe intake.
[2] AI.
[3] 7–12 months 10ug, AI.
[4] > 70 years, 20ug.

TABLE A6 Dietary reference values for vitamin E (mg/day, α-tocopherol)

Age	UK safe intake	USA RDA/AI	Europe AI
0–6 months	0.4 mg/g PUFA	4[1]	–
7–12 months	0.4 mg/g PUFA	5[1]	5
1–3 years	0.4 mg/g PUFA	6	6
4–10 years		7	9
Males			
11–50+ years	>4 mg/day	11–15	13

TABLE A6 *Continued*

Age	UK safe intake	USA RDA/AI	Europe AI
Females			
11–50+ years	>3 mg/day	11–15	11
Pregnancy		15	11
Lactation		19	11

PUFA, polyunsaturated fatty acid.

[1] AI.

No specific recommendations concerning vitamin E requirements were made by FAO/WHO in their 2001 Report (FAO/WHO (2001) *Human Vitamin and Mineral requirements*. Report of a joint FAO/WHO expert consultation. FAO, Bangkok, Thailand, Rome.

TABLE A7 Dietary reference values for vitamin K (µg/day)

Age	UK safe intake	USA AI	FAO/WHO RNI	Europe AI[1]
0–6 months	10	2.0	5[2]	–
7–12 months	10	2.5	10	10
1–3 years		30	15	12
4–6 years		55	20	20
7–10 years		60	25	30
Males				
11–14 years	1 µg/kg body weight	60	35–55	45
15–18 years	1 µg/kg body weight	75	35–55	65
19–24 years	1 µg/kg body weight	120	65	70
25+ years	1 µg/kg body weight	120	65	70
Females				
11–14 years	1 µg/kg body weight	60	35–55	45
15–18 years	1 µg/kg body weight	75	35–55	65
19–24 years	1 µg/kg body weight	90	55	70
25+ years	1 µg/kg body weight	90	55	70
Pregnancy	1 µg/kg body weight	90	55	70
Lactation	1 µg/kg body weight	90	55	70

[1] AI is based on phylloquinone only.

[2] This intake cannot be met by infants who are exclusively breastfed. (FAO/WHO (2001) *Human Vitamin and Mineral Requirements*. Report of a joint FAO/WHO expert consultation. FAO, Bangkok, Thailand, Rome.

TABLE A8 Dietary reference values for vitamin C (mg/day)

Age	UK			USA RDA/AI	FAO/WHO RNI	Europe PRI
	LRNI	EAR	RNI			
0–6 months	6	15	25	40[1]	25	–
7–12 months	6	15	25	50[1]	30	20

TABLE A8 *Continued*

Age	UK			USA RDA/AI	FAO/WHO RNI	Europe PRI
	LRNI	EAR	RNI			
1–3 years	8	20	30	15	30	20
4–6 years	8	20	30	25	30	30
7–10 years	8	20	30	25	35	45
Males						
11–14 years	9	22	35	45	40	70
15–50+ years	10	25	40	90	45	110
Females						
11–14 years	9	22	35	45	40	70
15–50+ years	10	25	40	75	45	95
Pregnancy			+10	85	55	+10
Lactation			+30	120	70	+60

[1] AI.

TABLE A9 Dietary reference values for thiamin

Age	UK				USA RDA/AI (mg/day)	FAO/WHO RNI (mg/day)	Europe PRI[1] (mg/MJ)
	LRNI (µg/1000 kcal)	EAR (µg/1000 kcal)	RNI (µg/1000 kcal)	RNI (mg/day)			
0–6 months	0.2	0.23	0.3	0.2	0.2[2]	0.2	–
7–12 months	0.2	0.23	0.3	0.3	0.3[2]	0.3	0.1
1–3 years	0.23	0.3	0.4	0.5	0.5	0.5	0.1
4–6 years	0.23	0.3	0.4	0.7	0.6	0.6	0.1
7–10 years	0.23	0.3	0.4	0.7	0.6	0.9	0.1
Males							
11–14 years	0.23	0.3	0.4	0.9	0.9	1.2	0.1
15–50+ years	0.23	0.3	0.4	0.9	1.2	1.2	0.1
Females							
11–14 years	0.23	0.3	0.4	0.7	0.9	1.1	0.1
15–50+ years	0.23	0.3	0.4	0.8	1.1	1.1	0.1
Pregnancy	0.23	0.3	0.4	+0.12	1.4	1.4	0.1
Lactation	0.23	0.3	0.4	+0.2	1.4	1.5	0.1

[1] EFSA expresses thiamin requirement in mg/MJ, assuming that thiamin requirement is related to energy requirement.
[2] AI.

TABLE A10 Dietary reference values for riboflavin (mg/day)

Age	UK			USA RDA/AI	FAO/WHO RNI	Europe PRI
	LRNI	EAR	RNI			
0–6 months	0.2	0.3	0.4	0.3[1]	0.3	–
7–12 months	0.2	0.3	0.4	0.4[1]	0.4	–
1–3 years	0.3	0.5	0.6	0.5	0.5	0.6
4–6 years	0.4	0.6	0.8	0.6	0.6	0.7
7–10 years	0.5	0.8	1.0	0.6	0.9	1.0
Males						
11–14 years	0.8	1.0	1.2	0.9	1.3	1.4
15–18 years	0.8	1.0	1.3	1.3	1.3	1.6
19–50 years	0.8	1.0	1.3	1.3	1.3	1.6
50+ years	0.8	1.0	1.3	1.3	1.3	1.6
Females						
11–14 years	0.8	0.9	1.1	0.9	1.0	1.4
15–50+ years	0.8	0.9	1.1	1.1	1.1	1.6
Pregnancy			+0.3	1.4	1.4	1.9
Lactation			+0.5	1.6	1.6	2.0

[1] AI.

TABLE A11 Dietary reference values for niacin (niacin equivalent (NE))

AGE	UK				USA RDA/AI (mg/day)	FAO/WHO RNI (NE/day)	Europe PRI (mg NE/MJ)
	LRNI (mg/1000 kcal)	EAR (mg/1000 kcal)	RNI (mg/1000 kcal)	RNI (mg/day)			
0–6 months	4.4	5.5	6.6	3	2[1]	2	–
7–12 months	4.4	5.5	6.6	5	4[1]	4	1.6
1–3 years	4.4	5.5	6.6	8	6	6	1.6
4–6 years	4.4	5.5	6.6	11	8	8	1.6
7–10 years	4.4	5.5	6.6	12	8	12	1.6
Males							
11–14 years	4.4	5.5	6.6	15	12	16	1.6
15–18 years	4.4	5.5	6.6	18	16	16	1.6
19–50 years	4.4	5.5	6.6	17	16	16	1.6
50+ years	4.4	5.5	6.6	16	16	16	1.6

766　Appendix　Dietary reference values

TABLE A11 *Continued*

AGE	UK				USA RDA/AI (mg/day)	FAO/WHO RNI (NE/day)	Europe PRI (mg NE/MJ)
	LRNI (mg/1000 kcal)	EAR (mg/1000 kcal)	RNI (mg/1000 kcal)	RNI (mg/day)			
Females							
11–14 years	4.4	5.5	6.6	12	12	16	1.6
15–18 years	4.4	5.5	6.6	14	14	16	1.6
19–50 years	4.4	5.5	6.6	13	14	14	1.6
50+ years	4.4	5.5	6.6	12	14	14	1.6
Pregnancy	*	*	*	*	18	18	*
Lactation	*	*	+2.3 mg/day	+2	17	17	*

* No increment.

[1] AI.

TABLE A12 Dietary reference values for vitamin B_6

Age	UK				USA RDA/AI (mg/day)	FAO/WHO RNI (mg/day)	Europe PRI (mg/day)
	LRNI (μg/g protein)	EAR (μg/g protein)	RNI (μg/g protein)	RNI (mg/day)			
0–6 months	3.5	6	8	0.2	0.1[1]	0.1	–
7–9 months	6	8	10	0.3	0.3[1]	0.3	–
10–12 months	8	10	13	0.4	0.3[1]	0.3	–
1–3 years	8	10	13	0.7	0.5	0.5	0.6
4–6 years	8	10	13	0.9	0.6	0.6	0.7
7–10 years	8	10	13	1.0	0.6	1.0	1.0
Males							
11–14 years	11	13	15	1.2	1.0	1.3	1.4
15–18 years	11	13	15	1.5	1.3	1.3	1.7
19–50+ years	11	13	15	1.4	1.7	1.3[2]	1.7
Females							
11–14 years	11	13	15	1.0	1.0	1.2	1.4
15–18 years	11	13	15	1.2	1.2	1.2	1.6
19–50+ years	11	13	15	1.2	1.5	1.3[3]	1.6
Pregnancy	*	*	*	*	1.9	1.9	1.8
Lactation	*	*	*	*	2.0	2.0	1.7

* No increment.

[1] AI.

[2] >50 years, 1.7 mg/day.

[3] >50 years, 1.5 mg/day'

TABLE A13 Dietary reference values for folate (µg/day)

Age	UK			USA[1] RDA/AI	FAO/WHO[1] RNI	Europe PRI[1]
	LRNI	EAR	RNI			
0–3 months	30	40	50	65[2]	80	–
4–6 months	30	40	50	65[2]	80	–
7–12 months	30	40	50	80[2]	80	–
1–3 years	35	50	70	150	160	120
4–6 years	50	75	100	200	200	140
7–10 years	75	110	150	200	300	200
Males						
11–14 years	100	150	200	300	400	270
15–50+ years	100	150	200	400	400	330
Females						
11–14 years	100	150	200	300	400	270
15–50+ years	100	150	200	400	400	330
Pregnancy	+100	600	600	600	600	600[2]
Lactation	+60	500	500	500	500	500

[1] Dietary folate equivalent (DFE) = µg food folate + (1.7 × µg folic acid).
[2] AI.

TABLE A14 Dietary reference values for vitamin B_{12} (µg/day)

Age	UK LRNI	EAR	RNI	USA RDA/AI	FAO/WHO RNI	Europe AI
0–6 months	0.1	0.25	0.3	0.4[1]	0.4	–
7–12 months	0.25	0.35	0.4	0.5[1]	0.5	1.5
1–3 years	0.3	0.4	0.5	0.9	0.9	1.5
4–6 years	0.5	0.7	0.8	1.2	1.2	1.5
7–10 years	0.6	0.8	1.0	1.2	1.8	2.5
Males						
11–14 years	0.8	1.0	1.2	1.8	2.4	3.5
15–50+ years	1.0	1.25	1.5	2.4	2.4	4
Females						
11–14 years	0.8	1.0	1.2	1.8	2.4	3.5
15–50+ years	1.0	1.25	1.5	2.4	2.4	4
Pregnancy			*	2.6	2.6	4.5
Lactation			+0.5	2.8	2.8	5

* No increment.
[1] AI.

TABLE A15 Dietary reference values for biotin (µg/day)

Age	UK safe intake	USA AI	FAO/WHO RNI	Europe AI
0–6 months		5	5	–
7–12 months		6	6	6
1–3 years		8	8	20
4–10 years		12–20	12–20	25
Males and females				
11–50+ years	10–20	30	30	40

TABLE A16 Dietary reference values for pantothenic acid (mg/day)

Age	UK safe intake	USA AI	FAO/WHO RNI	Europe AI
0–6 months	1.7	1.7	1.7	–
7–12 months	1.7	1.8	1.8	3
1–3 years	1.7	2	2.0	4
4–10 years	3–7	3–4	3–4	4
Males and females				
11–50+ years	3–7	5	5	5

TABLE A17 Dietary reference values for calcium (mg/day)

Age	UK LRNI	UK EAR	UK RNI	USA RDA/AI	FAO/WHO RNI	Europe PRI
0–6 months	240	400	525	200[1]	300	–
6–12 months	240	400	525	260[1]	400	–
1–3 years	200	275	350	700	500	450
4–6 years	275	350	450	1000	600	800
7–10 years	325	425	550	1000	700	800
Males						
11–14 years	450	750	1000	1300	1300	1150
15–18 years	450	750	1000	1300	1300	1150
19–24 years	400	525	700	1000	1000	1000
25–50 years	400	525	700	1000	1000	950
50+ years	400	525	700	1200	1000	950
Females						
11–14 years	480	625	800	1300	1300	1150
15–18 years	480	625	800	1300	1300	1150
19–24 years	400	525	700	1000	1000	1000
25–50 years	400	525	700	1000	1000	950
50+ years	400	525	700	1200	1300	950
Pregnancy	*	*	*	1000[2]	1000[3]	*
Lactation	*	*	*	1000[2]	1000	*[4]

* No increment.

[1] AI.

[2] Pregnant or lactating 14–18yr 1300mg/day.

[3] Last trimester, 1200 mg/day.

[4] Lactating women 18–24 years, 1000 mg/day.

TABLE A18 Dietary reference values for phosphorus (mg/day)

Age	UK RNI[1]	USA RDA/AI	Europe AI
0–6 months	400	100[2]	–
7–12 months	400	275[2]	160
1–3 years	270	460	250
4–10 years	350	500	440
Males			
11–18 years	775	1250	640
19–24 years	550	700	550
25–50 years	550	700	550
50+ years	550	700	550
Females			
11–18 years	625	1250	640
19–24 years	550	700	550
25–50+ years	550	700	550
Pregnancy	*	*	*
Lactation	+440	*	*

* No increment.

[1] Phosphorus RNI is set equal to calcium in molar terms.

[2] AI.

No FAO/WHO values available.

TABLE A19 Dietary reference values for magnesium (mg/day)

Age	UK			USA RDA/AI	FAO/WHO RNI	Europe AI
	LRNI	EAR	RNI			
0–3 months	30	40	55	30[1]	26–36[2]	–
4–6 months	40	50	60	30[1]	26–36[2]	–
7–9 months	45	60	75	75[1]	54	80
10–12 months	45	60	80	75	54	80
1–3 years	50	65	85	80	60	170[3]
4–6 years	70	90	120	130	76	230
7–10 years	115	150	200	130	100	230
Males						
11–14 years	180	230	280	240	230	300
15–18 years	190	250	280	410	230	300
19–50+ years	190	250	75[1]	420	260	350

TABLE A19 *Continued*

Age	UK			USA RDA/AI	FAO/WHO RNI	Europe AI
	LRNI	**EAR**	**RNI**			
Females						
11–14 years	180	230	280	240	220	250
15–18 years	190	250	300	360	220	250
19–50 years	190	250	300	320	220	300
50+ years	150	200	270	320	220	300
Pregnancy	*	*	*	350	*	*
Lactation	*	*	*	310	*	*

* No increment.

[1] AI.

[2] According to whether breast or formula-fed respectively?

[3] 3 years, 230mg/day.

TABLE A20 Dietary reference values for sodium (mg/day)

Age	UK		USA AI	Europe Safe and adequate intake
	LRNI	**RNI**		
0–3 months	140	210	120	
4–6 months	140	280	120	
7–9 months	200	320	370	
10–12 months	200	350	370	
1–3 years	200	500	1000	1100
4–6 years	280	700	1200	1300
7–10 years	350	1200	1500	1700
11–14 years	460	1600	1500	2000
15–50+ years	575	1600	1500	2000
Pregnancy	*	*	*	*
Lactation	*	*		*

* No increment.

TABLE A21 Dietary reference values for potassium (mg/day)

Age	UK		USA AI	Europe AI
	LRNI	**RNI**		
0–3 months	400	800	400	
4–6 months	400	850	400	
7–9 months	400	700	700	750
10–12 months	450	700	700	750

TABLE A21 *Continued*

Age	UK		USA AI	Europe AI
	LRNI	RNI		
1–3 years	450	800	3000	800
4–6 years	600	1100	3800	1100
7–10 years	950	2200	4500	1800
11–14 years	1600	3100	4700	2700
15–50+ years	2000	3500	4700	3500
Pregnancy	*	*	*	*
Lactation	*	*	5100	4000

* No increment.

TABLE A22 Dietary reference values for chloride (mg/day)

Age	UK RNI[1]	USA AI	Europe Safe and adequate intakes
0–3 months	320	180	
4–6 months	400	180	
7–9 months	500	570	
10–12 months	500	570	
1–3 years	800	1500	1700
4–6 years	1100	1900	2000
7–10 years	1100	2300	2600
Males and females			
11–50+ years	2500	3600	3100
Pregnancy	*	*	*
Lactation	*	*	*

* No increment.
No FAO/WHO values available.

TABLE A23 Dietary reference values for iron[1] (mg/day)

Age	UK			USA RDA/AI	FAO/WHO[2] RNI	Europe PRI
	LRNI	EAR	RNI			
0–3 months	0.9	1.3	1.7	0.27[3]	–	
4–6 months	2.3	3.3	4.3	0.27[3]	–	
7–12 months	4.2	6.0	7.8	11	6.2	11
1–3 years	3.7	5.3	6.9	7	3.9	7

TABLE A23 *Continued*

Age	UK			USA RDA/AI	FAO/WHO[2] RNI	Europe PRI
	LRNI	EAR	RNI			
4–6 years	3.3	4.7	6.1	10	4.2	7
7–10 years	4.7	6.7	8.7	10	5.9	11
Males						
11–14 years	6.1	8.7	11.3	8	9.7	11
15–18 years	6.1	8.7	11.3	11	12.5	11
19–50+ years	4.7	6.7	8.7	8	9.1	11
Females						
11–14 years	8.0	11.4	14.8[4]	8	9.3[5]	13
15–50 years	8.0	11.4	14.8[4]	15	19.6–20.7	16
50+ years	4.7	6.7	8.7	8	7.5[6]	11
Pregnancy	*	*	*	27	*	16
Lactation	*	*	*	9	10	16

[1] 1 mmol iron = 55.9 mg.

[2] Recommended nutrient intake based on 15% dietary bioavailability.

[3] AI.

[4] Insufficient for women with high menstrual losses, who may need iron supplements.

[5] Non-menstruating.

[6] Post-menopausal.

TABLE A24 Dietary reference values for zinc (mg/day)

Age	UK			USA RDA/AI	FAO/WHO RNI[1]	Europe PRI
	LRNI	EAR	RNI			
0–3 months	2.6	3.3	4.0	2.0[2]	2.8	
4–6 months	2.6	3.3	4.0	2.0[2]	2.8	
7–12 months	3.0	3.8	5.0	3	4.1	2.9
1–3 years	3.0	3.8	5.0	3	4.1	4.3
4–6 years	4.0	5.0	6.5	5	4.8	5.5
7–10 years	4.0	5.4	7.0	5	5.6	7.4
Males						
11–14 years	5.3	7.0	9.0	8	8.6	10.7
15–18 years	5.5	7.3	9.5	11	8.6	14.2
19–50+ years	5.5	7.3	9.5	11	7.0	9.4–16.3[3]

TABLE A24 *Continued*

Age	UK			USA RDA/AI	FAO/WHO RNI[1]	Europe PRI
	LRNI	EAR	RNI			
Females						
11–14 years	5.3	7.0	9.0	8	7.2	10.7
15–18 years	4.0	5.5	7.0	9	7.2	11.9
19–50+ years	4.0	5.5	7.0	8	4.9	7.5–12.7[3]
Pregnancy	*	*	*	11	4.9–10	(+)1.6
Lactation						(+)2.9
0–4 months			+6.0	12.0	9.5	
4+ months			+2.5	12.0	8.8	

* No increment.

[1] Assuming a diet with moderate bioavailability.

[2] AI.

[3] Depending on phytate intake.

TABLE A25 Dietary reference values for copper[1] (mg/day)

Age	UK RNI	USA RDA/AI	Europe AI
0–3 months	0.3	0.2[2]	
4–6 months	0.3	0.2[2]	
7–12 months	0.3	0.2	0.4
1–3 years	0.4	0.34	0.7
4–6 years	0.6	0.44	1.0
7–10 years	0.7	0.44	1.0
Males			
11–14 years	0.8	0.7	1.3
15–18 years	1.0	0.89	1.3
19–50+ years	1.2	0.90	1.6
Females			
11–14 years	0.8	0.7	1.1
15–18 years	1.0	0.89	1.1
19–50+ years	1.2	0.90	1.3
Pregnancy	1.0	1.0	1.5
Lactation	+0.3	1.3	1.5

[1] 1 μmol copper = 63.5 μg.

[2] AI.

TABLE A26 Dietary reference values for selenium (µg/day)[1]

Age	UK		USA RDA/AI	FAO/WHO RNI	Europe AI
	LRNI	RNI			
0–3 months	4	10	15[2]	6	
4–6 months	5	13	15[2]	6	
7–9 months	5	10	20[2]	10	15
10–12 months	6	10	20[2]	10	15
1–3 years	7	15	20	17	15
4–6 years	10	20	30	22	20
7–10 years	16	30	30	21	35
Males					
11–14 years	25	45	40	32	55
15–18 years	40	70	55	32	70
19–50+ years	40	75	55	34	70
Females					
11–14 years	25	45	40	26	55
15–18 years	40	60	55	26	70
19–50+ years	40	60	55	26	70
Pregnancy	*	*	60	28–30	70
Lactation	115	115	70	35–42	85

* No increment.

[1] 1 µmol selenium = 79 µg.

[2] AI.

TABLE A27 Dietary reference values for iodine (µg/day)

Age	UK		USA RDA/AI	FAO/WHO RDI µg/kg/day	Europe AI
	LRNI	RNI			
0–3 months	40	50	110[1]	15	
4–6 months	40	60	110[1]	15	
7–12 months	40	60	130[1]	15	70
1–3 years	40	70	90	6	90
4–6 years	50	100	90	6	90
7–10 years	55	110	90	4	90
Males and females					
11–14 years	65	130	120	4	120
15–18 years	70	140	150	2	130
19–50+ years	70	140	150	2	150
Pregnancy	*	*	220	3.5	200
Lactation	*	*	290	3.5	200

* No increment.

[1] AI.

Index

Tables, figures, and boxes are indicated by an italic *t*, *f*, and *b* following the page number.